P9-CLH-041

HARRIS COUNTY PUBLIC LIBRARY
3 4028 05296 0862

Grzimek's
Animal Life Encyclopedia

Second Edition

●●●●

Grzimek's Animal Life Encyclopedia

Second Edition

Volume 15
Mammals IV

Devra G. Kleiman, Advisory Editor
Valerius Geist, Advisory Editor
Melissa C. McDade, Project Editor

Joseph E. Trumpey, Chief Scientific Illustrator

Michael Hutchins, Series Editor
In association with the American Zoo and Aquarium Association

Detroit • New York • San Diego • San Francisco • Cleveland • New Haven, Conn. • Waterville, Maine • London • Munich

Grzimek's Animal Life Encyclopedia, Second Edition

Volume 15: Mammals IV

Project Editor
Melissa C. McDade

Editorial
Stacey Blachford, Deirdre S. Blanchfield, Madeline Harris, Christine Jeryan, Kate Kretschmann, Mark Springer, Ryan Thomason

Indexing Services
Synapse, the Knowledge Link Corporation

Permissions
Margaret Chamberlain

Imaging and Multimedia
Randy Bassett, Mary K. Grimes, Lezlie Light, Christine O'Bryan, Barbara Yarrow, Robyn V. Young

Product Design
Tracey Rowens, Jennifer Wahi

Manufacturing
Wendy Blurton, Dorothy Maki, Evi Seoud, Mary Beth Trimper

For more information contact
The Gale Group, Inc.
27500 Drake Rd.
Farmington Hills, MI 48331-3535
Or you can visit our Internet site at
http://www.gale.com

Cover photo of humpback whale (*Megaptera novaeangliae*) by John Hyde, Bruce Coleman, Inc. Back cover photos of sea anemone by AP/Wide World Photos/University of Wisconsin-Superior; land snail, lionfish, golden frog, and green python by JLM Visuals; red-legged locust © 2001 Susan Sam; hornbill by Margaret F. Kinnaird; and tiger by Jeff Lepore/Photo Researchers. All reproduced by permission.

ISBN 0-7876-5362-4 (vols. 1–17 set)
0-7876-6573-8 (vols. 12–16 set)
0-7876-5788-3 (vol. 12)
0-7876-5789-1 (vol. 13)
0-7876-5790-5 (vol. 14)
0-7876-5791-3 (vol. 15)
0-7876-5792-1 (vol. 16)

This title is also available as an e-book.
ISBN 0-7876-7750-7 (17-vol set)
Contact your Gale sales representative for ordering information.

LIBRARY OF CONGRESS CATALOGING-IN-PUBLICATION DATA

Grzimek, Bernhard.
[Tierleben. English]
Grzimek's animal life encyclopedia.— 2nd ed.
v. cm.
Includes bibliographical references.
Contents: v. 1. Lower metazoans and lesser deuterosomes / Neil Schlager, editor — v. 2. Protostomes / Neil Schlager, editor — v. 3. Insects / Neil Schlager, editor — v. 4-5. Fishes I-II / Neil Schlager, editor — vv. 6. Amphibians / Neil Schlager, editor — v. 7. Reptiles / Neil Schlager, editor — v. 8-11. Birds I-IV / Donna Olendorf, editor — v. 12-16. Mammals I-V / Melissa C. McDade, editor — v. 17. Cumulative index / Melissa C. McDade, editor.
ISBN 0-7876-5362-4 (set hardcover : alk. paper)
1. Zoology—Encyclopedias. I. Title: Animal life encyclopedia. II. Schlager, Neil, 1966- III. Olendorf, Donna IV. McDade, Melissa C. V. American Zoo and Aquarium Association. VI. Title.
QL7 .G7813 2004
590'.3—dc21
2002003351

Printed in Canada
10 9 8 7 6 5 4 3 2 1

Recommended citation: *Grzimek's Animal Life Encyclopedia,* 2nd edition. Volumes 12–16, *Mammals I–V,* edited by Michael Hutchins, Devra G. Kleiman, Valerius Geist, and Melissa C. McDade. Farmington Hills, MI: Gale Group, 2003.

Contents

Foreword ix
How to use this book xii
Advisory boards xiv
Contributing writers xvi
Contributing illustrators xx

Volume 12: Mammals I
What is a mammal? 3
Ice Age giants 17
Contributions of molecular genetics to phylogenetics 26
Structure and function 36
Adaptations for flight 52
Adaptations for aquatic life 62
Adaptations for subterranean life 69
Sensory systems, including echolocation 79
Life history and reproduction 89
Reproductive processes 101
Ecology 113
Nutritional adaptations 120
Distribution and biogeography 129
Behavior 140
Cognition and intelligence 149
Migration 164
Mammals and humans: Domestication and commensals 171
Mammals and humans: Mammalian invasives and pests 182
Mammals and humans: Field techniques for studying mammals 194
Mammals and humans: Mammals in zoos 203
Conservation 213

Order MONOTREMATA
Monotremes 227
  Family: Echidnas 235
  Family: Duck-billed platypus 243

Order DIDELPHIMORPHIA
New World opossums
  Family: New World opossums 249

Order PAUCITUBERCULATA
Shrew opossums
  Family: Shrew opossums 267

Order MICROBIOTHERIA
Monitos del monte
  Family: Monitos del monte 273

Order DASYUROMORPHIA
Australasian carnivorous marsupials 277
  Family: Marsupial mice and cats, Tasmanian devil 287
  Family: Numbat 303
  Family: Tasmanian wolves 307

For further reading 311
Organizations 316
Contributors to the first edition 318
Glossary 325
Mammals species list 330
Geologic time scale 364
Index 365

Volume 13: Mammals II
Order PERAMELEMORPHIA
Bandicoots and bilbies 1
  Family: Bandicoots 9
  Subfamily: Bilbies 19

Order NOTORYCTEMORPHIA
Marsupial moles
  Family: Marsupial moles 25

Order DIPROTODONTIA
Koala, wombats, possums, wallabies, and kangaroos 31
  Family: Koalas 43
  Family: Wombats 51
  Family: Possums and cuscuses 57
  Family: Musky rat-kangaroos 69
  Family: Rat-kangaroos 73
  Family: Wallabies and kangaroos 83
  Family: Pygmy possums 105
  Family: Ringtail and greater gliding possums 113
  Family: Gliding and striped possums 125

Family: Honey possums 135
Family: Feather-tailed possums 139

Order XENARTHRA
Sloths, anteaters, and armadillos 147
Family: West Indian sloths and two-toed tree sloths 155
Family: Three-toed tree sloths 161
Family: Anteaters 171
Family: Armadillos 181

Order INSECTIVORA
Insectivores 193
Family: Gymnures and hedgehogs 203
Family: Golden moles 215
Family: Tenrecs 225
Family: Solenodons 237
Family: Extinct West Indian shrews 243
Family: Shrews
I: Red-toothed shrews 247
II: White-toothed shrews 265
Family: Moles, shrew moles, and desmans 279

Order SCANDENTIA
Tree shrews
Family: Tree shrews 289

Order DERMOPTERA
Colugos
Family: Colugos 299

Order CHIROPTERA
Bats 307
Family: Old World fruit bats
I: *Pteropus* 319
II: All other genera 333
Family: Mouse-tailed bats 351
Family: Sac-winged bats, sheath-tailed bats, and ghost bats 355
Family: Kitti's hog-nosed bats 367
Family: Slit-faced bats 371
Family: False vampire bats 379
Family: Horseshoe bats 387
Family: Old World leaf-nosed bats 401
Family: American leaf-nosed bats 413
Family: Moustached bats 435
Family: Bulldog bats 443
Family: New Zealand short-tailed bats 453
Family: Funnel-eared bats 459
Family: Smoky bats 467
Family: Disk-winged bats 473
Family: Old World sucker-footed bats 479
Family: Free-tailed bats and mastiff bats 483
Family: Vespertilionid bats
I: Vespertilioninae 497
II: Other subfamilies 519

For further reading 527
Organizations 532
Contributors to the first edition 534
Glossary 541
Mammals species list 546
Geologic time scale 580
Index 581

Volume 14: Mammals III
Order PRIMATES
Primates 1
Family: Lorises and pottos 13
Family: Bushbabies 23
Family: Dwarf lemurs and mouse lemurs 35
Family: Lemurs 47
Family: Avahis, sifakas, and indris 63
Family: Sportive lemurs 73
Family: Aye-ayes 85
Family: Tarsiers 91
Family: New World monkeys
I: Squirrel monkeys and capuchins 101
II: Marmosets, tamarins, and Goeldi's monkey 115
Family: Night monkeys 135
Family: Sakis, titis, and uakaris 143
Family: Howler monkeys and spider monkeys 155
Family: Old World monkeys
I: Colobinae 171
II: Cercopithecinae 187
Family: Gibbons 207
Family: Great apes and humans
I: Great apes 225
II: Humans 241

Order CARNIVORA
Land and marine carnivores 255
Family: Dogs, wolves, coyotes, jackals, and foxes 265
Dogs and cats 287
Family: Bears 295
Family: Raccoons and relatives 309
Family: Weasels, badgers, skunks, and otters 319
Family: Civets, genets, and linsangs 335
Family: Mongooses and fossa 347
Family: Aardwolf and hyenas 359
Family: Cats 369
Family: Eared seals, fur seals, and sea lions 393
Family: Walruses 409
Family: True seals 417

For further reading 437
Organizations 442
Contributors to the first edition 444
Glossary 451
Mammals species list 456
Geologic time scale 490
Index 491

Volume 15: Mammals IV
Order CETACEA
Whales, dolphins, and porpoises 1
Family: Ganges and Indus dolphins 13

Family: Baijis 19
Family: Franciscana dolphins 23
Family: Botos 27
Family: Porpoises 33
Family: Dolphins 41
Family: Beaked whales 59
Family: Sperm whales 73
Family: Belugas and narwhals 81
Family: Gray whales 93
Family: Pygmy right whales 103
Family: Right whales and bowhead whales 107
Family: Rorquals 119

The ungulates 131
Ungulate domestication 145

Order TUBULIDENTATA
Aardvarks
Family: Aardvarks 155

Order PROBOSCIDEA
Elephants
Family: Elephants 161

Order HYRACOIDEA
Hyraxes
Family: Hyraxes 177

Order SIRENIA
Dugongs, sea cows, and manatees 191
Family: Dugongs and sea cows 199
Family: Manatees 205

Order PERISSODACTYLA
Odd-toed ungulates 215
Family: Horses, zebras, and asses 225
Family: Tapirs 237
Family: Rhinoceroses 249

Order ARTIODACTYLA
Even-toed ungulates 263
Family: Pigs 275
Family: Peccaries 291
Family: Hippopotamuses 301
Family: Camels, guanacos, llamas, alpacas, and vicuñas 313
Family: Chevrotains 325
Family: Deer
Subfamily: Musk deer 335
Subfamily: Muntjacs 343
Subfamily: Old World deer 357
Subfamily: Chinese water deer 373
Subfamily: New World deer 379
Family: Okapis and giraffes 399
Family: Pronghorn 411

For further reading 419
Organizations 424
Contributors to the first edition 426
Glossary 433
Mammals species list 438
Geologic time scale 472
Index 473

Volume 16: Mammals V
Family: Antelopes, cattle, bison, buffaloes, goats, and sheep 1
I: Kudus, buffaloes, and bison 11
II: Hartebeests, wildebeests, gemsboks, oryx, and reedbucks 27
III: Gazelles, springboks, and saiga antelopes 45
IV: Dikdiks, beiras, grysboks, and steenboks 59
V: Duikers 73
VI: Sheep, goats, and relatives 87

Order PHOLIDOTA
Pangolins
Family: Pangolins 107

Order RODENTIA
Rodents 121
Family: Mountain beavers 131
Family: Squirrels and relatives
I: Flying squirrels 135
II: Ground squirrels 143
III: Tree squirrels 163
Family: Beavers 177
Family: Pocket gophers 185
Family: Pocket mice, kangaroo rats, and kangaroo mice 199
Family: Birch mice, jumping mice, and jerboas 211
Family: Rats, mice, and relatives
I: Voles and lemmings 225
II: Hamsters 239
III: Old World rats and mice 249
IV: South American rats and mice 263
V: All others 281
Family: Scaly-tailed squirrels 299
Family: Springhares 307
Family: Gundis 311
Family: Dormice 317
Family: Dassie rats 329
Family: Cane rats 333
Family: African mole-rats 339
Family: Old World porcupines 351
Family: New World porcupines 365
Family: Viscachas and chinchillas 377
Family: Pacaranas 385
Family: Cavies and maras 389
Family: Capybaras 401
Family: Agoutis 407
Family: Pacas 417
Family: Tuco-tucos 425
Family: Octodonts 433
Family: Chinchilla rats 443
Family: Spiny rats 449
Family: Hutias 461
Family: Giant hutias 469
Family: Coypus 473

Order LAGOMORPHA
Pikas, rabbits, and hares 479
Family: Pikas 491
Family: Hares and rabbits 505

Order MACROSCELIDEA
Sengis
Family: Sengis 517

For further reading 533
Organizations 538
Contributors to the first edition 540
Glossary 547
Mammals species list 552
Geologic time scale 586
Index 587

• • • • •

Foreword

Earth is teeming with life. No one knows exactly how many distinct organisms inhabit our planet, but more than 5 million different species of animals and plants could exist, ranging from microscopic algae and bacteria to gigantic elephants, redwood trees and blue whales. Yet, throughout this wonderful tapestry of living creatures, there runs a single thread: Deoxyribonucleic acid or DNA. The existence of DNA, an elegant, twisted organic molecule that is the building block of all life, is perhaps the best evidence that all living organisms on this planet share a common ancestry. Our ancient connection to the living world may drive our curiosity, and perhaps also explain our seemingly insatiable desire for information about animals and nature. Noted zoologist, E. O. Wilson, recently coined the term "biophilia" to describe this phenomenon. The term is derived from the Greek *bios* meaning "life" and *philos* meaning "love." Wilson argues that we are human because of our innate affinity to and interest in the other organisms with which we share our planet. They are, as he says, "the matrix in which the human mind originated and is permanently rooted." To put it simply and metaphorically, our love for nature flows in our blood and is deeply engrained in both our psyche and cultural traditions.

Our own personal awakenings to the natural world are as diverse as humanity itself. I spent my early childhood in rural Iowa where nature was an integral part of my life. My father and I spent many hours collecting, identifying and studying local insects, amphibians and reptiles. These experiences had a significant impact on my early intellectual and even spiritual development. One event I can recall most vividly. I had collected a cocoon in a field near my home in early spring. The large, silky capsule was attached to a stick. I brought the cocoon back to my room and placed it in a jar on top of my dresser. I remember waking one morning and, there, perched on the tip of the stick was a large moth, slowly moving its delicate, light green wings in the early morning sunlight. It took my breath away. To my inexperienced eyes, it was one of the most beautiful things I had ever seen. I knew it was a moth, but did not know which species. Upon closer examination, I noticed two moon-like markings on the wings and also noted that the wings had long "tails", much like the ubiquitous tiger swallow-tail butterflies that visited the lilac bush in our backyard. Not wanting to suffer my ignorance any longer, I reached immediately for my *Golden Guide to North American Insects* and searched through the section on moths and butterflies. It was a luna moth! My heart was pounding with the excitement of new knowledge as I ran to share the discovery with my parents.

I consider myself very fortunate to have made a living as a professional biologist and conservationist for the past 20 years. I've traveled to over 30 countries and six continents to study and photograph wildlife or to attend related conferences and meetings. Yet, each time I encounter a new and unusual animal or habitat my heart still races with the same excitement of my youth. If this is biophilia, then I certainly possess it, and it is my hope that others will experience it too. I am therefore extremely proud to have served as the series editor for the Gale Group's rewrite of *Grzimek's Animal Life Encyclopedia*, one of the best known and widely used reference works on the animal world. *Grzimek's* is a celebration of animals, a snapshot of our current knowledge of the Earth's incredible range of biological diversity. Although many other animal encyclopedias exist, *Grzimek's Animal Life Encyclopedia* remains unparalleled in its size and in the breadth of topics and organisms it covers.

The revision of these volumes could not come at a more opportune time. In fact, there is a desperate need for a deeper understanding and appreciation of our natural world. Many species are classified as threatened or endangered, and the situation is expected to get much worse before it gets better. Species extinction has always been part of the evolutionary history of life; some organisms adapt to changing circumstances and some do not. However, the current rate of species loss is now estimated to be 1,000–10,000 times the normal "background" rate of extinction since life began on Earth some 4 billion years ago. The primary factor responsible for this decline in biological diversity is the exponential growth of human populations, combined with peoples' unsustainable appetite for natural resources, such as land, water, minerals, oil, and timber. The world's human population now exceeds 6 billion, and even though the average birth rate has begun to decline, most demographers believe that the global human population will reach 8–10 billion in the next 50 years. Much of this projected growth will occur in developing countries in Central and South America, Asia and Africa—regions that are rich in unique biological diversity.

Finding solutions to conservation challenges will not be easy in today's human-dominated world. A growing number of people live in urban settings and are becoming increasingly isolated from nature. They "hunt" in supermarkets and malls, live in apartments and houses, spend their time watching television and searching the World Wide Web. Children and adults must be taught to value biological diversity and the habitats that support it. Education is of prime importance now while we still have time to respond to the impending crisis. There still exist in many parts of the world large numbers of biological "hotspots"—places that are relatively unaffected by humans and which still contain a rich store of their original animal and plant life. These living repositories, along with selected populations of animals and plants held in professionally managed zoos, aquariums and botanical gardens, could provide the basis for restoring the planet's biological wealth and ecological health. This encyclopedia and the collective knowledge it represents can assist in educating people about animals and their ecological and cultural significance. Perhaps it will also assist others in making deeper connections to nature and spreading biophilia. Information on the conservation status, threats and efforts to preserve various species have been integrated into this revision. We have also included information on the cultural significance of animals, including their roles in art and religion.

It was over 30 years ago that Dr. Bernhard Grzimek, then director of the Frankfurt Zoo in Frankfurt, Germany, edited the first edition of *Grzimek's Animal Life Encyclopedia*. Dr. Grzimek was among the world's best known zoo directors and conservationists. He was a prolific author, publishing nine books. Among his contributions were: *Serengeti Shall Not Die*, *Rhinos Belong to Everybody* and *He and I and the Elephants*. Dr. Grzimek's career was remarkable. He was one of the first modern zoo or aquarium directors to understand the importance of zoo involvement in *in situ* conservation, that is, of their role in preserving wildlife in nature. During his tenure, Frankfurt Zoo became one of the leading western advocates and supporters of wildlife conservation in East Africa. Dr. Grzimek served as a Trustee of the National Parks Board of Uganda and Tanzania and assisted in the development of several protected areas. The film he made with his son Michael, *Serengeti Shall Not Die*, won the 1959 Oscar for best documentary.

Professor Grzimek has recently been criticized by some for his failure to consider the human element in wildlife conservation. He once wrote: "A national park must remain a primordial wilderness to be effective. No men, not even native ones, should live inside its borders." Such ideas, although considered politically incorrect by many, may in retrospect actually prove to be true. Human populations throughout Africa continue to grow exponentially, forcing wildlife into small islands of natural habitat surrounded by a sea of humanity. The illegal commercial bushmeat trade—the hunting of endangered wild animals for large scale human consumption—is pushing many species, including our closest relatives, the gorillas, bonobos and chimpanzees, to the brink of extinction. The trade is driven by widespread poverty and lack of economic alternatives. In order for some species to survive it will be necessary, as Grzimek suggested, to establish and enforce a system of protected areas where wildlife can roam free from exploitation of any kind.

While it is clear that modern conservation must take the needs of both wildlife and people into consideration, what will the quality of human life be if the collective impact of short-term economic decisions is allowed to drive wildlife populations into irreversible extinction? Many rural populations living in areas of high biodiversity are dependent on wild animals as their major source of protein. In addition, wildlife tourism is the primary source of foreign currency in many developing countries and is critical to their financial and social stability. When this source of protein and income is gone, what will become of the local people? The loss of species is not only a conservation disaster; it also has the potential to be a human tragedy of immense proportions. Protected areas, such as national parks, and regulated hunting in areas outside of parks are the only solutions. What critics do not realize is that the fate of wildlife and people in developing countries is closely intertwined. Forests and savannas emptied of wildlife will result in hungry, desperate people, and will, in the long-term lead to extreme poverty and social instability. Dr. Grzimek's early contributions to conservation should be recognized, not only as benefiting wildlife, but as benefiting local people as well.

Dr. Grzimek's hope in publishing his *Animal Life Encyclopedia* was that it would "...disseminate knowledge of the animals and love for them", so that future generations would "...have an opportunity to live together with the great diversity of these magnificent creatures." As stated above, our goals in producing this updated and revised edition are similar. However, our challenges in producing this encyclopedia were more formidable. The volume of knowledge to be summarized is certainly much greater in the twenty-first century than it was in the 1970's and 80's. Scientists, both professional and amateur, have learned and published a great deal about the animal kingdom in the past three decades, and our understanding of biological and ecological theory has also progressed. Perhaps our greatest hurdle in producing this revision was to include the new information, while at the same time retaining some of the characteristics that have made *Grzimek's Animal Life Encyclopedia* so popular. We have therefore strived to retain the series' narrative style, while giving the information more organizational structure. Unlike the original *Grzimek's*, this updated version organizes information under specific topic areas, such as reproduction, behavior, ecology and so forth. In addition, the basic organizational structure is generally consistent from one volume to the next, regardless of the animal groups covered. This should make it easier for users to locate information more quickly and efficiently. Like the original Grzimek's, we have done our best to avoid any overly technical language that would make the work difficult to understand by non-biologists. When certain technical expressions were necessary, we have included explanations or clarifications.

Considering the vast array of knowledge that such a work represents, it would be impossible for any one zoologist to have completed these volumes. We have therefore sought specialists from various disciplines to write the sections with which they are most familiar. As with the original *Grzimek's*, we have engaged the best scholars available to serve as topic editors, writers, and consultants. There were some complaints about inaccuracies in the original English version that may have been due to mistakes or misinterpretation during the complicated translation process. However, unlike the original *Grzimek's*, which was translated from German, this revision has been completely re-written by English-speaking scientists. This work was truly a cooperative endeavor, and I thank all of those dedicated individuals who have written, edited, consulted, drawn, photographed, or contributed to its production in any way. The names of the topic editors, authors, and illustrators are presented in the list of contributors in each individual volume.

The overall structure of this reference work is based on the classification of animals into naturally related groups, a discipline known as taxonomy or biosystematics. Taxonomy is the science through which various organisms are discovered, identified, described, named, classified and catalogued. It should be noted that in preparing this volume we adopted what might be termed a conservative approach, relying primarily on traditional animal classification schemes. Taxonomy has always been a volatile field, with frequent arguments over the naming of or evolutionary relationships between various organisms. The advent of DNA fingerprinting and other advanced biochemical techniques has revolutionized the field and, not unexpectedly, has produced both advances and confusion. In producing these volumes, we have consulted with specialists to obtain the most up-to-date information possible, but knowing that new findings may result in changes at any time. When scientific controversy over the classification of a particular animal or group of animals existed, we did our best to point this out in the text.

Readers should note that it was impossible to include as much detail on some animal groups as was provided on others. For example, the marine and freshwater fish, with vast numbers of orders, families, and species, did not receive as detailed a treatment as did the birds and mammals. Due to practical and financial considerations, the publishers could provide only so much space for each animal group. In such cases, it was impossible to provide more than a broad overview and to feature a few selected examples for the purposes of illustration. To help compensate, we have provided a few key bibliographic references in each section to aid those interested in learning more. This is a common limitation in all reference works, but *Grzimek's Encyclopedia of Animal Life* is still the most comprehensive work of its kind.

I am indebted to the Gale Group, Inc. and Senior Editor Donna Olendorf for selecting me as Series Editor for this project. It was an honor to follow in the footsteps of Dr. Grzimek and to play a key role in the revision that still bears his name. *Grzimek's Animal Life Encyclopedia* is being published by the Gale Group, Inc. in affiliation with my employer, the American Zoo and Aquarium Association (AZA), and I would like to thank AZA Executive Director, Sydney J. Butler; AZA Past-President Ted Beattie (John G. Shedd Aquarium, Chicago, IL); and current AZA President, John Lewis (John Ball Zoological Garden, Grand Rapids, MI), for approving my participation. I would also like to thank AZA Conservation and Science Department Program Assistant, Michael Souza, for his assistance during the project. The AZA is a professional membership association, representing 215 accredited zoological parks and aquariums in North America. As Director/William Conway Chair, AZA Department of Conservation and Science, I feel that I am a philosophical descendant of Dr. Grzimek, whose many works I have collected and read. The zoo and aquarium profession has come a long way since the 1970s, due, in part, to innovative thinkers such as Dr. Grzimek. I hope this latest revision of his work will continue his extraordinary legacy.

Silver Spring, Maryland, 2001
Michael Hutchins
Series Editor

How to use this book

Grzimek's Animal Life Encyclopedia is an internationally prominent scientific reference compilation, first published in German in the late 1960s, under the editorship of zoologist Bernhard Grzimek (1909-1987). In a cooperative effort between Gale and the American Zoo and Aquarium Association, the series is being completely revised and updated for the first time in over 30 years. Gale is expanding the series from 13 to 17 volumes, commissioning new color images, and updating the information while also making the set easier to use. The order of revisions is:

Vol 8–11: Birds I–IV
Vol 6: Amphibians
Vol 7: Reptiles
Vol 4–5: Fishes I–II
Vol 12–16: Mammals I–V
Vol 1: Lower Metazoans and Lesser Deuterostomes
Vol 2: Protostomes
Vol 3: Insects
Vol 17: Cumulative Index

Organized by taxonomy

The overall structure of this reference work is based on the classification of animals into naturally related groups, a discipline known as taxonomy—the science through which various organisms are discovered, identified, described, named, classified, and catalogued. Starting with the simplest life forms, the lower metazoans and lesser deuterostomes, in volume 1, the series progresses through the more complex animal classes, culminating with the mammals in volumes 12–16. Volume 17 is a stand-alone cumulative index.

Organization of chapters within each volume reinforces the taxonomic hierarchy. In the case of the Mammals volumes, introductory chapters describe general characteristics of all organisms in these groups, followed by taxonomic chapters dedicated to Order, Family, or Subfamily. Species accounts appear at the end of the Family and Subfamily chapters To help the reader grasp the scientific arrangement, each type of chapter has a distinctive color and symbol:

● =Order Chapter (blue background)

◓ =Monotypic Order Chapter (green background)

▲ =Family Chapter (yellow background)

△ =Subfamily Chapter (yellow background)

Introductory chapters have a loose structure, reminiscent of the first edition. While not strictly formatted, Order chapters are carefully structured to cover basic information about member families. Monotypic orders, comprised of a single family, utilize family chapter organization. Family and subfamily chapters are most tightly structured, following a prescribed format of standard rubrics that make information easy to find and understand. Family chapters typically include:

Thumbnail introduction
- Common name
- Scientific name
- Class
- Order
- Suborder
- Family
- Thumbnail description
- Size
- Number of genera, species
- Habitat
- Conservation status

Main essay
- Evolution and systematics
- Physical characteristics
- Distribution
- Habitat
- Behavior
- Feeding ecology and diet
- Reproductive biology
- Conservation status
- Significance to humans

Species accounts
- Common name
- Scientific name
- Subfamily
- Taxonomy
- Other common names
- Physical characteristics
- Distribution
- Habitat
- Behavior

Feeding ecology and diet
Reproductive biology
Conservation status
Significance to humans
Resources
Books
Periodicals
Organizations
Other

Color graphics enhance understanding

Grzimek's features approximately 3,000 color photos, including approximately 1,560 in five Mammals volumes; 3,500 total color maps, including nearly 550 in the Mammals volumes; and approximately 5,500 total color illustrations, including approximately 930 in the Mammals volumes. Each featured species of animal is accompanied by both a distribution map and an illustration.

All maps in *Grzimek's* were created specifically for the project by XNR Productions. Distribution information was provided by expert contributors and, if necessary, further researched at the University of Michigan Zoological Museum library. Maps are intended to show broad distribution, not definitive ranges.

All the color illustrations in *Grzimek's* were created specifically for the project by Michigan Science Art. Expert contributors recommended the species to be illustrated and provided feedback to the artists, who supplemented this information with authoritative references and animal skins from University of Michgan Zoological Museum library. In addition to species illustrations, *Grzimek's* features conceptual drawings that illustrate characteristic traits and behaviors.

About the contributors

The essays were written by scientists, professors, and other professionals. *Grzimek's* subject advisors reviewed the completed essays to insure consistency and accuracy.

Standards employed

In preparing these volumes, the editors adopted a conservative approach to taxonomy, relying on Wilson and Reeder's *Mammal Species of the World: a Taxonomic and Geographic Reference* (1993) as a guide. Systematics is a dynamic discipline in that new species are being discovered continuously, and new techniques (e.g., DNA sequencing) frequently result in changes in the hypothesized evolutionary relationships among various organisms. Consequently, controversy often exists regarding classification of a particular animal or group of animals; such differences are mentioned in the text.

Grzimek's has been designed with ready reference in mind and the editors have standardized information wherever feasible. For **Conservation status,** *Grzimek's* follows the IUCN Red List system, developed by its Species Survival Commission. The Red List provides the world's most comprehensive inventory of the global conservation status of plants and animals. Using a set of criteria to evaluate extinction risk, the IUCN recognizes the following categories: Extinct, Extinct in the Wild, Critically Endangered, Endangered, Vulnerable, Conservation Dependent, Near Threatened, Least Concern, and Data Deficient. For a complete explanation of each category, visit the IUCN web page at <http://www.iucn.org/>.

Advisory boards

Contributing writers

Mammals I–V

Clarence L. Abercrombie, PhD
Wofford College
Spartanburg, South Carolina

Cleber J. R. Alho, PhD
Departamento de Ecologia (retired)
Universidade de Brasília
Brasília, Brazil

Carlos Altuna, Lic
Sección Etología
Facultad de Ciencias
Universidad de la República Oriental del Uruguay
Montevideo, Uruguay

Anders Angerbjörn, PhD
Department of Zoology
Stockholm University
Stockholm, Sweden

William Arthur Atkins
Atkins Research and Consulting
Normal, Illinois

Adrian A. Barnett, PhD
Centre for Research in Evolutionary Anthropology
School of Life Sciences
University of Surrey Roehampton
West Will, London
United Kingdom

Leonid Baskin, PhD
Institute of Ecology and Evolution
Moscow, Russia

Paul J. J. Bates, PhD
Harrison Institute
Sevenoaks, Kent
United Kingdom

Amy-Jane Beer, PhD
Origin Natural Science
York, United Kingdom

Cynthia Berger, MS
National Association of Science Writers

Richard E. Bodmer, PhD
Durrell Institute of Conservation and Ecology
University of Kent
Canterbury, Kent
United Kingdom

Daryl J. Boness, PhD
National Zoological Park
Smithsonian Institution
Washington, DC

Justin S. Brashares, PhD
Centre for Biodiversity Research
University of British Columbia
Vancouver, British Columbia
Canada

Hynek Burda, PhD
Department of General Zoology Faculty of Bio- and Geosciences
University of Essen
Essen, Germany

Susan Cachel, PhD
Department of Anthropology
Rutgers University
New Brunswick, New Jersey

Alena Cervená, PhD
Department of Zoology
National Museum Prague
Czech Republic

Jaroslav Cerveny, PhD
Institute of Vertebrate Biology
Czech Academy of Sciences
Brno, Czech Republic

David J. Chivers, MA, PhD, ScD
Head, Wildlife Research Group
Department of Anatomy
University of Cambridge
Cambridge, United Kingdom

Jasmin Chua, MS
Freelance Writer

Lee Curtis, MA
Director of Promotions
Far North Queensland Wildlife Rescue Association
Far North Queensland, Australia

Guillermo D'Elía, PhD
Departamento de Biología Animal
Facultad de Ciencias
Universidad de la República
Montevideo, Uruguay

Tanya Dewey
University of Michigan Museum of Zoology
Ann Arbor, Michigan

Craig C. Downer, PhD
Andean Tapir Fund
Minden, Nevada

Amy E. Dunham
Department of Ecology and Evolution
State University of New York at Stony Brook
Stony Brook, New York

Stewart K. Eltringham, PhD
Department of Zoology
University of Cambridge
Cambridge, United Kingdom.

Melville Brockett Fenton, PhD
Department of Biology
University of Western Ontario
London, Ontario
Canada

Kevin F. Fitzgerald, BS
Freelance Science Writer
South Windsor, Connecticut

Theodore H. Fleming, PhD
Department of Biology
University of Miami
Coral Gables, Florida

Gabriel Francescoli, PhD
Sección Etología
Facultad de Ciencias
Universidad de la República Oriental del Uruguay
Montevideo, Uruguay

Udo Gansloßer, PhD
Department of Zoology
Lehrstuhl I
University of Erlangen-Nürnberg
Fürth, Germany

Valerius Geist, PhD
Professor Emeritus of Environmental Science
University of Calgary
Calgary, Alberta
Canada

Roger Gentry, PhD
NOAA Fisheries
Marine Mammal Division
Silver Spring, Maryland

Kenneth C. Gold, PhD
Chicago, Illinois

Steve Goodman, PhD
Field Museum of Natural History
Chicago, Illinois and
WWF Madagascar
Programme Office
Antananarivo, Madagascar

Nicole L. Gottdenker
St. Louis Zoo
University of Missouri
St. Louis, Missouri and The Charles Darwin Research Station
Galápagos Islands, Ecuador

Brian W. Grafton, PhD
Department of Biological Sciences
Kent State University
Kent, Ohio

Joel H. Grossman
Freelance Writer
Santa Monica, California

Mark S. Hafner, PhD
Lowery Professor and Curator of Mammals
Museum of Natural Science and Department of Biological Sciences
Louisiana State University
Baton Rouge, Louisiana

Alton S. Harestad, PhD
Faculty of Science
Simon Fraser University Burnaby
Vancouver, British Columbia
Canada

Robin L. Hayes
Bat Conservation of Michigan

Kristofer M. Helgen
School of Earth and Environmental Sciences
University of Adelaide
Adelaide, Australia

Eckhard W. Heymann, PhD
Department of Ethology and Ecology
German Primate Center
Göttingen, Germany

Hannah Hoag, MS
Science Journalist

Hendrik Hoeck, PhD
Max-Planck- Institut für Verhaltensphysiologie
Seewiesen, Germany

David Holzman, BA
Freelance Writer
Journal Highlights Editor
American Society for Microbiology

Rodney L. Honeycutt, PhD
Departments of Wildlife and Fisheries Sciences and Biology and Faculty of Genetics
Texas A&M University
College Station, Texas

Ivan Horácek, Prof. RNDr, PhD
Head of Vertebrate Zoology
Charles University Prague
Praha, Czech Republic

Brian Douglas Hoyle, PhD
President, Square Rainbow Limited
Bedford, Nova Scotia
Canada

Graciela Izquierdo, PhD
Sección Etología
Facultad de Ciencias
Universidad de la República Oriental del Uruguay
Montevideo, Uruguay

Jennifer U. M. Jarvis, PhD
Zoology Department
University of Cape Town
Rondebosch, South Africa

Christopher Johnson, PhD
Department of Zoology and Tropical Ecology
James Cook University
Townsville, Queensland
Australia

Menna Jones, PhD
University of Tasmania School of Zoology
Hobart, Tasmania
Australia

Mike J. R. Jordan, PhD
Curator of Higher Vertebrates
North of England Zoological Society
Chester Zoo
Upton, Chester
United Kingdom

Corliss Karasov
Science Writer
Madison, Wisconsin

Tim Karels, PhD
Department of Biological Sciences
Auburn University
Auburn, Alabama

Serge Larivière, PhD
Delta Waterfowl Foundation
Manitoba, Canada

Adrian Lister
University College London
London, United Kingdom

W. J. Loughry, PhD
Department of Biology
Valdosta State University
Valdosta, Georgia

Geoff Lundie-Jenkins, PhD
Queensland Parks and Wildlife Service
Queensland, Australia

Peter W. W. Lurz, PhD
Centre for Life Sciences Modelling
School of Biology
University of Newcastle
Newcastle upon Tyne, United Kingdom

Colin D. MacLeod, PhD
School of Biological Sciences (Zoology)
University of Aberdeen
Aberdeen, United Kingdom

James Malcolm, PhD
Department of Biology
University of Redlands
Redlands, California

David P. Mallon, PhD
Glossop
Derbyshire, United Kingdom

Robert D. Martin, BA (Hons), DPhil, DSc
Provost and Vice President
Academic Affairs
The Field Museum
Chicago, Illinois

Gary F. McCracken, PhD
Department of Ecology and Evolutionary Biology
University of Tennessee
Knoxville, Tennessee

Colleen M. McDonough, PhD
Department of Biology
Valdosta State University
Valdosta, Georgia

William J. McShea, PhD
Department of Conservation Biology
Conservation and Research Center
Smithsonian National Zoological Park
Washington, DC

Rodrigo A. Medellín, PhD
Instituto de Ecología
Universidad Nacional Autónoma de México
Mexico City, Mexico

Leslie Ann Mertz, PhD
Fish Lake Biological Program
Wayne State University
Detroit, Michigan

Gus Mills, PhD
SAN Parks/Head
Carnivore Conservation Group, EWT
Skukuza, South Africa

Patricia D. Moehlman, PhD
IUCN Equid Specialist Group

Paula Moreno, MS
Texas A&M University at Galveston
Marine Mammal Research Program
Galveston, Texas

Virginia L. Naples, PhD
Department of Biological Sciences
Northern Illinois University
DeKalb, Illinois

Ken B. Naugher, BS
Conservation and Enrichment Programs Manager
Montgomery Zoo
Montgomery, Alabama

Derek William Niemann, BA
Royal Society for the Protection of Birds
Sandy, Bedfordshire
United Kingdom

Carsten Niemitz, PhD
Professor of Human Biology
Department of Human Biology and Anthropology
Freie Universität Berlin
Berlin, Germany

Daniel K. Odell, PhD
Senior Research Biologist
Hubbs-SeaWorld Research Institute
Orlando, Florida

Bart O'Gara, PhD
University of Montana (adjunct retired professor)
Director, Conservation Force

Norman Owen-Smith, PhD
Research Professor in African Ecology
School of Animal, Plant and Environmental Sciences
University of the Witwatersrand
Johannesburg, South Africa

Malcolm Pearch, PhD
Harrison Institute
Sevenoaks, Kent
United Kingdom

Kimberley A. Phillips, PhD
Hiram College
Hiram, Ohio

David M. Powell, PhD
Research Associate
Department of Conservation Biology
Conservation and Research Center
Smithsonian National Zoological Park
Washington, DC

Jan A. Randall, PhD
Department of Biology
San Francisco State University
San Francisco, California

Randall Reeves, PhD
Okapi Wildlife Associates
Hudson, Quebec
Canada

Peggy Rismiller, PhD
Visiting Research Fellow
Department of Anatomical Sciences
University of Adelaide
Adelaide, Australia

Konstantin A. Rogovin, PhD
A.N. Severtsov Institute of Ecology and Evolution RAS
Moscow, Russia

Randolph W. Rose, PhD
School of Zoology
University of Tasmania
Hobart, Tasmania
Australia

Frank Rosell
Telemark University College
Telemark, Norway

Gretel H. Schueller
Science and Environmental Writer
Burlington, Vermont

Bruce A. Schulte, PhD
Department of Biology
Georgia Southern University
Statesboro, Georgia

John H. Seebeck, BSc, MSc, FAMS
Australia

Melody Serena, PhD
Conservation Biologist
Australian Platypus Conservancy
Whittlesea, Australia

David M. Shackleton, PhD
Faculty of Agricultural of Sciences
University of British Columbia
Vancouver, British Columbia
Canada

Robert W. Shumaker, PhD
Iowa Primate Learning Sanctuary
Des Moines, Iowa and Krasnow Institute at George Mason University
Fairfax, Virginia

Andrew T. Smith, PhD
School of Life Sciences
Arizona State University
Phoenix, Arizona

Karen B. Strier, PhD
Department of Anthropology
University of Wisconsin
Madison, Wisconsin

Karyl B. Swartz, PhD
Department of Psychology
Lehman College of The City University of New York
Bronx, New York

Bettina Tassino, MSc
Sección Etología

Facultad de Ciencias
Universidad de la República Oriental del Uruguay
Montevideo, Uruguay

Barry Taylor, PhD
University of Natal
Pietermaritzburg, South Africa

Jeanette Thomas, PhD
Department of Biological Sciences
Western Illinois University-Quad Cities
Moline, Illinois

Ann Toon
Arnside, Cumbria
United Kingdom

Stephen B. Toon
Arnside, Cumbria
United Kingdom

Hernán Torres, PhD
Santiago, Chile

Rudi van Aarde, BSc (Hons), MSc, PhD
Director and Chair of Conservation Ecology Research Unit
University of Pretoria
Pretoria, South Africa

Mac van der Merwe, PhD
Mammal Research Institute
University of Pretoria
Pretoria, South Africa

Christian C. Voigt, PhD
Research Group Evolutionary Ecology
Leibniz-Institute for Zoo and Wildlife Research
Berlin, Germany

Sue Wallace
Freelance Writer
Santa Rosa, California

Lindy Weilgart, PhD
Department of Biology
Dalhousie University
Halifax, Nova Scotia
Canada

Randall S. Wells, PhD
Chicago Zoological Society
Mote Marine Laboratory
Sarasota, Florida

Nathan S. Welton
Freelance Science Writer
Santa Barbara, California

Patricia Wright, PhD
State University of New York at Stony Brook
Stony Brook, New York

Marcus Young Owl, PhD
Department of Anthropology and Department of Biological Sciences
California State University
Long Beach, California

Jan Zima, PhD
Institute of Vertebrate Biology
Academy of Sciences of the Czech Republic
Brno, Czech Republic

Contributing illustrators

Drawings by Michigan Science Art

Joseph E. Trumpey, Director, AB, MFA
Science Illustration, School of Art and Design, University of Michigan

Wendy Baker, ADN, BFA

Ryan Burkhalter, BFA, MFA

Brian Cressman, BFA, MFA

Emily S. Damstra, BFA, MFA

Maggie Dongvillo, BFA

Barbara Duperron, BFA, MFA

Jarrod Erdody, BA, MFA

Dan Erickson, BA, MS

Patricia Ferrer, AB, BFA, MFA

George Starr Hammond, BA, MS, PhD

Gillian Harris, BA

Jonathan Higgins, BFA, MFA

Amanda Humphrey, BFA

Emilia Kwiatkowski, BS, BFA

Jacqueline Mahannah, BFA, MFA

John Megahan, BA, BS, MS

Michelle L. Meneghini, BFA, MFA

Katie Nealis, BFA

Laura E. Pabst, BFA

Amanda Smith, BFA, MFA

Christina St.Clair, BFA

Bruce D. Worden, BFA

Kristen Workman, BFA, MFA

Thanks are due to the University of Michigan, Museum of Zoology, which provided specimens that served as models for the images.

Maps by XNR Productions

Paul Exner, Chief cartographer
XNR Productions, Madison, WI

Tanya Buckingham

Jon Daugherity

Laura Exner

Andy Grosvold

Cory Johnson

Paula Robbins

Cetacea

(Whales, dolphins, and porpoises)

Class Mammalia

Order Cetacea

Number of families 14

Number of genera, species
40 genera; 86 species

Photo: A spinner dolphin (*Stenella longirostris*) leaping in Hawaiian waters. (Photo by Animals Animals ©James Watt. Reproduced by permission.)

Introduction

Linnaeus originally assigned the name Cete to the order of mammals consisting of whales, dolphins, and porpoises. The term is derived from the classical noun *cetos*, meaning a large sea creature. Linnaeus conceived Cete to be the sole member of the group Mutica, one of his three primary subdivisions of placental mammals. The term Cetacea is the plural of *cetos* and was coined by Brisson in 1762. The study of cetaceans has come to be known as cetology, those who practice it as cetologists.

The lines of demarcation between the living cetaceans and other orders of mammals are firmly drawn, and there is no ambiguity. Similarly, the two living suborders of Cetacea are unequivocally distinct from each other, but also monophyletic; that is, derived from a common ancestor. The Mysticeti, or baleen whales, and Odontoceti, or toothed whales, differ fundamentally in the ways that the bones of their skulls have become "telescoped." The mysticete skull features a large, bony, broad, and flat upper jaw, which thrusts back under the eye region. In contrast, the main bones of the odontocete upper jaw thrust back and upward over the eye sockets, extending across the front of the braincase. Mysticetes have baleen and no teeth as adults, and they have paired blowholes (nostrils). Odontocetes, in contrast, have teeth and no baleen (in some species, many or most of the teeth are unerupted and non-functional, however), and a single blowhole. A major additional factor in the anatomical divergence of the two groups is the development in odontocetes of a sophisticated echolocation system, which has required various unique anatomical specializations for producing, receiving, and processing sound. Mysticetes generally lack the enlarged facial muscles and nasal sacs that characterize odontocetes.

Below the level of suborder, many different approaches to classification have been proposed, involving varying numbers and combinations of infraorders, superfamilies, families, and subfamilies. For simplicity here and in what follows, only families, genera, and species are considered. The present-day consensus among cetologists is that there are four extant families, six genera, and at least 14 species of mysticetes, and ten families, 34 genera, and about 72 species of odontocetes. These numbers will inevitably change as larger samples become available and as more sophisticated analytical methods are applied. It is instructive that no less than five "new" species of cetaceans have been described over the past 15 years, including two mysticetes (Antarctic minke whale, *Balaenoptera bonaerensis*, and pygmy Bryde's whale, *Balaenoptera edeni*) and three odontocetes (pygmy beaked whale, *Mesoplodon peruvianus*, spade-toothed whale, *Mesoplodon traversii*, and Perrin's beaked whale, *Mesoplodon perrini*). Some of these represent the formal recognition and description of species long known to exist, but others are genuine discoveries. More of both types of developments are to be expected.

Vernacular uses of the terms whale, dolphin, and porpoise have always been complicated and, occasionally, confusing. All baleen-bearing cetaceans are considered whales, but any of the three terms can be applied to toothed cetaceans, depending upon a number of factors. Body size is a useful, but not definitive, basis for distinguishing whales from dolphins and porpoises. In general, cetaceans with adult lengths greater than about 9 ft (2.8 m) are called whales, but some "whales" (e.g., dwarf sperm and melon-headed; *Kogia sima* and *Peponocephala electra*, respectively) do not grow that large and some dolphins (e.g., Risso's and common bottlenosed; *Grampus griseus* and *Tursiops truncatus*, respectively) can grow larger. There is considerable overlap in body size between dolphins and porpoises as well. Strictly speaking, the term porpoise should be reserved for members of the family Phocoenidae, all of which are relatively small (maximum length less than 8

The Atlantic spotted dolphin (*Stenella frontalis*) is very active at the water's surface. (Photo by François Gohier/Photo Researchers, Inc. Reproduced by permission.)

ft [2.5 m]) and have numerous small, spatulate (spade-shaped) teeth. The proclivity of seafarers and fishers to apply the term "porpoise" (singular and plural) to any small cetacean that they encounter has led to its rather loose application to marine dolphins by scientists as well. It is occasionally suggested that porpoises can be distinguished from dolphins by their lack of a pronounced beak (the elongated anterior portion of the skull that includes both the upper and lower jaw), but a number of dolphins are at least as blunt-headed as any porpoise. In fact, there is no strict definition of "dolphin," as the term is equally valid for species as diverse as the very long-beaked, bizarre-looking river dolphins (superfamily Platanistoidea), the round-headed "blackfish" (pilot, false killer, and pygmy killer whales; *Globicephala* spp., *Pseudorca crassidens*, and *Feresa attenuata*, respectively), and the archetypal bottlenosed and common dolphins (*Tursiops* spp. and *Delphinus* spp., respectively). One other variant that often finds its way into the popular lexicon is "great whales." In most contexts, those who use this term mean it to refer to all of the baleen whales plus the sperm whale (*Physeter macrocephalus*). In essence, the great whales are those that had great commercial value and therefore were seriously depleted by the whaling industry.

Evolution and systematics

Cetaceans are related to the hoofed mammals, or ungulates, and their ancestry is linked more or less closely to that of cows, horses, and hippopotamuses. Current thinking is that they are highly derived artiodactyls, with a particularly close evolutionary relationship to the hippos. The fossil record of cetacean ancestry dates back more than 50 million years to the early Eocene epoch. Most paleontologists agree that cetaceans arose from the Mesonychidae, an extinct family of primitive terrestrial mammals that inhabited North America, Europe, and Asia. Mesonychids can generally be described as cursorial (adapted for running) carrion feeders with large heads, powerful jaws, and five-toed feet with hoof-like claws. The transition from a wholly terrestrial to an amphibious existence is believed to have taken place initially in the Tethys Sea, a large, shallow, near-tropical seaway that extended from the present-day Mediterranean eastward to beyond the South Asian subcontinent. Most of the fossil evidence for this initial radiation of the stem or basal Cetacea, the extinct suborder Archaeoceti, has come from Eocene Tethys sediments in India, Pakistan, and Egypt, although some archaeocete material has also been found in Nigeria and Alabama (United States). The archaeocetes diversified between 45 and 53 million years ago (mya), and the group had spread into mid-temperate waters by 40 mya, toward the end of the middle Eocene. More than 35 different species have been identified for the interval 35–53 mya, during which time archaic cetaceans had expanded from riverine and near-shore habitats and become adapted to occupy oceanic settings as well. Their eyes and kidneys had probably become capable of tolerating different salt balances, they may have lost much of their hair and begun to acquire blubber for insulation and fat storage, their underwater hearing capability had become enhanced, and they had probably developed nasal plugs to close the nostrils when diving. Presumably, they had also begun to move their tails in an up-and-down, rather than side-to-side, fashion for more efficient swimming.

Archaeocetes exhibited many features typical of living cetaceans, including an elongate upper jaw with bony nostrils set back from the tip, a broad shelf of bone above the eye, anteroposteriorly aligned incisors, and an enlarged mandibular canal on the inner side of the lower jaw. They had a dense outer ear bone, or tympanic bulla, and later forms had an expanded basicranial air sinus similar to that of modern cetaceans. A major difference between archaeocetes and the more derived cetaceans is that the archaeocete skull was not telescoped; that is, it did not have overlapping bony elements. Most, and possibly all, archaeocetes had external hind limbs. In some instances at least, they probably used all four limbs for locomotion both in water and on land. Although they are often depicted as having sinuous, almost eel-like bodies, the basic skeletal structures of most archaeocetes would have supported bodies not much different in overall design to those of living cetaceans.

Five families of Archaeoceti are recognized: Pakicetidae, the amphibious earliest cetaceans; Ambulocetidae, the walking whales; Remingtonocetidae, the gavial-convergent cetaceans (the gavial is a long-snouted, freshwater, fish-eating crocodilian of the south Asian subcontinent); Protocetidae, the first pelagic cetaceans; and Basilosauridae, the so-called zeuglodonts, referring to their complex, many-cusped teeth (the Greek *zugotos* means yoked or joined, and *odous*, of course, tooth). The most primitive archaeocete identified to date was *Nalacetus*, known mainly from isolated teeth. *Pakicetus*, another small, very early archaeocete, had eyes on top of its head, drank only fresh water (confirmed from oxygen isotope ratios in its tooth enamel), and was predominantly wolf- or hyena-like in appearance. The other families of archaeocetes had been largely supplanted by the zeuglodonts during the late Eocene.

Probably the best-known zeuglodont was *Basilosaurus*, or the "king lizard" (from the Greek *basileus* for king and *sauros* for lizard). This animal could be almost 70 ft (21 m) long and weighed at least 11,000 lb (5,000 kg). Its small head in relation to the long body made it appear truly serpentine. The front appendages had been modified into short, broad paddles, but were still hinged at the elbow; and the rear appendages had atrophied to nothing more than stumps. Basilosaurids may have had dorsal fins and horizontal tail flukes, and they were likely hairless, or nearly so. In short, *Basilosaurus* was well along the path to becoming what cetologists now think of as a whale.

The archaeocetes are replaced in the fossil record by odontocetes and mysticetes beginning in the Oligocene, about 38 mya. By approximately the middle of that epoch, the archaeocetes appear to have died out completely. The oldest known cetacean in the mysticete clade is *Llanocetus denticrenatus*, found in late Eocene rocks on the Antarctic Peninsula. This species' most characteristic feature was its series of lobed, widely spaced teeth, which were somewhat reminiscent of the teeth of the crabeater seal (*Lobodon carcinophagus*). Like the crabeater seal, *L. denticrenatus* was probably a filter feeder on krill-like invertebrates or possibly small schooling fish. At least four families of tooth-bearing mysticetes have been described from the Oligocene (24–38 mya). The transition leading to rudimentary baleen plates in the spaces between teeth probably occurred about 30 mya with the emergence of the Cetotheriidae, or primitive baleen-bearing mysticetes. It is a slight misconception to say that the presence of teeth is a diagnostic feature of Odontoceti, the so-called toothed whales, because all archaeocetes and some of the primitive fossil mysticetes also had teeth. Further, all of the modern baleen-bearing mysticetes have teeth in the early fetal stages of their development.

Odontocetes also radiated rapidly and widely during the Oligocene, by the end of which there were more than 13 families and 50 species of cetaceans in the world's oceans. This diversity was probably driven by changes in foraging opportunities related to breakup of the southern supercontinent of Gondwana, opening of the Southern Ocean, and the consequent polar cooling and sharpening of latitudinal temperature gradients. Several of the early odontocete lineages failed to survive beyond the Miocene (5–23 mya). The shark-toothed dolphins (Squalodontidae), with their sharp, triangular, serrated teeth, were likely active carnivores, while the very long-beaked Eurhinodelphinidae, with their overhanging upper jaws and many small, conical teeth, were more like the dolphins that cetologists know today. Both of these groups had vanished from the fossil record, and others had dwindled to mere remnants, by the end of the Miocene.

The cetotheres radiated further during the Miocene (5–23 mya), with more than 20 genera in which the blowholes were positioned about as far back on the top of the head as they

A pilot whale (*Globicephala* sp.) in Roatan, Honduras. (Photo by Corp. F. Stuart Westmorland/Photo Researchers, Inc. Reproduced by permission.)

are in living mysticetes. Also, by the early Miocene, the two main branches of cetotheres were evident, one leading to the modern right whales (Balaenidae) and the other to the rorquals (Balaenopteridae) and gray whale (Eschrichtiidae). Gray whales do not appear in the fossil record until only about 100,000 years ago, and their ancestry is therefore particularly problematic. For their part, the odontocetes also experienced a major Miocene radiation. Beaked whale (Ziphiidae) fossils are common in marine sediments worldwide by 5–10 mya, and these include animals belonging to the modern genus *Mesoplodon*. Sperm whales in the family Physeteridae, similar in some important ways to the living species, were present by 22 mya.

Dolphins and porpoises as cetologists know them today also emerged in the Miocene, perhaps about 12 mya. The large, speciose odontocete family Delphinidae is one of the least resolved of the 14 extant cetacean families. In spite of fairly blatant external morphological differences among genera within the family, such as the globe-headed (pilot whales) versus long-beaked (common dolphins) dichotomy, the family's validity is supported by several lines of evidence. For ex-

A killer whale (*Orcinus orca*) shows its teeth. (Photo by Bruce Frisch/Photo Researchers, Inc. Reproduced by permission.)

ample, intergeneric hybrids have been observed for many delphinids both in captivity and in the wild, and all 17 included genera share the same basic skull architecture. Most of the morphological diversification within the family is related to body size and foraging structures such as rostral length and width, and the number, size, and form of the teeth. A recent phylogenetic analysis of the delphinids based on full cytochrome b gene sequences has revealed that certain of the genera may represent artificial assemblages of species and that extensive revision is needed at both the genus and subfamily levels.

One of the more high-profile and controversial issues in cetacean systematics that has arisen in recent years is the contention by some molecular biologists that sperm whales are more closely related to the baleen whales than to other odontocetes. However, this view has been refuted, contradicting as it does a host of morphological, paleontological, and even some other molecular evidence confirming that the odontocetes are a monophyletic group. As one expert summarized it, the proposed split linking sperm whales with mysticetes "would require morphological convergences and reversals of a magnitude that defies credibility."

Physical characteristics

The absolute range in body size is vastly greater for cetaceans than for any other mammalian order, from scarcely 5 ft (1.5 m) in length and 120 lb (55 kg) in weight for some dolphin and porpoise species to at least 110 ft (33 m) in length and 400,000 lb (180,000 kg) in weight for the Antarctic blue whale (*Balaenoptera musculus*). There is also considerable variation in morphology. Several species completely lack a dorsal fin (right whales and right whale dolphins, Balaenidae and *Lissodelphis* spp., respectively), others have only a hump or ridge (gray whale and Ganges river dolphin, *Eschrichtius robustus* and *Platanista gangetica*, respectively), and still others have a tall, prominent, even outsized dorsal fin (male killer whales and spectacled porpoises, *Orcinus orca* and *Phocoena dioptrica*, respectively). The very long, flexible pectoral flippers of the humpback whale (*Megaptera novaeangliae*) are in stark contrast to the small, rounded flippers of beaked whales (Ziphiidae) that fit into molded depressions on the sides of the body, so-called "flipper pockets." A cetacean's dorsal fin, like its tail flukes, has no bony support. The stiffness of these structures comes from tough fibrous tissue and, in the case of the flukes, tendons. The flippers, in contrast, are modified front limbs and therefore contain a full complement of arm and hand bones, which, however, are greatly compressed in length.

Body streamlining is obviously an essential feature of the cetacean form. The eyes are on the sides of the head and the blowhole, or blowholes, are on top. The paired blowholes on all living mysticetes are positioned in approximately the same place—at the back and in the center of the rostrum. The single blowhole of odontocetes can vary in both its appearance and placement, but in all species it is skewed to the left of the midline, thereby reflecting the sinistral skew of the underlying cranium. A sperm whale's blowhole is a deep slit at the very front of the top of the head, which makes its blow cant forward and to the left, allowing an observer to identify the species at a considerable distance. In most dolphins, the blowhole is much farther back on the head, approximately even with the eyes, and it appears as a round hole. However, the blowhole of the Ganges river dolphin is a longitudinal slit well back on the top of the head. Another extraordinary feature of this species is its vestigial eyes, which are tiny and effectively non-functional. Cetaceans have no external ear appendages, and all reproductive and excretory organs are concealed within the body. Both males and females have a navel, genital slit, and anus along the ventral midline, and females normally have, in addition, a small mammary slit on each side of the genital slit. Two small, rudimentary pelvic bones embedded in muscle are the only vestiges of hind limbs.

Cetaceans have compensated for their lack of fur or hair by acquiring an adipose-rich hypodermis, a dense endodermal layer of fat, called "blubber," which functions not only as extremely efficient insulation (a core body temperature of about 98.6°F [37°C] is maintained regardless of ambient conditions), but also as an energy depot. They also have a highly developed counter-current heat exchange system, with arteries completely surrounded by bundles of veins. This system is configured so that heat loss and retention are controlled largely through blood flow to the flippers, flukes, and dorsal fin, none of which has a thick layer of insulative blubber.

Distribution

Cetaceans inhabit all marine waters throughout the world, as well as several large rivers and associated freshwater systems in Asia and South America. Their distribution is limited at the poles only by solid ice coverage. Land, ice massifs, and more subtle features such as depth and temperature gradients, current boundaries, and zones of low productivity constitute the biogeographical barriers that separate species and populations. Competitive interactions have probably also helped to shape the global pattern of cetacean distribution. It is worth emphasizing that cetaceans even occur in all large semi-enclosed seas and gulfs, such as the Black, Red, Baltic, and Japan Seas, the Arabian Gulf, and Hudson Bay.

It is important to recognize that human activities have played a major role in determining the present-day global distribution of cetaceans. Although human actions are not known to have exterminated any cetacean species entirely, they have at least reduced certain species to levels at which they no longer play a significant role in the ecosystem. For example, bowhead whales (*Balaena mysticetus*) were conspicuous members of the marine fauna of the eastern Atlantic Arctic (Greenland and Barents Seas) before European commercial whalers arrived at the end of the sixteenth century. By the early twentieth century, only scattered individual bowheads remained. Gray whales were present in the North Atlantic Ocean until at least as recently as the seventeenth century but have been extinct there for more than 150 years and now occur only in the North Pacific Ocean. The disappearance of river dolphins from large segments of their range in the Indian subcontinent, Southeast Asia, and China is a well-documented result of deliberate killing, incidental mortality in fishing gear, and dam construction. Moreover, in the Antarctic and no doubt elsewhere, the severe depletion of blue, fin (*Balaenoptera physalus*), and humpback whales have probably changed the species composition and relative abundance of other high-order consumers. Although difficult to test, the hypothesis that minke whales (as well as crabeater seals and perhaps even some seabirds) increased and expanded their range as the larger krill-consuming whales were eliminated is at least plausible. Some scientists have also argued that sei whales (*Balaenoptera borealis*), as copepod specialists, were given a competitive advantage and thus proliferated in temperate regions as the numbers of copepod-eating right whales (*Eubalaena* spp.) were decimated. Again, this hypothesis is all but impossible to prove or disprove.

Generally speaking, human agency has not been responsible for the introduction of cetaceans into new areas of distribution; that is, made them into "alien invaders." However, a few relevant incidents have been documented. It was recently reported that one or more Indo-Pacific humpback dolphins (*Sousa chinensis*) had breached the Suez Canal, moving from the Red Sea into the Mediterranean Sea—a transoceanic switch facilitated by canal construction. On a few occasions, captive bottlenosed dolphins that originated in one ocean basin have escaped or been released into another basin, opening the possibility that an invasive species or genetic variant could become established accidentally. Thus far, there has been no report of movement through the Panama Canal by a cetacean, but manatees (*Trichechus* spp.) have negotiated this route from the Atlantic to the Pacific during the last few decades of the twentieth century and into the early years of the twenty-first century.

The bottlenosed dolphin (*Tursiops truncatus*) is found worldwide, except in the polar regions of the world. (Photo by Tom Brakefield. Bruce Coleman, Inc. Reproduced by permission.)

Habitat

Three living families of cetaceans, Lipotidae, Iniidae, and Platanistidae, consist of dolphins that are obligate inhabitants of freshwater environments. The Iniidae, in particular, exhibit a remarkable ability to survive, indeed flourish, in habitat that seems unlikely for a cetacean. Amazon River dolphins, or botos (*Inia geoffrensis*), occupy both the large, turbid, "white-water" rivers and the "black-water" streams and lake systems of Amazonia and Orinoquia, seasonally entering the flooded rainforest to forage among roots and vines. Some platanistids in the upper reaches of the Ganges River system live in relatively cool, clear, fast-flowing streams, while their relatives downriver occupy the wide, brown, slower-flowing channels

The false killer whale (*Pseudorca crassidens*) can be found in groups of up to several hundred individuals. (Photo by J. T. Wright. Bruce Coleman, Inc. Reproduced by permission.)

of the Gangetic plain. All river dolphins tend to be most abundant in counter-current eddies, where prey is more easily available and less energy is needed to maintain position.

Some delphinids (e.g., the tucuxi, *Sotalia fluviatilis*, and Irrawaddy dolphin, *Orcaella brevirostris*) and one species of porpoise (the finless porpoise, *Neophocaena phocaenoides*) are called "facultative" freshwater cetaceans because they have populations that live not only far up rivers and in freshwater lake systems, but also in marine coastal waters. Some of the other coastal small cetaceans, notably the humpback dolphins and the franciscana (*Sousa* spp. and *Pontoporia blainvillei*, respectively), tend to exist in greatest densities in portions of coastline with high volumes of continental runoff, that is, in and near large river mouths. Such areas are typically very productive.

Numerous cetacean species are best characterized as inhabitants of the continental shelf, and they are found mainly inside the 660 ft (200 m) depth contour. Among these, several of the great whales are strongly migratory, going from winter calving and breeding grounds in tropical waters to high-latitude feeding grounds in summer. Gray whales, for example, congregate in warm, shallow lagoons along the Pacific coast of Mexico's Baja California peninsula in winter, and many then travel close along the western North American coast for 4,600–6,200 mi (7,500–10,000 km) to shallow feeding grounds in the Bering and Chukchi Seas, only to return south again by approximately the same route to Mexico during the following autumn. Humpback whales are also long-distance migrators, congregating on shallow banks and reefs in tropical latitudes to give birth, nurse their young, and breed in winter, and moving to productive subpolar and polar waters to feed in summer. Some humpbacks cover 10,000 mi (16,000 km) in their annual round-trip migration. Unlike gray whales, they often strike out across expanses of deep water to get from one segment of habitat to another.

Still other cetacean species are pelagic, or "blue-water," animals, living along the steep contours of continental slopes, near the edges of offshore banks and seamounts, or in canyon areas where sharp depth gradients create beneficial foraging conditions. Some pelagic species forage in the deep scattering layer, a complex of organisms that migrate vertically in the water column, approaching to within about 650 ft (200 m) of the surface at night and descending to depths of 1,000 ft (300 m) during the day. Dolphins that are not especially deep divers take advantage of this phenomenon by resting and socializing during the day and foraging at night. The spinner dolphin (*Stenella longirostris*), for example, is one of the most widespread warm-water species of cetaceans. Many spinner populations centered on offshore islands or atolls move inshore to bays or reef-fringed lagoons during the day, then offshore at night to feed.

Behavior

The behavior of cetaceans, like so many other aspects of this diverse order, spans a wide range of characteristics. When at the surface, porpoises, beaked whales, and pygmy and dwarf sperm whales (*Kogia breviceps* and *K. sima*, respectively) are cryptic and undemonstrative. In contrast, some dolphin species are energetic and conspicuous, leaping high above the surface, spinning, somersaulting, and churning the water. Bow-riding species charm seafarers as they race toward a fast-moving boat and "hitch a ride" in the pressure wave. Some species live in small groups of 10 or fewer individuals and can be considered almost solitary, while others are among the most gregarious mammals. At both extremes, however, it is important to consider that appearances may not reveal the entire story. Given the fact that most cetacean communication is acoustic, not visual, it is possible that individuals and small groups maintain contact over large distances. Thus, the level of social integration may be much greater than an apparently "scattered" pattern of distribution implies. In this regard, the low-frequency calls of blue and fin whales can be heard at distances of hundreds of miles when entrained in deep sound channels.

Remarkably, even many of the earliest odontocetes appear to have been capable of echolocation; that is, able to use sound echoes for detection and navigation as a supplement to, or substitute for, vision. High-frequency clicks produced by the movement of recycled air within the diverticula, sacs, and valves of the nasal passages are projected into the environment via the melon (the lump of fatty tissue that forms an odontocete's "forehead"). These sounds reflect off objects and bounce back. The echoes are transmitted to the ears via the side of the face and pass through the thin wall of the mandible before reaching the ear region. The ear bones, iso-

A killer whale (*Orcinus orca*) spy-hopping in Tysfjord, Norway. (Photo by François Gohier/Photo Researchers, Inc. Reproduced by permission.)

lated in fat bodies, receive a given sound at different times, thus facilitating directional hearing. Although proven experimentally for only a few species, it is likely that all odontocetes echolocate. Mysticetes, in contrast, do not echolocate, although it has been speculated that bowhead whales may "read" the undersurface of sea ice, and thus assess the dimensions of a floe, for example, by listening to the reverberations of their calls. This would be a crude form of "echo-sensing." Besides their echolocation clicks, many odontocetes produce high-frequency whistles that are used to communicate. Some mysticetes produce patterned sequences of sounds that constitute "song" in a technical sense, and that are believed to function as sexual advertisement during the mating season.

The social structure of several odontocete species has been studied in detail. Killer whales, for example, have a society centered on matrilineal groups that coalesce to form pods of up to about 60 individuals. Pods are organized into clans, which are collections of pods with similar vocal dialects. Sperm whale social structure has been likened to that of elephants, with adult males roving between stable matrilineal pods on the tropical breeding grounds and becoming essentially solitary while on their high-latitude feeding grounds. Bottlenose dolphins live in fission-fusion societies in which group composition changes frequently as individuals join and leave. Nevertheless, calves stay with their mothers for several years, and in some areas males establish pair bonds that last for decades. The social systems of baleen whales are generally thought to be less complex and structured than those of toothed cetaceans.

Although it is widely assumed that whales are "gentle giants," there is considerable evidence of aggressive behavior in some species. Quite apart from the fact that killer whales regularly kill and eat mammalian prey, male Indo-Pacific bottlenosed dolphins (*Tursiops aduncus*) form coalitions to fight with other males and aggressively herd females; common bottlenosed dolphins occasionally kill harbor porpoises (*Phocoena phocoena*) for reasons not readily apparent; adult male beaked whales and narwhals (*Monodon monoceros*) engage in combat that results in extensive body scarring; and male humpback whales, while competing for access to an adult female on the breeding grounds, may engage in bouts of slashing and scraping that result in bleeding or abrasion of a competitor's head knobs and dorsal fin.

The diving abilities of cetaceans vary in relation to their ecology, distribution, and diet. Sperm whales can dive to depths in excess of 6,080 ft (1,853 m). Both they and bottlenosed whales (*Hyperoodon* spp.) can remain submerged for well over an hour at a time, and they are known to feed near the bottom in very deep water. Mysticetes generally do not dive as deep, or for as long, although some are capable of staying down for half an hour or longer.

The boto (*Inia geoffrensis*), or Amazon River dolphin, is the largest of the river dolphins. (Photo by Gergory Ochocki/Photo Researchers, Inc. Reproduced by permission.)

Feeding ecology and diet

Cetaceans are generally regarded as apex predators, and even the baleen whales, which in many respects feed more like grazers than predators, are positioned relatively high on the trophic pyramid. With their specialized feeding apparatus, the baleen whales are all filter feeders although their actual strategies for collecting prey vary. The balaenids and the sei whale are skim feeders, meaning that they tend to swim steadily through the water, mouth open, allowing prey organisms (usually zooplankton) to be continuously filtered against the mat of baleen fringes on the inside of the mouth. At the end of a feeding run, the whale uses its massive tongue to sweep the food into the throat. It then resumes the food-gathering process. Balaenopterids other than the sei whale are gulp feeders, meaning that they take large volumes of seawater into the mouth, normally causing substantial distention of the throat (ventral grooves), then close the mouth and squeeze the water out through the baleen, trapping the prey inside the mouth and swallowing it. Skim feeders tend to have supple, finely fringed baleen, while gulp feeders have stiffer, coarser baleen. The diets of baleen whales range from the stenophagous habits of the blue whale, a krill (euphausiid) specialist, to the more euryphagous habits of the minke, humpback, and fin whales, which take zooplankton, schooling fish, and occasionally even squid.

Toothed cetaceans also prey upon a very broad spectrum of organisms that includes fish of many sizes, from small (herring, capelin, sand lance) to medium (cod, salmon, halibut) to large (sharks and tuna), cephalopods (especially squid but also cuttlefish and octopus), shrimp, and crabs. Killer whales are the only cetaceans known to prey upon warm-blooded animals on a regular basis. Their diet can include everything from seabirds and sea turtles to seals, sea lions, sea otters, and fellow cetaceans. While the baleen whales often consume thousands or even millions of animals in a single feeding bout, odontocetes mainly catch one creature at a time. Those species with reduced dentition, notably most of the beaked whales (Ziphiidae), Risso's dolphin, the pilot whales, and narwhal, probably use suction to capture their prey, which are mostly squid. For the most part, prey is swallowed whole, although groups of rough-toothed dolphins (*Steno bredanensis*), for example, have been seen tearing chunks from large fish that they had apparently captured cooperatively. Killer whales obviously must bite pieces of flesh from their larger prey. In fact, when they kill a baleen whale, they typically consume the tongue, lips, and throat region first. One odontocete species, the boto, has differentiated dentition. Its rear teeth are flanged and molar-like, presumably so that hard-bodied prey such as armored catfish can be crushed before swallowing.

Reproductive biology

The reproductive and excretory organs are all concealed within the body. The male's retractile penis, similar anatomically to that of the bull, contains a great deal of tough, fibrous tissue. Erections apparently result at least in part from the elasticity of that tissue, which comes into play when the retractor muscles relax. The elongated testes lie within the abdominal cavity just behind the kidneys, rather than in an external scrotum. Female reproductive anatomy is basically similar to that of most other mammals, with the two ovaries in the same position as the male's testes. The ovaries of odontocetes are elongated and somewhat egg-shaped, while those of mysticetes are much more irregular in shape, studded with rounded protuberances. A unique aspect of cetacean reproductive anatomy is that the corpora albicantia; that is, the degenerated corpora lutea that follow ovulation remain evident throughout a female's life. This means that the ovaries provide a complete and permanent record of the animal's reproductive history, allowing scientists to count the number of times that ovulation (but not necessarily pregnancy) has occurred.

The reproductive strategies of cetaceans are generally typical of *K*-selected species; that is, ones that grow slowly, have relatively few offspring, live for a long time, and exhibit substantial parental involvement in the rearing of young. Even the harbor porpoise and franciscana, two of the fastest-maturing species, take at least several years to achieve sexual maturity, and they give birth to only one calf per year when in their prime. Some of the longer-lived social odontocetes take at least 10 years to mature, and they give birth at intervals of at least three years. The gestation period of sperm whales is 14–16 months, and although the calf may begin taking solid food before the end of its first year, it may continue to be suckled for at least five more years. The reproductive parameters of most odontocetes fall between those of the harbor porpoise and the sperm whale. Baleen whales generally mature before 10 years of age, have a gestation period of 10–14 months, a lactation period of six months to one year, and give birth at intervals of two to five years. Most species are migratory to a greater or lesser extent, and give birth and breed during the winter months in relatively low latitudes.

Humpback whale (*Megaptera novaeangliae*) spy-hopping in Alaska. (Photo by John Hyde. Bruce Coleman, Inc. Reproduced by permission.)

Conservation

Cetacean conservation emerged during the late twentieth century as one of the world's most highly publicized environmental issues. International focus on the decimation of the stocks of great whales portrayed the human capacity for greed and wanton destruction of natural resources like few other issues could have. The collapse of blue and fin whale stocks in the Antarctic, following as it did the sequential destruction of the stocks of right, bowhead, humpback, and gray whales in other oceans, finally brought serious international regulation to the commercial whaling industry. Having closed the fisheries for one species and stock after another, the International Whaling Commission (IWC) finally agreed in the 1980s to impose a global moratorium on commercial whaling, which remains in effect. Controversy continues, however, over Norway's ongoing commercial hunts for minke whales in the North Atlantic, and Japan's hunts for an expanding variety of species in the western North Pacific and Antarctic. The hunts by Norway are legal because that country exercised its sovereign right to object to the moratorium in the first instance, and the Japanese hunts are justified through a loophole in the whaling convention that allows member states to issue national permits for "scientific" catches regardless of prohibitions in the IWC schedule.

The deliberate killing of whales, dolphins, and porpoises for meat and other products continues in many parts of the world, including Japan, where tens of thousands of small cetaceans are taken annually in addition to the "scientific" catch of minke and larger whales; the Faeroe Islands, where many hundreds of long-finned pilot whales and Atlantic white-sided dolphins (*Globicephala melas* and *Lagenorhynchus acutus*, respectively) are killed in most years; Greenland, where 160–180 minke whales and 10–15 fin whales are taken annually under the IWC's exemption for "aboriginal subsistence" whaling, as well as many hundreds of harbor porpoises, narwhals, and belugas (*Delphinapterus leucas*); and Canada, the United States (Alaska), and Russia (Chukotka), where thousands of belugas and narwhals, plus several hundred bowhead and gray whales, are killed each year in what are considered

The Pacific white-sided dolphin (*Lagenorhynchus obliquidens*) can be found in groups of several hundred. (Photo by JACANA Scientific Control/Jean Philippe Varin/Photo Researchers, Inc. Reproduced by permission.)

traditional hunts for "subsistence." While it is true that the absolute scale of the killing of great whales has declined with regulation over the last few decades of the twentieth century and into the early years of the twenty-first century, serious problems remain as many of the hunts for small cetaceans are inadequately regulated to ensure sustainability or permit recovery from depletion.

During the past several decades of the twentieth century and into the early twenty-first century, incidental mortality in fishing gear (so-called bycatch), especially in large-mesh gillnets, has become of paramount importance as a threat factor for cetaceans. Some species, notably the Critically Endangered vaquita (*Phocoena sinus*) and baiji (*Lipotes vexillifer*), have been driven close to extinction, and numerous populations of other cetacean species have been greatly depleted, as a result of interactions with fisheries. Efforts to reduce the scale of incidental mortality have centered on development, testing, and mandatory use of acoustic pingers to deter the animals from approaching nets; time and area fishery closures; and establishment of protected areas where high-risk fishing is forbidden. Another threat factor for some populations, and particularly for the Endangered North Atlantic right whale (*Eubalaena glacialis*) population off the North American east coast, is mortality from collisions with ships. Thus far, mitigation measures have consisted of reconfiguring the ship channels in southeastern Canada to reduce traffic in areas where right whales congregate during summer, and implementation of early-warning systems in portions of the U.S. East Coast where right whales and heavy ship traffic overlap.

Several other factors are of increasing concern: underwater noise, chemical contaminants, and climate change. The possibility that whales are disturbed by industrial noise (e.g., seismic testing, ocean drilling for oil and gas) has been a source of concern for decades, but recent evidence suggests that under certain circumstances, high-energy artificial sounds can actually cause lethal injuries to beaked whales. Pollution of the world's waterways and oceans has become recognized as a serious threat to many forms of life. Cetaceans and other marine mammals are no exception. Because they store large amounts of fat in their bodies, they tend to accumulate very high levels of lipophilic contaminants such as the organochlorines (e.g., PCB, DDT). Interestingly, heavy doses of these toxic chemicals are transmitted to first-born calves through the placenta and milk, which means that this age-class within a cetacean population may be especially at risk. Finally, the rapid ongoing change in global climate is certain to have implications for cetaceans, as for other wildlife. Those species that live in high latitudes could be affected the most. Thinning of sea ice and melting of glaciers will certainly influence productivity and change the character of habitat in the Arctic and Antarctic. While some wild species could benefit, others are likely to be harmed.

Significance to humans

Cetaceans have been of great significance to humans for millennia, beginning when primitive coast-dwellers scavenged stranded carcasses for meat, blubber oil, and bone material. The flesh was eaten by people but also fed to domestic animals, most importantly sled dogs in the Arctic and Subarctic. Whale oil was burned to illuminate homes and footpaths, and in lamps to provide warmth. Bones of whales were used in the construction of dwellings and to manufacture tools and appliances. Baleen had many uses as well. Ironically, some early whalers in the Arctic fashioned sea anchors from woven baleen and attached them to harpoon lines to provide resistance for a harpooned whale trying to escape; they thus used a product obtained from one whale to help them capture another. Although the widespread, critical reliance upon whales for food, oil, and other products no longer applies, some aboriginal communities in the Arctic still consider whale hunting central to their identity and sustenance.

As early maritime communities in more temperate regions ventured into coastal waters and learned to capture cetaceans, they established markets to distribute and sell the oil and baleen (whalebone), giving rise to the global whaling industry, as mentioned earlier. The pursuit of whales was a motivating force in exploration and in the development of many remote regions. Whalers brought trade goods, diseases, firearms, and employment to the people they visited as they scoured the planet for their prey. They also enlisted crewmembers from island outposts like the Azores, Cape Verde Islands, and Hawaii, facilitating a diaspora of sorts. Even if unintended, the consequences of activities of whalers were often disastrous to local societies. An obvious example is the degree to which commercial whalers destroyed the stocks of whales, in some instances literally depriving indigenous people of an essential natural resource.

Whales and dolphins are popular, but high-maintenance and controversial, performers in captivity. Bottlenosed dolphins, belugas whales, and killer whales are the most common species in oceanaria, but numerous other species have been trained to perform as well. The captive display industry played a key role in raising awareness about these animals and in getting people to view them as both sentient and vulnerable. In fact, the killer whale's reputation was completely transformed once people had been exposed to several captive individuals. Ironically, oceanaria have now themselves become targets of protest by campaigners who view the keeping of cetaceans as unethical. Dolphins and small whales have also been the subjects of ex situ research of various kinds, including one program in Hawaii that focuses on developing ways for humans and dolphins to communicate with one another. Some success has been reported in efforts to treat autism by allowing patients to interact with captive dolphins, and luxury hotels in a number of tropical holiday destinations keep animals in sea pens and offer "swim-with-the-dolphin" options for guests. Finally, the U.S. Navy has, for decades, used trained dolphins and small toothed whales to locate and recover objects from the sea floor and participate in at-sea research of various kinds. There were reports during the 2003 invasion of Iraq that dolphins were being used by American forces to detect and help destroy mines in the Persian Gulf. The captive population of common bottlenosed dolphins in the United States is considered by some experts to be self-sustaining; that is, capable of replenishing itself without the need for more captures from the wild. In some respects, the domestication of this species may be at hand.

Resources

Books

Dizon, Andrew E., Susan J. Chivers, and William F. Perrin, eds. *Molecular Genetics of Marine Mammals.* Lawrence, KS: Society for Marine Mammalogy, 1997.

Evans, Peter G. H., and Juan Antonio Raga, eds. *Marine Mammals: Biology and Conservation.* New York: Kluwer Academic/Plenum, 2001.

Harrison, Richard, and M. M. Bryden, eds. *Whales, Dolphins and Porpoises.*New York: Facts on File, 1988.

Hoelzel, A. Rus, ed. *Marine Mammal Biology: An Evolutionary Approach.* Oxford, U.K.: Blackwell Science, 2002.

Mann, Janet, Richard C. Connor, Peter L. Tyack, and Hal Whitehead, eds. *Cetacean Societies: Field Studies of Dolphins and Whales.* Chicago: University of Chicago Press, 2000.

Perrin, William F., Bernd Würsig, and J. G. M. Thewissen, eds. *Encyclopedia of Marine Mammals.* San Diego: Academic Press, 2002.

Reeves, Randall R., Brent S. Stewart, Phillip J. Clapham, and James A. Powell. *National Audubon Society Guide to Marine Mammals of the World.* New York: Alfred A. Knopf, 2002.

Reeves, Randall R., Brian D. Smith, Enrique A. Crespo, and Giuseppe Notarbartolo di Sciara. *Dolphins, Whales, and Porpoises: 2002–2010 Conservation Action Plan for the World's Cetaceans.* Gland, Switzerland: International Union for Conservation of Nature and Natural Resources, 2003.

J. E., Reynolds, III, and Sentiel A. Rommel, eds. *Biology of Marine Mammals.* Washington, DC: Smithsonian Institution Press, 1999.

Rice, Dale W. *Marine Mammals of the World: Systematics and Distribution.* Lawrence, KS: Society for Marine Mammalogy, 1998.

Ridgway, Sam H., and Richard Harrison, eds. *Handbook of Marine Mammals.* Vol. 3, *The Sirenians and Baleen Whales.* London: Academic Press, 1985.

———. *Handbook of Marine Mammals,* Vol. 4, *The Sirenians and Baleen Whales.* London: Academic Press, 1985.

———. *Handbook of Marine Mammals,* Vol. 5, *The Sirenians and Baleen Whales.* London: Academic Press, 1985.

———. *Handbook of Marine Mammals,* Vol. 6, *The Sirenians and Baleen Whales.* London: Academic Press, 1985.

Twiss, John R. Jr., and Randall R. Reeves, eds. *Conservation and Management of Marine Mammals.* Washington, DC: Smithsonian Institution Press, 1999.

Randall Reeves, PhD

Ganges and Indus dolphins

(Platanistidae)

Class Mammalia
Order Cetacea
Suborder Odontoceti
Family Platanistidae

Thumbnail description
Small gray dolphin with long beak, exposed interlocking teeth and tiny eyes; broad flippers and flukes

Size
Females 8.2 ft (2.5 m) and males 6.6 ft (2.0 m); 185.2 lb (84 kg)

Number of genera, species
1 genus; 1 species

Habitat
Rivers and tributaries

Conservation status
Endangered

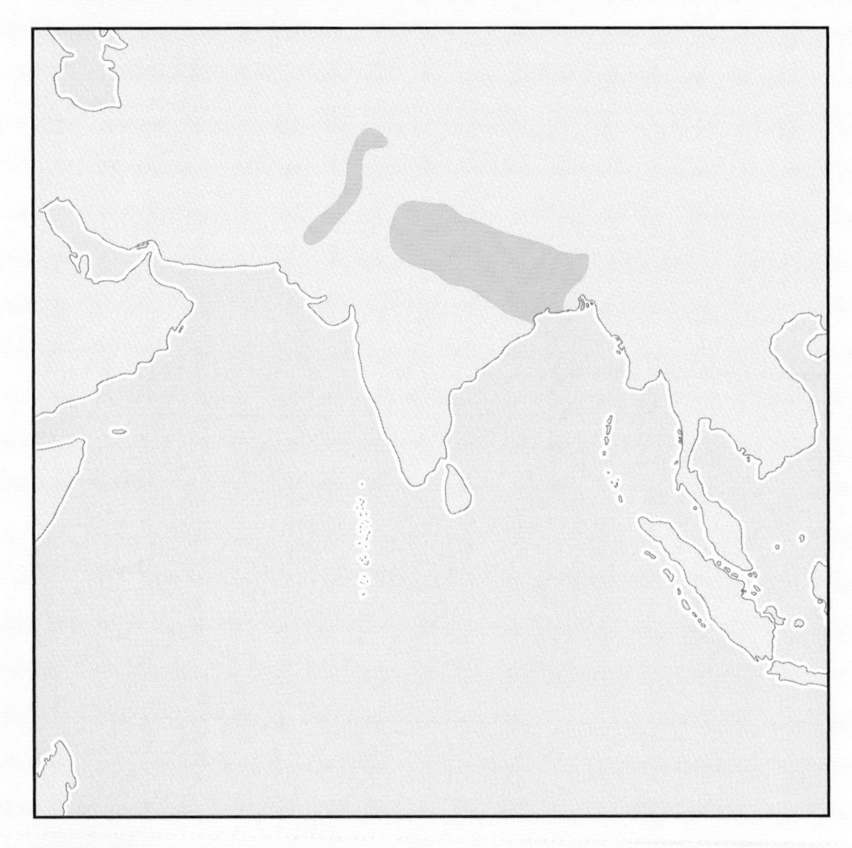

Distribution
Indus River in Pakistan; Ganges River drainage in India, Bangladesh, and Nepal

Evolution and systematics

This is the most primitive group of all river dolphins. It is closely related to five extinct families that were widely distributed during the Oligocene (34–24 million years ago [mya]) and Miocene (24–5 mya). Two fossils exist from the middle to late Miocene. *Zarhachis* and *Pomatodelphis* were found in marine environments in North America and Europe, but paleontological data are too scarce to establish when these marine ancestors first entered rivers. It is hypothesized that they inhabited the estuarine regions created during the rise of the sea level in the middle Miocene and survived in rivers as the waters regressed in the late Miocene. Different hypotheses have been advanced about their phylogeny. Placement between other river dolphins and Ziphiidae (beaked whales) or between Ziphiidae and Physeteridae (sperm whales) has gained considerable support from genetic and morphologic data.

Classification of this species at the family level is the least controversial of all river dolphins. It is the only species of family Platanistidae. It is grouped under superfamily Plastanistoidae with five fossil families: Prosqualondotidae, Squalondotidae, Squalodelphinidae, Waipatiidae, and Dalpiazinidae.

Although *Platanista* from the Indus and Ganges drainages have been proposed to be different species, namely *Platanista gangetica* and *Platanista minor* (or *indi*), based on morphologic and biochemical analysis, currently they are considered a single species, *Platanista gangetica*. While differences in tail length between the Indus and Ganges dolphins has led some authors to consider two subspecies, *P. g. gangetica* and *P. g. minor*, genetic analysis has not resolved this issue.

The taxonomy for this species is *Plantanista gangetica* (Roxburgh, 1801), Hooghly River, Ganges River Delta. Other common names include: English: Blind river dolphin, susu; French: Plataniste du Gange, plataniste de l'Indus, sousou; Spanish: Delfín del Ganges, delfín del Indo.

Physical characteristics

The primitive appearance of the Ganges and Indus dolphin is unlike that of any other dolphin, even other river dolphins. The snout is elongated, about one-fifth of the body, and widens towards the tip. The anterior teeth are larger and exposed, especially close to the tip. The dorsal fin is merely a small hump close to the rear of the bulky body. However, the flippers and

The Ganges and Indus dolphin (*Platanista gangetica*) is a solitary mammal and only uses about 5% of its sounds for communication. (Photo by Toby Sinclair/Naturepl.com. Reproduced by permission.)

fluke are relatively large. On top of the head, there is a longitudinal ridge. The blowhole is a longitudinal slit in contrast to a horizontal opening typical in dolphins. A wattle, forming several folds, adds to the species' ungainly appearance. Uniquely, the external ear sits below eye level. The eyes are tiny, smaller than the ear opening. The optical apparatus is underdeveloped and is thought to perceive only shades rather than images; hence, the name blind river dolphin. The skull is extremely asymmetrical compared to most odontocetes and has prominent facial inflections unseen in other dolphins. The neck is very long and, because of unfused vertebrae, flexible. The brain has the simplest cerebral cortex among odontocetes. Coloration is gray or brown, occasionally with a pinkish belly.

Distribution

Currently, the species is found in the Ganges/Brahmaputra/Megna and Karnapuli River systems and their tributaries in India, Bangladesh, and Nepal, and in Pakistan in the Indus River system. Previously, its range extended further upstream into several tributaries. In the Ganges River, the species no longer occurs beyond the Bijnor Barrage (gated dam), completed in 1984 with a loss of a 62-mi (100-km) segment of their habitat. In the Indus River, it does not inhabit the tributaries above Chasma, Trimmu, Sidhnai, and Islam Barrages, built from 1927 to 1971. Its southern range also has shrunk; the lower limit in the Indus River is the Kotri Barrage. Reduced precipitation may drastically affect their distribution, forcing the dolphins to leave smaller tributaries during the dry season.

Habitat

These dolphins occupy rivers and tributaries that run through hills (up to 820 ft [250 m] above sea level in Nepal) and plains, some with turbulent rapids and sharp meanders. River bends, mid-channel islands, or convergences of tributaries create eddy countercurrents, a preferred habitat for dolphins. Dolphins are found both in shallow and deep water and appear to favor 10–30 ft (3–9 m) depths. Water temperatures are 46.4–91.4°F (8–33°C). Dolphins have been seen at the mouths of the rivers that flow into the Bay of Bengal and are thought to disperse between the Ganges/Bramahputra/Meghna and Karnaphuli/Sangua systems along the coast. This may occur during the monsoon when a freshwater plume from the river extends into coastal waters.

Behavior

A remarkable behavior is the dolphins' side swimming, with their tail slightly higher than the head, thought to be an adaptation to very shallow waters. Aerial behaviors are uncommon, leaping being performed mainly by calves. Surfacing usually occurs beak first, followed by the melon. Only the front of the body is exposed. This is a very vocal species, which produces pulsed sounds rather than whistles. For navigation and foraging, they use echolocation in place of vision. Mostly solitary, their mean group size is fewer than three individuals, although groups of 25–30 have been observed.

Feeding ecology and diet

They feed predominantly on benthic species, including reports of catfish, herring, carp, gobies, and mahseers. Invertebrates such as prawns and clams have also been found in their

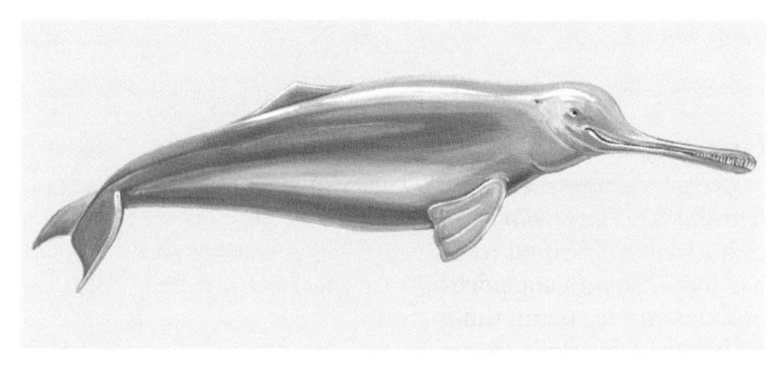

Ganges and Indus dolphin (*Platanista gangetica*). (Illustration by Patricia Ferrer)

A Ganges and Indus river dolphin (*Platanista gangetica*) and fisherman, in India. (Photo by © Roland Seitre/Seapics.com. Reproduced by permission.)

stomach contents. In captivity, individual daily consumption varies from 1–3.3 lb (500–1,500 g). In the Brahmaputra River, they often feed in association with the river tern (*Sterna aurantia*), sightings of which may be used to help locate dolphins.

Reproductive biology

Sexual maturity is estimated at 10 years. Estimates of gestation range 8–11 months. Neonate length is estimated at 3 ft (1 m). Lactation may last from two months up to one year. Calving appears to occur throughout the year. Information on other reproductive parameters and mating behavior is scarce.

Conservation status

River dolphins are among the world's most threatened mammals. *P. gangetica* is the second most vulnerable river dolphin, being classified as Endangered. In the Indus River, it has lost a significant portion of its historical range. Subpopulations in Nepal and the Karnaphuli River in Bangladesh are believed to be close to extinction. In the Indus and Ganges River systems, respectively, it is estimated that only a few hundred and several thousand occur. Perhaps the worst threat is posed by nearly 100 water development projects such as dams, barrages, embankments, and dikes. The dams reduce downstream flow and, hence, eliminate periodic enrichment during flooding, reducing riverine productivity. Dams also disrupt seasonal migrations and spawning habitat of fishes. Over-fishing further aggravates this loss of prey. In addition, dams split dolphins into smaller groups, potentially reducing genetic diversity and compromising the long-term viability of populations.

Hunting of dolphins is another threat. Tribal people in the Bramahputra River, Nepal, and in parts of Bangladesh continue to hunt dolphins for meat and oil. Although the direct hunt has decreased following implementation of protective regulations in 1972, enforcement is ineffective and many fishermen are unaware of the laws. By-catch occurs mainly in gill nets and mosquito nets or *kapda jal* (very fine-meshed nets that are illegal). It is estimated that 90–160 dolphins are caught annually in monofilament gillnets in Sirajganj, a town near the Jamuna River. It is unclear whether these catches truly are accidental since the meat and oil are used as fish attractants.

The high human population density in this region, combined with poverty, also causes acute pollution problems from untreated sewage and agricultural run-off. There are few tox-

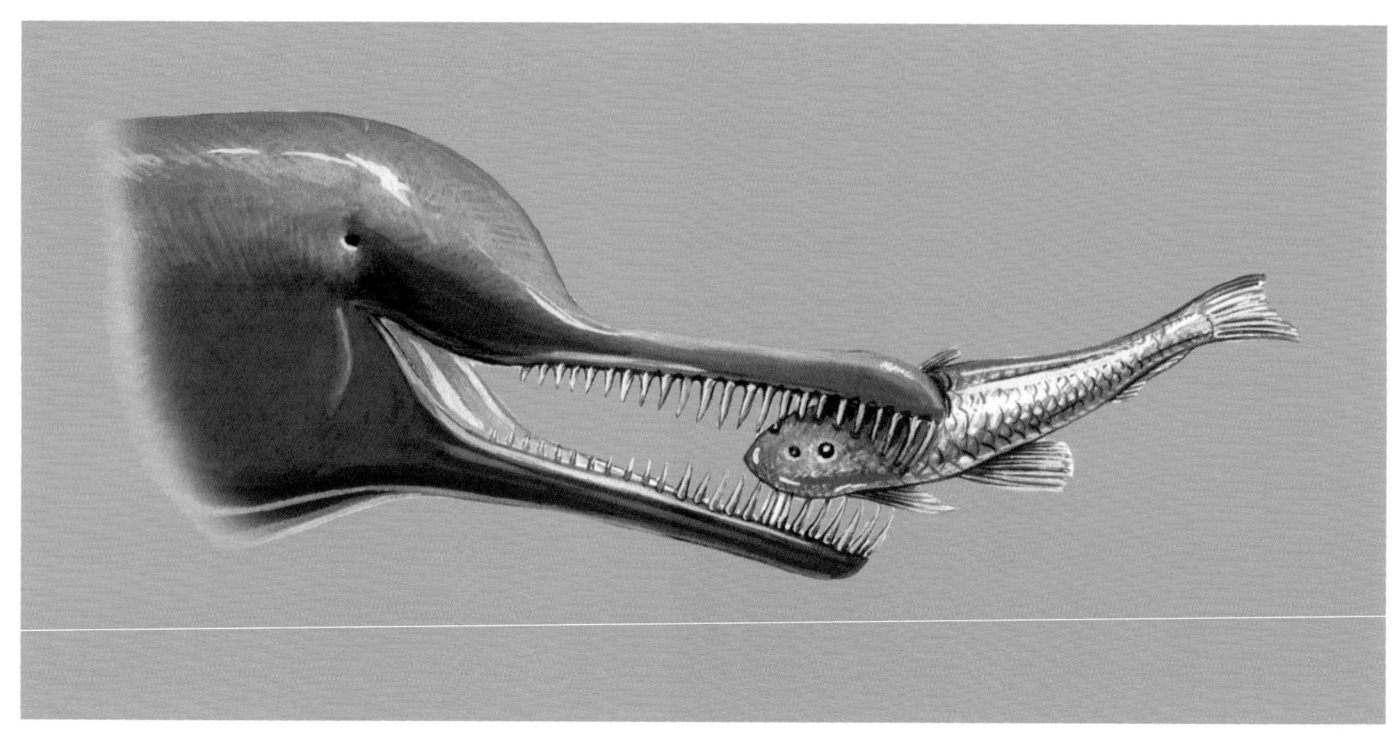

Ganges and Indus dolphins use their teeth to capture prey. (Illustration by Patricia Ferrer)

icological studies, but high concentrations of heavy metals were found in a river dolphin neonate from Bangladesh.

The Asian River Dolphin Committee has proposed better consideration of dam locations and monitoring of their impacts. It also recommended creation of artificial eddy countercurrents and "Managed Resource Protected Areas," where fisheries are conducted in a sustainable manner.

Significance to humans

Its oil has been valued as medicine for a variety of diseases (e.g., arthritis, rheumatism), as an aphrodisiac, and as an ointment for humans and livestock in India, Bangladesh, and Pakistan. It is also used as a fish attractant, and the meat is consumed in some regions. A common practice of fishermen in the Ganges and Bramahputra Rivers is to hang pieces of dolphin meat on the side of the boat and sprinkle the water with a mixture of oil and minced meat. In a site where 15–20 boats target dolphins for oil, it was estimated that about 20 dolphins are required annually for a fishery that operates only two months per year. There is a considerable demand for dolphin oil, especially in the catfish fishery in northeast India. Recent research shows, however, that fish scraps, freely available to fishermen, are equally effective as catfish bait. Thus, educating fishermen to use fish oil rather than dolphin oil may be a promising conservation measure.

Resources

Books

Berta, Annaliesa, and James L. Sumich. *Marine Mammals Evolutionary Biology.* San Diego: Academic Press, 1999.

Klinowska, Margaret. *Dolphins, Porpoises and Whales of the World.* Switzerland: International Union for the Conservation of Nature and Natural Resources (IUCN), 1991.

Perrin, William F., Bernd Würsig, and J. G. M. Thewissen. *Encyclopedia of Marine Mammals.* San Diego: Academic Press, 2002.

Pilleri, Giorgio. *Die Geheimnisse der Blinden Delphine*. Bern and Stuttgart: Hallwag Verlag, 1975.

Ridgway, Sam H., and Richard J. Harrison. *Handbook of Marine Mammals.* Vol. 4. London: Academic Press, 1989.

Periodicals

Cassens, I., et al. "Independent Adaptation to Riverine Habitats Allowed Survival of Ancient Cetacean Lineages." *Proceedings of the National Academy of Sciences* 97, no. 21 (2000): 11343–11347.

Pilleri, Giorgio. "Ethology, Bioacoustics, and Behavior of *Platanista indi* in Captivity." *Investigations on Cetacea* VI (1976): 13–69.

Sinha, Ravindra K. "An Alternative to Dolphin Oil as a Fish Attractant in the Ganges River System: Conservation of the Ganges River Dolphin." *Biological Conservation* 107 (2002): 253–257.

Resources

Smith, Brian D. "River Dolphin in Bangladesh: Conservation and the Effects of Water Development." *Environmental Management* 22, no. 3 (1998): 323–335.

Yang, G., K. Zhou, W. Ren, G. Ji, and S. Liu. "Molecular Systematics of River Dolphins Inferred From Complete Mitochondrial Cytochrome-B Gene Sequence." *Marine Mammal Science* 18, no. 1 (2002): 20–29.

Organizations

The World Conservation Union (IUCN). Rue Mauverney 28, Gland, 1196 Switzerland. Phone: 41 (22) 999-0000. Fax: 41 (22) 999-0000. E-mail: mail@iucn.org Web site: <http://www.iucn.org>

Other

Reeves, Randall R., Brian D. Smith, and Toshio Kasuya. "Biology and Conservation of Freshwater Cetaceans in Asia." *The IUCN Species Survival Commission.* Switzerland and Cambridge: International Union for the Conservation of Nature and Natural Resources (IUCN), 2000.

Reeves, Randall R., Stephen Leatherwood, and R. S. Lal Mohan. "A Future for Asian River Dolphins." *Report from a Seminar on the Conservation of River Dolphins in the Indian Subcontinent.* Bath, England: Whale and Dolphin Conservation Society, 1993.

Paula Moreno, MS

Baijis
(Lipotidae)

Class Mammalia
Order Cetacea
Suborder Odontoceti
Family Lipotidae

Thumbnail description
Light-colored dolphin with robust body, small bluff head, tiny eyes set high on sides of head, long narrow beak slightly upturned at tip, blunt-peaked triangular dorsal fin, and broad flippers

Size
7.5–8.5 ft (2.3–2.5 m); 290–370 lb (130–170 kg)

Number of genera, species
1 genus; 1 species

Habitat
Freshwater, rivers, and lakes

Conservation status
Critically Endangered

Distribution
Yangtze River of China, from Three Gorges to the sea, including tributary lake systems

Evolution and systematics

Although the genus *Prolipotes* was assigned to a mandible fragment from the Miocene of China, Fordyce and Muizon considered this fossil specimen to be non-diagnostic and therefore incertae sedis. The only good fossil cranial material for a lipotid, belonging to the extremely long-beaked genus Parapontoporia, comes from the latest Miocene (6–8 million years ago [mya]) to Late Pliocene (2–4 mya) of Mexico and California. Based on the fact that lipotids are known only from the Northern Hemisphere, and there only from China (the living baiji) and western North America (the long-extinct *Parapontoporia*), it is provisionally assumed that the evolutionary history of Lipotidae took place in the North Pacific.

The genus *Lipotes* was traditionally classified in either of two families of long-beaked river dolphins—Platanistidae or Iniidae. In 1978 Zhou et al. proposed that it be assigned to a separate family, Lipotidae, on the basis of osteology and stomach anatomy. Although Barnes later placed *Lipotes* in a subfamily of Pontoporiidae, the current consensus supports placement of *Lipotes* and *Parapontoporia* in their own family, Lipotidae. Until recently, the four living genera of long-beaked "river dolphins"—*Platanista*, *Inia*, *Lipotes*, and *Pontoporia*—were lumped together in Simpson's superfamily Platanistoidea. However, it is now recognized that only *Platanista*, the Ganges and Indus dolphin of the south Asian subcontinent, belongs in that superfamily. Muizon has assigned *Lipotes* and *Parapontoporia* to the monofamilial superfamily Lipotoidea.

The taxonomy of this species is *Lipotes vexillifer* Miller, 1918, Tung Ting Lake, about 600 mi (965 km) up the Yangtze River, China. Other common names include: English: Chinese lake dolphin, white fin dolphin, French: Baiji, dauphin fluvia de Chine; Spanish: Baiji, delfín de China.

Baiji (*Lipotes vexillifer*). (Illustration by Barbara Duperron)

Physical characteristics

The baiji (*Lipotes vexillifer*) has a spindle-shaped, robust body, with a rounded, rather bluff melon (forehead) and a very long, narrow beak. The beak is often slightly upturned at the tip. There are 30–34 teeth in each of the upper jaws and 32–36 in the lower jaws. The eyes are small, regressed, and dark, situated high on the sides of the head. The oval-shaped blowhole is oriented longitudinally on top of the head, slightly left of the midline. The baiji's dorsal fin is low and triangular, its flippers broad and rounded at the tips.

The baiji's coloration is a subtle blend of gray, bluish gray, and white. Basically, the dorsal surfaces are gray or bluish gray, the ventral surfaces white or ashy white. A broad, irregular white stripe sweeps up onto each side ahead of the flipper, and two more brush strokes of white intrude onto the gray sides of the tail stock.

Distribution

The baiji is endemic to the Yangtze River of China. Its historical distribution extended for approximately 995 mi (1,600 km), from the Yangtze estuary upstream to the Three Gorges above Yichang (655 ft [200 m] above sea level). During floods, dolphins also entered the two large tributary lakes of the Yangtze—Dongting and Poyang. During the great flood of 1955, a few specimens were reported in the Fuchun River, which flows into the East China Sea to the south of the Yangtze mouth. In recent years, there have been no observations upstream of Shashi, which is about 93 mi (150 km) below the Gezhouba Dam, which in turn is about 30 mi (50 km) downstream of the Three Gorges.

Habitat

Within the Yangtze system, the baiji shows a strong preference for eddy countercurrents that form below meanders and channel convergences. Therefore, prime areas for finding these dolphins tend to be near sandbanks, just below islands, and where tributary streams enter or lakes connect with the main channel.

A captive baiji (*Lipotes vexillifer*). (Photo by WANG Xiaoqiang and WANG Ding. Reproduced by permission.)

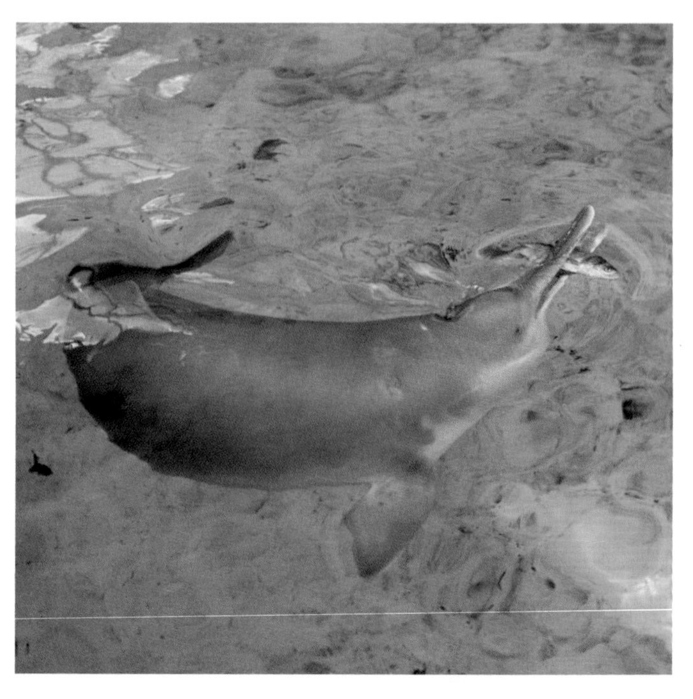

A baiji (*Lipotes vexillifer*) eating fish, in China. (Photo by © Roland Seitre/Seapics.com. Reproduced by permission.)

Behavior

There has been little opportunity to study the baiji's behavior in the wild, particularly over the last decade or two when just finding a few animals has been a major challenge. Group size ranges from two to seven; groups occasionally form temporary aggregations of 15–20. Although baiji generally do not breach or exhibit aerial activity of any sort, they typically expose the head and beak on the first surfacing after a dive. Dives can last one to two minutes. These dolphins are strong swimmers; several animals were observed to move 60 mi (100 km) upriver against the Yangtze's current in just three days.

Feeding ecology and diet

Based on stomach contents of wild dolphins as well as the behavior of captives, the baiji's diet it believed to consist entirely of small fish. It consumes a large variety of species, the only limitation appearing to be the size of its mouth and throat. Most fish eaten are less than 2.6 in (6.5 cm) long and weigh less than 9 oz (250 g). Fish are ingested whole and headfirst.

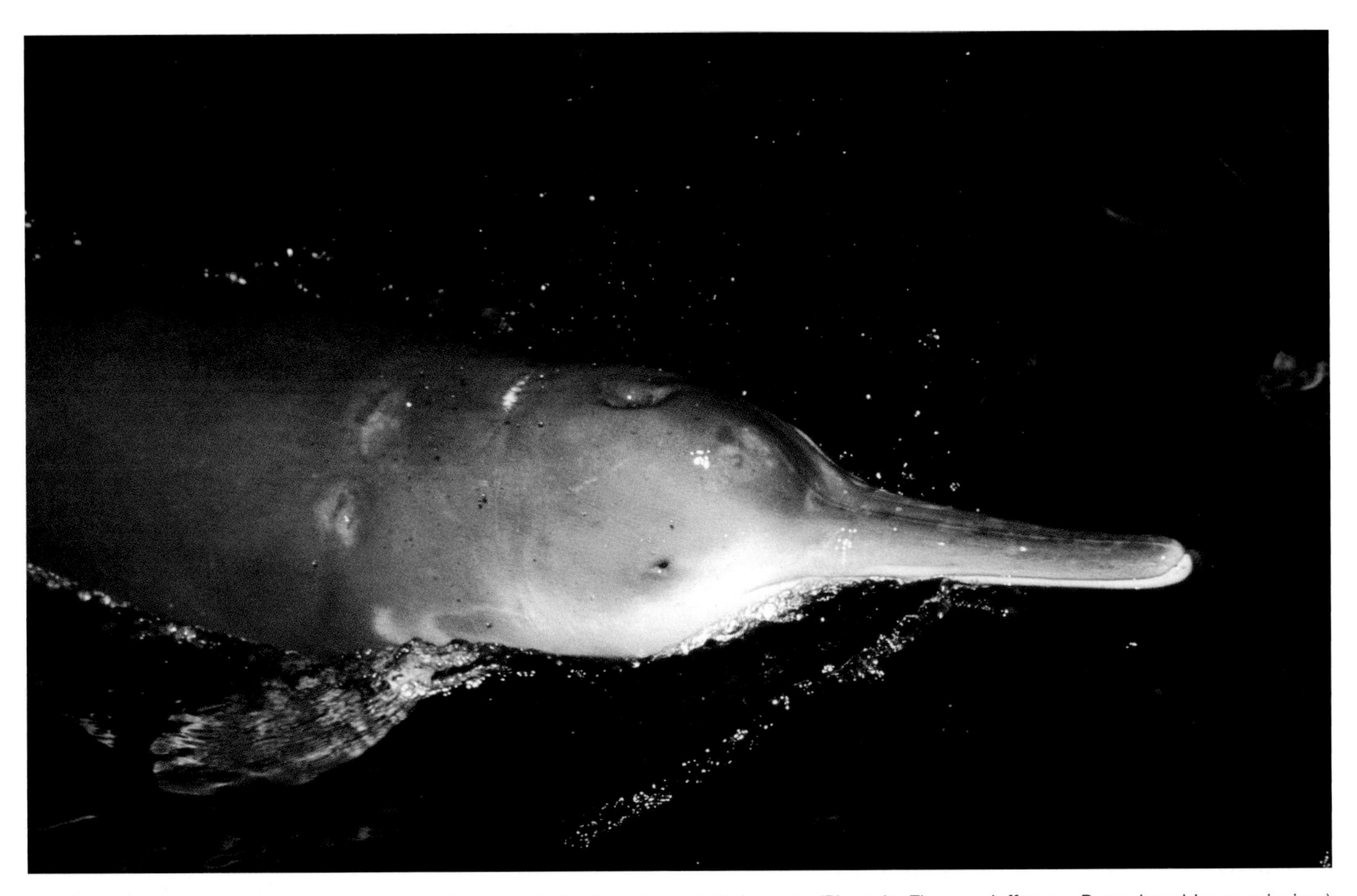

The baiji (*Lipotes vexillifer*) uses its long snout to unearth food at the water's bottom. (Photo by Thomas Jefferson. Reproduced by permission.)

Reproductive biology

Little is known because no observational research on baiji reproduction has been conducted. All that is known about the species' reproductive biology has come from examinations of specimens collected opportunistically, most of them killed incidentally in fishing gear. Females apparently become sexually mature at a body length greater than 6.5 ft (2 m). Males of approximately that length have mature, active testes. Single calves, about 3 ft (91 cm) long, are born mainly in spring, following gestation period of probably 10–11 months. Age at sexual maturation is about six (females) or seven (males) years.

Conservation status

The baiji is the most endangered species of cetacean, numbering only a few tens of individuals. It has probably been declining in abundance and range for a very long time, but there is little reliable information on absolute abundance or trends for any time period. Dolphins apparently were still common and widely distributed in the Yangtze when China's Great Leap Forward began in 1958. Intensive hunting for meat, oil, and leather ensued. Purchasing stations along the river received dead cetaceans from fishermen and supplied them to a central leather factory where bags and gloves were produced from baiji skin. A few hundred animals are believed to have survived as recently as the late 1970s, but the main threats—incidental mortality in fisheries, heavy vessel traffic, declining prey resources, and pollution—have continued unabated.

Since 1986, efforts have been made within China to develop "semi-natural reserves," with the intention of providing safe refuges for dolphins. These reserves were expected to provide opportunities for captive breeding and eventual restocking of the river. However, only one animal was captured—an adult female translocated to the Shishou Baiji Semi-natural Reserve near Wuhan in December 1995. She survived for six months, and during that time no effort was made to place her with the other captive baiji, a male that had been salvaged after becoming hooked and entangled in fishing line in 1980. This male died in 2002.

Despite full legal protection from deliberate harm since 1983, the baiji appears doomed. Its habitat has become thoroughly dominated by humans, and there is abundant evidence that intensive human use of the Yangtze is incompatible with the dolphin's survival.

Significance to humans

The baiji is characterized in Chinese folklore as "Goddess of the Yangtze." Legends and myths portray the dolphin as a friendly and beneficent creature, and it was long revered by fishing people along the Yangtze. Thus, the wanton killing of the late 1950s and 1960s went against traditional cultural norms and probably can be viewed as an aberration.

"Qi Qi," the male baiji held at the Wuhan Institute of Hydrobiology from 1980 to 2002, was a symbol of hope for the species. Most published baiji photographs and video footage depict "Qi Qi" in his tank. The symbolic importance of the baiji to aquatic conservation in China may be likened to that of the giant panda (*Ailuropoda melanoleuca*) to forest conservation. It appears, however, that the baiji will become extinct long before the giant panda, if for no other reason than because it has proven impossible to find, capture, and maintain significant numbers of these dolphins in captivity.

Resources

Books

Chen, P. "Baiji *Lipotes vexillifer* Miller, 1918." In *Handbook of Marine Mammals.* Vol. 4, *River Dolphins and the Larger Toothed Whales,* edited by S. H. Ridgway and R. Harrison. London: Academic Press, 1989.

de Muizon, C. "River Dolphins, Evolutionary History." In *Encyclopedia of Marine Mammals,* edited by W.F. Perrin, B. Würsig, and J. G. M. Thewissen. San Diego: Academic Press, 2002.

Perrin, W. F., R. L. Brownell Jr., K. Zhou, and J. Liu, eds. *Biology and Conservation of the River Dolphins: Occasional Papers of the IUCN Species Survival Commission No. 3.* Gland, Switzerland: IUCN, 1989.

Reeves, R. R., B. D. Smith, and T. Kasuya, eds. *Biology and Conservation of Freshwater Cetaceans in Asia: Occasional Papers of the IUCN Species Survival Commission No. 23.* Gland, Switzerland: IUCN, 2000.

Reeves, R. R., B. S. Stewart, P. J. Clapham, and J. A. Powell. *National Audubon Society Guide to Marine Mammals of the World.* New York: Alfred A. Knopf, 2002.

Zhou, K. "Baiji *Lipotes vexillifer.*" In *Encyclopedia of Marine Mammals,* edited by W. F. Perrin, B. Würsig, and J.G.M. Thewissen. San Diego: Academic Press, 2002.

Zhou, K., and Zhang, X. *Baiji, the Yangtze River Dolphin and other Endangered Animals of China.* Washington: Stone Wall Press, 1991.

Periodicals

Zhou, K., J. Sun, A. Gao, and B. Würsig. "Baiji (*Lipotes vexillifer*) in the Lower Yangtze River: Movements, Numbers, Threats and Conservation Needs." *Aquatic Mammals* 24 (1998): 123–132.

Zhou, K., W. Qian, and Y. Li. "Recent Advances in the Study of the Baiji, *Lipotes vexillifer.*" *Journal of Nanjing Normal College (Natural Sciences)* 1 (1978): 8–13.

Randall Reeves, PhD

▲

Franciscana dolphins

(Pontoporiidae)

Class Mammalia

Order Cetacea

Suborder Odontoceti

Family Pontoporiidae

Thumbnail description
Small gray or brown dolphin with very long and slender beak and prominent forehead; the flippers are broad with a squared trailing edge and the dorsal fin is triangular with a rounded tip

Size
Females 4.4–5.7 ft (1.34–1.74 m), males 4.1–5.2 ft (1.25–1.58 m); mature females weigh 75–117 lb (34–53 kg), males 64–94.8 lb (29–43 kg)

Number of genera, species
1 genus; 1 species

Habitat
Temperate coastal waters and estuaries

Conservation status
Data Deficient

Distribution
Central Atlantic waters of South America from Espírito Santo, Brazil (18°25′S) to Peninsula Valdés (42°35′S), Argentina

Evolution and systematics

Two fossils (*Pliopontes* and *Brachydelphis*), dating from early Pliocene and middle Miocene, were recovered from Peru. Uncertainty continues to surround the fossil *Parapontoporia*; some workers argue that it is similar to *Lipotes* (baiji) and include it in superfamily Lipotoidae.

Phylogenetic relationships continue to be debated. A consensus suggests a close association between the franciscana (*Pontoporia blainvillei*) and boto, or the Amazon River dolphin (*Inia geoffrensis*), forming a sister group of Delphinoidea (porpoises, monodontids, and marine dolphins). Thus, *P. blainvillei* is seen as distant from other river dolphins such as the Ganges and Indus river dolphin, *Platanista gangetica*.

Pontoporia blainvillei is the single member of the family Pontoporiidae. Together with *I. geoffrensis*, it forms the superfamily Inioidea. Classification of this species is still controversial. By some, it has been grouped with *Lipotes* in the family Pontoporidae, while other researchers combined the three species under the family Iniidae.

The taxonomy for this species is *Pontoporia blainvillei* (Gervais and d'Orbigny, 1844), mouth of the Rio de La Plata near Montevideo, Uruguay. Other common names include: English: La Plata river dolphin; French: Dauphin de la Plata; Spanish: Delfín de la Plata, tonina.

Physical characteristics

Franciscana is one of the smallest cetaceans, not exceeding 5.2 ft (1.58 m) in males and 5.7 ft (1.74 m) in females. The most distinct feature is the long and slender beak, which in adults reaches 15% of the total length. The mouth line is straight, curving slightly upward at the ends. The forehead is prominent, particularly in juveniles. The dorsal fin is triangular with a rounded tip. The flippers are broad and truncated, while the flukes are crescent-shaped with a medial notch. Like other river dolphins, all the cervical vertebrae are separated, providing great flexibility. Coloration is dark gray or brown, lighter ventrally and on the lower flanks.

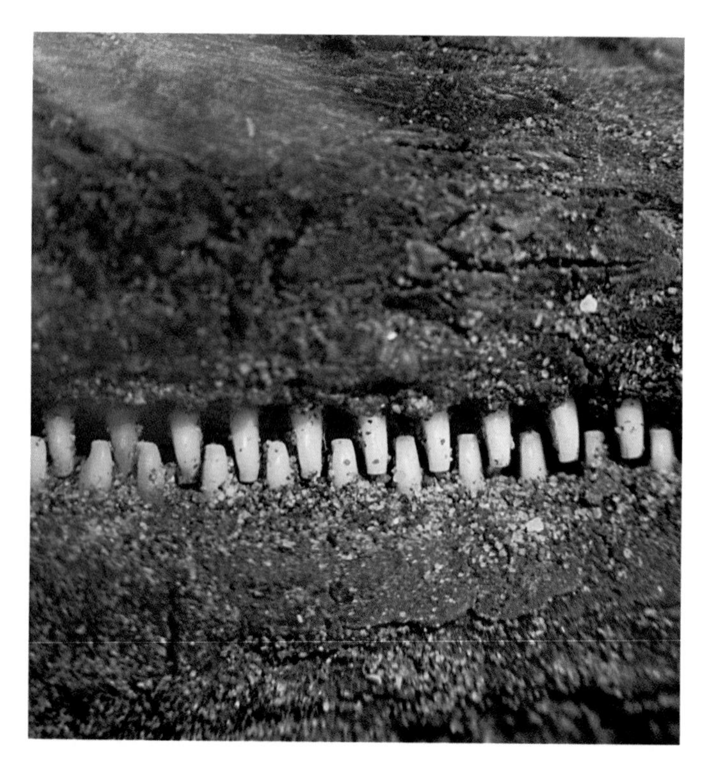

The franciscana dolphin (*Pontoporia blainvillei*) eats bottom-dwelling fish found with echolocation. Shown here is a close-up of its teeth. (Photo by Aníbal Parera. Reproduced by permission.)

The brain is the smallest among cetaceans (0.45–0.55 lb [205–250 g]).

Distribution

Franciscana occurs in the coastal central Atlantic waters of South America. Currently, it ranges from Espirito Santo in southeastern Brazil to Golfo Nuevo, Peninsula Valdés in Argentina. Although related to other river dolphin species, it is primarily a marine species. Franciscana is the only representative of the four river dolphins to inhabit marine waters. It is also known as La Plata River dolphin because the first described specimen was from the mouth of La Plata River in Uruguay.

Morphometrical and genetic differences between dolphins occurring south and north of Santa Catarina, Brazil, suggest the existence of at least two populations.

The range of franciscana overlaps, at least partially, with other small odontocete species such as the bottlenosed dolphin (*Tursiops truncatus*), dusky dolphin (*Lagenorhynchus obscurus*), tucuxi (*Sotalia fluviatilis*), and Burmeister's porpoise (*Phocoena spinipinnis*).

Habitat

Franciscana is found in coastal waters and estuaries, mainly in shallow waters less than 100 ft (30 m) deep and within 30 nautical miles (56 km) of shore. Occasionally, it is found further offshore in waters up to 200 ft (60 m) deep. In much of its range, the water is very turbid. Franciscanas are often found in areas of high turbulence such as countercurrents and eddies.

Unlike the case in Brazilian waters, seasonal movements of franciscanas have been reported in Argentine waters, in which dolphins move offshore during the winter. Seasonality in Argentina has been attributed to the marked variation in water temperature (42.8–69.8°F [6–21°C]), in contrast to the more constant temperature in Brazilian waters (68–75.2°F [20–24°C]).

Behavior

Franciscana are found in small groups of one to 15 dolphins, typically fewer than six animals. Their social organization remains unknown, partly because these dolphins are difficult to detect and observe. They seem to avoid boats and spend very little time (4%) at the surface. Moreover, their coloration closely matches that of the murky waters they inhabit, and their surfacing behavior is very inconspicuous. Typically, the beak emerges first, followed by the head, then the body arches forward, exposing little more than the dorsal fin.

Franciscanas produce low- to high-frequency clicks that are associated with echolocation used for navigation and foraging.

Sevengill (*Notorlynchus cepedianus*) and hammerhead sharks (*Sphyrna* spp.), and possibly killer whales (*Orcinus orca*), are believed to be their only predators. However, direct attacks by these species have not been documented.

In parts of its range, feeding and breeding is more prevalent near shore. Tides may influence activity patterns. For example, feeding increases during high tide.

Feeding ecology and diet

Franciscanas have a diversified diet, consisting of at least 24 species of fish, cephalopods (e.g., squid, octopus), and, less importantly, crustaceans such as shrimp. The target fish species vary across their range, but mostly consist of demersal (bottom-dwelling) species. Common prey fish species belong to the families Sciaenidae (croakers, drums), Engraulidae (anchovies), and Batrachoididea (toadfishes). Prey is mostly juvenile fish, less than 3.9 in (10 cm) long.

Franciscana may engage in cooperative feeding by swimming in a circle in a coordinated manner, thus concentrating the fish in the center.

Reproductive biology

Sexual maturity is reached between two to five years in females and two to three years in males. Researchers have reported differences across its range of about one year in the age of sexual maturity of females: Uruguay (2.8 years), Brazil (3.7 years), and Argentina (4.5 years). However, more current data are needed for Uruguay to allow reliable comparisons,

Franciscana dolphin (*Pontoporia blainvillei*). (Illustration by Barbara Duperron)

since estimates there are based on data collected more than 25 years ago.

After 11 months of gestation, females give birth to one calf, whose length may range 2–2.6 ft (0.6–0.8 m). Lactation may last nine months, but calves may start feeding on prey as early as three months of age. Females may give birth every year or every other year. Calving is seasonal in some areas, occurring from September to December. In northern Brazil, calving occurs throughout the year.

Compared to most odontocetes, franciscana has a lower age of sexual maturity, shorter calving intervals, and a very short life span (estimated at 15 years for females and 18–20 years for males). Their mating system is unknown.

Conservation status

Franciscana is endemic to the southwest Atlantic. Sadly, it is the most rare, and among the most poorly understood, of the South American dolphins. Its distribution makes it particularly vulnerable to entanglements in gillnets used in coastal fisheries. It is one of the most threatened small cetaceans in the southwest Atlantic due to substantial incidental takes in fisheries. Entanglements occur both in surface and bottom gillnets. Gillnet fisheries target sharks, anchovies, and sciaenids, depending on the region. Incidental catches occur throughout its range but are of particular concern on the coasts of southern Brazil (Rio Grande do Sul) and Uruguay. In these areas, density of franciscana is estimated to be 1.8 dolphins per mi^2 (0.7 dolphins per km^2), and annual catches reach 550–1,500 dolphins. These takes correspond to an annual removal rate of 1–3.5% of the stock. It is estimated that a 2% rate of removal may not be sustainable for this population.

Despite the growing pressure resulting from the rapid expansion of the coastal fisheries, franciscana is classified as Data Deficient in the IUCN Red List. This is because estimates of abundance and incidental mortality are unavailable for its entire range. A recent study using population viability analysis (PVA) to model the impact of incidental catches in southern Brazil found that this population is decreasing. The same model predicted that if the current incidental catch level persists, this population could plummet to as little as 10% of its current abundance within 25 years.

National regulations in all three countries of its range prohibit hunting.

Significance to humans

Blubber of dolphins caught incidentally is occasionally used as bait in the longline shark fishery. However, franciscanas are not targeted for this purpose. In Uruguay, the blubber oil has been used in the tanning industry, and carcasses are discarded or used as fish flour.

The fin of a franciscana dolphin (*Pontoporia blainvillei*). (Photo by Aníbal Parera. Reproduced by permission.)

Resources

Books

Klinowska, Margaret. *Dolphins, Porpoises and Whales of the World.* Switzerland: World Conservation Union (IUCN), 1991.

Ridgway, Sam H., and Richard J. Harrison. *Handbook of Marine Mammals.* Vol. 4. London: Academic Press, 1989.

Periodicals

Bordino, Pablo, Gustavo Thompson, and Miguel Iniguez. "Ecology and Behaviour of the Franciscana (*Pontoporia blainvillei*) in Bahia Anegada, Argentina." *Journal of Cetacean Research and Management* 1, no. 2 (1999): 213–222.

Di Beneditto, Anna P. M., and Renata M. A. Ramos. "Biology and Conservation of the Franciscana (*Pontoporia blainvillei*) in the North of Rio de Janeiro State, Brazil." *Journal of Cetacean Research and Management* 3, no. 2 (2001): 185–192.

Kinas, Paul G. "The Impact of Incidental Kills by Gillnets on the Franciscana Dolphin (*Pontoporia blainvillei*) in Southern Brazil." *Bulletin of Marine Science* 70, no. 2 (2002): 409–421.

Nikaido, M., et al. "Retroposon Analysis of Major Cetacean Lineages: The Monophyly of Toothed Whales and the Paraphyly of River Dolphins." *Proceedings of the National Academy of Sciences* 98, no. 13 (June 2001): 7384–7389.

Zhou, Kaiya. "Classification and Phylogeny of the Superfamily Platanistoidae, with Notes on Evidence of the Monophyly of the Cetacea." *Scientific Reports of the Whales Research Institute* 34 (1982): 93–108.

Organizations

Grupo de Estudos de Mamiferos Aquaticos do Rio Grande do Sul (GEMARS). Rua Felipe Neri, 382/203, Porto Alegre, RS 90440-150 Brazil. Phone: (51) 335-2886. Fax: (51) 267-1667. E-mail: gemars@zaz.com.br

United Nations Environmental Programme (UNEP). PO Box 30552, United Nations Avenue, Gigiri, Nairobi, Kenya. Phone: 254 (2) 621234. Fax: 254 (2) 624489. E-mail: eisinfo@unep.org Web site: <http://www.unep.org>

The Whale and Dolphin Conservation Society (WDCS). 38 St Paul Street, Chippenham, Wiltshire, SN15 1LY United Kingdom. Phone: 44 (0) 1249-449500. Fax: 44 (0) 1249-449501. E-mail: info@wdcs.org Web site: <http://www.wdcs.org>

Paula Moreno, MS

▲

Botos

(Iniidae)

Class Mammalia

Order Cetacea

Suborder Odontoceti

Family Iniidae

Thumbnail description
Pink or grayish dolphin with narrow, long beak and prominent forehead; long flippers and low dorsal crest

Size
Maximum length is 8.4 ft (2.55 m) for males and 6.6 ft (2 m) for females; weight is 345 lb (156.5 kg) for males and 217 lb (98.5 kg) for females

Number of genera, species
1 genus; 1 species

Habitat
Rivers, tributaries, floodplains, and inundated forests

Conservation status
Vulnerable

Distribution
South America in the Orinoco and Amazon river systems

Evolution and systematics

Ischyrorhyncus, a fossil from the late Miocene found in Argentina, is a confirmed ancestor of *Inia geoffrensis* that was present in freshwater. Precisely how ancestors of boto entered the Amazon remains unsolved. Two theories posit the timing and origin of colonization in freshwater: one suggests an earlier colonization from the Pacific Ocean about 15 million years ago (mya), while the second proposes a more recent one (1.8–5 mya) from the Atlantic Ocean. In spite of contentious issues of phylogeny and taxonomy, a growing body of morphologic, genetic, and fossil evidence position *I. geoffrensis* closest to franciscana (*Pontoporia blainvillei*), and suggest that these two also are strongly related to Delphinoidea. Compared to other river dolphins, *I. geoffrensis* is the most recent taxa. Presently, *I. geoffrensis* is divided into three subspecies: *I. g. geoffrensis* (mainstem Amazon River); *I. g. boliviensis* (Amazon in eastern Bolivia); and *I. g. humboldtiana* (Orinoco River). This analysis is supported by morphologic studies and, in the case of the first two subspecies, also by molecular evidence.

Boto (*Inia geoffrensis*). (Illustration by Patricia Ferrer)

The taxonomy for this species is *Inia geoffrensis* (Blainville, 1817), probably Upper Amazon. Other common names include: English: Amazon River dolphin, pink river dolphin; French: Dauphin de l'Amazone, inia; Spanish: Bufeo.

Physical characteristics

Bulky, with a prominent forehead and a long snout with sparse hair. The dorsal fin is reduced to a long dorsal crest. The flippers are large, paddle-shaped, and can rotate freely. The peduncle is compressed laterally and the flukes have a concave trailing edge with a medial notch. Botos are very flexible, largely due to free cervical vertebra.

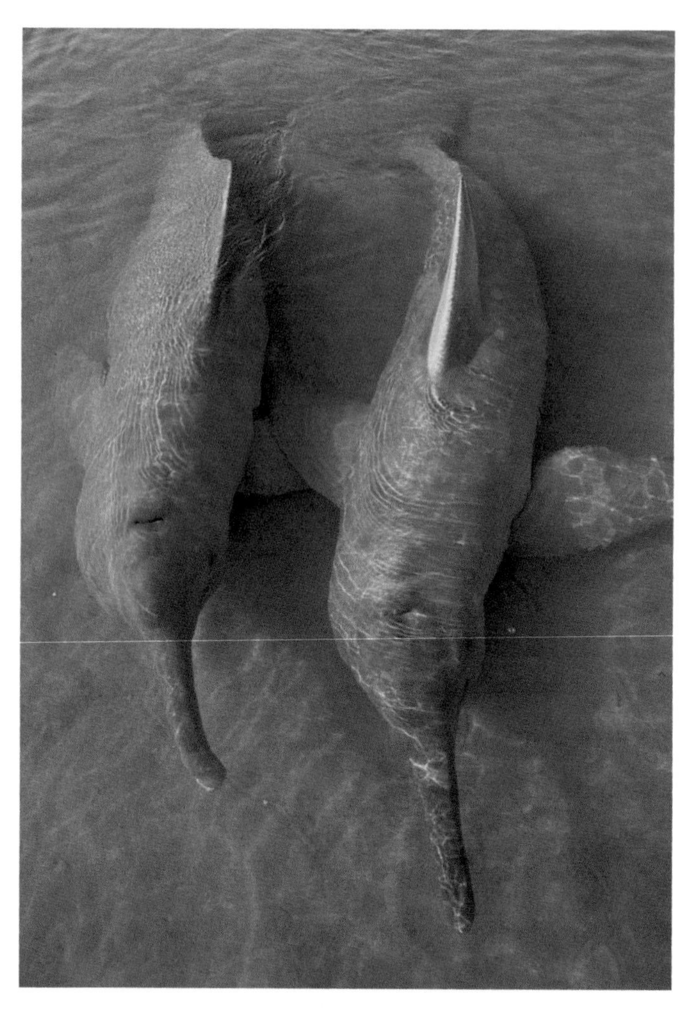

Botos (*Inia geoffrensis*) are able to swim in very shallow water. (Photo by Flip Nicklin/Minden Pictures. Reproduced by permission.)

Their eyes are very small but, unlike the Ganges and Indus dolphin, they provide good vision, both underwater and above it.

Coloration varies, apparently related to age and environmental factors such as water clarity. Calves are dark gray, while adults are pink. Dolphins living in "black waters" are generally darker than those from "white waters."

Distribution

Widely distributed in the main rivers of Amazon and Orinoco basins throughout Venezuela, Colombia, Ecuador, Guyana, Peru, Bolivia, and Brazil.

Fast-moving water such as Teotonio rapids in the Madeira River constitute barriers to the species' dispersal.

Current or historical abundance estimates for their entire range are unavailable. A growing number of surveys have provided local densities. This is the most abundant of the river dolphins, but population numbers are generally unknown.

Habitat

Exclusively a freshwater species, the boto occupies rivers, tributaries, and lakes, but does not tolerate the brackish waters of estuaries. It has an affinity for high turbulence zones, often where tributaries merge and where there are higher prey concentrations. The Amazon and Orinoco drainages undergo seasonal flood cycles, causing the water level to rise or fall up to 33 ft (10 m). The rainy season is from November to May and the dry season is from June to October. During flooding, dolphins disperse into floodplains and inundated forests where plentiful seeds and fruits attract fish. When the waters recede, the dolphins migrate back into the main channels of rivers or into deep pools. Water temperature where boto occurs ranges from 73.4 to 86°F (23–30°C). It shares its habitat with another river dolphin, *Sotalia fluviatilis* (tucuxi), which is not strictly a freshwater dolphin.

Behavior

Typically, sightings consist of single individuals, pairs, or a few individuals to a few dozen. Botos swim slowly but are extremely maneuverable, negotiating shallow waters and making their way around obstacles such as woody debris. A common mode of surfacing is ascending horizontally and showing only the top of the head. A much briefer surfacing, associated with deep-diving, occurs when a boto breaks the water with the snout and then exposes the dorsal crest by arching and rolling forward. Leaping out of the water is rare. Sometimes, the dolphins' presence can be detected even without seeing them because of the loud blowing sound produced when they exhale.

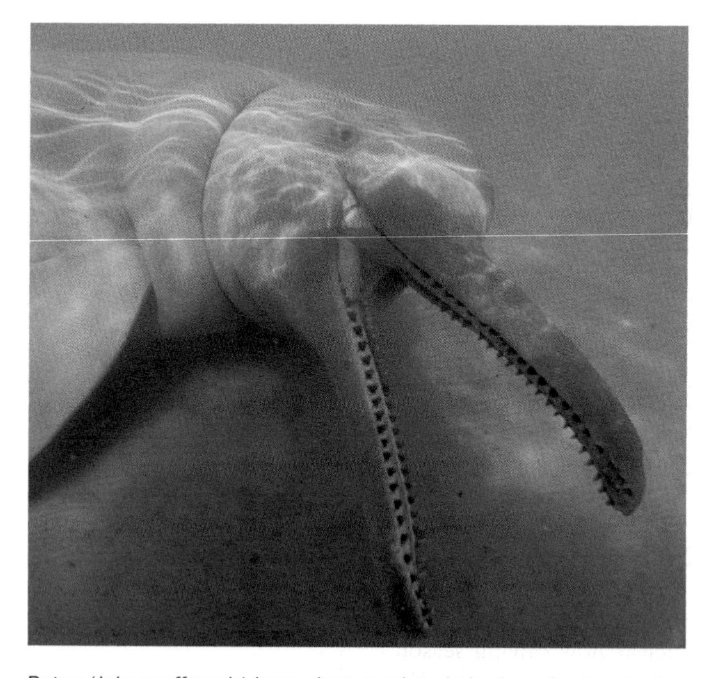

Botos (*Inia geoffrensis*) have sharp teeth to help them feed on turtles and crabs. (Photo by Flip Nicklin/Minden Pictures. Reproduced by permission.)

A boto (*Inia geoffrensis*), also known as a pink river dolphin, swimming in the Orinoco River. (Photo by Gregory Ochocki/Photo Researchers, Inc. Reproduced by permission.)

Botos are very curious and playful, using almost any object available (logs, turtles, or paddles) as toys. This phenomenon is also observed in captivity.

Nearly 40 dolphins were tracked with radio transmitters in Mamiraua System, Brazil, showing some individuals to be year-round residents. There is evidence that the boto uses echolocation for navigation and to capture prey. The sounds they produce are known as "clicks," which are predominantly between 85 and 100 kHz, well above the frequency audible to human beings. The boto also produces low-frequency sounds. Some researchers have recorded only pulsed sounds, while others have also recorded whistling. Further research is needed to determine whether the boto produces whistles and, if so, whether these sounds have similar functions to those made by oceanic species.

Feeding ecology and diet

Their agility allows the boto to pursue and capture fish in shallow or deep channels and in densely vegetated areas. Feeding tends to occur more often during early morning or in the afternoon. However, nocturnal fish species have also been found among stomach contents, suggesting that night feeding may occur.

The boto diet includes over 40 species of fish, mainly sciaenids, cichlids, and characins. The size of fishes varies 2–31 in (5–80 cm), but is typically 8 in (20 cm). Crustaceans, mollusks, and turtles have also been found in stomachs. Their teeth are differentiated into conical and molar-like teeth that facilitate masticating prey with hard exteriors. This dentition is unique among dolphins, which commonly have equal-sized conical teeth suited for grasping.

Reproductive biology

Reproduction is seasonal. Peak of births varies with river systems, but is generally between May and August. It is still unclear how calving season relates to the flood cycle of the rivers. Best and da Silva (1989) reported that in the Amazon, births appeared to be associated with seasonally receding waters and suggested that this timing would be favorable for pregnant and lactating females because prey fish would become more concentrated as surface waters diminished. However, in a tributary of the Orinoco River, McGuire and Winemiller (1998) observed calves when water levels began to rise, but never witnessed them during the period of falling waters. Lactation lasts more than one year, and females give birth every two to three years.

Sexual maturity (about five years old) is reached at total body lengths of 6.6 ft (2 m) and 5.2–5.9 ft (1.6–1.8 m) in males and females, respectively. Gestation is estimated at 10–11 months. The length at birth is approximately 2.6 ft (0.8 m).

The species' mating system is unknown. This is partly because their behavior and morphology hinder identification of individuals.

Conservation status

Although its range does not appear to be shrinking, the boto faces many serious threats to its habitat. Thus, it is classified as Vulnerable. National protective regulations are still incomplete, and most are recent. In Brazil, botos became protected in 1986.

Hydroelectric dams are common throughout the Orinoco and Amazon Rivers, and construction of many more are planned. Dams isolate dolphin populations into ever-smaller groups, making them increasingly vulnerable to environmental changes because of reduced genetic diversity. Moreover, dams and extensive deforestation reduce fish productivity. Pollution is another concern: mercury used in gold mining and pesticides, which tend to bio-accumulate in animals

A boto (*Inia geoffrensis*) leaping, near Venezuela. (Photo by © Fernando Trujillo/Seapics.com. Reproduced by permission.)

Botos (*Inia geoffrensis*) lack distal and dorsal fins. (Photo by Flip Nicklin/Minden Pictures. Reproduced by permission.)

higher on the food chain, are particularly detrimental to top predators like dolphins. Females transfer pollutants to their calves while nursing. Very few studies have reported on levels of toxic substances in fish and dolphins. Pesticides and high concentrations of mercury—close to levels considered toxic for humans—were detected in milk of botos. Pulp mills are also an important source of pollution. As the use of gill nets and seine nets increases, so does the incidence of entanglements. Although illegal, dynamite fishing still occurs and attracts dolphins to feed on stunned fish, exposing them to subsequent blasts. Preservation of boto habitat will require implementation of environmental impact assessment for new dams and monitoring pollution levels.

Significance to humans

During the 1960s and 1970s, about 100 dolphins were captured for exhibition at aquaria and oceanaria. Of these, about 70% went to the United States. However, high mortality occurred. Presently, only a handful are still exhibited worldwide. Many of these deaths resulted from aggression between the dolphins. Considering that botos do not generally occur in large groups, it is likely that confining several botos in small enclosures prevented them from maintaining minimum necessary spacing.

The boto once was a strong cultural influence on indigenous people—the focus of many folk tales and superstitions. It was both respected and feared for its supposed supernatural powers and its reputed ability to bring misfortune. Currently, dolphins interfere with fisheries, damaging nets while stealing fish. There are no reports of direct dolphin killings other than to stop destruction of fishing gear. However, when they are killed or found dead, genitalia and eyes are sometimes traded as love charms.

Resources

Books

Klinowska, Margaret. *Dolphins, Porpoises and Whales of the World.* Gland, Switzerland: World Conservation Union (IUCN), 1991.

Ridgway, Sam H., and Richard J. Harrison. *Handbook of Marine Mammals.* Vol. 4. London: Academic Press, 1989.

Periodicals

Ding, Wang, Bernd Würsig, and Stephen Leatherwood. "Whistles of Boto, *Inia geoffrensis*, and Tucuxi, *Sotalia fluviatilis.*" *Journal of the Acoustical Society of America* 109, no. 1 (January 2001): 407–411.

Hamilton, H., S. Caballero, A. G. Collins, and R. L. Brownell Jr. "Evolution of the River Dolphins." *Proceedings of the Royal Society of London* 268 (2001): 549–558.

Kamminga, C., M. T. Van Hove, F. J. Engelsma, and R. P. Terry. "Sonar X: A Comparative Analysis of Underwater Echolocation of *Inia* spp. and *Sotalia* spp." *Aquatic Mammals* 19, no.1 (1993): 31–43.

McGuire, Tamara L., and Kirk O. Winemiller. "Occurrence Patterns, Habitat Associations, and Potential Prey of the River Dolphin, *Inia geoffrensis*, in the Cinaruco River, Venezuela." *Biotropica* 30, no. 4 (1998): 625–638.

Messenger, Sharon L., and Jimmy A. McGuire. "Morphology, Molecules, and the Phylogenetics of Cetaceans." *Systematic Biology* 47, no. 1 (1998): 90–124.

Perrin, W. F., R. L. Brownell Jr., K. Zhou, and J. Liu. "Biology and Conservation of the River Dolphins." *Occasional Papers, IUCN Species Survival Commission*, no. 3 (1989).

Podos, Jeffrey, Vera M. F. da Silva, and Marcos R. Rossi-Santos. "Vocalizations of Amazon River Dolphins, *Inia geoffrensis*: Insights into the Evolutionary Origins of Delphinid Whistles." *Ethology* 108 (2002): 601–612.

Rosas, Fernando C. W., and Kesae K. Lehti. "Nutritional and Mercury Content of Milk of the Amazon River Dolphin, *Inia geoffrensis.*" *Comparative Biochemistry and Physiology* 115A, no. 2 (1996): 117–119.

Yang, G., K. Zhou, W. Ren, G. Ji, and S. Liu. "Molecular Systematics of River Dolphins Inferred From Complete Mitochondrial Cytochrome-B Gene Sequence." *Marine Mammal Science* 18, no. 1 (2002): 20–29.

Organizations

Instituto Nacional de Pesquisas da Amazonia (INPA). Alameda Cosme Ferreira 1796, Aleixo, Manaus, 69011-970 Brazil. Phone: (092) 643-3184. Fax: (092) 643-3292. E-mail: tucuxi@cr-am.rnp.br Web site: <http://www.cnpq.br>

Omacha Foundation. Web site: <http://www.omacha.org>

The World Conservation Union (IUCN), Species Survival Commission. Rue Mauverney 28, Gland, 1196 Switzerland. Phone: 41 (22) 999-0152. Fax: 41 (22) 999-0015. E-mail: ssc@iucn.org Web site: <http://www.iucn.org>

Paula Moreno, MS

Porpoises
(*Phocoenidae*)

Class Mammalia
Order Cetacea
Suborder Odontoceti
Family Phocoenidae

Thumbnail description
Small to medium dolphin-like aquatic carnivores with dark gray to black body covering and a pale gray underbelly, the absence of a distinct beak, compressed spatula-shaped teeth, fused neck vertebrae, small triangular-shaped dorsal fins, and a rounder shape than their dolphin relations

Size
4.8–6.5 ft (1.4–2.0 m); 90–485 lb (40–220 kg)

Number of genera, species
3 genera; 6 species

Habitat
Oceans, bays, harbors, estuaries, rivers; deep, shallow, benthic, and pelagic

Conservation status
Critically Endangered: 1 species; Threatened: 1 species; Lower Risk/Conservation Dependent: 1 species; Data Deficient: 3 species

Distribution
Globally distributed across the northern Pacific and Atlantic Oceans, along the coasts of South America, and the northern and southern Indian Ocean

Evolution and systematics

While dolphins and porpoises are both related to the squalodonts (earliest true-toothed whales) and the kentridontids (ancestral dolphins), porpoises have been distinct and set apart from dolphins for approximately 11 million years.

Porpoises started to appear in the fossil record during the Miocene epoch, around 10 to 12 million years ago (mya). Scientists conjecture that they probably resembled modern-day finless porpoises, *Neophocaena phocaenoides*, and lived in the same type of habitat—the warm, tropical waters of the Pacific Ocean.

According to the fossil record, the original Neophocaenid ancestor that later colonized the temperate and subarctic regions of the North Pacific from the late Miocene onwards, most likely gave rise to the genus *Phocoenoides*, to which the Dall's porpoise (*P. dalli*) also belongs. The Phocoenid porpoises emerged sometime in the middle of the Pliocene. By that time, the genus *Phocoena* appeared to have already colonized both the Northern and Southern Hemispheres. It has been hypothesized that their migratory movements most likely coincided with the equatorial inflow of cooler subtropical waters during the cooler periods of the Pliocene, or in another scenario, during the glacial intervals in the Pleistocene. The original ancestor in the north is believed to have led to the emergence of the harbor porpoise (*P. phocoena*), while its southern radiations resulted in at least two other species—Burmeister's porpoise (*P. spinipinnis*) and the spectacled porpoise (*P. dioptrica*).

The vaquita (*P. sinus*), believed to be the "youngest" species in the genus, appears to have descended from the modern Burmeister's porpoise after it migrated to the Northern Hemisphere during the cooler periods of the late Pleistocene but was later cut off in the upper Sea of Cortez when tropical waters warmed up and the last glaciers made their retreat.

Physical characteristics

Typical of most porpoises is the absence of a forehead or beak on their small rounded heads. The six neck vertebrae of a porpoise are extremely foreshortened and fused together to create an immobile neck, with the body tapering down to a pair of notched flukes. Contrary to the hooked or curved dorsal fins of dolphins, the ones found in most porpoises are well defined and triangular in shape and matched by small pointed or rounded flippers usually positioned near the head. The leading edge of the dorsal fin may also be lined with rows of tubercles, or circular bumps. Unlike the homodont, conical teeth of dolphins, porpoises possess spatulate (spade-shaped) teeth in both jaws.

While female porpoises are generally larger than the males—except in spectacled porpoises, where the reverse holds true—a considerable range of morphological variation exists between the different species of porpoises. Differences in stockiness, robustness, and coloration can sometimes occur even within individual species. For instance, three different color patterns have been observed in Dall's porpoise alone:

Dall's porpoise (*Phocoenoides dalli*) inhabits the cooler waters of the North Pacific. (Photo by © Robert L. Pitman/Seapics.com. Reproduced by permission.)

uniform black or uniform white; inter-mixed stripes of black and white across the length of the body; and a solid black dorsal portion with a white underside. The finless porpoise, which resembles a small Beluga whale (*Delphinapterus leucas*), has a light coloration which darkens slightly with age and quickly turns black after death. Eight stocks of Dall's porpoise, found in the Pacific waters of Japan, are recognized by the International Whaling Commission, with each stock having its own unique color morphotype.

Most species have significant characteristics that distinguish them from one another. The spectacled porpoise has eyes surrounded by a black circle edged in white, giving the illusion that the mammal is wearing eyeglasses. The finless porpoise has a lack of dorsal fin, while the harbor porpoise has a pronounced keel, or distinctive bulge on the tail stock near the flukes.

Distribution

It is commonly believed that porpoises originated in the North Pacific and then later spread to the Atlantic and southern waters, although they can be sighted today in rivers and estuaries as well. Harbor porpoises can be found in the temperate waters of the Northern Hemisphere and generally inhabit coastal waters with a depth of less than 500 ft (152 m). In fact, their common name is derived from regular appearance in bays and harbors. Spectacled porpoises are only found in the temperate and subantarctic waters of the Southern Hemisphere, while Burmeister's porpoises can only be located in the coastal waters of South America, around the southern coastline of Tierra del Fuego towards northern Peru.

Dall's porpoises are only found in the Pacific Ocean in the Northern Hemisphere and range across the entire North Pacific at latitudes greater than 32°N but not any further beyond the lower, deeper half of the Bering Sea. Generally oceanic in nature, they also appear to prefer cold waters and are not usually found in the southern extremes of their range during the summer months.

Vaquitas, the smallest of all porpoises and perhaps even all cetaceans, have the most restricted range of any marine cetacean as they appear to live only in the northern end of the Gulf of California. Most sightings of vaquita are in shallow water of less than 130 ft (40 m) and within 16 mi (25 km) of shore.

More widely distributed are the finless porpoises, which can be found in the coastal waters of Asia from the Persian Gulf, east and north to central Japan, and as far south as the northern coast of Java and the Strait of Sunda. Frequently sighted near the coast, they are described as a coastal, estuarine, or riverine species.

Studies of the harbor porpoise in Monterey Bay suggest a seasonal north-to-south migratory pattern along the coast of California, according to the seasonal availability of prey items. While this wide range of distribution suggests a significant population of harbor porpoises, specific numbers remain unknown.

Habitat

Because of their long-range migratory patterns, porpoises live in a variety of habitats. While the spectacled porpoise prefers the cold temperate waters and subantarctic waters of the Southern Hemisphere, specifically the coastal waters of eastern South America, for instance, the finless porpoise prefers saltwater and freshwater environs, such as estuaries, mangroves, and rivers. Porpoises may be both benthic or pelagic. For example, Burmeister's porpoises inhabit the shallow temperate waters of coastal South America, but Dall's

A harbor porpoise (*Phocoena phocoena*) in a kelp bed. (Photo by © Florian Graner/Seapics.com. Reproduced by permission.)

The finless porpoise (*Neophocaena phocaenoides*) can be found in both fresh and saltwater environments. (Photo by WANG Xiaoqiang and WANG Ding. Reproduced by permission.)

porpoises are found in the deeper waters of the northern North Pacific and Bering Sea.

Behavior

Frequently elusive, secretive, unapproachable, and wary of human presence, porpoises are rarely observed in groups of more than a few individuals. Vaquitas in particular intentionally seem to avoid boats. On the other hand, the Yangtze River populations of finless porpoises are atypically unfazed by boats or people, probably because they have become accustomed to the river's heavy traffic. They have also been observed in groups of five to 10 individuals and even in pods of 50 members, perhaps in order to take advantage of rich feeding grounds.

Feeding ecology and diet

Dietary preferences naturally follow the range of habitat that porpoises are situated in. Harbor porpoises are deep divers, capable of reaching depths in excess of 650 ft (200 m) and their diet tends to consist of herring, capelin, and gadoid fishes such as pollack and hake. Recently weaned porpoises eat euphausiid shrimp. Vaquitas, on the other hand, feed primarily on teleost fishes and squids that are commonly found in the demersal and benthic zones of the shallow waters of the upper Gulf of California, while Burmeister's porpoises feed primarily on anchovy and hake, although squid, mysid shrimp and euphasiids are also consumed. Burmeister's porpoises in Chilean waters also appear to eat mollusks. In the Pacific Ocean, Dall's porpoises feed on a wide variety of fish and cephalopods, most of which are deepwater or vertically migratory in nature. Almost nothing is known about the food preferences of spectacled porpoises although a single stranded animal found in Argentina had anchovy and small crustaceans in its stomach.

Reproductive biology

Porpoises, on average, become sexually mature between three and five years of age, after which the females produce one calf annually. Calving season for Dall's porpoises take place during the summer from June to September and gestation lasts about 11 months. Subsequently, mothers tend to nurse their young for approximately two years.

Knowledge of porpoise reproductive biology is not uniform for the various species. The vaquita, because of its elusiveness, is virtually undocumented, although it has been noted that their juveniles have white spots on the leading edge of the dorsal fin.

Conservation status

Many harbor porpoise populations around the world have been depleted through bycatch by fisheries, with chemical and noise pollution acting as a contributing factor. In North America alone, bottom-set gill nets catch hundreds of porpoises annually. This has led some countries to afford them special status. In Atlantic Canada, harbor porpoises are listed as Threatened by the Committee on the Status of Endangered Wildlife in Canada (COSEWIC). Northwest Atlantic harbor porpoises are designated as a Strategic Stock under the U. S. Marine Mammal Protection Act (MMPA) because current levels of killing exceed the estimated Potential Biological Removal (PBR) level for the population. Under the IUCN Red List of Threatened Species, harbor porpoises are listed as Vulnerable.

Very little is known about the abundance of the vaquita, although a 1997 abundance survey, jointly conducted by the National Fisheries Institute of Mexico and the U.S. National Marine Fisheries placed an estimate at 547 animals, with a 95% confidence interval of between 177 and 1,073 individuals. Vaquitas are currently listed on the IUCN Red List of Threatened Species as Critically Endangered. The greatest

The finless porpoise (*Neophocaena phocaenoides*) has a beakless head. (Photo by WANG Xiaoqiang and WANG Ding. Reproduced by permission.)

A group of finless porpoises (*Neophocaena phocaenoides*). (Photo by Thomas Jefferson. Reproduced by permission.)

threat to remaining populations of vaquita is incidental mortality in fishing gear since vaquitas are known to die in gill nets legally set for sharks, rays, mackerel, and chano, and illegal but occasionally permitted gill nets set for totoaba (*Totoaba macdonaldi*), an endangered species of fish.

There are no abundance estimates for the spectacled porpoises, Burmeister's porpoises, or finless porpoises, although bycatch and the occasional harpooning for bait or human consumption are also seen as the largest threats to their populations. The IUCN list them as Data Deficient under the Red List of Threatened Species. Threats to finless porpoises in the Yangtze River include incidental mortality from entanglement in passive fishing gear, electric fishing, collisions with powered vessels, and exposure to explosives used for harbor construction. Most of their habitat, according to the IUCN, has undergone severe degradation due to the damming of the Yangtze tributaries and the high volume of traffic in the river.

Significantly more abundant are the Dall's porpoises, which are listed as Lower Risk/Conservation Dependent by IUCN. One estimate places their numbers in the North Pacific and Bering Sea at 1,185,000. Despite their existing numbers, these porpoises have also been taken in large numbers in a variety of Asian-based pelagic drift net fisheries for salmon and squid. Directed fisheries may also pose a threat to their populations.

Several mitigation measures have been put in place to reduce the large bycatches of porpoises, including an acoustic deterrent device known as a "pinger," which alerts porpoises to echolocate in the presence of nets. However, it was discovered that harbor porpoises habituate to pingers, thus reducing their effectiveness over time.

Significance to humans

From the 1830s until the end of World War II, a major fishery for harbor porpoises in the Lille Bælt in Denmark took several hundred to more than a thousand animals annually. Before Turkish fisheries were suspended in 1983, 34,000–44,000 animals were taken per year between 1976 and 1981, with harbor porpoises making up about 80% of the total catch.

The harbor porpoise (*Phocoena phocoena*) must come up for air about every 25 seconds. (Photo by © Armin Maywald/Seapics.com. Reproduced by permission.)

1. Burmeister's porpoise (*Phocoena spinipinnis*); 2. Harbor porpoise (*Phocoena phocoena*). (Illustration by Michelle Meneghini)

Species accounts

Harbor porpoise

Phocoena phocoena

SUBFAMILY
Phocoenidae

TAXONOMY
Phocoena phocoena (Linneaus, 1758), Baltic Sea.

OTHER COMMON NAMES
English: Common porpoise; French: Marsouin commun.

PHYSICAL CHARACTERISTICS
Length 4.9–6.6 ft (1.5–2.0 m); weight 99–143 lb (45–65 kg). Forehead and beak are absent from the head; mouth is short and straight while curving slightly at the ends; dorsal side is often deep brown or gray; ventral side ranges from light gray to white; lips and chin are black; black lines extend from the jawline to the flippers.

DISTRIBUTION
Mainly in the North Atlantic and northern Pacific, with populations in western Europe and the North American coasts; a seasonal north-to-south migratory pattern along the coast of California. A 1995 ship survey in northern California found a specific distribution according to water depth—significantly more porpoises than expected occurred at depths of 65–200 ft (20–60 m) and fewer at depths of more than 200 ft (60 m). There is also a colony of harbor porpoises in the Mediterranean Sea.

HABITAT
Coastal temperate and subarctic waters of the North Atlantic and northern Pacific; bays, rivers, estuaries, and tidal channels in Western Europe and both coasts of North America.

BEHAVIOR
Secretive and seldom observed in the wild; can be identified by the distinct sound it produces, which is a puffing noise resembling a sneeze.

FEEDING ECOLOGY AND DIET
Mostly cephalopods and fishes; schooling non-spiny fishes such as herring, mackerel, and sardine.

REPRODUCTIVE BIOLOGY
Mating season occurs in the summer, from June to October after a gestation period of 11 months; age and length of sexual maturity is still being debated.

CONSERVATION STATUS
Listed as Threatened by COSEWIC; Strategic Stock under the U. S. MMPA; Vulnerable by the IUCN Red List of Threatened Species. Bycatches are known to be high in the North Sea, and in France, Spain, and Portugal, although the extent of these bycatches is unknown.

SIGNIFICANCE TO HUMANS
Direct fishing for food and oil. ◆

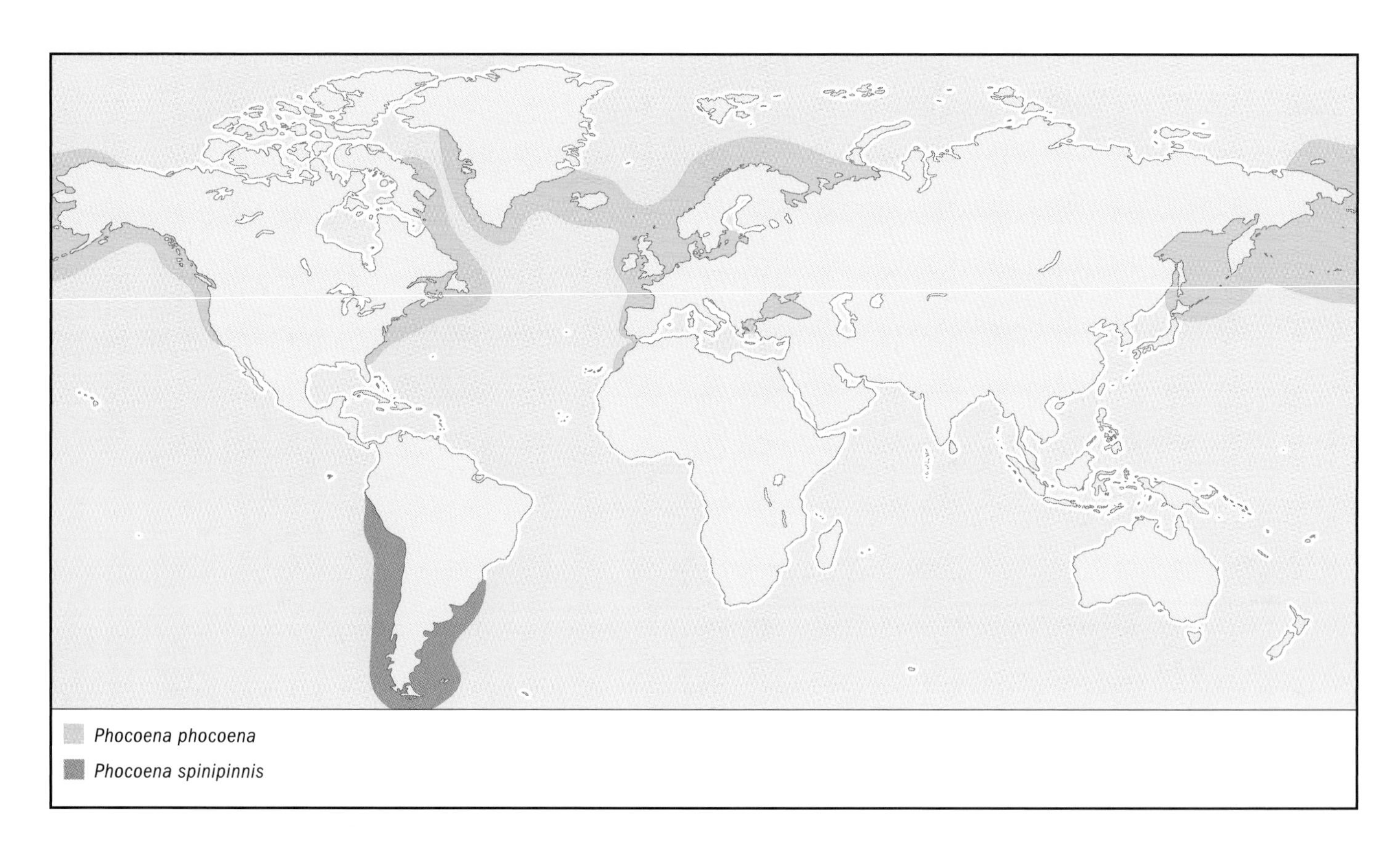

Burmeister's porpoise

Phocoena spinipinnis

SUBFAMILY
Phocoenidae

TAXONOMY
Phocoena spinipinnis Burmeister, 1865, Buenos Aires, Argentina.

OTHER COMMON NAMES
English: Black porpoise; French: Marsouin de Burmeister; Spanish: Marsopa espinosa.

PHYSICAL CHARACTERISTICS
Length 4.6–5.9 ft (1.4–1.8 m); weight 88–154 lb (40–70 kg); dorsal side is dark gray to black in color; ventral portion is slightly lighter; referred to as the "black porpoise" because it turns completely black after death.

DISTRIBUTION
Mostly in the Atlantic and Pacific Oceans, along the coastal waters of South America—the coast of Brazil, south along the coastlines of Tierra del Fuego, and the Falkland Islands north into the coastal Pacific waters of Peru.

HABITAT
Shallow waters of less than 500 ft (152 m) depth; rivers and estuaries; coasts bordering the Atlantic and Pacific Oceans.

BEHAVIOR
Travel is usually in small groups, although observers have rarely found more than eight individuals together at any one time. They swim in quick, jerky movements but remain fairly inconspicuous and barely break the surface of the water when they come up to breathe. Burmeister's porpoises are very timid and scatter rapidly when approached by boats, and can be identified on the surface by respiration sounds.

FEEDING ECOLOGY AND DIET
Primarily anchovies and hake. Also squid, euphasiids, mysid shrimp, up to nine species of fish, and mollusks.

REPRODUCTIVE BIOLOGY
Sexual maturity peaks at an average length of 61 in (155 cm) for males and 63 in (160 cm) for females; mating season occurs from June to September, with calving in May through August after a gestation of 10 months.

CONSERVATION STATUS
Listed as Data Deficient by the IUCN.

SIGNIFICANCE TO HUMANS
Like most cetaceans, Burmeister's porpoises are frequently inadvertent bycatch in fishing nets. In Peru and Chile, animals are shot or harpooned and then sold for their meat, either as bait in crab fisheries or for human consumption. After stricter legislation was implemented in 1994, purposeful catches have declined, although bycatches have not. ◆

Common name / Scientific name/ Other common names	Physical characteristics	Habitat and behavior	Distribution	Diet	Conservation status
Spectacled porpoise *Phocoena dioptrica* French: Marsouin de lahille, marsouin à lunettes; German: Brillenschweinswal; Spanish: Marsopa de anteojo	Coloration is black, ventral areas are light gray. Black coloration around eye, giving the appearance of an eye patch, usually outlined by white. Dorsal fins are triangular, head is small and rounded, little forehead present. Males are generally larger than females. Head and body length 4.9–6.6 ft (1.5–2.0 m), weight 132–185 lb (60–84 kg).	Prefer cold (41.9–49.1°F/5.5–9.5°C), open oceanic waters. Good swimmer, very shy. Little is known of reproductive cycles.	Temperate and subantarctic waters of the Southern Hemisphere.	Mainly anchovy and small crustaceans.	Data Deficient
Finless porpoise *Neophocaena phocaenoides* French: Marsouin aptère, marsouin sans nageoires; German: Glattschweinswal; Spanish: Marsopa negra	Coloration is blue-gray. Small ridge runs from blowhole to tail flukes. Small, curving mouth. Head and body length 6.2 ft (1.9 m), weight 66–99 lb (30–45 kg).	Found in shallow, warm waters. Group size is 1–4 individuals, though aggregations of 20–50 are not uncommon. They are known to spy-hop, and some mothers carry their calves upon their backs.	Coastal waters and all major rivers of the Indian and western Pacific Oceans.	Fish, shrimp, prawns, and octopus.	Data Deficient
Vaquita *Phocoena sinus* English: Gulf porpoise; French: Marsouin du Golfe de Californie; German: Kalifornischer Schweinswal; Spanish: Cochito	Coloration is gray, paler on sides, and gray or white belly. Dark patch around eyes and mouth. Triangular dorsal fin, bumps and whitish spots on leading edge. Head and body length 3.9–4.9 ft (1.2–1.5 m), weight up to 121 lb (55 kg).	Found in shallow, murky waters of the Gulf of California. Groups consists of 1–5 individuals, but primarily solitary. Multi-male breeding systems, sonar is used in communication.	Northern end of the Gulf of California.	Consists mainly of squid, grunt, and croaker.	Critically Endangered
Dall's porpoise *Phocoenoides dalli* French: Marsouin de Dall; German: Weißflankenschweinswal; Spanish: Marsopa de Dall	Narrow mouth, steeply sloping forehead, small flippers, triangular dorsal fin. White patch on belly and flanks. Head and body length 5.6–7.2 ft (1.7–2.2 m), weight 298–485 lb (135–220 kg).	Often found in water with a surface temperature between 37.4°F and 68°F (3–20°C) in open ocean, some in more coastal waters. Groups consist of 10–20 individuals, aggregations of several thousand are not uncommon. Forwardly directed splashes, known as a "rooster tail."	North Pacific Ocean and adjacent seas.	Fish and squid in the open ocean, and schooling fish in coastal areas.	Lower Risk/ Conservation Dependent

Resources

Books

Gaskin, D. E. *The Ecology of Whales and Dolphins.* London: Heinemann Educational Books Ltd., 1982.

Jefferson, T. A., S. Leatherwood, and M. A. Webber. *The Marine Mammals of the World. FAO Species Identification Guide.* Rome: United Nations Environmental Programme, 1993.

Klinowska, M. *Dolphins, Porpoises and Whales of the World. The IUCN Red Data Book.* Gland, Switzerland: IUCN, 1991.

Perrin, W. F., G. P. Donovan, and J. Barlow. *Gillnets and Cetaceans.* Cambridge: International Whaling Commission, 1994.

Ridgeway, S. H., and R. Harrison, eds. *Handbook of Marine Mammals.* Vol. 6, *The Second Book of Dolphins and Porpoises.* San Diego: Academic Press, 1999.

Tolley, K. A. *Population Structure and Phylogeography of Harbor Porpoises in the North Atlantic.* Bergen, Norway: Department of Fisheries and Marine Biology, University of Bergen, 2001.

Periodicals

Caretta, J. V, B. L. Taylor, and S. J. Chivers. "Abundance and Depth Distribution of Harbor Porpoise (*Phocoena phocoena*) in Northern California Determined From a 1995 Ship Survey." *Fish Bulletin* 99, no. 1 (2001): 29–39.

Cox, T. M., A. J. Read, A. Solow, and N. Trengenza. "Will Harbor Porpoises (*Phocoena phocoena*) Habituate to Pingers?" *Journal of Cetacean Research and Management* 3, no. 1 (2001): 81–86.

Frantzis, A., J. Gordon, G. Hassidis, and A. Komnenou. "The Enigma of Harbor Porpoise Presence in the Mediterranean Sea." *Marine Mammal Science* 17, no. 4 (2001): 937–944.

Law, T. C., and R. W. Blake. "Swimming Behaviors and Speeds of Wild Dall's Porpoises (*Phocoenoides dalli*)." *Marine Mammal Science* 10, no. 2 (1994): 208–213.

Sekiguchi, K. "Occurrence, Behavior and Feeding Habits of Harbor Porpoises (*Phocoena phocoena*) at Pajaro Dunes, Monterey Bay, California." *Aquatic Mammals* 21, no. 2 (1995): 91–103.

Jasmin Chua, MS

▲

Dolphins

(Delphinidae)

Class Mammalia
Order Cetacea
Suborder Odontoceti
Family Delphinidae

Thumbnail description
Medium to large fully aquatic carnivores, characterized by a fusiform body, a head with a projecting beak carrying homodont teeth, well-formed eyes, lacking external ears, a single blowhole on top for respiration, pectoral appendages reduced to flippers, loss of pelvic appendages, and a horizontal tail consisting of two flukes for propulsion

Size
4.5–30 ft (1.4–9.0 m); 117–12,000 lb (53–5,600 kg)

Number of genera, species
17 genera; 34 species

Habitat
Oceans, bays, estuaries, and rivers

Conservation status
Endangered: 1 species; Lower Risk/Conservation Dependent: 5 species; Data Deficient: 17 species

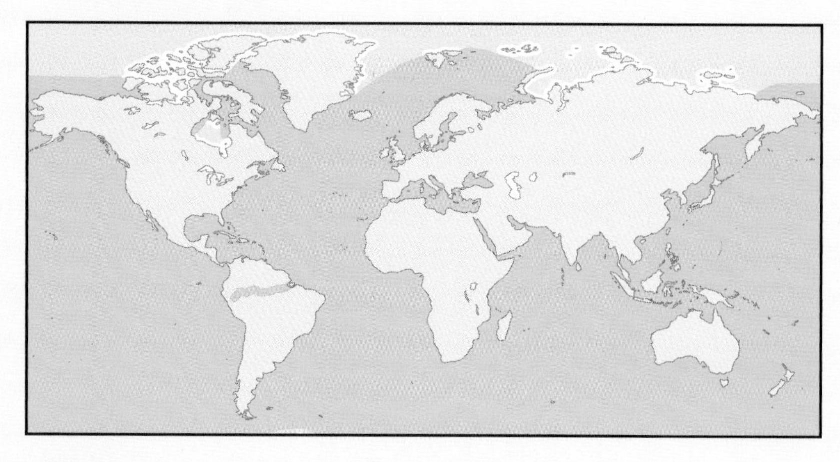

Distribution
Worldwide in all pelagic and coastal marine waters from the Arctic ice edge to the Antarctic ice edge and including a number of rivers

Evolution and systematics

The cetaceans appear to have been derived from ancient ungulate stock. The earliest whales are described as archaeocetes, which evolved from mesonychian condylarths. The mesonychids were ungulate ancestors that were primarily terrestrial. Fossil archaeocetes from about 52–42 million years ago (mya) have been found in Africa, North America, Pakistan, and India. Recent fossil findings have identified a "walking whale" of the genus *Ambulocetus* as a clear example of the transition from terrestrial to aquatic life. One of the more obvious evolutionary developments in the transition from archaeocetes to modern cetaceans was the movement of bones in the skull as the nasal openings migrated to the more effective position on top of the cetacean's head. This telescoping resulted in elongated premaxillary and maxillary bones of the skull, creating a rostrum or beak. Baleen whales (Mysticetes) and toothed whales (Odontocetes) appear to have diverged about 25–35 mya. Three families of archaic, extinct dolphins are known from the Miocene. These led to the current family Delphinidae, which is now the most diverse of the cetacean families. The first modern dolphins appeared in the fossil record from about 11 mya.

Comparisons of the fossil record, morphological features of existing animals, and genetic data have led to a variety of controversial descriptions of the phylogenetic relationships between cetaceans and other mammals, and within the odontocetes. While there is general agreement regarding ties between cetaceans and ungulates, the nature of this relationship remains unresolved. It has been suggested that artiodactyls (even-toed ungulates) such as the hippopotamus may be their closest living relatives, but available data are in conflict. Within the Cetacea, dolphins are distinguished by the loss of the posterior nasal sac, and a reduction of the posterior end of the premaxilla skull bone. Distinctions between dolphins and other groups of small cetaceans such as river dolphins and porpoises involve comparison of a variety of skull features. In more general terms, dolphins differ from porpoises in that the dolphins tend to have a larger and more falcate dorsal fin as compared to the lower, more triangular porpoise fin; dolphins tend to have a longer, more clearly demarcated beak; and dolphin teeth are conical in shape as compared to the spatulate teeth of porpoises.

The classification of dolphins is undergoing much revision with the advent of genetic analysis techniques and the increased efforts by scientists to collect small genetic samples from specimens from around the world. In 2003, 17 genera, 34 species, and 16 subspecies of dolphins were recognized. Though subfamily designations must be considered tentative pending additional study, one proposed scheme identifies five subfamilies: Delphininae (*Delphinus*, *Lagenodelphis*, *Lagenorhynchus*, *Sousa*, *Stenella*, and *Tursiops*), Globicephalinae (*Feresa*, *Grampus*, *Globicephala*, *Pseudorca*, and *Peponocephala*), Lissodelphinae (*Cephalorhynchus* and *Lissodelphis*), Orcininae (*Orcinus* and *Orcaella*), and Stenoninae (*Sotalia* and *Steno*). A

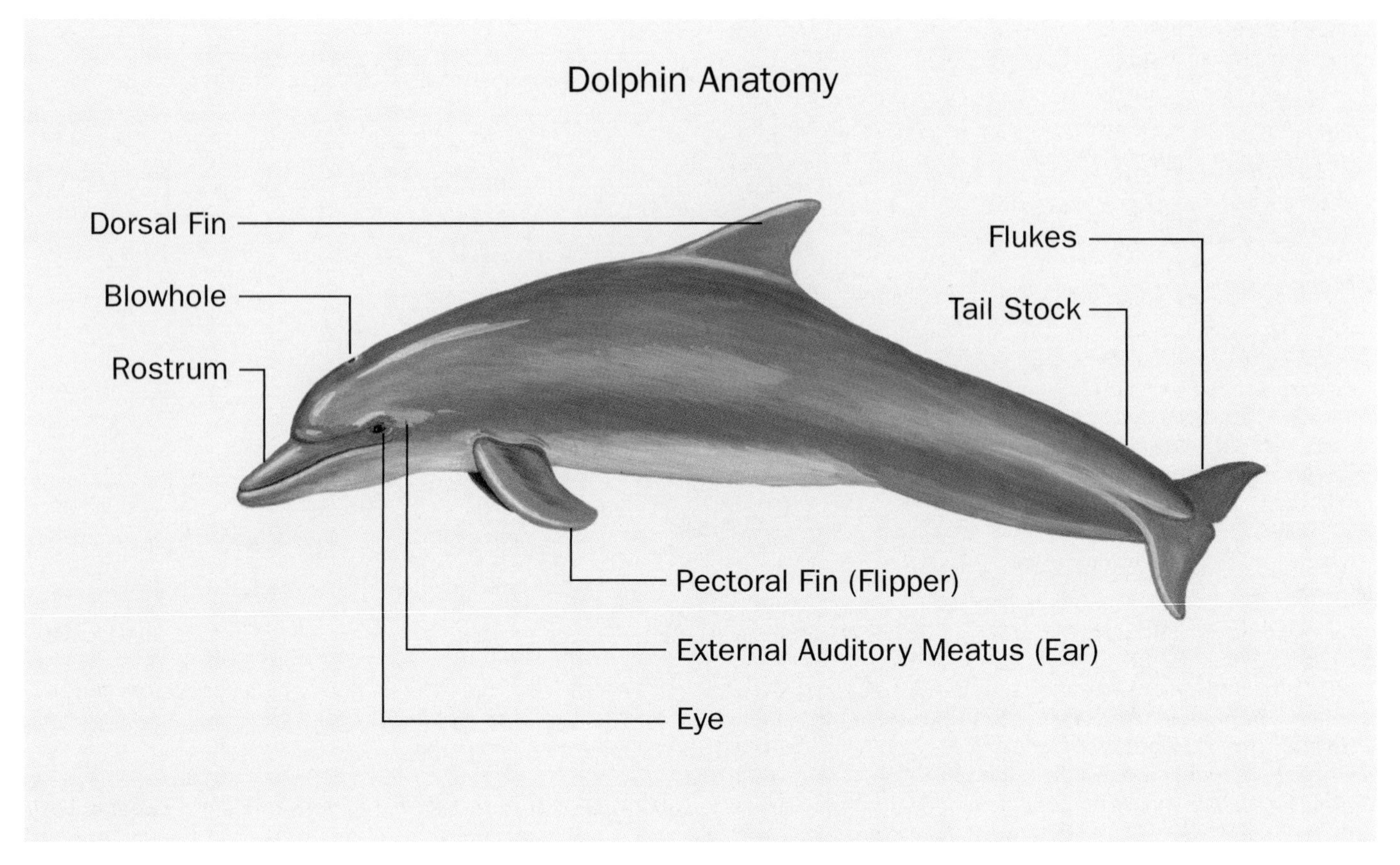

Dolphin anatomy. (Illustration by Michelle Meneghini)

number of hybrid dolphins have been identified in captive breeding situations and in the wild, serving to further blur the distinctions between species. Some of these hybrids have been both viable and fertile.

Physical characteristics

All dolphins have essentially the same fusiform body shape, streamlined for efficient movement through a dense water medium. The dolphins possess flippers as their pectoral appendages, with the bones of the hand and arm modified into a solid wing-like structure, articulating with the shoulder, and serving as control surfaces for maneuvering. Pelvic appendages are essentially nonexistent, reduced to small, internal pelvic bones. The vertebral column, with its variably fused cervical (neck) vertebrae and prominent processes for attachment of the strong musculature used for propulsion, tapers toward the tail, where a pair of fibrous horizontal fins form flukes. Contrary to the baleen whales, which have two nostrils, dolphins have a single blowhole on top of their heads for respiration. Variation exists among the dolphins in terms of robustness, the presence of a clearly demarcated projecting beak, numbers and size of teeth, and the presence of a dorsal fin. In general, body shape of the smaller dolphins seems to roughly grade from more slender to more robust, with greater mid-body girth, moving away from the equator. Dolphin teeth are homodont, meaning that all of the teeth in a dolphin's jaws are alike in structure. The pointy teeth are designed for grasping individual prey items, rather than for chewing. The cone-shaped teeth differ in size and number, depending on the prey of the different dolphin species. The number and size of the teeth, in turn, influence the size of the beak and shape of the mouth of each species. For example, common dolphins (*Delphinus delphis*) may have more than 250 small teeth in long, forceps-like jaws for capturing small schooling fish and invertebrates, whereas killer whales (*Orcinus orca*) have about 50 large teeth for capturing large fish and removing large pieces from a variety of marine mammals. A fibrous dorsal fin located near the middle of the back is found on all but two species of dolphins, the northern and southern right whale dolphins (*Lissodelphis* spp.). Dorsal fins vary from species to species in height and shape, but in general they serve to stabilize the swimming dolphins, as radiators for cooling the internal reproductive organs, and as weapons. As a secondary benefit to researchers, the dorsal fins of many dolphins are individually distinctive based on shape and natural notch patterns, providing a means of reliable identification for observing individuals over time. A few small hairs occur on dolphin beaks at the time of birth, but these are soon lost. Males and females look quite similar in most dolphin species, though there may be differences in body and/or appendage size in some. Female dolphins typically have a genital and an anal opening in a single ventral (belly) groove, with a nipple located in a mammary slit on each side of the genital opening. Males typically have a genital opening in a ventral groove anterior to the separate groove for the anus.

Dolphins are medium- to large-sized aquatic mammals. They range from the tiny and endangered Hector's dolphin (*Cephalorhynchus hectori*) at 4.5 ft. (1.4 m) and 117 lb (53 kg) to adult male killer whales at 30 ft (9 m) and 12,000 lb (5,600 kg). Gender differences in adult body size (sexual size dimorphism) occur in some dolphin species. In general, the females of some of the smaller species (*Cephalorhynchus* spp.; the tucuxi, *Sotalia fluviatilis*) are slightly larger than the males, whereas the males of the largest species (killer whales, false killer whales [*Pseudorca crassidens*], pilot whales [*Globicephala* spp.]) tend to be much larger than the females. In some cases, features that might affect performance in battle or chasing females, such as dorsal fin height (for example, killer whales), tail stock height, or fluke span are disproportionately larger for males than for females. In most of the dolphin species, males and females are of similar size or males are somewhat larger.

Dolphins exhibit a tremendous range of species-specific color patterns, including shades of black, white, gray, brown, orange, and pink. Countershading occurs for most dolphin species. Presumably this pattern of light bellies and dark backs provides camouflage for dolphins avoiding predators or approaching prey. Dolphins living in clear water tend to have more striking and complex color patterns on their sides than do dolphins living in the murky waters of estuaries or rivers. One hypothesis for this difference is that dolphins inhabiting clearer waters can use flashing of color patterns in much the same way as birds in coordinated flocks to signal changes in direction, or for other social displays.

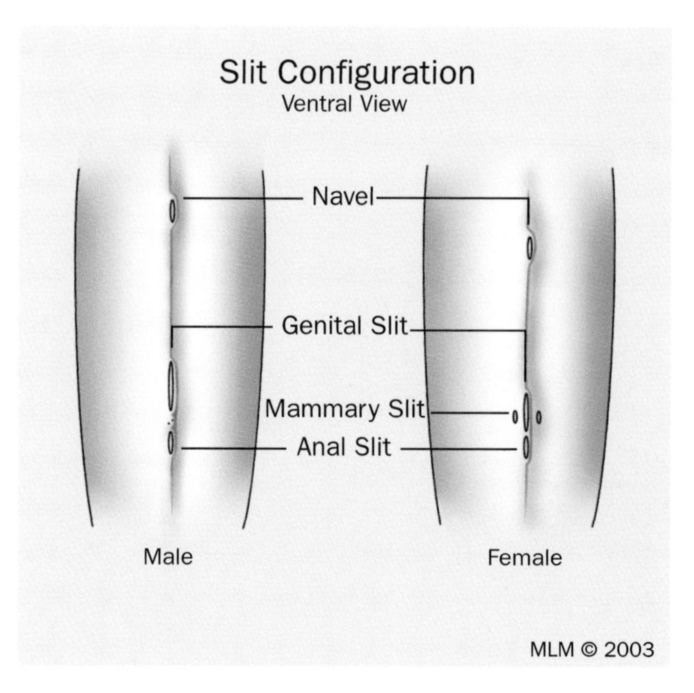

Differences in slit configurations between male and female dolphins. (Illustration by Michelle Meneghini)

Distribution

Dolphins are distributed worldwide through marine waters. Killer whales have been termed cosmopolitan in their distribution because they are found along the ice edge in both arctic and antarctic waters, and in many areas in between. Though most genera have representatives in each of the major ocean systems of the world, individual species tend to be more limited in their distributions. Tucuxi are found only in coastal waters and rivers of South and Central America, the Irrawaddy dolphin (*Orcaella brevirostris*) only in inshore and riverine Indo-Pacific waters, and all four species of *Cephalorhynchus* are found only in the southern hemisphere.

A single Atlantic spotted dolphin (*Stenella frontalis*). (Photo ©Tony Wu/www.silent-symphony.com. Reproduced by permission.)

Habitat

Dolphins are entirely aquatic, meaning that they must find food, mate, produce and rear young, and avoid predators in water. Access to the water's surface for air, availability of prey, predator abundance, and temperature constraints are among the major factors determining habitat use by dolphins.

Dolphins can be found in all available marine habitats. These habitats are truly three-dimensional, and water depth and physiography are important habitat features. Dolphins occur in greatest abundance where resources are most available. In open-ocean or pelagic habitats, this tends to be near islands or seamounts, where nutrient-rich waters carried in deep currents are brought to the surface and support an extensive ecosystem, or where different water masses meet. Similarly, upwelling is a wind- and current-driven phenomenon that creates highly productive areas along the continental slope. Estuaries, where rivers meet marine waters, are extremely productive, and support large numbers of dolphins.

Suites of adaptations are associated with patterns of habitat use. For example, oceanic dolphins of the genera *Stenella* and *Delphinus* are among the most streamlined of the dol-

Atlantic spotted dolphins (*Stenella frontalis*) swimming in a pod. (Photo ©Tony Wu/www.silent-symphony.com. Reproduced by permission.)

phins. With small appendages and many fused cervical (neck) vertebrae, they are designed to maintain a rigid body and move at high speed through pelagic waters free of obstacles. Their long beaks and numerous tiny teeth facilitate the capture of small prey found in rich patches in the open ocean habitat. In contrast, some of the more coastal species such as bottlenosed dolphins (*Tursiops* spp.) and Irrawaddy dolphins are more flexible and/or have larger appendages for maneuvering around bottom and shoreline features in more restrictive habitats. Most killer whales are found near shorelines, where their large appendages facilitate maneuvering around ice floes, shore features, and highly maneuverable, sometimes amphibious, prey (killer whales sometimes slide onto beaches to capture pinnipeds). Their large size helps them to chase and capture large, fast-swimming prey (for example, blue whales, *Balaenoptera musculus*) in more open habitats. The blackfish, including pilot whales (*Globicephala* spp.) and related species, are among the deepest diving dolphins, diving hundreds of feet (meters) in search of squid and fish. They lack projecting beaks, tend to be mostly black (the deepest parts of their habitat lack light), have rounded or bulbous heads, and heavily muscled tailstocks that would facilitate reaching depth and returning to the surface quickly after a prolonged foraging dive. The tucuxi and Irrawaddy dolphin frequent rivers; tucuxi are found several thousand miles (kilometers) up the Amazon River in Peru.

Behavior

Dolphins exhibit a wide range of sociality, reflecting the diversity of morphological variations and habitats of the family. Few of the species include solitary individuals as a common feature of the social system. Group size varies from species to species. Among the smaller dolphins, inshore and riverine species such as bottlenosed and tucuxi dolphins tend to form small groups, typically of fewer than 10–20 individuals. In more open habitats offshore, common dolphin and spinner and spotted dolphin (*Stenella* spp.) groups may number in the thousands. Such variability in the size of groups is likely related to the abundance and distribution of prey (rich, patchy prey offshore, more evenly distributed, predictable prey inshore), and to exposure to predators (dolphins in open habitats are likely more vulnerable to detection and attack by predators than in shallow inshore waters where predator-approach options are more limited). Spreading out in long formations, large groups of dolphins increase the probability of finding fish schools. Similarly, larger groups can work together to better detect and avoid or defend against predators. Larger dolphins such as the pilot whales and killer whales form groups of intermediate size, typically involving fewer than 100 individuals. Because they are larger than most potential predators, predation pressure on adults is minimal, and it is likely that available resources place the primary constraints on group size.

Group composition and cohesion are also quite variable among the dolphins. Many of the smaller dolphin species live in groups of fluid composition, a situation referred to as fission/fusion. Swimming associates may change from day to day, but repeated associations over time are common, especially involving individuals sharing a home range and that are of the same gender, and similar age and reproductive status. In some cases such as male bottlenosed dolphins, small numbers of individuals may be close associates for many years. Some of the larger dolphin species (pilot whales, killer whales) maintain much more stable groups, consistent with the term pod. In these species, the groups may consist of one or more closely associating maternal lineages of several generations, and many of these individuals may remain together for years or decades. In the case of killer whales, it has been suggested that this long-term stable, multi-generational grouping facilitates highly coordinated prey capture such as when the group works together to attack a large baleen whale.

Communication between dolphins remains incompletely understood. Dolphins may signal one another by changing their orientation to show more or less of their high-contrast color patterns to others, often by rolling slightly from one side to another. While visual signals may be important to dolphins living in clear water, many dolphins inhabit turbid waters where vision is limited to a few feet (meters) at best.

Acoustics play a premiere role in the lives of dolphins. Dolphins have exceptional hearing, and can hear frequencies nearly 10 times higher than humans. They are also capable of producing three classes of sounds within this broad range of hearing. Dolphins lack vocal chords. Instead, they produce sounds by cycling air past tissues in the nasal region. Dolphins often produce "burst pulse" sounds, which sound like squawks, when socializing with one another. They also produce sonar, or echolocation, clicks to investigate their environment. These broadband clicks bounce off objects in the environment, and the dolphin is capable of interpreting, with a high degree of discrimination, the echoes returning through the animal's lower jaw. The dolphin can adjust the rate of production of these rapidly repeated clicks to allow the echo to return between clicks. Clicks are very directional signals, projected forward in a narrow beam from the dolphin's melon. Dolphins also produce frequency-modulated tonal whistles, which are much less directional. Individual bottlenosed dolphins tend to produce one specific whistle more often than others. This is termed its signature whistle, and playback experiment results suggest that they are used at least as identifiers or contact calls, presumably to maintain group cohesion in murky or dark water. Dolphins responded much more strongly to recorded signature whistles of kin or close associates than to those of less familiar individuals. Variations in the pitch or other features of these whistles may carry addi-

Several Hector's dolphins (*Cephalorhynchus hectori*) swimming close to the water's surface. (Photo ©Tony Wu/www.silent-symphony.com. Reproduced by permission.)

A dusky dolphin (*Lagenorhynchus obscurus*) exhibits its well-known social behavior. (Photo ©Tony Wu/www.silent-symphony.com. Reproduced by permission.)

tional information about the emotional state of the producer. As a variation on this identifier theme, killer whales produce calls that are pod-specific, rather than individual-specific, and remain unchanged for several decades. Captive dolphins have been taught to understand artificial language, but they have yet to initiate acoustic responses or to clearly demonstrate the nature of a true language of their own.

Some dolphin communication may combine visual and acoustic modes, or involve other senses. For example, "jaw pops" involve opening and closing the mouth in a threatening manner and producing a loud popping sound. Tail slaps, when the flukes are slapped sharply on the water's surface also produce a loud report. Dolphins are very tactile animals, and many affiliative interactions involve physical contact between various body parts. Socio-sexual displays are also an important part of the dolphins' behavioral repertoire. Dolphins are apparently able to taste, but not smell. Though it remains to be demonstrated conclusively, it is possible that dolphins may communicate such things as reproductive readiness through production of chemicals that can be tasted by others.

Dolphins exhibit a variety of ranging patterns. Knowledge of the ranging patterns of dolphins inhabiting pelagic and continental shelf waters is limited by the inherent difficulties of conducting research in these regions. However, available information suggests that dolphins typically do not range through entire ocean basins, but instead occupy generally definable regions within basins. This pattern becomes more evident where geographical features help to define ranges. For example, scientists working from oceanic islands have been able to repeatedly identify individuals of a variety of dolphin species, including spinner dolphins and pilot whales, over periods of years in the waters near these islands. Along shorelines and in enclosed bays, residency is frequently noted. Some killer whale pods residing in Puget Sound and near Vancouver Island have been observed repeatedly in the same waters for nearly three decades; some of these pods have also been observed as far away as Alaska and California. Atlantic spotted dolphins and bottlenosed dolphins inhabiting the Bahama Banks have been identified repeatedly for nearly two decades. Tucuxi dolphins in bay waters of southern Brazil have been observed repeatedly for nearly a decade. Probably the best-known ranging patterns are those of inshore bottlenosed dolphins. In nearly every study around the world, at least a few, if not most, of the dolphins have been determined to exhibit residency in an area for at least part of the year and over multiple years. The longest-term study, in Sarasota Bay, Florida, continues to identify four generations of year-round residents after more than 33 years of observation. In places where the animals are living at the extremes of the species' range, seasonal migrations have been noted. For example, bottlenosed dolphins along the Atlantic seaboard of the United States move as far north as New Jersey in summer, but migrate at least as far south as North Carolina as waters cool. Territoriality, in terms of defended areas, has rarely been described for dolphins.

Dolphin activities tend to occur in bouts. Dolphins are nearly constantly on the move, surfacing to breathe, diving, and traveling from one location to the next or milling in one area. They intersperse bouts of travel with foraging, socializing, play, and/or rest, or some combination of these activities. As voluntary breathers, dolphins must remain conscious at all times. Thus, they do not sleep in the same way as humans. It is believed that they decrease their overall activity level and rest one hemisphere of the brain at a time. In some species, the degree of synchrony of group members can be very high. Different species of dolphins exhibit different levels of aerial activity. Spinner dolphins and right whale dolphins engage in frequent leaps, whereas other dolphins such as bottlenosed or tucuxi tend to be more subdued.

Feeding ecology and diet

Dolphins are carnivorous, with most eating fish and/or squid. In contrast to the batch-feeding baleen whales, dolphins typically capture prey one item at a time. Some dolphins may eat other invertebrates such as shrimp, and others, especially killer whales and false killer whales, may eat other marine mammals. Killer whales are known to prey upon sea otters, pinnipeds, porpoises, dolphins, and baleen whales. In the vicinity of Vancouver Island, different pods of killer whales specialize on different prey. The residents feed primarily on fish, while the transients emphasize marine mammals in their diet. Dolphins typically consume about 5% of their body weight in food each day.

Wild bottlenosed dolphin (*Tursiops truncatus*) plays with an octopus in the Red Sea. (Photo by Jeff Rotman/Photo Researchers, Inc. Reproduced by permission.)

Dolphins use a variety of techniques to find and capture prey, ranging from individual hunting to coordinated, cooperative efforts involving entire dolphin groups. Dolphins find prey visually and acoustically. Some leaps performed by dusky dolphins (*Lagenorhynchus obscurus*) prior to feeding are believed to be for the dolphins to locate bird flocks that may be flying over fish schools. Dolphins use both passive and active acoustics to find prey. Many fish and marine mammals produce sounds that can be heard by the dolphins, and allow them to locate the prey. Dolphins also use their echolocation to actively search for and zero in on prey. The idea has been proposed that dolphins can produce echolocation clicks of sufficient strength to stun prey, but little evidence exists to suggest that this approach is used widely. Inshore dolphins often hunt prey individually, especially when feeding on nonschooling fish inhabiting seagrass meadows, reefs, or other seafloor features. In open ocean habitats, large schools of foraging dolphins such as common dolphins may spread across broad areas in a line-abreast formation to search for schools of fish or squid, and then converge on the schools once they are found, presumably as a result of some acoustic cue. Working cooperatively, dolphins will circle prey schools, condensing them and driving them to the surface where that barrier further limits their escape. Dolphins then pass through the densely packed prey and grab individuals. The extreme case of cooperative feeding involves killer whale pod members working together to subdue large baleen whales, much like wolf packs attacking large ungulates.

Dolphins exhibit a variety of specialized foraging behaviors. Some dolphins in Australia place sponges on their rostra presumably to aid in prey capture. Killer whales in Argentina and bottlenosed dolphins in several locations engage in strand feeding, in which they slide onto beaches after prey. Bottlenosed dolphins also engage in "kerplunking," a behavior that involves driving their flukes and tailstock through the water's surface, creating a large splash and bubbles that may flush prey from cover. They also engage in "fish-whacking," which involves striking fish with flukes, often sending them soaring, stunned, through the air. Atlantic spotted (*Stenella frontalis*) and bottlenosed dolphins use a behavior known as "crater feeding" to dig into sandy seafloors in search of buried prey. Dolphins around the world have also learned to take advantage of human fishing efforts, including obtaining fish lost or discarded from trawlers and seiners, and working to drive schools of fish toward artisanal fishermen working with cast or seine nets from shore. In the latter case, the barriers provided by the fishermen and the confusion from their fishing activity may enhance the dolphins' prey-capture efficiency. Many of these specialized foraging and feeding behaviors are believed to provide evidence for the cultural transmission of knowledge through dolphins' societies.

Typically, dolphin prey are eaten intact, swallowed head first. If a fish is too large to take in this way, or if it has dangerous spines that could injure the dolphin if ingested, the dolphin may break it into smaller pieces by tossing it, rubbing it on the seafloor, or in some cases, working with another dolphin to tear the fish apart. When feeding on large marine mammals, killer whales often work together to restrain the large prey while they bite off pieces of the animal.

Reproductive biology

Behaviors for courtship and mating for reproduction are not well understood for most dolphins. Sexual behaviors are used in several contexts, including developing and maintaining dominance and other social relationships, in addition to reproduction. Males and females of all ages, from the time they are several weeks old, engage in sexual behavior with members of the opposite and the same gender, and sometimes with their own close relatives. The vast majority of these sex-

A killer whale (*Orcinus orca*) hunts South American sea lions on a beach in Argentina. (Photo by François Gohier/Photo Researchers, Inc. Reproduced by permission.)

Spinner dolphins (*Stenella longirostris*) are found in the tropical and subtropical waters of the Atlantic, Indian, and Pacific Oceans. (Photo by François Gohier/Photo Researchers, Inc. Reproduced by permission.)

ual interactions are social in nature, but they confuse the issue of research that tries to identify those involved specifically in reproduction. Sexual behaviors include stroking with flippers and flukes, rubbing the genital region, inserting fin tips or beaks into the genital slits, and intromission of the fibro-elastic penis. Dolphins typically mate belly-to-belly, but mounting may involve males approaching at a different angle to the female's body. Copulations generally are brief, lasting less than a minute, but they may be repeated. Dolphins have large testes and exceptionally high sperm counts, facilitating multiple copulations.

Data for defining mating systems are difficult to collect for dolphins, but genetic studies are now allowing some of the first dolphin paternity testing, and continued work should clarify understanding. Available evidence suggests that monogamy is not a practice in which dolphins engage. Bottlenosed dolphin paternity tests indicate that females may use different sires for subsequent calves. For the better-studied dolphins, associations between breeding males and females tend to be brief, lasting days to weeks, and one male or male coalitions may associate with one receptive female at a time, sometimes battling with other males for access to the female. Males may move between female groups during a breeding season. This pattern has been referred to as serial polygyny or promiscuity.

Dolphin reproduction can occur anywhere within the animals' range, but calf rearing may lead to shifts in habitat use to more protected or productive areas, or to the creation of nursery subgroups of mothers with calves inside of larger dolphin groups. Reproductive seasonality may dictate where reproduction occurs for species that move over large areas or migrate seasonally. Seasonality tends to be most evident in environments that experience strong seasonal variations in water temperature or other environmental factors such as flooding cycles for riverine species. For example, most bottlenosed dolphins have well defined breeding seasons that vary with latitude, but births, after a 12-month gestation, tend to occur as temperatures warm and food is abundant. Similarly, tucuxi dolphins give birth during flood stage in the Amazon River, when food is most abundant, after a 10-month gestation.

Sexual maturity occurs between five and 16 years of age for dolphins, depending on the species. Larger species tend to mature later, and females tend to mature before males. Dolphins usually produce a single calf after gestation periods of 10–15 months, again depending on the species. The calf is born tail-first, and once the mother snaps the umbilical cord, the calf swims to the surface for its first breath. Following expulsion of the placenta, the calf begins nursing from the nipples on each side of the mother's genital slit. Nursing bouts are brief, but repeated often, as the calf receives milk that is very rich in fat. Calves may begin to capture small prey on their own when they are only a few months old, but may not be fully weaned from milk until they are up to 3.5 years old, and bottlenosed dolphins and pilot whales more than seven years old have been found with lactating mothers or with milk in their stomachs. The association between mother and calf may last for one or more years beyond nutritional weaning, suggesting the importance of learning and protection. Killer whales carry this pattern to an extreme, with adult female and male offspring remaining within the mother's pod. For all dolphins examined to date, it seems that mothers are fully responsible for calf rearing; paternal investment is limited to insemination.

Conservation status

The conservation status of dolphins varies by species, subspecies, and population. According to the IUCN Red List, only Hector's dolphin is considered Endangered, due primarily to incidental mortality in fishing gear. Of the 34 dolphin species, 17 are Data Deficient and five are Lower Risk/Conservation Dependent. Given the lack of information available for 67% of the taxa, one should not draw too much solace from the current listing of only a single species as endangered. Data on population sizes and numbers of losses from specific human activities are lacking for many species.

Dolphins face many threats from human activities. The degree of risk from these threats varies from site to site. Few countries actively hunt dolphins in directed fisheries. Japan continues to harvest striped dolphins (*Stenella coeruleoalba*) in fisheries that involve driving schools ashore, harpooning, or crossbows. The dolphin population is in decline, and continued hunting is unsustainable. As numbers of striped dolphins decline, 10 other species of dolphins are being hunted in their place. International efforts to halt this fishery have not been

A killer whale (*Orcinus orca*) breaching. (Photo by © Brandon D. Cole/Corbis. Reproduced by permission.)

successful. In Peru, Sri Lanka, the Philippines, and elsewhere, incidental catches of dolphins in fishing gear have led to the development of directed fisheries for thousands of dolphins for meat, using purse seines, harpoons, gillnets, and explosives. Efforts to regulate some of these fisheries have had little effect on takes due to difficulties with monitoring and enforcement.

Other directed dolphin fisheries around the world involve commercial collection of bottlenosed dolphins and other species for oceanarium displays, research, and military activities. The vast majority of dolphins collected from the wild are bottlenosed, but smaller numbers of killer whales, false killer whales, Pacific white-sided dolphins (*Lagenorhynchus obliquidens*), Commerson's dolphins (*Cephalorhynchus commersonii*), and rough-toothed dolphins (*Steno bredanensis*) have also been collected for public display and interactive programs. These fisheries typically involve relatively small numbers of individuals, but when removals are concentrated in small areas and emphasize young females, locally resident populations can be placed at risk. Data indicate that removal of individual bottlenosed dolphins from stable communities can adversely impact the remaining community members in terms of calf survivorship or availability of appropriate social associates. Modeling efforts have shown that the existing captive population of bottlenosed dolphins in North America, if breeding is managed appropriately, is sufficient to sustain itself with appropriate genetic diversity for decades without the need for new genes from the wild. Commercial collection of dolphins has been banned or restricted in a number of countries, but Cuba, Japan, and some African nations, among others, still engage in this activity.

Of much greater concern is the issue of dolphins killed incidentally in nets set for fish. One of the best known examples involved hundreds of thousands of pantropical spotted (*Stenella attenuata*) and spinner dolphins killed each year in the tuna seine net fishery in the eastern tropical Pacific Ocean from the 1950s into the 1990s. Currently, fewer than several thousand dolphins are killed each year in this fishery. This turnaround occurred as a result of changes in equipment and approaches, and increased efforts by the fishing crews to reduce mortalities, driven in large part by public outrage and pressure over the situation. Dolphins continue to be killed in large numbers in fishing nets in many parts of the world. Trawls, seines, and gillnets are responsible for dolphin mortalities. Efforts to make nets more reflective to dolphin echolocation, or the use of noisemaking "pingers" to alert dolphins to the presence of nets have met with mixed success. In the southeastern United States, most states have banned the use of large gillnets in inshore waters, but North Carolina continues to allow such fishing, leading to numbers of deaths

A pod of spinner dolphins (*Stenella longirostris*). (Photo ©Tony Wu/www.silent-symphony.com. Reproduced by permission.)

of bottlenosed dolphins in excess of what scientists consider to be sustainable. Efforts are underway through a federal "take reduction team" negotiation process to bring stakeholders together to arrive at a solution that considers the needs of the animals as well as those of the humans involved.

Other kinds of fishing activity also impact dolphins. Baited hooks on long-lines set for swordfish and other species kill or injure dolphins such as pilot whales. Monofilament fishing line and lures and hooks used in recreational fishing kill or injure dolphins through entanglement and ingestion. Crab trap float lines also entangle and drown dolphins.

Among the most insidious, worldwide threats to dolphins are environmental contaminants. More than 10,000 chemicals have entered the environment as a result of human activities. Many of these persistent toxic chemicals enter the dolphins' environment through airborne deposition and runoff, and work their way up through the food chain into dolphins through the fish they eat. Many of these chemicals bind with lipids, and as a result, they accumulate in fatty tissues such as blubber and are transferred to young through fat-rich milk. Chemicals such as PCBs and DDT metabolites and other pesticides have been found in some species at levels of great concern. Based on recent population declines correlated with measured levels of contaminants in blubber samples, killer whales in Puget Sound have recently come under scrutiny for endangered species status. Similarly, concentrations of contaminants in bottlenosed dolphin blubber from the southeastern United States have been measured at levels in excess of those of concern for human health, and evidence is mounting for the role of these contaminants in first-born calf mortality and in the decline of immune system function. While some chemicals have been banned as knowledge of their effects on health and reproduction become known, some will persist in the environment for decades to come; additionally, other chemicals are now emerging as potential sources of concern.

Habitat degradation and loss can take many forms for dolphins. Shoreline alteration and dredge-and-fill operations can decrease productivity of the prey that supports dolphin populations. For riverine species such as Irrawaddy dolphins and tucuxi, dams across rivers can isolate subpopulations, or prevent access to critical resources. Pollution from sewage, garbage, and other wastes can threaten dolphin health. Heavy fishing can lead to competition with fisheries for prey. Marine construction and vessel operations introduce noise that can lead to disturbance of normal activities, or interference with communication or biologically important sounds for the dolphins. High-energy pressure waves from explosives or military activities may injure or kill dolphins. Boat collisions also cause dolphin deaths and injuries in areas where high levels of boat traffic occur.

Significance to humans

Dolphins have figured prominently in human lives for thousands of years. They are depicted in ancient Greek and Roman artwork, are incorporated into mythology, and appear in early writings. Their relationships to humans have long been considered special, with numerous ancient and contemporary accounts of dolphins saving humans lost at sea. The dolphins' smile (actually a fixed fact of anatomy rather than an expression of emotion), their endearing (mostly trained) antics in oceanaria, films, and television, reports of complex social behavior, the ease with which they can be trained to perform complex behaviors, their problem-solving capabilities, and early (false) claims that they could likely communicate in English have supported the idea in many peoples' minds that these creatures should be considered superior to other animals, on an elevated plane with humans.

As more is learned about dolphins, people have begun to develop an appreciation for the animals because of how well they are adapted to life in the sea, without the need to credit them with supernatural powers. Today, millions of people worldwide visit marine parks to see, feed, and/or swim with captive dolphins, or view wild dolphins on dolphin-watching tours, creating a multi-billion dollar industry. It has been argued that this increased familiarity with the animals encourages action to protect them. Others argue that holding dolphins in captivity is cruel exploitation. Increased interest in dolphins is creating conservation issues in some cases, as people begin to feed, swim with, and otherwise disturb wild dolphins.

Dolphins have been and are now used by humans for other purposes besides entertainment and education. In some places, they assist artisanal fishermen, or are harvested as an inexpensive source of protein. They are studied to understand their exceptional sonar and diving capabilities, and to evaluate their cognitive abilities. Dolphins have also been used by the military in the United States and the former Soviet Union to search for weapons, assist divers, and perform surveillance tasks.

1. Male spinner dolphin (*Stenella longirostris*); 2. Common bottlenosed dolphin (*Tursiops truncatus*); 3. Female killer whale (*Orcinus orca*). (Illustration by Michelle Meneghini)

Species accounts

Killer whale

Orcinus orca

SUBFAMILY
Orcininae

TAXONOMY
Orcinus orca (Linnaeus, 1758), eastern North Atlantic.

OTHER COMMON NAMES
English: Orca; French: Orque.

PHYSICAL CHARACTERISTICS
Length 30 ft (9 m); weight 12,000 lb (5,600 kg). Mostly black bodies with white eye patches, saddles, and bellies. Large flippers and prominent dorsal fin, reaching height of 6 ft (2 m) on adult males.

DISTRIBUTION
Cosmopolitan, found in marine waters worldwide, with highest densities at higher latitudes.

HABITAT
Pelagic and coastal habitats, and along ice edges, concentrated where prey is abundant.

BEHAVIOR
Live in long-term, multi-generational stable pods, work cooperatively to capture prey.

FEEDING ECOLOGY AND DIET
Includes fish, pinnipeds, sea otters, dolphins, porpoises, and baleen whales, though different pods may specialize on mammal or fish prey.

REPRODUCTIVE BIOLOGY
Single calf born after 15-month gestation period. Calves produced every three to eight years. Believed to be polygamous.

CONSERVATION STATUS
According to the IUCN Red List, in most parts of the species' range, killer whales are considered Lower Risk/Conservation Dependent. Because of declines in population size and findings of exceptionally high concentrations of environmental contaminants in blubber, killer whales inhabiting Puget Sound have been considered for endangered species status under the U.S. Endangered Species Act.

SIGNIFICANCE TO HUMANS
These are perhaps the highest-profile animals on public display at marine parks at several sites around the world, and form the basis of dolphin-watching businesses where they can be found regularly near shore. ◆

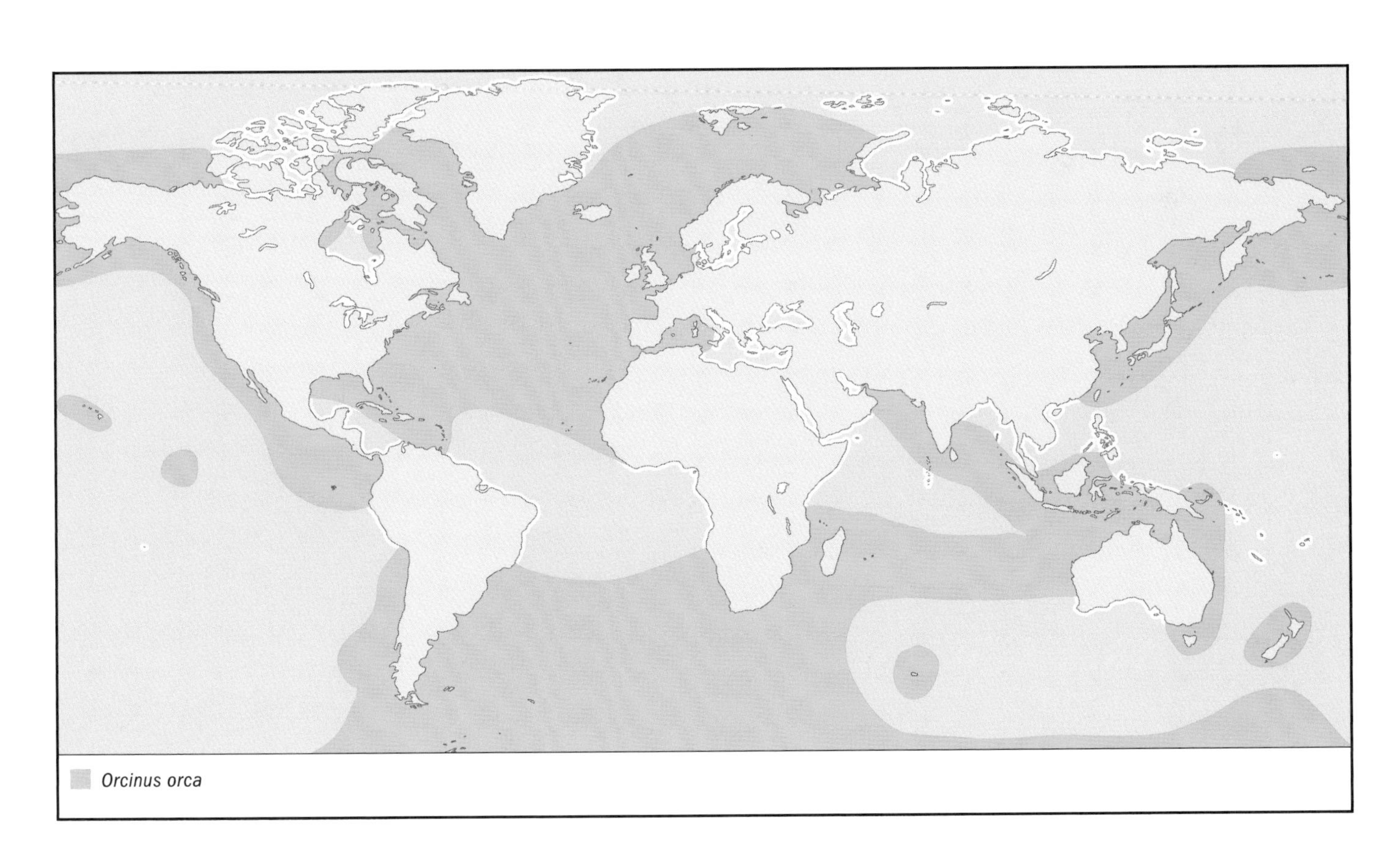

Common bottlenosed dolphin

Tursiops truncatus

SUBFAMILY
Delphininae

TAXONOMY
Tursiops truncates (Montagu, 1821), Devonshire, England, UK.

OTHER COMMON NAMES
English: Atlantic bottlenosed dolphin; French: Souffleur; German: Grosser tümmler; Spanish: Tonina.

PHYSICAL CHARACTERISTICS
Length 8–12.5 ft (2.5–3.8 m); weight 500–1,100 lb (227–500 kg). Geographic variability. Gray or brown backs, shading lighter laterally, light gray or white belly.

DISTRIBUTION
Worldwide in tropical to temperate waters.

HABITAT
Density highest near shorelines and in bays and estuaries, but also found in pelagic waters.

BEHAVIOR
Form small groups of somewhat fluid composition, often occupying a long-term home range. Some associations may last for decades. Seasonally migratory at extremes of species range.

FEEDING ECOLOGY AND DIET
Includes fish and invertebrates, especially squid. They feed individually or cooperatively, and they often take advantage of human fishing activities to supplement their own foraging efforts. A variety of specialized feeding patterns have been reported, specific to particular habitats.

REPRODUCTIVE BIOLOGY
A single calf is born after a 12-month gestation period. The calf is reared by mother for three to six years, until the birth of her next calf. Studies have indicated a promiscuous, polygynous mating system with male-male competition.

CONSERVATION STATUS
The IUCN Red List indicates that this species is Data Deficient. As the most common coastal dolphin in much of the world, it is exposed to a high level of human activity, including pollution, commercial and recreational fisheries, boat traffic, marine construction, commercial collection, and habitat loss. The effects of these impacts remain to be fully evaluated, but data suggest serious concerns in parts of the species range.

SIGNIFICANCE TO HUMANS
Represented in the works of ancient Greeks and Romans, bottlenose dolphins today are among the most familiar dolphins, from television, film, and marine parks around the world where they are displayed. They are hunted in Japan, Peru, the Philippines, Chile, Venezuela, Sri Lanka, and the West Indies. They have been used in military operations by several countries, and for research. ◆

Spinner dolphin

Stenella longirostris

SUBFAMILY
Delphininae

TAXONOMY
Stenella longirostris (Gray, 1828), type locality unknown.

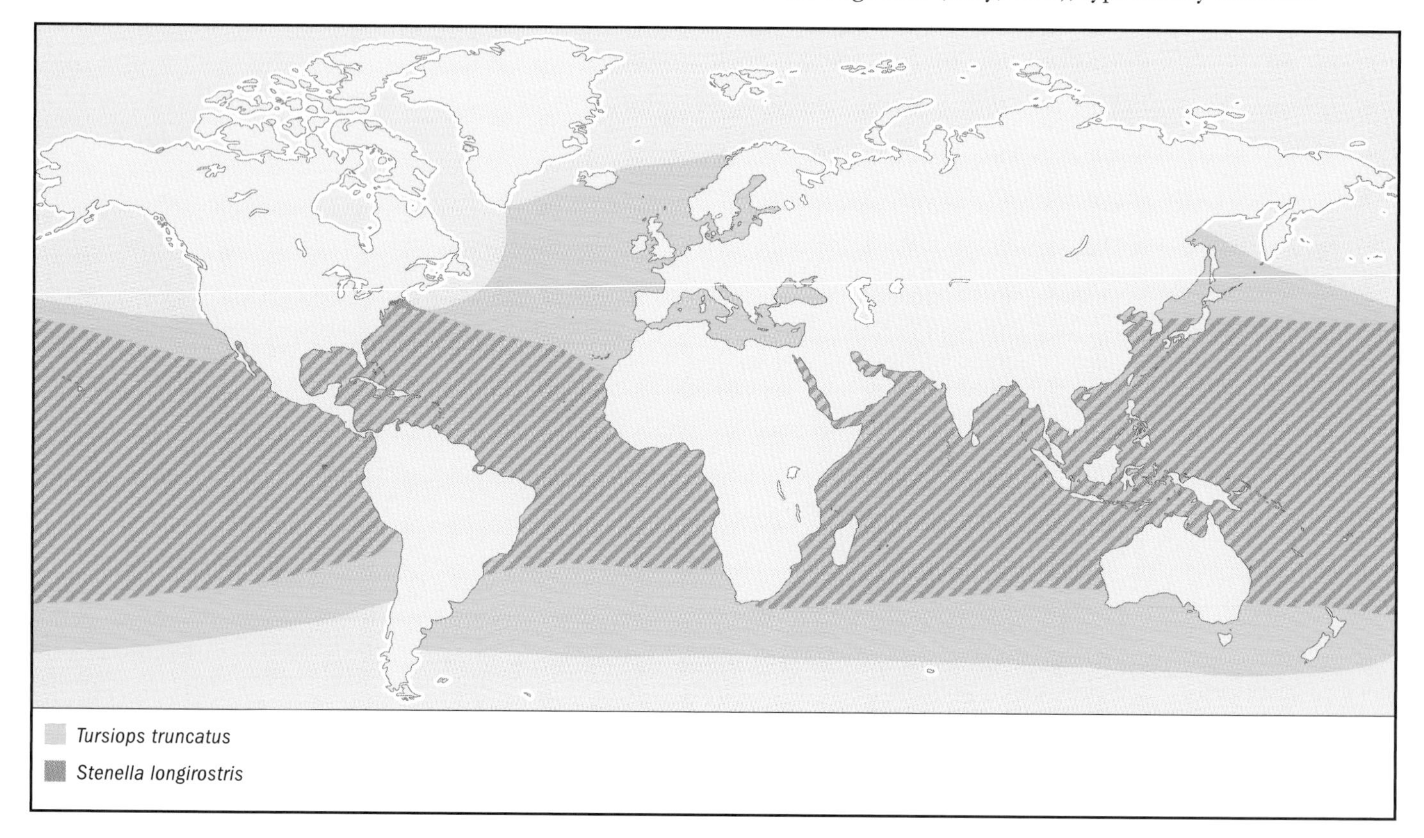

OTHER COMMON NAMES
English: Long-snouted dolphin; French: Dauphin à ventre rose; Spanish: Delfín tornillon.

PHYSICAL CHARACTERISTICS
Length 7.7 ft (2.3 m); weight 170 lb (78 kg). Geographic variability. Some forms are nearly uniformly gray, while others are strikingly marked with black backs, gray sides, and white bellies, with black flippers, beaks, and a connecting stripe.

DISTRIBUTION
Found in tropical to warm temperate waters worldwide.

HABITAT
Live in pelagic habitats, in open ocean regions as well as near oceanic island shorelines. They feed over escarpments, and rest in nearby shallow bays and inside atolls where these are available to them.

BEHAVIOR
Spinner dolphins were named after their characteristic behavior of leaping from the water and spinning one or more times around their longitudinal axis before returning to the water. They swim in large groups of fluid composition. Near oceanic islands they are known to occupy long-term home ranges.

FEEDING ECOLOGY AND DIET
Includes small fish and invertebrates associated with the deep scattering layer. They feed at night when this layer moves up to within 600 ft (200 m) of the surface.

REPRODUCTIVE BIOLOGY
Give birth to a single calf after a 10.5-month gestation period. Calves are produced about every three years on average. Thought to be polygynous.

CONSERVATION STATUS
Spinner dolphins are listed as Lower Risk/Conservation Dependent. Some populations were reduced almost by half by tuna seine net fisheries during the 1950s to early 1990s. Today, some spinner dolphins are killed in nets in the tropics for bait, meat, or incidentally, and exposure to harassment from dolphin watching is of concern in other areas.

SIGNIFICANCE TO HUMANS
Used by tuna fishermen to indicate the presence of tuna below. They also form the basis for a dolphin-watching industry in several locations. ◆

Common name / Scientific name/ Other common names	Physical characteristics	Habitat and behavior	Distribution	Diet	Conservation status
Commerson's dolphin *Cephalorhynchus commersonii* German: Commerson-Delfin; Spanish: Marsopa de anteojos	Coloration is black and white. Black color extends from head to behind blowhole, down the side (including the flippers, dorsal fin, and tailstock). Rest of the body is pure white. Head and body length 25.2–57.7 in (64–146.5 cm), weight 57.3–189.6 lb (26–86 kg).	Coastal regions near the mouths of bays and estuaries, or over the wide shallow continental shelf, where the tidal range is great. Birthing season is between early spring and late summer, from October to March. Groups are made up of one to three individuals.	Indian Ocean and Atlantic Ocean.	Consists of mysid shrimp, three species of fish, squid, algae, isopods, and other benthic invertebrates.	Not threatened
Black dolphin *Cephalorhynchus eutropia* English: Chilean dolphin; German: Weißbauchdelfin; Spanish: Delfín chileno	Coloration is black, except for white coloration of belly, chin, sides, and anterior of the back. No beak, a number of teeth. Females slightly larger than males. Head and body length 31.5–43.3 in (80–110 cm), weight 57.3–163.1 lb (26–74 kg).	Areas of strong tidal flow above a steeply dropping continental shelf. Mate in the early winter and bear their young in the spring. Females have one calf every two years. Very shy species, groups of 8 to 14 individuals.	Pacific Ocean, off the coast of Chile and as far south as Tierra del Fuego.	Feed on sea bottom, mainly on fish, squid, and crustaceans.	Not threatened
Short-beaked saddleback dolphin *Delphinus delphis* English: Common dolphin; German: Gemeiner Delfin; Spanish: Delfín común	Coloration of back is black or dark brown, underside is white or cream colored. Dark streak runs from the lower jaw to the flipper. Eyes are encircled with black markings, beak is black. Crisscross pattern runs across the side. Head and body length 60–96 in (152.4–243.8 cm), weight 220.4–300 lb (100–136 kg).	Offshore and occasionally inshore. Groups consist of 10–500, largest groups in eastern tropical Pacific. May often be found in large, active schools.	Atlantic Ocean, Pacific Ocean, and Indian Ocean; abundant in the Mediterranean Sea, as well as in the Black Sea, the Gulf of Mexico, and the Red Sea.	Fish and squid or octopus.	Not threatened
Pygmy killer whale *Feresa attenuata* German: Zwerggrindwal	Coloration is dark gray to black with paler underside, some white on belly. Rounded head, no beak, white lips, white patch on tip of lower jaw. Large and conical teeth. Sub-triangular, high dorsal fin that points backwards. Head and body length 82.7–102.4 in (210–260 cm), weight 242.5–374.8 lb (110–170 kg).	Tropical waters, though it has been spotted in cooler waters off the west coast of southern Africa and Peru. Prefers subtropical and tropical waters usually in deep water in the open oceans and is rarely found in closed water. Believed to be non-migratory. Aggressive and extremely acrobatic. Calves are born during summer months.	Around Japan and Hawaii and in the warmer eastern areas of the North Pacific Ocean. Also in the west Indian Ocean, around tropical western Africa in the Atlantic Ocean, and in the Gulf of Mexico.	Mainly squid, octopus, and large fish, e.g., tuna and dolphinfish.	Data Deficient
Short-finned pilot whale *Globicephala macrorhynchus* French: Globicéphale tropical; German: Indischer Grindwal; Spanish: Calderón negro	Coloration is very dark brown, or gray-black. Light gray or white patch in the shape of an anchor on the throat and chest. Streak behind eye and behind dorsal. Bulbous forehead, small mouth, pointed flippers, rounded dorsal fin. Head and body length: male 23.6 ft (7.2 m), female 16.7 ft (5.1 m); maximum body weight: male 4.35 tons (3.95 tonnes), female 1.5 tons (1.4 tonnes).	Tropical, subtropical, and warm temperate waters where the surface has a temperature of 46.4–77°F (8–25°C). Will enter coastal and shallow waters in search of food. Families consist of up to 40 individuals and may even reach 100 individuals. Extremely social, not acrobatic, vocalize with clicks and whistles.	All tropical, subtropical, and warm temperate oceans.	Mainly squid and octopus, otherwise fish.	Lower Risk/ Conservation Dependent
Long-finned pilot whale *Globicephala melas* German: Gewöhnlicher Grindwal	Coloration is black, white belly, spherical head, long tapering trunk. Dorsal fin has a long base. Tail is long and high, flipper is one-fifth of the body. Average head and body length: male 19.7–26 ft (6–8 m), female 16.4–19.7 ft (5–6 m).	Cooler waters. There are two population, northern and southern, which do not meet, mix, and interbreed because they are separated by a wide band of warm tropical water. Travels in small groups of 4 to 6 individuals. Occasionally seen in herds of 50 or more and, on some occasions, they will gather by the hundreds. Inborn fear of killer whale.	Nearctic area of Atlantic Ocean.	Mainly squid and cod, but also a variety of other fish.	Not threatened

[continued]

Common name / Scientific name/ Other common names	Physical characteristics	Habitat and behavior	Distribution	Diet	Conservation status
Risso's dolphin *Grampus griseus* French: Dauphin de Risso, grampus; German: Rundkopfdelfin; Spanish: Delfín de Risso, fabo calderón	Coloration may vary from blue-gray, gray-brown, to almost white. Large, blunt head, no beak, males may have scars. Mouth slants upwards, dorsal fin is tall and curved, tail stock is thick. Head and body length 9.2–12.6 ft (2.8–3.85 m), weight up to 1,100 lb (500 kg).	Warm temperate and tropical offshore waters and are seen close to shore only when the continental shelf is narrow. Groups consist of 3 to 50 individuals. Occasional aggregations of up to 4,000 are also seen, and they often mix with other dolphin species. Make a variety of sounds, including signature whistle. Mass and individual strandings not uncommon.	Deep tropical and warm temperate waters in both the Northern and Southern Hemispheres.	Mainly squid and octopus, but also other varieties of fish.	Data Deficient
Fraser's dolphin *Lagenodelphis hosei* French: Dauphin de Fraser; German: Borneo-Delfin, Fraser-Delfin; Spanish: Delfín de Borneo	Robust, small, and pointed flippers and dorsal fin. Coloration is dark blue-gray or gray-brown. Short beak, dark upper jaw. Chin, throat, and belly are white. Gray line runs from melon to flanks. Head and body length 6.6–8.7 ft (2–2.65 m), weight around 440 lb (200 kg).	Only in tropical and subtropical waters. Breeding appears to be year-round with a possible peak in the summer months. Group sizes consist of 100–2,500 individuals.	Atlantic Ocean, Pacific Ocean, and Indian Ocean.	Primarily eat fish, but they also feed on squid, cuttlefish, and shrimp.	Data Deficient
Atlantic white-sided dolphin *Lagenorhynchus acutus* German: Weißseitendelfin	Back is black or dark purple to gray, ventral side is yellow to tan. A blaze runs from each side to the tail stock. Pale gray stripe also runs along the length of the body. Belly is pale, yellowish. Dark stripe runs from the corner of the jaw to the insertion point of flippers. Large, dark eye patches. Head and body length 95.7–8.4 in (243–250 cm), weight 401.2–515.9 lb (182–234 kg).	Cool waters, average 44.6–53.6°F (7–12°C), of the North Atlantic Ocean. Calving season is in summer. Uses clicks and whistles to communicate. Fast swimmer, feeds in small groups.	North Atlantic Ocean.	Mainly small schooling fish and squid, but also herring, smelt, silver hake, and shrimp.	Not listed by IUCN
Atlantic humpback dolphin *Sousa teuszii* French: Dauphin à bosse de l'atlantique; German: Kamerunfluss-Delfin; Spanish: Bufeo africano, delfín blanco africano	Younger individuals are cream colored, graying as they age. Unusual dorsal fin that is curved and humped. Second, smaller hump exists on dorsal side. Head and body length 47.2–98.4 in (120–250 cm), average weight 165.3–330.7 lb (75–150 kg).	Tropical waters close to the West African shoreline. They are not thought to venture more than 0.6–1.2 mi (1–2 km) away from the shore in an effort to avoid killer whales. Solitary, often travel and feed alone. Groups may range from 2 to 10 individuals. One offspring is born at a time.	Atlantic Ocean waters off the coast of western Africa, from Mauritania south to Angola.	Fish such as sardines and mullet.	Data Deficient
Striped dolphin *Stenella coeruleoalba* French: Dauphin bleu et blanc, dauphin rayé; German: Blau-Weißer Delfin; Spanish: Delfín blanco y azul	Fusiform body, tall dorsal fins, long and narrow flippers, prominent beak, distinctive color, and stripe pattern on body. Coloration is bluish gray with dark dorsal cape and light ventral coloration. Head and body length 86.6–92.9 in (220–236 cm), weight up to 198.4–330.7 lb (90–150 kg).	Offshore and inshore warm temperate and tropical waters. Mating season is in winter and early summer. Group size ranges from a few to over 1,000 individuals. Three different kinds of schools often occur: juvenile, breeding adults, and nonbreeding adults.	Atlantic Ocean, Pacific Ocean, and Indian Ocean in warm temperate and tropical seas throughout the world.	Mainly cephalopods, crustaceans, and bony fishes.	Lower Risk/ Conservation Dependent

Resources

Books

Berta, A., and J. L. Sumich. *Marine Mammals: Evolutionary Biology*. San Diego: Academic Press, 1999.

Leatherwood, S., and R. R. Reeves. *The Sierra Club Handbook of Whales and Dolphins*. San Francisco: Sierra Club Books, 1983.

Le Duc, R. G. *A Systematic Study of the Delphinidae (Mammalia: Cetacea) using Cytochrome b Sequences*. PhD dissertation, University of California, San Diego, 1997.

Norris, K. S., B. Würsig, R. S. Wells, and M. Würsig. *The Hawaiian Spinner Dolphin*. Berkeley: University of California Press, 1994.

Reeves, R. R., B. D. Smith, E. A. Crespo, and G. Notarbartolo di Sciara. *Dolphins, Whales, and Porpoises: 2002–2010 Conservation Action Plan for the World's Cetaceans*. Gland, Switzerland: IUCN, 2003.

Reeves, R. R., B. S. Stewart, P. J. Clapham, and J. A. Powell. *Guide to Marine Mammals of the World*. New York: Alfred A. Knopf, 2002.

Reynolds, J. E. III, and S. A. Rommel, eds. *Biology of Marine Mammals*. Washington, DC: Smithsonian Institution Press, 1999.

Reynolds, J. E. III, R. S. Wells, and S. D. Eide. *The Bottlenose Dolphin: Biology and Conservation*. Gainesville: University Press of Florida, 2000.

Rice, D. W. *Marine Mammals of the World: Systematics and Distribution*. Special Publication No. 4. Society for Marine Mammalogy, 1998.

Ridgway, S. H., and R. Harrison, eds. *Handbook of Marine Mammals*. Vol. 5. San Diego: Academic Press, 1994.

Ridgway, S. H., and R. Harrison, eds. *Handbook of Marine Mammals*. Vol. 6. San Diego: Academic Press, 1999.

Twiss, J. R. Jr., and R. R. Reeves, eds. *Conservation and Management of Marine Mammals*. Washington, DC: Smithsonian Institution Press, 1999.

Randall S. Wells, PhD

Beaked whales
(*Ziphiidae*)

Class Mammalia
Order Cetacea
Suborder Odontoceti
Family Ziphiidae

Thumbnail description
Small- to medium-sized whales, characterized by a noticeable beak with no crease between it and the forehead, a single pair of throat grooves, a robust cigar-shaped body, small dorsal fin positioned two-thirds of the way along the body, large flukes lacking a central notch, and very reduced dentition with only one or two pairs of teeth that form tusks in most species

Size
13–42 ft (3.9–12.8 m)

Number of genera, species
6 genera; 21 species

Habitat
Deep ocean, particularly around seabed features such as canyons, seamounts, and escarpments

Conservation status
Lower Risk/Conservation Dependent: 4 species; Data Deficient: 15 species

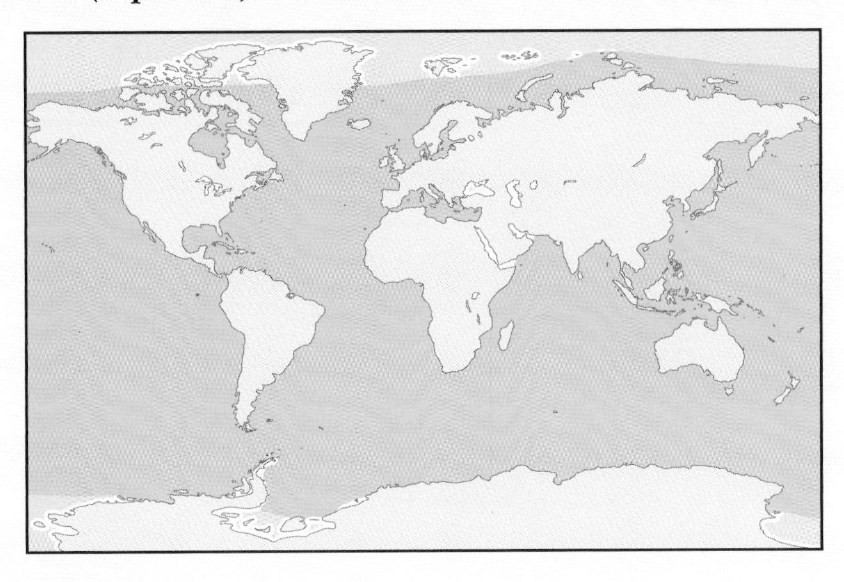

Distribution
All oceanic waters deeper than 660 ft (200 m), from the ice edge to the equator

Evolution and systematics

The beaked whales were one of the first lineages to split from the ancestral group of toothed whales, and now form the second largest family of living species in the order Cetacea. Although some fossil beaked whales are known, it is not clear how these relate to modern species. Fossils may represent extinct lineages or ancestors to modern species. The genera *Ziphius*, *Tasmacetus*, and *Indopacetus* all contain a single species each, while the genera *Berardius* and *Hyperoodon* both contain two species. In contrast, the genus *Mesoplodon* contains 14 species. Currently, no beaked whale species is separated into subspecies.

Physical characteristics

Adult beaked whales range in body length from 13 ft (3.9 m) in the pygmy beaked whale (*Mesoplodon peruvianus*) to 42 ft (12.8 m) in Baird's beaked whale (*Berardius bairdii*). With the exception of Cuvier's beaked whale (*Ziphius cavirostris*), all beaked whales have a noticeable beak, and in none of the species is there a crease between the beak and the forehead, which is found in other toothed whales. Between the lower jaws is a single pair of throat grooves that are used in feeding. The body is cigar-shaped, with the greatest girth occurring in the middle. The dorsal fin is relatively small, sub-triangular in shape, and is set two-thirds of the way along the body. The pectoral fins are also relatively small and can be held against the body in recessed flipper pockets. The tail flukes are broad and are unique in cetaceans in lacking a central notch. The dentition is greatly reduced in all species, except Shepherd's beaked whale (*Tasmacetus shepherdi*), which has only one or two pairs of teeth remaining in the lower jaw and none remaining in the upper jaw. In all species apart from those of the genus *Berardius*, these teeth erupt in adult males to form tusks, though they never erupt in adult females. The position and shape of the tusks vary between beaked whale

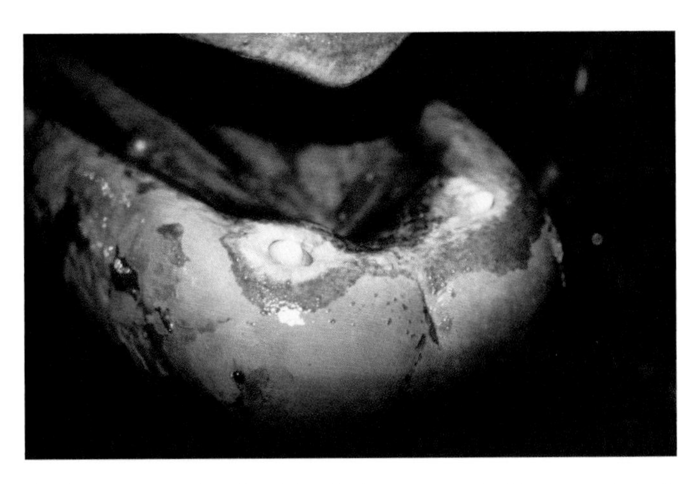

The lower jaw and teeth of Baird's beaked whale (*Berardius bairdii*), seen in Wadaura, Boso Peninsula, Japan. (Photo by © Mako Hirose/Seapics.com. Reproduced by permission.)

Before diving, Cuvier's beaked whale (*Ziphius cavirostris*) takes about a 20 second breath. (Photo by © Todd Pusser/Seapics.com. Reproduced by permission.)

An adult male Blainville's beaked whale (*Mesoplodon densirostris*), Hawaii. (Photo by © Michael S. Nolan/Seapics.com. Reproduced by permission.)

Cuvier's beaked whale (*Ziphius cavirostris*), seen in the Azores Islands, Portugal. (Photo by © Whale Watch Azores/Seapics.com. Reproduced by permission.)

A northern bottlenosed whale (*Hyperoodon ampullatus*) diving. (Photo by Flip Nicklin/Minden Pictures. Reproduced by permission.)

The northern bottlenosed whale (*Hyperoodon ampullatus*) has a short, bottle-shaped beak. (Photo by Flip Nicklin/Minden Pictures. Reproduced by permission.)

species and are the primary features used to identify species. The color of beaked whales varies from light brown to slate gray and black. In some species there are also areas of contrasting light colors, which can be sexually dimorphic. Adult males of many species are covered in long pale lines that are scars caused by the tusks of other males.

Distribution

Beaked whales have been recorded from all oceanic waters beyond the continental shelf edge and from the ice-edge at the poles to the equator. However, individual species generally have a more restricted distribution. Only two species, Cuvier's beaked whale and Blainville's beaked whale (*Mesoplodon densirostris*), have been recorded from more than two oceans, and some species appear to have a very restricted distribution, such as the newly described Perrin's beaked whale (*Mesoplodon perrini*), which to date has only been recorded from the waters along the coast of California.

Habitat

Beaked whales occupy deep oceanic waters of greater than 660 ft (200 m) beyond the edges of the continental shelf, with most sightings occurring in water depths between 3,300 and 9,900 ft (1,000–3,000 m). Sightings of living beaked whales are often concentrated around marine features such as canyons, seamounts, shelf edges, and escarpments.

Behavior

Beaked whales are generally seen in small groups of fewer than 10 animals, but group size and structure vary between species. Where studies exist, individual animals have been resighted within and between years in the same area, indicating some level of site fidelity. Some species of beaked whales show evidence of regular migrations.

Feeding ecology and diet

Beaked whales primarily feed on deepwater squid, fish, and occasionally crustaceans, ranging in sizes from a few ounces (grams) to several pounds (kilograms). The presence of bottom-living prey species as well as small stones in the stomachs of beaked whales suggests that they are capable of diving to the seabed in water depths of 3,300 ft (1,000 m) or more to forage. Prey capture is thought to occur by suction feeding, with animals using their piston-like tongues and expandable throat-grooves to suck prey into the mouth.

Reproductive biology

Single calves are the norm in beaked whales and these calves remain close to their mothers for at least the first year before being weaned. The presence of large sexually dimorphic weapons (the tusks) and the restriction of intraspecific scarring to adult males suggests that most species are polygamous, with adult males actively competing aggressively for access to receptive females.

Conservation status

Four beaked whale species are listed as Lower Risk/Conservation Dependent by the IUCN, and 15 are listed as Data Deficient, which reflects the lack of studies of these species.

Baird's beaked whale (*Berardius bairdii*) has the ability to stay underwater for over an hour. (Photo by Randall S. Wells. Reproduced by permission.)

An adult male Blainville's beaked whale (*Mesoplodon densirostris*), seen in the Canary Islands, Spain. (Photo by © Fabian Ritter/Seapics.com. Reproduced by permission.)

The northern bottlenose whale was hunted commercially in the past, but in 1977 the population was considered to be depleted by the International Whaling Commission (IWC) and the fishery was closed.

Significance to humans

In general, beaked whales have little significance to humans. Three species of beaked whales have been commercially hunted by humans: the northern bottlenosed whale in the North Atlantic and Cuvier's beaked whale and Baird's beaked whale by the Japanese in the North Pacific. Other beaked whale species are occasionally taken in several parts of the world either as bycatch in fishing nets or from whale fisheries directed at other species.

1. Male northern bottlenosed whale (*Hyperoodon ampullatus*); 2. Female Longman's beaked whale (*Indopacetus pacificus*); 3. Male Shepherd's beaked whale (*Tasmacetus shepherdi*); 4. Male Cuvier's beaked whale (*Ziphius cavirostris*); 5. Male Blainville's beaked whale (*Mesoplodon densirostris*); 6. Male Baird's beaked whale (*Berardius bairdii*). (Illustration by Bruce Worden)

Species accounts

Northern bottlenosed whale
Hyperoodon ampullatus

TAXONOMY
Hyperoodon ampullatus (Forster, 1770), Maldon, England.

OTHER COMMON NAMES
English: Altantic bottlenosed whale, flathead, bottlehead; French: Grand souffleur à bec d'oie; German: Butskof; Spanish: Hocico de botella.

PHYSICAL CHARACTERISTICS
Adult males reach up to 33 ft (10 m) in length, while adult females are noticeably smaller, reaching only 28 ft (8.5 m). The body is robust, with a large bulbous head and short beak. The forehead becomes more bulbous in adult males due to the growth of two large bony crests on the skull. The single pair of teeth found in the lower jaw remains relatively small and barely erupted even in the oldest males. Dark on the back, with a paler belly, although adult males may also have a pale area on the forehead.

DISTRIBUTION
Endemic to the polar to warm-temperate waters of the North Atlantic. They are regularly recorded from the Norwegian Sea, around Greenland, and off Labrador south to the Bay of Biscay, the Azores, and Nova Scotia, as well as further south.

HABITAT
Generally found in deep waters beyond the edge of the continental shelves and is usually seen in water depths of greater than 3,300 ft (1,000 m). Off the coasts of Nova Scotia, one population is regularly sighted over a deep marine canyon known as the Gully, however, northern bottlenose whales are also recorded over continental slopes and plateaus.

BEHAVIOR
Usually occur in groups of about four individuals, although larger groups of more than 10 have been sighted. Adult males have been seen aggressively head-butting each other, using the large bony crests in their foreheads as battering rams, and such combat may relate to competition for females. Some populations may migrate, moving northward in late winter and spring and southward in late summer and autumn. However, other populations such as in the Gully off Nova Scotia are apparently nonmigratory.

FEEDING ECOLOGY AND DIET
In the northern North Atlantic, squid of the genus *Gonatus* are the most commonly consumed prey, although other squid species and deepwater fish are eaten in some locations. Northern bottlenose whales usually dive close to the seabed when foraging and can dive to depths of over 5,000 ft (1,500 m) for up to 80 minutes or more.

REPRODUCTIVE BIOLOGY
Calves are born in late spring and summer after a gestation of approximately 12 months and will nurse for approximately another 12 months. Both males and females mature around the age of seven to 10 years. Adult males are larger than adult fe-

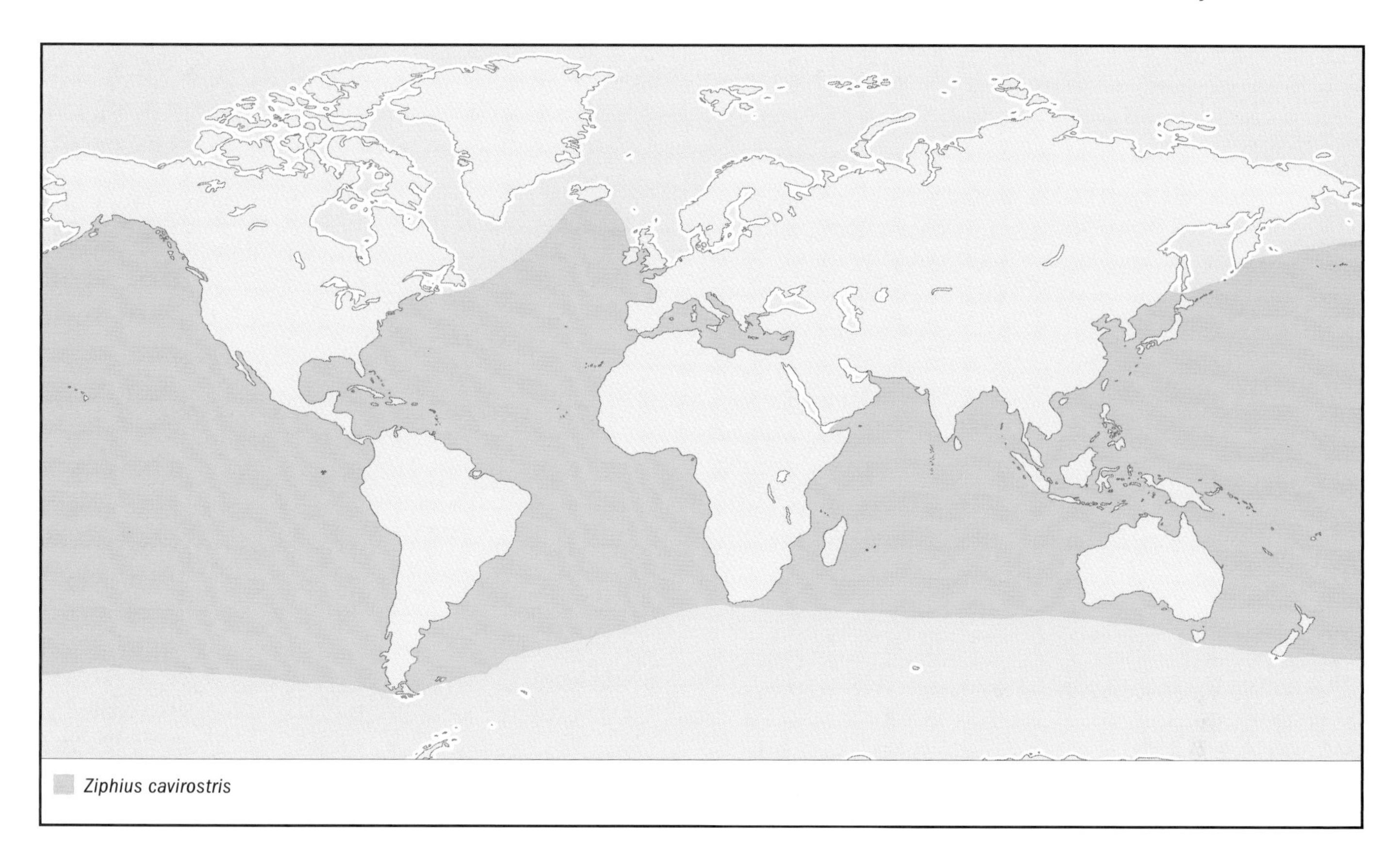

males, and this species may well be polygamous, however the exact mating system remains unclear.

CONSERVATION STATUS
In 1977, northern bottlenosed whales were declared depleted by the International Whaling Commission, and commercial whaling for this species ceased. It is currently classified as Data Deficient by the IUCN.

SIGNIFICANCE TO HUMANS
None known. ◆

Baird's beaked whale
Berardius bairdii

TAXONOMY
Berardius bairdii (Stejneger, 1882), Commander Islands, Russia.

OTHER COMMON NAMES
English: Northern four-toothed whale, giant bottlenosed whale.

PHYSICAL CHARACTERISTICS
Reaches up to 39 ft (12 m) in males and 42 ft (12.8 m) in females. They have a noticeable beak with two pairs of teeth in the lower jaw and a relatively steep forehead. The teeth erupt to become functional in both males and females as they mature. Their bodies are generally dark gray or black all over, with adults gaining pale linear scars as they grow older.

DISTRIBUTION
Endemic to the North Pacific and found from as far south as Japan and Mexico, and north to Siberia, the Aleutian Islands, and Alaska.

HABITAT
Generally found in deep waters beyond the edge of the continental shelves and usually seen in water depths of greater than 3,300 ft (1,000 m). Most sightings occur over marine features such as seamounts, escarpments, and continental slopes.

BEHAVIOR
Usually found in groups of up to 10 animals, although larger groups of up to 30 have been recorded; groups may contain several mature adults of both sexes. They typically dive for about 30 minutes or more and may dive to more than 3,300 ft (1,000 m) to feed.

FEEDING ECOLOGY AND DIET
Consumes deepwater squid and fish species.

REPRODUCTIVE BIOLOGY
Despite being hunted, reproduction of Baird's beaked whales remains poorly known. The length of gestation is unclear, with some speculating that it lasts 10 months while others think it lasts 17 months; females may only produce a calf every three or so years. Males mature before females and apparently live longer. The mating system is unknown.

CONSERVATION STATUS
Classified as Data Deficient by the IUCN.

SIGNIFICANCE TO HUMANS
The Japanese hunt Baird's beaked whales off the coasts of Japan and in the western Pacific. Other than this, this species has little significance to humans. ◆

Cuvier's beaked whale

Ziphius cavirostris

TAXONOMY

Ziphius cavirostris G. Cuvier, 1823, France.

OTHER COMMON NAMES

English: Goose-beaked whale; French: Ziphius de Cuvier; German: Cuvier-Schnabelwal; Spanish: Ballena de Cuvier.

PHYSICAL CHARACTERISTICS

Up to 23 ft (7 m) in length and brown to tan in color. Adult males have pale heads and backs, and can be covered in large numbers of long pale scars caused by the tusks of other males. Their tusks are conical in shape, situated at the tip of the lower jaw, and only erupt to become functional in adult males.

DISTRIBUTION

The widest distribution of any beaked whale species, being found in tropical to cold-temperate waters of all the world's oceans.

HABITAT

Generally found in deep waters beyond the edge of the continental shelves and usually seen in water depths of greater than 3,300 ft (1,000 m). Sightings are often reported over marine features such as canyons, escarpments, and shelf edges.

BEHAVIOR

Recorded in groups of up to seven animals. While at the surface, the most characteristic behavior is a three-quarter lunge out of the water, often in response to a close approach by a boat. Dives may last for more than 30 minutes and animals may reach depths of over 3,300 ft (1,000 m).

FEEDING ECOLOGY AND DIET

Consumes deepwater squid, fish, and crustaceans, and dives to great depths to capture them.

REPRODUCTIVE BIOLOGY

Almost nothing is known about reproduction. Adult males are larger than females and have high levels of scarring caused by the tusks of other males. These features are consistent with a polygamous mating system.

CONSERVATION STATUS

Classified as Data Deficient by the IUCN.

SIGNIFICANCE TO HUMANS

Taken in small numbers by Japanese whalers in the North Pacific. Other than this, Cuvier's beaked whales are of little significance to humans. ◆

Blainville's beaked whale

Mesoplodon densirostris

TAXONOMY

Mesoplodon densirostris (Blainville, 1817), type locality unknown.

OTHER COMMON NAMES

English: Dense beaked whale; German: Blainville-Schnabelwal.

PHYSICAL CHARACTERISTICS

Between 15 and 16 ft long (4.5–5 m), with a relatively long beak and a noticeable arch midway along the lower jaw. In

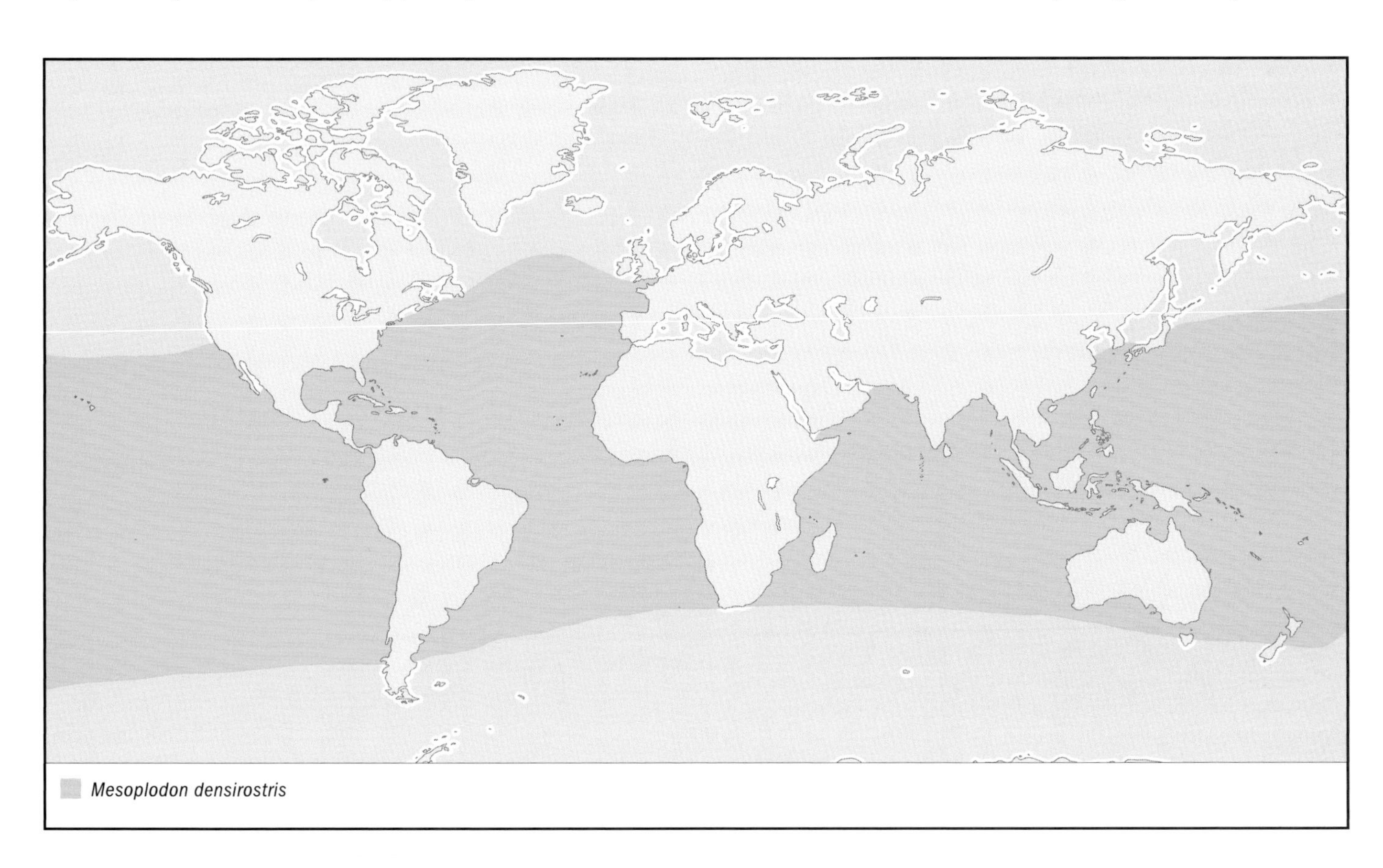

adult males, this arch becomes enlarged and two large tusks erupt from the top. Younger animals are gray or dark gray on top and pale underneath, however adult animals are often brown, gray, or dark all over. In adult males, the head and back can be covered in a very large number of white scars caused by the tusks of other males.

DISTRIBUTION
Found in all tropical to warm-temperate waters around the globe, although they are most commonly recorded around oceanic tropical islands.

HABITAT
This species is recorded in shallower waters than most other beaked whale species and can be seen in water depths of as little as 330 ft (100 m). Around oceanic islands, most sightings are in water depths of less than 3,300 ft (1,000 m).

BEHAVIOR
Relatively little is known about behavior other than the fact that they are deep-divers and often dive for up to 30 minutes or more. However, the heavy scarring found on adult males indicates that they engage in aggressive combat, presumably over access to females.

FEEDING ECOLOGY AND DIET
Deepwater squid and fish have been recorded in the stomachs of Blainville's beaked whales, and they are thought to forage at or close to the seabed.

REPRODUCTIVE BIOLOGY
Apparently polygamous, with adult males competing aggressively with each other for access to females. Groups with more than one adult male are rarely recorded. However, males do not appear to remain with a single female group and may rove between them looking for receptive females. Calves remain close to their mothers for the first year, with weaning occurring after 12 months. Both sexes mature at around 10 years of age.

CONSERVATION STATUS
Classified as Data Deficient by the IUCN. Although occasionally killed by fishermen, either purposefully or accidentally in fishing nets set for other species, this species has never been hunted commercially.

SIGNIFICANCE TO HUMANS
Generally unknown to most humans and, when encountered, few people even identify them correctly, most thinking they are large dolphins. ◆

Shepherd's beaked whale
Tasmacetus shepherdi

TAXONOMY
Tasmacetus shepherdi Oliver, 1937, New Zealand.

OTHER COMMON NAMES
English: Tasman beaked whale.

PHYSICAL CHARACTERISTICS
Up to 23 ft (7 m) in length, with a long beak and rounded forehead. They are dark on the back, with lighter sides and belly. Has numerous small, peg-like teeth in both the upper and lower jaws as well as a pair of large sexually dimorphic tusks at the tip of the lower jaw.

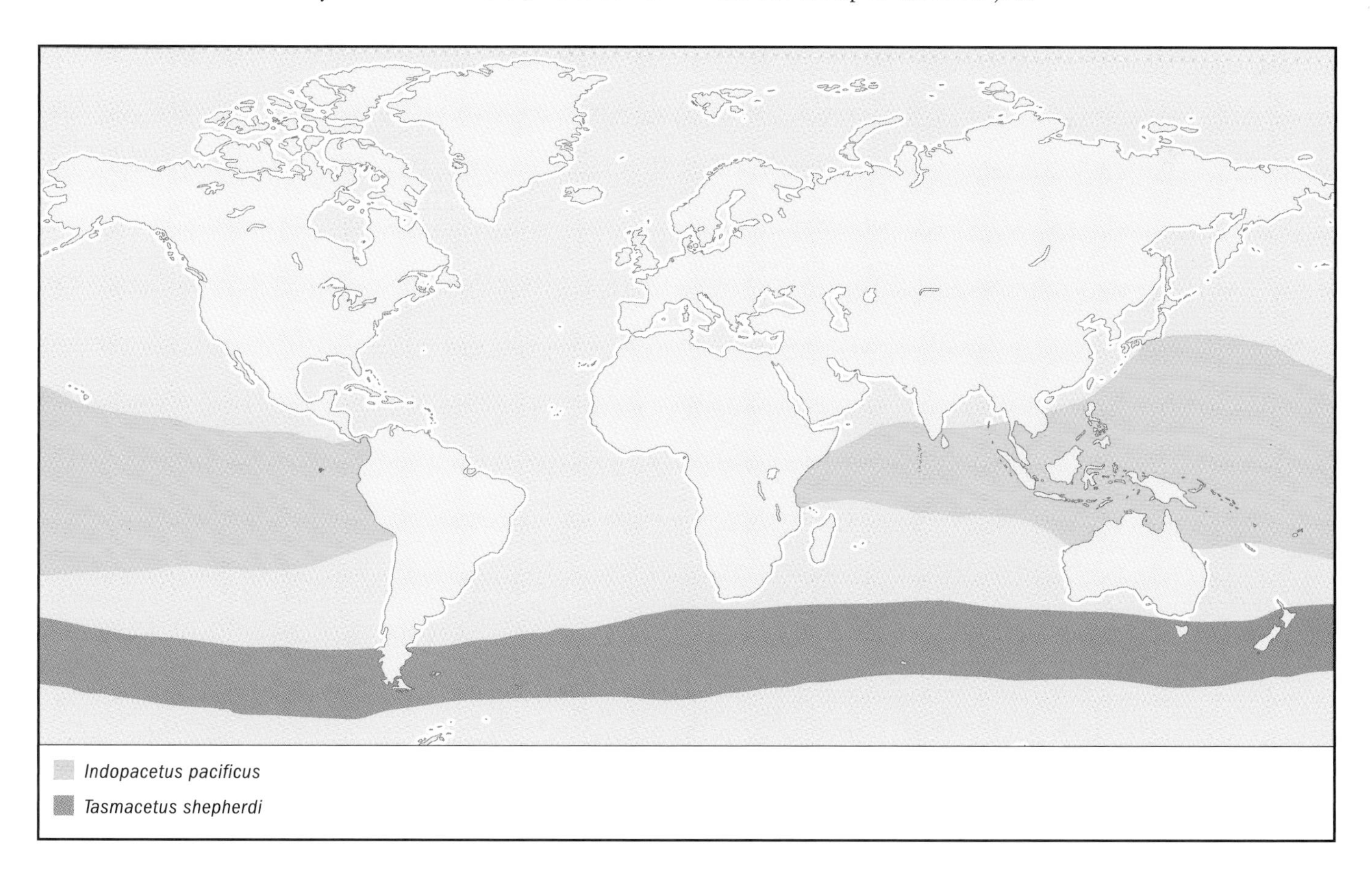

DISTRIBUTION
All records come from the colder waters of the Southern Hemisphere; the majority of records come from New Zealand, with additional records from southern Australia, the Chatham Islands, Juan Fernandez Islands, Argentina, and South Africa.

HABITAT
With no confirmed sightings at sea, nothing is currently known about its habitat preferences.

BEHAVIOR
Little is known about its behavior.

FEEDING ECOLOGY AND DIET
It is thought that deepwater fish may make up a greater proportion of the diet of Shepherd's beaked whale than other beaked whales, although this is based on stomach contents analysis from a single animal.

REPRODUCTIVE BIOLOGY
Nothing is known about its reproductive biology.

CONSERVATION STATUS
Classified as Data Deficient by the IUCN.

SIGNIFICANCE TO HUMANS
Shepherd's beaked whale was only discovered in 1937, and has had little significance to humans. ◆

Longman's beaked whale
Indopacetus pacificus

TAXONOMY
Indopacetus pacificus (Longman, 1926), Queensland, Australia.

OTHER COMMON NAMES
English: Indopacific beaked whale.

PHYSICAL CHARACTERISTICS
The only two definite records consist of two skulls. From this, it has been estimated that this species is about 23 ft (7 m) long. Possible sightings are of animals about 23–26 ft (7–8 m) long, with a moderately long beak and bulbous forehead and varying in color from tan to grayish brown.

DISTRIBUTION
The two definite records of Longman's beaked whale were found on a beach in Queensland, Australia, and in a fertilizer factory in Somalia. However, possible sightings have been recorded from the tropical waters of the Pacific and Indian Oceans.

HABITAT
This species' habitat is not known.

BEHAVIOR
Sightings of possible Longman's beaked whales are of groups of animals usually between 15 and 20 animals, although some groups number up to 100.

FEEDING ECOLOGY AND DIET
Nothing is known, as it has never been seen in the wild.

REPRODUCTIVE BIOLOGY
Nothing known.

CONSERVATION STATUS
Classified as Data Deficient by the IUCN.

SIGNIFICANCE TO HUMANS
None known. ◆

Common name / Scientific name/ Other common names	Physical characteristics	Habitat and behavior	Distribution	Diet	Conservation status
Arnoux's beaked whale *Berardius arnuxii* English: Southern four-toothed whale, southern giant bottlenosed whale; French: Bérardien d'Arnoux; Spanish: Ballena de pico de Arnoux	Dark or dark brown, often covered with many long, thin white scars. Long beak and steeply sloping forehead. Both males and females have only two pairs of teeth, found in the lower jaw. These erupt to become functional only in adults. Body length 25–30 ft (7.8–9.7 m).	Groups of up to 80 individuals. Found in deep open waters, along the ice edge, and in open areas within seasonal ice fields.	Southern oceans from Brazil, South Africa, and Australia south to the Antarctic ice edge.	Deepwater squid and fish.	Lower Risk/ Conservation Dependent
Southern bottlenosed whale *Hyperoodon planifrons* English: Antarctic bottlenosed whale, flathead; French: Hyperoodon austral; Spanish: Ballena hocico de botella del sur	Dark brown, dark or slate gray. Short beak and bulbous forehead, particularly in adult males. Reduced dentition with only one pair of teeth at the tip of the lower jaw that erupt beyond the gums only in adult males. Body length 19–25 ft (6–7.5 m).	Generally seen in groups of one or two. Found in deep waters. May dive to great depths to feed.	Southern oceans from Brazil, South Africa, and Australia south to the Antarctic ice edge.	Deepwater squid and fish.	Lower Risk/ Conservation Dependent
Andrew's beaked whale *Mesoplodon bowdoini* English: Splay-toothed beaked whale, Bowdoin's beaked whale; French: Mesoplodon de Bowdoin; Spanish: Ballena de pico de Andrew	Dark blue-black or brown with white tip to beak, males may have long, thin, white scars on body. Short beak and gently sloping forehead, adult males have two large tusks in middle of lower jaw. Body length 13–16 ft (4–4.7 m).	There are no confirmed sightings of this species in the wild, so habitat use and behavior remain unknown.	Most of the 35 known records are from New Zealand, with additional records from the southern coast of Australia, the Falkland Islands and Tristan da Cunha. Apparently restricted to southern temperate waters south to the Antarctic Convergence.	Unknown.	Data Deficient
Sowerby's beaked whale *Mesoplodon bidens* English: North Sea beaked whale; French: Mesoplodon de Sowerby; Spanish: Ballena de pico de Sowerby	Dark gray or brown on top, paler underneath. Long, narrow beak, gently sloping forehead. Reduced dentition with only a single pair of teeth in the middle of the lower jaw that form tusks in adult males. Body length 14.8–18.2 ft (4.5–5.5 m).	Generally seen in groups of fewer than ten animals. Found in deep waters of several thousand feet (meters) or more and may dive to great depths to feed	Temperate to subpolar North Atlantic from Norway, Iceland, and Labrador south to the Azores and northern United States.	Deepwater fish and squid.	Data Deficient
Hubbs' beaked whale *Mesoplodon carlhubbsi* English: Arch-beaked whale; French: Mesoplodon de Hubbs; Spanish: Ballena de pico de Hubbs	Dark gray or black with a white tip to the beak and a white cap, males may have high levels of thin white scars all over the body. Relatively short beak with a noticeable arch in the middle of the lower jaw. Reduced dentition with only a single pair of teeth in the middle of the lower jaw that form tusks in adult males. Body length 16.5–18.2 ft (5.0–5.5 m).	Rarely seen in the wild; habitat use and behavior remain unknown.	Temperate waters of the North Pacific along the coast of Japan and continental North America from Mexico to southern Canada.	Deepwater squid and fish.	Data Deficient
Gervais' beaked whale *Mesoplodon europaeus* English: Antillean beaked whale, Gulf Stream beaked whale, European beaked whale; French: Mesoplodon de Gervais; Spanish: Ballena de pico de Gervais	Dark gray or brown on top, paler underneath. Relatively short beak and steep forehead. Reduced dentition with only a single pair of teeth in the one third of the jaw length from the tip of the lower jaw. Body length 14.8–17.2 ft (4.5–5.2 m).	Rarely seen in the wild; habitat use and behavior remain unknown.	Warm temperate to tropical waters of the Atlantic on both sides of the equator.	Deepwater squid, fish, and crustaceans.	Data Deficient
Ginkgo-toothed beaked whale *Mesoplodon ginkgodens* English: Japanese beaked whale; French: Mesoplodon de Nishiwaki	Dark gray or brown all over. Relatively short beak and smoothly sloping forehead. Reduced dentition with only a single pair of teeth in the middle of the lower jaw that only barely erupt in adult males. Body length 14.8–17.2 ft (4.5–5.0 m).	Rarely seen in the wild; habitat use and behavior remain unknown.	Recorded from Mexico, southwestern United States, Galápagos Islands, Japan, Taiwan, Sri Lanka, Malaysia, New South Wales, Australia, and New Zealand.	Unknown.	Data Deficient

[continued]

Common name / Scientific name/ Other common names	Physical characteristics	Habitat and behavior	Distribution	Diet	Conservation status
Gray's beaked whale *Mesoplodon grayi* English: Southern beaked whale; French: Mesoplodon de Gray; Spanish: Ballena de pico de Gray	Dark gray or brown with a white beak and face in some individuals. Long, narrow beak, gently sloping forehead. Reduced dentition with only a single pair of teeth in the middle of the lower jaw that form tusks in adult males. Body length 14.8–18.2 ft (4.5–5.5 m).	Groups of four or five have been sighted, but one stranding had about 28 animals. Generally seen in deep water far from shore.	Temperate to subpolar waters of the Southern Hemisphere.	Deepwater squid.	Data Deficient
Hector's beaked whale *Mesoplodon hectori* English: Skew-beaked whale; French: Mesoplodon d'Hector; Spanish: Ballena de pico de Héctor	Dark gray or brown on top, paler underneath with a white beak. Short beak with only two triangular teeth close to the tip of the lower jaw that erupt only in adult males. Body length 13.2–14.9 ft (4–4.5 m).	Rarely seen in the wild; habitat use and behavior remain unknown.	Very small number of records from Tasmania, New Zealand, South Africa, Argentina, and the Falkland Islands.	Unknown, but probably deepwater squid and fish.	Data Deficient
Strap-toothed whale *Mesoplodon layardii* English: Layard's beaked whale; French: Mesoplodon de Layard; Spanish: Ballena de pico de Layard	Dark gray or black with white areas on belly, throat and on the back behind the head. Long, narrow beak, gently sloping forehead. Adult males have two tusks that emerge from the middle of the lower jaw and cross over the upper jaw. Body length 16.5–20.5 ft (5.0–6.2 m).	Occasionally sighted in deep waters far from shore in small groups.	Throughout the Southern Hemisphere from 30°S to the Antarctic Convergence.	Deepwater squid.	Data Deficient
True's beaked whale *Mesoplodon mirus* French: Mesoplodon de True; Spanish: Ballena de pico de True	Dark gray or brown on top, paler underneath with a dark patch around the eyes. Relatively short beak and steeply sloping forehead beak. Reduced dentition with only a single pair of teeth of conical teeth at the tip of the lower jaw that form tusks in adult males. Body length 16.2–17.5 ft (4.9–5.3 m).	Few, if any, positive sightings at sea; habitat use and behavior remain unknown.	Temperate North Atlantic from Ireland to the Canaries and the eastern seaboard of the United States and Nova Scotia, Canada, as well as records in South Africa and Australia, indicating that a separate Southern Hemisphere population also exists.	Deepwater squid.	Data Deficient
Perrin's beaked whale *Mesoplodon perrini*	Dark gray or brown on top, paler underneath with a white beak. Short beak with only two triangular teeth close to the tip of the lower jaw that only erupt in adult males. Body length 13.2–14.9 ft (4–4.5 m).	Known only from five stranded animals and two possible sightings; habitat use and behavior remain unknown.	All currently known records come from the waters of California, United States, but the species may have a wider distribution.	Unknown.	Not listed by IUCN
Pygmy beaked whale *Mesoplodon peruvianus* English: Peruvian beaked whale lesser beaked whale; French: Mesoplodon pygmée	Dark gray or brown on top, paler underneath. Males may have a pale flank and back as well as numerous pale white scars. Short beak and gently sloping forehead. Slight curve in lower jaw and only a single pair of teeth set on a slightly raised area of the lower jaw. Body length up to 12.8 ft (3.9 m).	Groups of one to five animals have been seen in deep oceanic waters.	Most records come from the eastern tropical Pacific from Peru to Mexico out to about 126°W, with an additional record from New Zealand.	Deepwater fish.	Data Deficient
Stejneger's beaked whale *Mesoplodon stejnegeri* English: Bering Sea beaked whale; French: Mesoplodon de Stejneger; Spanish: Ballena de pico de Stejneger	Dark brown, gray, or black all over with the exception of white on the beak and underside of the tail. Males may have many narrow, pale scars on the back. Long, narrow beak grading into a gently sloping forehead. Single pair of teeth set in the middle of the lower jaw that are visible only in adult males. Body length 16.5–17.5 ft (5–5.5 m).	Groups of up to 15 animals have been seen in deep waters ranging from 2,500 to 5,000 ft (750–1500 m) deep over continental slope areas	Cold temperate to subpolar North Pacific from as far south as California and Japan to Siberia, the Aleutian Islands, and Alaska.	Deepwater squid.	Data Deficient
Spade-toothed whale *Mesoplodon traversii* French: Zifo de travers; Spanish: Baleine a bec de Travers	Known only from three skulls. The most distinguishing feature is a single pair of spade-shaped teeth set in the middle of the lower jaw.	Habitat use and behavior remain unknown.	Of the three currently known records, two are from New Zealand and one is from Robinson Crusoe Island, Chile.	Unknown.	Not listed by IUCN

Resources

Books

Balcomb, Kenneth. "Baird's Beaked Whale *Berardius bairdii*, Stejneger, 1883; Arnoux's Beaked Whale *Berardius arnuxii*." In *Handbook of Marine Mammals*. Vol. 4, *River Dolphins and Larger Toothed Whales*, edited by Samuel Ridgway and Richard Harrison. London: Academic Press, 1989.

Heyning, John. "Cuvier's Beaked Whale *Ziphius cavirostris*." In *Handbook of Marine Mammals*. Vol. 4, *River Dolphins and Larger Toothed Whales*, edited by Samuel Ridgway and Richard Harrison. London: Academic Press, 1989.

Mead, James. "Beaked Whales of the Genus *Mesoplodon*." In *Handbook of Marine Mammals*. Vol. 4, *River Dolphins and Larger Toothed Whales*, edited by Samuel Ridgway and Richard Harrison. London: Academic Press, 1989.

———. "Bottlenose Whales *Hyperoodon ampullatus* (Forster, 1770) and *Hyperoodon planifrons* (Flower, 1882)." In *Handbook of Marine Mammals*. Vol. 4, *River Dolphins and Larger Toothed Whales*, edited by Samuel Ridgway and Richard Harrison. London: Academic Press, 1989.

———. "Shepherd's Beaked Whale *Tasmacetus shepherdi* Olivier, 1937." In *Handbook of Marine Mammals*. Vol. 4, *River Dolphins and Larger Toothed Whales*, edited by Samuel Ridgway and Richard Harrison. London: Academic Press, 1989.

Reeves, Randall, Brent Stewart, Phillip Clapham, and James Powell. *Sea Mammals of the World*. London: A&C Black, 2002.

Colin D. MacLeod, PhD

▲

Sperm whales

(Physeteridae)

Class Mammalia

Order Cetacea

Suborder Odontoceti

Family Physeteridae

Thumbnail description
Small to large whales, with distinctive, barrel-shaped heads, blowholes left of center, narrow, underslung lower jaws with uniform teeth, and paddle-shaped flippers

Size
Sperm whale: 34–60 ft (10.4–18.3 m), 26,000–125,000 lb (12,000–57,000 kg); pygmy sperm whale: 11 ft (3.4 m), 900 lb (400 kg); dwarf sperm whale: 9 ft (2.7 m), 600 lb (270 kg)

Number of genera, species
2 genera; 3 species

Habitat
Mainly deep waters off edge of continental shelf, but also continental shelf and slope

Conservation status
Vulnerable: 1 species

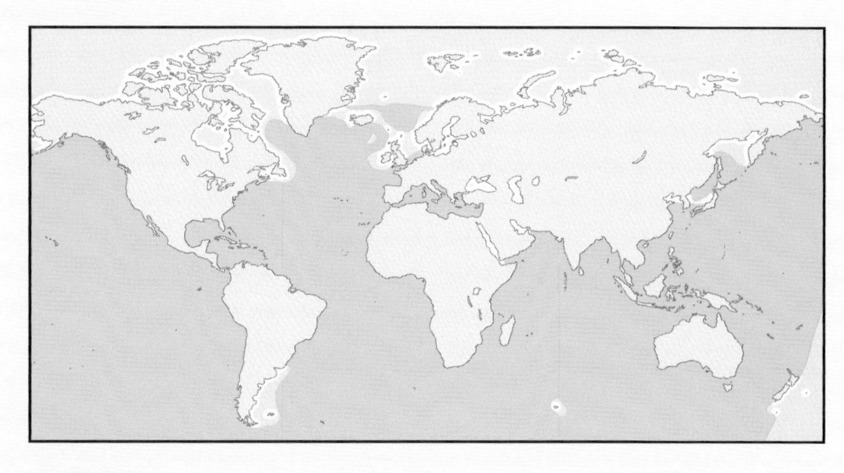

Distribution
Worldwide in tropical and temperate waters reaching latitudes of about 40°; mature male sperm whales up to edges of pack ice

Evolution and systematics

The family Physeteridae is about 30 million years old and among the oldest of living families of whales and dolphins. They are thought to have separated quite early from the main odontocete (toothed whale) line, and retain many of the characteristics that are considered primitive in odontocetes. Some of their features are highly derived, however. There is little similarity to the primitive Eocene cetaceans from which they descended.

The earliest physeterid (*Ferecetotherium*), appearing in the late Oligocene, was small with a reduced head size. It represents the earliest record of a living odontocete family. By the middle of the Miocene, the physeterids were fairly diverse, and the family is moderately well documented in the fossil record. The sperm whale (*Physeter macrocephalus*) is the only surviving species of the genus (*Physeter*) and is the most phylogenetically distinct of all the species of living odontocete. The pygmy and dwarf sperm whales (*Kogia* species) are often placed in a separate family, Kogiidae, as they emerged much later, about 7 million years ago in the late Miocene.

Physical characteristics

All three species are characterized by a highly asymmetrical rostrum, but *Kogia* has a lesser developed melon (fat-filled forehead), which makes up a much smaller proportion of their body length than in the sperm whale. The blowhole is also more posterior, the head more conical, and the rostrum is shorter in *Kogia* (indeed, this genus has the shortest rostrum of all the living cetaceans). At least two structures contained within the forehead are unique to the Physeteridae: the spermaceti organ and the *museau de singe*. The spermaceti organ is an elongated structure made up of spongy tissue containing a liquid or semi-liquid waxy oil, and the *museau de singe* is a valve-like clapper system that lies at the end of the right nasal passage. Both are believed to be involved in sound production. The right nasal passage is much smaller than the left; the former is thought to be used for sound production, and the latter, for respiration.

Members of this family are dark gray in color, with a lighter-colored belly. None of the three species has teeth in their upper jaw, though small, unerupted teeth (around 10 on each side of the sperm whale jaw) remain in the upper jaw during their lifetime. The teeth of the lower jaw all erupt around the time of sexual maturity. The sperm whale has 17–29 teeth on each side of its lower jaw; the pygmy sperm whale, 12–16; and the dwarf sperm whale, eight to 11. *Physeter* teeth are large and conical. In contrast, the teeth of *Kogia* are thin, very sharp and curved, and lack enamel.

The sperm whale is the largest odontocete and the most sexually dimorphic one. Adult females are about 36 ft (11 m) in length and weigh 33,000 lb (15,000 kg), but can reach 41 ft (12.5 m) and 53,000 lb (24,000 kg). Physically mature males are roughly 52.5 ft (16 m) long and 100,000 lb (45,000 kg), though maximum sizes are 60 ft (18.3 m) and 125,000 lb (57,000 kg). In the *Kogia* spp., the sexes are roughly the same size.

Side profile of a sperm whale (*Physeter macrocephalus*). (Photo ©Tony Wu/www.silent-symphony.com. Reproduced by permission.)

Distribution

The distributions of the Physeteridae family members are cosmopolitan. Few animals in the world have such wide distributions as the sperm whale, occurring in all oceans, from the pack ice to the Equator. The ranges of the pygmy and dwarf sperm whales are perhaps somewhat smaller than that of the sperm whale, but they are found worldwide in warm-temperate and tropical waters of the Atlantic, Pacific, and Indian Oceans. The dwarf sperm whale seems to prefer warmer waters than the pygmy sperm whale. Knowledge of *Kogia* distributions is sketchy, however, as most records are based on stranded animals.

Habitat

The Physeteridae occupy mainly deep, oceanic waters over 3,300 ft (1,000 m) in depth, (off the edge of the continental shelf. However, the *Kogia* spp., particularly the dwarf sperm whale, live in shallower water over the continental shelves and shelf edges.

Behavior

All three species are known to strand, but the pygmy sperm whale is one of the most commonly stranded cetacean species. Much of what is known of *Kogia* is based solely on information from strandings. When surfacing for air, all of the Physeteridae produce a low, relatively inconspicuous blow.

Very little is known about the behavior of pygmy or dwarf sperm whales, as there have been no comprehensive behavioral studies. Group sizes in the *Kogia* spp. range from single individuals to a maximum of six (pygmy sperm whale) or 10 (dwarf sperm whale) animals of varying age and sex composition. Groups of dwarf sperm whales can be composed of adults of both sexes with calves, females with calves, or immatures only. The *Kogia* spp. rise slowly and deliberately when they are surfacing and dive without showing their flukes. They spend a considerable amount of time lying motionless at the surface, with their tail hanging limply down and the back of their head exposed. The *Kogia* spp. are reported to have an interesting adaptation, presumably used in predator defense. When startled or distressed, they discharge a quantity of reddish brown intestinal fluid. This may function like octopus ink to confuse predators, enabling their escape. Neither *Kogia* is known to be highly vocal, though they do use echolocative, directional clicks.

The sperm whale, in contrast, is much more social and vocal than *Kogia*. Indeed, female sperm whales are exceptionally colonial. Sperm whale society is complex and built around the long-term unit, which is made up of around 10 females and their young. These units are generally matrilineal, but consist of more than one matriline. Most females probably spend their lives in the same unit, surrounded by close female relatives like aunts, cousins, and grandmothers. Two or more units may join together for several days at a time, forming a group of 20–30 animals.

A sperm whale (*Physeter macrocephalus*) spy-hopping in the Gulf of California, Mexico. (Photo by © Michael S. Nolan/Seapics.com. Reproduced by permission.)

Female sperm whales regularly gather at the surface to socialize or rest for several hours a day, or 25% of their time. Whales can be tightly aggregated, lying quietly and parallel to one another, a behavior called "logging," or they may vigorously twist and turn about one another, often touching one another. Breaches (leaps from the water), lobtails (hitting the water with tail flukes), and spyhops (raising the head vertically out of the water) can also be observed during these social times.

These are also the circumstances in which sperm whales typically arc cither silent or emit "codas," which is a patterned series of around three to 20 clicks that are somewhat reminiscent of the Morse code. Codas are less than two seconds long and can often be heard as exchanges between individuals; codas clearly represent communication in *Physeter*. Groups vary in their usage of different coda types, thus exhibiting dialects, and these dialects are stable over time. Moreover, units can be clustered into "vocal clans" based on the similarity of their dialects. Variation in vocal behavior between clans is likely cultural, passed down from mother and clan to offspring.

The most commonly heard sounds from sperm whales, though, are the usual clicks, which are long trains of regularly-spaced clicks, produced at rates of about two clicks per second. These very loud, highly directional clicks most likely represent echolocation used in food-finding behavior. Slow clicks, or clangs, are distinctively ringing clicks produced by large males, generally on the breeding grounds. They are emitted every six seconds or so, and may advertise a breeding male's presence and/or fitness.

Male sperm whales leave their natal units at an age of about six years. They then form bachelor schools, which are loose aggregations of males of about the same size and age. As males grow, they are found in progressively smaller schools, with the largest males being mostly solitary.

Feeding ecology and diet

All of the Physeteridae feed principally on mid- and deep-water squid, although they also eat some fish and octopus. Their anatomy suggests that they use powerful suction feeding. The sperm whale eats mainly deep-ocean squid, 0.2–15.5 lb (0.1–7 kg) in weight, but occasionally, it will prey on giant and jumbo squid, over 50 ft (15 m) in length. Scars from the squids' sucker marks can be found on sperm whales' heads as proof of these undersea battles. Males are more likely to eat fish, but will also feed on larger species and larger individuals of the same species of squid that females eat.

Kogia have an anterior-ventrally flattened snout, which points to a tendency to bottom-feed at least some of the time. Indeed, bottom-dwelling fish and crabs have been found in their stomachs. The *Kogia* spp. eat some of the same species as the sperm whale, but because they also inhabit the continental shelf region, they feed on shelf-living squids and octopods.

The Physeteridae are deep divers, with the sperm whale being the champion in this department. It can dive to depths

Underwater view of a sperm whale (*Physeter macrocephalus*). (Photo by François Gohier/Photo Researchers, Inc. Reproduced by permission.)

A freediver with a sperm whale (*Physeter macrocephalus*). (Photo by © 2003 Jonathan Bird/Seapics.com. Reproduced by permission.)

of 3,300–6,500 ft (1,000–2,000 m), possibly even 10,000 ft (3,000 m). More typical are dives to 1,000–2,600 ft (300–800 m). Dives are usually 30–45 minutes in length with seven to 10 minutes at the surface between dives to breathe. Dives can last over an hour, however. Females spend about 75% of their time foraging. The pygmy and dwarf sperm whales are also probably capable of diving to greater than 1,600 ft (500 m), judging by the prey they ingest.

Reproductive biology

In the sperm whale, the sexes show an unusual pattern of distribution, where females and immatures inhabit warmer waters at latitudes below 40° and males are found at higher latitudes. As males mature, they move to higher and higher latitudes, with the largest males found close to the edge of the pack ice. Mature males in their late twenties and older make the long migrations to the tropics to mate. The sperm whale is polygynous. On the breeding grounds, large males rove between groups of females, most likely searching for receptive females. They spend only minutes to hours with each female group. Rare fights occur between mature males.

Very little is known of the mating system of the pygmy or dwarf sperm whales. Little or no sexual dimorphism is apparent in either species, so they may have a different reproductive strategy from the highly sexually dimorphic sperm whale.

Sperm whales give birth about once every five years. Calves can ingest solid food by the age of one year, but continue to suckle for two or more years. This species exhibits one of the longest periods of parental care among marine mammals, as calves can be suckled for as long as 13–15 years. The calves are the probable reason behind female sperm whale sociality. Young calves seem unable to make the prolonged dives to the depths their mothers do to feed. Left alone at the surface, they would be vulnerable to attacks by killer whales or sharks. Thus, calves are "babysat" by other members of the group re-

Sperm whale (*Physeter macrocephalus*) showing the bulbous spermaceli organ at the front of its head. (Photo by Tui De Roy. Bruce Coleman, Inc. Reproduced by permission.)

maining at the surface. Groups containing calves stagger their dives, leaving some adults available to stand guard at the surface at all times, while groups without calves dive more synchronously. In addition to this communal care for the young, there is strong evidence for females suckling calves that are not their own.

Gestation is 14–16 months for sperm whales; 11 months for pygmy sperm whales; and nine months for dwarf sperm whales.

Conservation status

All three species are reported to ingest ocean debris such as plastic bags, causing death on occasion. This family is also vulnerable to ship strikes. Population sizes for *Kogia* are unknown, though they are apparently not common. The global population of sperm whales is very roughly 360,000. The sperm whale is listed as Endangered under the U.S. Endangered Species Act and Vulnerable by the IUCN. All three species are listed on CITES Appendix I or II.

Sperm whales were heavily whaled around the world in the eighteenth and nineteenth centuries, but especially by the New Englanders. In the twentieth century, a second wave of whaling took place, using mechanized catcher vessels and explosive harpoons. Up to 30,000 sperm whales were killed

Eye of baby sperm whale (*Physeter macrocephalus*). (Photo by © Doug Perrine/Seapics.com. Reproduced by permission.)

Sperm whales (*Physeter macrocephalus*) surfacing for air. (Photo by François Gohier/Photo Researchers, Inc. Reproduced by permission.)

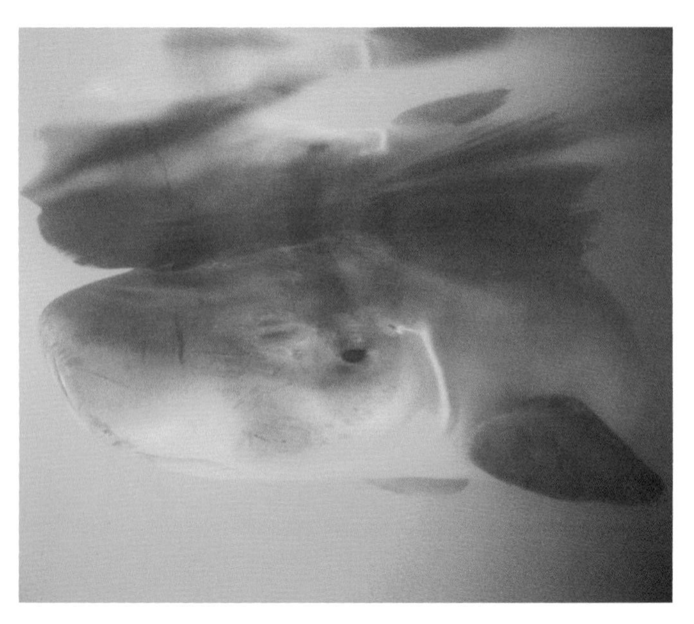

A pygmy sperm whale (*Kogia breviceps*) baby, about two weeks old. (Photo by © Doug Perrine/Seapics.com. Reproduced by permission.)

every year, particularly the large males. Commercial whaling more or less ceased in 1988 with the International Whaling Commission's moratorium.

The sperm whale, with a maximal rate of increase of just 1%, is slow to recover from past whaling. Additional threats include chemical pollution, evident in its blubber, and noise pollution, because of its dependence on sound for all aspects of its life.

Significance to humans

The sperm whale is memorialized in Herman Melville's novel, *Moby Dick*. Japan has restarted sperm whaling in 2000, taking five to eight sperm whales a year. Whale-watching for sperm whales is a profitable business around the world, bringing in substantial revenue from tourists. Pygmy and dwarf sperm whales are occasionally taken in commercial harpoon fisheries in the Indian and Pacific Oceans, and the Caribbean.

Species accounts

Sperm whale
Physeter macrocephalus

TAXONOMY
Physeter macrocephalus Linnaeus, 1758, "Oceano Europaeo."

OTHER COMMON NAMES
French: Cachalot; German: Pottwal; Spanish: Cachalote.

PHYSICAL CHARACTERISTICS
Square forehead makes up one quarter to one third the body length. Body dark gray, but mouth has bright, white lining. Skin corrugated, except for head and flukes.

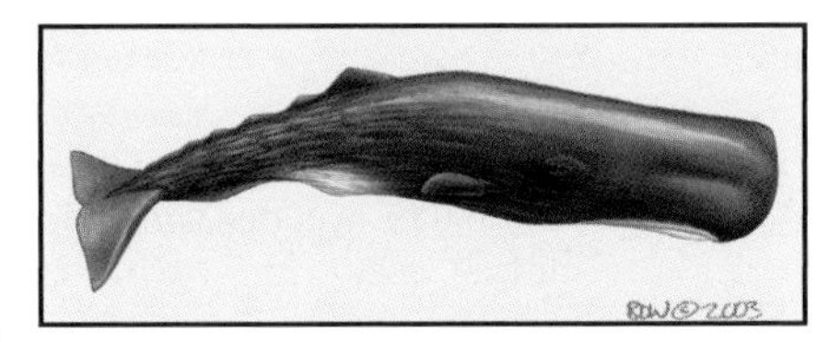

Physeter macrocephalus

DISTRIBUTION
Cosmopolitan.

HABITAT
Deep water near productive areas; often near steep drop-offs or strong oceanographic features. Males more likely in shallower waters.

BEHAVIOR
Females highly social, males solitary. Very vocal.

FEEDING ECOLOGY AND DIET
Eats mainly deep-ocean squid. Dives vertically, raising flukes in air. Females forage in rank abreast of each other.

REPRODUCTIVE BIOLOGY
Polygynous. Breeding season in northern hemisphere, January–August; in southern hemisphere, July–March.

CONSERVATION STATUS
Listed as Vulnerable by the IUCN. In 2002, global population size was about 32% of pre-whaling numbers. Large, breeding males scarce, and calving rates below sustainability in southeast Pacific. Still widespread, however.

SIGNIFICANCE TO HUMANS
In the past, whalers used the valuable oil from the spermaceti organ and blubber to fuel the industrial revolution. Minimal sperm whaling since 2002. Sperm whale-watching operations off New Zealand, West Indies, Norway, Madeira, the Azores, etc. ◆

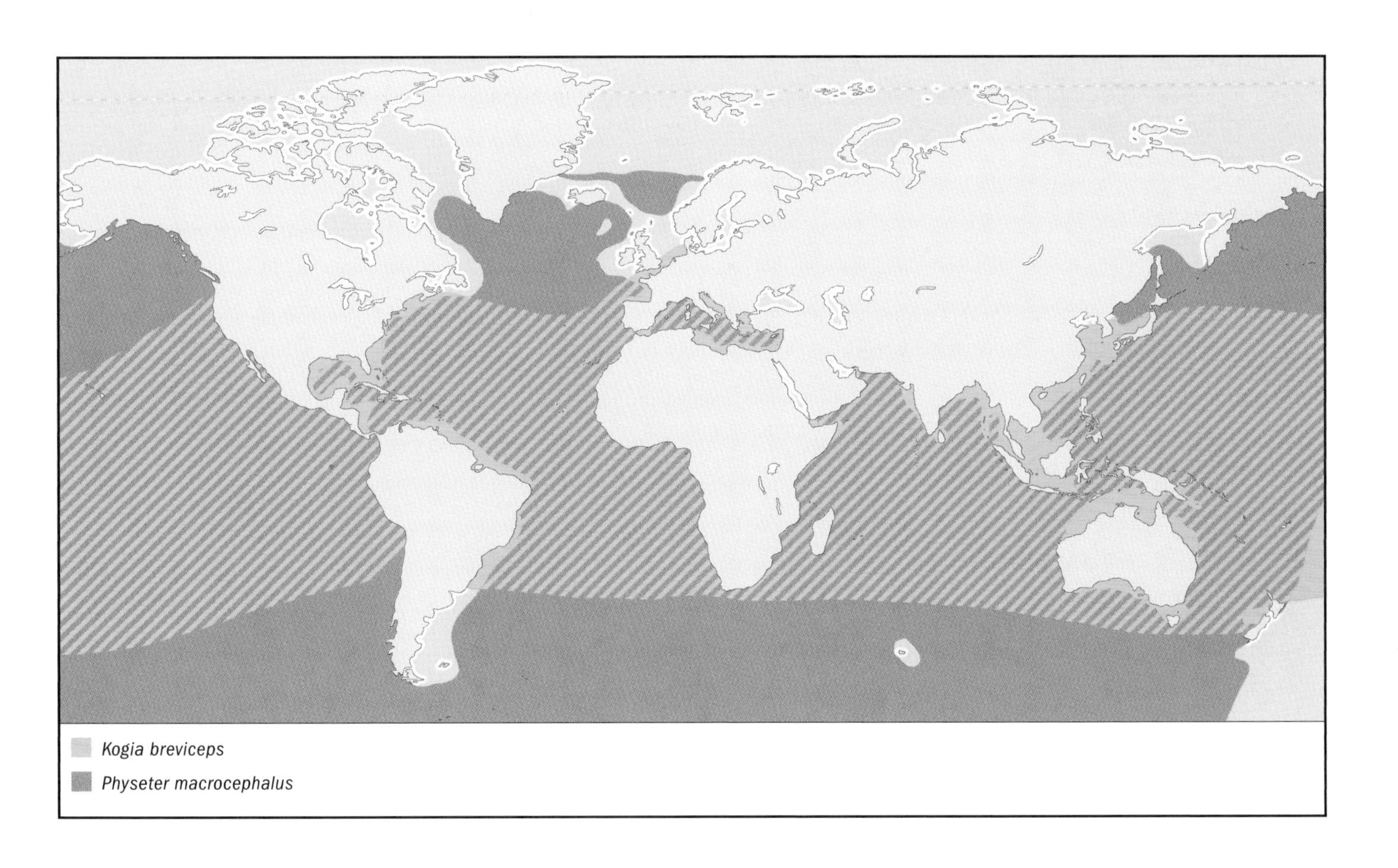

Pygmy sperm whale

Kogia breviceps

TAXONOMY

Kogia breviceps (Blainville, 1838), Cape Province, South Africa.

OTHER COMMON NAMES

French: Cachalot pygmée; German: Zwergpottwal; Spanish: Cachalote pigmeo.

PHYSICAL CHARACTERISTICS

Bluish steel gray on back. Shark-like appearance. Bracket-shaped mark on side of head, resembling gill slit.

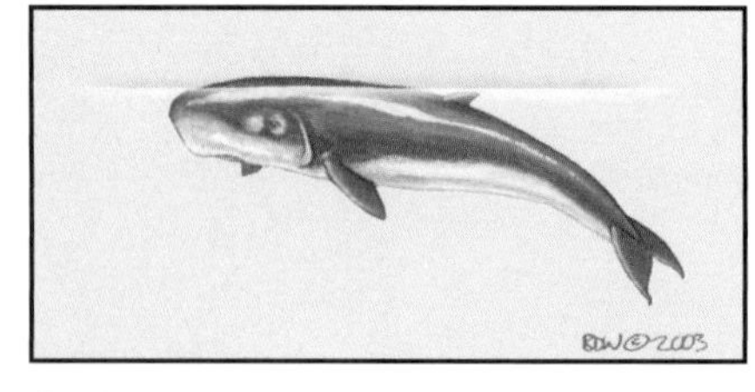

Kogia breviceps

DISTRIBUTION

Worldwide in warm-temperate and tropical waters.

HABITAT

Deep, oceanic waters and over the continental shelf.

BEHAVIOR

Difficult to observe at sea. Easily approached, timid and slow-moving.

FEEDING ECOLOGY AND DIET

Deep-ocean and shelf-dwelling squid and octopus; fish and crabs.

REPRODUCTIVE BIOLOGY

May give birth two years in succession, i.e., can be pregnant and nursing simultaneously. Calves nurse for about one year. Mating believed to occur in summer, births in spring. Mating system unknown.

CONSERVATION STATUS

Population size unknown, though not common. Not listed by the IUCN.

SIGNIFICANCE TO HUMANS

Sometimes taken in commercial harpoon fisheries. ◆

Resources

Books

Berta, A., and J. L. Sumich. *Marine Mammals: Evolutionary Biology.* San Diego: Academic Press, 1999.

Perrin, W. F., B. Würsig, and J. G. M. Thewissen, eds. *Encyclopedia of Marine Mammals.* San Diego: Academic Press, 2002.

Rice, D. W. *Marine Mammals of the World: Systematics and Distribution.* Lawrence, Kansas: Allen Press, 1998.

Ridgway, S., and R. J. Harrison, eds. *Handbook of Marine Mammals.* Vol. 4, *River Dolphins and the Larger Toothed Whales.* London: Academic Press, 1989.

Whitehead, H. *Sperm Whales: Social Evolution in the Ocean.* Chicago: University of Chicago Press, 2003.

Periodicals

Gosho, M. E., D. W. Rice, and J. M. Breiwick. "The Sperm Whale (*Physeter macrocephalus*)." *Marine Fisheries Review* 46, 4 (1984): 54–64.

Weilgart, L., H. Whitehead, and K. Payne. "A Colossal Convergence." *American Scientist* 84 (1996): 278–287.

Whitehead, H. "Estimates of the Current Global Population Size and Historical Trajectory for Sperm Whales." *Marine Ecology Progress Series* 242 (2002): 295–304.

Organizations

American Cetacean Society. P.O. Box 1391, San Pedro, CA 90733-1391 United States. Phone: (310) 548-6279. Fax: (310) 548-6950. E-mail: info@acsonline.org Web site: <http://www.acsonline.org>

Cetacean Society International. P.O. Box 953, Georgetown, CT 06829 United States. Phone: (203) 431-1606. Fax: (203) 431-1606. E-mail: rossiter@csiwhalesalive.org Web site: <http://csiwhalesalive.org>

European Cetacean Society. Web site: <http://web.inter.nl.net/users/J.W.Broekema/ecs>

Society for Marine Mammalogy. Web site: <http://www.marinemammalogy.org>

Whale and Dolphin Conservation Society. P.O. Box 232, Melksham, Wiltshire SN12 7SB United Kingdom. Phone: 44(0)1249 449500. Fax: 44(0)1249 449501. E-mail: info@wdcs.org Web site: <http://www.wdcs.org>

Other

The Quest for Moby Dick. Videotape. Dokumente des Meeres, 1992.

Audiotape of Sperm Whale Sounds. Cornell Laboratory of Ornithology, Library of Natural Sounds. 159 Sapsucker Woods Road, Ithaca, NY 14850. Phone: (607) 254-2405.

Lindy Weilgart, PhD

▲

Belugas and narwhals

(Monodontidae)

Class Mammalia
Order Cetacea
Suborder Odontoceti
Family Monodontidae

Thumbnail description
Medium-sized toothed whales with robust bodies, small bulbous heads, no dorsal fins, short rounded flippers, and elegantly butterfly-shaped tail flukes

Size
13–16 ft (4–4.9 m); 1,500–3,500 lb (680–1,590 kg)

Number of genera, species
2 genera; 2 species

Habitat
Marine, estuaries, deep, shallow, benthic, pelagic

Conservation status
Vulnerable: 1 species; Data Deficient: 1 species

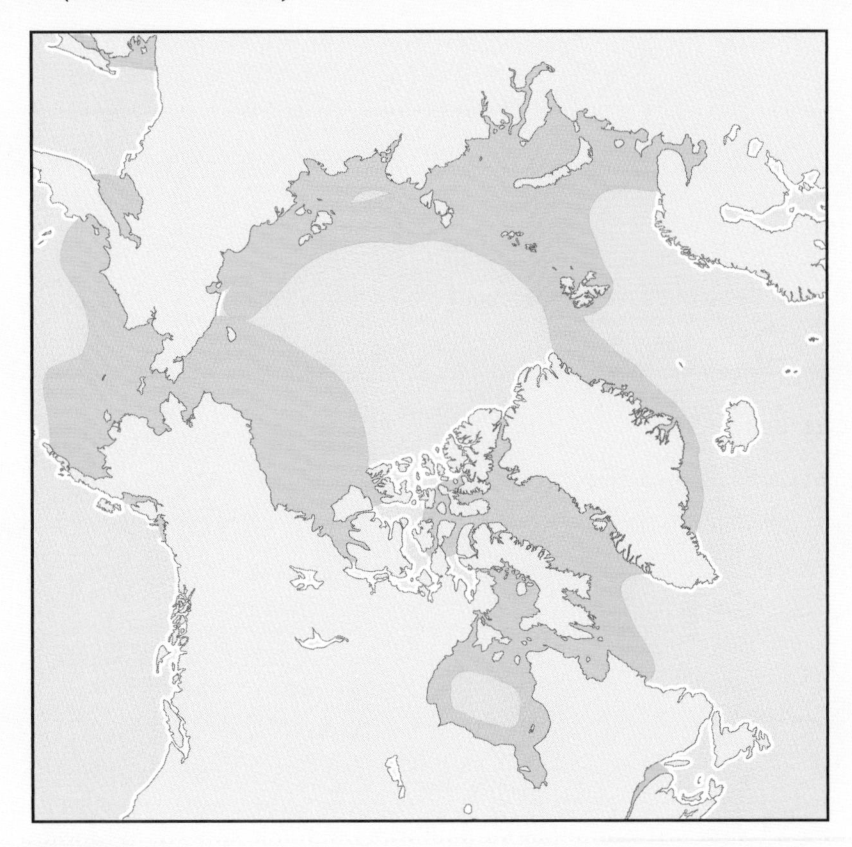

Distribution
Arctic and subarctic, with isolated remnant populations of belugas in a few cold temperate areas

Evolution and systematics

Only a single fossil monodontid is known, and it came from the late Miocene of Baja California, Mexico. Fossils found in Pleistocene clays of eastern North America indicate that belugas underwent large range extensions and contractions in response to glacial processes. In the case of narwhals, the low level of nucleotide diversity found in animals from the eastern Canadian Arctic, west Greenland, and east Greenland has been interpreted as indicating a rapid and recent expansion from a small founding population ("recent" meaning, in this context, perhaps several tens of thousands of years ago).

Past formulations of monodontid systematics have included: two subfamilies (Delphinapterinae and Monodontinae) either under Delphinidae or sequestered into the family Monodontidae; two genera in family Delphinapteridae; Monodontidae as a family within its own superfamily Monodontoidea; or assignment of the two genera to separate families, Monodontidae and Delphinapteridae, with the latter including the genus *Orcaella*. Most recent authorities agree that the monodontids comprise a single family in the superfamily Delphinoidea. No subfamilies or subspecies are recognized.

The two genera are well differentiated and have long been recognized as separate taxa. However, at least one example of a narwhal-beluga hybrid has been documented from west Greenland.

Physical characteristics

The two living species in the family Monodontidae are medium-sized toothed whales (odontocetes) with almost no beak and a small head. Their melon (forehead) is rounded and can appear bulbous. The cervical vertebrae are unfused, allowing considerable lateral and vertical flexibility—a characteristic readily noticed by anyone observing a beluga in captivity. Another interesting feature of the beluga (*Delphinapterus leucas*) is the malleability of its rostral bulge, or melon—described by one scientist as reminiscent of a balloon filled with warm lard. The flippers of both species are broad, short, and rounded at the tips, and their outer margins tend to curl upward in adult males. There is no dorsal fin, and this lack of a dorsal fin is exceptional among the cetaceans. Both species have a low, fleshy ridge along the back in the area where the dorsal fin would normally be situated. The tail

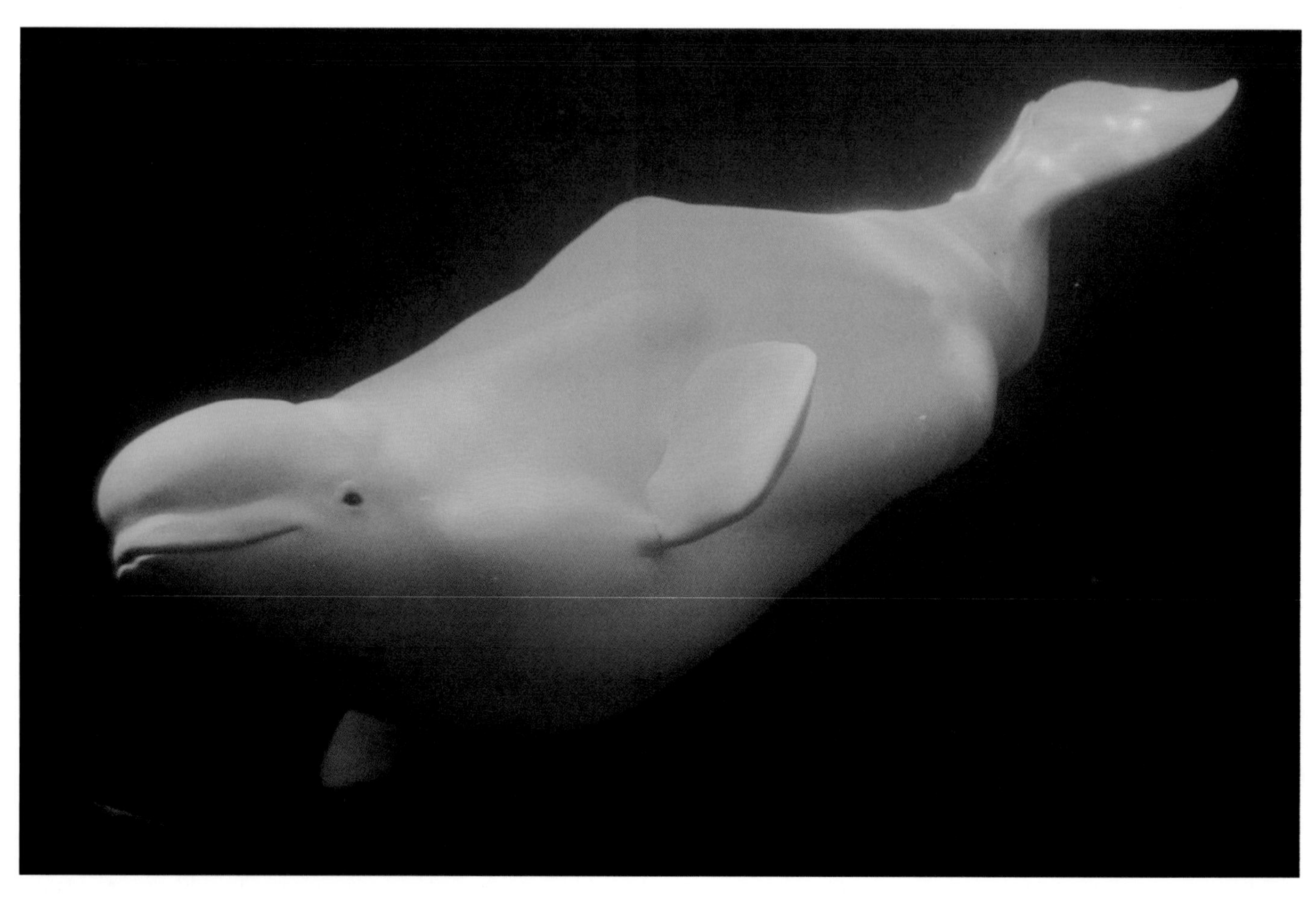

Underwater viewing of a beluga whale (*Delphinapterus leucas*) at the Vancouver Aquarium in British Columbia, Canada. (Photo by Terry Whittaker/ Photo Researchers, Inc. Reproduced by permission.)

flukes are relatively small in relation to the rest of the body. In young animals the rear margins are straight, but as the animals age these become strongly convex, giving the flukes a butterfly shape in dorsal view. In old animals, the flukes may overlap in the region of the notch that separates them.

Both species exhibit marked ontogenetic changes in coloration. Newborn animals are uniformly gray or brownish gray. Belugas become progressively lighter as they age and are pure white by about seven (females) to nine (males) years of age. Narwhals (*Monodon monoceros*) become black and then begin to acquire white patches and streaks on the belly and sides. Adults are spotted or mottled dorsally, white ventrally, with black areas persisting on the head and nape. Very old male narwhals are almost completely white.

A major difference between the two species is their dentition. Belugas have rows of eight or nine simple, peg-like teeth in both upper and lower jaws, while narwhals have no erupted teeth within the buccal cavity. Rather, they have only two pairs of maxillary teeth, all of which remain concealed within the jaws of females. In males, one of the teeth in the left side of the upper jaw erupts and protrudes forward from the front of the head, as the leftward-spiraled "unicorn" tusk for which the narwhal is famous. The tusk can be nearly 10 ft (3 m) long and weigh more than 20 lb (10.5 kg). Some males have two tusks ("double-tuskers") but the left member of the pair is often larger.

Distribution

Monodontids occur only in high latitudes of the Northern Hemisphere, i.e., entirely north of 45°N and mostly north of 55°N. Although their total distribution can be considered circumpolar, gaps exist in areas with heavy year-round ice cover. Numerous stocks have been identified on the basis of differences in distribution and migration patterns, morphology (including body size), tissue contaminant profiles, trends in abundance (e.g., depletion by whaling in one area with no observed decline in other areas), and genetics. There are more than 20 such stocks of belugas and at least three of narwhals, which are less well studied. Although the two species generally do not occur in mixed schools, they are broadly sympatric in portions of their range. For example, both species become concentrated along the pack ice blocking westward movement into Lancaster Sound during spring; their routes then diverge as the ice breaks up and they move to their respective summering grounds.

The principal areas of beluga distribution are: Alaska, USA (Cook Inlet, Bristol Bay, Norton Sound and Yukon Delta, Kotzebue Sound, and North Slope coast); Canada (Mackenzie Delta, Beaufort Sea, Peel Sound, Barrow Strait, Prince Regent Inlet, Lancaster Sound, Jones Sound, Cumberland Sound, Frobisher Bay, Ungava Bay, Hudson and James Bays, Foxe Basin, and St. Lawrence River); Greenland (all along west coast); Norway (mainly Barents Sea coast of Svalbard); Russia (Barents Sea: Franz Josef Land coast; Kara Sea: Ob and Yenisey Gulfs; White Sea: Onezhsky, Dvinsky, and Mezhensky Bays; southwestern Laptev Sea; ice edge of Chukchi and East Siberian Seas, Anadyr Gulf; Okhotsk Sea: Shelikov and Shantar Bays, Amur Lagoon and River).

The principal areas of narwhal distribution are limited to the Nearctic between the eastern Canadian Arctic and the western Russian Arctic. Animals that winter in the pack ice of Davis Strait and Baffin Bay move through Lancaster and Jones Sounds into the Canadian Arctic archipelago for the summer, while those that winter in and at the eastern entrance of Hudson Strait move west into northern Hudson Bay and Repulse Bay in summer. Similarly, at least some of those that winter in the pack ice of the Greenland Sea move inshore along the east coast of Greenland in summer, particularly into Scoresby Sound and Kangerlussuaq.

A beluga whale (*Delphinapterus leucas*) with its head out of the water. (Photo by Tim Davis/Photo Researchers, Inc. Reproduced by permission.)

Narwhals (*Monodon monoceros*) spouting at the surface of the Arctic Ocean. (Photo by © Goran Ehlme/Seapics.com. Reproduced by permission.)

Wandering belugas have been observed as far south as Japan and Washington State in the Pacific, and New Jersey and France in the Atlantic. Narwhals observed in the Beaufort, Chukchi, and Bering Seas are essentially strays, as no permanent population is known to inhabit those seas. Both narwhals and belugas have been recorded in the Arctic Ocean in latitudes as high as 80 to 85°N.

Habitat

Monodontids are ice-adapted and typically spend at least several months, if not most of the year, in close proximity to pack ice. Narwhals, in particular, are often encountered in heavy offshore ice where the only access to air is in shifting cracks and leads between floes. In general, the beluga is a more inshore, coastal species, tending to congregate in very large numbers (hundreds to thousands of individuals) in estuaries to molt. Nevertheless, some belugas have been radio-tracked moving up to 680 mi (1,100 km) from shore and penetrating 435 mi (700 km) into the dense polar ice cap where more than 90% of the surface is ice-covered. These

Hundreds of whales may be confined to small breathing holes in solid ice cover, making them vulnerable to hunters, polar bears, or suffocation. (Illustration by Patricia Ferrer)

same individuals may have spent time earlier in the same season in turbid, shallow estuaries or lagoons. As migratory animals with strong diving abilities, monodontids range across varied habitat in the course of a given year.

As a rule, belugas arrive in summering areas by June and July, and move out to wintering areas by the end of September before freeze-up. In early spring they can be seen in the narrow leads along the edge of the fast-ice. Their presence can be detected by the breathing holes that they make when the ice is thin enough. During periods of maximum ice cover, pods of whales frequent isolated areas of open water, called "polynyas," and are usually separated from one another by large areas of impenetrable ice. Individuals that become trapped in ice fields try to break through the ice by ramming it from the underside and the cushion located on top of the head lessens the shock.

Behavior

Belugas and narwhals usually occur in pods of two to 10 individuals, but these pods are often traveling or milling in close proximity so that they appear as large schools or herds. A pod may consist only of adult males or only of females and young males. Monodontids are not particularly fast swimmers, and their traveling speeds average only about 3 mph (5 km/hr).

Some populations of monodontids migrate long distances each year in response to seasonal changes in ice cover. Their migrations are not necessarily latitudinal; often the movement is inshore-offshore, or involves passing in spring through connecting corridors to reach particular estuaries (belugas) or fjords (narwhals), then returning in autumn along the same route as ice formation drives them away from the summering grounds. Both species are vulnerable to entrapment when wind-driven or fast-forming ice blocks them from moving to seasonal refugia. Ice-entrapped whales become easy prey for polar bears or human hunters. If not discovered, however, they can survive for weeks or months on fat reserves as long as they are able to maintain breathing holes in the ice.

Like other toothed cetaceans, belugas and narwhals are vocal animals and probably rely more upon sound than any other sense to detect and capture prey, to communicate, and to navigate in a harsh (and often dark) environment. Both emit pulsed series of high-frequency clicks for echolocation, and belugas, in particular, have a varied repertoire of pure tones and modulated whistles that appear to be used for communication.

Feeding ecology and diet

Both belugas and narwhals dive to the sea bottom, even in areas deeper than 3,300 ft (1,000 m). They can remain submerged for 25 minutes but usually do not stay down for longer than about 20 minutes. There is some overlap in their diets, as both species prey upon shrimp, squid, and schooling pelagic fish, such as arctic cod (*Boreogadus saida*). They forage in the water column, at times on organisms associated with the undersurface of sea ice, but also on demersal and benthic species. Narwhals in some areas may feed exclusively on the squid *Gonatus fabricii.* They also consume deepwater species such as Greenland halibut (*Reinhardtius hipposlossoides*) and redfish (*Sebastes marinus*). The list of prey items of belugas is considerably longer that that of narwhals, perhaps because they are more widely distributed and occupy a greater range of habitat types. In some areas, belugas take advantage of seasonal concentrations of anadromous and coastal fish, such as salmon (*Oncorhynchus* spp.), herring (*Clupea harengus*), and capelin (*Mallotus villosus*).

Reproductive biology

Like other whales, monodontids give birth to single young at relatively long intervals, and the period of calf dependence is prolonged. The newborn calf is about 5 ft (1.5–1.6 m) long and weighs 175–220 lb (80–100 kg). Lactation lasts for one to two years. Mating takes place in late winter and early spring when the whales are generally inaccessible for observation. As a result, little is known about the social organization and behavior associated with conception. The gestation period is estimated at between 13 and 16 months. Thus, the average calving interval for both belugas and narwhals is thought to be about three years.

Fluke of a beluga whale (*Delphinapterus leucas*) in Cunningham Inlet, Somerset Island, Canada. (Photo by Animals Animals © Stefano Nicolini. Reproduced by permission.)

The tusk of the narwhal (*Monodon monoceros*) can be up to 10 ft (3 m) long. (Photo by © John K.B. Ford/Ursus/Seapics.com. Reproduced by permssion.)

Belugas give birth between March and August, with a peak occurring in June and July. Calving is believed to take place in warm shallow rivers. Some well-known calving areas are the Mackenzie Delta at Inuvik, Cunningham Inlet on the north coast of Somerset Island, and the Seal and Nelson rivers in western Hudson Bay. Calves are delivered in bays and estuaries, where the water is relatively warm: about 50 to 60°F (10 to 15°C). They are born either tail-first or head-first. Observers of wild beluga populations have estimated that beluga calves average 5.2 ft (1.6 m) and weigh about 176 lb (80 kg). Beluga calves are generally dark gray to bluish or brownish gray, and darken about one month after birth. Like other whales, they can swim at birth. As other mammals do, mother belugas nurse their calves. A calf suckles below the water from nipples concealed in abdominal mammary slits. The calf may begin nursing several hours after birth and then nurses at hourly intervals thereafter. Beluga calves depend upon nursing for their first year, until their teeth come out. They then supplement their diets with shrimps and small fish. Most calves nurse on average for 20 to 24 months.

Sexual maturation, defined as first pregnancy, is attained in belugas at 4–7 years of age. Males become sexually mature at 7–9 years but may not be socially adept, and thus capable of successful reproduction, until somewhat older. Less is known about age-length relationships in narwhals because their embedded teeth do not provide a complete record of annual growth layers, as beluga teeth do (although uncertainty remains as to whether one or two growth-layer-groups are formed annually in beluga teeth). Male narwhals are adolescent when their tusk begins to erupt at a body length of 8.5 ft (2.6 m), and they are sexually mature by the time the tusk has reached a length of about 5 ft (1.5 m). Body lengths at sexual maturity in narwhals are around 11.9 ft (3.6 m) and 13.8 ft (4.2 m) for females and males, respectively. The mating systems for both belugas and narwhals are unknown, although it is suspected that beluga males mate with multiple females.

Conservation status

The IUCN listing of the beluga as Vulnerable reflects the fact that many populations have been depleted by overhunting. Although the aggregate world population of the species is well over 100,000, several stocks are close to extinction. For example, the Ungava Bay stock, once numbering more than 1,000, is represented by only a few scattered survivors. Many thousands of belugas used to assemble each summer in the mouths of the Great Whale and Little Whale Rivers of eastern Hudson Bay, but large concentrations no longer occur in these areas. In fact, there may be no more than about 1,000 belugas remaining along the entire east coast of Hudson Bay. Similar major declines, also due to overhunting, have been documented in Cook Inlet and off west Greenland. The population in the St. Lawrence River numbered at least 5,000 in the late nineteenth century, whereas today there are no more than about 1,200 there. Tissue concentrations of contaminants in St. Lawrence belugas are extremely high, and environmental conditions have changed considerably over the past 100 years.

Narwhals have also been intensively hunted in many parts of their range, and it is reasonable to assume that their numbers have been reduced considerably as a result. However,

Underwater pod of beluga whales (*Delphinapterus leucas*). (Photo by François Gohier/Photo Researchers, Inc. Reproduced by permission.)

Beluga whales (*Delphinapterus leucas*) in the shallows of Somerset Island, Canada. (Photo by Animals Animals ©D. Allan, OSF. Reproduced by permission.)

much less is known about stock identity (population structure), numbers, and trends for narwhals than for belugas. Abundance estimates for areas that have been surveyed total about 40,000. Allowing for negative bias in the estimation procedures, there are likely at least 50,000 narwhals in the waters bordering Canada and Greenland, plus unknown numbers in the Eurasian Arctic.

Many of the rivers that formerly provided estuarine habitat for belugas have been dammed. Although it has not been possible to establish direct cause-and-effect links between such development and beluga declines, the changed ecological conditions downstream of dams are likely to have made at least some of the estuaries and nearby waters less hospitable to the whales. Another major concern for belugas and narwhals is climate change. Given their close association with sea ice, effects of some kind can be expected.

Significance to humans

Belugas and narwhals have played, and continue to play, a prominent role in the subsistence economy of Inuit. Whale skin, called muktuk or mattak, is a northern delicacy. The desire to obtain this valued food drives the continued hunting of these animals in much of their range. In the case of narwhals, the cash value of the ivory tusk is an added incentive. Narwhal tusks are sold as curios in international trade, and the ivory is used extensively to make carved jewelry and ornaments.

Belugas have been taken extensively in commercial drive fisheries by non-indigenous hunters, especially in Russia and Canada. Such operations were responsible for large and rapid

A beluga whale (*Delphinapterus leucas*) in Arctic waters. (Photo by Animals Animals ©Zig Leszczynski. Reproduced by permission.)

A narwhal (*Monodon monoceros*) mother with pup near Baffin Island, Canada. (Photo by Animals Animals ©D. Allan, OSF. Reproduced by permission.)

declines in abundance. In contrast, narwhals were taken only occasionally by commercial whalers and have not been subjected to industrial exploitation in the same way as belugas.

For more than a century, belugas have been popular animals in captive displays. They adapt relatively well to capture, handling, and confinement, and they can be trained to perform in shows or to engage in research tasks. There have been numerous captive births of belugas. Narwhals, in contrast, have been brought into captivity on only a few occasions, and their survival was poor.

A limited amount of tourism has focused on both species in specific areas, notably the St. Lawrence River in southeastern Canada and the Churchill River in western Hudson Bay for belugas, and the Pond Inlet area of northern Baffin Island (Canada) for narwhals.

Beluga whales (*Delphinapterus leucas*) have seven vertebrae that are not fused, giving it a flexible neck. (Photo by Bruce Coleman, Inc. Reproduced by permission.)

1. Beluga (*Delphinapterus leucas*); 2. Narwhal (*Monodon monoceros*). (Illustration by Patricia Ferrer)

Species accounts

Beluga
Delphinapterus leucas

TAXONOMY
Delphinus leucas (Pallas, 1776), mouth of Ob River, northeastern Siberia, Russia.

OTHER COMMON NAMES
French: Belouga, marsouin blanc; German: Weissfisch; Spanish: Beluga.

PHYSICAL CHARACTERISTICS
Length 13–16 ft (3.9–4.9 m); weight 1,500–3,500 lb (700–1,600 kg).

DISTRIBUTION
Circumpolar in Arctic and subarctic; relict populations in St. Lawrence River, Canada; Cook Inlet, Alaska.

HABITAT
Marine and estuarine waters of almost any depth, depending on season and circumstance. Concentrate in shallow estuaries to molt, but also move into deep trenches where they dive to depths in excess of 3,300 ft (1,000 m). Occasionally ascend rivers. Tend to stay in polynyas and large coastal expanses of open water in winter, but can also be found in cracks and lanes in dense pack ice.

BEHAVIOR
Usually occur in pods of two to 10 animals, often with several associated pods. Swim slowly and roll at surface, usually without lifting head or flukes clear of water. Pure whiteness of adults makes them conspicuous, but also can make it difficult to tell them apart from whitecaps and small ice floes. Varied vocal repertoire; known to some whalers as "sea canary."

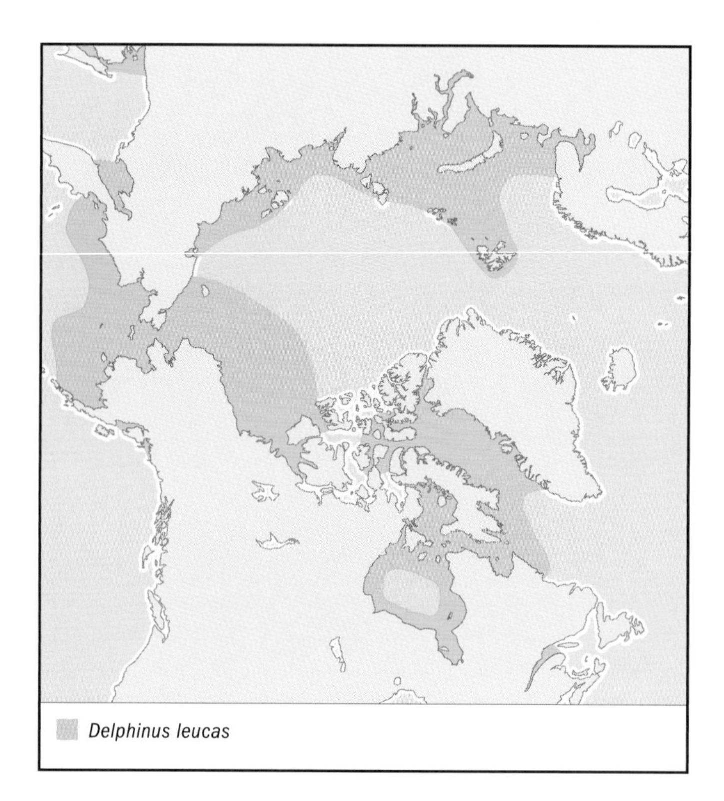

FEEDING ECOLOGY AND DIET
The diet includes shrimp, squid, octopus, marine worms, and many species of fish.

REPRODUCTIVE BIOLOGY
Single calf born in late spring or early summer, following gestation period of 14 to 14.5 months. Calves nurse for up to two years. Inter-birth interval averages three years. Female likely to bear first calf at age six or seven, male likely to mate successfully at age seven or older. Mating system is unknown.

CONSERVATION STATUS
Overall, still relatively abundant (over 100,000 individuals), but many populations reduced from past and continuing hunting pressure. Damming of northern rivers for hydroelectric power; industrial pollution of riverine, estuarine, and coastal habitat; and rapid climatic warming likely to have discrete and cumulative effects on populations.

SIGNIFICANCE TO HUMANS
Subsistence use of skin, plus some blubber and meat, is important to maritime Inuit of Canada, Alaska, and Greenland. Live-capture continues to supply animals for display; most new captive stock in recent years has come from Sea of Okhotsk, Russia. Limited amount of nature tourism in specific areas. ◆

Narwhal
Monodon monoceros

TAXONOMY
Monodon monoceros Linnaeus, 1758, northern seas of Europe and America.

OTHER COMMON NAMES
French: Narval; German: Narwal; Spanish: Narval.

PHYSICAL CHARACTERISTICS
Length 14–15.5 ft (4.2–4.7 m); weight 2,200–3,500 lb (1,000–1,600 kg).

DISTRIBUTION
Disjunct Arctic circumpolar; main concentrations in Greenland Sea, Davis Strait/Baffin Bay, Hudson Strait, northern Hudson Bay, and their adjacent sounds and inlets.

HABITAT
Deep marine waters, including inshore fjords and sounds in summer and offshore heavy pack-ice zone in winter.

BEHAVIOR
Roll at surface showing back but generally not head or flukes; same-sex pods (e.g., groups containing only males with large tusks); strongly migratory, moving in large groups of associated pods, totaling hundreds of animals. Reports of males crossing their tusks above the surface ("fencing") are difficult to interpret. Scars on head region and high incidence of broken tusk tips imply aggressive tusk use, perhaps in dominance interactions.

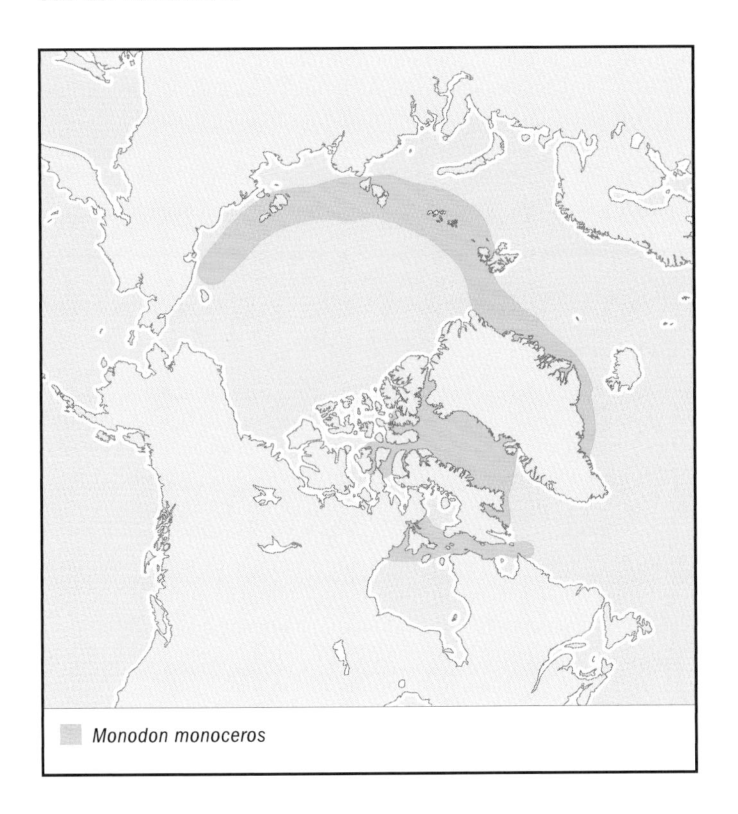

FEEDING ECOLOGY AND DIET
Deep divers that feed on shrimp, squid, schooling pelagic fish, and deepwater species such as halibut and redfish.

REPRODUCTIVE BIOLOGY
Single calf born in summer after gestation period of 13 to 16 months; lactation one to two years; inter-birth interval three years, on average. Mating system is unknown.

CONSERVATION STATUS
Present and historical abundance uncertain, but recent estimates for main areas of distribution in Canada and Greenland total close to 40,000, without adjusting for negative bias. Continued hunting, with no quotas, growing human populations in hunting districts, and no expected reduction in demand for products, signals the need for closer monitoring and management to prevent further depletion.

SIGNIFICANCE TO HUMANS
"Unicorn" tusks of adult males give species a special place in history and mythology. While intact, tusks continue to have high commercial value as curiosities in international trade, and the ivory is also used for carving, especially in Greenland. Skin a much-valued human food among Inuit. ◆

Resources

Books

Hay, K. A., and A. W. Mansfield. "Narwhal *Monodon monoceros* Linnaeus, 1758." In *Handbook of Marine Mammals.* Vol. 4, *River Dolphins and the Larger Toothed Whales,* edited by S. H. Ridgway and R. Harrison. London: Academic Press, 1989.

Heide-Jørgensen, M. P. "Narwhal *Monodon monoceros.*" In *Encyclopedia of Marine Mammals,* edited by W. F. Perrin, B. Würsig, and J. G. M. Thewissen. San Diego: Academic Press, 2002.

O'Corry-Crowe, G. M. "Beluga Whale *Delphinapterus leucas.*" In *Encyclopedia of Marine Mammals,* edited by W. F. Perrin, B. Würsig, and J. G. M. Thewissen. San Diego: Academic Press, 2002.

Reeves, R. R., B. S. Stewart, P. J. Clapham, and J. A. Powell. *National Audubon Society Guide to Marine Mammals of the World.* New York: Alfred A. Knopf, 2002.

Periodicals

Born, E. W., R. Dietz, and R. R. Reeves, eds. "Studies of White Whales (*Delphinapterus leucas*) and Narwhals (*Monodon monoceros*) in Greenland and Adjacent Waters." *Meddelelser om Grønland, Bioscience* 39 (1994): 1-259.

International Whaling Commission. "Report of the Sub-committee on Small Cetaceans." *Journal of Cetacean Research and Management* 2 Suppl. (2000): 235–263.

Reeves, R. R., and D. J. St. Aubin, eds. "Belugas and Narwhals: Application of New Technology to Whale Science in the Arctic." *Arctic* 54, 3 (2001): 207–353.

Smith, T. G., D. J. St. Aubin, and J. R. Geraci, eds. "Advances in Research on the Beluga Whale, *Delphinapterus leucas.*" *Canadian Bulletin of Fisheries and Aquatic Sciences* 224 (1990): 1–206.

Stewart, B. E., and R. E. A. Stewart. "*Delphinapterus leucas.*" *Mammalian Species* 336 (1989): 1–8.

Randall Reeves, PhD

Gray whales
(*Eschrichtiidae*)

Class Mammalia
Order Cetacea
Suborder Mysticeti
Family Eschrichtiidae

Thumbnail description
Medium-sized, bottom-feeding baleen whales with black or slate-gray skin, much blotched, mottled and encrusted with barnacles; gray whales are distinguished by their short, coarse baleen plates and by having a dorsal ridge instead of a dorsal fin

Size
43–46 ft (13–14.1 m); 44,000–81,500 lb (20,000–37,000 kg)

Number of genera, species
1 genus; 1 species

Habitat
Shallow coastal waters

Conservation status
Lower Risk/Conservation Dependent

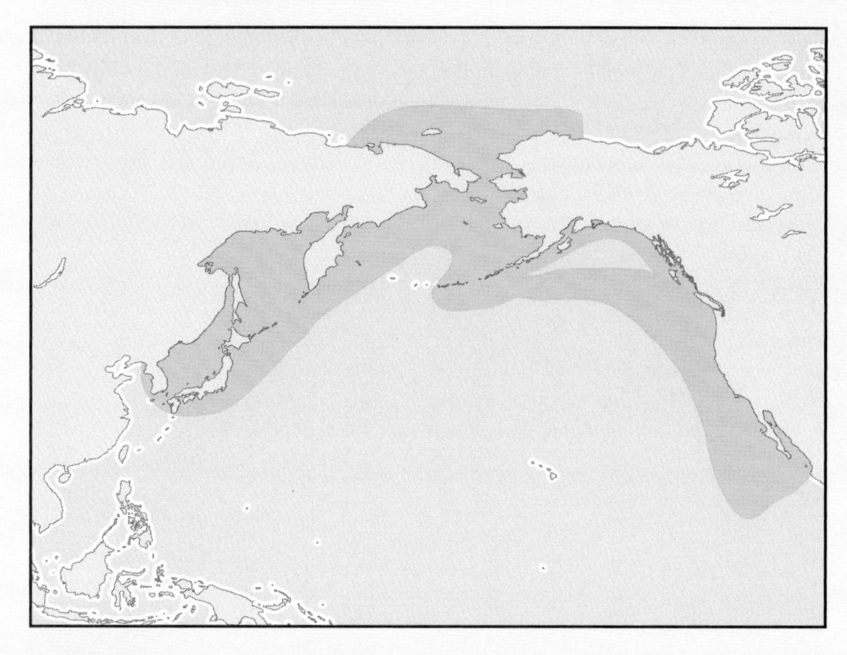

Distribution
Extant populations migrate seasonally between Arctic and warm temperate waters of the North Pacific; western population summers in the Sea of Okhotsk and winters off the coasts of South Korea and Japan; eastern population summers in the Bering, Chukchi, and Beaufort Seas off northeastern Alaska, then migrates south along the west coast of North America to winter on calving and breeding grounds in coastal Baja California and western Mexico

Evolution and systematics

The gray whale, *Eschrichtius robustus*, is the only species in the family Eschrichtiidae (formerly Rhachianectidae). Escrichtidae is one of four families in the suborder Mysticeti (the other three are Balaenidae, the right whales; Neobalaenidae, the pygmy right whale; and Balaenopteridae, the rorquals). Of these groups, Eschrichtiidae is considered to be the most primitive. Evidence suggests that gray whales, rorquals, and right whales diverged from a common ancestor during the Miocene (26 to 7 million years ago). Concerning the fossil record, only a single fossil gray whale specimen, dating to the Pleistocene (50,000 to 120,000 years ago) has been found in southern California.

The taxonomy for this species is *Eschrichtius robustus* (Lilljeborg, 1861), Sweden. Other common names include: English: devil-fish, desert whale, the friendly whale; French: Baleine grise; Spanish: Ballenna gris.

Physical characteristics

Although they are called "gray," these whales are actually black—at least, they are at birth. The skin color changes with time, primarily because of scarring caused by barnacles. These sedentary marine crustaceans attach to whale skin after it has been scraped during feeding bouts; they mostly cluster on the head but may occur anywhere on the body. A large whale may carry hundreds of pounds of barnacles. Although scarring is what changes the skin color, the presence of so many barnacles and also sea lice (orange or yellow in color and which also routinely infest gray whales) contributes to a gray whale's overall light, mottled appearance.

Gray whales have been described as "living fossils" and as "the most primitive of the great whales" (Nollman 1999) because of their short, coarse baleen plates and their lack of a dorsal fin. On the back, toward the tail, where most other baleen whales have a dorsal fin, gray whales have a series of 8 to 14 bumps that form a "dorsal ridge". The short baleen plates are less than 20 in (50 cm) long (compare this to bowhead whales, which have the longest baleen, 14 ft [4.3 m] long). The plates are ivory to yellow in color; thicker than the baleen plates of other baleen whales; and arranged in two groups of 130 to 180 plates each on either side of the mouth; the two rows of baleen plates do not meet at the front of the mouth, as they do in members of the closely related family Balaenopteridae. The mouth itself is slightly arched and the outside of the jaw is studded with small, sensitive hairs. Gray whales have two or three (rarely four) shallow furrows on their throats; these "throat pleats" are shorter and less numerous

Gray whales have a unique heart-shaped spray when they blow. (Illustration by Michelle Meneghini)

than those of many other baleen whales (for example, Bryde's whales have as many as 70).

Baleen whales as a group include large species and small; the largest living animal on earth, the blue whale (*Balaenoptera musculus*), is 90 ft (27.5 m) long, whereas the minke whale measures a mere 25–30 ft (7.9 m). Compared to these relatives, gray whales are considered medium-sized. Adult females average 46 ft (14 m) in length with an average weight of almost 70,000 lb (32,000 kg); adult males are somewhat smaller than females, averaging 43 ft (13 m) in length. Gray whales are comparatively slender for their size, with narrow heads that are small in relation to the total body length. The pelvis is relatively large, however.

Other distinguishing traits include two blowholes; 56 vertebrae (with the neck vertebrae being separate, not fused); and flippers with only four digits (the first finger [thumb] is absent; some baleen whales have five digits in the flippers). Male gray whales have very large testes for their body size, and are assumed to produce large volumes of sperm.

Distribution

Historically, gray whales occurred both in the North Atlantic and North Pacific basins; however Atlantic populations were extinct by the mid-1700s due to whaling. The eastern Atlantic population is thought to have spent the summer in the Baltic Sea and wintered off the Atlantic and Mediterranean coasts of southern Europe and also North Africa. The western Atlantic population of gray whales may also have summered in the Baltic sea; this population migrated south along the eastern coast of North America to breed and bear young in shallow lagoons and bays off southeastern Florida.

The Pacific basin continues to support two genetically distinct gray whale populations, one in the western Pacific and one in the eastern Pacific. Members of the eastern population spend the summer on feeding grounds in the Bering, Beaufort, and Chukchi Seas between northern Alaska and Siberia (although some whales do feed farther south in summer, off the coasts of southeast Alaska, British Columbia, Washington, Oregon, and California). Starting in October and continuing through January, whales in the Arctic move south along the east side of the Bering Sea and through Unimak Pass in the Aleutian Islands, then continue down the western coast of North America. By January or February a significant portion of the population has arrived at breeding and calving grounds in warm, shallow, nearly landlocked lagoons and bays along the west coast of Baja California and the eastern side of the Gulf of California in western Mexico. (Some members of the population do pass the winter farther north along the coast.)

Members of the western or Korean gray whale population are thought to spend summer months feeding in the sea of Okhotsk and then move south to breed in winter somewhere along the coast of southern Korea or Japan. (A western population that bred in the Inland Sea of Japan was extinct by the

A gray whale (*Eschrichtius robustus*) spy-hopping. (Photo by Tui De Roy. Bruce Coleman, Inc. Reproduced by permission.)

A gray whale (*Eschrichtius robustus*) calf. (Photo by François Gohier/Photo Researchers, Inc. Reproduced by permission.)

beginning of the twentieth century; a population that bred off southern Korea was all but wiped out by the 1930s.)

Habitat

Gray whales are notable for their habit of migrating and feeding in very shallow water; they are typically found closer to shore than any other large whale—usually within 1.9 to 3.1 mi (3–5 km) of land.

This species is sometimes referred to as "the desert whale" because it leaves the open sea to breed and calve in shallow desert lagoons—the only whale to do this routinely. As of 2003, the gray whales' most important winter breeding habitat consisted of just four areas on the coast of Baja California: Guerrero Negro Lagoon, Scammon's Lagoon, San Ignacio Lagoon, and Magdalena Bay.

Behavior

Gray whales make the longest known migration of any whale—and indeed of any mammal. Members of the eastern Pacific population travel as much as 10,000–12,000 mi (16,000–19,300 km) round-trip each year. In contrast to their relatives, the humpbacks and blue whales, which migrate across the open ocean, gray whales migrate exclusively in coastal waters.

Gray whales swim slowly, averaging 4.4–5.6 mph (7–9 km/hr), but can speed up to 8 mph (13 km/hr) when pursued. The "spout" or "blow" (exhalation) is described as low and spreading; it rises only about 118 in or about 10 ft (300 cm) above the ocean surface (in comparison, the blow of a blue whale rises more than twice as high).

Swimming whales follow a characteristic pattern; they make a series of short surface dives followed by a longer, deeper dive; observers will see three to five blows, 30 to 50 seconds apart, followed by a dive that typically lasts for 4 to 5 minutes; the whale blows three to five times when it surfaces. Gray whales do not arch their backs before diving the way humpbacks (*Megaptera novaeangliae*) do; their tail flukes rise above the surface of the water at the start of a deep dive but not a shallow one. The deepest documented dive by a gray whale was to a depth of 248 ft (75.6 m). To put this in context, sperm whales—the deepest-diving whales—are known to reach depths of 6,560 ft (2,000 m) when they feed.

Whales on migration dive to shallower depths than do feeding whales. Diving whales sometimes engage in a behav-

A gray whale (*Eschrichtius robustus*) snout covered in barnacles. (Photo by Animals Animals ©Bob Cranston. Reproduced by permission.)

ior called "bubble-blasting" —they submerge, then release air underwater, so that it bubbles to the surface. It is not clear whether this behavior is a stress response or a mechanism for regulating buoyancy.

Although female gray whales accompany their calves on the northward migration, migrating gray whales do not otherwise appear to travel in family groups. Whales on migration may swim alone, in groups of two or three, or in pods of up to 16 animals; a typical migration group has about nine animals. The composition of these groups is not stable; instead the group make-up changes constantly (shifting social alliances are typical of baleen whales).

Of all migrating gray whales, female whales and their calves often swim the farthest. On their northward migration, they continue past the Bering Sea—where the majority of gray whales congregate to feed—until they reach the Chukchi Sea. Here, food may not be quite so abundant, but neither are predatory killer whales, or orcas (*Orcinus orca*).

Gray whales seem to be able to detect their orca enemies from a distance, by their vocalizations. In one study conducted along the coast of California, researchers broadcast killer whale vocalizations underwater; migrating gray whales avoided the sounds by rapidly changing course, swimming toward shore, and moving into concealing beds of kelp. Females with calves defend their young fiercely—not just from killer whales but from any perceived threat. This behavior caused nineteenth-century whalers working the breeding lagoons off Baja California to give gray whales the nickname of "devil-fish." They were described as "a cross between a sea-serpent and an alligator," as American whaleboat captain Charles Scammon wrote in his landmark 1874 book on whale behavior, adding, "The casualties from coast and kelp whaling are nothing to be compared with the accidents that have been experienced by those engaged in taking the females in the lagoons. Hardly a day passes but that there is upsetting or staving of boats, the crews receiving bruises, cuts, and, in many instances having limbs broken; and repeated accidents have happened in which men have been instantly killed, or received mortal injury."

Though fierce when threatened, gray whales that feel secure will allow humans to approach remarkably close. Since the latter half of the twentieth century, whale-watching has been a popular activity in the lagoons of Baja California. In 1975, tour operators first noticed that some whales would actually approach whale-watch boats and allow themselves to be petted. This behavior has earned them a new nickname: "the friendly whale."

Gray whales display such common cetacean behaviors as breaching (in which the whale leaps above the water's surface, then falls back into the water, landing on its back or side) and spy-hopping (positioning the body vertically in the water, with the head raised above the sea surface, sometimes while turning slowly). Gray whales rarely breach outside of their southern breeding lagoons, however, leading some researchers to hypothesize that this behavior is a component of courtship. Alternately, breaching may represent an effort to get rid of itchy parasites, or it may be an expression of stress, or a form of play. The function of spy-hopping also is not known. One idea is that it helps whales to orient while on migration; however whales sometimes spyhop while allowing their eyes to remain underwater.

Among the baleen whales, humpback whales are known for their long, complex courtship songs. Gray whales do not make such sustained or complex vocalizations. Most vocalizations are at a frequency range of less than 1500 Hz. Researchers have described several types of calls, including "pulses" that sound like clangs, pops, and croaks; low-frequency "moans" that, to human ears, sound like a cow mooing; "rapid upsweeps"; and grunting and groaning sounds, like a zipper being pulled open. Bubble blats and bubble trails are also categorized as vocalizations. Field observations suggest females use pulses to communicate with their calves.

Feeding ecology and diet

Like all mysticete whales, gray whales use their comb-like baleen plates the way humans use a tea-strainer or colander—to collect small food items from the water. Most baleen whales strain free-floating plankton out of the water column, and gray whales are capable of doing this too—but usually, they take their prey from the sea floor. A feeding whale dives to the bottom, rolls onto its side—usually the right side—then shoves its body forward and upward, taking in a mouthful of soft sediment and water. Then, with its large, muscular tongue—which is the size of a compact car—the whale pushes the muddy mouthful forward against its baleen. Mud and water filter through, leaving edible components caught against the baleen strands. This unusual bottom-feeding behavior

A gray whale (*Eschrichtius robustus*) spouting near Baja California, Mexico. (Photo by Joe McDonald. Bruce Coleman, Inc. Reproduced by permission.)

A gray whale (*Eschrichtius robustus*) calf surfacing for air. (Photo by François Gohier/Photo Researchers, Inc. Reproduced by permission.)

earned gray whales the nicknames of "hard head" and "mud digger" from nineteenth-century whalers.

Amphipods make up the major part of a gray whale's diet. These small crustaceans, which are related to shrimp, live in burrows on the ocean floor. Amphipods are particularly plentiful in the cold waters off the northern Pacific Coast, where whales spend the summer; a single species, *Ampelisca macrocephala*, can account for 95% of the whales' intake on their Arctic feeding grounds. Gray whales do eat various other small, bottom-dwelling invertebrates, however, including clams, crabs, and marine worms; a study of whales feeding in Clayquot Sound, British Columbia, documented mysid shrimp, pelagic porcelain crab larvae, and benthic ghost shrimp as part of the diet. Other studies of gut contents reveal that gray whales sometimes consume small fish, as well as kelp and other marine vegetation; however it is not clear whether the whales eat algae on purpose or accidentally, simply because it happens to be growing on the bottom.

Gray whales do most of their feeding between May and November, while they are in Arctic waters. This means the insides of their mouths are constantly exposed to freezing water; however a network of blood vessels at the base of the tongue functions as a highly effective countercurrent exchange mechanism to reduce the loss of body heat. In one study, scientists measuring heat loss by a captive gray whale calf were surprised to discover that the animal lost more heat through the thick, insulating layer of blubber covering its entire body than it did through its tongue.

A lone adult whale may eat as much as 65 tons (59 tonnes) of food per year—except that feeding is not spread out over an entire year but is concentrated in the five months spent on the summer feeding grounds in the Arctic. The whales accumulate a thick layer of blubber during the summer, then eat very little on migration and while at the breeding lagoons, living off the stored energy in their blubber. Pregnant female whales store enough calories that they can make the long southward migration, give birth, nurse a calf, and swim north again—all while fasting.

Reproductive biology

Gray whales are one of the three species of baleen whales (right whales [*Eubalaena* spp.] and humpback whales are the other two) that form "breeding aggregations," in other words, gather in large numbers to mate and give birth. The other eight mysticete species do not gather to breed.

Gray whales use different habitats for mating than for calving. Mating typically occurs at the entrances to the lagoons, or just outside (although courtship and mating can also occur while on migration); while this frenzied courtship goes on outside, the females and their calves stay well inside the lagoons. Although gray whales are not particularly social, there are some reports of whales supporting laboring females or injured animals so they can stay at the surface and breathe.

Courting gray whales are very active; they can be seen rolling and breaching and swimming in line. These whales are not monogamous; observations suggest each female probably mates with a number of different males. This, along with the fact that males have large testes and produce large amounts of sperm, suggests that sperm competition occurs.

Gray whales are sexually mature at the age of six to eight years; however the average age of females when they first give birth is nine years. The breeding cycle takes two years, with gestation lasting for 13.5 months of that period. Pregnant females are the first members of the population to leave the breeding lagoons; they depart in mid-February. During the more than year-long gestation period, females make the long trip north to the feeding grounds and then back to the breeding lagoons.

Most gray whale calves are born in the Baja lagoons between early January and mid-February (a few newborn calves have been also sighted along the California coast, suggesting that some births occur outside of the lagoons). A typical newborn calf may be 15 ft (4.6 m) long and weigh about 1,100 lb (500 kg). It drinks 50 gallons of milk (190 l) a day and grows rapidly. By the end of winter most calves reach a length of 18–19 ft (5.5–6 m).

If pregnant females leave the breeding lagoons early, mothers with calves are the last to leave, sometimes heading north as late as May or June. By spending the maximum possible amount of time in protected habitat, they give their calves

A gray whale (*Eschrichtius robustus*) passes through kelp. (Photo by Animals Animals ©Bob Cranston. Reproduced by permission.)

A gray whale (*Eschrichtius robustus*) breaching. (Photo by François Gohier/Photo Researchers, Inc. Reproduced by permission.)

time to grow large enough and strong enough to evade killer whales on the trip north. (Killer whales will not enter the shallow lagoons.) Like most female baleen whales, gray whales nurse their calves for about six to seven months. This means the calves are nursing throughout the northbound migration. They are weaned in late summer, on the Arctic feeding grounds where food is plentiful. Maximum longevity is reported at 70 years.

Conservation status

The eastern population of Atlantic gray whales is thought to have become extinct as early as A.D. 500; the western Atlantic population, which migrated seasonally along the eastern coast of North America, became extinct in the eighteenth century. As of 2003, the western Pacific or Korean population of gray whales is critically endangered, with fewer than 100 individuals remaining.

In contrast to these sad stories of extinction, the eastern Pacific population of gray whales represents one of the few success stories in whale conservation—indeed in the conservation of any endangered species. The exploitation that came with the European settlement of North America led to near-extinction of the eastern Pacific population by the middle of the twentieth century, but the species made a remarkable and rapid recovery once it was protected, with populations returning to pre-exploitation levels by the mid-1990s. In 1946, the International Whaling Commission banned the commercial take of gray whales; they were later additionally protected under the U. S. Endangered Species Act in 1973. With these protections gray whales made a remarkable recovery. In 1994, when the population numbered about 22,000 individuals, the U. S. Fish and Wildlife Service removed the gray whale from the U. S. Endangered species list. Experts hypothesize that the gray whale's habit of forming breeding aggregations helped the species to make this remarkable recovery; in species that breed solitarily, such as blue whales, it's thought that individuals have a hard time finding a mate when populations are very small.

The western Pacific population of gray whales does remain very small, despite having been protected at the same time as the eastern population. The probable reason this population failed to recover was that illegal whaling continued. The World Conservation Union (IUCN) listed the western gray whale as Critically Endangered in 2000.

Habitat quality is of concern. The Mexican government and Mitsubishi Corporation have proposed joint development of a salt-production facility in San Ignacio Lagoon, which is

Gray whale (*Eschrichtius robustus*) flukes. (Photo by George D. Lepp/Photo Researchers, Inc. Reproduced by permission.)

one of the whales' key calving grounds. However experience suggests the noise and activity associated with this development would be harmful to the whales; when dredging for a salt works was conducted in Guerrero Negro Lagoon for several years, breeding whales deserted it. As of 2003, a coalition of Mexican and U. S. environmental groups has been able to prevent development in San Ignacio.

As whale populations have rebounded, animals have started to move back into the waters around such major cities as San Diego, San Francisco, and Seattle, where they historically occurred, and this has led to more concerns; decades of industrial discharge, discharge from sewage treatment facilities, and agricultural runoff have left the sediments in these coastal areas highly polluted—a potential problem for a bottom-feeding whale that wallows in sediment. Offshore oil exploration and oil production are other issues. In controlled experiments, gray whales actively avoided the noise from these activities and, as a result, sometimes moved into very shallow water, risking stranding.

Even though gray whales were removed from the Endangered Species list in 1994, researchers have continued to monitor populations. A significant increase in gray whale strandings was noted in 1999 and 2000. The population estimate in 2001–2002 was 17,500 individuals—significantly down from a high of 26,635 in 1997–98; however, researchers estimate the total carrying capacity for the western Pacific at 20,000 to 24,000 whales and say the most recent decline is probably within normal fluctuation parameters—and possibly due to a natural climate cycle that has temporarily put food in short supply. Population monitoring continues.

Significance to humans

Indigenous peoples of northwestern North American and eastern Siberia hunted gray whales for oil, meat, hide, and baleen for hundreds if not thousands of years; indeed, indigenous whaling was the major economic activity along the Chukchi, Bering, and Okhotsk seas before Europeans came on the scene and began taking large numbers of whales.

Indigenous peoples of Europe and Japan probably hunted gray whales as well. The indigenous peoples of eastern North America and Baja California are not thought to have hunted the whales, but evidence shows they did take advantage of stranded whales as a source of food and other materials. Images of gray whales can easily be identified in cave and rock paintings made by indigenous people northeast of San Ignacio Lagoon.

In Baja California, Magdelena Bay was the center of American whaling from 1845 to 1874. In the United States and Europe at this time, whale oil was a valuable commodity, burned in lamps for household illumination and used to lubricate machinery.

With the recovery of eastern Pacific populations, whale watching has replaced whaling as a major money-making industry in Baja California and along North America's West Coast. Millions of people watch these whales on migration each year. Magdelena Bay has sponsored a major gray whale festival since 1994.

Gray whale (*Eschrichtius robustus*). (Illustration by Michelle Meneghini)

Resources

Books

Dedina, Serge. *Saving the Gray Whale: People, Politics, and Conservation in Baja California.* Tucson: University of Arizona Press, 2000.

Mann, Janet, Richard C. Connor, Peter L. Tyack, and Hal Whitehead. *Cetacean Societies: Field Studies of Dolphins and Whales.* Chicago: University of Chicago Press, 2000.

Mead, James G. *Whales and Dolphins in Question.* Washington, DC: The Smithsonian Institution, 2002.

Nollman, Jim. *The Charged Border: Where Whales and Humans Meet.* New York: Henry Holt and Company, 1999.

Simmons, Mark P., and Judith D. Hutchinson. *The Conservation of Whales and Dolphins: Science and Practice.* New York: John Wiley & Sons, 1996.

Periodicals

Buckland, S. T., and J. M. Breiwick. "Estimated Trends of Eastern Pacific Gray Whales from Shore Counts (1967/68 to 1995/96)." *Journal of Cetacean Research and Management* 4, no. 1 (2002): 41–48.

Dunham, Jason S., and David A. Duffus. "Foraging Patterns of Gray Whales in Central Clayquot Sound, British Columbia, Canada." *Marine Ecology Progress Series* 223 (2001): 299–310.

Heyning, John E. "Thermoregulation in Feeding Baleen Whales: Morphological and Physiological Evidence." *Aquatic Mammals* 27, no. 3 (2001): 284–288.

Moore, Sue E., and Janet T. Clarke. "Potential Impact of Offshore Human Activities on Gray Whales *(Eschrichtius robustus).*" *Journal of Cetacean Research and Management* 4, no. 1 (2002): 19–25.

Weller, David W., Alexander M. Burdin, Bernd Wursig, Barbara L. Taylor, and Robert L. Brownell, Jr. "The Western Gray Whale: A Review of Past Exploitation, Current Status, and Potential Threats." *Journal of Cetacean Research and Management* 4, no. 1 (2002): 7–12.

Wolff, Wim J. "The South-eastern North Sea: Losses of Vertebrate Fauna During the Past 2000 Years." *Biological Conservation* 95, no. 2 (2000): 209–217.

Other

Gray Whales (*Eschrictius robustus*): Eastern North Pacific Stock. *National Marine Fisheries Service.* December 10, 2000. <http://www.nmfs.noaa.gov.prot_res/PR2/Stock_Assessment_Program/Cetaceans/Gray_Whale_(Eastern_N._Pacific)/AK 00gray whale_E.N.Pacific.pdf>.

Cynthia Berger, MS

Pygmy right whales
(*Neobalaenidae*)

Class Mammalia
Order Cetacea
Suborder Mysticeti
Family Neobalaenidae

Thumbnail description
Smallest baleen whale, with sickle-shaped dorsal fin, arched rostrum, inconspicuous blow, and coloration ranging from black or gray above to pale below

Size
16.4–21.3 ft (5–6.5 m); maximum weight is 7,562 lb (3,430 kg)

Number of genera, species
1 genera; 1 species

Habitat
Coastal and pelagic waters of the Southern Hemisphere, 31–52° S of the equator

Conservation status
Not listed by the IUCN

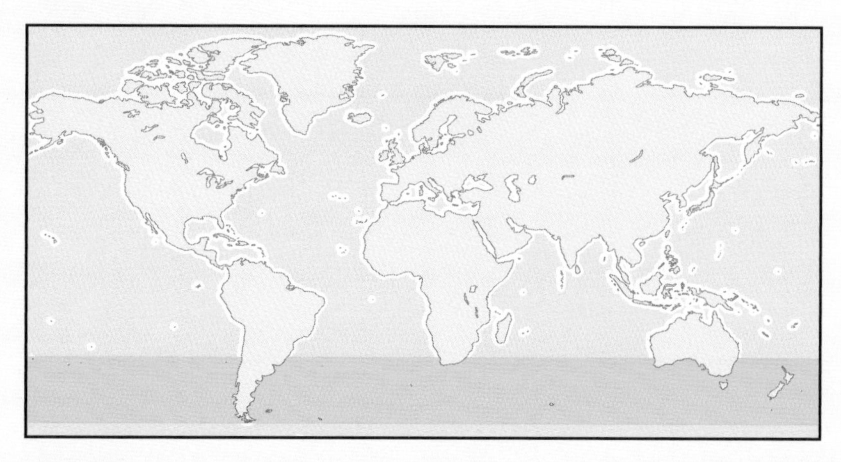

Distribution
Circumpolar, from about 30–55° S of the equator

Evolution and systematics

The pygmy right whale, *Caperea marginata*, is the smallest baleen whale and the only member of family Neobalaenidae. In the past, it was placed among the right and bowhead whales (Balaenidae). It is now considered to be a closer relative to the rorquals (Balaenopteridae) and gray whales (Eschrichtiidae). DNA research from 1992 and 1993 by Úlfur Árnason and colleagues, as well as morphological evidence, have supported its membership in a separate family.

Caperea marginata differs from other right whales in having a proportionally smaller head and humerus, a different type of baleen, a dorsal fin, four digits instead of five, and 44 or fewer vertebrae. The skull is also very different from those of the Balaenidae. The occipital shield is larger, and projects farther to the fore. The rostrum is shorter, wider, and although it forms less of an arch, the arch becomes more pronounced with age. The supraorbital processes are shorter. The nasal bones are smaller. The glenoid fossae and the orbits are less ventrally placed than on the Balaenidae.

No studies of geographic variation have been conducted, and there are no recognized subspecies. No fossils are known.

The taxonomy for this species is *Caperea marginata* (Gray, 1846), temperate waters of the Southern Hemisphere.

Physical characteristics

Caperea marginata is the smallest of the baleen whales. The greatest length and weight ever recorded are 21.3 ft (6.5 m) and 7,562 lb (3,430 kg).

Caperea marginata is streamlined. Adults are heftier than rorquals but not as wide as bowhead or right whales. The relatively large head comprises about one-fourth of total length. The jawline arches and then reverses direction just northwest of the eye, somewhat like a scythe or a question mark that has been laid over to the left and the sharp angle smoothed into a gentle curve. The flippers, located just aft of and below the posterior end of the jawline, are small and narrow.

Features that distinguish *C. marginata* from other right whales include a small and strongly hooked dorsal fin 25–30% of body length forward from the tip of the flukes, and two throat grooves.

The color is dark gray or black on the back, shading to paler on the belly. The inside of the mouth and tongue are ivory colored, as is the gum at the base of the baleen. The baleen is long and narrow, with very fine bristles, and it is creamy white in color except for the margin, which varies from brown to black. Each side of the jaw has 213 to 230 baleen plates, which measure up to 33 in (840–850 mm). The area of the filtering apparatus is relatively large for the body size.

The skeleton and skull are different from those of all other cetaceans. *Caperea marginata* has the most ribs (34–36) and the least number of vertebrae (40–44) of any cetacean. The ribs, wide and flat, become more so posteriorly, presumably to protect internal organs, and they extend farther aft than those of any other genera, leaving only two ribless vertebrae anterior to the tail. The seven cervical vertebrae are all fused.

Pygmy right whale (*Caperea marginata*). (Illustration by Brian Cressman)

C. marginata can easily be confused with the minke whale (*Balaenoptera acutorostrata*) when seen from the rear.

Distribution

The pygmy right whale appears to inhabit the Southern Hemisphere between latitudes 31° and 52°, in both coastal and pelagic waters of the Atlantic, Pacific, and Indian Oceans. More sightings—more than one-third of the total—have taken place off the coast of southeastern Australia, primarily off Tasmania, than anywhere else. Sightings and strandings have also occurred in South Africa, southern South America, New Zealand, The Falkland Islands, the Crozet Islands, and the south Atlantic Ocean.

Habitat

Caperea marginata inhabits temperate and sub-antarctic waters in coastal and pelagic zones, between the 5° and 20° isotherms.

A pygmy right whale (*Caperea marginata*) off New Zealand. (Photo by © Robert L. Pitman/Seapics.com. Reproduced by permission.)

The dorsal fin of the pygmy right whale (*Caperea marginata*) distinguishes it from true right whales. (Photo by © Robert L. Pitman/Seapics.com. Reproduced by permission.)

Behavior

Little is known about this small, obscure species, because living animals have been observed only rarely. There have been fewer than 20 observations of individuals or groups at sea, partly due to the animal's inconspicuous behavior, and to the paucity of activity at the surface. Pods of up to ten animals have been seen, and in one instance, roughly 80 individuals were spotted together in pelagic waters, but most live pygmy right whales have been observed singly or in pairs.

Caperea marginata have been viewed in the company of dolphins and pilot (*Globicephala melas*), sei (*B. borealis*), and minke whales.

Caperea marginata usually swims slowly, around three to five knots, but can accelerate and swim very quickly if necessary, propagating a notable wake. It swims by flexing its body laterally, in waves. It does not appear to jump, and the blow is inconspicuous. *Caperea marginata* has been observed to dive for only up to four minutes at a time, surfacing briefly between dives.

Vocalizations are characterized as sounding like intense thumps or tone bursts, the volume of each quickly rising, then slowly falling, while the frequency drops. They come in cycles of 11–19, each with a mean duration of 180 m/sec, separated from each other by a mean of 460 m/sec.

Strandings can occur at any time of year, suggesting that pygmy right whales may not migrate seasonally. However, increased sightings and strandings in the waters off the Cape Peninsula of South Africa during December, January, and February have suggested a seasonal migration during these months.

Feeding ecology and diet

Diet appears to consist almost entirely of copepods, and possibly other plankton. Observations and comparisons with other copepod eaters (sei and right whales) suggest that *C. marginata* may use a surface-skimming technique for feeding.

The pygmy right whale (*Caperea marginata*) surfacing. (Photo by S. Whiteside/ANTPhoto.com.au. Reproduced by permission.)

A stranded pygmy right whale (*Caperea marginata*). (Photo by Tui De Roy/Minden Pictures. Reproduced by permission.)

As mentioned, increased sightings and strandings in the waters off the Cape Peninsula of South Africa from December to February have suggested a seasonal inward migration during these months, which coincides with an increase in the biomass of zooplankton that occurs during the Southern Hemisphere's spring and summer.

Reproductive biology

Little is known about reproductive biology. The mating season, the mating system, the gestation period, and the calving interval are all unknown. Some researchers believe that calving may take place year-round. Newborns are around 6.5 ft (2 m) and reach 9–11.5 ft (3.0–3.5 m) at weaning. Sexual maturity is reached at 16–20 ft (5–6 m).

Conservation status

Caperea marginata is now listed on appendix I of the CITES, meaning it is considered "Threatened, though it is not listed by the IUCN." Though it is the only baleen whale to escape large-scale commercial whaling, it is thought to be comparatively rare. It may be at risk due to the difficulty of distinguishing it from the Antarctic minke whale (*B. bonaerensis*), 440 of which the Japanese harvested annually in the Southern Ocean. The International Whaling Commission has estimated population size due to lack of data. *Caperea marginata* is thought to be threatened by global climate change, but not by toxic contamination.

Significance to humans

Though *C. marginata* have never been caught commercially, individuals have been caught deliberately by inshore fisheries, and accidentally in fishing nets.

Resources

Books

Jefferson, T. A., S. Leatherwood, and M. A. Webber, eds. *Marine Mammals of the World.* Heidelberg: Springer-Verlag, 1993.

Kemper, Catherine M. "Pygmy Right Whale *Caperea marginata*." In *Encyclopedia of Marine Mammals*, edited by William F. Perrin, Bernd Würsig, and J. G. M. Thewissen. San Diego: Academic Press, 2000.

Resources

Periodicals

Árnason, Úlfur, Sólveig Grétarsdóttir, and Bengt Widegren. "Mysticete (Baleen Whale) Relationships Based upon the Sequence of the Common Cetacean DNA Satellite." *Molecular Biology Evolution* 9 (1992): 1018–1028.

Árnason, Úlfur, Anette Gullberg, and Bengt Widegren. "Cetacean Mitochondrial DNA Control Region: Sequences of All Extant Baleen Whales and Two Sperm Whale Species." *Molecular Biology Evolution* 10 (1993): 960–970.

Dawbin, William H., and Douglas H. Cato. "Sounds of a Pygmy Right Whale (*Caperea marginata*)." *Marine Mammal Science* 8, no. 3 (July 1992): 213–219.

Kemper, Catherine M. "Sightings and Strandings of the Pygmy Right Whale *Caperea marginata* Near Port Lincoln, South Australia and a Review of other Australasian Sightings." *Transactions of the Royal Society of South Australia* 121 (1997): 79–82.

———. "Distribution of the Pygmy Right Whale, *Caperea marginata*, in the Australasian Region." *Marine Mammal Science* 18, no. 1 (2002): 99–111.

———. "New Information on the Feeding Habits and Baleen Morphology of the Pygmy Right Whale *Caperea marginata*." *Marine Mammal Science* 8, no. 3 (July 1992): 288–293.

Other

"Pygmy Right Whale." In *Walker's Mammals of the World Online 5.1*, by Ronald M. Nowak. Baltimore: Johns Hopkins University Press, 1997.

David Holzman, BA

▲

Right whales and bowhead whales

(Balaenidae)

Class Mammalia
Order Cetacea
Suborder Mysticeti
Family Balaenidae

Thumbnail description
Large, mainly black, baleen whales with proportionally large heads, narrow rostra, strongly arched mouthlines, broad flippers, and no dorsal fins

Size
43–65 ft (13–20 m); 168,000–224,000 lb (76,200–101,600 kg)

Number of genera, species
2 genera; 4 species

Habitat
Marine, coastal, pelagic, shallow, and deep waters

Conservation status
Endangered: 1 species; Lower Risk/ Conservation Dependent: 2 species

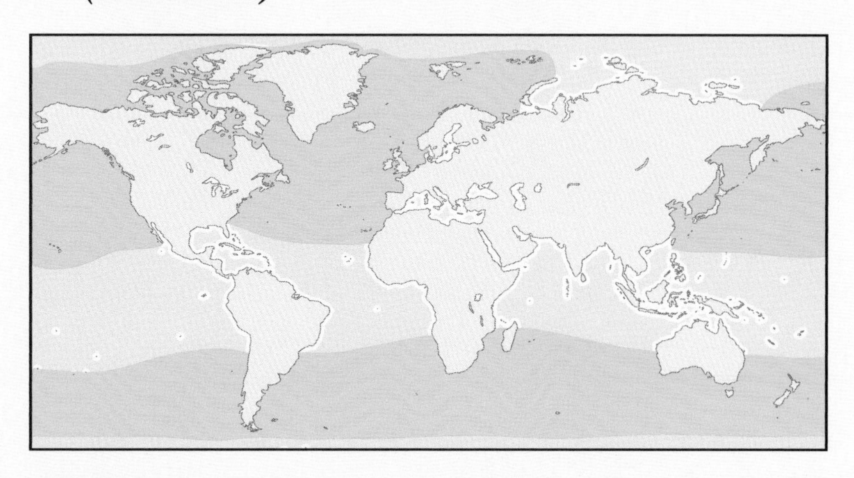

Distribution
Arctic and subarctic (bowhead), temperate Northern and Southern Hemispheres, with movement into Antarctic and subarctic waters in summer (right whales); largely absent from tropical belt

Evolution and systematics

The balaenids diverged from the other mysticetes relatively early, possibly around 30 million years ago (mya). The oldest fossil is *Morenocetus parvus*, a primitive balaenid from the early Miocene of Argentina (23 mya). Numerous fossils of more derived balaenids have been reported from deposits of late Miocene and Pliocene ages, especially in Europe. In fact, species in the extinct genus *Balaenula* have more highly derived crania than either of the living genera.

Balaena was the only genus in the suborder Mysticeti recognized by Linnaeus, and early classification systems generally placed all baleen whales within it. For more than 100 years, however, there has been a firm consensus that bowhead and right whales belong to a separate family of mysticetes, the Balaenidae. The pygmy right whale (*Caperea marginata*) was assigned to a family of its own, Neobalaenidae, in the 1920s.

There has been considerable disagreement until recently about whether bowheads and right whales should be assigned to one genus, *Balaena*, or instead the right whales should be in a separate genus, *Eubalaena*. In addition, there has been difficulty agreeing on the number of right whale species. Two had long been recognized: southern, *Eubalaena australis*, and northern, *E. glacialis*. Current convention is to recognize two genera, one containing only *B. mysticetus*, or the bowhead whale; the other containing three species of right whales: North Atlantic, *E. glacialis*, North Pacific, *E. japonica*, and southern, *E. australis*, although there is some debate on whether *E. japonica* is a separate species.

Referring to the balaenids generically as right whales causes some confusion. As recently as the nineteenth century, whalemen and scientists referred to the bowhead as the Greenland or Arctic right whale, and to the other species in the family as black right whales. Given the difficulties, it seems more appropriate, or at least less confusing, to refer to the family as a whole as balaenids, and to reserve the term "right whales" for the genus *Eubalaena*.

Physical characteristics

The most obvious distinguishing characteristics of the family Balaenidae are the large head (up to one third of the total body length), the narrow, arched rostrum, the complete absence of a dorsal fin, and the broad flippers. The body is rotund, and the mouth-line greatly arched to accommodate long baleen plates (to 9 ft [2.8 m] in right whales, 13 ft [4 m] in bowheads) that hang from the roof of the mouth. There is a space at the front of the upper jaw separating the rows of baleen plates into a left and a right series. The blows of bowheads and right whales are more consistently V-shaped than those of other whales.

The body color is basically black. Bowheads have a white chin patch and a light gray-to-white band around the base of the tail, sometimes extending onto the flukes. Right whales often have irregular white ventral patches.

The most conspicuous difference in appearance between bowheads and right whales is that the latter have callosities

Close-up of a southern right whale (*Eubalaena australis*) head. (Photo by Jen & Des Bartlett. Bruce Coleman, Inc. Reproduced by permission.)

The North Atlantic right whale (*Eubalaena glacialis*) showing its V-shaped spout. (Photo by Sam Fried/Photo Researchers, Inc. Reproduced by permission.)

A southern right whale (*Eubalaena australis*) feeding at dusk with baleen exposed. (Photo by © Doug Perrine/Seapics.com. Reproduced by permission.)

The fluke of a North Atlantic right whale (*Eubalaena glacialis*). (Photo by François Gohier/Photo Researchers, Inc. Reproduced by permission.)

on their heads, especially on the rostrum, lower lips, and chin, and above the eyes. These irregular, thickened patches of hard skin are colonized by small amphipod crustaceans, called cyamids, or "whale lice." The skin of both bowhead and right whales is generally free of barnacle infestation, although the callosities of southern right whales are colonized by the barnacle genus *Tubicinella*.

Distribution

Balaenids have a nearly cosmopolitan distribution in marine waters, or at least they did before being drastically depleted by commercial whaling. They are absent only between the tropics of Cancer and Capricorn and in the far southern reaches of the Antarctic. The only large marine area poleward of 30° of latitude where they are not known ever to have been common is the Mediterranean Sea, although there have been a few records there. Bowheads occurred historically throughout the Arctic, including the Sea of Okhotsk, Hudson Bay, and Gulf of St. Lawrence. Ice massifs along the northern coasts of Asia and North America periodically blocked their passage between ocean basins.

Right whales were widely distributed across the North Atlantic and North Pacific north of 30°N, but are now almost entirely absent in the eastern portions of those basins. In the Southern Hemisphere, right whales occur in summer throughout most of the sub-Antarctic zone between 35–40°S and 55–60°S. Their wintering grounds are centered in discrete areas of coastline along the South American, African, and Australian continents as well as around certain oceanic islands, including St. Paul and Tristan da Cunha.

A North Atlantic right whale (*Eubalaena glacialis*) surfacing for air. (Photo by François Gohier/Photo Researchers, Inc. Reproduced by permission.)

Habitat

The defining characteristic of balaenid feeding habitat is zooplankton productivity, as these filter-feeding whales need very high-density concentrations of prey on which to forage efficiently. Thus, their summer distribution centers on coastal and offshore areas where physical processes, involving bottom topography, water column structure, and currents, aggregate plankton.

All balaenid populations appear to have large ranges and to migrate over fairly long distances. Bowheads are exceptionally adapted to coping with the annual formation and disintegration of sea ice in high latitudes. They travel through areas where cracks and pools of open water are widely and irregularly spaced, and they can break through new ice 9 in (22 cm) thick.

Known right whale calving grounds tend to be in warm temperate bays and shallow coastal regions. While such areas have been identified in the North Atlantic and Southern Hemisphere, none have been specifically located for North Pacific right whales or bowheads.

Behavior

The slow swimming speeds of balaenids made them vulnerable to capture by early whalers who could approach in small open boats powered by hand or sail. They were, nevertheless, dangerous quarry because of their powerful tails.

A southern right whale (*Eubalaena australis*) breaching. (Photo by Tom Brakefield. Bruce Coleman, Inc. Reproduced by permission.)

Numerous whaleboats were "stove," that is, broken and splintered by the thrashing tail of a harpooned bowhead or right whale. The tail is also used to defend against attempts at predation by killer whales (*Orcinus orca*). Besides fighting back, bowheads move into heavy ice to elude killer whales. The general distribution and migration patterns of bowheads may have been strongly influenced, via selection pressure, by the need to reduce frequency of encounters with killer whales.

These whales typically raise their flukes above the surface at the beginning of a long dive. They probably do not dive deeper than a few hundred feet (meters), and most dives do not last longer than 10–20 minutes. Bowheads under duress (e.g., when harpooned, or perhaps when transiting long distances under solid ice) can remain submerged for much longer, possibly up to an hour.

Balaenids are fairly vocal, producing low-frequency moans, grunts, belches, and pulses. Some bowhead calls have been described as growls, roars, trumpet sounds, or wild complex screams. They occasionally make high-frequency whines or squeals. During spring, bowheads produce songs that typically consist of one to three themes, composed of one- to five-note phrases. These songs are believed to serve a reproductive function, perhaps to attract females or dominate rival males. Loud, sharp sounds reminiscent of gunshots are sometimes heard from southern right whales and bowheads. It has been suggested that reverberations from their calls are used by bowheads to sense the undersides of ice floes and thereby navigate under sea ice.

Feeding ecology and diet

Balaenids prey exclusively on zooplankton, mainly copepods and euphausiids. Their long, finely fringed baleen and capacious mouths are well adapted to filter huge quantities of very small organisms. As skim-feeders, they swim forward with the mouth open, allowing water to flow in through the front of the mouth and pass out through the baleen filter, trapping food organisms on the inside fringed surfaces. The prey are then swept off the baleen and into a narrow digestive tract by the massive tongue.

Reproductive biology

Like other cetaceans, balaenids give birth to single young. Gestation takes at least a year and calves are nursed for six months or longer. The inter-birth interval is generally three years for prime-aged females, and probably increases as they approach senescence. Female right whales generally do not

A North Atlantic right whale (*Eubalaena glacialis*) courting group. (Photo by François Gohier/Photo Researchers, Inc. Reproduced by permission.)

give birth to their first calf until they are eight or nine years old. One photo-identified individual gave birth to seven calves over a 29-year period. While right whales are known to be capable of living for close to 70 years, bowheads live even longer, and thought to reach ages of well over 100.

Right whales have the largest testes of any animal (6 ft [2 m] long and nearly 2,000 lb [900 kg]). It is therefore assumed that sperm competition is a central feature of their reproductive strategy. There is no evidence to suggest long-term pairing between males and females. The mating system is polygamous, with multiple males engaging in what behaviorists call "scramble competition" for opportunities to copulate with a focal female.

Conservation status

All balaenids have been legally protected from commercial whaling since the 1930s, but aboriginal people in the Arctic are exempted and continue to hunt bowheads. Most ongoing whaling is subject to management by the International Whaling Commission (IWC) in cooperation with national and local authorities, and the limits on removals appear adequate to ensure continued population recovery. The hunting in eastern Canada, however, takes place outside the IWC's purview, and there is no assurance that the populations hunted there will be allowed to recover. Although the bowhead is still reasonably abundant in the western Arctic, its numbers in the eastern Arctic are much less than 5% of what they were when commercial whaling began.

Several populations of southern right whales are making strong recoveries from the depletion caused by commercial whaling, but aggregate abundance is still only about 10% of what it was in the late sixteenth century. The situation is much less hopeful in the Northern Hemisphere. A few hundred right whales remain in the western portions of the North Pacific and North Atlantic, but the populations on the east sides of these basins are all but extirpated, consisting of only scattered individuals and small groups. Even with best-case reasoning, one cannot escape the fact that there are less than 5% as many Northern Hemisphere right whales alive today than there were when commercial whaling began.

The clearest threat to right whales today is incidental mortality, caused mainly by collisions with ships and by entanglements in fishing gear such as set or drifting gillnets and lines connecting surface buoys with bottom traps for crustaceans (crabs, lobsters, etc.). There is also uncertainty whether some very small populations, such as bowheads around Svalbard and right whales in the eastern North Pacific and eastern North Atlantic, have the intrinsic capacity to recover, given their loss of genetic and demographic diversity, changes in their ecological circumstances, and the pos-

A southern right whale (*Eubalaena australis*) mother and calf. (Photo by Jen & Des Bartlett. Bruce Coleman, Inc. Reproduced by permission.)

sibility that they or their prey are being affected by chemical pollution.

The IUCN lists the bowhead and one of its populations as Lower Risk/Conservation Dependent, one of its populations as Critically Endangered, two as Endangered, and one as Vulnerable. The southern right whale is listed as Lower Risk/Conservation Dependent and the northern right whale and its populations are listed as Endangered.

Significance to humans

Northern people traditionally used bowhead baleen to construct toboggans, baskets, and traps and snares for catching birds and mammals. They also depended on bowheads for food, for oil to produce light and heat, and for construction materials. Bowhead bones supported the walls and roofs of ancient Thule-culture dwellings across the Arctic. In the present day, Eskimos in Alaska continue to organize much of their cultural life around the annual bowhead hunt, and bowhead meat and blubber remain staples in their diet.

Right whales and bowheads were the chief targets of early Basque, and eventually other European and American whalers. The tough, flexible baleen, known to whalemen as whalebone or bone, had great commercial value. It was used as a stiffener for hoop skirts, shirt collars, and corsets, and to make horse whips and umbrella ribs. Brooms and brushes were fashioned from the fibrous fringes. Discovery of spring steel, and later plastics, obviated the need for baleen, which was, in any event, in short supply by the end of the nineteenth century due to the depletion of balaenid stocks. The blubber oil of right whales and bowheads lighted streets and homes in Europe and North America, and its value did not decline significantly until the advent of electricity and petroleum in modern times.

Today, right whales are the focus of intensive research in North and South America, South Africa, Australia, and New Zealand. Special legislation exists to ensure their protection, and regulations have been implemented to prevent ship strikes and entanglements, especially off the eastern United States and Canada. The accessibility of right whales near shore during winter makes them popular tourist attractions in South Africa, Argentina, and Australia, and their summer presence in eastern Canada's Bay of Fundy supports a number of local tour enterprises.

A southern right whale (*Eubalaena australis*) breaching. (Photo by Jen & Des Bartlett. Bruce Coleman, Inc. Reproduced by permission.)

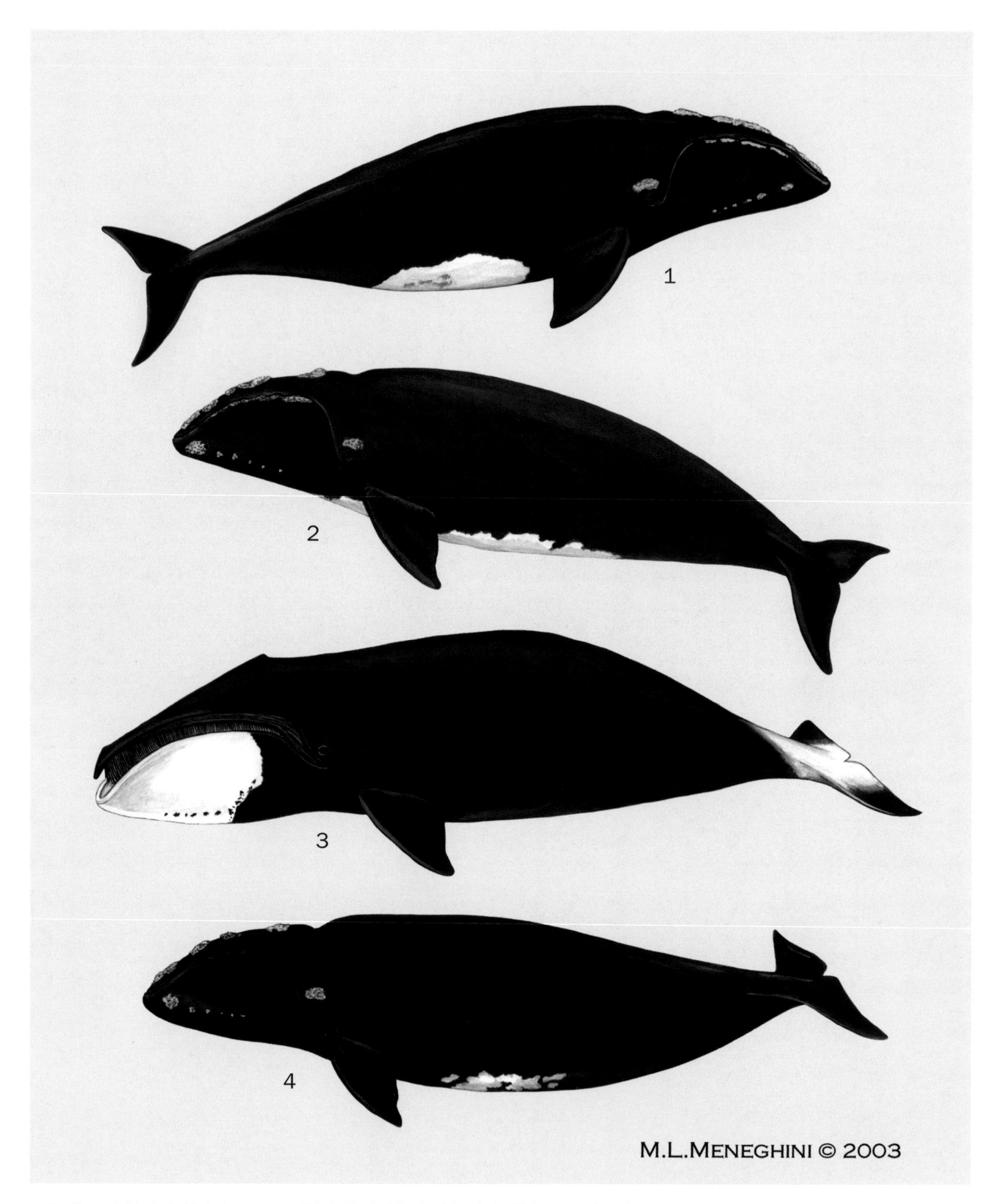

1. Southern right whale (*Eubalaena australis*); 2. North Atlantic right whale (*Eubalaena glacialis*); 3. Bowhead whale (*Balaena mysticetus*); 4. North Pacific right whale (*Eubalaena japonica*). (Illustration by Michelle Meneghini)

Species accounts

Bowhead whale

Balaena mysticetus

TAXONOMY

Balaena mysticetus Linnaeus, 1758, Greenland Sea.

OTHER COMMON NAMES

French: Baleine du Groenland; German: Grönlandwal; Spanish: Ballena polar, ballena de Groenlandia.

PHYSICAL CHARACTERISTICS

Length 46–65 ft (14–20 m); weight 168,000–224,000 lb (76,200–101,600 kg). Rotund shape, but with a distinct "neck" region. No dorsal fin or ridge, very broad back. Flippers have blunt tips and flukes wide with smooth contours. Muscular bulge (the stack) in the blowhole area. Predominantly black; a white patch at the front of the lower jaw may have several dark gray to black spots indicating chin hair. Light gray to white band around tail stock, just in front of the flukes. 250 to 350 baleen plates in each side of the jaw up to 17 ft (5.2 m) long, longest of all whales.

DISTRIBUTION

Arctic circumpolar; largely separate populations (stocks) centered in Sea of Okhotsk, Bering-Chukchi-Beaufort Seas, Hudson Bay-Foxe Basin, Davis Strait, Baffin Bay, and Greenland-Barents Seas; waters bordering northern Russia, United States (Alaska), northern Canada, Greenland, and Norway (Svalbard).

HABITAT

Marine waters of any depth in high northern latitudes, often associated with pack ice, including very dense (greater than 90%) ice coverage, but also found in open water during summer.

BEHAVIOR

Strongly migratory in response to ice formation and disintegration; slow-swimming; generally found alone or in small groups that converge on feeding areas and when several males are attempting to mate with a female.

FEEDING ECOLOGY AND DIET

Forage at surface, in water column, and on sea floor; 60 different species have been identified in stomach contents; copepods and euphausiids are preferred prey; mysids and gammarid amphipods also eaten.

REPRODUCTIVE BIOLOGY

Mating season late winter and spring, calving season spring or early summer, gestation 13–14 months, lactation less than a year. Single calves are born at intervals of three to four years. Females believed to reach sexual maturity at roughly 15 years of age.

CONSERVATION STATUS

About 10,000 bowheads still exist, most in the western Arctic population. Numbers in the other stocks are in the hundreds or less. The once large (25,000) Svalbard stock may number only tens and is considered Critically Endangered. There is concern that habitat deterioration caused by climate change in the Arctic will impair recovery. Also, resumed hunting by Inuit in eastern

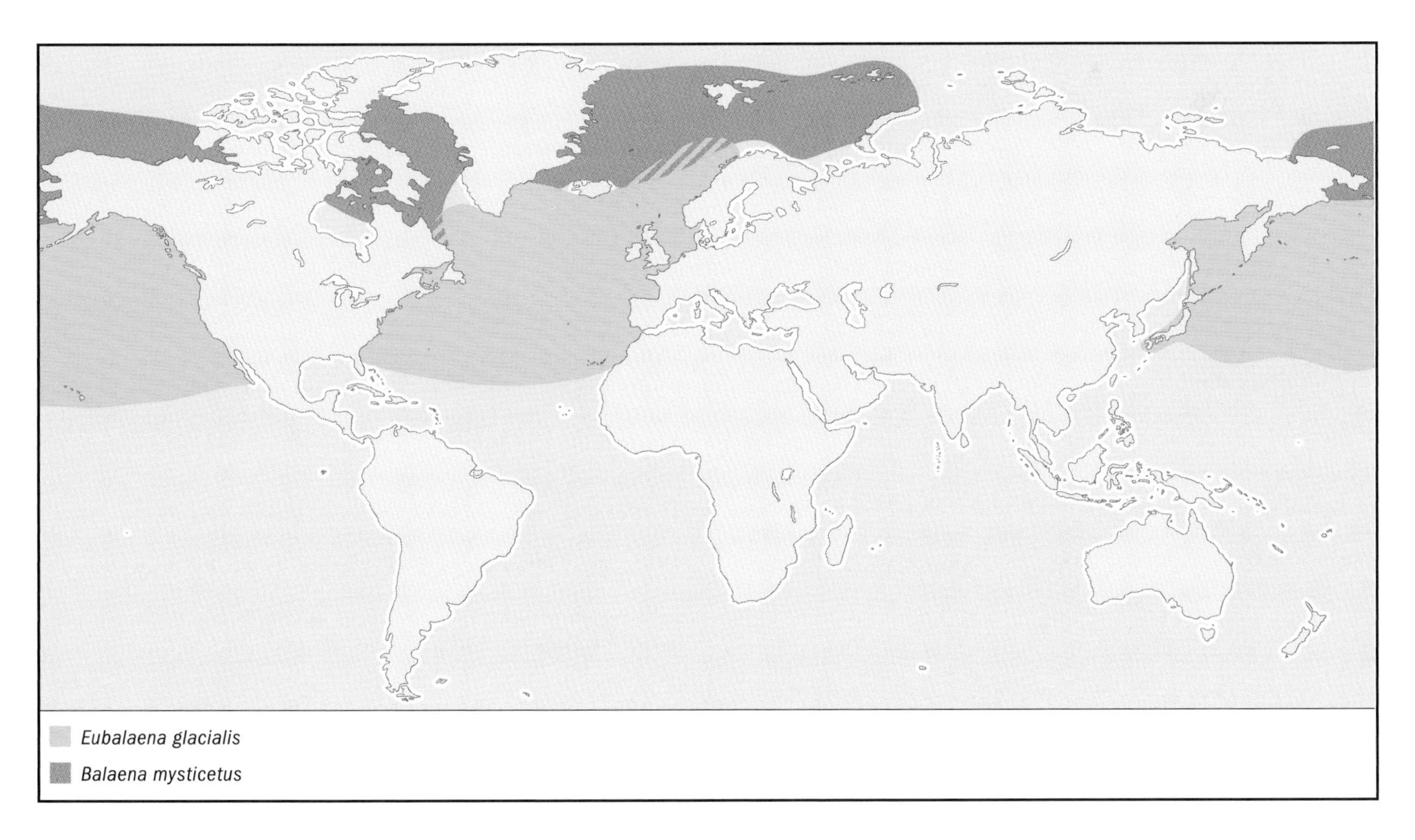

Canada is controversial because bowhead numbers there are a small fraction of what they were historically.

SIGNIFICANCE TO HUMANS
Hunting of bowheads probably influenced early human settlement patterns and was a major impetus for Arctic exploration. European countries competed for ascendancy on the Spitsbergen and Davis Strait grounds, while the American fleet dominated in Hudson Bay and the western Arctic. The species continues to be a cultural icon in some Arctic communities with a whaling tradition, and contributes to the Eskimo diet in Alaska and, to a much smaller degree, in Russia and Canada. ◆

North Atlantic right whale
Eubalaena glacialis

TAXONOMY
Eubalaena glacialis (Müller, 1776), North Cape, Norway.

OTHER COMMON NAMES
French: Baleine de Biscaye; German: Nordkaper; Spanish: Ballena franca del Atlántico Norte.

PHYSICAL CHARACTERISTICS
Length 43–53 ft (13–16 m); weight 200,000 lb (90,000 kg). One of the stockiest of all whales. Flippers are broad and tend to be more fan-shaped than for most other cetaceans. No dorsal fin or ridge on the broad back. Flukes are very wide and smoothly tapered, with a smooth trailing edge and a deep notch. Most predominantly black, but large white splotches on the belly and chin may be present. Head is covered with areas of roughened skin (callosities) to which whale lice and sometimes barnacles attach, the largest of which is called the bonnet. 200 to 270 baleen plates up to nearly 10 ft (3 m) long.

DISTRIBUTION
Originally across rim of North Atlantic from Florida in west to northwestern Africa in east, with large numbers on feeding grounds in Gulf of Maine, Gulf of St. Lawrence, off southeastern Greenland, Iceland, and northwestern Europe. Presently seen regularly only off eastern North America in coastal waters from northern Florida to Nova Scotia.

HABITAT
Shallow coastal waters in low latitudes used for calving and nursing in winter months; migratory routes partly coastal but individuals travel long distances offshore as well; summer feeding habitat in cooler northern waters with dense plankton concentrations.

BEHAVIOR
Courtship groups can involve more than 20 males boisterously competing for access to an adult female, amid much pushing, nudging, and rolling at the surface. In resting state, whales log at surface for long periods, broad backs exposed. Feeding dives last 10–20 minutes.

FEEDING ECOLOGY AND DIET
Large calanoid copepods are primary prey; also eat smaller copepods, euphausiids, pteropods, barnacle larvae, and salps. Mouth open and baleen visible while surface skim-feeding. Mud on head after surfacing from extended dive implies foraging near sea bottom in some instances.

REPRODUCTIVE BIOLOGY
Females occasionally bear first calf at five years of age, but average closer to nine to 10 years. Most calves born in winter, weaned by about one year of age. Normal calving interval about three years, but this has been increasing, which raises concern about possible reproductive dysfunction in this population.

CONSERVATION STATUS
Endangered. Only about 300–350 survive, compared with many thousands, and possibly tens of thousands historically. High incidence of ship strikes and entanglement in fishing gear along east coast of North America is preventing recovery. Other concerns include chemical pollution and genetic or demographic effects of small population size.

SIGNIFICANCE TO HUMANS
Basque and possibly Norse whalers began whaling for this species about a thousand years ago. Oil and baleen (whalebone) from these whales were valuable commodities, and shore whalers in New York and North Carolina continued to hunt them until the early twentieth century. Given its low numbers and lack of recovery, the small remnant population in the western North Atlantic now commands multi-million dollar annual investments by government agencies and conservation groups. Right whales support economically significant whale-watching in Nova Scotia and New Brunswick, eastern Canada. ◆

North Pacific right whale
Eubalaena japonica

TAXONOMY
Eubalaena japonica Lacépède, 1818, Japan.

OTHER COMMON NAMES
French: Baleine japonaise; German: Pazifischer Nordkaper; Spanish: Ballena franca del Pacífico Norte.

PHYSICAL CHARACTERISTICS
Length 46–59 ft (14–18 m); weight 220,000 lb (100,000 kg). One of the stockiest of all whales. Flippers are broad and tend to be more fan-shaped than for most other cetaceans. No dorsal fin or ridge on the broad back. Flukes are very wide and smoothly tapered, with a smooth trailing edge and a deep notch. Most predominantly black, but large white splotches on the belly and chin may be present. Head is covered with areas of roughened skin (callosities) to which whale lice and sometimes barnacles attach, the largest of which is called the bonnet. 200 to 270 baleen plates up to nearly 10 ft (3 m) long.

DISTRIBUTION
Throughout temperate and subarctic North Pacific; to central Bering Sea in north, to Baja California (Mexico) in east, to Taiwan and Bonin (Ogasawara) Islands (Japan) in west, occasionally south to Hawaiian Islands in central Pacific. Most sightings in recent years have been in southeastern Bering Sea (outer Bristol Bay) and southern Okhotsk Sea.

HABITAT
Unlike for other species of *Eubalaena*, specific near-shore calving areas have not been identified. General distribution appears to extend all across North Pacific basin, with major feeding areas (at least historically) in Gulf of Alaska, Bristol Bay, south-

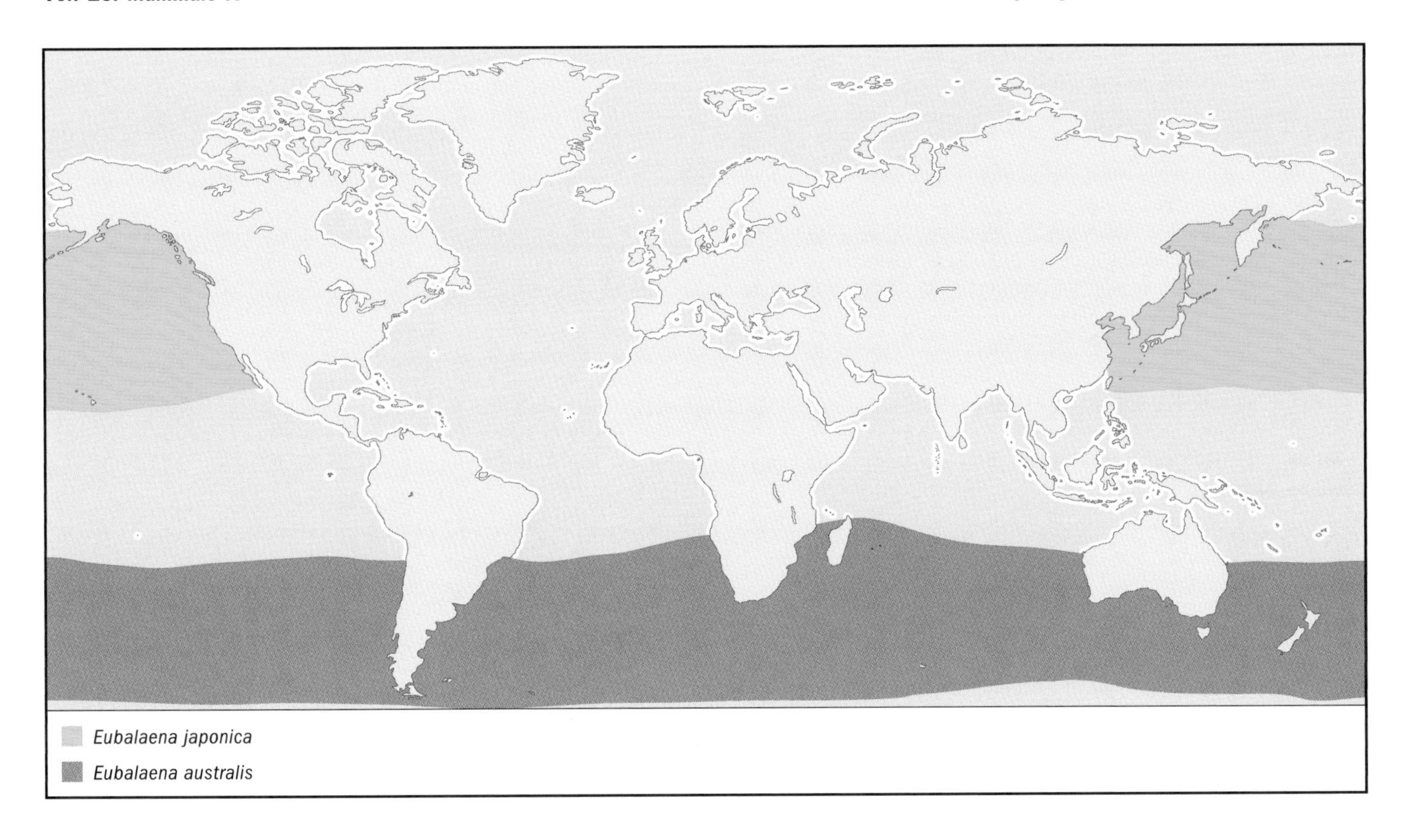

ern Okhotsk Sea, and around Aleutian and Commander Islands.

BEHAVIOR
Similar to other right whales. The few recent observations have been of small groups and lone individuals, the latter sometimes associated with humpback whales.

FEEDING ECOLOGY AND DIET
Diet dominated by calanoid copepods and larval stages of some euphausiids.

REPRODUCTIVE BIOLOGY
Similar to other *Eubalaena* species.

CONSERVATION STATUS
Commercial whalers killed 21,000–30,000 between 1840 and 1849, amounting to about 80% of the total killed between 1839 and 1909. Eastern population almost extinct, having been dealt a final blow by illegal Soviet whaling during 1960s, when more than 370 were killed in the Bering Sea and Gulf of Alaska. Now threatened by effects of small population size. Western population may still be viable, numbering in hundreds. Entanglement in fishing gear a continuing threat.

SIGNIFICANCE TO HUMANS
As true of all balaenids, hunted to very low levels by end of nineteenth century for oil and baleen. Now only significant as a focus of conservation and recovery efforts. ◆

Southern right whale
Eubalaena australis

TAXONOMY
Eubalaena australis Desmoulins, 1822, Algoa Bay, South Africa.

OTHER COMMON NAMES
French: Baleine du Cap; German: Südkaper; Spanish: Ballena franca del sur.

PHYSICAL CHARACTERISTICS
Length 43–53 ft (13–16 m); weight 200,000 lb (90,000 kg). One of the stockiest of all whales. Flippers are broad and tend to be more fan-shaped than for most other cetaceans. No dorsal fin or ridge on the broad back. Flukes are very wide and smoothly tapered, with a smooth trailing edge and a deep notch. Largely black, but may have white patches on the belly or back. Some blue-black, light brown, and nearly white individuals have been noted. Callosities on the head are present, as for all right whales. Whale lice are common in creases and folds of the body. 200 to 270 baleen plates up to nearly 10 ft (3 m) long.

DISTRIBUTION
Circumpolar in temperate to subpolar Southern Hemisphere; South Africa, Namibia, southern Mozambique, Madagascar (formerly at least), southern Angola, western and southern Australia, New Zealand (formerly at least), Chile, Argentina, southern Brazil, around numerous oceanic islands such as Crozet, Kerguelen, Amsterdam, and St. Paul (France), Prince Edward (South Africa), Auckland and Campbell (New Zealand), Falkland, South Georgia, Tristan, and Gough (United Kingdom).

HABITAT
Very widely distributed from near-shore waters to pelagic zone, and from temperate waters south to Antarctic. Main determinant of offshore habitat appears to be availability of dense concentrations of zooplankton. Near-shore wintering grounds typically have a gently sloping sandy bottom, relatively sheltered. Avoidance of unwanted attention by male suitors may help explain female's choice of nursery.

BEHAVIOR
Migratory, moving south and offshore in summer, north and inshore in winter, with at least portions of population congregating in coastal calving areas. Generally occur in small groups with no obvious social structure apart from close affiliation between mothers and calves. Aggregations form in productive feeding areas and when numerous males attempt to mate with focal female. Breaching (leaping clear of surface) and lobtailing (slapping water with flukes) are common on wintering grounds. Behavior called "tail-sailing" observed off Patagonia, with flukes high above surface acting as a sail to propel whale horizontally. Playful and curious behavior toward buoys, tide gauges, and kelp fronds.

FEEDING ECOLOGY AND DIET
Zooplankton, mainly copepods and euphausiids, must be found in dense concentrations to allow right whales to feed efficiently. Whales surface skim-feeding on krill (*Euphausia superba*), which are relatively fast-swimming and adept at predator avoidance, can engage in high-speed bursts (8 knots) and create considerable turbulence.

REPRODUCTIVE BIOLOGY
Average age at first calving about nine years, some individuals giving birth at six, others not until 13 years old. Single calves are born at intervals of three years. Calving occurs over a period of about four months during austral winter. Calves closely associate with mothers for at least several months but are weaned by one year of age. Gestation assumed to last about 12 months.

CONSERVATION STATUS
Lower Risk/Conservation Dependent. Although severely depleted by commercial whaling throughout range, strong recoveries underway in some areas, notably southern Africa, Argentina, and Australia. Numbers there total approximately 7,000, with annual increase rates of 7–8%. In other areas such as New Zealand, Chile, and Madagascar, there is little or no evidence of recovery. An important factor limiting recovery was unreported and only recently disclosed: illegal killing of more than 3,200 southern right whales by Soviet factory ships between 1951 and 1970, a period during which the species was legally protected.

SIGNIFICANCE TO HUMANS
Like other right whales, hunted relentlessly for oil and baleen. For last 30 years, interest in whale-watching has grown rapidly in South Africa, Argentina, and Australia. Marine-protected areas exist in all three countries to protect winter concentrations of right whales and facilitate exploitation as objects of tourism and study. ◆

Resources

Books

Best, P. B., J. L. Bannister, R. L. Brownell Jr., and G. P. Donovan, eds. "Right Whales: Worldwide Status." *Journal of Cetacean Research and Management*, Special Issue 2. Cambridge, UK: International Whaling Commission, 2001.

Burns, J. J., J. J. Montague, and C. J. Cowles, eds. *The Bowhead Whale*. Lawrence, KS: Special Publication No. 2. Society for Marine Mammalogy, 1993.

Brownell, R. L. Jr., P. B. Best, and J. H. Prescott, eds. "Right Whales: Past and Present Status." *Report of the International Whaling Commission*, Special Issue 10. Cambridge, UK: International Whaling Commission, 1986.

Kenney, R. D. "North Atlantic, North Pacific, and Southern Right Whales *Eubalaena glacialis, E. japonica*, and *E. australis*." In *Encyclopedia of Marine Mammals*, edited by W. F. Perrin, B. Würsig, and J. G. M. Thewissen. San Diego: Academic Press, 2002.

McLeod, S. A., F. C. Whitmore Jr., and L. G. Barnes. "Evolutionary Relationships and Classification." In *The Bowhead Whale*, edited by J. J. Burns, J. J. Montague, and C. J. Cowles. Lawrence, KS: Special Publication No. 2, Society for Marine Mammalogy, 1993.

Reeves, R. R., and R. L. Brownell Jr. "Baleen Whales *Eubalaena glacialis* and Allies." In *Wild Mammals of North America: Biology, Management, and Economics*, edited by J. A. Chapman, and G. A. Feldhamers. Baltimore: Johns Hopkins University Press, 1982.

Reeves, R. R., and R. D. Kenney. "Baleen Whales: The Right Whales, *Eubalaena* spp., and Allies." In *Wild Mammals of North America: Biology, Management, and Conservation*, 2nd ed., edited by G. A. Feldhamer, B. C. Thompson, and J. A. Chapman. Baltimore: Johns Hopkins University Press, in press.

Rugh, D. J., and K. E. W. Shelden. "Bowhead Whale *Balaena mysticetus*." In *Encyclopedia of Marine Mammals*, edited by W. F. Perrin, B. Würsig, and J. G. M. Thewissen. San Diego: Academic Press, 2002.

Randall Reeves, PhD

Rorquals

(Balaenopteridae)

Class Mammalia
Order Cetacea
Suborder Mysticeti
Family Balaenopteridae

Thumbnail description
Large-sized whales with furrows or pleats on their undersides, and baleen instead of teeth. Unlike other baleen whales, rorquals have pointed dorsal fins, longer, streamlined bodies, and relatively small heads

Size
32–102 ft (10–31 m); 22,000–400,000 lb (9,980–181,440 kg)

Number of genera, species
2 genera; 7 or 8 species

Habitat
Marine, estuaries, aquatic, deep, benthic, and pelagic

Conservation status
Endangered: 3 species; Vulnerable: 1 species; Lower Risk/Conservation Dependent: 1 species; Lower Risk/Near Threatened: 1 species; Data Deficient: 1 species

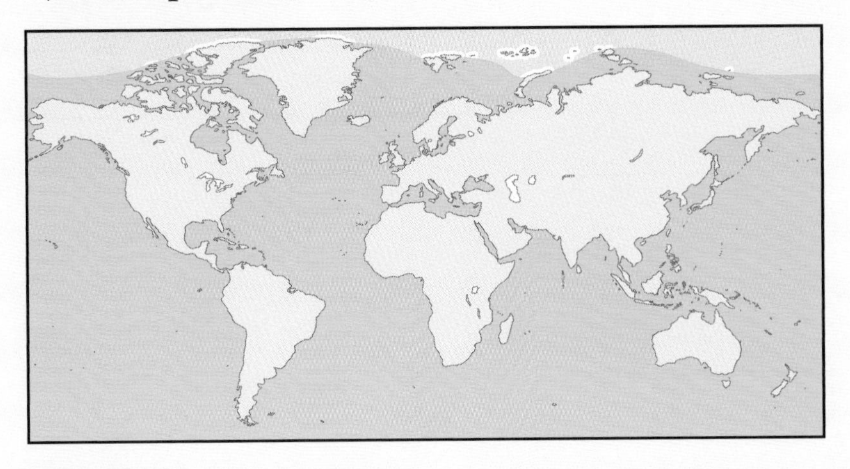

Distribution
All oceans and adjoining seas of the world

Evolution and systematics

The oldest fossil whales are often grouped together in a taxon known as the archaeocetes. They exhibit several features that modern whales lack, such as teeth of different types and nostrils near the tip of the nose. For many years, paleontologists thought that whales evolved from a group of now-extinct, wolf-like hoofed mammals called mesonychians. The similarities in the shape and construction of the skull and in the shape of the teeth were the best evidence for this. However, several phylogenetic studies of gene sequences of living mammals have argued that whales are most closely related to artiodactyls, which are the hoofed mammals with an even number of toes, such as cows, pigs, camels, deer, and hippopotamus. In fact, the genetic evidence suggests that whales are most closely related to hippos and are thus actually a subgroup of artiodactyls. One of the main synapomorphies (a shared character that originated in their last common ancestor) linking all living and extinct artiodactyls is an anklebone that has rounded, pulley-shaped joints on each end. This anklebone allows only front-to-back motion between the leg and ankle, and the ankle and toes. The front-and-back motion is well suited for efficient running. Recent descriptions of the ankle bones of Eocene whale species from Pakistan and India show that whales have the double-pulley anklebone of artiodactyls, suggesting that the ancestor of whales and hippos may have ventured into the water more than 55 million years ago.

In 1978, the earliest known well-preserved cetacean, a 52-million-year-old skull, was discovered in Pakistan. The new bones, dubbed *Pakicetus*, proved to have key features that were transitional between terrestrial mammals and the earliest true whales, including an ear that was modified for directional hearing underwater. An amphibious animal, *Pakicetus* was found in near-shore marine sediments. *Basilosurus* is another Eocene archaeocete, from the Gulf Coast, which retained tiny hind limbs that projected from the body, although there was no joint between the pelvic bones and the vertebrae.

By the late Oligocene, the two modern lineages of cetaceans had evolved from archaeocete ancestors. The late Oligocene whale, *Aetiocetus* from Oregon, has skull and jaw features typical of baleen whales and is considered the earliest mysticete, yet it also bares a full set of teeth. By the late Miocene, rorquals were relatively common fossils in many marine deposits.

Gray created this family (the Balaenopteridae) in 1864 to include all the rorqual whales—defined as those that have a number of throat grooves. All except the humpback whale (*Megaptera novaeangliae*) share a strong family resemblance.

Physical characteristics

Rorqual whales are relatively streamlined in appearance and have pointed heads and small pointed fins. They can be distinguished from other whales by many (25–200) deep groves along their throats that expand when they feed. The tongue is soft and fleshy, well adapted for licking food off

A fin whale (*Balaenoptera physalus*) expelling through its blowhole. (Photo by François Gohier/Photo Researchers, Inc. Reproduced by permission.)

their baleen. There are eight species of rorqual whales: humpback whale (*Megaptera novaeangliae*), fin whale (*Balaenoptera physalus*), Bryde's whale (*Balaenoptera edeni*), blue whale (*Balaenoptera musculus*), northern minke (*Balaenoptera acutorostrata*), Antarctic minke (*Balaenoptera bonaerensis*), and sei whale (*Balaenoptera borealis*). *Balaenoptera edeni* may represent two species, *B. edeni* and *B. brydei*, though this is not yet determined.

Rorquals range in size from 32 to 102 ft (10–31 m) and weigh 22,000–400,000 lb (9,980–181,440 kg). Females grow slightly larger than males.

Major distinctive features of the rorquals are a white right lower lip and a white edge on the upper jaw in fin whales; a single rostral ridge extending from the base of the blowhole in sei whales; three prominent ridges on the rostrum (upper jaw or snout) in Bryde's whale; and a triangular-shaped rostrum with a single prominent ridge in the minke whales. The blue whale, the largest of all whales, possesses a small dorsal fin, a flat rostrum that appears U-shaped when viewed from above, and a tall, dense spout. In the Antarctic, a yellowish film of diatoms is often present on the ventral and lateral surfaces of these whales, prompting the whaler's term "sulfur-bottom."

Distribution

Bryde's whale is found in tropical and temperate waters around the world. They are especially abundant in areas of high food productivity. In the western Pacific, Bryde's whale occurs from Japan to New Zealand, and in the eastern Pacific, from Baja California to Chile. In the northeast Pacific, they move between Bonin Islands and the coast of Japan, west Kyushu, and further north. In the Atlantic, the species is reported from Virginia to the Gulf of Mexico and the Caribbean, south to Brazil. In the east, reports range from the Canaries and Morocco south to the Cape of Good Hope. In the Indian Ocean, their north-south range is from the Persian Gulf to the Cape of Good Hope, and from Myanmar to Australia.

Minke whales, the most widespread of the rorquals, are found in tropical, temperate, and polar waters of both hemispheres. The species is frequently seen in inshore northern and western coastal waters of the United Kingdom, and occasional records have been reported from the channel coast of mainland Europe, the Mediterranean, and the Azores and Portugal. In the Pacific, they range from the tropics (Vietnam, Baja California) to the Bering Sea. During summer, minke whales are found from temperate waters all the way up

A blue whale (*Balaenoptera musculus*) calf breaching. (Photo by François Gohier/Photo Researchers, Inc. Reproduced by permission.)

to the ice pack. Their winter movements are poorly known; some may stay in temperate waters year-round, and there is recent acoustic evidence that some minke whales in the North Atlantic may move into tropical waters in the Caribbean during winter. Within their range, they are widely distributed, and are found over a more widespread area than their larger relatives.

Fin whales migrate to polar waters in summer for feeding and return to warmer seas in winter for breeding. Photo-identification work indicates that fin whales in the North Atlantic have been detected to move throughout the New England/Nova Scotia region, but have never been sighted off of Newfoundland or the Gulf of St. Lawrence. Similarly, genetic work indicates that fin whales in the Mediterranean, on the other side of the North Atlantic, are a separate population. No wintering concentration area is known anywhere in the world; the speculation is that these animals go to deep waters and disperse. There is a year-round resident group in the Gulf of California in Mexico.

In the eastern North Pacific, fin whales winter from at least central California southward, and summer from central Baja California into the Chukchi Sea. In the western North Pacific they winter in the Philippine Sea, including concentrations in the East China Sea and the Sea of Japan. In the western North Atlantic, they winter from the ice-edge south to Florida and the Greater Antilles, and into the Gulf of Mexico, primarily in offshore waters. They summer from below the latitude of Cape Cod to the Arctic Circle. They are present in the Mediterranean Sea and in the eastern North Atlantic from the Strait of Gibraltar to southwestern Norway in winter. Although fin whales are in the Mediterranean Sea year-round, they apparently migrate to more northerly waters along the eastern European coasts. In the southern hemisphere, fin whales migrate from summering grounds in the Antarctic past New Zealand into the southwestern Pacific, along South America to Peru on the west coast and Brazil on the east coast, to the central Atlantic off the west coast of Africa, and to the southern Indian Ocean.

Sei whales are largely oceanic, and widely distributed in temperate and polar waters of both hemispheres. Blue whales are found closer to shore, often near deep coastal canyons. Humpback whales migrate between their breeding and feeding grounds, moving mainly along the continental coasts in the Northern and Southern Hemispheres.

Habitat

Rorquals are found in open seas, but mainly over continental shelves, and sometimes in bays, inlets, and estuaries.

A minke whale (*Balaenoptera acutorostrata*) surfacing. (Photo by François Gohier/Photo Researchers, Inc. Reproduced by permission.)

Behavior

The social organization of rorquals is still poorly understood. Groups generally only include two to five individuals, although larger, temporary aggregations occur on rich feeding grounds and during the breeding season. Rorquals are not deep divers, generally feeding within 330 ft (100 m) of the surface.

Because of the loud, low-frequency sounds made by fin whales, animals may remain in vocal contact over long distances, making it difficult to know when whales are or are not associated. The fin whale is sometimes called the "greyhound of the sea" because of its fast swimming speed; it can swim up to 23 mph (37 km/hr) in short bursts.

Some minke whales undertake lengthy migration, totaling 5,590 mile (9,000 km) or more, but others may move little. There are many regions where minkes are found year-round. However, seasonal variation in abundance and distribution suggests that the whales probably do undergo some migration, from higher latitudes in summer to lower latitudes in winter. Pregnant females seem to move farther north in summer than lactating and immature females, but in some temperate waters these animals are present year-round.

Long migrations are probably not typical of Bryde's whales, although there are indications that some animals may shift towards the equator in winter and toward more temperate waters in summer. They are not, however, one of the species that frequents the Antarctic Ocean each year. Bryde's whales mainly make local seasonal movements, and may form resident populations in some regions. In certain areas, for example, off South Africa, two forms of the species are found: one is resident year-round, and found within 25 miles (40 km) of the coast, and the other generally occurs about 62 miles (100 km) from the shore and appears there in autumn and spring. The offshore form undertakes north-south migrations as it follows shoals of fish throughout the year. In the northwest Pacific, Bryde's whale moves from the Bonin Islands north to the coast of Japan, a seasonal migration of only about 342 miles (550 km). In the Gulf of California, Bryde's whales probably make limited north-south migrations on a local scale, following sardine and herring concentrations. They seem to be relatively resident in the area year-round.

Minke whales are more likely to be seen close-up than other rorquals, as they often approach boats, especially stationary vessels, and are notoriously inquisitive. They are fast moving and may swim at speeds in excess of 13 miles (20 km) per hour. The surfacing and blow rates of minke whales tend to be less regular than those of the large baleen whales, and may be affected by the presence of vessels, time of day, activity of the animal, and/or the environmental conditions. A typical dive sequence is five to eight blows, at intervals less than one minute, followed by a dive, lasting from two to six minutes, although minkes can stay underwater for 20 minutes or longer. They are also known to breach more often than other baleen whales, leaping clear of the surface and reentering the water head-first or with a splash.

Bryde's whales also seem to breach often and, when feeding, often change direction and splash and roll around at the surface. They generally take four or five short breaths before starting a long dive. They may dive for 20 minutes or so, and rarely show their tail flukes as they dive. They often surface steeply, like the fin whale, with the blow becoming visible well before the dorsal fin is exposed. They have been observed to exhale underwater, surfacing with little or no visible blow. They are known to be inquisitive and sometimes approach boats, circling them or swimming alongside. Its lifespan is approximately 50 years.

A minke whale (*Balaenoptera acutorostrata*) near Great Barrier Reef, Australia. (Photo by François Gohier/Photo Researchers, Inc. Reproduced by permission.)

A humpback whale (*Megaptera novaeangliae*) bubble-net feeding (cooperative feeding) in Alaska. (Photo by François Gohier/Photo Researchers, Inc. Reproduced by permission.)

Rorquals produce four types of sounds: low-frequency moans, including the "songs" of the humpback whale; grunt-like thumps and knocks of short duration; high-frequency chirps, cries, and whistles; and low-frequency clicks or pulses. Though few sounds are known to be linked with specific behaviors, it is thought that the sounds are social, for greeting, courtship, threat, individual identification, and other purposes. The origin of these sounds is suspected to be the larynx, although whales have no vocal cords.

Humpback whale "songs" are sung only by solitary males, and they all sing the same song during each breeding season. The specific function of the song is not known, but likely communicates information to other male and female humpbacks. When the solitary males join social groups they no longer sing their song.

Feeding ecology and diet

Rorquals feed with the help of their baleen, curtains of horny fronds hanging from the top of their giant, bowed upper jaws. Most rorquals drop their pleated lower jaws and engulf schools of small fish or invertebrates. Closing their mouths, they ram their tongues against the baleen, squeezing out the water through their lips, while leaving the food behind. Dives for food rarely last longer than 10 minutes, to depths less than 660 ft (200 m).

The main food of rorquals is various species of krill (euphasiids). Rorquals also eat animals such as small squid, lantern fish (Myctophidae), and certain amphipods. Sei whale prefer copepods if available. Bryde's whales feed on krill in pelagic waters and fish in coastal areas. Blue whales feed almost exclusively on swarms of krill. However, off Baja California, Mexico, they also eat seagoing crabs in the winter. In the Antarctic, daily food consumption for a single blue whale is up to 8 tons (7.3 tonnes) of krill.

In certain feeding areas, one or two humpbacks swim in an upward spiral around swarms of krill found on or below the surface. As they circle the krill, they expel a chain of bubbles from their blowholes. The rising bubbles form a "bubble net" that forces the krill to mass to the center of the bubbles.

Because of their relatively small size, and lowered energetic needs, minke whales consume a wider variety of fish than the larger fin and humpback whales. At times, they may even take single larger fish rather than large quantities of smaller fish. Feeding minke whales are often seen near the surface chasing fish. The species has been reported to feed in one of two ways: lunge feeding or "bird association" feeding (depending mainly on the feeding areas). Most individuals seem

A humpback whale (*Megaptera novaeangliae*) family feeding, Inside Passage, southeasten Alaska, USA. (Photo by John Hyde. Bruce Coleman, Inc. Reproduced by permission.)

to specialize in just one of these methods. Bird-associated foraging exploits the concentration of fish fry below flocks of feeding gulls and auks, while lunge feeding consists of the whale actively concentrating the prey against the air-water interface with no feeding birds involved. The minke whale is often seen turning on its side when lunge feeding. In the North Pacific, the minke whale feeds on krill and sand lance.

Bryde's whale often exploits the activities of other predators, swimming through and engulfing the fish they have herded. Therefore, it is frequently found in areas of high fish abundance, along with seabirds, seals, sharks, and other cetaceans. When feeding, Bryde's whales often roll onto their sides or churn the water at the surface by pinwheeling or halfheartedly breaching. They also frequently accelerate and change direction suddenly. So their movements when feeding look more like those of feeding dolphins than those of the other large whales. Bryde's whales feed actively year-round.

Reproductive biology

The entire reproductive cycle is correlated with the migrations of the whales between rich feeding areas and breeding/calving grounds. Mating usually occurs with the pair swimming on their side, belly to belly. The male's testes are retained permanently inside the abdominal cavity. Most rorquals have a polygamous mating system, with both males and females being promiscuous.

The gestation period is a year or slightly longer in all rorquals, except in minkes, which is about 10 months. The mean length at birth ranges from 9 ft (2.7 m) for the minke whale to about 23 ft (7m) for the blue whale. Only a single young is born. The teats of the mammary glands are found within paired slits on either side of the female reproductive opening. Contact with the teats during suckling causes the milk to spurt freely into the mouth of the calf. The milk of rorquals is unusually high in fat content: 30–53%. The high-fat content likely accounts for the rapid growth of the calf during the suckling period (usually one year), during which it can increase its body weight five to eight times. The larger, mature rorquals breed every two to three years; smaller minke whales breed almost every year. Sexual maturity is attained in both sexes between five and 15 years. Whales in depleted populations attain sexual maturity at an earlier age than those in populations at their carrying capacity. Rorquals average a life span of 50 to 80 years.

Conservation status

In 1946, 20 whaling nations set up the International Whaling Commission (IWC) in an attempt to regulate whale hunt-

An adult humpback whale (*Megaptera novaeangliae*) near Cierva Cove, Antarctic Peninsula. (Photo by Rod Planck/Photo Researchers, Inc. Reproduced by permission.)

A humpback whale (*Megaptera novaeangliae*) breaching, Frederick Sound, Inside Passage, southeast Alaska, USA. (Photo by John Hyde. Bruce Coleman, Inc. Reproduced by permission.)

ing to stop over-fishing. It collected data on the number of whales, though the numbers came mostly from the whalers themselves. The commission set annual quotas for the number of whales to be killed. These quotas, however, were nonbinding and could not be enforced. Furthermore, some whaling nations did not belong to the IWC. The blue whale, for example, was not completely protected by the IWC until the 1965–1966 season, long after its numbers had been drastically reduced. And even under the protection of the IWC, blue whales were hunted at least until 1971 by the fleets of countries that did not belong to the IWC.

Under mounting pressure from conservationists, the IWC gradually banned the hunting of other whales. The United States Congress separately passed the Marine Mammal Protection Act of 1972, which bans the hunting of all marine mammals (except in the traditional fisheries of Alaskan natives) and the importation of their products. By 1974, the IWC had included the blue whale and the humpback whale under its protection. Minke whales, sei whales, and fin whales were still being hunted in large numbers, but worldwide catches began to dwindle. Catches fell from 64,418 in 1965 to 6,623 in 1985. A moratorium on all commercial whaling was finally declared by the IWC in 1985, a move considered long overdue by conservation groups. Japan, Iceland, and Norway, however, opted in 1988 to continue to hunt minke, fin, and sei whales, a fishery permitted by the IWC under the controversial guise of "scientific whaling."

The fin whale, blue whale, and sei whale are listed as Endangered by the IUCN; the humpback whale is Vulnerable; the northern minke whale is Lower Risk/Near Threatened; the Antarctic minke whale as Lower Risk/Conservation Dependent; and Bryde's whale as Data Deficient.

Significance to humans

Historically, demand for whale blubber, mostly oil used in the manufacture of margarine, soaps, and lubricants, and as a substitute for kerosene, was high. By the 1980s, artificial substitutes had been found for whale oil. Whale meat, however, remains valued as both a pet food and as human food, mostly in the *kujiraya*, or whale-meat bars, of Japan. About 70% of the cetacean products from market surveys in Japan and South Korea have proven to be from minke whales (about 19% *Balaenoptera acutorostrata* and 51% *B. bonaerensis*). The Japanese surveys included collections organized by Earthtrust, Whale and Dolphin Conservation Society, TRAFFIC Japan, and Greenpeace Germany from 1999 and 1998.

Blue whales, the largest of them all, were especially sought for their blubber. A large specimen yielded more than 9,000 gal (34,000 l) of oil. It has been estimated that more than 200,000 blue whales were taken worldwide between 1924 and 1971, close to 30,000 during the 1930–1931 whaling season alone. Soon catches pushed way above optimal yield level. As many as 80% of all blue whales caught by 1963 were sexually immature, meaning that there were even less individuals in the ocean to perpetuate the species.

Fin whales, the second largest of all whales, became the next major target as blue whales became scarcer. The 1950s and early 1960s saw annual catches of 20,000–32,000 fin whales per year, mostly from Antarctica. As their stocks dwindled, whalers shifted once again in the mid-1960s, this time to the smaller sei whale.

Whale watching has become, in many cases, an economically beneficial alternative to hunting. In 2000, it attracted some nine million enthusiasts in 87 countries, and generated a record-breaking $1 billion in revenue, according to the World Wildlife Fund. The income earned by the industry has doubled in only six years. In Iceland, whale-watching passenger numbers have grown from just 100 in 1991 to 44,000 in 2000.

1. Northern minke whale (*Balaenoptera acutorostrata*); 2. Fin whale (*Balaenoptera physalus*); 3. Bryde's whale (*Balaenoptera edeni*). (Illustration by Brian Cressman)

Species accounts

Fin whale
Balaenoptera physalus

SUBFAMILY
Balaenopterinae

TAXONOMY
Balaenoptera physalus (Linnaeus, 1758), Spitsbergen Sea, near Svalbard, Norway.

OTHER COMMON NAMES
English: Finback, herring whale, razorback; French: Baleine fin, rorqual commun; Spanish: Ballena aleta, ballena boba.

PHYSICAL CHARACTERISTICS
Grows to 78 ft (24 m) and 88.5 ft (27 m) in the Northern and Southern Hemispheres, respectively. A full-grown adult weighs about 30–80 tons (27–73 t). Females are generally larger than the males. The dorsal fin—which often slopes backwards—is set about two-thirds back along the body. The flukes are broad and triangular; the head is pointed. It is dark gray to brownish black, with white undersides. Has an asymmetrical head; the bottom lip is dark on the left side and white on the right side. There are 520–950 baleen plates per animal, the largest of which is 35 in (90 cm) in length, and 50–200 pleats on the lower jaw that expand during feeding.

DISTRIBUTION
Distributed worldwide, with three major distinct populations: the North Atlantic, North Pacific, and southern oceans.

HABITAT
Rare in tropical waters and among pack ice; is rarely seen inshore.

BEHAVIOR
Gregarious, and are usually found either in pairs (as in mother and calf) or in groups of six to 10 animals. Although individuals are also common, congregations of approximately 100 can be found on the feeding grounds. Dives to a maximum of about 984 ft (300 m) and communicates via moans, pulses, clicks, and grunts, as well as breaching.

FEEDING ECOLOGY AND DIET
A wide variety of small fish, with some krill (their primary diet in the southern hemisphere). Some fish, such as herring and capelin, as well as squid, are also taken as food. It is unknown whether this species fasts through the winter months.

REPRODUCTIVE BIOLOGY
Born during the winter at 15–18 ft (4.6–5.5 m) and approximately 3,000 lb (1,360 kg) after a 12-month gestation. The calf stays with its mother for six to eight months. Maturity is believed to take place at six to eight years of age, and females produce a single calf every two to five years. May live 60–100 years.

CONSERVATION STATUS
Fin whales were killed extensively once whalers had virtually extinguished blue whales. Between the 1930s and the 1960s,

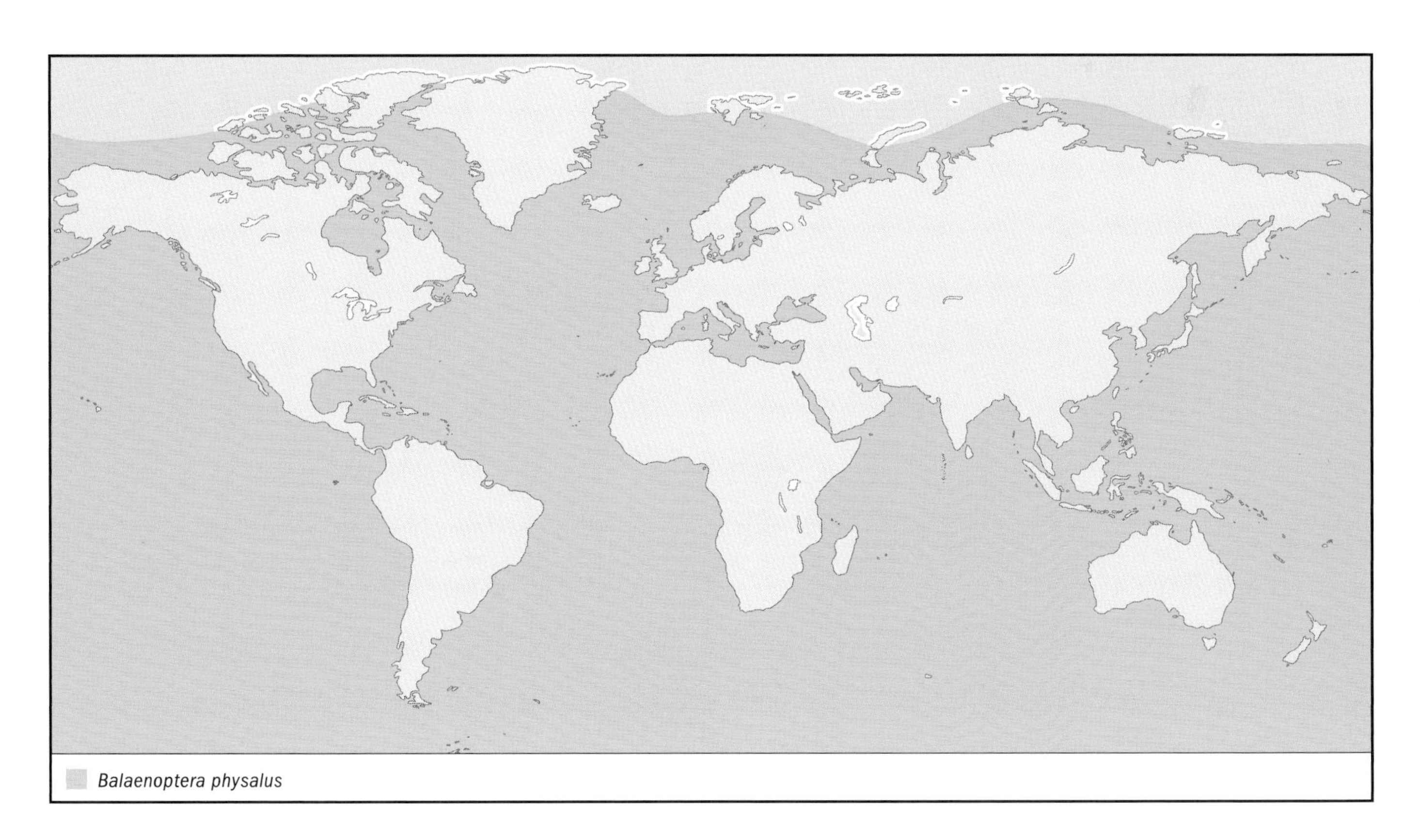

more than 500,000 fin whales were killed worldwide, mostly in the Antarctic. Although whaling for fin whales took place as recently as 1989, kills were highly limited after 1970. Now protected worldwide, fin whales are estimated to number 50,000–90,000, and are listed as Endangered on the IUCN Red List. Currently, the largest threats to fin whales are development and habitat destruction, entanglement, and the interest in several countries for resumed whaling.

SIGNIFICANCE TO HUMANS
None known. ◆

Northern minke whale
Balaenoptera acutorostrata

SUBFAMILY
Balaenopterinae

TAXONOMY
Balaenoptera acutorostrata Lacépède, 1804, Manche, France. A subspecies was identified by Burmeister in 1867.

OTHER COMMON NAMES
English: Piked whale, finner, lesser rorqual; French: Petit rorqual; Spanish: Ballena minke.

PHYSICAL CHARACTERISTICS
Smallest of the rorqual whales, ranging 26–33 ft (8–10 m) as adults and weighing about 5–8 tons (4.5–7.2 t), they are sleek, small, and dolphin-like; a streamlined body and a tall, falcate dorsal fin; the rostrum is very narrow and pointed, with a single ridge from the blowhole. They sport a white stripe across each flipper; sometimes also have a light chevron on the back, behind the head, and two regions of light gray on each side. The broad tail flukes may be pale gray, blue-gray or white on the underside, usually with a dark margin. The baleen plates (between 230 and 330 pairs) are white, gray, or cream. Between 50 and 70 thin ventral pleats. Generally, they are black, gray, or brown dorsally and light ventrally.

DISTRIBUTION
Distributed from the tropics to the ice edges worldwide, with two major distinct populations: the North Atlantic and North Pacific. They range from Florida to Labrador and Greenland and from North Africa to north of Spitsbergen.

HABITAT
Of all baleen whales, they are found closest to the edge of the polar ice, sometimes entering the ice fields. In general, they approach close to shore and often enter bays, inlets, and estuaries.

BEHAVIOR
Almost always seen by themselves or in pairs or threes, although they appear to aggregate in concentrations that can number up 50 in productive food areas. While true side-by-side associations are unusual, they may work in small bands where individuals stay in each other's general vicinity. Have feeding strategies that are specialized within the locality.

FEEDING ECOLOGY AND DIET
Feed on whatever food source is most abundant in a given area, primarily krill and small schooling fish, but occasionally larger fish such as mature Arctic cod and haddock. In the North Atlantic, the northern minke whale is known to feed on sand lance, sand eel, krill, salmon, capelin, mackerel, cod, herring, and a number of other fish species.

REPRODUCTIVE BIOLOGY
Very little is known. Breeding may occur throughout the year, but there seems to be a calving peak in winter. The gestation

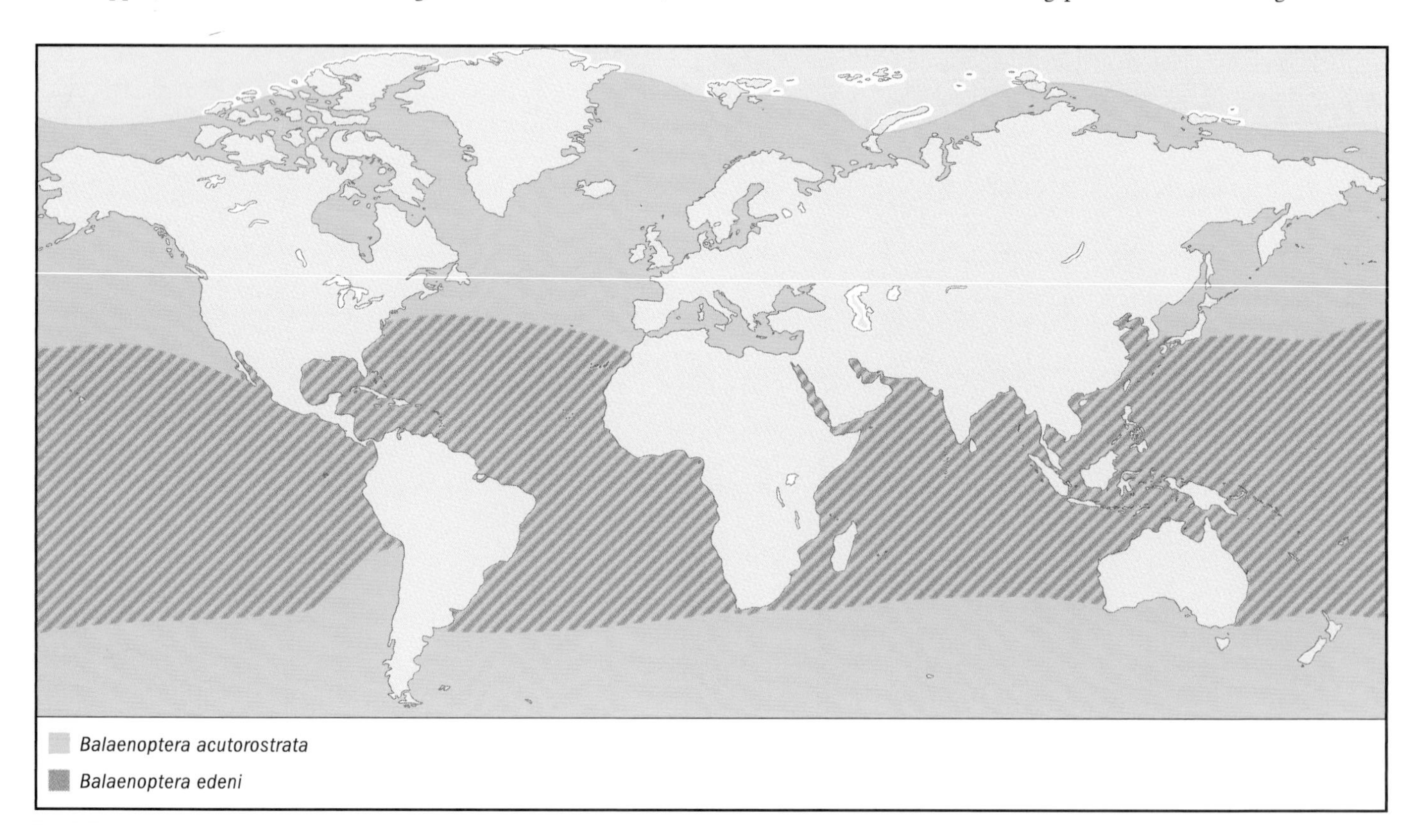

period is thought to be about 10 months and lactation probably occurs for three to six months. The newborn calf is only about 8.5 ft (2.6 m) long and stays with its mother for about two years. In the Pacific, females are thought to give birth to one calf at a time once every one to two years. Females become sexually mature at ages six to eight and males at five to eight years. In the North Atlantic, females may give birth every year. The age at sexual maturity has been estimated at 7.1 years in females and six years in males. They live to approximately 60 years.

CONSERVATION STATUS
The most abundant whale in the world today, numbering between 103,000–204,000 animals. They are listed as Low Risk/Conservation Dependent on the IUCN's Red List. They are still commercially hunted by Norway, where about 500 are killed per year, and the Japanese take up to 800 minke whales per year. Current threats include expansion of the current hunt (which is being promoted by Japan, Norway, and Iceland), entanglements in fishing gear, and degradation of their habitat from pollution.

SIGNIFICANCE TO HUMANS
Despite the IWC's Moratorium on Commercial Whaling that came into effect in 1986, meat from these whales—and many other species—still ends up on many butchers' slabs. The meat is considered a delicacy in Japan, where it sells for several hundred dollars per pound. The Korean fishery is now the largest coastal minke whale fishery in the world. Individuals are also taken sporadically for food by Eskimos on St. Lawrence Island, Alaska. ◆

Bryde's whale
Balaenoptera edeni

SUBFAMILY
Balaenopterinae

TAXONOMY
Balaenoptera edeni Anderson, 1879, Burma. (Pronounced "brude's.")

OTHER COMMON NAMES
English: Tropical whale; French: Rorqual de Bryde; Spanish: Ballena de Bryde.

PHYSICAL CHARACTERISTICS
Length 43–50 ft (13–15.3 m); females are slightly larger than males. The color is variable, but usually the dorsal side is bluish black and the ventral side white or yellowish. A dark bluish gray area extends from the throat to the flippers. The flippers are slender and somewhat pointed. The dorsal fin is pointed and falcate. The ventral grooves extend to the umbilicus. Unique to the Bryde's whale is the presence of two lateral ridges that run from the tip of the snout to the blowholes. The baleen is about 7.5 in (19 cm) wide and about 20 in (50 cm) long. The inner margin is concave. They usually have 250–280 fully developed baleen plates. In some populations, whitish gray, oblong spots occur over much of the body surface, which may be scars from parasites or sharks.

DISTRIBUTION
Found in the tropical and temperate areas of the southern Pacific, Atlantic, and Indian Oceans. On the northern hemisphere, this species can be found in the tropical and temperate areas of the Pacific and the western Atlantic, as well as in the Indian Ocean.

HABITAT
Tropical to warm temperate inshore and offshore waters, following food.

BEHAVIOR
Either swim alone or in pairs; the largest group sizes of 10–23 animals are usually in loose congregations when feeding. Dives to a maximum of 980 ft (300 m) and communicates via moans, pulses, clicks, and grunts, as well as breaching. Its life span is approximately 50 years.

FEEDING ECOLOGY AND DIET
Feeds predominantly on krill, schooling fish such as pilchards, anchovies, herring, and mackerel; also bonito, shark, and squid.

REPRODUCTIVE BIOLOGY
Females become sexually mature at 10 years of age. Males become sexually mature at the age of 9–13 years. Breed throughout the year; gestation lasts about one year. Calves are weaned at about six months. Females probably give birth less than once every two years.

CONSERVATION STATUS
The IUCN recognizes the species as Data Deficient. Population size is estimated between 40,000–80,000 animals.

SIGNIFICANCE TO HUMANS
Bryde's whales are regularly captured in the artisanal whale hunt of the Philippines and in Indonesia. They have only been systematically exploited in this region of the world and this ceased when the IWC's Moratorium on Commercial Whaling was introduced in 1986. Therefore, Bryde's whale is not believed to be in danger nor at depleted levels. ◆

Common name / Scientific name/ Other common names	Physical characteristics	Habitat and behavior	Distribution	Diet	Conservation status
Blue whale *Balaenoptera musculus* French: Baleine bleue, rorqual bleu; German: Blauwal; Spanish: Ballena azul, rorcual azul	Largest mammal. Coloration is slate or grayish blue, mottled with light spots. Underparts may acquire yellowish coating of microorganisms. About 90 ventral grooves extend to navel. Head and body length 73.8–78.7 ft (22.5–24 m).	Found in temperate and subtropical zones during the winter, and around the poles in the spring. Does not eat for a period of up to 8 months and lives off stored fat. Mating and calving take place in late spring and summer.	All oceans and adjoining seas.	Consists almost entirely of shrimp-like crustaceans of the family Euphausiidae.	Endangered
Antarctic minke whale *Balaenoptera bonaerensis* French: Petit rorqual austral; German: Südlicher Zwergwal, Antarktischer Zwergwal; Spanish: Ballena minke Antárctica	Coloration of back is dark gray, belly and area under flippers is white. There is a white diagonal band on each flipper. May be pale chevron on back behind head or pale gray bracket marks above each flipper. Row of about 300 baleen plates on each side of the upper part of the mouth, mostly yellowish white in color. Head and body length, males 24 ft (7.3 m), females 25.9 ft (7.9 m). Females are slightly larger than males.	Found within 100 mi (160 km) of coastline, often in bays and estuaries. Moves far into the polar ice fields. Fast swimmer, very acrobatic. Generally solitary or in groups of two to four individuals. Usually one offspring produced each year.	All oceans and adjoining seas.	Mainly plankton, but also squid, herring cod, sardines, and various other kinds of small fish.	Lower Risk/ Conservation Dependent
Sei whale *Balaenoptera borealis* French: Baleinoptere de Rudolphi; German: Seiwal; Spanish: Rorcual norteño	Coloration is typically dark steel gray with irregular white markings ventrally. Ventrum has 38–56 deep grooves, side of upperpart of mouth contains 300–380 baleen plates. Head and body length 40–50 ft (12.2–15.2 m).	These whales are found far from shore. They are among the fastest cetaceans and can travel up to speeds of 31 mph (50 kph). Typical groups consist of 2 to 5 individuals. Mating occurs during winter months.	Atlantic, Pacific, and Indian Oceans.	Consists of copepods, amphipods, euphausiids, and small fish.	Endangered
Humpback whale *Megaptera novaeangliae* French: Baleine à bosse, mégaptére; German: Buckelwal; Spanish: Ballena jorobada, gubarte	Coloration is black, white on ventral part, flippers, and throat. Small dorsal fin on hump. Head, jaw, and flippers are covered with bumps. Head and body length 37.7–49.2 ft (11.5–15 m), weight 27.6–33.1 tons (25–30 tonnes). Males are smaller.	Groups consist of 2 to 5 individuals. Breeding takes place in tropical waters in the winter, usually once every two years. Known for males' detailed songs. May live up to 77 years.	Atlantic and Pacific Oceans.	Consists mainly of fish caught through baleen.	Vulnerable

Resources

Books

Ellis, Richard. *The Book of Whales.* New York: Knopf, 1985.

Evans, Peter G. H. *The Natural History of Whales and Dolphins.* New York: Facts On File, 1987.

Periodicals

Baskin, Yvonne. "Blue Whale Population May Be Increasing Off California." *Science* 260, 5106 (1993): 287.

Olsen, E., and O. Grahl-Nielsen. "Blubber Fatty Acids of Minke Whales: Stratification, Population Identification and Relation to Diet." *Marine Biology* 142 (2003): 13.

Tennesen, Michael. "Biggest Animal that Ever Lived. Scientists Attempt to Solve the Many Riddles Posed by the Blue Whale." *International Wildlife* 31, no. 2 (2001): 22–29.

Young, Emma. "Minke Whales Are Out for the Count." *New Scientist* 170, 2295 (2001): 12.

Organizations

International Whaling Commission. Website: <http://www iwcoffice.org>

Society for Marine Mammalogy. Web site: <http://www .marinemammalogy.org>

The Whale Center of New England. Web site: <http://www .whalecenter.org>

WhaleNet. Internet Marine Mammal Resource List. Web site: <http://whale.wheelock.edu/whalenet-stuff/interwhale.html>

Gretel H. Schueller

• • • • •

The ungulates

(Hoofed mammals)

Introduction

Ungulates or hoofed mammals are a large, diverse, and highly successful group of terrestrial mammals classified into a series of superorders and orders. Originally the term "ungulate" referred to plant-eating animals with hooves on the terminal digits (toes) of their legs. This foot structure is their unique morphological adaptation and defining characteristic. In all species, the size of at least one toe has decreased, and in many, the number of toes has been reduced through natural selection. In addition, as their common name indicates, most have evolved hooves, which are modified claws or nails at the tips of the toes. There are two basic specialized foot plans in the ungulates: mesaxonic, in which the main weight is borne on the third digit (toe) as in horses, for example; and paraxonic, in which the weight is borne equally by the third and fourth digits as seen in cattle.

Ungulates are not only the most successful and widespread group of large mammals in the world today, but they are also the single most important group of animal species directly beneficial to humans. Almost all the important domestic animals are ungulates, including horses, pigs, cattle, sheep, goats, camels, and water buffalo. From these species, humans obtain meat, milk, hides, fibers, draft animal power, and much more. These animals, along with domestic plants, fostered development of modern civilizations and continue to support humankind across the globe.

Evolution and systematics

The evolutionary story of the ungulates is complex and not always taxonomically clear. It revolves around changes towards herbivory and accompanying changes to morphology of skulls, teeth and digestive systems. The story also involved evolution of the limb structure in response to predation pressure such that the limbs became increasingly adapted for fast running (cursorial locomotion). The early evolution is particularly challenging to understand, so it is difficult to provide a clear account of their history. Many of the mammals included in the grand order Ungulata are not necessarily closely related. However, even widely separated forms show many cases of parallel evolution, often more than once and often to a remarkable degree of similarity. The evolutionary story is also problematic because ancestral ungulates are poorly represented in the fossil record resulting in an incomplete history; thus it is sometimes difficult to determine with confidence whether or not a particular fossil was an ancestral form. The taxonomy of ungulates will always change as new fossils are discovered and new evidence based on DNA and other molecular analyses help reconstruct their evolutionary history.

Traditionally, five superorders of ungulates have been recognized based on morphological evidence from fossil remains, encompassing up to 16 orders. The superorders consisted of: the Protoungulata, comprised of the primitive ungulates within the Condylarthra, the tillodonts, the litopterns, notoungulates and the Astrapotheria, as well as the Tubulidentata; the Amblypoda with three or four orders, the Dinocerata, Pyrotheria, and Desmostylia, and possibly the Pantodonta; the Paenungulata or near-ungulates, which include the Sirenia, Proboscidea, and Hyracoidea, along with extinct and distantly related Embrithopoda and Desmostylia; the Paraxonia containing the Artiodactyla; and the Mesaxonia with the Perissodactyla. Of these 16 orders, only six are extant: Tubulidentata (aardvarks), Sirenia (manatees and dugongs), Proboscidea (elephants), Hyracoidea (hyraxes), and the two modern ungulate orders, the Artiodactyla and Perissodactyla. Recent interpretations of the grand order Ungulata include the order Cetacea (whales and dolphins), while the Pantodonta, formerly included in the Amblypoda, are probably not ungulate. Of the remaining previous members of the amblypods, the uintatheres (Dinocerata) are thought to be more closely related to the paenungulates and perissodactyls, and the desmostylians are placed closer to the proboscidean root.

The following synopsis of the history of modern ungulates is best understood within the context of broad groups that included other extant orders and their common ancestry. Based almost solely on the number of digits, a large portion of modern ungulates can be placed in two morphological groups: the even-toed Artiodactyla (e.g., sheep and deer) and the odd-toed Perissodactyla (e.g., horse and rhinoceros). Although even-toed and odd-toed ungulates are about as closely related as are the rodents and primates, both ungulate orders most probably arose from within the Condylarthra. Artiodactyla were previously considered to have evolved from the mesonychids, which were themselves thought to be derived from the

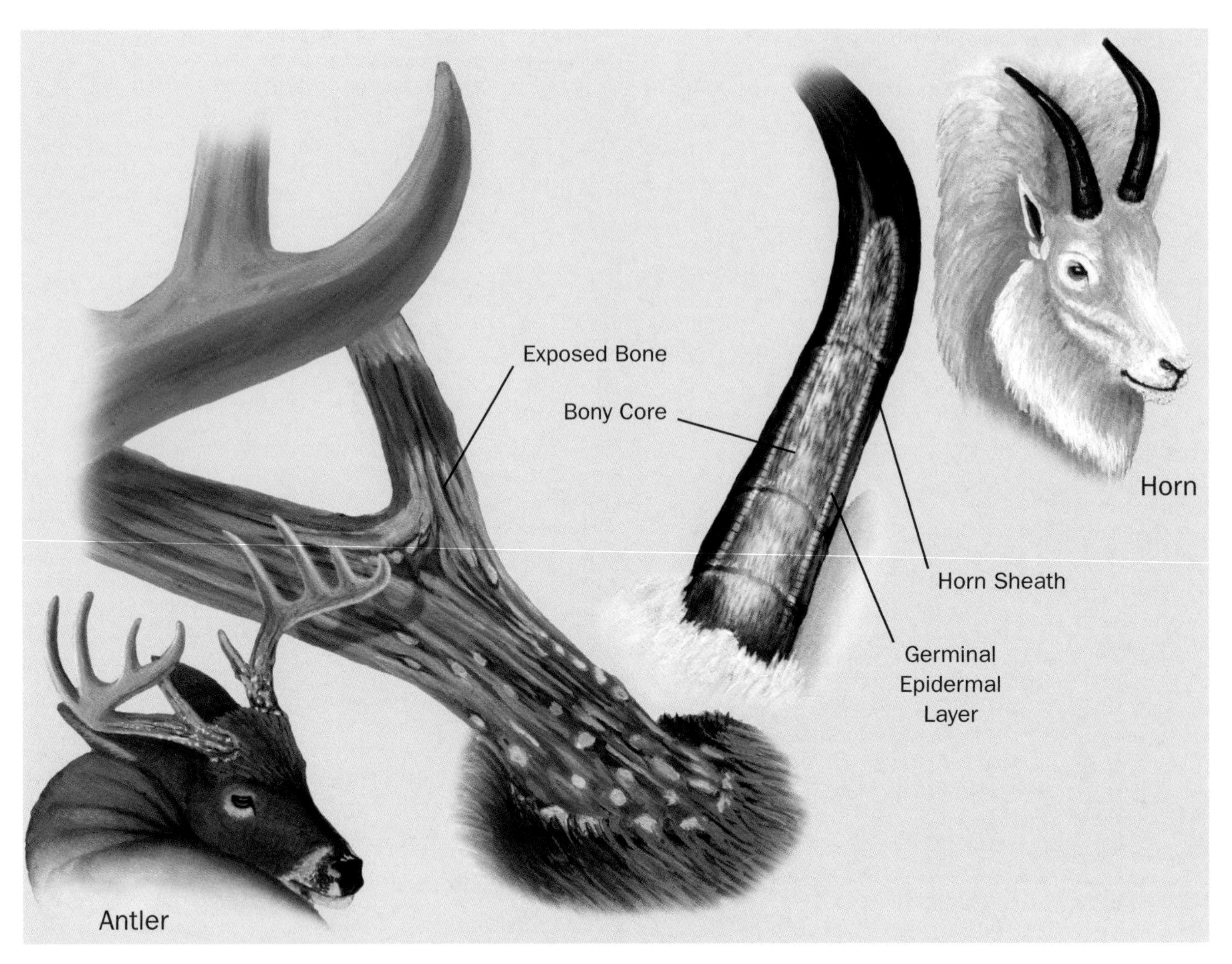

Antlers are exposed bony outgrowths, while horns have a covering over their bony core. (Illustration by Jarrod Erdody)

raccoon-like arctocyonids. Artiodactyls are now considered to have arisen from an arctocyonid ancestor, and though the most successful and numerous of the living ungulates, this order is also probably one of the most "primitive" in terms of its origins. The arctocyonids are some of the oldest of the condylarths that first appeared in the fossil record of the late Upper Cretaceous, and eventually dispersed throughout North America and Eurasia. These primitive mammals were probably omnivores based on their teeth structure which resembled that of modern-day bears. The Perissodactyla are generally thought to have evolved from the Phenacodontidae. These were early condylarths ranging in size from modern foxes to tapirs, probably with claws on their terminal phallanges and dental adaptations towards herbivory, but otherwise showing typically primitive mammalian characteristics. The earliest phenacodontid considered leading to the perissodactyl lineage was *Tetraclaenodon*. This fox-sized mammal appeared in middle Paleocene and gave rise to three main groups, one of which is presumed to be a proto-perissodactyl. However, so far, fossil intermediates have not been found between *Tetraclaenodon* and what Leonard Radinsky considers to the earliest known primitive perissodactyl, *Hyracotherium* of the early Eocene.

The earliest ancestors of ungulates are poorly represented in Paleocene strata. The generally accepted picture is that ancestral ungulates evolved from a group of primitive, small-bodied mammals belonging to the order Condylarthra, the most primitive of which was probably *Protungulatum*. Condylarths themselves first appeared in the late Cretaceous but are one of the most characteristic groups of mammals of the Paleocene. Unlike earlier insectivorous mammals, most condylarths were probably omnivorous having bunodont cheek teeth suitable for crushing and grinding, differentiated canines, and elongated skulls. The name "condylarth" was originally given to the earliest fossil herbivores considered ancestral to ungulates by the American paleontologist E. D. Cope (1840–1897). Later it came to encompass a wide array of ancient mammals, leading some paleontologists to question the order's validity. It contains a diverse, paraphyletic as-

A moose (*Alces alces*) cow feeds on a pond's bottom in the evening. (Photo by YVA Momatiuk & John Eastcott/Photo Researchers, Inc. Reproduced by permission.)

semblage even amongst extant taxa, including the ancestors of artiodactyls, perissodactyls, cetaceans, sirenians, proboscideans, and hyraxes.

Later in the Paleocene, the Paenungulata, also called primitive or sub-ungulates, diverged from the Condylarthra. Except for the embrithopods and desmostylians, the other three orders within this superorder persist to this day, forming an interesting and diverse grouping; the aquatic Sirenia (dugong, sea cow and manatees), the Proboscidea (elephants), and the Hyracoidea (hyraxes or dassies). The relationship of cetaceans (whales) within the ungulates is somewhat controversial. Until the application of molecular techniques, morphological and fossil evidence suggested that whales had probably evolved from the mesonychids, an offshoot of the condylarthran arctocyonids considered to have given rise also to the early ungulates. In 2000, the discovery of early fossil whales in Eocene deposits from Pakistan, show that they had a double-pulley astragalus (anklebone) suggesting that cetaceans were closely related to the artiodactyls. Molecular evidence also pointed to the cetaceans being most closely related to the predecessor of the modern hippopotamus. This has led some to suggest that cetaceans should be included in the Artiodactyla and the name changed to the Cetartiodactyla. However, the earliest fossil hippos do not appear until much later than these early cetaceans, and computer analyses using morphological data from living and fossil forms also cast doubt on the hippo-whale connection. No matter whether morphological or molecular analyses are used, whales are still most closely related to artiodactyls and belong in the Ungulata.

Besides providing the ancestors of the modern ungulates, paenungulates and cetaceans, the Condylarthra also probably gave rise to several other ungulate groups including the tillodonts, litopterns, notoungulates, tubulidents and astrapotherians. The Tillodontia is an order with uncertain affinities that is found in deposits from Asia, Europe and North America. It probably evolved from the condylarths and its members had a pair of enlarged rodent-like incisors. The condylarths also gave rise to three orders that were restricted almost entirely to South America. The Litopterna, ecologically and morphologically reminiscent of modern camels and horses, had a mesaxonic limb with the weight borne by the third digit. This order included two major families: the proterotherids which evolved one- and three-toed species, illustrating another case of parallel evolution this time with the equids; and the macrauchenids, which also showed a reduction to three toes, and in some later species, the suggestions of an elongated, trunk-like upper lip that would have been used for feeding. A second order was the diverse Notoungulata, containing some of the oldest South American mammals. Notoungulates evolved a great diversity of species and form from small, rabbit-sized creatures to the giant rhinoceros-like toxodonts. This was a long-lasting order that spanned into the Pleistocene until artiodactyls, perissodactyls and proboscideans migrated into South America and out-competed the resident notoungulates. The third South American order of primitive ungulates was the Astrapotheria represented by the rhinoceros-sized *Astrapotherium*, which lived in the

A male Thomson's gazelle (*Gazella thomsonii*) drinking. (Photo by Leonard Lee Rue III. Bruce Coleman, Inc. Reproduced by permission.)

Mountain goats (*Oreamnos americanus*) on the slopes near the Athabasca River in Jasper National Park, Canada. (Photo by Animals Animals ©Victoria Mc Cormick. Reproduced by permission.)

Miocene, and based on cranial anatomy, most probably had a trunk.

The last two groups, Notoungulata and Astrapotheria, are in many ways so unlike the other early ungulates that they were not always classed with them. Although its origins are unknown, the order Tubulidentata (aardvarks) is included within the primitive ungulates based on various anatomical features. At one time, its specialized food habits, enamel-free teeth, tubular skull and long tongue, led taxonomists to class aardvarks together with the New World edentates (anteaters), but this was not supported by other key anatomical features. Much of their specialized morphology for termite feeding had been acquired by the Miocene. Besides the Tubulidentata, a second order, the Periptychoidea, differed significantly from other primitive ungulates in their dentition, brain morphology, and structure of the limbs. They sometimes were classed with pantodonts. Although some members resembled tubulidentates, the periptychoids were herbivores, lacking the characteristic skull and teeth of the termite-eating aardvark. In North America, they were the dominant condylarths in the early Paleocene ranging in size from rat- to sheep-sized forms.

The generally accepted interpretation of the phylogenetic relationships of ungulates described above was based on morphological evidence provided by fossil remains. This is now being challenged by new evidence based on genetic distances, which present a quite different view of mammalian phylogeny. Genetic data suggest that there are four superorders of mammals, Afrotheria, Xenarthra, Eurarchintoglires, and Laurasiatheria. The Sirenia, elephants, hyraxes and aardvarks are separated from the rest of the ungulates and placed into the Afrotheria, while all other modern ungulates would be placed in the superorder Laurasiatheria.

The evolutionary history of the ungulates has taken place against a backdrop of major climatic and environmental changes. These changes and their effects on plant life and habitats have driven the evolution of ungulates and other mammals. Starting in late Cretaceous and early Paleocene when primitive mammals considered ancestors of ungulates appeared, the world's temperature began increasing with only a slight cooling phase around 58 million years ago (mya) towards the end of the Paleocene. The warming trend continued, with mean annual temperatures reaching a maximum about 50 mya between the early and middle Eocene. After this, temperatures declined, at first steadily but around 35 million years ago, they dropped dramatically at the Eocene-Oligocene boundary. This was probably the result of major changes in the patterns of oceanic circulation, brought about by continental drift that had separated Greenland from Norway, and Australia from Antarctica. As vertebrate paleontologist Professor Christine Janis described, it was a "transition from the Mesozoic 'hothouse' world to the 'ice house' world" of the Neogene (Miocene and Pliocene). A much slower warming of global temperatures followed, peaking in the late Middle Miocene, but at a temperature only about half that of

Guanacos (*Lama guanicoe*) can run at speeds of up to 35 mph (56 kph). (Photo by Hans Reinhard. Bruce Coleman, Inc. Reproduced by permission.)

A giraffe (*Giraffa camelopardalis*) with its young. (Photo by St. Myers/OKAPIA/Photo Researchers, Inc. Reproduced by permission.)

the Eocene maxima. Thereafter, global temperatures once again began to cool, and from the Pliocene to present they have continued to fluctuate, but with a significantly lower amplitude and with lower maxima than in the previous epochs. Such fluctuations exerted changes to terrestrial plant communities and acted as selective pressures upon the ungulates that exploited those communities in terms of adaptations of their teeth and digestive systems to cope with the problems of eating plants.

In the northern continents, the Middle Paleocene saw the appearance of small herbivorous condylarths (phenacodontids, meniscotheriids), and also larger herbivores such as the pantodonts (including *Barylamda* estimated to weigh over 1,323 lb or 600 kg). Even larger were the rhinoceros-like and possibly semi-aquatic uintatheres, of which *Uintatherium* was estimated to weigh as much as 9,900 lb (4,500 kg). Condylarths, pantodonts, and uintatheres were present in both Asia and North America, while Tillodonts migrated into North America from Asia in the late Paleocene. In South America, the Paleocene ungulate fauna included small condylarths, the dilododontids, rhinoceros-like astrapotheres, xenungulates, litopterns, and notoungulates. By the middle Paleocene, new ungulates had appeared, some with mesodont cheek teeth, and by the late Paleocene, early South America ungulates had evolved cheek teeth that showed dental adaptations to a more fibrous diet, indicating a probable shift to more temperate, open forested habitats as the world climate became warmer. At this time, palaeanodonts, uintatheres and arctostylopids (hyrax-like notoungulates), might have dispersed from South to North America.

The warming trend continued from the Paleocene into the beginning of Eocene, and faunas indicate generally tropical habitats in the north. Conditions appeared to shift towards drier environments than in the Paleocene, creating more diverse understory habitats in which browsing ungulates could flourish. The first artiodactyls and perissodactyls appear relatively abruptly in fossil record at the beginning of the Eocene in strata from Asia, Europe, and North America. This is a ma-

A male steenbok (*Raphicerus campestris*). (Photo by Jen & Des Bartlett. Bruce Coleman, Inc. Reproduced by permission.)

jor mystery in mammalian evolutionary history. It is speculated that that their "sudden" appearance in Holarctic strata might have been because they originated elsewhere, such as in Africa, Central or South America, or India, and that the disappearance of physical barriers or the warming climate in the early Eocene allowed them to disperse. A significant faunal exchange did take place between North America and Europe starting in the early Eocene and continuing until the middle Eocene, but no evidence has yet been found for any faunal interchange with South America during the Eocene. In South America at this time, ungulates and condylarths became rare and xenungulates disappeared, while small to medium-sized litopterns and notoungulates and larger astrapotheres and pyrotheres all diversified.

Ungulates with browsing and frugivorous feeding habits also appeared, and became established by the late Eocene. At this time in North America, there was an increase in ungulates with cursorial adaptations and hypsodont cheek teeth with high crowns and short roots. Similarly in South America, a change to more fibrous vegetation is suggested by notoungulate cheek teeth because they show a greater degree of hypsodonty.

Towards the end of the Eocene, major extinctions occurred within the North American fauna, almost certainly the result of the climate change that brought about a pronounced seasonality in plant production, and hence in food supply. Many archaic ungulate groups were lost including condylarths, uintatheres, and protoungulates such as tillodonts, as well as some artiodactyls and perissodactyls. Body size increased in surviving ungulates, and the chewing surface of their cheek teeth became more complex, both probably reflecting a diet that was more fibrous, hard to digest, and less nutritious. This was the time when the artiodactyls began to flourish and diversify with the establishment of three major lineages: suines, tylopods, and ruminants. However, the perissodactyls still dominated despite having generally declined through the late Eocene in both diversity and abundance. In Europe, the impacts of climate change were not expressed until into the Oligocene, but when the climate change occurred, it too was accompanied by significant faunal extinctions. Asia, on the other hand, seemed to miss the Eocene-Oligocene extinctions and archaic forms such as condylarths, along with the more advanced pantodonts, brontotheres and tapiroids, which lingered until the mid-Oligocene.

The Oligocene was drier and cooler than the preceding Paleocene and Eocene epochs, especially in North America. The extent of tropical forests was restricted and large areas of open habitats had yet to develop, while in the northern latitudes, the broad-leaved deciduous forests in the polar region were replaced with temperate broad-leaved deciduous woodlands. Conditions were relatively stable through much of the Oligocene, although there was a gradual shift towards more seasonal vegetation types. This probably accounts for the replacement of many Eocene perissodactyl herbivores by folivorous artiodactyls (eating mainly leaves), which might have been able to feed more selectively. The most common artiodactyls were goat-sized browsers resembling modern capybaras. Hippo- and equid-like rhinocerotoids were found throughout Asia, Europe and North America, while true rhinos were small to medium-sized, either lacking horns or having horns with bony cores unlike extant species. Llama-sized camels and browsing, pony-sized equids were common in North America. Other early ungulates included anthracotheres and small, hornless pecoran ruminants in Eurasia, and oreodonts in North America. Most Oligiocene ungulates with their shorter limbs were not as well-adapted to cursorial locomotion than are modern forms. It was not until about 26 mya at the beginning of the Late Oligocene, that a second period of mammalian extinctions began. For the ungulates, this was marked mainly by the disappearance of Eocene forms including archaic rhinocerotoids, and anthracotheres in North America and paleotheres and tylopod artiodactyls in Europe.

The Miocene climate became much drier and warmer. Subtropical and tropical forest zones expanded once more, and a dry thorn-scrub developed on the western edges of the continents. In general, mammalian faunas were somewhat comparable to modern-day savanna-woodlands. Lowered sea-levels towards the end of the Early Miocene, allowed faunal exchanges between North America and Eurasia and between Africa and Eurasia. Ungulate migrations included browsing horses from North America to Eurasia, cervoid percoran ruminants and rhinos from the New to the Old World, and proboscidians from Africa to Europe. Bovids, cervids, giraffids, and giraffe-like palaeomerycids appeared in the early Miocene of Europe. These relatively large animals, which replaced the earlier, smaller ruminant forms, were notable for the diversity of their species-specific horns or antlers. Suids diversified at this time to include browser-types with tapir-like incisors. While the traguloid ruminants declined, other artiodactyls became more diverse by the middle Miocene and the bovids developed hypsodont cheek teeth. Size also increased, especially in the bovids and giraffids. In North America, camel diversity increased and the earliest equine *Merychippus* appeared, both groups later developing hypsodont dentition. Later in the Miocene of North America, the antilocaprids diversified as did the deer-like dromomerycids. Although there was a trend to more open grassland savannas in North America during the Miocene, sufficient productive woodlands remained for browsing peccaries and tapirs to survive. In South America, woodlands gave way to more open habitats as indicated by the trend towards increased hypsodonty (high-crowned teeth), which characterized Miocene herbivores on this continent. Litopterns adapted to these open habitats evolving single-toed forms, and although the notoungulates showed similar dental adaptations, their diversity declined during this epoch. In mid-Miocene Africa, woodland ungulates diversified including rhinos, bovids, giraffoids, suids, and traguloids.

The steady cooling trend and further drying in the late Miocene was accompanied by an expansion of savannas and a decrease in mammalian faunal diversity. Grasses had been gradually spreading since their appearance in the Eocene, but open savannas were probably not extensive until the Late Miocene, which was about the time that C_4 grasses (photosynthesis takes place in chloroplasts) are first recorded. While C_4 grasses were taking over from C_3 grasses (photosynthesis occurs within the mesophyll) in Asia, the dominance of C_4 grasses in temperate regions occurred later in the Pliocene. True grasslands did not appear until much later, probably in the Pleistocene, but savanna grasslands were extensive in North and South America. The development of Late Miocene open savanna-like habitats in Eurasia was accompanied by hipparionid horses developing hypsodont dentition and three-toed feet, and by hypsodont bovids. Interestingly, however, the hypsodont dentition was not indicative of predominantly grazing diets. Examination of the micro-wear patterns on the teeth of these and other herbivorous ungulates suggest that feeding habits were predominantly mixed grazing-browsing. Presumably the habitat was of an open wooded savanna-type, rather than being primarily grassland. The rise of C_4 grasses and the increasing aridity of habitats are suspected to be a main selective force towards full hypsodonty. C_4 grasses contain up to three times as much silica as do C_3 species and hence teeth of animals feeding on them will wear down more quickly. In Africa around this time, more modern forms of bovids giraffes and hippos evolved, while North American hipparionid horses entered the continent via Asia and Europe.

Oryx (*Oryx gazella*) calf and nursing mother in Samburu, Kenya. (Photo by K & K Ammann. Bruce Coleman, Inc. Reproduced by permission.)

The close of the Miocene was accompanied by major extinctions, especially of savanna-dwelling mammalian fauna, across Eurasia and North America in association with colder climates and habitat changes. Tundra and taiga ecosystems developed in the northern latitudes, and extensive dry grasslands to the south, and by the beginning of the Pliocene an ice cap may have existed in the Arctic. Global temperatures had begun to rise before the beginning of the Pliocene and continued until about 2.5 mya in the Late Pliocene, leading to the development of desert and semi-desert habitats. Modern *Equus* took the place of hipparionid horses in Eurasia, and along with camels, both originating in North America, entered Africa around 2.5 mya.

Around the beinning of the Pliocene, 3 to 2.5 mya, continental processes and possibly a lowering of the sea level created the Panamanian land bridge between North and South America. Animals were once more able to move between the two continents that had been separated for over 100 million years. What followed is referred to by paleontologists as the "Great American Interchange." These faunal exchanges of large mammals, which peaked in the middle Pleistocene, involved equids, tapirs, peccaries, llamas, and deer moving from North to South America along with gompotheres (related to elephants) and some large carnivores. Of the South American ungulates only a few notoungulates moved north, although two other large herbivores, glyptodonts and giant ground sloths, also entered North America. Among the ungulates that emigrated to South America, the equids eventually became extinct, as they did in North America, although these perissodactyls did survive in Africa and Eurasia.

By the end of the Pliocene global temperatures were dropping, and the following Pleistocene was a time of alternating glacial and interglacial periods repeated on a roughly 41,000-year cycle. Apart from Antarctica, most of the ice accumulation on land was in the northern hemisphere especially in North America. As this ice formed huge glaciers and ice caps, sea level fell and a land bridge formed once more between

Takin (*Budorcas taxicolor*) family group in Bhutan. (Photo by Harald Schütz. Reproduced by permission.)

Eurasia and North America. The interchange between Eurasia and North America was not symmetrical, with more species moving into the New World than moved in the other direction. Bovids such as bison, mountain sheep and mountain goats, mammoth along with other mammals such as wolves and cave lions, migrated across the Bering land bridge from Siberia into Alaska, and when the ice masses retreated, made their way south into the rest of the continent. Humans also entered North America around this time over the land bridge and perhaps also down the west coast of the continent. Modern horses and camels were the two ungulates that moved to the Old World and survived to the present day.

Towards the end of the Pleistocene, many species of large mammals became extinct, including many large herbivores. Eurasia lost mammoths and the woolly rhinoceros; North America lost mammoths, all but one member of the Antilocapridae, woodland musk ox, horses, and camels; and North and South America lost ground sloths, mastodons and glyptodonts. In South America, the litopterns—the last of the indigenous ungulates on this continent—finally vanished. Other species of mammals besides ungulates disappeared, and in North America, a total of between 35 and 40 large mammal species vanished between 12,000 and 9,000 years ago. Three main hypotheses have been proposed to explain these megafaunal extinctions: climate, overkill by humans, and disease (epizootic). Each has supporting evidence and some scientists suspect that the extinctions might have resulted from a combination, at least of the first two factors. Certainly climate change and the appearance of humans in North America, coincide with extinctions of large mammals.

Physical characteristics

The largest land mammal that has ever existed was an ungulate, *Paraceratherium Indricotherium transouralicum*. This was a Perissodactyl that belonged to the rhinoceros family and it roamed central Asia about 35 mya during the late Eocene. Although not as big as the largest dinosaurs, *Paraceratherium* stood over 15 ft (5 m) at the shoulder with a skull 4 ft (1.3 m) in length. It could probably browse tall vegetation to a height of 25 ft (8 m). *Paraceratherium* was estimated to have weighed about 20 tons (18 tonnes), or about four times heavier than the largest weight recorded for modern African elephants. This feature of relatively large body size extends to modern ungulates although some species are small. Weighing around 1.6 tons (1.5 tonnes), the hippopotamus is the heaviest of the living ungulates, while the smallest are the mouse deer or chevrotains (Tragulidae), which can weigh less than 2 lb (1 kg). The tallest living mammal, the giraffe, which stands almost 19 ft (6 m) tall, is also an ungulate. Living perissodactyls are all large animals from 440 to 7,700 lb (200–3,500 kg). The largest artiodactyls are heavier and their weights span a greater range, from 2 to 10,000 lb (less than 1 kg to about 5,000 kg).

As stated earlier, hooves surrounding the terminal toe bones or phalanges of their feet, are unique at least to modern ungulates. The hooves are composed of a hard protein material called keratin that creates a tough outer sheath protecting the terminal toe bones. Immediately under the hoof is the softer sub-unguis, and behind this is the pad. The hard outer sheath is technically the *unguis* and derives its name from the Latin *ungula* for claw or nail. The word "ungulate," the alternative name for hoofed mammals, also originates from the Latin word *ungula*, meaning hoof. Hooves are modified claws or nails found in most other groups of mammals.

Besides their feet, other characteristic adaptations of ungulates include their limbs, eyes, digestive systems, and teeth. They also show a wide diversity of weapons and head ornamentation such as tusks, horns, and antlers, although amongst the modern perissodactyls, only the various species of horned rhinoceroses have such specialized weapons. All of these morphological adaptations evolved to help ungulates avoid predators, gather and process food, and to interact with other members of their species for survival and reproduction.

Two of the biggest challenges facing ungulates are to avoid being killed by predators and to extract sufficient nutrients from the plants they eat. Ungulates are the main prey of large mammalian predators, and the resulting intense and constant selection pressure has led to several adaptations that help ungulates reduce the risk of being killed by a predator. The position and structure of their eyes are adapted for predator detection, their pelage and other features can make them more difficult to detect by predators, and finally, their limbs are adapted for running fast which helps them escape from predators.

Disruptive or camouflaged coats that help hide from predators are common in forest-dwelling ungulate species. Coat color can blend with the surroundings, while disruptive color patterns such as light stripes against a dark background, help break up body form in the dappled light of the tropical forest or savanna bushlands; both make detecting potential prey more difficult. Disruptive striping is seen in many tragelaphins such as the bongo (*Tragelaphus eurycerus*), and in the tiny forest-dwelling chevrotains (Tragulidae). Small size can also make it easier to hide and some species, such as duikers (Cephalophinae), literally dive into dense vegetation to hide from predators. This behavior has given rise to their common name which, is derived from the Dutch word for diver.

It is more difficult for large species, especially those living in open habitats, to hide. Instead, they rely primarily on vision to detect predators well in advance, and then on speed to outrun them. Ungulate vision is adapted for detecting movement over a wide field of view. The basic anatomy of their eye is similar to that of other mammals, but in the ungulate there is no central focusing spot or fovea, which allows discrimination of fine details. Also, the ungulate pupil is elliptical rather than round, and oriented horizontally, while the eyes are located on the sides, rather than on the front of the head. Eyes in this position, together with elliptical pupils, give most ungulates a field of view over 200°, and some can probably see almost as well behind as in front or to the side. So while ungulates might not easily distinguish fine details, they readily detect movements almost anywhere around them and so have a good chance of detecting approaching predators. Whether ungulates have color vision is uncertain, but based on the internal structure of their eye, they probably can distinguish some colors, but not as well as do humans, for example. It is also suggested that in some forms, the elongated skull (especially the muzzle region) not only aids with feeding, but at the same time allows an ungulate to keep its eyes above the vegetation and wary for predators. Extreme examples of elongated skulls and high placement of the eyes is seen in the hartebeests (*Beatragus*, *Damaliscus*, *Alcelaphus*) of Africa.

Very few species of ungulates are large enough, compared to their common predators, to be able to defend themselves physically. This is especially true if the predator normally attacks as a group. African buffalo, a large-bodied species with heavy horns, will sometimes cooperatively defend themselves against a predator such as a lion. Similarly, musk oxen will form a defensive ring against a wolf pack, shielding the young calves between their bodies. Defense, however, is the least common anti-predator strategy used by ungulates.

Lone red hartebeest (*Alcelaphus buselaphus*) pronking to attract the attention of a nearby herd of springboks (*Antidorcas marsupialis*) in Kalahari Gemsbok Park, South Africa. (Photo by Animals Animals ©Francois Savigny. Reproduced by permission.)

Instead of defense, most ungulates flee from their predators. Their limbs are adapted to allow them to run fast. The evolution of ungulate limbs is one of a gradual change towards longer, lighter limbs, with smaller, lighter feet, and specialized limb joints. Long limb bones provide an obvious advantage for speed because they provide a longer lever action, and thus longer stride length. The greatest increase in leg length came through elongation of the metapodials, the bones between the wrist/ankle and the fingers/toes. Lighter limbs also help improve speed because they require less effort to overcome the inertia of locomotion. Lightness results from musculature in the lower extremities being replaced with tendons and ligaments, and by smaller feet resulting from a decrease or loss of the lateral toes. Although the rhinoceros and hippopotamus do not show the extreme limb adaptations common in other modern ungulates, their limbs still illustrate the same basic patterns and they are capable of surprising speed.

Ancestral ungulates had the basic vertebrate plan of five digits, but during their evolution between one and four outer toes were lost, and often the outer metapodial bones were also lost or reduced in size. The number of functional toes remaining is key to whether the ungulate is classified as perissodactyl or as artiodactyl. Artiodactyls have lost the first toe and three metapodials. What remains are the two central metapodials fused into a single unit, two functioning toes (the third and fourth), and two greatly reduced outer toes (the second and fifth) called lateral hooves or dew claws. All the toes including the dew claws are covered by hooves. A similar reduction in the number of toe bones occurred in perissodactyls, with the most extreme found in the modern horses. Zebras, asses, and horses have only the third metapodial, the cannon bone, and the associated single toe has been retained, though vestigial metapodials or splint bones may persist. In rhinos, the number of toes has been reduced to three. The result of all these reductions in the number of toes is that most modern ungulates walk with their weight on their hooves or tips of their toes. This type of gait is referred as unguligrade, and it helps the animal to move quickly over hard ground. This contrasts with plantigrade found in bears and humans which walk on the soles of their feet, and digigrade found in cats and dogs and many other species where the weight of the body is taken by the entire digits not just by the tips. The very heavy-bodied ungulates which rely less on speed, such as the rhinoceros and hippopotamus, have feet that are rather short and wide, with splayed digits that are needed to support their mass.

Increased speed and efficiency for cursorial locomotion have also been achieved through adaptations of the limb joints, and the way that they attach to the vertebral column. Many ungulate limb joints restrict movement to a strong and efficient forward-backward movement, which means that power is transferred more effectively to forward movement. For example, the ungulate astragalus is grooved, on the distal end in perissodactyls and at both ends in artiodactyls, to minimize lateral movement of the ankle joint during articulation.

The ungulate hind legs are firmly attached to the sacral section of the vertebral column via a strong ball-and-socket joint between the femur and the pelvis. In this way, the force of the limb motion is transferred directly to the body through the backbone, propelling the animal forward with each powerful stroke of the hind limbs. The front limbs in ungulates are not firmly attached to the rest of the skeleton. As in most cursorial mammals, the last upper bone of the forelimb, the scapula or shoulder blade, is not attached to the backbone, but is attached only by muscles to the upper thorax. This cushions the running animal when its forelimbs hit the ground. Additionally, flexibility is gained because ungulates, and other cursorial mammals, have lost the clavicle or collar-bone. Its absence allows the shoulder blade to move relatively freely when the forelimbs swing forward, increasing stride length and thus speed.

Besides evading and outrunning predators, ungulates also have to acquire energy and because most are herbivores, this energy must be extracted from plants. Plants are difficult to digest, and to help maximize the nutritional value of their food, ungulates not only have specialized teeth to crop and chew plants, but also unique digestive systems. In some species, even their body shape helps them feed more effectively (e.g., giraffes and gerenuks). Last, ungulates are generally social species forming small groups to large herds that lead to a range of communication systems, mating strategies, and fighting styles.

At a broad evolutionary level, herbivores can choose one of two approaches: they can feed on low-quality but abundant forage, or they can feed on high-quality but uncommon forage. These different approaches to the exploitation of plants have morphological consequences that affect feeding efficiency and thus energy acquisition. High-quality forage is typically sparse, hence many ungulates depend heavily on grasses and shrubs, which are usually abundant but generally low in nutritional quality. To help them gather sufficient quantities of food, the largest species have wide mouths and spatulate cropping incisors, allowing them to take large bunches of forage into their mouths. Many smaller species have narrower mouths that permit them to feed more selectively on the low-abundance but high-quality forage.

Ungulates, like most mammals, have heterodont dentition: incisors, canines, premolars, and molars. Except for the Suidae and Tayassuidae, which have less specialized teeth, most ungulates have highly specialized dentition that reflects their wholly herbivorous diet. To meet their energy requirements, these obligate herbivores must eat a lot of vegetation relative to their body size. Large browsing and grazing species such as the giraffe or bison use their tongue, wrapping it around a clump of vegetation to pull it into their mouth. Grazing Perissodactyls such as the Equids have retained the upper incisors and use these with the lower ones to crop grasses close to the ground. By contrast, grazing Artiodactyls lack upper incisors and instead grasp the food between their modified spatulate lower incisors and their hardened upper palate, and by jerking their mouth upward tear off a mouthful of forage. Browsers generally have narrower muzzles and incisor tooth rows than do grazing species. The former's narrow muzzle allows them to select the more nutritious parts of plants such as leaves and tips of twigs.

In most ungulates, there is a characteristic space between the lower canines and the first of the premolars, called the diastema. This might aid browsers to strip off leaves or simply be a result of an elongated jaw that evolved to allow greater feeding selectivity, and perhaps helps keep their eyes above the vegetation while feeding so they can watch for predators.

Besides the muzzle and incisors, the cheek teeth of herbivorous ungulates also show diet-related adaptations. Plants not only need to be chewed into small pieces to help digestion, but some plants such as grasses are highly abrasive. The crown of a mammalian tooth is covered in a layer of enamel, which is the hardest and most wear-resistant part of the tooth. The cheek teeth, or premolars and molars, are used for chewing, and these teeth show the greatest specialization. The molars of Suidae and Tayassuidae have bunodont teeth in which the crown consists of low conical cusps covered by a layer of enamel. These teeth are suited to masticating their diverse and generally softer, less abrasive foods. In the obligate herbivorous ungulates, however, the crown enamel of the cheek teeth is highly modified and formed into lateral or vertical folds. The tips of these enamel folds wear off quickly, creating several hard cutting edges juxtaposed to layers of softer dentine. Differential wear allows these self-maintaining ridges of enamel to create a rough surface that helps grind plants into small pieces. When the grinding surface is examined, two general enamel patterns are seen in the herbivorous ungulates: the selenodont or crescent-shaped pattern, and the lophodont or convoluted pattern. The selenodont pattern is found in artiodactyls but not in modern perissodactyls. The lophodont pattern, though typical of modern perissodactyls, is also found in premolars of some artiodactyls.

Tooth wear, especially in herbivores, is a major determinant of an animal's life span. Abrasive foods naturally wear teeth more quickly, thus potentially shortening the time that cheek teeth function to grind vegetation. Browsers (e.g., moose) consume mainly leaves, twigs, fruits, buds, and young shoots of woody plants—all foods that are much softer than tooth enamel. Species with these food habits have brachydont or low-crowned teeth, with selenodont or lophodont enamel patterns. Grazers (e.g., gnu), on the other hand, eat primarily grasses and forbs, and most grasses have phytoliths in their epidermal cells. These phytoliths are comprised of silica that abrade even the hard tooth enamel. This abrasion problem is compounded for ungulate species living in relatively dry areas, because plants growing there are invariably coated with a fine layer of silica dust that further abrades the teeth.

Grazers have evolved two main adaptations to resist tooth wear. First, the diameter of the premolars and molars is enlarged, increasing the grinding surface so that the tooth takes longer to wear away. Second, in the most highly specialized grazers, the molars and sometimes premolars are hypsodont or high-crowned, having tall crowns and short roots, so again these teeth take longer to wear away, thus extending the life of the animal.

Feeding ecology

Pigs and peccaries are not obligate herbivores and eat foods other than plants. Although they mainly eat vegetation, they consume everything from roots, bulbs, and fruits to bird's eggs and insects. Of greater relevance to the evolution and radiation of ungulates are the specializations in the exploitation of plants. All other artiodactyls and perissodactyls feed almost exclusively on plants, although the types of plants eaten vary greatly. These obligate herbivores can be divided into either grazers, browsers, or mixed grazer-browsers.

The leaves and twigs of plants contain a great deal of potential energy, but they are difficult to digest because their cell walls contain cellulose and sometimes lignin. This energy is not directly accessible because no ungulate (nor any other mammal) has enzymes that can digest these two constituents. Ungulates evolved adaptations to overcome this problem by using microorganisms and fermentation to digest cellulose and some lignins. The mainly bacterial microorganisms produce the enzyme cellulase that breaks down cellulose, while the fermentation process continues the breakdown of cellulose into simpler compounds called volatile fatty acids. The main fatty acids, acetic and propionic acid, are absorbed directly into blood vessels in the gut wall, transported to the liver, and metabolized as the ungulate's primary energy source.

Artiodactyl and perissodactyl ungulates evolved distinctly different strategies for where this symbiotic microbial fermentation occurs in their digestive systems. In artiodactyls, it takes place at the front of the digestive system, referred to as fore-gut fermentation. Depending on species, there are between one and three chambers or false stomachs located before the true stomach. In the first of these, in the fore-stomach or rumen, the microbes ferment ingested plants. The most successful of the artiodactyls have evolved an added adaptation: regurgitating food (cud) to rechew it into smaller food particles, thus increasing the surface area on which digestion by microorganisms can occur.

In perissodactyls, microbial fermentation does not occur until the food has passed through the stomach, along the small intestine, and has reached the enlarged cecum located towards the end of the digestive system. Not surprisingly, this approach is referred to as hind-gut fermentation. Perissodactyls only chew their food once, and have a single stomach where digestive enzymes are released. Food is further digested in the intestines where proteins are broken down to amino acids, and sugars and carbohydrates to glucose, before being absorbed. The remaining undigested food reaches the cecum and there it is further digested by bacterial fermentation which

Bongos (*Boocercus euryceros*) have distinctive white spots on the sides of their faces. (Photo by David M. Maylen III. Reproduced by permission.)

breaks down cellulose and other plants components into volatile fatty acids. As in the ruminant, these microbial byproducts are then absorbed and used for energy.

These two fermentation systems differ in their relative benefits and costs. There are at least two and possibly three significant benefits enjoyed by the ruminant artiodactyls. First and probably most important is the ability to regurgitate and rechew food. This increases digestive efficiency simply because it more effectively breaks the consumed plant material into very small particles, increasing the surface area on which microbes can operate and thus facilitating digestion. A second benefit comes from having the fermentation occur at the beginning of the digestive system. It means that the ruminant can obtain a valuable protein source by digesting the microorganisms themselves when they are transported out of the rumen with the digestia. Artiodactyls with greatly enlarged fore-stomachs gain a third benefit: they can harvest a large amount of food and then move to safer areas to digest it. The main cost of the fore-gut fermentor system occurs when food is of low quality. Rumen fermentation is slowed when low-quality food is ingested because crude protein in the diet limits microbe population growth and thus microbial digestion is impeded. This slower fermentation diminishes the amount of energy available to the ruminant. The ruminant is faced with being more selective in its feeding. In constrast, the hind-gut fermentors, such as some perissodactyls (e.g., modern horses), are able to increase the rate that food passes through their gut, so they extract only the most readily digestible fraction of the food and excrete the undigestible material. As a result, although they must feed almost continuously, they can be much less selective in what they eat. This allows horses to survive on poorer-quality food than artiodactyls are able to do.

Body size and feeding strategy

To survive, any animal must balance its energy budget by meeting its energy requirements through food intake and by limiting energy losses or expenditures. There are three principal constraints to meeting energy requirements through food intake: 1) the quality of available food, 2) the animal's metabolic requirements, and 3) the animal's physical capacity to eat (primarily regulated by its mouth size and stomach capacity).

The energy requirements of any mammal are approximately proportional to its body weight to the power 0.75 ($W^{0.75}$), and are closely linked to heat transfer and hence to surface area of the body. This general relationship has been shown to hold for comparisons among species, but not necessarily within species (e.g., between sexes, or between juveniles and adults). This relationship is partly explained by considering that heat transfer is a function of surface area and so larger animals have a smaller surface area-to-volume ratio than do small animals. Also, not all parts of an animal metabolize energy (heat) at the same rate. In larger animals, a higher proportion of their body is made up of structural components with relative low metabolic rates. As a consequence, small animals have a greater maintenance cost per unit body weight than do large animals.

Many aspects of the body relate to metabolic size and energy requirements, but at some point they are limited by the animal's physical size. This is true of the gastrointestinal capacity of the digestive system. One consequence is that smaller herbivores cannot develop a sufficiently large gastrointestinal tract to match their needs. Ideally, animals evolve a gastrointestinal capacity that match their metabolic requirements, but because of the link between gastrointestinal capacity and physical size, small herbivores cannot use this strategy. Small herbivores need more energy per unit of body size because of higher metabolic requirements, but their small gastrointestinal capacity limits the amount of food they can process. There are least three ways to meet this need, but because of their relatively high metabolic requirement and small gastrointestinal capacity, small herbivores (including ungulates) gain the greatest benefit from consuming higher-quality diets than do larger herbivores. This relationship between selection for high-quality foods and small body size is apparent within the Artiodactyla, especially the Bovidae. Small species of antelope, such as dik-dik, feed primarily on plants and plant parts with high nutrient quality, whereas large species such as African buffalo have broader diets and consume large quantities of low-quality forage such as grasses.

Social behavior

Ungulates are social mammals; most live in groups that can range from a pair to several thousands, although a few can be relatively solitary. Group size seems generally related to the visual density of the habitat; large groups form in open areas, and smaller ones in more closed habitats. In most species, males and female are sexually segregated and live in separate groups for most of the year, coming together during the mating season. However, in some species (e.g., equids, vicuñas) an adult male will live with a group of females and young throughout the year. The most likely reason for sexual segregation is that the sexes have different requirements for food and security habitat. Adult females need areas for raising their young that are relatively safe from predators, even if this means feeding in areas with poorer forage conditions. Males require abundant, high-quality food so they can maximize their growth and body condition for competing with other males for females. Often these different requirements can only be met in different habitats, leading to sexual segregation.

Ungulates enjoy at least two main benefits from living in groups: reduced predation and greater feeding efficiency. Reducing the chances of being killed by a predator is the main reason ungulates live in groups. Group living means that an individual can use other group members to hide behind when a predator attacks, and it also means greater vigilance because there are many pairs of eyes to keep watch for predators. Because there will always be several group members scanning for predators, other individuals can spend more time feeding than if they were alone. In species that are large compared their common predators, they might have a better chance to defend themselves collectively as a group than if they tried to defend themselves alone. However, the most important benefit ungulates gain from group living is the dilution effect. This arises simply because for each group member, the probability of being killed when a predator attacks, is inversely proportional to group size. Even in a group of two, when attacked by a single predator, the probability of an individual being killed is reduced by 50%.

Many ungulates use chemicals to communicate with each other, often depositing them on the ground, bushes or other places where conspecifics will encounter them. This means that an animal can make its presence known without actually having to be present, which can be useful for species defending large territories. Urine and feces are commonly used chemical signals, but many species also possess glands. Many of these are called epithelial glands because they are modifications of the skin or epithelium. Glands produce odoriferous chemical secretions as either volatile chemicals or waxy material, and they use these chemicals to communicate with each other for a variety of purposes. The glands themselves can be found at different locations on the body depending on species. Typical paired glands are ant- or pre-orbital glands seen as depressions, pits or slits just in front of the eye. Others are found between the hooves of some artiodactyls (pedal glands), while others can be associated with the tail (caudal glands) and hind legs.

Reproduction and mating systems

Most wild ungulates breed only once each year and the timing of birth is the main determining factor, which in turn is related to the annual cycle of plant growth. The seasonal pattern of plant growth is governed primarily by moisture and temperature. In temperate regions, this means that plants begin to grow in spring when warmer days lead to higher soil temperatures and snow melt or rainfall is also sufficient. In warmer arid areas, plant growth begins with the rainy season. Newborn ungulates need to grow rapidly so they can minimize the chances of falling prey to predators, and consequently they need to be born as early in the plant-growing season as possible. Hence, they are born either in early spring or at the beginning of the rainy season. Gestation period and birth season effectively determine when mating takes place. In some species, this is up to nine months before births and often when the adults are in their best physical condition.

Ungulates have evolved a variety of mating systems, almost all based on polygynous mating in which one male mates with several females. Basically males either defend and court a single female (tending pair), or defend a group of females (harem), mating with each as they come into heat (estrus). Only a few species form pair bonds.

The mating period in ungulates is often referred to as the rut, and is the period when males seek females coming into heat, and when females try to select a suitable mate. Courtship helps both genders achieve their goals. Professor Niko Tinbergen, one of the founders of modern ethology (animal behavior), suggested that there are up to four main functions in courtship in animals: orientation, persuasion, synchronization, and reproductive isolation. In ungulates, persuasion is probably the most important, the rest have minor roles. Although male ungulates seem to be the most active using often elaborate courtship patterns and rituals, the apparently passive females play an important role.

When a female comes into heat, chemicals in her urine act as signals that males use as cues to her reproductive condition. Males seeking females coming into estrus usually approach in a submissive or non-aggressive posture, so they can get close to investigate her. Quite often the female urinates when the male approaches and the male then tests the urine using a behavior called flehmen or lip curl. It appears that by doing this, the male is using his paired vomeronasal organs located in his upper palate to test the urine. These organs are sensitive to the chemical cues found in an estrous female's urine. Once a female coming into heat is located, the male begins to court her with species-specific courtship patterns that, if she chooses, leads to copulation. The female's role in selection of her mate, while not overt, is critical to the evolution of the mating system.

Sexual dimorphism and elaborate weapons are both usually indications of competition for mates or for attracting mates, commonly associated with polygynous mating systems. In polygynous mating, a male will usually mate with more than one female, so only a few males in a population will have the opportunity to breed. This can lead to intense competition among males, often involving ritual displays and fighting. Charles Darwin was one of the first to recognize that animal weapons function primarily for intraspecific competition (competition between members of the same species), and are used only occasionally for defense against predators (interspecific interactions between different species). Besides their use in fighting, ungulates also frequently use their weapons for displays.

Four basic weapons systems are recognized in ungulates. The simplest are the hard, often sharp hooves, though obviously these did not evolve primarily for fighting. Some species have long canine teeth that they use for fighting. Suoids have sharply pointed upper and lower canines (tusks), camels have smaller but sharp canine teeth, and some cervoids have relatively long, dagger-like upper canines. A third weapon type are antlers typical of modern cervid deer, though not all species have them. Finally, there are horns, which are found in four living families, each with its own unique type; the rhinoceroses, the giraffes, the bovids, and the pronghorn "antelope."

The various aspects of ungulate's lives are intertwined. The habitats a species uses influence food selection and impose constraints on body and group size. In turn these affect antipredator strategies as well as mating systems. The intricacies of these various ecological processes provide the rich morphological and behavioral differences characteristic of ungulates in modern times. Also, it is these diverse and large herbivores that sustain large predator species. The ungulate herbivores and carnivores define the mammalian assemblages that form communities throughout the world except Australia. It is from these diverse ungulates that humans have drawn sustenance and domestic animals that have supported their development.

Resources

Books

Bubenik, G. A., and A. B. Bubenik, eds. *Horns, Pronghorns, and Antlers: Evolution, Morphology, Physiology, and Social Significance.* New York: Springer-Verlag, 1990.

Feldhamer, G. A., L. C. Drickamer, S. H. Vessey, and J. F. Merritt. *Mammalogy: Adaptation, Diversity and Ecology.* Boston: WCB/McGraw-Hill, 1999.

Geist, V. *Mountain Sheep: A Study in Behavior and Ecology.* Chicago: The University of Chicago Press, 1971.

Gosling, L. M. "The even-toed ungulates: order Artiodactyla—sources, behavioural context, and function of chemical signals." In *Social Odors in Mammals*, Vol. 2, edited by R. T. E. Brown and D. W. Macdonald, pp 550-618. Oxford: Clarendon Press, 1985.

Krebs, J. R., and N. B. Davies. *An Introduction to Behavioural Ecology.* 3rd edition. Oxford: Blackwell Scientific Publications, 1993.

Macdonald, D., ed. *The Encyclopaedia of Mammals.* New York: Facts On File Publications; and Oxford: Equinox (Oxford) Ltd., 1984.

Nowak, R. M. *Walker's Mammals of the World*, 5th edition, Vol. 2. Baltimore and London: The John Hopkins University Press, 1991.

Pough, F. H., C. M. Janis, and J. B. Heiser. *Vertebrate Life*, 6th edition. Upper Saddle River, NJ: Prentice Hall, 2002.

Van Soest, P. J. *Nutritional Ecology of the Ruminant.* 2nd Edition. Ithaca: Cornell University Press, 1994.

Vaughan, T. A. *Mammalogy.* 3rd edition. New York: Saunders College Publishing, 1986.

Walther, F. R. *Communication and Expression in Hoofed Mammals.* Bloomington: Indiana University Press, 1984.

Periodicals

Bleich, V. C., R. T. Bowyer, and J. D. Wehausen. "Sexual segregation in mountain sheep: Resources or predation?" *Wildlife Monographs* 134 (1997): 1–50.

Fortelius, M., J. Eronen, J. Jernvall, L. Liu, D. Pushkina, J. Rinne, A. Tesakov, I. Vislobokova, Z. Zhang, and L. Zhou. "Fossil mammals resolve regional patterns of Eurasian climate change over 20 million years." *Evolutionary Ecology Research* 4 (2002): 1005–1016.

Gore, R. "The rise of mammals." *National Geographic* 203 (2003): 2–37.

Graur, D. and D. G. Higgins. "Molecular evidence for the inclusion of the Cetaceans within the order Artiodactyla." *Molecular Biology and Evolution* 11 (1994): 357–364.

Janis, C. M. "New ideas in ungulate phylogeny and evolution." *Trends in Ecology and Evolution* 3 (1988): 291–297.

———. "Tertiary mammal evolution in then context of changing climates, vegetation, and tectonic events." *Annual Review of Ecology and Systematics* 24 (1993): 467–500.

MacFadden, B. J. "Cenozoic mammalian herbivores from the Americas: Reconstructing ancient diets and terrestrial communities." *Annual Review of Ecology and Systematics* 31 (2000): 33–59.

Main, M. B., F. W. Weckerly, and V. C. Bleich. "Sexual segregation in ungulates: New directions for research." *Journal of Mammalogy* 77 (1996): 449–461.

Montgelard, C., F. M. Catzeflis, and E. Douzery. "Phylogenetic relationships of artiodactyls and cetaceans as deduced from the comparison of cytochrome b and 12S rRNA mitochondrial sequences." *Molecular Biology and Evolution* 14 (1997): 550–559.

Nikaido, M., A. P. Rooney, and N. Okada. "Phylogenetic relationships among cetartiodactyls based on insertions of short and long interpersed elements: Hippopotamuses are the closest extant relatives of whales." *Evolution* 96 (1999): 10261–10266.

Radinsky, L. B. "The adaptive radiation of the phenacodontid condylarths and the origin of the Perissodactyla." *Evolution* 20 (1966): 408–417.

———. "The early evolution of the Perissodactyla." *Evolution* 23 (1968): 308–328.

Rose, K. D. "On the origin of the order Artiodactyla." *Proceeding of the National Academy of Sciences, U.S.A.* 93 (1996): 1705–1709.

David M. Shackleton, PhD
Alton S. Harestad, PhD

Ungulate domestication

What is domestication?

The domestication of animals, especially the hoofed mammals, has been practiced for centuries and has profoundly shaped the course of human history in the last 13,000 years. It contributed to the rise of civilization, single-handedly transformed global demography, and provides a sizable amount of our food and clothing today.

A prime example are horses, which were domesticated in southeast Europe 6,000 years ago. While their original purpose was riding, they also provided meat, milk, and long-distance transportation and warfare capabilities that led to the spread of Indo-European languages and culture and the transformation of ancient social orders.

In 1984, Price defined domestication as "that process by which a population of animals becomes adapted to man and to the captive environment by some combination of genetic changes occurring over generations and environmentally induced developmental events recurring over each generation." Price himself admits that this rather simplified definition of domestication does not allow for the possibility that genes and the environment may operate as independent factors that acted additively on each other. Lickliter and Ness in 1990 pointed out that the development of domestic physical characteristics can only be understood in terms of the complex interplay of organic and environmental factors during embryo development.

Both these definitions assume that the animals have been plucked from their ancestral wild environments and placed into a different, captive one. It also supposes that certain animal management and housing practices have been applied over time in rearing and maintaining each species in captivity.

How animals were selected for domestication

Both cultural and practical reasons have determined which animals were domesticated and where or when their domestication occurred. With the possible exception of the dog, domestication occurred when humans were faced a specific need and requirement that could only be fulfilled through corralling and breeding a certain animal population.

Although food supplies were available through hunting, fishing, and gathering in tropical regions, animals in those regions were domesticated as work animals or to provide fiber, as well as for food. The water buffalo (*Bubalus bubalis*), for example, was used primarily as a beast of burden and not for meat or milk.

Another issue that comes into play is whether the animal can be domesticated or not. Most wild animals avoid close human contact unless they have been habituated to the presence of people. The interaction of captive animals with humans thus plays a key factor in the domestication process. The degree of *tameness* in individual animals determines, to a large extent, the nature of that interaction.

Tameability, like tameness, is a desirable trait in animals undergoing domestication. We can surmise that some species, breeds, and individuals are more tameable than others, and that tameability is a heritable trait. The component of tameness, some argue, is not transmitted from mother to offspring.

In 1974, Blaxter determined that the offspring of hand-reared (very tame) ungulates, when exposed to people in the absence of members of their own species, exhibit similar flight distances from humans as the mother-reared (relatively untamed). Still, relatively untamed animals often show less fear of people in the presence of tame members of the same species. Lyons, Price, and Moberg reported that relatively untamed mother-reared dairy goats (*Capra hircus*) exhibited shorter flight distances from humans when exposed to people in the presence of tame herd-males.

The question of tameability may provide an answer to why, in a world that has provided 148 species weighing 99 lb (45 kg) or more, only 14 of those species have actually been domesticated. There are many cases in which only one out of a closely related group of species was successfully domesticated. Horses and donkeys, for example, were one of the first domesticates, but none of the four zebra species that are able to interbreed with them have been actually tamed. At the same time, while five of the most valuable domestic mammals—the goat, sheep, cow, pig, and horse—had all been domesticated repeatedly by 4,000 B.C., the sole addition within the last millennium has been the reindeer.

Ongoing efforts at domesticating other large wild mammals have often resulted in failure, as in the cases of the eland, elk, moose, musk ox, and zebra, or at best, have led to ranched

The donkey (*Equus asinus*) was domesticated about 6,000 years ago and is used throughout the world. (Photo by © Hanan Isachar/Corbis. Reproduced by permission.)

animals such as the deer and American bison that still cannot be herded and that have trivial economic value in comparison to the five most valuable domesticated mammals.

According to Diamond, the obstacle lies with the species itself, not with the local people. In the case of zebras, European horse breeders who settled in South Africa in the 1600s, like the African herders of previous millennia, gave up trying to domesticate zebras after several centuries. Zebras are exceptionally and incurably vicious, for one thing. Diamond recounts their nasty habit of biting a handler and not letting go until the handler is dead. They also have better peripheral vision than horses, which makes them impossible to lasso since they can see the rope coming and will flick away their head.

Other examples of species characteristics that stand in the way of domestication include: the tendency of gazelle and deer to grow territorial and violent against their keepers; the absence of follow-the-leader dominance hierarchies in antelope; and the fierce dispositions of rhinoceroses. Though elephants have been tamed and domesticated, their slow growth rate and long birth spacing provides an impediment to large-scale domestication.

Not all wild animals were recognized for their domestic value, however. The quagga (*Equus quagga*), a small, horse-like animal from South Africa, was regarded as a pest and hunted to extinction despite its docile nature, potential ease of tameability, and resistance to diseases that plagued imported horses of European descent.

Domestication of the goat and sheep

Goats were first domesticated in the highlands of western Iran 10,000 years ago and were probably the first ruminants to be domesticated because of the presence of wild goats in regions where agriculture was developing. Archaeological evidence and carbon dating show a distinct shift toward the selective harvesting of young male goats. This probably marks initial human management and the transition from hunting to the herding of the species.

The ancestry of the domestic goat (*Capra hircus*) can be traced to wild goats (*C. aegagrus*) and the Nubian ibex (*C. ibex*) native to the canyon system. As it is difficult to distinguish between the bones of these two species, the exact percentages of each species in today's domestic goats are difficult to determine. According to Luckart and his co-workers, the earliest unambiguous fossil evidence of domestic goats was found in southwest Iran dating 9,000 years ago and in the Iranian plateau dating 10,000 years ago.

Sheep are bred for their quality of wool. (Photo by © Chris Lisle/Corbis. Reproduced by permission.)

A domestic goat (*Capra hircus*) nursing her young. (Photo by Hans Reinhard. Bruce Coleman, Inc. Reproduced by permission.)

In the case of domestic sheep (*Ovis aries*), DNA sequencing and analyses carried out in 2002 provided strong evidence that they originated from two subspecies of the mouflon (*O. musimon*). The mouflon is a rare breed of primitive domestic sheep whose population is shrinking on the Mediterranean islands of Sardinia, Corsica, and Cyprus. However, it has been successfully introduced into central Europe, including Germany, Austria, Czech Republic, Slovak Republic, and Romania.

Many morphological, physiological, and behavioral traits characterize and distinguish domestic sheep and goats from their wild ancestors, such as diminished sexual dimorphism, decreased brain, body, and horn size, changes in horn shape, and even in the color of the coat. In 1998, Zohary and his colleagues proposed that these differences were shaped largely by unconscious selection. Once the founder herd had been assembled and controlled by humans, the transfer of these animals from their wild environments into the markedly different human-made husbandry system resulted in drastic changes in selection pressures. Several adaptations that were vital for survival in the wild then lost their fitness under the new conditions, subsequently breaking down through lack of use. New traits, which now characterize domestic sheep and

The United States produces over 46 million lb (20 million kg) of wool a year from domesticated sheep. (Photo by © Stephanie Maze/Corbis. Reproduced by permission.)

The cow (*Bos taurus*) can be used for dairy production, meat, or to move heavy loads. (Photo by © Dave G. Houser/Corbis. Reproduced by permission.)

goats, were selected for instead. Protection from predators, culling of young males, protection from the elements, and changes in land use and in food and water supplies are considered the main ecological factors introduced by humans at the start of sheep and goat domestication.

Variation in the behavior patterns of different breeds of sheep, including level of tameness, has also been attributed to breeding under the control of humans as well as selection for productivity. Selection of rams for domestic behavior, for instance, often leads to hereditary changes of behavior diversity. When Lankin studied the domestic behavior of 11 breeds of sheep (*O. aries*) in 1997, he came to the conclusion that breeds subjected to intensive selection for commercial purposes tended to be tamer toward humans than "wilder" breeds. The East Fresian breed, which has been intensively selected for meat and milk production, for instance, is particularly tame toward humans. A few years later, Lankin found that "wild" ewes showed increased levels in the stress hormone corticosteriod when they were isolated, transported, or competing for feed. On the other hand, "domestic" ewes showed the greatest stress hormone response to being paired with another sheep in a cage.

Domestication of the cow

It is generally believed that cattle were first domesticated in southwest Asia, particularly Anatolia, or in southeast Europe, where their remains have been found in several sites dated between 9,000 and 8,000 years ago. Large bovid bones discovered at several small sites in the Western Desert of Egypt have been identified as belonging to domestic cattle. The bones were radiocarbon dated to between 9,500 and 8,000 years ago, raising the possibility that there was a separate, independent center for cattle domestication in northeast Africa. There remains, however, some contention about the identity and therefore, the validity of the bones.

A 2003 study suggests that Britons were harvesting milk as early as 6,000 years ago. Evershed and his University of Bristol colleagues examined over 950 broken pieces of crockery from 14 archaeological sites in Britain that date to the Neolithic, Bronze Age, and Iron Age. They found evidence of dairy consumption, although the ages of the milk-spotted pieces of pottery varied from site to site. The authors concluded that animals were already being exploited for milk at the time farming arrived in Britain in the late fifth millennium B.C.

Domestication of the pig

Domestication of the pig is likely to have occurred first in the Middle East about 9,000 years ago and may have occurred repeatedly from local populations of wild boars. The wild boar population, with at least 16 different subspecies proposed, is widespread in Eurasia and occurs in Northwest Africa. His-

Oxen are used to pull heavy loads, such as this cart loaded with sugar cane. (Photo by © Tony Arruza/Corbis. Reproduced by permission.)

torical records indicate that Asian pigs were introduced into Europe during the eighteenth and early nineteenth centuries.

Modern domestic pigs show marked morphological differences when compared to their wild ancestors. However, it has not yet been established whether domestic pigs have a single or multiple origin.

In 1868, Darwin described two major forms of domestic pigs—a European and an Asian form. The former was assumed to originate from the European wild boar, while the wild ancestor of the latter was unknown. Darwin considered the two forms as distinct species on the basis of profound phenotypic differences. It is well documented that Asian pigs were used to improve European pig breeds during the eighteenth and early nineteenth centuries but to what extent Asian pigs have contributed genetically to different European pig breeds is unknown.

A 1998 study using microsatellite markers estimated the divergence between major European breeds and the Chinese Meishan breed at about 2,000 years. Some mitochondrial DNA studies during that time also indicated genetic differences between European and Asian pigs, but provide no estimate of the time the breeds diverged.

Finally, in 1999, an investigation aiming to provide a more comprehensive molecular analysis of the origin of domestic pigs evaluated wild and domestic pig populations from Asia and Europe. Giuffra and his fellow researchers came to the conclusion that the European and Asian subspecies of the wild boar were independently domesticated, and estimated the time since divergence of the ancestors for European and Chinese Meishan domestic pigs at 500,000 years ago. Their data showed that European domestic pigs and Chinese Meishan pigs are closely related to existing subspecies of the Eurasian wild boar (*S. scrofa*).

Domestication of the horse

The domestication of the horse facilitated the development of human civilization by providing the means of effective transport, agriculture, industry, and warfare.

Wild horses were widely distributed throughout the Eurasian steppe during the Upper Paleolithic around 35,000 to 10,000 years ago, but in many regions, they disappeared from the fossil record about 10,000 years ago. Today, only one supposed wild population, the Przewalski's horse (*Equus caballus przewalskii*), remains. Przewalski's horse is Extinct in the wild, but has lived in captivity and efforts are underway to reintroduce it into the wild.

Horse remains are common in archaeological sites of the Eurasian grassland steppe dating from about 6,000 years ago,

The domestic pig (*Sus scrofa*) has a wide variety of uses for humans, the main use being for food. (Photo by © Macduff Everton/Corbis. Reproduced by permission.)

suggesting the time and place of their first domestication. Evidence from bit wear patterns on the teeth suggests that some horses could have been ridden.

There are two different hypotheses for the origin of the domestic horse (*E. caballus*) from wild populations. The "restricted origin" hypothesis suggests that the domestic horse was selectively and multigenerationally bred from a limited wild stock from a few domestication centers. The domestic horses were then distributed to other regions.

Another school of thought suggests that domestication involved a large number of founder animals recruited over an extended period of time throughout the extensive Eurasian range of the horse. In this "multiple origins" scenario, horses may have been captured from diverse wild populations and then increasingly bred in captivity as wild populations declined.

In 2001, an analysis of mitochondrial DNA from 191 domestic horses revealed a high diversity of matrilines, or lines of descent traced exclusively through female members from one founding female ancestor. This suggests that wild horses from a large number of populations were utilized as founders of the domestic horse. A single, geographically restricted population would have limited the founding lineages. The results also showed a bias toward females in ancient breeding and trade. This is consistent with modern breeding practices in which select studs are used for a much larger population of females.

Domestication of the camelids

Information on the presence and utility of camels in India from the proto-historic period to the present is supported by archaeological evidence, literature, and the arts. The camel in India has been an animal of utility from early Harappan level of civilization (c. 3000–1800 B.C.). One hypothesis suggests that the single-humped camel, or dromedary (*Camelus dromedarius*), was independently domesticated by the Indus people, while others are of the opinion that the domesticated double-humped camel was the species present in the Indus Valley during the third millennium B.C.

Because of the absence of detailed analysis of bones excavated from archaeological sites, scientists and archaeologists are hard-pressed to reach any conclusive identification regarding the species level of the ancient Indian camel. Today, the domestic bactrian camel (*C. bactrianus*) is on the verge of extinction in India, while the dromedary, boasting population

The camel (*Camelus* sp.) can go many days without food or water, which made it a viable resource for desert travelers. (Photo by Corbis. Reproduced by permission.)

numbers of 1.1 million, is an important domesticate widely distributed throughout the northwestern parts of India.

Some debate rages over the origins of the guanaco (*Lama guanicoe*) and llama (*L. glama*) and if faunal remains from Andean archaeological sites can resolve this issue. Changes in incisor morphology during the domestication process suggest that the alpaca may be descended from the vicuña (*Vicugna vicugna*). A comparison of fiber production characteristics in preconquest and extant llama and alpaca breeds, however, indicates that extensive hybridization between the two species is likely to have occurred since European contact.

Consequences of domestication

The transition from the hunter-gatherer lifestyle to food production beginning around 8,500 B.C. enabled people to adopt a more sedentary lifestyle, rather than migrating to follow seasonal shifts in wild food supplies. (Some movement was still necessary with herds of domestic animals.) The production of food led to a human population explosion, since this lifestyle allowed shorter birth intervals.

Animal production did not come without a cost as it meant that humans had to share disease-causing microbes with the animals. Molecular studies of microbes have revealed the presence of their closest relatives in the domesticated animal species. The measles virus may have evolved from the rinderpest virus found in cattle. Smallpox virus may have evolved from cowpox virus. It has also been postulated that human influenza virus may have arisen from a mixture of influenza viruses in ducks and pigs.

The future of domestication?

In his 2002 paper on the past, present, and future of domestication, Diamond wondered aloud if the rise of molecular biology, genetics, and improved understanding of animal behavior might not allow the domestication of species that have proven undomesticable in the past. On the other hand, challenges exist with our existing varieties of domesticates, even as science and technology becomes increasingly sophisticated.

A study published online by the journal *Nature Biotechnology* provides a sneak peek into the possible future of dairy farming. Brophy and her coworkers at the Ruakura Research Center in New Zealand, took cells from female dairy cows and altered them to include additional copies of two genes that are instrumental in the production of the milk protein casein. The cells were then fused with donor eggs and implanted into surrogate mothers to bring to term. Out of the 11 cloned cows that managed to survive, only nine yielded milk with elevated levels of two casein molecules. Their milk had 8–20% more beta-casein and nearly twice as much kappa-casein as milk from regular dairy cows.

These enhanced properties of the milk should speed up the cheese manufacturing process and increase the cows' productivity. This was the first time that scientists have modified cow's milk solely to improve its quality instead of altering it to manufacturing proteins of pharmaceutical interest. The authors concluded that "the magnitude of the observed changes highlights the potential of transgenic technology to tailor milk composition in dairy cows."

Horses (*Equus caballus caballus*) are some of the most important domesticated animals. (Photo by © Gunter Marx Photography/Corbis. Reproduced by permission.)

Resources

Books

Anthony, D.W. *Horses Through Time*, edited S. L. Olsen. Boulder, CO: Roberts Rinehart for the Carnegie Museum of Natural History, 1996.

Bökönyi, S. *History of Domestic Mammals in Central and Eastern Europe*. Budapest: Akademiai Kiado, 1974.

Clutton-Brock, J. *A Natural History of Domesticated Mammals*. 2nd ed. Cambridge: Cambridge University Press, 1999.

Darwin, C. *The Variation of Animals and Plants under Domestication*. London: John Murray, 1868.

Hale, E. B. "Domestication and the Evolution of Behavior." In *The Behavior of Domestic Animals*, edited by H. S. E. Hafez. London: Bailliére, Tindall, and Cox, 1969.

Isaac, E. *Geography of Domestication*. Englewood Cliffs, NJ: Prentice-Hall, 1970.

Price, E. O. *Animal Domestication and Behavior*. Cambridge, MA: CAB International, 2002.

Ridgeway, W. *The History and Influence of Thoroughbred Horses*. Cambridge: Cambridge Biological Series, 1905.

Ruvinsky, A., and M. F. Rothschild, eds. *The Genetics of the Pig*. Wallingford, Oxon, UK: CAB International, 1998.

Zeuner, F. E. *A History of Domesticated Animals*. New York: Harper and Row, 1963.

Periodicals

Anthony, D. W. "The Kurgan Culture. Indo-European Origins, and the Domestication of the Horse: A Reconsideration." *Current Anthropology* 27 (1986): 291.

Blaxter, K. L. "Deer Farming." *Mammal Review* 4 (1974): 119–122.

Brisbin, L. L., Jr. "The Ecology of Animal Domestication: Its Relevance to Man's Domestic Crisis—Past, Present and Future." *Association of Southeastern Biologists Bulletin* 21 (1974): 3–8.

Diamond, J. "Evolution, Consequences and Future of Plant and Animal Domestication." *Nature* 21, no. 418 (2002): 700–707.

Downs, J.F. "Domestication: An Examining of the Changing Social Relationships Between Man and Animals." *University of California, Berkeley, Anthropological Society Papers* 22 (1960): 18–67.

Giuffra, E., J. M. H. Kijas, V. Amarger, Ö Carlborg, J.-T. Jeon, and L. Andersson. "The Origin of Domestic Pigs: Independent Domestication and Subsequent Introgression" *Genetics* 154 (2000): 1785–1791.

Hediger, H. "Tierpsychologie und Haustierforschung." *Zeitschrift für Tierpsychologie* (1938): 29–46.

Hiendleder, S., B. Kaupe, R. Wassmuth, and A. Janke "Molecular Analysis of Wild and Domestic Sheep Questions Current Nomenclature and Provides Evidence for Domestication from Two Different Subspecies." *Proceedings of the Royal Society of London* 269, no. 1494 (2002): 893–904.

Khanna, N. D. "Camels in India from Protohistoric to the Present Times." *Indian Journal of Animal Sciences* 60, no. 9 (1990): 1093–1101.

Lankin, V. "Factors of Diversity of Domestic Behavior in Sheep." *Genetics Selection Evolution* 29, no.1 (1997): 73–92.

———. "Domesticated Behavior in Sheep. Role of Behavioral Polymorphism in the Regulation of Stress Reactions in Sheep." *Genetics Selection Evolution* 35, no. 8 (1999): 1109–1117.

Lickliter, R., and J. W. Ness. "Domestication and Comparative Animal Psychology: Status and Strategy." *Journal of Comparative Pyschology* 104 (1990): 211–218.

Luckart, G., L. Gielly, L. Excoffier, V. Curry, N. Pidancier, J. Bouvet, and P. Taberlet. "Domestication Origins and Phylogenetic History of Domestic Goats." *Proceedings of the 7th International Conference of Goats in Tours* 104 (May 15–18, 2002).

Lyons, D. M., E. O. Price, and G. P. Moberg. "Individual Differences in Temperament of Dairy Goats: Constancy and Change." *Animal Behavior* 36 (1988a): 1323–1333.

———. "Social Modulation of Pituitary-Adrenal Responsiveness and Individual Differences in Behavior of Young Domestic Goats." *Physiology and Behavior* 43 (1988b): 451–458.

Paszek, A. A., G. H. Fluckinger, L. Fontanesi, C. W. Beattie, and G. A. Rohrer. "Evaluating Evolutionary Divergence with Microsatellites." *Journal of Molecular Evolution* 46 (1998): 121–126.

Price, E. O. "Behavioral Aspects of Animal Domestication." *Quarterly Review of Biology* 59 (1984): 1–32.

Vilà, C., J. A. Leonard, A. Götherström, S. Marklund, K. Sandberg, K. Lidèn, R. K. Wayne, and H. Ellegren. "Widespread Origins of Domestic Horse Lineages." *Science* 291 (2001): 474–477.

Watanabe, T., Y. Hayahsi, J. Kimura, Y. Yasuda, and N. Saitou. "Pig Mitochondrial DNA; Polymorphism, Restriction Map Orientation, and Sequence Data." *Biochemical Genetics* 24 (1986): 385–396.

Wendorf, F., and R. Schild. "Are the Early Holocene Cattle in the Eastern Sahara Domestic or Wild?" *Evolutionary Anthropology* 3 (1994): 118–128.

Wheeler, J. C. "Evolution and Present Situation of the South American Camelidae." *Biological Journal of the Linnean Society* 54, no. 3 (1995): 271–295.

Yuichi, T. "The Roles of Domesticated Animals in the Cultural History of the Humans." *Asian-Australasian Journal of Animal Sciences* 14 (2001): 13–18.

Zeder, M. A., and B. Hesse. "The Initial Domestication of Goats (*Capra hircus*) in the Zagros Mountains 10,000 Years Ago." *Science* 287, no. 5461 (2000): 2254–2257.

Zohary, D., E. Tchernov, and L. K. Horwitz. "The Role of Unconscious Selection in the Domestication of Sheep and Goats" *Journal of Zoology* 245, no. 2 (1998): 129–135.

Other

Graham, S. "Dairy Farming Old and New." *Scientific American Online*, January 28, 2003. <http://www.sciam.com/article.cfm?articleID=0008EBF9-AB8B-FC809EC5880000>.

Brophy, B., G. Smolenski, T. Wheeler, D. Wells, P. L'Huiller, and Götz Laible. "Cloned Transgenic Cattle Produce Milk with Higher Levels of β-casein and κ-casein." *Nature Biotechnology Online*, January 27, 2003. <http://www.nature.com/cgi-taf/DynaPage.taf?file=/nbt/journal/v21/n2/full/nbt783.html>

Jasmin Chua, MS

Tubulidentata
Aardvarks
(Orycteropodidae)

Class Mammalia
Order Tubulidentata
Family Orycteropodidae
Number of families 1

Thumbnail description
Small- to medium-sized stocky anteater with short, powerful limbs (front shorter than back) with claws used for digging; tapering, muscular tail, pig-like snout, elongated head, and pale, yellowish gray sparsely haired bodies; long, tubular ears are held upright

Size
Length 67–79 in (170–200 cm); tail length 18–25 in (45–63 cm); weight 99–139 lb (45–65 kg)

Number of genera, species
1 genus; 1 species

Habitat
Savanna and woodland

Conservation status
Vulnerable

Distribution
Sub-Saharan Africa

Evolution and systematics

The Tubulidentata are the last living group of primitive ungulates. The only extant species in the order, the aardvark, was once thought to be an edentate. However, it has no phylogenetic ties with this group, and the similarities between the aardvark and the South American anteater result from convergent evolution since both animals feed on termites and ants. The Tubulidentata arose in Africa, and the modern aardvark existed as early as the middle Tertiary (Miocene). In the Pliocene, there were aardvarks in southern Europe and western Asia, and during the Pleistocene they were on Madagascar as well. All modern aardvarks belong to one species, *Orycteropus afer*. None of the 18 subspecies are considered of taxonomic importance any longer.

The taxonomy for this species is *Orycteropus afer* (Pallas, 1766), Cape of Good Hope, South Africa.

Physical characteristics

The head is elongated and the snout is long and pig-like. The ears are tubular, and the yellow-gray colored body has little hair. The fore feet have four toes and well-developed claws, while the hind feet have five toes. The tail is well developed and resembles the tail of a kangaroo.

The name aardvark, Afrikaans for earth pig, is derived from the pig-like snout and digging behavior. The aardvark is uniquely shaped: the short neck joins the massive body with an elongated head and a narrow, rounded snout. The nasal openings can be closed. The muscular tail has a circumference of 18 in (40 cm) at its base. The legs are short, with the hind legs longer than the fore legs. The fore feet have toes adapted for digging; the claws of the five toes on the hind feet are somewhat shorter and weaker. Although the hind feet have soles, aardvarks always move on their toes. The soles do rest on the ground when the aardvark assumes its characteristic stooped position as it digs up termite hills.

There are many embryonic teeth, which rest in tooth cavities, but adults have teeth only in the rear of the jaw. They are columnar and rootless, and have hexagonal prisms of dentine. The tooth formula is 0 0 2 3/0 0 2 3. The teeth are unusual in that they grow throughout life and are not coated in

Aardvark (*Orycteropus afer*). (Illustration by Joseph E. Trumpey)

enamel. Instead, each tooth has numerous hexagonal prisms of dentine surrounding tubular pulp cavities. The largest tooth, the second molar, is composed of 1,500 such hexagonal prisms. An aardvark typically has five functional molars in each jaw half, though sometimes as few as four or as many as seven. The canines are usually missing.

Distribution

Aardvarks have a wide distribution on the continent of Africa alone, but are not common anywhere and, as a result of their secretive nocturnal habits, are rarely seen. They occur across Africa south of the Sahara, but generally avoid true forests (although they have been recorded in the northeastern parts of the Congo Basin forests) and extremely arid areas.

Habitat

Aardvarks are found in a variety of habitats in their range, although their local occurrence is determined by the availability of food and the distribution of sandy soils. They are also capable of utilizing heavier soils, but will avoid rocky terrain, preferring more open areas.

Behavior

Aardvarks usually live alone and are never found in large numbers. Being nocturnal animals, they are rarely seen in the wild. They normally emerge from their burrows shortly after nightfall, though they may emerge late in the afternoon during winter. They forage on both dark and bright moonlit nights, but may take shelter in one of several burrow systems within their home range during spells of adverse weather or when disturbed. They forage over distances varying 1.2–3 mi (2–5 km) per night at a speed of about 1,640 ft (500 m) per hour. In the arid Karoo (South Africa), home ranges vary from 321–988 ac (130–400 ha). Although the ranges of neighboring aardvark overlap, individuals spend about half their time in a core area represented by one-quarter to one-third of their home range.

An aardvark (*Orycteropus afer*) burrowing at night. (Photo by Rudi van Aarde. Reproduced by permission.)

Aardvarks can dig with astonishing speed in suitable soil. The burrow can be 6.5–9.8 ft (2–3 m) long, on a 45° angle, and with a diameter of about 16 in (40 cm). It terminates in a rounded chamber where the aardvark sleeps coiled up and where the female bears her young. There is typically just one entrance, but some aardvark burrows form tunnel systems with numerous entrances; the main burrow may have several side tunnels. Abandoned aardvark burrows are used as dens by creatures such as warthogs, porcupines, wild dogs, viverrids, jackals, hyenas, birds (such as ant thrushes), and the bat *Nycteris thebaica*.

An aardvark (*Orycteropus afer*) in its burrow. (Photo by Alan Root/OKAPIA/Photo Researchers, Inc. Reproduced by permission.)

An aardvark (*Orycteropus afer*) foraging for ants and termites in Tussen de Riviere Reserve, South Africa. (Photo by Nigel J. Dennis/Photo Researchers, Inc. Reproduced by permission.)

Enemies of aardvarks, besides man (some do eat aardvark meat), include lions, hyenas, and leopards. Pythons occasionally enter aardvark burrows and may eat the young. An aardvark can defend itself only with the claws on its fore feet. When it is threatened, it lies on its back, raises up all four legs, and threatens with its claws. When it is pursued, it starts running in leaps and bounds to gain speed, then continues at a trot.

An aardvark (*Orycteropus afer*) resting during the heat of the day. (Photo by N. Myers. Bruce Coleman, Inc. Reproduced by permission.)

An aardvark (*Orycteropus afer*) emerging from its burrow in South Africa. (Photo by Animals Animals ©Anthony Bannister. Reproduced by pemission.)

Feeding ecology and diet

In the wild, practically the only food for aardvarks is termites and ants. In the savanna, they feed chiefly on the termite genera *Trinervitermes*, *Cubitermes*, and *Macrotermes*. In the semi-arid Karoo (South Africa), the ant *Anoplolepis custodiens* is the main food item throughout the year, followed by a termite (*T. trinervoides*). In the Karoo, termites are fed on more often in winter than in summer, coinciding with a decrease in the availability of ants. Aardvarks cannot fully satisfy their hunger by breaking open termite hills. The only way they can get enough termites is to find termite colonies making mass movements on the ground. Swarms of harvester termites (*Trinervitermes*) contain thousands of individuals marching in armies with columns 33–130 ft (10–40 m) long; *Macrotermes* and *Hodotermes* form similar nocturnal marches. Aardvarks usually dig up the termite hills at the base, using the sharp claws of the fore feet. As soon as termites appear at the surface, the aardvark extends its long tongue and licks the termites, which stick to the tongue. Aardvarks even press their disc-shaped snout against termite hills and suck the termites in.

An aardvark (*Orycteropus afer*) entering a burrow, showing its tapered tail. (Photo by Rudi van Aarde. Reproduced by permission.)

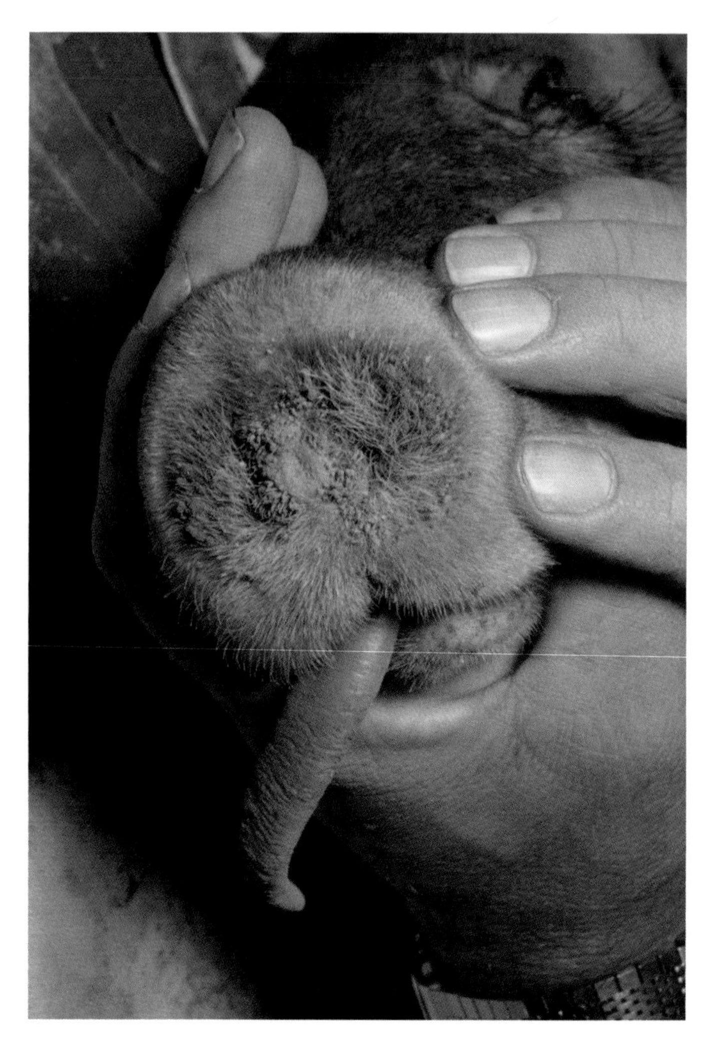

A close-up of an aardvark's (*Orycteropus afer*) snout, showing its circular nostrils and tapered tongue. (Photo by © Anthony Bannister; Gallo Images/Corbis. Reproduced by permission.)

Aardvarks occasionally consume the underground fruits of the cucumber species, *Cucumis humifructus*, a plant that in South Africa is known as the "aardvark pumpkin" or the "aardvark cucumber." The plant fruits underground and occasionally thrives in the vicinity of abandoned aardvark burrows. The Kung San people know the plant as "aardvark dung." The aardvark buries its feces outside the burrow, and thereby also the seeds of this plant, enabling the plant to reproduce.

Reproductive biology

Little is known about aardvark reproduction or raising of the young. The gestation period is about seven months. Usually one young of less than 4.4 lb (2 kg) is born at a time. In the southern Congo region, aardvarks mate between April and May, and the young are born in October or November. Females in Ethiopia bear their young in May or June. Anecdotal accounts from Africa suggest that aardvarks do not construct a birth nest in the burrow. A newborn aardvark is naked and has tender, flesh-colored skin. After about two weeks the young aardvark begins accompanying its mother on forays. Once the young aardvark is six months old it digs its own burrow, just a few feet (meters) from the mother's, although it continues hunting with her. During the next breeding season, the young male aardvark leaves its mother, but the young female stays with her until after the next offspring is born. The males roam about, associating with females only during the short mating period. Due to the fact that only the females keep a consistant home range, aardvarks are thought to be polygamous.

Conservation status

In being a specialized feeder, aardvarks are extremely vulnerable to habitat changes. While intensive crop farming over vast areas may reduce their numbers, increased cattle herding, whose trampling creates better conditions for termites, may increase their numbers. However, although they are widely distributed, aardvarks are not common anywhere. Though not listed by the IUCN, the aardvark is considered worthy of protection, according to the South Africa Red Data Book. There are no known conservation efforts directed primarily at aardvarks, but they do occur in most large conservation areas in Africa.

Significance to humans

Apart from aardvark flesh, which is said to taste like pork, various parts of the aardvark's body are prized. Its teeth are worn on necklaces by some tribes of the Democratic Republic of the Congo (Zaire) to prevent illness and as a good-luck charm. Its bristly hair is sometimes reduced to powder and, when added to local beer, regarded as a potent poison. It is also believed that the harvest will be increased when aardvark claws are put into baskets used to collect flying termites for food.

Resources

Books

Lindsey, Andrew P. *The Feeding Ecology and Habitat Use of the Aardvark* Orycteropus afer. Pretoria: MSc thesis, University of Pretoria, 1999.

Skinner, John D., and Reay H. N. Smithers. *The Mammals of the Southern African Sub-region.* Pretoria: University of Pretoria, 1990.

Taylor, Andrew. *The Ecology of the Aardvark* Orycteropus afer *(Tubulidentata-Orycteropodidae).* Pretoria: MSc thesis, University of Pretoria, 1999.

van Aarde, Rudolph J. "Aardvark." In *The Complete Book of Southern African Mammals,* edited by Gus Mills and Lex Hes. Cape Town: Struik, 1997.

Periodicals

Melton, D. A. "The Aardvark at Night." *Animals' Magazine* 16 (1974): 108–110.

———. "The Biology of the Aardvark (Tubulidentata-Orycteropodidae)." *Mammal Review* 6 (1976): 75–88.

Melton, D. A., and C. Daniels. "A Note on the Ecology of the Aardvark *Orycteropus afer.*" *South African Journal of Wildlife Research* 16 (1986): 112–114.

Taylor, Andrew, Peter A. Lindsey, and John D. Skinner. "The Feeding Ecology of the Aardvark *Orycteropus afer.*" *Journal of Arid Environments* 50 (2002): 135–152.

van Aarde, Rudolph J., et al. "Range Utilization by the Aardvark, *Orycteropus afer* (Pallas, 1766) in the Karoo, South Africa." *Journal of Arid Environments* 22 (1992): 387–394.

Rudi van Aarde, PhD

Proboscidea
Elephants
(Elephantidae)

Class Mammalia
Order Proboscidea
Family Elephantidae
Number of families 1

Thumbnail description
The largest living land animals, entirely herbivorous, characterized by the presence of a proboscis (trunk) and greatly elongated incisor teeth (tusks)

Size
Height at shoulder 6.5–13 ft (2–4 m); weight 2.2–7.7 tons (2–7 tonnes)

Number of genera, species
2 genera; 2 species (3 according to some authorities)

Habitat
Forest, savanna, and semi-desert

Conservation status
Endangered: 2 species

Distribution
Southern and Southeast Asia; Africa south of the Sahara

Evolution and systematics

The living elephant species are contained within a single family, the Elephantidae, and are the sole remaining representatives of the mammalian order Proboscidea. The name Proboscidea derives from the proboscis, or trunk; the name elephant derives from the Greek words for a large arch, referring to the elephant's arched back supported by pillar-like legs.

Both DNA and anatomical data indicate that the closest living relatives of elephants are the Sirenia—dugongs and seacows. Recently, it has become clear that elephants and sirenians fall within a larger grouping, including hyraxes, tenrecs, golden moles, elephant shrews (whose long nose is, however, independently acquired from that of the elephant), and the aardvark. Together, this diverse assemblage of mammals has been named "Afrotheria," since all are believed to have arisen in Africa from a common ancestor, 70 million or more years ago.

Approximately 165 fossil species of Proboscidea have been identified. The majority of early proboscidean fossils, 60–40 million years old, have been found in North Africa, and several of them appear to have been amphibious in habit. Moeritherium probably lived on a diet of aquatic plants, rather like a small modern day hippopotamus, which it resembled in build. Recent research has suggested that elephants retain some features reflecting a distant semi-aquatic ancestry. These include internal testicles, primitive embryonic kidney structures called nephrostomes, and the arrangement of the embryonic respiratory system. However, the suggestion that the trunk arose as a snorkeling device is probably not correct, as no living aquatic mammal has developed a "snorkel," and the trunk was only fully developed in later, fully terrestrial proboscideans.

In early Oligocene times, about 36–30 million years ago (mya), *Palaeomastoden* and *Phiomia* show the beginnings of an elephant-like appearance, with enlarged body size (up to about 6.5 ft [2m] shoulder height), distinct upper and lower tusks, and a short trunk. They may have been woodland browsers. The peculiar adaptive complex of features characterizing elephants developed during the course of proboscidean evolution. The increase in body size to massive proportions supported a huge gut for fermentation of large quantities of poor-quality forage. The massive molar teeth and heavy tusks are supported within a very large head and held on a very short neck, easing the task of raising and lowering it. But with a head high above the ground, short neck, and cumbersome tusks, feeding with the mouth (especially on low-growing plants) would be difficult; hence, the evolution of the trunk.

Among the diverse fossil proboscidean groups that evolved from about 25 million years onwards, the stegodons are the best known and regarded as the sister-group (closest relatives) of the elephants. Common fossils of Africa and Asia from

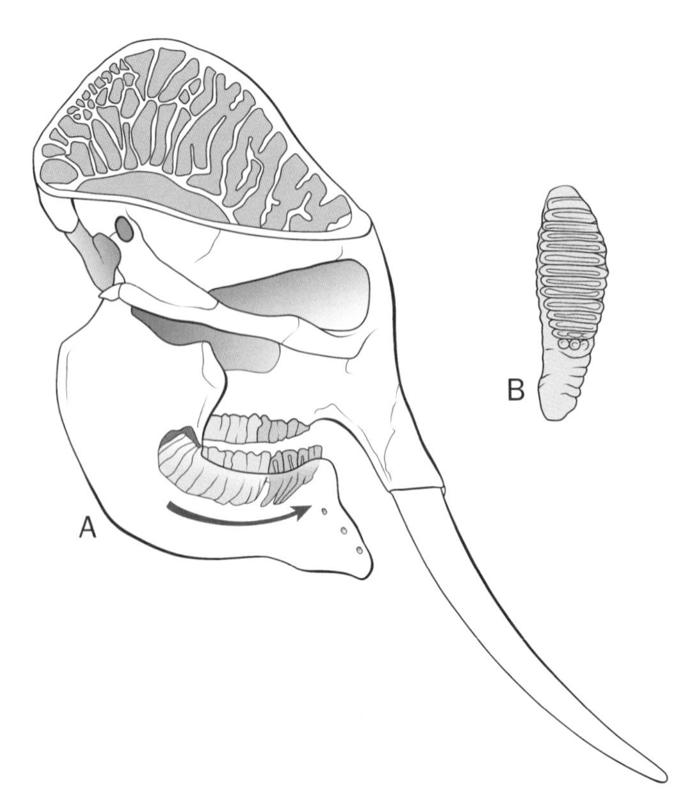

A. An elephant skull. The upper shaded section shows the air spaces that lighten the skull. Also shown are two upper and lower teeth, the front ones wear out and the back ones move forward to take their place. B. The chewing surface of an elephant molar. (Illustration by Patricia Ferrer.)

about 7–0.5 mya, the stegodon achieved tremendous size (11.5 ft; 3.5 m shoulder height) and sported extremely long, parallel, closely-spaced tusks.

True elephants (Elephantidae) are differentiated from their ancestors by detailed skull characters as well as by the loss of enamel covering the tusks, multiplication of enamel ridges on the molars, and heightening of the molar tooth crown. Since the earliest fossil representatives, 7–4 million years old, are found almost exclusively in Africa, it is reasonable to assume that the group originated there. *Primelephas* was the earliest member of the subfamily Elephantinae. It stood about as tall as a female Asian elephant, had upper and lower tusks, and is believed to have lived in an open wooded savanna.

Three great branches of the elephant family can be recognized in the fossil record of the last 4 million years or so. These are *Elephas* and its relatives (including the living Asian elephant), *Loxodonta* (including the living African elephants), and *Mammuthus* (including the woolly mammoth, not to be confused with the very distantly related but similarly named proboscidean *Mammut*, the American mastodont).

The earliest *Loxodonta* appears at 7.3–5.4 mya in Kenya and Uganda. It subsequently divided into two apparently coexisting species, *L. adaurora* and *L. exoptata*. The living species, *L. africana* presumably derived from one of these forms, and initially coexisted with a further species, *L. atlantica*, with grazing-adapted dentition and small tusks. The relationships among all these forms are unclear, but recent genetic evidence has cast new light on the history of the African elephant. The ancestors of the living species almost certainly lived in the forests of central Africa. Between about 3.5 and 2.5 mya, drying of the climate led to the development of savanna-adapted populations in the south and east. This division led to the modern subspecies, *L. africana cyclotis* (the forest elephant) and *L. africana africana* (savanna elephant). *L. a. cyclotis* is today the more primitive of the subspecies; its skull is similar to that of *L. adaurora*.

Today, the bush or savanna elephant, *L. a. africana*, is distributed in eastern and southern Africa, while the forest elephant, *L. a. cyclotis*, occupies much of central and western Africa. The physical differences between them are very marked. In *L. a. africana*, the body size is larger and rangier, the ears are very large and triangular, the tusks are massive and curve outwards

An African bull elephant (*Loxodonta africana*) estimated at 55 to 60 years of age. The tattered ears are probably a file of history of conflict with other elephant bulls. Older elephant bulls roam widely, and at times come into conflict with other bulls. In general they are not aggressive except when in musth. (Photo by Rudi van Aarde. Reproduced by permission.)

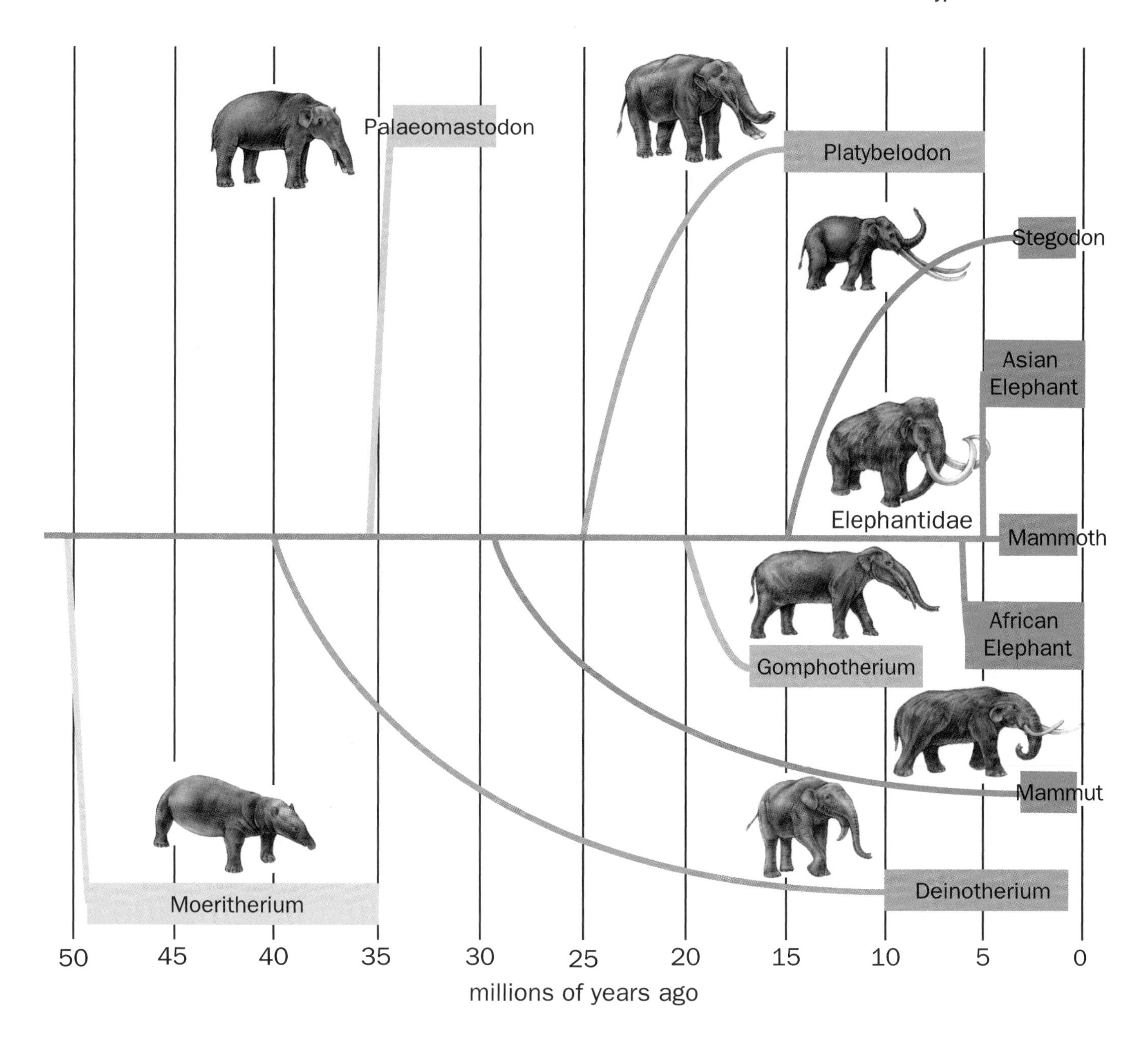

Elephant evolution. (Illustration by Patricia Ferrer)

and forwards, and the back is distinctly saddle-shaped. In *L. africana cyclotis*, the body is distinctly smaller and more compact, the ears are smaller and rounded, the tusks are narrow, long, and downward-pointing, and the back is straighter.

These forms have been treated as subspecies of *L. africana*, but recent research has raised the possibility that they are separate species, *L. africana* and *L. cyclotis*. There are pronounced differences in skull and mandible anatomy, while DNA sequence studies found that forest elephants from central Africa were genetically quite distinct from those of the east and south African savannas, the genetic distance between the two groups being more than half as great as that between them and the Asian elephant. Finally, there are clear differences in habitat, feeding, social behavior, and communication. Hybridization, if it occurs at all, is limited and has not destroyed the integrity of the two forms.

However, the precise geographical limits of the two forms have yet to be established. Moreover, the forest elephants of West Africa, although they appear visually similar to *L. a. cyclotis* of central Africa, have been found to differ from them genetically. For these reasons, the IUCN's African Elephant Specialist Group decided in 2002 to provisionally retain the designations at subspecies level, pending further research, and this policy will be followed here. Thus, the elephants are named *L. africana africana* for the savanna elephants of southern and eastern Africa and the West African sahel, and *L.*

Two male African elephants (*Loxodonta africana*) sparring. (Photo by Peter Oxford/Naturepl.com. Reproduced by permission.)

africana cyclotis for the forest elephants of central and West Africa. Records of the so-called pygmy elephant, *L. pumilio*, are almost certainly *L. a. cyclotis*.

The genus *Elephas*, leading ultimately to the *E. maximus*, the living Asian elephant, first appears as a fossil in Ethiopia 6.7–5.2 mya. This lineage produced a diversity of species in Africa, Europe, and Asia. The African *Elephas ekorensis*, from around 4.5–4 mya, appears to be close to the common ancestry of this radiation. Elephants entered Asia about 3 mya. One species, *Elephas hysudricus*, inhabited northern India and Myanmar between about 2 and 1 mya, and is believed to be close to the ancestry of *E. maximus*. *E. hysudricus* was of large size, massive tusks, and, like *E. maximus*, had a well-developed double head dome. There are remains of the *hysudricus-maximus* lineage about half a million years old in the Middle East, and by 120,000 years ago, *E. maximus* is recorded on Java. An earlier form on Java, *E. hysudrindicus*, lived from about 1 to 0.5 mya, but its skull and dental anatomy appear too specialized to have given rise to the living species.

A recent study of DNA sequences has identified two main genetic groups among Asian elephants. Although now widely dispersed and co-occurring in many areas, these may have originated in separate populations, one in Indonesia, and one on the mainland of Asia, which subsequently intermingled. Since the two genotypes are different enough to have separated a million or more years ago, researchers speculate that these two populations may be those identified in the fossil record as *E. hysudricus* (continental) and *E. hysudrindicus* (island Indonesia). This interesting theory must, however, be weighed against the anatomical differences between *E. hysudrindicus* and *E. maximus* as well as the observation that *maximus* replaced *hysudrindicus* on Java as part of a wave of colonization from the mainland.

There are three currently recognized subspecies of *Elephas maximus*: *E. m. maximus* of Sri Lanka and southern India, *E. m. sumatrensis* of Sumatra, and *E. m. indicus* throughout the rest of the range. The differences are a matter of degree and are expressed as gradual changes across the range. Elephants from Sri Lanka are the largest, have the darkest skin color, the largest ears, and are most prone to pink depigmentation of the skin on the face, trunk, and ears. Animals from Sumatra are the smallest, lightest in color, and least prone to depigmentation. Those in-between generally show

An African elelphant (*Loxodonta africana*) family. (Photo by David M. Maylen III. Reproduced by permission.)

intermediate characters. However, there are exceptions: for example, elephants from western Nepal are perhaps the largest living anywhere today. And there are other particularities of the populations: most male elephants in Sri Lanka today are tusk-less, while the Sumatran subspecies is said to possess an extra pair of ribs (20 instead of 19 in the other subspecies).

Studies of Asian elephant genetics have not so far provided much insight into the systematics of the modern subspecies, probably due to genetic mixing resulting from the extensive historical movements of animals in captivity. In mitochondrial DNA, at least, there appears to be nothing distinctive about *E. m. maximus* of Sri Lanka, although the current population of Sumatra is genetically distinctive and may have been isolated for a hundred thousand years or more.

The mammoth was an elephant, and not the ancestor of living elephants but their cousin: the earliest mammoth fossils date from 4.5 to 4 mya in southern Africa and Ethiopia. Some time after that date, populations migrated north, arriving in Europe about 3 mya. They ultimately spread throughout northern Eurasia and North America, producing several species. The woolly mammoth, *M. primigenius*, probably arose in northern Siberia about 750,000 years ago in response to ice age conditions, later spreading to Europe and North America. Its adaptations included a furry coat, small ears and tail to prevent heat loss and frostbite, and teeth adapted to a grass-dominated diet.

There has been debate over the relationships among *Elephas*, *Loxodonta*, and *Mammuthus*. Anatomical studies of the skulls and other features initially suggested that *Elephas* and *Mammuthus* were more closely related to each other, and this was supported by some DNA sequence data. However, further studies of anatomy and DNA have cast doubt on this conclusion. Shared features of *Elephas* and *Mammuthus*, such as the higher skull and increased number of molar enamel bands, may have arisen independently, and the DNA data suggest that all three elephants may have diverged relatively close together some time in the interval 7–5 mya, a conclusion not inconsistent with the fossil record.

Physical characteristics

Elephants weigh 200–265 lb (90–120 kg) at birth. Unlike other mammals, they continue to grow well into adult life. Females cease growth at 25–30 years, males at 35–45. Adult weights range typically from 3.3 tons (3 tonnes) in a female Asian elephant, to 7.7 tons (7 tonnes) in a large African savanna bull; typical respective shoulder heights are 7.2–11.8 ft (2.2–3.6 m).

The elephant's head is proportionately very large, weighing up to half a ton/tonne; the neck is short. The body is supported on four extremely strong pillar-like legs. The elephant has five splayed toes buried within its foot, and stands on tiptoe; the first visible joint, some distance above the ground, is not the elbow or knee, but the wrist or ankle. The foot contains a pad of springy tissue that causes the elephant's foot to

A forest elephant (*Loxodonta cyclotis*) with young in Dzanga Ndoki National Park in central Africa. (Photo by Animals Animals ©Peter Weismann. Reproduced by permission.)

swell sideways when it bears the animal's weight. The tail is long, extending to below the knee, and ends in a tuft of very coarse hairs. Otherwise, the body is sparsely covered by short hair, more pronounced in very young animals. As far as is known, there are no sweat glands. The ears are very large and thin, except for a thicker supporting ridge along the top. They are richly supplied with blood vessels for heat loss, and are flapped mainly for this purpose. The skin of both species is a uniform gray. Elephants may take on brown or other hues after wallowing in mud.

The elephant's trunk is, anatomically, a fusion between its nose and upper lip. The trunk is remarkably sensitive, flexible, and maneuverable, as well as being immensely strong. It contains no bone or cartilage, but is principally composed of muscle, in eight main sets (four on each side) comprising a total of about 150,000 separately moveable muscle units. Two nostrils run the entire length of the trunk for breathing.

The tusks are, anatomically, greatly expanded lateral incisor teeth. They are comprised almost entirely of dentine. About a third of their length is buried within a socket in the animal's skull. The tusks are solid, except the upper part within the socket, where there is a pulp cavity. The tusks grow by addition of dentine there, pushing them out by up to 6 in (15 cm) a year. The tusks of a large bull can extend 79 in (200 cm) in total length and weigh 110 lb (50 kg) each, although such figures are rare nowadays.

Distribution

The current range of the African elephant is Africa, south of the Sahara; it formerly extended into North Africa up to the Mediterranean coast. The Asian elephant currently occupies the Indian subcontinent and Southeast Asia, formerly extending from the Near East to the Pacific coast of China.

Habitat

Although occupying exclusively tropical and subtropical zones, elephants live in a wide range of habitats, including evergreen rainforests, dry deciduous forests, savannas (mixed woodland and grassland mosaics), and semi-deserts. They are essentially mixed feeders, so accessibility to a wide range of plants, and to water within one day's walk, are essential prerequisites.

Estimates of natural animal density are hard to make. The carrying capacity will also vary enormously with the environment. In general, an area of about 2 mi^2 (5 km^2) per animal is probably typical in the wild, although the figure may be as high as 7.7 mi^2 (20 km^2) in rainforest habitats.

In many areas where they live, elephants are the dominant mammalian species in terms of biomass, and have a major ecological role. Their massive dung production recycles nutrients back into the soil. They can disperse seeds and fruits over wide distances. Elephants seem quite resistant to the tannins present, for example, in acacia bark and, by consuming a wide variety of species, they limit the intake of toxic defensive compounds specific to particular plant types.

Dominant African elephant (*Loxodonta africana*) bull threat display. (Photo by Harald Schütz. Reproduced by permission.)

An African elephant (*Loxodonta africana*) bull feeding on tree branches. (Photo by Tom & Pat Leeson/Photo Researchers, Inc. Reproduced by permission.)

The elephants' habit of destroying trees has led to debate about their role in changing their own environment. In some parts of Africa, elephants have transformed wooded areas into open grassland. However, it is likely that in the formerly natural situation, such phenomena formed part of a natural cycle, with long-term balance between different habitats. If a high number of elephants in one area caused a reduction in the tree density, either the elephant population would limit its own reproduction, or the animals would migrate to another area, allowing regeneration of woodland. In some areas even today, vegetation regeneration seems to keep pace with elephant feeding; it is primarily in savanna habitats, and particularly where elephants have been constrained within the boundaries of reserves, that problems arise, and in the present situation these are certainly important issues for conservation. Many other factors such as fire and climate change also contribute to the balance between elephants and their habitats. In the severe drought of 1970–71, thousands of elephants died in Africa as a result of food and water shortage.

Behavior

The understanding of the complex social lives of elephants has been built up thanks to long-term studies over 20 or 30 years by dedicated field researchers, notably Cynthia Moss and colleagues in East Africa, and Raman Sukumar in India. By learning to recognize individual animals, much has been learned about social organization and the factors influencing the status and success of families and individuals.

Elephant society has a structure that has been termed matriarchal. The core element is the family unit, a group of 3–25 individuals, comprising related adult females and their young. Females within the family unit are closely bonded for life. By contrast, adult males tend to be solitary, or may form temporary associations of two or three unrelated bulls. They leave the family of their birth at 12–15 years of age and, after that time, although they may frequently associate with female groups for feeding or mating, they have no long-term bonds with them, or with each other.

Within the female groups, a few older individuals, and in particular the lead individual, termed the matriarch, are instrumental in deciding the group's pattern of movement, in defending the group against danger, and in monitoring and responding to other approaching elephants. Calves, especially when very young, stay close to their mother, but all females in the group will aid in its upbringing. At the approach of a predator, adult females wheel round to face the source of danger, protecting the calves that stay close behind. The members of a family unit may separate for short intervals during the day, but will soon regroup. Family units also form looser associations or "bond groups," with more distantly related families. Occasionally, very large herds of 500 or even 1,000 elephants can be seen, primarily during migration. Even then, within the mass of animals, individual family groups maintain their integrity.

Elephants are highly intelligent animals with a complex repertoire of social interactions. Within a family group, individuals of all ages greet, and maintain bonding, by touching their trunk tips to each other's bodies, rubbing together, and with sound communication and scent. In calves, play is a dominant behavior. They mock charge, chase each other, or wrestle with their trunks. Males, from an early age, engage in mock sparring matches. They are also more independent of their mothers than the females, a trend that increases as they get older.

There is a dominance hierarchy among bulls, generally related to their age, size, and power. If two bulls of roughly equal size meet, they may assess each other through intertwining trunks, pushing and pulling, or lightly engaging their tusks. Rarely, sparring may lead to a full-scale fight, sometimes (but not always) for access to an oestrus female. The combatants will charge each other with ears outstretched, or cross tusks and attempt to twist the opponent off-balance, all accompanied by loud vocalizations. Each tries ultimately to gore the other with his tusks, sometimes resulting in fatal wounds by deep penetration of the head or chest. Broken tusks may result from twisting with the full body weight. The fight

Mother and adolescent African elephants (*Loxodonta africana*) assisting a young elephant climbing up the bank of Mara River in Kenya. (Photo by Charles V. Angelo/Photo Researchers, Inc. Reproduced by permission.)

will end either by withdrawal of the weaker animal, or with death.

Male elephants enter a periodic state called "musth." The temporal gland, located on the side of the head between the ear and the eye, produces a dark musky fluid (temporin) with a strong, musky odor. Musth males also intermittently dribble urine. A male elephant generally enters musth once a year, for a period of anything up to a month, the time of year varying with the animal. Musth bulls have heightened levels of testosterone and are very aggressive, especially toward other bulls. Musth is associated with heightened sexual activity, although non-musth bulls also mate. Females also have a temporal gland, which can occasionally be seen to ooze secretion, and elephants have been observed rubbing their cheeks against trees, so temporin may have broader communication functions. Recent research has indicated that subordinate bulls produce a different chemical signal, with a sweet aroma, which may be used to signal submissiveness to the dominant bulls and so avoid attack.

Elephants have relatively poor vision, but highly developed senses of taste and smell. They obtain chemical cues by using their trunks to touch each other's genitals, mouths, temporal glands, dung, and urine. They also often lift their trunks and rotate the open tips, testing the air for the scent of other animals in the vicinity. It is very likely that they can identify different individual elephants from these cues.

Elephants also have acute hearing and communicate through a wide variety of vocalizations. At least 25 different calls, audible to the human ear, have been identified in African elephants, 15 of them in a low-frequency group termed rumbles. Some of them are known to be associated with different events such as musth in a bull and oestrus or copulation in a female. In addition, a range of infrasound vocalizations extends down to 5 Hz, well below the frequency of human hearing. Low-frequency sound is less subject to environmental attenuation, and elephant rumbles and infrasound are audible to other elephants over a range of up to 3 mi (5 km). It has also been suggested that elephants may communicate over even longer distances as they stamp their feet on the ground; this theory remains to be tested.

An elephant can live to around 60 years; many die before this age, from disease, injury, starvation, drought, or predation (though the latter is rare for healthy adult animals). A remarkable aspect of elephant behavior is their response to injured, sick, and dead members of their species. Many accounts have been recorded: adult females immediately circling around a wounded animal to prevent further attack; lifting a wounded animal to its feet and shouldering it away to safety;

jumping into water where a wounded animal has fallen, and heaving it out again; pulling and pushing a calf out of mud where it had become stuck; standing guard over a stricken, but living, animal lying on the ground; covering the body of the relative with grass and leaves as soon as it had died; returning to the carcass or even skeleton of a dead relative; and tasting, picking up, and moving the remains with their trunks.

The idea of an elephant graveyard, a place where elephants go to die, is a myth. Sick and dying elephants often go to a lakeside or river, where there is a ready supply of food and water within easy reach, and several might die in one area for that reason. In times of drought, animals congregate around water holes and many may die there.

Elephants are not territorial. Although individuals or family units have home ranges, those of different animals overlap and are not defended as such. There are daily and seasonal activity patterns within the home range. They sleep lying down, usually for two to four hours in the early morning. They may also, in the hottest part of the day, stand motionless in the shade, but even when the eyes are closed, they are most likely dozing rather than sleeping.

Seasonal movements, especially in open country, may see large aggregations of hundreds of animals. In other situations, particularly in forest environments, matriarchs lead their families along the same paths that have been used for generations; these elephant trails, trampled, barren ground 3–6 ft (1–2 m) wide, can extend for tens of miles (kilometers).

Elephants walk or amble, but cannot canter or gallop. A charging animal can attain 16 ft (5 m) per second, or 12.4 mph (20 kph), while walking speed ranges 1.6–8.2 ft (0.5–2.5 m) per second, or 1.2–6.2 mph (2–10 kph). Elephants walk cautiously, appearing to place each foot with care to avoid ground that is too soft or cobbled, for example. Even so, they can maneuver very dense terrain and can climb up and down remarkably steep, slippery slopes. They are also adept swimmers, paddling with all four feet and using the trunk as a snorkel.

Feeding ecology and diet

Elephants consume a huge range of plant types, including grasses, herbs (forbs), shrubs, broadleaved trees, palms, and vines. Depending on the plant, they can take every conceivable part, including leaves, shoots, twigs, branches, bark, flowers, fruit, pods, roots, tubers, and bulbs. The range of plants taken by an elephant can be anything between 100–500 species, although in a given time and place the animals may concentrate on a few species.

Patterns of consumption change with the seasons. In the savanna-woodland habitats of Africa, and in the dry forests of Asia, new growth grasses are favored in the rainy season, comprising 50–60% of the diet, but as these become tough in the dry season, the elephants switch to browse, so that the leaves and fruit of trees and shrubs now comprise 70% of the intake. In the forests of Asia, bamboo is an important component throughout the year. For elephants in rainforest habitats such as those of central Africa and Malaysia, for example, the year-round supply of succulent leaves and fruits ensures that grass plays a lesser part in their diet.

Young male Asian elephant (*Elephas maximus*) digging for salt in India. Photo by Animals Animals ©A. Desai, OSF. Reproduced by permission.)

Tree bark is eaten because it provides essential minerals and fatty acids, as well as roughage. Elephants also frequent salt licks, those patches of soil or exposed rock high in minerals such as sodium.

Food consumption is 220–660 lb (100–300 kg) per day. Elephants spend 12–18 hours per day eating, most intensively in the morning and in the late afternoon to evening. In food-rich forest areas, elephants will typically move slowly through the day, browsing on a variety of plants, and eventually covering several miles (kilometers). In many areas, there are daily rhythms: where both woodland and open grassland are available, for example, the elephants may spend the morning and early afternoon browsing in the woodland, emerging in the cool of the late afternoon to graze. Fluid consumption can be 53 gal (200 l) of water per day in hot weather. When water is scarce, elephants will dig holes in dry stream or lake beds, using their feet, trunk, and tusks, until water seeps in and can be sucked up.

When plants become ready at particular times of year, such as fruits or new shoots, elephants will gravitate towards them, using both smell and a memory from past years. Generally speaking, the poorer the quality, abundance, or predictability of food and water, the greater the distances elephants must travel to find it. Home ranges, measured by radio-collaring individuals, vary from 23 mi^2 (60 km^2) in a rich rainforest habitat in Malaysia, to 1,158 mi^2 (3,000 km^2) in the Namib Desert, where individuals can easily walk 50 mi (80 km) in a day. In many areas, migrations are seasonal. Where water is a key issue, elephants tend to accumulate in the dry season in areas it can be found, dispersing more widely when this constraint is lifted during the wet season.

An Asiatic elephant (*Elephas maximus*) chewing on vegetation. (Photo by Harald Schütz. Reproduced by permission.)

Small items can be plucked or picked up with the terminal "fingers" of the trunk, the larger items, such as branches, by curling the trunk around them and pulling or twisting. Elephants are highly inventive and can be seen, for example, kicking up sods of dry turf with their feet, picking up the resulting grassy clump with the trunk, banging it against their leg to shake off the earth, and putting in the mouth. To reach high branches where young, succulent leaves are to be found or up in acacia trees, which have fewer thorns, they can rear up on their back legs, giving a total reach of up to 26 ft (8 m). They will also uproot or push over trees. Finally, the trunk is important in drinking; water is not sucked all the way up into the nose like a drinking straw, but is sucked into the lower part of the trunk, then the trunk is arched and water squirted into the mouth. The capacity of an Asian elephant's trunk has been measured at 2.2 gal (8.5 l). The only time in its life when an elephant feeds directly with its mouth is when suckling, the mouth being pressed directly against the breast with the trunk curled up out of the way.

The tusks are used to strip bark from trees, which is then eaten; to dig for roots or for water in the dry season; and to scrape or hack salt and other minerals from the soil or exposed rock.

The molar teeth display a series of long, thin enamel ridges running side to side; for this reason, an elephant chews by swinging its lower jaw fore and aft, so that the enamel ridges on the upper and lower teeth cut past each other, shearing the food. The tremendous wear caused by feeding long hours every day on abrasive food causes the teeth to grind down to the root, and elephants not only have high-crowned teeth, but replace their teeth five time through their life, making six sets in all. Each set, however, comprises only four massive teeth, lower and upper, left and right. As one tooth wears out, it moves forward in the jaw and is gradually replaced by another from behind.

The majority of an elephant's digestion is accomplished with the aid of cellulose-digesting microorganisms inhabiting its large intestine, especially a large blind sac opening from it, the cecum. This is a relatively inefficient method of digestion—only 40% or so of food, by weight, is utilized—but it does allow the animal to process large quantities of relatively low-nutrient food. The intestine is up to 115 ft (35 m) long and may weigh up to a ton (0.9 tonnes) when full of food, releasing an average of 220 lb (100 kg) of dung per day.

Reproductive biology

Elephant reproduction is slow; a female gives birth only every four to five years or so, and usually to one calf at a time, though twinning occurs in roughly one in 100 births. Growth to adulthood is also a long process. In consequence, a tremendous amount of time and energy is expended in the rearing of the young, a task that falls entirely to the females.

Elephant cows become sexually mature at the age of 12–14, and begin to reproduce soon after that date. Bulls start producing sperm around the same time, but in practice rarely father any calves until they are approaching 30.

Female elephants come into estrus about every 16 weeks, and are sexually receptive for only a day or so during this period, so male and female behavior must be tightly attuned. A male, especially when in musth, will visit female groups, testing for estrus females by touching their vulvas with the tip of his trunk. He then touches his trunk tip on a specialized taste gland, the Jacobson's organ, on the roof of his mouth. It has recently been discovered that the females' urine contains a pheromone indicating that she is in oestrus. She will also signal her readiness by behavioral cues. Copulation begins when the male reaches over the female's shoulder with his trunk from behind. The female exerts some choice in the matter, and may run off even at this stage. Otherwise, the bull mounts, placing most of his weight on his back legs. The penis is S-shaped, up to 3.2 ft (1 m) long, and highly muscular, finding and entering the vulva without pelvic movement. The testes are internal (unusual in mammals) and situated near the kidneys; up to 1 qt (1 l) of ejaculate is produced. The bull will remain with the female for anything from a few hours to a few days, mating with her occasionally and guarding her from the advances of rival males.

Pregnancy lasts about 22 months, and birth, accomplished with the mother squatting or lying, is assisted by other females of the group. The two mammary glands are situated between the front legs (unusual for quadrupedal mammals). Calves suckle until the second or third year or even longer, depending on when the next calf is born. Male calves suckle more frequently than females and, after the first few years, the difference in size between them becomes apparent. Female calves will remain in their family unit for life, eventually taking over its leadership, while males leave at sexual maturity, often aided by increasing impatience of the mother.

In drought years, cows are unlikely to come into oestrus, naturally regulating their reproduction. Otherwise, they can conceive at any time of year, but in seasonal environments, a definite peak has been observed some weeks after the onset of the rains. With a 22-month gestation, this ensures that the calf will be born when rainy-season greening has begun two years later, providing the mother with a rich food supply for lactation.

An African elephant (*Loxodonta africana*) baby holds the tail of another elephant. Photo by Animals Animals ©Anup Shah. Reproduced by permission.)

Conservation status

Elephants are faced with a dual threat to their survival: the destruction of their habitat, and hunting. The former is common to many species; the latter is due to the elephant's possession of a precious commodity: ivory.

Habitat destruction has both reduced the total range of elephants, and has greatly fragmented it within human settlement and agriculture. The principal cause is human population growth, but also activities such as logging for financial gain. Over much of the range, the remaining habitats correspond to national parks, nature reserves, and the like. Many of these fragments retain less than 100 individuals and prospects for their long-term survival are not good. If there is no exchange of individuals with other populations, inbreeding reduces the genetic health of the population. If climatic fluctuations produce a series of stressful years, the population will suffer increased mortality and reduced birth rate, and may not recover. In West Africa through the 1980s, elephant populations in habitat fragments of less than 96 mi^2 (250 km^2) had only a 20% chance of surviving the decade, while those in areas of more than 290 mi^2 (750 km^2) had almost a 100% chance of survival.

The hunting of elephants for meat has been practiced since prehistoric times, but only with the use of firearms has the thirst for ivory posed a threat to the very survival of the species. By 1800, the elephant populations of southern and West Africa had already been seriously depleted. A century later, the trade from Africa alone had increased to 1,100 tons (1,000 tonnes) per year. The 1970s and 1980s proved critical: the total African population fell from an estimated 1.3 million animals in 1979, to just over 400,000 in 1987. Asian elephant populations have also suffered at the hand of humans, both through ivory hunting and the gathering of wild animals for domestic use. The effect of ivory hunting on the two species is somewhat different, since in the African elephant both males and females carry tusks and are hunted, while in the Asian species only the males have ivory. This has led to a situation in some parts of Asia where the natural female-to-male ratio of 2:1 has risen to anything from 5:1 in the best-protected areas, to 100:1 in the worst; in the latter cases, the survival of even sizeable populations is threatened because of lowered reproductive rate.

From its foundation in the 1970s, CITES placed Asian elephants on its Appendix I and African elephants on Appendix II. In 1989, however, the African elephant was raised to Appendix I, effectively banning all trade in elephant ivory. The policy worked: ivory prices fell, and many countries reported a drastic reduction in poaching. However, in 1997, some southern African countries with healthy elephant populations won from CITES the permission to sell ivory stocks. The market was stimulated, and in subsequent years, increased poaching has been reported by a number of African countries. Nonetheless, in 2002, CITES agreed to allow further sales of stockpiled ivory by these countries, despite almost universal opposition from conservation organizations. Combating the ivory trade is a complex issue that requires the enforcement not only of bans against hunting, but international action to trace both the organizers of poaching, the middle men, and the ultimate consumers.

The management and protection of elephant habitats is also a major goal, especially in Asia. International support enabling poor countries to maintain existing wildlife reserves, or to create new ones, is crucial. Properly managed ecotourism can be beneficial, as it provides an income underscoring the value of the reserve. Yet small reserves, even when protected, may not support enough animals to give a viable population. Raman Sukumar has suggested that 50 breeding individuals, translating into 125–150 animals, is a minimum goal, with 10 times that number an ideal. One solution to this problem is to create corridors of habitat, allowing animals to migrate between parks, so that populations are effectively merged into one, viable unit.

Elephant-human conflict is a serious issue in some areas. Elephants enter agricultural areas and can destroy the entire crop of a smallholding in a single night. They also damage buildings and annually kill dozens of villagers in Asia. Traditional countermeasures include lighting flares, throwing rocks, employing domestic elephants to chase away the marauders, or digging trenches around fields. The latter are of some use but elephants learn how to fill them with earth or logs. Electric fences are employed by rich landowners, but are too expensive to bound large national parks or small private holdings. Other measures include not planting crops favored by elephants in the area around their habitat, and relocating farms and villages (with compensation paid to the farmers). The latter may also be necessary when extending reserves or creating habitat corridors.

In some African countries, elephant populations in wildlife parks have been held in check by government-approved culling. The stated rationale is to prevent the populations increasing to the point where they turn woodland into grassland, reducing biodiversity, and leading to elephant mortality when drought hits, as happened in Tsavo National Park, Kenya, in the 1960s and 1970s. Opponents counter that culling (sometimes of entire family groups) is inhumane and causes stress to surviving animals; is a temptation for illicit ivory dealing; interferes with natural cycles; and depresses tourism. Possible alternatives include relocating animals to areas of low density and subcutaneous implants of birth-control hormones.

Current estimates of world population size are between 34,000 and 54,000 wild Asian elephants, with roughly 13,000–16,000 in captivity. For the African elephant, the latest estimate is between 300,000 and 500,000 animals.

Significance to humans

Elephants, especially in Asia, have a long history of interaction with people, and an important place in many cultures. Ivory carving has been practiced since at least 30,000 years ago, when Palaeolithic people in Europe made tools and ornaments out of mammoth tusk. Ivory is hard, fine-grained, and has an elasticity that makes it excellent for carving: skilled craftspeople can produce objects of great beauty. Countless functional objects have been made throughout history: in recent centuries, piano keys and billiard balls were one of the main uses in the West, and in recent decades, the ornamental "signature seal" of Japan has become a major end-product.

The earliest evidence of elephant domestication is in the third millennium B.C. in the Indus Valley of India. The initial domestication was probably for purposes of traction, tree felling, and portage; this usage continues today in parts of Southeast Asia, although it is declining. Elephants were formerly captured from the wild, either singly in pits, or as family units in stockades; now they are bred and trained from calves. An elephant can recognize and respond to 30 or more commands issued by its *mahout*, or driver.

Soon after their domestication, elephants were pressed into military service. In 326 B.C., the Indian king Porus, with 200 elephants in his army, was famously defeated by Alexander. A typical battle formation of the Vedas included 45 elephants, which were the first to charge, throwing the enemy into disorder and knocking down stockades. Kings and princes hunted from elephant-back, a practice taken over with enthusiasm by European colonizers. In general, elephants came to embody royalty, largely because of the high price of their capture and maintenance.

The elephant also plays a prominent part in the Hindu pantheon. Airavata was the elephant mount of Lord Brahma, creator of the universe. Two elephants were the massive pillars of the world and bore the earth on their enormous heads. Ganesh, the elephant god, is one of the best loved of all Hindu deities: as the Remover of Obstacles and Lord of Beginnings, he is invoked at the start of any undertaking. The worship of Lord Ganesh originated in the third or fourth century A.D. and created a strong ethos against the killing of elephants. In Buddhist countries, especially in Indochina, the very rare white elephant was revered as an incarnation of the Buddha; when captured, it was ministered to with the utmost care.

The comparative rarity of domestication in African elephants appears to be for human cultural reasons rather than any innate inability of the species to be domesticated. The Carthaginians fought the Romans with them, and Hannibal's famous crossing of the Alps was probably with the African species. In modern times, Belgian colonizers domesticated elephants for traction and other uses in Central Africa.

1. African elephant (*Loxodonta africana*); 2. Asian elephant (*Elephas maximus*). (Illustration by Patricia Ferrer)

Species accounts

Asian elephant
Elephas maximus

SUBFAMILY
Elephantinae

TAXONOMY
Elephas maximus Linnaeus, 1758, Sri Lanka.

OTHER COMMON NAMES
English: Indian elephant; French: Eléphant d'Asie, eléphant asiatique; German: Asiatische Elefant; Spanish: Elefante asiático.

PHYSICAL CHARACTERISTICS
Weight 3.3–5.5 T (3–5 t), shoulder height 6.6–9.8 ft (2–3 m), back convex, high double head domes, ears smaller than African species, fold forwards at top, one finger at tip of trunk. Pigment loss with age, resulting in pink speckling of the ears and eventually of the face and trunk: particularly noticeable in *E. m. maximus* of Sri Lanka. Hairier than African species. Only males bear tusks, although females frequently possess tiny tusks called "tushes," which can just be seen protruding form the lip, especially when the trunk is raised. A percentage (currently increasing) of males congenitally lack tusks: known as "mukhnas," these animals are thought to compensate by being especially strongly-built, especially in the upper trunk region.

DISTRIBUTION
Principally in Burma, Cambodia, India, Indonesia (Sumatra), Laos, Malaysia, Sri Lanka, Thailand, and Vietnam, with small populations (fewer than 500 individuals) in Bangladesh, Bhutan, southwest China, Indonesia (Kalimantan), and Nepal. About half the world population is in India, and half of that in the southwest of the country. The principal characteristic of the global distribution is its fragmentation.

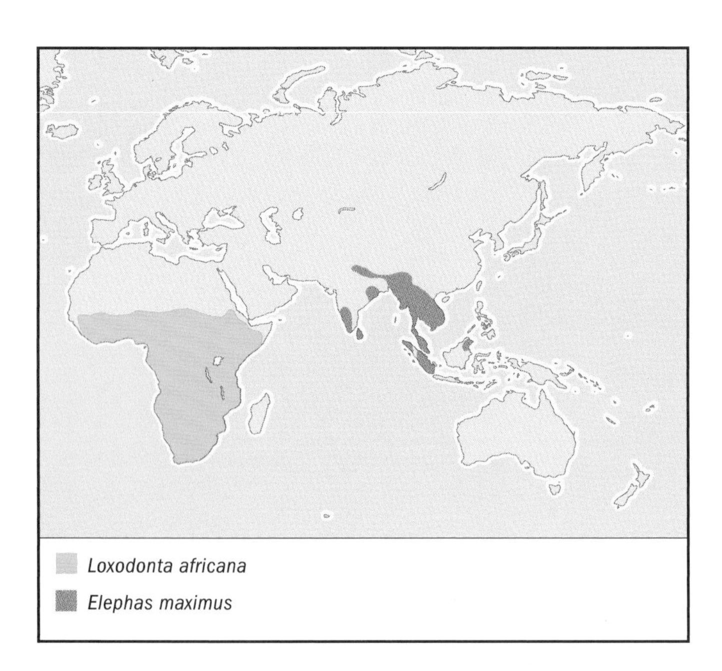

HABITAT
Wet evergreen forest, montane evergreen forest and grassland, semi-evergreen forest, moist deciduous forest, dry deciduous forest, savanna woodland, bamboo forest, dry scrub, and swampy floodplain grassland.

BEHAVIOR
May spend as much as 18–20 hours a day feeding. Very social, with a matriarchal structure: the oldest female is instrumental in deciding the group's movements. There is a dominance hierarchy among bulls; they tend to be solitary. This species is not territorial.

FEEDING ECOLOGY AND DIET
Adult food consumption is 220–440 lb (100–200 kg) per day. In mixed habitats, averaged over the season, approximately 50/50 browse and graze taken. Molars with greater number of enamel ridges than African elephant.

REPRODUCTIVE BIOLOGY
Pregnancy slightly less than 22 months; birth weight 198 lb (90 kg). Courtship behavior includes female standing face to face and intertwining trunks with the male. Males are competitive.

CONSERVATION STATUS
Listed as Endangered by the IUCN, and on Appendix I of CITES.

SIGNIFICANCE TO HUMANS
Highly important in cultures of southern Asia. Revered in religion, though captured for domestic work and warfare. ◆

African elephant
Loxodonta africana

SUBFAMILY
Elephantinae

TAXONOMY
Loxodonta africana (Blumenbach, 1797), Orange River, South Africa. Two subspecies.

OTHER COMMON NAMES
French: Eléphant d'Afrique, eléphant africain; German: Der Afrikanische Elefant; Spanish: Elefante Africano.

PHYSICAL CHARACTERISTICS
L. a. africana: weight 4.4–7.7 T (4–7 t), shoulder height 8.2–13 ft (2.5–4 m), back concave; *L. a. cyclotis*: weight 2.2–4.4 T (2–4 t), shoulder height 6–9.8 ft (1.8–3 m), back straighter. Head less high than Asian species and single domed, ears larger and fold back at top, two fingers at end of trunk. Both sexes possess tusks, those of the female being relatively smaller.

DISTRIBUTION
Occurs in about 35 African states. In West Africa, only thinly scattered, small populations remain, mostly *L. a. cyclotis*; northernmost Sahel population (Mali) probably a western extension of *L. a. africana*. Central African rainforests: still substantial, largely continuous, populations of *L. a. cyclotis*. East and southern African savannas down to northernmost Namibia,

Botswana, Zimbabwe, and South Africa (with large hole in central Angola and neighboring areas): *L. a. africana*.

HABITAT
L. a. africana: woodland, shrub and tree savanna, floodplain grassland, and desert; *L. a. cyclotis*: evergreen rainforest, moist semi-deciduous forest, woodland, and forest/grassland margins.

BEHAVIOR
L. a. africana, at least historically, had a tendency to aggregate in large herds, especially for seasonal migration. Social matriarchal society.

FEEDING ECOLOGY AND DIET
Daily adult food consumption 220–660 lb (100–300 kg). Molar teeth with fewer, lozenge-shaped enamel bands.

REPRODUCTIVE BIOLOGY
Female signals oestrus by a special walk, with the head held high while looking back over her shoulder, as well as loud vocalizations. In courtship, female's movements include spinning round and leading the male on a mock chase. Males are competitive. Gestation period is 22 months. Birth weight 265 lb (120 kg).

CONSERVATION STATUS
Listed as Endangered by the IUCN, and on CITES Appendix I, except some southern African countries; moved to Appendix II in 2002.

SIGNIFICANCE TO HUMANS
Hunted for meat and ivory. Rarely domesticated. ◆

Resources

Books

Buss, Irven O. *Elephant Life: Fifteen Years of High Population Density.* Ames IA: Iowa State University Press, 1990.

Delort, Robert. *The Life and Lore of the Elephant.* London: Thames and Hudson; New York: Harry N. Abrams, 1992.

Eltringham, S.K., ed. *The Illustrated Encyclopaedia of Elephants.* New York: Crescent Books, 1991.

Lister, Adrian, and Paul Bahn. *Mammoths: Giants of the Ice Age.* London: Marshall Editions, 2000.

Moss, Cynthia. *Elephant Memories: Thirteen Years in the Life of an Elephant Family.* University of Chicago Press, 2000.

Payne, Katy. *Silent Thunder: The Hidden Voice of Elephants.* Phoenix: Wiedenfeld and Nicholson, 1999.

Shoshani, Jeheskel, and Pascal Tassy, eds. *The Proboscidea: Evolution and Palaeoecology of Elephants and their Relatives.* Oxford: Oxford University Press, 1996.

Shoshani, Jeheskel, ed. *Elephants.* London: Simon & Schuster, 1992.

Periodicals

Eggert, Lori S., et al. "The Evolution and Phylogeography of the African Elephant Inferred from Mitochondrial DNA Sequence and Nuclear Microsatellite markers." *Proceedings of the Royal Society of London* B 269 (2002): 1993–2006.

Fleischer, Robert C., et al. "Phylogeography of the Asian Elephant (*Elephas maximus*) Based on Mitochondrial DNA." *Evolution* 55 (2001): 1882–1892.

Grubb, Peter, et al. "Living African Elephants Belong to Two Species: *Loxodonta africana* (Blumenbach, 1797) and *Loxodonta cyclotis* (Matschie, 1900)." *Elephant* 2, no.4 (2000): 1–4.

Maglio, Vincent J. "Origin and Evolution of the Elephantidae." *Transactions of the American Philosophical Society* 63, no. 2 (1973): 1–149.

Thomas, M. G., et al. "Molecular and Morphological Evidence on the Phylogeny of the Elephantidae." *Proceedings of the Royal Society of London* B 267 (2000): 2493–2500.

Other

Elephant Information Repository. <http://elephant.elehost.com>

African Elephant Specialist Group (AfESG) of the IUCN. <http://iucn.org/themes/ssc/sgs/afesg>

African Elephant (*Loxodonta africana*). Eleventh Meeting of the Conference of Parties to CITES. Nairobi, 10–20 April 2000. <http://www.panda.org/resources/publications/species/cites/fs_afeleph.html>

Elefriends Campaign. Born Free Foundation. <http://www.bornfree.org.uk/elefriends>

Adrian Lister

Hyracoidea
Hyraxes
(Procaviidae)

Class Mammalia
Order Hyracoidea
Family Procaviidae
Number of families 1

Thumbnail description
Hyraxes are small- to medium-sized herbivores, with short legs, a rudimentary tail, and round ears

Size
Head-body length 17–21 in (44–54 cm); weight 4–12 lb (1.8–5.4 kg)

Number of genera, species
3 genera; 5–11 species

Habitat
Forests, woodlands, and rock boulders in vegetation zones from arid to alpine

Conservation status
Vulnerable: 3 species

Distribution
Southwest and northeast Africa, Sinai to Lebanon, and southeast Arabian Peninsula

Evolution and systematics

Hyraxes were probably the most important medium-sized grazing and browsing ungulates in Africa, as investigations on the 40-million-year-old fossil beds in the Fayûm, Egypt, have shown. During this period, there were at least six genera, ranging in size from that of contemporary hyraxes to that of a hippopotamus. During the Miocene (about 25 million years ago [mya]), at the time of the first radiation of the bovids, hyrax diversity was greatly reduced, with species persisting only among rocks and in trees—habitats that were not invaded by bovids.

Fossil and morphological evidence shows that hyraxes share many features with elephants and seacows. Recent research using mitochondrial DNA provides additional support for the association of the paenungulates (elephants, hyraxes, and seacows), which, together with sengis (elephant shrews), aardvarks, tenrecs, and golden moles, are called the Afrotheria, a supraordinal grouping of mammals whose radiation is rooted in Africa.

Contemporary hyraxes retain several primitive features, notably the feeding mechanism, which involves cropping with the molars instead of the incisors as with most modern hoofed mammals, imperfect endothermy, and short legs and feet.

Hyraxes are members of the order Hyracoidea, family Procaviidae. Three living genera contain five species superficially similar in size and appearance. The word hyrax derives from the Greek word *hyrak*, which means shrew.

The genus *Procavia* contains two extinct and one living species (*Procavia capensis*) of Africa and southwest Asia. However, several authors have described four species for this genus: Cape hyrax (*P. capensis*), Abyssinian hyrax (*P. habessinica*), Johnston's hyrax (*P. johnstoni*), and western hyrax (*P. ruficeps*), while Bothma, in 1966, added a fifth species, the Kaokoveld hyrax (*P. welwitschii*). More recent studies on the geographic variation in mitochondrial DNA in South Africa indicate that at least two species in what conventionally has been regarded as *P. capensis*. Therefore, the monospecificity of the genus *Pro-*

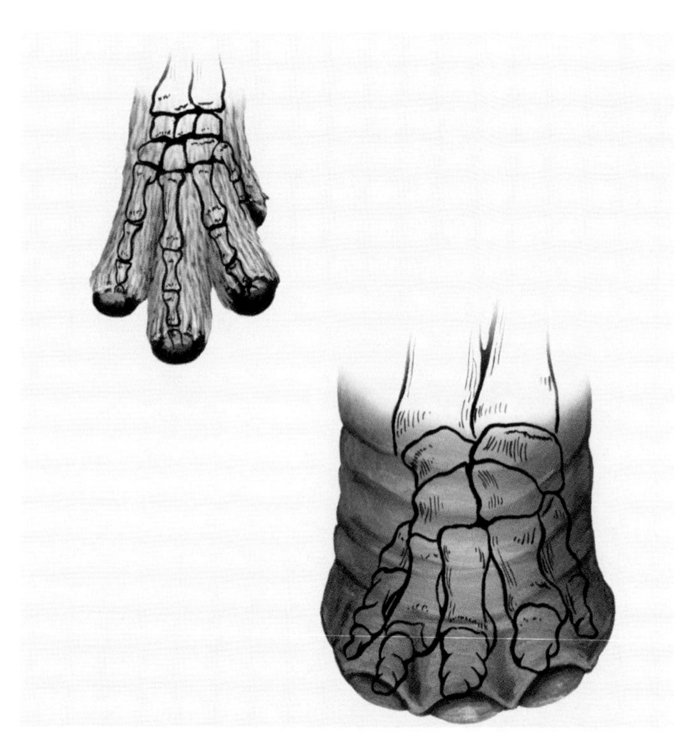

Foot comparison of hyrax (top) and elephant (bottom). (Illustration by Jarrod Erdody)

cavia by Olds and Shoshani in 1982 is debatable and only further research on the genetics will provide clarification.

For the bush hyrax, genus *Heterohyrax*, a single species (*H. brucei*) and 25 subspecies have been described though some consider *H. atineae* and *H chapini* as separate species.

Tree hyraxes in the genus *Dendrohyrax* include three living species endemic to Africa: western tree hyrax (*D. dorsalis*), southern tree hyrax (*D. arboreus*), and eastern tree hyrax (*D. validus*).

As for rock hyraxes, the species level classification for bush and tree hyraxes is still uncertain; there may be more species than currently accepted. Further research on their anatomy, genetics, behavior, and bioacoustics is necessary.

Physical characteristics

Hyraxes are small- to medium-sized herbivores, with short legs, a rudimentary tail, and round ears. They have a rabbit-like appearance, hence the vernacular name, rock rabbit. Males and females are approximately the same size. The average size of the adult rock hyrax varies greatly across Africa, and it seems to be closely linked to precipitation. The odd appearance of the hyrax has caused some confusion. Their superficial similarity to rodents led Storr, in 1780, to mistakenly link them with guinea pigs of the genus *Cavia*, and he thus gave them the family name of Procaviidae, or "before the guinea pigs." Later, the mistake was discovered and the group was given the equally misleading name of hyrax, which means "shrew mouse."

The feet have rubbery pads containing numerous sweat glands, and are ill equipped for digging. The feet sweat while the animal is running, which greatly enhances its climbing ability. Bush and tree hyraxes are very are agile climbers and good jumpers; they can ascend a smooth tree trunk up to 20 in (50 cm) in diameter. Feet are flexible and can be turned sole upwards. The forefoot has four digits, while the hindfoot has three. All digits have flat, hoof-like nails, except for the second digit of the hind foot, which has a long, curved claw for grooming.

Species living in arid and warm zones have short fur, but tree hyraxes and species in alpine areas have thick, soft fur. The pelage is dense, with short, thick underfur, and black guard hairs (tactile hairs or vibrissae) up to 1.2 in (30 mm) long are widely distributed over their bodies, probably for orientation in dark fissures and holes. Vibrissae are evident on the snout, above the eyes, under the chin, along the back and sides, on the abdomen, and on fore- and hind limbs.

A dorsal gland lies beneath a raised patch of skin approximately 0.6 in (1.5 cm) long that is surrounded by the dorsal spot of erectile hairs. These hairs are erected during mating behavior, when animals are aroused, and may function as an alarm or threat signal to other hyraxes. The gland in sexually active, mature adults consists of lobules of glandular tissue, is odiferous, and may function in mating and recognition of the mother by young. During courtship, the male erects the hairs of the dorsal spot, exposing the dorsal gland. This probably provides olfactory stimulation and allows dissemination of scent that may communicate the identity and status of the individual. The erectile hairs around the dorsal gland is a light cream to yellow-colored circle of hairs in all species, with the exception of the rock hyrax of southern Africa, in which it is not conspicuous.

Coat color of rock hyrax is light to dark brown. Bush hyraxes have a light gray to brown fur, ventral color is white or creamy, in distinct contrast to the sides, back, head, and rump. Eyebrows are strikingly white to creamy and conspicuous at a distance. In tree hyraxes, the coat is thick, coarse, dark brown, gray-brown, or black, and with scattered lighter yellowish hairs. *D. dorsalis* can be distinguished from other members of the genus by a shorter, coarser hair, longer dorsal patch, naked rostrum, and white spot beneath the chin.

Eyes are bulging, the head is flat dorsally, and the muzzle is skunk-like. A pair of upper incisors is tusk-like, ridged, or triangular in cross-section in males. The faces of these incisors are rounded in females. A space (diastema) 0.39–0.47 in (10–12 mm) long in adults precedes the molar teeth that bear transverse cusps adapted for a herbivorous diet.

The pupil of the eye houses a shield (umbraculum) that allows a basking individual to stare into the sun to detect aerial predators.

The digestive system is complex, with three separate areas of microbial fermentation for the food: the fore stomach, the caecum, and the paired colonic appendages.

Body temperature typically ranges from 95 to 98.6°F (35–37°C), but fluctuates up to 12°F (7°C) with air temperature. At air temperatures above 77°F (25°C) body temperature is maintained by evaporative water loss from the nostrils, soles of the feet, panting, salivating, and grooming. Little free water is consumed because of the low metabolic rate, low urine volume, and thermal lability.

Hyraxes conserve energy by having a low metabolic rate and a labile body temperature. The metabolic rate is 30% lower than that predicted on a weight basis, while the labile body temperature suggests a strategy adopted by larger animals such as an eland. The labile body temperature is activated by acclimatization and not by a rhythmic daily drop in body temperature.

As hyraxes have the habit of urinating in the same place, crystallized calcium carbonate forms deposits that whiten the cliff faces below latrines. The precipitated calcium oxylate where urine soaked through the dung heaps and then crystallized where it seeped out was used as medicine (hyraceum) by several South African tribes and by European settlers.

Testes are permanently abdominal and the uterus is duplex. Average distance between anus and penis is 3.1 in (8 cm) in bush hyraxes, 1.3 in (3.5 cm) in rock hyraxes, and 0.7 in (2 cm) in tree hyraxes. Anatomy of the bush hyrax penis is complex, and the penis measures less than 2.3 in (6 cm) when fully erected.

Female rock and bush hyraxes have one pair of pectoral and two pairs of inguinal mammae. Some tree hyraxes have the same number, while other subspecies have one pair pectoral and one pair inguinal.

Individuals dust-bathe to remove external parasites. In addition to the modified claw on the second digit of the hind foot, the four lower incisors are comb-like for grooming the fur.

The adult males of all three genera produce distinctive and loud calls (territorial calls), which can be used as a simple and confident method of locating and recognizing different species. In the tree hyrax, the differences in the characteristics of calls can even be used to differentiate between subspecies.

Ears are small and rounded and may be tipped with white. The tail does not extend past end of body.

Distribution

Hyraxes are endemic to Africa with the exception of bush hyrax found in Sinai and the rock hyrax from Lebanon to Saudi Arabia. Rock and bush hyraxes are dependent on the presence of suitable refuges in rocky outcrops (kopjes), piles of large boulders, and cliffs. These rock outcrops can provide a constant, moderate temperature 62.6–77°F (17–25°C) and humidity (32–40%), and protection from fire.

Rock hyraxes live in a wide range of habitats, from arid deserts to rainforests, and from sea level to the alpine zone of Mt. Kenya (10,500–13,800 ft [3,200–4,200 m]). Bush hyrax is found in parts of north and northeast Africa and the Sinai, east and south from Egypt to South Africa, and from Namibia to Congo. *H. b. antineae* is isolated in Algeria.

Bush hyraxes (*Heterohyrax brucei*) with young, in Wankie National Park, Zimbabwe. (Photo by Animals Animals ©Bertram G. Murray, Jr. Reproduced by permission.)

Tree hyraxes are found in arboreal habitats, but in the alpine areas (up to 14,763 ft [4,500 m]) of the Ruwenzori Mountains in Uganda and Congo, they are also rock dwellers. The eastern tree hyrax (*D. validus*) might be the earliest type of forest-living tree hyrax, being a member of the primitive fauna and flora of the islands of Zanzibar and Pemba in East Africa.

Habitat

Hyraxes adapt to any shelter that provides adequate protection from predators and the elements. Their occurrence, distribution, and numbers depend upon a combination of several abiotic factors such as rainfall and availability of holes and hiding places and biotic factors such as interspecific and intraspecific competition for food, predation, and parasites.

Rock hyraxes are dependent on the presence of suitable refuges in mountain cliffs and rocky outcrops. In several parts of Africa, bush and rock hyraxes occur together and live in close associations. For example, in the Serengeti National Park, rock and bush hyraxes are the most important resident herbivores of the kopjes (rock outcrops). Their numbers depend on the size of the kopje. The population density of rock hyraxes range 5–56 individuals, for bush hyraxes from 20–81 animals per 2.5 ac (1 ha) of kopje, and group size varies 5–34 for the former and 2–26 for the latter. In the Matobo National Park in Zimbabwe, density of *P. capensis* was estimated at 0.5–1.1 individuals/2.5 ac (1 ha) (1.2–2.6/2.5 ac [ha] of kopje) over a five-year period. This population consisted of 19.4–27.5% juveniles, 7.2–13.1% sub-adults, and 62.9–73.7% adults. Juvenile mortality was estimated at 52.4–61.3%.

These long-term observations have shown that in the Serengeti and Matobo, hyrax populations fluctuate and small colonies are prone to extinction.

Rock hyraxes (*Procavia capensis*) live in colonies of about 50 in natural crevices of rocks or bolders. (Photo by David M. Maylen III. Reproduced by permission.)

In Karoo National Park in South Africa, a strong link between drought and increased predation seems to be the causative agents for hyrax decline. The erratic rainfall is directly responsible for annual variation in hyrax recruitment, and thus population structure. This relationship may have been more responsible for irregularities in the age distribution than the variable mortality, which can affect all age classes. As rainfall precedes the birth season, it shows a close relationship with recruitment. Recent rains have an effect on the moisture of vegetation, and thus on the quality and quantity of milk of lactating females.

Tree hyraxes are found in moist forests, moist savannas, riverine vegetation, and montane habitats. At higher elevations, they can live among rock formations and are partly diurnal. Individuals maintain territories, but population densities and structure are poorly known. In the Mau Forest in Kenya, the density of *D. arboreus* varied between 1.3–6.2 animals/2.5 ac (1 ha), depending on the type of forest. In Mt. Kilimanjaro, the density of *D. validus* at three different sites was of 13, 23, and 70 animals/2.5 ac (ha).

External parasites such as ticks, lice, mites, and fleas, and internal parasites such as nematodes, cestodes, and anthrax play a role in hyrax mortality. In the Serengeti, the sarcoptic mite, that causes mange, is an important cause of mortality for rock hyraxes. In West Africa, tree hyraxes have been found to have nematode parasites (*Crossophorus collaris*, *Libyostrongylus alberti*, *Hoplodontophorus flagellum*, and *Theileriana brachylaima*).

The most important avian predator for rock and bush hyraxes in certain parts of Africa is the black, or Verreaux's, eagle (*Aquila verreauxii*), which feeds almost exclusively on hyraxes. This eagle preys on hyrax despite low availability. They remove substantial numbers, limit their population growth, and exert hard selection on adult hyraxes as observed in Karoo National Park. Other predators are martial and tawny eagles, leopards, lions, jackals, spotted hyena, and several snake species. In South Africa, the caracal is the second most important predator of hyrax, which can comprise more than 50% of its food.

Predators of the tree hyrax are the African crowned eagles (*Stephanoaetus coronatus*), leopards (*Panthera pardus*), and possibly also larger eagle owls (*Bubo* spp.) or hawk-eagles (*Hieraaetus* spp.). Chimpanzees have also been documented capturing and killing adult *D. dorsalis*, but have not been seen to eat them.

Behavior

Rock and bush hyraxes are diurnal and gregarious, but tree hyraxes are mainly nocturnal and usually solitary, although groups of two to three animals can be found.

The basic social unit of rock and bush hyraxes is a cohesive and stable polygynous harem, with a territorial adult male, up to 17 adult females, and juveniles. The territorial male repels all intruding males from an area largely encompassing the females' core area. The adult sex ratio is skewed in favor of females, but the sex ratio of newborns is 1:1.

In the Serengeti bush hyraxes, there are four classes of mature males: territorial, peripheral, and early and late dispersers. Territorial males are the most dominant. Their aggressive behavior toward other adult males escalates in the mating season when the weight of their testes increases 20-fold. These males monopolize receptive females and show a preference for copulating with females over 28 months of age. A territorial male monopolizes "his" female group year round and repels other males from sleeping holes, basking places, and feeding areas. Males can fight to the death, although this is quite rare. While his group members feed, a territorial male will often stand guard on a high rock and be the first to call in case of danger. Males utter the territorial call all year round.

A young rock hyrax (*Procavia capensis*). (Photo by Ann & Steve Toon Wildlife Photography. Reproduced by permission.)

A tree hyrax (*Dendrohyrax arboreus*) in the Serengeti National Park in Tanzania. (Photo by Animals Animals ©Bruce Davidson. Reproduced by permission.)

Peripheral males are those unable to settle on small kopjes, but on large kopjes can occupy areas on the periphery of the territorial males' territories. They are solitary, and the highest ranking among them takes over a female group when a territorial male disappears. These males show no seasonality in aggression, but call only in the mating season. Most of their mating attempts and copulations are with females younger than 28 months. The majority of juvenile males—the early dispersers—leave their birth sites at 16–24 months old, soon after reaching sexual maturity. The late dispersers leave a year later, but before they are 30 months old. Before leaving their birth sites, both early and late dispersers have ranges that overlap their mothers' home ranges. They disperse in the mating season to become peripheral males. Almost no threat, submissive, or fleeing behavior has been observed between territorial males and late dispersers.

Larger kopjes may support several family groups, each occupying a traditional range. The females' home ranges are not defended and may overlap. Rarely, an adult female from outside a group will be incorporated into the family group. In the bush hyrax, these immigrants are responsible for bringing new alleles into local populations, preventing inbreeding, and consequently reduce the risk of local extinction.

Individuals of rock and bush hyraxes were observed to disperse over a distance of at least 1.2 mi (2 km). However, the further a dispersing animal has to travel across the open grass plains, where there is little cover and few hiding places, the greater are its chances of death, either through predation or as a result of its inability to cope with temperature stress. First results on DNA analysis of rock hyraxes in the Serengeti show almost no genetic variation between colonies.

Body temperature is maintained mainly by gregarious huddling, long periods of inactivity, and basking. Although their physiology allows them to exist in very dry areas and use food of relatively poor quality, they are dependent on shelters (boulders and tree cavities) that provide relatively constant temperature and humidity.

Where both species live together, they huddle together in the early mornings after spending the night in the same holes. They also use the same urination and defecation sites. Parturition tends to be synchronous, and the two species cooperate. Newborns are greeted and sniffed intensively by members of both species, and they form a nursery group and play together. Most of their vocalizations are also similar. However, bush and rock hyraxes do differ in key behavior patterns. They do not interbreed because their mating behavior is different, and they have different reproductive anatomy. The male territorial call, which might function as a "keep out" sign, is also different. Finally, the bush hyrax browses on leaves, but the rock hyrax feeds mainly on grass. The latter is probably the main factor that allows both species to live together.

Tree hyraxes live primarily solitary, but groups of two and three can be found (likely mother and sub-adult young). They have small home ranges, with each defended male territory overlapping those of several smaller female ranges. Individuals in captivity rubbed dorsal glands, probably used in the wild to mark territory boundaries and individual identification. Individuals use latrines, defecating repeatedly at the bases of trees. Largely inactive, they emerge regularly at dusk and have another period of activity before daylight. Tree hyraxes produce loud and distinct calls, characterized by long cries, repeated between 22 and 42 times at gradually increasing amplitude and intervals, reaching a loud crescendo at the end. They call throughout the night, but with marked peaks in late evening and early morning corresponding with the activity patterns. They are also heard to call during the day, normally after being disturbed.

A rock hyrax (*Procavia capensis*) in its rock crevice home. (Photo by Animals Animals ©Werner Layer. Reproduced by permission.)

Young tree hyrax (*Dendrohyrax arboreus*) emerging from nest in Democratic Republic of the Congo (Zaire). (Photo by Animals Animals ©Bruce Davidson. Reproduced by permission.)

Feeding ecology and diet

Hyraxes are herbivorous, consuming mostly leaves, twigs, fruit, and bark. Hyraxes do not ruminate. Their gut is complex, with three separate areas of microbial digestion, and their ability to digest fiber efficiently is similar to that of ruminants. Their efficient kidneys allow them to exist on minimal moisture intake. In addition, they have a high capacity for concentrating urea and electrolytes and excreting large amounts of undissolved calcium carbonate.

Rock hyraxes in the Serengeti were observed feeding on 79 plant species. The animals have a high seasonal adaptability: in the wet season, they showed a high preference for grasses (78%), but in the dry season when grasses became parched and poor in quality, they browsed (57%) extensively, and more or less in proportion to the foliage density of each vegetation class. As rock hyraxes feed mainly on grass, which is a relatively coarse material because of phytoliths (plant opal), their molars and premolars are hypsodont, i.e., they have high crowns with relatively shorter roots.

Bush and tree hyraxes are obligate browsers. In the Serengeti, bush hyraxes were observed feeding on 64 plant species, but two to 11 species formed 90% of the animal's staple diet. They browsed leaves, buds, flowers, and fruits of trees, bushes, and herbs predominantly in the wet (81%) and dry (92%) season. Browse material is softer than grass. This difference is shown in the brachydont dentition (short crowns with relatively long roots) of bush and tree hyraxes.

Examination of ^{13}C:^{12}C ratios of carbonate and collagen fractions of bone and microwear patterns of the molariform teeth confirmed that the bush hyrax is a browser and the rock hyrax switches between grazing and browsing.

Most feeding occurs between 7:30 and 11 A.M. and 3:30 and 6 P.M., but occasionally to 9 P.M. Individuals may feed alone or in a group. Group feeding can occur up to 164 ft (50 m) from the center of the colony, although casual feeding rarely occurs at distances greater than 65.6 ft (20 m) from the den site. Feeding bouts average 20 minutes and last no longer than 35 minutes. Individuals can climb vertical trunks of trees and balance on thin branches to strip the vegetation of leaves.

Territorial male rock and bush hyraxes usually show sentinel behavior by sitting on a high rock or tree branch while the family is group feeding. This is the time when hyraxes are most vulnerable to predation because they venture furthest from shelter. The guarding animals are often the first to give a warning or alarm call in case of a sudden danger, whereupon the feeding animals take cover immediately. At group feeding times, individuals of both species may guard simultaneously; the warning or alarm call of either species is acted upon by all animals.

Reproductive biology

Females have six pairs of teats, one pair pectoral and two pairs inguinal. Females become receptive once a year, and a peak in births seems to coincide with rainfall. Before mating, a bush hyrax male emits a shrill cry while approaching the female, and she erects her dorsal spot hairs. The male sniffs the female's vulva, rests his chin on her rump, then slides onto her back as he makes thrusting movements followed by in-

A rock hyrax (*Procavia capensis*) foraging near burrow in Kenya, Africa. (Photo by Animals Animals ©Joe Mc Donald. Reproduced by permission.)

A rock hyrax (*Procavia capensis*) colony sunning in Tsitsikanna National Park, South Africa. (Photo by Animals Animals ©Anthony Bannister. Reproduced by permission.)

tromission in three to five minutes. A second copulation may occur in one to three hours.

Territorial bush hyrax males copulate more often than peripheral males and mate preferentially with females younger than 28 months of age. Peripheral males exhibit a dominance hierarchy and mate more often with young females in a polygynous system.

Gestation is between 26 and 30 weeks. Within a family group, the pregnant females all give birth within a period of about three weeks. The number of young per female bush hyraxes ranges from one to three, and in rock hyraxes from one to four. In tree hyraxes, one to two young are born. Litter size is smaller in *Dendrohyrax* than other hyrax genera. In southern Africa, a female rock hyrax with six embryos has been collected. Numbers depend on the size (age) of the mother; first breeders have only one to two.

The young are precocial, being fully developed at birth, and weigh 6.35–13.4 oz (180–380 g). Mothers suckle only their own as the young assume a strict teat order. Weaning occurs at one to five months, and both sexes reach sexual maturity at about 16–17 months of age. Upon sexual maturity, females usually join the adult female group, while males disperse before they reach 30 months. Adult females live significantly longer than adult males and may reach an age of more than 11 years.

A study on shot samples in the Karoo indicated a relatively high incidence of adult males in the study population, which appears to be an effect of drought. Rock hyrax males pre dominated in age classes that were born during dry conditions, while females predominated in age classes born during wet conditions. This difference might be due to late fetal reabsorption or mummification in the uterus. The ability to reabsorb fetuses at a late stage would be highly adaptive for hyraxes in an unpredictable environment.

Play behavior of the young consists of nipping, biting, climbing, pushing, fighting, chasing, and mounting. When rock and bush hyraxes live together, young of both species will play. Young in nurseries are attended by their own mothers, mothers of other young, non-maternal conspecifics, or even individuals from the other species.

In tree hyrax, both mating and birth peaks tend to coincide with the dry season, but offspring may be born throughout the year. Females excrete cinnamon-smelling oil from their dorsal gland prior to mating. Young reach sexual maturity around 16 months of age. Lifespan is poorly known, although captive animals have been reported to live up to 12 years.

A rock hyrax (*Procavia capensis*) family in Kyle National Park, Zimbabwe, Africa. (Photo by Animals Animals ©Bertram G. Murray, Jr. Reproduced by permission.)

Conservation status

Dendrohyrax validus, *Heterohyrax antinea*, and *H. chapini* are currently categorized as Vulnerable in the IUCN Red List. Others have no special status for IUCN, CITES, or U.S. ESA. All three of the tree hyrax species are probably sensitive to habitat degradation, as they are mainly confined to primary forests in Africa. They are killed for their fur and for food, but apparently are widespread and common in large forest tracks. According to the African Mammals Database, only about 6% of geographical range of *D. dorsalis* is protected. *D. validus* is heavily hunted for its fur in the forest belt around Mt. Kilimanjaro.

Significance to humans

The rock hyrax is mentioned several times in the Bible: Solomon says they are "wise" (Proverbs 30:26) because "the conies are a feeble folk, yet they make their houses in the rocks." And, "The high mountains are for the wild goats; the rocks are a refuge for the conies" (Psalms 104:18).

In Phoenician and Hebrew, hyraxes are known as *shaphan*, meaning "the hidden one." Some 3,000 years ago, Phoenician seamen explored the Mediterranean, sailing westward from their homeland on the coast of Syria. They found land where they saw small mammals, which they thought were hyraxes, and so they called the place "I-shaphan-im"—Island of the Hyrax. The Romans later modified the island's name to Hispania.

Several African tribes hunt, snare, or trap hyrax as a food source and for the skin. The Hadza or Watindiga, a Bushman tribe in Tanzania, hunt rock and bush hyraxes. Hadza boys catch a newborn hyrax and, when caused some pain, the young animal emits a loud bird-like chirrup distress call, which incites adult females and males to leave the safe holes or cracks to help and are then shot with arrows. The most common principle for catching *D. arboreus* in the Mau Forest in Kenya was to dislodge the animals from the trees and then kill them on the ground. The meat is eaten and the skin collected. It is estimated that the total off-take for the southwest Mau area per year is 16,000 hyraxes, representing about a quarter of the annual population increment. In the forest belt around Mt. Kilimanjaro, the eastern tree hyrax is heavily hunted for its skin; 48 animals yield one rug.

The forest dwelling people of the southwest Mau use tree hyrax in the traditional medicine as a means of prevention and to cure a number of ailments. The principal medicinal use was to cure deep coughing by drinking the ash of burnt hairs mixed with water or honey. In rituals, hyrax also played a role. Some clans traditionally bless their newborn babies by wrapping them in hyrax skins to ensure good health. Hyraxes are also regarded as an omen.

In Kenya and Ethiopia, rock and tree hyraxes might be an important reservoir for the parasitic disease cutaneous leishmaniasis, which can also affect humans.

1. Eastern tree hyrax (*Dendrohyrax validus*); 2. Southern tree hyrax (*Dendrohyrax arboreus*); 3. Bush hyrax (*Heterohyrax brucei*); 4. Western tree hyrax (*Dendrohyrax dorsalis*); 5. Rock hyrax (*Procavia capensis*). (Illustration by Joseph E. Trumpey)

Species accounts

Southern tree hyrax
Dendrohyrax arboreus

TAXONOMY
Dendrohyrax arboreus (A. Smith, 1827), Cape of Good Hope, South Africa. Eight subspecies have been described.

OTHER COMMON NAMES
English: Tree dassie; French: Daman d'arbre; German: Baumschliefer; Spanish: Daman de árbol.

PHYSICAL CHARACTERISTICS
Head and body length 12.5–24 in (32–60 cm); weight 3.7–9.9 lb (1.7–4.5 kg). Males and females are approximately the same size. Coat is long, soft, and dark brown; dorsal spot light to dark yellow, from 0.9–1.2 in (23–30 mm) long. Number of mammae variable. Longevity over 10 years.

DISTRIBUTION
Found in the evergreen forests of the eastern Cape Province and the Natal midlands of South Africa; northwestern Zambia; northeastern and eastern Zambia; eastern Congo; northwestern Tanzania; Burundi and Rwanda; western Uganda; central and southern Kenya.

HABITAT
Evergreen forests up to about 13,500 ft (4,500 m). In the Ruwenzori, they live also among rock boulders.

BEHAVIOR
Nocturnal and live primarily solitary, but groups of two and three can be found (likely mother and subadult young). Individuals maintain territories, but population densities and structure poorly known.

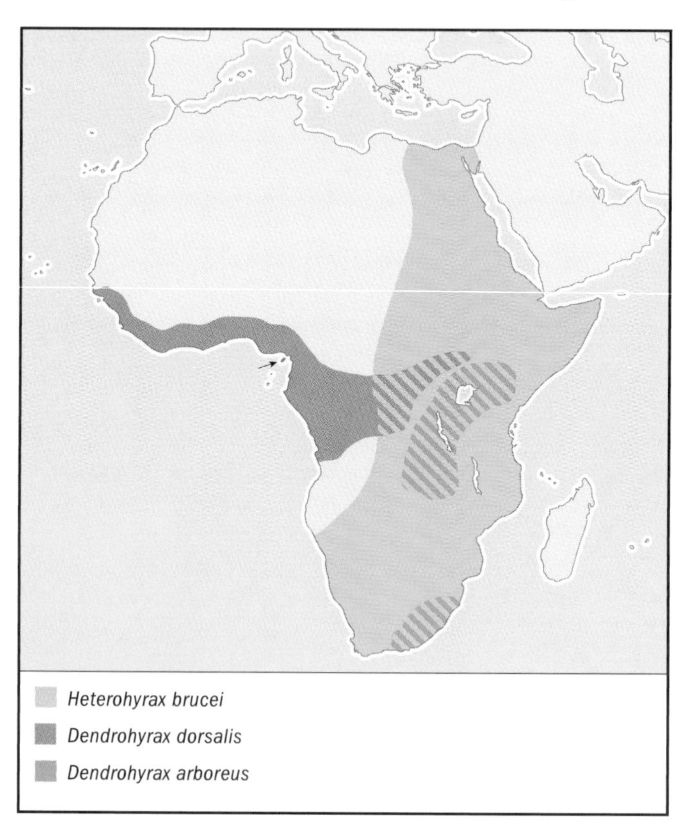

FEEDING ECOLOGY AND DIET
Herbivorous, browsing leaves, buds, twigs, fruits from forbs and trees all year-round.

REPRODUCTIVE BIOLOGY
Not well known. Gestation: 220–240 days; one to two young per female.

CONSERVATION STATUS
Not threatened.

SIGNIFICANCE TO HUMANS
Certain African tribes hunt hyrax as a source of food, to collect skins, and as medicine. For some tribes it is also important in their spiritual traditions. ◆

Western tree hyrax
Dendrohyrax dorsalis

TAXONOMY
Dendrohyrax dorsalis (Fraser, 1854), Bioko, Equatorial Guinea. Six subspecies have been described.

OTHER COMMON NAMES
English: Tree dassie; French: Daman d'arbre; German: Baumschliefer; Spanish: Daman de árbol.

PHYSICAL CHARACTERISTICS
Head and body length 12.5–24 in (32–60 cm); weight 3.7–9.9 lb (1.7–4.5 kg). Males and females are approximately the same size. Coat is long, soft, and dark brown; dorsal spot light to dark yellow, from 1.6 to 2.9 in (40–75 mm) long. One pair of inguinal mammae only.

DISTRIBUTION
Found on Fernando Po; the West African coastal forests from Gambia to Angola; central and northeastern Congo, and northern Uganda.

HABITAT
Found in moist forests up to about 12,000 ft (3,650 m), moist savannas, and montane habitats.

BEHAVIOR
Nocturnal and live primarily solitary, but groups of two and three can be found (likely mother and subadult young). Individuals maintain territories, but population densities and structure poorly known.

FEEDING ECOLOGY AND DIET
Herbivorous, browsing leaves, buds, twigs, fruits from forbs and trees all year-round.

REPRODUCTIVE BIOLOGY
In the tree hyrax both mating and birth peaks tend to coincide with the dry season, but offspring may be born throughout the year. Gestation: 220–240 days; one to two young per female.

CONSERVATION STATUS
Not threatened.

SIGNIFICANCE TO HUMANS
Certain African tribes hunt hyrax for food and for their skin. ◆

Eastern tree hyrax
Dendrohyrax validus

TAXONOMY
Dendrohyrax validus True, 1890, Mt. Kilimanjaro, Tanzania. Two subspecies have been described: *D. v. validus* on the continent and *D. v. neumanni* on the islands of Pemba, Zanzibar and Tumbatu.

OTHER COMMON NAMES
English: Tree dassie; French: Daman d'arbre; German: Baumschliefer; Spanish: Daman de árbol; Kiswahili: Perere.

PHYSICAL CHARACTERISTICS
Head and body length 12.5–24 in (32–60 cm); weight 1.7–4.0 kg (3.7–8.0 lb). Males and females are approximately the same size. Coat is long, soft and very dark brown, dorsal spot light to dark yellow, from 0.8 to 1.6 in (20–40 mm) long. One pair of inguinal mammae. Longevity unknown.

DISTRIBUTION
Kilimanjaro, Mt. Meru, Usambara, Zanzibar, Pemba, and the relict forests of the Kenyan coast.

HABITAT
Evergreen forests up to about 11,500 ft (3,500 m). On the Kenyan coast they live in the fossil reef area.

BEHAVIOR
Tree hyraxes are nocturnal and live primarily solitary but groups of two and three can be found (likely mother and subadult young). Population densities in the Kilimanjaro varies 7–23 animals/2.5 acres (1 ha).

FEEDING ECOLOGY AND DIET
herbivorous, browsing leaves, buds, twigs, fruits from forbs and trees all year round.

REPRODUCTIVE BIOLOGY
Not known. Gestation: 220–240 days; one to two young per female.

CONSERVATION STATUS
Listed as Vulnerable on the IUCN Red List.

SIGNIFICANCE TO HUMANS
African tribes hunt hyraxes as a source of food. In the Kilimanjaro area, they are hunted extensively for their skins, which have commercial value. ◆

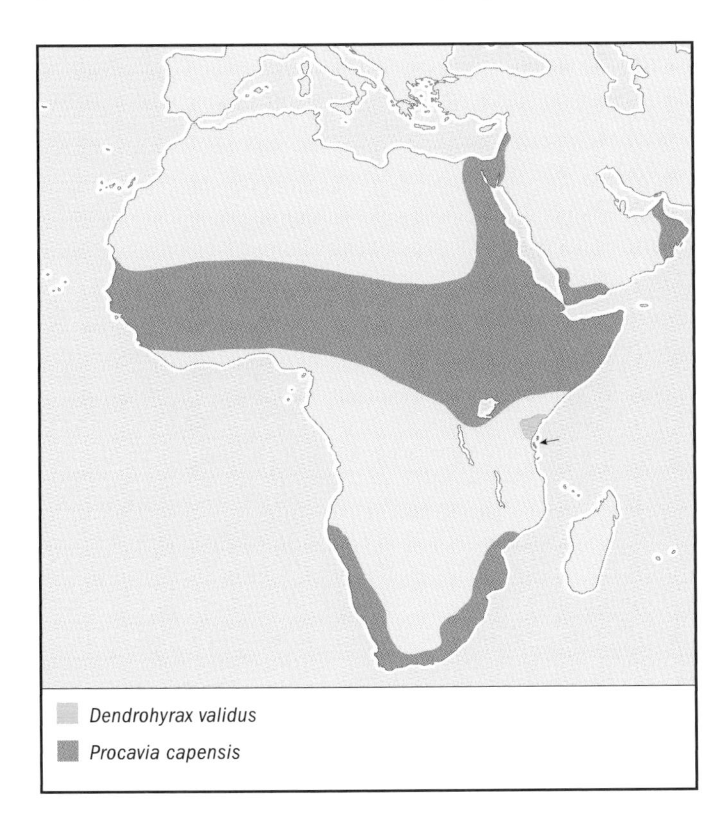

Bush hyrax
Heterohyrax brucei

TAXONOMY
Heterohyrax brucei (Gray, 1868), Ethiopia. Twenty-five subspecies have been described.

OTHER COMMON NAMES
English: Bush dassie, yellow-spotted rock hyrax, yellow-spotted hyrax, klipdassie; French: Daman d'arbuste; German: Buschschliefer; Spanish: Daman de arbusto.

PHYSICAL CHARACTERISTICS
Head and body length 12.5–18.5 in (32–47 cm); weight 2.9–5.3 lb (1.3–2.4 kg). Males and females are approximately the same size. Average distance between anus and penis is 3.1 in (8 cm). Coat is light gray to brown; ventral color is white or creamy in distinct contrast to the sides, back, head, and rump. Eyebrows are strikingly white to creamy and conspicuous at a distance.

DISTRIBUTION
Found in parts of northeast Africa and the Sinai; east to south, from Egypt to South Africa and Namibia to Congo. *H. b. antineae* might occur isolated in Algeria and central Sahara.

HABITAT
Adapt to any shelter that provides adequate protection from predators and the elements. Rock boulders and outcrops in different vegetation zones in Africa, sometimes in hollow trees.

BEHAVIOR
Diurnal and gregarious, the social unit being a polygynous harem, with a territorial adult male, several adult females, and juveniles.

FEEDING ECOLOGY AND DIET
Herbivorous, browsing leaves, buds, twigs, and fruits all year-round.

REPRODUCTIVE BIOLOGY
Females become receptive once a year, and a peak in births seems to coincide with rainfall. Polygynous. Gestation: 212–240 days; one to three young per female. Weaning at one to five months, and both sexes reach sexual maturity at about 16–17 months of age. Longevity: 9–12 years. Adult females live significantly longer than adult males.

CONSERVATION STATUS
Not threatened.

SIGNIFICANCE TO HUMANS
Certain African tribes hunt hyrax for food. ◆

Rock hyrax
Procavia capensis

TAXONOMY
Procavia capensis (Pallas, 1766), Cape of Good Hope, South Africa. Four subspecies identified.

OTHER COMMON NAMES
English: Klipdassie; French: Daman les roches; German: Klippschliefer; Spanish: Damán de rocas.

PHYSICAL CHARACTERISTICS
Head and body length 17–21 in (44–54 cm); weight 4–12 lb (1.8–5.4 kg). Males and females are approximately the same size. Mean distance between anus and penis 1.3 in (3.5 cm). Coat is light to dark brown; the dorsal spot is light creamy to yellow-colored in all species, with the exception of the rock hyrax of southern Africa, where it is not conspicuous.

DISTRIBUTION
Southwest and northeast Africa, Sinai to Lebanon, and southeast Arabian Peninsula.

HABITAT
Dependent on the presence of suitable refuges in mountain cliffs and rocky outcrops. Live in a wide range of habitats, from arid deserts to forests, and from sea level to the alpine zone of Mt. Kenya (10,500–13,800 ft [3,200–4,200 m]).

BEHAVIOR
Diurnal and gregarious, the social unit being a polygynous harem, with a territorial adult male, several adult females, and juveniles.

FEEDING ECOLOGY AND DIET
Herbivorous, consuming mostly leaves, twigs, fruit, and bark. The animals have a high seasonal adaptability: during the rainy season, they have a high preference for grasses, but in the dry season, they browse extensively.

REPRODUCTIVE BIOLOGY
Females become receptive once a year, and a peak in births seems to coincide with rainfall. Polygynous. Gestation: 212–240 days; one to four young per female. Weaning at one to five months, and both sexes reach sexual maturity at about 16–17 months of age. Longevity: 9–12 years. Adult females live significantly longer than adult males.

CONSERVATION STATUS
Not threatened.

SIGNIFICANCE TO HUMANS
Certain African tribes hunt hyraxes for food. Mentioned several times in the Bible as "conie." ◆

Common name / Scientific name/ Other common names	Physical characteristics	Habitat and behavior	Distribution	Diet	Conservation status
Ahaggar hyrax *Heterohyrax antineae* English: Hoggar hyrax	Coloration is brownish or grayish, sometimes suffused with black, under-parts are white. Patch of yellow, red, or white in middle of back, indicating gland. Head and body length 12.6–22 in (32–56 cm), weight 2.9–9.9 lb (1.3–4.5 kg).	Rocky kopjes, rocky hillsides, krantzes, and piles of loose boulders, particularly where there is a cover of trees and bushes on which it can feed, from sea level to at least 12,470 ft (3,800 m). Very keen and quite aggressive. Diurnal. Groups of 5–34 individuals.	Ahaggar Mountains of southern Algeria.	Many kinds of bushes and trees.	Vulnerable
Yellow-spotted hyrax *Heterohyrax chapini* French: Daman de steppe ou gris	Coloration is brownish or grayish, sometimes suffused with black, underparts are white. Patch of yellow, red, or white in middle of back, indicating gland. Head and body length 12.6–22 in (32–56 cm), weight 2.9–9.9 lb (1.3–4.5 kg).	Rocky kopjes, rocky hillsides, krantzes, and piles of loose boulders, particularly where there is a cover of trees and bushes on which it can feed, from sea level to at least 12,470 ft (3,800 m). Very keen and quite aggressive. Diurnal. Groups of 5–4 individuals.	Much of eastern Africa from western Egypt to northern South Africa, extending as far west as southern Angola.	Many kinds of bushes and trees, even those that are poisonous to most other mammals.	Vulnerable
Kaokoveld hyrax *Procavia welwitschii*	Small, brown body covered with thick, coarse hair. Small snout, short tail. Head and body length 9.8–11.8 in (25–30 cm), weight 5.5–7.7 lb (2.5–3.5 kg).	Montane grasslands and shrubs, usually taking shelter in rocky outcrops. Females come into heat once a year. Peak in births generally coincides with rainy season.	Southwestern Angola, and Namibia.	A wide variety of plants with emphasis on grasses.	Not threatened
Red-headed rock hyrax *Procavia ruficeps* English: Western hyrax	Small, brown body covered with thick, coarse hair. Small snout, short tail. Head and body length 9.8–11.8 in (25–30 cm), weight 5.5–7.7 lb (2.5–3.5 kg).	Rocky outcrops in arid regions. Females come into heat once a year, birthing peak follows rainy season. Average 2.4 offspring per litter.	Southern Algeria and Senegal to Central African Republic.	A wide variety of plants with emphasis on grasses.	Not threatened
Johnston's hyrax *Procavia johnstoni*	Small, brown body covered with thick, coarse hair. Small snout, short tail. Head and body length 9.8–11.8 in (25–30 cm), weight 5.5–7.7 lb (2.5–3.5 kg).	Rock outcroppings in arid zones. Birthing season correlates with rainy season. Average 2.4 offspring produced per litter.	Northeastern Zaire and central Kenya to Malawai.	A wide variety of plants with emphasis on grasses.	Not threatened
Syrian hyrax *Procavia syriacus*	Small, brown body covered with thick, coarse hair. Small snout, short tail. Head and body length 9.8–11.8 in (25–30 cm), weight 5.5–7.7 lb (2.5–3.5 kg).	Rocky outcrops in arid zones. Active during daylight. Tends to be solitary. Makes a variety of whistles, chatters, and other sounds.	Egypt to Kenya, and the southwest Asian portion of the range of the genus.	Consists of leaves, bark, and grasses, and they also eat some insects.	Not threatened

Resources

Books

Barry, R., and H. N. Hoeck. "*Heterohyrax brucei*." In *The Mammals of Africa: A Comprehensive Synthesis*, edited by Jonathan Kingdon, David Happold, and Thomas Butynski. London: Academic Press, 2003.

Bothma, J. P. "Du Hyracoidea." In *Preliminary Identification Manual of African Mammals*, edited by P. J. Meesters. Washington DC: Smithsonian Institution, 1966.

Davies, R. A. G. "Black Eagle (*Aquila verrauxii*) Predation on Rock Hyrax (*Procavia capensis*) and Other Prey in the Karoo." Unpublished PhD Dissertation. University of Pretoria, Pretoria, South Africa. 1994.

Fischer, M. S. "Hyracoidea." *Handbuch der Zoologie. Band VIII Mammalia.* Berlin and New York: Walter de Gruyter, 1992.

Gargett, V. *The Black Eagle: A Study.* Randburg, South Africa: Acorn Books, 1990.

Hoeck, H. N. *Systematics of the Hyracoidea: Towards a Clarification.* Pittsburgh: Bulletin Carnegie Museum of Natural History, 1978.

———. "*Procavia capensis*." In *The Mammals of Africa: A Comprehensive Synthesis*, edited by Jonathan Kingdon, David Happold, and Thomas Butynski. London: Academic Press, 2003.

Kingdon, J. *East African Mammals. An Atlas of Evolution in Africa.* London: Academic Press, 1971.

Roberts, D., E. Topp-Jorgensen, and D. Moyer. "*Dendrohyrax validus*." In *The Mammals of Africa: A Comprehensive Synthesis*, edited by Jonathan Kingdon, David Happold, and Thomas Butynski. London: Academic Press, 2003.

Resources

Schultz, D., and D. Roberts. "*Dendrohyrax dorsalis*." In *The Mammals of Africa: A Comprehensive Synthesis*, edited by Jonathan Kingdon, David Happold, and Thomas Butynski. London: Academic Press, 2003.

Periodicals

Bartholomew, G., and M. Rainy. "Regulation of Body Temperature in the Rock Hyrax (*Heterohyrax brucei*)." *Journal of Mammalogy* 52, 1994: 81–95.

Barry, R. E. "Synchronous Parturition of *Procavia capensis* and *Heterohyrax brucei* during Drought in Zimbabwe." *South African Journal of Wildlife Research* 24 (1994): 1–5.

Barry, R. E., and P. J. Mundy. "Population Dynamics of Two Species of Hyraxes in the Matobo National Park, Zimbabwe." *African Journal of Ecology* 36, (1998): 221–233.

Barry, R. E., and J. Shoshani. "*Heterohyrax brucei*." *Mammalian Species* no. 645 (2000): 1–7.

Coetzee, C. "The Relative Position of the Penis in Southern African Dassies (Hyracoidea) as a Character of Taxonomic Importance." *Zoologica Africana* 2, (1966): 223–224.

De Niro, M. J., and S. Epstein. "Carbon Isotopic Evidence for Different Feeding Patterns in Two Hyrax Species Occupying the Same Habitat." *Science* 201, no. 4359 (1978): 906–908.

Gerlach, G., and H. N. Hoeck. "Island on the Plains: Metapopulation Dynamics and Female Based Dispersal in Hyraxes (Hyracoidea) in the Serengeti National Park." *Molecular Ecology* 10 (2001): 2307–2317.

Hoeck, H. N. "Demography and Competition in Hyrax: A 17-year Study." *Oecologia* 79 (1989): 353–360.

———. "Differential Feeding Behaviour of the Sympatric Hyrax *Procavia johnstoni* and *Heterohyrax brucei*." *Oecologia* 22 (1975): 15–47.

———. "Population Dynamics, Dispersal and Genetic Isolation in Two Species of Hyrax (*Heterohyrax brucei* and *Procavia johnstoni*) on Habitat Islands in the Serengeti." *Zeitschrift für Tierpsychologie* 59 (1982): 177–210.

———. "Teat Order in Hyrax (*P. johnstoni* and *H. brucei*)." *Zeitschrift für Säugetierkunde* 42, (1977): 112–115.

Hoeck, H. N., H. Klein, and P. Hoeck. "Flexible Social Organization in Hyrax." *Zeitschrift für Tierpsychologie* 59 (1982): 265–298.

Janis, C. M. "New Ideas in Ungulate Phylogeny and Evolution." *Trends in Ecology and Evolution* 3, no. 11 (1988).

Jones, C. "*Dendrohyrax dorsalis*." *Mammalian Species* 113 (1978): 1–4.

Klein, R. G., and K. Cruz-Uribe. "Size Variation in the Rock Hyrax (*Procavia capensis*) and Late Quaternary Climatic Change in South Africa." *Quaternary Research* 46 (1996): 193–207.

Milner, J. "Relationships between the Forest Dwelling People of South-West Mau and the Tree Hyrax *Dendrohyrax arboreus*." *Journal of East African Natural History* 83 (1994): 17–29.

Olds, N., and J. Shoshani. "*Procavia capensis*." *Mammalian Species* 171 (1982): 1–7.

Prinsloo, P., and T. J. Robinson. "Geographic Mitochondrial DNA Variation in the Rock Hyrax, *Procavia capensis*." *Molecular Biology and Evolution* 9 (1992): 447–456.

Rübsamen K., I. D. Hume, and W. V. Engelhardt. "Physiology of the Rock Hyrax." *Comparative Biochemical Physiology* 72A (1982): 271–277.

Seibt, U., H. N. Hoeck, and W. Wickler. "*Dendrohyrax validus*, True, 1890 in Kenia." *Zeitschrift für Säugetierkunde* 42 (1977): 115–118.

Springer, M. S., et al. "Endemic African Mammals Shake the Phylogenetic Tree." *Nature* 388 (1997): 61–64.

Walker A., H. N. Hoeck, and L. Perez. "Microwear of Mammalian Teeth as an Indicator of Diet." *Science* 201 (1978): 908–910.

Yang, F., et al. "Reciprocal Chromosome Painting among Human, Aardvark, and Elephant (Superorder Afrotheria) Reveals the Likely Eutherian Ancestral Karyotype." *Proceedings of the National Academy of Sciences* 100, no. 3 (2003): 1062–1066.

Other

Afrotheria Specialist Group. [April 2003]. <http://www.calacademy.org/research/bmammals/afrotheria/ASG.html>.

Hoeck, H. N. "Ethologie von Busch- und Klippschliefer." Film D 1338 des Institut für den Wissenschaftlichen Film, Göttingen 1980. Publikation von H. N. Hoeck, Publikation Wissenschaftlicher Film, Sektion Biologie, Serie 15, Number 32/D1338 (1982), 24 Seiten.

———. "Nahrungsökologie bei Busch- und Klippschliefern." Sympatrische Lebensweise. Film D 1371 des Institut für den Wissenschaftlichen Film, Göttingen 1980. Publikation von H. N. Hoeck, Publikation Wissenschaftlicher Film, Sektion Biologie, Serie 15, Nr. 32/D1371 (1982), 19 S.

Hyrax: Behavioural Ecology of Two Species. [April 2003]. <http://www.mpi-seewiesen.mpg.de/knauer/hoeck/klip.html>.

Hendrik Hoeck, PhD

Sirenia

(Dugongs, sea cows, and manatees)

Class Mammalia
Order Sirenia
Number of families 2
Number of genera, species 3 genera; 5 species

Photo: A profile of a dugong (*Dugong dugon*). (Photo by © David B. Fleetham/Seapics.com. Reproduced by permission.)

Introduction

The sirenians are unique among mammals and marine mammals in that they are the only fully aquatic, herbivorous marine mammals. Their complete life cycle occurs in the water as do the life cycles of the cetaceans. Early explorers may have mistaken them for mermaids. Since sirenians eat water plants exclusively, they occur relatively close to coastlines where humans frequently are found in great numbers. Humans have exploited and over-exploited the sirenians range-wide and are responsible for the extinction of the Steller's sea cow (*Hydrodamalis gigas*). While the remaining sirenians (dugong and manatees) are protected range-wide by a variety of laws, humans continue to have a severe impact on them, primarily through habitat destruction. Given their incredible uniqueness we still know very little about their biology. So little, in fact, that it is difficult to develop detailed conservation strategies. The Florida manatee is probably the most intensively studied of all the living sirenians, yet the information gap is massive. For example, in 1979 Daniel Hartman estimated that the Florida manatee had a gestation period of "about a year." Yet nearly 25 years later (2003) that estimate is still the only estimate. The information gap for other sirenians is greater.

Evolution and systematics

The oldest sirenian fossils date back some 50 million years to the early Eocene along the Old World shores of the Tethys Sea. They probably evolved from primitive hoofed mammals with an ancestral line vastly different from that of the cetaceans. Their closest living relatives are the elephants (Proboscidea) with distant links to the hyraxes (Hyracoidea). While the seat of sirenian evolution was likely along what is now the eastern side of the Atlantic Ocean, the oldest fossil sirenians are from the western side of the Atlantic, namely Jamaica. The Dugongidae (dugongs and sea cows) evolved from the cosmopolitan subfamily Halitheriinae from the Eocene to the Pliocene. The subfamily Hydrodamalinae arose in the Miocene and was endemic to the North Pacific. The Dugonginae arose in the Pliocene and occurred in both Atlantic and Pacific regions. Apparently the manatees were able to out-compete the dugongs in the Atlantic. The modern manatees evolved from early dugongids in freshwater regions of northern South America in the Miocene. Expansion to Africa and North America occurred in the Pliocene or Pleistocene. The living and recent members of the order Sirenia are divided into two families, Trichechidae and Dugongidae. The Trichechidae contains three species in the genus *Trichechus: T. manatus*, the West Indian manatee; *T. inunguis*, the Amazonian manatee; and *T. senegalensis*, the West African manatee. The West Indian manatee is further divided into two subspecies: *T. m. latirostris*, the Florida manatee and *T. m. manatus*, the Antillean manatee. The Dugongidae has two subfamilies: Dugonginae and Hydrodamalinae. Subfamily Dugonginae has a single living species, the dugong (*Dugong dugon*). Likewise, the Hydrodamalinae contains a

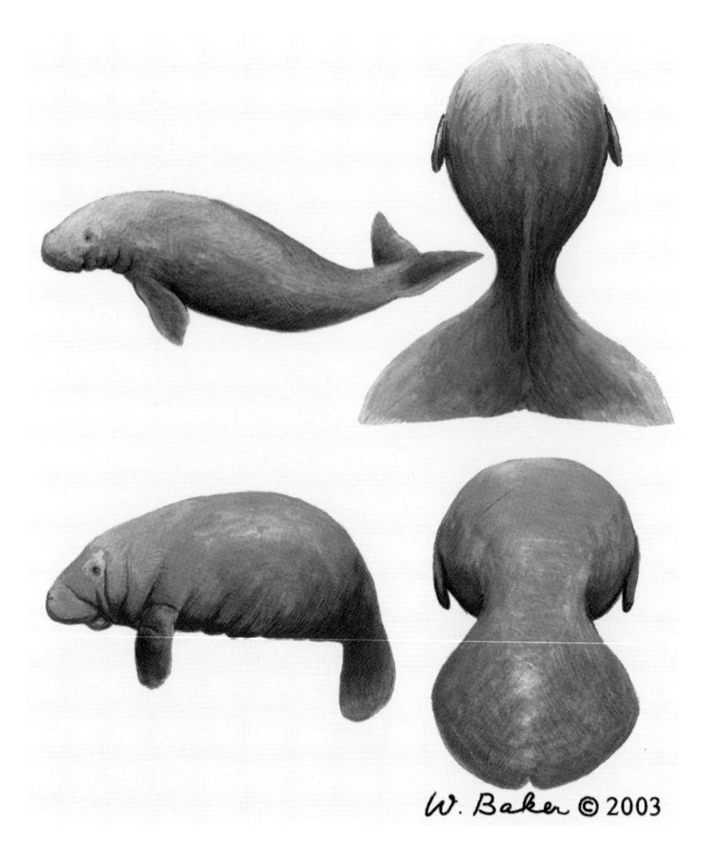

Comparison of dugong (top) and manatee (bottom) tails. (Illustration by Wendy Baker)

single species, the Steller's sea cow, which is recently (1768) extinct.

Physical characteristics

The sirenians range in length from about 9.8 ft (3 m) for the Amazonian manatee to as much as 32.8 ft (10 m) for the extinct Steller's sea cow. Maximum body mass ranges from about 992 lb (450 kg) for the Amazonian manatee to more than 9,920 lb (4,500 kg) for the sea cow. All sirenians are nearly hairless. Their skin varies from relatively smooth (dugong) to rough (manatees) to rugose (sea cow). All lack hind limbs and vestigial pelvic bones can be found in the deep pelvic musculature. The forelimbs are short and flexible. The West Indian and West African manatees have nails, but they are absent in the Amazonian manatee and the dugong. The sea cow lacked phalanges (finger bones) and seems to have had some sort of hooked structure at the end of the flippers to provide a grasping surface. The tail is paddle-shaped in the manatees and fluke-shaped in the dugong and sea cow. External ears are absent and the eyes are small. Color varies from gray to brown and often depends on ectobionts (like algae) that may be growing on the skin. The manatees lack incisors and canine teeth, but have molars in both upper and lower jaws. These molars are replaced continually during the life of the animal. They erupt at the rear of each jaw quadrant, move forward as a result of forces generated during chewing, and fall out the front of the row as the roots are resorbed. They have been called "marching molars." The dugongs have tusks (incisors) and several peg-like molars in each row that are not replaced. Tusks generally erupt in mature males but rarely in females. Growth layer groups in dugong tusks can be used to estimate age. In manatees, growth layer groups in earbones are used for age estimation. The sea cow was toothless. All four species have "heavily whiskered faces" due to the stiff, thickened vibrissae (tactile hairs) around the lips. Sirenian bones are dense (osteosclerotic) and thick (pachyostotic) bones, particularly the ribs, that probably function in buoyancy regulation.

Distribution

The recent sirenians are distributed, with one notable exception, in tropical, subtropical, and temperate regions of the world. The exception is the recently extinct (1765) Steller's sea cow, which was found only in the frigid waters around Bering and Medney Islands in the northwestern Pacific Ocean. The manatees occur on both sides of the Atlantic Ocean. On they western side, they occur from Rhode Island (USA) southward along the U. S. coast to the Gulf of Mexico, Central America, many Caribbean islands, and along the South American coast to northeastern Brazil. The Amazonian manatee occurs only in the Amazon River basin, whereas the West Indian manatee is found throughout the described range but does not enter the Amazon basin. On the eastern side of the Atlantic the West African manatee occurs from Senegal southward to Angola. The dugong is distributed from the southeast coast of Africa northeastward along the Indian Ocean, Red Sea, Persian Gulf, the west coast of India, including Sri Lanka, and eastward to Indonesia, the Philippines, Papua New Guinea, the Caroline Islands, and Australia. The northern limit is around Okinawa and the Ryukyu Islands.

Habitat

The manatees and dugong occupy relatively shallow, warm coastal waters dominated by vegetation. The West Indian and West African manatees occur in coastal, estuarine, and riverine habitats, while the Amazonian manatee is found only in freshwater. There are good indications that that at least the West Indian manatee requires access to freshwater for drinking. Certainly Florida manatees appear to drink from various freshwater sources. The dugong occupies, to the best of our knowledge, a strictly saltwater habitat. In contrast to the manatees and dugong, the sea cow occupied an extremely cold, relatively inhospitable (compared to the tropics!) environment around a few islands in the northwest Pacific Ocean. This intertidal and subtidal habitat was dominated by macroalgae (kelp) upon which the sea cows fed. There is also some indication in the old literature that the sea cow had a preference for sites where freshwater streams entered the ocean.

Behavior

The sirenians, in general and to the best of our current knowledge (2003), may be considered as semi-social; the primary social unit is the female and her most recent calf. Breeding behavior is discussed below. Dugongs may be

West Indian manatees (*Trichechus manatus*) are endangered animals. (Photo by Brandon D. Cole. Bruce Coleman, Inc. Reproduced by permission.)

found in foraging herds numbering in the tens or hundreds of individuals. Some dugongs have been documented to travel tens to hundreds of kilometers in the matter of a few days. Florida manatees also make relatively rapid, long distance movements. Many Florida manatees typically have a north-south migratory pattern in response to water temperature. Florida manatees also congregate in natural and artificial warm water refugia in the winter. Artificial refugia are those such as the outfalls of the cooling systems of electrical generating and other industrial activities. Some Antillean manatees may make seasonal movement to and from freshwater rivers. Sirenians make sounds with fundamental frequencies in the 3–10 kHz range. These sounds appear to function, in part, to maintain the cow-calf bond. Male dugongs on leks may use sound to attract females. Sirenians are not known to use echolocation and the full extent of their hearing capabilities is not known. Little is known about the behavior of Steller's sea cow.

Feeding ecology and diet

The manatees eat a rather wide variety of submerged, floating, emergent, shoreline, and overhanging vegetation. The dugong is restricted primarily to sea grasses. However, both manatees and dugongs have been noted to consume benthic invertebrates and some manatees have even eaten fish. The West Indian and West African manatees feed from the bottom to the surface and above while the Amazonian manatee is primarily a surface feeder. In contrast, the dugong is exclusively a bottom feeder. Manatees use their flippers and oral vibrissae to manipulate vegetation. Similarly, dugongs feed on all parts of the plants they consume, often uprooting the entire plant. Both manatees and the dugong have cornified plates on the anterior portions of their upper and lower jaws to crush vegetation and move it back for further crushing by the molars. The role of dugong tusks in feeding is unclear, since they only erupt in adult males. Manatees and dugongs have a single stomach compartment with an associated digestive gland followed by an equally capacious enlargement of the upper small intestine with its associated duodenal ampullae. The small and large intestines may reach lengths of 65.6 ft (20 m) each and are joined by a relatively small caecum. The sirenians are hind-gut digesters like the horse and use a variety of anaerobic microorganisms in the large intestine to break cellulose down into volatile fatty acids that can be absorbed by the gut. Food passage time is about one week. The sea cow was toothless and presumably crushed ingested kelp using the cornified plates on its upper and lower jaws. It apparently fed only on the surface and on algae growing on rocks exposed at low tide. Manatees studied in aquaria consume about 10% of their body weight in vegetation (wet weight) per day. In the field, both manatees and dugongs spend a large portion of their time feeding.

Reproductive biology

Little information is known about sirenian reproductive biology in general, but those facts that are available are outlined here. Manatees mature at two to 11 years of age. Females may be seasonally polyestrous. Gestation is thought to be about 12 months, but this has not been confirmed. Typically, a single calf is born that is 39.4–59 in (100–150 cm) long. Twins are rare. The typical calving interval is 2.5–3

The depth of this propeller in the water shows how easily a manatee can be injured. (Photo by Douglas Faulkner/Photo Researchers, Inc. Reproduced by permission.)

A manatee (*Trichechus manatus*) eating a hyacinth. (Photo by J. Foott. Bruce Coleman, Inc. Reproduced by permission.)

years. The mating system has been described as "scramble competition polygamy or polyandry" or "scramble promiscuity." Individual estrous females are pursued by as many as 20 or more males. While males may mature at three to five years of age, they may not be able to secure mating rights until they are physically larger. In this case, size does matter. There is no pair bonding. Males play no role in care of the young. Calves may be born at any time of the year, but there may be seasonal peaks in parts of the range.

Typically, dugongs produce single calves after a gestation period of about a year; the calf remains with the cow for more than a year. Male dugongs are not known to provide any parental care. Steller's sea cow reproductive biology is speculative. Steller wrote of family groups and suggested male-female pair bonding, which contrasts with the reproductive behavior of other living sirenians. Calves appear to have been seen at all times of the year, but may have been more common in autumn. This would suggest that mating occurred at most times of the year. Gestation period seems to have been at least one year, but may have been longer. Apparently only single calves were born. There are no data on age/size at sexual maturity or the degree of parental care. Steller's sea cows are thought to have been monogamous, but dugongs exhibit a variety of reproductive behavior, from scramble competition polygamy to lekking.

Conservation

The manatees are listed variously as endangered, threatened, or Vulnerable under international and national legislation. The Convention on Trade in Endangered Species of Flora and Fauna (CITES) lists the Amazonian manatee and

The dugong (*Dugong dugon*) is an endangered animal. (Photo by © Doug Perrine/Seapics.com. Reproduced by permission.)

West Indian manatee (*Trichechus manatus*) in the Homosassa River, Florida, USA. (Photo by Animals Animals ©James Watt. Reproduced by permission.)

A dugong (*Dugong dugon*) feeding on sea grass (*Halophila ovalis*) accompanied by pilot jacks (*Gnathanodon speciosus*). (Photo by © Doug Perrine/Seapics.com. Reproduced by permission.)

both subspecies of the West Indian manatee in Appendix I and the West African manatee in Appendix II. The IUCN lists all three species, including the subspecies, as Vulnerable. The U. S. Endangered Species Act (ESA) of 1973 lists both subspecies of the West Indian manatee and the Amazonian manatee as "endangered" and the West African manatee as "threatened." The African Convention for the Conservation of Nature and Natural Resources lists the West African manatee as "protected" under Class A. The Florida manatee is protected by the Florida Manatee Sanctuary Act in the State of Florida, USA. The dugong is listed in CITES Appendix I and is considered Vulnerable under IUCN criteria. It is listed as "endangered" under the U. S. Endangered Species Act. Steller's sea cow is extinct. In general, range-wide population estimates are not available for manatees or the dugong. In Florida, the Florida manatee population in 2003 was about 3,500 individuals. In Australia, the dugong population has been estimated to be over 80,000 individuals. Similarly, with the possible exception of Florida, population trends are unknown. While hunting of manatees and dugongs still occurs, it is probably at a very low level. However, other human activities are likely having significant impacts on manatee and dugong populations. Among these, habitat destruction is probably the most pervasive. Dredging, siltation from inland runoff, and human activities that result in salinity changes have drastic effects on sea grass beds. In Florida (and increasingly in other areas inhabited by manatees and dugongs), other human activities are directly responsible for killing manatees. Dozens of manatees are killed each year as a result of watercraft collisions, and they also are killed in flood control dams and navigation locks. Manatees become entangled in various kinds of fishing gear (crab pot float lines, monofilament fishing line, shrimp trawls) and are killed. Other fatalities have occurred when these animals have ingested fishing gear and various kinds of plastic. Dugongs in Australia have been killed as a result of entanglement in nets set to protect swimmers from sharks. The management of human population growth and associated activities in coastal and certain inland areas is essential for the protection of manatee and dugong habitat and for the long-term survival of the four species.

Significance to humans

Historically, manatees, dugongs, and the sea cow have been hunted by humans for food, hides, and bone. Hunting resulted in the extinction of Steller's sea cow, making it virtually the only marine mammal species eliminated by human activities. However, other sirenian species are close to ex-

A manatee (*Trichechus manatus*) nursing calves. (Photo by © Douglas Faulkner/Corbis. Reproduced by permission.)

West Indian manatee (*Trichechus manatus*) surfacing in Crystal River, Florida, USA. (Photo by William Goulet. Bruce Coleman, Inc. Reproduced by permission.)

tinction, and this dubious record may not hold for very long. Manatees and dugongs are of cultural significance to indigenous peoples in various parts of the world. The ecological significance of manatees to humans is not clear. Manatees and dugongs recycle nutrients in sea grass beds and keep the plants in a continual state of regrowth. Many other species of value to humans (e.g. sea turtles, fish, shrimp) rely on sea grass beds for shelter. This aspect of "what good are they" is often overlooked when trying to convince people to protect manatees and dugongs. A substantial ecotourism industry has developed at Crystal River, Florida, where a large number of manatees spend the winter in the clear, warm waters of natural springs. Manatee observation platforms have been set up near the effluents of power plant cooling canals where manatees congregate.

Resources

Books

Anderson, P. K., and D. P. Domning. "Steller's Sea Cow." In *Encyclopedia of Marine Mammals*, edited by William F. Perrin, Bernd Würsig, and J. G. M. Thewissen. San Diego: Academic Press, 2002.

Australian Parks and Wildlife Service. *Management Program for the Dugong* (Dugong dugon) *in the Northern Territory of Australia: 2003-2008*. Palmerston, Australia: Department of Infrastructure, Planning and Environment, 2003.

Dierauf, L. A., and F. M. D. Gulland, eds. *CRC Handbook of Marine Mammal Medicine*, 2nd ed. Boca Raton, FL: CRC Press, 2001.

Domning, D. P. "Desmostylia." In *Encyclopedia of Marine Mammals*, edited by William F. Perrin, Bernd Würsig, and J. G. M. Thewissen. San Diego: Academic Press, 2002.

———. "Sirenian Evolution." In *Encyclopedia of Marine Mammals*, edited by William F. Perrin, Bernd Würsig, and J. G. M. Thewissen. San Diego: Academic Press, 2002.

Hartman, D. S. *Ecology and Behavior of the Manatee* (Trichechus manatus) *in Florida*. Lawrence, KS: American Society of Mammalogists, 1979.

Kaiser, H. E. *Morphology of the Sirenia*. Basel: S. Karger, 1974.

Marmontel, M., D. K. Odell, and J. E. Reynolds III. "Reproductive Biology of South American Manatees." In *Reproductive Biology of South American Vertebrates*, edited by William C. Hamlett. New York: Springer-Verlag, 1992.

Marsh, H. "Dugong (*Dugong dugon*)." In *Encyclopedia of Marine Mammals*, edited by William F. Perrin, Bernd Würsig, and J. G. M. Thewissen. San Diego: Academic Press, 2002.

Odell, D. K. "Sirenian Life History." In *Encyclopedia of Marine Mammals*, edited by William F. Perrin, Bernd Würsig, and J. G. M. Thewissen. San Diego: Academic Press, 2002.

O'Shea, T. J., B. B. Ackerman, and H. F. Percival. *Population Biology of the Florida Manatee*. Washington, DC: U.S. Department of the Interior, National Biological Service, 1995.

Reynolds, J. E. III, and D. K. Odell. *Manatees and Dugongs*. New York: Facts On File, 1991.

Reynolds, J. E. III, and J. A. Powell. "Manatees." In *Encyclopedia of Marine Mammals*, edited by William F. Perrin, Bernd Würsig, and J. G. M. Thewissen. San Diego: Academic Press, 2002.

Reynolds, J. E. III, and S. A. Rommel, eds. *Biology of Marine Mammals*. Washington, DC: Smithsonian Institution Press, 1999.

Rommel, S. A., D. A. Pabst, and W. A. McLellan. "Skull Anatomy." In *Encyclopedia of Marine Mammals*, edited by William F. Perrin, Bernd Würsig, and J. G. M. Thewissen. San Diego: Academic Press, 2002.

Rommel, S. A., and J. E. Reynolds, III. "Skeletal Anatomy." In *Encyclopedia of Marine Mammals*, edited by William F. Perrin, Bernd Würsig, and J. G. M. Thewissen. San Diego: Academic Press, 2002.

Rice, D. W. *Marine Mammals of the World: Systematics and Distribution*. Lawrence, KS: The Society For Marine Mammalogy, Special Publication Number 4, 1998.

Twiss, J. R., and R. R. Reeves, eds. *Conservation and Management of Marine Mammals*. Washington, DC: Smithsonian Institution Press, 1999.

Periodicals

Anderson, P. K. "Habitat, Niche, and Evolution of Sirenian Mating Systems." *Journal of Mammalian Evolution* 9, nos. 1-2 (2003): 55–98.

Deutsch, C. J., J. P. Reid, R. K. Bonde, D. E. Easton, H. I. Kochman, and T. J. O'Shea. "Seasonal Movements, Migratory Behavior, and Site Fidelity of West Indian Manatees Along the Atlantic Coast of the United States." *Wildlife Monographs* 151 (2003): 1–77.

Domning, D. P. "Bibliography and Index of the Sirenia and Desmostylia." *Smithsonian Contributions in Paleobiology* 80 (1996): 1–611.

O'Shea, T. J. "Manatees." *Scientific American* 271, no. 1 (1994): 50–56.

Organizations

Save the Manatee Club. 500 North Maitland Avenue, Maitland, FL 32751 United States. Phone: (407) 539-0990. Fax: (407) 539-0871. E-mail: education@savethemanatee.org Web site: <http://www.savethemanatee.org>

Sirenian International, Inc.. 200 Stonewall Drive, Fredericksburg, VA 22401 United States. E-mail: caryn@sirenian.org Web site: <http://www.sirenian.org>

Daniel K. Odell, PhD

▲

Dugongs and sea cows

(Dugongidae)

Class Mammalia

Order Sirenia

Family Dugongidae

Thumbnail description
Large, fully aquatic marine herbivorous mammals

Size
9.8–32.8 ft (3–10 m); 881–13,000 lb (400–5,900 kg)

Number of genera, species
2 genera; 2 species

Habitat
Steller's sea cow: coastal around Bering and Medney Islands; dugong: Indo-Pacific tropical and subtropical coastal waters where seagrasses occur

Conservation status
Extinct: 1 species; Vulnerable: 1 species

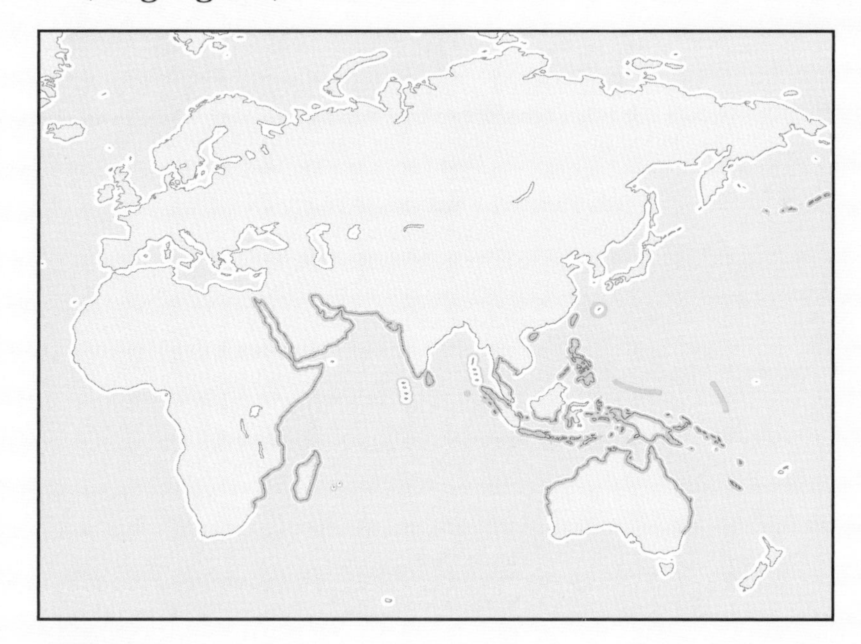

Distribution
Steller's sea cow: western North Pacific around the Bering and Medney Islands; dugong: tropical and subtropical, coastal, Indo-Pacific

Evolution and systematics

The oldest sirenians are from the early Eocene and are related to elephants, hyraces, and the extinct desmostylids. The Dugongidae appeared at the end of the Eocene. The subfamily Dugoninae probably appeared in the early to middle Oligocene, but the subfamily Hydrodamalinae did not appear until the early Miocene and was endemic to the North Pacific.

Physical characteristics

The dugongids have a streamlined, nearly hairless body reaching a length of 9.8 ft (3 m) for the dugong (*Dugong dugon*) and 23–33 ft (7–10 m) for Steller's sea cow (*Hydrodamalis gigas*). They lack hind limbs, but do have vestigial pelvic bones embedded in the pelvic musculature. The tail is forked, similar to that of a cetacean's flukes. Forelimbs are shortened and flexible without nails. Body color is variable gray-brown; color is unknown for the sea cow. The dugong may weigh more than 881 lb (400 kg) and the Steller's sea cow is estimated to have weighed in excess of 9,920 lb (4,500 kg).

Distribution

The recent dugongids are distributed disjunctively in the tropical and subtropical Indo-Pacific (dugong) and western North Pacific (Steller's sea cow).

A dugong (*Dugong dugon*) swimming near Indonesia. (Photo by Animals Animals ©D. Fleetham, OSF. Reproduced by permission.)

A dugong (*Dugong dugon*) forages on the water's bottom. (Photo by © Ingrid Visser/Seapics.com. Reproduced by permission.)

Habitat

The dugongids occupy a coastal marine habitat dictated by the presence of sea grasses and macroalgae (sea cow only).

Behavior

In general, the dugongids may be described as "semi-social," with the primary social unit being the female and her calf. Little is known about the behavior of Steller's sea cow. The dugong may occur in herds numbering in the tens or hundreds in areas of abundant sea grass. Satellite-tracked dugongs have been shown to make regular, short distance (9–25 mi [15–40 km]) round-trip movements between feeding areas and warmer coastal areas. In tropical Australia, trips of 62–373 mi (100–600 km) have been documented over the course of a few days. Dugongs form mating herds with several males attempting to mate with a single female; males in a part of Western Australia establish and defend territories and display behaviors to attract females, a behavior known as lekking.

Feeding ecology and diet

The dugongids are marine herbivores feeding almost exclusively on sea grasses (the dugong) and on macroalgae (Steller's sea cow). Steller's sea cows apparently fed on the surface and were not known to dive. Being toothless and the kelp being avascular, the sea cows crushed the kelp fronds between horny, keratinized plates at the front of their upper and lower jaws. If their digestive process were similar to those of living sirenians, they had a simple stomach, long intestines, and digestion occurred in the large intestine (hindgut) with the aid of anerobic microorganisms. The dugong is strictly a bottom feeder and, in some parts of its range, they ingest numbers of bottom-dwelling invertebrates. In some areas, at low tide, dugong-feeding trails are observed in the exposed sea grass beds. As with other living sirenians, the dugong is a hindgut digestor with long intestines and likely uses anerobic microorganisms to digest cellulose. Details are lacking.

A dugong (*Dugong dugon*) with its calf. (Photo by © Doug Perrine/Seapics.com. Reproduced by permission.)

A typical dive for a dugong (*Dugong dugon*) lasts about two to three minutes. (Photo by © Doug Perrine/Seapics.com. Reproduced by permission.)

Reproductive biology

Little is known about sirenian reproductive biology in general, and the dugongids are no exception. Typically, single calves are born after a gestation period of about a year and remain with the cow for more than a year. Male dugongs are not known to provide any parental care. Steller's sea cow reproductive biology is speculative. Steller wrote of family groups and suggested male-female pair bonding, which contrasts with the reproductive behavior of other living sirenians. Calves appear to have been seen at all times of the year, but may have been more common in autumn. This would suggest that mating occurred at most times of the year. Gestation period seems to have been at least one year, but may have been longer. Apparently only single calves were born. There are no data on age/size at sexual maturity or the degree of parental care. Steller's sea cows are thought to have been monogamous but dugongs exhibit a variety of reproductive behavior, from scramble competition polygamy to lekking.

Conservation status

Steller's sea cow is Extinct, while the dugong is Vulnerable to extinction under IUCN criteria. CITES Appendix I lists the dugong as Endangered under the United States Endangered Species Act. Increasing human activity (boating, fishing, inland habitat destruction) in coastal dugong habitats has the potential for directly injuring dugongs (boat collisions, fishing gear entanglement) or for sea grass destruction from silt in river outflows. The largest dugong population is in Australia where incomplete population estimates have been 80,000 or more animals.

Significance to humans

The dugong has been hunted for meat, hides, and bone, and is of cultural significance to some indigenous peoples in the Indo-Pacific region. The Steller's sea cow was hunted to extinction. This event, in the mid-sixteenth century, shows how much pressure humans can exert on the environment.

1. Steller's sea cow (*Hydrodamalis gigas*); 2. Dugong (*Dugong dugon*). (Illustration by Wendy Baker)

Species accounts

Steller's sea cow
Hydrodamalis gigas

SUBFAMILY
Hydrodamalinae

TAXONOMY
Hydrodamalis gigas (Zimmerman, 1780), Bering Island, Bering Sea.

OTHER COMMON NAMES
English: Great northern sea cow, rhytina.

PHYSICAL CHARACTERISTICS
Extremely large (23–33 ft [7–10 m] body length; 9,920–13,000 lb [4,500–5,900 kg] body mass), fully aquatic mammal; tail fluke-like; hind limbs absent; forelimbs lacking phalanges (finger bones); head relatively small; teeth absent.

DISTRIBUTION
The Steller's sea cow is unique among sirenians in that it inhabited the extremely cold waters around Medney and Bering Islands in the northwestern Pacific Ocean.

HABITAT
Coastal areas where macroalgae (kelp) grew.

BEHAVIOR
Seemed to be strictly coastal where the kelp grew and were reported as having some preference for the mouths of freshwater creeks. Apparently, they remained near the islands year-round and did not migrate.

FEEDING ECOLOGY AND DIET
The Steller's sea cow fed exclusively on several species of macroalgae (kelp) growing in the intertidal and subtidal areas around Bering and Medney Islands.

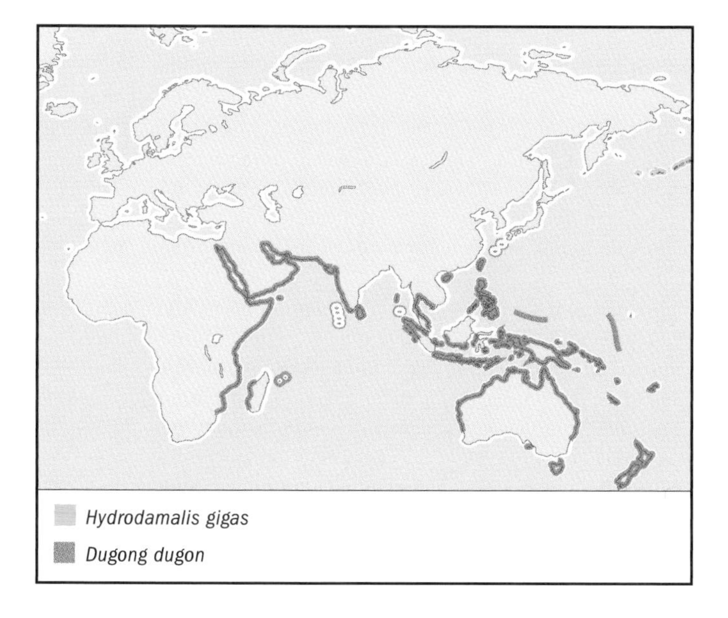

REPRODUCTIVE BIOLOGY
Virtually nothing is known about the sea cow's reproductive biology, and the published information is speculative at best.

CONSERVATION STATUS
Discovered 1741 and hunted to extinction by 1767.

SIGNIFICANCE TO HUMANS
Provided a ready source of meat for the stranded sailors who discovered the species and for subsequent explorers who visited the islands. This is the only recent sirenian that has been sent to extinction by human activities. ◆

Dugong
Dugong dugon

SUBFAMILY
Dugonginae

TAXONOMY
Dugong dugon (Müller, 1776), Cape of Good Hope to the Philippines.

OTHER COMMON NAMES
English: Seacow; French: Dugong: Spanish: Dugong, dugon.

PHYSICAL CHARACTERISTICS
Has gray-colored, nearly hairless skin; reaches a body length of about 9.8 ft (3 m) and a body mass of about 880 lb (400 kg); body streamlined and laterally compressed to some extent; tail fluke-like; hind limbs absent; forelimbs shortened and flexible; both sexes possess tusks, but they rarely erupt in the females.

DISTRIBUTION
Tropical and subtropical Indo-Pacific margin from eastern Africa to the Philippines and the South China and East China Seas where sea grasses may occur.

HABITAT
Coastal, strictly saltwater, and relatively shallow depths (up to about 98 ft [30 m]) as determined by the presence of sea grass beds.

BEHAVIOR
The primary social unit is the female and her calf. May form large (tens to hundreds of individuals) foraging herds.

FEEDING ECOLOGY AND DIET
A bottom feeder that eats the leaves and roots of a number of species of sea grasses.

REPRODUCTIVE BIOLOGY
Female dugongs typically give birth to a single calf after a gestation period of about one year. Males compete for mating rights. Animals of both sexes mature at over 10 years of age. Calves nurse for about 18 months, but start eating sea grass within weeks of birth. Calving interval estimates range from three to seven years. Maximum longevity is 50–70 years.

CONSERVATION STATUS
Listed as Vulnerable by the IUCN. Protected by various international and national regulations, but hunted by some indigenous peoples. There is no range-wide population estimate.

SIGNIFICANCE TO HUMANS
Dugongs have been hunted for meat, fat, hides, and bones, and are of cultural significance to many indigenous peoples in the Indo-Pacific region. ◆

Resources

Books

Anderson, P. K., and D. P. Domning. "Steller's Sea Cow." In *Encyclopedia of Marine Mammals*, edited by William F. Perrin, Bernd Würsig, and J. G. M. Thewissen. San Diego: Academic Press, 2002.

Domning, D. P. "Desmostylia." In *Encyclopedia of Marine Mammals*, edited by William F. Perrin, Bernd Würsig, and J. G. M. Thewissen. San Diego: Academic Press, 2002.

———. "Sirenian Evolution." In *Encyclopedia of Marine Mammals*, edited by William F. Perrin, Bernd Würsig, and J. G. M. Thewissen. San Diego: Academic Press, 2002.

Kaiser, H. E. *Morphology of the Sirenia.* Basel: S. Karger, 1974.

Marsh, H. "Dugong (*Dugong dugon*)." In *Encyclopedia of Marine Mammals*, edited by William F. Perrin, Bernd Würsig, and J. G. M. Thewissen. San Diego: Academic Press, 2002.

Odell, D. K. "Sirenian Life History." In *Encyclopedia of Marine Mammals*, edited by William F. Perrin, Bernd Würsig, and J. G. M. Thewissen. San Diego: Academic Press, 2002.

Reynolds, J. E. III, and D. K. Odell. *Manatees and Dugongs.* New York: Facts On File, 1991.

Reynolds, J. E. III, and S. A. Rommel, eds. *Biology of Marine Mammals.* Washington, DC: Smithsonian Institution Press, 1999.

Rommel, S. A., D. A. Pabst, and W. A. McLellan. "Skull Anatomy." In *Encyclopedia of Marine Mammals*, edited by William F. Perrin, Bernd Würsig, and J. G. M. Thewissen. San Diego: Academic Press, 2002.

Rommel, S. A., and J. E. Reynolds III. "Skeletal Anatomy." In *Encyclopedia of Marine Mammals*, edited by William F. Perrin, Bernd Würsig, and J. G. M. Thewissen. San Diego: Academic Press, 2002.

Rice, D. W. *Marine Mammals of the World: Systematics and Distribution.* Lawrence, KS: The Society For Marine Mammalogy, 1998.

Twiss, J. R., and R. R. Reeves, eds. *Conservation and Management of Marine Mammals.* Washington, DC: Smithsonian Institution Press, 1999.

Periodicals

Anderson, P. K. "Habitat, Niche, and Evolution of Sirenian Mating Systems." *Journal of Mammalian Evolution* 9, 1-2 (2003): 55–98.

Australian Parks and Wildlife Service. "Management Program for the Dugong (*Dugong dugon*) in the Northern Territory of Australia: 2003–2008." *Department of Infrastructure, Planning and Environment, Parks and Wildlife Service, Palmerston, Northern Territory, Australia* (2003).

Domning, D. P. "Bibliography and Index of the Sirenia and Desmostylia." *Smithsonian Contributions in Paleobiology* 80 (1996): 1–611.

Organizations

Sirenian International, Inc.. 200 Stonewall Drive, Fredericksburt, VA 22401 United States. E-mail: caryn@sirenian.org Web site: <http://www.sirenian.org>

Daniel K. Odell, PhD

▲

Manatees

(Trichechidae)

Class Mammalia

Order Sirenia

Family Trichechidae

Thumbnail description
Large, fully aquatic, nearly hairless, herbivorous marine and freshwater mammals

Size
9–13 ft (3–4 m) total length; 1,100–3,300 lb (500–1500 kg) body mass

Number of genera, species
1 genus; 3 species

Habitat
Tropical and subtropical Atlantic; coastal, estuarine and riverine

Conservation status
Vulnerable: 3 species

Distribution
West coast of Africa from Senegal south to Angola; Southeastern United States, throughout the Caribbean to southeastern Brazil; Amazon River basin

Evolution and systematics

The oldest sirenians are from the early Eocene and are related to elephants, hyraxes, and the extinct desmostylids. The trichechids may have arisen from dugongids in the late Eocene or early Oligocene. The subfamily Trichechinae (modern manatees) first appeared in freshwater Miocene deposits in Colombia. It is likely that much of the early developmental history occurred in South America. The spread to North America and Africa was likely in the Pliocene or Pleistocene.

Physical characteristics

Manatees have a standard body length of 9–13 ft (3–4 m) and weigh 1,100–3,300 lb (500–1,500 kg) depending on the species. The body is nearly hairless, robust and oval in cross section. Hind limbs are absent but vestigial pelvic bones remain embedded in the pelvic musculature. The tail is a broad, rounded paddle. Forelimbs are short, flexible, and have tree to four nails except in the Amazonian manatee. Color is gray to brownish and, in the field, may depend on the epiphytes (algae, etc.) that are growing on the skin. The eyes are very small and there is no external ear (pinna). The external ear canal opening is very small and difficult to see. Testes are internal and the male genital opening is anterioventral just posterior to the umbilicus. Mammary glands are paired and there is a single nipple in each axilla. The upper lips are split, covered with stiff vibrissae, and have been described as "prehensile" from the manatee's ability to manipulate vegetation.

Distribution

The manatees are found on both sides of the Atlantic Ocean in tropical and subtropical regions. In West Africa, they range from Senegal southward to Angola. Along the eastern Atlantic seaboard, they range from the southeastern United States (primarily Florida) southward throughout the Caribbean region to southeastern Brazil plus the Amazon River basin.

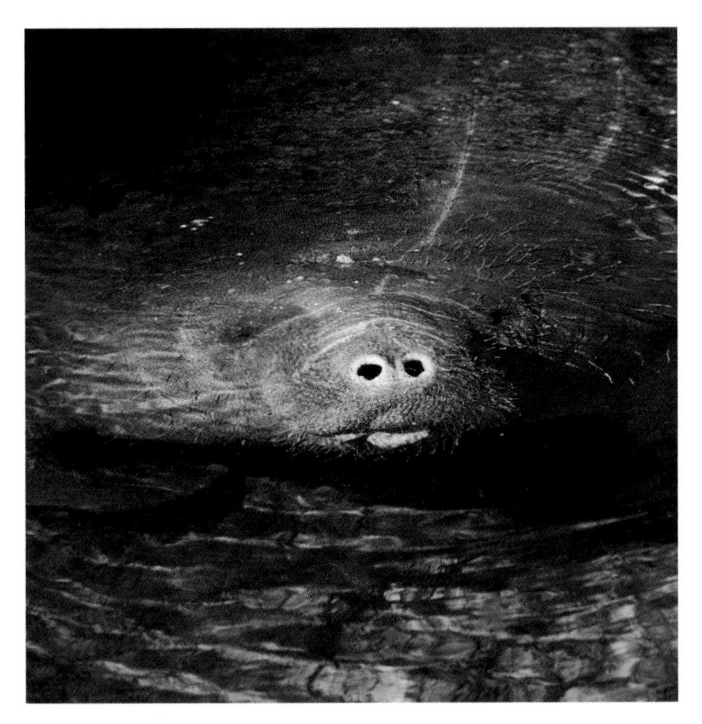

A manatee (*Trichechus manatus*) surfacing in Crystal River, Florida, USA. (Photo by M. H. Sharp/Photo Researchers, Inc. Reproduced by permission.)

A manatee (*Trichechus manatus*) eating hydrilla (*Hydrilla verticillata*) in Crystal River, Florida, USA. (Photo by Douglas Faulkner/Photo Researchers, Inc. Reproduced by permission.)

Habitat

Manatees occur in coastal, estuarine and freshwater/riverine habitats. Since two of the species depend extensively on marine vegetation (seagrasses), they rarely venture into deep waters.

A manatee (*Trichechus manatus*) floats on the water's surface at the Chassahowitzka National Wildlife Refuge in Florida, USA. (Photo by M. H. Sharp/Photo Researchers, Inc. Reproduced by permission.)

Behavior

Manatees are typically semi-social. The primary social group is the cow-calf pair that may remain intact for two years or more. There is no pair bonding between males and females. Their reproductive behavior has been described as "scramble promiscuity" wherein several males compete for mating rights with a single estrous female. Manatees may undertake local migrations in response to water temperature, water depth, or to the presence/absence of freshwater. Manatees communicate using sound, vision, taste and touch. In turbid waters, acoustic communication is important for maintaining the cow-calf bond. Most manatee sounds are in the 3–5 kHz range but have not been well studied.

Feeding ecology and diet

The manatees are the only marine mammals that are herbivorous. However, they are not obligate herbivores and will consume fish and invertebrates in some areas if they are available. The typical manatee diet consists of a wide variety of marine and freshwater vascular plants and algae as well as terrestrial vegetation that may be accessible on shorelines, overhanging and touching the water or floating such as red mangrove propagules. Manatees may feed on bottom, mid-

A manatee (*Trichechus manatus*) yawning in Crystal River, Florida, USA. (Photo by Douglas Faulkner/Photo Researchers, Inc. Reproduced by permission.)

Two manatees (*Trichechus manatus*) playing in Crystal River, Florida, USA. (Photo by Douglas Faulkner/Photo Researchers, Inc. Reproduced by permission.)

A group of manatees (*Trichechus manatus*) in Crystal River, Florida, USA. (Photo by Douglas Faulkner. Bruce Coleman, Inc. Reproduced by permission.)

A West Indian manatee (*Trichechus manatus*) nursing at the sandy bottom of a river in Florida, USA. (Photo by Animals Animals © James Watt. Reproduced by permission.)

water, or floating vegetation and they may climb partially out of the water to access shoreline vegetation.

Reproductive biology

Manatees mature at two to 11 years of age. Females may be seasonally polyestrus. Gestation is about 12 months but has not been confirmed. Typically, a single calf is born and is 3.3–4.9 ft (100–150 cm) long. Twins are rare. The mating system has been described as "scramble competition polygamy or polyandry" or "scramble promiscuity." Individual estrous females are pursued by as many as 20 or more males. While males may mature at three to five years of age, they may not be able to secure mating rights until they are physically larger. The typical calving interval is two-and-a-half to three years. There is no pair bonding. Males play no role in care of the young. Calves may be born at any time of the year but there may be seasonal peaks in parts of the range.

Conservation status

The manatees are listed variously as endangered, threatened, or vulnerable under international and national legislation. The Convention on Trade in Endangered Species of

A manatee (*Trichechus manatus*) in Crystal River, Florida, USA. (Photo by E & P Bauer. Bruce Coleman, Inc. Reproduced by permission.)

Two manatees (*Trichechus manatus*) surfacing for air. (Photo by Andrew J. Martinez/Photo Researchers, Inc. Reproduced by permission.)

Flora and Fauna (CITES) lists the Amazonian manatee and both subspecies of the West Indian manatee in Appendix I and the West African manatee in Appendix II. The World Conservation Union (IUCN) lists all three species, including the subspecies, as Vulnerable. The United States Endangered Species Act (ESA) of 1973 lists both subspecies of the West Indian manatee and the Amazonian manatee as "endangered" and the West African manatee as "threatened." The African Convention for the Conservation of Nature and Natural Resources lists the West African manatee as "protected" under Class A. The Florida manatee is protected by the Florida Manatee Sanctuary Act in the State of Florida, United States.

Significance to humans

Historically, manatees have provided humans with a meat for food and bones and hides for tools, implements, and leather. Some indigenous peoples may have used manatee parts for medicinal and aphrodisiac purposes. In Guyana and Florida, manatees have been used to clear vegetation-choked canals and waterways. As of 2003, while manatees are protected range-wide, they are still hunted for food in many areas. Ecologically, manatees may benefit humans indirectly by recycling nutrients in seagrass beds, keeping the vegetation in a constantly regenerating state and maintaining habitat for fish and invertebrates used by humans. In some regions, especially in Florida, the manatee is the basis for ecotourism.

The Amazonian manatee (top) does not have nails on its pectoral flippers, but West Indian and west African manatees do (bottom). (Illustration by Wendy Baker)

1. West African manatee (*Trichechus senegalensis*); 2. Amazonian manatee (*Trichechus inunguis*); 3. West Indian manatee (*Trichechus manatus*). (Illustration by Wendy Baker)

Species accounts

West Indian manatee

Trichechus manatus

TAXONOMY

Trichechus manatus Linnaeus, 1758, West Indies.

OTHER COMMON NAMES

English: Caribbean manatee, Antillean manatee, Florida manatee; French: Lamantin; German: Seekuh, manati; Spanish: Vaca marina, manatí.

PHYSICAL CHARACTERISTICS

13 ft (4 m) body length; 3,300 lb (1,500 kg) body mass; color gray; body nearly hairless and slightly compressed dorso-ventrally; no hind limbs; tail broad and spatulate; forelimbs short, flexible and with three to four nails.

DISTRIBUTION

Southeastern United States, Caribbean, to Bahia, Brazil. The Florida manatee occurs primarily in Florida and southeastern Georgia but is known from as far north as Rhode Island and as far west as Texas. A few Florida manatees have apparently crossed the Gulf Stream to the northern Bahamas. The Antillean manatee ranges from Mexico, the Caribbean, to northeastern Brazil. It apparently does not enter the Amazon River.

HABITAT

Coastal and estuarine areas, freshwater rivers connected to the coast.

BEHAVIOR

Generally considered solitary or semi-social except for mating herds or winter congregations in warm water refugia. The only long-term social unit is the female-calf pair which may last two or more years. Florida manatees migrate along a north-south axis in response to air and water temperature. In other parts of the range, manatee movements may be dictated by wet and dry seasons.

FEEDING ECOLOGY AND DIET

The West Indian manatee feeds on submerged, mid-water, floating, overhanging and bank vegetation. Fish and invertebrates are ingested on occasion. Individuals are estimated to consume 10% body weight in vegetation per day.

REPRODUCTIVE BIOLOGY

Males and females mature at 2.5–6 years. Calving interval is 2.5–3 years. Females are seasonally polyestrous. Both males and females are promiscuous and polygamous. Estrous females may be pursued by 20 or more males for up to a month. There is no pair bonding and males play no role in care of the young. Typically, a single calf is born but twins may account for 1–2% of pregnancies. Calving is broadly seasonal in some areas and both males and females may show seasonal patterns in sexual activity. Some Florida manatees may live for 50 years or more and reproductive senility is not known.

CONSERVATION STATUS

Protected throughout the range but laws are difficult to enforce. Unknown numbers are killed illegally each year for food.

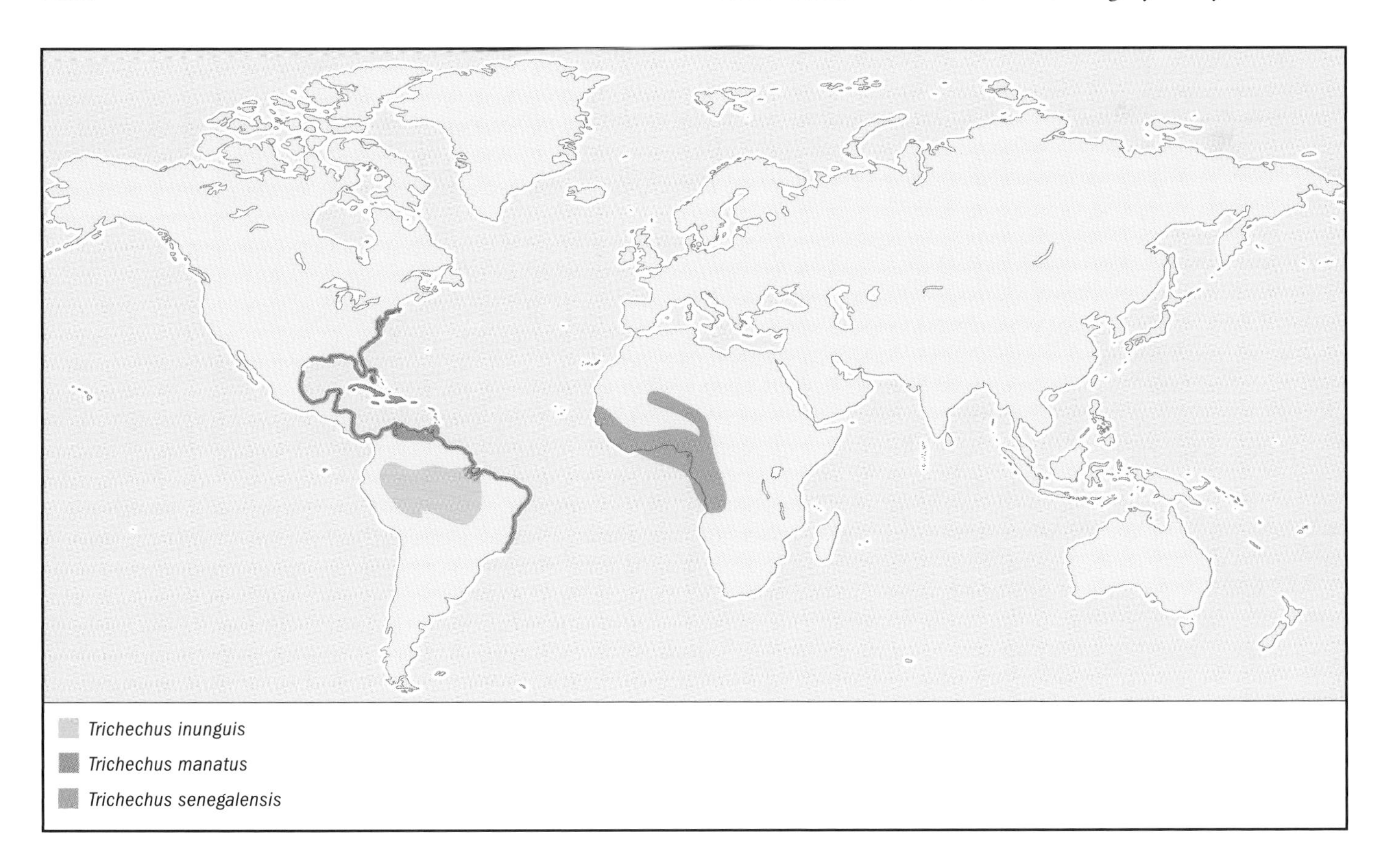

In Florida, significant numbers are killed each year from collisions with watercraft, entanglement in fishing gear, ingestion of plastic and recreational fishing gear, or drowning/crushing in canal locks and flood gates. Actual and potential habitat (seagrass) destruction is a significant conservation factor throughout the range. Runoff of anthropogenic chemicals (pesticides, etc) may be a problem range-wide. Natural and artificial warm water refugia in Florida are key habitat components in Florida. The Florida manatee has a minimum population of 3,000–3,500 (as of 2003). Population estimates for the Antillean manatee are not available.

SIGNIFICANCE TO HUMANS
Historically, a source of meat, fat, oil, hide, bone. Some use in clearing waterway vegetation in Florida and Georgetown, Guyana. ◆

West African manatee
Trichechus senegalensis

TAXONOMY
Trichechus senegalensis Link, 1795, Senegal.

OTHER COMMON NAMES
English: African manatee; French: Lamantin; German: Seekuh, manati.

PHYSICAL CHARACTERISTICS
Similar in body shape and size to the West Indian manatee but with a blunter snout, protruding eyes, and a slimmer body. Fingernails are present. Good data are lacking.

DISTRIBUTION
West Africa from Senegal southward to Angola.

HABITAT
Coastal, estuarine, riverine, and lacustrine (lakes).

BEHAVIOR
Similar to the West Indian manatee but details are lacking. The major social unit is the cow-calf pair.

FEEDING ECOLOGY AND DIET
Feeds on submerged, mid-water, floating, overhanging, and bank vegetation. Diet may include some invertebrates.

REPRODUCTIVE BIOLOGY
Details not well known. Thought to be similar to the West Indian manatee.

CONSERVATION STATUS
Animals are still hunted throughout their range. Habitat degredation is a potential problem. Range-wide data are lacking and there are no reliable population estimates.

SIGNIFICANCE TO HUMANS
Historically hunted for meat, fat, oil, hides and bones. Culturally significant in some areas and some hunting continues to this day for food and reduce damage to fishing gear and rice fields. In some areas of Cameroon, local people fear manatees and do not hunt them. ◆

Amazonian manatee
Trichechus inunguis

TAXONOMY
Trichechus inunguis (Natterer, 1883), Rio Madeira, Brazil.

OTHER COMMON NAMES
French: Lamantin de l'amazone; Spanish: Vaca marina amazónica.

PHYSICAL CHARACTERISTICS
The smallest of the three species of manatee. Maximum body length is 9 ft (3 m) or less and body mass is less than 1,100 lb (500 kg). Color generally gray and most individuals have a white or pink abdominal patch of variable size. Fingernails are absent.

DISTRIBUTION
Throughout the Amazon River basin. No reliable population estimates are available.

HABITAT
All accessible freshwater rivers and lakes in the Amazon basin. Does not enter saltwater.

BEHAVIOR
Considered to be semi-social with the cow-calf pair being the dominant social grouping.

FEEDING ECOLOGY AND DIET
Feeds on a variety of floating, overhanging, and bank vegetation. In the dry season, some Amazonian manatees are thought to fast for several months when water levels in lakes are low and vegetation is not accessible.

REPRODUCTIVE BIOLOGY
Thought to be similar to the West Indian manatee in general terms, but good details are lacking. Gestation is about one year. Calves are 2.5–3.5 ft (85–105 cm) long and weigh 22–33 lb (10–15 kg) at birth.

CONSERVATION STATUS
Still hunted to an unknown extent. Actual and potential habitat destruction and degradation are probably the biggest threats.

SIGNIFICANCE TO HUMANS
Historically hunted for meat, fat, oil, hides and bone. Middle ear bones (stapes) thought to have aphrodisiac powers. ◆

Resources

Books

Dierauf, L. A. and F. M. D. Gulland, eds. *CRC Handbook of Marine Mammal Medicine*, 2nd ed. Boca Raton, Florida: CRC Press, 2001.

Domning, D. P. "Desmostylia." In *Encyclopedia of Marine Mammals*, edited by William F. Perrin, Bernd Würsig and J. G. M. Thewissen. San Diego: Academic Press, 2002.

Resources

———. "Sirenian Evolution." In *Encyclopedia of Marine Mammals*, edited by William F. Perrin, Bernd Würsig and J. G. M. Thewissen. San Diego: Academic Press, 2002.

Hartman, D. S. *Ecology and Behavior of the Manatee (Trichechus manatus) in Florida.* Lawrence, KS: American Society of Mammalogists, 1979.

Kaiser, H. E. *Morphology of the Sirenia.* Basel: S. Karger, 1974.

Marmontel, M., D. K. Odell, and J. E. Reynolds III. "Reproductive Biology of South American Manatees." In *Reproductive Biology of South American Vertebrates*, edited by William C. Hamlett. New York: Springer-Verlag, 1992.

Odell, D. K. "Sirenian Life History." In *Encyclopedia of Marine Mammals*, edited by William F. Perrin, Bernd Würsig and J. G. M. Thewissen. San Diego: Academic Press, 2002.

O'Shea, T. J., B. B. Ackerman, and H. F. Percival. *Population Biology of the Florida Manatee.* Washington, DC: U. S. Department of the Interior, National Biological Service, 1995.

Reynolds, J. E. III, and D. K. Odell. *Manatees and Dugongs.* New York: Facts On File, 1991.

Reynolds, J. E. III, and J. A. Powell. "Manatees." In *Encyclopedia of Marine Mammals*, edited by William F. Perrin, Bernd Würsig and J. G. M. Thewissen. San Diego: Academic Press, 2002.

Reynolds, J. E. III, and S. A. Rommel, eds. *Biology of Marine Mammals.* Washington, DC: Smithsonian Institution Press, 1999.

Rice, D. W. *Marine Mammals of the World: Systematics and Distribution.* Lawrence, KS: The Society For Marine Mammalogy, Special Publication Number 4. 1998.

Rommel, S. A., D. A. Pabst, and W. A. McLellan. "Skull Anatomy." In *Encyclopedia of Marine Mammals*, edited by William F. Perrin, Bernd Würsig and J. G. M. Thewissen. San Diego: Academic Press, 2002.

Rommel, S. A., and J. E. Reynolds III. "Skeletal Anatomy." In *Encyclopedia of Marine Mammals*, edited by William F. Perrin, Bernd Würsig and J. G. M. Thewissen. San Diego: Academic Press, 2002.

Twiss, J. R., and R. R. Reeves, eds. *Conservation and Management of Marine Mammals.* Washington, DC: Smithsonian Institution Press, 1999.

Periodicals

Anderson, P. K. "Habitat, niche, and evolution of sirenian mating systems." *Journal of Mammalian Evolution* 9, nos. 1–2 (2003): 55–98.

Deutsch, C. J., J. P. Reid, R. K. Bonde, D. E. Easton, H. I. Kochman, and T. J. O'Shea. "Seasonal movements, migratory behavior, and site fidelity of West Indian manatees along the Atlantic coast of the United States." *Wildlife Monographs* 151 (2003):1–77.

Domning, D. P. "Bibliography and index of the Sirenia and Desmostylia." *Smithsonian Contributions in Paleobiology* 80 (1996): 1–611.

O'Shea, T. J. "Manatees." *Scientific American* 271, no. 1 (1994): 50–56.

Organizations

Save the Manatee Club. 500 North Maitland Avenue, Maitland, FL 32751 United States. Phone: (407);539-0990. Fax: (407) 539-0871. E-mail: education@savethemanatee.org Web site: <http://www.savethemanatee.org>

Sirenian International, Inc.. 200 Stonewall Drive, Fredericksburt, VA 22401 United States. E-mail: caryn@sirenian.org Web site: <http://www.sirenian.org>

Daniel K. Odell, PhD

Perissodactyla
(Odd-toed ungulates)

Class Mammalia
Order Perissodactyla
Number of families 3
Number of genera, species 6 genera; 16 species

Photo: Grevy's zebra (*Equus grevyi*) grazing. (Photo by K. & K. Ammann. Bruce Coleman, Inc. Reproduced by permission.)

Introduction

Perissodactyla are exclusively large terrestrial herbivores. Also commonly known as perrisodactyls, or odd-toed ungulates, this order is extremely diverse; from the robust, horned white rhinoceros (*Ceratotherium simum*) to the sleek, slender, and striped mountain zebra (*Equus zebra*).

Traditionally, there are three families within the order Perissodactyla: the Tapiridae (tapirs), the Rhinocerotidae (rhinoceroses), and Equidae (asses, horses, and zebras). These three families include six genera and 16 species.

Evolution and systematics

Despite excellent fossil records, the phylogeny of Perissodactyla is not well understood in terms of both the relationship within the order and the position among other orders of mammal. The perissodactyls as well as artiodactyls originated from the Condylarthra, the dominant mammalian herbivores of the early Paleocene (about 65 million years ago [mya]). Condylarths are considered to be ancestors of many of the other lineages of large mammals. Despite the superficial similarities between horses and cows, rhinos and hippos, tapirs and pigs, the former of each pair belongs to the Perissodactyla,

A Clydesdale horse and a miniature horse side by side. (Photo by R. Van Nostrand/Photo Researchers, Inc. Reproduced by permission.)

and the later to the Artiodactyla. The similarities between them have largely come about due to convergent evolution. However, mitochondrial genomes studies suggest that the order Perissodactyla is part of one eutherian clade, comprising also Pholidota, Carnivora, and Cetertiodactyla (Artiodactyla and Cetacea). The oldest identifiable perissodactyl fossils are from the early Eocene (about 50 mya). By this time, 14 radiated families were evident. During this epoch, perissodactyls were dominant ungulates, far outnumbering the artiodactyls. By the end of Oligocene (25 mya), eight families were extinct. By the early Miocene epoch, only the tapirids, rhinocerotids, equids, and Chalicotheriidae remained. This last family included unusual ungulates with large forelimbs and short hind limbs adapted for standing semi-erect to feed on tall trees. As the perissodactyls declined, there also seems to have been some definite ecological replacement of them by artiodactyls.

Originating in the early Eocene epoch of North America, tapirs migrated into Asia and Central and South America. Tapirs were extirpated throughout most of North America by the late Pleistocene epoch. A combination of migration and extirpation resulted in a discontinuous distribution today. The current genus *Tapirus* dates from 20 mya in the Miocene epoch. There are four extant species in the single genus *Tapirus*. Tapirs belong to among the most primitive large mammals in the world.

Fossil evidence of rhinocerotids dates from the late Eocene in Asia and North America. Most of today's genera date from the Miocene (10–25 mya). They were extinct in North America by the end of the Pliocene (2 mya). Rhinocerotids were abundant and widespread in the Old World until the late Pleistocene epoch (about 60,000 years ago). The largest land mammal that ever lived was a rhinocerotid, *Indricotherium transouralicum* (*Baluchitherium grangeri*), which was at least 16.5 ft (5 m) high at the shoulder and to 44,000 lb (20,000 kg) of body mass. Mitochondrial analysis identified a basal divergence between the African and the Asian species about 26 mya. There are four extant genera (*Diceros*, *Rhinoceros*, *Dicerorhinus*, and *Ceratotherium*), with five living species.

The fossil history of equids is one of the best documented for any mammalian family. This history shows increasing body size or skull proportion and reduction of the number of digits. However, the evolution of equids was not a directed progressive process, but a complex radiation of numerous divergent and overlapping lineages. Equids passed most of their evolution in North America, with migration to Eurasia and Africa during the Miocene and to Central and South America in the Pliocene and Pleistocene epochs. The earliest of the horse-like ancestors, *Hyracotherium*, appeared in the Eocene, about 54 mya. It was a small dog-size mammal that browsed on low shrubs of forest floor. When grasses extended in the Miocene, equids began to radiate. Overall body size increased, which reduced relative nutritional demands. By the early Pleistocene (2 mya), the one-toed equids had spawned the genus *Equus*, which rapidly spread. As environment changed, populations became isolated, giving rise to the living species. The first to split off from the equid stream was the Grevy's zebra (*Equus grevyi*), which, despite its stripes, is only distantly related to the other two zebra species. However, the ancestor of all equines was probably striped. The horses became extinct only about 10,000 years ago in the New World, but horses were reintroduced by the Spanish conquistador Hernando Cortes in 1519. The number of extant equid species is open to debate: seven to 10 species, all in the genus *Equus*, are recognized. Many subspecies and regional forms (mainly in zebras) are known.

Physical characteristics

The unifying characteristic of Perissodactyla is their single toe (or three toes together) bearing the weight of the animal, with the axis of each limb passing through the enlarged third digit. Tapirs have four digits on the forefeet and three digits on the hind feet, whereas rhinos have three digits on all feet. The single third digit is the only one remaining in equids. This is a condition of maximum specialization for running. The functional toes in both Perissodactyla and Artiodactyla end in hoofs, but the structure of the foot is different. The median metacarpals and metatarsal bones are not fused into a cannon bone as in artiodactyls. Perissodactyls have a deep pulley-like groove in the proximal surface of the anklebone (astralagus), which limits the limbs to forward-backward movements. The fibula does not articulate with the heel bone (calcaneum).

The perissodactyls show great range in body size and shape. The smallest species is the mountain tapir (*Tapirus pinchaque*), which weighs to 485 lb (220 kg). The largest species, the white rhinoceros (*Ceratotherium simum*), can weigh over 7,700 lb (3,500 kg). In most species, adult males are notably larger than females, while in some tapirs, the females are bigger.

The general form of tapirs is heavy, with short and stout limbs, a short tail, medium-sized oval ears extending out and upwards, small eyes flush with the head, and short fleshy proboscis formed by the upper lip and nose, with the nostrils located at the tip. The back of tapirs is arched, with the hind legs about 4 in (10 cm) higher than the forelegs, so the hind feet support the majority of their weight. This may be related to the fourth toe being absent on the hind feet. The compact, streamlined shape of the body is ideal for pushing through the dense undergrowth of the forest floor. Neotropical tapirs have short bristly manes extending along the back of the neck, protecting the most vulnerable part of the body against predators. Tapir skin is tough and covered with sparse hairs; only the mountain tapir has a thick coat, which protects it from the cold.

Rhinoceroses are large, heavy animals with short, stout legs. The eyes are small and pig-like, located on the sides of head. Vision is not especially acute. The ears are erect, tubular, tufted with hair, and fairly large, and hearing is excellent. The skin is thinly covered with hairs in *Dicerorhinus* and is nearly naked in the other species; in *Rhinoceros*, the skin is deeply folded across the back. One horn in *Rhinoceros* or two horns in all other rhinos have no bony core or keratinized sheath, but a dermal mass of agglutinated hair rest. The horns are developed upon a rough vascular cushion of bone on the midline of the nasal bones for the anterior horn and on the midline of the frontal bones for the posterior horn, when present. The horns can reach 70 in (175 cm) in length in the white rhinoceros (*Ceratotherium simum*). Rhinoceros horns grow throughout the animal's life, and are re-grown if lost. A further peculiarity of rhinos is that, as in elephants, the testes do not descend into a scrotum. The penis, when retracted, points backwards so that the urine is directed to the rear by both sexes.

All equids are medium sized with long heads and necks and slender legs. The ears are moderately long and erect, but can be moved to localize sounds. An erect mane covers the neck, but in the domestic horse, it falls to the side. All equids have long tails, with flowing hair in horses or with short hair only at the tip in ass and zebras. All African equids are partially striped on the legs (ass) or wholly black and white striped (zebra). The coat of Asian asses and horses are more uniform in color: dun in Przewalski's horse (*Equus caballus przewalskii*), from tan to gray in asses.

The skull in all species is elongated by the lengthening of the rostrum. The cheek teeth of all perissodactyls are brachydont (low-crowned), hypsodont (high-crowned), or lophodont (with elongated ridges), with 24–44 teeth. Three upper incisors are retained in the tapirids and equids; in the rhinos, they are reduced in numbers, or absent altogether. In tapirs, the canines are sharp, conical, and relatively short, separated by a diastema (space) from the cheek teeth. Perissodactyls have to consume large quantities of tough fibrous food, so the lower jaw is very deep and the masseter muscle is very large. The lips are thick and freely movable in all species.

In the Perissodactyla, the stomach is simple, and the cecum is enlarged to form a chamber in which microorganisms live and digest plant cellulose (hindgut fermentation). Food passes through the digestive system about twice as fast as through that of ruminating artiodactyls. Because food is retained for less time, digestion is less efficient. For example, the digestive efficiency of a horse is only about 70% of that of a cow. Perissodactyls compensate for reduced efficiency by consuming more food per unit of body weight. The enlarged cecum and colon provide storage and surface area for absorbtion of nutrients. In white rhinoceros (*Ceratotherium simum*), the cecum consists of a small chamber, whereas the colon is enlarged. The cecum may be functionally replaced by the well-developed colon, which may act as the main fermentation tank. Perissodactyls have a bicornuate uterus, diffuse placentation, and no clavicle or penis bone (baculum). The gallbladder is missing, and milk nipples are placed in groins.

Distribution

Perissodactyla have limited distribution in Africa, Asia, and America, but some hundred years ago they occupied much bigger areas of these continents, as well as Europe. The tapirs have a discontinuous distribution in the Neotropical Central and South America and in southeastern Asia. The rhinoceroses are distributed in sub-Saharan Central and East Africa and in the Indomalayan region of tropical Asia. The natural distribution of equids includes eastern and southern Africa and Asia from Near East to Mongolia. The wild horses became extinct in Europe during the nineteenth century. Only the domestic horse is found worldwide, and it has spawned feral populations in North America on the western plains and on east coast barrier islands, and in the mountains of Western Australia.

Habitat

Perissodactyls are able to occupy different habitats from the deserts to tropical rainforests. Tapirs usually do not occur far from permanent water and are associated with a variety of moist tropical forested habitat, including dry deciduous

Przewalski's wild horse (*Equus caballus przewalskii*) spends the majority of the day foraging for food. (Photo by Tom McHugh/Photo Researchers, Inc. Reproduced by permission.)

A dwarf pony at the Frankfurt Zoo, Germany. (Photo by Tom McHugh/Photo Researchers, Inc. Reproduced by permission.)

forest as well as multistratal tropical evergreen forests. Only one species, mountain tapir, is adapted to higher elevations to 14,700 ft (4,500 m) in the Andes and predominantly inhabit the paramos and dwarf forest. The other species occur in lowland and premontane forests, or in swampy grasslands. The lowland tapir (*T. terrestris*) is also common in dry parts of Paraguayan and Argentine Chaco.

Depending on the species, rhinoceroses occupy tropical rainforests, floodplains, grasslands, and scrublands. All species are dependent on a permanent water supply for frequent drinking and bathing. African rhinoceroses occupy drier savannas and arid scrublands as well as mountain rainforest. They mostly prefer edges of thickets and savannas with areas of short woody regrowth and numerous shrubs and herbs. Asian rhinos inhabit both the swamps and dense rainforests to the elevation 6,600 ft (2,000 m). Fossil forms has been found as high as 16,100 ft (4,900 m) in the Himalayas.

Equids inhabit short grasslands and desert scrublands. Only the plains zebra (*Equus burchellii*) and the mountain zebra (*Equus zebra*) occupy lusher grasslands and savannas where the vegetation is more abundant. The remaining species live in the more arid environments with sparsely distributed vegetation.

Behavior

Some perissodactyls are highly social mammals, but others live more or less solitary. Tapirs tend to move singly except for female and her dependent young. They often spend part of the day resting in mud wallows or standing water or lying in shaded thickets. Tapirs frequently use the same trail to and from wallowing and feeding sites. They mark their territories and daily routes with urine. Tapirs defecate in water or at special places near water. Communal use of trails and defecation sites permits a loose communication system within a population. Tapirs are mostly nocturnal and only partly diurnal animals. They confine much of their activity to the darkness. In the Central American tapir (*Tapirus bairdii*) in Costa Rica, diurnal movements could encompass as little as 7.5 acres (3 ha) or as much as 67 acres (27 ha). Nocturnal movements are larger, with activity included to 445 acres (180 ha). An adult and subadult of mountain tapir in central Colombia moved within the range of about 4.3 mi^2 (11 km^2). Tapirs are excellent swimmers and their existence is centered close to water, which offers lush feeding grounds and escape routes from predators. Tapirs can cross deep streams by walking on the bottom. They may seek refuge in water and can stay submerged for several minutes. Tapirs have an acute sense of smell and good hearing, but seem to have poor eyesight. Tapirs walk with the proboscis close to the ground, so they can detect the scent of predators. When threatened, they usually crash off into thick brush or into the water or defend themselves by biting. Tapirs are usually silent, but communicate over distance with long whistling calls. Grunts, hiccups, and whimpers are used at close range. A low snort may be given in alarm. Density of tapirs can reached 2.1 animals per mi^2 (0.8 animals per km^2). Males may fight ferociously to gain mating rights with sexually receptive females. Remarkable associations between tapir and a male coati exist. Tapirs can become infested with ticks and coati fed on the blood-engorged ticks by gleaning them from the tapirs' bodies.

Aside from mother and offspring pairs, rhinos generally are solitary. Small group of immature individuals may form in Indian (*Rhinoceros unicornis*) and white rhinos. Several individuals may come together mainly in favored wallowing areas or saltlicks. Mated pairs may form couples for a short time. The tolerance of dominant individuals, particularly males, varies by species, by region, and in response to changes in density. Rhinos display territorial behavior. Displays between males are very conspicuous: there is much strutting, broadsides with lowered head, flattened ears, and rolling eyes. Various snorts during encounters seem to have different shades of meaning. Home ranges of males can cover 50 mi^2 (130 km^2), but some are as small as 0.4 mi^2 (1 km^2). Females and their immediate offspring occupy home ranges 1.5–6 mi^2 (4–15 km^2), overlapping with one or more female neighbors. When territories are maintained by resident males, their border patrolling and scent marking leave foot-scuffs, dung middens, urine spray, rubbing posts, and horned vegetation along boundaries. Males of all species sometimes fight viciously, inflicting gaping wounds. Both African species fight by jabbing one another with upward blows of their front horns. In contrast, the Asian rhinos attack by jabbing open-mouthed with lower incisor tusks, or in the case of Sumatran rhinoceros (*Dicerorhinus sumatrensis*), with the lower canines. Black rhinoceroses (*Diceros bicornis*) have a reputation for unprovoked aggression, but very often their charges are merely blind rushes designed to get rid of the intruder. Some rhinos may travel 9.5–12.5 mi (15–20 km) within 24 hours, and some individuals of the Sumatran rhinoceroses have been observed swimming in the ocean. Temperature control, digestion, and scent communication all depend upon water, and rhinos are unable to survive extreme droughts. The wallowing of rhinoceros is probably necessary to help control body temperature and to reduce insect harassment. The sense of smell is

A Persian onager (*Equus hemionus onager*). (Photo by Tom Brakefield/OKAPIA/Photo Researchers, Inc. Reproduced by permission.)

well developed. Rhinos have poor vision and are unable to detect a motionless person at a distance of more than 35 yd (30 m). Social systems can be related to differences in the density and distribution of food resources. Rhinos rarely exceed local densities of 0.4–2 animals per mi^2 (1–5 per km^2).

Equids are highly social mammals that exhibit two basic patterns of social organization. Zebras are high-density, tropical grassland horses; asses are low-density, desert-adapted forms. The basic social units of zebras is the family group (or harems), generally 10–15 individuals made up of a highly territorial male, several females, and their offspring. Nonbreeding males, and occasionally young females, form small unstable bachelor groups. Each harem has a home range, which overlaps with those of neighbors. Home ranges vary, depending on the quality of the habitat, from 31–232 mi^2 (80–600 km^2). In the plains zebra, temporary aggregations of 100,000 individuals may form, depending on ecological conditions. Zebras very often associate with other ungulates. Social contacts are primarily by sounds, and zebras are extremely vocal. The adult males are particularly noisy during nocturnal movements. For each social unit, the stallion's individual song (a glottal-barking bray) becomes the focal point for all harem members. The second social system, typified by the asses and Grevy's zebra, involves more ephemeral adult associations, rarely lasting longer than a few months. Temporary aggregations of one or both sexes are common, but most adult males live alone within territories from 0.8–8 mi^2 (2–20 km^2). Within this territory, owners obtain exclusive mating access to receptive females that wander through them. Other males are tolerated within the territory, but the territory holder monopolizes all access to females. The preferred range of females with offspring is often within vast territories held by mature males. Dung piles mark boundaries of territories. Estrus stimulates frequent loud braying in males. Dominance is asserted by a proud posture, with arched neck and high-stepping gait. Submission is signaled by a lower head and raised tail. The stripes of zebra appear to serve as a visual bonding device. Black and white stripes stimulate visual neurones very strongly and appear to make them attractive to each other. The widespread theory that their strips are camouflage is therefore contradicted by the zebra behavior. The aggressively antisocial behavior of territorial animals would thus appear to be counteracted to some extent by visual attractiveness. Notwithstanding this, both males and females are mutually antagonistic and even young foals prefer members of the opposite sex. Equids are most active when the weather is cooler: at dusk, dawn, and during the night. Equid eyes are set far back in the head, giving a wide field of view. Their only blind spot lies directly behind the head, and they even have binocular vision in front. They probably can see color and, although their daylight vision is most acute, their night vision ranks with that of dogs or owls. Moods are often indicated visually by changes in ear, mouth, and tail position. Males use the flehmen, or lip-curl, response to assess the sexual states of females, and the womeronasal, or Jacobson organ, which is used for this is well developed. Reproductive competition among males for receptive females is keen. This begins with pushing contests, or ritualized bouts of defecting and sniffing, biting at necks, tearing at knees, or thrusting hind legs towards faces and chests. In contrast, amicable activities such as mutual grooming cement relationships among the females. All equids keep their coat in condition by dust bathing, rolling, rubbing, and some mutual nibbling.

Feeding ecology and diet

All perissodactyls are strictly terrestrial herbivores. The tapirs are browsers and frugivores and selectively feed on

A domesticated ass (*Equus asinus*). (Photo by St. Meyers/OKAPIA/Photo Researchers, Inc. Reproduced by permission.)

leaves, twigs, various fallen fruit, grass, aquatic vegetation, and occasionally on cultivated crops, but prefer green shoots. They follow a zigzag course in feeding, moving continuously and taking only a few leaves from any one plant. Although they may crush many seeds of fleshy fruit, the tapirs are an important disperser of seeds, since some pass through its gut unharmed. In Mexico and South America, tapirs sometimes damage young maize and other grain crops, and in Malaya, they are reputed to raid young rubber plantation.

Rhinos forage on woody or grassy vegetation and occasionally fruits, but prefer leafy material when available. They need a large daily intake of food to support their great bulk. The main rhino specialty is an ability to feed on coarse plant material. Because of large size and hindgut fermentation, rhinos can tolerate relatively high contents of fiber in their diet. Rhinos are more selective than elephants, but less so that most antelopes. In black rhinoceroses, there are special seasonal preferences for legumes, while certain other plants are always avoided. Salt is a major attraction. All rhinos are basically dependent upon water, drinking almost daily. But under arid conditions, both African species can survive up to five days without water, if their food is moist. In white rhinoceroses, a preference for short grass areas and seasonal movements to avoid waterlogged long grass confirm that this species evolved within a larger ungulate community that maintained short swards. In areas of high density, the rhinos themselves maintain grazing lawns. Linear pathways may form between grazing and water points. Frequent alterations between grazing and resting changes to long midday rest at the height of the dry season. Asian rhinos are browsers and during the course of feeding, branches up to 1 in (2 cm) thick are torn off, stems are broken or decorticated, and trees up to 6 in (15 cm) in diameter are uprooted when they lean against the tree. The rhino thus modifies the environment to meet dietary needs.

Plains zebra (*Equus burchellii*) by Mara River, Masai Mara, Kenya. (Photo by Gregory G. Dimijian/Photo Researchers, Inc. Reproduced by permission.)

All equids forage primarily on fibrous foods. Although horses and zebras feed mainly on grasses and sedges, they will consume bark, leaves, buds, fruits, and roots, which are common for asses. Wild asses are well adapted to graze the harshest desert grasses. They used their incisors and hooves to break open tussocks. Equids can sustain themselves in more marginal habitats and on diets of lower quality than can ruminants. Equids spend most of a day and night foraging (about 60–80% of 24 hours activity). Equids are able to go without water for about three days, but zebras are totally dependent on frequent drinking. Some local populations can dig waterbeds with the hooves. Mountain zebra often occupy separate summer and winter ranges in distances up to 75 mi (120 km) apart. They move between pastures and water sources on well-worn traditional paths. The timing and intensity of grazing is strongly influenced by temperature and season, with animals taking shelter and becoming inactive during the middle of the day in summer.

Reproductive biology

Perissodactyls are long-lived, very slow breeders with low recruitment rate. Sexual maturity in tapirs is reached at two to four years of age. Breeding occurs at any time during the year and females can be sexually receptive every two months. Mating is preceded by a noisy courtship. Male and female stand head to tail, sniffing their partner's sexual parts and moving around in a circle at an increasing speed. They nip each other's feet, ears, and flanks, and prod their bellies. Usually, a single precocial young (twins are rare) is born after a gestation of 383–395 days. The average neonatal weight is about 5 lb (2.3 kg). Young have a reddish brown coat with camouflage of white spots and lines. For the first week after birth, it remains in a secluded spot while the mother feeds, and she returns periodically to nurse. The calves actively follow the mother at 10 days of age and stay with her for six to eight months, by which time the juvenile pelage is replaced with adult pelage and it is nearly adult size. A female in her prime probably produces a calf every second year. Some animals are known to live for 30 years.

Rhino females become sexually mature at three to five years of age and bear their first calves when six to eight years of age. Females are probably polyestrous, coming into heat every 46–48 days. In the African rhinos, a birth peak occurs from the end of the rainy season through the middle of the dry season. Gestation is 7–8 months in the Sumatran rhi-

Grevy's zebra (*Equus grevyi*) on the left and plains zebra (*Equus burchellii*) on the right. (Photo by K. & K. Ammann. Bruce Coleman, Inc. Reproduced by permission.)

noceros, and 15–16 months in the other species. Mating in rhinos can be a prolonged business with several hours' foreplay, and copulation often lasting for one hour. Births, usually a single non-horned calf, occur at intervals of two to five years. The body weight of newborn calves varies 55–145 lb (25–65 kg), according the species, and is about 4% of the mother's weight. Mothers produce to 5.5 gal (25 l) of milk a day and the calves gain 5.5 lb (2.5 kg) a day. Young nurse for one to four years, although the white rhinoceros begins to eat solid food by one week of age. Calves suckle milk for up one or two years and only begin to drink water after four or five months. Males first become sexually potent at seven to eight years of age, but generally do not breed before 10 years old. Nonetheless, rhinos formerly suffered low levels of natural mortality, a trait that goes with long life up to 50 years. In spite of a long lifespan, this is one of slowest recruitment rates of any large mammal.

Both young male and female equids become sexually mature at about two years age. Males do not breed until they leave the family group and gain access to other females, which is at five years of age. A single foal is born after a gestation period about a year. Only in the Grevy's zebra does gestation last slightly longer (about 400 days). Births and subsequent mating 7–10 days later occur during the wet season, when vegetation is more abundant. Neonates are precocial. Young are up and about within an hour of birth. Initially, the foal is passive and fearless and will remain alone for long periods while the mother seeks water to maintain her lactation. The offspring begin to graze at one month of age and are weaned at 6–13 months of age. Some animals are known to live for 40 years.

Conservation

There are only 16 extant species of Perissodactyla. Of these, 13 (more than 81%) are threatened with extinction and listed as endangered species (IUCN Red Data Books, CITES). A few species like the quagga (*Equus quagga*) have become extinct (the last animal died at the Amsterdam Zoo in 1883), but many other subspecies, forms, or populations also disappeared. The principal threats to perrisodactyls are the degradation of their natural habitat and human activities, mainly poaching as well as legal over-hunting. Local wars also contribute to the decline of the animals in various areas. The Special Survival Commission of IUCN—The World Conservation Union is interested in the perrisodactyls and has Specialist groups for equids, Asian rhinos, tapirs, and Afrotherian conservation.

Although tapirs have survived for millions of years, their future is not secure. All species of tapirs suffer from loss of

A black rhinoceros (*Diceros bicornis*) baby. (Photo by Christian Grzimek/OKAPIA/Photo Researchers, Inc. Reproduced by permission.)

habitat due to huge deforestation for agriculture. Most populations are declining in number and distribution. The Malayan tapir (*Tapirus indicus*) and mountain tapir are considered Endangered or threatened. Tapir meat is much prized, and tapirs are easy to locate with dogs or calls and thus vulnerable to local extinction. Tapirs avoid people and leave areas of human activities, even when hunting is controlled. The best management for survival of tapirs is forest reserves like Taman Negara in Malaya.

Populations of all rhinoceroses have declined during the last 150 years and drastic declines will probably continue. By 1970, some 70,000 black rhinoceroses were estimated to have survived in Africa. This was a fraction of the number of existing rhinos at the turn of the century. By 1990, the total number of this species living within 38 officially protected conservation areas was about 3,300. Today, the estimate is about 2,700 individuals. All species are considered to be Endangered, with the Asian species near extinction. The population of Javan rhinoceros (*Rhinoceros sondiacus*) is now confined to a remnant of 60 animals in the Ujung Kulon Reserve in western Java and in National Park Cat Tien in Vietnam. The Sumatran rhinoceros is now restricted to perhaps 150 animals, and the Indian rhinoceros to a few reserves in Assam, west Bengal, and Nepal, with a total number of about 1,500 animals. Rhinos are among the world's most endangered mammals. The natural habitat destruction due to the enormous increase of human population, as well as the destruction of forest by elephants, plays a significant part in the threats to the survival of these animals. Rhinos have been brought to extinction because of the very large amounts of money that people are prepared to pay for rhinoceros products. The best protection the animals have is to remain in well-managed national parks and game reserves. White rhinoceroses have been successfully translocated to parts of their former range in southern Africa and its abundance has increased from about 20 animals to 13,000. Meanwhile, strict international legislation is intended to prevent trade with rhinoceros products and to punish those profiteers who deal in them. In many of the countries where rhinoceroses live, heavily armed patrols of game wardens battle with equally determined bands of poachers. Until cultural and environmental attitudes have changed, survival of rhinos will continue to depend upon captive or closely managed populations.

Despite the proliferation of domestic horses, their wild relatives are in a precarious situation. Several species are Endangered. These include Przewalski's horse and African ass (*E. africanus*), both of which are probably Extinct in the Wild. Their only immediate salvation lies in captive breeding and release back to available habitats in the wild. Many Przewalski's horse individuals were produced at Prague and Hellabrun zoos, and many African asses at Basel Zoo and Hai Bar Reserve in the Negev Desert. The release of Przewalski's horse to China in 1988 (from Hellabrun) and to the Mongolian desert (from Prague) is a very good example of species protection. In Europe, the last tarpans (*E. caballus*) died in 1806 in Poland and in 1879 in Ukraine. Almost all subspecies of Asiatic asses are seriously threatened: the Persian onager (*E. hemionus onager*), sometimes considered a distinct species, is Endangered; the kulan (*E. h. kulan*) occurs only in a reserve on the Barsakelmes Island in the Aral Sea; the khur (*E. h. khur*) occurs only in the Danghandra Reserve in India; and the Syrian onager (*E. h. hemippus*) is now Extinct. The true plains zebra (*E. burchelli burchelli*) is also Extinct since 1910. Populations of both subspecies of the mountain zebra, the Cape mountain zebra (*E. zebra zebra*) and Hartmann's mountain zebra (*E. z. hartmanni*), are small and are protected in national parks or reserves, but those of Grevy's zebra have been drastically reduced, as their beautiful coats fetch high prices. However, the primary threat to all equids has come from increased livestock farming, leading to their exclusion from traditional pastures and watering places. It would be tragic if these wonderful creatures were to go the way of the quagga.

Significance to humans

The important position of the perissodactyls in most communities of large herbivores, their use of habitat, size, social organization, and ecology, even their horns, have all exposed them to damaging interactions with humans; yet, those same characteristics have been the salvation of some. Humans are partly responsible for the dwindling numbers of perissodactyls since the Pleistocene. From the glacial and interglacial epochs in Eurasia comes a wealth of evidence from societies culturally dependent upon the hunting of perissodactyls and other ungulates. Cave paintings, drawings, and carvings made by ancient people in many areas of world bear witness to the relationship that has existed between man and perissodactyls since the dawn of human history. The crucial event in this relationship was the domestication of some equids. Despite the strong protection of most of extant wild species of perissodactyls, some of them still have a great importance for some local native nations as a food. Limited species are legally hunted for meat or for trophy.

Tapirs are hunted for food, sport, and for their thick skins, which provide good quality leather, much prized for whips

Female Indian rhinoceros (*Rhinoceros unicornis*) with its young. (Photo by Tom McHugh/Photo Researchers, Inc. Reproduced by permission.)

and bridles. The lowland tapir is quite easy to tame, and early colonists used them for field work.

Rhinos are illegally harvested for their horns, which, with other body parts (hide, hoof, teeth, various organs, blood, urine), are valued in traditional Asian medicine for supposed aphrodisiac and medicinal properties. While ground rhino horn is used as an aphrodisiac in parts of north India, its main use in China and neighboring counties of the Far East is as a fever-reducing agent. It is also used for headaches, heart and liver troubles, and for skin diseases. Rhino horns have also been used traditionally for making handles of daggers (*jambia*) worn by men in the Middle East as a sign of status. Between 1969 and 1977, horns representing the deaths of nearly 8,000 rhinos were imported into north Yemen alone. Until recent time, the hide and horn of both African species were fashioned into shields for use in battle or tribal ceremonies. As the number of rhinoceroses declines, the market value of their products naturally increases, so that today, horns are often hoarded as a good investment for the future. Under such circumstances, it is hardly surprising that rhinoceros are hunted wherever they live.

Horses and donkeys were the last of common big animals to be domesticated, and they have been the least affected by human activity. The first domestic horses appeared during the late Neolithic at about 6,000 years ago, when they may have initially been used for food. Surprisingly late, about 4,000 years ago, the use of the horse as a means of transport occurred. This event caused a revolution in the human mobility race and the development of modern techniques of warfare. All species of equids can interbreed, including the zebras, and hybrids can produce more or less viable offspring. Different species do not normally interbreed in nature and it usually requires human guile and expertise to bring it about. Hybridization between horses and donkeys occurs relatively often. The production of mules (offspring of a male donkey with a female horse) is easier than the production of hinnies (offspring of female donkey and a male horse). Today, the practical importance of domestic equids is much higher for humans than in extant wild species.

Many localities where tapirs, rhinoceroses, and wild equids occur now attract humans to observe these marvelous mammals in their mystery and beauty.

Resources

Books

Corbet, G. B., and J. E. Hill. *The Mammals of the Indomalayan Region: A Systematic Review.* New York and Oxford: Oxford University Press, 1992.

Eisenberg, J. F., and K. H. Redford, eds. *Mammals of the Neotropics. The Central Neotropic. Volume 3. Ecuador, Peru, Bolivia, Brazil.* Chicago and London: The University of Chicago Press, 1999.

Estes, R. D., and D. Otte. *The Behavior Guide to African Mammals.* Berkeley: University of California Press, 1990.

Feldhamer A., L. C. Drickamer, S. H. Vessey, and J. F. Merritt. *Mammalogy: Adaptation, Diversity, and Ecology.* New York: McGraw-Hill,1999.

Kingdon, J. *The Kingdon Field Guide to African Mammals.* London: Academic Press, 1997.

Reid, F. A. *A Field Guide to Mammals of Central America & Southeast Mexico.* New York and Oxford: Oxford University Press, 1997.

Wilson, D. E., and D. M. Reeder. *Mammal Species of the World.* Washington, DC: Smithsonian Institution Press, 1992.

Periodicals

Arnason, U., and A. Janke. "Mitogenomic Analyses of Eutherian Relationship." *Cytogenetic Genome Research* 96 (2002): 20–32.

Eisenmann, V. "Browsers and Grazers: Symphyseal Shapes in Equids and Tapirs (Perissodactyla, Mammalia)." *Geobios, Lyon* 31 (1998): 113–123.

Froehlich, D. "Quo Vadis Eohippus? The Systematic and Taxonomy of the Early Eocene Equids (Perrisodactyla)." *Zoological Journal of the Linnean Society, London* 134 (2002): 141–256.

Norman, J. V., and M. V. Ashley. "Phylogenetics of Perissodactyla and Tests of the Molecular Clock." *Journal of Molecular Evolution* 50 (2000): 11–21

Robinson-Rechavi, M., and Graur D. "Usage Optimalization of Unevenly Sampled Data through the Combination of Quartet Trees: A Eutherian Draft Phylogeny Based on 640 Nuclear and Mitochondrial Proteins." *Israel Journal of Zoology* 47 (2002): 259–270.

Organizations

The World Conservation Union (IUCN)/Species Survival Commission. Rue Mauverney 28, Gland, 1196 Switzerland. Phone: 41 (22) 999-0152. Fax: 41 (22) 999-0015. E-mail: ssc@iucn.org Web site: <http://www.iucn.org>

Jaroslav Cerveny, PhD

Horses, zebras, and asses

(Equidae)

Class Mammalia
Order Perissodactyla
Family Equidae

Thumbnail description
Medium-sized herbivore with long legs, hard, single-toed hooves, erect mane, long tail, and short coat; some species have a striped coat color

Size
Head and body length 77–118 in (195–300 cm); shoulder height 45–63 in (115–160 cm); weight 440–990 lb (200–450 kg)

Number of genera, species
1 genus; 7 species

Habitat
Savanna grassland, shrub-land, semiarid grassland, and desert grassland

Conservation status
Extinct in the Wild: 1 species; Critically Endangered: 1 species; Endangered: 2 species; Vulnerable: 1 species; Lower Risk/Least Concern: 2 species

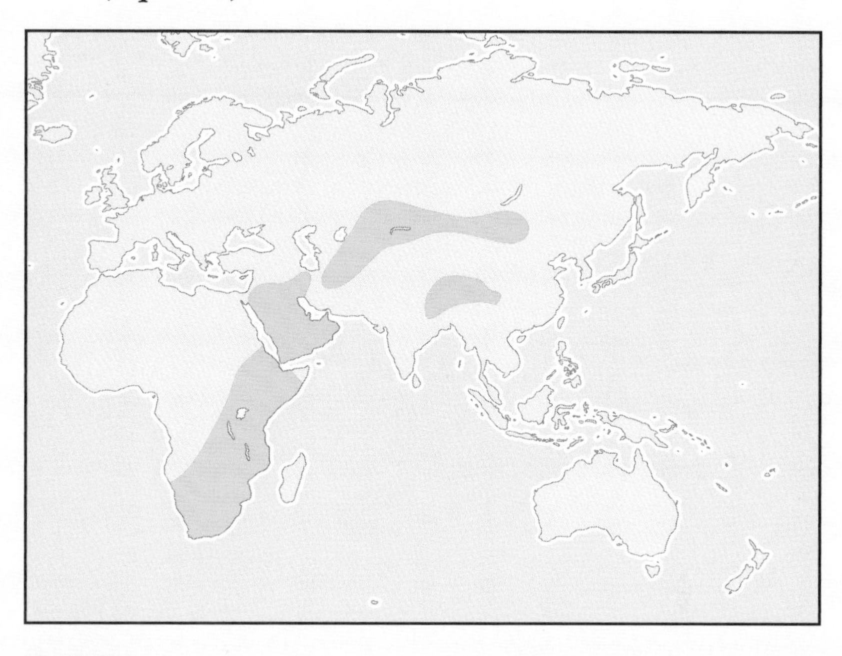

Distribution
Sub-Saharan Africa, Middle East, Arabia, Central Asia, and Mongolia

Evolution and systematics

The fossil record of the family Equidae begins 55 million years ago in the Eocene with the small "dawn horse," *Eohippus*. The main trends in the evolution of Equidae were an increase in body size, lengthening of the legs, reduction in the number of toes culminating with the single-toed hoof, increasing height and complexity of crown patterns in the cheek teeth, lengthening and deepening of the skull to accommodate the high-crowned cheek teeth, and an increase in the size and complexity of the brain. The mainstream of equid evolution occurred in North America and during the Pleistocene, when the modern genus *Equus* appeared and spread to Eurasia and Africa. In North America, there were approximately eight equid species. However, during the late Pleistocene, there was a mass extinction of mammals in North America and all the equid species disappeared. In Eurasia and Africa, seven species of equids survived.

Grevy's zebra (*Equus grevyi*) is taller and heavier than the Burchell's zebra. (Photo by David M. Maylen III. Reproduced by permission.)

Equidae is in the order Perissodactyla, the odd-toed ungulates. This order also includes Rhinocerotidae and Tapiridae. Equidae has one genus, *Equus*. Current taxonomy identifies seven species in this genus. However, taxonomy is an evolving science and questions remain concerning evolutionary and genetic relationships and whether some species should be split or combined.

Physical characteristics

Equids are medium-sized odd-toed ungulates. Anatomically, they are adapted for moving quickly and efficiently (long legs), feeding on higher-fiber grasses (high-crowned premolars and molars), quicker digestion of high-fiber forage (single stomach and rear-gut fermentation), and moving over hard and rocky substrate (single-toed hoof).

A young plains zebra (*Equus burchellii*) showing submissive behavior. (Photo by Rudi van Aarde. Reproduced by permission.)

The three species of zebra each have different stripe patterns. Within a species, it is possible to identify individuals by their unique rump and side stripe patterns. The African wild ass (*Equus africanus*) has stripe patterns on its legs that make it possible to distinguish individuals. The Asiatic wild ass (*E. hemionus*) and the kiang (*E. kiang*) do not have stripes and the identification of individuals is more difficult. Przewalski's horse (*E. caballus przewalskii*) may have leg stripes. All equid species have a short coat, although in temperate latitudes (e.g., the Przewalski's horse) they may grow a thicker, longer coat. Male and female may differ by less than 10% in terms of height and weight. There is little sexual dimorphism in equid species.

Distribution

Equids are found in sub-Saharan Africa, the Middle East, Arabia, Central Asia, and Mongolia. Although most of equid evolution occurred in North America, no wild equids are found there today. The radiation of the genus *Equus* occurred in Eurasia and into Africa. Four species are found in Africa and the remaining three species occur in the Middle East, Arabia, central Asia, and Mongolia.

Habitat

Equids are mainly grazers and the different species can thrive in habitats from below sea level in extreme deserts (e.g., the Danakil Desert) to mesic grasslands (e.g., the Serengeti Plains) to dry mountain grassland and shrubland.

Behavior

Social organization and reproductive strategies in equids are a complex interaction of individual feeding requirements, access to females, and defense against predation. Individuals will form groups when benefits exceed costs in terms of feeding, predation, disease, and reproduction. The density, continuity of distribution, and biomass of forage are key factors in the stability of association and the spacing of equids. However, water distribution and predation pressure are also important factors. When forage and water availability allows females to be gregarious and form stable groups, then a male can attempt to control access to these females. Among equids, these female groups form strong bonds, and if a male is removed, the females often maintain a stable group. Age and fighting ability and the adult sex ratio can also affect a harem male's success in defending his female group against bachelor males. The harem-defense type of social organization is characterized by long-term, stable non-territorial groups composed of one adult male and one or more females and their offspring (e.g., plains zebra, *Equus burchellii*; mountain zebra, *E. zebra*; Przewalski's horse. Other males live in what are often termed bachelor groups. The adult females remain in the social group through time, but another male might displace the current harem male. The group male has exclusive mating rights. Occasionally, there is a subordinate adult male within the group that is reproductively active, but has lower reproductive success. Foals born into a group usually remain with it for two to three years or longer. Dispersal time of young males and females and the causation varies. In plains zebra, a young female might be abducted during her first estrus, but young males tended to leave on their own. In mountain zebra, young females might be driven away by their mothers and young males might be chased away by the dominant male. Among equid populations that have a harem mating system (female-defense polygyny), the following have been observed:

- Multi-male harem groups in which all males defend the females and the dominant male achieves the most copulations.
- Harem males that form alliances and cooperatively defend their harems.
- Populations in which adult male sex ratio is significantly low and single-male harems encounter less intrusion/harassment from bachelor males.

Harassment involving chases and copulations can negatively affect a female's feeding rate and may even result in abortion or involve infanticide.

A prerequisite for the cost-effective viability of female-defense polygyny (stable family/harem groups) is a spatial and temporal patterning of resource availability such that it is possible for females to feed in close proximity. In more mesic habitats, forage for ungulates tends to be more abundant with a more continuous distribution. Large stable groups are more likely to form when one individual's foraging does not adversely affect a conspecifics' foraging. Consequently, closer spacing and larger aggregations are possible when food is abundant. Conversely, food shortages will tend to limit group size and stability.

Predation pressure on large mammals, like equids, should increase the tendency to form groups in order to improve detection of and/or defense against predators. The potential for polygyny among equids is further enhanced because females are able to provide nutritional care for their young, and females do not come into estrus synchronously, which enables a male to mate with several females. Thus, in a mesic habitat, a male can control access to multiple females by virtue of their gregariousness and their non-synchronous estrus. From the female point of view, abundant food allows

A herd of Grevy's zebras (*Equus grevyi*) grazing. (Photo by K & K Ammann. Bruce Coleman, Inc. Reproduced by permission.)

Plains zebras (*Equus burchellii*) drinking in Etosha National Park, Namibia. (Photo by Dr. Eckart Pott. Bruce Coleman, Inc. Reproduced by permission.)

closer spacing with other females and gregariousness enhances predator detection. In addition, the presence of a dominant male precludes harassment by other males in the population.

In more arid environments, limited food availability (both spatially and temporally) usually does not permit females to forage in close proximity and/or to be associated consistently. In dry habitats, equids exhibit the same nutritional and reproductive characteristics (e.g., females provide nutrition and females tend to come into estrus asynchronously), which allow males to attempt multiple matings, but indirectly control access to the females. In most cases, they actually control access to a critical resource, i.e., water. In the resource-defense type of social organization, the only stable groups are a female and her offspring. No permanent bonds persist between adult individuals (African wild asses, Grevy's zebra, *E. grevyi*; and Asiatic wild asses). Some males are territorial, dominate their areas for years, and have exclusive mating rights within their territories. Conspecifics of both sexes are tolerated in these territories.

Feeding ecology and diet

Equids are primarily grazers and have dental adaptations for feeding on grasses. Their high-crowned molars with complex ridges allow them to effectively grind grasses with higher-fiber content. Though individuals will select the most nutritious and lower-fiber forage, they can process senescent and higher fiber grasses. Equids also have a single stomach and hindgut fermentation. This allows them to digest and assimilate larger amounts of forage during a 24-hour period. By contrast, ruminants with a four-chambered stomach are limited in the volume of forage that can be digested in a 24-hour period. Equids are more effective in assimilating forage and can tolerate and survive on a greater breadth of diet in terms of relative forage quality/nutrition.

Reproductive biology

Equids are polyestrous and their estrous cycles are 19–35 days long. They breed seasonally and will cycle until conception or the end of the season. In the temperate zone, the breeding season is in the spring with the appearance of better forage and weather. In the tropics, breeding usually occurs during the rainy season. The mating system of equids tends to vary, depending on environmental conditions. Equid gestation is 11–12 months in duration. Soon after the birth of the foal, the female will come into postpartum estrus (7–18 days). This means that a female has the capability of producing a foal every year at approximately a 12-month interval. However, wild equids rarely produce a surviving foal every year. Normally, they will have a foal every other year if nutritional conditions permit. Natality in the African wild ass correlates significantly with rainfall during the previous 12 months, e.g., the period of gestation.

Wild equids have been observed to reach puberty at one to two years of age. However, wild equids normally produce their first foal at three to five years of age. In the more arid habitats, age of first reproduction may be five years. There is limited information on natality and survivorship in wild equids. Natality can be 0.0–1.0 in African wild ass, but with

Grevy's zebras (*Equus grevyi*) in a territorial fight. (Photo by K & K Ammann. Bruce Coleman, Inc. Reproduced by permission.)

Khurs (*Equus hemionus khur*) in Tann, India. (Photo by Animals Animals ©Anup Shah. Reproduced by permission.)

many wild equids, the average hovers around 0.5. Data on foal and yearling survival are equally sparse. More is known about adult survival in the plains zebra. Based on aging of skulls, adult female annual survival was 0.9–1.0 and most females died by age 16. Plains zebras live in mesic grasslands, and during most years, nutrition may be less of a limiting factor. Wild equids living in more extreme arid environments may have lower survival rates. There is useful information from domestic and feral horses and donkeys. However, these data need to be used with caution as these populations have a long history of domestic breeding and have been introduced to their current habitats.

Data on recently introduced wild equids (e.g., Asiatic wild ass) indicate that, with good environmental conditions, they can exhibit a rapid growth rate. But most wild equids either have stable populations (plains zebra in the Serengeti) or exhibit severe declines due to severe winters and severe drought. Predation can also impact population growth rates and stability. Once again, there are insufficient data as to what age and sex classes are most affected. Predation normally has the most impact on the survival of foals. Disease can also be a source of major mortality in equid populations. Predation by humans, transmission of disease from domestic livestock, and competition for forage and water are major threats to the continued existence of wild equids.

Conservation status

Family Equidae contains one genus, *Equus*, and seven species. Of these, one is Critically Endangered, two are Endangered, one is Vulnerable, and one is Extinct in the Wild. Only two species have large enough populations to be considered Lower Risk/Least Concern. But one of these, the Asiatic wild ass, illustrates the vulnerability of all wild equids. Kulan, the subspecies of Asiatic wild ass in Turkmenistan, were reduced from a population of approximately 6,000 to

A feral donkey (*Equus* sp.) in Custer State Park, South Dakota, USA. (Photo by Animals Animals ©Mickey Gibson. Reproduced by permission.)

Wild horses in the Amargosa Desert of Nevada. (Photo by Mark Newman. Bruce Coleman, Inc. Reproduced by permission.)

roughly 700 individuals in a few years time. This was attributed to human hunting.

Wild equids in their native habitats are threatened by hunting for food and medicine, competition with livestock and people for access to water and forage, fragmentation and reduction of habitat, small population size, and inter-breeding with domestic horses and donkeys.

Significance to humans

Since Paleolithic times, wild equids have been a source of inspiration to artists on cave walls and canvas. Their beauty and speed have been the personification of independence and freedom. Their domestic relatives have had major significance in the social and agricultural history of man. Less well understood is the significant role of wild equids in the ecology of multiple-grazer/browser ecosystems.

Mustangs fighting. (Photo by Jonathan Wright. Bruce Coleman, Inc. Reproduced by permission.)

A dwarf shetland pony with a foal. (Photo by Animals Animals ©Robert Maier. Reproduced by permission.)

1. Asiatic wild ass (*Equus hemionus*); 2. Grevy's zebra (*Equus grevyi*); 3. African wild ass (*Equus africanus*); 4. Plains zebra (*Equus burchellii*); 5. Przewalski's horse (*Equus caballus przewalskii*); 6. Kiang (*Equus kiang*); 7. Mountain zebra (*Equus zebra*). (Illustration by Barbara Duperron)

Species accounts

African wild ass
Equus africanus

TAXONOMY
Equus africanus Heuglin and Fitzinger, 1866, Atbara River, Sudan.

OTHER COMMON NAMES
English: Somali wild ass, Nubian wild ass, Abyssinian wild ass; French: Ane sauvage; German: Wildesel.

PHYSICAL CHARACTERISTICS
Body length 78.7 in (200 cm); shoulder height 49.2 in (125 cm); weight 615 lb (280 kg). Is a medium-sized, long-eared, long-legged, hoofed ungulate, with a short shiny coat that is tan to gray in color with white belly and chest. The mane is erect, pale in color with a dark edge. The muzzle is white with gray between and around the nostrils and on the lips. There is a stripe down the back. The Somali subspecies, *E. africanus somaliensis*, has leg stripes and occasionally a shoulder stripe. The Nubian subspecies, *E. africanus africanus*, has a shoulder stripe, but no leg stripes.

DISTRIBUTION
Within the last 20 years its historic range has been reduced by more than 90%. They are currently found in low density in Eritrea and Ethiopia. Small populations may persist in Somalia and Sudan.

HABITAT
They live in extreme desert conditions (less than 7.8 in [less than 200 mm] of rainfall), mostly in the rift valley of the Horn of Africa. They range from below sea level to approximately 2,000 ft (700 m). The substrate can vary from sandy soil to lava rock. Like all wild equids, they need to have access to water and it is estimated that during the dry season they stay within 18.6 mi (30 km) of permanent water sources.

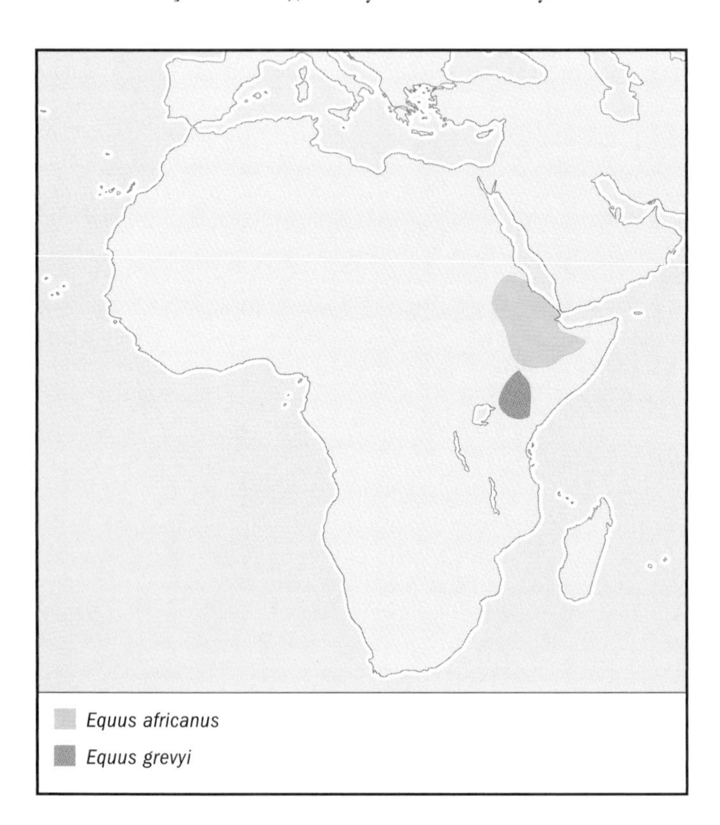

BEHAVIOR
Lives in small temporary groups that are typically composed of fewer than five individuals. The only stable groups are composed of a female and her offspring. In temporary groups, the sex- and age-group structure varies from single-sex adult groups to mixed groups of males and females of all ages. Adult males typically are solitary, but on occasion associate with other males. Adult females were usually associated with their foal and/or yearling. Some adult males are territorial and only territorial males have been observed copulating with estrous females.

FEEDING ECOLOGY AND DIET
They feed primarily on available grasses, but will also utilize browse.

REPRODUCTIVE BIOLOGY
Probably polygamous. Polyestrous, and most foals are born during the rainy season from October–February. A few females have produced foals every other year, but other females have surviving foals at longer inter-birth intervals. Females have not been observed to produce their first foal before the age of five years. Natality is strongly correlated with rainfall during the previous 12-month period. Gestation is approximately 12 months.

CONSERVATION STATUS
Critically Endangered, with a 90% reduction in range in the last 20 years. Major threats are hunting for food and medicine, potential competition for water and forage with domestic livestock, and possible interbreeding with domestic donkeys.

SIGNIFICANCE TO HUMANS
In some areas where they occur, they are used for meat and medicine. In Eritrea, they are conserved by the local Afar pastoralists as an important part of the natural environment. ◆

Grevy's zebra
Equus grevyi

TAXONOMY
Equus grevyi Oustalet, 1882, Galla country, Ethiopia.

OTHER COMMON NAMES
French: Zebra de Grevy; German: Grevyzebra.

PHYSICAL CHARACTERISTICS
Body length 118.1 in (300 cm); shoulder height 63 in (160 cm); weight 990 lb (450 kg). Is the largest wild equid; a medium-sized, long-legged, hoofed ungulate with large rounded ears and a short coat that is striped black and white. The belly is white and the mane is erect and striped. The muzzle is white with gray between and around the nostrils and on the lips. There is a dark stripe with white margins down the back.

DISTRIBUTION
Currently found in low density in Kenya from the Laikipia Plateau to the Ethiopian border. They are also found in southern Ethiopia in the Chalbi and Borana reserves. The northernmost population is found in the Alledeghi Wildlife Reserve of Ethiopia. A small population may persist in southeastern Sudan.

HABITAT
Live in arid and semiarid grasslands. They need to have access to water. Lactating females need access to water every one to two days.

BEHAVIOR
Has a territorial mating system; territorial males dominate on large resource territories, usually in the vicinity of permanent water sources. Lactating females with young foals (less than three months of age) tend to stay on these territories for daily access to water. Thus, when they come into postpartum estrus, the territorial male has better access to reproductive females. Live in small temporary groups and the only stable social group is composed of a female and her offspring. In temporary groups, the sex and age-group structure varies from single-sex adult groups to mixed groups of males and females of all ages. Adult males typically are solitary, but on occasion associate with other males.

FEEDING ECOLOGY AND DIET
Feed primarily on available grasses, but will also utilize browse during drought periods.

REPRODUCTIVE BIOLOGY
Polygamous. Polyestrous, and most foals are born after periods of good forage availability. During droughts, females will be anoestrous. Gestation is approximately 13 months, and age of puberty at three to four years.

CONSERVATION STATUS
Endangered, with a 70% reduction in population size in the last 30 years. Major threats are hunting for food, medicine, and hides, competition for water and forage with people and domestic livestock, and loss of habitat.

SIGNIFICANCE TO HUMANS
In some areas where they occur, they are used for meat and medicine. In recent times, they were one of the most important herbivores in the arid and semiarid grasslands of Kenya and Ethiopia. Due to their severe decline in numbers, they no longer play an important role in the biodiversity of these grassland ecosystems. ◆

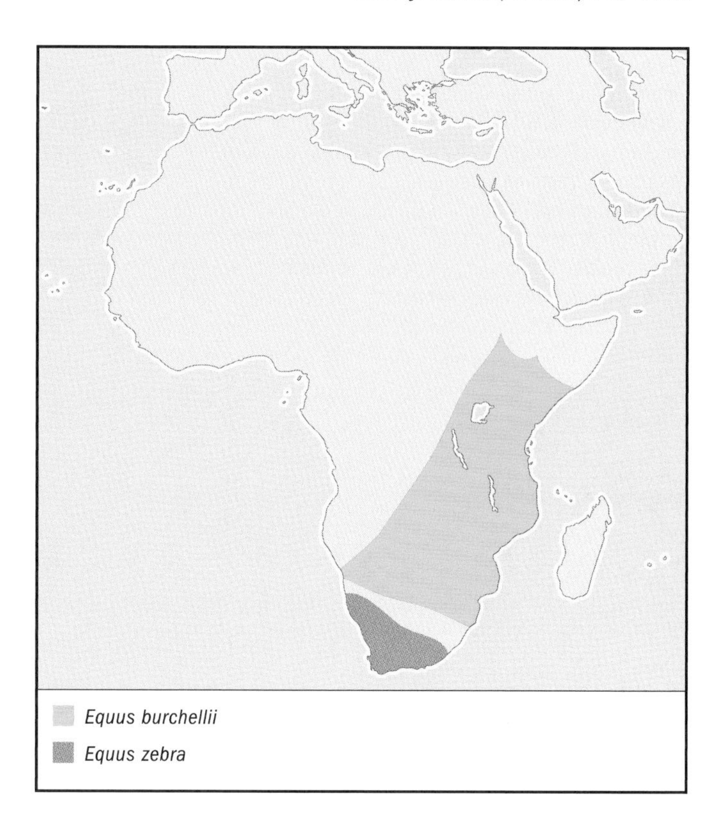

Mountain zebra
Equus zebra

TAXONOMY
Equus zebra Linnaeus, 1758, Paardeburg near Malmesbury, southwest Cape Province, South Africa.

OTHER COMMON NAMES
English: Cape mountain zebra, Hartmann's mountain zebra; French: Zebre de montagne; German: Bergezebra.

PHYSICAL CHARACTERISTICS
Body length 102 in (260 cm); shoulder height 59.1 in (150 cm); weight 750 lb (340 kg). Is a medium-sized, long-legged, hoofed ungulate, with a short coat that is striped black and white with wider stripes on the rump and a white belly. A dewlap gives a distinctive appearance. The mane is erect, and also striped. The muzzle is tan to dark gray between the nostrils and on the lips.

DISTRIBUTION
Occurs in small relict populations in Cape Province of South Africa. The largest populations occur in the Mountain Zebra National Park and the Karoo National Park. The Hartmann's mountain zebra occurs in small numbers in northwestern South Africa. The main population is in Namibia and occupies most of its historic range.

HABITAT
Live in semiarid mountainous grassland and shrubland. During the hotter months, Cape mountain zebra use the more open grasslands, and move to the ravine and wooded hills in the cold months.

BEHAVIOR
Has stable family (harem) groups composed of one male and one to five females and their offspring. Bachelor males are usually less than five years of age and travel in less stable groups. In the Mountain Zebra National Park family groups have overlapping home ranges.

FEEDING ECOLOGY AND DIET
Very selective in their grazing, e.g., more leaf than stalk. However, when forage quality decreases, they can feed on higher-fiber, more senescent grasses. They will also feed on browse when grass availability declines.

REPRODUCTIVE BIOLOGY
Polygamous. Polyestrous, and age at puberty may range 13–30 months. Males normally do not attain harem male status until they are five years old. Gestation is about 12 months.

CONSERVATION STATUS
Endangered, due to small population size. Major threats are fragmented and small populations, droughts and reduced access to water and forage, and interbreeding between the two subspecies.

SIGNIFICANCE TO HUMANS
They are a strikingly beautiful animal and are an important ecological component of their grassland and shrubland ecosystems. In Namibia, they are used for meat and the sale of skins. ◆

Plains zebra
Equus burchellii

TAXONOMY
Equus burchellii Gray, 1824, north Cape Province, South Africa.

OTHER COMMON NAMES
English: Common zebra, Burchell's zebra, painted quagga; French: Zebre de steppe; German: Steppenzebra.

PHYSICAL CHARACTERISTICS
Body length 98 in (250 cm); shoulder height 55 in (140 cm); weight 772 lb (350 kg). Is a medium-sized, long-legged, hoofed ungulate, with a short coat that is striped black and white. The stripe patterns vary with subspecies and geographic location. The mane is erect and striped.

DISTRIBUTION
Occurs in eastern sub-Saharan Africa from Sudan and Ethiopia to Namibia, Botswana, and South Africa. The largest populations occur in Kenya and Tanzania.

HABITAT
Live in mesic grasslands and are capable of migrating long distances. Their ability to utilize coarser vegetation of poorer quality means that they can thrive and survive in a range of habitats.

BEHAVIOR
Lives in stable family (harem) groups composed of a male and one to six females and their offspring. Both male and female offspring disperse from the natal group. Young males and deposed harem males live in bachelor groups that are loose aggregations. Family groups and bachelor groups will often form into larger aggregations/herds.

FEEDING ECOLOGY AND DIET
Feed primarily on grasses. They are selective in their feeding, but can feed on higher-fiber grasses when there is no choice. They must have access to water.

REPRODUCTIVE BIOLOGY
Polygamous. Polyestrous, and most foals are born during the rainy season. Age of puberty has been estimated at 15–22 months in the female. Males can reach puberty at 24 months, but rarely take over harems until they are older. Gestation is 12 months. Normally, less than 50% of the females had foals in any year. Throughout the species range, mortality rate can vary in the first year, from 19–47%. Adult mortality is lower and varies from 3–17%.

CONSERVATION STATUS
Lower Risk/Least Concern; however, one subspecies, the quagga, is already extinct and three subspecies are Data Deficient. Threats to this species include loss of habitat and overhunting.

SIGNIFICANCE TO HUMANS
In most parts of their range, they are used for meat and their hides are sold commercially. This utilization is potentially sustainable, but needs to be studied and closely monitored. The plains zebra is a symbol of the African savannahs and wildlife and is important to photographic tourism. This medium-sized herbivore plays an important role in grassland ecosystems. ◆

Asiatic wild ass
Equus hemionus

TAXONOMY
Equus hemionus Pallas, 1775, Transbaikalia, Russia. There are six subspecies currently recognized.

OTHER COMMON NAMES
English: Kulan, onager, khulan, khur, dzigettai.

PHYSICAL CHARACTERISTICS
Shoulder height 43–50 in (108–126 cm); weight 441–573 lb (200–260 kg). Is a medium-sized, long-eared, long-legged, hoofed ungulate, with a short coat that is tan to reddish in color with a white belly and chest. They also have distinctive white markings on the posterior portion of the shoulder and the anterior portion of the rump. The mane is erect, and dark in color. The muzzle is white with gray between and around the nostrils and on the lips. There is a stripe down the back that is edged in white.

DISTRIBUTION
The five remaining subspecies are found in restricted ranges in Mongolia, Kazakhstan, Uzbekistan, Turkmenistan, Iran, and India. They have been reintroduced in Israel and Saudi Arabia.

HABITAT
Live in habitats ranging from semiarid grasslands to extreme salt desert conditions.

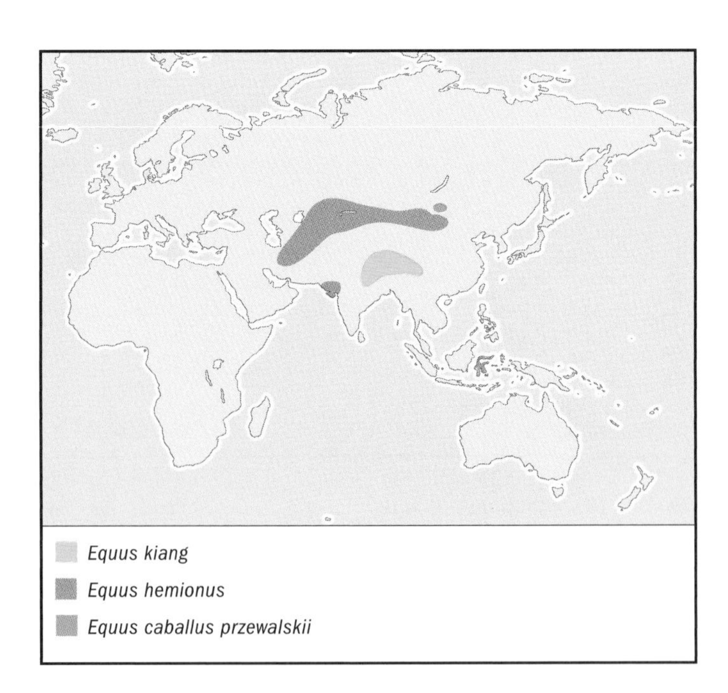

BEHAVIOR
Females live in small temporary groups that are typically composed of fewer than five individuals. The only stable groups are composed of a female and her offspring. Males may form small seasonal harem groups and hold seasonal territories near permanent water sources. Some females may stay on one territory during the entire breeding season.

FEEDING ECOLOGY AND DIET
Little is known about the feeding ecology of the Asiatic wild ass. When grass is abundant, they are primarily grazers. During drier seasons or in drier habitats, they will also feed on browse.

REPRODUCTIVE BIOLOGY
Monogamous. Polyestrous, breeding is seasonal, and gestation is 11 months. Most foals are born during a two- to three-month period. Foaling and mating occurs during the rainy season when vegetation is most abundant.

CONSERVATION STATUS
Listed as Vulnerable. Only one subspecies, the khulan, has a substantial population size. Major threats are over-hunting, loss of habitat, and competition for forage and water.

SIGNIFICANCE TO HUMANS
In some areas where they occur, they are used for meat. In all areas where they occur, they are an important component of the grassland and desert ecosystems. ◆

Kiang
Equus kiang

TAXONOMY
Equus kiang Moorcroft, 1841, Ladak, Kashmir, India.

OTHER COMMON NAMES
English: Tibetan wild ass.

PHYSICAL CHARACTERISTICS
Shoulder height 39–56 in (100–142 cm). Is a medium-sized, long-legged, hoofed ungulate; the color of the coat changes seasonally and is red-chestnut in summer and dark brown in the winter. During the winter, the length of the hair doubles to approximately 1.2 in (3 cm). Has a white belly, whitish legs, and distinctive patches of white on the neck, chest, and shoulder. The mane is erect, short, and dark in color. The muzzle is white with gray between and around the nostrils and on the lips.

DISTRIBUTION
Found in China, India, Nepal, and Pakistan.

HABITAT
Live in alpine grasslands and arid steppes. They are found at high altitudes ranging 8,800–17,400 ft (2,700–5,300 m).

BEHAVIOR
Has a social organization similar to other arid habitat equids. Information is limited, but during the breeding season males are solitary and their spacing suggests that they are territorial.

FEEDING ECOLOGY AND DIET
Feed primarily on available grasses, in particular *Stipa* spp.

REPRODUCTIVE BIOLOGY
Gestation is approximately 12 months. The peak in foaling and breeding occurs from June–September when forage is most abundant. Mating system not known.

CONSERVATION STATUS
Listed as Lower Risk/Least Concern. The total kiang population is estimated to be 60,000–70,000. Major threats are commercial hunting, loss of habitat, and competition for forage and water.

SIGNIFICANCE TO HUMANS
In some areas, they are hunted as a source of meat. In the Chang Tang Reserve in the Tibet Autonomous Region, they are one of the largest herbivores and their large herds have the potential to draw admiring tourists from throughout the world. ◆

Przewalski's horse
Equus caballus przewalskii

TAXONOMY
Equus ferus przewalskii (Groves, 1986).

OTHER COMMON NAMES
English: Przewalski's wild horse, Asiatic wild horse, Mongolian wild horse, tahki.

PHYSICAL CHARACTERISTICS
Shoulder height 49–58 in (124–146 cm); weight 772 lb (350 kg). Is a medium-sized, long-legged, hoofed ungulate. During the summer months, its coat is short and reddish-brown. During the colder winter months, hairs grow thicker and longer and provide good insulation. They have erect manes and the top of the tail has short hairs. The muzzle is white with dark gray around the nostrils and on the lips.

DISTRIBUTION
Extinct in the Wild; it has been reintroduced to Hustain Nuruu and Takhin Tal in Mongolia.

HABITAT
Last observed in the arid cold steppes of the Gobi desert. It may have also lived in the more mesic Eurasian steppes.

BEHAVIOR
Social organization and mating system is a stable family (harem) group that is composed of one male and several females and their offspring. Young males form bachelor groups. Dispersal of young males and females from their natal groups occurs 12.5–30.5 months.

FEEDING ECOLOGY AND DIET
Feed primarily on available grasses.

REPRODUCTIVE BIOLOGY
Polygamous. Reaches puberty at two to three years of age. Foaling is seasonal from April–August, with a significant peak in May. The age of first foaling is 4.5 years. In females over the age of five years, natality ranges 56–100%.

CONSERVATION STATUS
Extinct in the Wild; they are being reintroduced at several sites in Mongolia. Threats that contributed to their extinction were hunting for food, reduced access to water sources, loss of habitat and competition for water and forage with domestic livestock, and zoological capture expeditions. Currently, a major threat is loss of genetic diversity.

SIGNIFICANCE TO HUMANS
The Przewalski's horse is of great significance to the government and people of Mongolia. Concerned individuals and governments from Holland, Switzerland, and France are donating significant time, energy, and funds to make the re-introduction of the Przewalski's horse into the wild a reality. ◆

Common name / Scientific name/ Other common names	Physical characteristics	Habitat and behavior	Distribution	Diet	Conservation status
Domestic horse *Equus caballus caballus* English: Mustang; Spanish: Caballo	Stocky with short legs, short neck, massive head, long face, and powerful jaw. Eyes are set far back in skull, ears are long and erect. Stiff, erect, black mane, slender legs. Coloration is reddish brown, belly is yellowish white. Tail hairs are of graduated lengths. Head and body length 84 in (210 cm), tail length 36 in (90 cm).	Originally found in grassy deserts and plains in Western Mongolia, and reported to have lived at elevations of up to 8,000 ft (2,438 m). Not territorial. Social groups consist of only stallions or stallions and one or more mares. One offspring produced every two years.	Domestic horses occur worldwide.	Consumes mainly grass, plants and fruit. It sometimes eats bark, leaves and buds.	Not threatened
Quagga *Equus quagga*	Coloration is made up of various dark bands covering the body. However, less pronounced than in other zebras, and at times found with no stripes. Head and body length 78.7–94.5 in (200–240 cm), tail length 18.5–22.4 in (47–57 cm), weight 771.6 lb (350 kg).	Could often be found in arid to temperate grasslands, and sometimes wetter pastures. Lived in herds that contained life-long family members. Mainly diurnal, very family- and safety-oriented, always having a lookout, keeping track of members, and keeping pace with the slowest in the pack.	Formerly found in South Africa.	Frequently ate from tall grass vegetation or possibly wet pastures.	Extinct

Resources

Books

Duncan, P. *Horses and Grasses.* New York: Springer-Verlag Inc., 1991.

Moehlman, P. D., ed. *Equids: Zebras, Asses and Horses. Status Survey and Conservation Action Plan.* Gland, Switzerland and Cambridge, UK: IUCN/SSC Equid Specialist Group, IUCN, 2002.

Nowak, R. M. and J. L. Paradiso. *Walker's Mammals of the World,* 4th ed. Baltimore: John Hopkins University Press, 1983.

Wilson, D. E., and D. M. Reeder. *Mammal Species of the World: A Taxonomic and Geographic Reference,* 2nd ed. Washington DC: Smithsonian Institution Press, 1993.

Organizations

IUCN Species Survival Commission, Equid Specialist Group. Box 2031, Arusha, Tanzania. E-mail: tan.guides@habari.co.tz

Patricia D. Moehlman, PhD

Tapirs
(Tapiridae)

Class Mammalia
Order Perissodactyla
Family Tapiridae

Thumbnail description
Large herbivores with tapered, muscular bodies; stocky, medium-sized neck; short head with flat sides, short proboscis, relatively small, deep-set eyes and round, mobile ears; short legs; tiny tail; feet with spade-like hoofs: four on front and three on rear; low-crowned teeth

Size
5.9–8.2 ft (1.8–2.5 m); 330–660 lb (150–300 kg)

Number of genera, species
1 genus; 4 species

Habitat
Rainforest, bushland, and paramo; lowlands and mountains

Conservation status
Endangered: 3; Vulnerable: 1

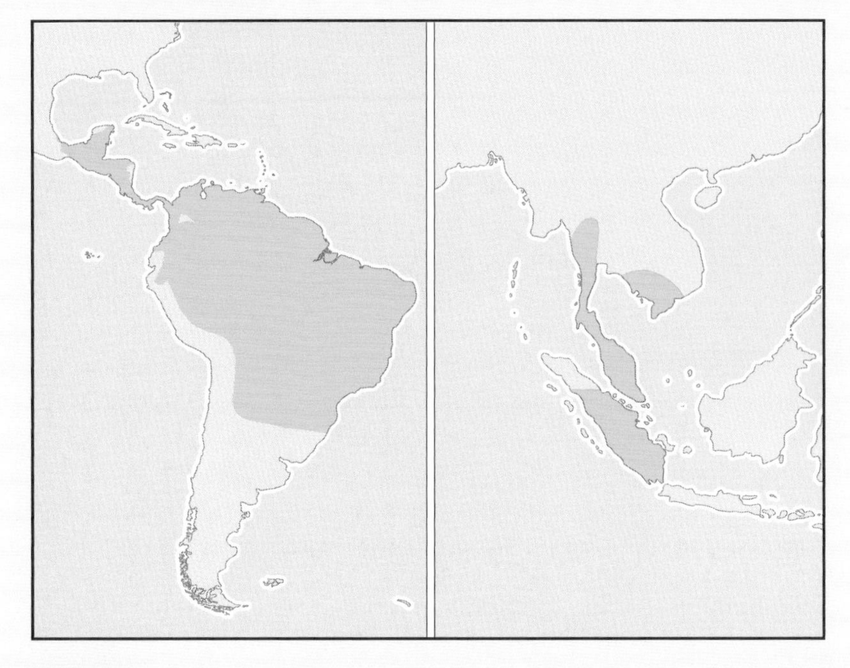

Distribution
Primarily tropical; southern Mexico, Central and South America, and Southeast Asia

Evolution and systematics

During the Tertiary period, the Ceratomorpha were a large group with many divergent forms. Today there are only a few survivors in the Tapiridae and Rhinocerotidae families. The various prehistorical tapir-like forms cannot be encompassed by one family alone and are considered members of the superfamily Tapiroidea.

Today it is known that the tapir, the rhinoceros, and the horse are related in the order Perissodactyla, whose members became the dominant herbivores in North America by the end of the Eocene. Among these surviving families, tapirs most adhere to the tropical moist forests and are considered the least altered from the root ancestors of the order. In relation to other Perissodactyls, their teeth are less specialized, their limbs have more digits and hooves, and they are shorter and more plump.

Tapirs are sometimes called "living fossils," survivors from past eras, as widely separated areas of past and present distribution indicate that they originated from ancient times. In the past, the Tapiridae and the Tapiroidea were much more widespread, with a more impressive variety of species. Numerous fossil discoveries trace tapiroid history and prove the existence of several long extinct families. For example, the Isectolophidae, belonging to the geologically oldest tapiroids, hails from the Eocene of North America and Asia. True Tapiridae in North America also date from the early Eocene. Together with the Lophiodontidae of Europe and the Helaletidae of North America and Asia, these tapiroids lived approximately 50 million years ago (mya) and resembled the ancient ancestors of all ungulates within the order Condylarthra, which date from the Cretaceous and were common during the Paleocene. Condylarths made a transition between insectivorous and herbivorous lifestyles.

Anatomical modifications encountered in tapir phylogeny affect mainly the skull. The set of teeth and the limbs have been transformed only very insignificantly. Early tapiroids lacked the movable extension of the upper lip and the nose that forms a true proboscis, like an elephant's trunk. This proboscis characterizes the true Tapiridae and is considered by paleontologists to be key to their evolutionary success. All animals with a proboscis, including elephants, have an especially large nasal cavity that reaches far to the back of their skulls. Among Perissodactyls, this nasal cavity occurs only in genuine Tapiridae and is first found in the family's oldest forms in the genus *Protapirus*, which lived in Europe during the Oligocene, approximately 40–25 mya. Similar forms in similar timeframes, probably of the same genus, have been found in North America. These genuine Tapiridae possibly derived from the Helaletidae during the early Eocene of North America. Fossils of the giant *Megatapirus* from the Sichuan province of China represent the only extinct tapir from the Old World Pleistocene. It developed temporarily in Asia at the same time that tapirs became extinct in Europe. It dwarfed in size today's tapirs, including the Malayan; its large size may have been an adaptation against large predators and cold. Tapirs disappeared from all but southernmost North America about 11,000 years ago,

A four-month-old female mountain tapir (*Tapirus pinchaque*) displays the black and white stripes and spots used for camouflage. (Photo by Craig C. Downer. Reproduced by permission.)

coinciding with the extinction of many other large animal species and the arrival of ever-larger numbers of human hunters over a period of 6,000 years, as indicated by the number of hearths discovered.

In 1996, a DNA analysis of the four extant tapir species revealed three diverging lines. The Malayan tapir lineage diverged from the neotropical tapirs about 21–25 mya, while that of the Central American tapir diverged from the common ancestor of the two South American tapirs about 19–20 mya. A common ancestor of the lowland and the mountain tapirs probably immigrated to South America over the Isthmus of Panama upon its connection about 3 mya. The separation of these two species probably relates to the geologically abrupt rise of the Andes.

Physical characteristics

Slightly higher at the rump than at the shoulder, tapirs possess tapered muscular bodies that can easily wedge their way through jungle tangles. Possessing a thicker neck at maturity, the males in most cases are somewhat smaller than females. On a stocky medium-sized neck, the head is relatively short, with rather flat sides, and is slightly arched upward, with a short, movable trunk. The latter is actually an extendible proboscis, or muscular hydrostat, used in smelling, selective feeding, etc. The eyes are proportionately small. The ears are round and mobile. The rump is bulky and slightly flat at the sides. Relatively more slender than in rhinos, the legs are powerful, short, and graceful.

Concerning the forelegs, the first digit is missing. In all four feet, the third digit is extremely strong, carrying the main weight, as is typical in all Perissodactyls. The second and fourth digits gradate to a lesser size; and the fifth digit is very short, touching down only on soft surfaces such as mud or sand. Though similar to the forefeet, the hind feet differ in being three-toed, with the third and central digit the strongest and longest. The spade-shaped hooves, sharing calloused, yet quite sensitive, sole balls, are relatively stronger than the homologues in rhinoceros. There is a bald callosity at the forearm that corresponds to a horse's "chestnut" (a vestige of the first inside digit). The teeth are those of a more generalized herbivore and are not specialized for the grazing of grass as is the case with the hypsodont dentition of equids.

Fur is short in three of the extant, mainly lowland-dwelling species, but long and woolly in the mountain tapir. Coat color varies from reddish brown to whitish gray to coal black to black-and-white two-tone. In the Central American and lowland tapirs, a short mane protects vital nervous centers. Newborn tapirs always have horizontal stripes and dots of a yellow-white color for about one year.

Distribution

In South America, tapirs occur in the Amazon and Orinoco basins, northern Andes, northern Caribbean coast, also northern Argentina, the northwestern Pacific coast, and the western slope of the Andes. In Central America, they occur from southern Mexico to Panama. In Southeast Asia, they are found in parts of Myanmar, Thailand, Cambodia, Vietnam, and Sumatra.

Habitat

All but the mountain tapir (*Tapirus pinchaque*) inhabit lowland rainforest, and lower to mid- and occasionally high-montane moist forests. Mid to high cloud forests and treeless paramo are inhabited by mountain tapirs. Lowland tapirs occupy drier scrub, woodlands, and grasslands at lower elevations in southern South America. All tapirs swim in rivers and lakes and require mineral procurement sites. Females often require secluded forests during parturition and raising of young.

A lowland tapir (*Tapirus terrestris*) baby nursing. (Photo by Erwin and Peggy Bauer. Bruce Coleman, Inc. Reproduced by permission.)

The lowland tapir (*Tapirus terrestris*) is a good swimmer. (Photo by Tom Brakefield. Bruce Coleman, Inc. Reproduced by permission.)

Behavior

At first glance, a tapir's movements do not seem so similar to those of its relatives, the rhinoceros and the horse. In a slow walk, it usually keeps its head lowered, though it may bob it up and down. In a trot, it lifts its head and moves its legs in an elastic manner. The amazingly fast gallop, during which all four feet may leave the ground, is seen in flight, playing, or when it is extremely excited. The first instinct when approached by an enemy is to remain still and to rely on its camouflage to avoid detection. Some hunters consider tapirs to be deaf for this reason, but this is far from being the case, given all tapirs' acute sense of hearing. Tapirs can climb remarkably well, regardless of their bulk. Even quite steep slopes may not present obstacles. Individuals have been observed to climb slopes of 70°. Tapirs climb or jump vertical fences or walls 9.8 ft (3 m) or more, rising on their hind legs, or springing down then leaping up, and in both cases grappling with their forefeet. In zoos, they are able to squeeze themselves through unbelievably narrow gaps and between bars, and to slip out under lower bars with their backs arched. This nimbleness is of advantage in the wild when they wander through jungles of bamboo and reed, tortuous shrubbery, trees, and roots, often thus eluding their enemies.

Getting up and lying down is similar in tapirs and rhinoceros. These two movements are very instructive phylogenetically. Tapirs do not support themselves on their wrists, as do horses, but in the process of lying down they briefly assume a sitting position. Female tapirs often give birth as well as suckle young while laying down on their sides.

Tapirs are very shy and retiring animals who rely greatly on concealment for safety and prefer to stay in the vicinity of water, where bathing helps to regulate their body temperature. They are excellent swimmers and cross even wide streams without great effort, including the Amazon River. In order to feed on aquatic plants or to escape when pursued, they can dive quite well, remaining several minutes or more under water without breathing. In the presence of an enemy, a tapir may clandestinely take air with the tip of its proboscis barely emerged above water and where sealable nostrils are located. Similar to the hippopotamus, tapirs can walk on the bottom of streams and lakes for brief periods. During the hottest time of the day, they frequently extend their daily baths. Not only does this cool them off, it also protects them from biting insects.

In order to remove dandruff, hair, or insects, tapirs scrub themselves on objects, or scratch their chests and front legs with their hind feet. Coatis have also been observed gleaning ticks from the Central American tapir (*Tapirus bairdii*). While lying down and resting, they sometimes lick their front legs. They also protect the surface of their body against insect bites by rolling in the mud; the dry mud remains as a thin protective film on the skin. This may also serve in predator avoidance by disguising their scent, as has been observed in the mountain tapir in Colombia's Ucumari Regional Park.

Being genuine forest animals, tapirs have an excellent sense of smell. Their eyesight, on the other hand, is less acute, but this shortsighted vision still serves in orientation, feeding, etc. The trunk, whose tip is equipped with hair-like, tactile bristles, plays an important role in exploring a new area. The trunk is also used during courtship.

All species of tapir share similar habits. Most are predominantly nocturnal in activity, though favoring the crepuscular, dawn and dusk periods, as do many animals. However, the mountain tapir is equally active both day and night. Tapirs are relatively unsocial, cautious creatures of the forest that usually avoid open territory, especially during daylight, and often depend on water for escape when threatened. They frequently use lakes, rivers, and streams to travel to separate areas within their home range, which may vary from a few to over a dozen square miles (kilometers). All tapirs are true amphibians, being excellent swimmers and waders, as well as walkers, runners, and climbers.

In more densely settled areas, lowland tapirs (*Tapirus terrestris*) become strictly nocturnal in activity. Under these circumstances, one hardly ever sees them unless they are routed out of their hiding places by dogs.

Well-established tapirs remain fairly close to one central locality and adhere to regular paths. These may eventually become tunnels in the vegetation. In a frightened state, a tapir generally clings to its path. When pursued by a predator or

A young Malayan tapir (*Tapirus indicus*) shows its changing, maturing coat. (Photo by Terry Whittaker/Photo Researchers, Inc. Reproduced by permission.)

A Malayan tapir (*Tapirus indicus*) wading in the river in Myanmar. This species of tapir is endangered. (Photo by E. & P. Bauer. Bruce Coleman, Inc. Reproduced by permission.)

suddenly confronted by hunters, it can dash with considerable speed through the forest or scrub with its head lowered. Its impetus may create a new path through the densest vegetation. In some cases, it can, thus, brush off a predator clinging to its back. Tapirs also thus manage to extract leeches. A tapir may also loosen the deadly grip of a predator by submerging under water.

Tapir paths often connect animal resting areas in dense forest with river banks or other clearings where more abundant food plants may be obtained. Major trails may also indicate migratory routes between higher, more open elevations, which are occupied during the dry season, and more densely forested lower elevations sites occupied during the wet season.

Tapir territories are frequently marked by urinated and rubbed signposts and dung piles. These discourage fellow males, but males welcome females more readily into their home ranges. A female's urine conveys general fitness, serving both to warn competitive males and to advertise to reproductive females. Social interaction occurs at salt licks and stream banks and is more common during the dry season and at full moons. Here, contesting for mates occurs between males and may entail dramatic fights, including rearing up and gnashing. After pairing, a couple engages in rambunctious courtship displays. These involve a characteristic head-to-tail juxtaposition, nipping at the feet, and a carousel-like whirling. Grunting and squealing sounds may be emitted during this process. A pregnant female has a gestation of around 13 months and secludes herself in a safe and thickly vegetated area before and after parturition.

Although considered unsociable loners, tapirs are harmoniously in tune with the other species. Except during the mating season and for mother with young, an adult tapir in the wild is usually alone. However, for this very reason, it may relate more to other species that form its ecological community. It is rare to see more than three animals together. Maintaining them in pairs or even in family groups the year round, as is usual in zoological gardens, is quite atypical. Yet, even in relatively small enclosures, tapirs get along well with one another. Serious squabbles occur rarely. There also seems to be no rank order within the group.

Orienting themselves more by smell than visually, tapirs' visible expressions are relatively few. Also with their ears and mouths, they transmit signals. The ears can be moved forward or backward to detect sounds and as an expression of mood. The white-fringed ears and lips of the mountain tapir provide a gestalt pattern clearly recognizable by conspecifics, either in the dark forest shadows or at night. When two tapirs meet, their white-rimmed ears point forward. When sniffed at or enraged, their ears point backward. A type of facial expression is seen in the dramatic *flehmen* (lip-curling), often after they have sniffed or licked the urine or feces of a conspecific. The tapir then rolls its tongue inside its open mouth and around the trunk several times, thus activating the Jacobson's organ located in the upper palate.

Only occasionally do tapirs demonstrate a threatening gesture. Then their rarely used weapons are exhibited as they pull their lips apart to expose their teeth. Though these teeth are not too frightening in appearance, if a threatening tapir does attack with them, it can seriously injure an opponent. Instances are known where seemingly harmless tapirs suddenly went wild and attacked everything within their reach with their teeth. They occasionally must use these against their natural enemies, the large cats, etc., but they usually manage to escape these through flight. Shrill whistles may also issue from adults as a means of warning a mate or young of an intruder or as issued by a juvenile as a cry for help. Clicking sounds are also known to communicate location among tapirs.

Feeding ecology and diet

Tapirs feed to a large degree on leaves, fresh sprouts, and small branches, plucking and biting them from bushes and low trees with their proboscis and teeth. In addition, they eat many varieties of fruits, fallen on the ground or on a plant, grasses, and aquatic plants. They feed in a way that is widely dispersed and low impacting to the plant community as a whole. On mountains, they are observed to zigzag across long slopes, only selecting a little from each plant they visit. In folivorous tapirs, the proboscis is especially well developed for harvesting leaves. It is easily moved in all directions and may be stretched out or contracted.

When tapirs are not able to reach a favorite food, they will rise on their hind legs while planting their front feet firmly against natural objects. Thus, mountain tapirs feed from high leaves and fruits of trees and even push these over. Malayan tapirs snap off the trunks of saplings up to 4.6 ft (1.4 m) in height in order to procure favored tender leaves and sprouts. By lowering trees, bushes, and saplings, this tapir makes food accessible for the sambar, the barking deer, and two species of chevrotains. The mountain tapir similarly benefits sympatric herbivores, including white-tailed and red brocket deer and the diminutive pudu. There have been reports that tapirs there occasionally eat fish. Stranded fish are reported to be eaten by lowland tapirs, according to people who live in the Amazon.

Shared with other Perissodactyls, the post-gastric, or caecal, digestive system of tapirs allows it to feed on coarse leaves and other vegetation without expending as much metabolic energy as would a ruminant digester. This system also allows it to pass many of the seeds it consumes intact and capable of germination. This relates to the important role mobile tapirs play as seed dispersers; their decomposing feces give a nutritional advantage to these seeds and help to build humus-rich soils. Tapirs' trailing and foraging patterns also serve to open up clear areas for new forest to grow.

Reproductive biology

Tapirs tend to be monogamous during any given breeding season, but may change mating partners over the course of their individual lifetimes. Some tapir populations pair up and breed during the dry season. In zoological gardens, female tapirs become pubescent at the age of three to four years. The gestation period is around 13 months; only one young is usually born, though twins do rarely appear. Generally, female tapirs seem to come into heat every 50–80 days. Heat normally lasts two days, but can last much longer.

Before copulation, male and female are highly excited, frequently uttering short, wheezing sounds or shrill, piercing whistles, and spraying large quantities of urine. Usually a female ready for mating will be pursued by the male, although it has been noted that a Malayan female tapir may, in the beginning, pursue the male. The animals walk toward, and stand parallel to, each other, but face in opposite directions so as to smell each other's anal region. From this position, frequent circling may develop. The male tries to push with his head under the female's underside or to snap at her hind legs. The female, in turn, snaps at the male's hind legs. Simultaneously, each animal tries to get its hind legs out of the other's reach, while the partner follows with the head. This whirling behavior is exhibited only during the early phases of courtship and, phylogenetically, may have been derived from primitive fighting positions, which still occur in horses. After copulation, which may be repeated several times in short succession, the female may become aggressive and ward the male off by biting.

Central American tapir (*Tapirus bairdii*) adult and baby in Central America. (Photo by Gail M. Shumwav. Bruce Coleman, Inc. Reproduced by permission.)

Conservation status

All four tapir species are listed as Endangered or Vulnerable by the IUCN. Throughout their occupied habitats in South and Central America and Southeast Asia, forested land is being steadily cleared and cultivated, and the number of tapirs is decreasing steadily. People kill them for their skin, meat, and for medicinal uses. Indigenous South Americans use poisoned arrows and occasionally chase them with dogs, a technique commonly employed by mestizo (a person of mixed blood) and white hunters. When pursued, the animals often plunge into the water where they can be killed from a boat with spears and knives. The lowland tapir is very susceptible

An adult mountain tapir (*Tapirus pinchaque*) rests by a pond at the Los Angeles Zoo, USA. All tapirs require moist habitats with plenty of water in the form of streams or lakes. (Photo by Craig C. Downer. Reproduced by permission.)

Mountain tapirs (*Tapirus pinchaque*) mating. (Photo by R. Mittermeier. Bruce Coleman, Inc. Reproduced by permission.)

Significance to humans

Already known to the Mayans for centuries, the Central American tapir served as their model for a human figure with a trunk, as is found on several of their temples. One of their distinctive glyphs is called the "laughing nose." The first explorers to America, Columbus, Pinzon, and Cabral, probably did not encounter tapirs. However, toward the end of the year 1500, an animal was described as, "the size of an ox, of the color of cattle, which had an elephant's trunk and hooves like a horse, but which was neither cattle, nor elephant, nor horse." Although initially oblivious as to which order it belonged, later explorers soon became familiar with the lowland tapir, whose name "tapir" derives from the word *tapyra* of the Brazilian Tupi language. Although the Malayan tapir had already been mentioned in old Chinese text, the Europeans, strangely enough, learned about this species much later than the lowland tapir. The mountain tapir was discovered to European culture in 1829 by French naturalist, X. Roulin, who first observed it in the high, frigid Páramo de Sumapaz, south of Bogota, Colombia. Some indigenous South American cultures (e.g., Calima) venerated tapirs, and they figured in their religious conceptions. In the northwestern part of the continent, indigenous people believed that tapirs gave the inhabitants of the world a needed impetus to dance to creation's music. In this connection, the steppe-dwelling Andean tapir could well be regarded as a master of balance.

Though the mountain tapir has a thinner skin than the other three extant species (probably due to the fact that it sprouts thick fur), all four extant tapirs possess a rather coarse, leathery skin. They are all frequently hunted in their native countries; people tan their skins and cut them into long straps

A lowland tapir (*Tapirus terrestris*) in the Brazilian rainforest. (Photo by Erwin and Peggy Bauer. Bruce Coleman, Inc. Reproduced by permission.)

to over-hunting, which is increasing alarmingly as civilization encroaches upon its remaining rainforest redoubts. Many of the original, shifting indigenous populations did not pose a real threat to tapir survival, and some even prohibited the killing of tapirs for religious reasons, as did Muslims in Southeast Asia. Often Caucasian or mestizo settlers kill these harmless vegetarians just for sport. In many villages, one can find young pet tapirs whose mothers have been killed. They become as tame as dogs, can be petted, and even let children ride on their backs. However, as adults, they frequently turn wild during the mating season and can do great damage in effecting their escape from captivity. For this and other reasons, tapirs are not well suited for domestication. They are best preserved in those sufficiently vast wilderness stretches that will support truly viable populations of 1,000 or more inter-breeding adults. A viable population area of 725,000 acres (293,500 ha) has been recommended for the mountain tapir. Key to conserving all species of tapirs is fervently dedicated public education, coupled with the implementation of ecologically compatible lifestyles in place of ones inherently destructive of tropical habitats. In 1997, an action plan for each of the four surviving tapirs was published by the World Conservation Union, Species Survival Commission, Tapir Specialist Group.

for reins, whips, sandals, etc. People eat their flesh and use their hooves, snouts, and other parts as folk remedies. Placing a monetary value on their dead bodies as well as for captive tapirs for zoos greatly increases their chances of extinction, and mounting evidence of international trade in tapirs, such as has recently come to light in northern Peru, is a call to conservation action to organizations such as the IUCN and CITES.

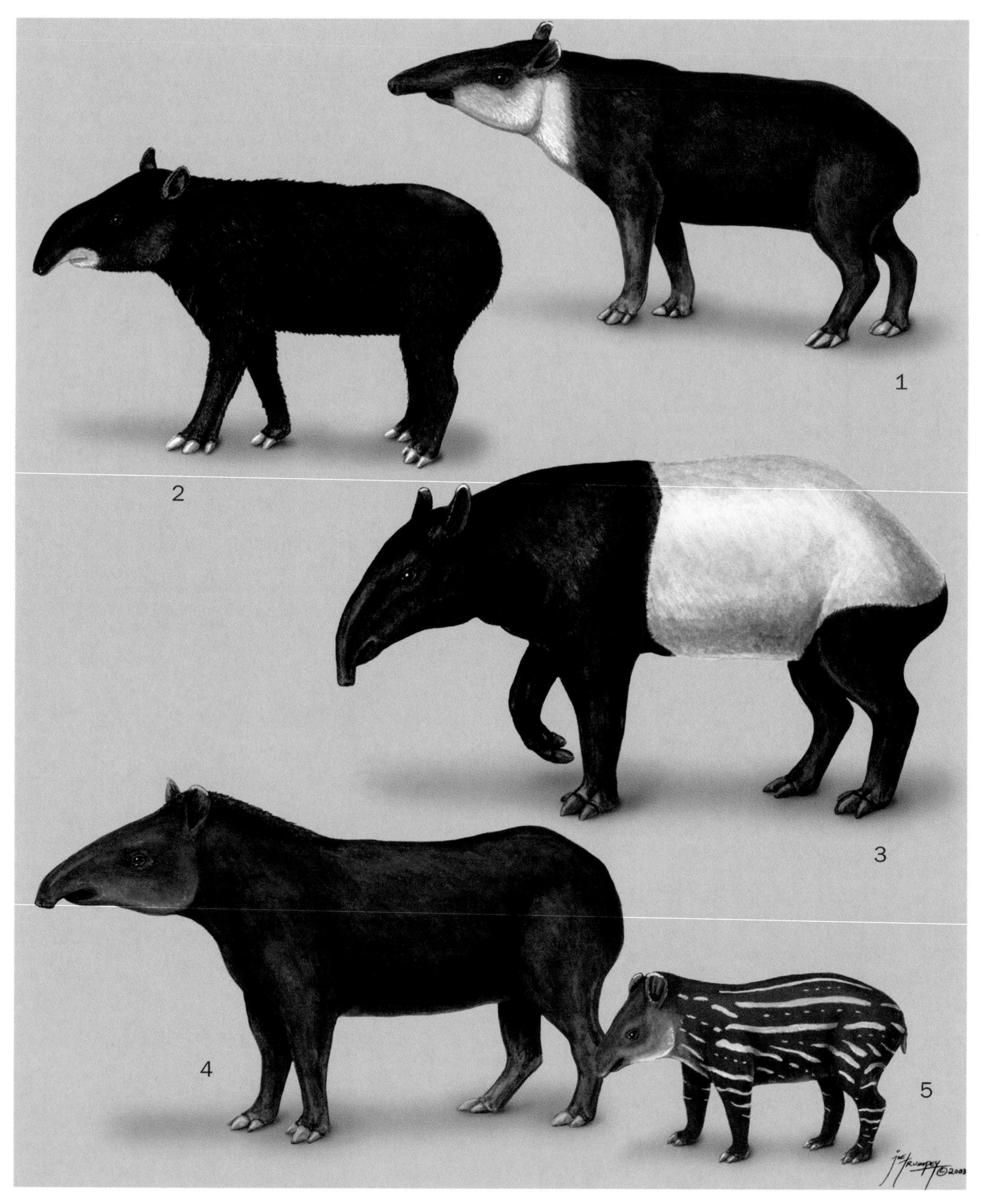

1. Central American tapir (*Tapirus bairdii*); 2. Mountain tapir (*Tapirus pinchaque*); 3. Malayan tapir (*Tapirus indicus*); 4. Lowland tapir (*Tapirus terrestris*). 5. Lowland tapir baby. (Illustration by Joseph E. Trumpey)

Species accounts

Lowland tapir
Tapirus terrestris

TAXONOMY
Tapirus terrestris (Linnaeus, 1758), Brazil. Six subspecies.

OTHER COMMON NAMES
English: South American tapir, Amazonian tapir, Brazilian tapir, bush cow; French: Tapir d'Amerique du Sud, tapir commun; German: Flachlandtapir; Spanish: Tapir de tierra baja, danta, gran bestia.

PHYSICAL CHARACTERISTICS
Head and body length: 6–7 ft (1.8–2.2 m); tail length: 2–4 in (5–10 cm); shoulder height: 2.5–3.5 ft (77–110 cm); weight: 396–660 lb (180–300 kg). Tan to black or reddish coloring. Undersides and legs typically dark; cheeks, throat, and ear edges lighter in color. Black mane from forehead to midback. Tall crest on head. Young are dark brown and have white spots and stripes.

DISTRIBUTION
East of Andes from northern Venezuela and Colombia to southern Brazil and northern Argentina. Also crosses eastern Andes of Colombia and Venezuela to inhabit Sierra Nevada de Santa Marta region and has been reported from the Colombian side of the Darien near Panama.

HABITAT
Lowland rainforest and montane cloud forest from sea level to about 4,265–4,920 ft (1,300-1,500 m) in Ecuador; 5,575 ft (1,700 m) or more in other locations.

BEHAVIOR
Congregate around mineral seeps where they interact socially. They use trails to link major habitat components and are generally solitary. They emit loud squealing sounds when alarmed, and often stand still to avoid detection. They are excellent swimmers.

FEEDING ECOLOGY AND DIET
Eats many species of trees, bushes, and herbs, successfully disperses the seeds of many of these (particularly documented in palms) and, in general, opens up clearings important for forest renewal. Also eats aquatic plants and may walk on river bottoms, which produces beneficial mutualistic effects.

REPRODUCTIVE BIOLOGY
Gestation of 385–412 days. Generally single birth, occasionally twins. In captivity may possibly reproduce every 15 months. Estrous lasts two or three days. Monogamous through breeding season.

CONSERVATION STATUS
Upgraded to Vulnerable from Extinction. Dire threats include wholesale forest destruction and indiscriminate hunting concomitant with rampant colonization and petroleum and mineral exploitation. Studies indicate that a lowland tapir population begins to decline after 20% of its numbers has been eliminated through hunting, and that mining pollution and obliteration of traditional salt licks have a devastating effect upon them.

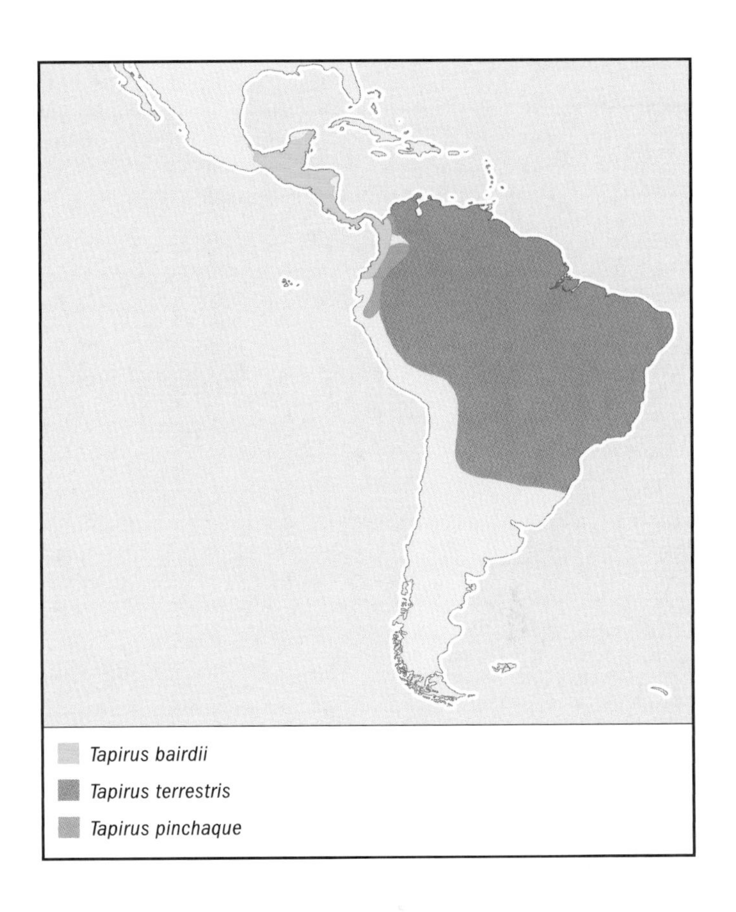

SIGNIFICANCE TO HUMANS
Figures in native religions as an animal with magical powers. Important for maintenance of biodiversity as a seed disperser and opener of clearings. Hunted for its meat, leather, and medicinal parts. Ecotourist attraction. ◆

Central American tapir
Tapirus bairdii

TAXONOMY
Tapirus bairdii (Gill, 1865), Panama.

OTHER COMMON NAMES
English: Baird's tapir, mountain cow; French: Tapir de Baird, tapir d'Amerique Central; German: Baird-Tapir, Mittel amerikanischer Tapir; Spanish: Tapir de Baird, tapir de America Central, danta, anteburro, macho de monte.

PHYSICAL CHARACTERISTICS
Head and body length: 6.5–6.7 ft (198–202 cm); tail length: 3–5 in (7–13 cm); shoulder height: about 4 ft (1.22 m); weight: 330–660 lb (150–300 kg). Dark brown coloring. Young are reddish-brown with white streaks and spots. Throat and cheeks light grayish-yellow; dark spot situated below and behind the eye. Edges of the ears are white. No head crest.

DISTRIBUTION
Yucatan Peninsula and Chiapas in southern Mexico; through all nations of Central America, except El Salvador where it is exterminated. South along Pacific coastal forests and swamps at least to mid elevations of western Andes, in Colombia to northern Ecuador.

HABITAT
Mainly inhabits lowland forests and swamps, but may ascend to mountainous mid elevations, and even has been reported above the tree line. It occupies a wide variety of habitats from sea level to over 6,560 ft (2,000 m) with reports as high as 11,800 ft (3,600 m). These include lowland forest, deciduous forest, montane cloud forest, swamp, marsh, mangrove, and alpine grassland. Requires plentiful water.

BEHAVIOR
This largely solitary, nocturnal tapir evinces a beautiful ecological harmony by rotating usage of portions of its annual home range. It can walk under water and has been observed completely submerged for 15 minutes. A high whistle is used to locate fellow tapirs when far apart.

FEEDING ECOLOGY AND DIET
Prefers browsing in flood plains, palm swamps, tree-fall gaps, and secondary lowland forests. Browses sparingly on selected plants, moving in zig-zag fashion. It relies more on fallen fruit during the dry season. Important seed disperser for native trees, etc.

REPRODUCTIVE BIOLOGY
Gestation period: 390–410 days. One, or rarely two, offspring. Sexual maturity occurs around two years of age. Inter-birth intervals are 18 months or more in captive females. Neonates may be hidden for brief periods while the mother is away feeding between nursing bouts. When a young reaches 10 days of age, it closely follows its mother. Association of mother with its young lasts up to two years. Monogamous through breeding season.

CONSERVATION STATUS
Upgraded to Endangered status. Most countries where it occurs list it as Endangered. Primarily habitat destruction threatens the survival of this species, but localized hunting is also a very serious menace, and a single hunter can have a devastating effect upon a population. Studies indicate that if forest settlements were spaced so as to allow more than 950 ft (290 m) of intact forest between farms, any tapir population could avoid becoming overly fragmented and entering into decline.

SIGNIFICANCE TO HUMANS
Figures in Mayan and other tribal religious beliefs. Traditionally, its meat and leather have been used, as well as parts, e.g., hoofs and snout, for folk medicine. This species aids economically important, native trees to reproduce by helping to disperse their seeds. Is also a major attraction for ecotourists. ◆

Mountain tapir
Tapirus pinchaque

TAXONOMY
Tapirus pinchaque (Roulin, 1829), Páramo de Sumapaz, Colombia.

OTHER COMMON NAMES
English: Andean tapir, woolly tapir, Roulin's tapir; French: Tapir pinchaque, tapir de Roulin, tapir des Andes, le pinchaque; German: Bergtapir, Wolltapir; Spanish: Danta de montaña, danta de Páramo, danta lanuda, danta cordillerana, danta negra, tapir de altura, gran bestia, bestia negra, pinchaque.

PHYSICAL CHARACTERISTICS
Head and body length: 6 ft (1.8 m); tail length: 2–4 in (5–10 cm); shoulder height: 30–32 in (75–80 cm); weight: 330–550 lb (150–250 kg). Coal black to dark reddish-brown coloring. Cheeks may be lighter. Young have white stripes and spots. Lips and edges of ears are white.

DISTRIBUTION
Lives from the northern Andes of Peru, including the Cordillera de Lagunillas and del Condor, through the eastern Andes of Ecuador and parts of western Andes in the north; thence, further north into Colombia in fragmented populations from the eastern, central, and, perhaps, the western Andes, to mid Colombia. Formerly occurred further north along both the central and eastern Andes of Colombia and was much more common in Ecuador and Peru as well as western Venezuela, particularly in the area of El Tama National Park near San Cristobal. Particularly the eastern flank of the eastern Andes is this endangered species' most important redoubt. Inhabit and are integral to the highland watersheds serving much of Amazonia and are associated with a global hot spot of biodiversity along the eastern Andes.

HABITAT
Montane cloud forest and paramo, and scrub ecotone from 4,920 to 15,420 ft (1,500–4,700 m) elevation, more common from 6,560 to 14,735 ft (2,000–4,500 m) elevation.

BEHAVIOR
More difficult to keep in captivity than other tapirs. Its tracks may be found up to the snow line. Adult home ranges of 1,360–2,175 acres (550–880 ha), divided between cloud forest and treeless paramo. The mountain tapir is especially active during crepuscular hours and is active half of the time during the night and half of the time during daylight hours. Shows increased nocturnal activity during the full moon.

FEEDING ECOLOGY AND DIET
Eats many different trees, shrubs, herbs, fern fronds, and horsetails, and also seeks out nitrogen-fixers, e.g., lupins and (*Gunnera* spp.). A highly significant correlation between frequency of seed germination from feces and dietary frequency indicates a mutualistic coevolution of the mountain tapir with the northern Andean flora.

REPRODUCTIVE BIOLOGY
Gestation is 390–400 days. One, rarely two, offspring, weighing 9–13 lb (4–6 kg) at birth. Young may stay with mother between one and two years. Monogamous through breeding season.

CONSERVATION STATUS
Remains Endangered with extinction. Estimated 2,500 mountain tapirs for the nation of Colombia. In captivity, they are very susceptible to disease, displaying little ability to adjust to lowlands. Global warming imposes a grave threat to them as cold-adapted, montane ecosystems are increasingly displaced. The mountain tapir is the most endangered of all tapir species due to its small numbers, its restricted global distribution, and the human onslaught against its remaining habitats and populations. It is the inhospitable cold and rain and the steepness of terrain that most preserve the mountain tapirs.

SIGNIFICANCE TO HUMANS
Hunted for meat, furry hide, and leather, and for parts such as hoofs and snout, which are used in folk medicine. Significant in Amerindian religious concepts. Important seed disperser for economically valuable trees and bushes and for maintaining bio-diversity and well-functioning of Andean ecosystems vital as watersheds. ◆

Malayan tapir
Tapirus indicus

SUBFAMILY
Ceratomorpha

TAXONOMY
Tapirus indicus Desmarest, 1819, Malay Peninsula, Malaysia.

OTHER COMMON NAMES
English: Asiatic tapir, Asian tapir, saddleback tapir, Indian tapir; French: Tapir des Indes; German: Schabrackentapir; Spanish: Tapir de Malasia.

PHYSICAL CHARACTERISTICS
Head and body length: 6–10 ft (185–250 cm); tail length: 2–4 in (5–10 cm); shoulder height: 35–41 in (90–105 cm); weight: 550–825 lb (250–375 kg). This sole Old World tapir species is considerably larger than the American tapirs. The short, smooth, and slick coat is strikingly colored with the rear half above the legs being white, while the rest of the coat is black, except for white ear fringes.

DISTRIBUTION
Its range has been greatly reduced in Myanmar (Burma), Thailand, Cambodia, and Sumatra. Current populations suffer from extreme fragmentation and are found in southern Vietnam, southern Cambodia, parts of southern Myanmar, the Tak Province in Thailand, and through all the states of the Malay Peninsula to Sumatra south of the Toba highlands, a faunal boundary for many species on the island. In the Pliocene, tapirs very similar to the Malayan lived in India and Myanmar.

HABITAT
Forests of lowland, swamp, montane, and hill types from sea level to about 6,560 ft (2,000 m) elevation. In parts of Indonesia, they dwell in the lowlands during the dry season and occupy mountains during the wet season. They need streams, lakes, swamps, and other habitat types with abundant water. Virgin swamps and lowland forests with well-drained soils support highest population levels.

BEHAVIOR
Nocturnally active. Rests in thick vegetation during the day. Excellent swimmer, as well as mountain climber. Produces shrill whistles when alarmed, or to placate young. Seeks salt licks avidly. Follows paths of its own creation, often with head down, sniffing. The male marks his path with urine, indicating possible territoriality. Average straight-line distance traveled by a male in a day was 0.20 mi (0.32 km).

FEEDING ECOLOGY AND DIET
A selective browser, it favors tender leaves and branchlets of certain trees, bushes, and succulents. It eats club moss, *Selaginella willdenonii*, and a variety of fruits. Disperses its feeding over a wide area. In a Thailand study, it preferred 39 plant species of which 86.5% were consumed as leaves, 8.1% as fruit, and 5.4% as twigs with leaves. Evidence exists for considerable successful seed dispersal through feces.

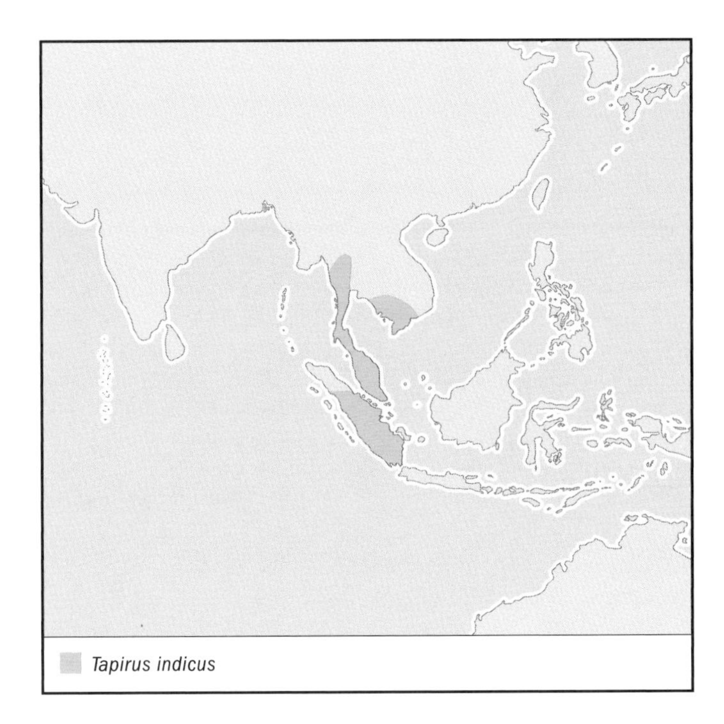
Tapirus indicus

REPRODUCTIVE BIOLOGY
Gestation: 390–407 days. Earliest age for mating was three years for males and average 2.8 for females in a zoo. Monogamous through breeding season. One, rarely two, offspring produced generally every two years. As with the other three tapir species, a young Malayan tapir has horizontal stripes and dots on its coat as part of an ancient "hider" strategy for survival. Yet young tapirs also strongly exhibit the "follower" strategy. In southern Sumatra, a crude density of 0.3 (undisturbed swamp) to 0.44 (lowland forest with well-drained soils) tapirs per square kilometer was estimated.

CONSERVATION STATUS
Upgraded to Endangered. Its forest habitat is quickly being eliminated. Additionally, although Asian countries have laws protecting the species, it is often killed by market suppliers and its meat sold under the name of *mu-nam*. Its survival status should be monitored by measuring browse usage of its favorite dietary plants.

SIGNIFICANCE TO HUMANS
Hunted in non-Muslim areas of southeastern Asia for meat and other products. Frequently tracked down and killed after it invades crops. Illegally traded for parts and live individuals. A fair number of these large tapirs live in the zoos and can reach up to 29 years. Wherever they occur, they act as important seed dispersers, as their post-gastric digestive system does not degrade seeds as much as a ruminant's digestive system. Tapirs act as forest architects. Their influence is both ancient and evolving and is known and felt at their grazing centers, on their trails and wallows they fashion, at their sites of social congregation, dung depots, etc. ◆

Resources

Books

Alvarez del Toro, M. *Los Mamiferos de Chiapas*. Tuxtla Gutierrez, Mexico: Universidad Autonoma de Chiapas, 1977.

Blouch, R. A. *Current Status of the Sumatran Rhino and Other Large Mammals in Southern Sumatra*. Bogor, Indonesia: WWF Indonesia Programme, 1984.

Brooks, D. M, R. E. Bodmer, and S. Matola, eds. *Status Survey and Conservation Action Plan: Tapirs*. Gland, Switzerland, and Cambridge, UK: IUCN/SSC Tapir Specialist Group, 1997.

Gade, D. W. *Nature and Culture in the Andes*. Madison: University of Wisconsin Press, 1999.

Lee, A. R. *Management Guidelines for Welfare of Zoo Animals: Tapirs (*Tapirus *spp.)*. London: The Federation of Zoological Gardens of Great Britain and Ireland, 1993.

Leigh, E. G., A. S. Rand, and D. M. Windsor, eds. *The Ecology of a Tropical Forest: Seasonal Rhythms and Long-Term Changes*. Washington, DC: Smithsonian Institution Press, 1982.

Mares, M. A., and D. J. Schmidly, eds. *Latin American Mammalogy: History, Biodiversity, and Conservation*. Norman, OK, and London: University of Oklahoma Press, 1991.

Prothero, D. R., and R. M. Schoch. *The Evolution of Perissodactyls*. New York and Oxford: Clarendon Press and Oxford University Press: 1989.

Robinson, J. G., and K. H. Redford. *Neotropical Wildlife Use and Conservation*. Chicago: University of Chicago Press, 1991.

Periodicals

Agenbroad, L. D., and W. R. Downs. "A Robust Tapir from Northern Arizona." *Journal Arizona-Nevada Academy of Science* 19 (1984): 91–99.

Ashley, M. V., J. E. Norman, and L. Stross. "Phylogenetic Analysis of the Perissodactylan Family Tapiridae Using Mitochondrial Cytochrome C Oxidase (COII) Sequence." *Journal of Mammalian Evolution* 3 (1996): 315–326.

Bodmer, R. E. "Fruit Patch Size and Frugivory in the Lowland Tapir (*Tapirus terrestris*)." *Journal of Zoology* 222 (1990): 121–128.

Bodmer, R. E., T. G. Fang, L. Moya-I., and R. Gill. "Managing Wildlife to Conserve Amazonian Rainforests: Population Biology and Economic Considerations of Game Hunting." *Biological Conservation* 67 (1993): 1–7.

Cohn, J. P. "On the Tapir's Tapering Trail." *Americas* 52 (2000): 40–47.

Downer, C. C. "The Mountain Tapir, Endangered "Flagship" Species of the High Andes." *Oryx* 30, no. 1 (1996): 45–58.

———. "Observations on the Diet and Habitat of the Mountain Tapir (*Tapirus pinchaque*)." *Journal of Zoology* 254 (2001): 279–291.

Flesher, K., and E. Ley. "A Frontier Model For Landscape Ecology: The Tapir in Honduras." *Environmental and Ecological Statistics* 3, no. 2 (1996): 119–125.

Fragoso, J. M. V. "Tapir-Generated Seed Shadows: Scale-Dependent Patchiness in the Amazon Rain Forest." *Journal of Ecology* 85 (1997): 519–529.

Fragoso, J. M. V., and J. M. Huffman. "Seed-Dispersal and Seedling Recruitment Patterns by the Last Neotropical Megafaunal Element in Amazonia, the Tapir." *Journal of Tropical Ecology* 16, no. 3 (2000): 369–385.

Henry, O., F. Feer, and D. Sabatier. "Diet of the Lowland Tapir (*Tapirus terrestris* L.) In French Guiana." *Biotropica* 32, no. 2 (2000): 364–368.

Holbrook, L. T. "The Phylogeny and Classification of Tapiromorph Perissodactyls (Mammalia)." *Cladistics* 15 (1999): 331–350.

Hunsaker, D., and T. Hahn. "Vocalization of the South American Tapir (*T. terrestris*)." *Animal Behavior* (1965): 69–74.

Lizcano, D. J., and J. Cavelier. "Daily and Seasonal Activity of the Mountain Tapir (*Tapirus pinchaque*) in the Central Andes of Colombia." *Journal of Zoology* 252 (2000): 429–435.

Lizcano, D. J., V. Pizarro, J. Cavelier, and J. Carmona. "Geographic Distribution and Population Size of the Mountain Tapir (*Tapirus pinchaque*) in Colombia." *Journal of Biogeography* 29 (2002): 7–15.

Novaro, A. J., K. H. Redford, and R. E. Bodmer. "Effect of Hunting in Source-Sink Systems in the Neotropics." *Conservation Biology* 14 (3) (2000): 713–721.

Overall, K. L. "Coatis, Tapirs and Ticks: A Case of Mammalian Interspecific Grooming." *Biotropica* 12 (2) (1980): 158.

Rodriguez, M., F. Olmos, and M. Galetti. "Seed Dispersal by Tapir in Southeastern Brazil." *Mammalia* 57 (1993): 460–461.

Santiopillai, C., and W. Sukohadi-Ramono. "The Status and Conservation of the Malayan Tapir in Sumatra, Indonesia." *Tigerpaper* 17, no. 4 (1990): 6–11.

Schauenberg, Paul. "Contribution a l'etude du Tapir pinchaque, *Tapirus pinchaque*, Roulin, 1829." *Revue Suisse de Zoologie* 76, no. 1 (1969): 211–256.

Terwilliger, V. J. "Natural History of Baird's Tapir on Barro Colorado Island, Panama Canal Zone." *Biotropica* 10, no. 3 (1978): 211–220.

Organizations

Andean Tapir Fund. P.O. Box 456, Minden, NV 89423 United States. E-mail: ccdowner@yahoo.com Web site: <http://www.dexlen.com/Tapir/andean_tapir.html>

IUCN Species Survival Commission, Tapir Specialist Group. E-mail: epmedici@uol.com.br Web site: <http://www.tapirback.com/tapirgal/iucn-ssc/tsg>

Tapir Preservation Fund. P.O. Box 118, Astoria, OR 97103 United States. Phone: (503) 325-3179; (503) 338-8646. Fax: (503) 325-3179. E-mail: tapir@tapirback.com Web site: <http://www.tapirback.com>

Craig C. Downer, PhD

▲

Rhinoceroses

(Rhinocerotidae)

Class Mammalia

Order Perissodactyla

Family Rhinocerotidae

Thumbnail description
Large, heavily built ungulates with three toes on each limb, one or two horns on the snout, and skin mostly devoid of hairs

Size
Shoulder height: 54–73 in (135–185 cm); head and body length: 100–150 in (250–380 cm); body mass: 1,750–5,000 lb (800–2,300 kg)

Number of genera, species
4 genera; 5 species

Habitat
From rainforest through savanna to semidesert

Conservation status
Critically Endangered: 3 species; Endangered: 1 species; Lower Risk/Near Threatened: 1 species

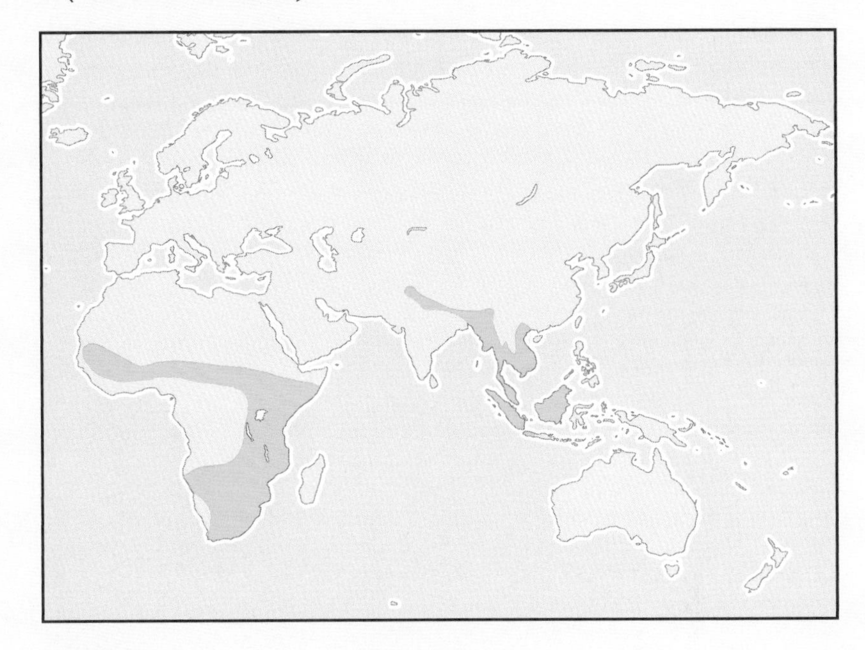

Distribution
Africa and tropical Asia; formerly also Eurasia

Evolution and systematics

The rhinoceros lineage split from the tapirs and equids in the late Eocene. The family was far more abundant and species-rich during the later Tertiary period than today. Among the Oligocene rhinoceroses, *Indricotherium asiaticum*, standing 16.4 ft (5 m) tall at the shoulder, was the largest land mammal ever. *Teleoceros* was a squat North American form with a single small horn on the end of the nose, while the *Diceratheres* had two horns side by side on the snout. *Elasmotherium sibiricum* was a Pleistocene giant with a huge single horn in the frontal region. The five extant species of rhinoceros fall into three distinct subfamilies. The Asian two-horned rhinos, or Dicerorhinae, may be traced back 40 million years to *Dicerorhinus tagicus*, an animal the size of a small tapir. One of its descendants was the woolly rhinoceros, *Coelodonta antiquitatis*, which was widespread through northern Eurasia during the Pleistocene ice ages. This species was primarily a grazer, as attested by its lengthened head, lack of incisors and canine teeth, and high-crowned cheek teeth. Two other rhinos from this subfamily occurred in Europe during the Pleistocene: the steppe rhinoceros, *Dicerorhinus hemitoechus*, and Merck's rhinoceros, *Dicerorhinus kirchbergensis*, which was more a forest inhabitant. The sole surviving species, the Sumatran rhinoceros, *Dicerorhinus sumatrensis*, has changed little from the Oligocene form.

The Asian one-horned rhinoceroses in the subfamily Rhinocerotinae can be traced back to *Gaindatherium browni* from mid-Miocene deposits in India. Of the two surviving species, the Javan rhinoceros, *Rhinoceros sondaicus*, is the more ancient, having changed little during the late Pleistocene in the last two million years.

The two African rhinoceroses represent the subfamily Dicerotinae. The earliest form was *Paradiceros mukiri*, which was found in Miocene deposits at Fort Ternan in Kenya and Beni Mellal in Algeria, dated to 12 million years ago. Rhinoceroses from this subfamily were found from Spain to Turkey and Iran during the late Miocene. The genus *Ceratotherium* first appears in late Pliocene deposits at Langebaanweg in the Cape and elsewhere. The modern species

White rhinoceros (*Ceratotherium simum*) mother and calf. (Photo by Harald Schütz. Reproduced by permission.)

A newborn Sumatran rhinoceros (*Dicerorhinus sumatrensis*) rests in its pen at the Cincinnati Zoo Thursday, Sept. 13, 2001. The rhino is the first Sumatran rhinoceros to be born in captivity in 112 years. Officials at the Cincinnati Zoo and Botanical Garden hailed the birth from the only breeding pair in the United States as a historic event that could help save a species. (Photo by AP Photo/Cincinnati Zoo. Reproduced by permission.)

A black rhinoceros (*Diceros bicornis*) mother chasing away a lion. (Photo by K. & K. Ammann. Bruce Coleman, Inc. Reproduced by permission.)

A black rhinoceros (*Diceros bicornis*) charging. (Photo by Tom Brakefield. Bruce Coleman, Inc. Reproduced by permission.)

Indian rhinoceros (*Rhinoceros unicornis*) checking for enemies. (Photo by Tom & Pat Leeson/Photo Researchers, Inc. Reproduced by permission.)

Ceratotherium simum is especially numerous in Pleistocene deposits at Olduvai Gorge in Tanzania. Some workers regard the distinctions between *Ceratotherium* and *Diceros* as insufficient to warrant the generic distinction.

Physical characteristics

Together with the elephants and the hippopotamus, rhinoceroses constitute the "megaherbivores," those species weighing over 2,200 lb (1,000 kg) as adults. Rhinos are large, graviportal animals with relatively short limbs and barrel-shaped bodies. The three toes on each foot leave a track resembling the ace-of-clubs. The skull includes enlarged nasal bones and an extended occipital crest, with the eyes perched strangely on the sides of the head. The cavity occupied by the nasal sinuses exceeds that of the brain. The chewing teeth comprise three molar-like premolars and three true molars in each half-jaw. In grazers like the white rhino and Indian rhino (*Rhinoceros unicornis*), these are high crowned with complex grinding surfaces, while in browsers the teeth are lower crowned with prominent cusps. The African species completely lack incisor and canine teeth, while the Indian and Javan rhinos retain a pair of tusk-like incisors in the lower jaw, and the Sumatran rhino (*Dicerorhinus sumatrensis*) has tusk-like lower canines as well as upper incisors. The skeleton is massively constructed to support the heavy body weight, with the vertebral spines greatly extended in the shoulder region and in the posterior thoracic region. The neck is short, as is the tail.

Rhino horns lack the bony core that is typical of the horns of cattle, goats, and antelope, but consist of the same proteinaceous substance, keratin, that forms the outer material of such horns, as well as the material of hooves, fingernails, and hairs. They are made up of tubular filamentous rods, resembling a mass of adherent hairs. Rather than being part of the skull, the horns adhere to roughened areas of bone. If knocked off during fighting, or through some other accident, the horns re-grow. Indeed, they continue growing throughout the life of a rhino, with increase in length counteracted by wear from the tip.

The skin thickness varies between 0.5 and 1.8 in (13–45 mm) over different regions of the body, and among species.

Black rhinoceros (*Diceros bicornis*) with raised upper lip demonstrating scenting of a tree in Masai Mara, Kenya. (Photo by Mary Beth Angelo/Photo Researchers, Inc. Reproduced by permission.)

White rhinoceros (*Ceratotherium simum*) male, female, and calf in Zululand, South Africa. (Photo by Nigel J. Dennis/Photo Researchers, Inc. Reproduced by permission.)

Black rhinoceroses (*Diceros bicornis*) mating in Ngorongoro Crater, Tanzania. (Photo by Joe McDonald. Bruce Coleman, Inc. Reproduced by permission.)

However, the outer epidermis is quite thin (about 0.04 in, or 1 mm), and well supplied with blood vessels, so that biting flies have only to penetrate this distance to draw blood. Instead of the usual type of sweat gland, rhinos have exceptionally large apocrine glands scattered over the skin, well designed for rapid and copious discharge of fluid.

The penis is muscular as in equids, and backwards pointing when retracted. There are laterally projecting lobes associated with extensions of the corpus cavernosum, and the tip terminates in an enlarged flattened flange. The testes are located close to the skin between the prepuce and the attenuated nipples, and there is no scrotum. Females possess two teats located between the hindlegs. Pedal glands are present in the genus *Rhinoceros*, but lacking in the African species. Preputial glands are present in the white rhinoceros only.

Distribution

Two species occur in Africa, and three species in Southeast Asia. All five species are much more restricted in their distribution today than they were in the past because of human impacts.

Habitat

Modern rhinoceroses occupy a diversity of habitats: dense rainforests for the Sumatran and Javan rhinos, swamplands

White rhinoceroses (*Ceratotherium simum*) form a defensive circle when a potential enemy threatens. (Photo by M. Reardon/Photo Researchers, Inc. Reproduced by permission.)

and adjoining meadows for the Indian rhino, grassy savannas for the white rhino, and dry bushland or semi-desert for the black rhino (*Diceros bicornis*).

Behavior

Rhinos are largely solitary animals, apart from the mother-offspring association, but the white rhino is more social and forms small groups. They have poor vision, and seem unable to recognize a stationary human observer at distances exceeding about 100 ft (30 m). Their hearing is good, with the ear pinnae moved independently to scan for sounds from different directions. Their sense of smell is acute, and rhinos can detect traces of human scent, and also follow the tracks of other rhinos, after many hours have passed. They can be frightening animals to encounter, because they often charge human intruders, or their vehicles, but when not threatened can be quite docile, and become very tame in captivity.

Feeding ecology and diet

The stomach is simple, and the capacious cecum and colon (or large intestine) in the hind-gut serve as the main sites of fermentation of plant food, with the help of bacteria. Large body size prolongs the period of retention, facilitating efficient digestion. Grass-feeders such as the white rhino and Indian rhino have a relatively longer colon than the other three rhino species, which are primarily browsers on the leaves and stems of woody plants.

Black rhinoceroses (*Diceros bicornis*) fighting. (Photo by Len Rue, Jr. Bruce Coleman, Inc. Reproduced by permission.)

A white rhinoceros (*Ceratotherium simum*) with muddy face and horns after digging for water during drought in Nakuru National Park, Kenya. Lake Nakuru (in background) is a soda lake. (Photo by Gregory G. Dimijian/Photo Researchers, Inc. Reproduced by permission.)

Reproductive biology

The gestation period is 15–16 months for all rhinoceros species, even the small Sumatran rhino, and the inter-birth interval is correspondingly between two and four years. Nursing generally continues for over a year, and the older calf is driven away by the mother around the time of birth of the next offspring. Estrous cycling begins while the mother is still nursing, and there is no narrow birth season. Rhinos are renowned for the extended duration of copulations, which last between 20 minutes and an hour or longer, with multiple ejaculations. They have proved surprisingly difficult to breed in zoos, with many strange features of their reproductive biology being revealed. For example, white rhinos show no reproductive activity if housed in pairs, and the presence of more than one male, or at least an exchange of males, seems necessary for females to show overt estrous behavior. For the Sumatran rhino, mating induces ovulation. Courtship behavior can be surprisingly aggressive and frequently results in injuries. Rhinos are polygynous, and generally males and females do not associate with each other outside of mating.

Conservation status

Both the Javan rhino and Sumatran rhino are Critically Endangered, with low numbers of animals persisting in just a few sanctuaries in Southeast Asia. Although more numerous, the black rhino is also classed as Critically Endangered, because no area harbors a large population and because it remains extremely vulnerable to poaching. The Indian rhino is marginally less insecure in numbers, and thus classed as merely Endangered. Only the white rhino is no longer seriously in danger, having recovered amazingly from its critically low numbers at the beginning of the twentieth century in southern Africa. Nevertheless, the distinct northern subspecies persists as just a relict of 30 animals in one park in northeast Congo. The threat to all of these species comes from illegal hunting driven by the high market value and legendary powers of their horns.

Significance to humans

Rhinoceroses have fascinated humans from early times, as is shown in cave art from the Early Stone Age in Europe, depicting the extinct woolly rhino, and in rock paintings and engravings spread across Africa, from the Cape to Algeria. The engraving of an Indian rhino by Durer in 1515, received as a gift to the Portuguese king from an Indian sultan (actually intended for the Pope), brought these animals to the attention of Europeans. However, Marco Polo described the Sumatran rhinos that he had seen during his travels through the Far East early in the fourteenth century, and the early Romans had imported some rhinos of unknown affinities. Rhinoceroses have become especially valued for their horns, used for the making of prestigious dagger handles in Yemen and adjoining parts of the Middle East, and in powdered form as a fever-reducing drug in China and aphrodisiac potion in India. The horns were also carved into cups, used by Indian and Far Eastern potentates to test whether beverages contained poison. The claimed medicinal power is without pharmaceutical foundation, since the substance of the horn is no different from that of hooves or fingernails. The legends about aphrodisiac properties seem to derive from the prolonged copulations typical of the family, perhaps supported by the phallic appearance of the anterior horn.

1. Black rhinoceros (*Diceros bicornis*); 2. Sumatran rhinoceros (*Dicerorhinus sumatrensis*); 3. Indian rhinoceros (*Rhinoceros unicornis*); 4. Javan rhinoceros (*Rhinoceros sondaicus*); 5. White rhinoceros (*Ceratotherium simum*). (Illustration by Joseph E. Trumpey)

Species accounts

Sumatran rhinoceros

Dicerorhinus sumatrensis

SUBFAMILY

Dicerorhinae

TAXONOMY

Dicerorhinus sumatrensis (Fischer, 1814), Sumatra.

OTHER COMMON NAMES

English: Asian two-horned rhinoceros; French: Rhinoceros de Sumatra; German: Sumatra-Nashorn; Spanish: Rinoceronte de Sumatra.

PHYSICAL CHARACTERISTICS

The smallest living rhinoceros, body mass to 1,750 lb (800 kg); shoulder height of 48–58 in (120–150 cm); and head and body length, 100–125 in (250–315 cm). Unusual among rhinos in having the body covered with short, stiff hairs. The hide is dark red-brown in color, the dark-colored hair quite dense in young animals, but thinning out in older animals. The ears are fringed with prominent hairs. The anterior horn length is up to 20 in (50 cm), but usually much less; the posterior one is maximally 6 in (15 cm), though commonly little more than a hump. Males have larger horns with a greater basal diameter than females, but otherwise there is little sexual dimorphism. The hide forms a thick horny cover over the front of the snout, and is also thickened around the eyes.

DISTRIBUTION

Formerly widely distributed from Assam and Burma (Myanmar) through Thailand into Indochina, and southwards through the Malay Peninsula to the islands of Sumatra and Borneo. Today thinly scattered through the southern Malay Peninsula, parts of Sumatra and Sarawak in north Borneo, plus possibly a few animals in Myanmar.

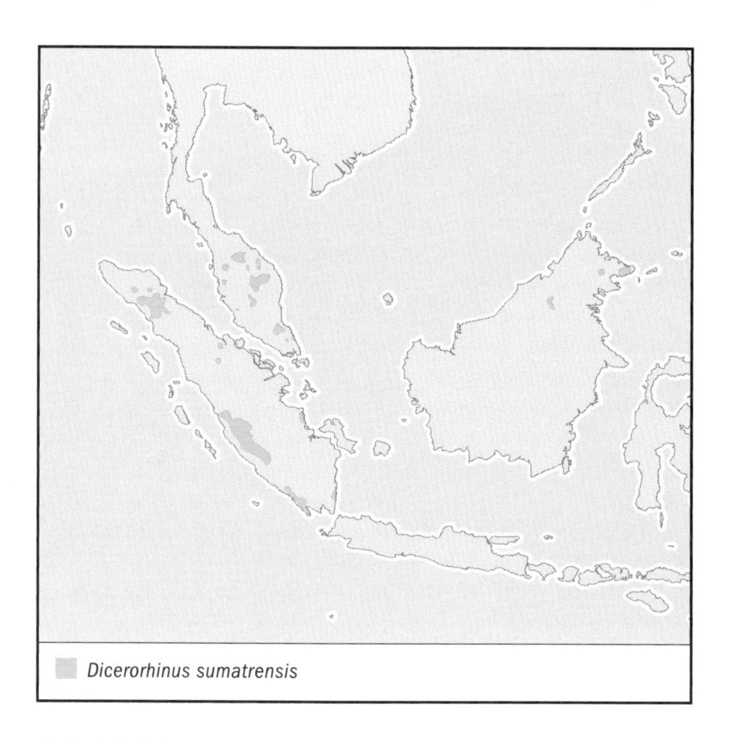
Dicerorhinus sumatrensis

HABITAT

Occurs in regions of broken mountainous forest, but in the past may have occupied lowland forest as well.

BEHAVIOR

Generally solitary, apart from calves accompanying mothers. Animals use a network of trails through the forest, and wallow frequently in muddy pools. The hair coat helps anchor a thick armor of mud to the body, which provides protection against biting insects, and also thorns. Salt licks are also visited frequently in the Gunung-Leuser Park in Sumatra. Females occupy home ranges covering 4–6 mi^2 (10–15 km^2), while adult males move over much larger areas up to 20 mi^2 (50 km^2). Male ranges have core areas, which may represent territories. Both sexes urinate in a spray, commonly over a bush, and leave scrape-marks, dung piles, and urine deposits along trails; mature males seem more active in placing such marks. Sharp dagger-like canines in the lower jaw are used as weapons in fights.

FEEDING ECOLOGY AND DIET

Feeds largely on the twigs and leaves of small trees or shrubs growing in the forest understory, and at times also on some fruits, herbs, and lianas of a wide variety of species. The preferred feeding habitat is dense undergrowth along streams and lower slopes, with the protected snout and eyes and mud-covered hide helping animals penetrate such vegetation. To reach higher shoots, animals bend or break saplings by walking over the plant and pressing down on the trunk with the body.

REPRODUCTIVE BIOLOGY

The species is an induced ovulator, and courtship is quite aggressive. Gestation period is 475 days, and birth weight 14 lb (33 kg). Females confine their movements to a small region close to a saltlick while nursing the calf. Calves separate from their mothers around 16–17 months of age, whereupon the mother returns to her non-breeding range, while the calf remains close to the saltlick. The usual birth interval seems to be close to four years, with a period of a year between weaning and the next pregnancy.

CONSERVATION STATUS

An estimated 300 animals still remained in the wild in 2002, but widely scattered among a number of parks in the Malay Peninsula, Sumatra, and north Borneo, with another 40 animals being managed in a captive breeding program. The first birth occurred in September 2001 in the Cincinnati Zoo in the United States. Poaching as well as encroachment by cultivators remain serious threats despite the formal establishment of several protected areas. Listed by the IUCN as Critically Endangered.

SIGNIFICANCE TO HUMANS

Perhaps the strangest of all the rhinoceros species, a living relict from far distant times. ◆

Javan rhinoceros
Rhinoceros sondaicus

SUBFAMILY
Rhinocerotinae

TAXONOMY
Rhinoceros sondaicus Desmarest, 1822, Java.

OTHER COMMON NAMES
French: Rhinoceros de al Sonde; German: Java-Nashorn; Spanish: Rinoceronte de Java.

PHYSICAL CHARACTERISTICS
A small version of the Indian rhinoceros, distinguished by minor differences in the folds of the skin, and by its browsing dentition. Body mass to 3,300 lb (1,500 kg); shoulder height 47–70 in (120–178 cm); horn length to 15 in (38 cm).

DISTRIBUTION
Formerly from Assam and Bangladesh through to Indochina and the islands of Sumatra and Java. Now restricted to the western tip of Java, plus a few animals in Vietnam.

HABITAT
Lowland rainforest, favoring transitional habitat between low secondary vegetation and the primal forest.

BEHAVIOR
Mostly solitary, although groupings of two to four animals are occasionally observed. Cows without calves move through overlapping ranges covering about 4 mi^2 (10 km^2), those with calves use a smaller area of about 1 mi^2 (2–3 km^2), while bulls traverse a larger area of 8 mi^2 (20 km^2). Animals often rest lying in ponds or mud wallows. The red-orange urine of male Javan rhinos is squirted alongside trails and in wallows. Dung deposits are widely spread and not broken up by the feet. A far-reaching whistling call may serve as a contact signal.

FEEDING ECOLOGY AND DIET
Feeds on the twigs and branches of saplings, shrubs, and lianas, using the finger-like upper lip to draw food towards the mouth. Small trees may be bent or broken to bring the upper foliage within reach.

REPRODUCTIVE BIOLOGY
Very little is known.

CONSERVATION STATUS
Among the rarest of mammal species, with a mere 50–60 animals surviving in the Ujung Kulon Reserve in Java, and an unknown, but very small, number in the Cat Loc Park in southern Vietnam. Although animals in Ujung Kulon are well protected, and there have been no cases of poaching, the population is not increasing. Listed as Critically Endangered by the IUCN.

SIGNIFICANCE TO HUMANS
The least known of the rhino species. ◆

Indian rhinoceros
Rhinoceros unicornis

SUBFAMILY
Rhinocerotinae

TAXONOMY
Rhinoceros unicornis Linnaeus, 1758, Assam Terai, India.

OTHER COMMON NAMES
English: Greater one-horned rhinoceros; French: Rhinoceros unicorne de l'Inde; German: Panzernashorn; Spanish: Rinoceronte unicornio indico.

PHYSICAL CHARACTERISTICS
Characterized by the armor-like plates formed by folds of skin on its sides and neck, and by the single horn perched on the snout. Males weigh to 4,600 lb (2,100 kg), females about 3,500 lb (1,600 kg); shoulder height in males to 71 in (180 cm), females 63 in (160 cm); head and body length in males 150 in (380 cm), females 135 in (340 cm). Maximum horn length is 18 in (45 cm) in both sexes. The hairless skin is gray in color, and shows flat bumps resembling rivets on a ship's hull. The two lower incisors are modified as short tusks used in fighting, up to 20 cm long, and thicker in males than in females. There are prominent glands on the edge of the sole of each foot.

DISTRIBUTION
Formerly from Industan through northern India and adjacent parts of Nepal and Bhutan into northern Burma. Now restricted to mostly isolated reserves in Nepal, Assam, and neighboring Bhutan.

HABITAT
Occupies floodplain and swampland habitats with tall cane-like grasses reaching heights of 13–20 ft (4–6 m), plus adjoining woodlands on drier ground.

BEHAVIOR
At Chitwan in Nepal, females occupy long, narrow home ranges bordering the river, covering up to 8 mi^2 (20 km^2), although most time is spent within a core area of about 1 mi^2 (2–4 km^2). Male core ranges are typically 1.1–1.5 mi^2 (3–4 km^2 in extent, and overlap without territorial exclusion. Indian rhinos are basically solitary, apart from associations being between mothers and their most recent offspring, and temporary associ-

ations among subadults. However, congregations of several animals may develop around wallows and bathing pools, and in feeding areas. Dominant males have a large head-on profile, plus prominent bibs formed by folds of skin, and perform squirt-urination and foot-dragging displays. Males respond aggressively when they meet strange intruding males, and violent fights can develop with fatal results. Submissive males may share the home ranges of dominant males, do not squirt-urinate, and run away when challenged. Dung deposits develop alongside trails and bordering feeding areas, and are used by animals of both sexes. They seem to serve as orientation points. Indian rhinos are highly vocal, and make a variety of sounds, from squeaks and grunts to loud roars. About 50–65% of their time is spent feeding, and the rest mainly lying down. Indian rhinos drink daily, and ingest mineral-rich soil when it is available.

FEEDING ECOLOGY AND DIET

Feeding is aided by the finger-like upper lip, used to grip grass stems and bushes. The lip is folded back when animals graze on short grasses. Tall cane-like grasses form the principal food source year-round, in particular species of *Saccharum*. Short grasses and herbs are favored during the monsoon period, and also aquatic plants. Woody browse becomes important during winter. Fruits of many species are also eaten, especially the hard green fruits of *Trewia nudiflora*, which fall to the ground in large numbers during the monsoon season. When feeding on tall grasses or shoots, animals often step over the plants, pulling the stems down between their legs and body so as to bite off the tips. The feeding activities of Indian rhinos have quite a large impact on their habitat, by trampling and breaking plants, and also by dispersing seeds in their dung. *Trewia* trees seem to be especially dependent on rhinos, with their hard seeds benefiting from passage through rhino gut, which increases their germination rate.

REPRODUCTIVE BIOLOGY

At Chitwan, there is a weak peak in the number of females in heat during late winter and the pre-monsoon period, with most births thus occurring during the monsoon period. A female in heat sprays urine, and makes rhythmical whistling sounds. Spectacular chases over 0.62–1.2 mi (1–2 km) are a feature of courtship, with the female making loud honking noises, and the male squeaky panting sounds. These chases attract the attention of other males, thus ensuring that the female mates with the strongest sire in the region. When the bull catches up with the cow, initial horn fighting may develop into biting, and gaping wounds can be inflicted. Several mounting attempts may be made before intromission is achieved. Copulation lasts around 60 minutes, with a maximum of 83 minutes recorded. There were as many as 56 ejaculations were recorded in one instance. Females may come on heat about 36–58 days later if conception fails.

The gestation period is 16 months. Mothers seek seclusion in thick grassland or forest to give birth, and are aggressive towards other rhinos while the calf is small. Calves weigh 140–150 lb (65–70 kg) at birth. Calves under two months of age may be left lying alone for periods of up to an hour while the mother forages several hundred feet (meters) away. Nursing continues until the offspring is about two years old. The offspring is driven away by the mother a week or two before the birth of the next calf. Young males tend to join up with other young males, and young females occasionally attach themselves to adult females, although such associations endure only a few days at most. Young females may remain within the maternal home range, while young males tend to disperse away from high-density areas.

Young males form shifting groups of from two to as many as 10 animals until over eight years of age, and may not achieve breeding status until 15 years of age. In the wild, females first give birth between six and eight years of age. In zoos, females can attain sexual maturity as early as three years of age, while males become sexually potent at seven years. The median inter-calving interval is 3.5–4 years in the wild. The shortest recorded calving interval is 18 months, which occurred after the newborn infant had been killed by a tiger. Birth intervals may lengthen in older animals. About 25% of all deaths of rhinos were as a result of fights, affecting especially young as well as adult males. Predation by tigers is an important cause of death among calves under eight months old. In zoos, a record longevity of 47 years has been recorded, but the oldest animal in the wild was estimated to be 30 years old, from counts of cementum lines in her teeth, when she died.

CONSERVATION STATUS

Listed as Endangered by the IUCN. As of 2002, the surviving population totaled about 2,400 free-ranging animals, including 1,550 rhinos in India's Kaziranga National Park, and 500 rhinos in the Royal Chitwan National Park in Nepal. Elsewhere in the Assam region, numbers have been declining because of social instability and associated poaching, as well as encroachment by agriculture and cattle grazing. Rhinos are vulnerable when they move out of parks to higher ground following flooding and when they raid the crops of surrounding villages. However, the situation seems to be improving, and animals have been translocated to establish populations in new parks in Nepal and India. A large stretch of the Brahmaputra River has been added to the Kaziranga National Park, which will allow the rhino access to fertile grazing on the islands in the river once cattle are excluded. This could also allow rhinos to move between Kaziranga and other nearby wildlife reserves.

SIGNIFICANCE TO HUMANS

Viewing Indian rhinos from the backs of elephants in sanctuaries like Kaziranga and Chitwan constitutes a considerable attraction to tourists. However, rhinos are costly to surrounding village people when they move out to feed on crop plants, including maize, rice, wheat, and potatoes in surrounding fields, generally at night. Fatal attacks on humans sometimes occur, with injuries inflicted by the lower incisor tusks. ◆

Black rhinoceros

Diceros bicornis

SUBFAMILY

Dicerotinae

TAXONOMY

Diceros bicornis (Linnaeus, 1758), South Africa. Four subspecies: *D. b. bicornis* in Namibia, re-introduced into former range in northern Cape; *D. b. minor* from eastern South Africa to southern Tanzania; *D. b. michaeli* from northern Tanzania through Kenya; and *D. b. longipes* from Cameroon and adjoining regions.

OTHER COMMON NAMES

English: Hook-lipped rhinoceros; French: Rhinoceros noir; German: Spitzmaul-Nashorn; Spanish: Rinoceronte negro.

PHYSICAL CHARACTERISTICS

Black rhinos attain a weight of 2,100–2,900 lb (950–1,300 kg); shoulder height of 56–63 in (143–160 cm); and head and body

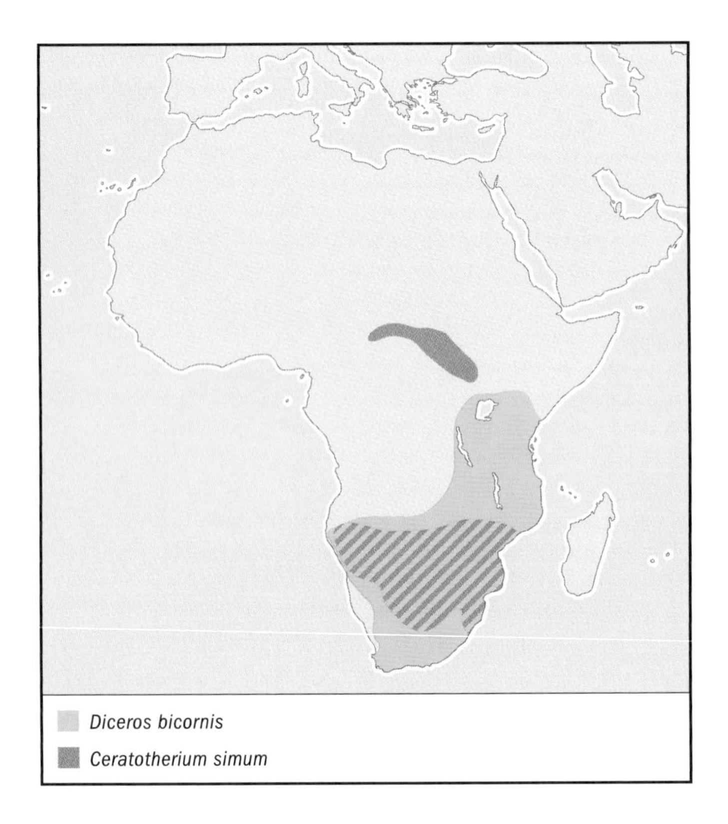

length of 112–120 in (286–305 cm). Male and female are of closely similar size. Despite name, the skin is gray to brownish gray in color, and devoid of hairs. The prehensile upper lip, used to grasp food, gives rise to alternate name. Anterior horn is 16.5–54 in (42–138 cm) in length, the posterior one 8–20 in (20–50 cm). Have a saddle-backed appearance, rounded ears, and tend to hold the head high, except when feeding on low vegetation.

DISTRIBUTION
Originally almost throughout Africa, from southwest Cape to Somaliland, and in the west to the border region between Cameroon and Ivory Coast, but absent from evergreen rainforest. Currently persists in fragmented remnants in parts of this range.

HABITAT
From various forms of savanna vegetation to arid shrub steppe.

BEHAVIOR
Home ranges vary enormously in size, from 1 mi^2 (2.5 km^2) in Ngorongoro Crater to 200 mi^2 (500 km^2) in the Namib Desert fringe in Namibia. Home ranges of males tend to be smaller than those of females, and in some populations may represent defended territories. May travel to water only at intervals of four to five days in dry areas, but drink more frequently when water is plentiful. Journeys to water usually take place during the late afternoon or early part of the night. Animals commonly lie in mud, and sometimes also in dust hollows. Saltlicks are also frequently visited. Somewhat more active at night than during the day.

Largely solitary, although groups of three to five animals may occasionally form. A cow and her calf comprise the basic social unit, and adult males are solitary, except when courting a female. Sub-adults join other rhinos more frequently, but still are more commonly alone than in groups. When adult males meet, a complex bull ceremony may take place, involving stiff-legged scraping, imposing postures, and short charges sometimes accompanied by screaming groans. Males mark the environment with long drag marks made by the legs, spray urine over bushes or other objects, and deposit feces on dung-heaps. Females use the same dung-heaps, or "middens," and animals of both sexes scatter their droppings with backwardly directed kicks.

Lions occasionally tackle black rhino calves, but can pay with their lives when the mother attacks back. Spotted hyenas are responsible for some mortality among small calves, which seem to be vulnerable to hyenas because of their habit of running behind the mother.

FEEDING ECOLOGY AND DIET
Predominantly low-level browsers, feeding on small saplings and shrubs under 5 ft (1.5 m) in height as well as a variety of herbs, and occasionally small amounts of grass. *Acacia* spp. are especially favored, as are various species of Euphorbiaceae, including succulent forms with milky sap reputed to be poisonous. The prehensile upper lip is used to pull twigs into the mouth, which are then bitten off with the cheek teeth. They crop branch tips up to 0.4 in (10 mm) in thickness and 4–9.8 in (100–250 mm) in length. The horns may be used to bend or break stems to reach higher branches. Bark may also be stripped from certain trees.

REPRODUCTIVE BIOLOGY
Matings and births can take place throughout the year; in South Africa, most matings take place during the early summer or wet season, with births subsequently peaking from late summer into the dry winter season. Males attach themselves to a female six to seven days prior to mating, and the courtship period may include attacks by the female on the male. Copulations last 20–40 minutes, and may be preceded by several mounting attempts before intromission is achieved. Females come on heat again after 35 days if the mating is not successful. Gestation lasts 15 months, and the birth weight is 9–11 lb (20–25 kg). The newborn infant may be left hidden for the first week after birth. When separated the mother calls with breathing pants, while the calf makes mewing squeals. Nursing continues until the calf is about 1.5 years old. Inter-birth intervals can vary widely, from as short as 2–2.5 years under favorable conditions, to four years in less favorable circumstances. Calves separate from their mothers around the time of birth of the next offspring, generally when aged 2.2–3.3 years. After leaving their mothers, sub-adults may join other sub-adults temporarily, but remain mostly solitary. Females first mate at 4.5 years of age, and give birth around six years of age, but in high-density populations the age at first parturition may be delayed until eight to 12 years. Males reach reproductive maturity around eight to nine years of age in the wild, but one captive male sired an offspring at the age of 3.5 years. The maximum longevity recorded in a zoo is 49 years, while the life span in the wild would probably not much exceed 35 years.

CONSERVATION STATUS
Listed as Critically Endangered by the IUCN. From being a common and widespread species, numbers plummeted to only 2,700 animals remaining in the wild in 2000. The largest segments are in the Kruger National Park (estimated 400), Hluhluwe-Umfolozi (about 350), and northern Namibia (around 700). Only about 400 remain in Kenya, once a major stronghold of the species. The West African subspecies, *D. b. longipes*, is Critically Endangered with only an estimated 10 animals persisting in Cameroon.

SIGNIFICANCE TO HUMANS
Renowned for their propensity to charge humans or their vehicles; many charges are merely curious advances. However, actual attacks may also result, leading to severe and sometimes fatal wounds, or damage to vehicles. ◆

White rhinoceros
Ceratotherium simum

SUBFAMILY
Dicerotinae

TAXONOMY
Ceratotherium simum (Burchell, 1817), Cape Province, South Africa. Two subspecies: *C. s. simum* in southern Africa and *C. s. cottoni* in northeast Africa.

OTHER COMMON NAMES
English: Square-lipped rhinoceros; French: Rhinoceros blanc; German: Breitmaul-Nashorn.

PHYSICAL CHARACTERISTICS
The largest rhinoceros, with males weighing to 5,000 lb (2,300 kg), females to 3,800 lb (1,700 kg); shoulder height for males 65–73 in (165–185 cm), females 61–70 in (155–177 cm); head and body length for males 140–150 in (360–380 cm), females 118–143 in (300–363 cm); anterior horn length males 20–47 in (50–120 cm), females 20–62 in (50–158 cm); posterior horn length 6.6–15 in (16–40 cm) in both sexes. The skin is battleship gray, with very sparse hairs on the body in the southern form, and none in the northern form, apart from fringes to the ears and the tip of the tail. The head is lengthened, and the lips broad, as adaptations for grazing. The ligament supporting the enormous weight of the head, and associated tissues, causes a hump on the back of the neck to form. The northern subspecies is slightly smaller, appears somewhat longer legged, and has the dorsal profile of the skull slightly less concave than the southern form.

DISTRIBUTION
Historically distributed in two discrete areas, separated by a gap of more than 1,240 mi (2,000 km). In southern Africa, they occurred south of the Zambezi to northern KwaZulu-Natal in the east, and westwards through Botswana and Northern Cape into the northern part of Namibia. The northern subspecies was distributed west of the Nile River, from northern Uganda into southern Sudan, and westwards through northeast Congo and the Central African Republic as far as the southern edge of Chad. However, teeth and rock art indicate that it formerly occurred through much of East Africa until quite recently, and extended as far north as Algeria. Hunting eliminated the southern subspecies over its entire range, except in the Hluhluwe-Umfolozi region, but subsequently animals have been reintroduced to parts of the former range. In northeast Africa, the species persists only in the Garamba National Park in the Democratic Republic of Congo.

HABITAT
Associated drier forms of savanna in southern Africa, but in the northern range occupies moist savanna, with tall grass prevalent except around termite mounds.

BEHAVIOR
Female home ranges extend over 8–16 mi^2 (10–20 km^2), including a smaller core area, in Hluhluwe-Umfolozi, but encompass 20 mi^2 (50 km^2) or more in low-density populations, or where habitat conditions are less favorable. Northern white rhinos cover 20–40 mi^2 (50–100 km^2) in Uganda and 80–200 mi^2 (200–500 km^2) in Garamba. Adult males restrict their movements to somewhat smaller areas, which constitute breeding territories. Active for about 50% of the time, both day and night, with most of this time taken up with feeding. There is generally a midday slumbering period, for which animals commonly resort to shady areas on ridge-crests. Wallowing in muddy hollows is a favorite activity, and they emerge coated with mud, which helps remove ticks and reduces the number of biting flies, and may also serve a cooling function. They sometimes lie in pools of water.

The typical group is a mother-offspring pair, but larger groups, including several subadults as well as one or more adult females, are also formed. Subadults almost invariably team up with one or more subadults of similar age, of the same or opposite sex, or with an adult female lacking a calf. Adult males are solitary, except when accompanying females.

Dominant males occupy clearly defined territories from which they exclude other dominant males, but share these with one or more subordinate adult males as well as with cows and sub-adults. These territories cover only 0.4–1 mi^2 (0.8–2.5 km^2) in the dense Hluhluwe-Umfolozi population, but may expand hugely when there is less pressure, with territories of 20–40 mi^2 (50-100 km^2) being patrolled in some sparsely populated localities. Males are dominant only within the boundaries of their own territory. Adult males manifest their subordinate status by uttering loud roars and shrieks when confronted by a territory holder, with curled tails indicating their nervousness. Adult females also use loud roars or snorts to deter a bull from a close approach. Occasional fights among males may lead to a change in territorial dominance. Interestingly, the defeated male may remain on in his former territory, provided he behaves submissively when challenged, and foregoes scent marking.

They largely ignore lions, even when a calf is present, although there are occasional records of lions preying on young white rhinos. Hyenas seem less a threat to calves than in the case of the black rhino, perhaps because white rhino calves run ahead of the mother and thus seem to be better protected. The longer but more slender horns of females seem designed to ward off predators.

FEEDING ECOLOGY AND DIET
The southern form is strictly a grazer, with herbs generally constituting no more than 1% of the diet, and only occasional records of munching on woody browse. Short grasses are the favored food source for most of the year. The grass can be cropped as short as 1 in (25 mm) above soil level. During the later dry season, animals turn to taller grasses, including buffalo grass (*Panicum maximum*) and red oats grass (*Themeda triandra*). The northern subspecies favors areas of short grass on termite mounds and after fires, but includes a range of medium-tall grass species in its diet.

REPRODUCTIVE BIOLOGY
Calves can be born throughout the year, but in Hluhluwe-Umfolozi there is a peak in the number of cows on heat following the first rains in early summer, and a corresponding peak in the number of calves born during the late summer/early winter period. Copulation lasts 15–30 minutes, with multiple ejaculations. The cow comes into heat again after about 30 days if the mating was not successful. Almost all matings are by territory holders. Cows seek seclusion before giving birth, either in dense bush or up on hillsides where few

other rhinos travel. The gestation period is 16 months, and the newborn infant weighs about 145 lb (65 kg). The older calf is driven away shortly before the birth. Weaning is completed by 15–24 months, and cows come on heat again while the calf is still being suckled. The mean inter-birth interval is 2.5 years for Hluhluwe-Umfolozi (range 1.9–3.5 years), although somewhat longer in some introduced populations. Subadults are itinerants, living in one area for a period, later shifting somewhere else. Females begin estrous cycling around four years of age, and the sub-adult period terminates with the birth of their first calf around 6.5–7.5 years of age. Young males start showing solitary tendencies around eight years of age, and reach the stage when they can challenge successfully for a territory around 10–12 years of age. The maximum life span for a white rhino is about 40 years.

CONSERVATION STATUS

By 2000, more than 10,000 white rhinos existed in the wild, including 1,700 in Hluhluwe-Umfolozi, over 3,000 in the Kruger National Park, and about 2,300 on private land in southern Africa. However, the ease with which animals can be tracked down, coupled with the high value of their horns, has led to the elimination of some reintroduced populations. Some hunting of the species is now allowed under strictly controlled conditions, as well as a limited trade in live white rhinos, but not in their horns. A strange new threat in some protected areas is the killing of white rhinos by young male elephants that have reached sexual maturity at an early stage in the absence of older bulls. The future of the northern subspecies is especially precarious, with a total population of only 30 animals in the wild, all in Garamba Park. The species is listed as Near Threatened by the IUCN.

SIGNIFICANCE TO HUMANS

A whiff of human scent, even at ranges of up to 2,625 ft (800 m), sends rhinos running away. The terror that humans inspire in these animals is a clear indication of the past hunting pressure that almost wiped out the species, not only during the era of guns but much earlier in East Africa. White rhino meat was highly regarded by early European hunters, and other body parts had many uses. Once guns became widely available across southern Africa, white rhinos changed from an abundant and widespread species to the brink of extinction within a few decades. The subsequent recovery of the southern subspecies, from a few score animals to the stage when populations could be reestablished in many parts of their former range, and even in East Africa, is one of the great success stories of conservation. ◆

Resources

Books

Anonymous. *Die Nashorner. Begegnung mit urzeitlichen Kolossen.* Furth: Filander Verlag, 1997.

Owen-Smith, Norman. *Megaherbivores. The Influence of Very Large Body Size on Ecology.* Cambridge: Cambridge University Press, 1988.

Van Strien, N. J. *The Sumatran Rhinoceros in the Gunung Leuser National Park, Sumatra, Indonesia: Its Distribution, Ecology and Conservation.* Berlin: Mammalia Depicta, Verlag, Paul, and Parey, 1986.

Periodicals

Dinerstein, E. "Effects of *Rhinoceros unicornis* on Riverine Forest Structure in Lowland Nepal." *Ecology* 73 (1992): 701–704.

Dinerstein, E., and L. Price. "Demography and Habitat Use by Greater One-horned Rhinoceros in Nepal." *Journal of Wildlife Management* 55 (1991): 401–411.

Dinerstein, E., and C. M. Wemmer. "Fruits Rhinoceros Eat: Dispersal of *Trewia nudiflora* (Euphorbiaceae) in Lowland Nepal." *Ecology* 69 (1988): 1768–1774.

Laurie, Andrew. "Behavioural Ecology of the Greater One-horned Rhinoceros." *Journal of Zoology, London* 196 (1982): 307–341.

Owen-Smith, Norman. "The Social Ethology of the White Rhinoceros." *Zeitschrift für Tierpsychologie* 38 (1975): 337–384.

Schenkel, R., and L. Schenkel-Hulliger. "The Javan Rhinoceros in Ujung Kulon Nature Reserve, Its Ecology and Behaviour." *Acta Tropica* 26 (1969): 98–135.

Van Gyseghem, R. "Observations on the Ecology and Behaviour of the Northern White Rhinoceros." *Zeitschrift fur Tierpsychologie* 49 (1984): 348–358.

Norman Owen-Smith, PhD

Artiodactyla

(Even-toed ungulates)

Class Mammalia

Order Artiodactyla

Number of families 10

Number of genera, species 82–84 genera; 221–227 species

Photo: A moose (*Alces alces*) shedding velvet in the Yukon Territory, Canada. (Photo by E. & P. Bauer. Bruce Coleman, Inc. Reproduced by permission.)

Introduction

Artiodactyls are one of the two living orders of terrestrial mammals that comprise the ungulates, or hoofed mammals. These orders are distinguished primarily by the animals' feet: the Artiodactyla are known as the even-toed ungulates in contrast to the Perissodactyla, or odd-toed ungulates. The name Artiodactyla comes from the Greek words *artios*, meaning entire or even numbered, and *dactylos* for finger or toe. Artiodactyls are a highly successful order and the most abundant large land mammals living today with more than 220 species worldwide. This order includes many familiar wild species such as antelopes, deer, bison, and giraffes, along with the familiar and important domestic species such as camels, cattle, goats, pigs, sheep, and water buffalo.

Although many artiodactyl species are relatively large and well known, scientists are still discovering new species. Since 1992, five new species of artiodactyls have been described, including one (*Pseudoryx*), and possibly another (*Megamuntiacus*), making two new genera. Each of the new species occurs in Southeast Asia (Laos, Cambodia, Vietnam). In addition, the Vietnam warty pig (*Sus bucculentus*) previously thought to have become extinct, was rediscovered, and there was also a new species of Bovidae discovered based on horns of the supposed "Linh Duong" (*Pseudonovibos spiralis*), although this may be a hoax as the horns of at least some specimens have turned out to be fashioned from domestic cattle horns.

Evolution and systematics

Understanding the evolutionary beginnings of the early artiodactyls, like that of the early ungulates, is hampered by an incomplete fossil record. Also, the artiodactyls appeared abruptly, along with early perissodactyls, without any clear intermediate forms between the early ungulates and the early artiodactyls. Some aspects of the evolutionary story are difficult to follow because the characteristics used to assign taxonomic position do not fossilize. For example, modern artiodactyls are divided into three suborders—non-ruminants, tylopods, and ruminants—based on the morphology of their digestive tracts, soft internal structures that are not preserved in fossils.

The oldest known fossils, clearly referable to artiodactyls, are in early Eocene deposits from Holarctica (Asia, Europe, and North America). These earliest artiodactyls were relatively abundant and widespread, and include *Diacodexis* and related genera in the Dichobunidae or Diacodexeidae. All were small mammals about the size of a rabbit or hare, weighing probably less than 11 lb (5 kg). They are considered to be early artiodactyls because they had a double-pulley astragalus (part of the ankle joint), which is a defining characteristic of this order, as well as other limb adaptations such as increased length for cursorial locomotion. However, the teeth of *Diacodexis* were still bunodont (low-crowned with rounded cusps), suggesting omnivorous food habits, and the skull shows no other traits diagnostic of artiodactyls. The sudden and widespread appearance of the early artiodactyls at the beginning of the Eocene about 55 million years ago (mya) suggests that they might have evolved elsewhere other than Holarctica. Perhaps they arose in Africa, India, or Central or South America and entered the northern continents only when physical or climatic barriers disappeared.

What is even less certain are the ancestors of these early artiodactyls from amongst the known fossil condylarths. So far, the closest condylarths to *Diacodexis* are the raccoon-like arctocyonids of the middle Paleocene. They were also small, being probably no more than 11 lb (5 kg) with long tails and teeth, suggesting an omnivorous diet. There is a tentatively

In ancient Egypt, the hooves of pigs (*Sus scrofa*) were used to create perfect holes in which to plant seeds. (Photo by Hans Reinhard/OKAPIA/Photo Researchers, Inc. Reproduced by permission.)

identified arctocyonid similar to *Chriacus*, a primitive oxyclaenid condylarth from the Paleocene, as being closest to the oldest artiodactyls so far discovered.

The family Dichobunidae, to which *Diacodexis* probably belonged, is the most primitive group of artiodactyls discovered so far. They are placed in the suborder Paleodonta, along with the closely related Leptochoeridae and the Entelodontidae. The entelodontids were much more advanced than either of the other two families and resembled giant pigs. In one genus, *Archaeotherium*, their elongated skulls had characteristic processes protruding from the jugal bone, as well as bony knobs on the lower jaw, reminiscent of the modern African warthog (*Phacochoerus africanus*). The incisors were blunt and heavy, while the canines were robust and capable of inflicting serious injury. The small molars were almost pig-like and, along with the premolars, were well spaced along the jaw. Their limbs had between two and three digits, with separate metapodials, although the ulna and radius were fused. Paleodonts have been found mainly in Europe, but also North America. They appeared in the early Eocene and became extinct by the end of the Miocene.

The suborder Ancodonta, another group of primitive, presumably non-ruminant artiodactyls, includes a rather loose grouping of three families: Anoplotheriidae, Anthracotheriidae, and Caenotheriidae. The anoplotheres were medium-sized ungulates that became extinct in the Oligocene. Anthracotheres, which probably evolved in the Eocene and flourished in the Oligocene, began as dog-sized mammals that evolved to hippo-size and are thought to have eaten soft plants and lived a semi-aquatic life in swamps. Their remains have been found in Africa, Asia, Europe, and North America, and representatives of this family lasted until the Pleistocene. In the early Miocene (about 18 mya), they are believed to have led to the modern hippopotamuses. The caenotheres were small, four-toed ungulates ranging in size from rabbits to small antelope that became extinct in the Miocene. Although no representatives of these three families exist today, genetic evidence supports the idea that hippos did not descend from the Old World representatives of Tyassuidae as previously suspected. They are more likely to have originated from within the suborder Ancodonta and, based on morphological characteristics, thought to have evolved from the anthracotheres.

Two families, the Agriochoeridae and Merycoidodontidae, are grouped together in the suborder Oreodontae. Their remains have been found only in deposits from Central and North America. The Merycoidodontidae were a very diverse group of small- and medium-sized, stocky-built herbivores, the largest of which was said to be about the size of a wild boar. They were highly successful having appeared in the late Eocene, flourished in Oligocene and Miocene, before becoming extinct in the Pliocene. Early forms had four prominent toes on the fore and hind feet, with a small almost vestigial fifth toe on the fore feet; in later forms, this fifth toe was lost. Their teeth showed interesting and characteristic modifications. There was no diastema (gap), but the lower canines had become incisor-like, while the first premolars had replaced the lower canine in form and function. In most forms, the orbit was closed and the skull often relatively large compared to the rest of the body so that they resembled modern peccaries in many ways. At least two forms showed skull characteristics suggestive of a proboscis. Most members of this family were plains-dwellers, but some were thought to have had an aquatic lifestyle similar to modern hippos. This is based on their skulls in which the eyes and nostrils are located high on the skull, similar to aquatic species. The Agriochoeridae, while less successful, are an equally interesting group of early oreodontid artiodactyls. They shared the same modifications of the lower canines and first premolars, but had a diastema after the upper canine and lower canine-like (caniniform) first premolars. Unlike the merycoidodontids, these animals had an open orbit and a defined saggital crest. Their lumbar region suggested an animal that could leap like a cat, and they also had a long heavy tail. They had five toes, although the first digit was much reduced, and rather than hooves, the toes terminated in claws. Despite this resemblance to carnivores, the teeth suggested that they were herbivorous. They also appeared in the late Eocene, but became extinct at the end of the Oligocene.

Toward the end of the Eocene, the world's climate began to change and by the beginning of the Oligocene epoch 38 mya, the Northern and Southern Hemispheres experienced definite seasons. Seasonality of climate resulted in significant and predictable variation in the growth and abundance of plants. Under these new conditions, both plants and the herbivores feeding on them evolved rapidly. Artiodactyls especially began to diversify and many large species evolved, with all but the pigs and peccaries becoming obligate herbivores. Molars of herbivorous artiodactyls evolved selenodont (crescent-shaped) enamel patterns adapted to grind plant food into small particles, to be followed much later in the Miocene by hypsodont (high-crowned) cheek teeth when the grasslands became established as an important terrestrial ecosystem.

When grasses first flourished about 20 mya in the Miocene, open savannas became a widespread ecosystem and the first specialized grazing ungulates began to appear. Browsing was not abandoned, but many species had mixed grazer-browser habits. Large pecoran ruminants, having either horns or antlers on their skulls, appeared in the early Miocene. These included cervids, bovids, giraffids, and the okapi-like palaeomerycids. By the end of the Miocene, all modern artiodactyl families were present.

In the mid Eocene, helohyids such as *Helohyus* appeared. These were primitive artiodactyls, somewhat larger than *Diacodexis* and with more robust limbs. These probably gave rise to the primitive pig-like *Propalaeochoerus*, which appeared in the late Eocene. By the beginning of the Oligocene, about 40 mya, the pig-like mammals had split into two families, the true Old World pigs (Suidae) and the New World peccaries (Tayassuidae). Both these modern groups have bunodont, relatively low-crowned molars, although in some ancestral forms, the molars are referred to as bunoselenodont because they show similar features to the crescent-shaped molars typical of ruminants. The earliest known peccary is *Perchoerus*, while the first recognized true pig is *Paleochoerus*. The ancestor of the modern *Sus* is probably *Hyotherium* from the early Miocene, which exhibited an elongated skull and tusks oriented laterally.

The earliest primitive tylopod was probably the small, four-toed *Poebrodon* of the late Eocene, followed in the mid Oligocene by *Poebrotherium*, a taller, longer-necked species with fused metapodials and each foot reduced to two toes. Subsequently, evolution in the camels saw the pads replacing hooves and a digitigrade rather than unguligrade posture. Miocene fossil footprints indicate that they had also developed a pacing rather than trotting gait. Miocene camels included *Protomeryx* in the early Miocene and *Procamelus* later in this epoch. *Lama* appeared to have split off at this time, but the other camels continued with the Pliocene (*Pliauchenia*) and then modern camels (*Camelus*) appeared in the late Pliocene. Camels probably originated in North America before dispersing to South America and Eurasia in the late Pliocene.

The late-Eocene Hypertragulidae showed signs of selenodont enamel and higher-crowned cheek teeth and so were probably primitive ruminants. Their upper incisors were reduced in size and, while the upper canine was little changed, the lower canine was reduced and incisiform (incisor-like). Although they had only four toes on the hind feet, they still retained five on the front, but the lateral toes were reduced in size. Their limbs were elongated, which together with their foot structure suggested they were able to run fast. The tragulids proper appeared in the Oligocene, showing a further reduction in the number of toes to four, with reduced lateral digits on both front and hind feet. The most advanced ruminants appeared sometime in the late Eocene–early Oligocene.

A Cape buffalo (*Syncerus caffer*) rolls in mud in Masai Mara, Kenya. (Photo by Erwin and Peggy Bauer. Bruce Coleman, Inc. Reproduced by permission.)

They had further modifications of the foot bones, and more complex cheek teeth.

One group of these early ruminants gave rise to the Giraffidae, Bovidae, Moschidae, Antilocapridae, and Cervidae. *Eumeryx* was a deer-like animal that appeared in the Oligocene and probably gave rise to the earliest giraffes and deer. At the beginning of the Miocene, two early giraffids appeared, *Climacoceras* and *Canthumeryx*, followed shortly after by *Paleomeryx* and *Palaeotragus*. The latter was short-necked, but possessed bony projections from the skull similar to the horn structures of modern giraffes. In the late Miocene, *Samotherium*, another short-necked giraffid appeared, after which the first members of the genus *Okapia* were recognized. It was not until the Pliocene that the long-necked giraffids (*Giraffa*) were seen.

Bovids first appeared in the fossil record in the late Oligocene, however, they were represented only by teeth typical of this family. The first, more complete representative of the early bovids was *Eotragus* from the middle Miocene. It was perhaps similar in size and habits to modern duikers (Cephalophinae), with small horns and slender limbs. Following *Eotragus*, the Tragocerinae evolved later in the Miocene. This was a primitive bovid group with a variety of horn shapes. The closest modern forms to these are probably the chousinga (*Tetracerus quadricornis*) and the nilgai (*Boselaphus tragocamelus*), both from India. The earliest sheep (*Oioceros*) and gazelles (*Gazella*) made their appearance in the mid to late Miocene about 14 mya. *Pachyporlax*, another genus from the late Miocene and a relative of the nilgai, is thought to be related to the earliest wild cattle such as *Leptobos* and *Parabos*, which appeared in the early Pliocene.

Some authors suggest that pronghorns (Antilocapridae) should be included within the Bovidae, but analysis of nuclear DNA suggests that they are a unique lineage. The antilocaprids arose from the Merycodontinae in North America, with the earliest form, *Paracosoryx prodromus*, being recognized from the early Miocene, about 20 mya. The Antilocaprinae followed in the late Miocene and appear to

Tule elk (*Cervus elaphus nannodes*) bulls sparring. (Photo by Tom & Pat Leeson/Photo Researchers, Inc. Reproduced by permission.)

have replaced the smaller Merycodontinae. The antilocaprins evolved into a highly successful and diversified group, especially in the Pliocene, and persisted until the Pleistocene, after which only one species remained, the living American pronghorn (*Antilocapra americana*). The antilocaprids are considered to be most closely related to the cervids rather than the bovids, as was previously thought. Hence, they have been placed together with the Cervidae into the Cervoidea. The deer-like *Eumeryx* of the Oligocene was followed in the early Miocene by *Dicrocerus*. This was the first true cervid to possess antlers, but they were short simple antlers and resembled those of modern-day muntjacs. Diversification of the Cervidae occurred later in the Miocene and also in the Pliocene.

By the late Pliocene, the Artiodactyla assumed more modern forms, and included many of the groups living today. The living mouse deer, or chevrotains (family Tragulidae), are the most primitive of the true ruminants. They are thought to have changed very little and, in fact, still resemble the early ancestors of this group. Analysis of nuclear DNA suggest that the Cervidae and Bovidae appear to be sister taxa, and the Giraffidae more primitive. The Antilocapridae are probably most closely related to the Cervidae and are placed together with the Moschidae in the Cervoidea.

Based on morphological considerations, the accepted taxonomy of modern artiodactyls recognizes three suborders: Suiformes (pigs, peccaries, and hippopotamuses), Tylopoda (camels), and Ruminantia (true ruminants). However, this may soon change. Cetaceans have always been considered closely related to ancestral artiodactyls based on fossil material, but until recently the link was not clear. Whales were considered to have arisen from within the mesonychids, a member of the Condylarthra, which are considered derived from arctocyonids. A key feature used to identify artiodactyls in the fossil record is their double-pulley astragalus. In the year 2000, two fossil early whales were discovered in 47-million-year-old Eocene deposits from Pakistan. One of these whales, *Artiocetus clavis*, possessed such an astragalus. This find, together with genetic evidence from other studies, suggests a clear evolutionary link between cetaceans and artiodactyls. In fact, some scientists contend that the name Artiodactyla should be replaced with Cetartiodactyla to indicate that whales belong to the same order. Based on the evidence to date, the closest ancestor of the cetaceans is believed to be a forerunner of the modern hippo, a notion that is not accepted by those who consider that there is strong fossil evidence showing whales are derived from the mesonychids. If the order Cetartiodactyla is accepted, it would be comprised of five rather than three modern suborders: the monophylic Suiformes (pigs and peccaries), the hippos in the Ancodonta, the Cetacea, Tylopoda, and Ruminantia.

Since the order Cetartiodactyle is not yet accepted, the Artiodactyla is considered to be divided into three suborders with 10 families. The suborder Suina contains three families: the Suidae (pigs), Tayassuidae (peccaries and javelinas), and Hippopotamidae (hippopotamuses). The Tylopoda contains only the family Camelidae (camels and llamas), while the suborder Ruminantia is comprised of the Tragulidae (mouse deer and chevrotains), Giraffidae (giraffe and okapi), Cervidae (deer), Antilocapridae (pronghorn), and Bovidae (antelopes, cattle, sheep, and goats). The modern Artiodactyla include a total of 79–81 genera and 217–223 species. A meaningful number of subspecies is difficult because many are disputed, but there are probably more than 800 recognized.

Physical characteristics

Artiodactyls vary greatly in physical characteristics. They range in size from the diminutive mouse deer (Tragulidae) of Southeast Asia, some of which weigh less than 2 lb (1 kg) and stand no more than 14 in (35 cm) at the shoulder, to the common hippopotamus (*Hippopotamus amphibius*) weighing almost 10,000 lb (up to 4,500 kg). The head varies in shape from a short to long facial structure, with hairy to naked muzzle that is either large or small. The laterally positioned eyes are often large with long lashes, and the ears, with either rounded or pointed tops, can be relatively large or small compared to the head. Neck length can be short to very long. In most species the hair covering the neck is relatively short, but some species, especially adult males, have longer hairs along the ventral edge forming a clearly defined ruff. In others, a flap of skin, or dewlap, hangs from the ventral surface. The back may be straight, or the shoulders higher than the rump as in bison and gaur, or the reverse as seen in the duikers. Tail length also varies widely, from very short to long. Hair length may be long over the entire tail or long only in a terminal tuft. The legs can be relatively long to short and relatively slender in most species, except in hippos, whose legs are quite stout. Hooves also vary from narrow to broad in width and short or long in length. The body pelage, made up of longer guard hairs and shorter underfur, shows a range from short and smooth to dense and long, although the longer hairs are present usually only in certain body regions. The longest guard hairs of any mammal are found on the musk oxen of the Arctic. Hair coat usually changes with the seasons, and in some species, coloration differs between sexes and among age classes.

The skull lacks an alisphenoid canal, and there is no posterior expansion of the nasal bones. The teeth are heterodont and highly specialized according to food habits, with most being adapted for a wholly herbivorous diet. The bovids, cervids, and giraffids have all lost the upper incisors. The lower canines are usually small and modified to function as incisors (incisiform). The upper canines in bovids and giraffids are absent. They are also absent in most cervids, but in some species in this family, they are present though small and blunt (e.g., *Cervus*). In the Tragulidae, muntjacs, and some of the small antlerless deer, upper canines are enlarged, while in the Suidae, Tayassuidae, Hippopotamidae, and Camelidae both upper and lower canines are well developed. There is almost always a space, or diastema, on the lower jaw, between the canine and first premolar. The dental formulae vary: (I0–3/3, C0–1/1, P2–4/2–4, M3/3) × 2 = 30–44.

Pere David's deer (*Elaphurus davidianus*) originally occurred in northeastern and east-central China, but became extinct in the wild over 1,000 years ago. (Photo by Animals Animals ©Darek Karp. Reproduced by permission.)

Female artiodactyls have two to four teats, except Suidae, which have six to 12. The gestation period lasts from five to 11 months, depending on species. Triplets are rare, except in suids. Physiological sexual maturity is typically reached at 18 months of age in both sexes, with females generally giving birth for the first time when they are two years old, In many species, males only begin to fully participate in mating several years later than females. Longevity varies considerably with species, ranging from 10–30 years, but average age at death (life expectancy) is much lower.

The defining characteristics of the Artiodactyla are their number of toes (digits) and the structure of their astragalus. Almost all have fewer toes than the five of the ancestral vertebrate plan. Except for two species (*Pecari* and *Tayassu*) in the Tayassuidae, for which there is conflicting evidence about the number of toes on the hind feet, all other artiodactyls have an even number of functional toes, either two or four, on each

A Valais goat (*Capra aegagrus hircus*) from Switzerland. (Photo by Tom McHugh/Photo Researchers, Inc. Reproduced by permission.)

foot. The first digit of the original mammalian plan of five digits has been lost during the evolution of artiodactyls. As a result, the symmetry of the artiodactyl foot passes between the middle two digits (third and fourth), creating a limb structure referred to as paraxonic, in which the weight is born on these two central elements. In artiodactyls with two main toes, the second and fifth digits are either reduced, vestigial, or absent; when present, they are referred to as lateral hooves or dew claws. The terminal phalanges of the two weight-bearing toes and the dewclaws are covered with keratin sheaths called hooves. In species with four functional (weight-bearing) toes, the toes form a spreading foot. Usually there are no hooves, but the nails at the end of the four toes are often enlarged. Most artiodactyls also have elongated metapodials that, except in pigs, are fused into a single functional unit.

Most artiodactyls share the same adaptations for speed as do other ungulates, including lighter structured feet and limbs, reduced lower limb musculature, strong attachment of the hind limbs and loose attachment of the front limbs to the vertebral column, and leg movements restricted to a fore-aft motion. Artiodactyls have another adaptation that affects limb movements and increases running efficiency. The astragalus, a tarsal (ankle) bone in their hind limb, has deeply arched grooves on both ends where it articulates with the corresponding limb bones. These grooved joints help resist lateral motion, and also create an efficient double-pulley system that increases the flexibility and springiness of the lower hind limb.

Generally, the hard hooves, coupled with small feet and long, light limbs, provide excellent adaptations for fast running. However, some species of artiodactyls show secondary foot adaptations to travelling over soft ground. In dromedary (*Camelus dromedarius*) and Bactrian (*C. bactrianus*) camels, for example, both toes on each foot are enlarged relative to the body size, thus providing a greater surface area that helps prevent the animal from sinking deeply into loose sand. Similarly, caribou and reindeer, which travel long distances over snow in winter, have relatively large main and lateral hooves. Together, these four toes serve to increase the foot's surface area, and in effect act as snowshoes to help the animals negotiate areas of deep snow. These deer also use their large feet to paw through the snow and uncover buried lichens.

Almost all species of artiodactyls have weapons of some kind, depending on family; for example, unbranched horns are present in all Bovidae and Giraffidae, forked horns in Antilocapridae, antlers in the Cervidae, and well-developed canines or tusks are typical of the Suidae, Tayassudiae, Tragulidae, and Moschinae. Well-developed upper canines are also found in some species of deer such as the antlerless water deer (*Hydropotes*), while the small-antlered muntjacs (*Muntiacus*) and tufted deer (*Elaphodus cephalophus*) have both antlers and large upper canines. New antlers grow each year and are shed, usually soon after the rut (mating season). In the bovids such as argali (*Ovis ammon*), horns are not shed but continue to grow throughout the animal's life. However, in the pronghorn (*Antilocapra americana*), which also possesses horns similar in structure to bovids, the keratinized horn sheath is shed each year.

In almost all cases, these weapons are largest in adult males and are either smaller or absent in females. Horns, antlers, and tusks are used primarily for intraspecific competition and not for defense against predators, as was first pointed out by Charles Darwin. He argued that the great variety of weapons, most of which were not particularly effective against predators, along with the fact that females either lacked or had much smaller weapons than males, and that males rarely defended females and young, supported an intra- rather than inter-specific function for these weapons. It is true, however, that under extreme circumstances, an animal might use its weapons against a predator, but this is usually a last resort and is not their primary function.

The coat or pelage consists of two parts: longer, stout guard hairs and shorter, usually finer, underfur. The guard hairs serve to shed rain and snow, and together with the underfur help control heat exchange. Most breeds of domestic sheep (*Ovis aries*) have been selectively bred so that they no longer have guard hairs, but have retained the underfur, or wool. Pelage color varies greatly across species, ranging from white to black, but most often varying shades of brown. The coat coloration of young artiodactyls in the first few months of life often is often distinctly different from the adults, and in some species, male coat color is related to age and social status. Species living in temperate and arctic regions shed their warm winter coat in early spring and grow a sleeker, shorter one for summer. Many species have a distinct color pattern. White spots or stripes against a dark coat can break up the animals outline, making it more difficult to see in the dappled sunlight filtering through forest vegetation. Other conspicuous coat patterns are used in social communication; for example, the erection of the hairs on a white rump patch signals danger to other members of the group.

Many species of artiodactyls have glands in different areas of the body, that they use for communication. Most are epithelial (skin) glands and are often paired, one on each side of the body. They secrete odoriferous chemical secretions either as volatile chemicals or waxy substances. Many territorial species use glandular secretions, sometimes along with urine and feces, to mark territorial boundaries. For example, adult male Thomson's gazelles (*Gazella thomsonii*) deposit secretions from their preorbital glands on twigs and tall grass stems along territorial boundaries. Some non-territorial species also use glandular secretions to mark objects in their environment, perhaps to advertise their presence. For example, male deer commonly rub their antlers against shrubs or small flexible trees, so species with glands on their heads probably leave olfactory signs, along with visual evidence of the scratched and broken vegetation. Muntjac use their large preorbital glands located in front of their eyes to mark conspecifics. Other species use their glands for self-marking, for communicating alarm, and possibly to advertise their physiological condition. Urination accompanied by wallowing is a common example of self-marking used by male artiodactyls in the mating season to advertise their presence. Other forms of self-marking include hock-rubbing, seen in some species of odocoilid deer, and urination spraying shown by caprid species such as alpine ibex (*Capra ibex*). Again, males in the rut most frequently perform these behaviors. Alarm signals may be given when the tail is raised in alarm, or from the tarsal glands when the hairs of the gland are erected on the hind legs. Interdigital or pedal glands found between the main hooves of many artiodactyls are thought to mark trails and bedding sites. In general, however, the role of most artiodactyl glands and their secretions is poorly understood.

Distribution

Artiodactyls are the most widespread of the ungulates. They are native to all continents, except for Antarctica and Australia, and are absent from the oceanic islands. However, introductions, primarily of domestic species, have been made around the world to areas outside their normal range, including to many small remote islands, where they have usually thrived.

Habitat

As Artiodactyls are widely distributed across much of the world, it is not surprising that they exhibit great variation in the habitats that they occupy. One factor that seems to define artiodactyl habitat is the presence of sufficient plant biomass to sustain their numbers. Depending on species, they inhabit most ecosystems and habitats from arctic tundra to tropical forest, including both hot and cold deserts, and they can be found at elevations ranging from valley floors to mountaintops. Four major patterns of habitat use occur. Some species (e.g., American pronghorn) specialize in exploiting open grasslands where they feed and at the same time use the excellent visibility to detect approaching predators. Rather than hiding, they simply move away relying on early detection and speed to avoid their predators. Other species (e.g., bighorn sheep, *Ovis canadensis*) specialize in exploiting open grasslands and meadows near the steep cliffs. This combination of habitats offers good foraging in the meadows with excellent security on narrow ledges and steep terrain. Other species (e.g., okapi, *Okapia johnstoni*) dwell in forest or shrubland. In this habitat, they feed on the great variety of available plants, but also obtaining concealment from predators among the dense vegetation. The fourth pattern is shown by species (e.g., roe deer, *Capreolus capreolus*, from Europe or axis deer, *Axis axis*, from India) that inhabit the ecozone between forest and open areas. In the complex landscapes comprised of patches of forest or thickets juxtaposed with open areas, these species move between forest and open habitats, using each for different resources. They might use forest for shelter from the sun and for security cover from predators during the day, often resting along the forest edge where the have a clear view of predators and ready access to the concealment offered in dense forest. During the night, or in the early morning and late evening, they move into open areas and along the forest edge where they feed on lush forbs and other vegetation.

A moose (*Alces alces*) cow feeding on aquatic vegetation. (Photo by Kerry Givens. Bruce Coleman, Inc. Reproduced by permission.)

The habitats used by artiodactyls are linked to features of their biology and there are clear trends with body size and taxonomy. Small- to medium-sized artiodactyls often use relatively tall, dense vegetation that provides both food and cover (e.g., dik-dik, *Madoqua* spp.). Most members of the Caprinae (wild sheep and goats) inhabit mountain regions where they find security in steep cliffs and feed in adjacent grasslands. The hippopotamus (*Hippopotamus amphibius*) provides another contrast because members of this African species feed primarily on land at night, and grazing on vegetation growing away from the rivers and pools that they return to during the day.

Behavior

Few artiodactyls are truly solitary, and even though they may consist of only two or three members, most species live in groups. Typically, however, the sexes remain separate for most of the year (sexual segregation), with adult males living apart from adult females and young. The sexes often use different habitat types, different parts of the range, or both.

Caribou (*Rangifer tarandus*) bulls grazing in Denali National Park, Alaska, USA. (Photo by Martin Grosnick. Bruce Coleman, Inc. Reproduced by permission.)

Group living seems driven primarily by predation pressure against which it confers several advantages, but probably the most important are dilution and vigilance. Sexual segregation is most likely driven by a trade-off between food requirements and predator avoidance. Males generally require more energy in order to grow large bodies and weapons for intraspecifc competition, and because of their larger body size and the fact that they have no vulnerable young to care for, they can use areas with higher predation risk than can females. Females are responsible for rearing the young, which are highly vulnerable to predators especially in the first few month of life. Females also need sufficient high-quality food for lactation, so they must balance predation risk for their offspring against their own food requirements. When the young are small, females tend to use areas with lower predation risk than do males, even if that means using areas with poorer quality or less abundant food.

Artiodactyls have a diverse array of weapons that they employ primarily in intraspecific interactions. The weapons often function not only for offence, but also have defensive functions. Antlers of deer are an excellent example of this dual function. Most cervids have antlers with several sharply pointed branches. Although such antlers are capable of inflicting serious injury, the forked branches also allow the animal to catch an opponent's antlers and so avoid injury. As a result of this dual offensive-defensive function, male deer fight head-to-head with antlers together as they try to twist each other off balance so they can stab the opponent with the antler points. Similar fighting strategies are seen in horned ruminants as well. Horns that grow in spirals, often with ridges, act in a similar way to branched antlers. The twists and ridges help to catch an opponent's horns, and the two often wrestle head-to-head, again trying to gain the upper hand and deliver an injurious blow.

Physical combat almost always carries a high risk for artiodactyls because, even though an individual might not be injured or killed, fights cost a lot of energy that could be used for mating or for feeding. Many artiodactyls use displays instead of actual fighting in an attempt to manipulate an opponent to withdraw. Displays are behavior patterns that are characteristically conspicuous and are usually directed at a conspecific. They include, for example, postures, vocalizations, and in some species, specialized morphological features have evolved as part of the display. Displays are used as threats, and as signals of dominance or submissiveness. Threat displays are typically aggressive intention movements, indicating a readiness to fight and almost always involve the weapons and stance of fighting style. For example, weapons such as antlers or horns may be pointed towards an opponent or be used to "attack" a nearby object such as a bush. Suids, which use their sharp canine teeth for fighting, threaten by grinding their teeth. Many artiodactyls also vocalize in aggressive situations (e.g., roaring by male bison, bellowing of domestic cattle, "bugling" of wapiti). When performing a dominance display, the animal tries to appear as large as possible and often they achieve this by standing sideways to an opponent (lateral or broadside display). Species such as chamois and kudu can raise the long hairs (piloerection) that run along the upper side of the neck and along the back, which in effect increases the apparent size of the displaying animal. Others such as some of the wild cattle (e.g., gaur, bison) have evolved elongated thoracic spines, which also increase the lateral profile. Submissive displays, on the other hand, appear to function by reducing aggression in an opponent, and it is common to see a submissive animal trying to make itself as small and non-threatening as possible. Frequently, subordinate males may even mimic the behavior of females.

Coloration patterns are also typically involved in displays and other social signals, but their connection to their "message" is not always obvious. For example, aggressive individuals will drop their ears as an indication of their behavioral state. In Nile lechwe (*Kobus megaceros*), the ears stand out noticeably against the black head and neck, further adding to the conspicuousness of this display. Other features of the pelage may have communication functions, and are mainly directed towards conspecifics. These characteristics are the result not only of different coat coloration, but also of different hair length. White-tailed deer (*Odocoileus virginianus*), for instance, raise their tails, exposing the long white hairs on the underside and over the rump, then bound off waving their tail from side to side as a warning signal to other deer that a predator is present. Other species such as pronghorn and Grant's gazelle (*Gazella granti*) have uncovered, prominent white rump patches that are permanent displays. They continuously signal among the group and are thought to help maintain group cohesiveness and alert members to disturbances. Also, coat characteristics may develop only in mature age and sex classes—in some Asiatic wild sheep and goats, only mature males develop large neck ruffs or beards, while in the Indian blackbuck antelope, males change from a reddish- to a black-colored coat on reaching maturity.

Feeding ecology and diet

Except for the Suidae and Tayassuidae, artiodactyls are obligate herbivores relying primarily on plants as their source of energy. The herbivorous diet of artiodactyls and the major adaptation to their anterior digestive system probably explains some of their success, because plants offer a diverse and abundant food source in most ecosystems. No mammal possesses enzymes capable of digesting cellulose or lignin, and so most obligate herbivores rely on microorganisms to breakdown these plant compounds. Within the artiodactyls, the most successful groups not only use microorganisms to help them breakdown plant tissues as do perissodactyls, but they also ruminate.

All artiodactyls have one or more chambers, or false stomachs, located just ahead of the true stomach, or abomasum. The pigs and peccaries have only one small chamber before the true stomach, hippos, camels, and tragulids have two, and Cervidae and Bovidae have three false stomachs (rumen, reticulum, and omasum) before the true stomach. Bacterial fermentation takes place in the first and largest chamber, the rumen, hence the alternative name for ruminants, the foregut fermentors. The large surface area of the cheek teeth with their selenodont enamel pattern, facilitate grinding of forage. The efficiency with which ruminant artiodactyls are able to digest plants is aided by their ability to ruminate, or chew their cud. In this process, the largest food particles move from the rumen to the reticulum where they are formed into a ball, or bolus, by the action of the honeycombed (reticulated) inner surface of this second chamber. The bolus is then regurgitated up the esophagus into the mouth, where it is once more chewed and mixed with saliva, before being swallowed once more. Together, these adaptations allow efficient digestion of vegetation and contribute to the success of artiodactyls as key components of ecosystems across much of the world.

Assisted by regrinding during rumination, the microorganisms are able to digest the smaller food particles, breaking them down into even smaller pieces. When reduced to a certain size, the particles move from the reticulum to the omasum through a small orifice that acts like a sieve and restricts the flow of food. In the omasum, most of the water is squeezed out of the food and reabsorbed through the folded and highly muscular walls. The now-drier food mass moves next into the abomasum, the true stomach where protein digestion begins. Most protein digestion, however, takes place in the intestines and the resulting amino acids are absorbed. The final step of digestion in ruminants occurs in the cecum, located towards the end of the gut where additional microbial fermentation takes place.

Ruminants are almost entirely dependant on the microorganisms for extracting nutrients from plants. Not only do the microorganisms allow the artiodactyls to extract energy from cellulose, but they are also the main protein source. The microorganisms use the plant protein to reproduce and, when they are transferred with undigested plant material from the rumen into the omasum and eventually the intestines, the ruminant digests the microbes, thereby recovering the protein that they contain. As a result, ruminants gain more valuable amino acids than if they relied on digesting plant proteins.

In addition to the benefits from the symbiotic relationship with the microorganisms for digesting their food, artiodactyls gain other advantages from rumination. They efficiently recycle nitrogen in the form of urea. During the fermentation process in the rumen, dietary protein is converted mainly to ammonia, which is either used by the bacteria or absorbed through the rumen wall and sent to the liver. Here, it is converted to urea, which is then returned to the rumen, either directly through the rumen wall or via saliva from the parotid glands. Another major advantage is that, after ruminants have filled their rumens, they can move to safer locations and habitats to digest their food. By selecting secure habitats to rest and digest, not only do they reduce predation risk, but also energy expenditures for locomotion are lowered.

A red goral (*Naemorhedus baileyi*) in Myanmar. (Photo by Tom McHugh/Photo Researchers, Inc. Reproduced by permission.)

While the majority of artiodactyls rely almost entirely on plants as a food source, there are well-documented records of this order, besides the Suidae, occasionally eating the eggs and young of ground-nesting birds as well as other sources of animal protein. These seem to be taken on a purely opportunistic basis and are a minor part of the normal diet, although it may be more frequent in the smaller-bodied artiodactyls.

There is a great range in body sizes among ruminants, and thus there are differences in both metabolic needs and the abilities to meet these needs. Small ruminants have relatively greater energy requirements than large species, and this influences the ways that they exploit plants. Although there can be exceptions, large species tend to be bulk feeders and thus less selective by gathering large quantities of low-quality vegetation. On the other hand, small species tend to be concentrate feeders and thus more selective by consuming plant species and parts that are high in nutrients and digestibility. Such feeding strategies facilitate division of resources and allow a diversity of artiodactyl species to coexist within the same ecosystem. This is especially evident in East Africa where dikdiks to African buffalo all form a species-rich community of herbivores.

Reproductive biology

All but the Suidae give birth to one, sometimes two young each year. Triplets are rare, but the pigs give birth typically to between four and eight young, with domestic pigs regularly giving birth to more than a dozen piglets per litter. Most species breed once each year, although in some tropical species there are two birth periods per year. The birth season is usually timed to coincide with onset of seasonal plant growth. As a result, most species give birth either in early spring in temperate and arctic regions or at the beginning of rainy season in tropics. The new flush of green nutritious vegetation at these times benefits milk production, which is the greatest physiological cost experienced by a healthy female. In addition, being born at the beginning of the plant-growing season allows the young a long period over which to grow while food is relatively abundant and of high quality. The faster the young grow, the lower their risk of predation.

The young of all artiodactyls are precocial, capable of walking and even running in some species within a few hours after birth. They can be classed into one of two broad types, depending on whether their mothers leave them during the day to feed elsewhere (hiders), or whether the young stay close to her (followers) during their first few weeks of life. Hider young have coat colors or patterns that help keep them concealed. Hider mothers lead their young to where they will leave them, and the young select the actual hiding site. The mother returns briefly once or twice during the day to nurse and clean her offspring. Later, when a hider young is older and more mobile, it accompanies its mother. Both are probably anti-predator tactics related to the visual density of the habitat and the size of the young. Species with the hider strategy tend to belong to smaller groups that inhabit habitats with suitable hiding places. Followers are typically larger species living in large groups and open habitats with few places to hide. Pigs show a variation of the hider strategy. The female just prior to birth builds a nest in which she has her litter of young. The piglets stay in the nest with the mother for a few days after parturition, then they are able to follow her.

The most common mating system is polygyny in which one male copulates with several females each mating season. In a few species such as the blue duiker (*Cephalophus monticola*), mated pairs may stay together most of the year, but such a mating system is the exception in artiodactyls. In many others, temporary mating pairs form and the male defends and courts a single female for as long as she is in estrous (e.g., *Bison*), or the male defends several females (harem) against other males (e.g., courting and mating with each female when she comes into heat). Both of these are forms of female defense polygyny. Another mating system, resource defense polygyny, is seen in many species such as the pronghorn (*Antilocapra americana*) or various African antelopes. In this system, a male defends a territory that attracts females because it contains valuable resources such as food or because it provides safety from predators. Females are also sometimes drawn to the male himself. Whether males hold territories during the rut depends on the economics of defense—whether the distribution and quality of resources that might attract females outweigh the costs of defending it. The most extreme form of mating territoriality is lekking, a behavior seen in fallow deer (*Dama dama*), topi (*Damaliscus lunatus*), and in several species of kob. In these species, large numbers of males congregate on a relatively small patch of ground where they defend territories that are no more than a few feet (meters) in diameter. When a female wanders onto one of these defended spaces, the males attempt to keep them there long enough to mate with them. Lekking does not occur in all populations of a species, but appears to depend, among other factors, on population density.

Conservation

A total of 168 artiodactyl species are listed in the IUCN Red List of threatened mammals. The listings statistics for this order are: Extinct: 7; Extinct in Wild: 2; Critically Endangered: 11; Endangered: 26; Vulnerable: 35; Near Threatened: 1; Lower Risk: 73; and Data Deficient: 13. Details of which species are threatened can be found at the IUCN Red List Web site, which is regularly updated.

Of the 155 species for which status has been determined, 53% (82 species) are in categories of conservation concern. Threats to artiodactyls range from over-harvesting (including poaching) to habitat loss and degradation (including deforestation, conversion to agricultural purposes), and competition with domestic livestock. Increased access such as roads and railways into to wildlife habitat exacerbates and precedes most of these threats. All, however, revolve around the central issue of increasing human demands for diminishing natural resources.

Significance to humans

Wild artiodactyls were the major large mammal prey for early hunter-gathers, and are still important today as sources of animal protein for many people who otherwise exist on subsistence agricultural production. Certain wild species (e.g., red deer, wild sheep and goats, kudu, antelopes, and African buffalo) are still much sought after by sport and trophy hunters. Rock and cave paintings and carvings attest to their long history of importance to humans; artiodactyls were the most common mammalian species depicted by the artists of the Paleolithic and later periods.

The most important domestic livestock species are cattle, sheep, goats, camels, pigs, all of which are artiodactyls. Sheep and goat were probably the first species to be domesticated after the dog, about 8,000–9,000 years ago. Domestic artiodactyls are key species in agriculture and food production, and as such are vitally important to all human societies throughout the world. Whether wild or domesticated, artiodactyls provide meat, fur, fiber, bones, medicinal products, antlers, and horns, while domestic species, in addition, are sources of milk, draught power, fertilizer, and wealth. Despite their usefulness, introductions of livestock, especially of domestic goats and pigs, to non-agricultural areas and islands (e.g., Isabela Island in the Galápagos) where they roamed freely, created feral populations, which, in turn, invariably lead to habitat degradation and conservation disasters.

Resources

Books

Byers, J. A. *American Pronghorn: Social Adaptations and the Ghosts of Predators Past.* Chicago and London: The University of Chicago Press, 1997.

Sowls, L. K. *Javelinas and Other Peccaries: Their Biology, Management, and Use.* 2nd ed. College Station: Texas A&M University Press, 1997.

Walther, F. R. *Communication and Expression in Hoofed Mammals.* Bloomington: Indiana University Press, 1984.

Periodicals

Gingerich, P. D., M. ul Haq, I. S. Zalmout, I. H. Khan, and M. S. Malkani. "Origin of Whales from Early Artiodactyls: Hands and Feet of Eocene Protocetidae from Pakistan." *Science* 293 (2001): 2239–2242.

Janis, C. M. "Evolution of Horns in Ungulates: Ecology and Paleoecology." *Biological Reviews* 57 (1982): 261–318.

Janis, C. M. "New Ideas in Ungulate Phylogeny and Evolution." *Trends in Ecology and Evolution* 3 (1988): 291–297.

Matthee, C. A., J. D. Burzlaff, J. F. Taylor, and S. K. Davis. "Mining the Mammalian Genome for Artiodactyl Systematics." *Systematic Biology* 50 (2001): 367–390.

Montgelard, C., F. M. Catzeflis, and E. Douzery. "Phylogenetic Relationships of Artiodactyls and Cetaceans as Deduced from the Comparison of Cytochrome b and 12S rRNA Mitochondrial Sequences." *Molecular Biology and Evolution* 14 (1997): 550–559.

Nikaido, M., A. P. Rooney, and N. Okada. "Phylogenetic Relationships among Cetartiodactyls Based on Insertions of Short and Long Interspersed Elements: Hippopotamuses Are the Closest Extant Relatives of Whales." *Evolution* 96 (1999): 10261–10266.

Other

IUCN Red List Web site. <http://www.redlist.org>.

David M. Shackleton, PhD
Alton A. Harestad, PhD

Pigs
(Suidae)

Class Mammalia
Order Artiodactyla
Suborder Suiformes
Family Suidae

Thumbnail description
Medium-sized omnivores, characterized by elongated head and discoid snout, bristly pelage, tusks, and short tail

Size
34–83 in (86–211 cm); 77–770 lb (35–350 kg)

Number of genera, species
5 genera; 16 species

Habitat
Grassland, forest, steppe, desert, swamp, and agricultural fields

Conservation status
Critically Endangered: 2 species; Endangered: 1 species; Vulnerable: 2 species; Data Deficient: 1 species

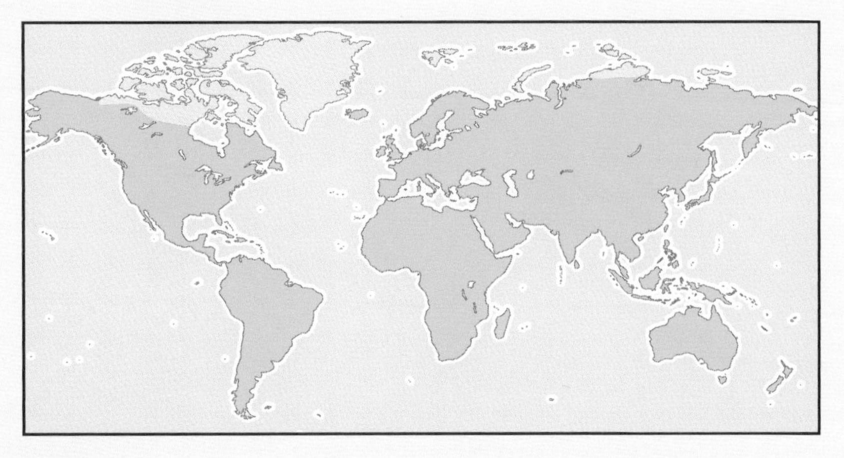

Distribution
Global, in association with humans; all continents, except Antarctica

Evolution and systematics

The family Suidac is one of three extant families belonging to the suborder Suiformes that has its origins about 48 million years ago (mya) in the middle to late Eocene. The Suiformes (pigs, peccaries, and hippopotamuses) are anatomically the most primitive among the Artiodactyls, being characterized by having simple stomachs, many low-crowned, bunodont teeth, and a less-advanced unguligrade limb structure.

During the Eocene epoch 60 mya, there were giant pigs (entelodonts) that may have been distant relatives of today's pig; however, there is no evidence of a direct lineage. Paleontological evidence suggests that the origin of the pig is in the Miocene about 40 mya. At this time, their distribution was restricted to the Old World, and there is not any evidence that North America was the site of origin for any modern genera. The African suids underwent a major radiation during the Pliocene and early Pleistocene epochs, with approximately 17 species known from the fossil record; most of this diversity had disappeared by the mid-Pleistocene. The African suid radiation has served as a useful tool in reconstructing hominid paleoecology because hominids and suids had some ecological characteristics in common. In addition, there are many suid taxa that appeared, evolved, and disappeared in a relatively short period of time in Africa, making them useful markers in biostratigraphy for aging archaeological sites. It has also been suggested that Pleistocene suids competed with *Homo erectus* for subterranean food sources.

The genus *Sus* is thought to have originated at the latest in the Miocene or near the Miocene/Pliocene boundary, 5 mya. Genetic studies suggest that the origin of the babirusa (*Babyrousa babyrussa*) is between two and 10 mya, and that of the warthog (*Phacochoerus africanus*), between five and 15 mya; these estimates are approximately 50% more recent than those suggested by paleontological and anatomical data. The only babirusa fossils to date are from the Pleistocene, and known fossils of the warthog are no older than two million years.

The 1993 action plan of the IUCN/SSC Pigs, Peccaries, and Hippo specialist group recognizes three subfamilies of Suidae: the Suinae, or true pigs, the Phacochoerinae (warthogs), and the Babirousinae (babirusa). The Suinae comprises three genera (*Sus*, *Potamochoerus*, and *Hylochoerus*) and 11 species. The Phacochoerinae includes two extant species (*P. aethiopicus* and *P. africanus*), and the Babirousinae includes one species (*Babyrousa babyrussa*). However, recent reports by Groves in 2001 and 2002 suggest that some previously identified subspecies may in fact warrant full species status. Anatomical studies suggest that there may be as many as 40 subspecies within the Suidae. Approximately 16 more or less distinct subspecies have been found in the *Sus scrofa* lineage.

Physical characteristics

The Suidae are medium-sized mammals with stocky, sometimes barrel-shaped, bodies. Body weight ranges from 77 to 770 lb (35–350 kg) and may be as high as 990 lb (450 kg) in some domesticated breeds. The length of the body is 34–83 in (86–211 cm), while body height is 21–43 in (53–109 cm). The exception is the pygmy hog (*Sus salvanius*), which is only 20–28 in (51–71 cm) in length, 10–12 in (25–30 cm)

A warthog (*Phacochoerus aethiopicus*) with young. (Photo by Nigel Dennis/Photo Researchers, Inc. Reproduced by permission.)

in height, and 14.5–21 lb (6.5–9.5 kg) in weight. The neck is short, whereas the head is long and pointed with a mobile snout. The tip of the snout is cartilaginous and discoid. The eyes are fairly small, and the ears are generally long, with a tassel of hair at the end in some species. The first digit is absent. Each foot has four digits, the middle two of which are flattened and bear hooves. The outer digits are higher up the leg and bear smaller hooves.

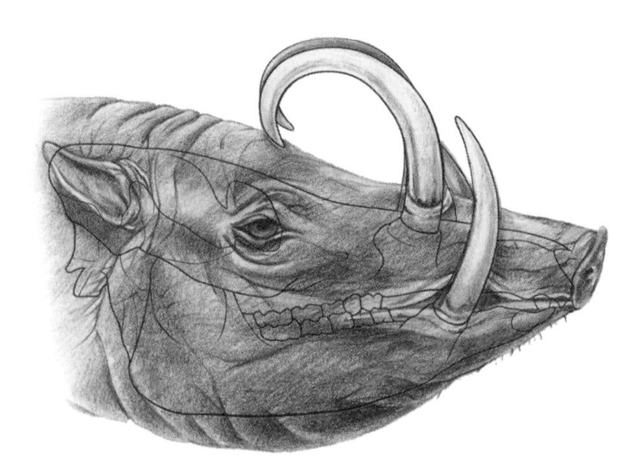

An adult male babirusa's tusks grow through the muzzle and curve back toward the face. (Illustration by Patricia Ferrer)

The complete dental formula is 44 teeth (3143/3143 per quadrant) in *Sus*, *Potamochoerus*, and *Hylochoerus*, while *Babyrousa* and *Phacochoerus* have 34 teeth (2123/3123 and 1133/3123, respectively). The upper canines are large and curve upward, protruding from the mouth to varying degrees. The exception is the babirusa, in which the upper canines protrude upwards through the skin of the head, never entering the mouth. The lower canines are generally sharper and may or may not be visible outside of the mouth. Suids have a simple, non-ruminating stomach composed of two chambers.

Skin color varies greatly in the suids, from brown to almost black, whereas the pelage may be dark gray or black to light red and of varying length. Some babirusa appear almost naked. Some species have a mane or dorsal crest of hair. The male Visayan warty pig (*S. cebifrons*) bears a long mane that is shed after the breeding season. In many species, the young bear stripes. Some members of the Suidae bear warts or fleshy ridges on the face. These are fleshy structures with no bony core or support. Females have from two to four pairs of mammae.

Distribution

Today, wild suids are found on every continent, except Antarctica, as well as on many oceanic islands. Their presence in the New World and on some islands (e.g., New Guinea, New Zealand, and possibly Madagascar) is due to introductions by humans. The domesticated pig has a global distribution due to its association with people, and has returned to a feral state in some areas (e.g., North America). The historical range of the Suidae was restricted to the Old World. Wild pigs occurred throughout much of sub-Saharan Africa, Europe, Asia, Asia Minor, India, and the East Indies as far southeast as the Philippines and Sulawesi. The ranges of many species have decreased in modern times because of expanding human populations and the associated loss of habitat and hunting pressure. In some cases, and for some species (e.g., wild boar), the conversion of land for agriculture has benefited local populations of wild pigs and allowed them to expand their range.

Habitat

The wild suids are found in a diverse array of habitats ranging from semi-arid environments and temperate woodlands to tropical rainforests and swamps. They also often take advantage of agricultural lands. They occupy altitudes ranging from sea level to over 13,000 ft (4,000 m). Habitat selection is dictated by the availability of energy-rich foods, climactic extremes, and predation pressure. Some species may occupy a range of habitat types as long as they have adequate food resources (e.g., wild boar), whereas others specialize on a particular habitat (e.g., pygmy hog). Forest type, elevation, and forest age have been found to influence habitat selection in some populations of wild boar. It has been suggested that wild pigs play an important role in forest diversity, regeneration, and structure through their depredation of seeds and young saplings. Studies of the effects of wild pigs on the environment have had varying results. Wild pigs may increase or decrease forest species' richness, negatively impact regeneration of trees, remove competitive vegetation such as weeds, and decrease soil macroinvertebrate populations. There is also evidence the Eurasian wild pigs will take advantage of acorn caches created by small mammals.

The white pig has been domesticated for large-scale human consumption. (Photo by Ernest A. James. Bruce Coleman, Inc. Reproduced by permission.)

A red river hog (*Potamochoerus porcus nyassae*) burrowing for roots. (Photo by Tom McHugh/Photo Researchers, Inc. Reproduced by permission.)

The suids can be divided into two groups in terms of territoriality or home range use. The African suids are generally more sedentary than other wild pigs, occupying small territories or home ranges that may overlap. The other suids are also more or less sedentary, but changing ecological conditions may cause these species to roam in search of better places to forage, sometimes over long distances. Whether territorial or not, areas occupied by wild suids tend to have several features in common. Resting places are one such feature, and these tend to take the form of nest sites or burrows. Most wild suids construct nests for farrowing, whereas others build nests for protection in bad weather (e.g., *P. larvatus*), but some construct nests all year round (*S. salvanius*). Most wild suids make nests of varying complexity and structure, using vegetation. Nests may be used year round or seasonally. Warthogs often occupy burrows dug by aardvarks (*Orycteropus afer*) and do not build their own nests. Wild suid home ranges also include tree trunks, rocks, or anthills that are used as spots for rubbing and scratching the body. These home ranges also include sources of shade and water, as well as mud wallows. These features are of critical importance as some suids do not have functional sweat glands for cooling the body. In forest hogs and warthogs, traditional defecation sites have been reported. Finally, all home ranges include sites for foraging. These features are often connected by a well-worn set of trails. Home range size fluctuates in response to food availability, reproductive condition, population density, age, and sex. Home ranges of 125–965 acres (52–390 ha) have been reported for wild boar.

The Vietnamese pot-bellied pig (*Sus* sp.) is one of many pig breeds. (Photo by Chris Sham/Photo Researchers, Inc. Reproduced by permission.)

Behavior

Suids are generally characterized by a loose social organization in which the basic unit is the mother-offspring pair. Group sizes vary from one to 15 individuals. Females may live alone or in groups with other females called "sounders." Offspring may remain in the natal group up to two years. Natal males almost always disperse, whereas natal females may sometimes remain in the group permanently. Males associate closely with females only during the breeding season; however, the African suids differ in this regard. Male forest hogs live with family groups year round, and male bush pigs associate with females longer to help rear the young. Male warthogs join females for breeding, leave, and then return after the young are born. Group size sometimes increases in proximity to fixed resources such as saltlicks or in response to availability of resources. Large groups may sometimes split into smaller ones if hunting pressure is high. Territorial or home range marking has not been well documented, but it has been suggested that the plowing behavior of babirusa may have a scent-marking function. The tusk gland has also been suggested to play a role in territorial marking. Phermones produced in the salivary glands of boars have been shown to induce standing in estrous females.

Suids communicate using a variety of sounds and displays. Vocalizations are used to convey fear, pain, comfort or well being, warning, or for breeding and establishing contact. Olfactory investigation of the snout and genitals are common features of greetings in wild pigs. Displays are often used in aggressive encounters but also during courtship and breeding. When displays fail to diffuse aggression, body contact is made. Several combat postures have been noted, including "boxing" on the hind limbs in babirusa. Wild pigs are fast runners and good swimmers. They will generally flee when threatened, but they will fight vigorously when they are wounded or cornered. The upper and lower tusks are formidable weapons in some species. There has been some suggestion of a dominance order both within and between the sexes among babirusa. Cannibalism and infanticide have been observed in some species. Solitary, social, and object play behavior have been documented in wild piglets.

The warthog is diurnal, but other wild pigs are often more active at night or at dawn and dusk. They generally become more nocturnal when hunting pressure is high. They may be active 40–65% of the time. There is a tendency for there to be two peaks in activity; however, this is affected by climate, reproductive condition, and resource availability.

Feeding ecology and diet

Wild pigs are usually omnivorous, consuming diets of leaves, grasses, young saplings, seeds, roots, tubers, fruits, fungi, eggs, invertebrates, carrion, and small vertebrates. The suid diet fluctuates considerably during the year to take advantage of energy-rich foods such as fruits and acorns. Most species also visit mineral licks where they ingest soil or water. Warthogs will chew animal bones, possibly as a source of calcium and other minerals. Wild pigs are known for their rooting behavior, but the extent to which they do so depends on the species and the soil conditions. The babirusa lacks a well-developed rostral bone and so can only root in very soft, moist soils. Warthogs,

The giant forest hog (*Hylochoerus meinertzhageni*) is hunted for its meat. (Photo by David Madison. Bruce Coleman, Inc. Reproduced by permission.)

on the other hand, have a very strong rhinarium and will go down on their wrists to dig in the hard soil to excavate roots and tubers. Pigs will forage alone or in small groups, though in some cases they will form very large groups to feed on ephemeral resources. Bearded pigs (*S. barbatus*) have been known to form groups of hundreds of individuals and make mass migrations following the fruiting of dipterocarps. These migrations occur over a period of months during which the animals may travel 150–400 mi (250–650 km). Aggregations of 30–60 red river hogs (*Potamochoerus porcus*) have been observed in Guinea and eastern Democratic Republic of the Congo (Zaire). Studies of Eurasian wild pig and bearded pig suggest a strong link between nutritional status and reproduction in these species. The populations of both of these species will increase dramatically following mass fruiting events. Alternatively, populations will decline significantly in drought years.

It has been argued that wild suids play a significant role in forest structure. A study in Queensland, Australia found that the germinability of mesquite seeds (*Porsopis pallida*) remained very high after passing through the gut of wild boar, suggesting that wild pigs may be important agents of seed dispersal. Exclosure studies in Malaysia have demonstrated that soil rooting and seed predation by Eurasian wild pigs decrease stem density and species richness; however, other studies have concluded that the foraging behavior of wild pigs may increase species richness in forests. The amount of rooting pigs do depends on food availability. In the temperate zone, most rooting occurs in mid-autumn to spring. After this period, rooting decreases as pigs switch to foraging on herbs and foliage.

A bearded pig (*Sus barbatus*) in a tropical rainforest in Bako Park, Sarawak, Borneo. (Photo by Fletcher & Baylis/Photo Researchers, Inc. Reproduced by permission.)

Reproductive biology

Suid reproduction differs from that of other ungulates in several ways: the gestation is relatively short, newborns are very small compared to the mother, suids are the only truly multiparous ungulates, and all show nest-building behavior prior to parturition. In some species, the young are unable to regulate their body temperature and, also in contrast to many other ungulates, mothers nurse from a recumbent position as opposed to standing. In most suids, males associate with females only during estrus. The mating system is generally polygynous, but there has been some suggestion of monogamy in warthogs.

Estrous females urinate more frequently than usual, and the voided urine is sniffed and sometimes licked by males. Courtship behavior includes broadside displays, chasing, vocalizations, female solicitation of the male, and male nuzzling of the sides and vulva of the female. Female estrus is accompanied by swelling of the labia and mucus discharge. Males also tend to salivate excessively during this time. There may also be mutual grooming of the genitals. There are multiple mounts before intromission, and copulation sessions may last from 15 to 30 minutes. Adults will copulate several times per day. The estrous cycle is 21–42 days in length, with estrus lasting one to four days. Gestation lasts 100–175 days, depending on the species. In some species, sexual maturity is reached as early as eight months whereas in others, maturity is attained at two to five years of age.

Prior to parturition, sows will separate from their sounders, if they live in one, and construct a nest in which to give birth. Nests are usually located in thick cover. Females may excavate a shallow area and line it with vegetation or create a bed of vegetation on which to farrow. Babirusa sows often farrow at night in captivity. In domestic pigs, the mother does not help free the young from membranes and usually does not eat the placenta. Litter sizes in wild pigs vary from one to 12 piglets, with litters of babirusa and Visayan warty pig having the smallest litters (one to three piglets). In domesticated pigs, litter sizes generally increase with age and depend on breed and may reach 18 piglets. Piglets will compete among themselves to establish a teat order shortly after birth and will nurse from the same teat throughout lactation. In domestic pigs, sows may nurse up to 20 times per day, and they wean their litters at eight to 14 weeks of age, whereas bearded piglets are weaned at five or six weeks, and babirusa piglets are weaned at 26–32 weeks. Piglets may leave the nest with their mothers as early as a few days postpartum. Males generally provide little, if any, parental care, but may defend the young in some species. Females will vigorously defend their young and will sometimes work with other females to repel predators.

Reproduction in the temperate zone is generally seasonal with parturition occurring in spring and mating in autumn or winter; however, these periods are variable and seem to be dictated in large part by food resources. Drought conditions tend to decrease the percent of adult females that breed in a given year. Similarly in the tropics, reproduction can

A herd of young wild boars (*Sus scrofa*). (Photo by Uwe Walz/Jacana/Photo Researchers, Inc. Reproduced by permission.)

occur year round when the availability of energy-rich foods is high. Studies in the Cape Provinces have shown that life history characteristics reflected differences in nutrient availability. Bush pigs (*P. larvatus*) in the eastern areas of the Province had a higher quality diet and showed a higher reproductive investment (e.g., small, young females had large, frequent litters with low survival rates). Southern bush pigs bred at a later age and larger size, and their litters were smaller and had higher survival rates. In Borneo, female bearded pigs appear to require a certain thickness of subcutaneous fat in order to be responsive to mating stimuli. In this species, young born to fat mothers grow and mature quickly and can breed within one year. Among domestic pigs, it has been shown that nutritional status can affect hormone secretion and fertility.

Conservation status

Though some species of wild pigs are very widespread and abundant, others are Critically Endangered with very limited distributions. According to the 1993 Survey and Action Plan of the IUCN/SSC Pigs, Peccaries, and Hippos Specialist Group, the pygmy hog is considered Critically Endangered. Indeed, its distribution is now restricted to only one national park in Assam. The common warthog is still considered to be secure, but the remaining desert warthog subspecies (*P. a. delamerei*) is considered Vulnerable. The forest hog is considered Rare to Endangered depending on the subspecies. Bush pigs and the red river hog are still considered widespread and abundant, or secure. The majority of subspecies of Eurasian wild pigs are also considered widespread and secure; however, the subspecies *S. s. riukiuanus* from the Ryukyu Islands in southern Japan is considered Vulnerable to Endangered. The bearded pig is considered potentially at risk or rare, as is the Philippine warty pig (*S. philippensis*). Javan warty pigs (*S. verrucosus*) are considered Vulnerable, and the Visayan Warty pig is Endangered, perhaps critically. The babirusa is Vulnerable or Endangered, whereas the other species sharing its range, the Sulawesi warty pig (*S. celebensis*), is still considered Secure. The Vietnam warty pig (*S. bucculentus*) is known from only a few recent skulls and may be extinct. The babirusa and pygmy hog are listed on Appendix I of CITES and are both considered Endangered by the United States Fish and Wildlife Service. Babirusa are protected under Indonesian law, and the pygmy hog is on Schedule 1 of the Indian Wildlife Protection Act of 1972. Forest hogs are listed on Class B of the African Convention on the Conservation of Nature and Natural Resources. The babirusa is endemic to the island of Sulawesi. In terms of species diversity and endemicity of wild pigs, the Philippines is the most important country in the world. Of the three species known in this area, two are endemic (*S. philippensis* and *S. cebifrons*). Seven subspecies are recognized, and six of these are endemic. More taxa await description as well.

The main threats to wild pigs are hunting and loss of habitat. For some species such as the Visayan warty pig, hybridization with local domesticated, feral, or wild stock of *S. scrofa* represents a major concern. The only way to ensure the survival of the Visayan warty pig at this point is captive breeding of pure stock. The conversion of forests to agricultural land has in some places benefited local populations of wild pigs, but in general the impact is negative, especially for forest specialists like the babirusa. Hunting generally take ones of three forms: subsistence hunting, commercial hunting, and hunting in reprisal for crop damage. Subsistence hunting is often not a significant threat for any of the wild pigs, but is being overrun in many places by a thriving market in bush meat. Some wild pigs are protected under law; however, enforcement is nonexistent in many cases. A disturbing development in the eradication of pigs as crop pests has been the use of "pig bombs," small vessels of gunpowder placed on or in the ground that explode when pigs root or chew them. Though a couple of species are listed by CITES, trade is not a significant threat for wild pigs. Population estimates for wild pigs are difficult to obtain given the types of habitats they prefer, their shy, reclusive nature, and nocturnal habits.

As of 2003, much of the conservation work being done on wild suids involves population surveys and habitat assessment as well as basic studies of behavior and ecology, though integrated conservation programs have been developed by the IUCN Pigs, Peccaries, and Hippos Specialist Group for both pygmy hogs and Visayan warty pigs under the aegis of formal agreements with the relevant national governmental authorities. There has also been a number of studies of local markets to assess the extent of the bush meat trade in wild pigs. For some wild pigs, population models have been developed to determine what sustainable levels of harvest may be. Public education campaigns have been launched in some areas to inform people that the wild pigs living around them are threatened with extinction. Some educational initiatives have also been aimed at reducing human-pig conflict. Captive breeding programs for pygmy hog, Visayan warty pigs, and babirusa have been initiated in their countries of origin; the latter two and some others are also being bred in the United States, Europe, and elsewhere. In some cases, these populations are very inbred because of small numbers of founders. Health regulations on imports of swine and other ungulates, coupled with historically little interest in suids as zoo exhibits, have contributed to this problem.

Male babirusa (*Babyrousa babyrussa*) reaching for vegetation on Sulawesi Island, Indonesia. (Photo by Kenneth W. Fink/Photo Researchers, Inc. Reproduced by permission.)

The red river hog (*Potamochoerus porcus*), rooting for vegetation. (Photo by Mark Newman. Bruce Coleman, Inc. Reproduced by permission.)

Significance to humans

Wild pigs represent either a pest or an important source of dietary protein for most humans. In places where they cooccur with people, pigs frequently do serious damage to agricultural lands through their consumption of crops and/or rooting behavior. They have also been known to do serious damage to timber plantations. Thus, they are often persecuted in reprisal for losses. However, even in areas dominated by Islam or other religions that forbid the consumption of pork, pigs are very frequently hunted for their flesh. Pigs represent a valuable resource for both subsistence and commercial hunting throughout much of their range in Asia and parts of Africa. In some areas wild pigs may be vectors of diseases that pose threats to domestic livestock. Some pig cultures still exist wherein pigs are valued as commodities, used as currency, or have ritualistic significance; however, these cultures are becoming more rare.

The earliest domestication of the pig appears to have taken place in Anatolia, Mesopotamia, and northern Iraq approximately 9,000 years ago. The domesticated pig came to Europe via the Caucasus and Balkans and is found in Aeneolithic and Bronze Age burials. The turbary is thought to be a transitional form between the wild boar and domestic pig. However, there is some evidence that other species besides *S. scrofa*

The Eurasian wild boar (*Sus scrofa*) wallowing in a pond. (Photo by Larry Ditto. Bruce Coleman, Inc. Reproduced by permission.)

were used as stock for domestication (e.g., *S. celebensis* in Southeast Asia). Local wild boar populations no doubt contributed to local breed development in Europe; however, Asia was the primary center of domestication, and many of today's European breeds are influenced by Asian stock. The skull of the domestic pig has changed from its wild form more than that of any other domesticated animal besides the dog. The skull is now broader, with a shortened anterior portion. The brain case is also higher. Other characteristics of this domestication include a larger body, smaller head, longer body, and shorter legs, the existence of flop ears, and a curly tail. In modern times, the major centers of domestic breed development have been England, China, and the United States, though several countries in Europe developed their own localized breeds, known as landraces. There are approximately 175 breeds and types recognized around the world. Some breeds are threatened with extinction or considered endangered, mainly because of declining popularity and changes in consumer demands (e.g., leaner pork). The American Livestock Breeds Conservancy lists nine suid breeds as critical or rare, and the Rare Breeds Survival Trust in Britain maintains populations of seven breeds of concern.

The pig is unique among domestic animals in that it has become widely used as an animal model in biomedical research given its similarity to humans in certain aspects of its anatomy, physiology, and even behavior. One of the most active areas of research involving pigs is xenotransplantation. Pigs represent a good source of organs owing to their size, availability, and limited risk of zoonosis. Potential donor organs for human recipients include porcine heart, kidney, liver, heart-lung, and pancreas tissue. Research is needed to overcome some of the immunological barriers before widespread use of porcine organs is possible; however, selective breeding and genetic engineering are also being explored as methods to reduce transplant rejection.

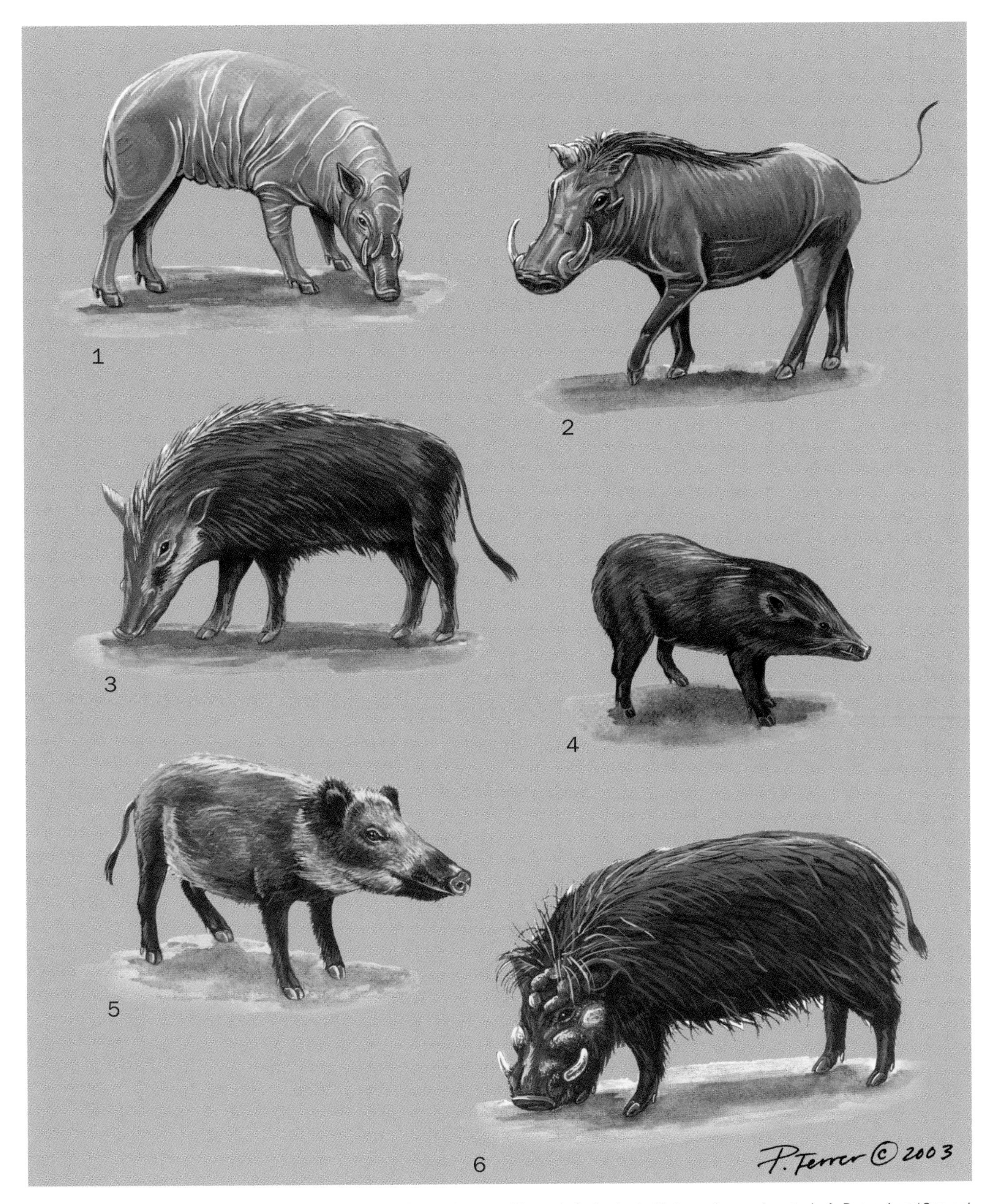

1. Babirusa (*Babyrousa babyrussa*); 2. Common warthog (*Phacochoerus africanus*); 3. Bush pig (*Potamochoerus larvatus*); 4. Pygmy hog (*Sus salvanius*); 5. Eurasian wild pig or boar (*Sus scrofa*); 6. Forest hog (*Hylochoerus meinertzhageni*). (Illustration by Patricia Ferrer)

Species accounts

Forest hog
Hylochoerus meinertzhageni

SUBFAMILY
Suinae

TAXONOMY
Hylochoerus meinertzhageni Thomas, 1904, Kenya. Three subspecies: West African forest hog (*H. m. ivoriensis*), Congo forest hog (*H. m. rimator*), and giant forest hog (*H. m. meinertzhageni*).

OTHER COMMON NAMES
French: Hylochére; German: Riesenwaldschwein; Spanish: Hilóquero.

PHYSICAL CHARACTERISTICS
Length 51–83 in (130–210 cm); height 30–43 in (76–110 cm); males 506.5 lb (230 kg) (range 319–606 lb [145–275 kg]), females 397 lb (180 kg) (range 286–449 lb [130–204 kg]). The pelage is sparse and composed of long, coarse, dark hairs. The skin is slate to blackish gray. Young are straw colored and lack stripes. Patches of skin anterior to the eyes are relatively hairless and form large protruding cheek pads in males. The snout is large and discoid in shape. Tusks are smaller than in the warthog (12 in [30 cm] or less). There are 32–34 teeth. Females have four pairs of mammae.

DISTRIBUTION
Found in a variety of forested habitats across western, central, and east tropical Africa.

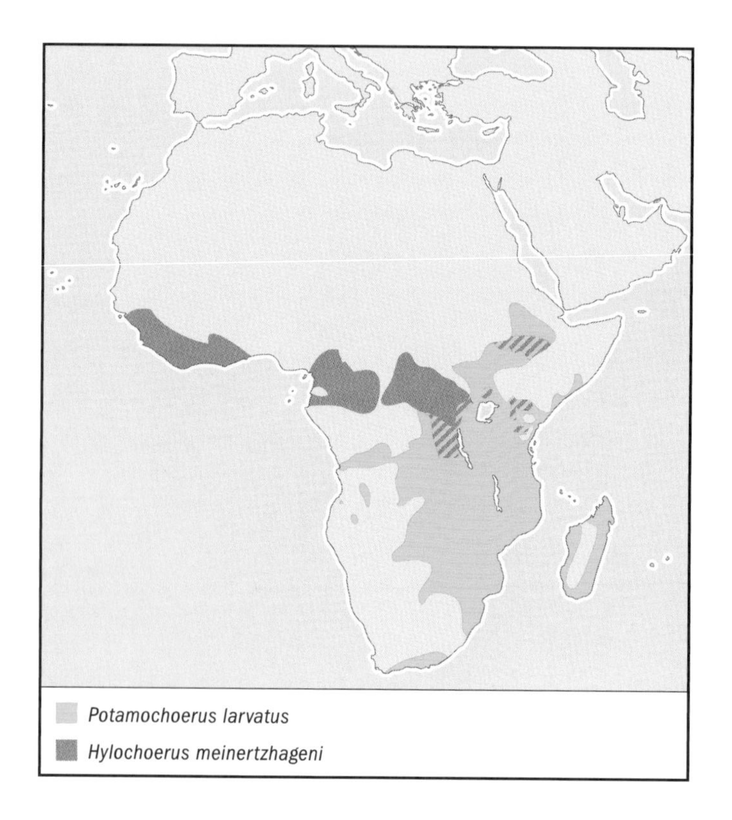

HABITAT
Elevations to 12,500 ft (3,800 m), subalpine forests, bamboo groves, montane, swamp, and gallery forests, savanna-forest mosaics, and wooded savannahs. More likely to occur in areas with permanent water sources, thick understorey vegetation, and a variety of vegetation types. Does well on habitat edges.

BEHAVIOR
Generally most active from dusk to midnight, but may also be diurnal in areas free of human persecution. The basic social group is a sounder containing one or more adult males, several adult females, and dependent juveniles or infants. These groups may sometimes form larger, temporary aggregations; however, males from different sounders are mutually intolerant of one another. Sounders have overlapping home ranges, with each range containing a network of well-worn trails to feeding sites, latrines, mineral licks, water holes for drinking and wallowing and sleeping sites. Home ranges may be up to 2.5 acres (1 ha) in size. Hyenas (*Crocuta crocuta*) and leopards (*Panthera pardus*) are the main predators.

FEEDING ECOLOGY AND DIET
Diet is mainly grass and dicotyledons in forested areas. Eggs and carrion may be eaten occasionally, while coprophagy and geophagy are common. Only root in soft soils.

REPRODUCTIVE BIOLOGY
Polygamous. Male-male fighting is less ritualized than in some other African suids and often leads to serious injury and sometimes death. Breeding is generally continuous, but mating is often most frequent towards the end of a rainy season. Two distinct breeding seasons have been identified in the Virunga National Park in eastern Democratic Republic of the Congo (Zaire). Copulation may last up to 10 minutes. Sows give birth in sheltered nests lined with dry grasses after a gestation of about 151 days. The litter size commonly ranges from two to four but may reach 11 piglets. Piglets remain in thick cover for about seven days, but then accompany the sow. Young are weaned at nine weeks and attain adult body size by approximately 18 months. Females may disperse from the natal group and conceive at one year of age, while males remain in the natal group until secondary sex characters such as cheek pads develop.

CONSERVATION STATUS
Main threats are excessive hunting and habitat destruction. Deforestation is particularly problematic in western Africa. Hunting pressure varies in intensity according to local religious beliefs or superstitions, but subsistence hunting is not thought to pose a serious problem. Commercial hunting is a serious problem in cities with surrounding forested areas. Forest hogs may also be eradicated in response to crop raiding. Population trends across its range are unclear in many cases. Not listed in CITES, but the West African forest hog is considered vulnerable by the IUCN.

SIGNIFICANCE TO HUMANS
Like other African suids, the forest hog is a significant source of meat for subsistence or commercial purposes. Certain tribes use the hides for war shields, while others superstitiously avoid killing them altogether in order to avoid trouble. ◆

Bush pig
Potamochoerus larvatus

SUBFAMILY
Suinae

TAXONOMY
Potamochoerus larvatus (Cuvier, 1822), Madagascar. Four subspecies: *P. l. hassama* from eastern Africa, *P. l. koiropotamus* from Angola and southeastern Africa, *P. l. larvatus* from Mayotte in the Comoro Islands and western Madagascar, and *P. l. hova* from eastern Magascar.

OTHER COMMON NAMES
French: Potamochère; German: Buschschwein; Spanish: Jabalí de río.

PHYSICAL CHARACTERISTICS
Length 39–59 in (100–150 cm); height 21.5–38 in (55–97 cm); weight 154.2 lb (70 kg) (range 101–286 lb [46–130 kg]). Snout is elongated. Animal ranges in color from light red to brown and gray to predominately black. Newborns are dark brown in color with rows of light spots. The tusks are least conspicuous among the African suids; the upper tusk averages 3 in (7.6 cm), whereas the lower tusks range from 3.5–6.5 in (9–16.5 cm) in length. Males have small warts in front of the eyes. There are six pairs of mammae.

DISTRIBUTION
Found from Somalia to east and southern Zaire down to the Cape Provinces in South Africa. It is also found on Madagascar and on Mayotte Island in the Comoros; however, it was most likely put here by humans.

HABITAT
Generally found in more moist habitats with significant vegetative cover and tend to avoid drier, more open areas. May occupy riverine, lowland, swamp, and montane forests as well as woodland savannas, scrub habitat, and cultivated areas.

BEHAVIOR
Mainly nocturnal, sleeping in self-excavated burrows or in heavy thickets of vegetation during the day. Those living in the southern Cape tend to be more diurnal during the colder months, suggesting temperature regulation may be an important factor influencing the activity rhythms in some populations. They live in single male family groups containing one or more females and their young (group size range four to 15 individuals), though both species have been observed in larger herds. Family groups are territorial, similar to other African suids. Territorial encounters are often characterized by ritualized displays and scent marking. Territory sizes vary with location and range from 0.05–2.5 acres (0.02–1 ha). Adult males play a role in rearing and defending the young. Males disperse from the natal group, whereas females may remain on the natal territory.

FEEDING ECOLOGY AND DIET
Omnivorous, consuming roots, bulbs, fungi, fruit, eggs, invertebrates, birds, small mammals, and carrion. They use their snouts to root in the ground and can do serious damage to crops in a short time. Field studies in Uganda have found that bush pigs may follow groups of monkeys as they forage and feed on the discarded fruits.

REPRODUCTIVE BIOLOGY
Polygamous. Reproduction is seasonal with young being commonly seen at the end of the dry season or beginning of the wet season. The average gestation is 120 days. Litter sizes range from one to six, with three to four piglets being the most common. Newborns weigh 24.5–28 oz (700–800 g). Females may start breeding at two to three years of age.

CONSERVATION STATUS
The IUCN has designated this species as widespread and locally abundant or relatively secure, depending on the subspecies. Though the bush pig is still relatively widespread, its distributions are patchy in certain regions. They have been reported to be rare outside of protected areas, and they are widely hunted either for subsistence or for commercial sale of the meat at local markets. In some countries, poaching inside of protected areas is a serious threat. The clearing of some forested areas and conversion to cropland has benefited this species in some areas.

SIGNIFICANCE TO HUMANS
Notorious crop raiders. They are also sometimes thought to pass diseases on to livestock. Eradication programs for these species have been largely unsuccessful because they breed quickly and are cryptic in nature. *Potamochoerus* does represent an important source of meat for food or revenue in some places, while in others the meat is avoided because it is thought to transmit epilepsy. ◆

Common warthog
Phacochoerus africanus

SUBFAMILY
Phacochoerinae

TAXONOMY
Phacochoerus africanus (Gmelin, 1788), Senegal. Four provisional subspecies: northern warthog (*P. a. africanus*), Eritrean warthog (*P. a. aeliani*), Central African warthog (*P. a. massaicus*), and southern warthog (*P. a. sundevallii*).

OTHER COMMON NAMES
French: Phacochère commun; German: Warzenschwein; Spanish: Jabalí verrugoso.

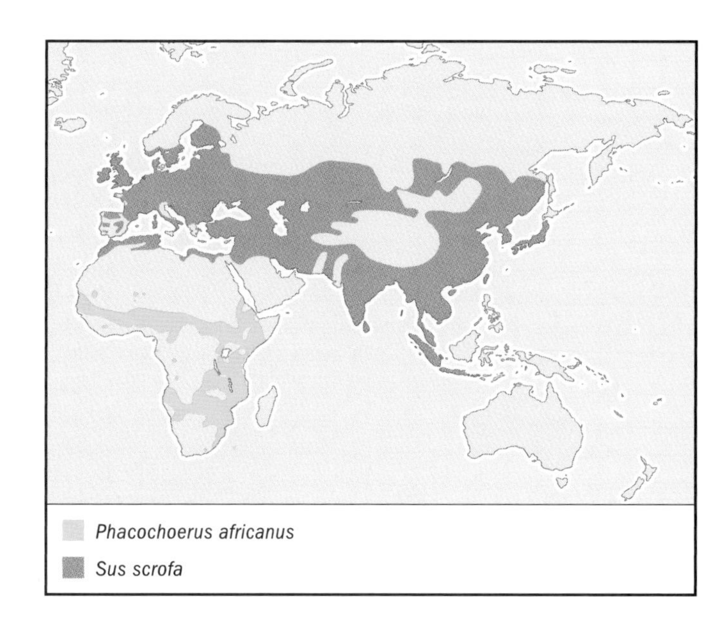

PHYSICAL CHARACTERISTICS
Barrel-shaped body 57–75 in (145–190 cm) in length, 25.5–33 in (65–84 cm) in height; males 180.5 lb (82 kg) (range 150–220 lb [68–100 kg]), females 143 lb (65 kg) (range 99–156 lb [45–71 kg]), mostly hairless except for a long, coarse mane of dark hair on the neck, shoulders, and rump, a tufted tail, white cheek whiskers, and some bristles scattered over the body. The skin is dark brown to black in color. Young are reddish brown. The head is large and flattened with one or two pair of warts on the face beneath the eyes and near the tusks. Both sexes have well-developed tusks. The upper tusks may measure 23.6 in (60 cm) in length, whereas the lower tusks are only 5.1 in (13 cm). There are 32–34 teeth, though old animals may be missing a number of teeth. There are four pairs of mammae.

DISTRIBUTION
Found in savannas in almost all sub-Saharan countries from Senegal and southern Mauritania to northern Ethiopia and Djibouti, south to Namibia, the Cape Province, and Natal.

HABITAT
Elevations to 9,840 ft (3,000 m), open and wooded savannas, grassy plains, and semi-arid areas.

BEHAVIOR
Mainly diurnal. Males tend to live alone or in small, bachelor groups with changing membership. Males are not territorial and only associate with females for mating. Females live in sounders consisting of one to three females and their offspring. Females may remain in the natal group for several breeding seasons, whereas males disperse from the group but remain on the natal home range (67–1,194 acres [27–483 ha]), depending on locality. Lone males, bachelor groups, and several sounders may use the home range simultaneously; these individuals are collectively known as a clan. Warthogs take shelter in subterranean burrows, often excavated by aardvarks, at night. The main diurnal predators are lions (*Panthera leo*), leopards, cheetah (*Acinonyx jubatus*), and wild dogs (*Lycaon pictus*), though leopards and cheetah tend to prey on younger warthogs. Warthogs communicate using scent marks deposited by both sexes, though more frequently by males, and a number of visual displays and vocalizations.

FEEDING ECOLOGY AND DIET
Selective grazers whose diet consists of grass, roots, fruits and berries, bark, and sometimes carrion. Usually drink on a daily basis, but may obtain sufficient water from succulent roots and bulbs when water is scarce.

REPRODUCTIVE BIOLOGY
Likely polygamous, but possibly monogamous. Males will fight among themselves for mating opportunities, sometimes inflicting serious wounds with their upper or lower canine tusks. Estrus in females is characterized by frequent urination and discharge from the swollen vulva. Copulation may last less than a minute. While warthogs may breed throughout the year in equatorial areas, breeding behavior and farrowing are seasonal in other areas and synchronized with the end and beginning of rainy periods, respectively. Females are seasonally polyestrous with estrus occurring about every six weeks and lasting approximately three days. The gestation ranges from 160–170 days, and the average litter is three piglets (range one to eight), each weighing 14–32 oz (400–900 g). Sows will leave their sounder temporarily to give birth in a hole in the ground. The sow rarely leaves the piglets during the first week and eats less than usual during the first several weeks postpartum. Piglets remain in the hole for six to seven weeks. The young may suckle 12–17 times per day, approximately every 40 minutes. The young may begin to graze on their own after two to three weeks and are fully weaned at six months. Both sexes reach puberty at 18 months, though males usually do not breed until about four years of age.

CONSERVATION STATUS
Anthropogenic threats include over-hunting for meat and persecution in response to crop raiding, which may be a serious problem in some areas. Also viewed as competitors for forage on cattle ranches. Hunting pressure varies by region, often according to prevailing religious beliefs, and can cause a serious decline in numbers. Religious taboos relating to consumption of pork generally benefit warthog in Muslim areas, but they may still be hunted and the meat sold to other areas. There is also a small trade in warthog tusks, however not at an international level. Despite these threats, warthogs are considered widespread and abundant and their populations relatively secure, according to the IUCN. They are not included in the CITES appendices.

SIGNIFICANCE TO HUMANS
Represent either a crop pest or source of meat for subsistence or commercial trade, depending on the region. Commercial ranching and harvesting occurs in some parts of Africa (Zimbabwe, South Africa, and Burkina Faso). ◆

Pygmy hog
Sus salvanius

SUBFAMILY
Suinae

TAXONOMY
Sus salvanius (Hodgson, 1847), Sikkim Terai, India.

OTHER COMMON NAMES
French: Sanglier nain; German: Zwergwildschwein; Spanish: Jabalí enano.

PHYSICAL CHARACTERISTICS
Length 20–28 in (50–71 cm); shoulder height 10–12 in (25–30 cm); weight 14.5–21 lb (6.6–9.7 kg). Aside from its smaller size, the only diagnostic characters of this species are its vestigial tail and the presence of only three pairs of mammae in females. This species has no facial warts. The hair is agouti-banded and medium brown along the sides of the animal, darkening along the back. Piglets are faintly striped at two to four weeks of age. The tusks of males are exposed, whereas those of females are not.

DISTRIBUTION
Now only found in a few isolated populations in Manas Tiger Reserve in northwest Assam, making it the rarest suid in the world.

HABITAT
Found only in tall grassland habitats associated with early successional riverine communities. Those grasslands dominated by *Narenga porphyrocoma*, *Saccharum spontaneum*, *S. bengalensis*, *S. munja*, *Erianthus ravennae*, *Imperata cylindrica*, and *Themeda* spp. are the most important for pygmy hogs.

BEHAVIOR
The main social groupings are sounders containing one or more adult females and their offspring and solitary males.

Unique among the suids in that it builds nests year round, which are made of grass leaves and built over shallow depressions in the ground. Both sexes and all age classes use nests on a nightly basis.

FEEDING ECOLOGY AND DIET
Omnivorous, feeding on roots, tubers, grass leaves, shoots, fruits, seeds, insects, worms, eggs, and small vertebrates.

REPRODUCTIVE BIOLOGY
Polygamous. Males will associate with estrous sows during the rut. Reproduction is highly seasonal with breeding beginning in late November and early December. Breeding may continue into the spring in captivity. Females are in estrus for three to four days during the 21-day cycle, and gestation lasts 120 days. In the wild, there is a clearly defined birth peak from April–June. Litters average around four piglets (range two to seven), weighing approximately 5.3 oz (150 g) at birth. Females are sexually mature at 20 months, whereas males reach maturity at two to three years of age. In captivity, both sexes reach sexual maturity at 10–11 months.

CONSERVATION STATUS
Considered Critically Endangered by the IUCN and is listed on Appendix I of CITES, as well as on Schedule 1 of the Indian Wildlife Protection Act of 1972. Also considered endangered by the U. S. Fish and Wildlife Service. The main threats are loss and degradation of habitat because of human settlements, agriculture, flood control schemes, and improper management. Tree planting in grasslands and the use of fire to open up grasslands for new growth have been detrimental. Hunting is becoming more of a threat. The survival is closely linked to the existence of the tall, wet grasslands. A conservation program was launched in 1995 as a cooperative effort among the IUCN/SSC Pigs, Peccaries, and Hippos Specialist group, the Durrell Wildlife Conservation Trust, the Forest Department (Government of Assam) and the Ministry of Environment and Forests (Government of India), and includes captive breeding, field research, and reintroduction.

SIGNIFICANCE TO HUMANS
Though they are not crop pests, hunting of pygmy hogs for meat is increasing. Their small size makes it a potentially very useful and valuable genetic resource for biomedical research and further domestication. ◆

Babirusa
Babyrousa babyrussa

SUBFAMILY
Babirusinae

TAXONOMY
Babyrousa babyrussa (Linnaeus, 1758), Borneo.

OTHER COMMON NAMES
French: Babiroussa; German: Hirscheber; Spanish: Babirusa.

PHYSICAL CHARACTERISTICS
May weigh 132–220 lb (60–100 kg), and measure 34–39 in (87–100 cm) in length and 25–32 in (65–80 cm) in height, depending on the subspecies. The length, thickness, and distribution of the pelage vary with location with some individuals appearing almost naked, while others have a long, coarse coat. Unlike most pigs, the young bear no stripes. The skin is brownish gray in color. The most striking features are the male's canines, which emerge vertically through the top of the snout and curve backwards toward the head. Females have two pair of mammae.

DISTRIBUTION
Endemic to the island of Sulawesi where it is still found in many areas. Also present on the Togian islands of Batudaka, Togian, and Talatakoh, the Sula islands of Mangole and Taliabu, as well as the island of Buru.

HABITAT
Are found in tropical rainforests, canebrakes, and on the banks of rivers and lakes where water vegetation is abundant.

BEHAVIOR
Diurnal with a peak in activity in the morning hours. Females may live in groups with one to five other adult females and their young. Males tend to be solitary. The most frequent type of grouping seen at a natural saltlick was solitary males followed by single females with young, but groups as large as 15 individuals have been observed. Construct nests for sleeping and shelter from the rain. Adult females are dominant to subadult males but subordinate to adult males. Larger females are dominant to smaller ones. Predators include pythons (*Python reticulatus* and *P. molurus*), and possibly civets (*Viverra* spp.). Agonistic behavior takes the form of displays, body pushing and rubbing, as well as boxing. The lower tusks may be used for attack or defense, whereas the upper, protruding canines seem to be used to protect the face from the lower canines of opponents. It has been suggested that ploughing behavior, wherein the individual pushes its head into the ground, slides forward, and rolls from side to side, has a scent-marking function.

FEEDING ECOLOGY AND DIET
Omnivorous, consuming a diet of fruit, nuts, leaves, roots, and some animal material. Also consume soil and rock fragments at saltlicks. Roots only in very soft, moist soil because it lacks a rostral bone in its snout. In captivity, both sexes have been known to opportunistically cannibalize young. A unique feature

of the digestive tract is a large area of mucous-producing cardiac glands that are thought to aid in fermentation.

REPRODUCTIVE BIOLOGY
Females will give birth year round in captivity, but may do so less frequently in the wild. Sexual maturity may be attained at five to 10 months of age. The estrous cycle is 28–42 days, with estrus lasting one to three days. Males may mount the female multiple times before intromission, and copulation bouts may last 15–30 minutes; average copulation lasts three minutes. The reproductive lifespan of females may begin at one year of age and continue through age 14. The gestation period is 155–175 days (mean 163 days) and the usual litter is one to two piglets. In captivity, females often give birth at night. Females separate from the group to give birth in nests. Young are weaned at 26–32 weeks, though they begin to nibble solid foods at one week of age. Nursing frequency is highest in the first month postpartum and declines thereafter. The average nursing bout length is 10 minutes, longer than most suids.

CONSERVATION STATUS
Considered Vulnerable or Endangered by the IUCN, depending on the subspecies, and it is on Appendix I of CITES. Designated as endangered by the United States Department of the Interior in 1980; given full protection under Indonesian law in 1931. The main threats are hunting, both commercial and subsistence, and loss of habitat. A 1997 census estimated that there were 5,000 babirusa remaining in the wild.

SIGNIFICANCE TO HUMANS
Most hunting activity is focused on the Sulawesi warty pig; however, babirusa are also taken as a source of protein. This species has not been reported to be a crop pest. Babirusa skulls are sold in local markets to tourists and in large department stores in Jakarta. ◆

Eurasian wild pig
Sus scrofa

SUBFAMILY
Suinae

TAXONOMY
Sus scrofa Linnaeus, 1758, Germany. Sixteen subspecies, which can be divided into four subspecies groupings: the western (Europe, Africa), eastern (Mongolia, Far East), Indian, and Indonesian races.

OTHER COMMON NAMES
French: Sanglier; German: Wildschwein; Spanish: Jabalí.

PHYSICAL CHARACTERISTICS
Body weight ranges 77–770 lb (35–350 kg); however, some domestic breeds of this species may reach 990 lb (450 kg). Height ranges 22–43 in (55–110 cm). The pelage is coarse and is typically composed of short bristles of varying color over the entire body. Animal has no facial warts. Males have canine tusks, whereas the canines of females are smaller.

DISTRIBUTION
Found on all continents, except Antarctica, in wild or in a barely modified feral form. It also inhabits many islands.

HABITAT
Found in many types of habitat from semi-arid areas to tropical rainforests, including temperate woodlands, grassland, steppe, broadleaf forests, and agricultural lands.

BEHAVIOR
Mainly diurnal, being most active in the morning and afternoon; however, they may shift to a more nocturnal activity pattern in disturbed areas or areas with significant hunting pressure. The basic social unit is a small group of females and their current litters. Adult males are solitary, and young from previous years are peripheral to the group. These groups may occasionally form larger (up to 100 individuals) aggregations. Home range size varies 48–4,950 acres (60–2,000 ha) across the species distribution in response to availability of resources, reproductive status, and hunting pressure. Active 40–65% of the time.

FEEDING ECOLOGY AND DIET
Omnivorous, with vegetable matter constituting about 90% of the diet. Consume roots, tubers, grasses, fruits, nuts, seeds, agricultural crops, soil, invertebrates, carrion, vertebrates, and mollusks. The composition of the diet shifts during the course of the year in many places in response to the fluctuating availability of energy-rich foods such as acorns. Animal has been known to migrate in response to food shortages.

REPRODUCTIVE BIOLOGY
Polygamous. Puberty occurs between eight and 24 months. Age of first breeding is generally 18 months for females and five years for males. The estrous cycle is 21 days, with estrus lasting two to three days. Gestation averages 112–120 days, and the average litter size is four to six piglets, eight to 12 in domestic pigs. Breeding is generally seasonal, occurring mostly in the fall, but reproduction appears to be heavily dependent on the availability of food and, thus, body condition in females. Most births are in the spring. Females separate from their group to give birth and return later with their young. Births tend to be synchronous within social groups; however, during bad conditions often only the dominant female will give birth. Weaning occurs at three to four months of age.

CONSERVATION STATUS
The IUCN has designated this species as widespread and abundant, known or believed relatively secure, or potentially at risk or rare, depending on the subspecies. The subspecies known from the Ryukyu Islands in southern Japan has been designated as Vulnerable by the IUCN. Most threatened by hunting and habitat destruction, but the survival of some subspecies is also threatened by hybridization with domesticated forms.

SIGNIFICANCE TO HUMANS
Consumed by humans more than any other species of suid and, in Asia, more than any other domesticated species. Represents a significant source of protein for subsistence and commercial hunters, and are persecuted in most of their range due to the significant damage they do to crops. Pig skulls and jaws are displayed as symbols of the hunter's ability and as protection from evil spirits. Some cultures use the intestines to read omens. In some island cultures, the domestic pig is treasured and has acquired significant cultural importance. Domesticated pigs may be used as currency for the payment of fines or fees for brides. ◆

Common name / Scientific name/ Other common names	Physical characteristics	Habitat and behavior	Distribution	Diet	Conservation status
Bearded pig *Sus barbatus* Spanish: Jabalí barbudo	Slim torso and long head. Two pairs of warts on face, first pair covered by beard. Large head, short neck, powerful and agile body. Coloration is dark brown-gray with white beard on face. Eyes are small, and long nose has set of tusks. Head and body length 3.3–5.5 ft (1–1.7 m), tail length 8–12 in (20.3–30.5 cm).	Rainforests, mangrove thickets, and secondary forests. Generally 2–8 offspring per litter. Pigs usually live in stable family groups throughout the year.	Malay Peninsula, Sumatra, Bangka, Borneo, Rhio Archipelago, and Palawan and Balabac Islands (Philippines).	Consume earthworms, roots, fruits, and gum tree seedlings.	Not threatened
Javan pig *Sus verrucosus* English: Javan warty pig; German: Pustelschwein; Spanish: Jabalí javanés	Coloration varies from reddish yellow to black, underparts are yellow. Long haired mane covers the nape of the neck. Slender legs, flat back, tail is long and simply tufted. Large ears, three pairs of warts on face. Shoulder height 27.6–35.4 in (70–90 cm), weight 77.2–330.7 lb (35–150 kg).	Secondary forests, predominantly teak forests. When threatened, the pig raises the long hairs that make up the mane. Uses shrill whistle as an alarm call. Groups consist of a sow and her current young, as adult males are usually solitary.	Java, Sulawesi, Molucca Islands, and Philippines.	Vegetation, including human crops.	Endangered
Red river hog *Potamochoerus porcus* Spanish: Jabalí de río	Coloration is predominantly reddish with white dorsal stripe. White facial masks are present. Ventrally pointed upper tusks on both sexes, long, white whiskers and ear tufts. Head and body length 3.3–4.9 ft (1–1.5 m), tail length 11.8–15.7 in (30–40 cm), average weight 101.4–286.6 lb (46–130 kg).	Primary and secondary forests, thickets in savanna, swamps, and steppes. They also congregate around human villages. Social animals, can live in groups of up to 11 individuals. Average litter contains four offspring.	West and central sub-Saharan Africa to northern South Africa and Madagascar.	May consume roots, fruit, seeds, water plants, nuts, grasses, crops, fungi, insects, bird eggs, snails, reptiles, carrion, and as piglets, goats, and sheep.	Not threatened
Warthog *Phacochoerus aethiopicus* English: Desert warthog; Spanish: Jabalí verrugoso	Long legs, large shovel-shaped head, broad muzzle with tusks. Females are smaller than males with shorter tusks. Coat is made up of sparse, bristly hair with a long mane running down the middle of the black. Few whiskers on lower jaw. Head and body length 3.5–4.5 ft (1.1–1.4 m), weight 110.2–330.7 lb (50–150 kg).	Grasslands and sparse forests. Prefer open plains of the savanna, with a nearby water source for drinking and wallowing. Gregarious animals. Live in small groups of 4–6 individuals. Not territorial, but competition does occur, followed by highly ritualized fighting.	Most open country of Africa south of the Sahara.	Graze on short grasses, feed on fruits and carrion, and also dig up bulbs, roots, and tubers.	Not threatened
Vietnam warty pig *Sus bucculentus* Spanish: Jabalí vietnamita	Coloration is brown or dark gray to black. The coat hair is coarse, bristle-like, and becomes heavier in winter for increased insulation. A narrow mane of long hair exists along the spine.	Habitat considered grasslands to forest. Piglets are usually seen in the dry season. The average number of piglets is 3–4 per litter.	Vietnam.	Graze on short grasses, feed on fruits and carrion, and also dig up bulbs, roots, and tubers.	Data Deficient
Cebu bearded pig *Sus cebifrons* English: Visayan warty pig; German: Visayas-Mähnenschwein	Long mane that extends to rump, large facial warts, white hairs on shoulders and sides. Skull length 11.7 in (29.9 cm). Little known of physical appearance.	Primary and secondary forest from sea level to mossy forest at 5,250 ft (1,600 m); now found only above 2,625 ft (800 m).	Recorded from the Visayan Islands of Cebu, Siquijor, Guimaras, Negros and Panay—identified as the Negros Faunal Region. However, already extinct in known, former range.	Consists of roots, fruits, leaves, shoots, carrion, and insects.	Critically Endangered
Celebes pig *Sus celebensis* English: Sulawesi warty pig; Spanish: Jabalí celebiano	Coloration is black, often with yellow or white intermixed. Reddish brown or yellow specimens have also been recorded. Underparts are light to creamy yellow. Dark dorsal stripe. Striking yellow band encircles snout. Head and body length 31.5–51.2 in (80–130 cm), weight 88–154 lb (40–70 kg).	Variety of habitats, from rainforest to swamp. Primarily diurnal, actively feeding during the day. Often interbreed with other pigs. Groups consist of family members and other individuals.	Presently throughout Indonesia.	Consists of roots, fruits, leaves, shoots, carrion, and insects.	Not threatened

[continued]

Common name / Scientific name/ Other common names	Physical characteristics	Habitat and behavior	Distribution	Diet	Conservation status
Philippine warty pig *Sus philippensis* German: Vietnam-Pustelschwein; Spanish: Jabalí filipino	Males typically develop three pairs of warts: on the cheek swellings, on the jaw angle, and above the canine root flanges. Coloration is dark brown to blackish brown. Very small body with whitish to yellowish stripes. Long trunk, small tusks.	Formerly abundant from sea level to at least 9,190 ft (2,800 m), in virtually all habitats, but is now only common in remote forests. Habitat considered grasslands to forest. Piglets are usually seen in the dry season (January–March in the Western Visayan Islands). The average number of piglets is probably about 3–4 per litter.	Luzon, Mindoro, Catanduanes, Mindanao and associated islands, and Polillo and Patnanungan.	Includes cultivated vegetables and fallen fruits.	Vulnerable
Timor wild boar *Sus timoriensis* English: Feral Celebes pig; Spanish: Jabalí de Timor	Coloration is black, often with yellow or white intermixed. Reddish brown or yellow specimens have also been recorded. Underparts are light to creamy yellow. Dark dorsal stripe. Striking yellow band encircles snout. Head and body length 31.5–51.2 in (80–130 cm), weight 88–154 lb (40–70 kg).	Variety of habitats, from rain-forest to swamp. Primarily diurnal, actively feeding during the day. Often interbreed with other pigs. Groups consist of family members and other individuals.	Lesser Sunda Island Chain in Indonesia.	Consists of roots, fruits, leaves, shoots, carrion, and insects.	Not threatened

Resources

Books

Estes, Richard D. *The Behavior Guide to African Mammals.* Berkeley: University of California Press, 1991.

Oliver, William L. R., ed. *Pigs, Peccaries, and Hippos: Status Survey and Conservation Action Plan.* Gland, Switzerland: International Union for Conservation of Nature and Natural Resources, 1993.

Pond, Wilson G., and Harry J. Mersmann, eds. *Biology of the Domestic Pig.* Ithaca, NY: Cornell University Press, 2001.

Periodicals

Cosgrove, J. R., S. T. Charlton, S. J. Cosgrove, L. J. Zak, and G. R. Foxcroft. "Interactions between Nutrition and Reproduction in the Pig." *Reproduction in Domestic Animals* 30 (1995): 193–200.

Groves, C. P. "Taxonomy of Wild Pigs (*Sus*) of the Philippines." *Zoological Journal of the Linnean Society* 120 (1997): 163–191.

Organizations

The American Livestock Breeds Conservancy. P.O. Box 477, Pittsboro, NC 27312 United States. Phone: (919) 542-5704. Fax: (919) 545-0022. E-mail: albc@albc-usa.org Web site: <http://www.albc-usa.org>

Rare Breeds Survival Trust. National Agricultural Centre, Stoneleigh Park, Warwickshire, CV82LG United Kingdom. Phone: 024 7669 6551. Fax: 024 7669 6706. E-mail: alderson@rbst.demon.co.uk Web site: <http://www.rare-breeds.com>

Other

American Zoo and Aquarium Association, Pig and Peccary Taxon Advisory Group. <http://www.pigpectag.org>.

IUCN Pigs, Peccaries, and Hippos Specialist Group. <http://www.iucn.org/themes/ssc/sgs/pphsg/home.htm>.

Asian Wild Pig Research and Conservation Group. <http://arts.anu.edu.au/awpn/>.

David M. Powell, PhD

▲

Peccaries

(Tayassuidae)

Class Mammalia

Order Artiodactyla

Suborder Suiformes

Family Tayassuidae

Thumbnail description
Pig-like mammals whose skin is covered in coarse bristles, with longer dorsal bristles and shorter side bristles; well-developed snout and nasal muscles; stout legs with a fused ulna and radius; relatively large, sharp, vertically directed canine teeth; a scent gland located approximately 5.9 in (15 cm) cranial to the base of the tail on the dorsal midline; a complex stomach with four sacculated compartments

Size
24.2–92.5 lb (11–42 kg)

Number of genera, species
3 genera; 3 species

Habitat
Desert, dry tropical forest, chaco, rainforest, Brazilian Atlantic forests, low Andean forests, llanos, pantanal, and cerrado

Conservation status
Endangered: 1 species

Distribution
Southwestern portion of North America; Central America; northern and central portions of South America

Evolution and systematics

The first peccaries were believed to have evolved during the Oligocene, with ancestral forms of peccary species migrating across the land bridge from North to South America around 2.5 million years ago. *Platygonus* is an example of an upper Pleistocene peccary genus that has been described in both North and South America, and *Catagonus wagneri* is proposed to be its closest extant relative. Out of the many ancestral peccary forms, there are three extant species: *Tayassu pecari* (white-lipped peccary), *Tayassu tajacu* (collared peccary), and *Catagonus wagneri* (Chacoan peccary).

Cladistic analysis suggest that white-lipped peccary and Chacoan peccary ancestors diverged in North America before coming to South America during the late Pliocene, and that white-lipped and Chacoan peccaries are more closely related to each other than white-lipped peccaries are to collared peccaries. Studies suggest that white-lipped and Chacoan peccaries are from a single clade that diverged about 1.7 million years ago (mya), with the white-lipped peccary clade separating from collared peccaries around 3.4–7.4.

Although common taxonomic nomenclature places white-lipped and collared peccaries in the genus *Tayassu*, and Chacoan peccaries in their own genus *Catagonus*, this classification is controversial based on evolutionary and genetic considerations. The "two-genera peccary theory" proposes that white-lipped and collared peccary are more similar to each other morphologically than to the Chacoan peccary, justifying the placement of white-lipped and collared peccaries in the genus *Tayassu* and the Chacoan peccary in *Catagonus*. The "three-genera peccary theory" breaks the Tayassuidae into three genera: *Tayassu* for white-lipped peccaries, *Dicotyles* for collared peccaries, and *Catagonus* for Chacoan peccaries. Genetic analysis of mitochondrial DNA supports the three-genera theory. Genetic studies place white-lipped and Chacoan peccaries in the same clade, separate from collared peccaries.

Physical characteristics

Tayassuids show few sexual dimorphisms, but skull measurements, particularly in the canine teeth and zygomatic processes, for male and female peccaries may differ. Unique

A collared peccary (*Tayassu tajacu*) chases away a mountain lion. (Photo by J & D. Bartlett. Bruce Coleman, Inc. Reproduced by permission.)

anatomical features of the Tayassuidae distinguishing them from other Suiformes include their skin, skull, and dentition, specialized digestive tract, dorsal scent gland, and hind leg structure. Peccary skin is covered in coarse bristles, with longer dorsal bristles and shorter side bristles. Color patterns differ among the three species, and there are intraspecific color variations in white-lipped and collared peccaries. Another unique feature of peccary skin is the dorsal scent gland, located approximately 5.9 in (15 cm) cranial to the base of the tail on the dorsal midline. This gland contributes to the very strong, musky odor that permeates peccary herds and meat. This dorsal gland appears to be used, in part, as a mechanism for social cohesion and herd territory definition. Additionally, peccaries do not have a thick layer of subcutaneous fat or dense fur and are poorly equipped to deal with low ambient temperatures.

Peccaries, like pigs, have well-developed snout and nasal muscles to facilitate rooting behavior. They have 38 teeth with a dental formula of I 2–2/3–3, C 1–1/1–1, P 3–3/3–3, M 3–3/3–3. Unlike swine, peccaries have interlocking canines with upper canines that point vertically downward, and a fused radius and ulna, as well as fused median metacarpals and metatarsals. Peccaries also possess a complex stomach with four sacculated compartments: two non-glandular blind sacs, a non-glandular gastric pouch, and a glandular hind stomach. This fore stomach comprises 85% of the total stomach volume of the peccary, 3–6% less than the fore stomach/stomach volume ratio of sheep or cattle. Because microbial fermentation with the formation of high volatile fatty acid levels have been observed in the peccary fore stomach, it has been proposed that their fore stomachs are a means to slow digestive passage and increase digestive efficiency for microbial fermentation of structural components of fruits, nuts, and herbaceous material. Peccaries do not have a gallbladder.

The legs are more highly evolved for running than those of the suids; the metatarsals are fused into a stout canonbone and the fifth digit is lost. The metacarpals III and IV are very stout, but unfused. The radius and ulna are co-ossified (fused together).

Distribution

The tayassuid family has a wide distribution, ranging from the southwestern portion of North America to Mexico and Central America; Pacific coasts of Colombia; tropical forests of northeastern and southeastern Peru; Venezuela and the Guianas; Amazon basin; southern Brazil; and Paraguay, northern Argentina, and lowland Bolivia. Large-bodied peccary species existed throughout the Pleistocene in North America, but vansihed with megafaunal extinction.

Habitat

Peccaries inhabit a wide range of habitat types, including desert regions of southwestern United States and northern

A collared peccary (*Tayassu tajacu*) youngster, tasting cactus. (Photo by Leonard Lee Rue III. Bruce Coleman, Inc. Reproduced by permission.)

Collared peccaries (*Tayassu tajacu*) fighting in south Texas. (Photo by G. C. Kelley/Photo Researchers, Inc. Reproduced by permission.)

Mexico; moist and dry tropical forests of Central America; Central American rainforests; choco; llanos; Amazon rainforest; wetlands and forests of the Pantanal region of South America; cerrado; Brazilian Atlantic forests; and the South American dry tropical thorn forests of Bolivia, Paraguay, and Argentina known as the Chaco.

Behavior

Peccaries are social herding animals with group sizes between three to more than 500, depending on the species. Peccaries show a classical kin-selected social structure, similar to chimpanzees.

In south Texas, collared peccary herd home ranges expand from 288–300 acres (116.5–121.4 ha) in dense chaparral to 550–757 acres (222.5–306.3 ha) in open mesquite habitat. In Costa Rica, a smaller home range may be attributed to higher habitat productivity or greater food quality. However, in Brazilian Amazonian forests, collared peccaries had home ranges that were relatively large. Peripheral areas of the home range of a collared peccary herd are often shared by other herds. Home ranges of collared peccary can also vary seasonally or in relation to forage conditions.

In general, collared peccaries are diurnal, with the greatest feeding activity in southwestern United States during the morning and late afternoon until dark. Peccaries in the Paraguayan Chaco rise at dawn, then rest during the hottest part of the day. In Arizona and Texas, collared peccaries are more nocturnal in summer and primarily diurnal in winter. They are very territorial and live in mixed sex groups of adult males and females, juveniles, and sub-adults. These groups are relatively cohesive and stable. Herds often are composed of subgroups, with individuals infrequently separating from the herds. When threatened by predators such as dogs, collared peccaries disperse and flee, often running into burrows or logs.

In order to define and mark their territory, collared peccaries use aggressive behavioral interactions and mark territorial boundaries on the ground or vertical structures with secretions from their dorsal scent glands, as well as by fecal deposition. Within their territory, collared peccaries paw the ground and create small dust beds where they often rest. When territory overlaps occur between collared peccary groups, males may display aggressive behavior in areas of overlap. Aggressive behavior includes vocalizations and gestures. Aggressive vocalizations include a continuous growl, tooth clicking, squeals, and alarm-like "woofs." Vocalizations are an important component of white-lipped peccary intragroup communication. In addition to aggressive vocalizations, white-lipped peccaries also have elaborate protocols of to-

A collared peccary (*Tayassu tajacu*) mother with two babies in Choke Canyon, Texas, USA. (Photo by G. C. Kelley/Photo Researchers, Inc. Reproduced by permission.)

getherness vocalizations. White-lipped peccaries tend to emit more aggressive vocalizations when they are spaced close together during feeding or resting as compared to when they are traveling in loosely spaced groups. When in danger, white-lipped peccaries clack their teeth, sniff loudly, and may let out a woof-like bark. Inter-individual interactions between peccaries can be quite aggressive, and consist of 14 reported agonistic behaviors, including teeth grinding and clacking, snorting, yawning or gaping, and grimacing. Fighting is relatively common in white-lipped peccary herds. In collared peccaries, the most common agonistic interaction is the squabble, which consists of two animals facing each other in a fighting position and chattering their teeth with their ears flattened back. When frightened, tayassuids bristle the hairs along their neck and back. When alarmed or when making social contact, white-lipped peccaries bristle their hairs, primarily along the dorsum and flank. Overall, intragroup aggression in collared peccary herds is less than in white-lipped peccary herds, and serious fighting is proposed to be rare within herds. Although there is relatively less aggression within collared peccary herds, there is a great deal of friendly interaction such as mutual grooming and rubbing. Similar to other species of peccaries, white-lipped peccaries exhibit allogrooming and mark each other extensively, often rubbing their head in the region of the dorsal gland of each respective partner (mutual marking).

White-lipped peccary females guard their young aggressively from other animals, including herd members, and it is very difficult to separate females from their newborns. Unlike white-lipped peccaries, collared peccary dams allow their young to nurse from other females, and young who have lost their mother may suckle other lactating females. Collared peccary newborns are also visited and touched by other members of the group. In the case of juvenile-adult interactions, a juvenile may pass underneath and between the hind legs of a dominant adult.

Feeding ecology and diet

Members of the Tayassuidae are primarily frugivorous and omnivorous. The ability of collared peccaries to subsist on a wide variety of food sources is likely an important factor in its ability to live in such a wide range of habitats. In southern Texas, prickly pear cladophils are a main staple of the collared peccary. In Texas and Arizona, *Opuntia engelmannii* is the most abundant food item eaten often during drought. Collared peccaries can also live in areas where there are no prickly pears. In the southwestern United States, collared peccaries need other plants for nutritional support, such as grass, forbs, acorns, pine nuts, and animal matter. They also eat plants and roots such as thistles (*Cirsium arvense*), gourds (*Cucurbita* spp.), and lechuguilla (*Agave lechuguilla*).

However, in Neotropical forests, collared peccaries are primarily frugivorous. Among fruits, the most common eaten by collared peccaries in the Peruvian Amazon are members of the Leguminosae, Sapotaceae, Menispermaceae, and the palms (*Iriartea* sp., *Jessenea* sp., and *Astrocaryum* sp.). As fruit availability decreases, collared peccaries shift to a more omnivorous diet.

When collared peccaries forage, they often root deeply into leaf litter, soil, base of trees, and mud of stream beds. Understanding the dietary composition of collared peccaries throughout their geographic range is very important, because collared peccary productivity is apparently very sensitive to their nutritional state.

White-lipped peccaries have efficient mechanisms for seed predation to exploit entire fruit resources including both the pulp and the seed. White-lipped peccaries are able to overcome the protective strategies of seeds by cracking seeds using their strong adductive jaw muscles, thick skull bones, strong resilient teeth, and interlocking canines. The strong adductive jaw muscles are attached to a large sagittal crest and

A four-day-old Chacoan peccary (*Catagonus wagneri*) infant explores its grassy habitat with its mother at the San Diego Zoo, USA. (Photo by AP Photo/San Diego Zoo. Reproduced by permission.)

large mandible, allowing the peccary to exert massive pressure on the seeds. The thick bones of the skull can withstand the pressure required to crack palm nuts. Premolar and molar teeth of peccaries are extremely strong and difficult to dislocate from the jaw. Interlocking canines are an adaptation to prevent the jaw from dislocating during the excretion of massive pressure.

The peccaries have a sacculated fore stomach that has microbial fermentation, similar to the deer rumen. Microbial fermentation in the stomach of peccaries occurs in three blind sacs, while gastric digestion occurs in the central main sac. These blind sacs are efficient at fermenting low-fiber plant products, but not plant material with high concentrations of cell wall. The blind sacs probably help soften and digest the hard palm seeds once they are cracked into small pieces by peccaries. However, since peccaries do not ruminate, they must crack the seeds into small pieces before ingesting them into the blind sacs.

Reproductive biology

The peccary reproductive tract is bicornuate, and each uterine horn is short compared to swine, and coiled caudally. Peccaries have diffuse epitheliochorial placentation. Regarding reproductive cyclicity, peccaries are polyestrous and can breed and give birth year-round. Males are also believed to be fertile year-round. Peccary litter sizes range from one to a maximum of four precocial young, but the litter size of most peccaries is one to two. A litter the size of one is much more common than a litter size of four for all peccary species. Mating in peccaries is not well known, but they may be promiscuous.

The white-lipped peccary (*Tayassu pecari*) is primarily nocturnal. (Photo by © Theo Allofs/Corbis. Reproduced by permission.)

A Chacoan peccary (*Catagonus wagneri*) in southeast Bolivia. (Photo by Kenneth W. Fink/Photo Researchers, Inc. Reproduced by permission.)

Conservation status

All Tayassuidae species are listed in CITES (Commission for International Trade of Endangered Species) appendices, and the Chacoan peccary is listed as Endangered by the IUCN. Hunting for bush meat is a major potential threat to peccaries. Determining sustainability of peccary harvests in hunted areas is crucial to future peccary conservation. Habitat loss, fragmentation, and degradation also pose severe threats to peccary populations. The World Conservation Union (IUCN) has established a Pig and Peccary Specialist Group. Group leaders convene and propose priority projects for study and of each species and harvest management. The IUCN Action Plan for human use of Tayassuidae advocates the development of rational and sustainable harvest management strategies by subsistence hunters, the improvement of local legislation and enforcement for harvest management, the outlaw of peccary hunting for solely commercial purposes, and to encourage the return of profits from the sale of sub-

A collared peccary (*Tayassu tajacu*) pair in a self-dug hole. (Photo by Kenneth Fink. Bruce Coleman, Inc. Reproduced by permission.)

sistence-byproduct peccary hides. This Action Plan also recommends further studies of peccary population dynamics, harvest monitoring and trade, and the development of management programs.

Significance to humans

Tayassuids are hunted throughout their range for sport, subsistence bush meat, commercial bush meat, and for commercial sale of their pelts. They are some of the most common species used for bush meat in Central and South America. In lowland South America, peccaries are the first ranked game in terms of biomass in seven out of eight indigenous groups and three out of four mestizo communities. In 12 out of 13 studies of hunting by rural inhabitants of Amazonia, peccaries rank first in terms of mammalian biomass hunted. Their meat is consumed locally in rural sectors and sold in city markets. Peccaries are also an important source of income for many rural inhabitants of the Neotropics. Game meat comprises more than 11% of the financial returns for rural inhabitants of the Peruvian Amazon.

In addition to meat, peccary skins are also a significant source of income for many rural inhabitants of the Neotropics. From 1946–1966, more than three million collared peccary skins and more than 800,000 white-lipped peccary skins were exported from Iquitos, Peru, to international leather markets. Currently, peccary skins are in demand in Europe, particularly Germany, for the manufacture of fine leather goods such as gloves. However, in the past decade, the peccary hide trade has declined in South America. Currently, Peru is the only country exporting peccary pelts. Since the peccary pelt trade became less profitable and more controlled, the exportation of pelts fell to the current level of around 35,000 peccary skins per year.

In accord with their economic and subsistence importance, peccaries also play an important cultural role in many indigenous societies. Peccaries are often associated with spiritual guardians of game animals in many indigenous communities of Amazonia. Many indigenous peoples in the Neotropics have myths and other spiritual beliefs relating to peccaries, and raise wild-caught juvenile peccaries as pets.

1. White-lipped peccary (*Tayassu pecari*); 2. Chacoan peccary (*Catagonus wagneri*); 3. Collared peccary (*Tayassu tajacu*). (Illustration by Patricia Ferrer)

Species accounts

Collared peccary
Tayassu tajacu

TAXONOMY
Tayassu tajacu (Linnaeus, 1758), Mexico.

OTHER COMMON NAMES
English: Javelina; French: Pecari a collier; German: Halsband-Pekari; Spanish: Sajino, chancho de monte, taitetu, jabalí, báquiro de collar, chácharo.

PHYSICAL CHARACTERISTICS
The smallest of the three peccary species, and adult animals range in size from approximately 24–88 lb (11–40 kg). Have a characteristic "collar" of pale hairs that cross behind the neck and extend bilaterally in front of the shoulders. The rest of the body hairs are gray or black with some whitish rings on individual hairs. The dorsal snout is relatively narrow.

DISTRIBUTION
The most wide-ranging of the Tayassuidae occurring from the southwestern United States through Central America, the entire Amazon forest region, the Pacific coasts of Colombia, Ecuador, and Peru, and the dry tropical thorn forests (Chaco) of Paraguay, Bolivia, Brazil, and northern Argentina.

HABITAT
Exploit the widest range of habitats of the Tayassuids, from dry, open deserts, mesquite bosques, and oak forests in the northern part of their range to moist and dry tropical forests in the southern part of their range. In the southwestern United States, they focus their activity in feeding areas, watering sites, and bedding grounds. In tropical forests of the Amazon, they use primarily moist terra firme habitat, and are uncommonly found in floodplain habitat.

Tayassu tajacu

BEHAVIOR
The herd size varies depending on their habitat, from 5–30 in the deserts of the southwestern United States and 2–15 in Neotropical forests. Territorial, with relatively well-defined home ranges that vary depending on the habitat; there is little exchange of animals between herds. Although they have relatively stable territories, herds in the southwestern United States often separate into subgroups more than 328 ft (100 m) apart from one another. Occasional solitary peccaries may be observed, but is relatively rare. Diurnal, with the greatest feeding activity during the morning and late afternoon until dark; they live in mixed sex groups of adult males and females, juveniles, and sub-adults.

FEEDING ECOLOGY AND DIET
Primarily grazers and rooters, with much of their diet comprised of spiny cacti, succulent plants, a few shrubs, forbs, grasses, tubers, beans, seeds, herbivores, as well as some small lizards and mammals.

REPRODUCTIVE BIOLOGY
Probably promiscuous. Ovarian cycles showed an average length of approximately 27.6 days. Behavioral estrus lasts for approximately of 5.7 days. Although lactational anoestrus does occur, an ovulatory postpartum oestrus has been observed. Males are fertile year-round, despite marked differences in rainfall between wet and dry seasons. Approximate age at first breeding is 16 months to two years. The average gestation period is 145 days. The average litter size is two fetuses/pregnant female.

CONSERVATION STATUS
Legally hunted in many Central and South American countries and the United States, and are listed on Appendix II of CITES, which permits international trade in their products so long as harvests do not overexploit natural populations.

SIGNIFICANCE TO HUMANS
The most frequently exploited Tayassuidae, they are hunted from southern North America through to Argentina. Peccary meat is an important resource and popular game meat for many rural people in northeastern Peru as a source of both food and monetary income. ◆

White-lipped peccary
Tayassu pecari

TAXONOMY
Tayassu pecari (Link, 1795), Cayenne, French Guiana.

OTHER COMMON NAMES
English: Wari; French: Pécari à Lévres blanches; German:

Weißbart-Pekari; Spanish: Huangana, pecari labiado, tropero, báquiro, puerco de monte, chancho de monte, chancho cariblanco.

PHYSICAL CHARACTERISTICS
Adult animals weigh 55–88 lb (25–40 kg). Has a characteristic growth of white hairs under the jaw and around the mouth. Body hairs range from gray to black, and younger animals often have a reddish tan hair color. Have a relatively broad snout, and have the strongest jaws of the Tayassuidae.

DISTRIBUTION
Restricted to Central and South America, occurring from eastern and southern Mexico to Panama, through the Amazon regions of Colombia, Venezuela, the Guianas, Suriname, Brazil, Boliva, and Peru, the Atlantic forests and Pantanal of Brazil, and south through the tropical dry thorn forests (Chaco) of Paraguay, Brazil, and northern Argentina.

HABITAT
In tropical rainforests, they use lowland moist terra firme, wet terra firme, stream beds, and floodplain habitats. Often visit water sources such as ponds or streams and saltlicks, palm swamps, and wallow in mud. Excellent swimmers that tend to forage near trees, roots, and other objects, and also appear to search for seeds scatter-hoarded by rodents.

BEHAVIOR
As herds move through their habitats, they give off a very strong odor, due to dorsal scent gland secretions. They make a great deal of noise by clacking their teeth, cracking palm nuts, snorting, grunting, and wheezing. They superficially graze the soil beneath the leaf litter as they forage. Herds are composed of males, females, and juveniles of various ages. They are diurnal, and forage and travel in the morning and afternoon, often resting near water sources and saltlick wallows during the midday heat.

FEEDING ECOLOGY AND DIET
Primarily frugivorous, but they also supplement their diet with insects, particularly insect larvae and worms, carrion, amphibians, bird, and reptile eggs, and occasionally fish.

REPRODUCTIVE BIOLOGY
May be polygynous. Ovarian cycles have an average length of 29.7 days. Behavioral oestrus has a mean of four days. Approximate age at first breeding is 1.5 years. The average gestation is 156–162 days. Breed and give birth year-around.

CONSERVATION STATUS
Legally hunted in many Central and South American countries, and are listed on Appendix II of CITES, which permits international trade in their products so long as harvests do not overexploit natural populations. However, in many regions, populations are rare or have been locally extirpated due to over-hunting or habitat loss. Although they are locally abundant in some regions, they are apparently susceptible to the effects of hunting. Because they often travel in large herds, hunters may kill many animals at once. There are historical and anecdotal reports of indigenous peoples and colonists killing large numbers while herds were swimming across rivers.

SIGNIFICANCE TO HUMANS
Socially and economically significant to many indigenous peoples and colonists throughout the Neotropics; meat is also an important source of protein for indigenous peoples, particularly in lowland, non-flooded tropical forests. Pelts are also sold in the peccary skin trade. They also play an important role in indigenous mythology and beliefs. Indigenous peoples in Central and South America occasionally capture juvenile white-lipped peccaries and keep them as pets. ◆

Chacoan peccary
Catagonus wagneri

TAXONOMY
Catagonus wagneri (Rusconi, 1930), Argentina. Believed to be a peccary species that became extinct during the Holocene, the Chacoan peccary was rediscovered by Western scientists in 1975. However, indigenous peoples in the Chaco have recognized the Chacoan peccary and other Tayassuids well before it was rediscovered. Chacoan peccaries are believed to be a relict species that survived in a thorn forest and scrub Pleistocene refugium.

OTHER COMMON NAMES
English: Giant peccary; French: Pecari du Chaco; German: Chaco-Pekari; Spanish: Tágua, pecari del Chaco, chanco quimilero.

PHYSICAL CHARACTERISTICS
Largest of the three peccary species, with adult animals weighing 66.1–94.7 lb (30–43 kg). They have relatively long, mule-like ears and large head, and well-developed sinuses, an apparent adaptation to dry, dusty environmental conditions. They have a whitish collar of hair that passes across the shoulder and extends under the chin. The rest of the body is covered in brown to gray hairs. They have relatively long bristles, giving them a shaggy appearance. Their legs are also relatively long.

DISTRIBUTION

Tayassu pecari
Catagonus wagneri

The most limited geographic range among the Tayassuids, ranging in thorn forest and steppe of the Gran Chaco in Northern Argentina, southeastern Bolivia, and Paraguay.

HABITAT
Preferentially range in thorn forest. They show an apparent selection for forested ecosystems during wet and dry seasons, but also use open savannas to a lesser extent. They require areas of low rainfall and high temperatures, making their range the smallest of the three peccary species.

BEHAVIOR
Live in stable, mixed-age and sex groups of approximately 4.5–4.6 animals; territorial. Often wallow in dust and/or mud. They are active from dawn to sunset, and move continally within their home range. Young have been observed to nurse from more than one female.

FEEDING ECOLOGY AND DIET
Cactus is a preferred food, likely due to its high water content. Although cactus is low in protein and high in oxalates, they can survive almost exclusively on this food source; they eat the giant cactus (*Opuntia quimilo*) known as *quimil*. The winter diet consists primarily of cactus (i.e., *Cleistocactus baumannii*, *Eriocereus* spp., *Opuntia* spp.), and flowers from *Stetsonia coryune*, *Cereus validus*, and *Quiabentia chacoensi*. They also eat a smaller amount of Acacia legumes, vine fruits, bromeliads, roots, and grubs, as well as soil from leaf-cutting ant mounds, likely due to its enriched mineral contents.

REPRODUCTIVE BIOLOGY
Probably promiscuous. Reach sexual maturity at over two years of age, and have between two and three young per litter. The gestation period is five months. The farrowing season is between July and December, with most young born in August–September.

CONSERVATION STATUS
The most endangered and least understood Tayassuid species, estimates are that the Paraguayan Chaco had a population of only 4,000 adult Chacoan peccaries in 1991. Unfortunately, little is known about Chacoan peccary population dynamics. The major threats are considered to be habitat loss to agricultural development, over-hunting, and disease. The species is listed as Endangered by the IUCN, and CITES classifies them as an Appendix I species, threatened by extinction. Appendix I species may not be traded between countries for commercial purposes. Chacoan peccaries may be exported and imported under highly regulated conditions for non-commercial purposes such as captive breeding in zoological collections. Recently, conservation groups and zoological parks have developed captive breeding programs for Chacoan peccaries in Paraguay, the United States, and Europe as a potential conservation strategy.

SIGNIFICANCE TO HUMANS
There is little information on the history of their socioeconomic and cultural significance. Despite their Endangered status, Chacoan peccaries are reported to be an occasional source of meat for local peoples and colonists in the Chaco. ◆

Resources

Books

Bodmer, R., R. Aquino, P. Puertas, C. Reyes, T. Fang, and N. Gottdenker. *Manejo y Uso Sustentable de Pecaríes in la Amazonía Peruana.* Quito and Geneva: IUCN Sur and CITES, 1997.

Bodmer, R. E., L. K. Sowls, and A. B.Taber. "Economic Importance and Human Utilization of Peccaries." In *Pigs, Peccaries, and Hippos, Status Survey and Conservation Action Plan*, edited by W. L. R. Oliver. Gland, Switzerland: IUCN, 1993.

Emmons, L. H. *Neotropical Rainforest Mammals, a Field Guide.* Chicago: University of Chicago Press, 1990.

Sowls, L. K. *The Peccaries.* College Station: Texas A&M Press, 1997.

Periodicals

Altrichter, M., J. C. Saenz, E. Carrillo, and T. K. Fuller "Seasonal Diet of *Tayassu pecari* (Artiodactyla:Tayassuidae)in Corcovado National Park, Costa Rica." *Revista de Biologia Tropical* 48 (2002): 689–701.

Barreto, G. R., O. E. Hernandez, and J. Ojasti. "Diet of Peccaries (*Tayassu tajacu* and *Tayassu pecari*) in a Dry Forest of Venezuela." *Journal of Zoology* 1 (1997): 241–256.

Carrillo, E., J. C. Saenz, and T. K. Fuller. "Movements and Activities of White-lipped Peccaries in Corcovado National Park, Costa Rica." *Biological Conservation* 108, no. 3 (2002): 317–324.

Fragoso, J. M. V. "Home Range and Movement Patterns of White-lipped Peccary (*Tayassu pecari*) Herds in the Northern Brazilian Amazon." *Biotropica* 30 (1998): 458–469.

Kiltie, R. W. "Bite Force as a Basis for Niche Differentiation between Rain Forest Peccaries (*Tayassu tajacu* and *Tayassu pecari*)." *Biotropica* 14 (1982): 188–195.

Taber, A. B, C. P. Doncaster, N. N. Neris, F. H. Colman. "Ranging Behavior and Population Dynamics of the Chacoan Peccary, *Catagonus wagneri*." *Journal of Mammalogy* 74, no. 2 (1993): 443–454.

Theimer, T. C., and P. C. Keim "Phylogenetic Relationships of Peccaries Based on Mitochondrial Cytochrome b DNA Sequences." *Journal of Mammalogy* 79, no. 2 (1998): 566–572.

Nicole Gottdenker
Richard Bodmer, PhD

▲

Hippopotamuses
(Hippopotamidae)

Class Mammalia
Order Artiodactyla
Suborder Suiformes
Family Hippopotamidae

Thumbnail description
Rotund, barrel-shaped body with short legs; large head with wide gape; eyes, ears, and nostrils high on the face; small tail flattened at the base

Size
Weight: female 500–3,000 lb (230–1,500 kg), male 600–4,000 lb (270–1,800 kg); length: female 58–106 in (150–270 cm), male 60–106 in (152–270 cm)

Number of genera, species
1 genus; 2 species

Habitat
Ponds, lakes, rivers, and wallows by day; grasslands and forests at night

Conservation status
Vulnerable: 1 species

Distribution
Sub-Saharan Africa

Evolution and systematics

The fossil record of hippos provides little evidence of their ancestry. All fossils can be readily assigned to one or other of the modern genera, except for a dwarf hippo from the Pleistocene of Cyprus, which has been placed in a separate genus, *Phanourios*, although it is considered to be of the genus *Hippopotamus* on the basis of its skull anatomy. Its limbs, however, are quite distinct and appear to be adapted for running over stony ground. It has been suggested that the differences are sufficiently great to justify its status as a separate genus.

There are numerous fossil *Hexaprotodon*, but none that can be equated with the modern pygmy hippo. Size is not a good criterion as there are dwarf *Hippopotamus* as well as full-sized *Hexaprotodon*. The former are usually found on islands, particularly those in the Mediterranean, and are probably examples of the phenomenon known as "island dwarfing," which occurs when herbivores are stranded without a predator on marine islands and subsequently shrink in body size. Simultaneously, they lose their abilities as saltors or cursors and evolve wide-ranging food habits as indicated, among others, by enlarging teeth to grind out what little nutrition there is from depleted and severely conditioned plant food supply. Island populations are inevitably small and, hence, are vulnerable to natural catastrophic events such as volcanic eruptions, as well as to human hunting. It is not surprising that dwarf hippos did not long survive the arrival of humans.

Threat display by a dominant bull (*Hippopotamus amphibius*). (Photo by Harald Schütz. Reproduced by permission.)

Hippopotamuses graze mostly at night at some distance from the water. (Illustration by Katie Nealis)

Most of the early fossil hippos of both genera are found in East Africa, suggesting that the family originated on that continent. But by the late Miocene, hippos had spread from Africa to most of Eurasia, though they did never reached Australia or the Americas. Some extinct species survived well into historical times. For example, *Hippopotamus lemerlei* was present in Madagascar as recently as A.D. 1000. As with the Mediterranean species, its extinction occurred soon after the arrival of man. In the case of Madagascar, this happened around 1,500 years ago.

The classical assumption is that hippos (Hippopotamidae) are related to pigs (Suidae) and peccaries (Tayassuidae), with the three families constituting the suborder Suiformes within the order Artiodactyla. Nine other families, containing only extinct forms, are recognized as belonging to the Suiformes, but none is obviously an ancestor to hippos. However, one superfamily, Anthracotheroidea, includes fossils that have been suggested as possible ancestors as they show some resemblance to hippos in their dentition. The anthracotherids resembled large pigs and were probably semi-aquatic. If they are ancestors, the hippo lineage branched off from them between the Oligocene and Miocene, about 25 million years ago. Hippos themselves date from the middle to late Miocene.

Some recent research points to a different evolutionary history as DNA analysis and gene sequencing suggest that the closest relatives to the hippo are whales (Cetacea). This

Fighting hippopotamuses (*Hippopotamus amphibius*) by the Rutshuru River, Uganda. (Photo by Animals Animals ©Bruce Davidson. Reproduced by permission.)

A mother hippopotamus (*Hippopotamus amphibius*) with a newborn baby in the Masai Mara National Reserve, Kenya. (Photo by Stephen J. Krasemann/Photo Researchers, Inc. Reproduced by permission.)

A hippopotomus (*Hippopotamus amphibius*) mother and baby in the river in Masai Mara National Reserve, Kenya. (Photo by Animals Animals ©Anup Shah. Reproduced by permission.)

Congregation behavior of the hippopotamus (*Hippopotamus amphibius*). (Photo by Harald Schütz. Reproduced by permission.)

Hippopotamuses scatter dung with their tails to mark territory. (Illustration by Katie Nealis)

relationship is not evident from comparative anatomy or from the fossil record, however, the evolutionary history of whales is about as poorly known as that of hippos. It is now generally accepted that whales evolved from ungulate stock. One 1994 study suggested a link with hippos from their sequencing of the mitochondrial DNA cytochrome *b* gene, though other research found no support for such a relationship from the mitochondrial DNA sequencing, although a link was demonstrated between cetaceans and artiodactyls. There is, therefore, no general agreement on a hippo/whale link, though if such a link exists, it is probably a weak one.

DNA analyses have also questioned the position of the Hippopotamidae within the Suiformes. Two separate studies did not find a close relationship between pigs and peccaries on the one hand and hippos on the other. The mitochondrial sequence comparisons support the relationship between pigs and peccaries, with hippos forming a separate evolutionary line. The conclusions from these genetic studies, however, differ from those of their morphological analyses, which support the monophyly of the Suiformes. With opposite conclusions from the two techniques, the question of the ancestry of hippos remains open.

Physical characteristics

Hippos have rotund bodies and short legs. Unlike other artiodactyls, which have only two functional hooves, there are four toes on each foot with slight webbing between them. The thick skin appears hairless apart from a few bristles around the mouth and on the tail, but there is a covering of

A hippopotamus (*Hippopotamus amphibius*) threatens a human intruder with a wide gape, Okavango Delta, Botswana. (Photo by Gregory G. Dimijian/Photo Researchers, Inc. Reproduced by permission.)

very fine hairs at low density over the whole of the body. There are no sweat glands as such, but there are large skin glands that secrete a viscous liquid that turns pink on exposure to air. The secretion probably acts as a sunscreen; it is also thought to have antiseptic properties. The skin needs to be kept wet, and cracks appear if the hippo is prevented from entering water.

The head is large with a wide mouth. The canines and incisors are tusk-like with open roots so that the teeth grow continuously, somewhat like the teeth of rodents. The lower canines are curved and are particularly large. They are kept sharp by rubbing against the upper canines. The incisors are straight and peg-like. The lower incisors point forward, but the shorter upper incisors point downward. There are six molariform (grinding) teeth in each half-jaw, with a pattern of cusps that is diagnostic of the species. The full dental formula for the family is (I3/3, C1/1, P4/4, M3/3) $\times$ 2 = 44. The nostrils, eyes, and ears are placed high on the face. Consequently, the animal can remain submerged with very little of its body in view, while at the same time being able to breath and sur-

A pygmy hippopotamus (*Hexaprotodon liberiensis*) courtship encounter in West Africa. (Photo by Rod Williams. Bruce Coleman, Inc. Reproduced by permission.)

A hippopotomus (*Hippopotamus amphibius*) with newborn and afterbirth, in Kazinga Channel, Zambia. (Photo by Animals Animals ©Anthony Bannister. Reproduced by permission.)

vey its surroundings. The ears are small but highly moveable and wag vigorously on surfacing.

The alimentary canal shows what is known as the pseudoruminant condition, i.e., digestion takes place through fermentation of food in a multi-chambered stomach, as in ruminants, but there is no chewing of the cud. The post-gastric gut is typical of a large herbivore. There is no cecum, which is not typical of large herbivores, but there is a gall bladder despite statements to the contrary.

The reproductive organs show some peculiarities. Unlike most mammals, the male lacks a scrotum as the testes do not fully descend at birth. The penis is normally retracted, so it is difficult to tell the sex of a hippo. The female reproductive tract shows two marked peculiarities, one being the series of ridges in the upper vagina and, the other, two large sacs that project from the vestibule. The functions of these organs remain obscure.

Distribution

The common hippo occurs in many sub-Saharan countries, but some of the populations are extremely small, especially in West Africa. The pygmy hippo is confined to West Africa with Liberia containing the majority of the population.

Habitat

The common hippo is found in shallow freshwater aquatic habitats during the day, but at night it emerges to graze some distance from water. The pygmy hippo is a forest animal, although it too spends the day in or near water.

Behavior

Although it is highly colonial in the water, the common hippo is not a social animal and the only social bond is that between a female and her dependent calves. Males are territorial in the water and maintain mating rights over the females within a defended length of shoreline. In addition to their aerial calls, common hippos vocalize under water. Some of the sounds are amphibious and pass through the air and water simultaneously. The pygmy hippo is usually found alone; although pairs occur; the duration of the pair bond is not known. There is no evidence that males are territorial.

Feeding ecology and diet

Hippos are herbivorous, although carnivory through scavenging has been reported in the common hippo on a few occasions. Grass is the principal food of the common hippo,

A pygmy hippopotamus (*Hexaprotodon liberiensis*) cow with calf. (Photo by Tom Brakefield. Bruce Coleman, Inc. Reproduced by permission.)

Hippopotamuses (*Hippopotamus amphibius*) protecting their young. (Photo by K & K Ammann. Bruce Coleman, Inc. Reproduced by permission.)

Two young male hippopotamuses (*Hippopotamus amphibius*) play-fight. (Photo by Erwin and Peggy Bauer. Bruce Coleman, Inc. Reproduced by permission.)

while the pygmy hippo has a much wider diet that includes fruits and ferns. Both species feed by nipping off the vegetation with their muscular lips.

Reproductive biology

The common hippo mates in the water as does the pygmy hippo, which also copulates on land. Births occur in the water in both species, although the pygmy hippo sometimes calves on land. Usually, only one calf is produced at a time, though twins do occur at a low rate. Births may take place throughout the year, but there are peaks associated with increased rainfall. Hippopotamuses are polygamous.

Conservation status

The common hippo is placed on Appendix II of CITES and the pygmy hippo on Appendix I. The common hippo is not in immediate danger over much of its range in southern and east Africa, but it is vulnerable to extinction in West Africa. The conservation status of the pygmy hippo is not clear, though the IUCN lists it as Vulnerable.

Significance to humans

The common hippo impinges on human affairs as it is a dangerous animal that also raids crops, particularly rice. The pygmy hippo is not a threat to people, although it does sometimes raid crops. Both species are hunted for meat.

1. Pygmy hippopotamus (*Hexaprotodon liberiensis*); 2. Common hippopotamus (*Hippopotamus amphibius*). (Illustration by Joseph E. Trumpey)

Species accounts

Common hippopotamus

Hippopotamus amphibius

TAXONOMY

Hippopotamus amphibius Linnaeus, 1758, Nile River, Egypt. Five subspecies have been described, though it is doubtful that they are valid. No external differences between them have been noted.

OTHER COMMON NAMES

English: River hippopotamus; French: Hippopotame; German: Grossflusspferd; Spanish: Hipopotamo.

PHYSICAL CHARACTERISTICS

The common hippo has a shoulder height of 54–60 in (137–152 cm), length up to 106 in (270 cm), and weighs up to 4,000 lb (1,800 kg). It has a rotund body with disproportionately short legs. The thick purple-brown skin appears hairless, with only a few bristles around the mouth and on the tail, but there is a covering of very fine hairs at low density over the whole of the body.

The head is large with a wide mouth that opens to a gape of nearly 180°. The number of teeth is slightly reduced in the common hippo to (I2/2, C1/1, P3/3, M3/3) × 2 = 36, although sometimes the fourth milk premolar is retained in the adult jaw.

DISTRIBUTION

The common hippopotamus occurs in some 35 sub-Saharan countries, but many populations are extremely small, especially in West Africa, where the hippos are fragmented into isolated groups each containing a few dozen at the most. Countries with substantial populations are situated in the eastern and southern parts of the continent of Africa and include Zambia, Democratic Republic of the Congo (Zaire), and Tanzania.

HABITAT

During the day, the hippo is found in shallow freshwater aquatic habitats, which may be rivers, lakes, wallows, or any wetland that keeps the skin moist. Wet mud is suitable, and hippos often occur in wallows that are drying out. At night, the hippo leaves the water to graze in grasslands that may be a mile (1.6 km) or more from its daytime retreat. It reaches the grazing areas by following well-worn trails.

BEHAVIOR

There have been few investigations into the behavior of the hippo as it is a difficult species to study and conclusions about its activities may not be of general applicability. The most detailed study, set in Uganda, reported that although the hippo is highly colonial in the water, it shows little in the way of a social life and is essentially solitary when grazing at night. The only social bond is that between a female and her dependent calves. Males are territorial in the water and maintain mating rights over the females within a defended length of shoreline. It is not a harem system, however, for the females do not necessarily return to the same territory every day. The territory may contain bachelor groups, which are tolerated by the territorial male provided they behave submissively on meeting. Fierce fighting may occur if a bachelor challenges the territory holder and the death of one of the combatants may follow. Hippos are not territorial on land. The spraying of bushes with feces occurs when the hippo vigorously wags its tail while defecating, thus spreading the droppings far and wide. The purpose is not clear but it may have a social function as a dominance display or possibly the deposits may serve to orientate the hippo during its nocturnal wanderings.

The bellows of hippos are well-known, though their role in communication is not clear. In addition to the aerial sounds, hippos vocalize under water, producing at least three types of call, including tonal whines, pulsed croaks, and clicks, all of which appear to be associated with communication. Some calls are amphibious and pass through the air and water simultaneously.

FEEDING ECOLOGY AND DIET

The common hippopotamus is predominantly a grazer. It plucks the grass with its lips, nipping it off close to the ground thereby creating short-grass patches known as hippo lawns. Relative to its weight, the hippo consumes less food than other herbivores, possibly reflecting its lower metabolic rate. The hippo is able to extend its grazing range in the wet season by making use of temporary wallows for daytime resting thus obviating the need to trek back to permanent water every day. However, it does not feed on aquatic vegetation to any extent. Carnivory, mainly through scavenging, has been reported several times, including a case of cannibalism.

REPRODUCTIVE BIOLOGY

Polygamous. The hippo mates in the water with the female remaining submerged other than when she breaks surface to

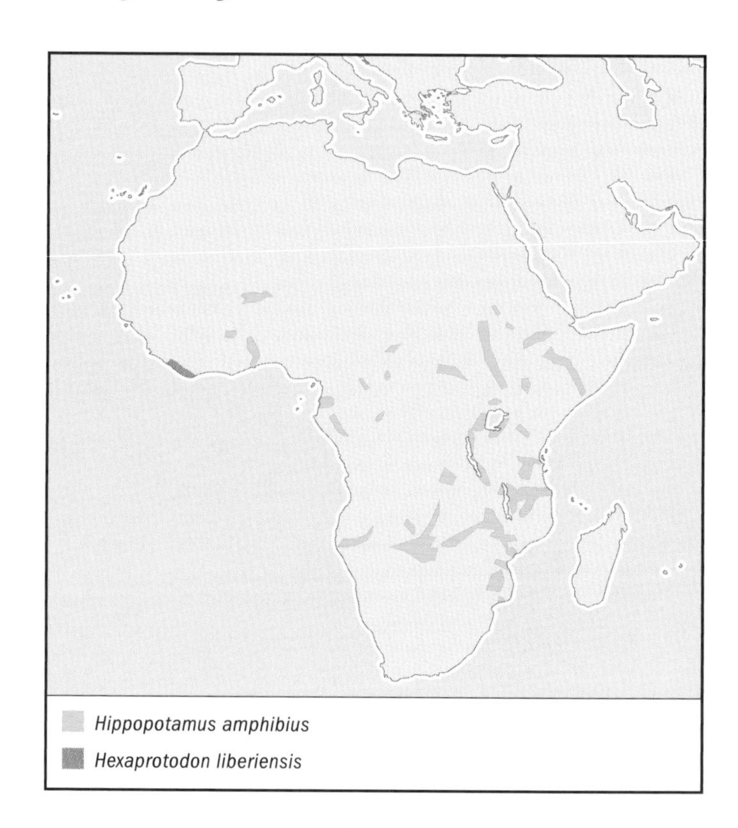

breathe. The gestation period of 240 days is short for such a large animal. The female withdraws from the herd before giving birth, which usually takes place in the water. The weight of the neonate is about 110 lb (50 kg). Normally, only one calf is produced at a time, though twins do occur at a low rate, which in Uganda was found to be 0.7%. The sex ratio at birth is 1:1.

Births occur throughout the year on the equator, but there are peaks associated with rainfall, with most births taking place at the beginning of the wet season. Away from the equator, where rainfall is more seasonal, there are some dry season months when there are no births at all.

The duration of lactation is unknown; it is probably of the order of one year. The calf remains with its mother after the second calf is born, but leaves before the onset of puberty, which in Uganda was estimated to be at about 7.5 years of age in the male and nine years in the female.

CONSERVATION STATUS

The common hippo is placed on Appendix II of CITES. It is not in immediate danger over much of its range in southern and east Africa, but the outlook in West Africa is bleak. The total number is low, and the remaining animals are scattered in small groups in which genetic problems resulting from inbreeding may be expected.

SIGNIFICANCE TO HUMANS

The hippo is a dangerous animal that is responsible for numerous human deaths each year, particularly of fishermen. It also raids crops, particularly rice. It is, in turn, hunted for meat and ivory; some of the human deaths occur during attempts to harpoon the animal. The hippo features in African folklore. ◆

Pygmy hippopotamus

Hexaprotodon liberiensis

TAXONOMY

Hippopotamus minor (Morton, 1849), St. Paul's River, Liberia. Two subspecies, *Hexaprotodon liberiensis liberiensis* and *H. l. heslopi*, but the latter is most probably extinct.

OTHER COMMON NAMES

French: Hippopotame nain; German: Zwergflusspferd; Spanish: Hipopotamo enano.

PHYSICAL CHARACTERISTICS

Length 5–6 ft (1.5–1.8 m); weight 350–600 lb (159–272 kg). The pygmy hippo is similar to its larger cousin in its general body shape; its head, however, is proportionately smaller with less protuberant eyes, which, with the ears and nostrils, are not so high on the head. Its limbs and neck are relatively longer, and the back slopes forward instead of being parallel with the ground as in the common hippo. The skin is darker than that of the larger species—a blackish color. The dental formula is (I2/1, C1/1, P3/3, M3/3) × 2 = 34.

DISTRIBUTION

The pygmy hippo is confined to West Africa with Liberia containing the majority of the population. Other countries with pygmy hippos are Guinea, Ivory Coast, and Sierra Leone.

HABITAT

A forest animal that spends the day in or near water, usually a river or stream, which it leaves at night to forage. Some, at least, occupy burrows in the banks of rivers; it is not clear whether the hippo excavates the burrow itself or merely enlarges an existing hole.

BEHAVIOR

Its behavior in the wild is insufficiently known. It is usually found alone, although pairs occur. It is not clear whether the species is monogamous or promiscuous. It follows game trails and spreads its feces with its tail.

FEEDING ECOLOGY AND DIET

Feeds with its lips; its diet is varied, with fruits and ferns figuring prominently. Grass is readily eaten when available.

REPRODUCTIVE BIOLOGY

Polygamous. Copulates either on land or in the water after a consorting period; it is not known how long the pair bond persists. Estimates of the gestation period vary from 188 to 210 days. Births occur in water or on land, and singletons are the rule. Birth weights of 35 calves in Basel Zoo averaged 12.61 lb (5.73 kg). Sexual maturity in zoo specimens occurs between three and five years of age; it is probably later in the wild. The calf does not follow its mother when she is foraging, remaining in hiding and visited at intervals to be suckled.

CONSERVATION STATUS

Listed on Appendix I of CITES, and the IUCN lists it as Vulnerable to extinction, mainly through loss of habitat. It adapts well to captivity and breeds readily so that it could, if necessary, be maintained indefinitely in zoos.

SIGNIFICANCE TO HUMANS

Not normally a threat to human beings, although it is a formidable animal and can cause injury to hunters. It occasionally raids crops. It is the subject of much folklore. ◆

Resources

Periodicals

Barklow, W. E. "Some Underwater Sounds of the Hippopotamus (*Hippopotamus amphibius*)." *Marine and Freshwater Behavioral Physiology* 29 (1997): 237–249.

Coryndon, S. C. "The Taxonomy and Nomenclature of the Hippopotamidae (Mammalia, Ardiodactyla) and a Description of Two New Fossil Species." *Proceedings of the Koninklijke Nederlandse Akademie Wetenschappen Series B* 80 (1977): 61–88.

Gansberger, K., and G. Forstenpointer. "On the Existence of a Gall Bladder in the Hippopotamus." *Wiener Tieraerztliche Monatsschrift* 82 (1995): 157–158.

Hasegawa, M., and J. Adachi. "Phylogenetic Position of Cetaceans Relative to Artiodactyls—Reanalysis of Mitochondrial and Nuclear Sequencies." *Molecular Biology and Evolution* 13 (1996): 710–713.

Houtekamer, J. L., and P. Y. Sondaar. "Osteology of the Forelimb of the Pleistocene Dwarf Hippopotamus from

Resources

Cyprus with Special Reference to Phylogeny and Function." *Proceedings of the Koninklijke Nederlandse Akademie Wetenschappen Series B* 82 (1979): 411–448.

Irwin, D. M., and U. Amason. "Cytochrome *b* Gene of Marine Mammals: Phylogeny and Evolution." *Journal of Mammalian Evolution* 2 (1994): 37–55.

Klingel, H. "The Social Organisation and Behaviour of *Hippopotamus amphibius*." *African Wildlife: Research and Management, International Council of Scientific Unions* (1991): 73–75.

Lang, E. M., M. vK. Hentschel, and W. Bulow. "Zwergflusspferde (Gattung *Choeropsis*)." *Grzimeks Enzyklopädie: Saugetiere* (1988): 62–64.

Laws, R. M., and G. Clough. "Observations on Reproduction in the Hippopotamus *Hippopotamus amphibius* Linn." *Symposia of the Zoological Society of London* 15 (1966): 117–140.

Montegelard, C., S. Ducrocq, and E. Douzery. "What is a Suiform (Artiodactyla)? Contribution of Cranioskeletal and Mitochondrial DNA Data." *Molecular Phylogenetics and Evolution* 9 (1998): 528–532.

Randi, E., V. Lucchini, and C. H. Diong. "Evolutionary Genetics of the Suiformes as Reconstructed Using mtDNA Sequencing." *Journal of Mammalian Evolution* 3 (1996): 163–194.

Stewart K. Eltringham, PhD

▲

Camels, guanacos, llamas, alpacas, and vicuñas

(Camelidae)

Class Mammalia
Order Artiodactyla
Suborder Tylopoda
Family Camelidae

Thumbnail description
Medium- to large-sized mammals with a long and thin neck, small head, and a slender snout with a cleft upper lip; they bend their legs beneath the body and rest on the stomach, the toes bear nails rather than hooves, and there are no hornlike structures on their low and elongated skull

Size
Bactrian and dromedary camels: average height 6–7.5 ft (183–229 cm); 1,000–1,800 lb (454–816 kg); vicuñas, guanacos, llamas, and alpacas: average height 3–4.3 ft (90–130 cm); 88.8–265.5 lb (40–120 kg)

Number of genera, species
3 genera; 6 species

Habitat
Semiarid to arid plains, grasslands, and deserts

Conservation status
Critically Endangered: 1 species; Vulnerable: 2 species

Distribution
Bactrian camels and dromedaries found in the arid plains and hills of Africa, Asia, and Australia; vicuñas, guanacos, llamas and alpacas located in the Andean high plateau and the arid plains of western and southern South America

Evolution and systematics

The family Camelidae originated in North America during the Eocene period, 45 million years ago. At the end of this period, camelids migrated to Africa and Asia through the Bering Strait, where after successive changes they evolved into the Camelini tribe, which includes the Bactrian, or two-humped camel, and the dromedary, or one-humped camel. The population also spread into South America through the Panama Isthmus, where it evolved into the Lamini tribe that gave way to the species now known as vicuña, guanaco, llama, and alpaca.

The camelidae family developed in North America from the late Eocene to the late Pleistocene, while in Europe it was present from the mid Pliocene to the Pleistocene. After many migrations and transformations, camelids became extinct in both. Members of the family are still present, however, in North Africa, where the family has developed from the late Pliocene until Recent in Asia, which it inhabits since the mid Pliocene until Recent, and in South America, where it has developed since the Pleistocene until Recent. There are currently six camelid species throughout these regions.

Two species belong to the genus *Camelus* (*Camelus bactrianus* and *Camelus dromedarius*); and three species to the genus *Lama* (*Lama guanicoe*, *Lama glama*, and *Lama pacos*). Only one species belongs to the genus *Vicugna* (*Vicugna vicugna*). Some authors place *Camelus* under the species *Camelus ferus bactrianus* and *Camelus ferus dromedarius*, but this taxonomic classification is no longer in use. Both species of the genus *Camelus* have been domesticated. Although there is still some controversy among several authors regarding when domestication started, it is estimated that it began some 6,000 or 5,000 years ago. The only known wild ancestor of *Camelus bactrianus* is a very small population found in the western Gobi Desert. However, taxonomists still hold the same genus and species for both wild and domestic populations in all its range. No wild ancestor is known for *Camelus dromedarius*.

The members of the Camelidae family found in South America include the wild guanaco (*Lama guanicoe*) and vicuña (*Vicugna vicugna*) as well as the domestic llama (*Lama glama*) and alpaca (*Lama pacos*). The origins of the alpaca are somewhat confusing, mainly due to a high level of hybridization and the lack of genetic analyses. However, recently, Kadwell et al. confirmed through mitochondrial and microsatellite DNA analysis that vicuña is the wild ancestor of alpaca. Therefore, this domestic species should be reclassified as *Vicugna pacos*. According to Kadwell et al., the domestication of this South American camelid began approximately some 7,000 to 6,000 years ago.

A similar situation occurred with the llama, which originated from the domestication of the guanaco, according to

The llama's (*Lama glama*) main use for humans today is for its wool. (Photo by © Corbis Images/PictureQuest. Reproduced by permission.)

A vicuña (*Lama vicugna*) in Altiplano, Peru. (Photo by Sullivan & Rogers. Bruce Coleman, Inc. Reproduced by permission.)

osteological remains from Andean archaeological sites approximately 6,000 years old.

Physical characteristics

There is a great difference in size between the genus *Camelus* in Africa and Asia, and *Lama* and *Vicugna* in South America. However, all three genera are characterized by a long and thin neck, a small head, and a slender snout with a cleft upper lip. Their mouths are tough, in order to allow them to eat thick grasses and thorny desert plants without hurting themselves.

The skull is long and elongated, and there are no horns or antlers. The usual dental formula in *Camelus* is: (I1/3, C1/1, P3/2, M3/3) × 2 = 34. In *Lama* and *Vicugna*, it is: (I1/3, C1/1, P2/1, M3/3) × 2 = 30. However, there is some variation depending on their age. The premaxilary bones of the skull bear the full number of upper incisors in the young, but only the outer incisors in the adult.

The hind part of the body is contracted and all of their limbs are long. The forelimbs have naked callosities in the guanaco, and prominent kneepads are present in *Camelus*, since they bend their legs beneath the body and rest on the stomach. In *Vicugna*, the knee joint is low in position, because of the long femur and its vertical placement. All camelids run with a swinging stride, as the front and hind legs move in unison on each side of the body.

Each foot has only two digits (the third and fourth). The proximal digital bones are expanded apart from the center. The other digital bones are small, not flattened on the inner surface, and not encased in hooves, bearing nails on the upper surface only. The digits are spread nearly flat on the ground. The feet are broad in *Camelus* and slender in *Lama* and *Vicugna*.

Both *Camelus* species have humps that help them survive in the desert, since they serve to store fat as a source of energy in times of need. When they are well fed, the hump is erect and plump, but when they do not have adequate food, the hump shrinks and often leans to one side. Dromedaries have only one hump, while Bactrian camels have two.

All Camelidae species have thick coats to protect them from the cold, but only the species of the genus *Camelus* shed

A Bactrian camel (*Camelus bactrianus*) yawning in the Gobi Desert. This species of camel is endangered in the wild. (Photo by Erwin and Peggy Bauer. Bruce Coleman, Inc. Reproduced by permission.)

Guanacos (*Lama guanicoe*) grazing in Chile. (Photo by Frank Krahmer. Bruce Coleman, Inc, Reproduced by permission.)

their hair when temperatures rise. Both Bactrians and dromedaries have special muscles that allow them to close their nostrils and lips tightly for long periods, in order to avoid breathing large amounts of sand or snow.

Distribution

Camelidae are found in the wild since ancient times, from the Arabian Peninsula to Mongolia and in western and southern South America. There has been a drastic reduction in the range of wild camelids, but domesticated members of the family have spread over much of the world.

The distribution of the wild Bactrian camel (*Camelus bactrianus*) in historic times extended from about the great bend of the Yellow River at 110°E westward across the deserts of southern Mongolia and northwestern China to central Kazakhstan. Populations of wild Bactrians are currently restricted to three reduced areas in Mongolia and China: the Taklimakan Desert, the deserts surrounding Lop Nur, and the region in and around Great Gobi Strict Protected Area.

The domesticated dromedary (*Camelus dromedarius*) is mainly found in the Sahara Desert, but its presence is also common in arid regions of the Middle East through northern India. In central Australia, there is a population composed of individuals that were introduced into the dry and arid regions and live in feral state.

The vicuña is distributed throughout the southern Andes of South America. In Peru, it is found in the Ancash, Ayacucho, and Arequipa Departments. In Chile, the species is located in the Tarapaca, Antofagasta, and Atacama regions. In Bolivia, vicuñas inhabit the La Paz, Oruro, Potosi, and Tarija Departments. In Argentina, the species is found in the San Juan, La Rioja, Catamarca, Salta, and Jujuy Provinces.

The llama is distributed in Argentina from the northeast, southward through the Pampa, to as far as Tierra del Fuego. In Chile, the species is irregularly distributed in the northern regions of Tarapaca, Copiapo, and La Serena, and the central O'Higgins region. However, the largest population is found in the southernmost areas of the Magallanes region. In Peru, the population is represented by small numbers in the departments of La Libertad, Arequipa, and Ayacucho. In Bolivia, individuals are located on the Mochara Range and in the Chaco region. In Paraguay, a small population is found in the northern section of the boreal Chaco in the Paulo Lagerenza area.

Alpacas and llamas are found throughout the southern Andes of South America, where they are kept in herds by the Andean local people. A significant number of both alpacas and llamas has been taken to the United States and Australia to be traded as pets.

Habitat

Wild camelids inhabit desert and semi-arid environments covered by sparse vegetation, some drought-resistant shrubs, and with a rigorous climate presenting a long dry season and a short rainy season.

Dromedary camels (*Camelus dromedarius*) drink from a stream in Tunisia. (Photo by Raymond Tercafs. Bruce Coleman, Inc. Reproduced by permission.)

Guanaco (*Lama guanicoe*) mother and young. (Photo by Wolfgang Bayer. Bruce Coleman, Inc. Reproduced by permission.)

The habitat of wild Bactrians is located in the arid continental areas in the temperate zone, where summer is hot and winter is severely cold, with large daily differences in temperature. Annual precipitation is less than 3.9 in (100 mm), and most of the distribution area presents aridity.

The dromedary inhabits desert environments where thorny plants, dry grasses, and saltbush are found.

The guanaco inhabits both warm and cold grasslands and shrublands, from sea level to approximately 13,120 ft (4,000 m). In some areas, the species inhabits forests during the winter.

The vicuña is more specific in its habitat and only occupies the grasslands of the Andes above 11,482 ft (3,500 m).

Behavior

All species are diurnal, adapted to harsh climates, and may spit, or occasionally kick, when threatened.

In the Gobi desert wild herds of Bactrians move widely in search of water. The animals tend to concentrate in and around the mountain areas where there are springs and snow on the slopes, which can provide the only moisture in winter. They also move to areas where a local shower has created a green spot. During October and November, large concentrations of Bactrians occur near the mountains. Most of them gather in herds of up to 30 individuals, while a few individuals of both sexes remain solitary.

In the Sahara, the domestic dromedary is left on its own for four or five months each year, coinciding with the mating season. The dromedaries form three types of herds during that time: bachelor males; adult females with their newborns; and up to 30 adult females, along with their one- and two-year-old offspring, each group led by a single adult male. Rival males that approach one another first employ dominance displays, including defacating, urinating, and slapping their tails on their backs. If neither male retreats from the display, the two animals fight by biting and thrashing with their forefeet.

The vicuña is a social animal. Territorial males maintain family groups consisting of the male adult and subadults, females, and young less than one year old. Adult males without territories form non-reproductive groups. Membership in these groups is limited to subadult males one to four years old, which have been expelled from their family groups, and of aging males, which have lost their territories.

There are three social groups in guanaco populations: family groups, male groups, and solitary males. Family groups consist of an adult male and one or several females with their

Mother vicuña (*Lama vicugna*) feeding a one-month-old baby in Pampa Galeras National Reserve, Peru. (Photo by Animals Animals ©Mark Jones. Reproduced by permission.)

year's offspring; young females may also be present. The number of animals found in this group may vary from a minimum of two to a maximum of 30 members. Male groups are entirely formed by males, both young and adult, whose number may amount to 50 individuals, but usually ranges from 5–20 animals. Solitary males are physically and sexually mature males prepared to form a family group and control a territory, or old, sick, or wounded males.

Alpaca and llama are raised as domestic stock and, as such, have lost their sense of social structure.

Feeding ecology and diet

Camelidae are grazers that feed on many kinds of grass and need very little water. They thrive on salty plants that are rejected by other grazing animals and need to eat halophytes, because although they are adapted for conservation of water, and since they hardly sweat or urinate, they will lose weight and strength if they go for long periods without drinking water. Dromedaries and guanacos have been seen drinking salty water, which no other animal could tolerate.

Both *Camelus* species eat practically any vegetation that grows in the desert or semiarid regions and are able to convert thorny desert shrubs and salty plants into highly nutritious food. They do not chew their food completely before swallowing it. After eating, they bring up the partly digested food, and then they re-chew it, swallow it again, and digest it.

Contrary to what is commonly believed, there is no evidence that *Camelus* store water in the stomach or in the humps. Humps are really masses of fat that nourish the animals when food is scarce. With this energy supply on their backs, they can go several days without eating. Both camel species can store up to 80 lb (36 kg) of fat in their humps. As camels use this fat, their humps shrink.

Reproductive biology

Females of the genus *Camelus* give birth to only one offspring after approximately 12–13 months of pregnancy. The newborn can stand shortly after its birth and can walk within a few hours. It stays with its mother until it is almost two years old, but it is not full-grown until the age of five.

Females of the genera *Lama* and *Vicugna* give birth to a single young after a gestation period of approximately 11 months. The offspring stay with their mothers until they are about one year old.

Because of their long gestation period and the high mortality rate of their newborns, females of all Camelidae species have an early estrous period that begins approximately one or two weeks after giving birth. Guanacos are polygamous, but the mating system for domesticated species is not well known.

Conservation status

The population of wild Bactrian camels has been reduced in Mongolia since the 1960s, due to heavy hunting and competition with domestic animals for water and grasses. This species is currently classified as Critically Endangered by the IUCN Red List.

Vicuña has shown a remarkable recuperation since its precarious existence during the 1960s, and guanaco shows the

A llama (*Llama glama*) herd in Lake Titicaca, Bolivia. (Photo by Norman Tomalin. Bruce Coleman, Inc. Reproduced by permission.)

Guanaco (*Lama guanicoe*) spitting at young in Patagonia, Chile. (Photo by Animals Animals ©Birgit Koch. Reproduced by permission.)

most stable population of all wild camelids. At present, the IUCN Red List has placed both species in the Vulnerable category.

Domestic Bactrian camel, dromedary, alpaca, and llama are not given a conservation status because they are grazing domestic animals, and unlikely to need protection.

Significance to humans

Camelids have been used for transportation as well as to obtain meat and fiber for clothes since approximately 7,000–6,000 years ago until today. Ancient civilizations in the Arabian Peninsula, Mongolia, and the South American Andes flourished thanks to these animals.

It is estimated that both bactrian and dromedary were domesticated at approximately the same time, some 6,000 years ago, in Mongolia and Arabia. This brought about an increase in trade that linked the Indus Valley civilization with Mesopotamian city-states along the Tigris and Euphrates rivers and even reached the waters of the Indian Ocean.

Human populations living in the hot deserts of North Africa and Asia have especially benefited from these species, using them to travel in caravans, since they may travel about 100 mi (161 km) without water and keep up a steady pace despite the heavy burdens they carry. Some desert peoples measure wealth by the number of camels a person owns. These species also provide their owners with food, clothing, and fuel. *Camelus* milk is thick and rich, and people sometimes eat camel meat. The fat contained in their humps can be melted and used for cooking. Also, when the animals shed their coats, their owners gather up the woolly hair and weave it into clothing, blankets, and tents.

Llama and alpaca domestication began thousands of years ago before the Inca Empire. Nevertheless, detailed knowledge of llama and alpaca breeding and pastoralism is available only from the Inca period, when the ownership of these animals was a symbol of wealth. Therefore, herds were owned by the state, the priests, the community, and certain individuals. The animals that were part of community herds pastured freely and mingled. To identify their owner, they were marked with a piece of wool tied to their ears, a practice that still exists today in the Andes.

Pastoralism specialists, who constituted a well-identified social level, controlled the herds belonging to the state and the priests and concentrated on the reproduction of animals of certain colors, which were later sacrificed to specific deities.

Unlike the llama and alpaca, the use of vicuña was limited to specific ritual practices. This restriction was very strict in the Inca Empire, therefore specialists improved upon a method of capturing animals, which had been used long before the Empire. This method was known as *chaku*, and consisted in rounding up the vicuñas toward the end of ravines or mountain slopes where there were stone corrals with enough space to capture a large number of animals without injuring them. Several hundred people participated in this ancestral management technique. After the animals were captured, they were sheared. Only a few were sacrificed in rituals, and the rest were set free to be used later on.

The guanaco's wide distribution allowed its existence without domestication. Many cultures in the southernmost areas of South America benefited from this species to obtain meat for food, hides for clothing and shelter, bezoar stones for medical purposes, fibers for sewing, domesticated juveniles for entertainment, a stimulus for the creation of myths, and several words to designate age, sex, color, etc.

An alpaca (*Llama pacos*) herd in Peru. (Photo by G. Gualco/F. Lane. Bruce Coleman, Inc. Reproduced by permission.)

1. Llama (*Lama glama*); 2. Dromedary camel (*Camelus dromedarius*); 3. Guanaco (*Lama guanicoe*); 4. Vicuña (*Vicugna vicugna*); 5. Alpaca (*Lama pacos*); 6. Bactrian camel (*Camelus bactrianus*). (Illustration by Joseph E. Trumpey)

Species accounts

Bactrian camel
Camelus bactrianus

TAXONOMY
Camelus bactrianus Linnaeus, 1758, Uzbekistan. No subspecies known.

OTHER COMMON NAMES
English: Two-humped camel; French: Chameau; German: Kamel; Spanish: Camello.

PHYSICAL CHARACTERISTICS
Measure an average of 7 ft (213 cm) tall at the hump and weigh about 1,800 lb (816 kg). The wild Bactrian camel has a sandy, gray-brown coat, rather than the predominately dark brown coat like that of the domestic Bactrian. Its body is small and slender, rather than large and bulky like that of the domestic Bactrian camel. The two humps of the wild Bactrian camel are small and pyramid-shaped, with a round base and a pointed end, while the two humps of the domestic Bactrian are distinctively large and irregular. Both Bactrians have very tough feet, especially adapted for crossing the rocky deserts of Asia.

DISTRIBUTION
Wild populations are restricted to three small areas in the Gobi desert of southwestern Mongolia and northwestern China. In contrast, the domestic Bactrian is widely bred in Mongolia and China.

HABITAT
Lives in arid plains and hills where there are scattered water sources and scarce vegetation. In these environments, temperatures may reach 100°F (38°C) in summer and -20°F (-30°C) in winter.

BEHAVIOR
Wild herds concentrate near mountains because it is where most springs are found, and snow on the slopes may at times be the only moisture available in winter.

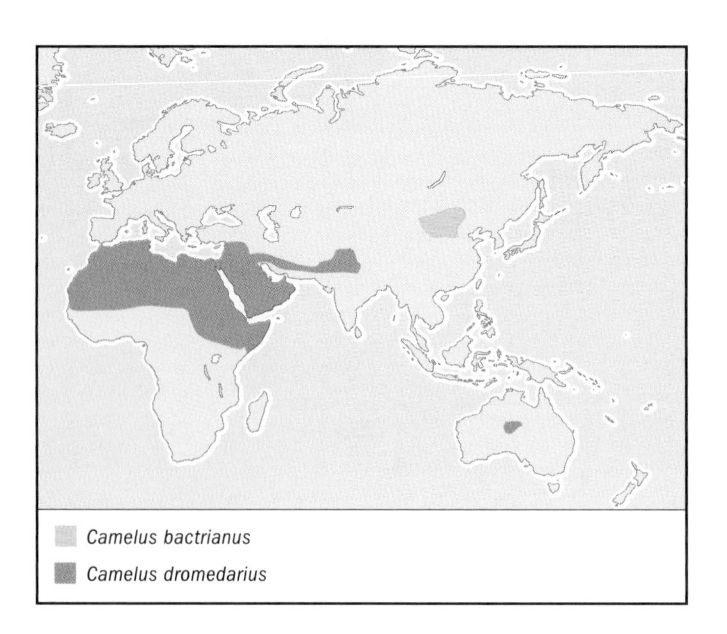

FEEDING ECOLOGY AND DIET
Consists of shrubs, grasses, and forbs. In winter, these plants provide enough moisture for bactrians to go without drinking for several weeks, although a thirsty individual can consume up to 30 gal (114 l) of water in just 10 minutes.

REPRODUCTIVE BIOLOGY
May be polygamous. Female Bactrians give birth at five years old. With a gestation period of approximately 406 days, the birth rate is two years between births and these usually take place during March and early April. Newborns can walk after two hours and can follow their mothers after 24 hours, but they reach independence when they are one year old. Life expectancy varies from 35–40 years.

CONSERVATION STATUS
Wild Bactrians are considered Critically Endangered, according to the IUCN Red List. The main threats it faces are heavy hunting and competition with domestic animals for water and pasture, as well as hybridization with domestic Bactrian stock. It is estimated that no more than 350 wild Bactrians survive in the Gobi desert. On the contrary, the domestic Bactrian has an estimated population of 2.5 million in Central Asia.

SIGNIFICANCE TO HUMANS
Starting some 6,000 years ago, people in Mongolia began using the Bactrian camel to transport themselves and to carry burdens, as well as a source for food, clothing, and fuel. ◆

Dromedary camel
Camelus dromedarius

TAXONOMY
Camelus dromedarius Linnaeus, 1758, Africa. No subspecies known.

OTHER COMMON NAMES
English: Arabian camel; French: Dromedaire; German: Dromedar; Spanish: Dromedario.

PHYSICAL CHARACTERISTICS
Stands 7 ft (213 cm) tall at the hump and weighs 1,600 lb (726 kg). The neck is long and curved, and they have a deep-narrow chest and a single hump. Their broad and thickly padded feet are especially adapted for traveling on sand. The hair is longer on the throat, shoulders, and hump area, and its color is usually caramel brown or sandy brown, although shades can range from almost black to nearly white. Its coating helps to block out the heat of the sun.

DISTRIBUTION
Found in arid regions of the Middle East through northern India, and arid regions in Africa, particularly the Sahara Desert. Introduced into Australia.

HABITAT
Desert environments with a long dry season and a short rainy season, where temperatures often rise above 120°F (49°C).

BEHAVIOR
Form a family group of usually two to 20 individuals, including one dominant male, females, juveniles and young. Males have

an inflatable soft palate that they use to attract females. The dominant male prevents contact between females and competitor males by driving them away. Confrontations consist in pushing each other with their whole body or lowered neck and head, snapping each other without biting, and occasionally spitting when they are hurt or excited.

FEEDING ECOLOGY AND DIET
Feed on thorny plants, dry grasses, and saltbush that grow in the desert, primarily browsing with shrubs and forbs that account for up to 70% of their diet. When foraging, they tend to spread over large areas and select only a few leaves from each plant to reduce the stress on the plant communities and ease competition with other region herbivores. They need six to eight times as much salt as other animals to absorb and store water, which they obtain mostly from halophytes. Since they do not easily sweat, they lose the moisture in their bodies slowly.

REPRODUCTIVE BIOLOGY
May be polygamous. Females reach sexual maturity at age three, while males do so at age six. Seasonal breeders in winter, though sometimes overlaps with the rainy season, depending on the group's geographic location. The gestation period can last up to 15 months.

CONSERVATION STATUS
Since the dromedary is domesticated, it has no special status in conservation. There are approximately 14 million dromedaries in its entire range of distribution.

SIGNIFICANCE TO HUMANS
Tribal peoples of the Arabian Peninsula hunted the native one-humped dromedaries for thousands of years, and started to use them mainly to carry people and things. It is also a source of milk, meat, wool, leather, and fuel from dried manure. Thus, it has become a key element for the survival of human populations in the seemingly inhabitable desert. ◆

Vicuña

Vicugna vicugna

TAXONOMY
Vicugna vicugna (Molina, 1782), Chile. One subspecies.

OTHER COMMON NAMES
French: Vigogne; German: Vikunja; Spanish: Vicuña.

PHYSICAL CHARACTERISTICS
Average height of 3 ft (90 cm) at the shoulder and weighs 99.2 lb (45 kg). Has a slender body, with a relatively long neck. The fur on the chest is long, of an off-white color, which serves to protect the animal when it is resting on the ground. The neck, back, and sides are a light brown color. The ventral and inner thigh surfaces are white. The head is relatively small, with prominent ears and eyes; the lower lip has a central crevice. The lower incisors of vicuña are unique among the Artiodactyla, because they are always growing and have the enamel on only one side. Sometimes, the canines are absent in the lower jaw. The front premolars are simple and usually separated from the other cheek teeth.

DISTRIBUTION
In the Andes of Peru, Bolivia, Chile, and Argentina.

HABITAT
Inhabits semiarid grasslands and plains at elevations ranging 11,480–18,860 ft (3,500–5,750 m) in the Andes.

BEHAVIOR
Territorial males maintain family groups consisting of the male adult and subadults, females, and young less than one year of age. Adult males without territories form non-reproductive groups, composed of subadult males from one to four years of age that have been expelled from their family groups and of aging males that have lost their territories. Establishes and defends a year-round feeding territory and a separate sleeping territory.

FEEDING ECOLOGY AND DIET
A grazer; its diet consists of almost all perennial grasses.

REPRODUCTIVE BIOLOGY
May be polygamous. Mating occurs in March and April, and births take place in February and March. The gestation period lasts 330–350 days, and a single young weighs 8.8–13.2 lb (4–6 kg) at birth. The young can stand and walk 15 minutes after being born. Most females mate at about two years, and some are reproductive until 19 years old.

CONSERVATION STATUS
The entire population remains Vulnerable, according to the IUCN Red List. Certain populations are on the CITES Convention, with the provision that only cloth woven from the sheared wool of a live vicuña may be traded. Otherwise, the vicuña is on Appendix I of the CITES Convention.

SIGNIFICANCE TO HUMANS
Vicuña produces one the finest wools in the world. Today, vicuña fiber is preferred for weaving fine cloaks and the cloths obtained from it are expensive in the international markets. At present, Peru and Chile have sustainable use programs based in the capturing, shearing, and release of these wild animals.

However, this sustainable use is allowed only on those populations placed in Appendix II of the CITES Convention. ◆

Guanaco
Lama guanicoe

TAXONOMY
Lama guanicoe (Müller, 1776), Chile. Three subspecies.

OTHER COMMON NAMES
French: Guanaco; German: Guanako; Spanish: Guanaco.

PHYSICAL CHARACTERISTICS
Average height of 3.7 ft (112 cm) and weighs 330.6 lb (150 kg); 5 ft (150 cm) long, including head and body; tail 9.8 in (25 cm) long. Slender body, with a relatively short wooly pelage that is a light brown with blackish tones on the head, while the area around the lips is whitish, as are the edge of the ears, the lower part of the body, and the inner side of the legs. The feet are brown, and there is a collar of white hair at the lower part of the neck.

DISTRIBUTION
Found throughout most of Argentina, the high Andes of Peru, and northeastern and southern latitudes of Chile; a small population is restricted to the Chaco region of Bolivia and Paraguay.

HABITAT
Grasslands and shrublands from sea level to over 11,482 ft (3,500 m). In southern latitudes, inhabits forests during the winter.

BEHAVIOR
Forms family groups of two to 30 individuals, consisting of an adult dominant male and females with their year's offspring. Young females may also be present. The dominant male defends the territory from other males. Males groups are made up of young and adult males whose number may amount to 50 individuals. Solitary males are physically and sexually mature males prepared to form a territory. Neither male groups nor solitary males are tolerated in family group areas and are violently expelled, so they are forced to cover a wider range in search of food.

FEEDING ECOLOGY AND DIET
Feeds on grasses, shrubs, epiphyte plants, lichens, fungi, and particularly halophyte plants.

REPRODUCTIVE BIOLOGY
Polygamous. Females reach sexual maturity when they are one year old, while males are sexually mature when they reach three to four years of age. After a gestation period that lasts 320–340 days, a single young is born, weighing 17.6–26.4 lb (8–12 kg). Conception generally takes place a week after the female has given birth. Within a month of age, the young starts grazing and is nursed by its mother until it is six to eight months old.

CONSERVATION STATUS
All populations remain Vulnerable, according to the IUCN Red List, and are on Appendix II of the CITES Convention. At present, there are more than 600,000 guanacos throughout its range of distribution.

SIGNIFICANCE TO HUMANS
Indigenous cultures in the southernmost areas of South America traditionally benefited from this species, obtaining meat for food, hides for clothing and shelter, bezoar stones for medical purposes, fibers for sewing, and domesticated juveniles for entertainment. At present, a commercial value is being assigned only to the pelts of young and adult hides. This commercial use is regulated in the CITES Convention. The three subspecies of guanaco are listed as Vulnerable or Endangered by the IUCN.◆

Alpaca
Lama pacos

TAXONOMY
Lama pacos (Linnaeus, 1758), Peru.

OTHER COMMON NAMES
French: Alpaga; German: Alpaka; Spanish: Alpaca.

PHYSICAL CHARACTERISTICS
Reaches 3 ft (90 cm) high and weighs 154.3 lb (70 kg). Has a small head, short ears with thin points, and a very long neck. The entire body, except the face and legs, is covered by long, thick, and soft wool. The legs are short and the hair can extend on the head, forming a tuft that, in males, covers the eyes. It presents a uniform color, generally dark chocolate or almost black.

DISTRIBUTION
Found in the Andes of Peru, Bolivia, and Chile at elevations ranging 9,840–15,750 ft (3,000–4,800 m). The largest populations are located in Peru.

HABITAT
Humid places of the Andean high plateaus or Altiplano known as *bofedales*, where tender grass can grow.

BEHAVIOR
Docile and gregarious, and show no social behavior.

FEEDING ECOLOGY AND DIET
Tender grasses.

REPRODUCTIVE BIOLOGY
Female alpacas reach maturity when they are two years old, while males do so when they are three years of age. One male usually copulates with 10 females, and the gestation period lasts 342–345 days.

CONSERVATION STATUS
Because it is domesticated, it has not been classified in any special conservation status. There are approximately 3.5 million alpacas throughout its entire geographic range.

SIGNIFICANCE TO HUMANS
Since before the arrival of the Incas, inhabitants of the High Andes have benefited from the alpaca's fine wool, hide, meat, and dung, which is used as fuel. During the Inca Empire, the species was bred specifically to utilize its soft fiber for fine textiles, and there was a great emphasis on the quality of the yarns, the fibers being carefully selected. Today, it is raised to provide fine wool that is commercialized on a large scale, especially in Peru and Bolivia. Most of the Andean people prefer alpaca meat to that of the llama. The dung is an important source of fuel in areas where there are no trees to supply wood. ◆

Llama
Lama glama

TAXONOMY
Lama glama (Linnaeus, 1758), Peru.

OTHER COMMON NAMES
French: Lama; German: Lama; Spanish: Llama.

PHYSICAL CHARACTERISTICS
Average height 3.8 ft (115 cm) and weighs 308.6 lb (140 kg). Its legs are long, and it also has inwardly curved tips of the ears. Presents a reddish brown color that is almost uniform over its entire body, with the face, ears, and legs tainted black, though they can also be black, white, or of mixed coloring.

DISTRIBUTION
Found in Bolivia, Peru, Argentina, Chile, and also in Ecuador and Colombia from 7,550–13,120 ft (2,300–4,000 m).

HABITAT
Lives in high, arid environments formed by grasses.

BEHAVIOR
Herds of up to 100 animals are driven by herders to grasslands every day.

FEEDING ECOLOGY AND DIET
Feeds on grasses and halophyte plants.

REPRODUCTIVE BIOLOGY
May be polygamous. Females become fertile when they reach between two and three years of age. One male can copulate with up to 30 females. The gestation period lasts an average of 11 months.

CONSERVATION STATUS
Since the llama is domesticated, it has not been classified with any special conservation status. There are approximately 2.5 million llamas throughout its entire geographic distribution.

SIGNIFICANCE TO HUMANS
Llama fiber is used to manufacture ropes and packing bags, the skin to produce leather goods, and the bones to make instruments for looms. The bezoaric stones, the fets, the blood, the fat, etc., are necessary elements in rituals to ensure the fertility and well being of the herds. The llama is an important source of meat, which may be consumed either fresh or salted and dried in the sun, a form known as *charqui* (jerky). ◆

Resources

Books

Bonavia, D. *Los Camélidos Sudamericanos (Una introducción a su estudio).* Lima, Peru: IFEA, UCH, Conservation International. 1996.

Gilmore, R. *Fauna and Ethnozoology of South America. Handbook of South American Indian.* Washington, DC: Smithsonian Institution, 1955.

Schaller, G. B. *Wildlife of the Tibetan Steppe.* Chicago: University of Chicago Press, 1998.

Tan, B. J. *Into the Wild—The Rare and Endangered Species of China.* Beijing, China: New World Press, 1996.

Torres, H. *South American Camelids. An Action Plan for their Conservation.* Gland, Switzerland: IUCN, 1992.

———. *Distribution and Conservation of the Guanaco.* Cambridge, Cambridge, UK: University Press, 1985.

———. *Distribution and Conservation of the Vicuña.* Gland, Switzerland: IUCN, 1983.

Periodicals

Afshar, A. "Camels at Persepolis." *Antiquity* LII (November 1978).

Kadwell, M., et al. "Genetic Analysis Reveals the Wild Ancestors of the Llama and the Alpaca." *The Royal Society Journal* 268 (2001): 2575–2584.

Wheeler, J., A. Russel, and H. Reden. "Llamas and Alpacas: Pre-conquest Breeds and Post-conquest Hybrids." *Journal of Archaeological Science* 22 (1995): 833–840.

Zarius, J. "The Camel in Ancient Arabia: A Further Note." *Antiquity* LII (March 1978).

Hernán Torres, PhD

Chevrotains
(Tragulidae)

Class Mammalia
Order Artiodactyla
Suborder Ruminantia
Family Tragulidae

Thumbnail description
The most ancient living representatives of early ruminants; rabbit-sized ungulates, among the smallest of the even-toed ruminants; resembles a diminutive hornless deer; slender legs, stocky body, and arching back; reddish brown to brown pelage (coat) may be striped or spotted

Size
Head and body length: 17–19 in (44–85 cm); weight 4.4–29 lb (2–13 kg)

Number of genera, species
2 genera; 4 species

Habitat
Rainforests, secondary forests, and mangrove forests and thickets

Conservation status
Data Deficient: 1 species

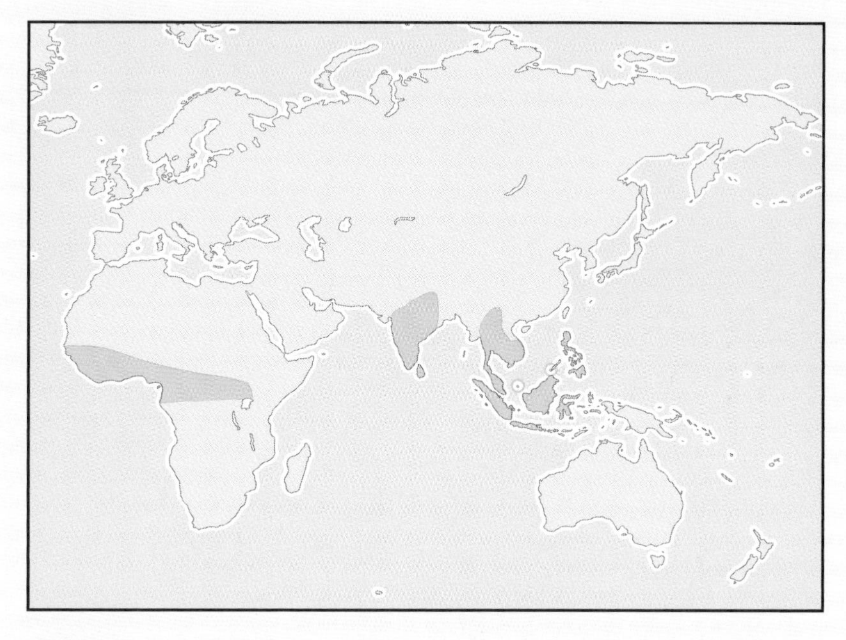

Distribution
Genus *Hyemoschus* (true chevrotain) indigenous to eastern Africa; genus *Tragulus* (mouse deer) indigenous to south and southeastern Asia

Evolution and systematics

Chevrotains (genera *Hyemoschus* and *Tragulus*) are regarded as living fossils, the most ancient living representatives of early ruminants. Traditional classifications place the family Tragulidae between non-ruminating ungulates such as pigs and hippos and ruminating artiodactyls like deer antelope and cattle. Two genera are recognized: true chevrotain is represented by one species in Africa (*Hyemoschus aquaticus*); the mouse deer by three species in Asia (*Tragulus napu*, *T. javanicus*, and *T. meminna*).

Throughout the Eocene, some 54–38 million years ago (mya), Artiodactyls radiated into diverse families. Today's chevrotains are descendants of early Tragulids (*Archaeotragulus krabiensis*) that appeared in Eocene. Based on fossil records, chevrotains similar to modern species did not appear until the Miocene (beginning 15 mya).

Like all ruminant artiodactyls, chevrotains evolved features that are adaptive for life on open grasslands. Global climate changes during the Miocene resulted in widespread cooling and drying, leading to the transition of many forest habitats to extensive grasslands. Ruminants such as chevrotains evolved specialized digestive systems and dentition that enabled them to use the nutrients in grasses and leaves—nutrients not available to non-ruminants that lack the ability to breakdown/digest cellulose and silica in the grasses and leaves. The multi-chambered ruminant gut is capable of digesting/breaking down cellulose and silica that are indigestible for non-ruminant mammals.

However, chevrotain feeding patterns and stomach structure differ from those of larger ruminants. Although they eat plants, chevrotains and other small ungulates tend to be more selective feeders than their larger grazing and browsing cousins. Since they do not need to gather large quantities of food daily, they can afford the time to carefully pick more digestible leaves, shoots, flowers, and fruits. Like all ungulates, they have a multi-chambered stomach in which the first two chambers, named the rumen and reticulum, mainly function as microbial fermentation vats for cellulose-rich plant cell-wall material that their own enzymes cannot breakdown. But, because more of their food is easy to digest, relative to the diet of the larger ruminants, their multi-chambered stomach structure permits easy-to-digest material to slip past the rumen and reticulum through a special groove and proceed directly into their conventional stomach, called the abomasum. In chevrotains, the chamber that connects the first two chambers to the abomasum, called the omasum, is absent.

A second adaptation that permits artiodactyls to thrive on grasslands is the evolution of molar teeth with crescent-shaped ridges (selenodont molar teeth) for more efficient grinding of plants. Like all Artiodactyl ruminants, chevrotains lack incisors and canines in the upper jaw as well. Their dentition resembles that of higher ruminants.

A water chevrotain (*Hyemoschus aquaticus*) resting among the vegetation. (Photo by Devez-CNRS/Jacana/Photo Researchers, Inc. Reproduced by permission.)

Modern-day chevrotains share many features with ancestral primitive ruminants. These small animals never developed antlers or horns. Instead, they grow tusk-like upper canines that are used by males in interspecific combat.

Of the four chevrotain species, the water chevrotain (*Hyemoschus aquaticus*) is regarded as the most primitive, due to several anatomical features that resemble pigs more than modern cervides. For example, the forelimbs lack a cannon bone and skin on the rump is especially tough, making it difficult for predators to bite into.

Physical characteristics

As their common name suggests, mouse deer (chevrotain) resemble diminutive hornless deer with their small heads, tapered snouts, pencil-thin legs, and stocky bodies. Yet, these rabbit-sized ungulates are in a separate family. The chevrotain family (Tragulidae) has all of the following distinguishing features, including short, slender legs; even toes; a small, pointed head; a tapered snout; large eyes; slit-like nostrils; and medium-sized rounded ears covered with a thin layer of hair. Their backs are rounded and rise toward the rear quarters. This slipper body-shape facilitates moving through dense forest undergrowth. Chevrotains also bear a strong physical resemblance to the South American agouti.

The pelage is short and thick, reddish brown to brown, with contrasting patterns of white and brown spots and stripes on the neck, chest, sides, and underbelly, depending on the species. In three of the four species, females are larger than males.

Unlike true deer, chevrotains lack horns or antlers. Male chevrotains fight with tusk-like teeth—enlarged upper canines that protrude downward from the mouth. These upper canines are only small studs in females. Beyond the canines, most chevrotain dentition resembles that of higher ruminants. They lack upper incisors; the lower canine resemble incisors; and the cheek teeth have crescent-shaped ridges (seledontic). The dentition pattern is: (I0/3; C1/1; P3/3; M3/3) $\times$ 2 = 34.

A greater Malay mouse deer (*Tragulus napu*) in forest undergrowth. (Photo by Kenneth W. Fink. Bruce Coleman, Inc. Reproduced by permission.)

Another distinct chevrotain feature is its less-specialized ruminant gut. The chevrotain has only three fully developed stomach chambers, as compared with the four-chambered stomach of larger ruminants. The omasm, the chamber that connects the first two chambers to the fourth chamber, is absent.

The female has four mammae, leading some researchers to suggest that chevrotains may be capable of larger litters than the small litters usually observed (one to two young). The chevrotain male's penis is spiral-shaped. Chevrotains have a gallbladder.

A greater Malay mouse deer (*Tragulus napu*) male with tusks. (Photo by Tom McHugh/Photo Researchers, Inc. Reproduced by permission.)

A greater Malay mouse deer (*Tragulus napu*) is about the same size as a rabbit. (Photo by Anthony Mercieca/Photo Researchers, Inc. Reproduced by permission.)

The skull is similar to true deer, except that the chevrotains have a unique ossified plate. There are four digits on each foot, but the second and fifth digits are short and slender.

Distribution

Distribution of the Tragulidae family was worldwide during the Oligocene and Miocene. At present, tragulids are restricted to the Old World. The three species of mouse deer (genus *Tragulus*) are endemic to Southeast Asia, while the single true chevrotain species (genus *Hyemoschus*) is found along east central Africa.

Habitat

Chevrotains inhabit rainforests, lowland forests, mangrove forests, and thickets. The three Asiatic species are found in dense vegetation during the day, occasionally frequenting open areas at night. These small mammals flee from disturbances by darting into dense vegetation or water. The African chevrotain is found in tropical rainforests, underbrush, and thick growth almost always along water courses where it can escape predators by diving into water.

Behavior

Chevrotains are extremely difficult to observe due to a timid behavior, have primarily nocturnal sleeping patterns and a preference for dense forests. These shy animals are flighty, easily excited, and prone to jumping in response to the slightest disturbance. They escape predators by darting into dense vegetation or water.

A spotted mouse deer (*Tragulus meminna*) foraging on the ground. (Photo by M. K. Ranjitsinh/Photo Researchers, Inc. Reproduced by permission.)

Chevrotains are regarded as extremely solitary, when compared with other forest species. Most chevrotains live alone, except during brief social periods when mating occurs or while raising young. One exception is the lesser Malay mouse deer, which is monogamous. Chevrotains are territorial. Even when

The lesser Malay mouse deer (*Tragulus javanicus*) is well camoflauged in the fallen leaves in Southeast Asia. (Photo by Art Wolfe, Inc./Photo Researchers, Inc. Reproduced by permission.)

Lesser Malay mouse deer (*Tragulus javanicus*) is 8–12 in (20–30 cm) in length. (Photo by Erwin & Peggy Bauer. Bruce Coleman, Inc. Reproduced by permission.)

home ranges are densely populated, chevrotains rarely come into contact with one another.

Chevrotains communicate through scent marks and vocalizations. When frightened, mouse deer make soft bleating sounds. Vocalizations are used to signal intent to approach, followed by answer calls.

Chevrotains mark their home territory with scent marks of urine, feces, or glandular secretions. The male water chevrotain has anal and preputial glands. Male water chevrotains and male and female greater Malay mouse deer mark objects by rubbing their chins over a leaf branch end or a tree root.

Males fight with sharp canines, although fighting between males is brief and infrequent. Fighting between females is rare. Some researchers suggest that the lack of more frequent fighting patterns may indicate a lack of social hierarchy.

Early observations suggest that all chevrotains are nocturnal, or crepuscular; however, more recent observations of the lesser Malay mouse deer indicate that some mouse deer may be diurnal.

Only mothers with young clean their offspring with their tongues, otherwise there is no mutual licking. Female chevrotains are more active than males. To rest, they sit on their hind legs or crouch with folded forelegs and hind legs.

Feeding ecology and diet

As ruminants, chevrotains are able to digest grasses and leaves that are indigestible to most non-ruminants. However, small ungulates such as chevrotains can afford to eat more selectively because they need less food than their larger ruminant cousins that must consume large quantities of food daily. As a result, the chevrotain diet tends to favor young shoots, forbs, fruits that have fallen to the ground, and seeds, in addition to occasional leaves and grasses. Some chevrotains have been observed eating arthropods and small animals.

A greater Malay mouse deer (*Tragulus napu*) resting among foliage. (Photo by © John White. Reproduced by permission.)

Reproductive biology

The reproductive biology of most chevrotain species is poorly known, though most are polygamous. Chevrotains reach sexual maturity sometime between five to 26 months. When a female enters estrus, males seek out and follow her while making cry-like vocalizations. In the case of water chevrotains, the male's cry causes the female to stop moving, allowing the male to lick her genital area. Among greater and lesser Malay mouse deer, males also stroke females with a special gland located between the rami of the male's lower jaw. After repeating a pattern of cries and physical contact, copulation takes place. Female greater and lesser Malay mouse deer can mate 85–155 minutes after giving birth; as a result, they are capable of almost continuous pregnancy through most of their adult lives.

Gestation lasts six to nine months, depending on the species, and females give birth to one young a year. The female has four mammae, leading some researchers to suggest that chevrotains may be capable of larger litters. Females ingest the placenta after giving birth. Offspring are precocial, capable of standing within an hour after birth, yet the young remain hidden on the forest floor. Females do not stay with young, except for brief feeding/suckling periods. Young are weaned at three to six months and disperse from the mother's home range when they reach sexual maturity between nine to 26 months. Individuals live to an age of 11–13 years.

Conservation status

Knowledge concerning the status of the four chevrotain is far from satisfactory. The combination of their shy, flighty behavior, small size, and their nocturnal activity patterns makes these diminutive ungulates especially difficult to study. All four chevrotain species are threatened by hunting and habitat destruction. The IUCN Red List classifies only one subspecies as Endangered, and the water chevrotain as Data Deficient.

Significance to humans

In all parts of its range, chevrotains are hunted by indigenous people for food. Although there is interest in using chevrotain as pets and for basic research on ungulates, most chevrotain are difficult to breed and care for in captivity. Zoos have had some success in breeding water chevrotain in captivity.

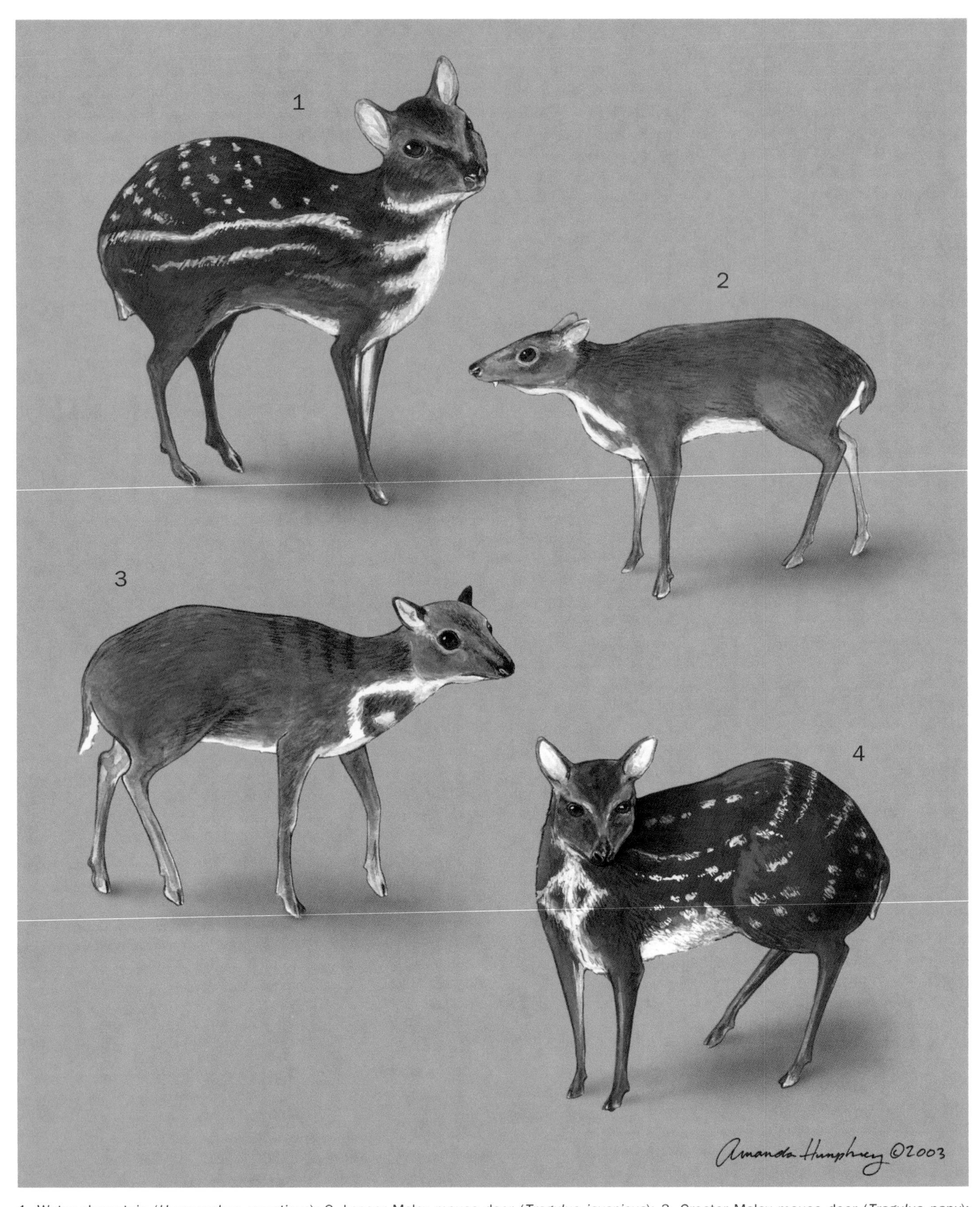

1. Water chevrotain (*Hyemoschus aquaticus*); 2. Lesser Malay mouse deer (*Tragulus javanicus*); 3. Greater Malay mouse deer (*Tragulus napu*); 4. Spotted mouse deer (*Tragulus meminna*). (Illustration by Amanda Humphrey)

Species accounts

Water chevrotain
Hyemoschus aquaticus

TAXONOMY
Hyemoschus aquaticus (Ogilby, 1841), Sierra Leone.

OTHER COMMON NAMES
English: African chevrotain; French: Chevrotain aquatique; German: Afrikanisches Hirschferkel; Spanish Antilope amizclero enano de agua; cervatillo almizclero acuatico.

PHYSICAL CHARACTERISTICS
Considerably larger than its Asian counterparts: head and body length: 28–31 in (70–80 cm); tail length: 4–5.5 in (10–14 cm); shoulder height: 14–16 in (35–40 cm); weight: 22–33 lb (10–15 kg). Males are slightly smaller than females. Males average 4.4 lb (9.7 kg); females average 5.1 lb (12 kg). The weight at birth is unknown.

The coat has an unmistakable pattern of spots and stripes that may provide camouflage in the shade of dense forests. The body is reddish brown with six or seven vertical rows of white spots along the back and a white line along each side from shoulder to rump. The head is covered with black and white bands.

DISTRIBUTION
Africa's single chevrotain species, it is endemic to lowland tropical forest zones of eastern Central Africa. While much of its range hugs the coast, this species has a disjointed distribution from Sierra Leone to western Uganda. The current range includes scattered regions of east Sierra Leone, Liberia, west Côte d'Ivoire, and south Ghana. There is a single record from southeast Nigeria.

HABITAT
Inhabit dense tropical rainforests and tropical scrub forests, always near water. These shy ungulates escape predators by diving into water. During the day, they hide in dense forest undergrowth; at night, they have been observed along exposed clearings and river banks.

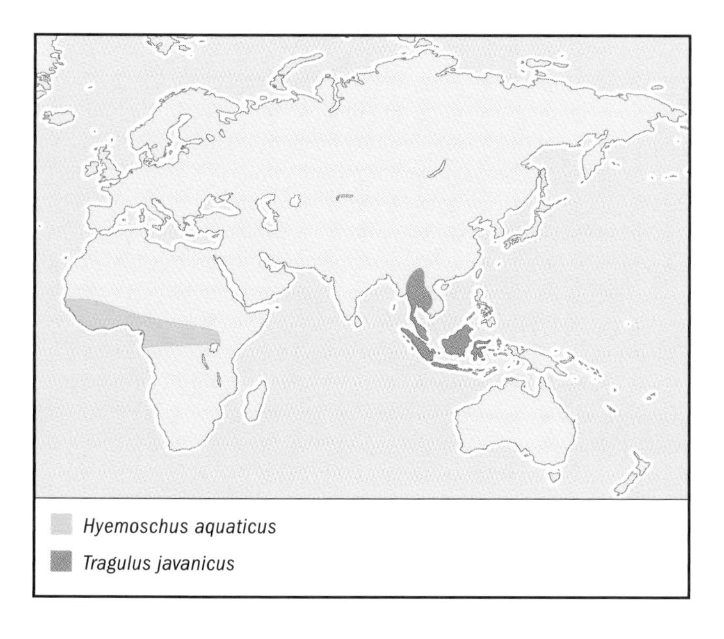

BEHAVIOR
May be the only exclusively nocturnal member of the chevrotain family. It hides in dense forest vegetation during the day and forages at night. As its name suggests, it is a capable swimmer although it may not be capable of swimming for extended periods of time. It always lives within 820 ft (250 m) of water. When disturbed, it escapes by plunging into the water.

These solitary forest animals are rarely seen together other than during mating and while females rear young. Few play behaviors have been observed. Males occasionally engage in brief fights consisting of short rushes and biting. Their population density ranges between 19.25–70 individuals per mi2 (7.7–28 individuals per km2). Females have small home ranges, approximately 32–34.5 acres (13–14 ha) in area, and may occupy the same range through their adult life. Male ranges often include the ranges of two or more females 49.4–74.1 ac (20–30 ha); males can be displaced from their home ranges several times in one year. Females are much more active than males. Lifespan is 10–14 years.

FEEDING ECOLOGY AND DIET
An herbivorous browser, primarily feeds on fallen fruit, along with occasional leaves, buds, trees, and shrubs. It has been observed eating insects, crustaceans, and small animals. Like all ruminants, it has a multi-chambered ruminating stomach that facilitates digestion of a low-nutrient diet.

REPRODUCTIVE BIOLOGY
Polygamous. When a female enters estrus, males follow her movements while making vocalizations. The cry of the male causes the female to stop moving and hold still, while he licks her genital area. The pattern of movement, cries, and licking is repeated until the male mounts the female and copulation takes place.

Gestation lasts six to nine months, after which the female gives birth to one or two young. The youngster is precocial, capable of standing within an hour after birth. Young remain hidden on the forest floor with the mother. Females do not stay with the young except for brief feeding/suckling sessions. Lactation lasts three to six months and the young disperse from the mother's home range when they reach sexual maturity (between nine and 26 months). Individuals live to an age of 11–13 years.

CONSERVATION STATUS
Overall numbers are currently decreasing due to hunting by humans and habitat destruction for timber resources. The 2000 IUCN Red List: Data Deficient. This is a change of status from 1996 Red List: Lower Risk/Near Threatened. This species is listed under Appendix III of CITES in Ghana.

SIGNIFICANCE TO HUMANS
Hunted by humans throughout its range. ◆

Lesser Malay mouse deer
Tragulus javanicus

TAXONOMY
Tragulus javanicus (Osbeck, 1765), west Java, Indonesia.

OTHER COMMON NAMES
French: Petit chrotain malais; German: Kleinkantschil.

PHYSICAL CHARACTERISTICS
The name mouse deer refers to the diminutive size of this deer-like ungulate, although it is not a true deer or mouse. The smallest living artiodactyl, it weighs only 4.4 lb (2 kg). Head and body length: 18–22 in (45–55 cm); tail length 2 in (5 cm); shoulder height 8–10 in (20–25 cm); weight 3.3–5.5 lb (1.5–2.5 kg).

It has a delicate black nose and large eyes surrounded by a lighter ring. The ears are sparsely covered with black hair. The upper coat is brown with an orange tinge, while the underside (including the belly, inner legs, and chin) is white. The neck has a series of white vertical markings. Females are slightly smaller than males.

DISTRIBUTION
Malaysia (Peninsular Malaysia), Range Brunei Darussalam, Cambodia, southwestern China (Yunan province), Indonesia, Borneo, Laos, People's Democratic Republic, Malaysia, Myanmar, Singapore, and Thailand.

HABITAT
Primary and secondary lowland forests. Found near water in dense vegetation, hollow trees, or among rocks.

BEHAVIOR
While it is more common than the other three chevrotain species, little has been published regarding its behavior and ecology. Previously it was assumed that it is nocturnal and solitary. However, recent studies suggest that at least some are diurnal and form monogamous pairs. Males are territorial, marking territory with urine feces and secretions from an intermandible gland under the chin; it fights with sharp canines, protecting itself and its mate against threatening rivals. When frightened, they beat their hooves on the ground as fast as seven times per second, creating a drum roll. Otherwise silent, a frightened mouse deer also makes a shrill cry.

FEEDING ECOLOGY AND DIET
Primarily herbivorous; its diet consists primarily of leaves, buds, grass, and fruits that have fallen from trees. In captivity, they have been observed eating arthropods.

REPRODUCTIVE BIOLOGY
Monogamous. Reach sexual maturity at five to six months. The female is able to conceive only 85–155 minutes after giving birth, so it has the potential to be constantly pregnant throughout its adult life. Gestation lasts 4.5–5 months and produces one fawn, occasionally two. Young are precocial, fully developed, and can stand within 30 minutes of birth. Fawns are exceptionally shy. Mothers nurse their young while standing on three legs. Offspring are weaned at 10–13 weeks. Lifespan is up to 12 years.

CONSERVATION STATUS
Since 1975, its range and density have increased, due to conservation efforts to rehabilitate native ecosystems. However, it remains threatened by hunting and habitat destruction. Predators are carnivorous mammals, large birds, and snakes. IUCN: not listed; ESA: Threatened; CITES: Appendix II.

SIGNIFICANCE TO HUMANS
Hunted and traded for its smooth skin. Due to the ease of taming mouse deer, it is sometimes used as a pet. ◆

Greater Malay mouse deer
Tragulus napu

TAXONOMY
Tragulus napu (F. Cuvier, 1822), south Sumatra, Indonesia.

OTHER COMMON NAMES
French: Grand chevrotain malais; German: Grosskantschil.

PHYSICAL CHARACTERISTICS
Similar to, but larger than its cousin, the lesser Malay mouse deer: head and body length: 2.3–2.5 ft (70–75 cm); tail length: 3.2–4 in (8–10 cm); shoulder height: 12–14 in (30–35 cm); weight: 11–17.6 lb (5–8 kg). Upper coat is brown to orange-brown, and the underparts are white. The underside of the chin is white with a series of white markings. The hindquarters are lightly grizzled with black.

DISTRIBUTION
Southern Thailand and Indochina, Malay Peninsula, and several nearby islands, Sumatra, Borneo, North and South Natuna island, Balabac Island, Brunei Darussalem, Cambodia, Indonesia, Myanmar, Philippines, Singapore, and Vietnam.

HABITAT
In dense undergrowth at the edge of dense lowland forests, usually close to water.

BEHAVIOR
Nocturnal and rarely seen, it travels through small tunnel-like trails through thick brush. Both males and females are territorial and regularly mark their territories with urine, feces, and

secretions from an intermandibular gland under the chin. Adult females with young occupy range of 32.5–35 ac (13–14 ha); adult males range is 50–75 ac (20–30 ha). When agitated, they drum on the ground with their hooves at a rate of four times per second.

FEEDING ECOLOGY AND DIET
Primarily herbivorous, choosing buds, leaves, and fruit that has fallen to the ground. Presumably, it also feeds on arthropods and other animals when available, based on observations of other chevrotains.

REPRODUCTIVE BIOLOGY
Polygamous. Gestation lasts for about five months, after which the female generally gives birth to one or two young. Life expectancy is unknown. A single deer has lived for more than 16 years in captivity.

CONSERVATION STATUS
Tragulus napu and its subspecies are threatened by habitat loss and hunting. Subspecies *T. n. nigricans* is classified as Endangered by the IUCN (2000). The subspecies occurs on Balabac Island in the southwest Philippines and is threatened by intense hunting pressure and habitat destruction.

SIGNIFICANCE TO HUMANS
Hunted for food. Although it is also easy to tame as a pet or for research, it may be too delicate to survive in captivity for any length of time. ◆

Spotted mouse deer
Tragulus meminna

TAXONOMY
Tragulus meminna Hodgson, 1843. Very little is known about the Indian or spotted mouse deer. On the basis of its anatomical and morphological differences from the two other Asian mouse deer, systematists have placed it into its own subgenus (*Moschiola*, Erxleben, 1777, Sri Lanka).

OTHER COMMON NAMES
English: Indian spotted chevrotain; French: Chevrotain tachete indien; German: Flecken kantschil.

PHYSICAL CHARACTERISTICS
Head and body length: 20–23 in (50–58 cm); weight: 6.6 lb (3 kg). As its name suggests, the upper side of the coat is finely speckled. The flanks and rump have a pattern of spots and stripes: five to seven white lines on the throat, and the underparts are whitish. The male has a black coat with a red or orange shoulder-patch; females heavily barred below.

DISTRIBUTION
Southern India, Nepal, and Sri Lanka.

HABITAT
Equatorial tropical rainforests, including rocky sites.

BEHAVIOR
Little information is available. This timid, nocturnal animal is extremely difficult to observe in the wild; even slight disturbances cause it to disappear into dense vegetation. Solitary, except for the mating period and when raising young.

FEEDING ECOLOGY AND DIET
Has a highly varied diet that includes both plants and small animals.

REPRODUCTIVE BIOLOGY
Polygamous. Gestation lasts for about five to six months, after which the female generally gives birth to one or two young.

CONSERVATION STATUS
Possibly endangered, though not listed by the IUCN. Both heavy hunting and habitat loss threaten its survival.

SIGNIFICANCE TO HUMANS
Hunted by indigenous people for food. ◆

Resources

Books

Geist, V., and F. Walther, eds. *The Behavior of Ungulates and Its Relation to Management.* Gland, Switzerland: IUCN, 1974.

Kingdon, Jonathan. *East African Mammals.* Vol. 3. New York: Academic Press, Inc., 1979.

Lydekker, R. *Catalogue of the Ungulate Mammals in the British Museum (Natural History).* London: Trustees of the British Museum of Natural History, 1913–1916.

Putman, R. *The Natural History of the Deer.* Ithaca, NY: Cornell University Press, 1988.

Robin, Klaus. "Chevrotains." In *Grzimek's Encyclopedia of Mammals.* New York: McGraw-Hill Publishing Company, 1990.

Van Soest, P. J. *Nutritional Ecology of the Ruminant.* 2nd ed. Ithaca, NY: Cornell University Press, 1994.

Vrba, E. S., and G. B. Schaller, eds. *Antelopes, Deer and Relatives: Fossil Record, Behavioral Ecology, Systematics and Conservation.* New Haven: Yale University Press, 2000.

Wemmer, C. M., ed. "The Comparative Behavior and Ecology of Chevrotains, Musk Deer, and Morphologically Conservative Deer." In *Biology and Management of Cervidae.* Washington, DC: Smithsonian Institution Press, 1987.

Wilson, D. E., and DeeAnn M. Reeder, eds. *Mammal Species of the World.* 2nd ed. Washington, DC: Smithsonian Institution Press, 1993.

Periodicals

Coley, P. D., J. P. Bryant, and F. S. Chapin. "Resource Availability and Plant Antiherbivore Defense." *Science* 230 (1985).

Dubost G. "Comparison of the Diets of Frugivorous Forest Ruminants of Gabon." *Journal of Mammalogy* 65, no. 2 (1984): 298–316.

Resources

Hofmann, R. "Evolutionary Steps of Ecophysiological Adaptation and Diversification of Ruminants: A Comparative View of Their Digestive System." *Oecologia* 78, no. 4 (1989).

Janis, C. M. "A Climatic Explanation for Patterns of Evolutionary Diversity in Ungulate Mammals." *Paleontology* 32 (1989): 463–481.

Matsubayashi, H., E. Bosi, and S. Kohshima. "Activity and Habitat Use of Lesser Mouse Deer (*Tragulus javanicus*)." *Journal of Mammalogy* 84, no. 1 (2003): 234–242.

Metais, G., Y. Chaimanee, and J. J. Jaeger, "New Remains of Primitive Ruminants from Thailand: Evidence of the Early Evolution of Ruminants." *Zoological Scripta* 30, no. 4 (2001): 231–248.

Murphy, W. J., et al. "Molecular Phylogenetics and the Origins of Placental Mammals. Home Range, Activity Patterns and Habitat Relations of Reeves Muntjacs in Taiwan." *Journal of Wildlife Management* (June 2002).

Vidyadaran, M. K., R. S. Sharma, S. Sumita, I. Zulkifli, and Mazlan Razeem. "Male Genital Organs and Accessory Glands of the Lesser Mouse Deer, *Tragulus javanicus*." *Journal of Mammalogy* 80, no. 1 (1999): 199–204.

Corliss Karasov

△

Musk deer

(Moschinae)

Class Mammalia

Order Artiodactyla

Family Cervidae

Subfamily Moschinae

Thumbnail description
Small-sized deer without antlers, coat color is grizzled brown with whitish yellow spots and stripes on the chest; both males and females have well-developed upper canines; in males, long and protruding as fangs, up to 3 in (7 cm) long; hind legs are longer than forelegs, thus the rump of the body is elevated and withers slope forward; animals move by jumps; males have a musk bag, externally visible near its reproductive organs

Size
Shoulder height: 20.8–31.4 in (53–80 cm); body length: 33.8–39.3 in (86–100 cm); tail length: 1.5–2.3 in (4–6 cm); weight: 22–39.6 lb (10–18 kg)

Number of genera, species
1 genus; 4 species

Habitat
Mountain forest

Conservation status
Vulnerable: 1 species; Lower Risk/Near Threatened: 3 species

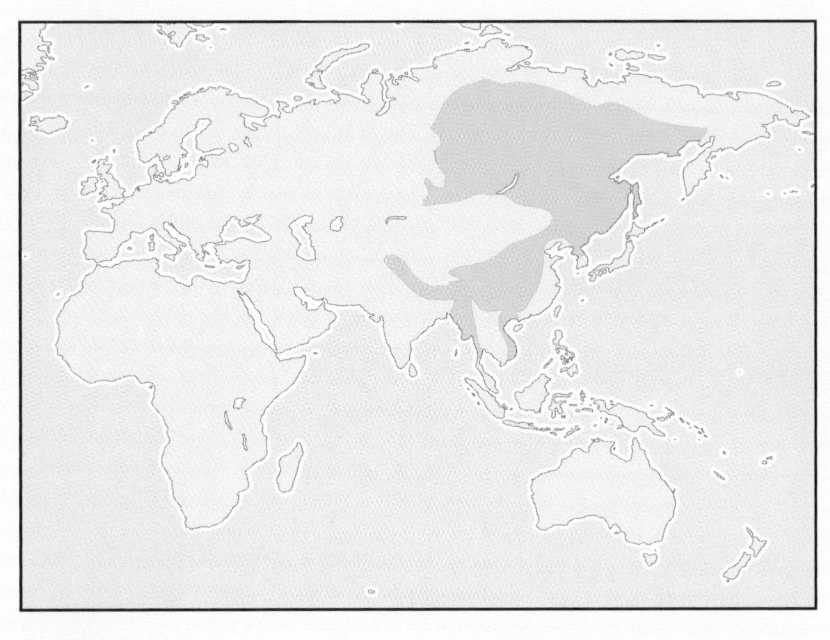

Distribution
Confined to the Old World: eastern Asia from the limits of forested zone at 70°N to Myanmar and Himalayas

Evolution and systematics

Musk deer are different from all other deer. They include a combination of primitive features (long tusks, lack of antlers) and advanced features (a four-chambered stomach). Unlike other Cervidae, musk deer have gallbladders. As for other ungulates that feed on concentrated fodder, musk deer are selective in their diet, sedentary, defend their home ranges, their diurnal rhythm of feeding and rest interchange up to 12 times, and they have a high reproductive rate. There are four species of musk deer: *Moschus chrysogaster*, *M. moschiferus*, *M. berezovskii*, and *M. fuscus*.

Physical characteristics

Musk deer are small-sized deer without antlers. Due to long and strong hind legs and shorter and weaker forelegs, musk deer look asymmetric, with heavy rump, banded back, and sloped withers. The distinctive body construction restricts the movement style of the animals: they walk or jump, and never run. The small head is adorned by a pair of big, sensible, hare-like ears. The muzzle around the black nostrils is hairless. Both sexes have long canines; in males, these tusks are protruding, in females, they are hidden in the mouth. Dewclaws on hooves are nearly the same size as central ones and their prints are visible in tracks.

The preputial gland is a protrusion of preputial skin with a separate opening; it has two layers of glands (inner and outer ones) around the mouth. Secretion starts as males reach the age of 8–9 months, and when males are around 15–16 months old, their sac is full of secretion. The maximum of secretion occurs from May–June, when the sac is filled. In addition, epithelium cells from the inner layer of the sac as well as masses of bacterium mix with the secretion. After short period of ripening, the sac is filled with a strongly odorous, granulated, reddish brown substance. This secretion of the preputial gland induces estrus in females and is very important in the course of mating.

Musk deer have many fragrant glands: on the nose mirror, pre-orbital, tarsal, metatarsal, circum anal, around the tail basement as well as on upper and side tail surfaces, and on the shanks of the hind legs. Secretion of all these glands is important in animal communication, in marking of home range, in individual distinguishing (mostly nasal glands), and in stimulation of a sexual partner.

The Siberian musk deer (*Moschus moschiferus*) uses its strong hind legs to jump instead of run. (Photo by Zoological Society of San Diego. Reproduced by permission.)

Distribution

The western edge of their distribution is the Altai Mountains. Eastward from there, musk deer are distributed in the mountains of southern Siberia. Musk deer range expands over the mountain crests of Siberia and eastern Siberia to the shores of the Japan Sea and the Okhotsk Sea. Musk deer are also distributed in China, Korea, Myanmar, and Vietnam; a wide area covers the Himalayas (Bhutan, China, India, and Nepal).

Habitat

Musk deer occur mostly on mountain slopes and on terraces, foothills, in mountain valleys, and on river bank escarpments. There are habitats at altitudes from 1,300–14,400 ft (400–4,400 m). Dense coniferous and broadleaved forests with rich undergrowths are common habitats of musk deer. Solitary rocks or rock pendants with very steep escarpments where musk deer, when necessary, can remain inaccessible to predators.

Behavior

Musk deer live solitarily, sedentary, and they keep strictly to their home ranges year-round, and never migrate. Home range borders are confined to natural margins like ridge crests and rivulets. There are feeding paths, watering places, and resting points, as well as defecation and urination points. Resting places are chosen for their access for their clear views: hilltops, anthill tops, ridge crests. In bad weather, animals take cover under branched trees or under protruding rocks.

They mark home range year-round with the secretion of their nasal and tail glands, by urine and pellets, and by scratching the ground with their hooves, which are also supplied with odorous glands to make scent-marks. Musk deer use the same latrines over and over, only for defecation (never to urinate). Home range includes rocky escarpments (usually small patches to 8–16 in [20–40 cm] wide) that are inaccessible to predators. Otherwise, animals hide in bushy thickets or under sloped trees and in piles of wooden trash. The home range of a male overlaps with home ranges of several does.

During mating season, the musk secretion issued by males in the urine is highly concentrated and marks snow with dark pink or red spots. During the rut (mating season), three or four animals make a group. Males start battles that are not especially fierce.

Musk deer have excellent vision and hearing, though their communication is predominantly by olfaction as they have an

The Siberian musk deer (*Moschus moschiferus*) is hunted for its glands, which are sold as remedies for impotence in Asia. This deer shows its drooping canine teeth. (Photo by Stanley Breeden/NGS Image Collection. Reproduced by permission.)

acute sense of smell. Once disturbed, musk deer freeze motionless, or jump to escape.

Feeding ecology and diet

Musk deer eat arboreal lichens, forbs, leaves, flowers, moss, needles of conifers shoots, twigs, and grass. They only nibble a small proportion of food at a time, as it minimizes the pressure on vegetation so that they can return to the same feeding place many times. Musk deer can also rise on hind legs or climb on bent trunks to reach leaves.

Reproductive biology

A polygamous group, musk deer have relatively a high reproductive rate, and twins and even triplets are not unusual. The mating season varies with locality and altitude, from November–January. After a gestation of 178–198 days, fawning takes place from May–June. Calves are born in hidden places, and within 25–30 minutes, will suckle its mother for the first time. Newborns will weigh 15.5–16.6 oz (440–470 g). Duration of

The Siberian musk deer (*Moschus moschiferus*) is found in Siberia, Mongolia, northeast China, North Korea, South Korea, and Sakhalin Island. (Photo by Kenneth Fink. Bruce Coleman, Inc. Reproduced by permission.)

nursing is three to four months. At the end of this period, a calf is nursed only once every five days. Calves younger than three months remain hidden and do not follow their mother. Young

The Siberian musk deer's (*Moschus moschiferus*) highly arched back separates it from other deer. (Photo by M. K. Ranjitsinh/Photo Researchers, Inc. Reproduced by permission.)

Male Siberian musk deer (*Moschus moschiferus*) have a musk pouch to attract females. (Photo by Silvestris/ANTPhoto.com.au. Reproduced by permission.)

grow quickly; females become sexually mature and capable of breeding in their first year.

Conservation status

Recently, the slaughter of the animal for its musk, which is used in medicine and perfume, has greatly reduced the numbers of Siberian musk deer (*Moschus moschiferus*), a species that used to be very numerous not so long ago. The IUCN lists it as Vulnerable. A high reproductive rate, its hidden mode of life, and breeding at farms has minimized the pressure of hunting. The other three musk deer species are Lower Risk/Near Threatened.

Significance to humans

Musk is highly valued in Chinese medicine, as it is used for the alleged improvement of health, treatment of inflammation, fever, and in the manufacture of soaps and perfumes. For cosmetic and alleged pharmaceutical properties, musk can fetch $24,000–45,000 per 2.2 lb (1 kg). Japan annually imports 220 –1,650 lb (100–750 kg) of musk; China 1,100–2,200 lb (500–1,000 kg); Taiwan 77 lb (35 kg); and the Republic of Korea 290 lb (130 kg). Since 1958, many farms have appeared in China where musk deer are bred and musk can be taken from the preputial sac without harm to the animal.

1. Himalayan musk deer (*Moschus chrysogaster*); 2. Siberian musk deer (*Moschus moschiferus*). (Illustration by Brian Cressman)

Species accounts

Himalayan musk deer
Moschus chrysogaster

TAXONOMY
Moschus chrysogaster (Hodgson, 1839), Nepal.

OTHER COMMON NAMES
English: Alpine musk deer; French: Porte-musk; German: Moschushirsch; Spanish: Ciervo almizclero de montana.

PHYSICAL CHARACTERISTICS
Shoulder height: 20–21 in (51–53 cm); body length: 2.8–3.3 ft (86–100 cm); tail length: 1.6–2.4 in (4–6 cm); weight: 24–40 lb (11–18 kg). General color is light grizzled brown; on the chest is a wide vertical whitish yellow stripe, which extends up the throat to the chin. Tail is hairless, but has a small tuff at the end. Ears are long.

DISTRIBUTION
Along Himalayas in Nepal, northern India, southern China, Afghanistan, Bhutan, and Pakistan.

HABITAT
Elevations of 6,600–14,100 ft (2,000–4,300 m). They use forest and shrub land, dwarf rhododendron, alpine woods, low shrubs on eastern and southern edges of Tibet, and the slopes of the Himalayas. They choose slopes that are not very steep in oak and fir woods with birch, pine, juniper, and bushes. Grasses and lichens in under story are very important for their habitats.

BEHAVIOR
Home range of a buck overlaps home ranges of several does; bucks fiercely defend their territories from rivals. Musk deer are active from dusk to dawn when they alternate feeding and rest; they are vigilant for predators. They dare to appear in clearings at night, though remain hidden in thickets during the day. When they hear a signal of danger, they make a loud double hiss, and flee.

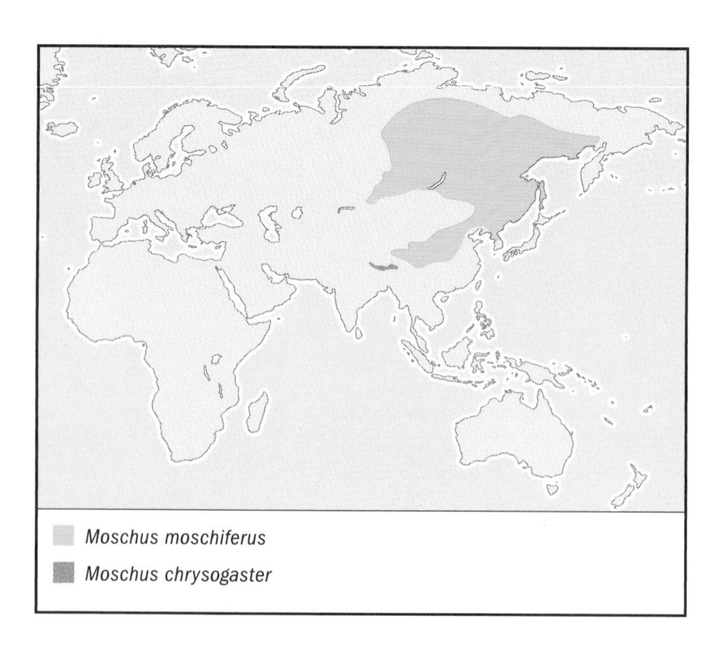

Musk deer stay in their home ranges the entire year, using an area of 2,200 acres (900 ha) for bucks and 740 acres (300 ha) for does. Home range comprises traditional trails, feeding places, watering points, and rocky promontories to escape from predators. Sometimes, several neighbors share the only steep outcrop in an area. Piles of wood and shrub thickets also serve as a cover from enemies. Main predators are yellow-throated marten, fox, wolf, and lynx.

FEEDING ECOLOGY AND DIET
In winter and autumn, they feed mostly on forbs, leaves of oak, gaultheria, and shrubs. In spring and summer, forbs, lichens, herbs, and moss are main food.

REPRODUCTIVE BIOLOGY
Polygamous. Gestation lasts 6.5 months, with one or two fawns per birth. Rut goes in December–January, calving in May–June. Fawns stay hidden in thickets where mother comes to nurse them. Weaning occurs at three to four months. Sexual maturity is reached at 1.5 to two years. Life expectancy high, 12–20 years, but actual lifespan is about three years in the wild and two to four years on farms.

CONSERVATION STATUS
Lower Risk/Near Threatened.

SIGNIFICANCE TO HUMANS
Commercial game species, mainly due to musk. ◆

Siberian musk deer
Moschus moschiferus

TAXONOMY
Moschus moschiferus Linnaeus, 1758, southwestern Siberia, Russia.

OTHER COMMON NAMES
French: Porte-Musc; German: Moschushirsch; Spanish: Ciervo almizclero.

PHYSICAL CHARACTERISTICS
Body length: 24–39 in (60–100 cm); tail length: 1.2–2.4 in (3–6 cm); weight: 18–36 lb (8–16 kg). General color of the coat varies from dark brown to grizzly brown. They have fuzzy whitish yellow spots on neck and chest, with rows of brighter spots on both sides of the body. Observers use these spot patterns to distinguish one animal from another. A light-colored band goes beneath neck to divide the chest. Newborns have thick pattern of yellowish spots. At the beginning of winter after shedding, calves obtain common color; though spots on their skin look brighter. Adult coat appears by their second winter.

DISTRIBUTION
Inhabit a wide area in eastern Asia from the border of the forest zone at the north (71°N) to Hindu Kush and Himalayan regions of Afghanistan, Nepal, Pakistan, and India to the south, and from the Altai Mountains eastward to the shores of Japan Sea and Okhotsk Sea.

HABITAT

Two factors are critical to their habitats: abundant tree lichens for fodder and shelter from predators. They inhabit dark coniferous forests with rich undergrowths and ground moss cover; in light coniferous forests (larch), and sometimes in coniferous-broadleaf forest. Solitary rocks or rock promontories are important. Sometimes, there is only one such point for all deer in the area.

In winter, they escape to areas with snow cover deeper than 23.6–27.5 in (60–70 cm). Due to their low weight and peculiar hoof structure (footing on all four fingers), they exert low pressure on snow, which is why they move easily on crusted snow surfaces. Deep and loose snow impedes their movement and causes mortality.

BEHAVIOR

Usually five to seven females, some with fawns, make a commune and their individual home ranges are overlapped by the home range of a dominating male. The stronger a female, the more central position in the commune area it occupies. As old animals perish, younger ones move closer to the center. Home ranges of males never overlap; the male marks his home range. There are seven to 10 latrines at each home range, each used many times by a host deer. Latrines serve also as important territory marks.

Musk deer are nocturnal, mostly active in twilight and at night. Daily home range reaches 4.9–24.7 acres (2–10 ha). To escape from predators, musk deer can jump 16.4–19.6 ft (5–6 m), landing to all four legs, as well as jump and turn in the air 90°. These deer are very vigilant and spend some 55% of feeding time listening for danger. Once approached, they rush away; when chased, they use many tricks to escape. Predators are yellow-throated marten, lynx, wolverine, less wolves and foxes.

FEEDING ECOLOGY AND DIET

Arboreal lichens and some terrestrial bushy lichens are their main food sources (more than 80% of the diet) in winter, as well as fir needles (either larch or pine needles, depending on type of coniferous forest), and twigs, leaves, dry cereals, berries, and mushrooms. Food is available on the snow surface and on tree branches; they can also dig in the snow for food. To reach lichens, an animal can stretch up to 55 in (140 cm). When snow is heavy, musk deer, otherwise sedentary, were observed to migrate up to 20 mi (35 km) for food. Lichens are constantly consumed in summer as a remedy to help digest green herbaceous plants. Also in summer, they feed on forbs, leaves, flowers, moss, shoots, twigs, and grass. Their daily input is 5.6–9.3 oz (160–265 g) of forage.

REPRODUCTIVE BIOLOGY

Polygamous. Rut occurs in November–December; calving in April–May in the Amur basin or in June in Yakutia.

CONSERVATION STATUS

Vulnerable. Subspecies (*M. m. sachalinensis*) inhabiting Sakhalin Island is in the most troublesome position. In Russia, *M. m. sachalinensis* is Endangered. According to 1997 estimations, there were about 50,000 *M. m. moschiferus*, 5,000 *M. m. parviceps*, and 300 *M. m. sachalinensis*.

SIGNIFICANCE TO HUMANS

An important hunting species. ◆

Resources

Books

Baskin, Leonid, and Kjell Danell. *Ecology of Ungulates. A Handbook of Species in Eastern Europe, Northern and Central Asia.* Heidelberg: Springer Verlag, 2003.

Bedi, Ramesh. *Wildlife of India.* New Delhi: Bridgebasi Printers Private Ltd., 1984.

Flerov, Konstantin K. *Musk Deer and Deer.* Moscow: Izdatelstvo Akademii Nauk SSSR, 1952.

Geist, Valerius. *Deer of the World: Their Evolution, Behavior, and Ecology.* Mechanicsburg, PA: Stackpole Books, 1998.

Hudson, Robert J., Karl R. Drew, and Leonid M. Baskin. *Wildlife Production Systems. Economic Utilization of Wild Ungulates.* Cambridge: Cambridge University Press, 1989.

Schaller, George B. *Wildlife of the Tibetan Steppe.* Chicago: University of Chicago Press, 1998.

Sokolov, Vladimir E., and Olga F. Chernova. *Skin Glands of Mammals.* Moscow: GEOS, 2001.

Sheng, Helin, and Lu Houji. *The Mammalian of China.* Beijing: China Forestry Publishing, 1999.

Leonid Baskin, PhD

△

Muntjacs
(*Muntiacinae*)

Class Mammalia
Order Artiodactyla
Family Cervidae
Subfamily Muntiacinae

Thumbnail description
Small grazing ungulates, described as Asian deer, characterized by short, two-tined antlers, and by upper canine teeth that are prolonged into tusks in the adult males; they lift their feet high when walking; always vigilant, and because they bark in response to predators and other disturbances, they are also commonly called barking deer

Size
Head and body length: 24–62 in (609.6–1,575 mm); tail length: 2.6–9.5 in (65–240 mm); shoulder height 15.8–30.7 in (401–780 mm); weight 24.3–110.2 lb (11–50 kg)

Number of genera, species
2 genera; 11 (recent) species

Habitat
Areas of dense vegetation, forests; tropical and subtropical

Conservation status
Vulnerable: 1 species; Data Deficient: 2 species

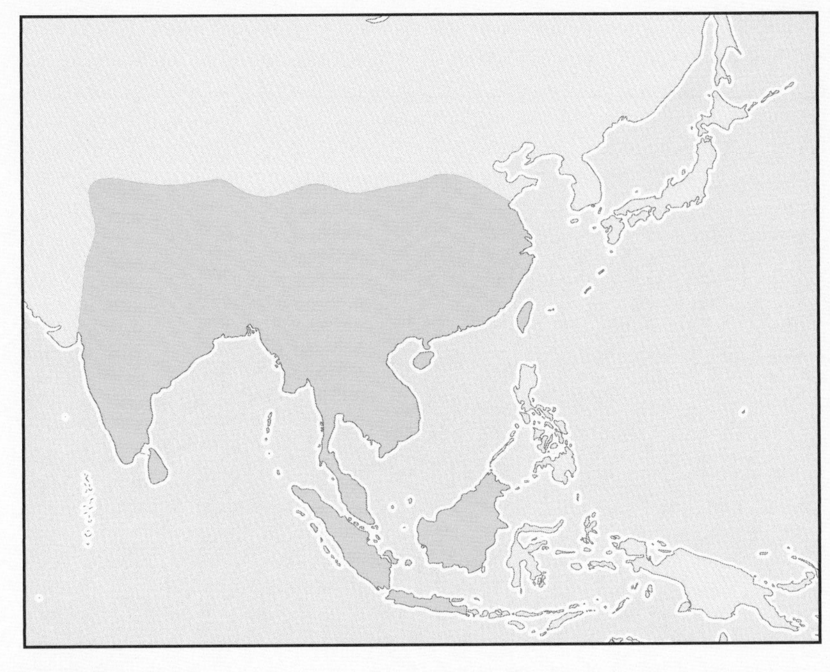

Distribution
East Asia

Evolution and systematics

The beginning of the Tertiary epoch marked the beginning of ungulates that came from order Condylarthra ancestors; this gave rise to the Eparctocyon line, which includes the recent order Artiodactyla. In late Eocene or early Oligocene epochs, the *Lophiomeryx* appeared. It was a ruminant, called "gelocid," which possessed an efficient and compact ankle, small side toes, complicated premolars, and almost completely covered mastoid bone. In late Eocene or early Oligocene epochs it split off into four families, one including *Dicrocerus* (early Miocene) with the first antlers (similar to living muntjacs). Cervids arose later from Palaeomerycid ancestry during the Oligocene epoch.

The subfamily Muntiacinae (muntjacs) belongs within the family Cervidae. The Cervidae is described as deer where the males possess bony antlers that molt annually, except in the Chinese water deer, and where the females lack antlers, except in the reindeer. They range in North and South America, Europe, Asia, and northern Africa. Included with the Muntiacinae are the other three subfamilies: Cervinae (deer and fallow deer), Hydropotinae (*Hydropotes inermis*, or Chinese water deer), and Capriolinae (such as moose and reindeer).

Previous phylogenetic studies for the Muntiacinae have shown that its species originated in the Pleistocene epoch. Specifically, the earliest fossil record of *Muntiacus reevesi* shows that it evolved in early Pleistocene (with paleontological specimens recovered from Nihewan, Xuanhua, and Hebei), *Muntiacus feae* in mid Pleistocene (with records from Yanjing and Sichuan), *Muntiacus rooseveltorum* in late Pleistocene (with records from Ziyang, Sichuan Province, Yuanmou, and Yunnan), *Muntiacus muntjak* in late Pleistocene (with records from Tongnam and Sichuan), and *Elaphodus cephalophus* from an indeterminate time in the Pleistocene epoch.

Physical characteristics

The body is covered with soft, short hairs, except for the ears that are barely covered. Coloration changes from dark brown to yellowish and grayish brown with markings that are whitish or creamy.

Antlers are possessed only by males and are shed annually in most species. The antlers rarely exceed 1–2 in (25–52 mm) in length and are usually positioned above long, bony, hair-covered pedicels. The females possess small, bony knobs and tufts of hair in the area where the antlers occur in the males.

The males possess upper canine teeth that are elongated into tusks, curving sharply outward from the lips. Such a configuration allows them to inflict serious injuries to small

An Indian muntjac (*Muntiacus muntjak*). (Photo by Anup Shah/Naturepl.com. Reproduced by permission.)

Reeve's muntjac or barking deer (*Muntiacus reevesi*) rubbing horns as part of the rutting behavior. (Photo by E & P Bauer. Bruce Coleman, Inc. Reproduced by permission.)

configuration allows them to inflict serious injuries to small animals.

Distribution

They are the most widespread but least known of all the Asian animals. Generally found throughout eastern Asia, they are specifically from southern to eastern China and extending north through into Tibet; Taiwan; from China southwest to Laos, Vietnam, Myanmar (formerly Burma), and Thailand; Indonesia, Malay Peninsula, Borneo, India, and Sri Lanka. They adapt well to captivity, and have been incorporated into zoos and private collections throughout the world.

Habitat

They are usually found in areas of dense vegetation and forests, from an altitude of sea level to medium elevations, up to 9,800 ft (3,000 m), in hilly country. They generally do not wander very far from water. The territory of males is usually exclusive of one another, but generally overlaps the territories of one or more females.

Behavior

The most obvious behavior is a deep, bark-like sound when danger is sensed from predators or when a stressful situation occurs. Barking may occur for an hour or more from a single incident. It also occurs most often when visibility is re-

The black muntjac (*Muntiacus crinifrons*) has upper tusk-like canines. (Photo by Helen Williams/Photo Researchers, Inc. Reproduced by permission.)

North Indian muntjacs (*Muntiacus muntjak vaginalis*) shed their short antlers annually. Photo by Animals Animals ©Michael Dick. Reproduced by permission.)

A Reeve's muntjac (*Muntiacus reevesi*) peering out from foliage. (Photo by © Nano Calvo/VW/The Image Works. Reproduced by permission.)

Reeve's muntjac (*Muntiacus reevesi*) is very territorial. (Photo by © George McCarthy/Corbis. Reproduced by permission.)

duced because of environmental conditions. In the past, barking was believed to be a means of communication during the mating season, but recent research shows this probably is untrue. Instead, distinctive barking most likely serves to identify different animals and to cause its predators to reveal themselves, to realize that they have been detected, and to provoke them to move away.

Both diurnal and nocturnal activities have been observed. Most sightings are of lone animals, with infrequent sightings of two to four individuals in a group. Males mark territory at intervals by lowering the head and rubbing the frontal gland on the ground and by scraping their hooves against the ground. They also mark trees by scraping the bark with the lower incisors and rubbing the base of their antlers.

Feeding ecology and diet

Muntjacs are herbivorous; their diet includes grasses, ivy, prickly bushes, low-growing leaves, bark, twigs, and tender shoots. They rarely feed in one place for very long, and prefer foods low in fiber, and rich in protein and nutrients.

A male Indian muntjac (*Muntiacus muntjak*). (Photo by David Kjaer/Naturepl.com. Reproduced by permission.)

Reproductive biology

Muntjacs are polygamous. Males often fight among themselves for possession of harem. Females reach sexual maturity sometime within the first year or two of life. There is no definite evidence of a specific breeding season, but in Thailand mating occurs frequently in December and January; general gestation period is 6–7 months. Females are polyestrous; the estrous cycle lasts 14–21 days, and estrus is about two days long. Females usually give birth to one young (infrequently two), generally in dense growth in order to remain hidden until it can move about with its mother. At birth young weigh 19.4–22.9 oz (550–650 g).

Conservation status

Muntiacus feae and *Muntiacus gongshanensis* are listed as Data Deficient, and *Muntiacus crinifrons* is considered Vulnerable, while *Muntiacus reevesi*, *Muntiacus muntjak*, *Muntiacus atherodes*, *Muntiacus gongshanensis*, *Elaphodus cephalophus*, *Muntiacus truongsonensis*, *Megamuntiacus vuquangensis*, and *Muntiacus putaoensis* are not listed by the IUCN. New muntjac species have been discovered since 1990, and it is likely that additional information about these species will help determine their conservation status in the future.

Significance to humans

They are hunted for their meat, skins, and sometimes antlers. They generally thrive in captivity and are found in many zoos. They play a vital ecological role in many ecosystems, and their economic importance in rural communities is significant. Phylogenetic relationships are of great interest because of their suitability for the study of evolutionary processes. They are also considered to be a nuisance in some areas, because they mutilate trees by tearing off the bark.

1. Black muntjac (*Muntiacus crinifrons*); 2. Indian muntjac (*Muntiacus muntjak*); 3. Gongshan muntjac (*Muntiacus gongshaniensis*); 4. Giant muntjac (*Megamuntiacus vuquangensis*); 5. Tufted deer (*Elaphodus cephalophus*). (Illustration by Brian Cressman)

1. Fea's muntjac (*Muntiacus feae*); 2. Leaf muntjac (*Muntiacus putaoensis*); 3. Truong Son muntjac (*Muntiacus truongsonensis*); 4. Roosevelt's muntjac (*Muntiacus rooseveltorum*); 5. Bornean yellow muntjac (*Muntiacus atherodes*); 6. Reeve's muntjac (*Muntiacus reevesi*). (Illustration by Brian Cressman)

Species accounts

Reeve's muntjac
Muntiacus reevesi

TAXONOMY
Muntiacus reevesi (Ogilby, 1839), Guangdong, China.

OTHER COMMON NAMES
English: Chinese muntjac.

PHYSICAL CHARACTERISTICS
Body length: 31.5–39 in (80–99 mm); tail length: 4.5–7 in (11–18 cm); shoulder height: 15.8–17.7 in (40–45 cm); weight: 24.3–35.3 lb (11–16 kg). Small, dainty, but fierce deer with rounded body, slender legs, and short tail. Adult male grows rudimentary, moderate-sized antlers about 2.4–3.2 in (60–80 mm) long; antlers are grown, averaging 2.75–3.2 in (69.9–81.3 mm), from bony, short pedicles that extend from the frontal bone on the skull and are shed annually; adult females have bony knobs on the forehead covered with tufts of hair. Tusks, which can grow up to 1 in (25 mm), are formed from the upper canine teeth; females have smaller tusks than males. They have a long tongue that is used to strip leaves from low bushes. Ranges in color from deep brown and reddish brown to yellowish or grayish brown with creamy markings; the short, soft coat is reddish brown in color; dorsal side is light red-brown with dorsal cervical stripe; undersides, including the lower legs and the ventral surface of the neck and chin, are creamy white. The nose and forehead are black, while the rest of the face is a pale tan; cranial coloring is dark red with a black stripe at center, extending to the neck.

DISTRIBUTION
Generally, eastern Asia; specifically, natural range is from southern to eastern (mainland) China (from Yunnan to Fujian) and extends north through Palaearctic China and Taiwan; introduced in Great Britain and the western region of Europe, especially France.

HABITAT
Especially temperate to tropical deciduous forest dweller because it requires large amounts of cover; ranges from areas of dense vegetation and hilly country from sea level to medium elevations; prefers to stay near water sources (especially streams). It does not hibernate in any way and remains active throughout the winter. It makes its home out of pine boughs and other large broken off branches.

BEHAVIOR
Limited amount information is available on its native habitats because of the difficulties in long-term observation doe to its timid nature; its ability to hide is primary means of defense. It is usually found alone or in groups of fewer than four individuals (it rarely forms herds). It is territorial, rarely leaving its home range; home ranges of males and females usually overlap. It prefers to stay under cover of vegetation, and is primarily nocturnal (and especially active during twilight), but may be active in the morning in quiet, undisturbed areas. During mating season, females emit a mewing noise. When competing for females, males fight with their canines (tusks), rather than their antlers, and make dog-like barking noises. Main predators are wolves, leopards, tigers, dhole (Asiatic wild dog), jackals, crocodiles, pythons, and birds of prey.

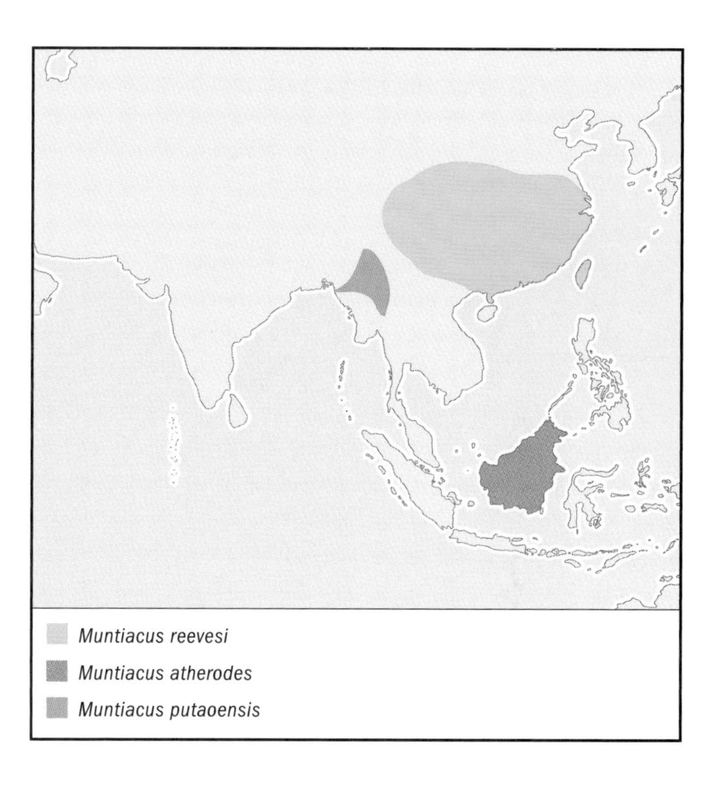

FEEDING ECOLOGY AND DIET
Herbivorous; usually forages for grasses, leaves (mostly of low-growing plants), tender shoots, fallen fruit and berries, and soft and hardwood tree bark.

REPRODUCTIVE BIOLOGY
Polygamous. Females sexually mature within their first year of life; mating can take place throughout the year, but primarily from January–March; gestation period is from 209–225 days (about seven months); birth is to one young (sometimes two) that weighs 19.4–23.8 oz (550–675 g); young fawns are usually born in dense jungle growth in order to hide until able to move around with mother, usually after two months; fawns have white spots on its coat for camouflage, which later disappear; Chinese populations number about 650,000. Lifespan can be up to 17 years.

CONSERVATION STATUS
Not considered threatened, and not listed by the IUCN. Humans have established wild populations throughout southern England, while some occur on private estates in France.

SIGNIFICANCE TO HUMANS
Its bark helps to alert humans as well as other muntjacs to potential danger, hunted for meat and skins; it is of interest because of its conservation and ungulate lineage importance. Some damage inflicted on agricultural lands, most damage onto young hardwood plantations in western Europe. ◆

Roosevelt's muntjac

Muntiacus rooseveltorum

TAXONOMY

Muntiacus rooseveltorum Osgood, 1932, Laos.

OTHER COMMON NAMES

None known.

PHYSICAL CHARACTERISTICS

Body length: 39–40 in (99–102 cm); shoulder height: approximately 21.1 in (53.5 cm). Relatively light colored brown-black stripes are present at the anterior section of the frontals that extend posterior to the ears, but do not converge. Body coloration is red-brown with very slight mottling. Hoofs are small. White bands on the posterior legs extend medially to the anterior side of the tarsal joint.

DISTRIBUTION

From western Yunnan and Yuanjiang in China southwest to Laos and Vietnam.

HABITAT

Very little is known about the animal's habitat. One of its few records was of one immature male specimen collected during Theodore Roosevelt's expedition to Indochina.

BEHAVIOR

Nothing is known.

FEEDING ECOLOGY AND DIET

Nothing is known.

REPRODUCTIVE BIOLOGY

Nothing is known, though other muntjacs are polygamous.

CONSERVATION STATUS

Generally assumed endangered, though not listed by the IUCN. Only recently rediscovered it is extremely low in population and borders on extinction. Endemic and near-endemic.

SIGNIFICANCE TO HUMANS

None known. ◆

Black muntjac

Muntiacus crinifrons

TAXONOMY

Muntiacus crinifrons (Sclater, 1885), Zhejiang, China.

OTHER COMMON NAMES

English: Hairy-fronted muntjac.

PHYSICAL CHARACTERISTICS

Length: 39.4 in (100 cm); height: 18.4 in (47 cm). Body color is totally black-brown. Its head is much lighter in color than the body. Antlers are short and small with tines of 2.6 (6.5 cm) in length. Pedicle is approximately 1.9–2.2 in (4.8–5.5 cm). Long, golden yellow tufts project off the top of the frontals. Dorsal tail is black.

DISTRIBUTION

Originally in southeastern China from the lower Yangtze River to Guangdong and eastern Yunnan; today restricted to only the lower reaches of the Yangtze River in China within the Anwei, Zhejiang, and Jiangxi Province region.

HABITAT

Occurs in mixed forests and scrub ground.

BEHAVIOR

Solitary.

FEEDING ECOLOGY AND DIET

Nothing is known.

REPRODUCTIVE BIOLOGY

Polygamous. Young are born throughout the year following a gestation period of about 210 days. Sexual maturity is reached at approximately one year.

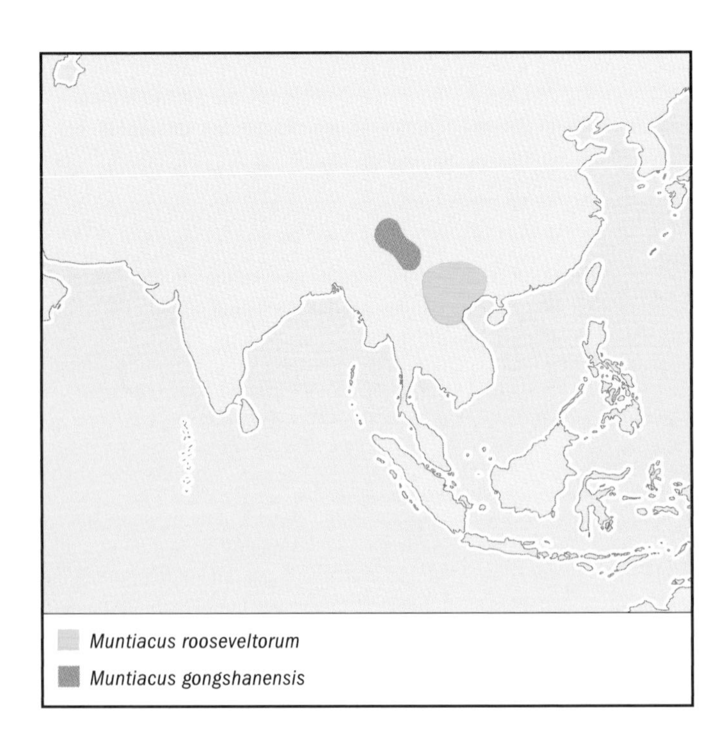

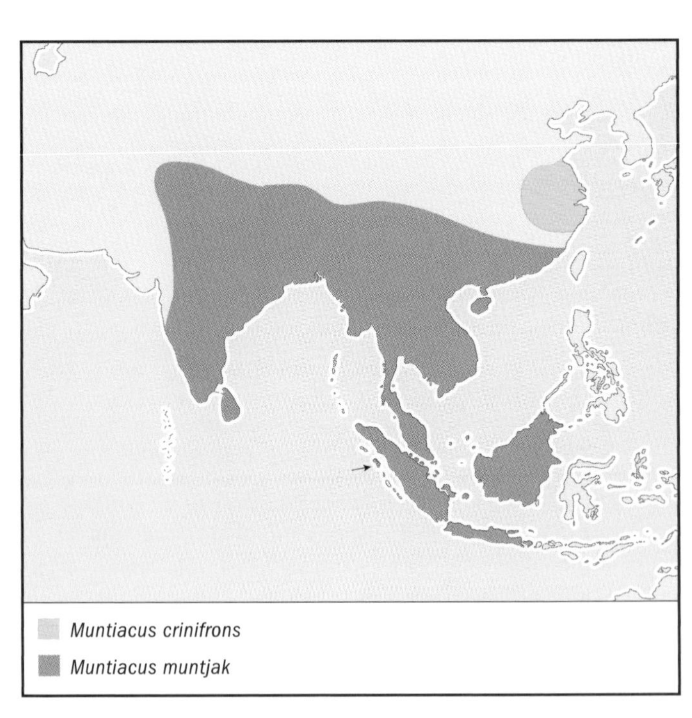

CONSERVATION STATUS
Listed as Vulnerable by the IUCN. Generally is declining in population because of habitat destruction and uncontrolled hunting. Once believed to be few in numbers, in the 1970s it was found to be living in four provinces of east Central China. Now, numbers are estimated to be about 10,000 individuals in an overall range of 29,500 mi^2 (76,500 km^2). Endemic in China.

SIGNIFICANCE TO HUMANS
Hunted for venison and skin. ◆

Fea's muntjac
Muntiacus feae

TAXONOMY
Muntiacus feae (Thomas and Doria, 1889), Tenasserim, Burma.

OTHER COMMON NAMES
English: Fea's rib-faced deer.

PHYSICAL CHARACTERISTICS
Moderate body size: body length: 39.4–40.9 in (100–104 cm); shoulder height: 23.2–26.8 in (59–68 cm); weight: 40–46 lb (18–21 kg). Tail is dark tan and frontal tufts are bright tan, and shorter than *Muntiacus crinifrons*. Antlers are short and small with tines equivalent to or shorter than the antler pedicles (there is distinct individual variation).

DISTRIBUTION
Yunnan (south Central China), Laos, eastern and peninsular Myanmar, Tenasserim, and Thailand.

HABITAT
Usually in evergreen forests in upland areas. In China, it has been found in mountainous forest comprised of a mixture of coniferous, broadleaf forest or shrub forest at an altitude to 8,200 ft (2,500 m).

BEHAVIOR
Diurnal and solitary.

FEEDING ECOLOGY AND DIET
Grasses, low-growing leaves, and tender shoots.

REPRODUCTIVE BIOLOGY
Polygamous. Gestation period is around 180 days. Young are usually born in dense growth, where they remain hidden until they can move about with the mother. It is now quite rare; total numbers are unknown, but are certainly small.

CONSERVATION STATUS
Generally considered highly endangered. Currently listed as Data Deficient by the IUCN, and listed on the U.S. Endangered Species Act. Restriction to a small roaming area and subjected to uncontrolled hunting by humans have led to its endangered status.

SIGNIFICANCE TO HUMANS
A nuisance in some areas because it destroys trees by ripping off the bark. ◆

Indian muntjac
Muntiacus muntjak

TAXONOMY
Muntiacus muntjak (Zimmermann, 1780), Java, Indonesia.

OTHER COMMON NAMES
English: Muntjac, barking deer.

PHYSICAL CHARACTERISTICS
Length: 35–53.2 in (89–135 cm); shoulder height: 15.7–25.6 in (40–65 cm); tail length: 5.2–9 in (13–23 cm); weight: 33.1–77.2 lb (15–35 kg); males usually larger than females. Males have small, simple antlers (about 6 in [150 mm]) with long burrs, only one branch with a broad and hairy pivot; females have tufts of hair and small bony knobs that are at the location of antlers in males; both have short coats of hair that can be thick and dense for those living in cooler climates, or thin and less dense for those living in warmer areas. Color of coat is golden tan on the dorsal side, white on the ventral side, and the limbs and face are dark brown to reddish brown. Ears have very little hair; also have tusk-like upper canine teeth measuring about 1 in (200 mm) in males.

DISTRIBUTION
Northeastern Pakistan, India, Sri Lanka, Nepal, southern China, Hainan (China), Vietnam, Malay Peninsula and some nearby islands, Riau Archipelago, Sumatra and Nias Island to the west, Bangka, Belitung Island, Java, Bali, and Borneo.

HABITAT
Tropical deciduous and tropical scrub forests, tropical rainforests, areas of dense vegetation, hilly country, savannas and grasslands, and monsoon forests; stays close to a water source; specifically, on the slopes of the Himalayas, they climb to more than 6,560 ft (2,000 m) of altitude.

BEHAVIOR
Adults are solitary (sometimes moving in pairs or small family groups), except during rutting when home ranges overlap for a short period of time; displays diurnal and nocturnal activity. When sensing detection of predator, they emit sounds similar to a barking dog that may last for more than an hour in order to make predator show itself or leave the area, and may bark

more frequently when its environment reduces its ability to see and when the male is rutting (small cries that may approach level of barking). Predators include pythons, jackals, tigers, leopards, and crocodiles.

FEEDING ECOLOGY AND DIET
Omnivorous, typically feeds on the edges of forests or in abandoned clearings, both as a browser and as a grazer; feeding on herbs, fruit, birds' eggs, small animals, sprouts, seeds, and grasses; typically feeds on the edges of forests or in abandoned clearings, they use their canine teeth to bite and their forelegs to deliver strong blows in order to catch small warm-blooded animals.

REPRODUCTIVE BIOLOGY
Polygamous. Females sexually mature in first year of life; they are polyestrous with the estrous cycle lasting 14–21 days and the estrus lasting about two days; no distinct breeding season occurs; usually bear one young at a time; gestation period is about 180 days and birth weight is usually 19.4–22.9 oz (550–650 g). Young leaves territory of mother when it is about six months old to find its own territory; adult may allow an immature male (without complete antlers so are not aggressive and unable to mate) into its territory.

CONSERVATION STATUS
Not threatened. Believed to number about 140,000–150,000 in China.

SIGNIFICANCE TO HUMANS
Indian pheasants hunters use muntjac barking noises as warning signals of approaching predators. The muntjac is hunted for its meat and skins. Heavy populations can destroy trees by tearing off bark, leading to loss of food sources and wood for shelter and fuel. ◆

Bornean yellow muntjac
Muntiacus atherodes

TAXONOMY
Muntiacus atherodes Groves and Grubb, 1982, Sabah, Malaysia.

OTHER COMMON NAMES
None known.

PHYSICAL CHARACTERISTICS
Length: 39.4 in (100 cm); height: 19.7 in (50 cm); weight: 29.76–39.02 lb (13.5–17.7 kg). Overall coat is a yellowish orange with a diffuse dark brown line running the length of the spine, being especially prominent on the nape of the neck. The underbelly is pale yellow, buff, or whitish in color. The tail is dark brown on its upper surface (a continuation of the dark dorsal stripe). The un-branched antlers are simply spikes, growing only 0.63–1.67 in (16.0–42.5 mm) long and are rarely, if ever, shed. They are positioned on top of slender forehead pedicels that are 2.6–3.5 in (6.5–8.7 cm) in length. It does not possess frontal tufts of hair.

DISTRIBUTION
Borneo.

HABITAT
Moist forests.

BEHAVIOR
Little is known. Primarily diurnal; usually appear as breeding pair or solitary. Males make the characteristic loud alarm bark, while adult females with young may make short, high-pitched mewing sounds.

FEEDING ECOLOGY AND DIET
Herbs, grasses, leaves, fallen fruit, and seeds.

REPRODUCTIVE BIOLOGY
Polygamous. Gestation most likely around seven months, with normally one young. Young are weaned within two months, and reach sexual maturity between 6–12 months. Population in the provinces of South Anhui and West Zhegang is estimated at 5,000–6,000 individuals. Unknown lifespan.

CONSERVATION STATUS
Not threatened. It is (assumed to be) fairly populated in its territory. Endemic.

SIGNIFICANCE TO HUMANS
Hunted for meat and skins. ◆

Gongshan muntjac
Muntiacus gongshanensis

TAXONOMY
Muntiacus gongshanensis Ma, 1990, Yunnan, China.

OTHER COMMON NAMES
None known.

PHYSICAL CHARACTERISTICS
Length: 39.4 in (100 cm); height: 19.7 in (50 cm).

DISTRIBUTION
Northwestern Yunnan (China) and adjacent (southeast) Tibet; northern Myanmar.

HABITAT
Forests.

BEHAVIOR
Diurnal.

FEEDING ECOLOGY AND DIET
Nothing is known.

REPRODUCTIVE BIOLOGY
Polygamous. Males and females become sexually mature at approximately two years. Mating season takes place during late fall and early winter. Gestation period lasts about six months, after which one or two fawns are born in late spring and early summer.

CONSERVATION STATUS
Listed by the IUCN as Data Deficient. Population has dropped due to habitat loss and hunting.

SIGNIFICANCE TO HUMANS
Extensively hunted by local people. ◆

Tufted deer
Elaphodus cephalophus

TAXONOMY
Elaphodus cephalophus Milne-Edwards, 1872, Sichuan, China.

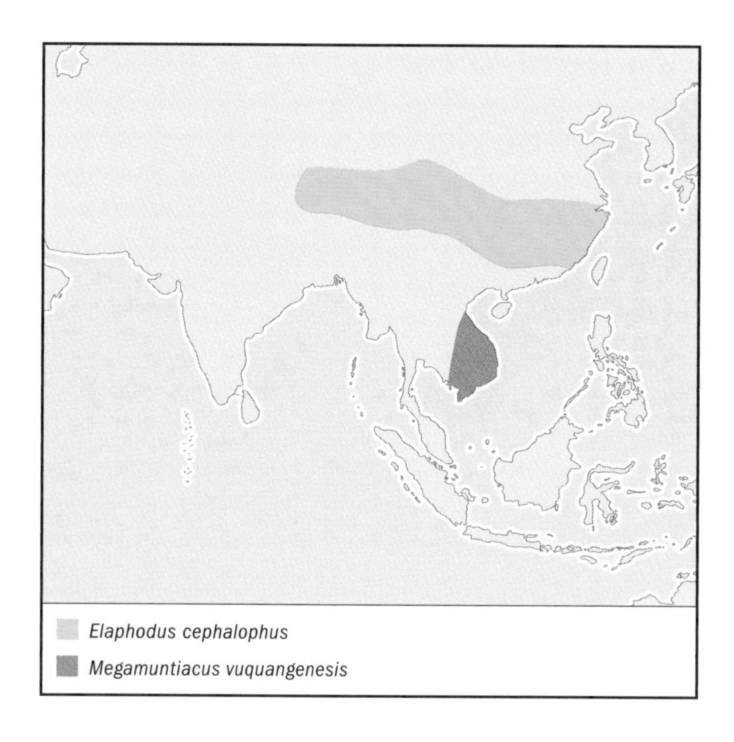

OTHER COMMON NAMES
English: Tibetan muntjac.

PHYSICAL CHARACTERISTICS
Largest muntjac species: head and body length: 43.2–62.4 in (110–160 cm); shoulder height: 19.7–27.6 in (50–70 cm); tail length: 2.8–5.9 in (7 15 cm); weight: 37.5–110.2 lb (17–50 kg). Body is covered with coarse, almost spine-like hairs that give it a shaggy look, general color of upper parts is deep chocolate brown, underbody is white, head and neck are gray; in some cases, a pale streak extends forward from the pedicel and above the eye. Tuft on forehead is blackish brown. Males have tuft of hair on the forehead at the base of the antlers; the antlers themselves are short and often almost hidden by the tuft. Bony pedicels are short and converge at tips.

DISTRIBUTION
Across southern China from eastern Tibet to Zhejiang and Fujian provinces, and northern Myanmar.

HABITAT
Mountainous forests, dense undergrowth at elevations of 980–14,700 ft (300–4,500 m); near water.

BEHAVIOR
Nocturnal, normally solitary, but occasionally travels in pairs. Both sexes bark when suddenly alarmed and during the mating season.

FEEDING ECOLOGY AND DIET
Feeds primarily on grass and other plant material; when feeding, it carries its tail high.

REPRODUCTIVE BIOLOGY
Polygamous. Females become sexually mature at about nine months. Mating occurs in late autumn and early winter; births occur in late spring and early summer; female gives birth to one young after a gestation period of about 180–210 days. Young is colored like adults, but has a row of spots along each side of the midline of the back.

CONSERVATION STATUS
Not threatened.

SIGNIFICANCE TO HUMANS
Hunted by local people. ◆

Truong Son muntjac
Muntiacus truongsonensis

TAXONOMY
Muntiacus truongsonensis Giao Tuoc et al, 1998, Vietnam.

OTHER COMMON NAMES
English: Dark Annamite muntjac.

PHYSICAL CHARACTERISTICS
Weighs 33 lb (15 kg), and half the size of the common muntjac.

DISTRIBUTION
Vietnam.

HABITAT
Lives in high ridges at altitudes ranging 1,300–3,280 ft (400–1,000 m), in forests with dense undergrowth (ferns and leaf litter), and in secondary wet evergreen forests; such places where its small size allows it to move freely.

BEHAVIOR
Diurnal and nocturnal.

FEEDING ECOLOGY AND DIET
Grasses, low-growing leaves, and tender shoots.

REPRODUCTIVE BIOLOGY
Polygamous. Generally, it has been found that young are born in dense jungle growth, where they remain hidden until they can move around with their mother.

CONSERVATION STATUS
Not threatened. Has not been identified previously because years of armed conflict, steep and rugged terrain, and remoteness have precluded scientific exploration of its territorial region until recently. Hunting may be a threat. As of 2002, the IUCN has not yet assessed. The Deer Specialist Group recommended in 1998 a rating of Data Deficient.

SIGNIFICANCE TO HUMANS
Although hunting is illegal in Vietnam, many people hunt, either for commercial or subsistence purposes. ◆

Giant muntjac
Megamuntiacus vuquangensis

TAXONOMY
Megamuntiacus vuquangensis Do Tuoc Vu Van Dung et al, 1994, Vietnam. DNA data suggest that the giant muntjac belongs with other muntjacs in the genus *Muntiacus* and not in a genus of its own.

OTHER COMMON NAMES
English: Large-antlered muntjac.

PHYSICAL CHARACTERISTICS
About the size of a large dog, weighs 66–100 lb (30–50 kg). Body color is gray-brown. Tail is short and broad.

DISTRIBUTION
Cambodia, Laos, and Vietnam.

HABITAT
Mostly in evergreen and semi-evergreen forests; apparently prefers primary forests, although it has been seen in second-growth areas and degraded forests. Ranges at altitudes 1,600–3,900 ft (500–1,200 m).

BEHAVIOR
Apparently solitary.

FEEDING ECOLOGY AND DIET
Nothing is known.

REPRODUCTIVE BIOLOGY
Only single young have been observed with females. Mating system not known.

CONSERVATION STATUS
Not listed by the IUCN. Threatened by heavy hunting pressure, as well as by habitat degradation due to logging and slash-and-burn agriculture.

SIGNIFICANCE TO HUMANS
None known. ◆

Leaf muntjac
Muntiacus putaoensis

TAXONOMY
Muntiacus putaoensis Amato Egan, Rabinowitz, 1999, northern Myanmar.

OTHER COMMON NAMES
English: Leaf deer.

PHYSICAL CHARACTERISTICS
Stands 24–30 in (60–80 cm) tall; shoulder height: 20 in (45 cm); weight: 25 lb (11 kg). Coat is a reddish brown color. Curved, thin, knife-like canine teeth hang from its mouth, with female canines just as long as male canines (very unusual in deer species). Antlers are just a single point, only 1.5 in (38 mm) long.

DISTRIBUTION
Found in northern portion of Myanmar.

HABITAT
Dense forest and high valley jungle habitats at an elevation of 1,500–2,000 ft (450–600 m). Almost always found close to water.

BEHAVIOR
Solitary; diurnal and nocturnal; females fight as much as males for territory.

FEEDING ECOLOGY AND DIET
Browsers and grazers, mostly grasses, fruit, and leaves.

REPRODUCTIVE BIOLOGY
Polygamous. Males and females become sexually mature at approximately two years of age, mating season takes place during late fall and early winter. Gestation period lasts about six months after which one or two fawns are born in late spring and early summer. Young are generally born in dense jungle growth and remain hidden until it can move around with its mother.

CONSERVATION STATUS
Not threatened.

SIGNIFICANCE TO HUMANS
Hunting with snares presumably reduced its numbers during the 1990s, even though it is not considered a particularly valuable species (for food, hides, and antlers) because of its small size. ◆

Resources

Books

Feldhemer, George A., Lee C. Drickamer, Stephen H. Vessey, and Joseph F. Merritt. *Mammalogy: Adaptation, Diversity, and Ecology.* Boston, MA: WCB McGraw-Hill, 1999.

Novak, Ronald M. *Walker's Mammals of the World*, Vol. II, 6th ed. Baltimore and London: The Johns Hopkins University Press, 1999.

Vaughan, Terry A., James M. Ryan, and Nicholas J. Czaplewski. *Mammalogy*, 4th ed. Philadelphia, PA: Saunders College Publishing, 2000.

Whitfield, Philip. *Macmillan Illustrated Animal Encyclopedia.* New York: Macmillan Publishing Company, 1984.

Wilson, Don E., and DeeAnn M. Reeder, eds. *Mammal Species of the World*, 2nd ed. Washington, D.C. and London: Smithsonian Institution Press, 1993.

Other

Artiodactyla—Sudokopytníci—Even-toed Ungulates.. Savci. June 5, 2002 [cited April 19,2003]. <http://savci.upol.cz/ce>.

Conova, Susan. "Science Finds Smallest Deer." *ABCNews.com*, [cited April 19,2003]. <http://abcnews.go.com/sections/science/DailyNews/deer990629.html>.

Datta, Aparajita, Japang Pansa, M. D. Madhusudan, and Charudutt Mishra. "Discovery of the Leaf Deer *Muntiacus putaoensis* in Arunachal Pradesh: An Addition to the Large Mammals of India." *Current Science*, vol. 84, no. 3. February 10, 2003 [cited April 19,2003]. <http://www.ncf-india.org/pubs/Datta%20et%20al%202003.pdf>.

Hunt, Kathleen. "Transitional Vertebrate Fossils FAQ, Part 2C." *The Talk.Origins Archives*, March 17, 1997 [cited April 17,2003]. <http://www.talkorigins.org/faqs/faq-transitional/part2c.html>.

Ma, Shilai, Yingxiang Wang, and Longhui Xu. "Taxonomic and Phylogenetic Studies on the Genus *Muntiacus*." *Acta Theriologica*, vol. VI, no. 3, (August 1986): 190–209. <http://www.nau.edu/qsp/will_downs/80.pdf>.

Massicot, Paul. *Animal Info—Giant Muntjac*, October 26, 2002 [cited April 19,2003]. <http://www.animalinfo.org/species/artiperi/megavuqu.htm>.

Resources

Massicot, Paul. *Animal Info—Leaf Muntjac*, October 26, 2002 [cited April 19,2003]. <http://www.animalinfo.org/species/artiperi/muntputa.htm>.

Massicot, Paul. *Animal Info—Truong Son Muntjac*, October 26, 2002 [cited April 17,2003]. <http://www.animalinfo.org/species/artiperi/munttruo.htm>.

"Microlivestock: Little-Known Small Animals with a Promising Economic Future." *The National Academies Press*, March 17, 1997 [cited April 17,2003]. <http://www.nap.edu/books/030904295X/html/299.html>.

Muntiacinae. National Center for Biotechnology Information, June 5, 2002 [cited April 19,2003]. <http://www.ncbi.nlm.nih.gov/Taxonomy/Browser/wwwtax.cgi?id=34877>.

"Muntjacs, or Barking Deer." *Walker's Mammals of the World Online*, [cited April 21, 2003]. <http://www.press.jhu.edu/books/walker/artiodactyla.cervidae.muntiacus.html>.

"Rare Endemic Mammals in the Pu Mat National Park." *MekongInfo*, [cited April 17,2003]. <http://www.mekonginfo.org/mrc_en/Contact.nsf/0/D62C8A67FE8831E447256C39002752F6/$FILE/Endemic_Mammals_e.html>.

"Saving Wild Places." *Northern Forest Complex Project in Myanmar*, Wildlife Conservation Society, [cited April 19,2003]. <http://www.savingwildplaces.com/swp-home/swp-protectedareas/61825>.

Schaller, George B., and Elisabeth S. Vrba. "Description of the Giant Muntjac (*Megamuntiacus vuquangensis*) in Laos." *Wildlife Conservation Society, Bronx, NY and Department of Geology and Geophysics, Yale University*, [cited April 19,2003]. <http://coombs.anu.edu.au/vern/species/schaller.html>.

"Taxonomy of Mammals, Birds, Amphibians and Reptiles." *Kenyalogy*, [cited April 21, 2003]. <http://www.kenyalogy.com/eng/fauna/taxonmam.html>.

William Arthur Atkins

△

Old World deer

(Cervinae)

Class Mammalia
Order Artiodactyla
Family Cervidae
Subfamily Cervinae

Thumbnail description
Only proximal parts of metacarpal bones of the second and fourth fingers (Plesiometacarpalian deer) are retained; vomer is short and does not divide posterior nasal holes; only males have antlers

Size
Large- and medium-sized deer

Number of genera, species
4 genera; 14 species

Habitat
Tend to stay in forests, woodlands, forest-steppe, partly in forested mountains; can adapt to varieties of habitats from marshlands to alpine meadows

Conservation status
Extinct: 1 species; Critically Endangered: 1 species; Endangered: 3 species; Vulnerable 3 species; Data Deficient: 2 species

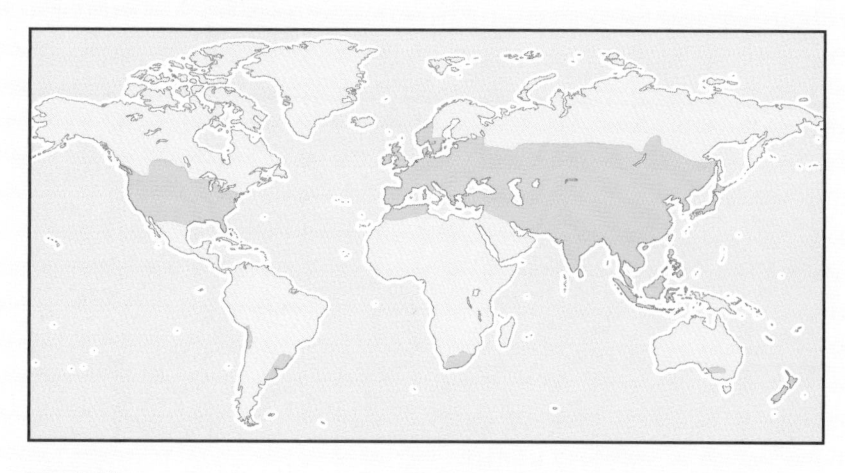

Distribution
Most of Old World, except Africa (only in northwest and introduced into South Africa); in North America, only one species (*Cervus elaphus*), which is a newcomer

Evolution and systematic

The most ancient deer forms, similar to the modern *Axis* deer, appeared in early Pleistocene, about two million years ago. A comparative study of extinct and existing forms revealed some trends in body construction and ecology, which helped distinguish between ancient and recent types of Cervinae. The most primitive of deer, they are distributed in India, China, and Indochina, and inhabit tropical forests, shrub lands, and grass thickets, often close to water basins. Species more advanced in evolution adapted to forests that interchanged with clearings so that their diet preferences are wider, and include leaves, tree and bush branches, forbs, and herbs. Some species such as Siberian maral (*Cervus elaphus sibiricus*) feed on grasses in high numbers, though they need a mixed diet that includes tree or bush leaves and twigs. None of the Cervinae became true grass and roughage eaters like sheep or cattle. A mixed diet facilitated the adaptation to a wide variety of habitats, including forest edges (with raids to neighboring grasslands), as did red deer, fallow deer, and barasingha. Teeth developed higher crowns and so are now adapted to partial feeding of herbal forages. Lower jaws grew longer to develop a significant gape between fangs and premolars. Incisors are still wide, with asymmetric outer edges, adapted to grazing softer vegetation and to browsing. Intestines in red deer and fallow deer are 15–17 times longer than body length, while in true grass and roughage eaters, the length of intestines exceeds body length by 25–30 times.

Fallow deer (*Dama dama*) have palmate, multi-pointed antlers. (Photo by St. Meyers/OKAPIA/Photo Researchers, Inc. Reproduced by permission.)

A barasingha, or swamp deer, (*Cervus duvaucelii*) stag. (Photo by Terry Whittaker/Photo Researchers, Inc. Reproduced by permission.)

Deer that inhabited northern areas differed from southern ones by the type of their coat. They had a thick undercoat, and hair in winter become air-filled and wavy, serving as an impermeable insulate for the animals.

Pere David's deer (*Elaphurus davidianus*) and barasingha (*Cervus duvaucelii*), as inhabitants of marshlands, had long hooves. The species that occupied mountainous areas, such as white-lipped deer (*Cervus albirostris*) had shorter and higher hooves, more adapted to harder ground. As well, the heels of the feet were enlarged, adding to their resilience on hard surfaces. When they moved from a continent to archipelago island, deer usually diminished in size, such as the *Axis* deer of the Indonesian archipelago.

Most deer fawns develop white spots, probably for camouflage. Most primitive deer are often spotty in the adult state, though they manifest no color differences in winter or summer coats, or between sexes. Multicolored ornate coats and different colors in bulls and hinds distinguish the evolutionary advanced species. They also develop a bright, white rump patch (a mirror) around the tail. These color differences all relate to more complicated social behavior. Both mirror and white underparts of the tail enable the herd to more easily follow a leader when danger threatens.

Antlers have become more and more significant as weapons, with the development of one or two brow tines that are important in defending themselves in rival fights. Three-pronged deer with long tails (hog deer, *Axis porcinus*; spotted deer, *Axis axis*; barasingha, *Cervus duvaucelii*; and sambar, *Cervus unicolor*) are found in the tropics; four-pronged deer in the warm, temperate zone (sika deer, *Cervus nippon*; and fallow deer, *Dama dama*), five- and six-pronged deer in the cold and alpine zones (izubr in Siberia, *Cervus elaphus xanthopygus*, elk in northern America, *C. e. canadensis*; and white-lipped deer, *Cervus albirostris*).

Cervinae (in particular the genus *Cervus*), compared with Odocoileinae, have lower reproductive rates, though they better utilize feeding resources and are more resistant to helminth diseases. The high degree of polymorphism is good for selection. The ability of interspecies (and even intergenera) hybridism is an important evolutionary feature of Cervinae. They are well adapted for intensive husbandry.

A sambar (*Cervus unicolor*) feeding. (Photo by Eric & David Hosking/Photo Researchers, Inc. Reproduced by permission.)

A red deer (*Cervus elaphus*) herd grazing. (Photo by Hans Reinhard. Bruce Coleman, Inc. Reproduced by permission.)

Physical characteristics

Deer vary in size from small to very large, standing high at the shoulders from 25–59 in (63–150 cm). Body is elongated, while legs are short or medium in length. A large bald spot partly covers the nostrils. Ears vary in size from short to long. Mane is typical for some specialized forms; primitive forms have no mane. The tail varies from extremely short and hidden by hairs, to very long. Antlers are rounded, with no less than three tines, and the brow tine is always developed. Some species wear antlers palmate at the top. Incisors are wide, with high crowns, while middle incisors have an elongated outer edge. Some species lack upper fangs. Molars always have high crowns.

Distribution

They occur in Europe and Asia, excluding areas to the north of 60°N, to northwest Africa. In North America, they occur from 60°N to Mexico. They are farmed in New Zealand, Australia, and the South African Republic.

Habitat

Deer inhabit predominantly ecotone habitats such as forest edges, and tend to move to more open habitats like alpine meadows, steppes, and farmed fields, rather than deep into forests. Some species inhabit tall shrub and grass thickets or marshlands.

Behavior

Most species live in small to moderately sized groups. In places of abundant forage, aggregations of tens and hundreds of deer gather. Bulls in these aggregations behave reasonably well, though they immediately establish hierarchy, which causes young males to stay at the periphery. In many species, bulls gather and defend harems of several does during the rut, trying to keep them on their breeding patches, and defend territories against rivals. Deer that inhabited thick groves, jungles, tall grass thickets in river or lake banks behave more like the Capriolinae: they live a solitary life, often are nocturnal, and are strictly linked to their home ranges.

Feeding ecology and diet

The type of feeding in these deer is mixed—they consume both concentrated forage (leaves, soft forest herbs, and fruits) and meadow grasses. This feature facilitates the farming and park breeding of them as they find the food supply (hay) to be acceptable.

The Javan rusa deer (*Cervus timorensis rusa*) is a native of Borneo and Java. (Photo by Andrew J. Martinez/Photo Researchers, Inc. Reproduced by permission.)

Red deer (*Cervus elaphus*) calling to attract mates. (Photo by Hans Reinhard. Bruce Coleman, Inc. Reproduced by permission.)

Reproductive biology

The reproductive rate of Cervinae is lower than in Odocoileinae. Does give birth to one fawn, and start mating at one and half years. Stags start mating at the age of five years, when they are in full physical maturity and are strong and heavy enough to dominate rivals. Most species are polygamous, with bulls gathering and defending a harem of several does.

Conservation status

For a long time, deer of these subgenera have been game animals, some since ancient times. As a result, many species have become threatened or scarce. Schomburgk's deer (*Cervus schomburgki*) may be Extinct. Pere David's deer (*Elaphurus davidianus*) is Critically Endangered; Calamian deer (*Axis calamianensis*), Bawean hog deer (*Axis kuhlii*), and Philippines spotted deer (*Cervus alfredi*) are Endangered. Many are con-

Red deer (*Cervus elaphus*) fleeing. (Photo by Hans Reinhard. Bruce Coleman, Inc. Reproduced by permission.)

sidered Vulnerable species, including barasingha (*Cervus duvaucelii*) and Eld's deer subspecies (*C. eldi eldi* from Manipur state in India and *C. e. siamensis* from Thailand, Vietnam, Kampuchea, Laos, and Hainan Island).

Significance to humans

Cervinae includes important game and farming deer due to venison (recently, it has become more highly esteemed due to its low-fat content) and skins (its raw material makes for the best suede). Many parts are valuable for Asian medicine (their prices have exploded since 1960), including velvet antlers, hard antlers, tail, bones, penis, heart, liver, sinews, placenta, and blood. The annual import of velvet antlers to Taiwan reaches 12 tons (11 tonnes) (including a minor percentage of reindeer velvet antlers); Korea also imports 12 tons (11 tonnes) annually and Thailand imports 1.1–3.3 tons (1–3 tonnes). The main antler manufacturer is China; its output is 44–55 tons (40–50 tonnes) per year (mostly from sika deer, and some from red deer). In Russia, which exports 13.2–15.4 tons (12–14 tonnes) per year, there are farms of red deer and sika deer. Antlers are also considered valuable trophies. Since medieval times, castles were adorned with deer heads with magnificent antlers.

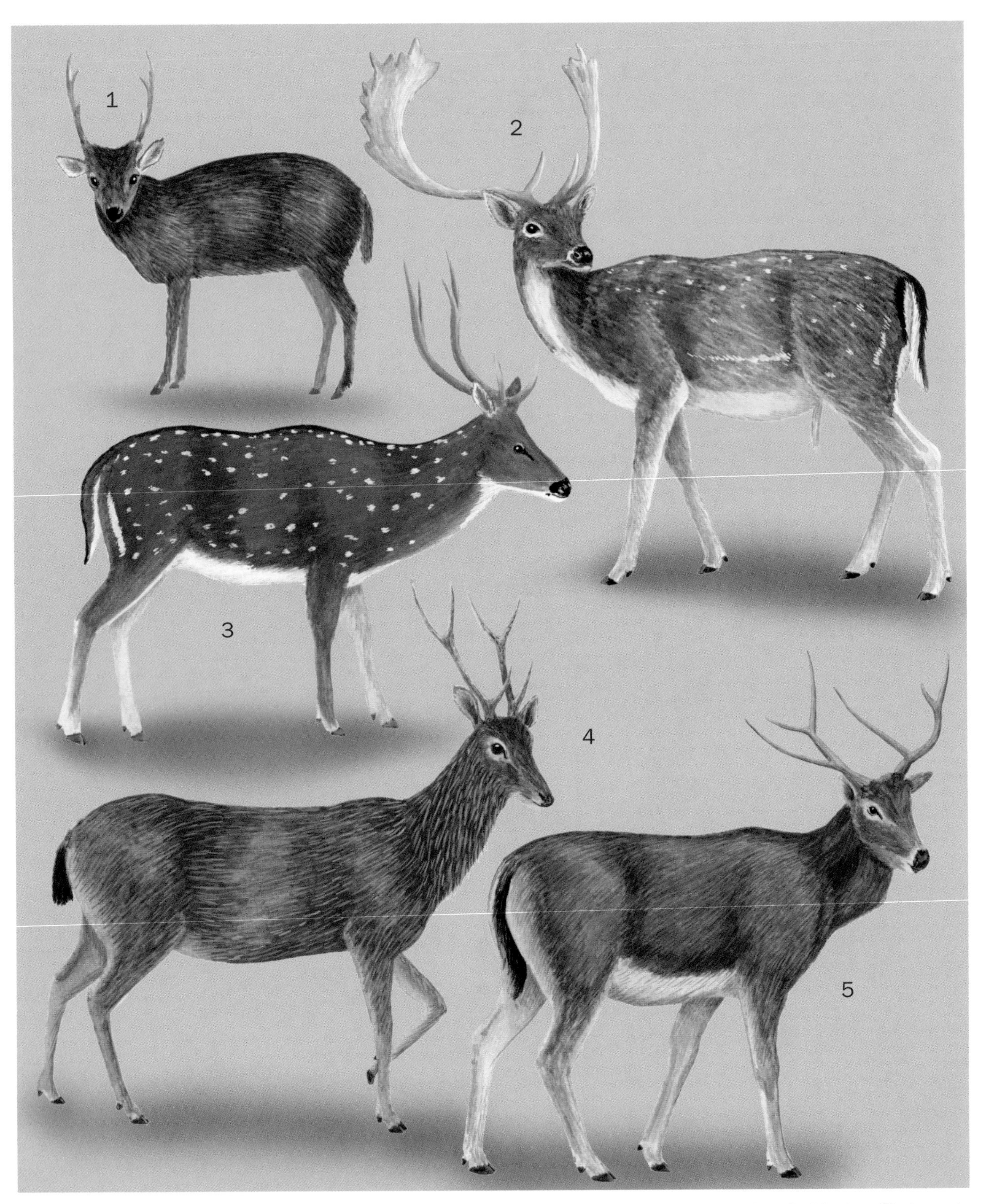

1. Hog deer (*Axis porcinus*); 2. Fallow deer (*Dama dama*); 3. Chital (*Axis axis*); 4. Sambar (*Cervus unicolor*); 5. Pere David's deer (*Elaphurus davidianus*). (Illustration by Dan Erickson)

1. Barasingha (*Cervus duvaucelii*); 2. Eld's deer (*Cervus eldi*); 3. Red deer (*Cervus elaphus*); 4. Sika deer (*Cervus nippon*); 5. White-lipped deer (*Cervus albirostris*). (Illustration by Dan Erickson)

Species accounts

Hog deer
Axis porcinus

TAXONOMY
Axis porcinus (Zimmermann, 1780), Bengal, India.

OTHER COMMON NAMES
French: Cerf-cochon; German: Schweinshirsch; Spanish: Ciervo porcino.

PHYSICAL CHARACTERISTICS
One of the smallest deer, looks strong despite its small size and short legs. Shoulder height: 24–30 in (60–75 cm); body length: 41–46 in (105–115 cm); tail length: 8 in (20 cm); weight: 79–110 lb (36–50 kg). Coat color is yellowish brown, darkened at the belly. Fawns develop white spots, while in adults spots are concealed. Bucks wear antlers 12 in (30 cm) long with three tines.

DISTRIBUTION
Originated in Pakistan, Indochina, and northern India; was introduced to Australia, New Zealand, the United States (Florida), and to the islands of Madagascar and Mauritius.

HABITAT
In Assam, they inhabit thick grasslands, with grasses taller than 4.9 ft (1.5 m); in shrub lands of Burma, they inhabit mangroves; in Manipur, they inhabit floating islets in Langmak Lake. Usually they keep to dark, poorly observed habitats. They use also clearings with moist pastures in big flood valleys.

BEHAVIOR
Called hog deer due to their special silhouette and behavior (they run away with the head bent downward) and to the peculiar manner in which they run through grass and shrub thickets, without jumping, like a wild hog; they also have a hog-like inclination to wallow in mud. They are crepuscular, live solitarily, so that only a doe and its fawn are paired. Stags are extremely aggressive in defending their home ranges, which they mark by scent gland secretion. In rut, pairs of stag and does are often together. In areas of abundant forage, 10–20 animals gather together. They can easily hide from hunters in grass thickets.

FEEDING ECOLOGY AND DIET
Feed on herbs, forbs, flowers, grasses, and fruits.

REPRODUCTIVE BIOLOGY
Polygamous. Bulls shed antlers in spring. Mating occurs mostly in July–October, and calving in January–April, though rut is possible year-round. Span between two births is about eight to 10 months, gestation period is 180–213 days, one fawn per birth is usual. Does become estrous a month after parturition and, if not mated, come into heat repeatedly. Life expectancy is 30 years.

CONSERVATION STATUS
Not listed by the IUCN, but the main threats are hunting and habitat loss.

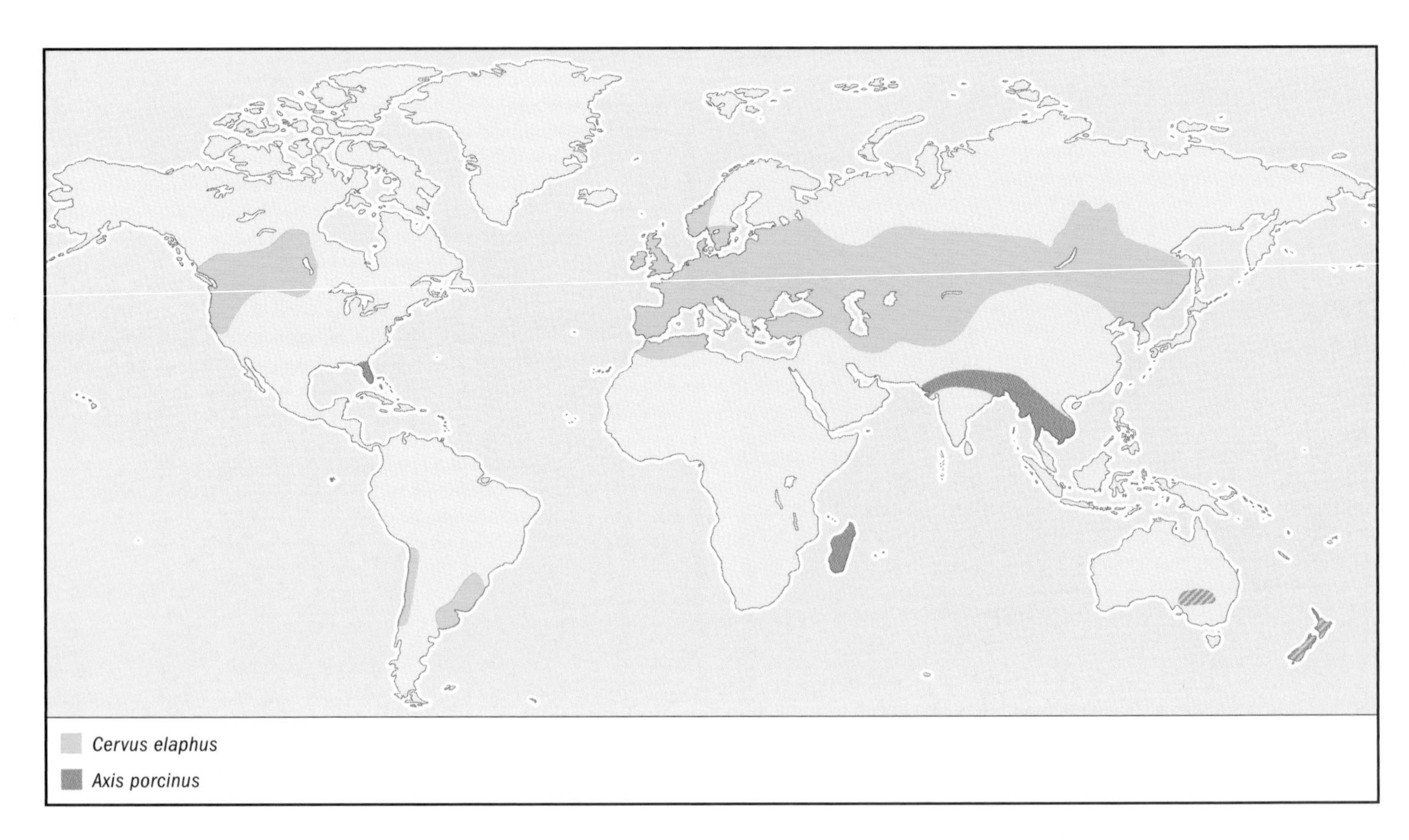

SIGNIFICANCE TO HUMANS
Game species hunted for meat and antler trophies. Some species become the predominate animals of game parks: red deer and sika deer. ◆

Chital
Axis axis

TAXONOMY
Axis axis (Erxleben, 1777), Bihar, India.

OTHER COMMON NAMES
English: Spotted deer; French: Chital; German: Axishirsch; Spanish: Chital.

PHYSICAL CHARACTERISTICS
A medium-sized animal. Shoulder height: 30–38 in (75–97 cm); body length: 43–55 in (110–140 cm); tail length: 8–12 in (20–30 cm); weight: 165–220 lb (75–100 kg). Stags are bigger than hinds. Regarded as the most beautiful of the Old World deer due to ornate coloration: brown with reddish or yellow tinge. Belly, inner parts of legs, and tail are white. Beautiful bright white spots decorate deer all yearlong. There are nearly no signs of sexual dimorphism in color, no mane. Antlers are lyre-like, widely spanned, of three points at each beam with a brow tine (found just above the base) and a forked main beam, inclined backside.

DISTRIBUTION
Originally inhabited thick forests of Hindustan and Sri Lanka, later was introduced by humans to many areas in Australia, New Zealand, the United States (Texas), South America, and Europe.

HABITAT
Usually live near water, in plains and hilly lands covered by monsoon deciduous forests, in thorny shrubs or bamboo forests, sometimes in dry pastures.

BEHAVIOR
Live in herds, up to hundreds individuals. Herds comprise animals of any sex and age, but old stags live solitary life. In large grazing herds, dominant stags, easily distinguished by size, gorgeous antlers, black neck, bright contrast color of muzzle, occupy a central position; they are surrounded by does, yearlings, and fawns. Peripheral ring is made up of weak stags, young or antler-less or with velvet antlers. Stags are not aggressive, they do not mark and defend their ranges, and rate rivals by evaluating body size, dimensions of antlers, behavior, and thus never launch useless combats. Only equal rivals fight. Chital are extremely settled, they stay in their ranges, even during deadly droughts, which are repeated in India and Sri Lanka every seven to 10 years. Then they die by hundreds and thousands. Chital active at dawn and sunset, they use hot middle part of day for rest. Deer approach human settlements, use arable lands.

FEEDING ECOLOGY AND DIET
Chital feed mainly on herbs and leaves.

REPRODUCTIVE BIOLOGY
Polygynous mating takes place year-round, though mainly during winter. In the same herd, there are animals of various reproductive status: stags with cleaned, hardened antlers ready to mate; does in heat; and animals not in rutting state. Gestation period is seven to eight months, a doe gives birth to one fawn, rarely two. Fawns as old as one and half months can eat herbs, but are usually nursed to an age of six months. Most hinds start reproduction at one and a half years. Stags, due to strong competition among males, participate in breeding from four to five

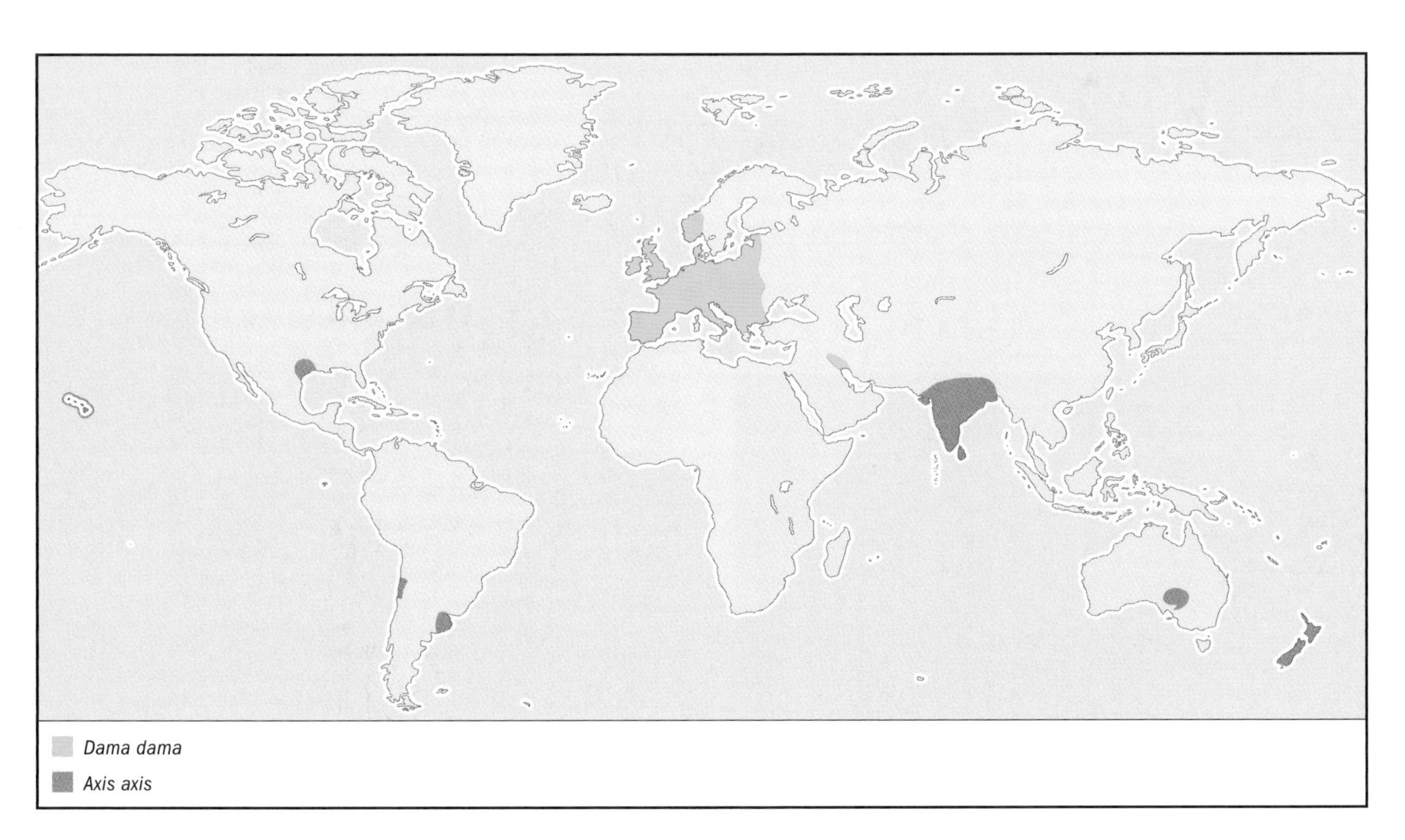

years when they reach full physical maturity. Life expectation in the wild is nine to 13 years; in parks, as old as 22 years.

CONSERVATION STATUS
Not threatened.

SIGNIFICANCE TO HUMANS
Chital is a popular game deer; once released from human pursuit and predators, its numbers soon exploded to make it a pest for cultivated lands. ◆

Fallow deer
Dama dama

TAXONOMY
Dama dama (Linnaeus, 1758), Sweden (introduced).

OTHER COMMON NAMES
English: Persian fallow deer; French: Daim; German: Damhirsch; Spanish: Gamo.

PHYSICAL CHARACTERISTICS
Shoulder height: 33–39 in (85–100 cm); body length: 51–63 in (130–160 cm); tail length: 6–7 in (16–19 cm); weight: males 176–186 lb (80–125 kg), females 132–187 lb (60–85 kg).

Coat color is variable; some are reddish, some dark brown, some nearly white. Most of animals develop whitish spots, bright in summer and poorly distinguished in winter. Spots often merge to white stripes, while in some animals a black line goes along the back to tail. Color variation might be caused by domestication; from the Roman Empire times, the deer were bred in game parks throughout Europe. Males sport large palmate antlers. At the age of one, they get spiky antlers; by age three or four, bucks grow three-pointed antlers and the third tine transforms into a wide palm with multiple small tines at the edge. Total length of antlers is up to 15 in (39 cm).

DISTRIBUTION
Primarily inhabited the Mediterranean, then were brought to Europe by Romans where they adapted to wilderness and became a preferred animal to breed in parks. Later introduced to many European countries, to New Zealand, and North and South America.

HABITAT
Prefer open plains and hilly grasslands for grazing; use shrub lands and mixed and deciduous forests for shelter, shade, calving.

BEHAVIOR
Yearlong home range of bulls rarely exceeds 740 acres (300 ha); that of does with calves is about 250 acres (100 ha). In North America, female herds use range to 15 mi^2 (40 km^2) in winter and spring, while in summer their home range decreases to 2–3 mi^2 (5–8 km^2). Insignificant shifts of home range sizes relate to abundance or availability of forage. During rut, bulls are strictly territorial, marking off a small patch, defending it against intrusion of rivals, keeping a harem of does and their offspring, and following each doe in heat until mating. Sounds made by stags in rut resemble snoring or hoarse coughs, and thus are quite distinct from those made by red deer. Stags often fight to establish hierarchy. As soon as rut is over, bulls cease defending activity and form bachelor groups. Segregated does with fawns make own groups.

In wilderness, they are very vigilant; it is difficult to approach them due to their excellent vision, hearing, and olfaction. Fleeing deer rise a tail, displaying a bright white patch bordered by black hairs.

FEEDING ECOLOGY AND DIET
Feed on herbs, forbs, and less on leaves and fruits.

REPRODUCTIVE BIOLOGY
Polygynous. Stags shed antlers in April–May, and regrow them in August. Rut occurs from mid September–November. Gestation period lasts seven and half months, does give birth to one fawn, rarely to twins. First 15–20 days a fawn hide, afterward follows doe in a herd. Fawn is nursed until it is six to nine months old, weaning precedes a new birth. In their second autumn, young doe reach sexual maturity, and by age two can participate in breeding. Males breed at age six to seven years, after reaching full physical maturity, though they are sexually mature at 14 months. Life expectation in captivity is up to 25 years, while usual lifespan in wild is 10–15 years.

CONSERVATION STATUS
Not threatened now. At the same time, subspecies *Dama dama mesopotamica* is considered to be a very rare, Endangered deer.

SIGNIFICANCE TO HUMANS
A significant game species, its meat and antlers considered valuable trophies. The deer most adapted to breeding in game parks. Europe's annual game harvest is up to 30,000 fallow deer. ◆

Pere David's deer
Elaphurus davidianus

TAXONOMY
Elaphurus davidianus Milne-Edwards, 1866, Chihli, China.

OTHER COMMON NAMES
English: Milu; French: Cerf du Pere David; German: Davidshirsche; Spanish: Ciervo del padre David.

PHYSICAL CHARACTERISTICS
Rather large animal. Shoulder height: 47 in (120 cm); body length: 6–6.3 ft (180–190 cm); tail length: 20 in (50 cm); weight: males 500 lb (220 kg); females 300 lb (135 kg).

They look odd, like a combination of deer, camel, cow, and donkey: very long and slender head, small ears, long legs with long and narrow (cow-like) hooves adapted to soft, boggy ground, very long tail ended by a black tuft. Unlike all other deer, the brow's first tine of antlers is forked into backward points. Also unlike other deer, Pere David's deer can grow two pair of antlers during a year, shedding summer antlers in November and growing a new pair in January (to shed it in some weeks). Bulls have a mane under the neck. Coat in summer is reddish rust, and retains long wavy axial hairs yearlong. In winter, the colors change to gray with bright creamy underneath, and dorsal dark stripe.

DISTRIBUTION
One thousand years ago, they inhabited northeastern and east central China. In 1939, the last wild deer was shot near the Yellow Sea.

HABITAT
It is believed that original habitats were marshlands.

BEHAVIOR
Observations in game parks show that the deer like water, readily stay in water for many hours, are good swimmers. Live in bull clans and doe herds. Bulls fight during rut, using antlers, boxing by forelegs, and kicking with hind legs.

FEEDING ECOLOGY AND DIET
Feed on herbs and probably on aquatic plants.

REPRODUCTIVE BIOLOGY
Polygynous. Gestation period is nine months, does give birth to one, rarely, two fawns, that wean in 10–11 months. Sexual maturity appears at 14 months. Life expectancy is to 23 years.

CONSERVATION STATUS
Critically Endangered. Pere David's deer conservation is a classic story of joint efforts of scientists, enthusiastic conservationists, and game managers. To prevent the extinction of the species, owners of European zoos gathered 18 animals, good reproducers, in the Woburn Abbey Park in 1914; there, a population of 90 Pere David's deer was maintained. In spite of foraging problems caused by World Wars I and II, by 1946 the population numbers were 300. Today, hundreds of the deer are in breeding centers throughout the world. In 1986, deer were brought to China to Park Nan Hai-tsu, the center of their distribution a century ago. Reintroduction to the wilderness in a forest reserve near the Yellow Sea is expected soon.

SIGNIFICANCE TO HUMANS
Cultivated as park animals. ◆

Sambar
Cervus unicolor

TAXONOMY
Cervus unicolor Kerr, 1792, Sri Lanka.

OTHER COMMON NAMES
French: Sambar; German: Indischer Pferdehirsch; Spanish: Sambar.

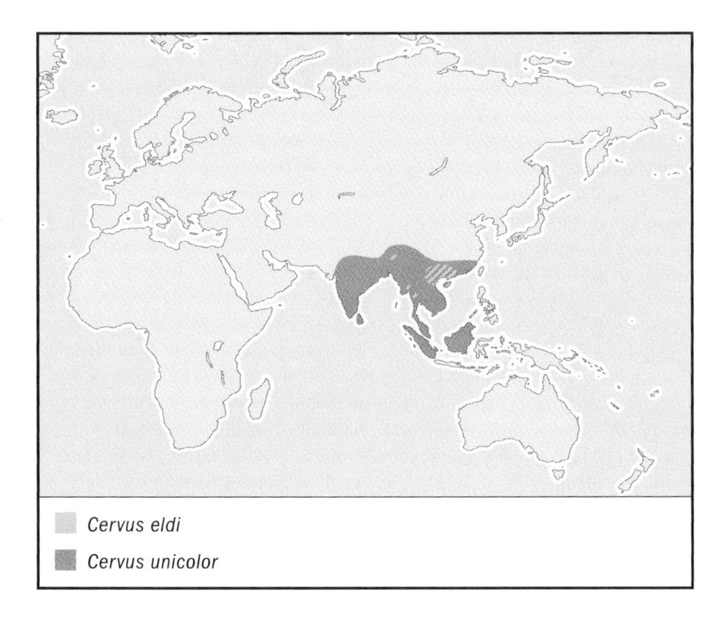

PHYSICAL CHARACTERISTICS
The largest deer in Southeast Asia. Shoulder height: 35–59 in (90–150 cm); body length: 6–9 ft (170–270 cm); tail length: 9–14 in (22–35 cm); weight: 220–300 lb (100–350 kg). Its colors are primitive, monotonous over entire body; sexual dimorphism is well displayed in some species. Antlers are simple, three pointed, later complicated by dividing.

DISTRIBUTION
India, Indochina, and Malaysia.

HABITAT
Inhabitant of moist tropical forests, visit cut clearings to feed on herbs.

BEHAVIOR
Polygynous. Live solitarily or in small family groups (does, yearlings, fawns). Stags keep strictly to their home ranges, which they mark by secretion of pre-antler glands on branches of trees and bushes. Both stags and does are aggressive; stags fight with antlers, does use teeth or start boxing matches with their rivals. They establish hierarchy only by fierce fights. In gatherings of 30–40 individuals, there is the dominating stag in the center, does around, subdominants, and young bachelors at periphery.

FEEDING ECOLOGY AND DIET
Feed on leaves, bark, fruits in woods, on grass in clearings, consuming a great variety of vegetation. Consume numbers of toxic plants, without adverse consequences due to special micro floras in intestines, big size of rumen, and many species of other plants consumed at the same time.

REPRODUCTIVE BIOLOGY
Stags grow antlers in January–April, wear hardened antlers from May–November. Rut occurs mostly in November–December. Gestation lasts six months, coinciding with rut time. Antler shedding occurs in December. Calving peak is April–May, though calving can take place nearly all year long.

CONSERVATION STATUS
Not threatened.

SIGNIFICANCE TO HUMANS
The most important game species in India and other countries. Sambar is bred in game parks of Australia (more than 5,000 deer) and elsewhere. ◆

Barasingha
Cervus duvaucelii

TAXONOMY
Cervus duvaucelii Cuvier, 1823, northern India.

OTHER COMMON NAMES
English: Swamp deer; French: Cerf de Duvaucel; German: Barasingha; Spanish: Ciervo de Duvaucel.

PHYSICAL CHARACTERISTICS
Shoulder height: 47–49 in (119–124 cm); body length: 71 in (180 cm); tail length: 5–8 in (12–20 cm); weight: 375–610 lb (170–280 kg). Its summer coat is short and pale creamy yellow. In winter, its coat is woolly and brown. Stags are darker with red shading.

DISTRIBUTION
Three subspecies are distinguished: *C. d. duvaucelii*, which inhabits Nepal and northern India; *C. d. branderi* of central India; and *C. d. ranjitsinghi* of eastern India.

HABITAT
Inhabit marshy flood plains, and move in winter to neighboring open grasslands, where deer survive on rough grasses and shrubs. Today, they also occupy broadleaved forests, both dry and moist, with under story of grasses, as well as evergreen thickets and mangroves.

BEHAVIOR
Live in small variable herds (four to 12 animals). The largest aggregations observed comprised 30 deer, though gathering only lasted a day. Breeding herds are mixed (bulls, does, young) and, at the peak of rut, are up to 50 animals. Bachelor groups were observed. Deer easily congregate or disperse, as there are no strong social bonds between animals. Only pairs of doe and a fawn younger than one year are linked.

FEEDING ECOLOGY AND DIET
Feed on grasses, less on aquatic plants in swamps.

REPRODUCTIVE BIOLOGY
Polygynous. Rut is during December–January. Females come into heat once a year, gestation period lasts 240–250 days. The breeding hierarchy is established by fighting among males, dominating males mating with females in heat.

CONSERVATION STATUS
C. d. branderi is Endangered, *C. d. ranjitsinghi* is Critically Endangered, and *C. d. duvaucelii* is Vulnerable. Decline in numbers is rapid due to poaching, drainage of wetlands, shooting to defend crops, and diseases contracted from livestock. The population range is fragmented in the limits of national parks in Nepal and India. However, in natural reserves, a steady revival and even increase in numbers occurs and numbers of *C. d. duvaucelii* currently expanded 3,500.

SIGNIFICANCE TO HUMANS
Game species. Low population numbers determine cultural significance today. Is preserved in parks. ◆

Eld's deer
Cervus eldii

TAXONOMY
Cervus eldi McClelland, 1842, Assam, India.

OTHER COMMON NAMES
English: Brow-antlered deer, thamin; French: Cerf D'Eld; German: Leierhirsch; Spanish: Ciervo de Elde.

PHYSICAL CHARACTERISTICS
Medium sized. Shoulder height: 48–51 in (120–130 cm); body length: 59–67 in (150–170 cm); tail length: 9–10 in (22–25 cm); weight: 210–330 lb (95–150 kg). Color in dorsal part is reddish brown in winter, lighter in summer; hinds lighter colored than stags. Whitish spots at back are noticeable. In adult males, thick and long hair forms a mane on the neck. Unlike other deer, the brow tine of antler forms a continuous curve with the beam (sometimes longer than 3.2 ft [1 m]) and resembles a bow from the side. Often, beam flattens at the end and forked to six to 12 tines. Hooves are adapted to boggy grounds.

DISTRIBUTION
India (Manipur state), Myanmar, Laos, and Vietnam.

HABITAT
Inhabit marshy lowlands with thick reed and grass. On Lake Longtak, wild population stays on a floating raft, formed by marshy soil, turf, and dead vegetation and overgrowth by reeds. During dry season, when the raft sinks, deer find shelter at lakeshores. Also use hilly islets in the lake for feeding and calving.

BEHAVIOR
Live in small groups of four to seven animals; gather in herds to 50 individuals. Adult stags live solitarily; join herd during rut. Wary, active at dusk, stay during heat of day resting at the edges of forests.

FEEDING ECOLOGY AND DIET
Sacharum latifoluum comprise main forage. They feed mostly on grass, while tree parts and fruits are subordinate food.

REPRODUCTIVE BIOLOGY
Polygynous. Shed antlers in June. New antlers are fully developed in three to four months, and are cleaned of velvet in November. Does reach sexual maturity at one or two years, are in heat February–September; estrus repeats each 17 days and lasts two days. Rut occurs in February–May, stags fight fiercely. Gestation period is 242 days, usually it is one young per birth. Weaning occurs in seven months.

CONSERVATION STATUS
C. e. eldii distributed in Manipur, India, in one national park is Critically Endangered, and *C. e. siamensis* distributed in Thailand and Kampuchea, total numbers estimated as 100–200 is Data Deficient. Population is threatened mainly by loss of habitat, and fawns are killed by poachers and wild dogs. Breeding in zoos followed by reintroduction to wilderness would be favorable for the species. *C. e. thamin* inhabiting Myanmar and Thailand are rather numerous, but listed as Lower Risk/Near Threatened. Poaching and loss of habitat due to farming and breeding of livestock cause its numbers to decrease.

SIGNIFICANCE TO HUMANS
Rare species. Many efforts by governments and social organizations are being made for its preservation. ◆

Sika deer

Cervus nippon

TAXONOMY

Cervus nippon Temminck, 1838, Japan.

OTHER COMMON NAMES

French: Sika; German: Sikahirsch; Spanish: Sika.

PHYSICAL CHARACTERISTICS

Medium sized. Shoulder height: 25–48 in (65–120 cm); male body length 66–74 in (168–187 cm); female: 59–69 in (149–176 cm); tail length: 6–7 in (17–19 cm); male weight: 230–310 lb (104–139 kg); female: 132–205 lb (60–93 kg). Males sport antlers to five tines, the second brow tine lacking or represented by a small prominence. Antlers forked in upper part or develop three-tined bush. Coat color in winter is brown-olive or reddish brown. Adult deer develop whitish spots on dorsal part of body and shoulders. Rump patch in winter is rather small. Tail adorned by wide black stripe above and is white underside. Rump hair rises to enlarge the mirror surface. In summer, coat is reddish to whitish below, with distinctive white spots on dorsal part and in stripes on sides. Spots better developed in young than in adult animals.

DISTRIBUTION

East China, Taiwan, Korean Peninsula, southern part of the Russian Far East, and Japan; introduced to New Zealand, European part of Russia, and other countries.

HABITAT

Prefer deciduous forests at seashores and surrounding mountain slopes, but escape coniferous and mixed coniferous-deciduous forests, marshy flood plains. Snow cover deeper than 15.7–19.6 in (40–50 cm) is limiting, and areas with snow cover lasting no more than 140 days are preferable.

BEHAVIOR

Live in small mixed herds, four to 20 individuals, but in spring and summer, females with fawns live in separate groups. During rut, a dominating male keeps some females on the home range. The dominating male banishes young males from a herd during rut, but they return back when rut is over. Sedentary; a summer home range of an individual is about 0.4–0.8 mi^2 (1–2 km^2), and groups ranging 1.5–1.9 mi^2 (4–5 km^2). Winter home range is more restricted, to 49–74 ac (20–30 ha). A male arranges six to seven rutting points: by trampling down vegetation, fraying trees with antlers, urinating into a pit, wallowing in mud. At rutting points, males roar.

Excellent runners, can jump to 19.6–26.2 ft (6–8 m) and cross a sea strait as wide as 6 mi (10 km). Animals are extremely vigilant.

FEEDING ECOLOGY AND DIET

Feed in winter on twigs of trees and shrubs, on bark, buds, leaves, and in some areas on acorns. In winter and autumn, herbs and fungi comprise a main part of the diet. Deer nip off very small pieces of each forage.

REPRODUCTIVE BIOLOGY

Polygynous. Shed antlers in March–April, first adult, then young males; growth of antlers starts soon after. Rut starts in October. Gestation period is 233–241 days. The first fawns appear in April, calving lasts till the end of May; there is usually one young per birth, rarely two. Both males and females reach sexual maturity early, but take part in breeding later: males at three to four years, females at two and half years.

CONSERVATION STATUS

Eleven subspecies of sika deer are listed by the IUCN: five as Critically Endangered, two as Endangered, and four as Data Deficient.

SIGNIFICANCE TO HUMANS

Important as game species. As trophies of sport, game meat, antlers, and skin are used. Velvet antlers are valued in Asian medicine. Velvet antler crop farming is significant in China, Thailand, and Korea. At the beginning of the 1980s, there were about 195,000 animals on Chinese farms, mostly *Cervus nippon hortulorum*. In Korea, mainly *Cervus nippon taiouanus* is farmed, to 80,000 by the end of the 1980s. Velvet antlers of a single male bring about $600 of pure profit. Also, meat, sinews, and tails from farms are used by local inhabitants. ◆

Red deer

Cervus elaphus

TAXONOMY

Cervus elaphus Linnaeus, 1758, Sweden.

OTHER COMMON NAMES

French: Cerf rouge; German: Edelhirsch; Spanish: Ciervo rojo.

PHYSICAL CHARACTERISTICS

Large sized. European red deer (*C. e. elaphus*) have shoulder height: 47–59 in (120–150 cm); male body length: 69–91 in (175–230 cm); female: 63–83 in (160–210 cm); tail length: 5–7 in (12–19 cm); male weight: 350–530 lb (160–240 kg); female: 264–374 lb (120–170 kg). North American wapiti (*C.e. canadensis*) have shoulder height: 47–59 in (120–150 cm); male body length: 83–110 in (210–280 cm); female: 70–105 in (180–270 cm); tail length 5–7 in (12–19 cm); male weight: 880 lb (400 kg); female: 570 lb (260 kg). Antlers develop at least five tines, with the second brow tine developed in most subspecies. Yearling males carry a set of long, single-point antlers, which are replaced in the subsequent year with a set of uneven, branched antlers with three or four points on each side. In their fourth year, bulls are fully matured and usually bear antlers with five or more points. Coat in adults is mostly monotonous or darkened at head, neck, lower part of body, and legs. Winter color is grayish brown; summer color is red. Rump patch is obvious, fringy.

DISTRIBUTION

Red deer in historical time inhabited temperate forests as well as plains in Europe, Asia, and North America. They were numerous in western Siberia, Kazakhstan, and the Urals. Extinguished by humans, deer survived in separated areas in mountain forests where there were more chances to survive in bushy thickets and high herbs in river valleys. Today, red deer is rapidly being restored to its distribution area, to occupy new grounds, even in very harsh climates (as in Yakutia).

HABITAT

An ecotone species whose habitat use is concentrated along relatively open areas that provide forage and densely forested areas that provide cover. About 95% of elk use of forage occurs within 650 ft (200 m) of a forage/cover edge. Patches of cover need to be at least 325 ft (100 m) and no more than 1,650 ft (500 m) wide to provide optimum elk habitat. Also inhabit shores of rivulets, open plains, hills, marshlands, and reed thickets in marshy river valleys, as well as mountain terraces and subalpine meadows. In middle Asia, they stay at altitudes to 7,500 ft (2,300 m) and use alpine meadows during many months. In vast monotonous coniferous (larch, less fir and spruce) forests of Siberia, they switch to more open fire-sites and cut clearings, where at the first year plenty of herbs grow and, in some years, overgrowth of broadleaf trees and regrowing of coniferous trees beneath occur. River valleys are especially important as they feed in willow shrubs and poplar groves.

In the most of their range, they meet snow problems. Animals move easily if depth of snow cover reaches 7.8–11.8 in (20–30 cm). Where it exceeds 19.6–23.6 in (50–60 cm), deer gather in limited areas with abundant forage (twigs, sprouts of trees, and shrubs) and move less. Snow depths of 27.5–29.5 in (70–75 cm) is crucial for does and calves, while strong stags survive in places with 39.3 in (100 cm) of snow.

BEHAVIOR

Live by singles (males, female with fawn), family groups (female, fawn, and yearling), and gatherings. During rut, there are either harems or bachelor groups; stags arrange, mark, and defend against invaders. Antler size is a key factor determining a bull's status and breeding privileges. When bulls are relatively equal in size, antlers are used in a pushing match. However, to avoid injuries and even death, physical combat is usually avoided in favor of visual displays to determine dominance.

Adult females often became leaders of family or mixed groups, while the biggest deer governs group of males. Leaders determine direction of movement, start of migration, raid to saltlicks, rhythm of grazing. If a predator attacks, large mixed groups often disperse in all directions.

During year, size of home range is 8–12 mi^2 (20–30 km^2). Animal uses trails, feeding and watering points, steep rock patches to escape from predators. In winter, especially in heavy snow, home range restricted. Observations show that a deer rests (and consequently feeds) five to eight times a day. It chooses rest areas of good observation (slope, forest edge).

In mountains, they make seasonal migrations: ascending in spring to upper forest border and then to alpine meadows. In autumn, animals move back to coniferous forests where snow cover is less deep.

Red deer are good walkers, trotters, and runners. Good swimmers, they can cross wide rivers and swim into an open sea for a significant distances.

Common prey for predators including humans, they are very vigilant. Animals look around 40 times per hour. They often defend themselves against predators by keeping to a rock prominence and using antlers or fore legs. Another way to escape is swimming across a wide and turbulent river or staying in water for a long time.

FEEDING ECOLOGY AND DIET

Feeding niche is wider than in other Cervidae. Main forage comprised of twigs, stems, and leaves of broadleaved trees and shrubs, needles and branches of larch and fir, herbs and sedges, forbs, horsetails, lichens, fruits, and fungi. Some preferable plants are willows, poplar, mountain ash, oak, cowberries, and blackberries.

In North America, they use western hemlock, fir, western red cedar, Oregon grape, Pacific ninebark, red elderberry, cowberries, willows, salal, ferns, sedges, bunchberry, salmonberry, twinflower, skunk cabbage, and wall lettuce.

In many areas, they visit saltlicks and use rivulets in summer and snow in winter for watering.

REPRODUCTIVE BIOLOGY

Polygynous. In July and August, stags begin to clean velvet from antlers. Rut starts at the end of August and the beginning of September and lasts till the end of September, October, or November. Harem of one stag includes one to five, rarely to 18, does. Harem sizes relate to population density. Elsewhere, bulls gather to leks, up to 50 individuals; they roar and display aggressive behavior.

After a gestation period of 210–255 days, does give birth to one, rarely two, fawns; at three or four weeks of life, fawns follow mothers. When a calf is three months old, mother grazes elsewhere and returns to it morning and evening to nurse. Lactation lasts till the next rut. Both males and females are sexual matured at one and a half years, though females mate at two and a half years and males at 3–5 years. Young bulls are not allowed to participate in rut by dominating rivals.

They keep a high reproductive rate, annually increasing in numbers to 30%. In a peak of reproduction, they extinguish forage resources, which leads to a population crash and brings trouble to other ungulates. Growth of population impeded by high mortality (some winters to 50%) due to diseases, affect of heavy and deep snow, predators, and poaching. The mortality among calves is the highest.

CONSERVATION STATUS

As game subjects, they can survive in densely populated countries due to skillful management and conservation measures. In Europe, they are considered endangered, as are red deer in Corsica and Sardinia. In North Africa, *C. e. barbarus* is Lower Risk/Near Threatened. The red deer inhabiting Central Asia and adjacent Afghanistan, the *C. e. hanglu*, are Vulnerable. In Sinkiang, the *C. e. yarkandensis* is Endangered and believed to be extinct. Four other subspecies are listed as Data Deficient.

SIGNIFICANCE TO HUMANS

Important sport game, and in New Zealand of commercial importance also. Since ancient times, value and appreciation of venison has increased, as well as for skins used for manufacturing the best suede. Since the end of 1960s, velvet antlers, tails, pizzles, and sinews are sold to Asia for traditional medicines. Red deer farm production developed so rapidly that they might be considered as a domestic animal. Today, there are more 65,000 deer at Chinese farms (*Cervus elaphus xanthopygus*, *C. e. songarius*). In Russia, mostly *C. e. sibiricus* are farmed. ◆

White-lipped deer

Cervus albirostris

TAXONOMY

Cervus albirostris Przewalski, 1883, Kansu, China.

OTHER COMMON NAMES

English: Thorold's deer; French: Cerf de Thorold; German: Weisslippenhirsch.

PHYSICAL CHARACTERISTICS

Large: shoulder height: 4–4.3 ft (120–130 cm); body length: 6.3–6.6 ft (190–200 cm); tail length: 4–5 in (10–12 cm); weight: 500 lb (230 kg). The deer muzzle, as well upper and lower lips, gorge, and orbit surroundings are pure white. General brown color of coat is ornate in summer with whitish spots, which nearly disappear in autumn and change to grayish brown with creamy under parts. Hairs at the rear back grow forward, giving an appearance of a saddle. Hooves are wide, tall, hard, adapted to mobility. Antlers span is to 4.2 ft (1.3 m), and carry five to six flattened tines, each antler reaching a weight to 15 lb (7 kg).

DISTRIBUTION
Tibetan Plateau in China.

HABITAT
Inhabits forests interchanged with clearings, shrub lands, keeps to alpine meadows at altitudes more than 11,500 ft (3,500 m), and to the border of vegetation that, in Tibet, is at 16,400 ft (5,000 m).

BEHAVIOR
Sedentary, undertake small vertical migrations. Does and yearlings live in groups to 40 individuals, while stags keep solitary or in small group (to eight animals). Social bonds within groups are strong. At rut, mixed herds unite stags, does, and calves in congregations to 200–300 deer (average number 50). Each mixed herd comprises one to eight stags, very aggressive towards each other. Deer easily climb and run across steep mountain slopes due to specific features of their hooves. Often prey of hunting, they are very wary and cannot be easily observed in the wild.

FEEDING ECOLOGY AND DIET
Grasses.

REPRODUCTIVE BIOLOGY
Polygynous. Rut occurs in September–November, gestation period is 7.5–8.3 months (according to other data, 270 days), calving takes place from the end of May to beginning of July, rarely to August. Usually it is one young per birth. Does start reproduction at age three years, stags first mate at five years of age. Life expectancy is to 19 years (in captivity).

CONSERVATION STATUS
Vulnerable. Previously distributed over eastern Tibet, but currently it occurs from the vicinity of Lhasa eastward into western Sichuan and in the eastern two-thirds of Qinghai and into Gansu, where total numbers are estimated at 50,000–100,000. Low density and strong fragmentation of the population reflect both rugged mountains and human impact such as poaching and competition with livestock for pastures.

SIGNIFICANCE TO HUMANS
Hunting for venison, as well as a growing demand for antlers and other parts of body in Asian medicine. ◆

Common name / Scientific name/ Other common names	Physical characteristics	Habitat and behavior	Distribution	Diet	Conservation status
Timor deer *Cervus (Rusa) timorensis* English: Sunda sambar; French: Sambar de Timor; German: Timor-Sambar; Spanish: Sambar de Sunda	Grayish brown, coat is rough and coarse in appearance. Ears are broad and slightly rounded. Short legs give stubby appearance. Males have three-tined antlers. Head and body length 55.9–72.8 in (142–185 cm), tail length 7.9 in (20 cm), weight 110.2–253.5 lb (50–115 kg).	Deciduous forests, plantations, and grasslands on the southern Indonesian islands. Family groups consist of previously single sex groups of 25–1,500 individuals. Primarily nocturnal. Breeding occurs throughout the year, females give birth to one to two offspring.	Java, Bali, Lesser Sunda Islands, Molucca Islands, Sulawesi and Timor (Indonesia). Australia, New Zealand, Caledonia and small islands in Indonesia and off the coast of Australia.	Primarily grasses, also leaves.	Not threatened
Philippine brown deer *Cervus (Rusa) mariannus* English: Philippine sambar; German: Mähnenhirsch; Spanish: Sambar filipino	Uniformly dark brown, being darker above and paler below and on the legs. Underside of the tail is white. Head and body length 39.4–59.4 in (100–151 cm), tail length 3.1–4.7 in (8–12 cm), shoulder height 121.3–27.6 in (55–70 cm), and weight 88.2–132.3 lb (40–60 kg).	Several habitats in the Philippines, including both lowlands and densely vegetated mountain slopes up to elevations of 9,510 ft (2,900 m). Females give birth to one offspring.	Philippines, on Luzon, Mindoro, Mindanao, and Basilan Islands. Introduced to Mariana, Caroline, and Bonin Islands (western Pacific Ocean).	Leaves, buds, grass, berries, and fallen fruit.	Data Deficient
Schomburgk's deer *Cervus schomburgki* Spanish: Ciervo de Schomburk	Upperparts are uniform brown, underparts are lighter. Ventral surface of tail is white, legs and crown have reddish tinge. Mane extends down front of the foreleg. Five-tine antlers. Shoulder height 39.4 in (100 cm), tail length 4.1 in (10.3 cm).	Swampy plains with long grass, cane, and shrubs. Avoids densely vegetated areas. Usually spent days resting in the shade and grazed during evening and night. Family groups generally consisted of a buck, a few does, and a few fawns.	Thailand.	Leaves, buds, grass, berries, and fallen fruit.	Extinct
Visayan spotted deer *Cervus alfredi* English: Philippine spotted deer; German: Prinz-Alfred-Hirsch	Dark brown, with a dark dorsal band bordered with faint dull ochre spots. Underparts and underside of tail is buff. Shoulder height 25.2 in (64 cm).	Habitat is tropical forests. Gestation period of 8 months. Offspring born from May to June. Very little known about behavior.	Philippines, on Masbate, Panay, and Negros Islands; formerly also Seguinjor, Guimares, Cebu, Bohol and perhaps other islands.	Leaves, buds, grass, berries, and fallen fruit.	Endangered

Resources

Books

Baskin, Leonid, and Kjell Danell. *Ecology of Ungulates. A Handbook of Species in Eastern Europe, Northern and Central Asia.* Heidelberg: Springer Verlag, 2003.

Bedi, Ramesh. *Wildlife of India.* New Delhi: Bridgebasi Printers Private Ltd., 1984.

Flerov, Konstantin K. *Musk Deer and Deer.* Moscow: Izdatelstvo Akademii Nauk SSSR, 1952.

Geist, Valerius. *Elk Country.* Minoqua, WI: NorthWord Press, 1991.

———. *Deer of the World: Their Evolution, Behavior, and Ecology.* Mechanicsburg, PA: Stackpole Books, 1998.

Hudson, Robert J., Karl R. Drew, and Leonid M. Baskin. *Wildlife Production Systems. Economic Utilization of Wild Ungulates.* Cambridge: Cambridge University Press, 1989.

Harrington, R. "Evolution and Distribution of the Cervidae." In *Biology of Deer Production*, edited by P. F. Fennessy and K. R. Drew. Wellington: The Royal Society of New Zealand, 1985.

Hofmann, R. R. "Evolution Digestive Physiology of the Deer—Their Morphophysiological Specialization and Adaptation." In *Biology of Deer Production*, edited by P. F. Fennessy and K. R. Drew. Wellington: The Royal Society of New Zealand, 1985.

Schaller, George B. *Wildlife of the Tibetan Steppe.* Chicago: University of Chicago Press, 1998.

Sheng, Helin, and Lu Houji. *The Mammalian of China.* Beijing, China Forestry Publishing, 1999.

Leonid Baskin, PhD

△

Chinese water deer

(Hydropotinae)

Class Mammalia
Order Artiodactyla
Family Cervidae

Thumbnail description
Small, dainty, antlerless deer with large ears and tusks; cold adapted and territorial; a saltor and hider in dense vegetation

Size
Height 19.6–21.6 in (50–55 cm) at the shoulder; weight 33 lb (15 kg)

Number of genera, species
1 genus; 1 species

Habitat
Riparian vegeation such as swamps, reedbeds, and grasslands

Conservation status
Lower Risk/Near Threatened

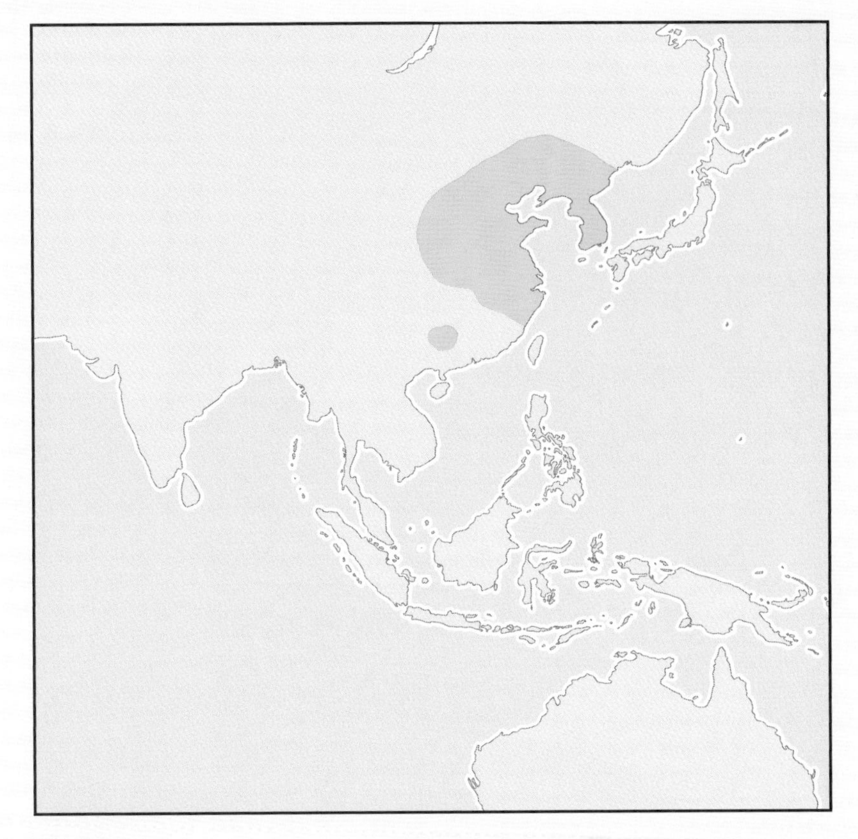

Distribution
Coasts and river valleys of eastern China, Korea; introduced in France and England

Evolution and systematics

In the structure of their skulls, brains, appendages, genetics, and the spotting pattern of their fawns, Chinese water deer are closely related to the European roe deer (*Capreolus capreolus*) and moose (*Alces*) of the New World deer subfamily. Males lack antlers, but possess large, protruding upper canines with which they fiercely defend resource territories in fertile flood plains. Together with small cheek teeth, small and simple brains, their selection of highly digestible plant foods and their escape behavior via rapid, long bounds followed by hiding, indicate a very primitive or ancestral condition among deer.

The water deer first appeared some 30 million years ago in the Oligocene. However, it may be secondarily primitive. It may have lost its antlers in favor of large tusks while reverting back to defending small, but resource-rich territories. A similar evolutionary reversal reducing antler size in favor of tusks has been identified in the muntjacs among the Old World deer. Here, phylogenetically young species defending small territories have small, simple, or rudimentary antlers compared to phylogenetically old species with large, complex antlers found on large territories in relatively infertile landscapes. The water deer is unusual in that its primitive adaptations are normally associated with warm climates. However, it is adapted to cold-temperate seasonal climates with frost and snow. It is characterized by an exceptionally high reproductive output and early sexual maturation, which matches the ecological opportunities and dangers in flat valleys with large, flooding rivers. The genus consists of a single species, *Hydropotes inermis*, and two subspecies: *Hydropotes inermis inermis*, found in eastern China, and *Hydropotes inermis argyropus* of Korea.

The taxonomy for this species is *Hydropotes inermis* Swinhoe, 1870, Kiangsu, China.

Physical characteristics

The water deer is a small, elegant animal with narrow pectoral and pelvic girdles, long legs, and a long, graceful neck. Its powerful hind legs are longer than its front legs, so that its haunches are carried higher than the shoulders. It runs with rabbit-like jumps. In the groin of each leg is an inguinal

The Chinese water deer (*Hydropotes inermis*) has a small scent gland in front of its eyes. (Photo by D. G. Huckaby/Mammal Images Library of the American Society of Mammalogists.)

gland used for scent marking; this deer is the only member of the Cervidae to possess such glands. The short tail is no more than 1.9–3.8 in (5–10 cm) in length and is almost invisible, except when it is held raised by the male during the rut. This deer characteristically stands alert, with its head held high and its ears erect.

The most striking features are the long, curved upper canines in the male, measuring 2.1 in (5.5 cm) on average in length. At the end of the male's first winter, his canines will be about half their full size; final length is reached after about 18 months. These canines are held loosely in their sockets, with their movement controlled by facial muscles. The male can draw them backwards out of the way when eating. In aggressive encounters, he thrusts his canines out and draws in his lower lip to pull his teeth closer together. He then presents an impressive two-pronged weapon to rival males. The female, by comparison, has tiny canines measuring just over 0.2 in (0.5 cm).

In the fall, this deer's red or reddish brown summer coat is gradually replaced by a thicker, coarse-haired winter coat that varies from light brown to grayish brown. Neither the head nor the tail poles are well differentiated as in gregarious deer; consequently, this deer's coat is little differentiated.

Distribution

Formerly widespread in eastern China, the Chinese subspecies is now largely restricted to the lower reaches of the Yangtze River, coastal Jiangsu province, and islands of Zhejiang. Feral populations dating to the mid-twentieth century are well established in southern England, and highly localized populations are found in France. The Korean subspecies is thought to have a wide distribution, but evidence is lacking.

Both sexes of Chinese water deer (*Hydropotes inermis*) have tusks that protrude from their upper jaw, but neither has antlers. (Photo by © Clive Druett/Papilio/Corbis. Reproduced by permission.)

Habitat

True to its name, the water deer occupies fertile river valleys as well as well-vegetated lake shores and coastal areas where tall reeds, rushes, sedges, and grasses provide it with cover. A proficient swimmer, it can swim several miles to make use of river islands.

Seasonal flooding of river deltas forces water deer to higher ground. They head up to hill grassland, where they venture into open fields, providing there is tall vegetation within easy reach. Other seasonal movements in search of better grazing may also be undertaken.

Behavior

Apart from during the rutting season, water deer are solitary animals, and males are highly territorial. Each male marks out his territory with urine and feces. Sometimes a small pit is dug and it is possible that in digging, the male releases scent from the interdigital glands on its feet. The male also scent-marks by holding a thin tree in his mouth behind the upper canines and rubbing his pre-orbital glands against it. Males may also bite off vegetation to delineate territorial boundaries.

Confrontations between males begin with the animals walking slowly and stiffly towards each other, before turning to walk in parallel 32–64 ft (10–20 m) apart, to assess each

A Chinese water deer (*Hydropotes inermis*) female with a newborn. (Photo by Michael Clark/FLPA–Images of Nature. Reproduced by permission.)

other. One male may then succeed in chasing off his rival, making clicking noises as he pursues his vanquished foe. However, if the conflict is not thus resolved, the males will fight. Each tries to wound the other on the head, shoulders, or back, by stabbing or tearing with his upper canines. Numerous long scars and torn ears seen on males indicate that fighting is frequent. Tufts of hair are most commonly found on the ground in November and December, showing that encounters are heavily concentrated around the rut.

Females do not seem to be territorial outside the breeding season and can be seen in small groups, although individual deer do not appear to be associated; they will disperse separately at any sign of danger. Females show aggression towards each other immediately before and after the birth of their young and will chase other females from their birth territories.

Communication between these solitary animals generally takes the form of low growling barks. They are certainly used as calls of alarm, but they are also directed repeatedly at other deer or humans for reasons that are unknown.

Feeding ecology and diet

This species spends roughly half its waking hours feeding, with 20 minute periods broken by spells of rest or rumination. Peak feeding activity is recorded around dawn and dusk. The water deer feeds at night too, although it is not known how much time is spent on nocturnal foraging. It is often observed grazing in the open, where it relies on good eyesight and smell to detect danger.

The water deer is a concentrate selector, avoiding low-grade food, but the degree of selectivity appears to vary between locations. One study in China of water deer rumens showed that the leaves of herbs made up 59% of the diet, with grasses and sedges comprising 24% and woody material 17%. Another study showed herbs constituted 93% of its diet.

Reproductive biology

During the annual rut in November and December, the male will seek out and follow females, giving soft squeaking contact calls and checking for signs of estrus by lowering his neck and rotating his head with ears flapping. Scent plays an important part in courtship, with both animals sniffing each other.

Mating among water deer is polygynous, with most females being mated inside the buck's own territory. After repeated mountings, copulation is brief. Gestation is normally around 180 days.

The Chinese water deer (*Hydropotes inermis*) is native to China. A male is shown here. (Photo by Rod Williams/Naturepl.com. Reproduced by permission.)

Water deer have been known to produce up to seven young, but two to three is normal for this species, the most prolific of all deer. The female often gives birth to her spotted young in the open, but they are quickly taken to concealing vegetation, where they will remain most of the time for up to a month. During these first few weeks, fawns come out play. Once driven from the natal territory in late summer, young deer sometimes continue to associate with each other, later separating to begin a solitary existence.

Conservation status

Although classified as a protected species in China, levels of hunting permitted during the 1990s did not appear to be sustainable. Population estimates at that time suggested a figure of no more than 10,000 individuals, yet this figure was given by the government as the annual tally hunted legally. Scientists also report at the beginning of the twenty-first century increased levels of poaching in eastern China. The IUCN lists the species as Lower Risk/Near Threatened.

Water deer are suffering increasing losses of habitat through conversion of wetlands to agriculture and aquaculture; significant losses have been recorded around the Yancheng coastal wetlands, one of the main strongholds. Construction of the Three Gorges Dam may have a huge impact on water deer habitats along the Yangtze River and associated waters. Ironically, numbers of this subspecies in southern England continue to expand following accidental introductions in the 1940s. It is feasible that numbers there may eventually exceed those in its native China.

Although the status of Siberian water deer (*Hydropotes inermis argyropus*) in Korea is not well documented (listed as Data Deficient), this subspecies is thought to be widespread and abundant. Proposals to establish a huge national park in the demilitarized zone between North and South Korea show that positive action that would protect this deer is being considered.

Significance to humans

This deer's tendency to remain standing still in the open makes it an easy target for hunters, both for meat and for its stomach colostrum, a prized ingredient in folk medicine. Formerly, it was viewed as a crop pest in China and slaughtered as such until rarity and legal protection ended this practice. Rapid growth, early maturity, and high fecundity make water deer considered for domestication. However, livestock farmers are deterred by this solitary species' reputation for territorial aggression.

Chinese water deer (*Hydropotes inermis*). (Illustration by Barbara Duperron)

Resources

Books

Cooke, A., and L. Farrell. *Chinese Water Deer.* Fordingbridge, UK: The Mammal Society and British Deer Society, 1998.

Geist, V. *Deer of the World.* Mechanicsburg, PA: Stackpole Books, 1998.

Macdonald, D. *The New Encyclopedia of Mammals.* Oxford: Oxford University Press, 2001.

Office of International Affairs. *Microlivestock: Little-known Small Animals with a Promising Economic Future.* Washington, DC: National Academy Press, 1991.

Putman, R. *The Natural History of Deer.* Ithaca, NY: Comstock Publishing Associates, 1989.

Organizations

British Deer Society. Burgate Manor, Fordingbridge, NH SP6 1EF England. Phone: 1425 655434. Fax: 1425 655433. E-mail: h.q@bds.org.uk Web site: <http://www.bds.org.uk>

Derek William Niemann, BA

△

New World deer

(Capriolinae)

Class Mammalia

Order Artiodactyla

Family Cervidae

Subfamily Capriolinae

Thumbnail description
New World deer have preserved the distal rudiments of the lateral metacarpal bones; the middle parts of the lateral metacarpae are reduced; and posterior portion of the nasal cavity is divided into two chambers by the vomer, which feature is retained in South American deer, but is lacking in moose and roe deer

Size
Vary in size from very small (pudu) to the largest among Cervidae (moose)

Number of genera, species
9 genera; 27 species

Habitat
Woodlands and shrublands, often forest edges; many populations of reindeer inhabit tundra or arctic desert yearlong, others migrate to openness of sub-arctic tundra for summer

Conservation status
Endangered: 1 species; Vulnerable: 2 species; Lower Risk/Near Threatened: 3 species; Data Deficient: 7 species

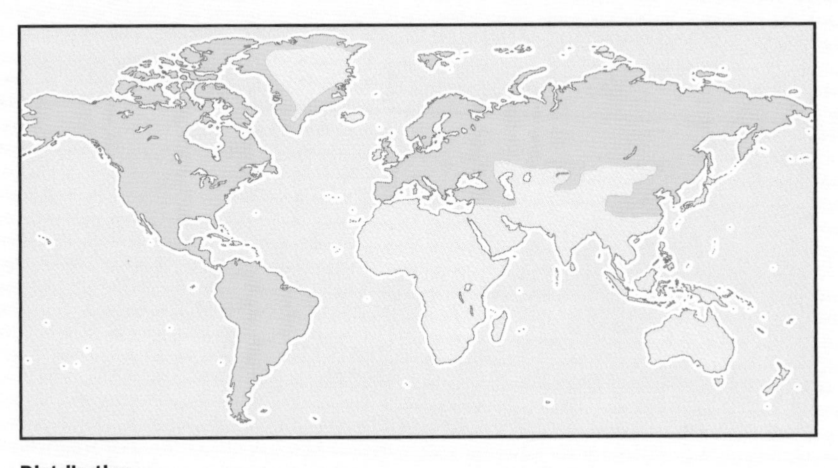

Distribution
North America, South America, Europe, and northern Asia

Evolution and systematics

Recent paleontological research in the high Arctic suggest that as late as mid-Pliocene times the Arctic portions of North America and Asia were covered by temperate forests. Together with the zoogeography and fossil record of New World deer species, this suggests that their origin lay in these northern land masses in late Tertiary times. From here they dispersed southward as climates cooled into periodic continental glaciations. The fossil record of New World deer is limited by the facts that huge continental glaciers destroyed Tertiary fossil deposits over enormous areas, while in unglaciated regions where rich plant fossil deposits are found, the acidic conditions favoring these dissolve bones. We have thus no idea what early New World deer looked like. When they first appear in areas south of the continental glaciations in North America or Eurasia, they are already well differentiated and close to modern genera. After they enter South America with the onset of major glaciations some 2 million years ago (mya), they evolved rapidly into a large array of diverse species, only some of which survived to the present. Here they evolved dwarfs, such as pudu (*Pudu pudu*, *P. mephistophiles*) and brocket deer (*Mazama* spp.); mountain climbing, short-legged specialists such as the huemul (*Hippocamelus antisensis*, *H. bisculus*); sophisticated swamp dwellers such as the marsh deer (*Blastocerus dichotomus*); gregarious plains dwellers such as pampas deer (*Ozotocerus bezoarticus*); as well as massively antlered, large-bodied steppe deer that are now extinct (*Morenelaphus*). In the north they evolved giants such as moose (*Alces*); herd forming reindeer or caribou (*Rangifer tarandus*); a large-bodied extinct form *Torontoceros*; forest and swamp dwellers such as the ancient white-tailed deer (*Odocoileus virginianus*), the oldest deer species in the world; rainforest and mountain dwellers such as black-tailed and mule deer (*O. hemionus*); as well as a large, short-legged, cliff adapted Rocky Mountain deer, *Navahoceros*, which vanished at the end of the Pleistocene. Closely related to these North Americans are the roe deer (*Capreolus capreolus*, *C. pygargus*) of Eurasia, which is a cold-climate specialist and the Chinese water deer (*Hydropotes inermis*). A sparse fossil record suggests that there were a few more species present in late Pliocene times. Compared to Old World deer, New World deer are more differentiated from and less related to on another. They thrived in the ecological turmoil of the Pleistocene and have often taken advantage of humanmade landscape changes in the Recent. They also suffered fewer extinctions than other groups of large mammals.

A characteristic of New World deer is that they appear in the fossil record in order of adaptation to cold climates. The

Black-tailed deer (*Odocoileus hemionus*) wading in the Pacific Ocean surf near the mouth of Ozette River, in Washington, USA. (Photo by Lee Rentz. Bruce Coleman, Inc. Reproduced by permission.)

species most tolerant of heat, and which therefore disperses into and colonizes tropical South America, is the white-tailed deer. A species very similar, if not identical, to it shows up in southern North America at the end of the Blancan period almost 4 mya. Moose appear about 2.6 mya. Roe deer appear about 2 mya, and reindeer, the most cold-adapted genus, appears about 1 mya. This genus is the sister genus to *Navahoceros*, suggesting that competition segregated these originally montane lineages, with reindeer exploiting the alpine and subalpine slopes while *Navahoceros* sought refuge in cliffs. Similar ecological division is seen in extant Asiatic goat (*Capra*) and sheep (*Ovis*).

Moose appear in the Pliocene fossil record first in western Eurasia. They are then large, plains-dwelling, long-legged runners (*Alces [Libralces] gallicus*), about the size of a red deer (*Cervus elaphus*), with a normal deer face and extraordinary long spoon-shaped antlers. There appear to be several species. Moose are next seen in mid-Pleistocene Eurasia during the major glaciations. It has grown into a massive giant (*Alces [Cervalces] latifrons*). The huge antlers are more palmate and have shorter beams than the Pliocene specimen. The skull is intermediate between that of a normal deer and contemporary moose. Such moose cross into North America, where they grow into long-legged trotters with complex large tri-lobed antlers, but retain the primitive deer-like skull (*Cervalces scotti*). This American stag-moose is narrowly associated with huge Pleistocene pro-glacial lakes, which were apparently its escape terrain from the many large predator species found in North America. In Siberia moose continue to evolve into the modern moose. These have even larger palms and much shorter antler beams, the muzzle became adapted to feeding on underwater vegetation, while the body changed from that of a cursor (runner) to that of a trotter. After post-Pleistocene megafaunal extinction in North America, Siberian moose colonize the northern half of that continent beginning about 10,000 years ago, where they are now widely distributed. Two fairly distinct modern moose evolved, one the west Siberian-European form, and the other the east Siberian-American form. Despite great physical differences, moose share diagnostic behaviors with white-tailed deer, mule deer, and caribou. They twin readily, and moose remained in cold climates throughout.

The Odocoileinae comprise four groups of species: North American *Odocoileus* deer and South American deer, reindeer, moose, and roe deer. They all branched early from an ancestral stem into different evolutionary radiations, but retain common features in body plan, behavior, and ecology. The *Odocoileus* and South American deer comprise one tribe, ac-

White-tailed buck (*Odocoileus virginianus*) with twigs and leaves caught in his antlers. He has been "fighting" brush, an activity related to rutting behavior. (Photo by Erwin and Peggy Bauer. Bruce Coleman, Inc. Reproduced by permission.)

cording to G. G. Simpson. These deer may be found from the Arctic Circle in Canada south to the glaciers of southern Chile, but the majority of species live in warm climates. Moose, reindeer, and roe deer are each placed into separate tribes. These deer live in the cold northern environments, including the high arctic, and have never colonized southern latitudes.

All Odocoileinae retain distal rudiments of the lateral metacarpal bones II and V, but which still retain hoof functions. Another feature is the division of the posterior portion of the nasal cavity into two chambers by the vomer. However, this feature is missing in moose and roe deer. Males have a pendular penis. Antlers are found in all genera. The dwarf deer of South America may be secondarily dwarfed, as they are exceedingly closely related to the older and more ubiquitous white-tailed deer. This species ranges from near the Arctic Circle in Canada to 18° south of the equator in South America. The short dagger-like antlers of the dwarf deer may be related to territorial defense. Reindeer and caribou have the largest antlers relative to body size among deer, while their

The pampas deer (*Ozotoceros bezoarticus*) is mostly sedentary. (Photo by François Gohier/Photo Researchers, Inc. Reproduced by permission.)

Black-tailed deer (*Odocoileus hemionus*) use their antlers to fight. (Photo by Mark Newman/Photo Researchers, Inc. Reproduced by permission.)

A key deer doe (*Odocoileus virginianus clavium*) with a cattle egret (*Bubulcus ibis*). (Photo by Claudine Laabs/Photo Researchers, Inc. Reproduced by permission.)

The Chilean pudu (*Pudu pudu*) is the smallest deer in the world. (Photo by Tom McHugh/Photo Researchers, Inc. Reproduced by permission.)

females also carry antlers except in some woodland populations. They also are the most cursorial and gregarious deer alive.

Physical characteristics

All Capriolinae retain distal rudiments of the lateral metacarpal bones II and V, though they are important to the functioning of lateral toes. Another feature, the posterior portion of the nasal cavity that is divided into two chambers by the vomer, is retained in South American deer, but is lacking in moose and roe deer. There are different varieties of antler structure in Odocoileinae: simple spiked (*Mazama*, *Pudu*), bifurcate (*Hippocamelus*), dichotomous (*Blastocerus*), or branched (*Odocoileus*, *Rangifer*). Moose often sport wide spade-like antlers. Normally, only males wear antlers, though reindeer females grow antlers as well.

Distribution

Six genera, including *Odocoileus*, *Ozotocerus*, *Blastocerus*, *Hippocamelus*, *Mazama*, and *Pudu*, live only in the New World. One genus, *Capreolus*, is known in Eurasia only. *Alces* and *Rangifer* inhabit both North America and Eurasia.

Habitat

Deer belonging to Capriolinae adapt to diverse habitats. Dwarf forms of deer with short antlers and long tails (brocket deer, *Mazama*) are inhabitants of tropical latitudes, while large

White-tailed deer (*Odocoileus virginianus*) fleeing. (Photo by Tom Brakefield. Bruce Coleman, Inc. Reproduced by permission.)

A reindeer (*Rangifer tarandus*) shedding velvet. (Photo by John Shaw. Bruce Coleman, Inc. Reproduced by permission.)

deer with dichotomous antlers (marsh deer, *Blastocerus dichotomus*) inhabit tropical and subtropical marshlands. Tropical savanna is the favored habitat for pampas deer (*Blastocerus campestris*). Subarctic dwellers are roe deer, white-tailed deer, black-tailed deer (*Odocoileus hemionus*), and moose and reindeer inhabit cold temperate subarctic forests, alpine, and subarctic tundra.

A mule deer (*Odocoileus hemionus*) lip curl seen during mating season. (Photo by E & P Bauer. Bruce Coleman, Inc. Reproduced by permission.)

Behavior

The dwarf and mountain deer of South and Central America are classical territory defenders and hiders. *Capreolus* males defend large territories and bond females that live within their territory. This species is a classical saltor (jumper) that relies on short runs over obstacles and then hides. *Odocoileus* may defend fawning territories, but otherwise related females form clans so that they move over a shared home range. Males form unstable fraternal groups but disperse and compete individually over females during the rut. These deer rely on sprinting or specialized locomotion to escape predators as well as on hiding. They may form a large selfish herd in open grasslands. Pampas deer also rely on large selfish herds to escape predators, while swamp deer use extensive wetlands for that purpose. South American deer, unlike North American species, are exceedingly sensitive to predation by feral dogs as there are no native wolf-sized canids in South America. Moose live dispersed much of the time, but bulls may form unstable fraternal groups in spring and after the rut in fall. Female moose are likely to join bulls socially only if they are neither pregnant nor have a calf at heel. Moose escape predators primarily by running over obstacles that are low relative to the moose, but high relative to the pursuing predators, who then must expend great amounts of energy to follow the fleeing moose. In deep snow moose turn and fight predators using both front and hind legs. Reindeer and caribou form highly gregarious selfish herds and excel at sustained high speed running to escape predators. In sedentary populations females disperse and give birth in hiding. In migratory populations they move northward in spring onto open terrain where they congregate into birthing herds. This overloads pursuing wolves and grizzly bears with calves. The calves are highly developed at birth and soon follow the female, who produces the richest milk among all deer, ensuring rapid growth of the calf to a survivable size. This is the most migratory species of large terrestrial mammal. Bulls advertise with antlers during the rut. Females use antlers to ward off young bulls in winter, who might otherwise parasitize the female's work of digging craters in deep snow to reach lichens. Both *Rangifer* and *Alces* are large Ice Age giants compared to other species in their family.

Feeding ecology and diet

The Capriolinae diet comprises highly nutritive forages of low-fiber content: forbs, flowers, and leaves, but rarely

The antlers of the Mexican red brocket deer (*Mazama americana temana*) point backwards. (Photo by Kenneth W. Fink/Photo Researchers, Inc. Reproduced by permission.)

The marsh deer (*Blastocerus dichotomus*) is the largest deer in South America. (Photo by Jany Sauvanet/Photo Researchers, Inc. Reproduced by permission.)

grass. The type of diet determines speciation of the digestive system: a large mouth enables the browsing of branches, while a long sensitive tongue helps to choose among forbs and foliage. Some species have a relatively small rumen, large salivary glands, and rapid digestion. Intestines are rather short—12 to 15 times longer than body. Consequently, interchanging periods of grazing and ruminating are short. If these deer, by necessity, feed on rough fibrous forage, there are fewer interchanging of grazing and ruminating periods during their diurnal activity (only five to six, rather than the usual eight to 12).

Reproductive biology

Capriolinae are polygynous, though the overall reproductive strategy of Capriolinae differs from that of Cervinae. They have a higher reproductive rate and mature sexually earlier; most genera produce two fawns per birth. Consequently, Capriolinae species have higher population densities, including mule deer, caribou, and moose. At the same time, the species are reproductively isolated: for instance, male hybrids between the European and Siberian roe deer as well as between white-tailed deer and mule deer are sterile. Hybridization between white-tailed and black-tailed deer is restricted. Capriolinae poorly resist diseases and parasites, and poorly adapt to new conditions.

Conservation status

Deer that inhabit Central and South America (pampas deer, *Ozotoceros bezoarticus*; marsh deer, *Blastocerus dichotomus*; huemul and pudu, *Pudu* spp.; brocket deer, *Mazama* spp.); and some subspecies from the southern part of North America have experienced severe pressure from hunters and have been on the brink of extinction. These species would greatly benefit from preservation and conservation efforts. Pampas deer, for instance, were harvested by the millions in the nineteenth century, and currently is Lower Risk/Near Threatened. The subfamily Capriolinae also includes the most numerous species on Earth, including white-tailed deer, black-tailed deer, reindeer, roe deer, and moose.

Most threatened are the Capriolinae in South America, where poor residents use subsistence hunting and do not distinguish between rare species that need to be preserved and flourishing species that need regular game management. In many South American countries, there is a fierce competition between deer and livestock for pastures. Also, a drastic decrease of habitats occurs due to drainage, farm development of grasslands, and forest cut. The Chilean huemul (*Hippocamelus bisulcus*) is considered Endangered. Marsh deer (*Blastocerus dichotomus*) and Chilean pudu (*Pudu pudu*) are considered Vulnerable. Less studied, Data Deficient are: Peruvian huemul (*Hippocamelus antisensis*), and six species (of seven known) of brocket deer (*Mazama* spp.).

Significance to humans

Roe deer, moose, and white-tailed deer benefit from successful, extremely productive game husbandry, though farming proved to be unsuccessful. Consequently, their future lies in game management.

Reindeer is the only deer species to become domesticated. It supports local cultures throughout northern Eurasia in this capacity. Wild migratory reindeer and caribou are also significant in northern economies. While moose can be tamed and used for riding, as beasts of burden, and for milking, their fickle feeding habits and susceptibility to livestock diseases makes them difficult to keep. They are a productive, highly appreciated meat source throughout their range. White-tailed, black-tailed, and mule deer, after recovery from severe depletion at the end of the nineteenth century, support today a rich hunting economy in North America. Roe deer fulfill a similar role in Europe.

1. Siberian roe deer (*Capreolus pygargus*); 2. European roe deer (*Capreolus capreolus*); 3. Marsh deer (*Blastocerus dichotomus*); 4. Black-tailed deer (*Odocoileus hemionus*); 5. White-tailed deer (*Odocoileus virginianus*). (Illustration by John Megahan)

1. Reindeer (*Rangifer tarandus*); 2. Chilean huemul (*Hippocamelus bisulcus*); 3. Red brocket (*Mazama americana*); 4. Southern pudu (*Pudu pudu*); 5. Moose (*Alces alces*). (Illustration by John Megahan)

Species accounts

European roe deer
Capreolus capreolus

TAXONOMY
Capreolus capreolus (Linnaeus, 1758), Sweden.

OTHER COMMON NAMES
French: Chevreuil; German: Reh; Spanish: El corzo.

PHYSICAL CHARACTERISTICS
Small animal. Shoulder height: 24–35 in (60–90 cm); body length: males 39–54 in (100–137 cm), females 37–54 in (94–136 cm); weight: males 46–75 lb (21–34 kg), females 42–71 lb (19–32 kg). Very slender, with long, slim legs and nearly invisible tail. Preorbital gland is rudimentary. There is large bald spot on the end of the muzzle between and around nostrils. Antlers are small, usually forked in three at the end, with a lack of brow tine. Small protuberances develop along beams. Coat is monotonously colored, gray in winter (with brownish or reddish tint) and red with lighter belly in summer. Rump patch is white. Fawns develop spots.

DISTRIBUTION
Throughout Europe, including Britain and Sicily (except in Corsica and Sardinia); none in close coniferous forests of Scandinavia and northern Russia.

HABITAT
Inhabit dry plains and mountains, in varieties of landscapes where forest islands interchange with steppe and meadows. Mature broadleaved forests also attract them, as well as shrubs and tall grass, which are used for shelter. In many countries of Central and Western Europe, they spend most time on farmed fields.

BEHAVIOR
Live solitarily or in small family groups (doe, fawn, yearling), except in rut season. Aggregations to tens of animals without social bonds appear in areas with abundant forage. Once disturbed, animals scatter. Males defend their territories against invaders: they mark it by horning trees, as well as leaving secretions from scent glands. Does keep to their home ranges and defend it from other does. Strong bonds to home ranges are a distinguishing feature of roe deer. A male home range about 990–2,100 ac (400–850 ha) usually overlap some ranges of does (490–700 ac [200–700 ha]).

Combats between bulls are frequent, even out of rut season; males even fight females. Good swimmers and jumpers. Diurnal activity is highest at dawn and dusk. Extremely wary; looking around takes up to 50% of active time. When disturbed, they make a typical bark, strike the ground with the front leg hooves, run away showing bright white tail mirror, and making signal jumps.

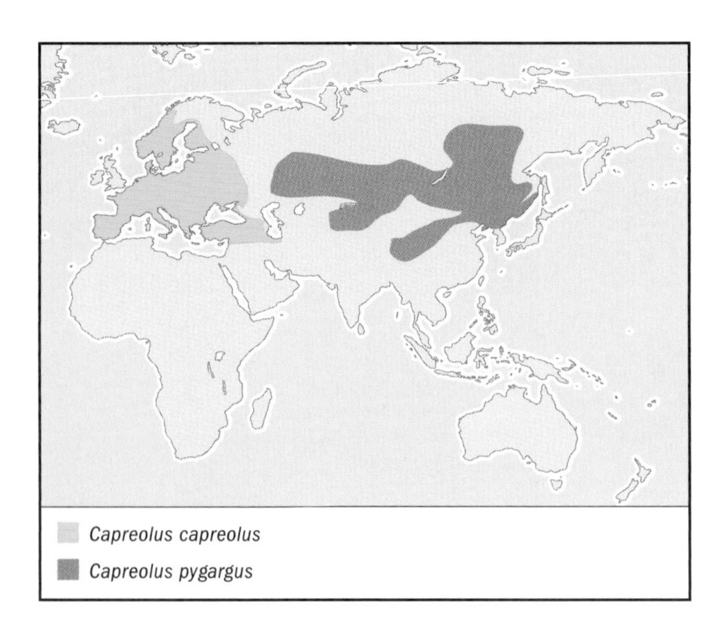

FEEDING ECOLOGY AND DIET
Feed mostly on herbs, less on leaves, buds, fruits, cereals, and sedges. In winter, twigs are preferable, as well as dry grasses, fallen leaves, mosses, and tree lichens; in summer, they browse green soft shoots of cereals and sedges. Fungi are eaten year-round. To reach upper branches or highly hanging fruits, they rise on hind legs. Animals even dig at snow or soil to get to roots in farmed fields. Can stay without water, using moisture in vegetation, and snow in winter.

REPRODUCTIVE BIOLOGY
Polygynous. First signs of rut appear in July into the beginning of August, or sometimes in May–June; peak of rut lasts from mid-July to the end of September (depending on latitudes). There are no harems; mating takes place when females stay in male's home range. Does are in heat for 4–5 days, gestation period lasts about 240 days. Calving takes place in herb or shrub thickets; one to two fawns per birth; fawns stay hidden for the first 6–8 days, rising only to be nursed; they later follow mothers, start feeding on herbs and leaves at one month, and are weaned at two months. Life expectancy is 11–12 years.

CONSERVATION STATUS
Not threatened.

SIGNIFICANCE TO HUMANS
Subject of game hunting, both venison and antlers are main attractions. ◆

Siberian roe deer
Capreolus pygargus

TAXONOMY
Capreolus pygargus (Pallas, 1771), Volga area, Russia.

OTHER COMMON NAMES
French: Chevreuil de Siberie; German: Reh von Sibirien; Spanish: Corzo siberiano.

PHYSICAL CHARACTERISTICS
Middle sized deer. Body length: males 47–61 in (120–156 cm), females 46–59 in (116–150 cm); weight: males 66–132 lb (30–60 kg), females 55–121 lb (28–55 kg). Coat in winter is grayish to

brownish on back, and creamy at belly and inside legs. Rump patch is white or cream. Summer coat is red at head and body. Fawns develop distinct spots arranged in four or five rows. Antlers to 16 in (40 cm) and long, pockmarked with bumps, some of which transform to protuberances and tines.

DISTRIBUTION
From the Volga to Russian Far East and northern China via northern Kazakhstan and north of Middle Asia.

HABITAT
Inhabit both plains and mountains to altitudes of 6,900 ft (2,100 m). The species adapt to deep snow to 20 in (50 cm) and to harsh winters, surviving in Yakutia and Transbaikal; also inhabit pine forests and mature coniferous-deciduous forests.

BEHAVIOR
In winter, they form groups of four to six, live solitary (does with fawns). Daily home range of 99 ac (40 ha) and annual range about 990 ac (500 ha). In Amur basin, home range in winter is to 34,500 ac (14,000 ha). Migrations (to avoid deep snow) to distances 62–250 mi (100–400 km) are frequent. In mountain ranges of the Caucasus, Altai, and the Urals, they make short migrations between altitudes 1,600–3,300 ft (500–1,000 m).

FEEDING ECOLOGY AND DIET
In winter, feed on tree and shrub branches, dry herbs, fallen leaves, mosses; in summer, mostly on sedges and grasses.

REPRODUCTIVE BIOLOGY
Polygynous. Rut peak is in August to the beginning of September, calving is May–June. Gestation period lasts 264–318 days. Does give birth to one, twins, or triples. Lifespan is about seven years; average age in wild is two and a half years. Strong hunting pressure and predators cause high mortality.

CONSERVATION STATUS
Not threatened.

SIGNIFICANCE TO HUMANS
Important game animal; in Russia, annual harvest is 5,000–10,000 deer, mostly in the Urals and in Amur valley. ◆

White-tailed deer
Odocoileus virginianus

TAXONOMY
Odocoileus virginianus (Zimmermann, 1780), Virginia, United States.

OTHER COMMON NAMES
French: Cerf de Virginie; German: Weisswedelhirsch; Spanish: Ciervo de Virginia.

PHYSICAL CHARACTERISTICS
Small. Shoulder height: males 39 in (100 cm), females 35 in (90 cm); body length: males 77 in (195 cm), females 67 in (170 cm); tail length: 11 in (27 cm); weight: males 128–300 lb (58–136 kg), females 110–175 lb (50–79 kg). Newborns weight 4–9 lb (1.8–4 kg); reach 55–86 lb (25–39 kg) by six months. In summer, coat is a foxy-red color, changing in autumn to tawny gray, with longer and thicker hairs. Newborns have reddish coat sprinkled with light spots. Under part of tail is bright white. Usually only bucks wear antlers. Glands producing strong scent are well developed on all four hooves.

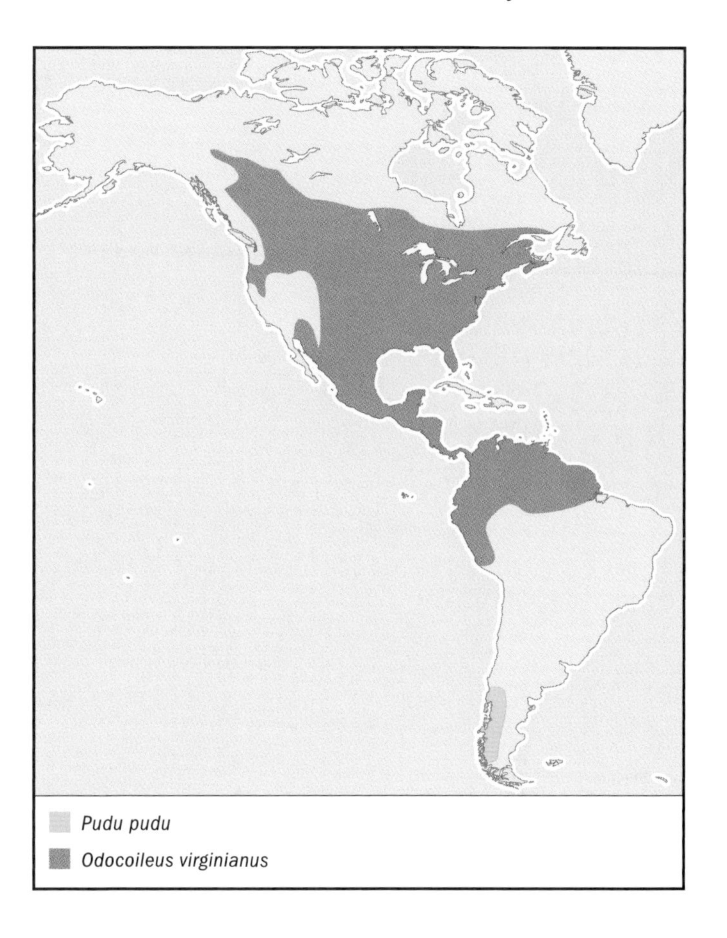

DISTRIBUTION
Distributed from Atlantic shore to Pacific shore of North America; from central Canada to Bolivia, Guiana, and northern Brazil.

HABITAT
Recently, they have come to inhabit edges of forest clearings: burned areas, logged sites, fields, and meadows. Clumps of broadleaved and coniferous forests in the middle of fields are their favored habitats. Also inhabit rugged river valleys, sandy hills covered by grass and trees, and sometimes stay in large forests. In summer, habitats are more diverse, including fields and meadows. In winter, deer keep to forests, especially coniferous stands, where they find shelter from harsh elements. Home range is from 0.1–1.5 mi^2 (0.2–4 km^2). Snow deeper than 17.7 in (45 cm), lasting more than two months, harsh weather, and ice crust are important elements of habitat. Use thick forest for cover from predators and hunters, as well as in tall grasses and bushes in prairies and riparian habitats.

BEHAVIOR
When deer run from danger, the tail bounces loosely from side to side, with white hairs erected; it serves as a signal of danger to all surrounding deer. It also helps a fawn following after the mother. They can run as fast as 30 mph (50 km/h), jump to 10 ft (3 m) high and to 30 ft (9 m) long.

FEEDING ECOLOGY AND DIET
Up to 70% of diet consists of tree and shrub leaves and twig ends; 19% is forbs; 11% is grasses. Agricultural crops accounted for 3% of yearlong diet. An adult needs 5–11 lb (2.5–5 kg) of forage daily. In winter, they can survive on two lb (1 kg)

of food daily. If succulent plants are available, they can manage without water for a long time, though watering places often are center of home range.

Preferable winter and autumn food: blackberry (*Rubus* spp.), dogwood (*Cornus* spp.), snowberry (*Symphoricarpos* spp.), chokecherry (*Prunus virginiana*), rose (*Rosa* spp.), Oregon grape (*Berberis repens*), ponderosa pine (*Pinus ponderosa*), elms (*Ulmus* spp.), and maples (*Acer* spp.). Nuts of oaks, hickories, beech, and walnuts and fungi are important.

REPRODUCTIVE BIOLOGY
Polygynous. Bucks shed antlers annually in winter; re-growth starts at the end of April to the beginning of May. Rut begins in October, to reach peak in mid November and end in December. At the beginning of rut, bucks get a swollen neck and they clean antlers of velvet; establish dominance hierarchy in October, displaying scraping activity: pawing the ground until leaf or grass litter is removed and bare ground is exposed. The animal then urinates on the bare spot. Does are in heat only one day, those that do not breed go back into heat in 28 days. Bulls keep no harems; stay with a doe a day or two, and after breeding, they search for another one, thus having 6–8 does mated during rut. Gestation period is 188–222 days, fawns (usually twins, rarely triplets) appeared in May to the beginning of June. Within the first hours, they can suckle and follow mother; nevertheless, stay hidden in bushes or tall grass first days of their life. In 10 days, they start nibbling green shoots. Calves participate in rut from their first year. There are 40–75% of pregnant does among very young ones. High productivity goes with high mortality in this species. Annual rate of mortality in the species is 30–50%. Life expectancy is six years; some individuals in wild live to 14 years, in captivity to 20 years.

CONSERVATION STATUS
Not threatened.

SIGNIFICANCE TO HUMANS
Important game species; the most numerous of big game animal in the world. Of 15 million white-tailed deer in North America, annual harvest is about three million animals. ◆

Black-tailed deer
Odocoileus hemionus

TAXONOMY
Odocoileus hemionus (Rafinesque, 1817), South Dakota, United States.

OTHER COMMON NAMES
English: Mule deer; French: Cerf-mulet; German: Schwarzwedelhirsch; Spanish: Ciervo mulo.

PHYSICAL CHARACTERISTICS
Medium size. Shoulder height: 37–39 in (95–100 cm); bulls weigh 220 lb (100 kg) with body length to 77 in (195 cm); does weigh 143 lb (65 kg) with body length to 62 in (160 cm); tail: 7 in (18 cm). Head is narrow, elongated, with big nasal bald patch. Hairs are longer at the top of neck, no mane. Adult have a mono-color coat, mostly black tail surrounded by a smaller white rump patch. Develop long ears, big antlers, and white tail with a black end. Fawns are reddish, lighter colored at under parts, with very long tail. Only bulls wear antlers, varying from simple spikes to forked, depending on bull's age.

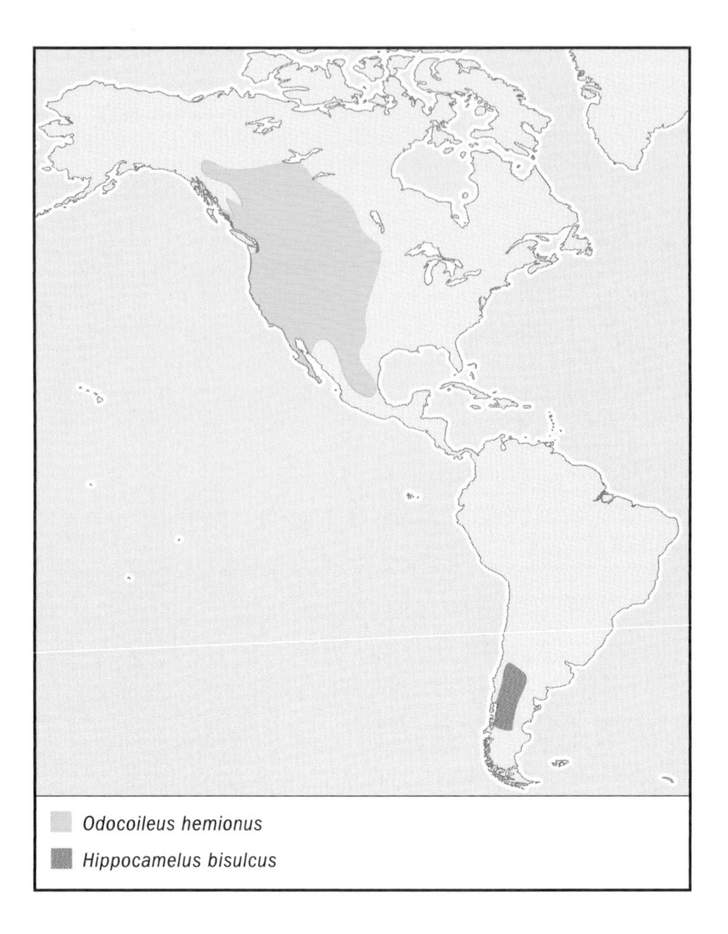

DISTRIBUTION
Western North America from the southernmost Yukon to Mexico.

HABITAT
Interspersion of food and cover is the major factor in habitats. Animals gather at cut clearings, rich in forage, and keep no farther than 330 ft (100 m) from edge of forest to escape when disturbed. In areas with deep snow, prefer southern slopes at altitudes less than 3,300 ft (1,000 m). Rich communities of herbs, ferns, and shrubs make the best spring ranges, usually at low elevations of less than 2,600 ft (800 m), southerly aspects, and moderate to steep slopes. Good spring range sometimes extends onto lower, gentler slopes where rich, moist soil produces more herbs and snowmelt earlier. Burned clear-cut sites often support abundant herbs—especially fireweed. Dense thickets of young cedar and hemlock provide security cover when they retain live branches within 7 ft (2 m) over the ground. Deep snow is critical for survival; though they move even through incrusted snow as deep as 20–25 in (50–60 cm), it takes too much energy.

BEHAVIOR
Live solitarily or in small groups (a mother and her young of the current and previous years) most of the year. Gather at places with abundant food or in shelters against harsh weather; no social bonds between individuals. Distance between summer and winter home range rarely exceeds 8 ft (2–3 m). Summer home range is 0.6–2.3 mi^2 (1.5–6 km^2); winter home range is 1.4–2.3 mi^2 (3.5–6 km^2).

In winter and spring, they live in groups comprised of both sexes and any age. As spring approaches, does wean yearlings and soon leave herd. In summer, does live isolated, nursing their fawns. Bulls in summer live in separate groups; young form temporary groups, easily scattered, and gather in new ones. At the approach of rut, in September–October, bulls become intolerant of each other. Does return to their winter groups, especially for feeding time. At the peak of rut, in November–December, bulls antagonistic toward each other, start antler fighting (mostly frequent among two-year-old males). Does in heat are driven from their groups by dominating bulls. At the end of rut, the estrous does rejoin their clans after breeding. Dominating bulls lose interest in does and search for good feeding places to graze in solitude. Later bulls return to their clans.

FEEDING ECOLOGY AND DIET
In winter, western red cedar, Douglas fir, western hemlock, blueberry, deer fern, bunchberry, salal, and the arboreal beard lichens are the most important forages. In spring, diet is Douglas fir, different species of *Rubus* (salmonberry, blackberry, thimbleberry, raspberry, bramble), salal, willows, bracken, *Pteridium acquilinum*, fireweed, horsetail, and pearly everlasting. There is a significant seasonal difference in nutritive value and digestibility of forage. In winter, they lose 20–25% of their autumn weight. During spring, most of the important nutrients in newly grown material are readily digestible.

REPRODUCTIVE BIOLOGY
Polygynous. Bucks shed antlers in January and re-grow during summer. Rut occurs from mid November to the beginning of December. Does are in heat repeatedly 22–29 days, with most does conceiving during their second ovulation. Three to four days before heat, a bull starts following a doe and stays with her three to four days after mating; during a rut, a bull can only mate three or four does. Gestation period lasts 200 days; fawns appear at the first half of June. Sometimes yearlings mate at their first year, and 45–80% become pregnant, while 90–95% of does older than two years become pregnant. By end of year, each 100 does bring 41–78 fawns. Along with high rate of productivity, display high mortality. Lifespan is six years. One-third of population perishes each year, especially after a harsh winter.

CONSERVATION STATUS
Not threatened.

SIGNIFICANCE TO HUMANS
Important game species. In North America, population numbers of black-tailed deer is about 1.5 million and annual game harvest about 0.5 million. ◆

Marsh deer
Blastocerus dichotomus

TAXONOMY
Blastocerus dichotomus (Illiger, 1815), Lake Ypoa, Paraguay.

OTHER COMMON NAMES
French: Cerf des marais, cerf de marecages; Spanish: Ciervo del los pantanos.

PHYSICAL CHARACTERISTICS
Largest of Cervidae in South America. Shoulder height: 3.6–4 ft (110–120 cm); head and body length: 70-80 in (180–200 cm); tail length: 4–6 in (10–15 cm); weight: 154–242 lb (70–110 kg), maximum 330 lb (50 kg). Male antlers: 24 in (60 cm) in length, dark yellow, doubly forked, each with four or sometimes five tines; weight of antlers 5.5 lb (2.5 kg). Harsh and longhaired coat is reddish brown to chestnut, lighter in lower parts, with

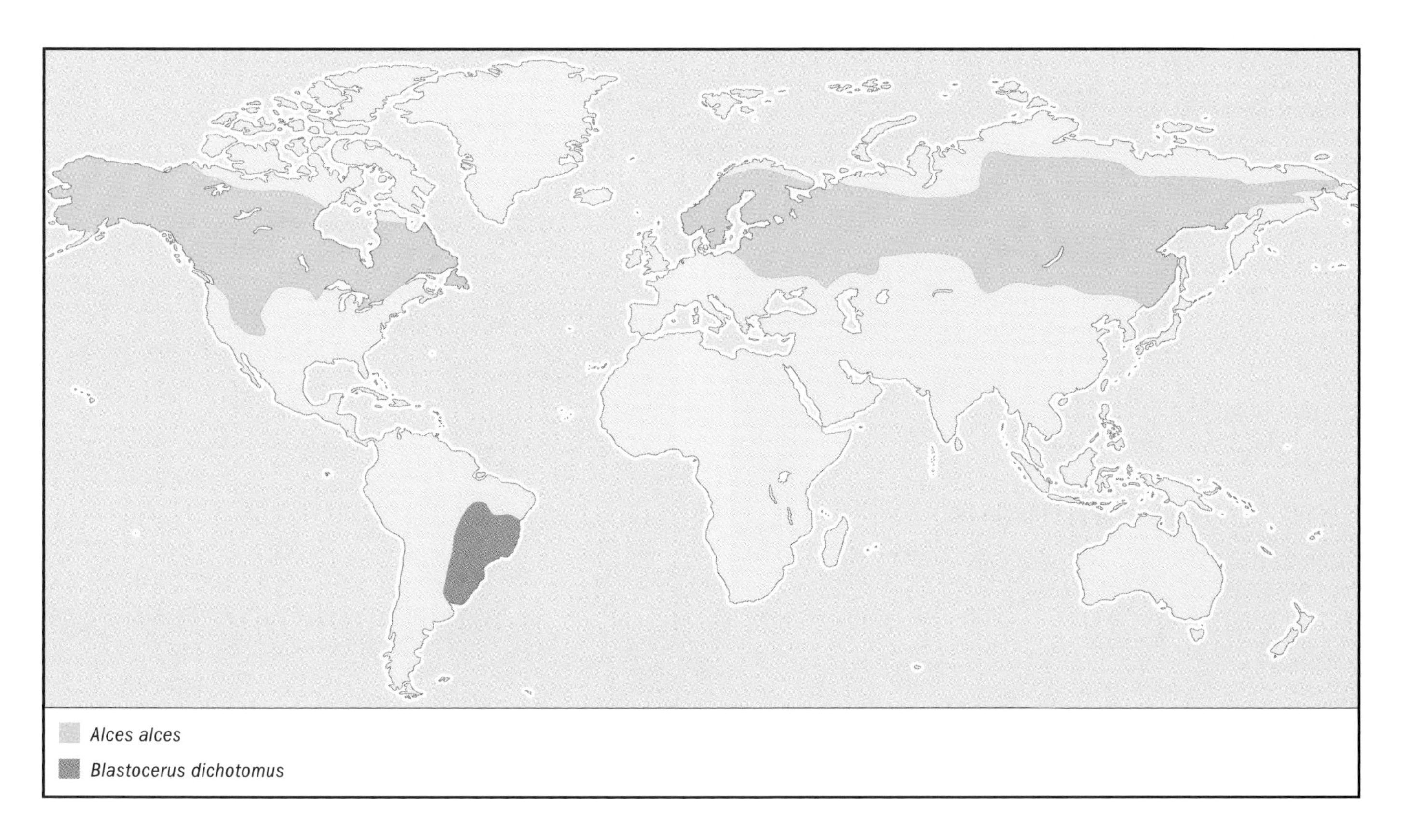

black muzzle, lips, and legs, and white orbital rounds. Large ears adorned by fluffy white hair. Tail is yellowish red above and dark brown to black underside. Fawns develop no spots, unlike other deer species. Hooves are well adapted to boggy habitats: with developed dewclaws, long hooves to 2.8–3.2 in (7–8 cm), with widely splayed middle "fingers."

DISTRIBUTION
Keeps to upper reaches of Rio Paraguay where the world's largest floodplain exists.

HABITAT
Inhabit floodplains and marshlands, sometimes interchanged with islands of wet savanna covered with high grass and small forest. Connected to riparian habitats, migrating near grasslands in flood time.

BEHAVIOR
Mostly nocturnal; start grazing at dusk at marshy clearings. They stay easily in water, though prefer shallow waters. During floods, they move to higher elevations. Live either solitarily or in small family groups (to six animals), including male, some females, and their offspring.

FEEDING ECOLOGY AND DIET
Feed on herbs, reeds, and aquatic plants.

REPRODUCTIVE BIOLOGY
Polygynous. No definite rut season or definite time of antler shed. Fawns can be found year-round. Gestation period is about 260 days; usually one young per birth. Fawn follows the mother until one year old, though weaning is usual in about five months. Both sexes reach sexual maturity at one year. Females mate again just after parturition.

CONSERVATION STATUS
Today considered an Endangered species, though in 1996 was regarded as Vulnerable. It is presumed that population of marsh deer is rapidly declining due to destruction of preferable habitats. Recent census shows that, in poorly accessible marshes in upper reaches of Rio Paraguay in south Brazil, still high density of marsh deer population. Status of marsh deer population is considered critical, as they vanished from Uruguay and became rare in Peru, Bolivia, Paraguay, and Argentina.

SIGNIFICANCE TO HUMANS
Livestock owners hunt for marsh deer to get rid of rival for livestock. ◆

Chilean huemul
Hippocamelus bisulcus

TAXONOMY
Hippocamelus bisulcus (Molina, 1782), Chilean Andes.

OTHER COMMON NAMES
English: Huemul, Chilean guemal; French: Cerf de andes méridionales, huémul des andes méridionales; Spanish: Ciervo andino meridional, huemul.

PHYSICAL CHARACTERISTICS
Small. Shoulder height: 31–35 in (77–90 cm); body length: 55–65 in (140–165 cm); tail length: 4.5–5.5 in (11–13 cm); weight: 100–145 lb (45–65 kg). Coat color of both sexes yearlong is monotonous, yellowish grizzly brown with whitish belly, a black Y-shaped stripe on muzzle, and a brown spot on rump. Bucks grow small two-pointed antlers. Both bucks and does have elongated canines, covered by a lip.

DISTRIBUTION
The Andes of central and southern Chile and Argentina.

HABITAT
Inhabit thick woods and bushes at altitudes 4,300–5,600 ft (1,300–1,700 m), on steep slopes, in rugged relief.

BEHAVIOR
Linked to constant home ranges to 90–200 ac (36–82 ha). Buck and adult does live together, sometimes in a small group of eight, comprising a buck and does. In rut, buck marks range by butting bushes, thus dispersing secretion of head scent glands. A buck defends does from rivals.

FEEDING ECOLOGY AND DIET
Grass eater, feeding on cereals and sedges.

REPRODUCTIVE BIOLOGY
Polygynous. Rut in winter (July–August). Gestation is eight months. Fawns appear at the end of rainy season February–April. Fawns stay hidden.

CONSERVATION STATUS
Now considered Endangered. Recently, deer inhabited high altitude plateaus in Andes. Hunting pressure, competition with cattle, and pursuit by wild dogs decreased population numbers to 1,300 animals living in some localities in southern Chile and Argentina.

SIGNIFICANCE TO HUMANS
Game species in the past, now preserved as rare species. ◆

Southern pudu
Pudu pudu

TAXONOMY
Pudu pudu (Molina, 1782), Chile.

OTHER COMMON NAMES
English: Chilean pudu; French: Pudou du sud; German: Pudu; Spanish: Ciervo enano, venadito chileno.

PHYSICAL CHARACTERISTICS
Smallest among Cervidae, with short legs and rounded body. Shoulder height: 14–18 in (35–45 cm); body length: 2.8 ft (85 cm); tail length: 3 in (8 cm); weight: 20–33 lb (9–15 kg). Thick bright coat is reddish brown, while lips and insides of ears are orangey. Fawns develop white spots. Males wear short spiked antlers, 2.8–4 in (7–10 cm) long.

DISTRIBUTION
Southern Chile.

HABITAT
Rainforests, bamboo groves, in mountains to the snow limit. Choose thickets to defend against human pursuit, though also prey of wild cats and foxes.

BEHAVIOR
Live solitarily or in pairs, rarely in small groups (to three). Home grounds to 40–65 ac (16–26 ha) are well arranged with trails connecting feeding grounds and resting points. Dung

piles often mark trails. Scent marks made by secretion of preorbital and frontal glands as well as urine spots on trails and tree branches play important role in pudu communications. Crepuscular and nocturnal, they are very cautious, regularly stop feeding to listen and sniff around. When in danger, they bark and flee in zigzag pattern through inaccessible thickets and steep rocks.

FEEDING ECOLOGY AND DIET
Feed on leaves, twigs, bark, buds, fruits, seeds, and rarely on herbs, To reach food, can stand on hind legs or scramble along fallen tree trunks.

REPRODUCTIVE BIOLOGY
Polygynous. Rut starts in autumn; fawning takes place the next spring (or in November–January in Southern Hemisphere). Gestation period is 210 days, usually it is one young each birth. Weaning occurs after two months. Females are sexually matured and can participate in reproduction during first year, males on their second autumn. Life expectancy is 8–10 years, to 15 years in zoos.

CONSERVATION STATUS
Endangered.

SIGNIFICANCE TO HUMANS
Game species. ◆

Red brocket
Mazama americana

TAXONOMY
Mazama americana (Erxleben, 1777), Cayenne, French Guiana.

OTHER COMMON NAMES
French: Daguet rouge; German: Grossen Roten Spiesshirsch; Spanish: Corzuela roja.

PHYSICAL CHARACTERISTICS
Small. Shoulder height: 14–30 in (35–75 cm); head and body length: 28–53 in (72–135 cm); tail length 3–6 in (8–15 cm); weight: 44 lb (20 kg). The hair on muzzle radiates in all directions from two whorls. Coat color is monotonous, light to dark brown; lighter in under parts and belly, tail end is white. The body is stout, limbs are slender, and the back is arched. Antlers are simple, spiked.

DISTRIBUTION
Mexico, Brazil, Argentina, Bolivia, and Paraguay.

HABITAT
Live in forests from sea level to altitudes of upper forest limit.

BEHAVIOR
Live solitary, sedentary life; active day and night. Wary, hide when disturbed, often remain unnoticed due to camouflage colors. Easy prey for predators.

FEEDING ECOLOGY AND DIET
Grasses, vines, and tender green shoots. Important part of diet are fruits, cereals, and fungi.

REPRODUCTIVE BIOLOGY
Polygynous. Bucks and does together only during rut, which is year long, though most of mating from July–September in main rainy season. Gestation period is 225 days, calving occurs in December–January, in short rainy period; usually one young per birth. Does participate in breeding from age one year. Life expectancy in captivity to 14 years.

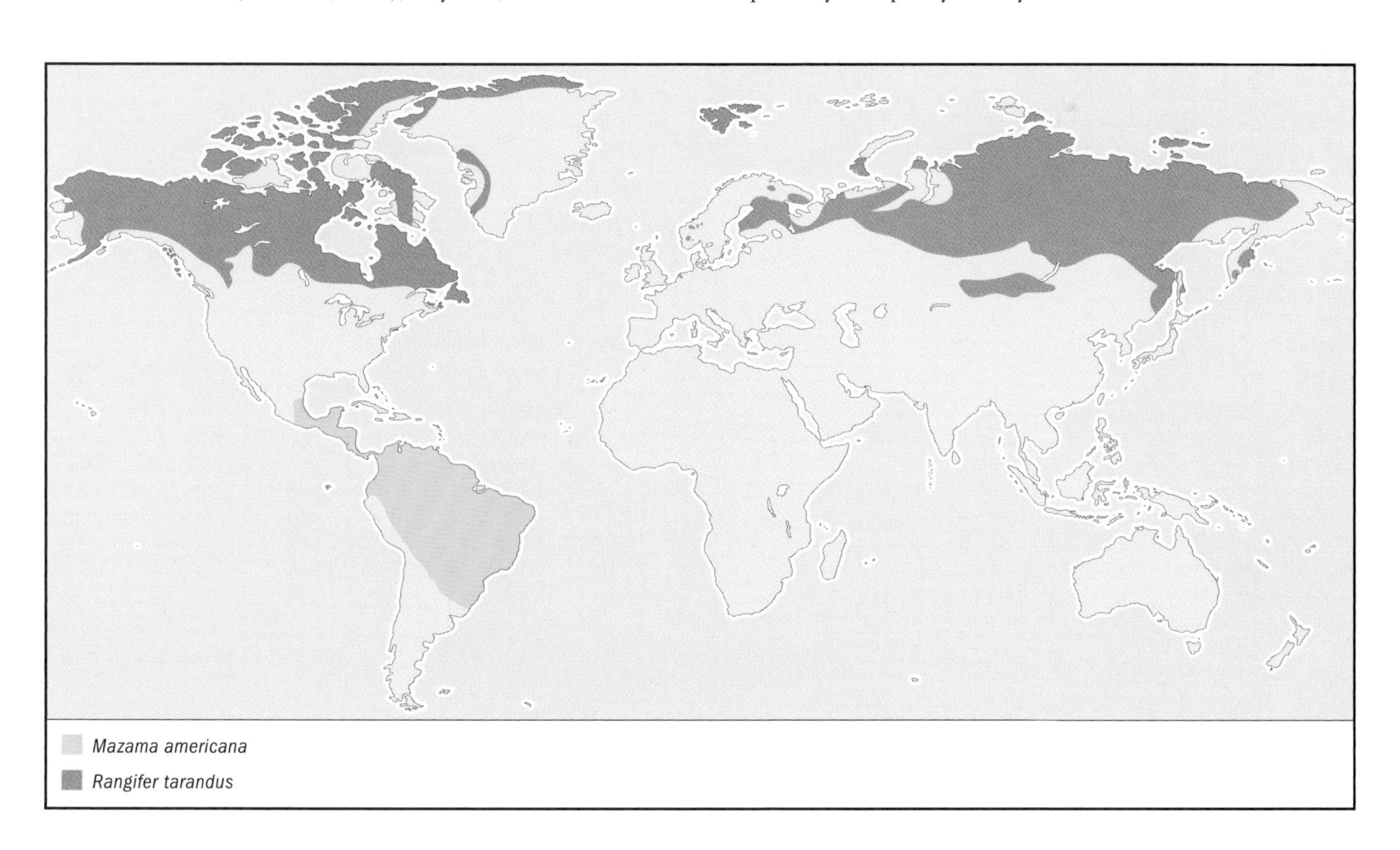

CONSERVATION STATUS
Data Deficient.

SIGNIFICANCE TO HUMANS
Experience severe hunting pressure and pursuit from farmers defending bean fields and corn crops. ◆

Moose
Alces alces

TAXONOMY
Alces alces (Linnaeus, 1758), Sweden.

OTHER COMMON NAMES
English: American moose; French: Elan; German: Elch; Spanish: Alce.

PHYSICAL CHARACTERISTICS
Largest of contemporary Cervidae. European moose: bull body length 87–110 in (220–280 cm), cow body length 87–106 in (220–270 cm); weight: bull 620–930 lb (280–420 kg), cows 600–770 lb (270–350 kg). North American moose: bull body length 140 in (350 cm), cow 125 in (315 cm); weight: bull 880–1,400 lb (400–630 kg), cows 1,200 lb (550 kg). Short body with hump-like withers, sloping rump, and very short tail, all mounted on long legs to 31 in (80 cm). Head is huge, long, and narrow, with a square upper lip hangs over the lower one. Muzzle is hairy, with a small bald spot between nostrils. The ears are large, oval, and vividly express all moods from fear to aggression. Short and thick neck furnished with mane, a skinny pendant (the bell) hanging from the throat; in North American moose, bell reaches 14 in (35 cm). Hooves are long and narrow. Dewclaws are also long, functional fingers surrounded by a strong stretchable membrane that reaches ground surface while walking. There is also a stretchable web between hoof fingers. Usually moose antlers are wide, palmate, though there are individuals with deer-like antlers in the same habitats. Coat color in adults is dark-brown with lighter legs. Moose have no rump patch. Calves are reddish brown, with no spots. Teeth are adapted to soft vegetable forage. Incisors are straight, chisel-like, good at nipping off wood bark. Moose cannot feed on harsh steppe grasses.

DISTRIBUTION
In North America, moose area extends from Alaskan tundra to Minnesota. Range of moose distribution in Eurasia spans from Scandinavia and eastern Poland to Pacific Ocean.

HABITAT
Inhabits nearly total forest zone of the northern hemisphere, and penetrates along forested or bushy ravines or river valleys far to the north to tundra, or to the south to steppes. Altitude of habitats varies from seashore plains to mountain forest limits. Moose adapted to climate with drastically changing temperatures. Snow cover deeper than 28–31 in (70–80 cm) impedes traveling, though they survives in areas with snow cover deeper than 7 ft (200 cm). Inhabit vast marshlands of western Siberia where they cross the swampiest patches crawling on their belly with forelegs stretched out in front.

In winter, they prefer mature coniferous and mixed deciduous-coniferous forests, young coniferous forests, forested banks of rivulets and lakes, burnouts, and cut clearings. In northern areas of America and Eurasia, moose come to shrub tundra and reach shores of the Arctic Ocean migrating along river valleys.

BEHAVIOR
Live solitarily or in small groups: mother, calf (calves), sometimes yearling(s). Two females with calves happen to make a congregation; sometimes a bull joins them. In areas rich in forages, there are sometimes large moose congregations, to 270 animals in Altai Mountains. No social bonds exist in those gatherings and, when disturbed, moose flee independently, without following any leader. Sedentary though summer and winter; home grounds might be separated. Migrate 18–24 mi (30–40 km) searching for more suitable habitat as seasons change. In winter, the deeper snow, the less desirable a home ground.

Active mostly in mornings and evenings, and switch to nocturnal life in summer, as insect harassment increases. Moose usually walk or trot (to 9.3 mph [15 km/h]) and can rush to gallop (18.6 mph [30 km/h]) for a short distance; they are good swimmers and divers.

FEEDING ECOLOGY AND DIET
Winter diet is wood and shrub bark and branches, while in summer they feed on leaves of trees and shrubs, on aquatic plants, forest under story grasses, and various herbs. It strongly prefers willow, poplar, aspen, mountain ash, blueberry, bird cherry tree, and buckthorn. An adult in winter consumes daily 22–30 lb (10–13 kg) of forage, and more than 66 lb (30 kg) during summer and spring. Aquatic plants make an important part of diet to supply animals with necessary nutritive components. Moose easily digest many toxic plants, the latter comprising up to 30% of diet. The highly unusual structure of the moose nose is apparently a specialization for feeding on aquatic vegetation. It evolved late in the Pleistocene in Eurasia.

REPRODUCTIVE BIOLOGY
Polygynous. Beginning of rut in August is signed by broken bushes, striped bark, as bulls are getting rid of antler velvet. Bulls produce typical sounds similar with groans. After one to two days together, they part and a bull starts searching for another female. The hairy skin flap under the jaw, the bell, has been identified as a scent distribution organ. It is splashed with urine when bulls dig rutting pits and serves to attract females, which are greatly attracted to bull moose scent. From end of August to mid October, during rut, females come to heat every 18–21 days. Gestation period varies 215–243 days, calving lasts from mid April to mid July. There are average 1.2-1.6 young per birth. Females from 3–7 years old often give birth to twins. Moose reach sexual maturity at one and a half years, bulls mate from an age of two and a half years.

CONSERVATION STATUS
Not threatened.

SIGNIFICANCE TO HUMANS
One of the most important game animal of the northern hemisphere. Many people use moose harvest as main source of food and skin. In Russian army of the eighteenth century, all horsemen wore trousers made of moose skin. Currently, moose became subject of sport game for sustainable use. Annual harvest in North America is more than 50,000; in Sweden 164,000; in Russia 80,000. ◆

Reindeer
Rangifer tarandus

TAXONOMY
Rangifer tarandus (Linnaeus, 1758), Swedish Lapland.

OTHER COMMON NAMES
English: Caribou; French: Renne; German: Rentier, Wildren; Spanish: Reno.

PHYSICAL CHARACTERISTICS
Medium-sized. Shoulder height: 33–59 in (85–150 cm); body length: males 70–84 in (180–214 cm), females 64–81 in (162–205 cm); tail length: 6–8 in (14–20 cm); weight: males 200–460 lb (92–210 kg), females 174–256 lb (79–116 kg). A dwarf subspecies (*Rangifer tarandus platyrhynchus*) of 31–37 in (80–95 cm) shoulder height inhabits Spitzbergen Islands. Reindeer have elongated body and short legs. Muzzle is covered by hair, and a thick mane along the lower part of neck adorns both sexes. Hooves are very large, wide, and flat, accompanied by wide and long dewclaws, with long coarse hairs between fingers.

Unlike all other deer, both males and females wear antlers. Antlers are forked, reaching 39 in (100 cm) width and 53 in (135 cm) length of beam in males. Coat color is nearly white to a shady light in winter and grayish brown in summer. Tail and rump patch are white. In winter, hairs are air-filled to supply thermal insulation. Construction of teeth is particularly adapted for grazing and to eat soft lichens; teeth are not fitted for browsing on woody vegetation.

DISTRIBUTION
Wide areas in Eurasia and North America from the Arctic Ocean and arctic islands southward to 50°N in Scandinavia, Finland, European part of Russia, and to 58°N in western Siberia.

HABITAT
Inhabit arctic deserts such as on Arctic Ocean islands, in lichen or lichen-mossy tundra, covered by dwarf bushes of birch, willow, alder; in high mountains, tundra with dwarf vegetation, in forest-tundra where clearings with dwarf vegetation interchange with larch clumps. Thin coniferous forests (pine, larch) with abundant woody lichens are also common habitat as well as vast marshlands (in western Siberia) and swamps amidst forests. Reindeer are perfectly adapted to life on seaside plains as well as in coniferous forests on hillsides and in open, woodless mountain plateaus to 8,850 ft (2,700 m).

BEHAVIOR
Reindeer live in families (female with calf), herds, and gatherings. Most typical are herds of 2,500–3,000 individuals coordinated in movements and following a single leader when disturbed. Gatherings during migrations reach 80,000–100,000 animals and more, consisting of several herds; each herd coordinated in movements. Forest herd is three to eight (maximum 55) animals, as visual contact between more animals is hard in a forest.

Following the leaders is a typical behavioral reaction of disturbed deer. Actually, experienced females and bulls are the first to leave a herd's protection to escape on their own.

Herd life increases competition for feeding in snow. Adult bulls shed antlers soon after rut, in the beginning of winter. Calves follow mothers during winter to feed in their crater. Barren females and young bulls shed antlers in March, so only pregnant females wear antlers in springtime.

Reindeer make seasonal migrations. Each population uses particular calving grounds where females return year after year. Summer pastures are also constant for each population, though position of those pastures might change, depending on natural conditions of the year. Reindeer use constant migration routes, especially points of river crossings. They move to points of most favorable combinations of snow cover and abundance of food.

Reindeer run with high speed. Good swimmers, they can cross sea straits that are 75 mi (120 km) wide. In areas under high hunting pressure, deer run 1,640–1,980 ft (500–600 m), but in places of low human pressure, they will allow a human to approach within a distance of 32–39 ft (80–100 m).

FEEDING ECOLOGY AND DIET
In Eurasia tundra, winter food consists of 20–90% lichens. Daily diet of an animal is about 11 lb (5 kg) of lichens. Also, winter diet is comprised of dry plants, green shoots of cotton grass, sedges, horse-tail, and green mosses. Important source of protein in autumn are mushrooms. Feeding in winter is mostly on lichens, which consist of 50–80% carbohydrates. At the same time, animals experience crucial lack of vitamins, proteins, and fat. They consume young leaves of birch and willow, twigs, buds, and flowers, cotton grass, and sedges.

REPRODUCTIVE BIOLOGY
Polygynous. Rut time coincides with autumn migrations in September–October. Fights between bulls are frequent. Winners, the strongest animals control 7–8 females. Rutting bulls eat rarely, and soon are exhausted, changing the hierarchy. Bulls younger than one and a half years do not take part in rut due to rivalry of elder ones. Most large females take part in rut; among those, more than 80% get pregnant. Gestation period is 192–246 days, mother gives birth to one calf. A newborn stands in one hour and first suckles in five hours. After 5–7 hours, mother can leave the birthing ground, followed by calf. Strong link between mother and calf lasts three months. Lifespan of bulls is 4–5 years, maximum to 14 years; of females, 6–7 years up to 19 years.

More than 40% of calves die in the first year and about 30% of the rest die at the second year. Wolves are the main predators, then brown bear, raven, golden eagle and sea eagle. Also, calves die during spring migration while crossing big rivers or from cold.

CONSERVATION STATUS
Not threatened.

SIGNIFICANCE TO HUMANS
Reindeer is the principal source of survival for indigenous people of the north. There are more than three million wild reindeer in North America and Eurasia and, also, about 2.5 million of tame reindeer. Reindeer meat is of specific quality. Deer pelts provide clothes necessary to survive in harsh northern climate. Velvet antlers are used in Asian medicines. ◆

Common name / Scientific name/ Other common names	Physical characteristics	Habitat and behavior	Distribution	Diet	Conservation status
Peruvian huemul *Hippocamelus antisensis* French: Cerf des Andes septentrionales, guémal péruvien; Spanish: Ciervo andino septentrional, guemal	Coloration is speckled yellowish gray-brown. Coat is coarse and brittle and longest on forehead and tail. Dark brow streak on face. Head and body length 55.1–65 in (140–165 cm), tail length 4.5–5.1 in (11.5–13 cm), shoulder height 30.5–35.4 in (77.5–90 cm), weight 99.2–143.3 lb (45–65 kg).	Inhabits mainly hills, rugged country, and steep mountain slopes, at elevations of 8,200–17,060 ft (2,500–5,200 m). Active during daylight. Groups form as segments, or subpopulations, of a larger group or population. Groups consist of adult males and females, accompanied by a few young. Solitary animals uncommon.	The Andes of Peru, western Bolivia, north-eastern Chile, and northwestern Argentina	Lichens, mosses, herbs, and grasses.	Data Deficient
Northern pudu *Pudu mephistophiles* French: Pudu du nord; Spanish: Pudu norteño, sachacabra, venadito de los páramos	Coloration is generally buffy, the middle of the back is dark brown, underparts are buffy to rufous. The face, outer ears, chin, and feet are dark brown to black. Head and body length 23.6–32.5 in (60–82.5 cm), tail length 1–1.8 in (2.5–4.5 cm), shoulder height 9.8–16.9 in (25–43 cm), weight 12.8–29.5 lb (5.8–13.4 kg).	Found in temperate zone forests and fringing grasslands at 6,560–13,120 ft (2,000–4,000 m). Has been found only alone or in pairs. Has a whistling vocalization.	Andes of Colombia, Ecuador, and Peru.	Herbaceous vegetation including bamboo, leaves, bark, twigs, buds, blossoms, fruit, and berries.	Lower Risk/Near Threatened
Brocket deer *Mazama bororo* English: Small red brocket; Spanish: Bororó de Sáo Paulo	General coloration is brown to dark brown. Head and body length 28.3–53.1 in (72–135 cm), tail length 2–7.9 in (5–20 cm), shoulder height 13.8–29.5 in (35–75 cm).	Usually found in woodlands and forests from sea level to elevations of 16,400 ft (5,000 m). Relatively sedentary, diurnal, and nocturnal. Extremely shy.	Presently found in Patagonia from about 39° to 45°S latitude.	Many kinds of plants, some of the preferred items include grasses, vines, and tender green shoots.	Data Deficient
Merioa brocket deer *Mazama bricenii* Spanish: Corzuela gris enana	Small, reddish coloration. Head and body length is 28.3–53.1 in (72–135 cm), tail length is 2–7.9 in (5–20 cm), and shoulder height is 13.8–29.5 in (35–75 cm).	Found in dimmed forests, evergreen forests, and deserts. Relatively sedentary, diurnal, and nocturnal. Extremely shy.	Western Venezuela and Colombia.	Many kinds of plants, some of the preferred items include grasses, vines, and tender green shoots.	Not threatened
Dwarf brocket deer *Mazama chunyi* French: Daguet gris nain; German: Kleiner Grauer Mazama; Spanish: Conzuela montera	Coloration is uniformly light to dark brown with a reddish tint. Head and body length 28.3–53.1 in (72–135 cm), tail length 2–7.9 in (5–20 cm), shoulder height 13.8–29.5 in (35–75 cm).	Usually found in woodlands and forests from sea level to elevations of 16,400 ft (5,000 m). Relatively sedentary, diurnal, and nocturnal. Extremely shy.	The Andes of southern Peru and Bolivia.	Many kinds of plants, some of the preferred items include grasses, vines, and tender green shoots.	Data Deficient
Gray brocket deer *Mazama gouazoubira* English: Brown brocket deer; Spanish: Corzuela gris	Coloration is grayish brown. Even, convex nasals. Males have slender, straight antlers that are ridged at the base. Shoulder height 17.7 in (45–61 cm), weight 37.5 lb (17 kg).	Usually solitary. Individuals of both sexes maintain stable home ranges for periods ranging from two to four years. Use urination, defecation, forehead rubbing, and thrashing to communicate and show territory.	San Jose Island (Panama); Peru, Ecuador, and Colombia east to Brazil and south to Bolivia, Paraguay, northern Argentina, and Uruguay.	Many kinds of plants, some of the preferred items include grasses, vines, and tender green shoots.	Data Deficient
Yucatán brown brocket deer *Mazama pandora*	Coloration is brown to gray-brown, underside is whitish. Males have long, divergent, and usually curved antlers. Large patch of long, dark, stiff hairs on forehead. Head and body length 28.3–53.1 in (72–135 cm), tail length 2–7.9 in (5–20 cm), shoulder height 13.8–29.5 in (35–75 cm).	Usually solitary. Individuals of both sexes maintain stable home ranges for periods ranging from two to four years. Use urination, defecation, forehead rubbing, and thrashing to communicate and show territory.	Yucatán Peninsula, Mexico.	Many kinds of plants, some of the preferred items include grasses, vines, and tender green shoots.	Data Deficient
Little red brocket deer *Mazama rufina* French: Daguet rouge nain; German: Kleiner Roter Mazama; Spanish: Conzua chica	Coloration is brown. Head and body length 28.7 in (73 cm), shoulder height 17.7 in (45 cm).	Reside in tropical forests. Relatively sedentary, diurnal, and nocturnal. Extremely shy.	Ecuador, southern Colombia.	Many kinds of plants, some of the preferred items include grasses, vines, and tender green shoots.	Lower Risk/Near Threatened
Pampas deer *Ozotoceros bezoarticus* French: Cerf des pampas; Spanish: Ciervo de las pampas; venado de campo	Coloration of upperparts is reddish brown or yellowish gray. Face, crown, and tail are darker. Head and body length 43.3–55.1 in (110–140 cm), shoulder height 27.6–29.5 in (70–75 cm), mass 66.1–88.2 lb (30–40 kg).	Reside in open savannas and cerrado which used to be found in most of the natural grasslands of South America south of the Amazon. Seasonal breeder. Largely sedentary. Generally solitary.	Brazil, northern Argentina, Paraguay, Uruguay, and southern Bolivia.	Herbivorous, but exact diet is unknown.	Lower Risk/Near Threatened

Resources

Books

Baskin, Leonid, and Kjell Danell. *Ecology of Ungulates. A Handbook of Species in Eastern Europe, Northern and Central Asia.* Heidelberg: Springer Verlag, 2003.

Bubenik, G. A., and Anthony B. Bubenik, eds. *Horns, Pronghorns, and Antlers. Evolution, Morphology, Physiology, and Social Significance.* Heidelberg: Springer-Verlag, 1990.

Flerov, Konstantin K. *Musk Deer and Deer.* Moscow: Izdatelstvo Akademii Nauk SSSR, 1952.

Geist, Valerius. *Deer of the World.* Mechanicsburg, PA: Stackpole Books, 1998.

Geist, Valerius, and Fritz Walther, eds. *The Behaviour of Ungulates and Its Relation to Management.* Gland: International Union Conservation Nature Publications, 1974.

Hudson, Robert J., Karl R. Grew, and Leonid M. Baskin, eds. *Wildlife Production Systems. Economic Utilization of Wild Ungulates.* Cambridge: Cambridge University Press, 1989.

Redform, K. H., and J. F. Eisenberg. *Mammals of the Neotropics: The Southern Cone.* Chicago and London: The University of Chicago Press, 1992.

Serret, A. *El Huemul. Fantasma de la Patagonia.* Buenos Aires: Zagier and Urruty Publishers, 2000.

Vislobokova, Inessa A. *Fossilized Deer of Eurasia.* Moscow: Nauka, 1990.

Whitehead, G. K. *The Whitehead Encyclopedia of Deer.* Stillwater, MN: Voyager Press, 1993.

Periodicals

Cowan, I. McT., and V. Geist. "Aggressive Behavior in Deer of the Genus *Odocoileus.*" *Journal of Mammalogy* 42, no. 4 (1961): 522.

Dusek, G. L. "Ecology of White-Tailed Deer in Upland Ponderosa Pine Habitat on Southern Montana." *Prairie Naturalist* 19, no. 1 (1987): 1.

Hershkovitz, P. "Neotropical Deer (Cervidae). Part 1. Pudus, Genus *Pudu* Gray." *Fieldiana Zoology* New Series 11 (1982): 1.

Pac, H. I., W. F. Kasworm, L. R. Irby, and R. J. Mackie. "Ecology of the Mule Deer, *Odocoileus hemionus*, along the Cast Front of the Rocky Mountains, Montana." *Canadian Field-Naturalist* 102, no. 2 (1988): 227.

Povilitis, A. "Characteristics and Conservation of a Fragmented Population of Huemul *Hippocamelus bisulcus* in Central Chile." *Biological Conservation* 86 (1998): 97.

Sher, Andrey V. "History and Evolution of Moose in USSR." *Swedish Wildlife Research "Viltrevy"* 1, no. 1 (1987): 71.

Leonid Baskin, PhD

▲

Okapis and giraffes

(Giraffidae)

Class Mammalia
Order Artiodactyla
Suborder Ruminantia
Family Giraffidae

Thumbnail description
Large to very large ruminants with high shoulders, sloping backs, long legs and neck, and two skin-covered horns. Very long, black, prehensile tongue; disruptive patterned coats for camouflage

Size
5–11 ft (1.5–3.3 m) shoulder height; up to 18 ft (5.5 m) to top of head; 460–4,250 lb (210–1,930 kg)

Number of genera, species
2 genera; 2 species

Habitat
Savanna and forest

Conservation status
Lower Risk/Conservation Dependent: 1 species; Lower Risk/Near Threatened: 1 species

Distribution
Sub-Saharan Africa

Evolution and systematics

Giraffes evolved relatively late, probably about 25 million years ago in the lower Miocene, from a branch of the even-toed ungulates, which was also to produce cattle, antelopes (Bovidae), and deer (Cervidae). Giraffids of various forms roamed Europe and Asia, benefiting from the climate change that saw subtropical woodland replaced by open savanna grasslands. This habitat change also allowed ruminants to spread and diversify into Africa. The most primitive form that can be distinctly classified as giraffes was Paleotraginae, which had a short neck and was about the size of red deer. They are generally considered the immediate ancestor of the modern giraffe. The okapi is very similar to this ancestral form of giraffe.

Physical characteristics

The most familiar of the two Giraffidae species, the giraffe (*Giraffa camelopardalis*), is the world's tallest mammal and largest ruminant. Males stand up to 11 ft (3.3 m) at the shoulder, but a very long neck of up to 8 ft (2.4 m) means the height of the top of their horns may be 18 ft (5.5 m) from the ground. The neck has only seven vertebrae, like most other mammals, but these are greatly elongated and strung with tendons and muscles anchored to a shoulder hump. A male giraffe weighs up to 4,350 lb (1,930 kg). Females are about 2–3 ft (0.7–1 m) shorter, and weigh up to 2,600 lb (1,180 kg). The giraffe's body is comparatively short, with high shoulders sloping steeply to the hindquarters. Legs are long, and the dinner-plate sized hooves are cloven. The giraffe's tail is hock length (up to 39 in [1 m]), with a long tassle.

The giraffe's head tapers to a point, eyes are large, and the tongue is black and very long (18 in [45 cm]). Both male and female giraffes have horns of solid bone covered with skin and up to 5 in (13.5 cm) long. Females' horns are thin and tufted, while males have thicker horns but the hair is worn smooth by sparring. Males also have a median horn and four or more smaller bumps. Giraffes are one of the few animals born with horns.

The giraffe's forest cousin, the okapi (*Okapia johnstoni*), has a number of features in common, including two skin-covered horns up to 6 in (15 cm) long (absent in females), a long,

A créche—a single adult watches over a number of giraffe calves while adults feed at a distance. (Illustration by Joseph E. Trumpey)

prehensile, black tongue, and lobed canine teeth. But while its neck is long relative to most ruminants, it is not nearly so well developed as the giraffe's. Superficially, the okapi more closely resembles a horse. Its shoulder height is no more than 5.6 ft (170 cm) and its weight no more than 550 lb (250 kg). The okapi's body also slopes to its hindquarters, but its head is more horse-like, with large, flexible ears.

Both giraffe and okapi have patterned coats that help camouflage them in their respective habitats. The giraffe's coat varies from pale brown to rich chestnut, dissected by a tapestry of creamy buff lighter hair. Patterns not only vary among subspecies, but each individual giraffe has a unique pattern. The giraffe's scientific name, *camelopardalis*, means "camel marked like a leopard." The okapi's coat is dark, chestnut to chocolate brown in color, with distinctive creamy white stripes on the upper legs, white stockings on the ankles, and dark garters at the leg joints. The head is lighter colored than the body, with a black muzzle.

The giraffe's unique evolutionary design has required adaptations of its circulatory and respiratory systems. Unusually elastic blood vessels and a series of valves help offset any sudden buildup of blood when the head is raised, lowered, or swung rapidly, keeping the animal from blacking out. Arteries near the feet are thick-walled and less elastic, with thick, tight skin around the lower leg, decreasing downward blood pressure and helping to avoid edema.

A newborn giraffe (*Giraffa camelopardalis*) can play and suckle 30 minutes after birth. (Photo by Connie Bransilver/Photo Researchers, Inc. Reproduced by permission.)

The giraffe's heart is enormous, 2 ft (60 cm) long and weighing 25 lb (11 kg). Beating up to 170 times a minute, it pumps the highest known blood pressure of any mammal, more than twice that of humans. Yet blood pressure in the giraffe's brain is similar to most other large mammals. Giraffes also breathe rapidly, around 20 times a minute, as the long neck contains partially inhaled and exhaled air that must be cleared.

Distribution

Both Giraffidae species are confined to sub-Saharan Africa. In historic times, the giraffe was found throughout arid and dry savanna zones, wherever trees occurred. That range has now been halved, and the giraffe has disappeared from most of West Africa (apart from a residual population in Niger). Elsewhere it remains fairly common, and is not confined to reserves. Okapis have a much more restricted range in rainforest in the Democratic Republic of the Congo (Zaire).

Habitat

Giraffes inhabit arid and dry savanna zones where there are adequate feed trees. Okapis, thought of as forest giraffes, require a quite different habitat of dense, moist, tropical lowland forest. They prefer secondary forest, near water, and are usually found at altitudes of 1,600–3,200 ft (500–1,000 m).

Behavior

Giraffes and okapis exhibit very different forms of social behavior. Giraffes are sociable, living in loose, open, unstable herds of up to 20 animals, with individuals joining and leaving the herd at will. Herds can be all female, all male, females with young, or of mixed genders and age. But in general, females are more sociable than males. A female giraffe is only likely to be out of sight of another female when calving, whereas mature males wander more widely in search of estrous females. Giraffes are non-territorial, but do have home ranges that can vary enormously in size, from 2 to 252 mi^2 (5–654 km^2), depending on food and water availability. Giraffes have been known to cover 580 mi^2 (1,500 km^2) in the Sahel of Niger looking for food in the dry season.

Okapis live mainly solitary lives, or in mother-offspring pairs, although ephemeral groupings are occasionally seen feeding together at prime food sites. Okapis have individual, overlapping home ranges of about 1–2 mi^2 (2.5–5 km^2), with estimated population densities of 0.25–1 animals per mi^2 (0.1–0.4 km^2).

Both species appear to have a dominance hierarchy among males. Sub-adult male giraffes spar to establish this hierarchy. Sparring may involve the animals standing stiff-legged in parallel, marching in step with necks horizontal and looking ahead, rubbing and intertwining necks and heads, and leaning against one another to assess each other's weight. Most dramatic is "necking,", in which two giraffes, standing side-by-side, swing their heads at one another, using their horns to aim blows at the rump, flanks, or neck of their opponent. The animals rock to dampen impacts, but a hard blow can knock down an opponent.

Reticulated giraffes (*Giraffa camelopardalis reticulata*) neck fighting. (Photo by K. & K. Ammann. Bruce Coleman, Inc. Reproduced by permission.)

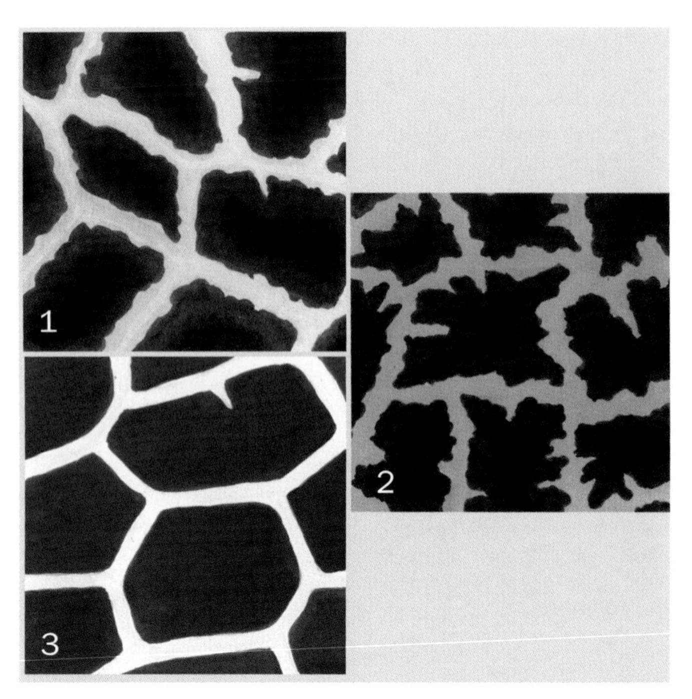

Spotting pattern differences between three giraffe subspecies. 1. Rothschild's giraffe (*Giraffa camelopardalis rothschildi*); 2. Masai giraffe (*G. c. tippelskirchi*); 3. Reticulated giraffe (*G. c. reticulata*). (Illustration by Joseph E. Trumpey)

Mature bulls that have established their dominance rarely resort to violent conflicts, but when they do these can be serious. The head of a giraffe bull gets heavier throughout its life, as it develops protective bony deposits, and can weigh 66 lb (30 kg) by age 20. A heavy blow to the underbelly of an opponent can cause considerable damage. Giraffes have a very thick hide as a first line of defense against the severe blows they can receive from other giraffes' armored heads.

Little is known of the okapi's behavior in the wild. However, captive studies suggest a dominance hierarchy. A dominant male displays by holding his head high and neck straight, while a subordinate signals submission by placing head and neck on the ground.

The fluid nature of giraffe society reflects their need to spend a lot of time feeding and to move independently between variable-sized trees. Their height and acute eyesight allow giraffes to maintain visual contact at long distances, and herds may disperse widely, only clustering at a particularly good food tree, or if bothered by lions.

Giraffe cows spend over half of their day browsing. Bulls spend less time feeding (43%), but more time walking (22% as against 13% for cows) in their perpetual search for females to mate with. Nights are mostly spent lying down, ruminating. Okapis are also most active during the daytime, but tend to follow well-worn paths through the jungle when feeding. In such a dense habitat, scent is believed to be important in locating breeding partners—okapis have glands on their feet, and have been observed urine-marking bushes.

Both species have a similar ambling gait, walking with their weight supported alternately on their left and right legs, like camels, while their necks move synchronously to maintain balance. The name giraffe may come from the Arabic *zarafa*, or *xirapha*, meaning "one who walks swiftly" or "graceful one." Giraffes can run at up to 35 mph (60 kph), with their hind legs swinging simultaneously ahead of and outside the fore legs, like a rabbit.

Lions are the main predator of giraffes, but hyena, leopard, and wild dog can also take young animals; up to 75% of calves die in their first year. Giraffes rely on their height and vision to spot predators and will defend themselves by kicking, but a pride of lions can take even a mature, healthy bull. Giraffe meat represents a substantial portion of lions' diet in some locations. Male giraffes are more vulnerable, because being more often alone they lack the benefits of group vigilance; in lion-rich areas, the giraffe population can be heavily skewed toward a majority of females. The okapi's main predator is the leopard.

Both giraffes and okapis are usually silent, but do have a range of vocalizations. Giraffe calves bleat or make mewing calls, and cows seeking lost calves bellow. Courting bulls may

A male giraffe establishes dominance over another male, Masai Mara Game Reservation in Kenya. (Photo by E. & P. Bauer. Bruce Coleman, Inc. Reproduced by permission.)

Giraffes have a prehensile tongue, making it easy to wrap it around leaves and pull them from the trees and into their mouths. (Photo by David M. Maylen, III. Reproduced by permission.)

cough raucously. Alarm snorts, moaning, snoring, hissing, and flutelike sounds have also been heard. Research suggests giraffes may communicate with infrasonic sound, as do elephants and blue whales—which suggests their social system may be more complex than once thought. Adult okapis may cough softly during rutting, and young okapis are noisier, with a repertoire of coughs, bleats, and whistles.

Feeding ecology and diet

The giraffe's unique physical form has evolved to exploit a 6 ft (2 m) band of foliage beyond the reach of any other browser except the elephant. A large bull giraffe, with a neck joint that allows its head to be extended vertically, can reach foliage 19 ft (5.8 m) high. Giraffes are almost pure browsers, feeding mainly on broad-leaved deciduous foliage in the rains and on evergreen species near watercourses at other seasons, and only occasionally eating grass, herbs, and fruit.

The giraffe's diet includes 100 species of tree but, in any one area, 40–60 will be utilized, with acacia and combretum being favored. They are selective feeders, using their narrow muzzles, flexible upper lip, and long, prehensile tongue to select the most nutritious leaves. The animal's molars crush acacia thorns, while smooth shoots are stripped of leaves by pulling them through a gap between the molars and canines. A male giraffe may consume 145 lb (66 kg) of food a day, but can survive periods of poor-quality fodder on as little as 15 lb (7kg). Giraffes are ruminants with a four-chambered stomach, and, unusually, they can chew cud while walking, which maximizes feeding opportunities. In mineral-deficient areas they sometimes chew on bones, and even carcasses, though this carries a risk of disease.

Okapi have also been recorded feeding on more than 100 species of plant, some of them poisonous to man. They also use their prehensile, 12-in (30-cm) tongue to pluck buds, leaves, and branches (used as well as for grooming). Okapis range about 0.6 miles (1 km) per day as they forage, and regularly eat sulfurous clay found along river banks to supplement their mineral intake.

Giraffes drink every two or three days if water is available, but extract much of their requirement from foliage. Their characteristic splay-legged, bent-kneed pose when drinking was thought to be necessary because of their long necks, but okapis adopt a similar pose, which suggests this is not the explanation. Drinking does make giraffes vulnerable to predators, and one animal will often keep watch while another drinks.

When high leafy vegetation is scarce, the giraffe (*Giraffa camelopardalis*) is able to eat from the ground. (Photo by St. Myers/OKAPIA/Photo Researchers, Inc. Reproduced by permission.)

Adult male okapi (*Okapia johnstoni*) in the forest. (Photo by Tom Brakefield. Bruce Coleman, Inc. Reproduced by permission.)

Reproductive biology

Giraffes are polygynous and breed year round, with a conception peak in the rainy season. Bulls always check the reproductive status of cows when they meet, by tasting their urine. Females reach sexual maturity at four to five years. Males compete for mating rights from age seven, but older, larger males have a strong advantage, and by the time a female in estrus is ready to mate, the local dominant male has usually asserted his rights without having to fight.

A male giraffe follows an estrous female closely, keeping other males away. The male's courtship behavior includes urine testing with a pronounced lip curl (called "flehmen"), rubbing his head on the female's rump and resting it on her back, licking her tail, and lifting his foreleg. After initially proving evasive, the female will circle, hold her tail out, and stand in a mating attitude to invite mounting.

Gestation lasts about 15 months, before a single calf (very rarely twins) is born, often in a favored calving ground. Giraffes have very long gestation periods compared to other ruminants, and bear—relative to the mother's body mass—very large young. These are very precocious, getting to their feet quite quickly. Giraffe milk is rich in fat, suggesting rapid growth of the calf. Mothers give birth standing up or even while walking, so the baby is dropped 6 ft (2 m) to the ground. A newborn calf is 6 ft (2 m) tall and weighs 110–120 lb (50–55 kg). In its first week, a calf lies out half the day and most of the night, guarded by its mother, which usually stays within 75 feet (25 m), but may leave to go to water. An absent mother always returns before dark to suckle the calf and stay with it overnight. After a few weeks, calves form nursery groups. Young start browsing after only one month, but are not weaned until one year, and may stay close to their mother until 22 months. Cows will usually conceive again about five months after calving. Young males remain in maternal herds until about three years old, after which they join bachelor herds and eventually leave the area. Females tend to remain in the range. Life expectancy is 20–25 years.

Okapi are believed to be polygynous, though courtship and mating rituals are known only from zoos. Females, sexually mature at two years old, remain in estrous for up to a month, advertising themselves by urine marking and calling. When a male is attracted, the two animals circle, sniff, and lick each other. The female is initially aggressive, while the male asserts dominance by extending his neck, tossing his head, and thrusting one leg forward. Male courtship displays include flehmen and showing his white throat patch. Mating then takes place.

A single calf is born between August and October after a gestation period of 14–15 months. It weighs 30–65 lb (14–30 kg) at birth, has a small head, short neck, long, thick legs, and a conspicuous mane. Mothers retreat into dense forest to give birth, after which the calf lies hidden for several weeks. It may spend 80% of its first two months in a hidden nest, growing

Giraffes have to bend a long way down to drink. When they're bent over, they are more vulnerable to predators, such as crocodiles or lions. Groups of giraffes take turns drinking, presumably for this reason. (Photo by David M. Maylen, III. Reproduced by permission.)

Reticulated giraffes (*Giraffa camelopardalis reticulata*) inspecting a newborn. (Photo by K. & K. Ammann. Bruce Coleman, Inc. Reproduced by permission.)

rapidly and avoiding predators. Calves are weaned after six months, but may continue to suckle for more than a year. They reach adult size after three years. Longevity is over 30 years in captivity.

Conservation status

Neither giraffes nor okapis are considered under immediate threat, but some giraffe subspecies are subject to intense pressure. The giraffe is classified as Lower Risk/Conservation Dependent by the IUCN. It is still common in east and southern Africa, but populations have fallen dramatically in West Africa, through over-hunting and habitat degradation. It used to be found from Senegal to Lake Chad, but is now only present in parts of Niger, where its conservation has been made a priority. Elsewhere, giraffes have survived where other large mammals have disappeared, perhaps because their height allows them to compete for food with domestic animals. The okapi is classified as Lower risk, Near Threatened, but accurate assessments of population size and trends are difficult because the animal lives in dense forest. Driven to extinction in Uganda in the early twentieth century, the okapi has been protected in the Democratic Republic of Congo since 1933. Habitat loss due to deforestation and poaching are its biggest threats.

Significance to humans

The giraffe has been revered by African cultures for millennia—it is depicted in early cave paintings and in an-

A female okapi (*Okapia johnstoni*) uses her long tongue to reach high leaves. (Photo by William Munoz/Photo Researchers, Inc. Reproduced by permission.)

Profile of an okapi (*Okapia johnstoni*). (Photo by William Munoz/Photo Researchers, Inc. Reproduced by permission.)

cient Egyptian art. A captive giraffe was recorded in Rome in 46 B.C. Giraffe tails were highly prized by the ancient Egyptians and later cultures, being used as good-luck bracelets, fly whisks, and thread for sewing. Tourist demand for giraffe-hair bracelets continues to account for giraffe poaching even today, though meat and hide are more common motives. Giraffe are an important tourist draw in game reserves of east and southern Africa, and in southern Africa this has encouraged private game ranchers to introduce the species to land previously farmed for cattle. Giraffe have also been reintroduced to their former range in the southern Kalahari by conservationists. They also do well in zoos throughout the world.

The okapi was unknown to Western science until 1900. Henry Stanley heard of the okapi living in the dense Congo forests in 1890, but conjectured it was some kind of African donkey. Explorer Sir Harry Johnston journeyed into the Congo in 1899 and also heard about the animal, assuming it to be a species of forest zebra. Examination of skin samples at the British Museum led to mistaken announcements of the discovery of a new species of zebra. After obtaining a complete skin and two skulls, it was finally ascertained that the new species was a forest giraffe. Excitement over this "new" animal, which quickly assumed unicorn-like status, led to a rush by zoos to obtain specimens but there was a high mortality rate during translocation, and it was not until air shipment was introduced in the second half of the twentieth century that okapis could be moved with reasonable success. Many zoos now breed okapi successfully.

1. Okapi (*Okapia johnstoni*); 2. Giraffe (*Giraffa camelopardalis*). (Illustration by Joseph E. Trumpey)

Species accounts

Giraffe

Giraffa camelopardalis

TAXONOMY

Cervus camelopardalis (Linneaus, 1758), Sudan. Up to nine subspecies, but taxonomy is not fully agreed, and some subspecies hybridize. The reticulated giraffe (*Giraffa c. reticulata*) of north Kenya is most distinctive with a latticework of thin pale lines separating large chestnut-colored square patches. The larger Baringo, or Rothschild's, giraffe (*Giraffa c. rothschildi*) of western Kenya and eastern Uganda has chestnut patches separated by broader white lines but no spotting below knees. The Masai giraffe (*Giraffa c. tippelskirchi*) of East Africa has the most irregular pattern of star-shaped brown or tan spots. Other races include the West African giraffe (*Giraffa c. peralta*) in Niger, Kordofan giraffe (*Giraffa c. antiquorum*) of western Sudan, Nubian giraffe (*Giraffa c. camelopardalis*) in eastern Sudan and Eritrea, Thornicroft giraffe (*Giraffa c. thornicrofti*) in Zambia, the Angolan giraffe (*Giraffa c. angolensis*) in southern Angola, northern Namibia, and western Botswana, and the southern giraffe (*Giraffa c. capensis*) of South Africa, Zimbabwe, and Mozambique. Some authorities synopsize Kordofan and West African, Nubian and Rothschild's, and Angolan and southern giraffes, respectively.

OTHER COMMON NAMES

French: Girafe; German: Giraffe; Spanish: Jirafa.

PHYSICAL CHARACTERISTICS

13–18 ft (3.9–5.5 m) tall; weight 1,200–4,350 lb (550–1,930 kg). Large eyes and long black tongue. Male and female have skin-covered horns. Coat is pale brown to chestnut patterned with lighter creamy buff.

DISTRIBUTION

Africa, south of the Sahara.

HABITAT

Arid and dry savanna with trees.

BEHAVIOR

Gregarious, in loose-knit herds of up to 20 animals. Not territorial, but males have dominance hierarchy based on seniority.

FEEDING ECOLOGY AND DIET

A browsing ruminant that feeds selectively on leaves of more than 100 trees and shrubs, especially acacia and combretum species.

REPRODUCTIVE BIOLOGY

Polygynous. Single calf born after 15 month gestation. Females pregnant in fourth year, then at least 16 months (usually 20) between births.

CONSERVATION STATUS

Lower Risk/Conservation Dependent (IUCN). Largely eliminated from former range in western Africa, but common in suitable habitat in eastern and southern Africa. Population estimated at 141,000 (IUCN, 1998).

SIGNIFICANCE TO HUMANS

Hunted, often illegally, for meat, skin, and good luck charms. Translocated to stock game farms and zoos. Darting and translocation carries high risk of heart attack in older animals, so younger specimens preferred. ◆

Giraffa camelopardalis

Okapia johnstoni

Okapi

Okapia johnstoni

TAXONOMY

Equus johnstoni (P. Sclater, 1901), Zaire.

OTHER COMMON NAMES

French: Okapi; German: Okapi; Spanish: Okapi.

PHYSICAL CHARACTERISTICS

Shoulder height 5–5.6 ft (150–170 cm); weight 462–550 lb (210–250 kg.). Females slightly taller than males. Long, black tongue. Male has skin-covered horns. Coat is dark chestnut to chocolate-brown with creamy white stripes on upper legs and white stockings.

DISTRIBUTION

Rainforest in northeastern Democratic Republic of the Congo.

HABITAT

Dense, moist tropical lowland forest near water, especially secondary forest.

BEHAVIOR

Mainly solitary, except mothers with calves or mating pairs. Rarely forming temporary groups when feeding. Overlapping ranges, not territorial, but male dominance hierarchy.

FEEDING ECOLOGY AND DIET
A browsing ruminant utilizing well-trodden paths linking favorite feeding areas.

REPRODUCTIVE BIOLOGY
Likely polygynous. Single calf born August–October after 14–15 months gestation. Females sexually mature at two years.

CONSERVATION STATUS
Lower Risk/Near Threatened. Estimated at 30,000 in wild. Populations are highly localized but relatively common where they occur.

SIGNIFICANCE TO HUMANS
Okapi is a corruption of the native name, *o'api*. Historically, subsistence hunting using noose traps, pitfalls, and (rarely) driving animals into nets was probably at sustainable levels. Okapis are now protected, but poaching for bush meat poses a threat to long-term survival. An okapi breeding reserve was established in Epulu, Democratic Republic of Congo, to supply okapi with fresh genes for zoos and breeding centers, but the program was disrupted by war. ◆

Resources

Books

Estes, Richard D. *The Safari Companion: A Guide to Watching African Mammals.* White River Junction, VT: Chelsea Green, 1999.

Lindsey, S. L., M. N. Green, and C. L. Bennett. *The Okapi: Mysterious Animal of Congo-Zaire.* Austin, TX: University of Texas Press, 1999.

Mills, G., and L. Hes. *Complete Book of Southern African Mammals.* Cape Town: Struik, 1997.

Periodicals

Bodmer, R. E., and G. Rabb. "*Okapia johnstoni*." *Mammalian Species* 422 (1992): 1–8.

Henk, P., H. van der Jeugd, and H. T. Prins. "Movements and Group Structure of Giraffe (*Giraffa camelopardalis*) in Lake Manyara National Park, Tanzania." *Journal of Zoology* 251, no. 1 (2000): 15–21.

Le Pendu, Y., I. Ciofolo, and A. Gosser. "The Social Organization of Giraffes in Niger." *African Journal of Ecology* 38, no.1 (2000): 78.

Ann and Stephen B. Toon

Pronghorn

(Antilocapridae)

Class Mammalia

Order Artiodactyla

Suborder Ruminantia

Family Antilocapridae

Thumbnail description
Small ruminant characterized by large head, elongated nose, large eyes and ears, heavily built body, long and slim but powerful limbs, and short tail

Size
Total length: 52.1–58.8 in (132.3–149.4 cm); shoulder height: 32.7–37 in (83.1–94 cm); for extant species: 87–129 lb (40–59 kg)

Number of genera, species
1 genus; 1 species

Habitat
Grassland, steppe, and desert plains

Conservation status
Not threatened

Distribution
Southeastern Alberta and southwestern Saskatchewan in Canada; in the 18 westernmost states (except Washington) of the United States; and in the states of Baja California Sur, Chihuahua, Coahuila, and Sonora, Mexico

Evolution and systematics

Two subfamilies, with five genera of Merycodontinae and 12 of Antilocaprinae, generally are recognized but researchers have not successfully divided them into species. Five subspecies have been named because of minor differences in color, size, and form. Mitochondrial DNA analyses since the early 1990s, plus intergrade zones among named subspecies, support the concept of clines within a wide-ranging species more than that of subspecies. Currently the subfamily Atilocapridae contains just one genus with one species.

The taxonomy for this species is *Antilocapra americana* (Ord, 1815), plains and highlands of the Missouri River, United States.

Physical characteristics

Extant antilocaprids (as were extinct antilocaprids) are long-legged runners. Merycodonts and early antilocaprins were small for ruminants: about 20 in (51 cm) high at the shoulder and weighing 26–44 lb (12–20 kg). Larger antilocaprins varied from that size to the size of pronghorn.

The pronghorn head, which varies from creamy white on the sides to various shades of brown on top, is marked in the male with brownish black patches that start just below the ears and extend downward 3–4 in (7.6–10.2 cm). Blackish markings sometimes occupy the entire face below a line connecting the horns. The throat is elegantly marked with a white crescent above a white shield; the point of the shield usually joins the white underparts at the base of the neck. Erectile hairs of the mane on the back of the neck are russet, tipped with black. A tan strip connects or nearly connects the upper surface of the short tail, about 4 in (10.2 cm) long, with the colored back. Rusty brown to tan hairs cover the back, most of the neck, and the outer sides of the limbs. White hair extends well up the sides of the body, forming a rectangular area between shoulder and hip. White hairs of twin rump patches are about 3 in (7.6 cm) long and almost gleam when erected, fanning beyond the normal contours of the body. The stalks of hairs are pithy with large central air cells that provide ex-

Pronghorn (*Antilocapra americana*) fleeing, in Arizona, USA. (Photo by Jack Couffer. Bruce Coleman, Inc. Reproduced by permission.)

cellent protection from wind and cold, and a pronghorn can regulate the insulating properties of its pelage by erecting or flattening hair.

The pronghorn is a plastic species with a variable number of chromosomes. Occasional individuals have extra horns, teats, or ribs. Females may or may not have horns, and some does have a different type of uterus than those possessed by most. The prongs that give pronghorn their name are unique, as is the annual shedding of horn sheaths. Normally, all adult bucks, and about 70% of adult does, have horns. The horns of does are small. In relation to its size, the pronghorn has the largest eyes of any North American ungulate. Pronghorn feet are long and slim and lack dewclaws common to most ruminants.

Distribution

A map of pronghorn distribution was developed in 1999, based on reports received from provincial and state wildlife agencies in Canada and the United States. Locations in Mexico were taken from a variety of publications. In 1909, E. T. Seton produced an outer delineation of what he called "primitive" range. That line agrees quite well with present knowledge concerning former distribution of the species. However, a good deal of the land within the outer delineation (high mountains, heavy forest, tall shrublands, etc.) was not pronghorn habitat.

Habitat

This small ruminant is a selective feeder adapted to succulent, high-protein vegetation. Thus, under most conditions, it does compete much with more generalized feeders such as bison (*Bison bison*) or cattle (*Bos taurus*). Excellent eyesight and great speed adapt the pronghorn to a "see and flee" existence on plains with short vegetation.

Pronghorn are most numerous in short-grass prairies where they can reach shrubs for forage when the ground is covered with snow. The next most important habitat is steppe, preferably with vegetation ranging 5–30 in (13–76 cm) in height. Deserts support less than 1% of the population.

Pronghorn are associated most frequently with treeless, undulating, or flat terrain. Herds range from near sea level to 11,000 ft (3,353 m) above sea level, but the greatest concentrations occupy landscapes 3,000–8,000 ft (914–2,438 m) above sea level.

Behavior

Pronghorn herd together for protection (especially of fawns) from predators, seasonal movements, and to feed on the best available forage during winters. Because the animals can maintain visual contact over considerable distances and come together quickly when threatened, pronghorn bands are well organized even when widely scattered. Erection of gleaming white rump patches and alarm snorts quickly alert the scattered animals.

Pronghorn produce a variety of vocalizations and mechanical sounds. Both sexes snort or blow through their noses

Pronghorn (*Antilocapra americana*) at water's edge in Yellowstone Park, USA. (Photo by Erwin and Peggy Bauer. Bruce Coleman, Inc. Reproduced by permission.)

Pronghorn (*Antilocapra americana*). (Illustration by Patricia Ferrer)

(sounding much like *cha-oo*) when they note something strange in their habitat. Fawns produce soft, mew-like bleats, which help does locate their concealed offspring. Distressed fawns and sometimes adults emit loud bleats. Does emit low grunts or clicking sounds when approaching their hidden fawns or when pursued by bucks. The buck roar sounds like a loud stomach growl and probably is caused by exhaling air. It is used principally during chases of either bucks or does. The snort-wheeze is a strictly male vocalization (a typical cha-oo, followed by a series of shorter bursts of sound descending in pitch and volume). Courting males make three distinct sounds: one is vocal, while the other two result from movements of the mouth. The pitch of the whine starts high and then smoothly decreases. If a doe moves away, the whine grades into the roar. During courtship approaches, bucks lip smack or tongue flick, both of which create a low, staccato, sucking sound.

Marking by territorial pronghorn bucks is accomplished during a linked sniff-paw-urinate-defecate (SPUD) sequence. Postural changes are extreme and each element of the sequence is performed in exaggerated form. A buck performing the SPUD precedes urination by pawing and, sometimes, by *flehmen* (lip curling). He then moves each front foot ahead to take an extended stance, urinates on the pawed spot, then moves his hind feet ahead, squats, and defecates. The feces fall on or close to the urine spot.

Nearly equal numbers of studies have described pronghorn bucks as territorial (defending a breeding area) and as having a harem-type mating system in which dominant bucks controlled and defended does without regard to specific locations. Breeding systems obviously have an ecological basis. When the best resources are clumped, pronghorn tend to be territorial, and bucks holding the best territories do most of the breeding. The breeding system shifts to harem formation as resources become more uniform, when population levels are low, or when sex ratios are unusually skewed (ratios of one male to 10 or more females). According to C. R. Maher, "Spacing systems are affected as much by the animals' internal state, i.e., physiology, as by external conditions, e.g., population density and food abundance. Physiological condition is reflected in metabolic processes and hormone levels, and hormones have important effects on reproduction and aggression."

Pronghorn sleep often, but without continuity or regularity, and usually for short periods. Activity is greatest just after sunrise and before sunset. Pronghorn spent the majority of their time feeding and the next greatest proportion of time reclining and ruminating. It has been observed that their time upright almost equals time reclining and greatly exceeds both feeding and reclining time for bucks during the rut. Time upright also exceeds that for reclining, and almost equals feeding for does during the rut.

Sizcs of home ranges seemingly should reflect amounts of activity. However, large variations in sizes of pronghorn home

Pronghorn (*Antilocapra americana*) one-week-old fawn hidden in the wildflowers of Big Hole River Valley, Montana, USA. (Photo by Erwin and Peggy Bauer. Bruce Coleman, Inc. Reproduced by permission.)

Pronghorn (*Antilocapra americana*) buck marking his territory. (Photo by Leonard Lee Rue III. Bruce Coleman, Inc. Reproduced by permission.)

ranges result from differing habitat quality, population and group sizes, density, past history of land use, and season. Thus, sizes of pronghorn home and seasonal ranges and distances moved daily or seasonally vary so much with these factors and weather conditions that results from studies in one area seldom have application to another area, or even to another year.

As reported in 1927, pronghorn of Jackson Hole traveled 150 mi (240 km) southward to the Red Desert. Those on the prairies of Saskatchewan moved 100 mi (160 km) south and west. Those on the plains near the Rockies went toward the foothills, and those on open prairies around the Black Hills flocked there from all directions. Observations of these movements between Jackson Hole and the Red Desert were later confirmed by radio-tracking.

Because environmental factors can influence or trigger moves, the timing of movements is variable. Fences, interstate highways, reservoirs, and railways complicate movements and reduce, often drastically, the carrying capacity in areas where pronghorn must move long distances to procure the year-round necessities of life. Few truly migratory herds remained by the beginning of the twenty-first century.

Feeding ecology and diet

Pronghorn prefer succulent forbs to other foods. An adequate supply of forbs during spring and early summer generally assure a good fawn crop. During snowy winters and periods of droughts, the animals depend on shrubs, mostly sagebrush (*Artemesia* spp.)

Feeding pronghorn are described as walking slowly, occasionally pausing and holding their noses just above or brushing through vegetation. Food items appear to be found by scent as well as sight. Individuals commonly pass their noses along the length of a forb and then remove one or two leaves from the plant. At other times, individuals temporarily specialize in single items such as flowers, walking briskly from plant to plant, obviously guided by vision. Pronghorn use their mobile and dexterous upper lips to select plant parts, draw them into their mouths, and hold them in place until cut free by the incisiform teeth, which are jerked forward and upward by movements of the head.

Reproductive biology

Pronghorn does usually breed for the first time when 16–17 months of age, but fawns only occasionally conceive. The gestation period of approximately 250 days is unusually long for so small a ruminant. About 98% of does on good habitat produce twins. Southern pronghorn have longer rutting periods that begin earlier than those of animals on northern prairies. Northern pronghorn and those in Oregon generally breed during mid- to late September. All breeding seasons are timed so that fawns are born when nutritional content of forage is best for the does.

Pronghorn are polygynous, and a few robust bucks often do most of the breeding. Breeding and birthing generally take place in prime habitat, where the most succulent forbs are available. After giving birth, a doe licks the entire fawn, but grooming of the anogenital region soon takes precedence over

Pronghorn (*Antilocapra americana*). (Photo by © Joe McDonald/Corbis. Reproduced by permission.)

After four weeks, pronghorn fawns group together for most of the day. One doe often stays near the young group. (Photo by Pat & Tom Leeson/Photo Researchers, Inc. Reproduced by permission.)

generalized licking and continues after each suckling bout until the fawn is two to three weeks old. During such grooming bouts, fawns assume a distinctive rump-up posture and eliminate, while the mother ingests the urine and feces. Fawns hide for the first three weeks of life and their dams feed them about three times a day. This is a predator-avoidance behavior involving both does and fawns.

Conservation status

Pronghorn are endemic to western North America and are common in suitable habitat. Numbers were reduced to less than 30,000 early in the twentieth century. The population rebuilt to more than a million animals by the early 1980s—perhaps the carrying capacity for the reduced habitat available.

The U. S. Fish and Wildlife Service has listed as Endangered two subspecies, the peninsular pronghorn in Baja California Sur, Mexico, and the Sonoran pronghorn in Arizona and Sonora, Mexico. The federal government of Mexico has listed all endemic subspecies in Mexico as Endangered. For the total pronghorn population, unrestricted hunting and competition with millions of domestic sheep (*Ovis aries*) reduced numbers. Managed hunting, protection, and reduction in sheep numbers helped restore the population.

Peninsular and Sonoran pronghorn persist in deserts where numbers were low even before the advent of European-Americans. Since then, desertification has put these animals in serious danger of extinction. The *mexicana* subspecies has

Pronghorn (*Antilocapra americana*) grazing in the snow in western North America. (Photo by © Joe McDonald/Corbis. Reproduced by permission.)

Pronghorn (*Antilocapra americana*) males engaging in rut jousting in Suster State Park, South Dakota, USA. (Photo by Stephen J. Krasemann/Photo Researchers, Inc. Reproduced by permission.)

suffered from habitat loss, desertification, and poaching. During the last 20 years of the twentieth century, the total number of pronghorn has varied from more than a million to about 700,000. Droughty summers and severe winters reduced numbers during some years.

A recovery plan for pronghorn in Mexico was completed in 1999. It contained an estimate of 1,200 pronghorn in all of Mexico. A 1998 U.S. Fish and Wildlife Service recovery plan for Sonoran pronghorn estimated fewer than 300 of that subspecies in the United States.

Significance to humans

For many Native Americans within historic times, pronghorn mainly personified speed, grace, vigilance, good fortune, and peace. Indian myths were used to explain how pronghorn originated, why they lived on the plains, why they had a gallbladder while the deer did not, etc.

Apparently because of their beauty, paintings of pronghorn adorned prehistoric pottery and kiva walls in what is now New Mexico. Later, cowboy artist, Charlie Russell, and others frequently featured pronghorn in their western paintings.

Pronghorn (*Antilocapra americana*) buck rubbing scent from subauricular gland on vegetation to mark territory. (Photo by Leonard Lee Rue III/Photo Researchers, Inc. Reproduced by permission.)

The importance of pronghorn in late prehistoric Native American economies depended on their abundance and the availability of other food resources. Over wide areas, the abundant bison was overwhelmingly the major food source of Plains Indians. However, pronghorn contributed to dietary diversity as well as providing hides, horns, etc., for clothing and tools. They were worthy of large cooperative hunting efforts yielding up to hundreds of animals at a time.

After the near demise of bison in the late nineteenth century when rifles became readily available, pronghorn were common fare for both red and white peoples throughout the West. Great numbers of pronghorn were slaughtered for their skins, which were dried and sold by the pound.

As pronghorn increased in numbers during the 1930s, Canadian provinces and U. S. states opened hunting seasons. By the end of the hunting season in 2002, nearly five million pronghorn had been legally harvested. The harvest provided many tons of meat and days of outdoor recreation besides millions of dollars of profit for businesses throughout pronghorn country.

Pronghorn occasionally damage crops, especially alfalfa (*Medicago sativa*), extensively. Wildlife agencies sometimes pay for such damage or increase hunting pressure to reduce the number of animals.

Resources

Books

Byers, J. A. *American Pronghorn: Social Adaptations and the Ghosts of Predators Past.* Chicago: University of Chicago Press, 1997.

Geist, V., and M. H. Francis. *Antelope Country, Pronghorns: The Last Americans.* Iola, WI: Krause Publications, 2001.

Hill, R. B. *Hanta Yo.* Garden City: Doubleday and Co., 1979.

O'Gara, B. W., and J. D. Yoakum. *Ecology and Management of Pronghorn.* Washington, DC: Wildlife Management Institute, in prep.

Seton, E. T. *Life-histories of Northern Animals.* New York: Charles Scribner's Sons, 1909.

———. *Lives of Game Animals.* Garden City: Doubleday, Doran and Co., Inc., 1927.

Turbak, G. *Portrait of the American Antelope.* Hong Kong: Northland Publishing, 1995.

Periodicals

Heffelfinger, J. R., B. W. O'Gara, C. M. Janis, and R. Babb. "A Bestiary of Ancestral Antilocaprids." *Proceedings of the Pronghorn Workshop* 20 (2002).

Maher, C. R. "Quantitative Variation in Ecological and Hormonal Variables Correlates with Spatial Organization of Pronghorn (*Antilocapra americana*) Males." *Behavioral Ecology and Sociobiology* 47, no. 5 (2000): 327–338.

Organizations

The Arizona Antelope Foundation. P.O. Box 15501, Phoenix, AZ 85060 United States. Phone: (602) 433-9077. E-mail: azantelope@aol.com Web site: <http://www.azantelope.org>

North American Pronghorn Foundation. P.O. Box 1383, Rawlins, WY 82301 United States. Phone: (307) 324-5238. E-mail: info@antelope.org Web site: <http://www.antelope.org>

Bart O'Gara, PhD

• • • • •

For further reading

Alcock, J. *Animal Behavior.* New York: Sinauer, 2001.

Alderton, D. *Rodents of the World.* New York: Facts on File, 1996.

Alterman, L., G. A. Doyle, and M. K. Izard, eds. *Creatures of the Dark: The Nocturnal Prosimians.* New York: Plenum Press, 1995.

Altringham, J. D. *Bats: Biology and Behaviour.* New York: Oxford University Press, 2001.

Anderson, D. F., and S. Eberhardt. *Understanding Flight.* New York: McGraw-Hill, 2001.

Anderson, S., and J. K. Jones Jr., eds. *Orders and Families of Recent Mammals of the World.* John Wiley & Sons, New York, 1984.

Apps, P. *Smithers' Mammals of Southern Africa.* Cape Town: Struik Publishers, 2000.

Attenborough, D. *The Life of Mammals.* London: BBC, 2002.

Au, W. W. L. *The Sonar of Dolphins.* New York: Springer-Verlag, 1993.

Austin, C. R., and R. V. Short, eds. *Reproduction in Mammals.* 4 vols. Cambridge: Cambridge University Press, 1972.

Avise, J. C. *Molecular Markers, Natural History and Evolution.* London: Chapman & Hall, 1994.

Barber, P. *Vampires, Burial, and Death: Folklore and Reality.* New Haven: Yale University Press, 1998.

Barnett, S. A. *The Story of Rats.* Crows Nest, Australia: Allen & Unwin, 2001.

Baskin, L., and K. Danell. *Ecology of Ungulates. A Handbook of Species in Eastern Europe, Northern and Central Asia.* Heidelberg: Springer-Verlag, 2003.

Bates, P. J. J., and D. L. Harrison. *Bats of the Indian Subcontinent.* Sevenoaks, U. K.: Harrison Zoological Museum, 1997.

Bekoff, M., C. Allen, and G. M. Burghardt, eds. *The Cognitive Animal.* Cambridge: MIT Press, 2002.

Bennett, N. C., and C. G. Faulkes. *African Mole-rats: Ecology and Eusociality.* Cambridge: Cambridge University Press, 2000.

Benton, M. J. *The Rise of the Mammals.* New York: Crescent Books, 1991.

Berta, A., and L. Sumich. *Marine Mammals: Evolutionary Biology.* San Diego: Academic Press, 1999.

Bonaccorso, F. J. *Bats of Papua New Guinea.* Washington, DC: Conservation International, 1998.

Bonnichsen, R, and K. L. Turnmire, eds. *Ice Age People of North America.* Corvallis: Oregon State University Press. 1999.

Bright, P. and P. Morris. *Dormice* London: The Mammal Society, 1992.

Broome, D., ed. *Coping with Challenge.* Berlin: Dahlem University Press, 2001.

Buchmann, S. L., and G. P. Nabhan. *The Forgotten Pollinators.* Washington, DC: Island Press, 1997.

Burnie, D., and D. E. Wilson, eds. *Animal.* Washington, DC: Smithsonian Institution, 2001.

Caro, T., ed. *Behavioral Ecology and Conservation Biology.* Oxford: Oxford University Press, 1998.

Carroll, R. L. *Vertebrate Paleontology and Evolution.* New York: W. H. Freeman and Co., 1998.

Cavalli-Sforza, L. L., P. Menozzi, and A. Piazza. *The History and Geography of Human Genes.* Princeton: Princeton University Press, 1994.

Chivers, R. E., and P. Lange. *The Digestive System in Mammals: Food, Form and Function.* New York: Cambridge University Press, 1994.

Clutton-Brock, J. *A Natural History of Domesticated Mammals.* 2nd ed. Cambridge: Cambridge University Press, 1999.

Conley, V. A. *The War Against the Beavers.* Minneapolis: University of Minnesota Press, 2003.

Cowlishaw, G., and R. Dunbar. *Primate Conservation Biology.* Chicago: University of Chicago Press, 2000.

Craighead, L. *Bears of the World* Blaine, WA: Voyager Press, 2000.

Crichton., E. G. and P. H. Krutzsch, eds. *Reproductive Biology of Bats.* New York: Academic Press, 2000.

Croft, D. B., and U. Gansloßer, eds. *Comparison of Marsupial and Placental Behavior.* Fürth, Germany: Filander, 1996.

Darwin, C. *The Autobiography of Charles Darwin 1809–1882 with original omissions restored.* Edited by Nora Barlow. London: Collins, 1958.

Darwin, C. *On The Origin of Species by Means of Natural Selection, or The Preservation of Favoured Races in the Struggle for Life.* London: John Murray, 1859.

Darwin, C. *The Zoology of the Voyage of HMS* Beagle *under the Command of Captain Robert FitzRoy RN During the Years 1832-1836.* London: Elder & Co., 1840.

Dawson, T. J. *Kangaroos: The Biology of the Largest Marsupials.* Kensington, Australia: University of New South Wales Press/Ithaca, 2002.

Duncan, P. *Horses and Grasses.* New York: Springer-Verlag Inc., 1991.

Easteal, S., C. Collett, and D. Betty. *The Mammalian Molecular Clock.* Austin, TX: R. G. Landes, 1995.

Eisenberg, J. F. *Mammals of the Neotropics.* Vol. 1, *The Northern Neotropics.* Chicago: University of Chicago Press, 1989.

Eisenberg, J. F., and K. H. Redford. *Mammals of the Neotropics.* Vol. 3, *The Central Neotropics.* Chicago: University of Chicago Press, 1999.

Ellis, R. *Aquagenesis.* New York: Viking, 2001.

Estes, R. D. *The Behavior Guide to African Mammals.* Berkeley: The University of California Press, 1991.

Estes, R. D. *The Safari Companion: A Guide to Watching African Mammals.* White River Junction, VT: Chelsea Green, 1999.

Evans, P. G. H., and J. A. Raga, eds. *Marine Mammals: Biology and Conservation.* New York: Kluwer Academic/Plenum, 2001.

Ewer, R. F. *The Carnivores.* Ithaca, NY: Comstock Publishing, 1998.

Feldhamer, G. A., L. C. Drickamer, A. H. Vessey, and J. F. Merritt. *Mammalogy. Adaptations, Diversity, and Ecology.* Boston: McGraw Hill, 1999.

Fenton, M. B. *Bats.* Rev. ed. New York: Facts On File Inc., 2001.

Findley, J. S. *Bats: A Community Perspective.* Cambridge: Cambridge University Press, 1993.

Flannery, T. F. *Mammals of New Guinea.* Ithaca: Cornell University Press, 1995.

Flannery, T. F. *Possums of the World: A Monograph of the Phalangeroidea.* Sydney: GEO Productions, 1994.

Fleagle, J. G. *Primate Adaptation and Evolution.* New York: Academic Press, 1999.

Frisancho, A. R. *Human Adaptation and Accommodation.* Ann Arbor: University of Michigan Press, 1993.

Garbutt, N. *Mammals of Madagascar.* New Haven: Yale University Press, 1999.

Geist, V. *Deer of the World: Their Evolution, Behavior, and Ecology.* Mechanicsburg, PA: Stackpole Books, 1998.

Geist, V. *Life Strategies, Human Evolution, Environmental Design.* New York: Springer Verlag, 1978.

Gillespie, J. H. *The Causes of Molecular Evolution.* Oxford: Oxford University Press, 1992.

Gittleman, J. L., ed. *Carnivore Behavior, Ecology and Evolution.* 2 vols. Chicago: University of Chicago Press, 1996.

Gittleman, J. L., S. M. Funk, D. Macdonald, and R. K. Wayne, eds. *Carnivore Conservation.* Cambridge: Cambridge University Press, 2001.

Givnish, T. I. and K. Sytsma. *Molecular Evolution and Adaptive Radiations.* Cambridge: Cambridge University Press, 1997.

Goldingay, R. L., and J. H. Scheibe, eds. *Biology of Gliding Mammals.* Fürth, Germany: Filander Verlag, 2000.

Goodman, S. M., and J. P. Benstead, eds. *The Natural History of Madagascar.* Chicago: The University of Chicago Press, 2003.

Gosling, L. M., and W. J. Sutherland, eds. *Behaviour and Conservation.* Cambridge: Cambridge University Press, 2000.

Gould, E., and G. McKay, eds. *Encyclopedia of Mammals.* 2nd ed. San Diego: Academic Press, 1998.

Groves, C. P. *Primate Taxonomy.* Washington, DC: Smithsonian Institute, 2001.

Guthrie, D. R. *Frozen Fauna of the Mammoth Steppe.* Chicago: University of Chicago Press. 1990.

Hall, L., and G. Richards. *Flying Foxes: Fruit and Blossom Bats of Australia.* Malabar, FL: Krieger Publishing Company, 2000.

Hancocks, D. *A Different Nature. The Paradoxical World of Zoos and Their Uncertain Future.* Berkeley: University of California Press, 2001.

Hartwig, W. C., ed. *The Primate Fossil Record.* New York: Cambridge University Press, 2002.

Hildebrand, M. *Analysis of Vertebrate Structure.* 4th ed. New York: John Wiley & Sons, 1994.

Hillis, D. M., and C. Moritz. *Molecular Systematics.* Sunderland, MA: Sinauer Associates, 1990.

Hoelzel, A. R., ed. *Marine Mammal Biology: An Evolutionary Approach.* Oxford: Blackwell Science, 2002.

Hunter, M. L., and A Sulzer. *Fundamentals of Conservation Biology.* Oxford, U. K.: Blackwell Science, Inc., 2001.

Jefferson, T. A., S. Leatherwood, and M. A. Webber, eds. *Marine Mammals of the World.* Heidelberg: Springer-Verlag, 1993.

Jensen, P., ed. *The Ethology of Domestic Animals: An Introductory Text.* Oxon, MD: CABI Publishing, 2002.

Jones, M. E., C. R. Dickman, and M. Archer. *Predators with Pouches: The Biology of Carnivorous Marsupials.* Melbourne: CSIRO Books, 2003.

Kardong, K. V. *Vertebrates: Comparative Anatomy, Function, Evolution.* Dubuque, IA: William C. Brown Publishers, 1995.

King, C. M. *The Handbook of New Zealand Mammals.* Auckland: Oxford University Press, 1990.

Kingdon, J. *The Kingdon Field Guide to African Mammals.* London: Academic Press, 1997.

Kingdon, J., D. Happold, and T. Butynski, eds. *The Mammals of Africa: A Comprehensive Synthesis.* London: Academic Press, 2003.

Kinzey, W. G., ed. *New World Primates: Ecology, Evolution, and Behavior.* New York: Aldine de Gruyter, 1997.

Kosco, M. *Mammalian Reproduction.* Eglin, PA: Allegheny Press, 2000.

Krebs, J. R., and N. B. Davies. *An Introduction to Behavioural Ecology.* 3rd ed. Oxford: Blackwell Scientific Publications, 1993.

Kunz, T. H., and M. B. Fenton, eds. *Bat Ecology.* Chicago: University of Chicago Press, 2003.

Lacey, E. A., J. L. Patton, and G. N. Cameron, eds. *Life Underground: The Biology of Subterranean Rodents.* Chicago: University of Chicago Press, 2000.

Lott, D. F. *American Bison: A Natural History.* Berkeley: University of California Press, 2002.

Macdonald, D. W. *European Mammals: Evolution and Behavior.* London: Collins, 1995.

Macdonald, D. W. *The New Encyclopedia of Mammals.* Oxford: Oxford University Press, 2001.

Macdonald, D. W. *The Velvet Claw: A Natural History of the Carnivores.* London: BBC Books, 1992.

Macdonald, D. W., and P. Barrett. *Mammals of Britain and Europe.* London: Collins, 1993.

Martin, R. E. *A Manual of Mammalogy: With Keys to Families of the World.* 3rd ed. Boston: McGraw-Hill, 2001.

Matsuzawa, T., ed. *Primate Origins of Human Cognition and Behavior.* Tokyo: Springer-Verlag, 2001.

Mayr, E. *What Evolution Is.* New York: Basic Books, 2001.

McCracken, G. F., A. Zubaid, and T. H. Kunz, eds. *Functional and Evolutionary Ecology of Bats.* Oxford: Oxford University Press, 2003.

McGrew, W. C., L. F. Marchant, and T. Nishida, eds. *Great Ape Societies.* Cambridge: Cambridge University Press, 1996.

Meffe, G. K., and C. R. Carroll. *Principles of Conservation Biology.* Sunderland, MA: Sinauer Associates, Inc., 1997.

Menkhorst, P. W. *A Field Guide to the Mammals of Australia.* Melbourne: Oxford University Press, 2001.

Mills, G., and M. Harvey. *African Predators.* Cape Town: Struik Publishers, 2001.

Mills, G., and L. Hes. *Complete Book of Southern African Mammals.* Cape Town: Struik, 1997.

Mitchell-Jones, A. J., et al. *The Atlas of European Mammals.* London: Poyser Natural History/Academic Press, 1999.

Neuweiler, G. *Biology of Bats.* Oxford: Oxford University Press, 2000.

Norton, B. G., et al. *Ethics on the Ark.* Washington, DC: Smithsonian Institution Press, 1995.

Nowak, R. M. *Walker's Bats of the World.* Baltimore: The Johns Hopkins University Press, 1994.

Nowak, R. M. *Walker's Mammals of the World.* 6th ed. Baltimore: Johns Hopkins University Press, 1999.

Nowak, R. M. *Walker's Primates of the World.* Baltimore: The Johns Hopkins University Press, 1999.

Payne, K. *Silent Thunder: The Hidden Voice of Elephants.* Phoenix: Wiedenfeld and Nicholson, 1999.

Pearce, J. D. *Animal Learning and Cognition.* New York: Lawrence Erlbaum, 1997.

Pereira, M. E., and L. A. Fairbanks, eds. *Juvenile Primates: Life History, Development, and Behavior.* New York: Oxford University Press, 1993.

Perrin, W. F., B. Würsig, and J. G. M. Thewissen. *Encyclopedia of Marine Mammals.* San Diego: Academic Press, 2002.

Popper, A. N., and R. R. Fay, eds. *Hearing by Bats.* New York: Springer-Verlag, 1995.

Pough, F. H., C. M. Janis, and J. B. Heiser. *Vertebrate Life.* 6th ed. Upper Saddle River, NJ: Prentice Hall, 2002.

Premack, D., and A. J. Premack. *Original Intelligence: The Architecture of the Human Mind.* New York: McGraw-Hill/Contemporary Books, 2002.

Price, E. O. *Animal Domestication and Behavior.* Cambridge, MA: CAB International, 2002.

Racey, P. A., and S. M. Swift, eds. *Ecology, Evolution and Behaviour of Bats.* Oxford: Clarendon Press, 1995.

Redford, K. H., and J. F. Eisenberg. *Mammals of the Neotropics.* Vol. 2, *The Southern Cone.* Chicago: University of Chicago Press, 1992.

Reeve, N. *Hedgehogs.* London: Poyser Natural History, 1994.

Reeves, R., B. Stewart, P. Clapham, and J. Powell. *Sea Mammals of the World.* London: A&C Black, 2002.

Reynolds, J. E. III, and D. K. Odell. *Manatees and Dugongs.* New York: Facts On File, 1991.

Reynolds, J. E. III, and S. A. Rommel, eds. *Biology of Marine Mammals.* Washington, DC: Smithsonian Institution Press, 1999.

Rice, D. W. *Marine Mammals of the World.* Lawrence, KS: Allen Press, 1998.

Ridgway, S. H., and R. Harrison, eds. *Handbook of Marine Mammals.* 6 vols. New York: Academic Press, 1985-1999.

Riedman, M. *The Pinnipeds.* Berkeley: University of California Press, 1990.

Rijksen, H., and E. Meijaard. *Our Vanishing Relative: The Status of Wild Orang-utans at the Close of the Twentieth Century.* Dordrecht: Kluwer Academic Publishers, 1999.

Robbins, C. T. *Wildlife Feeding and Nutrition.* San Diego: Academic Press, 1992.

Robbins, M. M., P. Sicotte, and K. J. Stewart, eds. *Mountain Gorillas: Three Decades of Research at Karisoke.* Cambridge: Cambridge University Press, 2001.

Roberts, W. A. *Principles of Animal Cognition.* New York: McGraw-Hill, 1998.

Schaller, G. B. *Wildlife of the Tibetan Steppe.* Chicago: University of Chicago Press, 1998.

Seebeck, J. H., P. R. Brown, R. L. Wallis, and C. M. Kemper, eds. *Bandicoots and Bilbies.* Chipping Norton, Australia: Surrey Beatty & Sons, 1990.

Shepherdson, D. J., J. D. Mellen, and M. Hutchins. *Second Nature: Environmental Enrichment for Captive Animals.* Washington, DC: Smithsonian Institution Press, 1998.

Sherman, P. W., J. U. M. Jarvis, and R. D. Alexander, eds. *The Biology of the Naked Mole-rat.* Princeton: Princeton University Press, 1991.

Shettleworth, S. J. *Cognition, Evolution, and Behavior.* Oxford: Oxford University Press, 1998.

Shoshani, J., ed. *Elephants.* London: Simon & Schuster, 1992.

Skinner, R., and R. H. N. Smithers. *The Mammals of the Southern African Subregion.* 2nd ed. Pretoria, South Africa: University of Pretoria, 1998.

Sowls, L. K. *The Peccaries.* College Station: Texas A&M Press, 1997.

Steele, M. A. and J. Koprowski. *North American Tree Squirrels.* Washington, DC: Smithsonian Institution Press, 2001.

Sunquist, M. and F. Sunquist. *Wild Cats of the World* Chicago: University of Chicago Press, 2002.

Sussman, R. W. *Primate Ecology and Social Structure.* 3 vols. Needham Heights, MA: Pearson Custom Publishing, 1999.

Szalay, F. S., M. J. Novacek, and M. C. McKenna, eds. *Mammalian Phylogeny.* New York: Springer-Verlag, 1992.

Thomas, J. A., C. A. Moss, and M. A. Vater, eds. *Echolocation in Bats and Dolphins.* Chicago: University of Chicago Press, 2003.

Thompson, H. V., and C. M. King, eds. *The European Rabbit: The History and Biology of a Successful Colonizer.* Oxford: Oxford University Press, 1994.

Tomasello, M., and J. Calli. *Primate Cognition.* Chicago: University of Chicago Press, 1997.

Twiss, J. R. Jr., and R. R. Reeves, eds. *Conservation and Management of Marine Mammals.* Washington, DC: Smithsonian Institution Press, 1999.

Van Soest, P. J. *Nutritional Ecology of the Ruminant.* 2nd ed. Ithaca, NY: Cornell University Press, 1994.

Vaughan, T., J. Ryan, and N. Czaplewski. *Mammalogy.* 4th ed. Philadelphia: Saunders College Publishing, 1999.

Vrba, E. S., G. H. Denton, T. C. Partridge, and L. H. Burckle, eds. *Paleoclimate and Evolution, with Emphasis on Human Origins.* New Haven: Yale University Press, 1995.

Vrba, E. S., and G. G. Schaller, eds. *Antelopes, Deer and Relatives: Fossil Record, Behavioral Ecology, Systematics and Conservation.* New Haven: Yale University Press, 2000.

Wallis, Janice, ed. *Primate Conservation: The Role of Zoological Parks.* New York: American Society of Primatologists, 1997.

Weibel, E. R., C. R. Taylor, and L. Bolis. *Principles of Animal Design.* New York: Cambridge University Press, 1998.

Wells, R. T., and P. A. Pridmore. *Wombats.* Sydney: Surrey Beatty & Sons, 1998.

Whitehead, G. K. *The Whitehead Encyclopedia of Deer.* Stillwater, MN: Voyager Press, 1993.

Wilson, D. E., and D. M. Reeder, eds. *Mammal Species of the World: a Taxonomic and Geographic Reference.* 2nd ed. Washington, DC: Smithsonian Institution Press, 1993.

Wilson, D. E., and S. Ruff, eds. *The Smithsonian Book of North American Mammals.* Washington, DC: Smithsonian Institution Press, 1999.

Wilson, E. O. *The Diversity of Life.* Cambridge: Harvard University Press, 1992.

Wójcik, J. M., and M. Wolsan, eds. *Evolution of Shrews.* Bialowieza, Poland: Mammal Research Institute, Polish Academy of Sciences, 1998.

Woodford, J. *The Secret Life of Wombats.* Melbourne: Text Publishing, 2001.

Wrangham, R. W., W. C. McGrew, F. B. M. de Waal, and P. G. Heltne, eds. *Chimpanzee Cultures.* Cambridge: Harvard University Press, 1994.

Wynne, C. D. L. *Animal Cognition.* Basingstoke, U. K.: Palgrave, 2001.

Organizations

African Wildlife Foundation
1400 16th Street, NW, Suite 120
Washington, DC 20036 USA
Phone: (202) 939-3333
Fax: (202) 939-3332
E-mail: africanwildlife@awf.org
<http://www.awf.org/>

The American Society of Mammalogists
<http://www.mammalsociety.org/>

American Zoo and Aquarium Association
8403 Colesville Road, Suite 710
Silver Spring, MD 20910 USA.
Phone: (301) 562-0777
Fax: (301) 562-0888
<http://www.aza.org/>

Australian Conservation Foundation Inc.
340 Gore Street
Fitzroy, Victoria 3065 Australia
Phone: (3) 9416 1166
<http://www.acfonline.org.au>

The Australian Mammal Society
<http://www.australianmammals.org.au/>

Australian Regional Association of Zoological Parks and Aquaria
PO Box 20
Mosman, NSW 2088
Australia
Phone: 61 (2) 9978-4797
Fax: 61 (2) 9978-4761
<http://www.arazpa.org>

Bat Conservation International
P.O. Box 162603
Austin, TX 78716 USA
Phone: (512) 327-9721
Fax: (512) 327-9724
<http://www.batcon.org/>

Center for Ecosystem Survival
699 Mississippi Street, Suite 106
San Francisco, 94107 USA
Phone: (415) 648-3392
Fax: (415) 648-3392
E-mail: info@savenature.org
<http://www.savenature.org/>

Conservation International
1919 M Street NW, Ste. 600
Washington, DC 20036
Phone: (202) 912-1000
<http://www.conservation.org>

The European Association for Aquatic Mammals
E-mail: info@eaam.org
<http://www.eaam.org/>

European Association of Zoos and Aquaria
PO Box 20164
1000 HD Amsterdam
The Netherlands
<http://www.eaza.net>

IUCN-The World Conservation Union
Rue Mauverney 28
Gland 1196 Switzerland
Phone: ++41(22) 999-0000
Fax: ++41(22) 999-0002
E-mail: mail@iucn.org
<http://www.iucn.org/>

The Mammal Society
2B, Inworth Street
London SW11 3EP United Kingdom
Phone: 020 7350 2200
Fax: 020 7350 2211
<http://www.abdn.ac.uk/mammal/>

Mammals Trust UK
15 Cloisters House
8 Battersea Park Road
London SW8 4BG United Kingdom
Phone: (+44) 020 7498 5262
Fax: (+44) 020 7498 4459
E-mail: enquiries@mtuk.org
<http://www.mtuk.org/>

The Marine Mammal Center
Marin Headlands
1065 Fort Cronkhite
Sausalito, CA 94965 USA
Phone: (415) 289-7325

Fax: (415) 289-7333
<http://www.marinemammalcenter.org/>

National Marine Mammal Laboratory
7600 Sand Point Way N.E. F/AKC3
Seattle, WA 98115-6349 USA
Phone: (206) 526-4045
Fax: (206) 526-6615
<http://nmml.afsc.noaa.gov/>

National Wildlife Federation
11100 Wildlife Center Drive
Reston, VA 20190-5362 USA
Phone: (703) 438-6000
<http://www.nwf.org/>

The Organization for Bat Conservation
39221 Woodward Avenue
Bloomfield Hills, MI 48303 USA
Phone: (248) 645-3232
E-mail: obcbats@aol.com
<http://www.batconservation.org/>

Scripps Institution of Oceanography
University of California-San Diego
9500 Gilman Drive
La Jolla, CA 92093 USA
<http://sio.ucsd.edu/gt;

Seal Conservation Society
7 Millin Bay Road
Tara, Portaferry
County Down BT22 1QD
United Kingdom
Phone: +44-(0)28-4272-8600
Fax: +44-(0)28-4272-8600
E-mail: info@pinnipeds.org
<http://www.pinnipeds.org>

The Society for Marine Mammalogy
<http://www.marinemammalogy.org/>

The Wildlife Conservation Society
2300 Southern Boulevard
Bronx, New York 10460
Phone: (718) 220-5100

Woods Hole Oceanographic Institution
Information Office
Co-op Building, MS #16
Woods Hole, MA 02543 USA
Phone: (508) 548-1400
Fax: (508) 457-2034
E-mail: information@whoi.edu
<http://www.whoi.edu/>

World Association of Zoos and Aquariums
PO Box 23
Liebefeld-Bern CH-3097
Switzerland
<http://www.waza.org>

World Wildlife Fund
1250 24th Street N.W.
Washington, DC 20037-1193 USA
Phone: (202) 293-4800
Fax: (202) 293-9211
<http://www.panda.org/>

Contributors to the first edition

The following individuals contributed chapters to the original edition of Grzimek's Animal Life Encyclopedia, *which was edited by Dr. Bernhard Grzimek, Professor, Justus Liebig University of Giessen, Germany; Director, Frankfurt Zoological Garden, Germany; and Trustee, Tanzanian National Parks, Tanzania.*

Dr. Michael Abs
Curator, Ruhr University
Bochum, Germany

Dr. Salim Ali
Bombay Natural History Society
Bombay, India

Dr. Rudolph Altevogt
Professor, Zoological Institute, University of Münster
Münster, Germany

Dr. Renate Angermann
Curator, Institute of Zoology, Humboldt University
Berlin, Germany

Edward A. Armstrong
Cambridge University
Cambridge, England

Dr. Peter Ax
Professor, Second Zoological Institute and Museum, University of Göttingen
Göttingen, Germany

Dr. Franz Bachmaier
Zoological Collection of the State of Bavaria
Munich, Germany

Dr. Pedru Banarescu
Academy of the Roumanian Socialist Republic, Trajan Savulescu Institute of Biology
Bucharest, Romania

Dr. A. G. Bannikow
Professor, Institute of Veterinary Medicine
Moscow, Russia

Dr. Hilde Baumgärtner
Zoological Collection of the State of Bavaria
Munich, Germany

C. W. Benson
Department of Zoology, Cambridge University
Cambridge, England

Dr. Andrew Berger
Chairman, Department of Zoology, University of Hawaii
Honolulu, Hawaii, U.S.A.

Dr. J. Berlioz
National Museum of Natural History
Paris, France

Dr. Rudolf Berndt
Director, Institute for Population Ecology, Hiligoland Ornithological Station
Braunschweig, Germany

Dieter Blume
Instructor of Biology, Freiherr-vom-Stein School
Gladenbach, Germany

Dr. Maximilian Boecker
Zoological Research Institute and A. Koenig Museum
Bonn, Germany

Dr. Carl-Heinz Brandes
Curator and Director, The Aquarium, Overseas Museum
Bremen, Germany

Dr. Donald G. Broadley
Curator, Umtali Museum
Mutare, Zimbabwe

Dr. Heinz Brüll
Director; Game, Forest, and Fields Research Station
Hartenholm, Germany

Dr. Herbert Bruns
Director, Institute of Zoology and the Protection of Life
Schlangenbad, Germany

Hans Bub
Heligoland Ornithological Station
Wilhelmshaven, Germany

A. H. Chisholm
Sydney, Australia

Herbert Thomas Condon
Curator of Birds, South Australian Museum
Adelaide, Australia

Dr. Eberhard Curio
Director, Laboratory of Ethology, Ruhr University
Bochum, Germany

Dr. Serge Daan
Laboratory of Animal Physiology, University of Amsterdam
Amsterdam, The Netherlands

Dr. Heinrich Dathe
Professor and Director, Animal Park and Zoological Research Station, German Academy of Sciences
Berlin, Germany

Dr. Wolfgang Dierl
Zoological Collection of the State of Bavaria
Munich, Germany

Dr. Fritz Dieterlen
Zoological Research Institute and A. Koenig Museum
Bonn, Germany

Dr. Rolf Dircksen
Professor, Pedagogical Institute
Bielefeld, Germany

Josef Donner
Instructor of Biology
Katzelsdorf, Austria

Dr. Jean Dorst
Professor, National Museum of Natural History
Paris, France

Dr. Gerti Dücker
Professor and Chief Curator, Zoological Institute, University of Münster
Münster, Germany

Dr. Michael Dzwillo
Zoological Institute and Museum, University of Hamburg
Hamburg, Germany

Dr. Irenäus Eibl-Eibesfeldt
Professor and Director, Institute of Human Ethology, Max Planck Institute for Behavioral Physiology
Percha/Starnberg, Germany

Dr. Martin Eisentraut
Professor and Director, Zoological Research Institute and A. Koenig Museum
Bonn, Germany

Dr. Eberhard Ernst
Swiss Tropical Institute
Basel, Switzerland

R. D. Etchecopar
Director, National Museum of Natural History
Paris, France

Dr. R. A. Falla
Director, Dominion Museum
Wellington, New Zealand

Dr. Hubert Fechter
Curator, Lower Animals, Zoological Collection of the State of Bavaria
Munich, Germany

Dr. Walter Fiedler
Docent, University of Vienna, and Director, Schönbrunn Zoo
Vienna, Austria

Wolfgang Fischer
Inspector of Animals, Animal Park
Berlin, Germany

Dr. C. A. Fleming
Geological Survey Department of Scientific and Industrial Research
Lower Hutt, New Zealand

Dr. Hans Frädrich
Zoological Garden
Berlin, Germany

Dr. Hans-Albrecht Freye
Professor and Director, Biological Institute of the Medical School
Halle a.d.S., Germany

Günther E. Freytag
Former Director, Reptile and Amphibian Collection, Museum of Cultural History in Magdeburg
Berlin, Germany

Dr. Herbert Friedmann
Director, Los Angeles County Museum of Natural History
Los Angeles, California, U.S.A.

Dr. H. Friedrich
Professor, Overseas Museum
Bremen, Germany

Dr. Jan Frijlink
Zoological Laboratory, University of Amsterdam
Amsterdam, The Netherlands

Dr. H. C. Karl Von Frisch
Professor Emeritus and former Director, Zoological Institute, University of Munich
Munich, Germany

Dr. H. J. Frith
C.S.I.R.O. Research Institute
Canberra, Australia

Dr. Ion E. Fuhn
Academy of the Roumanian Socialist Republic, Trajan Savulescu Institute of Biology
Bucharest, Romania

Dr. Carl Gans
Professor, Department of Biology, State University of New York at Buffalo
Buffalo, New York, U.S.A.

Dr. Rudolf Geigy
Professor and Director, Swiss Tropical Institute
Basel, Switzerland

Dr. Jacques Gery
St. Genies, France

Dr. Wolfgang Gewalt
Director, Animal Park
Duisburg, Germany

Dr. H. C. Viktor Goerttler
Professor Emeritus, University of Jena
Jena, Germany

Dr. Friedrich Goethe
Director, Institute of Ornithology, Heligoland Ornithological Station
Wilhelmshaven, Germany

Dr. Ulrich F. Gruber
Herpetological Section, Zoological Research Institute and A. Koenig Museum
Bonn, Germany

Dr. H. R. Haefelfinger
Museum of Natural History
Basel, Switzerland

Dr. Theodor Haltenorth
Director, Mammalology, Zoological Collection of the State of Bavaria
Munich, Germany

Barbara Harrisson
Sarawak Museum, Kuching, Borneo
Ithaca, New York, U.S.A.

Dr. Francois Haverschmidt
President, High Court (retired)
Paramaribo, Suriname

Dr. Heinz Heck
Director, Catskill Game Farm
Catskill, New York, U.S.A.

Dr. Lutz Heck
Professor (retired), and Director, Zoological Garden, Berlin
Wiesbaden, Germany

Dr. H. C. Heini Hediger
Director, Zoological Garden
Zurich, Switzerland

Dr. Dietrich Heinemann
Director, Zoological Garden, Münster
Dörnigheim, Germany

Dr. Helmut Hemmer
Institute for Physiological Zoology, University of Mainz
Mainz, Germany

Dr. W. G. Heptner
Professor, Zoological Museum, University of Moscow
Moscow, Russia

Dr. Konrad Herter
Professor Emeritus and Director (retired), Zoological Institute, Free University of Berlin
Berlin, Germany

Dr. Hans Rudolf Heusser
Zoological Museum, University of Zurich
Zurich, Switzerland

Dr. Emil Otto Höhn
Associate Professor of Physiology, University of Alberta
Edmonton, Canada

Dr. W. Hohorst
Professor and Director, Parasitological Institute, Farbwerke Hoechst A.G.
Frankfurt-Höchst, Germany

Dr. Folkhart Hückinghaus
Director, Senckenbergische Anatomy, University of Frankfurt a.M.
Frankfurt a.M., Germany

Francois Hüe
National Museum of Natural History
Paris, France

Dr. K. Immelmann
Professor, Zoological Institute, Technical University of Braunschweig
Braunschweig, Germany

Dr. Junichiro Itani
Kyoto University
Kyoto, Japan

Dr. Richard F. Johnston
Professor of Zoology, University of Kansas
Lawrence, Kansas, U.S.A.

Otto Jost
Oberstudienrat, Freiherr-vom-Stein Gymnasium
Fulda, Germany

Dr. Paul Kähsbauer
Curator, Fishes, Museum of Natural History
Vienna, Austria

Dr. Ludwig Karbe
Zoological State Institute and Museum
Hamburg, Germany

Dr. N. N. Kartaschew
Docent, Department of Biology, Lomonossow State University
Moscow, Russia

Dr. Werner Kästle
Oberstudienrat, Gisela Gymnasium
Munich, Germany

Dr. Reinhard Kaufmann
Field Station of the Tropical Institute, Justus Liebig University, Giessen, Germany
Santa Marta, Colombia

Dr. Masao Kawai
Primate Research Institute, Kyoto University
Kyoto, Japan

Dr. Ernst F. Kilian
Professor, Giessen University and Catedratico Universidad Australia, Valdivia-Chile
Giessen, Germany

Dr. Ragnar Kinzelbach
Institute for General Zoology, University of Mainz
Mainz, Germany

Dr. Heinrich Kirchner
Landwirtschaftsrat (retired)
Bad Oldesloe, Germany

Dr. Rosl Kirchshofer
Zoological Garden, University of Frankfurt a.M.
Frankfurt a.M., Germany

Dr. Wolfgang Klausewitz
Curator, Senckenberg Nature Museum and Research Institute
Frankfurt a.M., Germany

Dr. Konrad Klemmer
Curator, Senckenberg Nature Museum and Research Institute
Frankfurt a.M., Germany

Dr. Erich Klinghammer
Laboratory of Ethology, Purdue University
Lafayette, Indiana, U.S.A.

Dr. Heinz-Georg Klös
Professor and Director, Zoological Garden
Berlin, Germany

Ursula Klös
Zoological Garden
Berlin, Germany

Dr. Otto Koehler
Professor Emeritus, Zoological Institute, University of Freiburg
Freiburg i. BR., Germany

Dr. Kurt Kolar
Institute of Ethology, Austrian Academy of Sciences
Vienna, Austria

Dr. Claus König
State Ornithological Station of Baden-Württemberg
Ludwigsburg, Germany

Dr. Adriaan Kortlandt
Zoological Laboratory, University of Amsterdam
Amsterdam, The Netherlands

Dr. Helmut Kraft
Professor and Scientific Councillor, Medical Animal Clinic, University of Munich
Munich, Germany

Dr. Helmut Kramer
Zoological Research Institute and A. Koenig Museum
Bonn, Germany

Dr. Franz Krapp
Zoological Institute, University of Freiburg
Freiburg, Switzerland

Dr. Otto Kraus
Professor, University of Hamburg, and Director, Zoological Institute and Museum
Hamburg, Germany

Dr. Hans Krieg
Professor and First Director (retired), Scientific Collections of the State of Bavaria
Munich, Germany

Dr. Heinrich Kühl
Federal Research Institute for Fisheries, Cuxhaven Laboratory
Cuxhaven, Germany

Dr. Oskar Kuhn
Professor, formerly University Halle/Saale
Munich, Germany

Dr. Hans Kumerloeve
First Director (retired), State Scientific Museum, Vienna
Munich, Germany

Dr. Nagamichi Kuroda
Yamashina Ornithological Institute, Shibuya-Ku
Tokyo, Japan

Dr. Fred Kurt
Zoological Museum of Zurich University, Smithsonian Elephant Survey
Colombo, Ceylon

Dr. Werner Ladiges
Professor and Chief Curator, Zoological Institute and Museum, University of Hamburg
Hamburg, Germany

Leslie Laidlaw
Department of Animal Sciences, Purdue University
Lafayette, Indiana, U.S.A.

Dr. Ernst M. Lang
Director, Zoological Garden
Basel, Switzerland

Dr. Alfredo Langguth
Department of Zoology, Faculty of Humanities and Sciences, University of the Republic
Montevideo, Uruguay

Leo Lehtonen
Science Writer
Helsinki, Finland

Bernd Leisler
Second Zoological Institute, University of Vienna
Vienna, Austria

Dr. Kurt Lillelund
Professor and Director, Institute for Hydrobiology and Fishery Sciences, University of Hamburg
Hamburg, Germany

R. Liversidge
Alexander MacGregor Memorial Museum
Kimberley, South Africa

Dr. Konrad Lorenz
Professor and Director, Max Planck Institute for Behavioral Physiology
Seewiesen/Obb., Germany

Dr. Martin Lühmann
Federal Research Institute for the Breeding of Small Animals
Celle, Germany

Dr. Johannes Lüttschwager
Oberstudienrat (retired)
Heidelberg, Germany

Dr. Wolfgang Makatsch
Bautzen, Germany

Dr. Hubert Markl
Professor and Director, Zoological Institute, Technical University of Darmstadt
Darmstadt, Germany

Basil J. Marlow , BSc (Hons)
Curator, Australian Museum
Sydney, Australia

Dr. Theodor Mebs
Instructor of Biology
Weissenhaus/Ostsee, Germany

Dr. Gerlof Fokko Mees
Curator of Birds, Rijks Museum of Natural History
Leiden, The Netherlands

Hermann Meinken
Director, Fish Identification Institute, V.D.A.
Bremen, Germany

Dr. Wilhelm Meise
Chief Curator, Zoological Institute and Museum, University of Hamburg
Hamburg, Germany

Dr. Joachim Messtorff
Field Station of the Federal Fisheries Research Institute
Bremerhaven, Germany

Dr. Marian Mlynarski
Professor, Polish Academy of Sciences, Institute for Systematic and Experimental Zoology
Cracow, Poland

Dr. Walburga Moeller
Nature Museum
Hamburg, Germany

Dr. H. C. Erna Mohr
Curator (retired), Zoological State Institute and Museum
Hamburg, Germany

Dr. Karl-Heinz Moll
Waren/Müritz, Germany

Dr. Detlev Müller-Using
Professor, Institute for Game Management, University of Göttingen
Hannoversch-Münden, Germany

Werner Münster
Instructor of Biology
Ebersbach, Germany

Dr. Joachim Münzing
Altona Museum
Hamburg, Germany

Dr. Wilbert Neugebauer
Wilhelma Zoo
Stuttgart-Bad Cannstatt, Germany

Dr. Ian Newton
Senior Scientific Officer, The Nature Conservancy
Edinburgh, Scotland

Dr. Jürgen Nicolai
Max Planck Institute for Behavioral Physiology
Seewiesen/Obb., Germany

Dr. Günther Niethammer
Professor, Zoological Research Institute and A. Koenig Museum
Bonn, Germany

Dr. Bernhard Nievergelt
Zoological Museum, University of Zurich
Zurich, Switzerland

Dr. C. C. Olrog
Institut Miguel Lillo San Miguel de Tucumán
Tucumán, Argentina

Alwin Pedersen
Mammal Research and Arctic Explorer
Holte, Denmark

Dr. Dieter Stefan Peters
Nature Museum and Senckenberg Research Institute
Frankfurt a.M., Germany

Dr. Nicolaus Peters
Scientific Councillor and Docent, Institute of Hydrobiology and Fisheries, University of Hamburg
Hamburg, Germany

Dr. Hans-Günter Petzold
Assistant Director, Zoological Garden
Berlin, Germany

Dr. Rudolf Piechocki
Docent, Zoological Institute, University of Halle
Halle a.d.S., Germany

Dr. Ivo Poglayen-Neuwall
Director, Zoological Garden
Louisville, Kentucky, U.S.A.

Dr. Egon Popp
Zoological Collection of the State of Bavaria
Munich, Germany

Dr. H. C. Adolf Portmann
Professor Emeritus, Zoological Institute, University of Basel
Basel, Switzerland

Hans Psenner
Professor and Director, Alpine Zoo
Innsbruck, Austria

Dr. Heinz-Siburd Raethel
Oberveterinärrat
Berlin, Germany

Dr. Urs H. Rahm
Professor, Museum of Natural History
Basel, Switzerland

Dr. Werner Rathmayer
Biology Institute, University of Konstanz
Konstanz, Germany

Walter Reinhard
Biologist
Baden-Baden, Germany

Dr. H. H. Reinsch
Federal Fisheries Research Institute
Bremerhaven, Germany

Dr. Bernhard Rensch
Professor Emeritus, Zoological Institute, University of Münster
Münster, Germany

Dr. Vernon Reynolds
Docent, Department of Sociology, University of Bristol
Bristol, England

Dr. Rupert Riedl
Professor, Department of Zoology, University of North Carolina
Chapel Hill, North Carolina, U.S.A.

Dr. Peter Rietschel
Professor (retired), Zoological Institute, University of Frankfurt a.M.
Frankfurt a.M., Germany

Dr. Siegfried Rietschel
Docent, University of Frankfurt; Curator, Nature Museum and Research Institute Senckenberg
Frankfurt a.M., Germany

Herbert Ringleben
Institute of Ornithology, Heligoland Ornithological Station
Wilhelmshaven, Germany

Dr. K. Rohde
Institute for General Zoology, Ruhr University
Bochum, Germany

Dr. Peter Röben
Academic Councillor, Zoological Institute, Heidelberg University
Heidelberg, Germany

Dr. Anton E. M. De Roo
Royal Museum of Central Africa
Tervuren, South Africa

Dr. Hubert Saint Girons
Research Director, Center for National Scientific Research
Brunoy (Essonne), France

Dr. Luitfried Von Salvini-Plawen
First Zoological Institute, University of Vienna
Vienna, Austria

Dr. Kurt Sanft
Oberstudienrat, Diesterweg-Gymnasium
Berlin, Germany

Dr. E. G. Franz Sauer
Professor, Zoological Research Institute and A. Koenig Museum, University of Bonn
Bonn, Germany

Dr. Eleonore M. Sauer
Zoological Research Institute and A. Koenig Museum, University of Bonn
Bonn, Germany

Dr. Ernst Schäfer
Curator, State Museum of Lower Saxony
Hannover, Germany

Dr. Friedrich Schaller
Professor and Chairman, First Zoological Institute, University of Vienna
Vienna, Austria

Dr. George B. Schaller
Serengeti Research Institute, Michael Grzimek Laboratory
Seronera, Tanzania

Dr. Georg Scheer
Chief Curator and Director, Zoological Institute, State Museum of Hesse
Darmstadt, Germany

Dr. Christoph Scherpner
Zoological Garden
Frankfurt a.M., Germany

Dr. Herbert Schifter
Bird Collection, Museum of Natural History
Vienna, Austria

Dr. Marco Schnitter
Zoological Museum, Zurich University
Zurich, Switzerland

Dr. Kurt Schubert
Federal Fisheries Research Institute
Hamburg, Germany

Eugen Schuhmacher
Director, Animals Films, I.U.C.N.
Munich, Germany

Dr. Thomas Schultze-Westrum
Zoological Institute, University of Munich
Munich, Germany

Dr. Ernst Schüt
Professor and Director (retired), State Museum of Natural History
Stuttgart, Germany

Dr. Lester L. Short , Jr.
Associate Curator, American Museum of Natural History
New York, New York, U.S.A.

Dr. Helmut Sick
National Museum
Rio de Janeiro, Brazil

Dr. Alexander F. Skutch
Professor of Ornithology, University of Costa Rica
San Isidro del General, Costa Rica

Dr. Everhard J. Slijper
Professor, Zoological Laboratory, University of Amsterdam
Amsterdam, The Netherlands

Bertram E. Smythies
Curator (retired), Division of Forestry Management, Sarawak-Malaysia
Estepona, Spain

Dr. Kenneth E. Stager
Chief Curator, Los Angeles County Museum of Natural History
Los Angeles, California, U.S.A.

Dr. H. C. Georg H. W. Stein
Professor, Curator of Mammals, Institute of Zoology and Zoological Museum, Humboldt University
Berlin, Germany

Dr. Joachim Steinbacher
Curator, Nature Museum and Senckenberg Research Institute
Frankfurt a.M., Germany

Dr. Bernard Stonehouse
Canterbury University
Christchurch, New Zealand

Dr. Richard Zur Strassen
Curator, Nature Museum and Senckenberg Research Institute
Frandfurt a.M., Germany

Dr. Adelheid Studer-Thiersch
Zoological Garden
Basel, Switzerland

Dr. Ernst Sutter
Museum of Natural History
Basel, Switzerland

Dr. Fritz Terofal
Director, Fish Collection, Zoological Collection of the State of Bavaria
Munich, Germany

Dr. G. F. Van Tets
Wildlife Research
Canberra, Australia

Ellen Thaler-Kottek
Institute of Zoology, University of Innsbruck
Innsbruck, Austria

Dr. Erich Thenius
Professor and Director, Institute of Paleontolgy, University of Vienna
Vienna, Austria

Dr. Niko Tinbergen
Professor of Animal Behavior, Department of Zoology, Oxford University
Oxford, England

Alexander Tsurikov
Lecturer, University of Munich
Munich, Germany

Dr. Wolfgang Villwock
Zoological Institute and Museum, University of Hamburg
Hamburg, Germany

Zdenek Vogel
Director, Suchdol Herpetological Station
Prague, Czechoslovakia

Dieter Vogt
Schorndorf, Germany

Dr. Jiri Volf
Zoological Garden
Prague, Czechoslovakia

Otto Wadewitz
Leipzig, Germany

Dr. Helmut O. Wagner
Director (retired), Overseas Museum, Bremen
Mexico City, Mexico

Dr. Fritz Walther
Professor, Texas A & M University
College Station, Texas, U.S.A.

John Warham
Zoology Department, Canterbury University
Christchurch, New Zealand

Dr. Sherwood L. Washburn
University of California at Berkeley
Berkeley, California, U.S.A.

Contributors to the first edition

Eberhard Wawra
First Zoological Institute, University of Vienna
Vienna, Austria

Dr. Ingrid Weigel
Zoological Collection of the State of Bavaria
Munich, Germany

Dr. B. Weischer
Institute of Nematode Research, Federal Biological Institute
Münster/Westfalen, Germany

Herbert Wendt
Author, Natural History
Baden-Baden, Germany

Dr. Heinz Wermuth
Chief Curator, State Nature Museum, Stuttgart
Ludwigsburg, Germany

Dr. Wolfgang Von Westernhagen
Preetz/Holstein, Germany

Dr. Alexander Wetmore
United States National Museum, Smithsonian Institution
Washington, D.C., U.S.A.

Dr. Dietrich E. Wilcke
Röttgen, Germany

Dr. Helmut Wilkens
Professor and Director, Institute of Anatomy, School of Veterinary Medicine
Hannover, Germany

Dr. Michael L. Wolfe
Utah, U.S.A.

Hans Edmund Wolters
Zoological Research Institute and A. Koenig Museum
Bonn, Germany

Dr. Arnfrid Wünschmann
Research Associate, Zoological Garden
Berlin, Germany

Dr. Walter Wüst
Instructor, Wilhelms Gymnasium
Munich, Germany

Dr. Heinz Wundt
Zoological Collection of the State of Bavaria
Munich, Germany

Dr. Claus-Dieter Zander
Zoological Institute and Museum, University of Hamburg
Hamburg, Germany

Dr. Fritz Zumpt
Director, Entomology and Parasitology, South African Institute for Medical Research
Johannesburg, South Africa

Dr. Richard L. Zusi
Curator of Birds, United States National Museum, Smithsonian Institution
Washington, D.C., U.S.A.

Glossary

Adaptive radiation—Diversification of a species or single ancestral type into several forms that are each adaptively specialized to a specific niche.

Agonistic—Behavioral patterns that are aggressive in context.

Allopatric—Occurring in separate, nonoverlapping geographic areas.

Alpha breeder—The reproductively dominant member of a social unit.

Altricial—An adjective referring to a mammal that is born with little, if any, hair, is unable to feed itself, and initially has poor sensory and thermoregulatory abilities.

Amphibious—Refers to the ability of an animal to move both through water and on land.

Austral—May refer to "southern regions," typically meaning Southern Hemisphere. May also refer to the geographical region included within the Transition, Upper Austral, and Lower Austral Life Zones as defined by C. Hart Merriam in 1892–1898. These zones are often characterized by specific plant and animal communities and were originally defined by temperature gradients especially in the mountains of southwestern North America.

Bergmann's rule—Within a species or among closely related species of mammals, those individuals in colder environments often are larger in body size. Bergmann's rule is a generalization that reflects the ability of endothermic animals to more easily retain body heat (in cold climates) if they have a high body surface to body volume ratio, and to more easily dissipate excess body heat (in hot environments) if they have a low body surface to body volume ratio.

Bioacoustics—The study of biological sounds such as the sounds produced by bats or other mammals.

Biogeographic region—One of several major divisions of the earth defined by a distinctive assemblage of animals and plants. Sometimes referred to as "zoogeographic regions or realms" (for animals) or "phytogeographic regions or realms" (for plants). Such terminology dates from the late nineteenth century and varies considerably. Major biogeographic regions each have a somewhat distinctive flora and fauna. Those generally recognized include Nearctic, Neotropical, Palearctic, Ethiopian, Oriental, and Australian.

Blow—Cloud of vapor and sea water exhaled by cetaceans.

Boreal—Often used as an adjective meaning "northern"; also may refer to the northern climatic zone immediately south of the Arctic; may also include the Arctic, Hudsonian, and Canadian Life Zones described by C. Hart Merriam.

Brachiating ancestor—Ancestor that swung around by the arms.

Breaching—A whale behavior—leaping above the water's surface, then falling back into the water, landing on its back or side.

Cephalopod—Member of the group of mollusks such as squid and octopus.

Cladistic—Evolutionary relationships suggested as "tree" branches to indicate lines of common ancestry.

Cline—A gradient in a measurable characteristic, such as size and color, showing geographic differentiation. Various patterns of geographic variation are reflected as clines or clinal variation, and have been described as "ecogeographic rules."

Cloaca—A common opening for the digestive, urinary, and reproductive tracts found in monotreme mammals.

Colony—A group of mammals living in close proximity, interacting, and usually aiding in early warning of the presence of predators and in group defense.

Commensal—A relationship between species in which one benefits and the other is neither benefited nor harmed.

Congeneric—Descriptive of two or more species that belong to the same genus.

Conspecific—Descriptive of two or more individuals or populations that belong to the same species.

Contact call—Simple vocalization used to maintain communication or physical proximity among members of a social unit.

Convergent evolution—When two evolutionarily unrelated groups of organisms develop similar characteristics due to adaptation to similar aspects of their environment or niche.

Coprophagy—Reingestion of feces to obtain nutrients that were not ingested the first time through the digestive system.

Cosmopolitan—Adjective describing the distribution pattern of an animal found around the world in suitable habitats.

Crepuscular—Active at dawn and at dusk.

Critically Endangered—A technical category used by IUCN for a species that is at an extremely high risk of extinction in the wild in the immediate future.

Cryptic—Hidden or concealed; i.e., well-camouflaged patterning.

Dental formula—A method for describing the number of each type of tooth found in an animal's mouth: incisors (I), canines (C), premolars (P), and molars (M). The formula gives the number of each tooth found in an upper and lower quadrant of the mouth, and the total is multiplied by two for the total number of teeth. For example, the formula for humans is: I2/2 C1/1 P2/2 M3/3 (total, 16, times two is 32 teeth).

Dimorphic—Occurring in two distinct forms (e.g., in reference to the differences in size between males and females of a species).

Disjunct—A distribution pattern characterized by populations that are geographically separated from one another.

Diurnal—Active during the day.

DNA-DNA hybridization—A technique whereby the genetic similarity of different animal groups is determined based on the extent to which short stretches of their DNA, when mixed together in solution in the laboratory, are able to join with each other.

Dominance hierarchy—The social status of individuals in a group; each animal can usually dominate those animals below it in a hierarchy.

Dorso-ventrally—From back to front.

Duetting—Male and female singing and integrating their songs together.

Echolocation—A method of navigation used by some mammals (e.g., bats and marine mammals) to locate objects and investigate surroundings. The animals emit audible "clicks" and determine pathways by using the echo of the sound from structures in the area.

Ecotourism—Travel for the primary purpose of viewing nature. Ecotourism is now "big business" and is used as a non-consumptive but financially rewarding way to protect important areas for conservation.

Ectothermic—Using external energy and behavior to regulate body temperature. "Cold-blooded."

Endangered—A term used by IUCN and also under the Endangered Species Act of 1973 in the United States in reference to a species that is threatened with imminent extinction or extirpation over all or a significant portion of its range.

Endemic—Native to only one specific area.

Endothermic—Maintaining a constant body-temperature using metabolic energy. "Warm-blooded."

Eocene—Geological time period; subdivision of the Tertiary, from about 55.5 to 33.7 million years ago.

Ethology—The study of animal behavior.

Exotic—Not native.

Extant—Still in existence; not destroyed, lost, or extinct.

Extinct—Refers to a species that no longer survives anywhere.

Extirpated—Referring to a local extinction of a species that can still be found elsewhere.

Feral—A population of domesticated animal that lives in the wild.

Flehmen—Lip curling and head raising after sniffing a female's urine.

Forb—Any herb that is not a grass or grass-like.

Fossorial—Adapted for digging.

Frugivorous—Feeds on fruit.

Granivorous—Feeding on seeds.

Gravid—Pregnant.

Gregarious—Occuring in large groups.

Hibernation—A deep state of reduced metabolic activity and lowered body temperature that may last for weeks or months.

Holarctic—The Palearctic and Nearctic bigeographic regions combined.

Hybrid—The offspring resulting from a cross between two different species (or sometimes between distinctive subspecies).

Innate—An inherited characteristic.

Insectivorous—Technically refers to animals that eat insects; generally refers to animals that feed primarily on insects and other arthropods.

Introduced species—An animal or plant that has been introduced to an area where it normally does not occur.

Iteroparous—Breeds in multiple years.

Jacobson's organ—Olfactory organ found in the upper palate that first appeared in amphibians and is most developed in these and in reptiles, but is also found in some birds and mammals.

Kiva—A large chamber wholly or partly underground, and often used for religious ceremonies in Pueblo Indian villages.

Mandible—Technically an animal's lower jaw. The plural, mandibles, is used to refer to both the upper and lower jaw. The upper jaw is technically the maxilla, but often called the "upper mandible."

Marsupial—A mammal whose young complete their embryonic development outside of the mother's body, within a maternal pouch.

Matrilineal—Describing a social unit in which group members are descended from a single female.

Melon—The fat-filled forehead of aquatic mammals of the order Cetacea.

Metabolic rate—The rate of chemical processes in living organisms, resulting in energy expenditure and growth. Metabolic rate decreases when an animal is resting and increases during activity.

Migration—A two-way movement in some mammals, often dramatically seasonal. Typically latitudinal, though in some species is altitudinal or longitudinal. May be short-distance or long-distance.

Miocene—The geological time period that lasted from about 23.8 to 5.6 million years ago.

Molecular phylogenetics—The use of molecular (usually genetic) techniques to study evolutionary relationships between or among different groups of organisms.

Monestrous—Experiencing estrus just once each year or breeding season.

Monogamous—A breeding system in which a male and female mate only with one another.

Monophyletic—A group (or clade) that shares a common ancestor.

Monotypic—A taxonomic category that includes only one form (e.g., a genus that includes only one species; a species that includes no subspecies).

Montane—Of or inhabiting the biogeographic zone of relatively moist, cool upland slopes below timberline dominated by large coniferous trees.

Morphology—The form and structure of animals and plants.

Mutualism—Ecological relationship between two species in which both gain benefit.

Near Threatened—A category defined by the IUCN suggesting possible risk of extinction in the medium term (as opposed to long or short term) future.

Nearctic—The biogeographic region that includes temperate North America. faunal region.

Neotropical—The biogeographic region that includes South and Central America, the West Indies, and tropical Mexico.

New World—A general descriptive term encompassing the Nearctic and Neotropical biogeographic regions.

Niche—The role of an organism in its environment; multidimensional, with habitat and behavioral components.

Nocturnal—Active at night.

Old World—A general term that usually describes a species or group as being from Eurasia or Africa.

Oligocene—The geologic time period occurring from about 33.7 to 23.8 million years ago.

Omnivorous—Feeding on a broad range of foods, both plant and animal matter.

Palearctic—A biogeographic region that includes temperate Eurasia and Africa north of the Sahara.

Paleocene—Geological period, subdivision of the Tertiary, from 65 to 55.5 million years ago.

Pelage—Coat, skin, and hair.

Pelagic—An adjective used to indicate a relationship to the open sea.

Pestiferous—Troublesome or annoying; nuisance.

Phylogeny—A grouping of taxa based on evolutionary history.

Piscivorous—Fish-eating.

Placental—A mammal whose young complete their embryonic development within the mother's uterus, joined to her by a placenta.

Pleistocene—In general, the time of the great ice ages; geological period variously considered to include the last 1 to 1.8 million years.

Pliocene—The geological period preceding the Pleistocence; the last subdivision of what is known as the Tertiary; lasted from 5.5 to 1.8 million years ago.

Polyandry—A breeding system in which one female mates with two or more males.

Polygamy—A breeding system in which either or both male and female may have two or more mates.

Polygyny—A breeding system in which one male mates with two or more females.

Polyphyletic—A taxonomic group that is believed to have originated from more than one group of ancestors.

Post-gastric digestion—Refers to the type of fermentative digestion of vegetative matter found in tapirs and other animals by which microorganisms decompose food in a caecum. This is not as thorough a decomposition as occurs in ruminant digesters.

Precocial—An adjective used to describe animals that are born in an advanced state of development such that they generally can leave their birth area quickly and obtain their own food, although they are often led to food and guarded by a parent.

Proboscis—The prehensile trunk (a muscular hydrostat) found in tapirs, elephants, etc.

Quaternary—The geological period, from 1.8 million years ago to the present, usually including two subdivisions: the Pleistocene, and the Holocene.

Refugium (pl. refugia) —An area relatively unaltered during a time of climatic change, from which dispersion and speciation may occur after the climate readjusts.

Reproductive longevity—The length of an animal's life over which it is capable of reproduction.

Ruminant—An even-toed, hoofed mammal with a four-chambered stomach that eats rapidly to regurgitate its food and chew the cud later.

Scansorial—Specialized for climbing.

Seed dispersal—Refers to how tapirs and other animals transport viable seeds from their source to near or distant, suitable habitats where they can successfully germinate. Such dispersal may occur through the feces, through sputum, or as the seeds are attached and later released from fur, etc.

Semelparity—A short life span, in which a single instance of breeding is followed by death in the first year of life.

Sexual dimorphism—Male and female differ in morphology, such as size, feather size or shape, or bill size or shape.

Sibling species—Two or more species that are very closely related, presumably having differentiated from a common ancestor in the recent past; often difficult to distinguish, often interspecifically territorial.

Sonagram—A graphic representation of sound.

Speciation—The evolution of new species.

Spy-hopping—Positioning the body vertically in the water, with the head raised above the sea surface, sometimes while turning slowly.

Steppe—Arid land with vegetation that can thrive with very little moisture; found usually in regions of extreme temperature range.

Suspensory—Moving around or hanging by the arms.

Sympatric—Inhabiting the same range.

Systematist—A specialist in the classification of organisms; systematists strive to classify organisms on the basis of their evolutionary relationships.

Taxon (pl. taxa) —Any unit of scientific classification (e.g., species, genus, family, order).

Taxonomist—A specialist in the naming and classification of organisms. (See also Systematist. Taxonomy is the older science of naming things; identification of evolutionary relationships has not always been the goal of taxonomists. The modern science of systematics generally incorporates taxonomy with the search for evolutionary relationships.)

Taxonomy—The science of identifying, naming, and classifying organisms into groups.

Territoriality—Refers to an animal's defense of a certain portion of its habitat against other conspecifics. This is often undertaken by males in relation to one another and as a lure to females.

Territory—Any defended area. Territorial defense is typically male against male, female against female, and within a species or between sibling species. Area defended varies greatly among taxa, seasons, and habitats. A territory may include the entire home range, only the area immediately around a nest, or only a feeding area.

Tertiary—The geological period including most of the Cenozoic; from about 65 to 1.8 million years ago.

Thermoregulation—The ability to regulate body temperature; can be either behavioral or physiological.

Tribe—A unit of classification below the subfamily and above the genus.

Truncal erectness—Sitting, hanging, arm-swinging (brachiating), walking bipedally with the backbone held vertical.

Ungulate—A hoofed mammal.

Upper cone—The circle in which the arm can rotate when raised above the head.

Viable population—A population that is capable of maintaining itself over a period of time. One of the major conservation issues of the twenty-first century is determining what is a minimum viable population size. Population geneticists have generally come up with estimates of about 500 breeding pairs.

Vulnerable—A category defined by IUCN as a species that is not Critically Endangered or Endangered, but is still facing a threat of extinction.

• • • • •

Mammals species list

Monotremata [Order]

Tachyglossidae [Family]
Tachyglossus [Genus]
T. aculeatus [Species]
Zaglossus [Genus]
Z. bruijni [Species]

Ornithorhynchidae [Family]
Ornithorhynchus [Genus]
O. anatinus [Species]

Didelphimorphia [Order]

Didelphidae [Family]
Caluromys [Genus]
C. derbianus [Species]
C. lanatus
C. philander
Caluromysiops [Genus]
C. irrupta [Species]
Chironectes [Genus]
C. minimus [Species]
Didelphis [Genus]
D. albiventris [Species]
D. aurita
D. marsupialis
D. virginiana
Glironia [Genus]
G. venusta [Species]
Gracilinanus [Genus]
G. aceramarcae [Species]
G. agilis
G. dryas
G. emiliae
G. marica
G. microtarsus
Lestodelphys [Genus]
L. halli [Species]
Lutreolina [Genus]
L. crassicaudata [Species]
Marmosa [Genus]
M. andersoni [Species]
M. canescens
M. lepida
M. mexicana
M. murina
M. robinsoni
M. rubra
M. tyleriana
M. xerophila
Marmosops [Genus]
M. cracens [Species]
M. dorothea
M. fuscatus
M. handleyi
M. impavidus
M. incanus
M. invictus
M. noctivagus
M. parvidens
Metachirus [Genus]
M. nudicaudatus [Species]
Micoureus [Genus]
M. alstoni [Species]
M. constantiae
M. demerarae
M. regina
Monodelphis [Genus]
M. adusta [Species]
M. americana
M. brevicaudata
M. dimidiata
M. domestica
M. emiliae
M. iheringi
M. kunsi
M. maraxina
M. osgoodi
M. rubida
M. scalops
M. sorex
M. theresa
M. unistriata
Philander [Genus]
P. andersoni [Species]
P. opossum
Thylamys [Genus]
T. elegans [Species]
T. macrura
T. pallidior
T. pusilla
T. velutinus

Paucituberculata [Order]

Caenolestidae [Family]
Caenolestes [Genus]
C. caniventer [Species]
C. convelatus
C. fuliginosus
Lestoros [Genus]
L. inca [Species]
Rhyncholestes [Genus]
R. raphanurus [Species]

Microbiotheria [Order]

Microbiotheriidae [Family]
Dromiciops [Genus]
D. gliroides [Species]

Dasyuromorphia [Order]

Dasyuridae [Family]
Antechinus [Genus]
A. bellus [Species]
A. flavipes
A. godmani
A. leo
A. melanurus
A. minimus
A. naso
A. stuartii
A. swainsonii
A. wilhelmina
Dasycercus [Genus]
D. byrnei [Species]
D. cristicauda
Dasykaluta [Genus]
D. rosamondae [Species]
Dasyurus [Genus]
D. albopunctatus [Species]
D. geoffroii
D. hallucatus

D. maculatus
D. spartacus
D. viverrinus
Murexia [Genus]
M. longicaudata [Species]
M. rothschildi
Myoictis [Genus]
M. melas [Species]
Neophascogale [Genus]
N. lorentzi [Species]
Ningaui [Genus]
N. ridei [Species]
N. timealeyi
N. yvonnae
Parantechinus [Genus]
P. apicalis [Species]
P. bilarni
Phascogale [Genus]
P. calura [Species]
P. tapoatafa
Phascolosorex [Genus]
P. doriae [Species]
P. dorsalis
Planigale [Genus]
P. gilesi [Species]
P. ingrami
P. maculata
P. novaeguineae
P. tenuirostris
Pseudantechinus [Genus]
P. macdonnellensis [Species]
P. ningbing
P. woolleyae
Sarcophilus [Genus]
S. laniarius [Species]
Sminthopsis [Genus]
S. aitkeni [Species]
S. archeri
S. butleri
S. crassicaudata
S. dolichura
S. douglasi
S. fuliginosus
S. gilberti
S. granulipes
S. griseoventer
S. hirtipes
S. laniger
S. leucopus
S. longicaudata
S. macroura
S. murina
S. ooldea
S. psammophila
S. virginiae
S. youngsoni

Myrmecobiidae [Family]
Myrmecobius [Genus]
M. fasciatus [Species]

Thylacinidae [Family]
Thylacinus [Genus]
T. cynocephalus [Species]

Peramelemorphia [Order]

Peramelidae [Family]
Chaeropus [Genus]
C. ecaudatus [Species]
Isoodon [Genus]
I. auratus [Species]
I. macrourus
I. obesulus
Macrotis [Genus]
M. lagotis [Species]
M. leucura
Perameles [Genus]
P. bougainville [Species]
P. eremiana
P. gunnii
P. nasuta

Peroryctidae [Family]
Echymipera [Genus]
E. clara [Species]
E. davidi
E. echinista
E. kalubu
E. rufescens
Microperoryctes [Genus]
M. longicauda [Species]
M. murina
M. papuensis
Peroryctes [Genus]
P. broadbenti [Species]
P. raffrayana
Rhynchomeles [Genus]
R. prattorum [Species]

Notoryctemorphia [Order]

Notoryctidae [Family]
Notoryctes [Genus]
N. caurinus [Species]
N. typhlops

Diprotodontia [Order]

Phascolarctidae [Family]
Phascolarctos [Genus]
P. cinereus [Species]

Vombatidae [Family]
Lasiorhinus [Genus]
L. krefftii [Species]
L. latifrons
Vombatus [Genus]
V. ursinus [Species]

Phalangeridae [Family]
Ailurops [Genus]
A. ursinus [Species]
Phalanger [Genus]
P. carmelitae [Species]
P. lullulae
P. matanim
P. orientalis
P. ornatus
P. pelengensis
P. rothschildi
P. sericeus
P. vestitus
Spilocuscus [Genus]
S. maculatus [Species]
S. rufoniger
Strigocuscus [Genus]
S. celebensis [Species]
S. gymnotis
Trichosurus [Genus]
T. arnhemensis [Species]
T. caninus
T. vulpecula
Wyulda [Genus]
W. squamicaudata [Species]

Hypsiprymnodontidae [Family]
Hypsiprymnodon [Genus]
H. moschatus [Species]

Potoroidae [Family]
Aepyprymnus [Genus]
A. rufescens [Species]
Bettongia [Genus]
B. gaimardi [Species]
B. lesueur
B. penicillata
Caloprymnus [Genus]
C. campestris [Species]
Potorous [Genus]
P. longipes [Species]
P. platyops
P. tridactylus

Macropodidae [Family]
Dendrolagus [Genus]
D. bennettianus [Species]
D. dorianus
D. goodfellowi
D. inustus
D. lumholtzi
D. matschiei
D. scottae
D. spadix
D. ursinus
Dorcopsis [Genus]
D. atrata [Species]
D. hageni
D. luctuosa
D. muelleri
Dorcopsulus [Genus]
D. macleayi [Species]
D. vanheurni
Lagorchestes [Genus]

L. asomatus [Species]
L. conspicillatus
L. hirsutus
L. leporides
Lagostrophus [Genus]
L. fasciatus [Species]
Macropus [Genus]
M. agilis [Species]
M. antilopinus
M. bernardus
M. dorsalis
M. eugenii
M. fuliginosus
M. giganteus
M. greyi
M. irma
M. parma
M. parryi
M. robustus
M. rufogriseus
M. rufus
Onychogalea [Genus]
O. fraenata [Species]
O. lunata
O. unguifera
Petrogale [Genus]
P. assimilis [Species]
P. brachyotis
P. burbidgei
P. concinna
P. godmani
P. inornata
P. lateralis
P. penicillata
P. persephone
P. rothschildi
P. xanthopus
Setonix [Genus]
S. brachyurus [Species]
Thylogale [Genus]
T. billardierii [Species]
T. brunii
T. stigmatica
T. thetis
Wallabia [Genus]
W. bicolor [Species]

Burramyidae [Family]
Burramys [Genus]
B. parvus [Species]
Cercartetus [Genus]
C. caudatus [Species]
C. concinnus
C. lepidus
C. nanus

Pseudocheiridae [Family]
Hemibelideus [Genus]
H. lemuroides [Species]
Petauroides [Genus]
P. volans [Species]
Petropseudes [Genus]
P. dahli [Species]
Pseudocheirus [Genus]
P. canescens [Species]
P. caroli
P. forbesi
P. herbertensis
P. mayeri
P. peregrinus
P. schlegeli
Pseudochirops [Genus]
P. albertisii [Species]
P. archeri
P. corinnae
P. cupreus

Petauridae [Family]
Dactylopsila [Genus]
D. megalura [Species]
D. palpator
D. tatei
D. trivirgata
Gymnobelideus [Genus]
G. leadbeateri [Species]
Petaurus [Genus]
P. abidi [Species]
P. australis
P. breviceps
P. gracilis
P. norfolcensis

Tarsipedidae [Family]
Tarsipes [Genus]
T. rostratus [Species]

Acrobatidae [Family]
Acrobates [Genus]
A. pygmaeus [Species]
Distoechurus [Genus]
D. pennatus [Species]

Xenarthra [Order]

Megalonychidae [Family]
Choloepus [Genus]
C. didactylus [Species]
C. hoffmanni

Bradypodidae [Family]
Bradypus [Genus]
B. torquatus [Species]
B. tridactylus
B. variegatus

Myrmecophagidae [Family]
Cyclopes [Genus]
C. didactylus [Species]
Myrmecophaga [Genus]
M. tridactyla [Species]
Tamandua [Genus]
T. mexicana [Species]
T. tetradactyla

Dasypodidae [Family]
Chlamyphorus [Genus]
C. retusus [Species]
C. truncatus
Cabassous [Genus]
C. centralis [Species]
C. chacoensis
C. tatouay
C. unicinctus
Chaetophractus [Genus]
C. nationi [Species]
C. vellerosus
C. villosus
Dasypus [Genus]
D. hybridus [Species]
D. kappleri
D. novemcinctus
D. pilosus
D. sabanicola
D. septemcinctus
Euphractus [Genus]
E. sexcinctus [Species]
Priodontes [Genus]
P. maximus [Species]
Tolypeutes [Genus]
T. matacus [Species]
T. tricinctus
Zaedyus [Genus]
Z. pichiy [Species]

Insectivora [Order]

Erinaceidae [Family]
Atelerix [Genus]
A. albiventris [Species]
A. algirus
A. frontalis
A. sclateri
Erinaceus [Genus]
E. amurensis [Species]
E. concolor
E. europaeus
Hemiechinus [Genus]
H. aethiopicus [Species]
H. auritus
H. collaris
H. hypomelas
H. micropus
H. nudiventris
Mesechinus [Genus]
M. dauuricus [Species]
M. hughi
Echinosorex [Genus]
E. gymnura [Species]
Hylomys [Genus]
H. hainanensis [Species]
H. sinensis
H. suillus
Podogymnura [Genus]
P. aureospinula [Species]
P. truei

Chrysochloridae [Family]
Amblysomus [Genus]
A. gunningi [Species]
A. hottentotus
A. iris
A. julianae
Calcochloris [Genus]
C. obtusirostris [Species]
Chlorotalpa [Genus]
C. arendsi [Species]
C. duthieae
C. leucorhina
C. sclateri
C. tytonis
Chrysochloris [Genus]
C. asiatica [Species]
C. stuhlmanni
C. visagiei
Chrysospalax [Genus]
C. trevelyani [Species]
C. villosus
Cryptochloris [Genus]
C. wintoni [Species]
C. zyli
Eremitalpa [Genus]
E. granti [Species]

Tenrecidae [Family]
Echinops [Genus]
E. telfairi [Species]
Geogale [Genus]
G. aurita [Species]
Hemicentetes [Genus]
H. semispinosus [Species]
Limnogale [Genus]
L. mergulus [Species]
Microgale [Genus]
M. brevicaudata [Species]
M. cowani
M. dobsoni
M. dryas
M. gracilis
M. longicaudata
M. parvula
M. principula
M. pulla
M. pusilla
M. talazaci
M. thomasi
Micropotamogale [Genus]
M. lamottei [Species]
M. ruwenzorii
Oryzorictes [Genus]
O. hova [Species]
O. talpoides
O. tetradactylus
Potamogale [Genus]
P. velox [Species]
Setifer [Genus]
S. setosus [Species]
Tenrec [Genus]
T. ecaudatus [Species]

Solenodontidae [Family]
Solenodon [Genus]
S. cubanus [Species]
S. marcanoi
S. paradoxus

Nesophontidae [Family]
Nesophontes [Genus]
N. edithae [Species]
N. hypomicrus
N. longirostris
N. major
N. micrus
N. paramicrus
N. submicrus
N. zamicrus

Soricidae [Family]
Anourosorex [Genus]
A. squamipes [Species]
Blarina [Genus]
B. brevicauda [Species]
B. carolinensis
B. hylophaga
Blarinella [Genus]
B. quadraticauda [Species]
B. wardi
Chimarrogale [Genus]
C. hantu [Species]
C. himalayica
C. phaeura
C. platycephala
C. styani
C. sumatrana
Congosorex [Genus]
C. polli [Species]
Crocidura [Genus]
C. aleksandrisi [Species]
C. allex
C. andamanensis
C. ansellorum
C. arabica
C. armenica
C. attenuata
C. attila
C. baileyi
C. batesi
C. beatus
C. beccarii
C. bottegi
C. bottegoides
C. buettikoferi
C. caliginea
C. canariensis
C. cinderella
C. congobelgica
C. cossyrensis
C. crenata
C. crossei
C. cyanea
C. denti
C. desperata
C. dhofarensis
C. dolichura
C. douceti
C. dsinezumi
C. eisentrauti
C. elgonius
C. elongata
C. erica
C. fischeri
C. flavescens
C. floweri
C. foxi
C. fuliginosa
C. fulvastra
C. fumosa
C. fuscomurina
C. glassi
C. goliath
C. gracilipes
C. grandiceps
C. grandis
C. grassei
C. grayi
C. greenwoodi
C. gueldenstaedtii
C. harenna
C. hildegardeae
C. hirta
C. hispida
C. horsfieldii
C. jacksoni
C. jenkinsi
C. kivuana
C. lamottei
C. lanosa
C. lasiura
C. latona
C. lea
C. leucodon
C. levicula
C. littoralis
C. longipes
C. lucina
C. ludia
C. luna
C. lusitania
C. macarthuri
C. macmillani
C. macowi
C. malayana
C. manengubae
C. maquassiensis
C. mariquensis
C. maurisca
C. maxi
C. mindorus

C. minuta
C. miya
C. monax
C. monticola
C. montis
C. muricauda
C. mutesae
C. nana
C. nanilla
C. neglecta
C. negrina
C. nicobarica
C. nigeriae
C. nigricans
C. nigripes
C. nigrofusca
C. nimbae
C. niobe
C. obscurior
C. olivieri
C. orii
C. osorio
C. palawanensis
C. paradoxura
C. parvipes
C. pasha
C. pergrisea
C. phaeura
C. picea
C. pitmani
C. planiceps
C. poensis
C. polia
C. pullata
C. raineyi
C. religiosa
C. rhoditis
C. roosevelti
C. russula
C. selina
C. serezkyensis
C. sibirica
C. sicula
C. silacea
C. smithii
C. somalica
C. stenocephala
C. suaveolens
C. susiana
C. tansaniana
C. tarella
C. tarfayensis
C. telfordi
C. tenuis
C. thalia
C. theresae
C. thomensis
C. turba
C. ultima
C. usambarae
C. viaria
C. voi
C. whitakeri
C. wimmeri
C. xantippe
C. yankariensis
C. zaphiri
C. zarudnyi
C. zimmeri
C. zimmermanni
Cryptotis [Genus]
C. avia [Species]
C. endersi
C. goldmani
C. goodwini
C. gracilis
C. hondurensis
C. magna
C. meridensis
C. mexicana
C. montivaga
C. nigrescens
C. parva
C. squamipes
C. thomasi
Diplomesodon [Genus]
D. pulchellum [Species]
Feroculus [Genus]
F. feroculus [Species]
Megasorex [Genus]
M. gigas [Species]
Myosorex [Genus]
M. babaulti [Species]
M. blarina
M. cafer
M. eisentrauti
M. geata
M. longicaudatus
M. okuensis
M. rumpii
M. schalleri
M. sclateri
M. tenuis
M. varius
Nectogale [Genus]
N. elegans [Species]
Neomys [Genus]
N. anomalus [Species]
N. fodiens
N. schelkovnikovi
Notiosorex [Genus]
N. crawfordi [Species]
Paracrocidura [Genus]
P. graueri [Species]
P. maxima
P. schoutedeni
Ruwenzorisorex [Genus]
R. suncoides [Species]
Scutisorex [Genus]
S. somereni [Species]
Solisorex [Genus]
S. pearsoni [Species]
Sorex [Genus]
S. alaskanus [Species]
S. alpinus
S. araneus
S. arcticus
S. arizonae
S. asper
S. bairdii
S. bedfordiae
S. bendirii
S. buchariensis
S. caecutiens
S. camtschatica
S. cansulus
S. cinereus
S. coronatus
S. cylindricauda
S. daphaenodon
S. dispar
S. emarginatus
S. excelsus
S. fumeus
S. gaspensis
S. gracillimus
S. granarius
S. haydeni
S. hosonoi
S. hoyi
S. hydrodromus
S. isodon
S. jacksoni
S. kozlovi
S. leucogaster
S. longirostris
S. lyelli
S. macrodon
S. merriami
S. milleri
S. minutissimus
S. minutus
S. mirabilis
S. monticolus
S. nanus
S. oreopolus
S. ornatus
S. pacificus
S. palustris
S. planiceps
S. portenkoi
S. preblei
S. raddei
S. roboratus
S. sadonis
S. samniticus
S. satunini
S. saussurei
S. sclateri
S. shinto

S. sinalis
S. sonomae
S. stizodon
S. tenellus
S. thibetanus
S. trowbridgii
S. tundrensis
S. ugyunak
S. unguiculatus
S. vagrans
S. ventralis
S. veraepacis
S. volnuchini
Soriculus [Genus]
S. caudatus [Species]
S. fumidus
S. hypsibius
S. lamula
S. leucops
S. macrurus
S. nigrescens
S. parca
S. salenskii
S. smithii
Suncus [Genus]
S. ater [Species]
S. dayi
S. etruscus
S. fellowesgordoni
S. hosei
S. infinitesimus
S. lixus
S. madagascariensis
S. malayanus
S. mertensi
S. montanus
S. murinus
S. remyi
S. stoliczkanus
S. varilla
S. zeylanicus
Surdisorex [Genus]
S. norae [Species]
S. polulus
Sylvisorex [Genus]
S. granti [Species]
S. howelli
S. isabellae
S. johnstoni
S. lunaris
S. megalura
S. morio
S. ollula
S. oriundus
S. vulcanorum

Talpidae [Family]
Desmana [Genus]
D. moschata [Species]
Galemys [Genus]
G. pyrenaicus [Species]
Condylura [Genus]
C. cristata [Species]
Euroscaptor [Genus]
E. grandis [Species]
E. klossi
E. longirostris
E. micrura
E. mizura
E. parvidens
Mogera [Genus]
M. etigo [Species]
M. insularis
M. kobeae
M. minor
M. robusta
M. tokudae
M. wogura
Nesoscaptor [Genus]
N. uchidai [Species]
Neurotrichus [Genus]
N. gibbsii [Species]
Parascalops [Genus]
P. breweri [Species]
Parascaptor [Genus]
P. leucura [Species]
Scalopus [Genus]
S. aquaticus [Species]
Scapanulus [Genus]
S. oweni [Species]
Scapanus [Genus]
S. latimanus [Species]
S. orarius
S. townsendii
Scaptochirus [Genus]
S. moschatus [Species]
Scaptonyx [Genus]
S. fusicaudus [Species]
Talpa [Genus]
T. altaica [Species]
T. caeca
T. caucasica
T. europaea
T. levantis
T. occidentalis
T. romana
T. stankovici
T. streeti
Urotrichus [Genus]
U. pilirostris [Species]
U. talpoides
Uropsilus [Genus]
U. andersoni [Species]
U. gracilis
U. investigator
U. soricipes

Scandentia [Order]

Tupaiidae [Family]
Anathana [Genus]
A. ellioti [Species]
Dendrogale [Genus]
D. melanura [Species]
D. murina
Ptilocercus [Genus]
P. lowii [Species]
Tupaia [Genus]
T. belangeri [Species]
T. chrysogaster
T. dorsalis
T. glis
T. gracilis
T. javanica
T. longipes
T. minor
T. montana
T. nicobarica
T. palawanensis
T. picta
T. splendidula
T. tana
Urogale [Genus]
U. everetti [Species]

Dermoptera [Order]

Cynocephalidae [Family]
Cynocephalus [Genus]
C. variegatus [Species]
C. volans

Chiroptera [Order]

Pteropodidae [Family]
Acerodon [Genus]
A. celebensis [Species]
A. humilis
A. jubatus
A. leucotis
A. lucifer
A. mackloti
Aethalops [Genus]
A. alecto [Species]
Alionycteris [Genus]
A. paucidentata [Species]
Aproteles [Genus]
A. bulmerae [Species]
Balionycteris [Genus]
B. maculata [Species]
Boneia [Genus]
B. bidens [Species]
Casinycteris [Genus]
C. argynnis [Species]
Chironax [Genus]
C. melanocephalus [Species]
Cynopterus [Genus]
C. brachyotis [Species]
C. horsfieldi
C. nusatenggara
C. sphinx
C. titthaecheileus
Dobsonia [Genus]

D. beauforti [Species]
D. chapmani
D. emersa
D. exoleta
D. inermis
D. minor
D. moluccensis
D. pannietensis
D. peroni
D. praedatrix
D. viridis
Dyacopterus [Genus]
D. spadiceus [Species]
Eidolon [Genus]
E. dupreanum [Species]
E. helvum
Eonycteris [Genus]
E. major [Species]
E. spelaea
Epomophorus [Genus]
E. angolensis [Species]
E. gambianus
E. grandis
E. labiatus
E. minimus
E. wahlbergi
Epomops [Genus]
E. buettikoferi [Species]
E. dobsoni
E. franqueti
Haplonycteris [Genus]
H. fischeri [Species]
Harpyionycteris [Genus]
H. celebensis [Species]
H. whiteheadi
Hypsignathus [Genus]
H. monstrosus [Species]
Latidens [Genus]
L. salimalii [Species]
Macroglossus [Genus]
M. minimus [Species]
M. sobrinus
Megaerops [Genus]
M. ecaudatus [Species]
M. kusnotoi
M. niphanae
M. wetmorei
Megaloglossus [Genus]
M. woermanni [Species]
Melonycteris [Genus]
M. aurantius [Species]
M. melanops
M. woodfordi
Micropteropus [Genus]
M. intermedius [Species]
M. pusillus
Myonycteris [Genus]
M. brachycephala [Species]
M. relicta
M. torquata
Nanonycteris [Genus]
N. veldkampi [Species]
Neopteryx [Genus]
N. frosti [Species]
Notopteris [Genus]
N. macdonaldi [Species]
Nyctimene [Genus]
N. aello [Species]
N. albiventer
N. celaeno
N. cephalotes
N. certans
N. cyclotis
N. draconilla
N. major
N. malaitensis
N. masalai
N. minutus
N. rabori
N. robinsoni
N. sanctacrucis
N. vizcaccia
Otopteropus [Genus]
O. cartilagonodus [Species]
Paranyctimene [Genus]
P. raptor [Species]
Penthetor [Genus]
P. lucasi [Species]
Plerotes [Genus]
P. anchietai [Species]
Ptenochirus [Genus]
P. jagori [Species]
P. minor
Pteralopex [Genus]
P. acrodonta [Species]
P. anceps
P. atrata
P. pulchra
Pteropus [Genus]
P. admiralitatum [Species]
P. aldabrensis
P. alecto
P. anetianus
P. argentatus
P. brunneus
P. caniceps
P. chrysoproctus
P. conspicillatus
P. dasymallus
P. faunulus
P. fundatus
P. giganteus
P. gilliardi
P. griseus
P. howensis
P. hypomelanus
P. insularis
P. leucopterus
P. livingstonei
P. lombocensis
P. lylei
P. macrotis
P. mahaganus
P. mariannus
P. mearnsi
P. melanopogon
P. melanotus
P. molossinus
P. neohibernicus
P. niger
P. nitendiensis
P. ocularis
P. ornatus
P. personatus
P. phaeocephalus
P. pilosus
P. pohlei
P. poliocephalus
P. pselaphon
P. pumilus
P. rayneri
P. rodricensis
P. rufus
P. samoensis
P. sanctacrucis
P. scapulatus
P. seychellensis
P. speciosus
P. subniger
P. temmincki
P. tokudae
P. tonganus
P. tuberculatus
P. vampyrus
P. vetulus
P. voeltzkowi
P. woodfordi
Rousettus [Genus]
R. aegyptiacus [Species]
R. amplexicaudatus
R. angolensis
R. celebensis
R. lanosus
R. leschenaulti
R. madagascariensis
R. obliviosus
R. spinalatus
Scotonycteris [Genus]
S. ophiodon [Species]
S. zenkeri
Sphaerias [Genus]
S. blanfordi [Species]
Styloctenium [Genus]
S. wallacei [Species]
Syconycteris [Genus]
S. australis [Species]
S. carolinae
S. hobbit
Thoopterus [Genus]
T. nigrescens [Species]

Rhinopomatidae [Family]
Rhinopoma [Genus]
R. hardwickei [Species]
R. microphyllum
R. muscatellum

Emballonuridae [Family]
Balantiopteryx [Genus]
B. infusca [Species]
B. io
B. plicata
Centronycteris [Genus]
C. maximiliani [Species]
Coleura [Genus]
C. afra [Species]
C. seychellensis
Cormura [Genus]
C. brevirostris [Species]
Cyttarops [Genus]
C. alecto [Species]
Diclidurus [Genus]
D. albus [Species]
D. ingens
D. isabellus
D. scutatus
Emballonura [Genus]
E. alecto [Species]
E. atrata
E. beccarii
E. dianae
E. furax
E. monticola
E. raffrayana
E. semicaudata
Mosia [Genus]
M. nigrescens [Species]
Peropteryx [Genus]
P. kappleri [Species]
P. leucoptera
P. macrotis
Rhynchonycteris [Genus]
R. naso [Species]
Saccolaimus [Genus]
S. flaviventris [Species]
S. mixtus
S. peli
S. pluto
S. saccolaimus
Saccopteryx [Genus]
S. bilineata [Species]
S. canescens
S. gymnura
S. leptura
Taphozous [Genus]
T. australis [Species]
T. georgianus
T. hamiltoni
T. hildegardeae
T. hilli
T. kapalgensis
T. longimanus
T. mauritianus
T. melanopogon
T. nudiventris
T. perforatus
T. philippinensis
T. theobaldi

Craseonycteridae [Family]
Craseonycteris [Genus]
C. thonglongyai [Species]

Nycteridae [Family]
Nycteris [Genus]
N. arge [Species]
N. gambiensis
N. grandis
N. hispida
N. intermedia
N. javanica
N. macrotis
N. major
N. nana
N. thebaica
N. tragata
N. woodi

Megadermatidae [Family]
Cardioderma [Genus]
C. cor [Species]
Lavia [Genus]
L. frons [Species]
Macroderma [Genus]
M. gigas [Species]
Megaderma [Genus]
M. lyra [Species]
M. spasma

Rhinolophidae [Family]
Rhinolophus [Genus]
R. acuminatus [Species]
R. adami
R. affinis
R. alcyone
R. anderseni
R. arcuatus
R. blasii
R. borneensis
R. canuti
R. capensis
R. celebensis
R. clivosus
R. coelophyllus
R. cognatus
R. cornutus
R. creaghi
R. darlingi
R. deckenii
R. denti
R. eloquens
R. euryale
R. euryotis
R. ferrumequinum
R. fumigatus
R. guineensis
R. hildebrandti
R. hipposideros
R. imaizumii
R. inops
R. keyensis
R. landeri
R. lepidus
R. luctus
R. maclaudi
R. macrotis
R. malayanus
R. marshalli
R. megaphyllus
R. mehelyi
R. mitratus
R. monoceros
R. nereis
R. osgoodi
R. paradoxolophus
R. pearsoni
R. philippinensis
R. pusillus
R. rex
R. robinsoni
R. rouxi
R. rufus
R. sedulus
R. shameli
R. silvestris
R. simplex
R. simulator
R. stheno
R. subbadius
R. subrufus
R. swinnyi
R. thomasi
R. trifoliatus
R. virgo
R. yunanensis

Hipposideridae [Family]
Anthops [Genus]
A. ornatus [Species]
Asellia [Genus]
A. patrizii [Species]
A. tridens
Aselliscus [Genus]
A. stoliczkanus [Species]
A. tricuspidatus
Cloeotis [Genus]
C. percivali [Species]
Coelops [Genus]
C. frithi [Species]
C. hirsutus
C. robinsoni
Hipposideros [Genus]
H. abae [Species]
H. armiger

H. ater
H. beatus
H. bicolor
H. breviceps
H. caffer
H. calcaratus
H. camerunensis
H. cervinus
H. cineraceus
H. commersoni
H. coronatus
H. corynophyllus
H. coxi
H. crumeniferus
H. curtus
H. cyclops
H. diadema
H. dinops
H. doriae
H. dyacorum
H. fuliginosus
H. fulvus
H. galeritus
H. halophyllus
H. inexpectatus
H. jonesi
H. lamottei
H. lankadiva
H. larvatus
H. lekaguli
H. lylei
H. macrobullatus
H. maggietaylorae
H. marisae
H. megalotis
H. muscinus
H. nequam
H. obscurus
H. papua
H. pomona
H. pratti
H. pygmaeus
H. ridleyi
H. ruber
H. sabanus
H. schistaceus
H. semoni
H. speoris
H. stenotis
H. turpis
H. wollastoni
Paracoelops [Genus]
P. megalotis [Species]
Rhinonicteris [Genus]
R. aurantia [Species]
Triaenops [Genus]
T. furculus [Species]
T. persicus

Phyllostomidae [Family]
Ametrida [Genus]
A. centurio [Species]
Anoura [Genus]
A. caudifer [Species]
A. cultrata
A. geoffroyi
A. latidens
Ardops [Genus]
A. nichollsi [Species]
Ariteus [Genus]
A. flavescens [Species]
Artibeus [Genus]
A. amplus [Species]
A. anderseni
A. aztecus
A. cinereus
A. concolor
A. fimbriatus
A. fraterculus
A. glaucus
A. hartii
A. hirsutus
A. inopinatus
A. jamaicensis
A. lituratus
A. obscurus
A. phaeotis
A. planirostris
A. toltecus
Brachyphylla [Genus]
B. cavernarum [Species]
B. nana
Carollia [Genus]
C. brevicauda [Species]
C. castanea
C. perspicillata
C. subrufa
Centurio [Genus]
C. senex [Species]
Chiroderma [Genus]
C. doriae [Species]
C. improvisum
C. salvini
C. trinitatum
C. villosum
Choeroniscus [Genus]
C. godmani [Species]
C. intermedius
C. minor
C. periosus
Choeronycteris [Genus]
C. mexicana [Species]
Chrotopterus [Genus]
C. auritus [Species]
Desmodus [Genus]
D. rotundus [Species]
Diaemus [Genus]
D. youngi [Species]
Diphylla [Genus]
D. ecaudata [Species]
Ectophylla [Genus]
E. alba [Species]
Erophylla [Genus]
E. sezekorni [Species]
Glossophaga [Genus]
G. commissarisi [Species]
G. leachii
G. longirostris
G. morenoi
G. soricina
Hylonycteris [Genus]
H. underwoodi [Species]
Leptonycteris [Genus]
L. curasoae [Species]
L. nivalis
Lichonycteris [Genus]
L. obscura [Species]
Lionycteris [Genus]
L. spurrelli [Species]
Lonchophylla [Genus]
L. bokermanni [Species]
L. dekeyseri
L. handleyi
L. hesperia
L. mordax
L. robusta
L. thomasi
Lonchorhina [Genus]
L. aurita [Species]
L. fernandezi
L. marinkellei
L. orinocensis
Macrophyllum [Genus]
M. macrophyllum [Species]
Macrotus [Genus]
M. californicus [Species]
M. waterhousii
Mesophylla [Genus]
M. macconnelli [Species]
Micronycteris [Genus]
M. behnii [Species]
M. brachyotis
M. daviesi
M. hirsuta
M. megalotis
M. minuta
M. nicefori
M. pusilla
M. schmidtorum
M. sylvestris
Mimon [Genus]
M. bennettii [Species]
M. crenulatum
Monophyllus [Genus]
M. plethodon [Species]
M. redmani
Musonycteris [Genus]
M. harrisoni [Species]
Phylloderma [Genus]
P. stenops [Species]
Phyllonycteris [Genus]

P. aphylla [Species]
P. poeyi
Phyllops [Genus]
P. falcatus [Species]
Phyllostomus [Genus]
P. discolor [Species]
P. elongatus
P. hastatus
P. latifolius
Platalina [Genus]
P. genovensium [Species]
Platyrrhinus [Genus]
P. aurarius [Species]
P. brachycephalus
P. chocoensis
P. dorsalis
P. helleri
P. infuscus
P. lineatus
P. recifinus
P. umbratus
P. vittatus
Pygoderma [Genus]
P. bilabiatum [Species]
Rhinophylla [Genus]
R. alethina [Species]
R. fischerae
R. pumilio
Scleronycteris [Genus]
S. ega [Species]
Sphaeronycteris [Genus]
S. toxophyllum [Species]
Stenoderma [Genus]
S. rufum [Species]
Sturnira [Genus]
S. aratathomasi [Species]
S. bidens
S. bogotensis
S. erythromos
S. lilium
S. ludovici
S. luisi
S. magna
S. mordax
S. nana
S. thomasi
S. tildae
Tonatia [Genus]
T. bidens [Species]
T. brasiliense
T. carrikeri
T. evotis
T. schulzi
T. silvicola
Trachops [Genus]
T. cirrhosus [Species]
Uroderma [Genus]
U. bilobatum [Species]
U. magnirostrum
Vampyressa [Genus]
V. bidens [Species]
V. brocki
V. melissa
V. nymphaea
V. pusilla
Vampyrodes [Genus]
V. caraccioli [Species]
Vampyrum [Genus]
V. spectrum [Species]

Mormoopidae [Family]
Mormoops [Genus]
M. blainvillii [Species]
M. megalophylla
Pteronotus [Genus]
P. davyi [Species]
P. gymnonotus
P. macleayii
P. parnellii
P. personatus
P. quadridens

Noctilionidae [Family]
Noctilio [Genus]
N. albiventris [Species]
N. leporinus

Mystacinidae [Family]
Mystacina [Genus]
M. robusta [Species]
M. tuberculata

Natalidae [Family]
Natalus [Genus]
N. lepidus [Species]
N. micropus
N. stramineus
N. tumidifrons
N. tumidirostris

Furipteridae [Family]
Amorphochilus [Genus]
A. schnablii [Species]
Furipterus [Genus]
F. horrens [Species]

Thyropteridae [Family]
Thyroptera [Genus]
T. discifera [Species]
T. tricolor

Myzopodidae [Family]
Myzopoda [Genus]
M. aurita [Species]

Molossidae [Family]
Chaerephon [Genus]
C. aloysiisabaudiae [Species]
C. ansorgei
C. bemmeleni
C. bivittata
C. chapini
C. gallagheri
C. jobensis
C. johorensis
C. major
C. nigeriae
C. plicata
C. pumila
C. russata
Cheiromeles [Genus]
C. torquatus [Species]
Eumops [Genus]
E. auripendulus [Species]
E. bonariensis
E. dabbenei
E. glaucinus
E. hansae
E. maurus
E. perotis
E. underwoodi
Molossops [Genus]
M. abrasus [Species]
M. aequatorianus
M. greenhalli
M. mattogrossensis
M. neglectus
M. planirostris
M. temminckii
Molossus [Genus]
M. ater [Species]
M. bondae
M. molossus
M. pretiosus
M. sinaloae
Mops [Genus]
M. brachypterus [Species]
M. condylurus
M. congicus
M. demonstrator
M. midas
M. mops
M. nanulus
M. niangarae
M. niveiventer
M. petersoni
M. sarasinorum
M. spurrelli
M. thersites
M. trevori
Mormopterus [Genus]
M. acetabulosus [Species]
M. beccarii
M. doriae
M. jugularis
M. kalinowskii
M. minutus
M. norfolkensis
M. petrophilus
M. phrudus
M. planiceps
M. setiger
Myopterus [Genus]

M. daubentonii [Species]
M. whitleyi
Nyctinomops [Genus]
N. aurispinosus [Species]
N. femorosaccus
N. laticaudatus
N. macrotis
Otomops [Genus]
O. formosus [Species]
O. martiensseni
O. papuensis
O. secundus
O. wroughtoni
Promops [Genus]
P. centralis [Species]
P. nasutus
Tadarida [Genus]
T. aegyptiaca [Species]
T. australis
T. brasiliensis
T. espiritosantensis
T. fulminans
T. lobata
T. teniotis
T. ventralis

Vespertilionidae [Family]
Antrozous [Genus]
A. dubiaquercus [Species]
A. pallidus
Barbastella [Genus]
B. barbastellus [Species]
B. leucomelas
Chalinolobus [Genus]
C. alboguttatus [Species]
C. argentatus
C. beatrix
C. dwyeri
C. egeria
C. gleni
C. gouldii
C. kenyacola
C. morio
C. nigrogriseus
C. picatus
C. poensis
C. superbus
C. tuberculatus
C. variegatus
Eptesicus [Genus]
E. baverstocki [Species]
E. bobrinskoi
E. bottae
E. brasiliensis
E. brunneus
E. capensis
E. demissus
E. diminutus
E. douglasorum
E. flavescens
E. floweri
E. furinalis
E. fuscus
E. guadeloupensis
E. guineensis
E. hottentotus
E. innoxius
E. kobayashii
E. melckorum
E. nasutus
E. nilssoni
E. pachyotis
E. platyops
E. pumilus
E. regulus
E. rendalli
E. sagittula
E. serotinus
E. somalicus
E. tatei
E. tenuipinnis
E. vulturnus
Euderma [Genus]
E. maculatum [Species]
Eudiscopus [Genus]
E. denticulus [Species]
Glischropus [Genus]
G. javanus [Species]
G. tylopus
Harpiocephalus [Genus]
H. harpia [Species]
Hesperoptenus [Genus]
H. blanfordi [Species]
H. doriae
H. gaskelli
H. tickelli
H. tomesi
Histiotus [Genus]
H. alienus [Species]
H. macrotus
H. montanus
H. velatus
Ia [Genus]
I. io [Species]
Idionycteris [Genus]
I. phyllotis [Species]
Kerivoula [Genus]
K. aerosa [Species]
K. africana
K. agnella
K. argentata
K. atrox
K. cuprosa
K. eriophora
K. flora
K. hardwickei
K. intermedia
K. jagori
K. lanosa
K. minuta
K. muscina
K. myrella
K. papillosa
K. papuensis
K. pellucida
K. phalaena
K. picta
K. smithi
K. whiteheadi
Laephotis [Genus]
L. angolensis [Species]
L. botswanae
L. namibensis
L. wintoni
Lasionycteris [Genus]
L. noctivagans [Species]
Lasiurus [Genus]
L. borealis [Species]
L. castaneus
L. cinereus
L. ega
L. egregius
L. intermedius
L. seminolus
Mimetillus [Genus]
M. moloneyi [Species]
Miniopterus [Genus]
M. australis [Species]
M. fraterculus
M. fuscus
M. inflatus
M. magnater
M. minor
M. pusillus
M. robustior
M. schreibersi
M. tristis
Murina [Genus]
M. aenea [Species]
M. aurata
M. cyclotis
M. florium
M. fusca
M. grisea
M. huttoni
M. leucogaster
M. puta
M. rozendaali
M. silvatica
M. suilla
M. tenebrosa
M. tubinaris
M. ussuriensis
Myotis [Genus]
M. abei [Species]
M. adversus
M. aelleni
M. albescens
M. altarium
M. annectans
M. atacamensis

M. auriculus
M. australis
M. austroriparius
M. bechsteini
M. blythii
M. bocagei
M. bombinus
M. brandti
M. californicus
M. capaccinii
M. chiloensis
M. chinensis
M. cobanensis
M. dasycneme
M. daubentoni
M. dominicensis
M. elegans
M. emarginatus
M. evotis
M. findleyi
M. formosus
M. fortidens
M. frater
M. goudoti
M. grisescens
M. hasseltii
M. horsfieldii
M. hosonoi
M. ikonnikovi
M. insularum
M. keaysi
M. keenii
M. leibii
M. lesueuri
M. levis
M. longipes
M. lucifugus
M. macrodactylus
M. macrotarsus
M. martiniquensis
M. milleri
M. montivagus
M. morrisi
M. muricola
M. myotis
M. mystacinus
M. nattereri
M. nesopolus
M. nigricans
M. oreias
M. oxyotus
M. ozensis
M. peninsularis
M. pequinius
M. planiceps
M. pruinosus
M. ricketti
M. ridleyi
M. riparius
M. rosseti
M. ruber
M. schaubi
M. scotti
M. seabrai
M. sicarius
M. siligorensis
M. simus
M. sodalis
M. stalkeri
M. thysanodes
M. tricolor
M. velifer
M. vivesi
M. volans
M. welwitschii
M. yesoensis
M. yumanensis
Nyctalus [Genus]
N. aviator [Species]
N. azoreum
N. lasiopterus
N. leisleri
N. montanus
N. noctula
Nycticeius [Genus]
N. balstoni [Species]
N. greyii
N. humeralis
N. rueppellii
N. sanborni
N. schlieffeni
Nyctophilus [Genus]
N. arnhemensis [Species]
N. geoffroyi
N. gouldi
N. heran
N. microdon
N. microtis
N. timoriensis
N. walkeri
Otonycteris [Genus]
O. hemprichi [Species]
Pharotis [Genus]
P. imogene [Species]
Philetor [Genus]
P. brachypterus [Species]
Pipistrellus [Genus]
P. aegyptius [Species]
P. aero
P. affinis
P. anchietai
P. anthonyi
P. arabicus
P. ariel
P. babu
P. bodenheimeri
P. cadornae
P. ceylonicus
P. circumdatus
P. coromandra
P. crassulus
P. cuprosus
P. dormeri
P. eisentrauti
P. endoi
P. hesperus
P. imbricatus
P. inexspectatus
P. javanicus
P. joffrei
P. kitcheneri
P. kuhlii
P. lophurus
P. macrotis
P. maderensis
P. mimus
P. minahassae
P. mordax
P. musciculus
P. nanulus
P. nanus
P. nathusii
P. paterculus
P. peguensis
P. permixtus
P. petersi
P. pipistrellus
P. pulveratus
P. rueppelli
P. rusticus
P. savii
P. societatis
P. stenopterus
P. sturdeei
P. subflavus
P. tasmaniensis
P. tenuis
Plecotus [Genus]
P. auritus [Species]
P. austriacus
P. mexicanus
P. rafinesquii
P. taivanus
P. teneriffae
P. townsendii
Rhogeessa [Genus]
R. alleni [Species]
R. genowaysi
R. gracilis
R. minutilla
R. mira
R. parvula
R. tumida
Scotoecus [Genus]
S. albofuscus [Species]
S. hirundo
S. pallidus
Scotomanes [Genus]
S. emarginatus [Species]
S. ornatus

Scotophilus [Genus]
S. borbonicus [Species]
S. celebensis
S. dinganii
S. heathi
S. kuhlii
S. leucogaster
S. nigrita
S. nux
S. robustus
S. viridis
Tomopeas [Genus]
T. ravus [Species]
Tylonycteris [Genus]
T. pachypus [Species]
T. robustula
Vespertilio [Genus]
V. murinus [Species]
V. superans

Primates [Order]

Lorisidae [Family]
Arctocebus [Genus]
A. aureus [Species]
A. calabarensis
Loris [Genus]
L. tardigradus [Species]
Nycticebus [Genus]
N. coucang [Species]
N. pygmaeus
Perodicticus [Genus]
P. potto [Species]

Galagidae [Family]
Euoticus [Genus]
E. elegantulus [Species]
E. pallidus
Galago [Genus]
G. alleni [Species]
G. gallarum
G. matschiei
G. moholi
G. senegalensis
Galagoides [Genus]
G. demidoff [Species]
G. zanzibaricus
Otolemur [Genus]
O. crassicaudatus [Species]
O. garnettii

Cheirogaleidae [Family]
Allocebus [Genus]
A. trichotis [Species]
Cheirogaleus [Genus]
C. major [Species]
C. medius
Microcebus [Genus]
Microcebus coquereli [Species]
Microcebus murinus
Microcebus rufus
Phaner [Genus]
P. furcifer [Species]

Lemuridae [Family]
Eulemur [Genus]
E. coronatus [Species]
E. fulvus
E. macaco
E. mongoz
E. rubriventer
Hapalemur [Genus]
H. aureus [Species]
H. griseus
H. simus
Lemur [Genus]
L. catta [Species]
Varecia [Genus]
V. variegata [Species]

Indriidae [Family]
Avahi [Genus]
A. laniger [Species]
Indri [Genus]
I. indri [Species]
Propithecus [Genus]
P. diadema [Species]
P. tattersalli
P. verreauxi

Lepilemuridae [Family]
Lepilemur [Genus]
L. dorsalis [Species]
L. edwardsi
L. leucopus
L. microdon
L. mustelinus
L. ruficaudatus
L. septentrionalis

Daubentoniidae [Family]
Daubentonia [Genus]
D. madagascariensis [Species]

Tarsiidae [Family]
Tarsius [Genus]
T. bancanus [Species]
T. dianae
T. pumilus
T. spectrum
T. syrichta

Cebidae [Family]
Alouatta [Genus]
A. belzebul [Species]
A. caraya
A. coibensis
A. fusca
A. palliata
A. pigra
A. sara
A. seniculus
Callicebus [Genus]
C. brunneus [Species]
C. caligatus
C. cinerascens
C. cupreus
C. donacophilus
C. dubius
C. hoffmannsi
C. modestus
C. moloch
C. oenanthe
C. olallae
C. personatus
C. torquatus
Cebus [Genus]
C. albifrons [Species]
C. apella
C. capucinus
C. olivaceus
Saimiri [Genus]
S. boliviensis [Species]
S. oerstedii
S. sciureus
S. ustus
S. vanzolinii

Callitrichidae [Family]
Callimico [Genus]
C. goeldii [Species]
Callithrix [Genus]
C. argentata [Species]
C. aurita
C. flaviceps
C. geoffroyi
C. humeralifer
C. jacchus
C. kuhlii
C. penicillata
C. pygmaea
Leontopithecus [Genus]
L. caissara [Species]
L. chrysomela
L. chrysopygus
L. rosalia
Saguinus [Genus]
S. bicolor [Species]
S. fuscicollis
S. geoffroyi
S. imperator
S. inustus
S. labiatus
S. leucopus
S. midas
S. mystax
S. nigricollis
S. oedipus
S. tripartitus

Aotidae [Family]
Aotus [Genus]
A. azarai [Species]
A. brumbacki

A. hershkovitzi
A. infulatus
A. lemurinus
A. miconax
A. nancymaae
A. nigriceps
A. trivirgatus
A. vociferans

Pitheciidae [Family]
Cacajao [Genus]
C. calvus [Species]
C. melanocephalus
Chiropotes [Genus]
C. albinasus [Species]
C. satanas
Pithecia [Genus]
P. aequatorialis [Species]
P. albicans
P. irrorata
P. monachus
P. pithecia

Atelidae [Family]
Ateles [Genus]
A. belzebuth [Species]
A. chamek
A. fusciceps
A. geoffroyi
A. marginatus
A. paniscus
Brachyteles [Genus]
B. arachnoides [Species]
Lagothrix [Genus]
L. flavicauda [Species]
L. lagotricha

Cercopithecidae [Family]
Allenopithecus [Genus]
A. nigroviridis [Species]
Cercocebus [Genus]
C. agilis [Species]
C. galeritus
C. torquatus
Cercopithecus [Genus]
C. ascanius [Species]
C. campbelli
C. cephus
C. diana
C. dryas
C. erythrogaster
C. erythrotis
C. hamlyni
C. lhoesti
C. mitis
C. mona
C. neglectus
C. nictitans
C. petaurista
C. pogonias
C. preussi
C. sclateri
C. solatus
C. wolfi
Chlorocebus [Genus]
C. aethiops [Species]
Colobus [Genus]
C. angolensis [Species]
C. guereza
C. polykomos
C. satanas
Erythrocebus [Genus]
E. patas [Species]
Lophocebus [Genus]
L. albigena [Species]
Macaca [Genus]
M. arctoides [Species]
M. assamensis
M. cyclopis
M. fascicularis
M. fuscata
M. maura
M. mulatta
M. nemestrina
M. nigra
M. ochreata
M. radiata
M. silenus
M. sinica
M. sylvanus
M. thibetana
M. tonkeana
Mandrillus [Genus]
M. leucophaeus [Species]
M. sphinx
Miopithecus
M. talapoin
Nasalis [Genus]
N. concolor [Species]
N. larvatus
Papio [Genus]
P. hamadryas [Species]
Presbytis [Genus]
P. comata [Species]
P. femoralis
P. frontata
P. hosei
P. melalophos
P. potenziani
P. rubicunda
P. thomasi
Procolobus [Genus]
P. badius [Species]
P. pennantii
P. preussi
P. rufomitratus
P. verus
Pygathrix [Genus]
P. avunculus [Species]
P. bieti
P. brelichi
P. nemaeus
P. roxellana
Semnopithecus [Genus]
S. entellus [Species]
Theropithecus [Genus]
T. gelada [Species]
Trachypithecus [Genus]
T. auratus [Species]
T. cristatus
T. francoisi
T. geei
T. johnii
T. obscurus
T. phayrei
T. pileatus
T. vetulus

Hylobatidae [Family]
Hylobates [Genus]
H. agilis [Species]
H. concolor
H. gabriellae
H. hoolock
H. klossii
H. lar
H. leucogenys
H. moloch
H. muelleri
H. pileatus
H. syndactylus

Hominidae [Family]
Gorilla [Genus]
G. gorilla [Species]
Homo [Genus]
H. sapiens [Species]
Pan [Genus]
P. paniscus [Species]
P. troglodytes
Pongo [Genus]
P. pygmaeus [Species]

Carnivora [Order]

Canidae [Family]
Alopex [Genus]
A. lagopus [Species]
Atelocynus
A. microtis
Canis [Genus]
C. adustus [Species]
C. aureus
C. latrans
C. lupus
C. mesomelas
C. rufus
C. simensis
Cerdocyon [Genus]
C. thous [Species]
Chrysocyon [Genus]
C. brachyurus [Species]
Cuon [Genus]

C. *alpinus* [Species]
Dusicyon [Genus]
D. *australis* [Species]
Lycaon [Genus]
L. *pictus* [Species]
Nyctereutes [Genus]
N. *procyonoides* [Species]
Otocyon [Genus]
O. *megalotis* [Species]
Pseudalopex [Genus]
P. *culpaeus* [Species]
P. *griseus*
P. *gymnocercus*
P. *sechurae*
P. *vetulus*
Speothos [Genus]
S. *venaticus* [Species]
Urocyon [Genus]
U. *cinereoargenteus* [Species]
U. *littoralis*
Vulpes [Genus]
V. *bengalensis* [Species]
V. *cana*
V. *chama*
V. *corsac*
V. *ferrilata*
V. *pallida*
V. *rueppelli*
V. *velox*
V. *vulpes*
V. *zerda*

Ursidae [Family]
Ailuropoda [Genus]
A. *melanoleuca* [Species]
Ailurus [Genus]
A. *fulgens* [Species]
Helarctos [Genus]
H. *malayanus* [Species]
Melursus [Genus]
M. *ursinus* [Species]
Tremarctos [Genus]
T. *ornatus* [Species]
Ursus [Genus]
U. *americanus* [Species]
U. *arctos*
U. *maritimus*
U. *thibetanus*

Procyonidae [Family]
Bassaricyon [Genus]
B. *alleni* [Species]
B. *beddardi*
B. *gabbii*
B. *lasius*
B. *pauli*
Potos [Genus]
P. *flavus* [Species]
Bassariscus [Genus]
B. *astutus* [Species]
B. *sumichrasti*
Nasua [Genus]
N. *narica* [Species]
N. *nasua*
Nasuella [Genus]
N. *olivacea* [Species]
Procyon [Genus]
P. *cancrivorus* [Species]
P. *gloveralleni*
P. *insularis*
P. *lotor*
P. *maynardi*
P. *minor*
P. *pygmaeus*

Mustelidae [Family]
Amblonyx [Genus]
A. *cinereus* [Species]
Aonyx [Genus]
A. *capensis* [Species]
A. *congicus*
Arctonyx [Genus]
A. *collaris* [Species]
Conepatus [Genus]
C. *chinga* [Species]
C. *humboldtii*
C. *leuconotus*
C. *mesoleucus*
C. *semistriatus*
Eira [Genus]
E. *barbara* [Species]
Enhydra [Genus]
E. *lutris* [Species]
Galictis [Genus]
G. *cuja* [Species]
G. *vittata*
Gulo [Genus]
G. *gulo* [Species]
Ictonyx [Genus]
I. *libyca* [Species]
I. *striatus*
Lontra [Genus]
L. *canadensis* [Species]
L. *felina*
L. *longicaudis*
L. *provocax*
Lutra [Genus]
L. *lutra* [Species]
L. *maculicollis*
L. *sumatrana*
Lutrogale [Genus]
L. *perspicillata* [Species]
Lyncodon [Genus]
L. *patagonicus* [Species]
Martes [Genus]
M. *americana* [Species]
M. *flavigula*
M. *foina*
M. *gwatkinsii*
M. *martes*
M. *melampus*
M. *pennanti*
M. *zibellina*
Meles [Genus]
M. *meles* [Species]
Mellivora [Genus]
M. *capensis* [Species]
Melogale [Genus]
M. *everetti* [Species]
M. *moschata*
M. *orientalis*
M. *personata*
Mephitis [Genus]
M. *macroura* [Species]
M. *mephitis*
Mustela [Genus]
M. *africana* [Species]
M. *altaica*
M. *erminea*
M. *eversmannii*
M. *felipei*
M. *frenata*
M. *kathiah*
M. *lutreola*
M. *lutreolina*
M. *nigripes*
M. *nivalis*
M. *nudipes*
M. *putorius*
M. *sibirica*
M. *strigidorsa*
M. *vison*
Mydaus [Genus]
M. *javanensis* [Species]
M. *marchei*
Poecilogale [Genus]
P. *albinucha* [Species]
Pteronura [Genus]
P. *brasiliensis* [Species]
Spilogale [Genus]
S. *putorius* [Species]
S. *pygmaea*
Taxidea [Genus]
T. *taxus* [Species]
Vormela [Genus]
V. *peregusna* [Species]

Viverridae [Family]
Arctictis [Genus]
A. *binturong* [Species]
Arctogalidia [Genus]
A. *trivirgata* [Species]
Chrotogale [Genus]
C. *owstoni* [Species]
Civettictis [Genus]
C. *civetta* [Species]
Cryptoprocta [Genus]
C. *ferox* [Species]
Cynogale [Genus]
C. *bennettii* [Species]
Diplogale [Genus]
D. *hosei* [Species]
Eupleres [Genus]

E. *goudotii* [Species]
Fossa [Genus]
F. *fossana* [Species]
Genetta [Genus]
G. *abyssinica* [Species]
G. *angolensis*
G. *genetta*
G. *johnstoni*
G. *maculata*
G. *servalina*
G. *thierryi*
G. *tigrina*
G. *victoriae*
Hemigalus [Genus]
H. *derbyanus* [Species]
Nandinia [Genus]
N. *binotata* [Species]
Macrogalidia [Genus]
M. *musschenbroekii* [Species]
Paguma [Genus]
P. *larvata* [Species]
Paradoxurus [Genus]
P. *hermaphroditus* [Species]
P. *jerdoni*
P. *zeylonensis*
Osbornictis [Genus]
O. *piscivora* [Species]
Poiana [Genus]
P. *richardsonii* [Species]
Prionodon [Genus]
P. *linsang* [Species]
P. *pardicolor*
Viverra [Genus]
V. *civettina* [Species]
V. *megaspila*
V. *tangalunga*
V. *zibetha*
Viverricula [Genus]
V. *indica* [Species]

Herpestidae [Family]
Atilax [Genus]
A. *paludinosus* [Species]
Bdeogale [Genus]
B. *crassicauda* [Species]
B. *jacksoni*
B. *nigripes*
Crossarchus [Genus]
C. *alexandri* [Species]
C. *ansorgei*
C. *obscurus*
Cynictis [Genus]
C. *penicillata* [Species]
Dologale [Genus]
D. *dybowskii* [Species]
Galerella [Genus]
G. *flavescens* [Species]
G. *pulverulenta*
G. *sanguinea*
G. *swalius*
Galidia [Genus]
G. *elegans* [Species]
Galidictis [Genus]
G. *fasciata* [Species]
G. *grandidieri*
Helogale [Genus]
H. *hirtula* [Species]
H. *parvula*
Herpestes [Genus]
H. *brachyurus* [Species]
H. *edwardsii*
H. *ichneumon*
H. *javanicus*
H. *naso*
H. *palustris*
H. *semitorquatus*
H. *smithii*
H. *urva*
H. *vitticollis*
Ichneumia [Genus]
I. *albicauda* [Species]
Liberiictis [Genus]
L. *kuhni* [Species]
Mungos [Genus]
M. *gambianus* [Species]
M. *mungo*
Mungotictis [Genus]
M. *decemlineata* [Species]
Paracynictis [Genus]
P. *selousi* [Species]
Rhynchogale [Genus]
R. *melleri* [Species]
Salanoia [Genus]
S. *concolor* [Species]
Suricata [Genus]
S. *suricatta* [Species]

Hyaenidae [Family]
Crocuta [Genus]
C. *crocuta* [Species]
Hyaena [Genus]
H. *hyaena* [Species]
Parahyaena [Genus]
P. *brunnea* [Species]
Proteles [Genus]
P. *cristatus* [Species]

Felidae [Family]
Acinonyx [Genus]
A. *jubatus* [Species]
Caracal [Genus]
C. *caracal* [Species]
Catopuma [Genus]
C. *badia* [Species]
C. *temminckii*
Felis [Genus]
F. *bieti* [Species]
F. *chaus*
F. *margarita*
F. *nigripes*
F. *silvestris*
Herpailurus [Genus]
H. *yaguarondi* [Species]
Leopardus [Genus]
L. *pardalis* [Species]
L. *tigrinus*
L. *wiedii*
Leptailurus [Genus]
L. *serval* [Species]
Lynx [Genus]
L. *canadensis* [Species]
L. *lynx*
L. *pardinus*
L. *rufus*
Neofelis [Genus]
N. *nebulosa* [Species]
Oncifelis [Genus]
O. *colocolo* [Species]
O. *geoffroyi*
O. *guigna*
Oreailurus [Genus]
O. *jacobita* [Species]
Otocolobus [Genus]
O. *manul* [Species]
Panthera [Genus]
P. *leo* [Species]
P. *onca*
P. *pardus*
P. *tigris*
Pardofelis
P. *marmorata*
Prionailurus [Genus]
P. *bengalensis* [Species]
P. *planiceps*
P. *rubiginosus*
P. *viverrinus*
Profelis [Genus]
P. *aurata* [Species]
Puma [Genus]
P. *concolor* [Species]
Uncia [Genus]
U. *uncia* [Species]

Otariidae [Family]
Arctocephalus [Genus]
A. *australis* [Species]
A. *forsteri*
A. *galapagoensis*
A. *gazella*
A. *philippii*
A. *pusillus*
A. *townsendi*
A. *tropicalis*
Callorhinus [Genus]
C. *ursinus* [Species]
Eumetopias [Genus]
E. *jubatus* [Species]
Neophoca [Genus]
N. *cinerea* [Species]
Otaria [Genus]
O. *byronia* [Species]

Phocarctos [Genus]
P. hookeri [Species]
Zalophus [Genus]
Z. californianus [Species]

Odobenidae [Family]
Odobenus [Genus]
O. rosmarus [Species]

Phocidae [Family]
Cystophora [Genus]
C. cristata [Species]
Erignathus [Genus]
E. barbatus [Species]
Halichoerus [Genus]
H. grypus [Species]
Hydrurga [Genus]
H. leptonyx [Species]
Leptonychotes [Genus]
L. weddellii [Species]
Lobodon [Genus]
L. carcinophagus [Species]
Mirounga [Genus]
M. angustirostris [Species]
M. leonina
Monachus [Genus]
M. monachus [Species]
M. schauinslandi
M. tropicalis
Ommatophoca [Genus]
O. rossii [Species]
Phoca [Genus]
P. caspica [Species]
P. fasciata
P. groenlandica
P. hispida
P. largha
P. sibirica
P. vitulina

Cetacea [Order]

Platanistidae [Family]
Platanista [Genus]
P. gangetica [Species]
P. minor

Lipotidae [Family]
Lipotes [Genus]
L. vexillifer [Species]

Pontoporiidae [Family]
Pontoporia [Genus]
P. blainvillei [Species]

Iniidae [Family]
Inia [Genus]
I. geoffrensis [Species]

Phocoenidae [Family]
Australophocaena [Genus]
A. dioptrica [Species]
Neophocaena [Genus]
N. phocaenoides [Species]
Phocoena [Genus]
P. phocoena [Species]
P. sinus
P. spinipinnis
Phocoenoides [Genus]
P. dalli [Species]

Delphinidae [Family]
Cephalorhynchus [Genus]
C. commersonii [Species]
C. eutropia
C. heavisidii
C. hectori
Delphinus [Genus]
D. delphis [Species]
Feresa [Genus]
F. attenuata [Species]
Globicephala [Genus]
G. macrorhynchus [Species]
G. melas
Grampus [Genus]
G. griseus [Species]
Lagenodelphis [Genus]
L. hosei [Species]
Lagenorhynchus [Genus]
L. acutus [Species]
L. albirostris
L. australis
L. cruciger
L. obliquidens
L. obscurus
Lissodelphis [Genus]
L. borealis [Species]
L. peronii
Orcaella [Genus]
O. brevirostris [Species]
Orcinus [Genus]
O. orca [Species]
Peponocephala [Genus]
P. electra [Species]
Pseudorca [Genus]
P. crassidens [Species]
Sotalia [Genus]
S. fluviatilis [Species]
Sousa [Genus]
S. chinensis [Species]
S. teuszii
Stenella [Genus]
S. attenuata [Species]
S. clymene
S. coeruleoalba
S. frontalis
S. longirostris
Steno [Genus]
S. bredanensis [Species]
Tursiops [Genus]
T. truncatus [Species]

Ziphiidae [Family]
Berardius [Genus]
B. arnuxii [Species]
B. bairdii
Hyperoodon [Genus]
H. ampullatus [Species]
H. planifrons
Indopacetus [Genus]
I. pacificus [Species]
Mesoplodon [Genus]
M. bidens [Species]
M. bowdoini
M. carlhubbsi
M. densirostris
M. europaeus
M. ginkgodens
M. grayi
M. hectori
M. layardii
M. mirus
M. peruvianus
M. stejnegeri
Tasmacetus [Genus]
T. shepherdi [Species]
Ziphius [Genus]
Z. cavirostris [Species]

Physeteridae [Family]
Kogia [Genus]
K. breviceps [Species]
K. simus
Physeter [Genus]
P. catodon [Species]

Monodontidae [Family]
Delphinapterus [Genus]
D. leucas [Species]
Monodon [Genus]
M. monoceros [Species]

Eschrichtiidae [Family]
Eschrichtius [Genus]
E. robustus [Species]

Neobalaenidae [Family]
Caperea [Genus]
C. marginata [Species]

Balaenidae [Family]
Balaena [Genus]
B. mysticetus [Species]
Eubalaena [Genus]
E. australis [Species]
E. glacialis

Balaenopteridae [Family]
Balaenoptera [Genus]
B. acutorostrata [Species]
B. borealis
B. edeni
B. musculus
B. physalus
Megaptera [Genus]
M. novaeangliae [Species]

Tubulidentata [Order]

Orycteropodidae [Family]
Orycteropus [Genus]
O. afer [Species]

Proboscidea [Order]

Elephantidae [Family]
Elephas [Genus]
E. maximus [Species]
Loxodonta [Genus]
L. africana [Species]
L. cyclotis

Hyracoidea [Order]

Procaviidae [Family]
Dendrohyrax [Genus]
D. arboreus [Species]
D. dorsalis
D. validus
Heterohyrax [Genus]
H. antineae [Species]
H. brucei
Procavia [Genus]
P. capensis [Species]

Sirenia [Order]

Dugongidae [Family]
Dugong [Genus]
D. dugon [Species]
Hydrodamalis [Genus]
H. gigas [Species]

Trichechidae [Family]
Trichechus [Genus]
T. inunguis [Species]
T. manatus
T. senegalensis

Perissodactyla [Order]

Equidae [Family]
Equus [Genus]
E. asinus [Species]
E. burchellii
E. caballus
E. grevyi
E. hemionus
E. kiang
E. onager
E. quagga
E. zebra

Tapiridae [Family]
Tapirus [Genus]
T. bairdii [Species]
T. indicus
T. pinchaque
T. terrestris

Rhinocerotidae [Family]
Ceratotherium [Genus]
C. simum [Species]
Dicerorhinus [Genus]
D. sumatrensis [Species]
Diceros [Genus]
D. bicornis [Species]
Rhinoceros [Genus]
R. sondaicus [Species]
R. unicornis

Artiodactyla [Order]

Suidae [Family]
Babyrousa [Genus]
B. babyrussa [Species]
Phacochoerus [Genus]
P. aethiopicus [Species]
P. africanus
Hylochoerus [Genus]
H. meinertzhageni [Species]
Potamochoerus [Genus]
P. larvatus [Species]
P. porcus
Sus [Genus]
S. barbatus [Species]
S. bucculentus
S. cebifrons
S. celebensis
S. heureni
S. philippensis
S. salvanius
S. scrofa
S. timoriensis
S. verrucosus

Tayassuidae [Family]
Catagonus [Genus]
C. wagneri [Species]
Pecari [Genus]
P. tajacu [Species]
Tayassu [Genus]
T. pecari [Species]

Hippopotamidae [Family]
Hexaprotodon [Genus]
H. liberiensis [Species]
H. madagascariensis
Hippopotamus [Genus]
H. amphibius [Species]
H. lemerlei

Camelidae [Family]
Camelus [Genus]
C. bactrianus [Species]
C. dromedarius
Lama [Genus]
L. glama [Species]
L. guanicoe
L. pacos
Vicugna [Genus]
V. vicugna [Species]

Tragulidae [Family]
Hyemoschus [Genus]
H. aquaticus [Species]
Moschiola [Genus]
M. meminna [Species]
Tragulus [Genus]
T. javanicus [Species]
T. napu

Cervidae [Family]
Alces [Genus]
A. alces [Species]
Axis [Genus]
A. axis [Species]
A. calamianensis
A. kuhlii
A. porcinus
Blastocerus [Genus]
B. dichotomus [Species]
Capreolus [Genus]
C. capreolus [Species]
C. pygargus
Cervus [Genus]
C. albirostris [Species]
C. alfredi
C. duvaucelii
C. elaphus
C. eldii
C. mariannus
C. nippon
C. schomburgki
C. timorensis
C. unicolor
Dama [Genus]
D. dama [Species]
D. mesopotamica
Elaphodus [Genus]
E. cephalophus [Species]
Elaphurus [Genus]
E. davidianus [Species]
Hippocamelus [Genus]
H. antisensis [Species]
H. bisulcus
Hydropotes [Genus]
H. inermis [Species]
Mazama [Genus]
M. americana [Species]
M. bricenii
M. chunyi
M. gouazoupira
M. nana
M. rufina
Moschus [Genus]
M. berezovskii [Species]
M. chrysogaster
M. fuscus
M. moschiferus
Muntiacus [Genus]
M. atherodes [Species]
M. crinifrons
M. feae
M. gongshanensis

M. muntjak
M. reevesi
Odocoileus [Genus]
O. hemionus [Species]
O. virginianus
Ozotoceros [Genus]
O. bezoarticus [Species]
Pudu [Genus]
P. mephistophiles [Species]
P. puda
Rangifer [Genus]
R. tarandus [Species]

Giraffidae [Family]
Giraffa [Genus]
G. camelopardalis [Species]
Okapia [Genus]
O. johnstoni [Species]

Antilocapridae [Family]
Antilocapra [Genus]
A. americana [Species]

Bovidae [Family]
Addax [Genus]
A. nasomaculatus [Species]
Aepyceros [Genus]
A. melampus [Species]
Alcelaphus [Genus]
A. buselaphus [Species]
Ammodorcas [Genus]
A. clarkei [Species]
Ammotragus [Genus]
A. lervia [Species]
Antidorcas [Genus]
A. marsupialis [Species]
Antilope [Genus]
A. cervicapra [Species]
Bison [Genus]
B. bison [Species]
B. bonasus
Bos [Genus]
B. frontalis [Species]
B. grunniens
B. javanicus
B. sauveli
B. taurus
Boselaphus [Genus]
B. tragocamelus [Species]
Bubalus [Genus]
B. bubalis [Species]
B. depressicornis
B. mephistopheles
B. mindorensis
B. quarlesi
Budorcas [Genus]
B. taxicolor [Species]
Capra [Genus]
C. caucasica [Species]
C. cylindricornis
C. falconeri
C. hircus
C. ibex
C. nubiana
C. pyrenaica
C. sibirica
C. walie
Cephalophus [Genus]
C. adersi [Species]
C. callipygus
C. dorsalis
C. harveyi
C. jentinki
C. leucogaster
C. maxwellii
C. monticola
C. natalensis
C. niger
C. nigrifrons
C. ogilbyi
C. rubidus
C. rufilatus
C. silvicultor
C. spadix
C. weynsi
C. zebra
Connochaetes [Genus]
C. gnou [Species]
C. taurinus
Damaliscus [Genus]
D. hunteri [Species]
D. lunatus
D. pygargus
Dorcatragus [Genus]
D. megalotis [Species]
Gazella [Genus]
G. arabica [Species]
G. bennettii
G. bilkis
G. cuvieri
G. dama
G. dorcas
G. gazella
G. granti
G. leptoceros
G. rufifrons
G. rufina
G. saudiya
G. soemmerringii
G. spekei
G. subgutturosa
G. thomsonii
Hemitragus [Genus]
H. hylocrius [Species]
H. jayakari
H. jemlahicus
Hippotragus [Genus]
H. equinus [Species]
H. leucophaeus
H. niger
Kobus [Genus]
K. ellipsiprymnus [Species]
K. kob
K. leche
K. megaceros
K. vardonii
Litocranius [Genus]
L. walleri [Species]
Madoqua [Genus]
M. guentheri [Species]
M. kirkii
M. piacentinii
M. saltiana
Naemorhedus [Genus]
N. baileyi [Species]
N. caudatus
N. crispus
N. goral
N. sumatraensis
N. swinhoei
Neotragus [Genus]
N. batesi [Species]
N. moschatus
N. pygmaeus
Oreamnos [Genus]
O. americanus [Species]
Oreotragus [Genus]
O. oreotragus [Species]
Oryx [Genus]
O. dammah [Species]
O. gazella
O. leucoryx
Ourebia [Genus]
O. ourebi [Species]
Ovibos [Genus]
O. moschatus [Species]
Ovis [Genus]
O. ammon [Species]
O. aries
O. canadensis
O. dalli
O. nivicola
O. vignei
Pantholops [Genus]
P. hodgsonii [Species]
Pelea [Genus]
P. capreolus [Species]
Procapra [Genus]
P. gutturosa [Species]
P. picticaudata
P. przewalskii
Pseudois [Genus]
P. nayaur [Species]
P. schaeferi
Raphicerus [Genus]
R. campestris [Species]
R. melanotis
R. sharpei
Redunca [Genus]
R. arundinum [Species]
R. fulvorufula
R. redunca

Rupicapra [Genus]
R. pyrenaica [Species]
R. rupicapra
Saiga [Genus]
S. tatarica [Species]
Sigmoceros [Genus]
S. lichtensteinii [Species]
Sylvicapra [Genus]
S. grimmia [Species]
Syncerus [Genus]
S. caffer [Species]
Taurotragus [Genus]
T. derbianus [Species]
T. oryx
Tetracerus [Genus]
T. quadricornis [Species]
Tragelaphus [Genus]
T. angasii [Species]
T. buxtoni
T. eurycerus
T. imberbis
T. scriptus
T. spekii
T. strepsiceros

Pholidota [Order]

Manidae [Family]
Manis [Genus]
M. crassicaudata [Species]
M. gigantea
M. javanica
M. pentadactyla
M. temminckii
M. tetradactyla
M. tricuspis

Rodentia [Order]

Aplodontidae [Family]
Aplodontia [Genus]
A. rufa [Species]

Sciuridae [Family]
Aeretes [Genus]
A. melanopterus [Species]
Aeromys [Genus]
A. tephromelas [Species]
A. thomasi
Ammospermophilus [Genus]
A. harrisii [Species]
A. insularis
A. interpres
A. leucurus
A. nelsoni
Atlantoxerus [Genus]
A. getulus [Species]
Belomys [Genus]
B. pearsonii [Species]
Biswamoyopterus [Genus]
B. biswasi [Species]
Callosciurus [Genus]
C. adamsi [Species]
C. albescens
C. baluensis
C. caniceps
C. erythraeus
C. finlaysonii
C. inornatus
C. melanogaster
C. nigrovittatus
C. notatus
C. orestes
C. phayrei
C. prevostii
C. pygerythrus
C. quinquestriatus
Cynomys [Genus]
C. gunnisoni [Species]
C. leucurus
C. ludovicianus
C. mexicanus
C. parvidens
Dremomys [Genus]
D. everetti [Species]
D. lokriah
D. pernyi
D. pyrrhomerus
D. rufigenis
Epixerus [Genus]
E. ebii [Species]
E. wilsoni
Eupetaurus [Genus]
E. cinereus [Species]
Exilisciurus [Genus]
E. concinnus [Species]
E. exilis
E. whiteheadi
Funambulus [Genus]
F. layardi [Species]
F. palmarum
F. pennantii
F. sublineatus
F. tristriatus
Funisciurus [Genus]
F. anerythrus [Species]
F. bayonii
F. carruthersi
F. congicus
F. isabella
F. lemniscatus
F. leucogenys
F. pyrropus
F. substriatus
Glaucomys [Genus]
G. sabrinus [Species]
G. volans
Glyphotes [Genus]
G. simus [Species]
Heliosciurus [Genus]
H. gambianus [Species]
H. mutabilis
H. punctatus
H. rufobrachium
H. ruwenzorii
H. undulatus
Hylopetes [Genus]
H. alboniger [Species]
H. baberi
H. bartelsi
H. fimbriatus
H. lepidus
H. nigripes
H. phayrei
H. sipora
H. spadiceus
H. winstoni
Hyosciurus [Genus]
H. heinrichi [Species]
H. ileile
Iomys [Genus]
I. horsfieldi [Species]
I. sipora
Lariscus [Genus]
L. hosei [Species]
L. insignis
L. niobe
L. obscurus
Marmota [Genus]
M. baibacina [Species]
M. bobak
M. broweri
M. caligata
M. camtschatica
M. caudata
M. flaviventris
M. himalayana
M. marmota
M. menzbieri
M. monax
M. olympus
M. sibirica
M. vancouverensis
Menetes [Genus]
M. berdmorei [Species]
Microsciurus [Genus]
M. alfari [Species]
M. flaviventer
M. mimulus
M. santanderensis
Myosciurus [Genus]
M. pumilio [Species]
Nannosciurus [Genus]
N. melanotis [Species]
Paraxerus [Genus]
P. alexandri [Species]
P. boehmi
P. cepapi
P. cooperi
P. flavovittis
P. lucifer

P. ochraceus
P. palliatus
P. poensis
P. vexillarius
P. vincenti
Petaurillus [Genus]
P. emiliae [Species]
P. hosei
P. kinlochii
Petaurista [Genus]
P. alborufus [Species]
P. elegans
P. leucogenys
P. magnificus
P. nobilis
P. petaurista
P. philippensis
P. xanthotis
Petinomys [Genus]
P. crinitus [Species]
P. fuscocapillus
P. genibarbis
P. hageni
P. lugens
P. sagitta
P. setosus
P. vordermanni
Prosciurillus [Genus]
P. abstrusus [Species]
P. leucomus
P. murinus
P. weberi
Protoxerus [Genus]
P. aubinnii [Species]
P. stangeri
Pteromys [Genus]
P. momonga [Species]
P. volans
Pteromyscus [Genus]
P. pulverulentus [Species]
Ratufa [Genus]
R. affinis [Species]
R. bicolor
R. indica
R. macroura
Rheithrosciurus [Genus]
R. macrotis [Species]
Rhinosciurus [Genus]
R. laticaudatus [Species]
Rubrisciurus [Genus]
R. rubriventer [Species]
Sciurillus [Genus]
S. pusillus [Species]
Sciurotamias [Genus]
S. davidianus [Species]
S. forresti
Sciurus [Genus]
S. aberti [Species]
S. aestuans
S. alleni
S. anomalus
S. arizonensis
S. aureogaster
S. carolinensis
S. colliaei
S. deppei
S. flammifer
S. gilvigularis
S. granatensis
S. griseus
S. ignitus
S. igniventris
S. lis
S. nayaritensis
S. niger
S. oculatus
S. pucheranii
S. pyrrhinus
S. richmondi
S. sanborni
S. spadiceus
S. stramineus
S. variegatoides
S. vulgaris
S. yucatanensis
Spermophilopsis [Genus]
S. leptodactylus [Species]
Spermophilus [Genus]
S. adocetus [Species]
S. alashanicus
S. annulatus
S. armatus
S. atricapillus
S. beecheyi
S. beldingi
S. brunneus
S. canus
S. citellus
S. columbianus
S. dauricus
S. elegans
S. erythrogenys
S. franklinii
S. fulvus
S. lateralis
S. madrensis
S. major
S. mexicanus
S. mohavensis
S. mollis
S. musicus
S. parryii
S. perotensis
S. pygmaeus
S. relictus
S. richardsonii
S. saturatus
S. spilosoma
S. suslicus
S. tereticaudus
S. townsendii
S. tridecemlineatus
S. undulatus
S. variegatus
S. washingtoni
S. xanthoprymnus
Sundasciurus [Genus]
S. brookei [Species]
S. davensis
S. fraterculus
S. hippurus
S. hoogstraali
S. jentinki
S. juvencus
S. lowii
S. mindanensis
S. moellendorffi
S. philippinensis
S. rabori
S. samarensis
S. steerii
S. tenuis
Syntheosciurus [Genus]
S. brochus [Species]
Tamias [Genus]
T. alpinus [Species]
T. amoenus
T. bulleri
T. canipes
T. cinereicollis
T. dorsalis
T. durangae
T. merriami
T. minimus
T. obscurus
T. ochrogenys
T. palmeri
T. panamintinus
T. quadrimaculatus
T. quadrivittatus
T. ruficaudus
T. rufus
T. senex
T. sibiricus
T. siskiyou
T. sonomae
T. speciosus
T. striatus
T. townsendii
T. umbrinus
Tamiasciurus [Genus]
T. douglasii [Species]
T. hudsonicus
T. mearnsi
Tamiops [Genus]
T. macclellandi [Species]
T. maritimus
T. rodolphei
T. swinhoei
Trogopterus [Genus]

T. xanthipes [Species]
Xerus [Genus]
X. erythropus [Species]
X. inauris
X. princeps
X. rutilus

Castoridae [Family]
Castor [Genus]
C. canadensis [Species]
C. fiber

Geomyidae [Family]
Geomys [Genus]
G. arenarius [Species]
G. bursarius
G. personatus
G. pinetis
G. tropicalis
Orthogeomys [Genus]
O. cavator [Species]
O. cherriei
O. cuniculus
O. dariensis
O. grandis
O. heterodus
O. hispidus
O. lanius
O. matagalpae
O. thaeleri
O. underwoodi
Pappogeomys [Genus]
P. alcorni [Species]
P. bulleri
P. castanops
P. fumosus
P. gymnurus
P. merriami
P. neglectus
P. tylorhinus
P. zinseri
Thomomys [Genus]
T. bottae [Species]
T. bulbivorus
T. clusius
T. idahoensis
T. mazama
T. monticola
T. talpoides
T. townsendii
T. umbrinus
Zygogeomys [Genus]
Z. trichopus [Species]

Heteromyidae [Family]
Chaetodipus [Genus]
C. arenarius [Species]
C. artus
C. baileyi
C. californicus
C. fallax
C. formosus
C. goldmani
C. hispidus
C. intermedius
C. lineatus
C. nelsoni
C. penicillatus
C. pernix
C. spinatus
Dipodomys [Genus]
D. agilis [Species]
D. californicus
D. compactus
D. deserti
D. elator
D. elephantinus
D. gravipes
D. heermanni
D. ingens
D. insularis
D. margaritae
D. merriami
D. microps
D. nelsoni
D. nitratoides
D. ordii
D. panamintinus
D. phillipsii
D. spectabilis
D. stephensi
D. venustus
Microdipodops [Genus]
M. megacephalus [Species]
M. pallidus
Heteromys [Genus]
H. anomalus [Species]
H. australis
H. desmarestianus
H. gaumeri
H. goldmani
H. nelsoni
H. oresterus
Liomys [Genus]
L. adspersus [Species]
L. irroratus
L. pictus
L. salvini
L. spectabilis
Perognathus [Genus]
P. alticola [Species]
P. amplus
P. fasciatus
P. flavescens
P. flavus
P. inornatus
P. longimembris
P. merriami
P. parvus
P. xanthanotus

Dipodidae [Family]
Allactaga [Genus]
A. balikunica [Species]
A. bullata
A. elater
A. euphratica
A. firouzi
A. hotsoni
A. major
A. severtzovi
A. sibirica
A. tetradactyla
A. vinogradovi
Allactodipus [Genus]
A. bobrinskii [Species]
Cardiocranius [Genus]
C. paradoxus [Species]
Dipus [Genus]
D. sagitta [Species]
Eozapus [Genus]
E. setchuanus [Species]
Eremodipus [Genus]
E. lichtensteini [Species]
Euchoreutes [Genus]
E. naso [Species]
Jaculus [Genus]
J. blanfordi [Species]
J. jaculus
J. orientalis
J. turcmenicus
Napaeozapus [Genus]
N. insignis [Species]
Paradipus [Genus]
P. ctenodactylus [Species]
Pygeretmus [Genus]
P. platyurus [Species]
P. pumilio
P. shitkovi
Salpingotus [Genus]
S. crassicauda [Species]
S. heptneri
S. kozlovi
S. michaelis
S. pallidus
S. thomasi
Sicista [Genus]
S. armenica [Species]
S. betulina
S. caucasica
S. caudata
S. concolor
S. kazbegica
S. kluchorica
S. napaea
S. pseudonapaea
S. severtzovi
S. strandi
S. subtilis
S. tianshanica
Stylodipus [Genus]

S. andrewsi [Species]
S. sungorus
S. telum
Zapus [Genus]
Z. hudsonius [Species]
Z. princeps
Z. trinotatus

Muridae [Family]
Abditomys [Genus]
A. latidens [Species]
Abrawayaomys [Genus]
A. ruschii [Species]
Acomys [Genus]
A. cahirinus [Species]
A. cilicicus
A. cinerasceus
A. ignitus
A. kempi
A. louisae
A. minous
A. mullah
A. nesiotes
A. percivali
A. russatus
A. spinosissimus
A. subspinosus
A. wilsoni
Aepeomys [Genus]
A. fuscatus [Species]
A. lugens
Aethomys [Genus]
A. bocagei [Species]
A. chrysophilus
A. granti
A. hindei
A. kaiseri
A. namaquensis
A. nyikae
A. silindensis
A. stannarius
A. thomasi
Akodon [Genus]
A. aerosus [Species]
A. affinis
A. albiventer
A. azarae
A. bogotensis
A. boliviensis
A. budini
A. cursor
A. dayi
A. dolores
A. fumeus
A. hershkovitzi
A. illuteus
A. iniscatus
A. juninensis
A. kempi
A. kofordi
A. lanosus
A. latebricola
A. lindberghi
A. longipilis
A. mansoensis
A. markhami
A. mimus
A. molinae
A. mollis
A. neocenus
A. nigrita
A. olivaceus
A. orophilus
A. puer
A. sanborni
A. sanctipaulensis
A. serrensis
A. siberiae
A. simulator
A. spegazzinii
A. subfuscus
A. surdus
A. sylvanus
A. toba
A. torques
A. urichi
A. varius
A. xanthorhinus
Allocricetulus [Genus]
A. curtatus [Species]
A. eversmanni
Alticola [Genus]
A. albicauda [Species]
A. argentatus
A. barakshin
A. lemminus
A. macrotis
A. montosa
A. roylei
A. semicanus
A. stoliczkanus
A. stracheyi
A. strelzowi
A. tuvinicus
Ammodillus [Genus]
A. imbellis [Species]
Andalgalomys [Genus]
A. olrogi [Species]
A. pearsoni
Andinomys [Genus]
A. edax [Species]
Anisomys [Genus]
A. imitator [Species]
Anonymomys [Genus]
A. mindorensis [Species]
Anotomys [Genus]
A. leander [Species]
Apodemus [Genus]
A. agrarius [Species]
A. alpicola
A. argenteus
A. arianus
A. chevrieri
A. draco
A. flavicollis
A. fulvipectus
A. gurkha
A. hermonensis
A. hyrcanicus
A. latronum
A. mystacinus
A. peninsulae
A. ponticus
A. rusiges
A. semotus
A. speciosus
A. sylvaticus
A. uralensis
A. wardi
Apomys [Genus]
A. abrae [Species]
A. datae
A. hylocoetes
A. insignis
A. littoralis
A. microdon
A. musculus
A. sacobianus
Arborimus [Genus]
A. albipes [Species]
A. longicaudus
A. pomo
Archboldomys [Genus]
A. luzonensis [Species]
Arvicanthis [Genus]
A. abyssinicus [Species]
A. blicki
A. nairobae
A. niloticus
A. somalicus
Arvicola [Genus]
A. sapidus [Species]
A. terrestris
Auliscomys [Genus]
A. boliviensis [Species]
A. micropus
A. pictus
A. sublimis
Baiomys [Genus]
B. musculus [Species]
B. taylori
Bandicota [Genus]
B. bengalensis [Species]
B. indica
B. savilei
Batomys [Genus]
B. dentatus [Species]
B. granti
B. salomonseni
Beamys [Genus]
B. hindei [Species]

B. major
Berylmys [Genus]
B. berdmorei [Species]
B. bowersi
B. mackenziei
B. manipulus
Bibimys [Genus]
B. chacoensis [Species]
B. labiosus
B. torresi
Blanfordimys [Genus]
B. afghanus [Species]
B. bucharicus
Blarinomys [Genus]
B. breviceps [Species]
Bolomys [Genus]
B. amoenus [Species]
B. lactens
B. lasiurus
B. obscurus
B. punctulatus
B. temchuki
Brachiones [Genus]
B. przewalskii [Species]
Brachytarsomys [Genus]
B. albicauda [Species]
Brachyuromys [Genus]
B. betsileoensis [Species]
B. ramirohitra
Bullimus [Genus]
B. bagobus [Species]
B. luzonicus
Bunomys [Genus]
B. andrewsi [Species]
B. chrysocomus
B. coelestis
B. fratrorum
B. heinrichi
B. penitus
B. prolatus
Calomys [Genus]
C. boliviae [Species]
C. callidus
C. callosus
C. hummelincki
C. laucha
C. lepidus
C. musculinus
C. sorellus
C. tener
Calomyscus [Genus]
C. bailwardi [Species]
C. baluchi
C. hotsoni
C. mystax
C. tsolovi
C. urartensis
Canariomys [Genus]
C. tamarani [Species]
Cannomys [Genus]
C. badius [Species]
Cansumys [Genus]
C. canus [Species]
Carpomys [Genus]
C. melanurus [Species]
C. phaeurus
Celaenomys [Genus]
C. silaceus [Species]
Chelemys [Genus]
C. macronyx [Species]
C. megalonyx
Chibchanomys [Genus]
C. trichotis [Species]
Chilomys [Genus]
C. instans [Species]
Chiromyscus [Genus]
C. chiropus [Species]
Chinchillula [Genus]
C. sahamae [Species]
Chionomys [Genus]
C. gud [Species]
C. nivalis
C. roberti
Chiropodomys [Genus]
C. calamianensis [Species]
C. gliroides
C. karlkoopmani
C. major
C. muroides
C. pusillus
Chiruromys [Genus]
C. forbesi [Species]
C. lamia
C. vates
Chroeomys [Genus]
C. andinus [Species]
C. jelskii
Chrotomys [Genus]
C. gonzalesi [Species]
C. mindorensis
C. whiteheadi
Clethrionomys [Genus]
C. californicus [Species]
C. centralis
C. gapperi
C. glareolus
C. rufocanus
C. rutilus
C. sikotanensis
Coccymys [Genus]
C. albidens [Species]
C. ruemmleri
Colomys [Genus]
C. goslingi [Species]
Conilurus [Genus]
C. albipes [Species]
C. penicillatus
Coryphomys [Genus]
C. buhleri [Species]
Crateromys [Genus]
C. australis [Species]
C. paulus
C. schadenbergi
Cremnomys [Genus]
C. blanfordi [Species]
C. cutchicus
C. elvira
Cricetomys [Genus]
C. emini [Species]
C. gambianus
Cricetulus [Genus]
C. alticola [Species]
C. barabensis
C. kamensis
C. longicaudatus
C. migratorius
C. sokolovi
Cricetus [Genus]
C. cricetus [Species]
Crossomys [Genus]
C. moncktoni [Species]
Crunomys [Genus]
C. celebensis [Species]
C. fallax
C. melanius
C. rabori
Dacnomys [Genus]
D. millardi [Species]
Dasymys [Genus]
D. foxi [Species]
D. incomtus
D. montanus
D. nudipes
D. rufulus
Delanymys [Genus]
D. brooksi [Species]
Delomys [Genus]
D. dorsalis [Species]
D. sublineatus
Dendromus [Genus]
D. insignis [Species]
D. kahuziensis
D. kivu
D. lovati
D. melanotis
D. mesomelas
D. messorius
D. mystacalis
D. nyikae
D. oreas
D. vernayi
Dendroprionomys [Genus]
D. rousseloti [Species]
Deomys [Genus]
D. ferrugineus [Species]
Dephomys [Genus]
D. defua [Species]
D. eburnea
Desmodilliscus [Genus]
D. braueri [Species]

Desmodillus [Genus]
D. auricularis [Species]
Dicrostonyx [Genus]
D. exsul [Species]
D. groenlandicus
D. hudsonius
D. kilangmiutak
D. nelsoni
D. nunatakensis
D. richardsoni
D. rubricatus
D. torquatus
D. unalascensis
D. vinogradovi
Desmomys [Genus]
D. harringtoni [Species]
Dinaromys [Genus]
D. bogdanovi [Species]
Diomys [Genus]
D. crumpi [Species]
Diplothrix [Genus]
D. legatus [Species]
Echiothrix [Genus]
E. leucura [Species]
Eropeplus [Genus]
E. canus [Species]
Eligmodontia [Genus]
E. moreni [Species]
E. morgani
E. puerulus
E. typus
Eliurus [Genus]
E. majori [Species]
E. minor
E. myoxinus
E. penicillatus
E. tanala
E. webbi
Ellobius [Genus]
E. alaicus [Species]
E. fuscocapillus
E. lutescens
E. talpinus
E. tancrei
Eolagurus [Genus]
E. luteus [Species]
E. przewalskii
Eothenomys [Genus]
E. chinensis [Species]
E. custos
E. eva
E. inez
E. melanogaster
E. olitor
E. proditor
E. regulus
E. shanseius
Euneomys [Genus]
E. chinchilloides [Species]
E. fossor
E. mordax
E. petersoni
Galenomys [Genus]
G. garleppi [Species]
Geoxus [Genus]
G. valdivianus [Species]
Gerbillurus [Genus]
G. paeba [Species]
G. setzeri
G. tytonis
G. vallinus
Gerbillus [Genus]
G. acticola [Species]
G. allenbyi
G. andersoni
G. bilensis
G. bottai
G. burtoni
G. cheesmani
G. dalloni
G. diminutus
G. dunni
G. floweri
G. gerbillus
G. grobbeni
G. henleyi
G. hoogstraali
G. juliani
G. lowei
G. maghrebi
G. mesopotamiae
G. nancillus
G. nigeriae
G. percivali
G. poecilops
G. pulvinatus
G. pyramidum
G. riggenbachi
G. ruberrimus
G. somalicus
G. syrticus
G. vivax
Golunda [Genus]
G. ellioti [Species]
Grammomys [Genus]
G. aridulus [Species]
G. caniceps
G. dolichurus
G. gigas
G. macmillani
G. rutilans
Graomys [Genus]
G. domorum [Species]
G. griseoflavus
Gymnuromys [Genus]
G. roberti [Species]
Habromys [Genus]
H. chinanteco [Species]
H. lepturus
H. lophurus
H. simulatus
Hadromys [Genus]
H. humei [Species]
Haeromys [Genus]
H. margarettae [Species]
H. minahassae
H. pusillus
Hapalomys [Genus]
H. delacouri [Species]
H. longicaudatus
Heimyscus [Genus]
H. fumosus [Species]
Hodomys [Genus]
H. alleni [Species]
Holochilus [Genus]
H. brasiliensis [Species]
H. chacarius
H. magnus
H. sciureus
Hybomys [Genus]
H. basilii [Species]
H. eisentrauti
H. lunaris
H. planifrons
H. trivirgatus
H. univittatus
Hydromys [Genus]
H. chrysogaster [Species]
H. habbema
H. hussoni
H. neobrittanicus
H. shawmayeri
Hylomyscus [Genus]
H. aeta [Species]
H. alleni
H. baeri
H. carillus
H. denniae
H. parvus
H. stella
Hyomys [Genus]
H. dammermani [Species]
H. goliath
Hyperacrius [Genus]
H. fertilis [Species]
H. wynnei
Hypogeomys [Genus]
H. antimena [Species]
Ichthyomys [Genus]
I. hydrobates [Species]
I. pittieri
I. stolzmanni
I. tweedii
Irenomys [Genus]
I. tarsalis [Species]
Isthmomys [Genus]
I. flavidus [Species]
I. pirrensis
Juscelinomys [Genus]
J. candango [Species]

J. vulpinus
Kadarsanomys [Genus]
K. sodyi [Species]
Komodomys [Genus]
K. rintjanus [Species]
Kunsia [Genus]
K. fronto [Species]
K. tomentosus
Lagurus [Genus]
L. lagurus [Species]
Lamottemys [Genus]
L. okuensis [Species]
Lasiopodomys [Genus]
L. brandtii [Species]
L. fuscus
L. mandarinus
Leggadina [Genus]
L. forresti [Species]
L. lakedownensis
Leimacomys [Genus]
L. buettneri [Species]
Lemmiscus [Genus]
L. curtatus [Species]
Lemmus [Genus]
L. amurensis [Species]
L. lemmus
L. sibiricus
Lemniscomys [Genus]
L. barbarus [Species]
L. bellieri
L. griselda
L. hoogstraali
L. linulus
L. macculus
L. mittendorfi
L. rosalia
L. roseveari
L. striatus
Lenomys [Genus]
L. meyeri [Species]
Lenothrix [Genus]
L. canus [Species]
Lenoxus [Genus]
L. apicalis [Species]
Leopoldamys [Genus]
L. edwardsi [Species]
L. neilli
L. sabanus
L. siporanus
Leporillus [Genus]
L. apicalis [Species]
L. conditor
Leptomys [Genus]
L. elegans [Species]
L. ernstmayri
L. signatus
Limnomys [Genus]
L. sibuanus [Species]
Lophiomys [Genus]
L. imhausi [Species]
Lophuromys [Genus]
L. cinereus [Species]
L. flavopunctatus
L. luteogaster
L. medicaudatus
L. melanonyx
L. nudicaudus
L. rahmi
L. sikapusi
L. woosnami
Lorentzimys [Genus]
L. nouhuysi [Species]
Macrotarsomys [Genus]
M. bastardi [Species]
M. ingens
Macruromys [Genus]
M. elegans [Species]
M. major
Malacomys [Genus]
M. cansdalei [Species]
M. edwardsi
M. longipes
M. lukolelae
M. verschureni
Malacothrix [Genus]
M. typica [Species]
Mallomys [Genus]
M. aroaensis [Species]
M. gunung
M. istapantap
M. rothschildi
Malpaisomys [Genus]
M. insularis [Species]
Margaretamys [Genus]
M. beccarii [Species]
M. elegans
M. parvus
Mastomys [Genus]
M. angolensis [Species]
M. coucha
M. erythroleucus
M. hildebrandtii
M. natalensis
M. pernanus
M. shortridgei
M. verheyeni
Maxomys [Genus]
M. alticola [Species]
M. baeodon
M. bartelsii
M. dollmani
M. hellwaldii
M. hylomyoides
M. inas
M. inflatus
M. moi
M. musschenbroekii
M. ochraceiventer
M. pagensis
M. panglima
M. rajah
M. surifer
M. wattsi
M. whiteheadi
Mayermys [Genus]
M. ellermani [Species]
Megadendromus [Genus]
M. nikolausi [Species]
Megadontomys [Genus]
M. cryophilus [Species]
M. nelsoni
M. thomasi
Megalomys [Genus]
M. desmarestii [Species]
M. luciae
Melanomys [Genus]
M. caliginosus [Species]
M. robustulus
M. zunigae
Melasmothrix [Genus]
M. naso [Species]
Melomys [Genus]
M. aerosus [Species]
M. bougainville
M. burtoni
M. capensis
M. cervinipes
M. fellowsi
M. fraterculus
M. gracilis
M. lanosus
M. leucogaster
M. levipes
M. lorentzii
M. mollis
M. moncktoni
M. obiensis
M. platyops
M. rattoides
M. rubex
M. rubicola
M. rufescens
M. spechti
Meriones [Genus]
M. arimalius [Species]
M. chengi
M. crassus
M. dahli
M. hurrianae
M. libycus
M. meridianus
M. persicus
M. rex
M. sacramenti
M. shawi
M. tamariscinus
M. tristrami
M. unguiculatus
M. vinogradovi
M. zarudnyi

Mesembriomys [Genus]
M. gouldii [Species]
M. macrurus
Mesocricetus [Genus]
M. auratus [Species]
M. brandti
M. newtoni
M. raddei
Microdillus [Genus]
M. peeli [Species]
Microhydromys [Genus]
M. musseri [Species]
M. richardsoni
Micromys [Genus]
M. minutus [Species]
Microryzomys [Genus]
M. altissimus [Species]
M. minutus
Microtus [Genus]
M. abbreviatus [Species]
M. agrestis
M. arvalis
M. bavaricus
M. breweri
M. cabrerae
M. californicus
M. canicaudus
M. chrotorrhinus
M. daghestanicus
M. duodecimcostatus
M. evoronensis
M. felteni
M. fortis
M. gerbei
M. gregalis
M. guatemalensis
M. guentheri
M. hyperboreus
M. irani
M. irene
M. juldaschi
M. kermanensis
M. kirgisorum
M. leucurus
M. limnophilus
M. longicaudus
M. lusitanicus
M. majori
M. maximowiczii
M. mexicanus
M. middendorffi
M. miurus
M. mongolicus
M. montanus
M. montebelli
M. mujanensis
M. multiplex
M. nasarovi
M. oaxacensis
M. obscurus
M. ochrogaster
M. oeconomus
M. oregoni
M. pennsylvanicus
M. pinetorum
M. quasiater
M. richardsoni
M. rossiaemeridionalis
M. sachalinensis
M. savii
M. schelkovnikovi
M. sikimensis
M. socialis
M. subterraneus
M. tatricus
M. thomasi
M. townsendii
M. transcaspicus
M. umbrosus
M. xanthognathus
Millardia [Genus]
M. gleadowi [Species]
M. kathleenae
M. kondana
M. meltada
Muriculus [Genus]
M. imberbis [Species]
Mus [Genus]
M. baoulei [Species]
M. booduga
M. bufo
M. callewaerti
M. caroli
M. cervicolor
M. cookii
M. crociduroides
M. famulus
M. fernandoni
M. goundae
M. haussa
M. indutus
M. kasaicus
M. macedonicus
M. mahomet
M. mattheyi
M. mayori
M. minutoides
M. musculoides
M. musculus
M. neavei
M. orangiae
M. oubanguii
M. pahari
M. phillipsi
M. platythrix
M. saxicola
M. setulosus
M. setzeri
M. shortridgei
M. sorella
M. spicilegus
M. spretus
M. tenellus
M. terricolor
M. triton
M. vulcani
Mylomys [Genus]
M. dybowskii [Species]
Myomys [Genus]
M. albipes [Species]
M. daltoni
M. derooi
M. fumatus
M. ruppi
M. verreauxii
M. yemeni
Myopus [Genus]
M. schisticolor [Species]
Myospalax [Genus]
M. aspalax [Species]
M. epsilanus
M. fontanierii
M. myospalax
M. psilurus
M. rothschildi
M. smithii
Mystromys [Genus]
M. albicaudatus [Species]
Nannospalax [Genus]
N. ehrenbergi [Species]
N. leucodon
N. nehringi
Neacomys [Genus]
N. guianae [Species]
N. pictus
N. spinosus
N. tenuipes
Nectomys [Genus]
N. palmipes [Species]
N. parvipes
N. squamipes
Nelsonia [Genus]
N. goldmani [Species]
N. neotomodon
Neofiber [Genus]
N. alleni [Species]
Neohydromys [Genus]
N. fuscus [Species]
Neotoma [Genus]
N. albigula [Species]
N. angustapalata
N. anthonyi
N. bryanti
N. bunkeri
N. chrysomelas
N. cinerea
N. devia
N. floridana
N. fuscipes
N. goldmani

N. lepida
N. martinensis
N. mexicana
N. micropus
N. nelsoni
N. palatina
N. phenax
N. stephensi
N. varia
Neotomodon [Genus]
N. alstoni [Species]
Neotomys [Genus]
N. ebriosus [Species]
Nesomys [Genus]
N. rufus [Species]
Nesokia [Genus]
N. bunnii [Species]
N. indica
Nesoryzomys [Genus]
N. darwini [Species]
N. fernandinae
N. indefessus
N. swarthi
Neusticomys [Genus]
N. monticolus [Species]
N. mussoi
N. oyapocki
N. peruviensis
N. venezuelae
Niviventer [Genus]
N. andersoni [Species]
N. brahma
N. confucianus
N. coxingi
N. cremoriventer
N. culturatus
N. eha
N. excelsior
N. fulvescens
N. hinpoon
N. langbianis
N. lepturus
N. niviventer
N. rapit
N. tenaster
Notiomys [Genus]
N. edwardsii [Species]
Notomys [Genus]
N. alexis [Species]
N. amplus
N. aquilo
N. cervinus
N. fuscus
N. longicaudatus
N. macrotis
N. mitchellii
N. mordax
Nyctomys [Genus]
N. sumichrasti [Species]
Ochrotomys [Genus]
O. nuttalli [Species]
Oecomys [Genus]
O. bicolor [Species]
O. cleberi
O. concolor
O. flavicans
O. mamorae
O. paricola
O. phaeotis
O. rex
O. roberti
O. rutilus
O. speciosus
O. superans
O. trinitatis
Oenomys [Genus]
O. hypoxanthus [Species]
O. ornatus
Oligoryzomys [Genus]
O. andinus [Species]
O. arenalis
O. chacoensis
O. delticola
O. destructor
O. eliurus
O. flavescens
O. fulvescens
O. griseolus
O. longicaudatus
O. magellanicus
O. microtis
O. nigripes
O. vegetus
O. victus
Ondatra [Genus]
O. zibethicus [Species]
Onychomys [Genus]
O. arenicola [Species]
O. leucogaster
O. torridus
Oryzomys [Genus]
O. albigularis [Species]
O. alfaroi
O. auriventer
O. balneator
O. bolivaris
O. buccinatus
O. capito
O. chapmani
O. couesi
O. devius
O. dimidiatus
O. galapagoensis
O. gorgasi
O. hammondi
O. intectus
O. intermedius
O. keaysi
O. kelloggi
O. lamia
O. legatus
O. levipes
O. macconnelli
O. melanotis
O. nelsoni
O. nitidus
O. oniscus
O. palustris
O. polius
O. ratticeps
O. rhabdops
O. rostratus
O. saturatior
O. subflavus
O. talamancae
O. xantheolus
O. yunganus
Osgoodomys [Genus]
O. banderanus [Species]
Otomys [Genus]
O. anchietae [Species]
O. angoniensis
O. denti
O. irroratus
O. laminatus
O. maximus
O. occidentalis
O. saundersiae
O. sloggetti
O. tropicalis
O. typus
O. unisulcatus
Otonyctomys [Genus]
O. hatti [Species]
Ototylomys [Genus]
O. phyllotis [Species]
Oxymycterus [Genus]
O. akodontius [Species]
O. angularis
O. delator
O. hiska
O. hispidus
O. hucucha
O. iheringi
O. inca
O. nasutus
O. paramensis
O. roberti
O. rufus
Pachyuromys [Genus]
P. duprasi [Species]
Palawanomys [Genus]
P. furvus [Species]
Papagomys [Genus]
P. armandvillei [Species]
P. theodorverhoeveni
Parahydromys [Genus]
P. asper [Species]
Paraleptomys [Genus]
P. rufilatus [Species]

P. wilhelmina
Parotomys [Genus]
P. brantsii [Species]
P. littledalei
Paruromys [Genus]
P. dominator [Species]
P. ursinus
Paulamys [Genus]
P. naso [Species]
Pelomys [Genus]
P. campanae [Species]
P. fallax
P. hopkinsi
P. isseli
P. minor
Peromyscus [Genus]
P. attwateri [Species]
P. aztecus
P. boylii
P. bullatus
P. californicus
P. caniceps
P. crinitus
P. dickeyi
P. difficilis
P. eremicus
P. eva
P. furvus
P. gossypinus
P. grandis
P. gratus
P. guardia
P. guatemalensis
P. gymnotis
P. hooperi
P. interparietalis
P. leucopus
P. levipes
P. madrensis
P. maniculatus
P. mayensis
P. megalops
P. mekisturus
P. melanocarpus
P. melanophrys
P. melanotis
P. melanurus
P. merriami
P. mexicanus
P. nasutus
P. ochraventer
P. oreas
P. pectoralis
P. pembertoni
P. perfulvus
P. polionotus
P. polius
P. pseudocrinitus
P. sejugis
P. simulus
P. sitkensis
P. slevini
P. spicilegus
P. stephani
P. stirtoni
P. truei
P. winkelmanni
P. yucatanicus
P. zarhynchus
Petromyscus [Genus]
P. barbouri [Species]
P. collinus
P. monticularis
P. shortridgei
Phaenomys [Genus]
P. ferrugineus [Species]
Phaulomys [Genus]
P. andersoni [Species]
P. smithii
Phenacomys [Genus]
P. intermedius [Species]
P. ungava
Phloeomys [Genus]
P. cumingi [Species]
P. pallidus
Phyllotis [Genus]
P. amicus [Species]
P. andium
P. bonaeriensis
P. caprinus
P. darwini
P. definitus
P. gerbillus
P. haggardi
P. magister
P. osgoodi
P. osilae
P. wolffsohni
P. xanthopygus
Pithecheir [Genus]
P. melanurus [Species]
P. parvus
Phodopus [Genus]
P. campbelli [Species]
P. roborovskii
P. sungorus
Platacanthomys [Genus]
P. lasiurus [Species]
Podomys [Genus]
P. floridanus [Species]
Podoxymys [Genus]
P. roraimae [Species]
Pogonomelomys [Genus]
P. bruijni [Species]
P. mayeri
P. sevia
Pogonomys [Genus]
P. championi [Species]
P. loriae
P. macrourus
P. sylvestris
Praomys [Genus]
P. delectorum [Species]
P. hartwigi
P. jacksoni
P. minor
P. misonnei
P. morio
P. mutoni
P. rostratus
P. tullbergi
Prionomys [Genus]
P. batesi [Species]
Proedromys [Genus]
P. bedfordi [Species]
Prometheomys [Genus]
P. schaposchnikowi [Species]
Psammomys [Genus]
P. obesus [Species]
P. vexillaris
Pseudohydromys [Genus]
P. murinus [Species]
P. occidentalis
Pseudomys [Genus]
P. albocinereus [Species]
P. apodemoides
P. australis
P. bolami
P. chapmani
P. delicatulus
P. desertor
P. fieldi
P. fumeus
P. fuscus
P. glaucus
P. gouldii
P. gracilicaudatus
P. hermannsburgensis
P. higginsi
P. johnsoni
P. laborifex
P. nanus
P. novaehollandiae
P. occidentalis
P. oralis
P. patrius
P. pilligaensis
P. praeconis
P. shortridgei
Pseudoryzomys [Genus]
P. simplex [Species]
Punomys [Genus]
P. lemminus [Species]
Rattus [Genus]
R. adustus [Species]
R. annandalei
R. argentiventer
R. baluensis
R. bontanus
R. burrus

R. colletti
R. elaphinus
R. enganus
R. everetti
R. exulans
R. feliceus
R. foramineus
R. fuscipes
R. giluwensis
R. hainaldi
R. hoffmanni
R. hoogerwerfi
R. jobiensis
R. koopmani
R. korinchi
R. leucopus
R. losea
R. lugens
R. lutreolus
R. macleari
R. marmosurus
R. mindorensis
R. mollicomulus
R. montanus
R. mordax
R. morotaiensis
R. nativitatis
R. nitidus
R. norvegicus
R. novaeguineae
R. osgoodi
R. palmarum
R. pelurus
R. praetor
R. ranjiniae
R. rattus
R. sanila
R. sikkimensis
R. simalurensis
R. sordidus
R. steini
R. stoicus
R. tanezumi
R. tawitawiensis
R. timorensis
R. tiomanicus
R. tunneyi
R. turkestanicus
R. villosissimus
R. xanthurus
Reithrodon [Genus]
R. auritus [Species]
Reithrodontomys [Genus]
R. brevirostris [Species]
R. burti
R. chrysopsis
R. creper
R. darienensis
R. fulvescens
R. gracilis
R. hirsutus
R. humulis
R. megalotis
R. mexicanus
R. microdon
R. montanus
R. paradoxus
R. raviventris
R. rodriguezi
R. spectabilis
R. sumichrasti
R. tenuirostris
R. zacatecae
Rhabdomys [Genus]
R. pumilio [Species]
Rhagomys [Genus]
R. rufescens [Species]
Rheomys [Genus]
R. mexicanus [Species]
R. raptor
R. thomasi
R. underwoodi
Rhipidomys [Genus]
R. austrinus [Species]
R. caucensis
R. couesi
R. fulviventer
R. latimanus
R. leucodactylus
R. macconnelli
R. mastacalis
R. nitela
R. ochrogaster
R. scandens
R. venezuelae
R. venustus
R. wetzeli
Rhizomys [Genus]
R. pruinosus [Species]
R. sinensis
R. sumatrensis
Rhombomys [Genus]
R. opimus [Species]
Rhynchomys [Genus]
R. isarogensis [Species]
R. soricoides
Saccostomus [Genus]
S. campestris [Species]
S. mearnsi
Scapteromys [Genus]
S. tumidus [Species]
Scolomys [Genus]
S. melanops [Species]
S. ucayalensis
Scotinomys [Genus]
S. teguina [Species]
S. xerampelinus
Sekeetamys [Genus]
S. calurus [Species]
Sigmodon [Genus]
S. alleni [Species]
S. alstoni
S. arizonae
S. fulviventer
S. hispidus
S. inopinatus
S. leucotis
S. mascotensis
S. ochrognathus
Sigmodontomys [Genus]
S. alfari [Species]
S. aphrastus
Solomys [Genus]
S. ponceleti [Species]
S. salamonis
S. salebrosus
S. sapientis
S. spriggsarum
Spalax [Genus]
S. arenarius [Species]
S. giganteus
S. graecus
S. microphthalmus
S. zemni
Spelaeomys [Genus]
S. florensis [Species]
Srilankamys [Genus]
S. ohiensis [Species]
Stenocephalemys [Genus]
S. albocaudata [Species]
S. griseicauda
Steatomys [Genus]
S. caurinus [Species]
S. cuppedius
S. jacksoni
S. krebsii
S. parvus
S. pratensis
Stenomys [Genus]
S. ceramicus [Species]
S. niobe
S. richardsoni
S. vandeuseni
S. verecundus
Stochomys [Genus]
S. longicaudatus [Species]
Sundamys [Genus]
S. infraluteus [Species]
S. maxi
S. muelleri
Synaptomys [Genus]
S. borealis [Species]
S. cooperi
Tachyoryctes [Genus]
T. ankoliae [Species]
T. annectens
T. audax
T. daemon
T. macrocephalus
T. naivashae

T. rex
T. ruandae
T. ruddi
T. spalacinus
T. splendens
Taeromys [Genus]
T. arcuatus [Species]
T. callitrichus
T. celebensis
T. hamatus
T. punicans
T. taerae
Tarsomys [Genus]
T. apoensis [Species]
T. echinatus
Tateomys [Genus]
T. macrocercus [Species]
T. rhinogradoides
Tatera [Genus]
T. afra [Species]
T. boehmi
T. brantsii
T. guineae
T. inclusa
T. indica
T. kempi
T. leucogaster
T. nigricauda
T. phillipsi
T. robusta
T. valida
Taterillus [Genus]
T. arenarius [Species]
T. congicus
T. emini
T. gracilis
T. harringtoni
T. lacustris
T. petteri
T. pygargus
Tscherskia [Genus]
T. triton [Species]
Thallomys [Genus]
T. loringi [Species]
T. nigricauda
T. paedulcus
T. shortridgei
Thalpomys [Genus]
T. cerradensis [Species]
T. lasiotis
Thamnomys [Genus]
T. kempi [Species]
T. venustus
Thomasomys [Genus]
T. aureus [Species]
T. baeops
T. bombycinus
T. cinereiventer
T. cinereus
T. daphne
T. eleusis
T. gracilis
T. hylophilus
T. incanus
T. ischyurus
T. kalinowskii
T. ladewi
T. laniger
T. monochromos
T. niveipes
T. notatus
T. oreas
T. paramorum
T. pyrrhonotus
T. rhoadsi
T. rosalinda
T. silvestris
T. taczanowskii
T. vestitus
Tokudaia [Genus]
T. muenninki [Species]
T. osimensis
Tryphomys [Genus]
T. adustus [Species]
Tylomys [Genus]
T. bullaris [Species]
T. fulviventer
T. mirae
T. nudicaudus
T. panamensis
T. tumbalensis
T. watsoni
Typhlomys [Genus]
T. chapensis [Species]
T. cinereus
Uranomys [Genus]
U. ruddi [Species]
Uromys [Genus]
U. anak [Species]
U. caudimaculatus
U. hadrourus
U. imperator
U. neobritanicus
U. porculus
U. rex
Vandeleuria [Genus]
V. nolthenii [Species]
V. oleracea
Vernaya [Genus]
V. fulva [Species]
Volemys [Genus]
V. clarkei [Species]
V. kikuchii
V. millicens
V. musseri
Wiedomys [Genus]
W. pyrrhorhinos [Species]
Wilfredomys [Genus]
W. oenax [Species]
W. pictipes
Xenomys [Genus]
X. nelsoni [Species]
Xenuromys [Genus]
X. barbatus [Species]
Xeromys [Genus]
X. myoides [Species]
Zelotomys [Genus]
Z. hildegardeae [Species]
Z. woosnami
Zygodontomys [Genus]
Z. brevicauda [Species]
Z. brunneus
Zyzomys [Genus]
Z. argurus [Species]
Z. maini
Z. palatilis
Z. pedunculatus
Z. woodwardi

Anomaluridae [Family]
Anomalurus [Genus]
A. beecrofti [Species]
A. derbianus
A. pelii
A. pusillus
Idiurus [Genus]
I. macrotis [Species]
I. zenkeri
Zenkerella [Genus]
Z. insignis [Species]

Pedetidae [Family]
Pedetes [Genus]
P. capensis [Species]

Ctenodactylidae [Family]
Ctenodactylus [Genus]
C. gundi [Species]
C. vali
Felovia [Genus]
F. vae [Species]
Massoutiera [Genus]
M. mzabi [Species]
Pectinator [Genus]
P. spekei [Species]

Myoxidae [Family]
Dryomys [Genus]
D. laniger [Species]
D. nitedula
D. sichuanensis
Eliomys [Genus]
E. melanurus [Species]
E. quercinus
Glirulus [Genus]
G. japonicus [Species]
Graphiurus [Genus]
G. christyi [Species]
G. hueti
G. lorraineus
G. monardi
G. ocularis

G. parvus
G. rupicola
Muscardinus [Genus]
M. avellanarius [Species]
Myomimus [Genus]
M. personatus [Species]
M. roachi
M. setzeri
Myoxus [Genus]
M. glis [Species]
Selevinia [Genus]
S. betpakdalaensis [Species]

Petromuridae [Family]
Petromus [Genus]
P. typicus [Species]

Thryonomyidae [Family]
Thryonomys [Genus]
T. gregorianus [Species]
T. swinderianus

Bathyergidae [Family]
Bathyergus [Genus]
B. janetta [Species]
B. suillus
Cryptomys [Genus]
C. bocagei [Species]
C. damarensis
C. foxi
C. hottentotus
C. mechowi
C. ochraceocinereus
C. zechi
Georychus [Genus]
G. capensis [Species]
Heliophobius [Genus]
H. argenteocinereus
Heterocephalus [Genus]
H. glaber [Species]

Hystricidae [Family]
Atherurus [Genus]
A. africanus [Species]
A. macrourus
Hystrix [Genus]
H. africaeaustralis [Species]
H. brachyura
H. crassispinis
H. cristata
H. indica
H. javanica
H. pumila
H. sumatrae
Trichys [Genus]
T. fasciculata [Species]

Erethizontidae [Family]
Coendou [Genus]
C. bicolor [Species]
C. koopmani
C. prehensilis
C. rothschildi
Echinoprocta [Genus]
E. rufescens [Species]
Erethizon [Genus]
E. dorsatum [Species]
Sphiggurus [Genus]
S. insidiosus [Species]
S. mexicanus
S. pallidus
S. spinosus
S. vestitus
S. villosus

Chinchillidae [Family]
Chinchilla [Genus]
C. brevicaudata [Species]
C. lanigera
Lagidium [Genus]
L. peruanum [Species]
L. viscacia
L. wolffsohni
Lagostomus [Genus]
L. maximus [Species]

Dinomyidae [Family]
Dinomys [Genus]
D. branickii [Species]

Caviidae [Family]
Cavia [Genus]
C. aperea [Species]
C. fulgida
C. magna
C. porcellus
C. tschudii
Dolichotis [Genus]
D. patagonum [Species]
D. salinicola
Galea [Genus]
G. flavidens [Species]
G. spixii
Kerodon [Genus]
K. rupestris [Species]
Microcavia [Genus]
M. australis [Species]
M. niata
M. shiptoni

Hydrochaeridae [Family]
Hydrochaeris [Genus]
H. hydrochaeris [Species]

Dasyproctidae [Family]
Dasyprocta [Genus]
D. azarae [Species]
D. coibae
D. cristata
D. fuliginosa
D. guamara
D. kalinowskii
D. leporina
D. mexicana
D. prymnolopha
D. punctata
D. ruatanica
Myoprocta [Genus]
M. acouchy [Species]
M. exilis

Agoutidae [Family]
Agouti [Genus]
A. paca [Species]
A. taczanowskii

Ctenomyidae [Family]
Ctenomys [Genus]
C. argentinus [Species]
C. australis
C. azarae
C. boliviensis
C. bonettoi
C. brasiliensis
C. colburni
C. conoveri
C. dorsalis
C. emilianus
C. frater
C. fulvus
C. haigi
C. knighti
C. latro
C. leucodon
C. lewisi
C. magellanicus
C. maulinus
C. mendocinus
C. minutus
C. nattereri
C. occultus
C. opimus
C. pearsoni
C. perrensis
C. peruanus
C. pontifex
C. porteousi
C. saltarius
C. sericeus
C. sociabilis
C. steinbachi
C. talarum
C. torquatus
C. tuconax
C. tucumanus
C. validus

Octodontidae [Family]
Aconaemys [Genus]
A. fuscus [Species]
A. sagei
Octodon [Genus]
O. bridgesi [Species]
O. degus
O. lunatus

Octodontomys [Genus]
O. gliroides [Species]
Octomys [Genus]
O. mimax [Species]
Spalacopus [Genus]
S. cyanus [Species]
Tympanoctomys [Genus]
T. barrerae [Species]

Abrocomidae [Family]
Abrocoma [Genus]
A. bennetti [Species]
A. boliviensis
A. cinerea

Echimyidae [Family]
Boromys [Genus]
B. offella [Species]
B. torrei
Brotomys [Genus]
B. contractus [Species]
B. voratus
Carterodon [Genus]
C. sulcidens [Species]
Clyomys [Genus]
C. bishopi [Species]
C. laticeps
Chaetomys [Genus]
C. subspinosus [Species]
Dactylomys [Genus]
D. boliviensis [Species]
D. dactylinus
D. peruanus
Diplomys [Genus]
D. caniceps [Species]
D. labilis
D. rufodorsalis
Echimys [Genus]
E. blainvillei [Species]
E. braziliensis
E. chrysurus
E. dasythrix
E. grandis
E. lamarum
E. macrurus
E. nigrispinus
E. pictus
E. rhipidurus
E. saturnus
E. semivillosus
E. thomasi
E. unicolor
Euryzygomatomys [Genus]
E. spinosus [Species]
Heteropsomys [Genus]
H. antillensis [Species]
H. insulans
Hoplomys [Genus]
H. gymnurus [Species]
Isothrix [Genus]
I. bistriata [Species]
I. pagurus
Kannabateomys [Genus]
K. amblyonyx [Species]
Lonchothrix [Genus]
L. emiliae [Species]
Makalata [Genus]
M. armata [Species]
Mesomys [Genus]
M. didelphoides [Species]
M. hispidus
M. leniceps
M. obscurus
M. stimulax
Olallamys [Genus]
O. albicauda [Species]
O. edax
Proechimys [Genus]
P. albispinus [Species]
P. amphichoricus
P. bolivianus
P. brevicauda
P. canicollis
P. cayennensis
P. chrysaeolus
P. cuvieri
P. decumanus
P. dimidiatus
P. goeldii
P. gorgonae
P. guairae
P. gularis
P. hendeei
P. hoplomyoides
P. iheringi
P. longicaudatus
P. magdalenae
P. mincae
P. myosuros
P. oconnelli
P. oris
P. poliopus
P. quadruplicatus
P. semispinosus
P. setosus
P. simonsi
P. steerei
P. trinitatis
P. urichi
P. warreni
Puertoricomys [Genus]
P. corozalus [Species]
Thrichomys [Genus]
T. apereoides [Species]

Capromyidae [Family]
Capromys [Genus]
C. pilorides [Species]
Geocapromys [Genus]
G. brownii [Species]
G. thoracatus
Hexolobodon [Genus]
H. phenax [Species]
Isolobodon [Genus]
I. montanus [Species]
I. portoricensis
Mesocapromys [Genus]
M. angelcabrerai [Species]
M. auritus
M. nanus
M. sanfelipensis
Mysateles [Genus]
M. garridoi [Species]
M. gundlachi
M. melanurus
M. meridionalis
M. prehensilis
Plagiodontia [Genus]
P. aedium [Species]
P. araeum
P. ipnaeum
Rhizoplagiodontia [Genus]
R. lemkei [Species]

Heptaxodontidae [Family]
Amblyrhiza [Genus]
A. inundata [Species]
Clidomys [Genus]
C. osborni [Species]
C. parvus
Elasmodontomys [Genus]
E. obliquus [Species]
Quemisia [Genus]
Quemisia gravis [Species]

Myocastoridae [Family]
Myocastor [Genus]
M. coypus [Species]

Lagomorpha [Order]

Ochotonidae [Family]
Ochotona [Genus]
O. alpina [Species]
O. cansus
O. collaris
O. curzoniae
O. dauurica
O. erythrotis
O. forresti
O. gaoligongensis
O. gloveri
O. himalayana
O. hyperborea
O. iliensis
O. koslowi
O. ladacensis
O. macrotis
O. muliensis
O. nubrica
O. pallasi
O. princeps
O. pusilla
O. roylei

O. rufescens
O. rutila
O. thibetana
O. thomasi
Prolagus [Genus]
P. sardus [Species]

Leporidae [Family]
Brachylagus [Genus]
B. idahoensis [Species]
Bunolagus [Genus]
B. monticularis [Species]
Caprolagus [Genus]
C. hispidus [Species]
Lepus [Genus]
L. alleni [Species]
L. americanus
L. arcticus
L. brachyurus
L. californicus
L. callotis
L. capensis
L. castroviejoi
L. comus
L. coreanus
L. corsicanus
L. europaeus
L. fagani
L. flavigularis
L. granatensis
L. hainanus
L. insularis
L. mandshuricus
L. nigricollis
L. oiostolus
L. othus
L. pequensis
L. saxatilis
L. sinensis
L. starcki
L. timidus
L. tolai
L. townsendii
L. victoriae
L. yarkandensis
Nesolagus [Genus]
N. netscheri [Species]
Oryctolagus [Genus]
O. cuniculus [Species]
Pentalagus [Genus]
P. furnessi [Species]
Poelagus [Genus]
P. marjorita [Species]
Pronolagus [Genus]
P. crassicaudatus [Species]
P. randensis
P. rupestris
Romerolagus [Genus]
R. diazi [Species]
Sylvilagus [Genus]
S. aquaticus [Species]
S. audubonii
S. bachmani
S. brasiliensis
S. cunicularius
S. dicei
S. floridanus
S. graysoni
S. insonus
S. mansuetus
S. nuttallii
S. palustris
S. transitionalis

Macroscelidea [Order]

Macroscelididae [Family]
Elephantulus [Genus]
E. brachyrhynchus [Species]
E. edwardii
E. fuscipes
E. fuscus
E. intufi
E. myurus
E. revoili
E. rozeti
E. rufescens
E. rupestris
Macroscelides [Genus]
M. proboscideus [Species]
Petrodromus [Genus]
P. tetradactylus [Species]
Rhynchocyon [Genus]
R. chrysopygus [Species]
R. cirnei
R. petersi

A brief geologic history of animal life

A note about geologic time scales: A cursory look will reveal that the timing of various geological periods differs among textbooks. Is one right and the others wrong? Not necessarily. Scientists use different methods to estimate geological time—methods with a precision sometimes measured in tens of millions of years. There is, however, a general agreement on the magnitude and relative timing associated with modern time scales. The closer in geological time one comes to the present, the more accurate science can be—and sometimes the more disagreement there seems to be. The following account was compiled using the more widely accepted boundaries from a diverse selection of reputable scientific resources.

Geologic time scale

Era	Period	Epoch	Dates	Life forms
Proterozoic			2,500-544 mya*	First single-celled organisms, simple plants, and invertebrates (such as algae, amoebas, and jellyfish)
Paleozoic	Cambrian		544-490 mya	First crustaceans, mollusks, sponges, nautiloids, and annelids (worms)
	Ordovician		490-438 mya	Trilobites dominant. Also first fungi, jawless vertebrates, starfishes, sea scorpions, and urchins
	Silurian		438-408 mya	First terrestrial plants, sharks, and bony fishes
	Devonian		408-360 mya	First insects, arachnids (scorpions), and tetrapods
	Carboniferous	Mississippian	360-325 mya	Amphibians abundant. Also first spiders, land snails
		Pennsylvanian	325-286 mya	First reptiles and synapsids
	Permian		286-248 mya	Reptiles abundant. Extinction of trilobytes. Most modern insect orders
Mesozoic	Triassic		248-205 mya	Diversification of reptiles: turtles, crocodiles, therapsids (mammal-like reptiles), first dinosaurs, first flies
	Jurassic		205-145 mya	Insects abundant, dinosaurs dominant in later stage. First mammals, lizards, frogs, and birds
	Cretaceous		145-65 mya	First snakes and modern fish. Extinction of dinosaurs and ammonites, rise and fall of toothed birds
Cenozoic	Tertiary	Paleocene	65-55.5 mya	Diversification of mammals
		Eocene	55.5-33.7 mya	First horses, whales, monkeys, and leafminer insects
		Oligocene	33.7-23.8 mya	Diversification of birds. First anthropoids (higher primates)
		Miocene	23.8-5.6 mya	First hominids
		Pliocene	5.6-1.8 mya	First australopithecines
	Quaternary	Pleistocene	1.8 mya-8,000 ya	Mammoths, mastodons, and Neanderthals
		Holocene	8,000 ya-present	First modern humans

*Millions of years ago (mya)

Index

Bold page numbers indicate the primary discussion of a topic; page numbers in italics indicate illustrations; "t" indicates a table.

1080 (Sodium monofluoroacetate), 12:186

A

AAA (Animal-assisted activities), 14:293
Aardvarks, 12:*48*, 12:129, 12:135, 15:131, 15:134, **15:155–159**, 15:*156*, 15:*157*, 15:*158*
Aardwolves, **14:359–367**, 14:*360*, 14:*361*, 14:*362*, 14:*363*, 14:*364*
behavior, 14:259, 14:362
conservation status, 14:362
distribution, 14:362
evolution, 14:359–360
feeding ecology, 14:255, 14:260, 14:362
habitats, 14:362
physical characteristics, 14:360
reproduction, 14:261, 14:362
taxonomy, 14:359–360
AAT (Animal-assisted therapy), 14:293
Abbott's duikers, 16:84*t*
Aberdare shrews, 13:277*t*
Abert squirrels, 16:*167*, 16:*168*, 16:*171*
Abrawayaomys spp., 16:264, 16:265
Abrocoma spp., 16:443–444, 16:445
Abrocoma bennettii. *See* Bennett's chinchilla rats
Abrocoma boliviensis. *See* Bolivian chinchilla rats
Abrocoma cinerea. *See* Ashy chinchilla rats
Abrocomidae. *See* Chinchilla rats
Abrothricines, 16:263–264
Abrothrix longipilis, 16:266–267
Abrothrix olivaceus, 16:267, 16:269
Abyssinian genets, 14:338
Abyssinian hyraxes, 15:177
Abyssinian wild asses. *See* African wild asses
Abyssinians, 14:291
Acacia rats. *See* Tree rats
Acerodon jubatus. *See* Golden-crowned flying foxes
Acinonychinae. *See* Cheetahs
Acinonyx jubatus. *See* Cheetahs
Acomyinae, 16:283
Acomys cahirinus. *See* Egyptian spiny mice
Acomys nesiotes. *See* Cyprus spiny mice
Acomys russatus. *See* Golden spiny mice
Aconaemys spp., 16:433
Aconaemys fuscus. *See* Rock rats
Aconaemys fuscus fuscus, 16:433
Aconaemys fuscus porteri, 16:433
Aconaemys sagei. *See* Sage's rock rats
Acouchis, 16:124, 16:407, 16:408–409, 16:*411*, 16:*412*, 16:413–414
Acoustics. *See* Hearing
Acrobates spp., 13:140, 13:141, 13:142
Acrobates pygmaeus. *See* Pygmy gliders
Acrobatidae. *See* Feather-tailed possums
Adapiformes, 14:1, 14:3
Adapis spp., 14:1
Adaptive radiation, 12:12
Addax nasomaculatus. *See* Addaxes
Addaxes, 16:2, 16:28–34, 16:*36*, 16:*37*, 16:40
Adelobasileus spp., 12:11
Aders's duikers, 16:84*t*
Adipose. *See* Body fat
Admiralty flying foxes, 13:331*t*
Aegyptopithecus spp., 14:2
Aepyceros melampus. *See* Impalas
Aepycerotinae, 16:1, 16:27
Aepyprymnus spp., 13:73, 13:74
Aepyprymnus rufescens. *See* Rufous bettongs
Aeretes melanopterus. *See* North Chinese flying squirrels
Aeromys tephromelas. *See* Black flying squirrels
Aeromys thomasi. *See* Thomas's flying squirrels
Aetiocetus spp., 15:119
Afghan foxes. *See* Blanford's foxes
Afghan mouse-like hamsters, 16:243
Afghan pikas, 16:488, 16:*496*, 16:*500*
Afghans, 14:288
Africa. *See* Ethiopian region; Palaearctic region
African anomalurids, 12:14
African antelopes, 15:272, 16:4, 16:5
African banana bats, 13:311, 13:497, 13:498
African bats. *See* Butterfly bats
African brush-tailed porcupines, 16:*355*, 16:*361*, 16:362–363
African buffaloes, 15:*265*, 16:*3*, 16:*6*, 16:*16*, 16:*19*
behavior, 16:5
evolution, 16:11
feeding ecology, 15:142, 15:272, 16:*14*
humans and, 15:273, 16:9
physical characteristics, 15:139, 15:142, 16:12, 16:*13*
taxonomy, 16:11
See also Buffaloes
African chevrotains. *See* Water chevrotains
African civets, 14:*336*, 14:*338*, 14:*339*, 14:*340*
African climbing mice. *See* Gray climbing mice
African Convention for the Conservation of Nature and Natural Resources
forest hogs, 15:280
West African manatees, 15:195, 15:209
African dassie rats, 16:121
African dormice, 16:317, 16:318, 16:319, 16:*321*
African elephants, 12:*8*, 15:*165*, 15:*173*, 15:*174*–175
behavior, 15:*162*, 15:*164*, 15:*166*, 15:*168*, 15:*171*
distribution, 12:*135*
feeding ecology, 15:*167*
hearing, 12:82
humans and, 12:173, 15:172
physical characteristics, 12:11–12
reproduction, 12:107, 12:*108*, 12:209
taxonomy, 12:*29*
vibrations, 12:83
vision, 12:79
in zoos, 12:209
See also Elephants
African epauletted bats, 13:312
African gazelles, 16:7
African golden cats, 14:391*t*
African golden moles, 12:13
African grass rats, 16:261*t*
African ground squirrels, 16:143
African hedgehogs, 13:203
African leaf-monkeys. *See* Colobus monkeys
African linsangs, 14:337, 14:*339*, 14:*341*
African long-tongued fruit bats, 13:*341*, 13:*345*, 13:347–348
African manatees. *See* West African manatees
African mole-rats, **16:339–350**, 16:*346*
behavior, 12:147, 16:124, 16:343
conservation status, 16:345
distribution, 16:123, 16:*339*, 16:342
evolution, 16:123, 16:339
feeding ecology, 16:343–344
habitats, 16:342–343
humans and, 16:345
parasites and, 12:78
photoperiodicity, 12:74
physical characteristics, 12:72–75, 16:339–342
reproduction, 16:126, 16:344–345
seismic communication, 12:76
species of, 16:347–349, 16:349*t*–350*t*
taxonomy, 16:121, 16:339
vision in, 12:76–77
African native mice, 16:249
African otter shrews, 12:14
African palm civets, 14:335, 14:*339*, 14:341–*342*
African pouched mice. *See* Gambian rats
African pouched rats. *See* Gambian rats
African rhinoceroses, 15:218, 15:220, 15:249
African tree pangolins. *See* Tree pangolins

African warthogs. *See* Common warthogs
African wild asses, 12:176–177, 15:222, 15:226, 15:228, 15:*231*, 15:*232*
African wild dogs, 14:*259*, 14:*275*
behavior, 14:258, 14:268, 14:269
conservation status, 12:216, 14:272, 14:273
evolution, 14:265
feeding ecology, 14:260, 14:270
habitats, 14:267
physical characteristics, 14:266
reproduction, 12:107, 14:261, 14:262, 14:271, 14:272
Afro-Australian fur seals, 14:393, 14:407*t*
Afrotheria, 12:12, 12:26, 12:33, 15:134, 15:161, 15:177
Agile gibbons, 14:207, 14:208–209, 14:210, 14:211, 14:*217*, 14:219–*220*
Agile mangabeys, 14:194, 14:204*t*
Agile wallabies, 13:*88*, 13:*91*, 13:*93*–94
Agouti paca. *See* Pacas
Agouti taczanowskii. *See* Mountain pacas
Agoutidae. *See* Pacas
Agoutis, 12:132, **16:407–415,** 16:*408*, 16:*410*, 16:415*t*
Agriculture, 12:213, 12:215
See also Domestic cattle
Agriochoeridae, 15:264
Ahaggar hyraxes, 15:177, 15:189*t*
Ailuropoda melanoleuca. *See* Giant pandas
Ailurops spp. *See* Bear cuscuses
Ailurops melanotis. *See* Yellow bear cuscuses
Ailurops ursinus. *See* Sulawesi bear cuscuses
Ailurus spp., 14:309
Ailurus fulgens. *See* Red pandas
Air-breathers, 12:67–68
See also specific animals
Akodon azarae. *See* Azara's field mice
Akodon boliviensis, 16:270
Akodon cursor, 16:267
Akodon kofordi, 16:270
Akodon mollis, 16:270
Akodon montensis. *See* Forest mice
Akodon puer, 16:270
Akodon reigi, 16:266
Akodon subfuscus, 16:270
Akodon torques, 16:270
Akodon varius, 16:270
Akodontines, 16:263–264
Alaskan brown bears. *See* Brown bears
d'Albertis's ringtail possums, 13:122*t*
Albignac, 14:74–75
Alcelaphinae, 16:1
Alcelaphini, 16:27, 16:28, 16:29, 16:31, 16:34
Alcelaphus buselaphus. *See* Red hartebeests
Alcelaphus buselaphus buselaphus. *See* Bubal hartebeests
Alcelaphus buselaphus cokii. *See* Coke's hartebeests
Alcelaphus buselaphus swaynei, 16:33–34
Alcelaphus buselaphus tora, 16:33–34
Alcelaphus lichtensteini. *See* Lichtenstein's hartebeests
Alces spp., 15:379, 15:383, 15:384
Alces alces. *See* Moose
Alces (Cervalces) latifrons, 15:380
Alces (Libralces) gallicus, 15:380
Algerian hedgehogs, 13:212*t*
Allactaga balicunica, 16:212, 16:216
Allactaga bullata, 16:212, 16:216
Allactaga elater. *See* Little five-toed jerboas
Allactaga euphratica, 16:216
Allactaga firouzi, 16:216
Allactaga major, 16:213–216
Allactaga severtzovi, 16:212–216
Allactaga sibirica, 16:215, 16:216
Allactaga tetradactyla. *See* Four-toed jerboas
Allactaga vinogradovi, 16:216
Allactaginae. *See* Five-toed jerboas
Allactodipus bobrinskii. *See* Bobrinski's jerboas
Allantois, 12:93
Allenopithecus spp., 14:191
Allenopithecus nigroviridis. *See* Allen's swamp monkeys
Allen's big-eared bats, 13:*507*, 13:*508*, 13:511
Allen's chipmunks, 16:160*t*
Allen's olingos, 14:316*t*
Allen's squirrels, 16:174*t*
Allen's swamp guenons. *See* Allen's swamp monkeys
Allen's swamp monkeys, 14:194, 14:*196*, 14:*197*
Allen's woodrats, 16:277*t*
Allied rock-wallabies, 13:102*t*
Allocebus spp. *See* Hairy-eared mouse lemurs
Allocebus trichotis. *See* Hairy-eared mouse lemurs
Allocricetulus spp., 16:239
Allocricetulus curtatus. *See* Mongolian hamsters
Allocricetulus eversmanni. *See* Eversmann's hamsters
Alopex lagopus. *See* Arctic foxes
Alouatta spp. *See* Howler monkeys
Alouatta belzebul. *See* Red-handed howler monkeys
Alouatta caraya. *See* Black howler monkeys
Alouatta coibensis. *See* Coiba howler monkeys
Alouatta guariba. *See* Brown howler monkeys
Alouatta palliata. *See* Mantled howler monkeys
Alouatta palliata mexicana, 14:162
Alouatta pigra. *See* Mexican black howler monkeys
Alouatta sara. *See* Bolivian red howler monkeys
Alouatta seniculus. *See* Venezuelan red howler monkeys
Alouatta seniculus insulanus, 14:163
Alpacas, 12:180, **15:313–323,** 15:*318*, 15:*319*, 15:*321*
Alpine chamois. *See* Northern chamois
Alpine ibex, 12:139, 15:269, 16:87, 16:*89*, 16:91, 16:95, 16:104*t*
Alpine marmots, 12:*76*, 12:147, 12:148, 16:*144*, 16:*150*, 16:*155*, 16:157
Alpine musk deer. *See* Himalayan musk deer
Alpine pikas, 16:501*t*
Alpine shrews, 13:*254*, 13:*258*, 13:260
Alston's woolly mouse opossums, 12:*256*, 12:*261*–262
Altai argalis, 16:90
Alticola spp., 16:225
Alticola argentatus. *See* Silvery mountain voles
Altricial offspring, 12:95–96, 12:97, 12:106, 12:108
See also Reproduction
Alymlestes spp., 16:480
Alymlestes kielanae, 16:480
Amami rabbits, 12:135, 16:482; 16:487, 16:505, 16:509, 16:516*t*
Amazon bamboo rats, 16:*452*, 16:*454*, 16:455–456
Amazon River dolphins. *See* Botos
Amazonian manatees, 15:191–193, 15:*209*, 15:*210*, 15:*211*–212
Amazonian marmosets, 14:115, 14:116, 14:117, 14:120
Amazonian tapirs. *See* Lowland tapirs
Amblonyx cinereus. *See* Asian small-clawed otters
Amblypoda, 15:131
Amblyrhiza spp., 16:469, 16:470
Amblyrhiza inundata. *See* Anguilla-St. Martin giant hutias
Amblysominae, 13:215
Amblysomus spp. *See* Golden moles
Amblysomus gunningi. *See* Gunning's golden moles
Amblysomus hottentotus. *See* Hottentot golden moles
Amblysomus iris. *See* Zulu golden moles
Amblysomus julianae. *See* Juliana's golden moles
Ambon flying foxes, 13:331*t*
Ambulocetidae, 15:2
Ambulocetus spp., 15:41
American bison, 16:*12*, 16:*16*, 16:*22*–23
behavior, 16:5, 16:6, 16:*11*, 16:*13*–14
conservation status, 12:215, 16:*14*
distribution, 16:4
domestication, 15:145–146
evolution, 16:2
reproduction, 12:*82*
American black bears, 14:*296*, 14:*301*, 14:*302*–303
conservation status, 14:300
evolution, 14:295
habitats, 14:296–297
reproduction, 12:109, 12:110, 14:*299*
translocation of, 12:224
American Kennel Club, 14:288
American leaf-nosed bats, **13:413–434,** 13:*421*, 13:*422*
behavior, 13:416–417
conservation status, 13:420
distribution, 13:*413*, 13:415
evolution, 13:413
feeding ecology, 13:412, 13:417–418
habitats, 13:311, 13:415–416
humans and, 13:316–317, 13:420
physical characteristics, 13:413–415, 13:*414*
reproduction, 13:419–420
species of, 13:*423*–432, 13:433*t*–434*t*
taxonomy, 13:413
American least shrews, 13:*253*, 13:*257*–258
American Livestock Breed Conservancy, 15:282
American martens, 12:110, 12:132, 14:*320*, 14:*324*
American mink, 14:*326*, 14:*328*–329
conservation status, 14:324
distribution, 12:132
ecological niche, 12:117
feeding ecology, 14:323
physical characteristics, 14:321
reproduction, 12:105, 12:110, 14:324
in United Kingdom, 12:182–183
American Pet Products Manufacturers Association, 14:291
American pikas, 12:*134*, 16:*494*, 16:*496*, 16:*497*
behavior, 16:482, 16:*484*, 16:492, 16:*493*

American pikas *(continued)*
distribution, 16:491
feeding ecology, 16:*495*
reproduction, 16:486
American pocket gophers, 12:74
American pronghorns. *See* Pronghorns
American pygmy shrews, 13:199, 13:247, 13:*252*, 13:*254*, 13:*260*, 13:262–263
American shrew moles, 13:280, 13:*282*, 13:*283*, 13:*284*
American Sign Language (ASL), 12:160–162
American Society for the Prevention of Cruelty to Animals (ASPCA), 14:293
American stag-moose, 15:380
American water shrews, 13:195, 13:196, 13:247, 13:249, 13:*254*, 13:*261*, 13:263
American Zoo and Aquarium Association (AZA), 12:203, 12:204, 12:209
Americans with Disabilities Act of 1990, 14:293
Ameridelphia, 12:11
Amino acids, 12:26–27, 12:30–31
Ammodorcas clarkei. See Dibatags
Ammospermophilus spp., 16:143, 16:144
Ammospermophilus harrisii. See Harris's antelope squirrels
Ammospermophilus nelsoni. See Nelson's antelope squirrels
Ammospermophilus parryii. See Arctic ground squirrels
Ammotragus spp., 16:87
Ammotragus lervia. See Barbary sheep
Amnion, 12:6–10, 12:10, 12:92–93
See also Reproduction
Amorphochilus spp., 13:467–468
Amorphochilus schnablii. See Schnabeli's thumbless bats
Amphinectomys spp., 16:268
Amphipithecus spp., 14:2
Amsterdam Island fur seals. *See* Subantarctic fur seals
Amur tigers, 12:*49*, 14:370
Anal glands, 12:37
See also Physical characteristics
Anathana spp., 13:292
Anathana ellioti. See Indian tree shrews
Anchieta, José de, 16:404
Ancodonta, 15:264, 15:266
Andalgalomys spp., 16:265
Andaman horseshoe bats, 13:388–389, 13:392
Andean cats, 14:391*t*
Andean hairy armadillos, 13:184, 13:191*t*
Andean mice, 16:*271*, 16:*273*, 16:274–275
Andean mountain cavies, 16:399*t*
Andean night monkeys, 14:135, 14:138, 14:141*t*
Andean tapirs. *See* Mountain tapirs
Andean titis, 14:148
Andinomys spp., 16:265
Andinomys edax. See Andean mice
Andrew's beaked whales, 15:69*t*
Andrews's three-toed jerboas, 16:212, 16:216, 16:223*t*
Angolan colobus, 14:184*t*
Angolan dwarf guenons. *See* Angolan talapoins
Angolan talapoins, 14:191, 14:193, 14:*196*, 14:*198*–199
Angoni vlei rats, 16:*287*, 16:*292*–*293*
Angora goats, 16:*91*
Anguilla-St. Martin giant hutias, 16:471*t*
Angwantibos, 12:116, 14:13, 14:14, 14:16, 14:*17*, 14:*19*–20
Animal-assisted activities (AAA), 14:293
Animal-assisted therapy (AAT), 14:293
Animal husbandry, in zoos, 12:209–211
Animal Record Keeping System (ARKS), 12:206
Animal rights movement, 12:183, 12:212, 14:293–294
Animal Welfare and Conservation Organization, 12:*216*
Annamite striped rabbits, 16:487, 16:509, 16:*510*, 16:*512*, 16:513
Anoas, 12:137, 16:4, 16:11, 16:24*t*
Anomalocaris spp., 12:64
Anomaluridae. *See* Scaly-tailed squirrels
Anomalurus spp., 16:123
Anomalurus beecrofti. See Beecroft's anomalures
Anomalurus derbianus. See Lord Derby's anomalures
Anomalurus pelii. See Pel's anomalures
Anomalurus pusillus. See Lesser anomalures
Anoplotheriidae, 15:264
Anotomys spp. *See* Aquatic rats
Anotomys leander, 16:266
Anoura geoffroyi. See Geoffroy's tailless bats
Anourosorex spp., 13:248
Anourosorex squamipes. See Mole-shrews
Anourosoricini, 13:247
Ant bears. *See* Giant anteaters
Antarctic blue whales. *See* Blue whales
Antarctic bottlenosed whales. *See* Southern bottlenosed whales
Antarctic fur seals, 14:*393*, 14:*401*, 14:*402*, 14:403
distribution, 12:138, 14:394, 14:395
feeding ecology, 14:397
humans and, 12:*119*
physical characteristics, 12:*66*
reproduction, 14:398, 14:399
Antarctic minke whales, 15:1, 15:120, 15:125, 15:130*t*
Antarctic Treaty, Weddell seals, 14:431
Antarctica, 12:138
Anteaters, 12:39, 12:46, 12:94, 13:147–153, **13:171–179**, 13:*175*, 13:*176*
Antechinomys laniger. See Kultarrs
Antechinus flavipes. See Yellow-footed antechinuses
Antechinus minimus, 12:279
Antechinus stuartii. See Brown antechinuses
Antechinus swainsonii. See Dusky antechinuses
Antechinuses, 12:279–283, 12:290
brown, 12:282, 12:*293*, 12:*296*–*297*
dusky, 12:279, 12:299*t*
yellow-footed, 12:277
Antelope ground squirrels. *See* Nelson's antelope squirrels
Antelope jackrabbits, 16:487, 16:*506*, 16:515*t*
Antelope squirrels, 12:131, 16:143
Harris's, 16:*150*, 16:*154*, 16:156
Nelson's, 16:124, 16:145, 16:147, 16:160*t*
Antelopes, 15:263, 16:1–4
African, 15:272, 16:4, 16:5
blue, 16:28
domestication and, 15:146
feeding ecology, 15:142
giant sable, 16:33
humans and, 15:273
migrations, 12:87
royal, 16:60, 16:63, 16:71*t*
Tibetan, 12:134, 16:*9*
See also specific types of antelope
Anthops ornatus. See Flower-faced bats
Anthracotheriidae, 15:136, 15:264
Anthracotheroidea, 15:302
Anthropoidea, 14:1, 14:3
Antidorcas marsupialis. See Springboks
Antillean beaked whales. *See* Gervais' beaked whales
Antillean fruit-eating bats, 13:*422*, 13:*424*, 13:426
Antillean ghost-faced bats, 13:436, 13:442*t*
Antillean manatees, 15:191, 15:211
See also West Indian manatees
Antillothrix spp., 14:143
Antilocapra americana. See North American pronghorns
Antilocapra americana mexicana spp., 15:415–416
Antilocapridae. *See* Pronghorns
Antilocaprinae, 15:266–267, 15:411
Antilope cervicapra. See Blackbucks
Antilopinae, 16:1, **16:45–58,** 16:*49*, 16:56*t*–57*t*
Antilopine kangaroos. *See* Antilopine wallaroos
Antilopine wallaroos, 13:101*t*
Antipodean fur seals. *See* New Zealand fur seals
Antlerless deer, 15:267
See also Chinese water deer
Antlers, 12:*5*, 12:10, 12:19–20, 12:22, 12:23, 12:99, 15:*132*
See also Physical characteristics
Antrozous spp. *See* Pallid bats
Antrozous pallidus. See Pallid bats
Aonyx spp., 14:321
Aotidae. *See* Night monkeys
Aotus spp. *See* Night monkeys
Aotus azarai. See Azari's night monkeys
Aotus dindinensis, 14:135
Aotus hershkovitzi. See Hershkovitz's night monkeys
Aotus lemurinus. See Gray-bellied night monkeys
Aotus lemurinus griseimembra, 14:138
Aotus miconax. See Andean night monkeys
Aotus nancymaae. See Nancy Ma's night monkeys
Aotus nigriceps. See Black-headed night monkeys
Aotus trivirgatus. See Three-striped night monkeys
Aotus vociferans. See Noisy night monkeys
Aoudads. *See* Barbary sheep
Apennine chamois. *See* Southern chamois
Apes
behavior, 14:7, 14:8
encephalization quotient, 12:149
enculturation of, 12:162
evolution, 14:2
feeding ecology, 14:9
habitats, 14:6
language, 12:160–161
memory, 12:152–153
numbers and, 12:155

Apes *(continued)*
taxonomy, 14:1, 14:3
theory of mind, 12:159–160
See also Gibbons; Great apes
Apidium spp., 14:2
Aplodontia rufa. See Mountain beavers
Aplodontia rufa nigra, 16:133
Aplodontia rufa phaea, 16:133
Aplodontidae. *See* Mountain beavers
Apocrine sweat glands, 12:36, 12:37–38
See also Physical characteristics
Apodemus agrarius. See Striped field mice
Apodemus sylvaticus. See Long-tailed field mice
Appendicular skeleton, 12:41
See also Physical characteristics
Aproteles bulmerae. See Bulmer's fruit bats
Aquatic desmans, 13:196, 13:197, 13:198–199, 13:279, 13:280
Aquatic genets, 14:335, 14:336, 14:*337*, 14:*339*
Aquatic mammals, 12:14
adaptations, **12:62–68**
field studies, 12:201
hair, 12:3
hearing, 12:82
locomotion, 12:44
neonatal milk, 12:127
reproduction, 12:91
See also specific aquatic mammals
Aquatic moles, 13:197, 13:198–199
Aquatic rats, 16:265
Aquatic shrews, 13:197, 13:198–199
Aquatic tenrecs, 13:234*t*
Aquila chrysaetos. See Golden eagles
AR. *See* Aspect ratio
Arabia. *See* Ethiopian region
Arabian camels. *See* Dromedary camels
Arabian oryx, 12:139, 12:204, 16:28, 16:29, 16:32, 16:33, 16:34, 16:*36*, 16:40–*41*
Arabian Peninsula. *See* Palaearctic region
Arabian rousettes. *See* Egyptian rousettes
Arabian sand gazelles, 12:139
Arabian tahrs, 16:90, 16:92, 16:94, 16:95, 16:*97*, 16:*100*, 16:101
Arabuko-Sokoke Forest, 16:524
Arboreal anteaters, 13:147
Arboreal mammals, 12:14
field studies of, 12:202
locomotion, 12:43
vision, 12:79
See also Habitats; specific arboreal mammals
Arboreal pangolins, 16:107–113
Arboreal spiny rats, white-faced, 16:450, 16:*452*, 16:*455*
Arborimus spp., 16:225
Arborimus albipes. See White-footed voles
Arborimus longicaudus. See Red tree voles
Arch-beaked whales. *See* Hubb's beaked whales
Archaeoceti, 15:2–3
Archaeolemurinae. *See* Baboon lemurs
Archaeotherium spp., 15:264
Archaeotraguludus krabiensis, 15:325
Archaoindris spp., 14:63
Archeoindris fontoynonti, 14:63
Archeolemur edwardsi. See Baboon lemurs
Arctic foxes, 12:132, 14:*260*, 14:267, 14:*270*, 14:272, 14:283*t*
Arctic ground squirrels, 12:113, 16:144, 16:145, 16:*149*, 16:*151*–152
Arctic hares, 16:482, 16:*482*, 16:487, 16:*506*, 16:*508*, 16:515*t*
Arctic lemmings, 16:*229*
Arctic sousliks. *See* Arctic ground squirrels
Arctictis binturong. See Binturongs
Arctocebus spp. *See* Angwantibos
Arctocebus aureus. See Golden angwantibos
Arctocebus calabarensis. See Calabar angwantibos
Arctocephalinae. *See* Fur seals
Arctocephalus australis. See South American fur seals
Arctocephalus forsteri. See New Zealand fur seals
Arctocephalus galapagoensis. See Galápagos fur seals
Arctocephalus gazella. See Antarctic fur seals
Arctocephalus philipii. See Juan Fernández fur seals
Arctocephalus pusillus. See Afro-Australian fur seals
Arctocephalus pusillus doriferus. See Australian fur seals
Arctocephalus pusillus pusillus. See Cape fur seals
Arctocephalus townsendi. See Guadalupe fur seals
Arctocephalus tropicalis. See Subantarctic fur seals
Arctocyonids, 15:132, 15:133
Arctogalidia trivirgata. See Small-toothed palm civets
Arctosorex polaris, 13:251
Arctostylopids, 15:135
Ardipithecus ramidus, 14:242
Arend's golden moles, 13:215
Arfak ringtails, 13:122*t*
See also Ringtail possums
Argalis, 12:178, 16:104*t*
Altai, 16:90
behavior, 16:92
distribution, 16:91, 16:92
evolution, 16:87
physical characteristics, 15:268, 16:89, 16:90
reproduction, 16:94
Argentine hemorrhagic fever, 16:270
Argentine Society of Mammalogists, 16:430
Aripuanã marmosets, 14:117, 14:118, 14:121, 14:*125*, 14:*127*, 14:130–131
Aristotle
classification of animals, 12:149
zoos and, 12:203
Ariteus flavescens, 13:420
Arizona gray squirrels, 16:167, 16:*168*, 16:*171*–172
Arkhars. *See* Argalis
ARKS (Animal Record Keeping System), 12:206
Armadillos, 13:147–152, **13:181–192,** 13:*186*, 13:190*t*–192*t*
behavior, 13:150–151, 13:183–184
conservation status, 13:152–153, 13:185
distribution, 13:150, 13:*181*, 13:182–183
evolution, 13:147–149, 13:181–182
feeding ecology, 13:151–152, 13:184
habitats, 13:150, 13:183
humans and, 13:153, 13:185
physical characteristics, 13:149, 13:182
reproduction, 12:94, 12:103, 12:110–111, 13:152, 13:185
species of, 13:187–190, 13:190*t*–192*t*
See also specific types of armadillos
Armored rats, 16:*452*, 16:453–*454*
Armored shrews, 13:196, 13:200, 13:*270*, 13:*273*–274
Armored spiny rats, 16:458*t*
Armored xenarthrans. *See* Cingulata
Arms race, 12:63–64
Árnason, Úlfur, 15:103
Arnoux's beaked whales, 15:69*t*
Arrector pili, 12:38
See also Physical characteristics
Arredondo's solenodons, 13:237–238
Artibeus harti, 13:*416*
Artibeus jamaicensis. See Jamaican fruit-eating bats
Artibeus lituratus. See Great fruit-eating bats
Artibeus watsoni, 13:416
Artiocetus clavis, 15:266
Artiodactyla, **15:263–273**
behavior, 15:269–271
cetaceans and, 12:30, 15:2
conservation status, 15:272–273
distribution, 12:129, 12:132, 12:136, 15:269
evolution, 15:131–133, 15:135–138, 15:263–267
feeding ecology, 15:141–142, 15:271–272
habitats, 15:269
humans and, 15:273
physical characteristics, 12:40, 12:79, 15:138–140, 15:267–269
reproduction, 12:127, 15:272
ruminant, 12:10
taxonomy, 15:263–267
See also specific artiodactyls
Arvicanthis niloticus. See African grass rats
Arvicola spp. *See* Water voles
Arvicola terrestris. See Northern water voles
Arvicolinae, 16:124, **16:225–238,** 16:*231*, 16:*232*, 16:237*t*–238*t*, 16:270, 16:282, 16:283
Asano, Toshio, 12:161–162
Asellia tridens. See Trident leaf-nosed bats
Aselliscus tricuspidatus. See Temminck's trident bats
Ashaninka rats, 16:443
Ashy chinchilla rats, 16:443, 16:*444*, 16:445, 16:*446*, 16:*447*
Ashy roundleaf bats, 13:410*t*
Asia. *See* Oriental region; Palaearctic region
Asian asses. *See* Asiatic wild asses
Asian banded boars, 12:179
Asian brush-tailed porcupines. *See* Asiatic brush-tailed porcupines
Asian buffaloes. *See* Water buffaloes
Asian elephants, 15:*170*, 15:*173*, 15:*174*
conservation status, 15:171, 15:172
distribution, 15:166
evolution, 15:162–164
feeding ecology, 15:*169*
humans and, 12:173
physical characteristics, 15:165
reproduction, 12:209
in zoos, 12:209, 12:*211*
See also Elephants
Asian garden dormice, 16:327*t*
Asian hamster mice. *See* Mouse-like hamsters
Asian house shrews. *See* Musk shrews
Asian leaf-monkeys, 14:172

Asian pigs, 15:149
Asian rhinoceroses, 15:220, 15:221, 15:222
Asian small-clawed otters, 14:325
Asian tapirs. *See* Malayan tapirs
Asian two-horned rhinoceroses. *See* Sumatran rhinoceroses
Asiatic black bears, 12:195, 14:295, 14:296, 14:297, 14:298, 14:300, 14:306*t*
Asiatic brown bears. *See* Brown bears
Asiatic brush-tailed porcupines, 16:*355*, 16:*358*, 16:363
Asiatic cave swiftlets, 12:53
Asiatic golden cats, 14:390*t*
Asiatic ibex. *See* Siberian ibex
Asiatic long-tailed shrews. *See* Hodgson's brown-toothed shrews
Asiatic mouflons, 12:178
See also Urals
Asiatic shrew-moles. *See* Chinese shrew-moles
Asiatic tapirs. *See* Malayan tapirs
Asiatic water shrews, 13:195
Asiatic wild asses, 12:134, 12:177, 15:217, 15:222, 15:226, 15:228, 15:229, 15:*231*, 15:*234*–235
Asiatic wild dogs, 14:270
Asiatic wild horses. *See* Przewalski's horses
Asiatic wild sheep, 15:271
ASL (American Sign Language), 12:160–162
Asoriculus maghrebensis, 13:247
ASPCA (American Society for the Prevention of Cruelty to Animals), 14:293
Aspect ratio (AR), bat wings, 12:57–58
Assam rabbits. *See* Hispid hares
Asses, 12:176–177, 15:139, 15:215–223, **15:225–236,** 15:*231*
Astrapotheria, 12:11, 15:131, 15:133–136
Astrapotherium spp., 15:133–134
Atalaye nesophontes, 13:243
Atelerix spp. *See* African hedgehogs
Atelerix albiventris. *See* Central African hedgehogs
Atelerix algirus. *See* Algerian hedgehogs
Atelerix frontalis. *See* Southern African hedgehogs
Atelerix sclateri. *See* Somalian hedgehogs
Ateles spp. *See* Spider monkeys
Ateles belzebuth. *See* White-bellied spider monkeys
Ateles belzebuth hybridsus. *See* Variegated spider monkeys
Ateles chamek. *See* Peruvian spider monkeys
Ateles fusciceps. *See* Brown-headed spider monkeys
Ateles fusciceps robustus. *See* Colombian black spider monkeys
Ateles geoffroyi. *See* Geoffroy's spider monkeys
Ateles hybridus, 14:160
Ateles marginatus. *See* White-whiskered spider monkeys
Ateles paniscus. *See* Black spider monkeys
Atelidae, **14:155–169,** 14:*161*
behavior, 14:157
conservation status, 14:159–160
distribution, 14:*155*, 14:156
evolution, 14:155
feeding ecology, 14:158
habitats, 14:156–157
humans and, 14:160
physical characteristics, 14:155–156
reproduction, 14:158–159
species of, 14:*162*–166, 14:166*t*–167*t*
taxonomy, 14:155
Atelins, 14:155, 14:157, 14:158, 14:159
Atelocynus microtis. *See* Short-eared dogs
Atherurus africanus. *See* African brush-tailed porcupines
Atherurus macrourus. *See* Asiatic brush-tailed porcupines
Atilax spp., 14:347
Atilax paludinosus. *See* Marsh mongooses
Atlantic bottlenosed dolphins. *See* Common bottlenosed dolphins
Atlantic bottlenosed whales. *See* Northern bottlenosed whales
Atlantic gray seals. *See* Gray seals
Atlantic humpbacked dolphins, 15:57*t*
Atlantic seals. *See* Gray seals
Atlantic spotted dolphins, 15:*2*, 15:*43*, 15:*44*, 15:46, 15:47
Atlantic walruses, 14:409, 14:*410*, 14:*412*, 14:415
Atlantic white-sided dolphins, 15:9, 15:57*t*
Atlantoxerus spp., 16:143, 16:144
Atlantoxerus getulus. *See* Barbary ground squirrels
Atmosphere, subterranean, 12:72–73
Attenborough, David, 13:217
Auckland sea lions. *See* Hooker's sea lions
Auditory Neuroethology Lab, 12:54
Audubon's cottontail. *See* Desert cottontails
Auliscomys boliviensis, 16:267
Auricles, 12:9, 12:72
See also Physical characteristics
Aurochs, 12:177, 16:*17*, 16:21–*22*
Australasian carnivorous marsupials. *See* Dasyuromorphia
Australasian Monotreme/Marsupial Action Plan, 13:71
Australian cattle dogs, 14:292
Australian false vampire bats, 13:313, 13:*380*, 13:*381*, 13:*382*, 13:*383*, 13:*384*–385
Australian fur seals, 14:394
Australian hopping mice. *See* Australian jumping mice
Australian jumping mice, 16:*250*, 16:*254*, 16:*260*
Australian Koala Foundation, 13:49
Australian marsupial moles, 12:13
Australian region, 12:136–138
feral cats, 12:185–186
rabbit control, 12:186–188, 12:193
Australian sea lions, 14:*401*, 14:404
behavior, 12:*145*
distribution, 14:394, 14:*403*
habitats, 14:*395*
reproduction, 14:*396*, 14:398, 14:*399*
Australian stick nest rats. *See* Greater stick-nest rats
Australopithecines, 12:20, 14:247
Australopithecus spp., 14:242
Australopithecus afarensis, 14:*242*, 14:*243*
Australopithecus africanus, 14:242
Australopithecus bahrelghazali, 14:242
Australopithecus boisei, 14:242
Australopithecus gahri, 14:242
Australopithecus robustus, 14:242, 14:247, 14:250
Autapomorphies, 12:7
Avahi spp. *See* Avahis
Avahi laniger. *See* Eastern woolly lemurs
Avahi occidentalis, 14:63
Avahis, **14:63–69,** 14:*68*
behavior, 14:6, 14:8, 14:65–66
conservation status, 14:67
distribution, 14:65
evolution, 14:63
feeding ecology, 14:8, 14:66
habitats, 14:65
physical characteristics, 14:63–65
species of, 14:*69*
taxonomy, 14:63
Axis spp., 12:19, 15:269, 15:357, 15:358
Axis axis. *See* Chitals
Axis calamianensis. *See* Calamian deer
Axis deer. *See Axis* spp.
Axis kuhlii. *See* Bawean hog deer
Axis porcinus. *See* Hog deer
Aye-ayes, 12:46, 12:83, 12:136, 14:4, 14:9, 14:10–11, **14:85–89,** 14:*86*, 14:*87*, 14:*88*
Aye-ayes (extinct). *See Daubentonia robusta*
AZA. *See* American Zoo and Aquarium Association
Azara's agoutis, 16:407, 16:409, 16:415*t*
Azara's field mice, 16:266–270
Azara's tuco-tucos, 16:*426*, 16:*427*
Azari's night monkeys, 14:135, 14:*136*, 14:138, 14:141*t*

B

Babakotia spp., 14:63
Babies. *See* Neonates
Babirousinae. *See* Babirusas
Babirusas, 12:137, 15:275, 15:*276*, 15:278, 15:279, 15:280, 15:*281*, 15:*283*, 15:*287*–288
Baboon lemurs, 14:63, 14:*68*, 14:*69*, 14:72
Baboons, 14:6, 14:188, 14:189, 14:191, 14:193, 14:194
Babyrousa spp. *See* Babirusas
Babyrousa babyrussa. *See* Babirusas
Bacardi bats, 13:317
Bactrian camels, 15:*320*
behavior, 15:317
conservation status, 15:317, 15:318
distribution, 12:134, 15:*315*
domestication of, 12:179, 15:150
evolution, 15:313
habitats, 15:316
humans and, 15:*319*
physical characteristics, 15:268, 15:314–*315*
taxonomy, 15:313
Baculum, 12:91, 12:111
See also Physical characteristics
Badgers, **14:319–325,** 14:*322*, 14:*326*, **14:329–334**
behavior, 12:145, 14:258–259, 14:321–323
distribution, 14:321
Eurasian, 12:77, 14:324
evolution, 14:319
feeding ecology, 14:260, 14:323
habitats, 14:321
humans and, 14:324
physical characteristics, 14:319–321
reproduction, 14:324
species of, 14:*329*, 14:331, 14:332*t*–333*t*
taxonomy, 14:256, 14:319
torpor, 12:113
Bahaman raccoons, 14:309, 14:310, 14:315*t*

Bahamian funnel-eared bats, 13:461, 13:*462*, 13:*463*, 13:464
Bahamian hutias, 16:461, 16:*462*, 16:467*t*
Bahía hairy dwarf porcupines, 16:366, 16:368, 16:*370*, 16:*371*, 16:373
Baijis, 15:5, 15:10, **15:19–22,** 15:*20*, 15:*21*
Baikal seals, 12:138, 14:422, 14:*426*, 14:*427*, 14:430
Bailey's pocket mice, 16:208*t*
Baiomys taylori. See Pygmy mice
Baird's beaked whales, 15:*59*, 15:*61*, 15:62, 15:*63*, 15:*64*, 15:65
Baird's tapirs. *See* Central American tapirs
Balaena spp. *See* Bowhead whales
Balaena mysticetus. See Bowhead whales
Balaenidae, 15:3, 15:4, 15:103, **15:107–118,** *15:114*
Balaenoptera acutorostrata. See Northern minke whales
Balaenoptera bonaerensis. See Antarctic minke whales
Balaenoptera borealis. See Sei whales
Balaenoptera brydei, 15:120
Balaenoptera edeni. See Bryde's whales
Balaenoptera musculus. See Blue whales
Balaenoptera physalus. See Fin whales
Balaenopteridae. *See* Rorquals
Balaenula spp., 15:107
Balantiopteryx infusca, 13:358
Balantiopteryx plicata. See Gray sac-winged bats
Bald eagles, 12:192
Bald-headed uakaris. *See* Bald uakaris
Bald uakaris, 14:144, 14:145, 14:146, 14:*147*, 14:148, 14:*149*, 14:*150*, 14:151
Baleen whales, 12:14, 15:1–2
  behavior, 15:7
  conservation status, 12:215
  digestive system, 12:120–121
  echolocation, 12:86
  evolution, 15:3–4
  feeding ecology, 15:8
  migrations, 12:169
  physical characteristics, 12:67
  reproduction, 15:9
  See also specific types of baleen whales
Bali cattle, 12:178
Ballou, J. D., 12:207
Baltic gray seals. *See* Gray seals
Baluchitherium spp., 12:12
Baluchitherium grangeri. See Indricotherium transouralicum
Bamboo bats, 13:311, 13:497, 13:*504*, 13:*513*, 13:514
Bamboo rats, 16:123
  Amazon, 16:*452*, 16:*454*, 16:455–456
  East African, 12:74
  large, 16:*283*, 16:*287*, 16:*288*, 16:294–295
  montane, 16:450–451, 16:*452*, 16:*455*, 16:457
  southern, 16:450, 16:*452*, 16:*454*, 16:456–457
Banana bats, African, 13:497, 13:498
Banana pipistrelles, 13:311
Banded anteaters. *See* Numbats
Banded duikers. *See* Zebra duikers
Banded hare-wallabies, 13:*91*
Banded leaf-monkeys, 14:*173*, 14:175, 14:*176*, 14:179–*180*
Banded linsangs, 14:*336*, 14:345*t*
Banded mongooses, 14:259, 14:*349*, 14:350, 14:*351*, 14:357*t*
Banded palm civets, 14:*337*, 14:344*t*
Bandicoots, 12:137, **13:1–7,** 13:19
  dry-country, 13:1, **13:9–18**
  physical characteristics, 13:*10*
  rainforest, **13:9–18**
  reproduction, 12:93
Bandicota spp., 16:249–250
Bands, human, 14:251
Bank voles, 16:*228*, 16:*232*, 16:*234*, 16:236
Banks flying foxes, 13:331*t*
Banner-tailed kangaroo rats, 16:200, 16:*201*–202, 16:203, 16:*204*, 16:*207*
Bantengs, 12:178, 16:4, 16:24*t*
Barasinghas, 12:19, 15:357, 15:*358*, 15:361, 15:*363*, 15:367–*368*, 15:384
Barbados raccoons, 14:262, 14:309, 14:310
Barbary apes. *See* Barbary macaques
Barbary ground squirrels, 16:159*t*
Barbary lions, 14:380
Barbary macaques, 14:*195*, 14:200–201
  behavior, 12:*150*
  distribution, 14:5, 14:191, 14:*198*
  evolution, 14:189
  habitats, 14:191
  reproduction, 12:*49*, 14:10
Barbary sheep, 16:4, 16:87, 16:90–92, 16:94, 16:*96*, 16:*98*, 16:100
Barbastella barbastellus. See Western barbastelles
Bare-tailed woolly opossums, 12:*250*, 12:252, 12:*255*, 12:*257*
Barking deer. *See* Indian muntjacs
Barred bandicoots. *See* Eastern barred bandicoots
Barylamda spp., 15:135
Basilosauridae, 15:2
Basilosaurus spp., 15:3, 15:119
Bassaricyon spp. *See* Olingos
Bassaricyon alleni. See Allen's olingos
Bassaricyon beddardi. See Beddard's olingos
Bassaricyon gabbii. See Olingos
Bassaricyon lasius. See Harris's olingos
Bassaricyon pauli, 14:310
Bassariscus spp., 14:309, 14:310
Bassariscus astutus. See Ringtails
Bassariscus sumichrasti. See Cacomistles
Bat Conservation International, 13:420
Bat-eared foxes, 14:*275*, 14:281
  behavior, 14:*260*
  canine distemper and, 12:216
  distribution, 14:*280*
  evolution, 14:265
  physical characteristics, 14:266–267, 14:*267*
  reproduction, 14:271, 14:272
Bate, D. M. A., 16:333
Bate's dwarf antelopes. *See* Dwarf antelopes
Bate's pygmy antelopes. *See* Dwarf antelopes
Bate's slit-faced bats, 13:378*t*
Bathyergidae. *See* African mole-rats
Bathyerginae. *See* Dune mole-rats
Bathyergus spp. *See* Dune mole-rats
Bathyergus janetta. See Namaqua dune mole-rats
Bathyergus suillus. See Cape dune mole-rats
Batodon spp., 13:193
Batonodoides, 13:194
Bats, 12:12–13, **13:307–318**
  behavior, 13:312–313
  brains, 12:49
  conservation status, 13:315–316
  distribution, 12:129, 12:132, 12:136, 12:137, 12:138, 13:310–311
  evolution, 12:11, 13:308–309
  feeding ecology, 13:313–314
  field studies, 12:200–201
  flight adaptations in, 12:52–60, 12:*53–59*
  habitats, 13:311
  hearing, 12:82
  hearts, 12:45
  humans and, 13:316–317
  as keystone species, 12:216–217
  locomotion, 12:14, 12:43–44
  navigation, 12:87
  physical characteristics, 13:307–308, 13:310
  reproduction, 12:90–91, 12:94, 12:95, 12:103, 12:105–106, 12:110, 13:314–315
  taxonomy, 13:309
  See also Vespertilionidae; specific types of bats
Bawean hog deer, 15:360
Bay cats, 14:390*t*
Bay duikers, 16:*76*, 16:*78*, 16:*83*
Bdeogale spp., 14:347
Bdeogale crassicauda. See Bushy tailed mongooses
Beagles, 14:288
Beaked whales, **15:59–71,** 15:*63*, 15:69*t*–70*t*
  behavior, 15:6, 15:7, 15:61
  conservation status, 15:10, 15:61–62
  distribution, 15:*59*, 15:61
  evolution, 15:3, 15:59
  feeding ecology, 15:8, 15:61
  habitats, 15:61
  humans and, 15:62
  physical characteristics, 15:4, 15:59, 15:61
  reproduction, 15:61
  species of, 15:64–68, 15:69*t*–70*t*
  taxonomy, 15:3, 15:59
Bear cats. *See* Binturongs
Bear cuscuses, 13:57
Beard, K. Christopher, 14:91
Bearded pigs, 15:*279*, 15:280, 15:289*t*
Bearded sakis, 14:143–148, 14:*147*, 14:*149*, 14:*150*–151
Bearded seals, 14:435*t*
Bears, **14:295–307,** 14:*301*
  behavior, 14:258, 14:297–298
  conservation status, 14:262, 14:300
  distribution, 12:131, 12:136, 14:*295*, 14:296
  evolution, 14:295–296
  feeding ecology, 14:255, 14:260–261, 14:298–299
  field studies of, 12:*196*, 12:200
  gaits of, 12:*39*
  habitats, 14:296–297
  hibernation, 12:113
  humans and, 14:300
  Ice Age, 12:24
  physical characteristics, 14:296
  reproduction, 12:109, 12:110, 14:299
  species of, 14:*302*–305, 14:306*t*
  taxonomy, 14:256, 14:295
  in zoos, 12:*210*
Beavers, 12:14, **16:177–184,** 16:*182*
  behavior, 16:179
  conservation status, 16:180
  distribution, 16:124, 16:*177*, 16:179
  evolution, 16:122, 16:177

Beavers *(continued)*
feeding ecology, 16:179–180
habitats, 16:179
humans and, 16:127, 16:180
Ice Age, 12:24
North American, 12:111
physical characteristics, 16:123, 16:177–179
reproduction, 12:107, 16:126, 16:180
species of, 16:183–184
taxonomy, 16:122, 16:177
See also specific types of beavers
Beccari's mastiff bats, 13:*484*, 13:494*t*
Bechstein's bats, 12:85
Beddard's olingos, 14:310, 14:316*t*
Beecroft, John, 16:303
Beecroft's anomalures, 16:*301*, 16:303, 16:*305*
Begall, S., 16:435
Behavior, 12:9–10, **12:140–148**
aardvarks, 15:156–157
agoutis, 16:408, 16:411–414, 16:415*t*
anteaters, 13:172–173, 13:177–179
armadillos, 12:71, 13:183–184, 13:187–190, 13:190*t*–192*t*
Artiodactyla, 15:269–271
aye-ayes, 14:86–87
baijis, 15:20
bandicoots, 13:3
dry-country, 13:10–11, 13:14–16, 13:16*t*–17*t*
rainforest, 13:10–11, 13:14–16, 13:16*t*–17*t*
bats, 13:312–313
American leaf-nosed bats, 13:416–417, 13:423–432, 13:433*t*–434*t*
bulldog bats, 13:446, 13:449–450
disk-winged, 13:474, 13:476–477
Emballonuridae, 13:356–357, 13:360–363, 13:363*t*–364*t*
false vampire, 13:381, 13:384–385
funnel-eared, 13:460, 13:463–465
horseshoe, 13:390, 13:396–400
Kitti's hog-nosed, 13:369
Molossidae, 13:484–485, 13:490–493, 13:493*t*–495*t*
mouse-tailed, 13:351, 13:353
moustached, 13:435–436, 13:440–441, 13:442*t*
New Zealand short-tailed, 13:454, 13:457–458
Old World fruit, 13:312, 13:319–320, 13:325–330, 13:331*t*–332*t*, 13:335, 13:340–347, 13:348*t*–349*t*
Old World leaf-nosed, 13:402, 13:405, 13:407–409, 13:409*t*–410*t*
Old World sucker-footed, 13:480
slit-faced, 13:371, 13:373, 13:375, 13:376*t*–377*t*
smoky, 13:468, 13:470–471
Vespertilionidae, 13:498–500, 13:506–514, 13:515*t*–516*t*, 13:521, 13:524–525, 13:526*t*
bears, 14:258, 14:297–298, 14:302–305, 14:306*t*
beavers, 16:179, 16:183–184
bilbies, 13:20–21
botos, 15:28–29
Bovidae, 16:5–6
Antilopinae, 16:46, 16:48, 16:50–55, 16:56*t*–57*t*
Bovinae, 16:13–14, 16:18–23, 16:24*t*–25*t*
Caprinae, 16:92–93, 16:98–103, 16:103*t*–104*t*
duikers, 16:74, 16:80–84, 16:84*t*–85*t*
Hippotraginae, 16:30–31, 16:37–42, 16:42*t*–43*t*
Neotraginae, 16:60–62, 16:66–71, 16:71*t*
bushbabies, 14:25–26, 14:28–32, 14:32*t*–33*t*
Camelidae, 15:316–317, 15:320–323
Canidae, 12:148, 14:267–269, 14:276–283, 14:283*t*–284*t*
capybaras, 16:402–*404*
Carnivora, 12:117, 12:141, 12:145, 14:257–260
cats, 12:145, 14:371–372, 14:379–389, 14:390*t*–391*t*
Caviidae, 16:391–392, 16:395–398, 16:399*t*
Cetacea, 12:86, 15:6–7
Balaenidae, 15:109–110, 15:115–117
beaked whales, 15:61, 15:64–68, 15:69*t*–70*t*
dolphins, 15:5, 15:44–46, 15:53–55, 15:56*t*–57*t*
franciscana dolphins, 15:24
Ganges and Indus dolphins, 15:14
gray whales, 15:95–97
Monodontidae, 15:84–85, 15:90
porpoises, 15:35, 15:38–39, 15:39*t*
pygmy right whales, 15:104
rorquals, 15:6, 15:7, 15:122–123, 15:127–130, 15:130*t*
sperm whales, 15:74–75, 15:79–80
chevrotains, 15:327–328, 15:331–333
Chinchillidae, 16:378–379, 16:382–383, 16:383*t*
colugos, 13:302, 13:304
coypus, 16:474–476
cultural, 12:157–159
dasyurids, 12:289, 12:294–298, 12:299*t*–301*t*
Dasyuromorphia, 12:280–281
deer
Chinese water, 15:374–375
muntjacs, 15:344–345, 15:349–354
musk, 15:336–337, 15:340–341
New World, 15:383, 15:388–396*t*
Old World, 15:359, 15:364–371*t*
Dipodidae, 16:213–215, 16:219–222, 16:223*t*–224*t*
Diprotodontia, 13:35–36
domestication and, 12:174–175
dormice, 16:318–319, 16:323–326, 16:327*t*–328*t*
duck-billed platypuses, 12:230–231, 12:244–245
Dugongidae, 15:200, 15:203
echidnas, 12:237, 12:240*t*
elephants, 15:167–169, 15:174–175
Equidae, 15:226–228, 15:226*t*, 15:232–235
Erinaceidae, 13:204–205, 13:209–213, 13:212*t*–213*t*
giant hutias, 16:469, 16:471*t*
gibbons, 14:210–211, 14:218–222
Giraffidae, 12:167, 15:401–403, 15:408
great apes, 14:228–232, 14:237–240
gundis, 16:312–313, 16:314–315
Herpestidae, 14:349–351, 14:354–356, 14:357*t*
Heteromyidae, 16:200–202, 16:205–209, 16:208*t*–209*t*
hippopotamuses, 15:306, 15:310–311
humans, 14:249–252, 14:*251*, 14:*252*
hutias, 16:461–462, 16:467*t*
Hyaenidae, 14:362, 14:364–367
hyraxes, 15:180–181, 15:186–188, 15:189*t*
Indriidae, 14:*64*, 14:65–66, 14:69–72
Insectivora, 13:197–198
koalas, 13:45–46
Lagomorpha, 16:482–483
lemurs, 14:6, 14:7, 14:8, 14:*49*, 14:50–51, 14:55–60
Cheirogaleidae, 14:36–37, 14:41–44
sportive, 14:8, 14:75–76, 14:79–83
Leporidae, 16:506–507, 16:511–515, 16:515*t*–*516*t
Lorisidae, 14:14–15, 14:18–20, 14:21*t*
luxury organs and, 12:22
Macropodidae, 12:145, 13:36, 13:86–87, 13:93–100, 13:101*t*–102*t*
manatees, 15:206, 15:211–212
Megalonychidae, 13:156–157, 13:159
moles
golden, 13:198, 13:216–217, 13:220–222, 13:222*t*
marsupial, 13:26–27, 13:28
monitos del monte, 12:274
monkeys
Atelidae, 14:157, 14:162–166, 14:166*t*–167*t*
Callitrichidae, 14:116–120, 14:*123*, 14:127–131, 14:132*t*
Cebidae, 14:103–104, 14:108–112
cheek-pouched, 14:191–192, 14:197–204, 14:204*t*–205*t*
leaf-monkeys, 14:7, 14:174, 14:178–183, 14:184*t*–185*t*
night, 14:137, 14:141*t*
Pitheciidae, 14:145–146, 14:150–153, 14:153*t*
monotremes, 12:230–231
mountain beavers, 16:132
Muridae, 16:284, 16:288–295, 16:296*t*–297*t*
Arvicolinae, 16:226–227, 16:233–238, 16:237*t*–238*t*
hamsters, 16:240–241, 16:245–246, 16:247*t*
Murinae, 16:251, 16:255–260, 16:261*t*–262*t*
Sigmodontinae, 16:267–268, 16:272–276, 16:277*t*–278*t*
musky rat-kangaroos, 13:70–71
Mustelidae, 12:145, 12:*186*, 14:258–259, 14:321–323, 14:327–331, 14:332*t*–333*t*
numbats, 12:304
octodonts, 16:435, 16:438–440, 16:440*t*
opossums
New World, 12:251–252, 12:257–263, 12:264*t*–265*t*
shrew, 12:268, 12:270
Otariidae, 14:396–397, 14:402–406
pacaranas, 16:387
pacas, 16:418, 16:*419*, 16:423–424
pangolins, 16:110–112, 16:115–120
peccaries, 15:293–294, 15:298–300
Perissodactyla, 15:218–219
Petauridae, 13:126–127, 13:131–133, 13:133*t*
Phalangeridae, 13:60, 13:*64*–*67t*
pigs, 15:278, 15:284–288, 15:289*t*–290*t*
pikas, 16:491–493, 16:497–501, 16:501*t*–502*t*

Behavior *(continued)*
pocket gophers, 16:188–189, 16:192–194, 16:195*t*–197*t*
porcupines
New World, 16:367–368, 16:371–373, 16:374*t*
Old World, 16:352–353, 16:357–364
possums
feather-tailed, 13:141, 13:144
honey, 13:136–137
pygmy, 13:106–107, 13:110–111
primates, 14:6–8
Procyonidae, 14:310, 14:313–315, 14:315*t*–316*t*
pronghorns, 15:412–414
Pseudocheiridae, 13:115–116, 13:119–123, 13:122*t*–123*t*
rat-kangaroos, 13:74–75, 13:78–81
rats
African mole-rats, 12:71, 12:74, 12:76, 12:83, 12:147, 16:*340*, 16:343, 16:347–349, 16:349*t*–350*t*
cane, 16:334, 16:337–338
chinchilla, 16:444, 16:447–448
dassie, 16:*331*
spiny, 16:450, 16:453–457, 16:458*t*
rhinoceroses, 15:253, 15:257–261
rodents, 16:124–125
sengis, 16:519–521, 16:524–530, 16:531*t*
shrews
red-toothed, 13:248–250, 13:255–263, 13:263*t*
West Indian, 13:244
white-toothed, 13:267, 13:271–275, 13:276*t*–277*t*
Sirenia, 15:192–193
solenodons, 13:238–239, 13:240–241
springhares, 16:308–309
squirrels
flying, 16:136, 16:139–140, 16:141*t*–142*t*
ground, 12:*71*, 12:141, 16:124, 16:144–145, 16:151–158, 16:158*t*–160*t*
scaly-tailed, 16:300, 16:302–305
tree, 16:165–166, 16:169–174, 16:174*t*–175*t*
Talpidae, 13:280–281, 13:284–287, 13:287*t*–288*t*
tapirs, 15:239–240, 15:245–247
tarsiers, 14:94, 14:97–99
tenrecs, 13:227–228, 13:232–234, 13:234t
three-toed tree sloths, 13:163, 13:167–169
tree shrews, 13:291, 13:294–296, 13:297*t*–298*t*
true seals, 12:85, 12:87, 14:418–419, 14:427–428, 14:430–434, 14:435*t*
tuco-tucos, 16:426–427, 16:429–430, 16:430*t*–431*t*
ungulates, 15:142
Viverridae, 14:337, 14:340–343, 14:344*t*–345*t*
walruses, 14:412–414
wombats, 13:52, 13:55–56
Xenartha, 13:150–151
Behavioral ecology, 12:148
Beintema, J. J., 16:311
Beiras, **16:59–72,** 16:63, 16:71*t*
Beisa oryx, 16:32
Belanger's tree shrews, 13:292, 13:297*t*
Belding's ground squirrels, 16:124, 16:*147*, 16:160*t*
Bell-Jarman principle, 12:119
Belomys pearsonii. See Hairy-footed flying squirrels
Belugas, **15:81–91,** 15:*82*, 15:*83*, 15:*86*, 15:*89*, 15:*90*
behavior, 15:84–85
conservation status, 15:9, 15:86–87
distribution, 15:*81*, 15:82–83
echolocation, 12:*85*, 12:87
evolution, 15:81
feeding ecology, 15:85
habitats, 15:83–84, 15:*87*
humans and, 15:10–11, 15:87–88
physical characteristics, 15:81–82, 15:*85*, 15:*88*
reproduction, 15:85–86
taxonomy, 15:81
Bengal cats, 14:*372*
Bengal slow lorises, 14:16, 14:21*t*
Bengali water mongooses, 14:349, 14:352
Beni titis, 14:148
Bennett's chinchilla rats, 16:443, 16:444, 16:*446*, 16:*447*
Bennett's tree kangaroos, 13:*86*, 13:*91*, 13:*99*–100
Bennett's wallabies. *See* Red-necked wallabies
Bensonomys spp., 16:264
Berardius spp., 15:59
Berardius arnouxii. See Arnoux's beaked whales
Berardius bairdii. See Baird's beaked whales
Beremendiini, 13:247
Bergh, Henry, 14:293
Bergmann's Rule, 12:21, 14:245
Bering Sea beaked whales. *See* Stejneger's beaked whales
Bern Convention, 13:260
Bernard's wallaroos. *See* Black wallaroos
Bernoulli lift, 12:56
Bettongia spp. *See* Bettongs
Bettongia gaimardi. See Tasmanian bettongs
Bettongia lesueur. See Boodies
Bettongia penicillata. See Brush-tailed bettongs
Bettongia tropica. See Northern bettongs
Bettongs, 13:34, 13:73
brush-tailed, 13:39, 13:74, 13:*75*, 13:76, 13:77, 13:*78*–79
northern, 13:74, 13:76, 13:77, 13:*79*
rufous, 13:*74*, 13:*75*, 13:*76*, 13:77, 13:*79*, 13:80–81
Tasmanian, 13:*73*, 13:74, 13:76, 13:77, 13:*78*
Bezoar goats, 12:179
Bharals. *See* Blue sheep
Biases, in subterranean mammal research, 12:70–71
Bibimys spp., 16:266
Bichon frises, 14:289
Bicolor-spined porcupines, 16:365, 16:374*t*
Bicolored shrews, 13:*266*
Bicornuate uterus, 12:90–91, 12:*102*
Big brown bats, 13:*499*, 13:*505*, 13:507–*508*
distribution, 13:311, 13:498
echolocation, 12:87
reproduction, 13:315, 13:500–501
smell, 12:85
Big cats, 12:136
See also Cheetahs; Leopards; Lions
Big crested mastiff bats, 13:495*t*
Big cypress fox squirrels, 16:*164*
Big-eared bats
Allen's, 13:*505*, 13:*507*, 13:509
common, 13:433*t*
Big-eared climbing rats, 16:277*t*
Big-eared flying foxes, 13:*324*, 13:*329*, 13:330
Big-eared flying mice, 16:300, 16:*301*, 16:*304*–305
Big-eared free-tailed bats. *See* Giant mastiff bats
Big-eared mastiff bats, 13:495*t*
Big-footed bats, Rickett's, 13:313
Big naked-backed bats, 13:435, 13:442*t*
Big pocket gophers, 16:196*t*
Bighorn sheep, 12:178, 16:*97*, 16:*99*, 16:102
behavior, 12:*141*,16:5, 16:*92*
conservation status, 16:95
distribution, 12:131, 16:91, 16:*92*
evolution, 16:87
feeding ecology, 16:93
habitats, 15:269
Bilbies, 12:137, 13:2, 13:3, 13:5, **13:19–23,** 13:*20*
Bilenca, D. N, 16:267
Binaural hearing, 12:81
Bini free-tailed bats, 13:494*t*
Binocularity, 12:79
See also Physical characteristics
Binturongs, 14:335, 14:336, 14:*339*, 14:*342*–343
Biodiversity, 12:11–15
conservation biology and, 12:216–225
threats to, 12:213
zoo management and, 12:206–209
See also Conservation status
Biodiversity surveys, 12:194–198
Biogeography, **12:129–139,** 12:*131*, 12:220
Bioko Allen's bushbabies, 14:25, 14:26, 14:32*t*
Biological control, 12:193
mongooses for, 12:190–191
rabbits, 12:187–188
Biological diversity. *See* Biodiversity
Biology, conservation, 12:216–225
Bioparks, 12:205
Birch mice, **16:211–224,** 16:*218*, 16:223*t*–224*t*
Birds
vs. bats, 12:58–59
flight, 12:52
humans and, 12:184
mammal predation and, 12:189–191
vs. mammals, 12:6
Birth control, in zoos, 12:207
Bishop's fossorial spiny rats, 16:450, 16:458*t*
Bison, 12:*50*, 12:*104*, 12:*130*, 15:263, **16:11–25,** 16:*17*
behavior, 12:*146*, 15:271, 16:5, 16:13–14
conservation status, 16:14–15
distribution, 12:131, 16:*11*, 16:13
evolution, 15:138, 16:11
feeding ecology, 16:14, 16:24*t*–25*t*
habitats, 16:13, 16:24*t*–25*t*
humans and, 12:139, 16:8, 16:15
migrations, 12:164, 12:168
physical characteristics, 15:140, 15:267, 16:11–13
reproduction, 12:*104*, 15:272, 16:14
species of, 16:18–25*t*

Bison *(continued)*
taxonomy, 16:11
in zoos, 12:210
See also specific types of bison
Bison spp. *See* Bison
Bison bison. See American bison
Bison bonasus. See Wisents
Biswamoyopterus biswasi. See Namdapha flying squirrels
Black agoutis, 16:415*t*
See also Mexican black agoutis
Black-and-rufous sengis, 16:522, 16:524, 16:531*t*
Black and white bears. *See* Giant pandas
Black-and-white colobus, 14:7
Black-backed jackals. *See* Silverback jackals
Black-backed squirrels. *See* Peters' squirrels
Black-bearded flying foxes. *See* Big-eared flying foxes
Black bearded sakis. *See* Bearded sakis
Black-bearded tomb bats, 13:*357*, 13:364*t*
Black bears, 14:297, 14:298
See also American black bears; Asiatic black bears
Black-bellied hamsters, 16:127, 16:239, 16:*240*, 16:*241*, 16:*242*, 16:243, 16:*244*, 16:*245*–246
Black-capped capuchins, 14:101, 14:102, 14:104–105, 14:106, 14:*107*, 14:110–*111*
Black-capped marmots, 16:*146*, 16:159*t*
Black colobus, 14:174, 14:175, 14:184*t*
Black crested gibbons, 14:207, 14:210, 14:*216*, 14:*220*, 14:221
Black dolphins, 15:56*t*
Black duikers, 16:*75*, 16:*78*, 16:80–*81*
Black-eared flying foxes. *See* Blyth's flying foxes
Black-faced black spider monkeys. *See* Peruvian spider monkeys
Black-faced impalas, 16:34
Black-faced kangaroos. *See* Western gray kangaroos
Black-faced lion tamarins, 14:11, 14:124, 14:132*t*
Black flying foxes, 12:*5*, 13:*320*, 13:*324*, 13:328–*329*
Black flying squirrels, 16:141*t*
Black-footed cats, 14:370, 14:390*t*
Black-footed ferrets, 12:*115*, 12:204, 12:214, 12:223, 14:*261*, 14:262, 14:321, 14:323, 14:324
Black-footed squirrels. *See* Yucatán squirrels
Black four-eyed opossums, 12:265*t*
Black-fronted duikers, 16:84*t*
Black-fronted titis, 14:148
Black gibbons. *See* Black crested gibbons
Black-handed spider monkeys. *See* Geoffroy's spider monkeys
Black-headed night monkeys, 14:135, 14:141*t*
Black-headed squirrel monkeys. *See* Bolivian squirrel monkeys
Black-headed uacaris. *See* Black uakaris
Black howler monkeys, 14:*159*, 14:166*t*
Black kangaroos. *See* Black wallaroos
Black lechwes, 16:34
Black lemurs, 14:*50*, 14:*53*, 14:*57*–58
Black-lipped pikas. *See* Plateau pikas
Black mangabeys, 14:194, 14:205*t*
Black-mantled tamarins, 14:118
Black miniature donkeys, 12:*172*
Black muntjacs, 15:*344*, 15:346, 15:*347*, 15:*350*–351
Black-necked rock hyraxes, 15:189*t*
Black porpoises. *See* Burmeister's porpoises
Black rats, 12:*186*, 12:190, 16:*123*, 16:*126*, 16:127, 16:253, 16:*254*, 16:*256*, 16:270
Black rhinoceroses, 15:*256*, 15:259–261
behavior, 15:218, 15:*250*, 15:*251*, 15:*253*
conservation status, 15:*222*, 15:254
distribution, 15:*260*
feeding ecology, 12:*118*, 15:220
habitats, 15:*252*
physical characteristics, 12:*47*
reproduction, 15:*252*
Black-rumped agoutis, 16:407, 16:415*t*
Black sea lions. *See* California sea lions
Black-shouldered bats. *See* Evening bats
Black-shouldered opossums, 12:249–253, 12:264*t*
Black spider monkeys, 14:*157*, 14:167*t*
Black-spotted cuscuses, 13:39, 13:60 13:*62*, 13:*63*, 13:64
Black squirrel monkeys. *See* Blackish squirrel monkeys
Black-striped duikers. *See* Bay duikers
Black-striped wallabies, 13:101*t*
Black-tailed deer, 15:*380*, 15:*386*, 15:*390*–391
evolution, 15:379
guilds of, 12:118
habitats, 15:383
humans and, 15:385
migrations, 12:165
physical characteristics, 12:38, 15:*381*
reproduction, 15:*383*, 15:384
Black-tailed hairy dwarf porcupines, 16:365, 16:374*t*
Black-tailed hutias, 16:467*t*
Black-tailed jackrabbits, 16:484, 16:*485*, 16:487, 16:*508*, 16:515*t*
Black-tailed phascogales. *See* Brush-tailed phascogales
Black-tailed prairie dogs, 16:*144*, 16:*147*, 16:*149*, 16:154–155
behavior, 16:*145*, 16:*146*
conservation status, 16:148
distribution, 16:*153*
habitats, 12:*75*
Black tree kangaroos, 13:32
Black tufted-ear marmosets, 14:117
Black uakaris, 14:*144*, 14:145, 14:153*t*
Black wallabies. *See* Swamp wallabies
Black wallaroos, 13:101*t*
Black wildebeests, 16:27, 16:28, 16:29, 16:32, 16:33, 16:*35*, 16:*38*–*39*
Blackbacks, 14:228
Blackbucks, 12:136, 16:2, 16:5, 16:48, 16:56*t*
Blackfish, 15:2, 15:44
Blackish squirrel monkeys, 14:101, 14:102, 14:105–106, 14:*107*, 14:*109*–110
Bladdernose seals. *See* Hooded seals
Blainville's beaked whales, 15:*60*, 15:61, 15:*62*, 15:*63*, 15:*66*–67
Blanford's foxes, 14:*275*, 14:*280*
Blanford's fruit bats, 13:351*t*
Blarina spp., 13:248
Blarina brevicauda. See Northern short-tailed shrews
Blarina carolinensis. See Southern short-tailed shrews
Blarina hylophaga. See Elliot's short-tailed shrews
Blarinella spp., 13:248
Blarinella quadraticauda. See Chinese short-tailed shrews
Blarinellini, 13:247
Blarinini, 13:247
Blarinomys spp. *See* Shrew mice
Blarinomys breviceps. See Brazilian shrew mice
Blasius's horseshoe bats, 13:393, 13:*394*, 13:*398*–399
Blastocerus spp., 15:382
Blastocerus campestris. See Pampas deer
Blastocerus dichotomus. See Marsh deer
Blaxter, K. L., 15:145
Blesboks, 16:27, 16:29, 16:33, 16:*35*, 16:*39*–40
Blind mole-rats, 12:70, 12:73, 12:74, 12:115, 12:116, 16:123, 16:124
circadian rhythms and, 12:74–75
orientation, 12:75
vision, 12:76–77
Blind river dolphins. *See* Ganges and Indus dolphins
Blind tree mice. *See* Malabar spiny dormice
Blood platelets, 12:8
See also Physical characteristics
Blossom bats, 13:333, 13:*336*, 13:*339*, 13:*346*–347
Blotched genets, 14:*336*, 14:344*t*
Blowholes, 12:67
Blubber, 12:66
See also Physical characteristics; Whales
Blue and white bearded wildebeests. *See* Blue wildebeests
Blue antelopes, 16:28
Blue duikers, 15:272, 16:*74*, 16:84*t*
Blue-eyed lemurs, 14:*2*
Blue fliers. *See* Red kangaroos
Blue hares. *See* Mountain hares
Blue monkeys, 12:38, 14:205*t*
Blue sheep, 16:87, 16:92–93, 16:94, 16:*96*, 16:*99*, 16:100–101
Blue whales, 12:63, 15:130*t*
behavior, 15:6, 15:130*t*
conservation status, 15:9, 15:125, 15:130*t*
distribution, 15:5, 15:*121*, 15:130*t*
feeding ecology, 15:8, 15:123, 15:130*t*
habitats, 15:130*t*
humans and, 15:125
physical characteristics, 12:12, 12:63, 12:79, 12:82, 12:213, 15:4, 15:120, 15:130*t*
reproduction, 15:124
Blue wildebeests, 16:*29*, 16:42*t*
behavior, 16:5, 16:31
coevolution and, 12:216
conservation status, 12:220–221
distribution, 16:29
feeding ecology, 16:32
habitats, 16:29–30
humans and, 16:34
migrations, 12:*114*, 12:*166*, 12:*169*, 16:6
reproduction, 16:*28*
taxonomy, 16:27
Bluebucks, 16:27, 16:33, 16:41*t*
Blunt-eared bats. *See* Peruvian crevice-dwelling bats
Blyth's flying foxes, 13:*323*, 13:*326*
Blyth's horseshoe bats, 13:*388*, 13:389, 13:390, 13:391

Bobcats, 12:*50*, 14:370, 14:*373*, 14:*378*, 14:*384*, 14:389
Bobrinski's jerboas, 16:212, 16:213, 16:216, 16:*218*, 16:*219*, 16:222
Bocages mole-rats, 16:349*t*
Bodenheimer's pipistrelles, 13:313
Body fat, 12:120
in hominids, 12:23–24
in neonates, 12:125
storage of, 12:19, 12:174
See also Physical characteristics
Body mass, 12:11–12, 12:24–25
Body size, 12:41
bats, 12:58–59
blue whales, 12:213
competition and, 12:117
domestication and, 12:174
field studies and, 12:200–202
Ice Ages and, 12:17–25
life histories and, 12:97–98
nocturnal mammals and, 12:115
reproduction and, 12:15
sexual dimorphism and, 12:99
tropics and, 12:21
See also Physical characteristics
Body temperature
body size and, 12:21
sperm and, 12:91, 12:103
subterranean mammals, 12:73–74
See also Thermoregulation
Bogdanowicz, W., 13:387
Bohor reedbucks, 16:27, 16:29, 16:30, 16:34
Boinski, S., 14:104
Bolivian chinchilla rats, 16:443, 16:444, 16:445, 16:*446*, 16:*447*, 16:448
Bolivian hemorrhagic fever, 16:270
Bolivian red howler monkeys, 14:167*t*
Bolivian squirrel monkeys, 14:101, 14:102, 14:103, 14:*107*, 14:*108*
Bolivian tuco-tucos, 16:430*t*
Bones, 12:41
bats, 12:58–59
growth of, 12:10
See also Physical characteristics
Bongos, 15:138, 15:*141*, 16:4, 16:*5*, 16:11
Bonn Convention for the Conservation of Migratory Animals
Brazilian free-tailed bats, 13:493
free-tailed bats, 13:487
Bonnet macaques, 14:194
Bonobos, 14:*236*, 14:*239*–240
behavior, 14:228, 14:231–232
conservation status, 14:235
distribution, 14:227
evolution, 12:33, 14:225
feeding ecology, 14:232–233
habitats, 14:227
humans and, 14:235
language, 12:162
physical characteristics, 14:226–227
reproduction, 14:235
taxonomy, 14:225
Bonteboks, 16:27, 16:29, 16:34, 16:*35*, 16:*39*–40
Boocercus euryceros, 15:141
Boodies, 13:74, 13:77, 13:79–*80*
Boraker, D. K., 16:436
Border collies, 14:288–289
Borean long-tailed porcupines. *See* Long-tailed porcupines
Bornean bay cats. *See* Bay cats
Bornean gibbons. *See* Mueller's gibbons
Bornean orangutans, 12:*156*, 12:222, 14:225, 14:*236*, 14:*237*
Bornean smooth-tailed tree shrews, 13:292, 13:*293*, 13:*294–295*
Bornean tree shrews, 13:292, 13:297*t*
Bornean yellow muntjacs, 15:346, 15:*348*, 15:*349*, 15:352
Borneo short-tailed porcupines. *See* Thick-spined porcupines
Bos spp., 16:11, 16:13, 16:14
Bos domestica. *See* Bali cattle
Bos frontalis. *See* Gayals
Bos gaurus. *See* Gaurs
Bos grunniens. *See* Yaks
Bos indicus. *See* Brahma cattle
Bos javanicus. *See* Bantengs
Bos mutus. *See* Yaks
Bos primigenius. *See* Aurochs
Bos sauveli. *See* Koupreys
Bos taurus. *See* Aurochs; Domestic cattle
Boselaphini, 16:1, 16:11
Boselaphus tragocamelus. *See* Nilgais
Bothma, J. P., 15:171
Botos, 15:5, 15:*8*, **15:27–31**, 15:*28*, 15:*29*, 15:*30*
Botta's pocket gophers. *See* Valley pocket gophers
Bottleheads. *See* Northern bottlenosed whales
Bottleneck species, 12:221–222
Bottlenosed dolphins, 12:*27*, 12:*67*, 12:68, 15:2
behavior, 12:*143*, 15:44, 15:45, 15:46
chemoreception, 12:84
conservation status, 15:49–50
distribution, 15:5
echolocation, 12:*87*
feeding ecology, 15:*47*
habitats, 15:44
humans and, 15:10–11
language, 12:*161*
physical characteristics, 12:67–68
reproduction, 15:48
See also Common bottlenosed dolphins
Bottlenosed whales, 15:7
See also Northern bottlenosed whales; Southern bottlenosed whales
Bovidae, **16:1–106**
Antilopinae, 16:1, **16:45–58**, 16:*49*, 16:*56t*–*57t*
behavior, 12:145, 12:146–147, 16:5–6
Bovinae, 16:1, **16:11–25**, 16:*16*, 16:*17*, 16:24*t*–*25t*
conservation status, 16:8
digestive systems of, 12:14
distribution, 16:4
evolution, 15:137, 15:138, 15:265, 15:266, 15:267, 16:1–2
feeding ecology, 15:142, 15:271, 16:6–7
habitats, 16:4–5
Hippotraginae, 16:1, **16:27–43**, 16:*35*, 16:41*t*–*42t*
horns, 12:40
humans and, 16:8–9
Neotraginae, 16:1, **16:59–72**, 16:*65*, 16:71*t*
physical characteristics, 15:142, 15:267, 15:268, 16:2–4
reproduction, 15:143, 16:7–8
taxonomy, 16:1–2
See also Caprinae; Duikers
Bovinae, 16:1, **16:11–25**, 16:*16*, 16:*17*, 16:24*t*–*25t*
Bovini, 16:1, 16:11
Bowdoin's beaked whales. *See* Andrew's beaked whales
Bowhead whales, **15:107–118**, 15:*114*, 15:*115*–116
behavior, 15:109–110
conservation status, 15:9–10, 15:111–112
distribution, 15:5, 15:*107*, 15:109
evolution, 12:*65*, 15:107
feeding ecology, 15:110
habitats, 15:109
humans and, 15:112
migrations, 12:87, 12:170
physical characteristics, 15:107, 15:109
reproduction, 15:110–111
species of, 15:*115*–118
taxonomy, 15:107
Brachydelphis spp., 15:23
Brachylagus spp. *See* Pygmy rabbits
Brachylagus idahoensis. *See* Pygmy rabbits
Brachyphylla cavernarum. *See* Antillean fruit-eating bats
Brachyphyllinae, 13:413, 13:415
Brachyteles spp. *See* Muriquis
Brachyteles arachnoides. *See* Southern muriquis
Brachyteles hypoxanthus. *See* Northern muriquis
Brachyuromy betsileonensis. *See* Malagasy reed rats
Bradbury, J. W., 13:342
Bradypodidae. *See* Three-toed tree sloths
Bradypus spp. *See* Three-toed tree sloths
Bradypus infuscatus. *See* Three-toed tree sloths
Bradypus pygmaeus. *See* Monk sloths
Bradypus torquatus. *See* Maned sloths
Bradypus tridactylus. *See* Pale-throated three-toed sloths
Bradypus variegatus. *See* Brown-throated three-toed sloths
Brahma cattle, 16:*4*
Braincase, 12:8, 12:9
Brains, 12:6, 12:36, 12:49
enlargement of, 12:9
evolution of, 12:19
learning and, 12:141
life spans and, 12:97–98
Neanderthals, 12:24
placentation and, 12:94
size of, 12:149–150
See also Physical characteristics
Brandt's hamsters, 16:247*t*
Brandt's hedgehogs, 13:213*t*
Branick's giant rats. *See* Pacaranas
Braude, S., 12:71
Brazilian agoutis. *See* Red-rumped agoutis
Brazilian free-tailed bats, 13:*484*, 13:*485*, 13:*486*, 13:*489*, 13:*492*, 13:493
behavior, 13:312, 13:484–485
distribution, 13:310
feeding ecology, 13:485, 13:486
humans and, 13:487–488
maternal recognition, 12:85
reproduction, 13:315, 13:486–487
Brazilian guinea pigs, 16:*392*
Brazilian shrew mice, 16:*271*, 16:*273*, 16:274
Brazilian spiny tree rats, 16:451, 16:458*t*
Brazilian tapirs. *See* Lowland tapirs
Brazilian three-banded armadillos, 13:192*t*

Breeds, dog, 14:288–289
Brewer's moles. *See* Hairy-tailed moles
Bridges's degus, 16:433, 16:440*t*
Bridled nail-tailed wallabies, 13:*40*, 13:*92*, 13:*98*–99
Brindled bandicoots. *See* Northern brown bandicoots
Bristle-spined porcupines. *See* Thin-spined porcupines
Broad-faced potoroos, 13:39
Broad-footed moles, 13:287*t*
Broad-striped dasyures, 12:299*t*
Brocket deer, 15:379, 15:382, 15:384, 15:396*t*
Brontotheres, 15:136
Bronx Zoo, 12:205
Bronze quolls, 12:279
Broom, R., 16:329
Brophy, B., 15:151
Brow-antlered deer. *See* Eld's deer
Brown, J. H., 16:200
Brown agoutis, 16:*408*
Brown antechinuses, 12:282, 12:*293*, 12:*296*–297
Brown-bearded sheath-tail bats, 13:*356*
Brown bears, 12:*23*, 14:*301*, 14:*303*
behavior, 12:145, 14:258, 14:297, 14:298
conservation status, 14:300
distribution, 14:296
evolution, 14:*295*–*296*
feeding ecology, 12:*124*, 14:260, 14:299
habitats, 14:297
physical characteristics, 14:296
taxonomy, 14:295
Brown brocket deer. *See* Gray brocket deer
Brown capuchins. *See* Black-capped capuchins
Brown dorcopsises, 13:102*t*
Brown fat, 12:113
Brown four-eyed opossums, 12:250–253, 12:264*t*
Brown greater bushbabies, 14:*24*, 14:25, 14:26, 14:*27*, 14:*30*, 14:31
Brown hairy dwarf porcupines, 16:366, 16:374*t*
Brown hares. *See* European hares
Brown-headed spider monkeys, 14:167*t*
Brown howler monkeys, 14:159, 14:160, 14:166*t*
Brown hyenas, 14:*361*, 14:*362*, 14:*363*, 14:366
behavior, 14:259, 14:260, 14:362
distribution, 14:*365*
evolution, 14:359
feeding ecology, 14:260–261, 14:*360*, 14:362
taxonomy, 14:359–360
Brown lemmings, 16:*228*
Brown lemurs, 14:8, 14:52, 14:*54*, 14:*55*, 14:59
Brown long-eared bats, 13:*502*, 13:*504*, 13:*513*–514
Brown mouse lemurs. *See* Red mouse lemurs
Brown murine bats. *See* Brown tube-nosed bats
Brown rats, 12:105–106, 12:*184*, 12:*187*, 12:190, 16:250–251, 16:*253*, 16:*254*, 16:*256*, 16:257–258
Brown-throated three-toed sloths, 13:*149*, 13:*161*, 13:*163*, 13:*164*, 13:*166*, 13:*167*
Brown titis, 14:*145*, 14:153*t*
Brown tube-nosed bats, 13:*523*, 13:524
Brown University, 12:57
Brumbies, 12:176
Brush mice, 16:266
Brush-tailed bettongs, 13:39, 13:74, 13:*75*, 13:76, 13:77, 13:*78*–79
Brush-tailed phascogales, 12:*290*, 12:*292*, 12:297
Brush-tailed rock wallabies, 13:85, 13:*92*, 13:*96*–97
Brushtailed possums. *See* Common brushtail possums
Bryde's whales, 15:94, 15:120, 15:122–*126*, 15:*128*, 15:129
Bubal hartebeests, 16:33
Bubalus spp., 16:11
Bubalus bubalis. See Water buffaloes
Bubalus depressicornis. See Anoas
Bubalus mephistopheles. See Short-horned water buffaloes
Budorcas taxicolor. See Takins
Budorcas taxicolor bedfordi. See Golden takins
Buettner-Janusch, 14:75
Buff-cheeked gibbons. *See* Golden-cheeked gibbons
Buffaloes, **16:11–25,** 16:*17*, 16:24*t*–25*t*
See also African buffaloes; Water buffaloes
Buffenstein, Rochelle, 12:74
Buffy flower bats, 13:*424*, 13:*425*, 13:426–427
Buffy-headed marmosets, 14:124, 14:*125*, 14:*130*
Buffy tufted-ear marmosets, 14:117, 14:121
Bugtilemur spp., 14:35
Bukovin mole rats, 16:297*t*
Bulldog bats, 12:*53*, **13:443–451,** 13:*447*, 13:*448*
Buller's pocket gophers, 16:*191*, 16:*192*, 16:194–195
Bulmer's fruit bats, 13:348*t*
Bumblebee bats. *See* Kitti's hog-nosed bats
Bunolagus monticularis. See Riverine rabbits
Bunopithecus spp., 14:207
Bunyoro rabbits, 16:482, 16:505
Buoyancy, 12:68
Burchell's zebras. *See* Plains zebras
Burgess shale, 12:64
Burmeister's porpoises, 15:33–36, 15:*37*, 15:*38*, 15:39
Burramyidae. *See* Pygmy possums
Burramys spp. *See* Mountain pygmy possums
Burramys parvus. See Mountain pygmy possums
Burros, 12:182
Burrowing bettongs. *See* Boodies
Burrowing wombats, 13:32
Burunduks. *See* Siberian chipmunks
Bush cows. *See* Lowland tapirs
Bush dassies. *See* Bush hyraxes
Bush dogs, 14:266, 14:267, 14:269, 14:270, 14:*274*, 14:284*t*
Bush duikers, 16:*75*, 16:*76*, 16:77
Bush hyraxes, 15:177–*179*, 15:180–*185*, 15:*186*, 15:187–188
Bush pigs, 12:136, 15:278, 15:280, 15:*283*, 15:*284*–285
Bush rats, 16:251
Bushbabies, 14:1, **14:23–34,** 14:*27*
behavior, 14:25–26
conservation status, 14:26
distribution, 14:5, 14:*23*, 14:24
evolution, 14:2, 14:3, 14:23
feeding ecology, 14:9, 14:26
habitats, 14:6, 14:24
humans and, 14:26
physical characteristics, 14:23–24
reproduction, 14:26
species of, 14:28–32, 14:32*t*–33*t*
taxonomy, 14:4, 14:23
Bushbucks, 16:11, 16:14, 16:24*t*
Bushdogs. *See* Tayras
Bushveld gerbils, 16:296*t*
Bushveld horseshoe bats, 13:390
Bushveld sengis, 16:*518*, 16:519, 16:531*t*
Bushy-tailed gundis. *See* Speke's pectinators
Bushy-tailed hutias. *See* Black-tailed hutias
Bushy-tailed mongooses, 14:357*t*
Bushy-tailed opossums, 12:249–253, 12:*255*, 12:*259*
Bushy-tailed rats, western Malagasy, 16:*284*
Bushy-tailed woodrats, 16:*124*, 16:277*t*
Butler, Percy M., 13:289
Butterfly bats, 13:515*t*
See also Gervais' funnel-eared bats

C

Cabassous spp. *See* Naked-tailed armadillos
Cabassous centralis. See Northern naked-tailed armadillos
Cabassous chacoensis. See Chacoan naked-tail armadillos
Cabassous tatouay. See Greater naked-tailed armadillos
Cabassous unicinctus. See Southern naked-tailed armadillos
Cabral, Pedro, 15:242
Cacajao spp. *See* Uakaris
Cacajao calvus. See Bald uakaris
Cacajao calvus calvus. See White bald uakaris
Cacajao calvus novaesi. See Novae's bald uakaris
Cacajao calvus rubicundus. See Red bald uakaris
Cacajao calvus ucayalii. See Ucayali bald uakaris
Cacajao melanocephalus. See Black uakaris
Cacomistles, 14:310, 14:316*t*
Cactus mice, 16:*264*
Caenolestes spp., 12:267–268
Caenolestes fuliginosus. See Silky shrew opossums
Caenolestidae. *See* Shrew opossums
Caenotheriidae, 15:264
Caipora bambuiorum, 14:155
Calabar angwantibos, 14:16, 14:*17*, 14:*19*–20
Calamian deer, 15:360
Calcium, subterranean mammals and, 12:74
Calcochloris spp., 13:216
Calcochloris obtusirostris. See Yellow golden moles
Calicivirus, 12:187–188, 12:223
California gray squirrels. *See* Western gray squirrels
California leaf-nosed bats, 12:79, 13:315, 13:*417*, 13:419, 13:420, 13:*422*, 13:*423*
California meadow voles, 16:125–126
California mice, 16:125–126, 16:278*t*
California sea lions, 12:124, 14:394, 14:396, 14:*397*, 14:399, 14:*401*, 14:*404*, 14:405
Callicebinae. *See* Titis
Callicebus spp. *See* Titis
Callicebus barbarabrownae. See Northern Bahian blond titis
Callicebus brunneus. See Brown titis

INDEX

Callicebus coimbrai. See Coimbra's titis
Callicebus cupreus. See Red titis
Callicebus donacophilus, 14:143, 14:144
Callicebus medemi. See Medem's collared titis
Callicebus melanochir. See Southern Bahian masked titis
Callicebus modestus, 14:143
Callicebus moloch. See Dusky titi monkeys
Callicebus nigrifrons. See Black fronted titis
Callicebus oenathe. See Andean titis
Callicebus olallae. See Beni titis
Callicebus ornatus. See Ornate titis
Callicebus personatus. See Masked titis
Callicebus torquatus. See Collared titis
Callimico goeldii. See Goeldi's monkeys
Callithrix spp. *See* Eastern Brazilian marmosets
Callithrix argentata. See Silvery marmosets
Callithrix aurita. See Buffy tufted-ear marmosets
Callithrix flaviceps. See Buffy-headed marmosets
Callithrix humeralifera. See Tassel-eared marmosets
Callithrix jacchus. See Common marmosets
Callithrix penicillata. See Black tufted-ear marmosets
Callitrichidae, **14:115–133**, 14:*125*, 14:*126*
  behavior, 12:147, 14:116–120, 14:*123*
  claws, 12:39
  conservation status, 14:124
  distribution, 14:*115*, 14:116
  evolution, 14:115
  feeding ecology, 14:120–121
  habitats, 14:116
  humans and, 14:124
  physical characteristics, 14:115–116
  reproduction, 14:121–124
  species of, 14:*127*–131, 14:132*t*
  taxonomy, 14:4, 14:115
Callorhinus spp. *See* Northern fur seals
Callorhinus ursinus. See Northern fur seals
Callosciurus spp., 16:163
Callosciurus erythraeus. See Pallas's squirrels
Callosciurus pygerythrus, 16:167
Callosciurus quinquestriatus, 16:167
Calls. *See* Behavior; Vocalizations
Calomys spp., 16:269, 16:270
Calomys laucha, 16:267, 16:270
Calomys musculinus, 16:269, 16:270
Calomys venustus, 16:269
Calomyscinae, 16:282
Calomyscus bailwardi. See Mouse-like hamsters
Calomyscus hotsoni. See Hotson's mouse-like hamsters
Calomyscus mystax. See Afghan mouse-like hamsters
Calomyscus tsolovi. See Tsolov's mouse-like hamsters
Calomyscus urartensis. See Urartsk mouse-like hamsters
Caluromyinae, 12:249
Caluromys spp. *See* Woolly opossums
Caluromys philander. See Bare-tailed woolly opossums
Caluromysiops spp. *See* Black-shouldered opossums
Caluromysiops irrupta. See Black-shouldered opossums
Camas pocket gophers, 16:195*t*
Camelidae, **15:313–323**, 15:*319*
  behavior, 15:316–317
  conservation status, 15:317–318
  distribution, 12:132, 12:134, 15:*313*, 15:315
  domestication of, 15:150–151
  evolution, 15:313–314
  feeding ecology, 15:317
  habitats, 15:315–316
  humans and, 15:318
  physical characteristics, 15:314–315
  reproduction, 15:317
  species of, 15:*320–323*
  taxonomy, 15:267, 15:313–314
Camelini, 15:313
Camels, 15:*151*, 15:263, **15:313–323**
  behavior, 15:316–317
  conservation status, 15:317–318
  distribution, 15:*313*, 15:315
  evolution, 15:136, 15:137, 15:138, 15:265, 15:313–314
  feeding ecology, 15:271, 15:317
  habitats, 15:315–316
  humans and, 15:273, 15:318
  neonatal milk, 12:127
  physical characteristics, 15:314–315
  reproduction, 15:143, 15:317
  species of, 15:320–323
  taxonomy, 15:313–314
  thermoregulation, 12:113
Camelus spp. *See* Camels
Camelus bactrianus. See Bactrian camels
Camelus dromedarius. See Dromedary camels
Camelus ferus. See Bactrian camels
Camelus thomasi, 12:179
Cameras, in field studies, 12:194–200
Cameroon scaly-tails, 16:300, 16:*301*, 16:*305*
Campbell's monkeys, 14:204*t*
Campfire program (Zimbabwe), 12:219
Canada lynx, 14:371, 14:*378*, 14:*383*, 14:388–389
Canadian beavers. *See* North American beavers
Canadian caribou, 12:180
Canadian porcupines. *See* North American porcupines
Canary shrews, 13:276*t*
Cane mice, 16:270
Cane rats, 16:121, **16:333–338**, 16:*336*
Canidae, **14:265–285**, 14:*275*
  behavior, 12:145, 12:147, 12:148, 14:258, 14:267–269
  conservation status, 14:262, 14:272–273
  distribution, 12:131, 14:*265*, 14:267
  evolution, 14:265–266
  feeding ecology, 14:261, 14:269–271
  habitats, 14:267
  humans and, 14:262, 14:263, 14:273–274
  as keystone species, 12:217
  physical characteristics, 14:266–267
  reproduction, 14:261, 14:271–272
  species of, 14:*276*–283, 14:283*t*–284*t*
  taxonomy, 14:256, 14:265–266
Canines (teeth), 12:46, 12:99
  See also Teeth
Canis spp., 14:265
Canis adustus. See Side-striped jackals
Canis aureus. See Golden jackals
Canis dirus. See Dire wolves
Canis familiaris. See Domestic dogs
Canis familiaris dingo. See Dingos
Canis latrans. See Coyotes
Canis lupus. See Gray wolves
Canis lupus lupus. See Eurasian wolves
Canis lupus pallipes, Indian wolves
Canis mesomelas. See Silverback jackals
Canis rufus. See Red wolves
Canis simensis. See Ethiopian wolves
Cansumys spp. *See* Gansu hamsters
Canthumeryx spp., 15:265
Cape buffaloes. *See* African buffaloes
Cape dune mole-rats, 16:345, 16:*346*, 16:*347*
Cape fur seals, 12:*106*, 14:394
Cape golden moles, 13:197, 13:215, 13:217, 13:222*t*
Cape gray mongooses, 14:350
Cape ground squirrels. *See* South African ground squirrels
Cape grysboks, 16:63, 16:71*t*
Cape grysbucks. *See* Cape grysboks
Cape hares, 16:*483*
Cape horseshoe bats, 13:389, 13:392, 13:*395*, 13:*398*
Cape hunting dogs. *See* African wild dogs
Cape hyraxes. *See* Rock hyraxes
Cape mole-rats, 12:74, 16:339, 16:*340*, 16:342, 16:343, 16:344, 16:345, 16:*346*, 16:*347*
Cape mountain zebras, 15:222
Cape pangolins. *See* Ground pangolins
Cape porcupines. *See* South African porcupines
Cape rock sengis. *See* Cape sengis
Cape rousettes. *See* Egyptian rousettes
Cape sengis, 16:522, 16:531*t*
Caperea marginata. See Pygmy right whales
Capped gibbons. *See* Pileated gibbons
Capped langurs, 12:*119*, 14:175
Capra spp., 16:87, 16:88, 16:90, 16:91, 16:92, 16:95
Capra aegagrus. See Bezoar goats; Wild goats
Capra aegagrus hircus. See Valais goats
Capra caucasica. See West Caucasian turs
Capra cylindricornis. See East Caucasian turs
Capra falconeri. See Markhors
Capra falconeri heptneri. See Tajik markhors
Capra hircus. See Domestic goats
Capra ibex. See Alpine ibex; Nubian ibex
Capra prisca. See Angora goats
Capra pyrenaica. See Spanish ibex
Capra sibirica. See Siberian ibex
Capra walie. See Walia ibex
Capreolus spp. *See* Roe deer
Capreolus capreolus. See European roe deer
Capreolus pygargus. See Siberian roe deer
Capricornis spp., 16:2, 16:87
Capricornis crispus. See Japanese serows
Capricornis sumatraensis. See Serows
Capricornis swinhoei. See Formosan serows
Caprinae, 12:134, **16:87–105**, 16:*96*, 16:97
  behavior, 16:92–93
  conservation status, 16:94–95
  distribution, 16:*87*, 16:90–91
  evolution, 16:1, 16:87–89
  feeding ecology, 16:93–94
  habitats, 15:269, 16:91–92, 16:98–102
  humans and, 16:95
  physical characteristics, 16:89–90
  reproduction, 16:7, 16:8, 16:94
  species of, 16:98–104*t*
  taxonomy, 16:1, 16:87–89

Caprini, 16:87, 16:88–89, 16:90
Capriolinae. *See* New World deer
Caprolagus spp. *See* Hispid hares
Caprolagus hispidus. See Hispid hares
Capromyidae. *See* Hutias
Capromys spp., 16:462, 16:463
Capromys pilorides. See Cuban hutias
Captive breeding, 12:223–224
See also Domestication; Zoos
Capuchins, 14:4, **14:101–107,** 14:*107,* **14:110–112**
behavior, 14:7, 14:103–104, 14:250
conservation status, 14:11, 14:105, 14:106
distribution, 14:*101,* 14:103
evolution, 14:101
feeding ecology, 14:9, 14:104–105
habitats, 14:103
humans and, 12:*174,* 14:106
memory, 12:153
physical characteristics, 14:101–102
reproduction, 14:105
species of, 14:*110*–112
taxonomy, 14:101
Capybaras, 12:14, 16:123, 16:126, **16:401–406,** 16:*402,* 16:*403,* 16:*404,* 16:*405*
Caracal (Felis) caracal. See Caracals
Caracals, 14:372, 14:*377,* 14:*387*–388
Carbohydrates, 12:120
See also Feeding ecology
Carbon cycle, 12:213
Carbon dioxide, subterranean mammals and, 12:72–73, 12:114
Cardim, Fernão, 16:404
Cardiocraniinae, 16:211, 16:212
Cardiocranius paradoxus. See Five-toed pygmy jerboas
Cardioderma cor. See Heart-nosed bats
Cardiovascular system, 12:45–46, 12:60
See also Physical characteristics
Caribbean manatees. *See* West Indian manatees
Caribbean monk seals. *See* West Indian monk seals
Caribou. *See* Reindeer
Carleton, M. D., 16:282
Carlocebus spp., 14:143
Carnivora, 12:14, **14:255–263**
behavior, 12:141, 12:145, 14:257–260
color vision, 12:79
conservation status, 14:262
digestive system, 12:120, 12:123–124
distribution, 12:129, 12:131, 12:132, 12:136, 14:257
evolution, 14:255–256
feeding ecology, 14:260–261
guilds, 12:117
habitats, 14:257
hearing, 12:82
humans and, 14:262–263
locomotion, 12:14
neonatal milk, 12:126–127
physical characteristics, 14:256–257
reproduction, 12:90, 12:92, 12:94, 12:95, 12:98, 12:109, 12:110, 14:261–262
taxonomy, 14:255, 14:256
teeth of, 12:14
water balance, 12:113
See also specific carnivores
Carollia castanea. See Chestnut short-tailed bats
Carollia perspicillata. See Seba's short-tailed bats
Carolliinae, 13:413, 13:417–418, 13:420
Carpal glands, 12:37
See also Physical characteristics
Carpitalpa spp. *See* Arend's golden moles
Carpitalpa arendsi. See Arend's golden moles
Carr, Archie, 12:218, 12:223
Carrying capacity, of zoos, 12:207
Carterodon sulcidens. See Cerrado rats
Caseasauria, 12:10
Casinycteris argynnis. See Short-palate fruit bats
Caspian seals, 12:138, 14:422, 14:435*t*
Castellarini, F., 16:269
Castor spp. *See* Beavers
Castor canadensis. See North American beavers
Castor fiber. See Eurasian beavers
Castoridae. *See* Beavers
Castoroides spp., 16:177
Castro-Vaszuez, A., 16:269
Cat bears. *See* Giant pandas
Cat Fanciers' Association, 14:291
Cat squirrels. *See* Fox squirrels
Catagonus spp., 15:291
Catagonus wagneri. See Chacoan peccaries
Catamounts. *See* Pumas
Catarrhini, 14:1
Caterodon spp., 16:450
Caterodon sulcidens. See Cerrado rats
Cathemeral lemurs, 14:8
Cathemerality, 14:8
Catopuma badia. See Bay cats
Catopuma temminckii. See Asiatic golden cats
Cats, **14:369–392,** 14:*377,* 14:*378*
behavior, 12:145, 14:371–372
conservation status, 14:262, 14:374–375
digestive system, 12:120
distribution, 12:131, 12:136, 14:257, 14:*369,* 14:370–371
evolution, 14:369
feeding ecology, 14:260, 14:372
feral, 12:185–186, 14:257
habitats, 14:371
humans and, 14:375–376
learning set, 12:152
neonatal requirements, 12:126
physical characteristics, 14:370
reproduction, 12:103, 12:106, 12:109, 12:111, 14:372–373
self recognition, 12:159
species of, 14:*379*–389, 14:390*t*–391*t*
taxonomy, 14:256, 14:369
vision, 12:84
in zoos, 12:210
See also Domestic cats
Cattle, 14:*247*
aurochs, 12:177, 16:*17,* 16:*21*–*22*
behavior, 16:6
Brahma, 16:4
conservation status, 12:215
digestive system, 12:122–123
distribution, 16:4
evolution, 15:265, 16:1
feeding ecology, 12:118, 15:271
humans and, 16:8
physical characteristics, 16:2
vision, 12:79
Zebu, 16:6
See also Bovidae; Domestic cattle; Ungulates; specific types of cattle
Cattle dogs, Australian, 14:292
Cavalli-Sforza, L. L., 14:250
Cave bears, Ice Age, 12:24
Cave-dwelling bats. *See* Long-fingered bats
Cave fruit bats. *See* Dawn fruit bats
Cave-hyenas, Ice Age, 12:24
Cave-lions, Ice Age, 12:24
Cavia spp. *See* Guinea pigs
Cavia aperea. See Guinea pigs
Cavia fulgida. See Shiny guinea pigs
Cavia magna. See Greater guinea pigs
Cavia porcellus. See Domestic guinea pigs
Cavia tschudii. See Montane guinea pigs
Cavies, 12:72, **16:389–400,** 16:*394,* 16:399*t*
Caviidae, 12:72, 16:123, 16:125, **16:389–400,** 16:*394,* 16:399*t*
Caviinae, 16:389–390
Caviomorphs, 16:121–126
Cebidae, 14:4, **14:101–112,** 14:*107*
behavior, 14:7, 14:103–104
conservation status, 14:11, 14:105–106
distribution, 14:*101,* 14:102–103
evolution, 14:101
feeding ecology, 14:9, 14:104–105
habitats, 14:103
humans and, 12:*174,* 14:106
physical characteristics, 14:101–102
reproduction, 14:105
species of, 14:*108*–112
taxonomy, 14:101
Cebinae, 14:101–102, 14:103
Cebu bearded pigs, 15:276, 15:279, 15:280, 15:281, 15:289*t*
Cebuella pygmaea. See Pygmy marmosets
Cebupithecia spp., 14:143
Cebus spp. *See* Capuchins
Cebus albifrons. See White-fronted capuchins
Cebus apella. See Black-capped capuchins
Cebus apella robustus, 14:106
Cebus capucinus. See White-throated capuchins
Cebus libidinosus, 14:101
Cebus nigritus, 14:101
Cebus olivaceus. See Weeper capuchins
Cebus xanthosternos. See Yellow-breasted capuchins
Celebes cuscuses. *See* Sulawesi bear cuscuses
Celebes macaques, 14:*188,* 14:193–194
Celebes pigs, 15:280, 15:282, 15:289*t*
Cellulase, 12:122
Central African hedgehogs, 13:*196,* 13:*205,* 13:*206,* 13:*207,* 13:213*t*
Central America. *See* Neotropical region
Central American agoutis, 16:*408,* 16:*410,* 16:*411*
Central American tapirs, 15:218, 15:238, 15:*239,* 15:*241,* 15:242, 15:*244,* 15:*245*–246
Central Asian gazelles, 12:134
Central Asian ground squirrels, 16:143
Central Park Zoo, New York, 12:203
Centronycteris centralis. See Shaggy-haired bats
Centurio senex. See Wrinkle-faced bats
Cephalophinae. *See* Duikers
Cephalophus spp., 16:5, 16:73
Cephalophus adersi. See Aders's duikers
Cephalophus callipygus. See Peters's duikers
Cephalophus dorsalis. See Bay duikers
Cephalophus harveyi. See Harvey's duikers
Cephalophus jentinki. See Jentink's duikers

Cephalophus leucogaster. See White-bellied duikers
Cephalophus maxwelli. See Maxwell's duikers
Cephalophus monticola. See Blue duikers
Cephalophus natalensis. See Natal duikers
Cephalophus niger. See Black duikers
Cephalophus nigrifrons. See Black-fronted duikers
Cephalophus ogilbyi. See Ogilby's duikers
Cephalophus rufilatus. See Red-flanked duikers
Cephalophus silvicultor. See Yellow-backed duikers
Cephalophus spadix. See Abbott's duikers
Cephalophus weynsi. See Weyns's duikers
Cephalophus zebra. See Zebra duikers
Cephalorhynchus spp., 15:43
Cephalorhynchus commersonii. See Commerson's dolphins
Cephalorhynchus eutropia. See Black dolphins
Cephalorhynchus hectori. See Hector's dolphins
Ceratotherium spp., 15:216, 15:249, 15:251
Ceratotherium simum. See White rhinoceroses
Cercartetus spp. *See* Pygmy possums
Cercartetus caudatus. See Long-tailed pygmy possums
Cercartetus concinnus. See Western pygmy possums
Cercartetus lepidus. See Little pygmy possums
Cercartetus nanus. See Eastern pygmy possums
Cercocebus spp., 14:188, 14:191
Cercocebus albigena. See Gray-cheeked mangabeys
Cercocebus atys. See Sooty mangabeys
Cercocebus galeritus. See Agile mangabeys
Cercocebus torquatus. See Collared mangabeys
Cercopithecidae. *See* Old World monkeys
Cercopithecinae. *See* Cheek-pouched monkeys
Cercopithecini. *See* Guenons
Cercopithecoides spp., 14:172
Cercopithecus spp. *See* Guenons
Cercopithecus ascanius. See Red-tailed monkeys
Cercopithecus campbelli. See Campbell's monkeys
Cercopithecus cephus. See Moustached guenons
Cercopithecus diana. See Diana monkeys
Cercopithecus dryas, 14:194
Cercopithecus erythrogaster, 14:193–194
Cercopithecus erythrotis, 14:194
Cercopithecus hamlyni, 14:194
Cercopithecus lhoesti. See l'Hoest's guenons
Cercopithecus mitis. See Blue monkeys
Cercopithecus petaurista. See Lesser white-nosed monkeys
Cercopithecus preussi, 14:193–194
Cercopithecus sclateri, 14:193–194
Cercopithecus solatus, 14:194
Cerdocyon thous. See Crab-eating foxes
Cerebral cortex, 12:24, 12:49
  See also Physical characteristics
Cerrado rats, 16:450–451, 16:*452*, 16:454, 16:*455*
Ceruminous glands, 12:37
Cervalces scotti. See American stag-moose
Cervidae. *See* Deer
Cervinae. *See* Old World deer
Cervoidea, 15:266
Cervus spp., 15:358
Cervus albirostris. See White-lipped deer
Cervus alfredi. See Visayan spotted deer
Cervus duvaucelii. See Barasinghas
Cervus elaphus. See Red deer
Cervus elaphus canadensis. See Elk
Cervus elaphus nannodes. See Tule elk
Cervus elaphus siamensis, 15:.361
Cervus elaphus sibiricus. See Siberian marals
Cervus elaphus xanthopygus. See Izubrs
Cervus eldi. See Eld's deer
Cervus nippon. See Sika deer
Cervus (Rusa) mariannus. See Philippine brown deer
Cervus (Rusa) timorensis. See Timor deer
Cervus schomburgki. See Schomburgk's deer
Cervus timorensis rusa. See Javan rusa deer
Cervus unicolor. See Sambars
Cetacea, **15:1–11**
  behavior, 15:6–7
  conservation status, 15:9–10
  distribution, 12:129, 12:138, 15:5
  echolocation, 12:86
  evolution, 15:2–4
  feeding ecology, 15:8
  geomagnetic fields, 12:84
  habitats, 15:5–6
  hippopotamuses and, 12:30
  humans and, 15:10–11
  locomotion, 12:44
  physical characteristics, 12:67, 15:4–5
  reproduction, 12:92, 12:94, 12:103–104, 15:9
  taxonomy, 12:14, 15:2–4, 15:133, 15:266
  See also Whales
Cetartiodactyla, 12:14, 15:266–267
Cetotheriidae, 15:3
Chacma baboons, 14:*189*, 14:*190*, 14:191, 14:193
Chacoan hairy armadillos, 13:191*t*
Chacoan naked-tail armadillos, 13:190*t*
Chacoan peccaries, 15:291, 15:*294*, 15:*295*, 15:*297*, 15:*299*–300
Chaerephon spp., 13:483
Chaerephon gallagheri, 13:487
Chaerephon jobensis. See Northern mastiff bats
Chaerephon plicata. See Wrinkle-lipped free-tailed bats
Chaerephon pumila. See Lesser-crested mastiff bats
Chaeropus spp., 13:10
Chaeropus ecaudatus. See Pig-footed bandicoots
Chaetodipus spp. *See* Coarse-haired pocket mice
Chaetodipus baileyi. See Bailey's pocket mice
Chaetodipus hispidus. See Hispid pocket mice
Chaetodipus spinatus. See Spiny pocket mice
Chaetomys spp., 16:365, 16:366, 16:367
Chaetomys subspinosus. See Thin-spined porcupines
Chaetophractus nationi. See Andean hairy armadillos
Chaetophractus vellerosus. See Small hairy armadillos
Chaetophractus villosus. See Large hairy armadillos
Chagas' disease, 12:254
Chalinolobus spp., 13:497
Chalinolobus gouldii. See Gould's wattled bats
Chalinolobus variegatus. See Butterfly bats
Chambius kasserinensis, 16:517
Chamois, 12:137, 15:271, 16:87–90, 16:*91*, 16:92, 16:94
  northern, 16:6, 16:87, 16:*96*, 16:*100*
  southern, 16:87, 16:103*t*
Chapman, C. A., 14:163
Checkered sengis, 16:*519*, 16:*521*, 16:522, 16:*523*, 16:*524*, 16:525
Cheek-pouched monkeys, **14:187–206,** 14:*195*, 14:*196*
  behavior, 14:191–192
  conservation status, 14:194
  distribution, 14:*187*, 14:191
  evolution, 14:187–189
  feeding ecology, 14:192–193
  habitats, 14:191
  humans and, 14:194
  physical characteristics, 14:189–191
  reproduction, 14:193
  species of, 14:*197*–204, 14:204*t*–205*t*
  taxonomy, 14:187–188
Cheek teeth. *See* Molar teeth; Premolars
Cheetahs, 12:*10*, 12:*93*, 12:*206*, 14:*374*, 14:*377*
  behavior, 14:258, 14:371
  bottleneck species, 12:221
  conservation status, 12:216
  distribution, 14:370
  feeding ecology, 14:372
  humans and, 14:262, 14:290
  Ice Age, 12:24
  physical characteristics, 14:370
  running speed, 12:42, 12:*45*
  species of, 14:*381*–382
  taxonomy, 14:369
Cheirogaleidae. *See* Dwarf lemurs; Mouse lemurs
Cheirogaleus spp. *See* Dwarf lemurs
Cheirogaleus major. See Greater dwarf lemurs
Cheirogaleus medius. See Western fat-tailed dwarf lemurs
Cheiromeles spp., 13:483, 13:484, 13:486
Cheiromeles parvidens. See Lesser naked bats
Cheiromeles torquatus. See Naked bats
Chelemys macronyx, 16:268
Chemical compounds, 12:120
Chemoreception, 12:84
  See also Physical characteristics
Cherrie's pocket gophers, 16:196*t*
Chestnut sac-winged bats, 13:356, 13:364*t*
Chestnut short-tailed bats, 13:433*t*
Chevrotains, **15:325–334,** 15:*330*
  behavior, 15:327–328
  conservation status, 15:329
  distribution, 15:*325*, 15:327
  evolution, 15:137, 15:265–266, 15:325–326
  feeding ecology, 15:328
  habitats, 15:327
  humans and, 15:329
  physical characteristics, 15:138, 15:267–268, 15:271, 15:326–327
  reproduction, 15:328–329
  species of, 15:*331*–333
  taxonomy, 15:267
Chibchanomys orcesi. See Water mice
Chickarees. *See* North American red squirrels
Chiefdoms, human, 14:251–252
Chilean coruros. *See* Coruros
Chilean dolphins. *See* Black dolphins
Chilean guemals. *See* Chilean huemuls
Chilean huemuls, 15:379, 15:384, 15:*387*, 15:*390*, 15:392

Chilean mouse opossums. *See* Elegant fat-tailed opossums
Chilean pudus. *See* Southern pudus
Chilean Red List, on *Spalacopus cyanus maulinus*, 16:439
Chilean shrew opossums, 12:267, 12:268, 12:*269*, 12:*270*
Chilean tree mice, 16:266, 16:*271*, 16:*273*, 16:274
Chimarrogale spp., 13:247, 13:248, 13:249
Chimarrogale himalayica. *See* Himalayan water shrews
Chimarrogale phaeura. *See* Sunda water shrews
Chimpanzees, 12:*196*, 14:*226*, 14:*236*
behavior, 12:145, 12:*151*, 12:157, 14:7, 14:*229–231*
conservation status, 12:215, 12:*220*, 14:235
distribution, 14:5, 14:227
enculturation of, 12:162
evolution, 12:33
feeding ecology, 14:9, 14:233–234
habitats, 14:227
humans and, 14:235
language, 12:160–162
memory, 12:153–154
numbers and, 12:155–156, 12:162
physical characteristics, 14:226, 14:227
reproduction, 14:10, 14:235
self recognition, 12:159
social cognition, 12:160
species of, 14:*239*
taxonomy, 14:225
theory of mind, 12:159–160
tool use, 12:157
touch, 12:83
China. *See* Oriental region
Chinchilla spp., 16:377
Chinchilla brevicaudata. *See* Short-tailed chinchillas
Chinchilla lanigera. *See* Long-tailed chinchillas
Chinchilla mice, 16:265, 16:270
Chinchilla rats, **16:443–448**, 16:*446*
Chinchillas, 12:132, **16:377–384**, 16:*381*, 16:383*t*
Chinchillidae. *See* Chinchillas; Viscachas
Chinchillones. *See* Chinchilla rats
Chinchillula spp. *See* Chinchilla mice
Chinese desert cats, 14:390*t*
Chinese dormice, 16:*322*, 16:*324*
Chinese ferret badgers, 14:332*t*
Chinese gymnures. *See* Hainan gymnures
Chinese hamsters, 16:243
Chinese hedgehogs, 13:204, 13:212*t*
Chinese jumping mice, 16:213, 16:216, 16:*223t*
Chinese lake dolphins. *See* Baijis
Chinese Meishan pigs, 15:149
Chinese muntjacs. *See* Reeve's muntjacs
Chinese pangolins, 16:110, 16:113, 16:*114*, 16:*115*
Chinese pygmy dormice, 16:297*t*
Chinese red pikas, 16:501*t*
Chinese rock squirrels, 16:143, 16:144
Chinese short-tailed shrews, 13:*253*, 13:*255–256*
Chinese shrew-moles, 13:282, 13:288*t*
Chinese water deer, 15:343, **15:373–377**, 15:*374*, 15:*375*, 15:*376*
Chionomys spp., 16:228
Chipmunks, 12:131, 16:127, 16:143, 16:144, 16:145, 16:147
Allen's, 16:160*t*
eastern, 16:126, 16:*148*, 16:*149*, 16:153–*154*
least, 16:*149*, 16:*152*
red-tailed, 16:159*t*
Siberian, 16:144, 16:158*t*
Tamias palmeri, 16:147–148
Townsend's, 16:*146*
Chipping squirrels. *See* Eastern chipmunks
Chiriquá pocket gophers, 16:196*t*
Chiroderma improvisum, 13:420
Chiroderma villosum. *See* Hairy big-eyed bats
Chironectes spp. *See* Water opossums
Chironectes minimus. *See* Water opossums
Chiropotes spp. *See* Bearded sakis
Chiropotes albinasus. *See* White-nosed bearded sakis
Chiropotes satanas. *See* Bearded sakis
Chiropotes satanas satanas. *See* Southern bearded sakis
Chiropotes satanas utahicki. *See* Uta Hick's bearded sakis
Chiroptera. *See* Bats
Chirus. *See* Tibetan antelopes
Chisel-toothed kangaroo rats, 16:202, 16:203, 16:209*t*
Chitals, 15:269, 15:358, 15:*362*, 15:*365–366*
Chlamyphorus spp., 13:149, 13:182
Chlamyphorus retusus. *See* Chacoan hairy armadillos
Chlamyphorus truncatus. *See* Pink fairy armadillos
Chlamytheres. *See* Giant armadillos
Chlorocebus spp., 14:188, 14:191
Chlorocebus aethiops. *See* Grivets
Chlorotalpa spp., 13:215
Chlorotalpa duthieae. *See* Congo golden moles
Chlorotalpa sclateri. *See* Sclater's golden moles
Choeronycteris mexicana. *See* Mexican hog-nosed bats
Choeropsis liberiensis. *See* Pygmy hippopotamuses
Choloepus spp., 13:149, 13:151, 13:153, 13:155, 13:156, 13:157
Choloepus didactylus. *See* Southern two-toed sloths
Choloepus hoffmanni. *See* Hoffman's two-toed sloths
Chorioallantoic placenta, 12:7, 12:12, 12:93
See also Physical characteristics; Reproduction
Chorion, 12:92–94
See also Physical characteristics; Reproduction
Chousinghas, 12:136, 15:265, 16:2, 16:11, 16:12, 16:14, 16:*17*, 16:*18*, 16:23–24
See also Antelopes
Chow chows, 14:293
Chozchozes, 16:433, 16:435, 16:*437*, 16:*438*, 16:439
Chromosome multiplication, 12:30–31
Chrossarchus spp. *See* Cusimanses
Chrotogale owstoni. *See* Owston's palm civets
Chrotopterus auritus. *See* Woolly false vampire bats
Chrysochloridae. *See* Golden moles
Chrysochlorinae, 13:215
Chrysochloris spp., 13:197, 13:215
Chrysochloris asiatica. *See* Cape golden moles
Chrysochloris stuhlmanni. *See* Stuhlmann's golden moles
Chrysocyon brachyurus. *See* Maned wolves
Chrysospalax spp., 13:215
Chrysospalax trevelyani. *See* Large golden moles
Chrysospalax villosus. *See* Rough-haired golden moles
Chuditches, 12:*280*, 12:283, 12:288, 12:290–291, 12:*293*, 12:*294–295*
Cimmarons, 12:176
Cimolestes spp., 14:255–256
Cincinnati Zoo, 12:203
Cingulata, 13:147, 13:148, 13:182
Cinnamon bears. *See* American black bears
Circadian rhythms, subterranean mammals, 12:74–75
See also Physical characteristics
Circannual cycles, subterranean mammals, 12:75
See also Physical characteristics
Circulatory system. *See* Cardiovascular system
Citellus spp. *See* Ground squirrels
CITES. *See* Convention on International Trade in Endangered Species
Civet cats. *See* African civets
Civet oil, 14:338
Civets, **14:335–345**, 14:*339*
behavior, 14:258, 14:337
conservation status, 14:338
distribution, 12:136, 14:336
evolution, 14:335
feeding ecology, 14:337
humans and, 14:338
physical characteristics, 14:335–336
reproduction, 14:337
species of, 14:*340–342*, 14:344*t*345*t*
taxonomy, 14:256, 14:335
Civettictis spp., 14:335, 14:338
Civettictis civetta. *See* African civets
Cladistics. *See* Taxonomy
Clarke's gazelles. *See* Dibatags
Claws, 12:39
See also Physical characteristics
Clethrionomys spp. *See* Red-backed voles
Clethrionomys gapperi. *See* Southern red-backed voles
Clethrionomys glareolus. *See* Bank voles
Clever Hans (horse), 12:155
Clidomys spp., 16:469, 16:470
Climacoceras spp., 15:265
Climate
glaciers and, 12:17–19
mammalian evolution and, 12:11, 12:17–19
Climbing mice, gray, 16:*286*, 16:*290–291*
Climbing shrews. *See* Forest musk shrews
Clinton, Bill, 12:184
Cloeotis percivali. *See* Percival's trident bats
Clouded leopards, 14:370, 14:372, 14:377, 14:*385*, 14:387
Club-footed bats. *See* Bamboo bats
Clydesdale horses, 15:*216*
Clyomys spp., 16:450
Clyomys bishopi. *See* Bishop's fossorial spiny rats
Coarse-haired pocket mice, 16:199, 16:200, 16:203
Coarse-haired wombats. *See* Common wombats
Coastal brown bears. *See* Brown bears

Coat coloration, 12:99
Coatimundis. *See* White-nosed coatis
Coatis, 14:256, 14:310, 14:311, 14:316*t*, 15:239
Coelacanth, 12:62
Coelodonta antiquitatis. *See* Woolly rhinoceroses
Coelops frithi. *See* East Asian tailless leaf-nosed bats
Coendou spp. *See* Tree porcupines
Coendou bicolor. *See* Bicolor-spined porcupines
Coendou ichillus, 16:365
Coendou koopmani. *See* Koopman's porcupines
Coendou melanurus. *See* Black-tailed hairy dwarf porcupines
Coendou pallidus, 16:365
Coendou paragayensis, 16:365
Coendou prehensilis. *See* Prehensile-tailed porcupines
Coendou pruinosus. *See* Frosted hairy dwarf porcupines
Coendou roosmalenorum, 16:365
Coendou rothschildi. *See* Rothschild's porcupines
Coendou sneiderni. *See* White-fronted hairy dwarf porcupines
Coevolution, 12:216–217
  See also Evolution
Cognition, 12:140–142, **12:149–164**
  See also Behavior
Coiba howler monkeys, 14:160, 14:167*t*
Coiba Island agoutis, 16:409, 16:*410*, 16:*412*, 16:413
Coimbra's titis, 14:11, 14:148
Coke's hartebeests, 16:*31*
Colburn's tuco-tucos, 16:430*t*
Coleura seychellensis. *See* Seychelles sheath-tailed bats
Collared anteaters. *See* Northern tamanduas; Southern tamanduas
Collared hedgehogs, 13:213*t*
Collared lemmings, 16:226, 16:230
Collared mangabeys, 14:194, 14:*196*, 14:*198*, 14:202
Collared peccaries, 15:267, 15:291, 15:*292*, 15:*293*, 15:*294*, 15:*296*, 15:*297*, 15:*298*
Collared pikas, 16:*480*, 16:484, 16:491, 16:*493*, 16:*495*, 16:501*t*
Collared titis, 14:143, 14:144, 14:145, 14:*149*, 14:*152*
Collared tuco-tucos, 16:431*t*
Collies, 14:288–289, 14:292
Colobidae, 14:4
Colobinae. *See* Leaf-monkeys
Colobus spp., 14:172, 14:173
Colobus angolensis. *See* Angolan colobus
Colobus guereza. *See* Mantled guerezas
Colobus monkeys, 12:14, 14:5, 14:11, 14:172, 14:173
  Angolan, 14:184*t*
  black, 14:174, 14:175, 14:184*t*
  black-and-white, 14:7
  king, 14:175, 14:184*t*
  olive, 14:173, 14:175, 14:*176*, 14:*178*, 14:179
  Penant's red, 14:185*t*
  red, 14:7, 14:172–175
Colobus polykomos. *See* King colobus
Colobus satanas. *See* Black colobus
Colobus vellerosus, 14:175
Colombian black spider monkeys, 14:*159*
Colombian weasels, 14:324
Colombian woolly monkeys, 14:160, 14:*161*, 14:*165*, 14:166
Color vision, 12:79
  See also Vision
Coloration, 12:38, 12:173
  See also Physical characteristics
Colostrum, 12:126
  See also Reproduction
Colugos, 12:14, 12:43, 12:136, **13:299–305**, 13:*300*, 13:*303*
Columbian ground squirrels, 12:*71*, 16:*150*, 16:*154*, 16:157
Columbian mammoths, 12:139
Columbus, Christopher, 15:242
Comb-toed jerboas, 16:212–215, 16:*218*, 16:220, 16:*221*
Commensalism, **12:171–181**
Commerson's dolphins, 15:49, 15:56*t*
Commerson's leaf-nosed bats, 13:401, 13:*402*, 13:*404*, 13:410*t*
Committee on the Status of Endangered Wildlife in Canada
  black-tailed prairie dogs, 16:155
  harbor porpoises, 15:35, 15:38
  Vancouver Island marmots, 16:156
Common bentwing bats, 13:310, 13:315, 13:*520*, 13:*521*, 13:*523*, 13:524–*525*
Common big-eared bats, 13:433*t*
Common blossom bats. *See* Southern blossom bats
Common bottlenosed dolphins, 12:126, 15:1, 15:*5*, 15:7, 15:11, 15:46, 15:*47*, 15:*52*, 15:*54*
Common brushtail possums, 12:215, 13:*58*, 13:*62*, 13:*65*
  behavior, 13:37
  conservation status, 13:61
  distribution, 13:34, 13:59, 13:*65*
  feeding ecology, 13:38, 13:60
  habitats, 13:34, 13:59
  humans and, 13:61
  translocation of, 12:224
Common chimpanzees. *See* Chimpanzees
Common cuscuses, 13:60, 13:66*t*
Common dolphins. *See* Short-beaked saddleback dolphins
Common dormice. *See* Hazel dormice
Common Eurasian moles. *See* European moles
Common European white-toothed shrews, 13:*266*, 13:267, 13:*268*, 13:*270*, 13:*271*
Common foxes. *See* Crab-eating foxes
Common genets, 14:*335*, 14:*337*, 14:*339*, 14:*340*–341
Common gibbons. *See* Lar gibbons
Common hamsters. *See* Black-bellied hamsters
Common hippopotamuses, 15:267, 15:*301*, 15:*303*, 15:*304*, 15:*305*, 15:*306*, 15:*307*, 15:*308*, 15:*309*, 15:*310*–311
Common house rats. *See* Black rats
Common jackals. *See* Golden jackals
Common langurs. *See* Northern plains gray langurs
Common long-nosed armadillos. *See* Nine-banded armadillos
Common marmosets, 14:*116*, 14:117, 14:118, 14:121–124, 14:*125*, 14:129–*130*
  See also Marmosets
Common mole-rats, 12:*70*, 12:74, 16:124, 16:*341*, 16:*342*, 16:344, 16:*346*, 16:*347*, 16:348
  See also African mole-rats
Common moles. *See* European moles
Common mountain viscachas. *See* Mountain viscachas
Common nectar-feeding fruit bats. *See* Dawn fruit bats
Common opossums, 12:249–254
  See also Southern opossums
Common porcupines, 16:353, 16:*356*, 16:*358*
Common porpoises. *See* Harbor porpoises
Common reedbucks. *See* Southern reedbucks
Common ringtails, 13:113, 13:*114*–115, 13:116, 13:117, 13:*118*, 13:120–*121*
  See also Ringtail possums
Common seals. *See* Harbor seals
Common short-tailed porcupines. *See* Common porcupines
Common shrews, 13:*248*, 13:*254*, 13:*261*
  See also Shrews
Common spotted cuscuses, 13:*62*, 13:*63*, 13:64
Common squirrel monkeys, 14:101, 14:*102*, 14:103, 14:*105*, 14:*107*, 14:*109*
Common tenrecs, 12:*84*, 13:200, 13:226, 13:*227*, 13:228, 13:*229*, 13:*231*, 13:*232*
  See also Tenrecs
Common tree shrews, 13:*289*, 13:*290*, 13:*291*, 13:*293*, 13:*294*, 13:295
Common vampires. *See* Vampire bats
Common wallaroos, 13:101*t*
Common wambengers. *See* Brush-tailed phascogales
Common warthogs, 15:264, 15:275, 15:280, 15:*283*, 15:*285*–286
  See also Warthogs
Common wombats, 13:*35*, 13:37, 13:38, 13:40, 13:51, 13:*52*, 13:*53*, 13:*54*, 13:*55*
  See also Wombats
Common zebras. *See* Plains zebras
Commonwealth Biological Control Act (Australia), 12:188
Communication
  echolocation, 12:53–55, 12:*59*, 12:*85*–87
  seismic, 12:76
  senses and, 12:84
  subterranean mammals, 12:114
  symbolic, 12:161–162
  vibrations in, 12:83
  See also Behavior
Communities. *See* specific types of communities
Comoro black flying foxes. *See* Livingstone's fruit bats
Competition
  deer, 12:118
  inter- and intra-specific, 12:114–115
  sperm, 12:98–99, 12:105–106
  See also Behavior; Feeding ecology
Complex-toothed flying squirrels, 16:136
Concolor gibbons. *See* Black crested gibbons
Condoro Island flying foxes. *See* Island flying foxes
Condylarthra, 15:131–132, 15:133, 15:135, 15:136, 15:263, 15:266, 15:343
Condylura spp., 13:281
Condylura cristata. *See* Star-nosed moles
Condylurinae, 13:279

Conepatus spp., 14:319, 14:321
Conepatus mesoleucus. See Western hog-nosed skunks
Congo golden moles, 13:222*t*
Congo water civets. *See* Aquatic genets
Congosorex polli. See Poll's shrews
Conies. *See* American pikas
Connochaetes spp. *See* Wildebeests
Connochaetes gnou. See Black wildebeests
Connochaetes taurinus. See Blue wildebeests
Conservation Action Plan for Eurasian Insectivores
  alpine shrews, 13:260
  Chinese short-tailed shrews, 13:256
  common shrews, 13:261
  elegant water shrews, 13:259
  Eurasian pygmy shrews, 13:262
  Eurasian water shrews, 13:258
  giant shrews, 13:262
  Himalayan water shrews, 13:259
  Hodgson's brown-toothed shrews, 13:259
  mole-shrews, 13:255
Conservation biology, 12:216–225
Conservation Endowment Fund, 12:204
Conservation International, 12:218
Conservation medicine, 12:222–223
Conservation status, 12:15, **12:213–225**
  aardvarks, 15:158
  agoutis, 16:409, 16:411–414, 16:415*t*
  anteaters, 13:175, 13:177–179
  armadillos, 13:185, 13:187–190, 13:190*t*–192*t*
  Artiodactyla, 15:272–273
  aye-ayes, 14:88
  baijis, 15:21
  bandicoots, 13:4–6
    dry-country, 13:11–12, 13:14–16, 13:16*t*–17*t*
    rainforest, 13:11–12, 13:14–17*t*
  bats, 13:315–316
    American leaf-nosed bats, 13:420, 13:423–432, 13:433*t*–434*t*
    bulldog, 13:447, 13:449–450
    disk-winged, 13:474, 13:476–477
    false vampire, 13:382, 13:384–385
    funnel-eared, 13:461, 13:463–465
    horseshoe, 13:392–393, 13:396–400
    Kitti's hog-nosed, 13:369
    Molossidae, 13:487, 13:490–493, 13:493*t*–495*t*
    mouse-tailed, 13:352, 13:353
    moustached, 13:436, 13:440–441, 13:442*t*
    New Zealand short-tailed, 13:455, 13:457–459
    Old World fruit, 13:321–322, 13:325–330, 13:331*t*–332*t*, 13:337, 13:340–347, 13:348*t*–349*t*
    Old World leaf-nosed, 13:405, 13:407–409, 13:409*t*–410*t*
    Old World sucker-footed, 13:480
    slit-faced, 13:373, 13:375–376, 13:376*t*–377*t*
    smoky, 13:468, 13:470–471
    Vespertilionidae, 13:501–502, 13:506–510, 13:512–514, 13:515*t*–516*t*, 13:522, 13:524–525, 13:526*t*
  bears, 14:262, 14:300, 14:302–305, 14:306*t*
  beavers, 16:180, 16:183–184
  bilbies, 13:22
  botos, 15:29–30
  Bovidae, 16:8
    Antilopinae, 16:48, 16:50–56, 16:56*t*–57*t*
    Bovinae, 16:14–15, 16:18–23, 16:24*t*–25*t*
    Caprinae, 16:94–95, 16:98–103, 16:103*t*–104*t*
    duikers, 16:77, 16:80–84, 16:84*t*–85*t*
    Hippotraginae, 16:33–34, 16:37–42, 16:42*t*–43*t*
    Neotraginae, 16:63, 16:66–71, 16:71*t*
  bushbabies, 14:26, 14:28–32, 14:32*t*–33*t*
  Camelidae, 15:317–318, 15:320–323
  Canidae, 14:272–273, 14:277–283, 14:283*t*–284*t*
  capybaras, 16:405
  Carnivora, 14:262
  cats, 14:374–375, 14:380–389, 14:390*t*–391*t*
  Caviidae, 16:393, 16:395–398, 16:399*t*
  Cetacea, 15:9–10
    Balaenidae, 15:9–10, 15:111–112, 15:115–118
    beaked whales, 15:61–62, 15:69*t*–70*t*
    dolphins, 15:48–50, 15:53–55, 15:56*t*–57*t*
    Emballonuridae, 13:358, 13:360–363, 13:363*t*–364*t*
    franciscana dolphins, 15:25
    Ganges and Indus dolphins, 15:15–16
    gray whales, 15:99–100
    Monodontidae, 15:86–87, 15:90–91
    porpoises, 15:35–36, 15:38–39, 15:39*t*
    pygmy right whales, 15:105
    rorquals, 15:9, 15:127–130, 15:130*t*
    sperm whales, 15:77–78, 15:79–80
  chevrotains, 15:329, 15:331–333
  Chinchillidae, 16:380, 16:382–383, 16:383*t*
  cognition and, 12:150–151
  colugos, 13:302, 13:304
  coypus, 16:477
  dasyurids, 12:290–291, 12:294–298, 12:299*t*–301*t*
  Dasyuromorphia, 12:283
  deer
    Chinese water, 15:376
    muntjacs, 15:346, 15:349–354
    musk, 15:338, 15:340–341
    New World, 15:384, 15:388–396, 15:396*t*
    Old World, 15:360–361, 15:364–371, 15:371*t*
  Dipodidae, 16:216–217, 16:219–222, 16:223*t*–224*t*
  Diprotodontia, 13:39–40
  domestication and, 12:175
  dormice, 16:321, 16:323–326, 16:327*t*–328*t*
  duck-billed platypuses, 12:233–234, 12:246–247
  Dugongidae, 15:201, 15:203–204
  echidnas, 12:238, 12:240*t*
  elephants, 15:171–172, 15:174–175
  Equidae, 15:229–230, 15:232–235, 15:236*t*
  Erinaceidae, 13:207, 13:209–213, 13:212*t*–213*t*
  giant hutias, 16:470, 16:471*t*
  gibbons, 14:214–215, 14:218–222
  Giraffidae, 15:405, 15:408–409
  great apes, 14:235, 14:237–240
  gundis, 16:313, 16:314–315
  Herpestidae, 14:352, 14:354–356, 14:357*t*
  Heteromyidae, 16:203, 16:205–209, 16:208*t*–209*t*, 16:209*t*
  hippopotamuses, 15:308, 15:311
  hutias, 16:463, 16:467*t*
  Hyaenidae, 14:362, 14:365–367
  hyraxes, 15:184, 15:186–188, 15:189*t*
  Indriidae, 14:67, 14:69–72
  Insectivora, 13:199–200
  koalas, 13:48–49
  Lagomorpha, 16:487
  lemurs, 14:52, 14:56–60
    Cheirogaleidae, 14:39, 14:41–44
    sportive, 14:76, 14:79–83
  Leporidae, 16:509, 16:511–515, 16:515*t*–516*t*
  Lorisidae, 14:16, 14:18–20, 14:21*t*
  Macropodidae, 13:89, 13:93–100, 13:101*t*–102*t*
  manatees, 15:208–209, 15:211–212
  Megalonychidae, 13:158, 13:159
  moles
    golden, 13:217, 13:220–222, 13:222*t*
    marsupial, 13:27, 13:28
  monitos del monte, 12:274
  monkeys
    Atelidae, 14:159–160, 14:162–166, 14:166*t*–167*t*
    Callitrichidae, 14:124, 14:127–131, 14:132*t*
    Cebidae, 14:105–106, 14:108–112
    cheek-pouched, 14:194, 14:197–204, 14:204*t*–205*t*
    leaf-monkeys, 14:175, 14:178–183, 14:184*t*–185*t*
    night, 14:138, 14:141*t*
    Pitheciidae, 14:148, 14:150–153, 14:153*t*
  monotremes, 12:233–234
  mountain beavers, 16:133
  Muridae, 16:285, 16:288–295, 16:296*t*–297*t*
    Arvicolinae, 16:229, 16:233–238, 16:237*t*–238*t*
    hamsters, 16:243, 16:245–246, 16:247*t*
    Murinae, 16:252–253, 16:256–260, 16:261*t*–262*t*
    Sigmodontinae, 16:270, 16:272–276, 16:277*t*–278*t*
  musky rat-kangaroos, 13:71
  Mustelidae, 14:324–325, 14:327–331, 14:332*t*–333*t*
  numbats, 12:305
  octodonts, 16:436, 16:438–440, 16:440*t*
  opossums
    New World, 12:254, 12:257–263, 12:264*t*–265*t*
    shrew, 12:268, 12:270
  Otariidae, 14:399–400, 14:402–406
  pacaranas, 16:388
  pacas, 16:420, 16:423–424
  pangolins, 16:113, 16:115–120
  peccaries, 15:295–296, 15:298–300
  Perissodactyla, 15:221–222
  Petauridae, 13:129, 13:131–133, 13:133*t*
  Phalangeridae, 13:60, 13:*63–66t*
  pigs, 15:280–281, 15:284–288, 15:289*t*–290*t*
  pikas, 16:495, 16:497–501, 16:501*t*–502*t*
  pocket gophers, 16:190, 16:192–195*t*, 16:196*t*–197*t*
  porcupines
    New World, 16:369, 16:372–373, 16:374*t*
    Old World, 16:353–354, 16:357–364
  possums
    feather-tailed, 13:142, 13:144
    honey, 13:138
    pygmy, 13:108, 13:110–111

Conservation status *(continued)*
primates, 14:11
Procyonidae, 14:310–311, 14:313–315, 14:315*t*–316*t*
pronghorns, 15:415–416
Pseudocheiridae, 13:117, 13:119–123, 13:122*t*–123*t*
rat-kangaroos, 13:76, 13:78–81
rats
African mole-rats, 16:345, 16:347–349, 16:349*t*–350*t*
cane, 16:335, 16:337–338
chinchilla, 16:445, 16:447–448
dassie, 16:331
spiny, 16:450–451, 16:453–457, 16:458*t*
rhinoceroses, 15:254, 15:257–262
rodents, 16:126
sengis, 16:522, 16:524–530, 16:531*t*
shrews
red-toothed, 13:252, 13:255–263, 13:263*t*
West Indian, 13:244
white-toothed, 13:268, 13:271–275, 13:276*t*–277*t*
Sirenia, 15:191, 15:194–195
solenodons, 13:239, 13:241
springhares, 16:310
squirrels
flying, 16:136, 16:137, 16:141*t*–142*t*
ground, 16:147–148, 16:151–158, 16:158*t*–160*t*
scaly-tailed, 16:300, 16:303–305
tree, 16:167, 16:169–174*t*, 16:175*t*
Talpidae, 13:282, 13:284–287, 13:287*t*–288*t*
tapirs, 15:241–242, 15:245–247
tarsiers, 14:95, 14:97–99
tenrecs, 13:229, 13:232–234, 13:234*t*
three-toed tree sloths, 13:165, 13:167–169
tree shrews, 13:292, 13:294–296, 13:297*t*–298*t*
true seals, 14:262, 14:422, 14:427, 14:429–434, 14:435*t*
tuco-tucos, 16:427, 16:429–430, 16:430*t*–431*t*
Viverridae, 14:338, 14:340–343, 14:344*t*–345*t*
walruses, 14:415
wombats, 13:40, 13:53, 13:55–56
Xenartha, 13:152–153
zoos and, 12:204, 12:*208*–209
Contreras, L. C.
on coruros, 16:435, 16:436
on octodonts, 16:433
Convention on International Trade in Endangered Species, 14:374
addaxes, 16:39
African elephants, 15:175
Arabian oryx, 16:40
Asian elephants, 15:174
babirusas, 15:288
bald uakaris, 14:151
Beecroft's anomalures, 16:303
big-eared flying mice, 16:305
black-capped capuchins, 14:111
Bolivian squirrel monkeys, 14:108
brown-throated three-toed sloths, 13:167
capuchins, 14:105, 14:106
Chacoan peccaries, 15:300
collared peccaries, 15:298
collared titis, 14:152
common hippopotamuses, 15:311
common squirrel monkeys, 14:109
dugongs, 15:201
dusky titi monkeys, 14:152
elephants, 15:171
gray-backed sportive lemurs, 14:79
guanacos, 15:322
hippopotamuses, 15:308
Indian pangolins, 16:116
jaguars and, 14:386
lesser Malay mouse deer, 15:332
Lord Derby's anomalures, 16:303
manatees, 15:208–209
masked titis, 14:153
Milne-Edwards's sportive lemurs, 14:81
mouse lemurs, 14:39
northern sportive lemurs, 14:83
pacas, 16:423
peccaries, 15:295
Pel's anomalures, 16:303
Perissodactyla, 15:221–222
pigs, 15:280–281
Pitheciidae, 14:148
pygmy hippopotamuses, 15:311
pygmy hogs, 15:287
pygmy right whales, 15:105
red mouse lemurs, 14:43
red-tailed sportive lemurs, 14:80
Sirenia, 15:194–195
sloths, 13:153
small-toothed sportive lemurs, 14:83
southern African hedgehogs, 13:207, 13:210
sperm whales, 15:78
sportive lemurs, 14:76
squirrel monkeys, 14:105
tapirs, 15:243
vicuñas, 15:321
volcano rabbits, 16:514
water chevrotains, 15:331
weasel sportive lemurs, 14:82
weeper capuchins, 14:112
white-faced sakis, 14:150
white-footed sportive lemurs, 14:82
white-fronted capuchins, 14:110
white-lipped peccaries, 15:299
white-nosed bearded sakis, 14:153*t*
white-throated capuchins, 14:111
Convention on the Conservation of European Wildlife and Natural Habitats, 16:180, 16:183
Convergent evolution, 12:27
of marine mammals, 12:68
of subterranean mammals, 12:78
See also Evolution
Cook, Captain James, 12:191
Cook, W. M., 16:268
Coons. *See* Northern raccoons
Cope, E. D., 15:132
Copperheads. *See* Golden-mantled ground squirrels
Coppery ringtail possums, 13:*115*, 13:123*t*
Coprophagy, 12:48, 12:121, 16:485
See also Behavior
Copulation, 12:103–104
See also Reproduction
Copulatory plugs, 12:105–106
Coquerel's mouse lemurs, 14:36, 14:*37*, 14:39, 14:*40*, 14:*41*, 14:43–44
Coquerel's sifakas, 14:*65*
Corbet, G. B., 16:199, 16:211
Corgis, 14:292
Cormura brevirostris. *See* Chestnut sac-winged bats
Corneum, 12:36
See also Physical characteristics
Cororos, 16:123
Corpora quadrigemina, 12:6
See also Physical characteristics
Corpus luteum, 12:92
See also Physical characteristics
Corsac foxes, 14:284*t*
Cortes, Hernando, 12:203, 15:216
Cortisole, 12:148
Coruros, 12:74, 12:75, 16:434, 16:435, 16:436, 16:*437*, 16:*438*, 16:439
COSEWIC. *See* Committee on the Status of Endangered Wildlife in Canada
Cotton rats, 16:127, 16:128, 16:263, 16:268–270, 16:275–276
Cotton-top tamarins, 14:119, 14:*122*, 14:123, 14:124, 14:*126*, 14:*127*
Cottontails, 16:*507*, 16:*510*, 16:514–515
behavior, 16:482
evolution, 16:505
habitats, 16:481, 16:506
humans and, 16:488, 16:509
Cougars. *See* Pumas
Countershading, 12:38
See also Physical characteristics
Counting, 12:155
See also Cognition
Cows. *See* Cattle
Coxal bone, 12:41
See also Physical characteristics
Coyotes, 12:*136*, 12:*143*, 14:*258*, **14:265–273**, 14:*266*, 14:*273*
behavior, 12:148, 14:268
conservation status, 14:272
evolution, 14:265
feeding ecology, 14:270, 14:271
habitats, 14:267
humans and, 14:262, 14:273
physical characteristics, 14:266–267
taxonomy, 14:265
Coypus, 16:*122*, **16:473–478**, 16:*474*, 16:*475*, 16:*476*, 16:*478*
distribution, 16:123, 16:*473*, 16:474
habitats, 16:124, 16:474
humans and, 12:184, 16:126–127, 16:477
physical characteristics, 16:123, 16:473–474
Cozumel Island coatis, 14:316*t*
Cozumel Island raccoons, 14:310, 14:311, 14:315*t*
Crab-eater seals, 12:138, 14:260, 14:420, 14:*425*, 14:*433*–434
Crab-eating foxes, 14:*268*, 14:271, 14:*275*, 14:*279*, 14:282–283
Crab-eating raccoons, 14:315*t*
Craseonycteridae. *See* Kitti's hog-nosed bats
Craseonycteris thonglongyai. *See* Kitti's hog-nosed bats
Cratogeomys spp., 16:185
Cratogeomys castanops. *See* Yellow-faced pocket gophers
Cratogeomys fumosus. *See* Smoky pocket gophers
Cratogeomys merriami. *See* Merriam's pocket gophers
Cratogeomys neglectus. *See* Querétaro pocket gophers

Cratogeomys zinseri. See Zinser's pocket gophers
Creek rats, 16:*254*, 16:*255*, 16:258–259
Creodonta, 12:11
Crepuscular activity cycle, 12:55
See also Behavior
Crest-tailed marsupial mice. *See* Mulgaras
Crested agoutis, 16:409, 16:415*t*
Crested free-tailed bats. *See* Lesser-crested mastiff bats
Crested gibbons, 14:207, 14:210, 14:215, 14:*216*, 14:*220*, 14:221
Crested rats, 16:*287*, 16:*289*, 16:293
Crevice bats. *See* Peruvian crevice-dwelling bats
Cricetinae. *See* Hamsters
Cricetomyinae, 16:282
Cricetomys gambianus. See Gambian rats
Cricetulus spp. *See* Rat-like hamsters
Cricetulus barabensis. See Striped dwarf hamsters
Cricetulus migratorius. See Gray dwarf hamsters
Cricetus spp. *See* Black-bellied hamsters
Cricetus cricetus. See Black-bellied hamsters
Crocidura spp., 13:265, 13:266–267, 13:268
Crocidura attenuata. See Gray shrews
Crocidura canariensis. See Canary shrews
Crocidura dsinezumi. See Dsinezumi shrews
Crocidura fuliginosa. See Southeast Asian shrews
Crocidura horsfieldii. See Horsfield's shrews
Crocidura leucodon. See Bicolored shrews
Crocidura monticola. See Sunda shrews
Crocidura negrina. See Negros shrews
Crocidura russula. See Common European white-toothed shrews
Crocidura suaveolens. See Lesser white-toothed shrews
Crocidurinae. *See* White-toothed shrews
Crocuta crocuta. See Spotted hyenas
Cross foxes. *See* Red foxes
Cross River Allen's bushbabies, 14:32*t*
Crossarchus spp., 14:347
Crossarchus obscurus. See Western cusimanses
Crowned gibbons. *See* Pileated gibbons
Crowned lemurs, 14:*9*, 14:*51*, 14:*54*, 14:*57*, 14:59–60
Cryptic golden moles, 13:215
Cryptochloris spp., 13:215
Cryptochloris wintoni. See De Winton's golden moles
Cryptochloris zyli. See Van Zyl's golden moles
Cryptomys spp., 16:339, 16:342–343, 16:344, 16:345
See also African mole-rats
Cryptomys amatus, 16:345
Cryptomys anselli. See Zambian mole-rats
Cryptomys bocagei. See Bocages mole-rats
Cryptomys damarensis. See Damaraland mole-rats; Kalahari mole-rats
Cryptomys darlingi. See Mashona mole-rats
Cryptomys foxi. See Nigerian mole-rats
Cryptomys hottentotus. See Common mole-rats
Cryptomys hottentotus hottentotus. See Common mole-rats
Cryptomys hottentotus pretoriae. See Highveld mole-rats
Cryptomys mechowi. See Giant Zambian mole-rats
Cryptomys ochraceocinereus. See Ochre mole-rats
Cryptomys zechi. See Togo mole-rats
Cryptoprocta spp., 14:347, 14:348
Cryptoprocta ferox. See Fossa
Cryptotis spp., 13:198, 13:248, 13:250
Cryptotis meridensis. See Mérida small-eared shrews
Cryptotis parva. See American least shrews
Ctenodactylidae. *See* Gundis
Ctenodactylus spp., 16:311, 16:312, 16:313
Ctenodactylus gundi. See North African gundis
Ctenodactylus vali. See Desert gundis
Ctenomyidae. *See* Tuco-tucos
Ctenomys spp. *See* Tuco-tucos
Ctenomys australis. See Southern tuco-tucos
Ctenomys azarae. See Azara's tuco-tucos
Ctenomys boliviensis. See Bolivian tuco-tucos
Ctenomys colburni. See Colburn's tuco-tucos
Ctenomys conoveri, 16:425
Ctenomys emilianus. See Emily's tuco-tucos
Ctenomys latro. See Mottled tuco-tucos
Ctenomys magellanicus. See Magellanic tuco-tucos
Ctenomys mattereri, 16:427
Ctenomys mendocinus, 16:427
Ctenomys nattereri. See Natterer's tuco-tucos
Ctenomys opimus, 16:427
Ctenomys pearsoni. See Pearson's tuco-tucos
Ctenomys perrensis. See Goya tuco-tucos
Ctenomys pundti, 16:425
Ctenomys rionegrensis. See Rio Negro tuco-tucos
Ctenomys saltarius. See Salta tuco-tucos
Ctenomys sociabilis. See Social tuco-tucos
Ctenomys talarum. See Talas tuco-tucos
Ctenomys torquatus. See Collared tuco-tucos
Cuban flower bats, 13:435*t*
Cuban funnel-eared bats. *See* Small-footed funnel-eared bats
Cuban hutias, 16:461, 16:*462*, 16:463, 16:*464*, 16:*465*
Cuban solenodons, 13:237, 13:*238*
Cuis, 16:391–392, 16:*394*, 16:*395*, 16:396
Culpeos, 14:265, 14:284*t*
Culture, 12:157–159
See also Behavior
Cuon alpinus. See Dholes
Cursorial locomotion. *See* Running
Cuscomys spp., 16:443
Cuscomys ashaninka. See Ashaninka rats
Cuscomys oblativa, 16:443
Cuscuses, 12:137, 13:31, 13:32, 13:34, 13:35, 13:39, **13:57–67**, 13:*62*, 13:66*t*
Cusimanses, 14:349, 14:350
Cuvier's beaked whales, 15:*60*, 15:61, 15:62, 15:*63*, 15:*65*, 15:66
Cyclopes spp., 13:151
Cyclopes didactylus. See Silky anteaters
Cynictis spp., 14:347
Cynictis penicillata. See Yellow mongooses
Cynocephalidae. *See* Colugos
Cynocephalus spp. *See* Colugos
Cynocephalus variegatus. See Malayan colugos
Cynocephalus volans. See Philippine colugos
Cynodontia, 12:10, 12:11
Cynogale bennettii. See Otter civets
Cynomys spp. *See* Prairie dogs
Cynomys gunnisoni. See Gunnison's prairie dogs
Cynomys ludovicianus. See Black-tailed prairie dogs
Cynomys mexicanus, 16:147
Cynopterus spp. *See* Short-nosed fruit bats
Cynopterus brachyotis. See Lesser short-nosed fruit bats
Cynopterus sphinx. See Indian fruit bats
Cyomys ludovicianus. See Black-tailed prairie dogs
Cyprus spiny mice, 16:261*t*
Cystophora cristata. See Hooded seals
Cyttarops spp., 13:355

D

Dactylomys dactylinus. See Amazon bamboo rats
Dactylomys peruanus, 16:451
Dactylopsila spp., 13:127, 13:128
Dactylopsila megalura. See Great-tailed trioks
Dactylopsila palpator. See Long-fingered trioks
Dactylopsila tatei. See Tate's trioks
Dactylopsila trivirgata. See Striped possums
Dactylopsilinae, 13:125, 13:126
Dagestan turs. *See* East Caucasian turs
Dairy cattle. *See* Domestic cattle
Dairy goats. *See* Domestic goats
d'Albertis's ringtail possums, 13:122*t*
See also Ringtail possums
Dall's porpoises, 15:33–*34*, 15:35, 15:36, 15:39*t*
Dall's sheep, 12:*103*, 12:178
Dalmatians, 14:289, 14:292
Dalpiazinidae, 15:13
Dama dama. See Fallow deer
Dama gazelles, 16:48, 16:57*t*
Dama wallabies. *See* Tammar wallabies
Damaliscus hunteri. See Hunter's hartebeests
Damaliscus lunatus. See Topis
Damaliscus lunatus jimela, 16:33
Damaliscus lunatus korrigum. See Korrigums
Damaliscus lunatus lunatus. See Tsessebes
Damaliscus pygargus. See Blesboks; Bonteboks
Damara mole-rats. *See* Damaraland mole-rats
Damaraland dikdiks. *See* Kirk's dikdiks
Damaraland mole-rats, 16:124, 16:*342*, 16:344–345, 16:*346*, 16:*348*
Darién pocket gophers, 16:196*t*
Dark Annamite muntjacs. *See* Truong Son muntjacs
Dark-handed gibbons. *See* Agile gibbons
Dark kangaroo mice, 16:*201*, 16:209*t*
Darwin, Charles
on animal weapons, 15:268
on armadillos, 13:181
on Bovidae, 16:2
on *Calomys laucha*, 16:267
on domestic pigs, 15:149
on mind and behavior, 12:149, 12:150
on ungulates, 15:143
See also Evolution
Darwin's foxes, 14:273
Dassie rats, 16:121, **16:*329*–332**, 16:*330*
Dasycercus cristicauda. See Mulgaras
Dasykaluta rosamondae. See Little red kalutas
Dasypodidae. *See* Armadillos
Dasypodinae, 13:182
Dasyprocta spp., 16:407, 16:*408*
Dasyprocta azarae. See Azara's agoutis

Dasyprocta coibae. See Coiba Island agoutis
Dasyprocta cristata. See Crested agoutis
Dasyprocta fuliginosa. See Black agoutis
Dasyprocta guamara. See Onnoco agoutis
Dasyprocta kalinowskii. See Kalinowski's agoutis
Dasyprocta leporina. See Red-rumped agoutis
Dasyprocta mexicana. See Mexican black agoutis
Dasyprocta prymnolopha. See Black-rumped agoutis
Dasyprocta punctata. See Central American agoutis
Dasyprocta ruatanica. See Roatán Island agoutis
Dasyprocta variegata. See Brown agoutis
Dasyproctidae. *See* Agoutis
Dasypus spp., 13:152, 13:*183*, 13:185
Dasypus hybridus. See Southern long-nosed armadillos
Dasypus kappleri. See Great long-nosed armadillos
Dasypus novemcinctus. See Nine-banded armadillos
Dasypus pilosus. See Hairy long-nosed armadillos
Dasypus sabanicola. See Llanos long-nosed armadillos
Dasypus septemcinctus. See Seven-banded armadillos
Dasyuridae. *See* Dasyurids
Dasyurids, 12:114, 12:277–284, **12:287–301,** 12:*292*, 12:*293*
Dasyuroidea. *See* Dasyuromorphia
Dasyuromorphia, 12:137, **12:277–285**
Dasyurus albopunctatus. See New Guinean quolls
Dasyurus geoffroii. See Chuditches
Dasyurus hallucatus. See Northern quolls
Dasyurus maculatus. See Spotted-tailed quolls
Dasyurus spartacus. See Bronze quolls
Dasyurus viverrinus. See Eastern quolls
Daubentonia spp. *See* Aye-ayes
Daubentonia madagascariensis. See Aye-ayes
Daubentonia robusta, 14:85, 14:86, 14:88
Daubentoniidae. *See* Aye-ayes
Daubenton's bats, 13:310–311, 13:313, 13:315, 13:500, 13:*504*, 13:*511*, 13:514
Daurian hedgehogs, 13:*208*, 13:211
Daurian pikas, 16:484, 16:488, 16:*492*
David's echymiperas, 13:12
Davy's naked-backed bats, 13:435, 13:*437*, 13:442*t*
Dawn fruit bats, 13:307, 13:*339*, 13:*344*, 13:345
De Winton's golden moles, 13:217, 13:222*t*
Deception, intentional, 12:159–160
  See also Cognition
Declarative memory, 12:154
  See also Cognition
Deer, 15:145–146, 15:263, 15:*362*, 15:*363*
  antlers, 12:40
  behavior, 15:270, 15:271
  digestive systems, 12:14
  distribution, 12:131, 12:134, 12:136, 12:137
  evolution, 15:137, 15:265–266
  gigantism and, 12:19–20
  Ice Age evolution, 12:22–23
  mate selection, 12:102
  physical characteristics, 15:267, 15:268, 15:269
  population indexes, 12:195
  reproduction, 12:103, 15:143
  sexual dimorphism, 12:99
  See also specific types of deer
Deer mice, 12:*91*, 16:124, 16:128, 16:*265*, 16:266
Defassa waterbucks. *See* Waterbucks
Defense mechanisms
  evolution of, 12:18–19
  senses and, 12:84
  See also Behavior
Deforestation, 12:214
  See also Conservation status
Degus, 16:*122*, 16:*434*, 16:*435*, 16:436, 16:*437*, 16:*438*, 16:440
  Bridges's, 16:433, 16:440*t*
  moon-toothed, 16:433, 16:440*t*
Deinogalerix spp., 13:194
Delanymys brooksi. See Delany's swamp mice
Delany's swamp mice, 16:297*t*
Delayed fertilization, 12:109–110
  See also Reproduction
Delayed implantation, 12:109–110, 12:111
  See also Reproduction
Delmarva fox squirrels, 16:167
Delomys spp., 16:264
Delphinapterus leucas. See Belugas
Delphinidae. *See* Dolphins
Delphininae, 15:41
Delphinoidea, 15:81
Delphinus spp., 15:2, 15:43–44
Delphinus delphis. See Short-beaked saddleback dolphins
Deltamys kempi. See Kemp's grass mice
Demand and consumption economies, 12:223
Demidoff's bushbabies, 14:7, 14:*24*, 14:*25*, 14:27, 14:*29*, 14:31
Dendrogale spp., 13:292
Dendrogale melanura. See Bornean smooth-tailed tree shrews
Dendrogale murina. See Northern smooth-tailed tree shrews
Dendrohyrax spp., 15:178, 15:183
Dendrohyrax arboreus. See Southern tree hyraxes
Dendrohyrax dorsalis. See Western tree hyraxes
Dendrohyrax validus. See Eastern tree hyraxes
Dendrolagus spp., 13:83, 13:85, 13:86
Dendrolagus bennettianus. See Bennett's tree kangaroos
Dendrolagus dorianus. See Doria's tree kangaroos
Dendrolagus goodfellowi. See Goodfellow's tree kangaroos
Dendrolagus lumholtzi. See Lumholtz's tree kangaroos
Dendrolagus matschiei. See Matschie's tree kangaroos
Dendrolagus mbaiso. See Forbidden tree kangaroos
Dendrolagus scottae. See Black tree kangaroos
Dendromurinae, 16:282
Dendromus melanotis. See Gray climbing mice
Dendromus vernayi, 16:291
Dense beaked whales. *See* Blainville's beaked whales
Density estimates, 12:195, 12:196–198, 12:200–202
Dental formulas, 12:46
  See also Teeth
Dentary-squamosal joint, 12:9
Dentition. *See* Teeth
Dent's horseshoe bats, 13:*394*, 13:*398*
Deppe's squirrels, 16:175*t*
Depth perception, 12:79
  See also Vision
Derived similarities, 12:28
Dermis, 12:36
  See also Physical characteristics
Dermoptera. *See* Colugos
Desert bandicoots, 13:2, 13:5, 13:11, 13:17*t*
Desert bighorn sheep, 16:92
Desert cavies. *See* Mountain cavies
Desert cottontails, 16:*507*, 16:*510*, 16:*514*
Desert dormice, 16:317, 16:318, 16:320, 16:*322*, 16:*324*, 16:325
Desert golden moles. *See* Grant's desert golden moles
Desert gundis, 16:312, 16:313, 16:*314*, 16:315
Desert kangaroo rats, 16:200, 16:*201*–202, 16:*203*
Desert lions, 14:372
Desert lynx. *See* Caracals
Desert oryx, 16:30
Desert pocket gophers, 16:195*t*
Desert pocket mice, 16:199, 16:200
Desert shrews, 13:*248*, 13:*254*, 13:259–*260*
Desert warthogs. *See* Warthogs
Desert whales. *See* Gray whales
Deserts, 12:20–21, 12:113, 12:*117*
  See also Habitats
Desmana spp., 13:279
Desmana moschata. See Russian desmans
Desmaninae. *See* Aquatic desmans
Desmans, **13:279–288,** 13:*283*, 13:287*t*
  aquatic, 12:14, 13:196–199, 13:279, 13:280
  behavior, 13:198
  distribution, 12:134
  feeding ecology, 13:198–199
  habitats, 13:197
  physical characteristics, 13:195–196
  taxonomy, 13:193, 13:194
Desmarest's spiny pocket mice, 16:*204*, 16:*206*
Desmatolagus gobiensis, 16:480
Desmodontinae, 13:415, 13:417
Desmodus rotundus. See Vampire bats
Desmostylia, 15:131, 15:133
Devil-fishes. *See* Gray whales
Dholes, 14:258, 14:265, 14:266, 14:268, 14:270, 14:273, 14:283*t*
Diacodexeidae, 15:263
Diacodexis spp., 15:263, 15:264
Diadem roundleaf bats, 13:*402*, 13:*406*, 13:407–*408*
Diademed sifakas, 14:63, 14:*66*, 14:67
Diaemus youngii. See White-winged vampires
Diamond, J., 15:146, 15:151
Diana monkeys, 14:193, 14:204*t*
Dian's tarsiers, 14:93, 14:95, 14:*96*, 14:*97*, 14:98–99
Diapause, kangaroos, 12:109
Diaphragm, 12:8, 12:36, 12:45–46
  See also Physical characteristics
Diarthrognathus spp., 12:11
Dibatags, 16:48, 16:56*t*
Dibblers, 12:288
Diceratheres spp., 15:249
Dicerorhinus spp., 15:216, 15:217

Dicerorhinus hemitoechus. See Steppe rhinoceroses
Dicerorhinus kirchbergensis. See Merck's rhinoceroses
Dicerorhinus sumatrensis. See Sumatran rhinoceroses
Dicerorhinus tagicus, 15:249
Diceros spp., 15:216, 15:251
Diceros bicornis. See Black rhinoceroses
Diceros bicornis longipes, 15:260
Dicerotinae, 15:249
Dichobunidae, 15:263, 15:264
Diclidurinae, 13:355
Diclidurus spp. *See* Ghost bats
Diclidurus albus. See Northern ghost bats
Dicotyles spp., 15:291
Dicrocerus spp., 15:266, 15:343
Dicrostonyx spp., 16:226
Dicrostonyx groenlandicus. See Northern collared lemmings
Dicrostonyx torquatus. See Arctic lemmings
Dicrostonyx vinogradovi, 16:229
Dicynodontia, 12:10
Didelphidae. *See* New World opossums
Didelphimorphia. *See* New World opossums
Didelphinae, 12:249
Didelphis spp. *See* Common opossums
Didelphis aurita, 12:251
Didelphis marsupialis. See Southern opossums
Didelphis virginiana. See Virginia opossums
Didodinae, 16:211
Diet. *See* Feeding ecology
Digestive system, 12:46–48
  carnivores, 12:120
  herbivores, 12:14–15, 12:120–124
  See also Physical characteristics
Digitigrades, 12:*39*
Dijkgraf, Sven, 12:54
Dikdiks, **16:59–72**
  behavior, 16:5, 16:60–62
  conservation status, 16:63
  distribution, 16:*59*–60, 16:71*t*
  evolution, 16:59
  feeding ecology, 15:142, 15:272, 16:62
  habitats, 15:269, 16:60
  humans and, 16:64, 16:66–70
  physical characteristics, 15:142, 16:59, 16:66–70
  reproduction, 16:7, 16:62–63
  species of, 16:66–71*t*
  taxonomy, 16:59
Dilododontids, 15:135
Dinagat gymnures, 13:207, 13:213*t*
Dinagat moonrats. *See* Dinagat gymnures
Dinaromys spp., 16:228
Dingisos. *See* Forbidden tree kangaroos
Dingos, 12:136, 12:180, 14:257, 14:266, 14:*270,* 14:273, 14:292
Dinocerata, 12:11, 15:131
Dinomyidae. *See* Pacaranas
Dinomys branickii. See Pacaranas
Dinopithecus spp., 14:189
Dinosaurs, mammals and, 12:33
Diodorus Siculus, 14:290
Diphylla ecaudata. See Hairy-legged vampire bats
Diphyodonty, 12:9, 12:89
Diplogale hosei. See Hose's palm civets
Diplomesodon spp., 13:265, 13:266–267
Diplomesodon pulchellum. See Piebald shrews
Diplomys caniceps, 16:450–451
Diplomys labilis. See Rufous tree rats
Diplomys rufodorsalis, 16:450
Dipodidae, 12:134, 16:121, 16:123, **16:211–224,** 16:*218,* 16:*223t–224t,* 16:283
Dipodinae, 16:211, 16:212
Dipodoidea, 16:211
Dipodomyinae, 16:199
Dipodomys spp., 16:124, 16:199, 16:200
Dipodomys deserti. See Desert kangaroo rats
Dipodomys elator. See Texas kangaroo rats
Dipodomys elephantinus. See Elephant-eared kangaroo rats
Dipodomys heermanni. See Heermann's kangaroo rats
Dipodomys heermanni morroensis, 16:203
Dipodomys ingens. See Giant kangaroo rats
Dipodomys merriami, 16:200, 16:203
Dipodomys merriami parvus, 16:203
Dipodomys microps. See Chisel-toothed kangaroo rats
Dipodomys nelsoni. See Nelson's kangaroo rats
Dipodomys nitratoides. See San Joaquin Valley kangaroo rats
Dipodomys nitratoides exilis, 16:203
Dipodomys nitratoides tipton, 16:203
Dipodomys ordii. See Ord's kangaroo rats
Dipodomys spectabilis. See Banner-tailed kangaroo rats
Dipodomys stephensi. See Stephen's kangaroo rats
Diprotodontia, 12:137, **13:31–41,** 13:83
Dipus sagitta. See Hairy-footed jerboas
Dire wolves, 14:265
Direct exploitation, 12:215
  See also Humans
Direct observations, 12:198–199, 12:200–202
Diseases
  Argentine hemorrhagic fever, 16:270
  Bolivian hemorrhagic fever, 16:270
  calicivirus, 12:187–188, 12:223
  Giarda, 16:127
  Gilchrist's disease, 16:127
  hantaviruses, 16:128, 16:270
  HIV, 12:222
  infectious, 12:215–216, 12:222–223
  Lyme disease, 16:127
  myxoma virus, 12:187
  rabbit calicivirus disease, 12:187–188, 12:223
  rickets, 12:74
  tetanus, 12:77–78
  viral resistance, 12:187
Disk-winged bats, 13:311, **13:473–477,** 13:*475*
Disney's Animal Kingdom, 12:204
Dispersal
  distribution and, 12:130–131
  for inbreeding avoidance, 12:110
  See also Distribution
Dispersed social systems, 12:145–148
Disruptive coloration, 12:38
  See also Physical characteristics
Distance sampling, 12:198, 12:200–202
Distoechurus spp., 13:140, 13:141, 13:142
Distoechurus pennatus. See Feather-tailed possums
Distribution, **12:129–139**
  aardvarks, 15:*155,* 15:156
  agoutis, 16:*407,* 16:*411*–414, 16:415*t*
  anteaters, 13:*171,* 13:172, 13:*177–179*
  armadillos, 13:*181,* 13:182–183, 13:*187*–190, 13:190*t*–192*t*
Artiodactyla, 12:129, 12:132, 12:136, 15:269
aye-ayes, 14:*85,* 14:86
baijis, 15:*19,* 15:20
bandicoots, 13:2
  dry-country, 13:*9,* 13:10, 13:*14*–16, 13:16*t*–17*t*
  rainforest, 13:*9,* 13:10, 13:*14*–17*t*
bats, 13:310–311
  American leaf-nosed, 13:*413,* 13:415, 13:*423*–432, 13:433*t*–434*t*
  bulldog, 13:*443,* 13:445, 13:*449–450*
  disk-winged, 13:*473,* 13:*476*–477
  Emballonuridae, 13:*355,* 13:*360*–363, 13:363*t*–364*t*
  false vampire, 13:*379,* 13:*384*–385
  funnel-eared, 13:*459,* 13:460, 13:*463*–465
  horseshoe, 13:*387,* 13:388–389, 13:*397*–400
  Kitti's hog-nosed, 13:*367,* 13:368
  Molossidae, 13:*483,* 13:*490*–493, 13:493*t*–495*t*
  mouse-tailed, 13:351, 13:*351,* 13:*353*
  moustached, 13:435, 13:*435,* 13:*440*–441, 13:442*t*
  New Zealand short-tailed, 13:*453,* 13:454, 13:457
  Old World fruit, 13:*319,* 13:*325*–330, 13:331*t*–332*t,* 13:*333,* 13:334, 13:*340*–347, 13:348*t*–349*t*
  Old World leaf-nosed, 13:*401,* 13:402, 13:*407*–409, 13:409*t*–410*t*
  Old World sucker-footed, 13:*479,* 13:480
  slit-faced, 13:*371,* 13:*375,* 13:376*t*–377*t*
  smoky, 13:*467,* 13:*470*
  Vespertilionidae, 13:*497,* 13:498, 13:*507*–514, 13:515*t*–516*t,* 13:*519,* 13:520–521, 13:*524–525,* 13:526*t*
bears, 14:*295,* 14:296, 14:*302*–305, 14:306*t*
beavers, 16:*177,* 16:179, 16:*183*
bilbies, 13:*19*–20
botos, 15:*27,* 15:28
Bovidae, 16:4
  Antilopinae, 16:*45,* 16:46, 16:*50*–55, 16:56*t*–57*t*
  Bovinae, 16:*11,* 16:13, 16:*18*–23, 16:24*t*–25*t*
  Caprinae, 16:*87,* 16:90–91, 16:*98*–103, 16:103*t*–104*t*
  duikers, 16:*73,* 16:*80*–84, 16:84*t*–85*t,* 16:85*t*
  Hippotraginae, 16:*27,* 16:29, 16:*37*–42*t,* 16:43*t*
  Neotraginae, 16:*59*–60, 16:*66*–71*t*
bushbabies, 14:5, 14:*23,* 14:24, 14:*28*–32, 14:32*t*–33*t*
Camelidae, 15:*313,* 15:315, 15:*320*–323
cane rats, 16:*333,* 16:334, 16:*337–338*
Canidae, 14:*265,* 14:267, 14:*276*–283, 14:283*t*–284*t*
capybaras, 16:*401,* 16:402
Carnivora, 12:129, 12:131, 12:132, 12:136, 14:257
cats, 12:131, 12:136, 14:257, 14:*369,* 14:370–371, 14:*379*–389, 14:390*t*–391*t*
Caviidae, 16:*389,* 16:390, 16:*395*–398, 16:399*t*

Distribution *(continued)*
Cetacea, 12:129, 12:138, 15:5
Balaenidae, 15:5, 15:*107*, 15:109, 15:*115*–117
beaked whales, 15:*59*, 15:61, 15:*64*–68, 15:69*t*–70*t*
dolphins, 15:*41*, 15:43, 15:*53*–55, 15:56*t*–57*t*
franciscana dolphins, 15:*23*, 15:24
Ganges and Indus dolphins, 15:*13*, 15:14
gray whales, 15:*93*, 15:94–95
Monodontidae, 15:*81*, 15:82–83, 15:*90*–*91*
porpoises, 15:*33*, 15:34, 15:*38*–39, 15:39*t*
pygmy right whales, 15:*103*, 15:104
rorquals, 15:5, 15:*119*, 15:120–121, 15:*127*–130, 15:130*t*
sperm whales, 15:*73*, 15:74, 15:79–80
chevrotains, 15:*325*, 15:327, 15:*331*–333
Chinchillidae, 16:*377*, 16:*382*, 16:383, 16:383*t*
colugos, 13:*299*, 13:302, 13:*304*
coypus, 16:123, 16:*473*, 16:474
dasyurids, 12:*287*, 12:288, 12:*294*–298, 12:299*t*–301*t*
Dasyuromorphia, 12:279
deer
Chinese water, 15:*373*, 15:374
muntjacs, 15:*343*, 15:344, 15:*349*–354
musk, 15:*335*, 15:336, 15:*340*–341
New World, 15:*379*, 15:382, 15:*388*–396, 15:396*t*:
Old World, 15:*357*, 15:359, 15:*364*–371, 15:371*t*
Dipodidae, 16:*211*, 16:212, 16:*219*–*222*, 16:223*t*–224*t*
Diprotodontia, 13:34
dormice, 16:*317*–318, 16:*323*–326, 16:327*t*–328*t*
duck-billed platypuses, 12:137, 12:230, 12:*243*, 12:244
Dugongidae, 15:*199*, 15:*203*
echidnas, 12:*235*, 12:236, 12:240*t*
elephants, 12:*135*, 15:*161*, 15:166, 15:*174*–175
Equidae, 15:*225*, 15:226, 15:*232*–235, 15:236*t*
Erinaceidae, 13:*203*, 13:204, 13:*209*–213, 13:212*t*–213*t*
giant hutias, 16:469, 16:*469*, 16:471*t*
gibbons, 14:*207*, 14:210, 14:*218*–*222*
Giraffidae, 15:*399*, 15:401, 15:*408*
great apes, 14:*225*, 14:227, 14:*237*–240
gundis, 16:*311*, 16:312, 16:*314*–315
Herpestidae, 14:*347*, 14:348–349, 14:*354*–356, 14:357*t*
Heteromyidae, 16:*199*, 16:200, 16:*205*–209, 16:208*t*–209*t*
hippopotamuses, 15:*301*, 15:306, 15:*310*–311
humans, 14:*241*, 14:247
hutias, 16:*461*, 16:467*t*
Hyaenidae, 14:*359*, 14:362, 14:*364*–*366*
hyraxes, 15:*177*, 15:179, 15:*186*–188, 15:189*t*
Indriidae, 14:*63*, 14:65, 14:*69*–72
Insectivora, 13:197
koalas, 13:*43*, 13:44
Lagomorpha, 16:481
lemurs, 12:135–136, 14:5, 14:*47*, 14:49, 14:*55*–60
Cheirogaleidae, 14:*35*, 14:36, 14:*41*–44
sportive, 14:*73*, 14:75, 14:*79*–*83*
Leporidae, 16:*505*–506, 16:*511*–515, 16:515*t*–516*t*
Lorisidae, 14:*13*, 14:14, 14:*18*–20, 14:21*t*
Macropodidae, 12:137, 13:*83*, 13:84–85, 13:*93*–100, 13:101*t*–102*t*
manatees, 15:*205*, 15:*211*–212
Megalonchidae, 13:*159*
Megalonychidae, 13:*155*–156
moles
golden, 13:197, 13:*215*, 13:216, 13:*220*–*222*, 13:222*t*
marsupial, 13:*25*, 13:26, 13:*28*
monitos del monte, 12:*273*, 12:274
monkeys
Atelidae, 14:*155*, 14:156, 14:*162*–166, 14:166*t*–167*t*
Callitrichidae, 14:*115*, 14:116, 14:*127*–131, 14:132*t*
Cebidae, 14:*101*, 14:102–103, 14:*108*–112
cheek-pouched, 14:*187*, 14:191, 14:*197*–204, 14:204*t*–205*t*
leaf-monkeys, 14:5, 14:*171*, 14:173, 14:*178*–183, 14:184*t*–185*t*
night, 14:*135*, 14:137, 14:141*t*
Pitheciidae, 14:*143*, 14:144–145, 14:*150*–*152*, 14:153*t*
monotremes, 12:136–137, 12:230
mountain beavers, 16:*131*
Muridae, 12:134, 16:123, 16:*281*, 16:283, 16:*288*–295, 16:296*t*–297*t*
Arvicolinae, 16:*225*, 16:226, 16:*233*–238, 16:237*t*–238*t*
hamsters, 16:*239*–240, 16:*245*–247*t*
Murinae, 16:*249*, 16:250, 16:*255*–260
Sigmodontinae, 16:*263*, 16:265–266, 16:*272*–276, 16:277*t*–278*t*
musky rat-kangaroos, 13:*69*, 13:70
Mustelidae, 14:*319*, 14:*327*–331, 14:332*t*–333*t*
numbats, 12:*303*
octodonts, 16:*433*, 16:434, 16:*438*–440, 16:440*t*
opossums
New World, 12:*249*, 12:250–251, 12:*257*–263, 12:264*t*–265*t*
shrew, 12:*267*, 12:*270*
Otariidae, 14:*393*, 14:394–395, 14:*402*–406
pacaranas, 16:*385*, 16:386
pacas, 16:*417*, 16:418, 16:*423*–424
pangolins, 16:*107*, 16:110, 16:*115*–119
peccaries, 15:*291*, 15:292, 15:*298*–*299*
Perissodactyla, 15:217
Petauridae, 13:*125*, 13:*131*–133, 13:133*t*
Phalangeridae, 13:*57*, 13:59, 13:*63*–*66t*
pigs, 15:*275*, 15:276–277, 15:*284*–288, 15:289*t*–290*t*
pikas, 16:*491*, 16:*497*–501, 16:501*t*–502*t*
pocket gophers, 16:*185*, 16:187, 16:*192*–194, 16:195*t*–197*t*
porcupines
New World, 16:*365*, 16:367, 16:*371*–373, 16:374*t*
Old World, 16:*351*, 16:352, 16:*357*–364
possums
feather-tailed, 13:*139*, 13:140, 13:*144*
honey, 13:*135*, 13:136
pygmy, 13:*105*, 13:106, 13:*110*–111
primates, 14:5–6
Procyonidae, 14:*309*, 14:309–310, 14:*313*–*315*, 14:315*t*–316*t*
pronghorns, 15:*411*, 15:412
Pseudocheiridae, 13:*113*, 13:114, 13:*119*–123, 13:122*t*–123*t*
rat-kangaroos, 13:*73*, 13:74, 13:*78*–*80*
rats
African mole-rats, 16:*339*, 16:342, 16:*347*–349, 16:349*t*–350*t*
chinchilla, 16:*443*, 16:444, 16:*447*–448
dassie, 16:*329*, 16:330
spiny, 16:*449*, 16:450, 16:*453*–457, 16:458*t*
rhinoceroses, 15:*249*, 15:252, 15:*257*–261
rodents, 16:123
sengis, 16:*517*, 16:519, 16:*524*–530, 16:531*t*
shrews
red-toothed, 13:*247*, 13:248, 13:*255*–263, 13:263*t*
West Indian, 13:*243*, 13:244
white-toothed, 13:*265*, 13:266, 13:*271*–275, 13:276*t*–277*t*
Sirenia, 15:192
solenodons, 13:*237*, 13:238, 13:*240*
springhares, 16:*307*, 16:308
squirrels
flying, 16:*135*, 16:*139*–*140*, 16:141*t*–142*t*
ground, 12:131, 16:*143*, 16:144, 16:*151*–158, 16:158*t*–160*t*
scaly-tailed, 16:*299*, 16:300, 16:*302*–*305*
tree, 16:*163*–164, 16:*169*–174*t*, 16:175*t*
Talpidae, 13:*279*, 13:280, 13:*284*–287, 13:287*t*–288*t*
tapirs, 15:*237*, 15:238, 15:*245*–247
tarsiers, 14:*91*, 14:93, 14:*97*–99
tenrecs, 13:*225*, 13:226, 13:*232*–234, 13:234*t*
three-toed tree sloths, 13:*161*, 13:162, 13:*167*–169
tree shrews, 13:*289*, 13:291, 13:*294*–*296*, 13:297*t*–298*t*
true seals, 12:129, 12:138, 14:*417*–418, 14:*427*–434, 14:435*t*
tuco-tucos, 16:*425*, 16:*429*–430, 16:430*t*–431*t*
Viverridae, 14:*335*, 14:336, 14:*340*–343, 14:344*t*–345*t*
walruses, 12:138, 14:*409*, 14:411–412
wombats, 13:*51*, 13:*55*–*56*
Xenartha, 13:150
Diurnal mammals, 12:79, 12:80
See also specific species
Diversity. *See* Biodiversity
Diving mammals, 12:67–68
See also specific diving mammals
Diving shrews, 13:*249*
Djarthia murgonensis, 12:277
DNA, phylogenetics and, 12:26–33
Doberman pinschers, 14:289
Dobsonia spp. *See* Naked-backed fruit bats
Dobsonia magna. *See* Naked-backed fruit bats
Dobson's horseshoe bats, 13:*395*
Dobson's long-tongued dawn bats. *See* Dawn fruit bats
Dobson's shrew tenrecs, 13:*231*, 13:*233*
Dog-faced fruit bats. *See* Indian fruit bats
Dog family. *See* Canidae
Doglike bats. *See* Lesser dog-faced bats
Dogs, 12:*215*, **14:265–285**, 14:*275*
behavior, 14:258, 14:268–269

Dogs *(continued)*
conservation status, 14:272–273
distribution, 12:137, 14:267
evolution, 12:29, 14:265–266
feeding ecology, 14:260, 14:269–271
gaits of, 12:*39*
habitats, 14:267
humans and, 14:273
neonatal requirements, 12:126
physical characteristics, 14:266–267
reproduction, 12:106, 12:107, 14:271–272
self recognition, 12:159
smell, 12:*81*
species of, 14:*278*, 14:281–282, 14:283*t*, 14:284*t*
taxonomy, 14:256, 14:265–266
See also Domestic dogs
Dolichopithecus spp., 14:172, 14:189
Dolichotinae, 16:389
Dolichotis spp., 16:389, 16:390
Dolichotis patagonum. *See* Maras
Dologale spp., 14:347
Dolphins, 12:14, **15:41–58,** 15:*42*, 15:*43*, 15:*52*
behavior, 15:5, 15:44–46
conservation status, 15:9, 15:48–50
distributioin, 12:138
distribution, 15:*41*, 15:43
echolocation, 12:86, 12:87
encephalization quotient, 12:149
evolution, 15:3, 15:41–42
feeding ecology, 15:46–47
field studies of, 12:201
habitats, 15:5, 15:43–44
hearing, 12:82
humans and, 12:173, 15:10, 15:51
physical characteristics, 12:63, 12:66, 12:67, 15:4, 15:42–43
reproduction, 12:91, 12:94, 12:103, 15:47–48
self recognition, 12:159
smell, 12:80
species of, 15:53–55, 15:56*t*–57*t*
taxonomy, 15:41–42
touch, 12:83
See also specific types of dolphins
Domestic camels, 12:179–180, 15:150–151
See also Camels
Domestic cats, 12:180, **14:289–294,** 14:*290*
animal rights movement and, 14:293–294
behavior, 12:145
distribution, 12:137, 14:370
evolution, 14:290–291
feeding ecology, 12:*175*, 14:372
history of, 12:172
humans and, 14:291, 14:292, 14:375
introduced species, 12:215
popularity of, 14:291
reproduction, 14:291–292
vision, 12:79
wildlife and, 14:291
See also Cats
Domestic cattle, 12:177, 15:*148*, 15:263, 16:8
behavior, 15:271, 16:6
cloning, 15:151
dairy, 12:122, 15:151
domestication of, 12:172, 15:145
humans and, 15:273, 16:8, 16:15
neonatal milk, 12:127
stomach, 12:*124*
See also Cattle; Ungulates; specific types of cattle
Domestic dogs, 12:180, **14:287–294**
animal rights movement and, 14:293–294
behavior, 12:143, 12:*180*, 14:268, 14:269
breeds, 14:288–289
distribution, 14:267, 14:288
evolution, 12:29, 14:287
history of, 12:172
humans and, 14:262, 14:288, 14:292–293
popularity of, 14:291
reproduction, 14:292
See also Dogs
Domestic donkeys, 12:176–177, 15:*219*
Domestic elephants, 15:172
See also Elephants
Domestic goats, 12:*177*, 12:179, 15:145, 15:*147*, 15:263
distribution, 16:91
domestication of, 12:172, 15:145, 15:146–148
humans and, 15:273, 16:8, 16:9, 16:95
See also Goats
Domestic guinea pigs, 12:180–181, 16:*391*, 16:399*t*
See also Guinea pigs
Domestic horses, 12:175–176, 15:*151*, 15:236*t*
behavior, 15:*230*
distribution, 15:217
domestication of, 15:149–150
humans and, 15:223
reproduction, 15:229
smell, 12:80
See also Horses
Domestic llamas, 12:180, 15:151
Domestic pigs, 12:179, 15:263
conservation status, 15:281
distribution, 15:277
domestication of, 12:172, 15:145, 15:148–149
humans and, 15:*150*, 15:273, 15:281, 15:282
pot-bellied, 12:191, 15:*278*
reproduction, 15:272, 15:279, 15:280
Yorkshire, 12:*172*
See also Pigs
Domestic rabbits, 12:180
See also Rabbits
Domestic sheep, 12:*174*, 12:178–179, 15:263, 16:*94*
distribution, 16:91
domestication of, 15:145, 15:146–148
humans and, 15:*147*, 15:273, 16:8, 16:95
physical characteristics, 15:268
reproduction, 12:*90*
See also Sheep
Domesticated asses. *See* Domestic donkeys
Domestication, **12:171–181, 15:145–153**
See also Zoos; specific domestic animals
Dominance, 12:*141*
See also Behavior
Donkeys, 12:*172*, 12:176–177, 15:145, 15:*146*, 15:*219*, 15:223, 15:*229*, 15:230
Dorcas gazelles, 12:222, 16:48, 16:*49*, 16:*50*–51
Dorcatragus megalotis. *See* Beiras
Dorcopsis spp., 13:83, 13:85, 13:86
Dorcopsis hageni. *See* White-striped dorcopsises
Dorcopsis luctuosa. *See* Gray dorcopsises
Dorcopsis macleayi. *See* Papuan forest wallabies
Dorcopsis veterum. *See* Brown dorcopsises
Dorcopsises
brown, 16:102*t*
gray, 13:*91*, 13:*94*, 13:100
white-striped, 16:102*t*
Dorcopsulus spp., 13:83, 13:85, 13:86
Dorcopsulus vanheurni. *See* Lesser forest wallabies
Doria's tree kangaroos, 13:36, 13:102*t*
Dormancy. *See* Hibernation
Dormice, 12:134, 16:122, 16:125, **16:317–328,** 16:*322*, 16:327*t*–328*t*
Dormouse possums, 13:35
Double helix, 12:26
Douc langurs. *See* Red-shanked douc langurs
Douglas's squirrels, 16:*164*
Dowler, R. C., 16:270
Draft dogs, 14:293
Draft guards, 14:288
Draft horses, 12:*176*
Draft mammals, 12:15
See also specific draft mammals
Drepanididae. *See* Hawaiian honeycreepers
Drills, 14:188, 14:191, 14:193–194, 14:205*t*
Dromedary camels, 12:*122*, 15:*315*, 15:*319*, 15:*320*–321
behavior, 15:316
domestication of, 12:179, 15:150–151
habitats, 15:316
humans and, 15:318
physical characteristics, 15:268, 15:314
taxonomy, 15:313
Dromiciops australis. *See* Monitos del monte
Dromiciops gliroides. *See* Monitos del monte
Dromomerycids, 15:137
Dry-country bandicoots, 13:1, **13:9–18,** 13:*13*, 13:16*t*–17*t*
Dryland mouse opossums. *See* Orange mouse opossums
Dryomys spp. *See* Forest dormice
Dryomys laniger. *See* Woolly dormice
Dryomys nitedula. *See* Forest dormice
Dryomys sichuanensis. *See* Chinese dormice
Dsinezumi shrews, 13:276*t*
Duck-billed platypuses, 12:*231*, **12:243–248,** 12:*244*
behavior, 12:230–231, 12:244–245, 12:*245*, 12:*246*
conservation status, 12:233–234, 12:246–247
distribution, 12:137, 12:230, 12:*243*, 12:244
evolution, 12:228, 12:243
feeding ecology, 12:231, 12:245
habitats, 12:*229*, 12:230, 12:244, 12:*247*
humans and, 12:234, 12:247
physical characteristics, 12:228–230, 12:243–244, 12:*244*, 12:*246*
reproduction, 12:101, 12:106, 12:108, 12:*228*, 12:231–233, 12:*232*, 12:245–246
taxonomy, 12:243
Dugong dugon. *See* Dugongs
Dugongidae, 12:44, 12:91, 12:94, 15:191, 15:*199*, **15:199–204,** 15:*202*
See also Dugongs; Steller's sea cows
Dugongids. *See* Dugongidae
Dugonginae. *See* Dugongs
Dugongs, 15:*194*, **15:199–204,** 15:*200*, 15:*202*, 15:*203*
behavior, 15:192–193, 15:200
conservation status, 15:195, 15:201

INDEX

Dugongs *(continued)*
distribution, 15:*191*, 15:192, 15:*199*
evolution, 15:191, 15:199
feeding ecology, 15:193, 15:*195*, 15:200
habitats, 15:192
humans and, 15:195–196, 15:201
physical characteristics, 15:*192*, 15:199
reproduction, 12:103, 15:194, 15:201
taxonomy, 15:191
Duikers, 16:1, **16:73–85,** 16:*78*, 16:*79*
behavior, 16:5, 16:74
conservation status, 16:80–84, 16:84*t*–85*t*
distribution, 16:*73*
evolution, 15:265, 16:73
feeding ecology, 16:74, 16:76
habitats, 16:74
humans and, 16:77
physical characteristics, 15:138, 15:267, 16:2, 16:73
reproduction, 16:76–77
species of, 16:80–83, 16:84*t*–85*t*
taxonomy, 16:73
Dunce hinnies, 12:177
Dune mole-rats, 16:339, 16:342, 16:343, 16:344
Cape, 16:345, 16:*346*, 16:*347*
Namaqua, 16:*344*, 16:345, 16:349*t*
Dunnarts, 12:280–282, 12:289
Duplicendentata. *See* Lagomorpha
Durer, Albrecht, 15:254
Durrell Wildlife Conservation Trust, 15:287
Dusicyon australis. See Falkland Island wolves
Dusky antechinuses, 12:279, 12:299*t*
Dusky bushbabies, 14:26, 14:32*t*
Dusky dolphins, 12:124, 15:*46*, 15:47
Dusky flying foxes, 13:331*t*
Dusky-footed sengis, 16:531*t*
Dusky-footed woodrats, 16:277*t*
Dusky hopping mice, 16:261*t*
Dusky leaf-nosed bats, 13:*403*, 13:*404*
Dusky marsupials. *See* Dusky antechinuses
Dusky sengis, 16:531*t*
Dusky titi monkeys, 14:143, 14:144, 14:145, 14:*149*, 14:151–*152*
Dust, glacial. *See* Loess
Dusty tree kangaroos. *See* Bennett's tree kangaroos
Dwarf antelopes, 16:2, 16:5, 16:7, 16:*60*, 16:63, 16:*65*, 16:*66*, 16:68–69
Dwarf armadillos. *See* Pichi armadillos
Dwarf blue sheep, 16:87, 16:91, 16:94, 16:103*t*
Dwarf brocket deer, 15:396*t*
Dwarf bushbabies. *See* Demidoff's bushbabies
Dwarf chimpanzees. *See* Bonobos
Dwarf deer, 15:381, 15:384
Dwarf dormice. *See* African dormice
Dwarf duikers, 16:73, 16:77
Dwarf epauletted fruit bats, 13:*340*, 13:*345*–346
Dwarf flying mice. *See* Zenker's flying mice
Dwarf flying squirrels, 16:136, 16:*139*
Dwarf gibbons. *See* Kloss gibbons
Dwarf hamsters, 16:*125*, 16:239, 16:240
Dwarf hippopotamuses, 15:301
Dwarf hutias, 16:467*t*
Dwarf lemurs, 14:*4*, **14:35–45,** 14:*40*
behavior, 14:36–37
conservation status, 14:39
distribution, 14:*35*, 14:36
evolution, 14:35
feeding ecology, 14:37–38
habitats, 14:36
humans and, 14:39
physical characteristics, 14:35
reproduction, 14:38
species of, 14:*41*–42
taxonomy, 14:3, 14:35
Dwarf little fruit bats, 13:419, 13:*423*, 13:*432*
Dwarf minke whales, 12:*200*
Dwarf mongooses, 12:146, 12:148, 14:259, 14:261, 14:*350*, 14:*353*, 14:*354*, 14:356
Dwarf ponies, 15:*218*
Dwarf porcupines. *See* Hairy dwarf porcupines
Dwarf rabbits, 16:*507*
Dwarf scalytails. *See* Lesser anomalures
Dwarf sloths. *See* Monk sloths
Dwarf sperm whales, 15:1, 15:6, 15:73, 15:74, 15:76, 15:77, 15:78
Dyacopterus spadiceus. See Dyak fruit bats
Dyak fruit bats, 12:106, 13:*339*, 13:*341*, 13:344
Dycrostonix torquatus, 16:270
Dynamic lift, 12:56
Dzhungarian hamsters, 16:*240*, 16:*242*, 16:243, 16:247*t*
Dzigettais. *See* Asiatic wild asses

E

Eagles, 12:184, 12:192
Ealey's ningauis. *See* Pilbara ningauis
Eared hutias, 16:461, 16:467*t*
Eared seals, **14:393–408,** 14:*401*
behavior, 14:396–397
conservation status, 14:262, 14:399–400
distribution, 14:394–395
evolution, 14:393
feeding ecology, 14:397–398
habitats, 14:395–396
humans and, 14:400
physical characteristics, 14:393–394
reproduction, 14:262, 14:398–399
species of, 14:*402*–406, 14:407*t*–408*t*
taxonomy, 14:256, 14:393
Ears, 12:8–9, 12:*41*
bats, 12:54
subterranean mammals, 12:77
See also Physical characteristics
East African hedgehogs, 13:*198*
East African long-nosed sengis. *See* Rufous sengis
East African mole rats, 12:74, 16:297*t*
East Asian tailless leaf-nosed bats, 13:409*t*
East Caucasian turs, 16:91, 16:92, 16:104*t*
East Fresian sheep, 15:147
East Mediterranean blind mole-rats, 12:74, 12:76–77
Eastern barred bandicoots, 13:4, 13:5, 13:10, 13:*11*–12, 13:*13*, 13:*14*
Eastern Brazilian marmosets, 14:115, 14:116, 14:117, 14:120
Eastern chipmunks, 16:126, 16:*148*, 16:*149*, 16:153–*154*
Eastern cottontails, 16:481, 16:487, 16:*510*, 16:*512*, 16:514–515
Eastern European hedgehogs, 13:212*t*
Eastern fox squirrels, 16:*164*, 16:*166*, 16:*168*, 16:*169*, 16:171
Eastern gorillas, 14:225, 14:*228*, 14:232, 14:*236*, 14:*238–239*
Eastern gray kangaroos, 12:108–109, 12:148, 12:185, 13:*32*, 13:38, 13:*87*, 13:*91*, 13:*94*
Eastern gray squirrels. *See* Gray squirrels
Eastern horseshoe bats, 13:*388*, 13:390, 13:392, 13:*395*, 13:*396*
Eastern lowland gorillas, 14:225, 14:227, 14:*230*, 14:232
Eastern Malagasy ring-tailed mongooses, 14:*348*
Eastern moles, 12:*71*, 12:*187*, 13:*195*, 13:198, 13:280, 13:*281*, 13:*283*, 13:*284–285*
Eastern pipistrelles, 13:*506*, 13:*509*, 13:515
Eastern pocket gophers. *See* Plains pocket gophers
Eastern pygmy possums, 13:*106*, 13:107, 13:*108*, 13:*109*, 13:*110*
Eastern quolls, 12:279, 12:280, 12:283, 12:*287*, 12:299*t*
Eastern rock sengis, 16:*518*, 16:*523*, 16:*528*–529
Eastern short-tailed opossums. *See* Southern short-tailed opossums
Eastern spotted skunks, 14:*321*, 14:324
Eastern tarsiers. *See* Spectral tarsiers
Eastern tree hyraxes, 15:177, 15:179, 15:180, 15:184, 15:*185*, 15:*187*
Eastern tube-nosed fruit bats. *See* Queensland tube-nosed bats
Eastern woodrats, 16:125, 16:277*t*
Eastern woolly lemurs, 14:63–64, 14:*65*, 14:*68*, 14:*69*
Eccrine sweat glands, 12:36
Echidnas, **12:235–241,** 12:*239*, 12:240*t*
behavior, 12:230–231, 12:*237*
conservation status, 12:233–234, 12:238
distribution, 12:137, 12:230, 12:*235*, 12:236
evolution, 12:227–228, 12:235
feeding ecology, 12:231, 12:237
habitats, 12:230, 12:237
humans and, 12:234, 12:238
physical characteristics, 12:228–230, 12:235–236
reproduction, 12:106, 12:*107*, 12:108, 12:231–233, 12:237–238
taxonomy, 12:227, 12:235
Echigo Plain moles, 13:282
Echimyidae. *See* Spiny rats
Echimys spp. *See* Spiny rats
Echimys blainvillei, 16:450–451
Echimys chrysurus. See White-faced arboreal spiny rats
Echimys pictus, 16:451
Echimys rhipidurus, 16:451
Echimys thomasi, 16:450
Echinoprocta rufescens. See Short-tailed porcupines
Echinops spp., 13:225, 13:226, 13:227, 13:230
Echinops telfairi. See Lesser hedgehog tenrecs
Echinosorex gymnura. See Malayan moonrats
Echinosorinae, 13:203
Echolocation, 12:53–55, 12:*59*, 12:*85*–87
See also Behavior; Physical characteristics
Echymipera spp., 13:10
Echymipera clara. See Large-toothed bandicoots

Echymipera davidi. See David's echymiperas
Echymipera echinista. See Menzies' echymiperas
Echymipera kalubu. See Spiny bandicoots
Echymipera rufescens. See Rufous spiny bandicoots
Ecological niches, 12:114–115, 12:117, 12:149–150, 12:183–184
 See also Habitats
Ecology, **12:113–119**
 animal intelligence and, 12:151
 behavioral, 12:148
 conservation biology and, 12:216–218
 domestication and, 12:175
 introduced species and, 12:182–193
 learning behavior and, 12:156
Economics, in conservation biology, 12:219–220, 12:223
Ectophylla alba. See White bats
Ecuador shrew opossums. *See* Silky shrew opossums
Edentates, 12:11
 locomotion, 12:14
 reproduction, 12:90
 See also Armadillos; Sloths
Edible dormice, 16:317, 16:*318*, 16:*319*, 16:*320*, 16:321, 16:*322*, 16:*323*, 16:326
Education, zoos and, 12:204, 12:*205*
Eggs, mammalian, 12:91–92, 12:101, 12:102–104, 12:106
 delayed implantation, 12:110, 12:111
 marsupials, 12:108
 placentals, 12:109
 See also Ovulation; Reproduction
Egyptian free-tailed bats, 13:495*t*
Egyptian fruit bats. *See* Egyptian rousettes
Egyptian maus, 14:291
Egyptian mongooses, 14:351–352, 14:357*t*
Egyptian rousettes, 12:46, 13:*309*, 13:*340*, 13:*342*
Egyptian slit-faced bats, 13:*371*, 13:*372*, 13:*373*, 13:*374*, 13:*375–376*
Egyptian spiny mice, 16:*251*, 16:*254*, 16:*257*, 16:259
Egyptian tomb bats, 13:364*t*
Ehrlich, Paul R., 12:217
Eidolon dupreanum. See Madagascan fruit bats
Eidolon helvum. See Straw-colored fruit bats
EIDs (Emerging infectious diseases), 12:222–223
Eira barbara. See Tayras
Eisenberg, John F., 16:201, 16:434, 16:486
Eisentraut, M., 13:341
Ekaltadeta sina, 13:70
Elands, 15:145–146, 16:6, 16:*9*, 16:11, 16:13
Elaphodus cephalophus. See Tufted deer
Elaphurus davidianus. See Pere David's deer
Elasmodontomys spp., 16:469
Elasmodontomys obliquus. See Puerto Rican giant hutias
Elasmotherium sibiricum, 15:249
Eld's deer, 12:19, 12:218, 15:361, 15:*363*, 15:*367*, 15:368
Elegant fat-tailed opossums, 12:253, 12:264*t*
Elegant water shrews, 13:*253*, 13:*258*, 13:259
Elephant-eared kangaroo rats, 16:209*t*
Elephant seals
 distribution, 14:418
 feeding ecology, 14:420
 habitats, 14:257
 physical characteristics, 14:256, 14:417
 reproduction, 14:261
 sexual dimorphism, 12:99
 See also Northern elephant seals; Southern elephant seals
Elephant shrews. *See* Sengis
Elephantidae. *See* Elephants
Elephantinae, 15:162
Elephants, 12:*18*, 12:*48*, 12:*221*, 15:146, **15:161–175**, 15:*163*, 15:*173*
 behavior, 15:167–169
 conservation status, 15:171–172
 digestive systems of, 12:14–15
 distribution, 12:*135*, 12:136, 15:*161*, 15:166
 evolution, 15:131, 15:133, 15:134, 15:161–165
 feeding ecology, 15:169–170
 feet, 12:*38*
 field studies of, 12:*199*, 12:*200*
 habitats, 15:166–167
 humans and, 12:173, 15:172
 locomotion, 12:42
 metabolism, 12:21
 migrations, 12:167
 neonatal milk, 12:127
 physical characteristics, 12:11–12, 12:79, 12:82, 12:83, 15:165–166
 reproduction, 12:90, 12:91, 12:94, 12:95, 12:98, 12:103, 12:106–108, 12:209, 15:170–171
 species of, 15:174–175
 taxonomy, 12:*29*, 15:131, 15:133, 15:134, 15:161–165
 teeth, 12:46
 vibrations, 12:83
 in zoos, 12:209
Elephantulus spp., 16:518–519, 16:521
Elephantulus brachyrhynchus. See Short-snouted sengis
Elephantulus edwardii. See Cape sengis
Elephantulus fuscipes. See Dusky-footed sengis
Elephantulus fuscus. See Dusky sengis
Elephantulus intufi. See Bushveld sengis
Elephantulus myurus. See Eastern rock sengis
Elephantulus revoili. See Somali sengis
Elephantulus rozeti. See North African sengis
Elephantulus rufescens. See Rufous sengis
Elephantulus rupestris. See Western rock sengis
Elephas spp., 15:162, 15:164–165
Elephas ekorensis, 15:163
Elephas hysudricus, 15:163–164
Elephas hysudrindicus, 15:164
Elephas maximus. See Asian elephants
Elephas maximus indicus, 15:164
Elephas maximus maximus, 15:164, 15:165
Elephas maximus sumatrensis, 15:164
Eligmodontia morgani, 16:267
Eligmodontia typus, 16:268, 16:269
Eliomys spp. *See* Garden dormice
Eliomys melanurus. See Asian garden dormice
Eliomys quercinus. See Garden dormice
Eliurus myoxinus. See Western Malagasy bushy-tailed rats
Eliurus tanala. See Greater Malagasy bushy-tailed rats
Elizabeth II (Queen of England), 16:423
Elk, 12:80, 12:*115*, 12:164–165, 15:145–146, 15:358
Elliot's short-tailed shrews, 13:198
Ellobius spp., 16:225
Emballonura alecto. See Small Asian sheath-tailed bats
Emballonura monticola. See Lesser sheath-tailed bats
Emballonura semicaudata. See Pacific sheath-tailed bats
Emballonuridae, **13:355–365**, 13:*359*, 13:*363t–364t*
Emballonurinae, 13:355
Embrithopoda, 15:131, 15:133
Embryonic development, 12:12
 See also Reproduction
Embryonic membranes, 12:92–94
Embryos, 12:92–94, 12:*94*, 14:*248*
 See also Reproduction
Emerging infectious diseases (EIDs), 12:222–223
Emily's tuco-tucos, 16:430*t*
Emmons, L., 16:165, 16:266, 16:457
Emperor tamarins, 14:*117*, 14:119, 14:121, 14:132*t*
Emulation, 12:158
 See also Behavior
Encephalization quotient (EQ), 12:149
Encoding information, 12:152
Enculturation, of apes, 12:162
Endangered species, in zoos, 12:204, 12:208, 12:212
Endangered Species Act (U.S.), 12:184
 black-tailed prairie dogs, 16:155
 dugongs, 15:201
 Fea's muntjacs, 15:351
 gray-backed sportive lemurs, 14:79
 gray whales, 15:99
 gray wolves, 14:277
 Idaho ground squirrels, 16:158
 island foxes, 12:192
 lesser Malay mouse deer, 15:332
 manatees, 15:209
 Milne-Edwards's sportive lemurs, 14:81
 northern sportive lemurs, 14:83
 red mouse lemurs, 14:43
 red-tailed sportive lemurs, 14:80
 Sirenia, 15:195
 small-toothed sportive lemurs, 14:83
 sperm whales, 15:78
 sportive lemurs, 14:76
 tree squirrels, 16:167
 Vespertilioninae, 13:504
 weasel sportive lemurs, 14:82
 white-footed sportive lemurs, 14:82
Endometrium, 12:106
 See also Physical characteristics
Endotheliodothelium placentation, 12:93–94
Endothermy, 12:3, 12:40–41, 12:50, 12:113
 See also Physical characteristics
Energy, 12:59, 12:71–72, 12:120
 See also Physical characteristics
Enhydra lutris. See Sea otters
Enrichment programs, in zoos, 12:209–210
Entelodontidae, 15:264, 15:275
Entertainment, zoos and, 12:203–204
Environment Protection and Biodiversity Conservation Act 1999 (Commonwealth of Australia), 13:76
Eodendrogale spp., 13:291
Eonycteris spp. *See* Dawn fruit bats
Eonycteris spelaea. See Dawn fruit bats
Eosimias spp., 14:2

Eotragus spp., 15:265, 16:1
Eozapus spp., 16:212
Eozapus setchuanus. See Chinese jumping mice
Eparctocyon, 15:343
Epauletted fruit bats, 13:312
dwarf, 13:*338*, 13:*343*–344
Wahlberg's, 13:*311*, 13:*338*, 13:*343*
Epidermis, 12:36
See also Physical characteristics
Episodic memory, 12:154–155
See also Cognition
Epitheliochorial placentation, 12:93–94
Epixerus spp., 16:163
Epixerus ebii, 16:167
Epixerus wilsoni, 16:167
Epomophorus spp. *See* Epauletted fruit bats
Epomophorus gambianus. See Gambian epauletted fruit bats
Epomophorus wahlbergi. See Wahlberg's epauletted fruit bats
Epomops spp. *See* African epauletted bats
Epomops franqueti. See Singing fruit bats
Eptesicus spp., 13:497, 13:498
Eptesicus fuscus. See Big brown bats
Eptesicus nilssoni. See Northern bats
Eptesicus serotinus. See Serotine bats
EQ (Encephalization quotient), 12:149
Equidae, **15:225–236,** 15:*231*
behavior, 15:142, 15:219, 15:226–228
conservation status, 15:221, 15:222, 15:229–230
distribution, 12:130–131, 12:134, 15:217, 15:*225*, 15:226
evolution, 15:136, 15:137, 15:215, 15:216, 15:225
feeding ecology, 12:118–119, 15:220, 15:228
habitats, 15:218, 15:226
humans and, 15:222, 15:223, 15:230
physical characteristics, 15:139, 15:140, 15:216, 15:217, 15:225–226
reproduction, 15:221, 15:228–229
species of, 15:232–235, 15:236*t*
taxonomy, 15:225
See also Asses; Horses; Zebras
Equus spp. *See* Equidae
Equus africanus. See African wild asses
Equus africanus africanus. See Nubian wild asses
Equus africanus somaliensis. See Somali wild asses
Equus asinus. See Domestic donkeys
Equus burchellii. See Plains zebras
Equus burchellii burchelli. See True plains zebras
Equus caballus. See Tarpans
Equus caballus caballus. See Domestic horses
Equus caballus przewalskii. See Przewalski's horses
Equus caballus silvestris. See Forest tarpans
Equus grevyi. See Grevy's zebras
Equus hemionus. See Asiatic wild asses
Equus hemionus hemippus. See Syrian onagers
Equus hemionus khur. See Khurs
Equus hemionus kulan. See Kulans
Equus hemionus onager. See Persian onagers
Equus kiang. See Kiangs
Equus quagga. See Quaggas
Equus zebra. See Mountain zebras
Equus zebra hartmanni. See Hartmann's mountain zebras
Equus zebra zebra. See Cape mountain zebras
Eremitalpa spp. *See* Grant's desert golden moles
Eremitalpa granti. See Grant's desert golden moles
Eremodipus lichtensteini. See Lichtenstein's jerboas
Eremotheres, 13:148
Erethizon spp., 16:366
Erethizon dorsatum. See North American porcupines
Erethizontidae. *See* New World porcupines
Erignathus barbaus. See Bearded seals
Erinaceidae, 13:193, 13:194, **13:203–214,** 13:*208*, 13:212*t*–213*t*
Erinaceinae, 13:203, 13:204
Erinaceus albiventris. See East African hedgehogs
Erinaceus amurensis. See Chinese hedgehogs
Erinaceus concolor. See Eastern European hedgehogs
Erinaceus europaeus. See Western European hedgehogs
Ermines, 12:116, 14:*324*, 14:*326*, 14:*327*–328
Ernest, K. A., 16:269
Erophylla sezekorni. See Buffy flower bats
Erythrocebus spp. *See* Patas monkeys
Erythrocebus patas. See Patas monkeys
Erythrocytes, 12:8, 12:36, 12:45, 12:60
See also Physical characteristics
Eschrichtiidae. *See* Gray whales
Eschrichtius robustus. See Gray whales
Estrous periods, 12:109
See also Reproduction
Ethics
in conservation biology, 12:217–218
zoos and, 12:212
Ethiopian hedgehogs, 13:213*t*
Ethiopian region, 12:135–136
Ethiopian wolves, 14:262, 14:269, 14:273, 14:*275*, 14:*276*, 14:277–278
Euarchontoglires, 12:26, 12:*31*, 12:33
Eubalaena spp. *See* Right whales
Eubalaena australis. See Southern right whales
Eubalaena glacialis. See North Atlantic right whales
Eubalaena japonica. See North Pacific right whales
Euchoreutes spp., 16:211
Euchoreutes naso. See Long-eared jerboas
Euchoreutes setchuanus. See Chinese jumping mice
Euchoreutinae. *See* Long-eared jerboas
Euderma maculatum. See Spotted bats
Eulemur spp., 14:8
Eulemur coronatus. See Crowned lemurs
Eulemur macaco flavifrons. See Blue-eyed lemurs
Eulemur mongoz. See Mongoose lemurs
Eumeryx spp., 15:265, 15:266
Eumetopias jubatus. See Steller sea lions
Eumops glaucinus. See Wagner's bonneted bats
Eumops perotis. See Western bonneted bats
Euoticus spp., 14:23, 14:26
Euoticus elegantulus. See Southern needle-clawed bushbabies
Euoticus pallidus. See Northern needle-clawed bushbabies
Eupelycosauria, 12:10
Eupetaurus cinereus. See Woolly flying squirrels
Euphractinae, 13:182
Euphractus sexcinctus. See Yellow armadillos
Eupleres spp., 14:347
Eupleres goudotii. See Falanoucs
Euplerinae, 14:335
Euprimates, 14:1
Eurarchintoglires, 15:134
Eurasian badgers. *See* European badgers
Eurasian beavers, 16:177, 16:*182*, 16:*183*
Eurasian lynx, 14:*378*, 14:*379*, 14:388
Eurasian pygmy shrews, 13:*254*, 13:*256*, 13:261–262
Eurasian red squirrels. *See* Red squirrels
Eurasian water shrews, 13:197, 13:*249*, 13:*250*, 13:*253*, 13:*258*
Eurasian wild boars. *See* Eurasian wild pigs
Eurasian wild pigs, 12:179, 15:148, 15:149, 15:*283*, 15:*285*, 15:288
behavior, 12:146
conservation status, 15:*280*, 15:281
evolution, 15:*264*, 15:*275*
habitats, 15:277
humans and, 15:281–*282*
reproduction, 15:279
smell, 12:85
Eurasian wolves, 12:180
Europe. *See* Palaearctic region
European badgers, 12:77, 14:324, 14:*326*, 14:*329*, 14:331
behavior, 12:145, 14:258–259, 14:*319*, 14:321–323
feeding ecology, 14:260
physical characteristics, 14:321
European beaked whales. *See* Gervais' beaked whales
European beavers. *See* Eurasian beavers
European bison, 16:24*t*
European brown bears. *See* Brown bears
European Community Habitats Directive, 16:246
European free-tailed bats, 13:*486*, 13:495*t*
European genets. *See* Common genets
European ground squirrels, 16:147, 16:*150*, 16:*155*–156
European hamsters. *See* Black-bellied hamsters
European hares, 16:*487*, 16:*508*, 16:*510*, 16:*511*–512
behavior, 16:482, 16:483, 16:*486*
conservation status, 12:222
distribution, 16:481, 16:505
humans and, 16:488
reproduction, 16:485–486, 16:508
European hedgehogs, 13:206
European mink, 12:117, 12:132, 14:324
European moles, 12:70–71, 12:*73*, 13:200, 13:*280*, 13:*281*, 13:*283*, 13:*285*–286
communication, 12:76
European mouflons, 12:178
European otters, 12:132, 14:*326*, 14:*330*
European pigs, 12:191–192
European pine martens, 12:132, 14:332*t*
European rabbits, 16:*510*, 16:513, 16:*514*
behavior, 16:482, 16:483
distribution, 16:481, 16:505–506
habitats, 16:482
humans and, 16:488, 16:509

European rabbits *(continued)*
integrated rabbit control, 12:186–188, 12:193
physical characteristics, 16:507
reproduction, 16:485, 16:486–487, 16:507–508
taxonomy, 16:505
European red deer, 12:19
European red squirrels. *See* Red squirrels
European roe deer, 15:384, 15:*386*, 15:*388*
European sousliks. *See* European ground squirrels
European water shrews, 13:195
European wild boars, 15:149
Euros. *See* Common wallaroos
Euroscaptor mizura. See Japanese mountain moles
Euroscaptor parvidens. See Small-toothed moles
Eurymylids. *See* Rodents
Euryzygomatomys spinosus. See Guiaras
Eusocial species, 12:146–147
Eustis, Dorothy Harrison, 14:293
Eutamias spp., 16:143
Eutherians. *See* Placentals
Evaporite lenses, 12:18
Even-toed ungulates. *See* Artiodactyla
Evening bats, 13:*504*, 13:*508*, 13:512–513
Eversmann's hamsters, 16:247*t*
Evolution, 12:10–11
aardvarks, 15:155
agoutis, 16:407
anteaters, 13:171
armadillos, 13:181–182
Artiodactyla, 15:131–133, 15:135–138, 15:263–267
aye-ayes, 14:85
baijis, 15:19
bandicoots, 13:1
dry-country, 13:9
rainforest, 13:9
bats, 13:308–309
American leaf-nosed, 13:413
bulldog, 13:443
disk-winged, 13:473
Emballonuridae, 13:355
false vampire, 13:379
funnel-eared, 13:459
horseshoe, 13:387
Kitti's hog-nosed, 13:367
Molossidae, 13:483
mouse-tailed, 13:351
moustached, 13:435
New Zealand short-tailed, 13:453
Old World fruit, 13:319, 13:333
Old World leaf-nosed, 13:401
Old World sucker-footed, 13:479
slit-faced, 13:371
smoky, 13:467
Vespertilionidae, 13:497, 13:519
bears, 14:295–296
beavers, 16:177
bilbies, 13:19
botos, 15:27
Bovidae, 16:1–2
Antilopinae, 16:45
Bovinae, 16:11
Caprinae, 16:87–89
duikers, 16:73
Hippotraginae, 16:27–28
Neotraginae, 16:59
bushbabies, 14:2, 14:3, 14:23
Camelidae, 15:313–314
Canidae, 14:265–266
capybaras, 16:401
Carnivora, 14:255–256
cats, 14:369
Caviidae, 16:389
Cetacea, 15:2–4
Balaenidae, 15:3, 15:107
beaked whales, 15:59
dolphins, 15:41–42
franciscana dolphins, 15:23
Ganges and Indus dolphins, 15:13
gray whales, 15:93
Monodontidae, 15:81
porpoises, 15:33
pygmy right whales, 15:103
sperm whales, 15:73
Cetacea rorquals, 12:*66*, 15:119
chevrotains, 15:325–326
Chinchillidae, 16:377
coevolution, 12:216–217
colugos, 13:299–300
conservation biology and, 12:218
convergent, 12:27
coypus, 16:473
dasyurids, 12:287
Dasyuromorphia, 12:277
deer
Chinese water, 15:373
muntjacs, 15:343
musk, 15:335
New World, 15:379–382
Old World, 15:357–358
Dipodidae, 16:211
Diprotodontia, 13:31–32
dormice, 16:317
duck-billed platypuses, 12:228, 12:243
Dugongidae, 15: 199
echidnas, 12:235
elephants, 15:161–165
Equidae, 15:225
Erinaceidae, 13:203
giant hutias, 16:469
gibbons, 14:207–209
Giraffidae, 15:399
great apes, 12:33, 14:225
gundis, 16:311
Herpestidae, 14:347
Heteromyidae, 16:199
hippopotamuses, 15:301–302, 15:304
humans, 14:241–244, 14:*243*
hutias, 16:461
Hyaenidae, 14:359–360
hyraxes, 15:177–178
Ice Ages, 12:17–25, **12:17–25**
Indriidae, 14:63
Insectivora, 13:193–194
intelligence and, 12:149–150
koalas, 13:43
Lagomorpha, 16:479–480
lemurs, 14:47
Cheirogaleidae, 14:35
sportive, 14:73–75
Leporidae, 16:505
Lorisidae, 14:13
Macropodidae, 13:31–32, 13:83
manatees, 15: 205
marine mammals, **12:62–68**
Megalonychidae, 13:155
molecular genetics and phylogenetics, **12:26–35**
moles
golden, 13:215–216
marsupial, 13:25
monitos del monte, 12:273
monkeys
Atelidae, 14:155
Callitrichidae, 14:115
Cebidae, 14:101
cheek-pouched, 14:187–189
leaf-monkeys, 14:171–172
night, 14:135
Pitheciidae, 14:143
monotremes, 12:11, 12:33, 12:227–228
mountain beavers, 16:131
Muridae, 16:122, 16:281–282
Arvicolinae, 16:225
hamsters, 16:239
Murinae, 16:249
Sigmodontinae, 16:263–264
musky rat-kangaroos, 13:69–70
Mustelidae, 14:319
numbats, 12:303
nutritional adaptations, **12:120–128**
octodonts, 16:433
opossums
New World, 12:249–250
shrew, 12:267
Otariidae, 14:393
pacaranas, 16:385–386
pacas, 16:417
pangolins, 16:107
peccaries, 15:291
Perissodactyla, 15:215–216
Petauridae, 13:125
Phalangeridae, 13:57
pigs, 15:275
pikas, 16:491
pocket gophers, 16:185
porcupines
New World, 16:365–366
Old World, 16:351
possums
feather-tailed, 13:139–140
honey, 13:135
pygmy, 13:105–106
primates, 14:1–3
Procyonidae, 14:309
pronghorns, 15:411
Pseudocheiridae, 13:113
rat-kangaroos, 13:73
rats
African mole-rats, 16:339
cane, 16:333
chinchilla, 16:443
dassie, 16:329
spiny, 16:449
rhinoceroses, 15:249, 15:251
rodents, 16:121–122
sengis, 16:517–518
sexual selection and, 12:101–102
shrews
red-toothed, 13:247
West Indian, 13:243
white-toothed, 13:265
Sirenia, 15: 191–192
solenodons, 13:237

Evolution *(continued)*
springhares, 16:307
squirrels
flying, 16:135
ground, 16:143
scaly-tailed, 16:299
tree, 16:163
subterranean adaptive, **12:69–78**
Talpidae, 13:279
tapirs, 15:237–238
tarsiers, 14:91
tenrecs, 13:225
three-toed tree sloths, 13:161
tree shrews, 13:289–291
true seals, 14:417
tuco-tucos, 16:425
ungulates, 15:131–138
Viverridae, 14:335
walruses, 14:409
wombats, 13:51
Xenartha, 13:147–149
See also Convergent evolution
Ex situ conservation, 12:223–224
See also Conservation status
Exhibits, zoo, 12:204–205, 12:*210*
Exilisciurus spp., 16:163
Exotics, introduction of, 12:215
Exploitation, of wildlife, 12:215
See also Humans
Extinct species, 12:129, 15:138, 16:126
animal intelligence and, 12:151
Archaoindris spp., 14:63
Archeoindris fontoynonti, 14:63
Archeolemur spp., 14:63
Babakotia spp., 14:63
baboon lemurs, 14:63, 14:72
Barbados raccoons, 14:262, 14:309, 14:310
Barbary lions, 14:380
black-footed ferrets, 14:262, 14:324
broad-faced potorros, 13:39
bubal hartebeests, 16:33
Camelus thomasi, 12:179
Caribbean monk seals, 14:422
causes of, 12:214–216
Columbian mammoths, 12:139
Cuscomys oblativa, 16:443
Daubentonia robusta, 14:85, 14:86, 14:88
desert bandicoots, 13:5, 13:11, 13:17*t*
dusky flying foxes, 13:331*t*
Falkland Island wolves, 14:262, 14:272
giant hutias, 16:469–471, 16:*470*, 16:471*t*
gray whales, 15:99
greater New Zealand short-tailed bats, 13:454, 13:455, 13:458
greater sloth lemurs, 14:71
Guadalupe storm-petrels, 14:291
Guam flying foxes, 13:332*t*
Hadropithecus spp., 14:63
Hippopotamus lemerlei, 12:136
Hypnomys spp., 16:317–318
imposter hutias, 16:467*t*
island biogeography and, 12:220–221
Jamaican monkeys, 12:129
koala lemurs, 14:73–74, 14:75
lesser bilbies, 13:22
lesser Haitian ground sloths, 13:159
Megaladapis edwardsi, 14:73
Megaladapis grandidieri, 14:73
Megaladapis madagascariensis, 14:73
Megaroyzomys spp., 16:270
Mesopropithecus spp., 14:63
Miss Waldron's red colobus, 12:214
Natalus stramineus primus, 13:463
Nesophontes spp., 12:133
Nesoryzomys darwini, 16:270
Nesoryzomys indefessus, 16:270
Omomyidae, 14:91
Oryzomys galapagoensis, 16:270
Oryzomys nelsoni, 16:270
Paleopropithecus spp., 14:63
pig-footed bandicoots, 13:*5*, 13:11, 13:15
Piliocolobus badius waldronae, 14:179
quaggas, 15:221, 15:236*t*
red deer, 15:370
red gazelles, 12:129
Robert's lechwes, 16:33
saber-toothed tigers, 14:369
Samana hutias, 16:467*t*
Schomburgk's deer, 15:360
sea minks, 14:262
sloth lemurs, 14:63
Steiromys spp., 16:366
Steller's sea cows, 12:215, 15:191, 15:195–196, 15:201, 15:203
Syrian onagers, 15:222
tarpans, 15:222
Tasmanian wolves, 12:137, 12:307
Toolache wallabies, 13:39
true plains zebras, 15:222
West Indian monk seals, 12:133, 14:435*t*
West Indian shrews, 13:244
Zanzibar leopards, 14:385
Eyes, 12:5–6, 12:50, 12:77, 12:*80*
See also Physical characteristics; Vision

F

Faces, 12:4–6
See also Physical characteristics
Fairy armadillos, 13:149, 13:*185*, 13:*186*, 13:*189*, 13:190
Falabella horses, 12:176
Falanoucs, 14:335, 14:336, 14:338, 14:*339*, 14:*342*, 14:343
Falkland Island wolves, 14:262, 14:272, 14:284*t*
Fallow deer, 15:*362*, 15:*365*, 15:366
competition, 12:118
evolution, 12:19, 15:357
habitats, 15:358
humans and, 12:173
physical characteristics, 15:*357*
reproduction, 15:272
False killer whales, 12:87, 15:2, 15:*6*, 15:43, 15:46, 15:49
False pacas. *See* Pacaranas
False pottos, 14:*17*, 14:20
False vampire bats, **13:379–385**, 13:*383*
feeding ecology, 12:13, 12:*84*, 12:*85*, 13:313–314
humans and, 13:317
physical characteristics, 13:310
woolly, 13:419, 13:433*t*
False water rats, 16:262*t*
FAO. *See* Food and Agriculture Organization of the United Nations
Fat. *See* Body fat
Fat dormice. *See* Edible dormice
Fat mice, 16:296*t*
Fat-tailed dunnarts, 12:280, 12:288, 12:*289*, 12:300*t*
Fat-tailed gerbils, 16:296*t*
Fat-tailed mouse opossums, 12:250–253
Fat-tailed pseudantechinuses, 12:*283*, 12:288, 12:300*t*
Fattening, seasonal, 12:168–169
See also Feeding ecology
Fauna and Flora International (FFI), on black crested gibbons, 14:221
Fea's muntjacs, 15:343, 15:346, 15:*348*, 15:*351*
Fea's rib-faced deer. *See* Fea's muntjacs
Feather-footed jerboas. *See* Hairy-footed jerboas
Feather-tailed possums, **13:139–145**, 13:*143*, 13:*144*
behavior, 13:35, 13:37, 13:141
conservation status, 13:142
distribution, 13:*139*, 13:140
evolution, 13:31, 13:139–140
feeding ecology, 13:141
habitats, 13:34, 13:141
humans and, 13:142
physical characteristics, 13:140
reproduction, 13:39, 13:142
species of, 13:144
taxonomy, 13:139–140
Feathertail gliders. *See* Pygmy gliders
Feeding ecology, 12:12–15
aardvarks, 15:157–158
agoutis, 16:408–409, 16:411–414, 16:415*t*
anteaters, 13:173–174, 13:177–179
armadillos, 13:184, 13:187–190, 13:190*t*–192*t*
Artiodactyla, 15:141–142, 15:271–272
aye-ayes, 14:*86*, 14:*87*–88
baijis, 15:20
bandicoots, 13:3–4
dry-country, 13:11, 13:14–16, 13:16*t*–17*t*
rainforest, 13:11, 13:14–17*t*
bats, 12:12–13, 12:54–55, 12:58, 12:59, 13:313–314
American leaf-nosed, 13:314, 13:417–418, 13:423–432, 13:433*t*–434*t*
bulldog, 13:446–447, 13:449–450
disk-winged, 13:474, 13:476–477
Emballonuridae, 13:357, 13:360–363, 13:363*t*–364*t*
false vampire, 13:313, 13:381–382, 13:384–385
funnel-eared, 13:461, 13:463–465
horseshoe, 13:390–391, 13:396–400
Kitti's hog-nosed, 13:368
Molossidae, 13:485, 13:490–493, 13:493*t*–495*t*
mouse-tailed, 13:351–352, 13:353
moustached, 13:436, 13:440–441, 13:442*t*
New Zealand short-tailed, 13:454, 13:457–458
Old World fruit, 13:320, 13:325–330, 13:331*t*–332*t*, 13:335–336, 13:340–347, 13:348*t*–349*t*
Old World leaf-nosed, 13:405, 13:407–409, 13:409*t*–410*t*
Old World sucker-footed, 13:480
slit-faced, 13:373, 13:375–376, 13:376*t*–377*t*
smoky, 13:468, 13:470–471

Feeding ecology *(continued)*
Vespertilionidae, 13:500, 13:506–514, 13:515*t*–516*t*, 13:521, 13:524–525, 13:526*t*
bears, 14:255, 14:298–299, 14:302–305, 14:306*t*
beavers, 16:179–180, 16:183–184
bilbies, 13:21–22
botos, 15:29
Bovidae, 16:6–7
Antilopinae, 16:48, 16:50–56, 16:56*t*–57*t*
Bovinae, 16:14, 16:18–23, 16:24*t*–25*t*
Caprinae, 16:93–94, 16:98–103*t*, 16:104*t*
duikers, 16:74, 16:76, 16:80–84, 16:84*t*–85*t*
Hippotraginae, 16:31–32, 16:37–42, 16:42*t*–43*t*
Neotraginae, 16:62, 16:66–71, 16:71*t*
bushbabies, 14:9, 14:26, 14:28–32, 14:32*t*–33*t*
Camelidae, 15:317, 15:320–323
Canidae, 14:269–271, 14:277–283, 14:283*t*–284*t*
capybaras, 16:404, 16:*405*
Carnivora, 14:260–261
cats, 14:372, 14:379–389, 14:390*t*–391*t*
Caviidae, 16:392–393, 16:395–398, 16:399*t*
Cetacea, 15:8
Balaenidae, 15:110, 15:115–117
beaked whales, 15:61, 15:64–68, 15:69*t*–70*t*
dolphins, 15:46–47, 15:53–55, 15:56*t*–57*t*
franciscana dolphins, 15:24
Ganges and Indus dolphins, 15:14–15
gray whales, 15:97–98
Monodontidae, 15:85, 15:90–91
porpoises, 15:35, 15:38–39, 15:39*t*
pygmy right whales, 15:104–105
rorquals, 15:8, 15:123–124, 15:127–130, 15:130*t*
sperm whales, 15:75–76, 15:79–80
chevrotains, 15:328, 15:331–333
Chinchillidae, 16:379–380, 16:382–383, 16:383*t*
colugos, 13:302, 13:304
coypus, 16:476–477
dasyurids, 12:289, 12:294–298, 12:299*t*–301*t*
Dasyuromorphia, 12:281
deer
Chinese water, 15:375
muntjacs, 15:345, 15:349–354
musk, 15:337, 15:340–341
New World, 15:383–384, 15:388–396, 15:396*t*
Old World, 15:359, 15:364–371, 15:371*t*
Dipodidae, 16:215, 16:219–222, 16:223*t*–224*t*
Diprotodontia, 13:37–38
dormice, 16:319–320, 16:323–326, 16:327*t*–328*t*
duck-billed platypuses, 12:231, 12:245
Dugongidae, 15:200, 15:203
echidnas, 12:237, 12:240*t*
elephants, 15:*167*, 15:169–170, 15:174–175
Equidae, 12:118–119, 15:228, 15:232–235, 15:236*t*
Erinaceidae, 13:206, 13:209–213, 13:212*t*–213*t*
giant hutias, 16:469, 16:471*t*
gibbons, 14:211–213, 14:218–222
Giraffidae, 15:403–404, 15:408–409
great apes, 14:232–234, 14:237–240
gundis, 16:313, 16:314–315
Herpestidae, 14:351–352, 14:354–356, 14:357*t*
Heteromyidae, 16:202, 16:205–209, 16:208*t*–209*t*
hippopotamuses, 15:306, 15:308, 15:310–311
humans, 14:*245*, 14:247–248
hutias, 16:462, 16:467*t*
Hyaenidae, 14:362, 14:364–367
hyraxes, 15:182, 15:186–188, 15:189*t*
Indriidae, 14:66, 14:69–72
Insectivora, 13:198–199
koalas, 13:46–47
Lagomorpha, 16:483–485
lemurs, 14:9, 14:51, 14:56–60
Cheirogaleidae, 14:37–38, 14:41–44
sportive, 14:8, 14:76, 14:79–83
Leporidae, 16:507, 16:511–515, 16:515*t*–516*t*
Lorisidae, 14:15, 14:18–20, 14:21*t*
Macropodidae, 13:87–88, 13:93–100, 13:101*t*–102*t*
manatees, 15:206, 15:208, 15:211–212
Megalonychidae, 13:157, 13:159
migrations and, 12:167–170
moles
golden, 12:84, 13:217, 13:220–222, 13:222*t*
marsupial, 13:27, 13:28
monitos del monte, 12:274
monkeys
Atelidae, 14:158, 14:162–166, 14:166*t*–167*t*
Callitrichidae, 14:120–121, 14:127–131, 14:132*t*
Cebidae, 14:104–105, 14:108–112
cheek-pouched, 14:192–193, 14:197–204, 14:204*t*–205*t*
leaf-monkeys, 14:8, 14:9, 14:174, 14:178–183, 14:184*t*–185*t*
night, 14:138, 14:141*t*
Pitheciidae, 14:146–147, 14:150–153, 14:153*t*
monotremes, 12:231
mountain beavers, 16:132–133
Muridae, 16:284, 16:288–295, 16:296*t*–297*t*
Arvicolinae, 16:227–228, 16:233–238, 16:237*t*–238*t*
hamsters, 16:241–242, 16:245–246, 16:247*t*
Murinae, 16:251–252, 16:255–260, 16:261*t*–262*t*
Sigmodontinae, 16:268–269, 16:272–276, 16:277*t*–278*t*
musky rat-kangaroos, 13:71
Mustelidae, 12:192, 14:260, 14:323, 14:327–331, 14:332*t*–333*t*
numbats, 12:304
octodonts, 16:435–436, 16:438–440, 16:440*t*
opossums
New World, 12:252–253, 12:257–263, 12:264*t*–265*t*
shrew, 12:268, 12:270
Otariidae, 14:397–398, 14:402–406
pacaranas, 16:387
pacas, 16:*418*–419, 16:423–424
pangolins, 16:112–113, 16:115–120
peccaries, 15:294–295, 15:298–300
Perissodactyla, 15:219–220
Petauridae, 13:127–128, 13:131–133, 13:133*t*
Phalangeridae, 13:60, 13:*63–66t*
pigs, 15:278–279, 15:284–288, 15:289*t*–290*t*
pikas, 16:493–494, 16:497–501, 16:501*t*–502*t*
pocket gophers, 16:189–190, 16:192–195*t*, 16:196*t*–197*t*
porcupines
New World, 16:368, 16:371–373, 16:374*t*
Old World, 16:353, 16:357–364
possums
feather-tailed, 13:141, 13:144
honey, 13:137
pygmy, 13:107, 13:110–111
primates, 14:8–9
Procyonidae, 14:310, 14:313–315, 14:315*t*–316*t*
pronghorns, 15:414
Pseudocheiridae, 13:116, 13:119–123, 13:122*t*–123*t*
rat-kangaroos, 13:75, 13:78–81
rats
African mole-rats, 16:343–344, 16:347–349, 16:349*t*–350*t*
cane, 16:334, 16:337–338
chinchilla, 16:445, 16:447–448
dassie, 16:331
spiny, 16:450, 16:453–457, 16:458*t*
rhinoceroses, 15:253, 15:257–261
rodents, 16:125
sengis, 16:521, 16:524–530, 16:531*t*
shrews
red-toothed, 13:250–251, 13:255–263, 13:263*t*
West Indian s, 13:244
white-toothed, 13:267, 13:271–275, 13:276*t*–277*t*
Sirenia, 15:193
solenodons, 13:239, 13:241
springhares, 16:309
squirrels
flying, 16:136, 16:139–140, 16:141*t*–142*t*
ground, 16:145, 16:147, 16:151–158, 16:158*t*–160*t*
scaly-tailed, 16:300, 16:302–305
tree, 16:166, 16:169–174*t*, 16:175*t*
Talpidae, 13:281–282, 13:284–287, 13:287*t*–288*t*
tapirs, 15:240–241, 15:245–247
tarsiers, 14:*94*–95, 14:97–99
tenrecs, 13:228–229, 13:232–234, 13:234*t*
three-toed tree sloths, 13:164–165, 13:167–169
tree shrews, 13:291–292, 13:294–296, 13:297*t*–298*t*
true seals, 14:260, 14:*261*, 14:419–420, 14:427–434, 14:435*t*
tuco-tucos, 16:427, 16:429–430, 16:430*t*–431*t*
ungulates, 15:141–142
Viverridae, 14:337, 14:340–343, 14:344*t*–345*t*
walruses, 14:260, 14:414
wombats, 13:52–53, 13:55–56
Xenartha, 13:151–152
Feet, 12:*38*
See also Physical characteristics

Felid Taxonomic Advisory Group (TAG), 14:369
Felidae. *See* Cats
Felinae, 14:369
Felines. *See* Cats
Felis spp., 14:369
Felis bengalensis euptilura. See Bengal cats
Felis bieti. See Chinese desert cats
Felis catus. See Domestic cats; Feral cats
Felis chaus. See Jungle cats
Felis concolor. See Pumas
Felis concolor coryi. See Florida panthers
Felis margarita. See Sand cats
Felis nigripes. See Black-footed cats
Felis rufus. See Bobcats
Felis silvestris. See Wild cats
Felis silvestris libyca. See Libyan wild cats
Felis viverrina. See Fishing cats
Felou gundis, 16:312, 16:313
Felovia spp., 16:311
Felovia vae. See Felou gundis
Female reproductive system, 12:90–91
  See also Physical characteristics; Reproduction
Fennec foxes, 14:*258*, 14:273
Fennecus zerda. See Fennec foxes
Feral Cat Coalition of San Diego, California, 14:292
Feral cats, 12:185–186, 14:257
  See also Domestic cats; Wild cats
Feral Celebes pigs. *See* Timor wild boars
Feral muntjacs, 12:118
Feral pigs, 12:191–192
  See also Domestic pigs
Ferecetotherium spp., 15:73
Feresa attenuata. See Pygmy killer whales
Feroculus spp., 13:265
Feroculus feroculus. See Kelaart's long-clawed shrews
Ferrets, 12:141, 14:321, 14:323, 14:324, 14:325
  See also Black-footed ferrets
Fertility control agents, 12:188
  See also Reproduction
Fertilization, delayed, 12:109–110
  See also Reproduction
FFI (Fauna and Flora International), on black crested gibbons, 14:221
Fiber duikers, 16:73
Field hamsters. *See* Black-bellied hamsters
Field mice, 16:270
  Azara's, 16:266–270
  long-tailed, 16:*250*, 16:261*t*
  old, 16:125
  striped, 16:261*t*
Field studies, **12:194–202,** 12:*195–201*
Field voles, 16:227, 16:*228*
Fin whales, 12:67, 15:*120*, 15:*126*, 15:*127*–128
  behavior, 15:6, 15:122
  conservation status, 15:9, 15:125
  distribution, 15:5, 15:121
  feeding ecology, 15:8
Finbacks. *See* Fin whales
Finless porpoises, 15:6, 15:33, 15:34, 15:*35*, 15:*36*, 15:39*t*
Finners. *See* Northern minke whales
Fire control, 14:250
Fischer's pygmy fruit bats, 13:350*t*
Fishers, 12:132, 14:321, 14:323, 14:324
Fishing bats, 13:313–314
Fishing cats, 14:*370*
Fishing genets. *See* Aquatic genets
Five-toed dwarf jerboas. *See* Five-toed pygmy jerboas
Five-toed jerboas, 16:211, 16:212, 16:213, 16:214
Five-toed pygmy jerboas, 16:215, 16:216, 16:*218*, 16:*220*
Fjord seals. *See* Harp seals
Flannery, Tim, 13:9, 13:90
Flashjacks. *See* Bridled nail-tailed wallabies
Flatheads. *See* Northern bottlenosed whales; Southern bottlenosed whales
Flehman response, 12:80, 12:*103*
Fleming, T. H., 16:200
Flickertails. *See* Richardson's ground squirrels
Flight adaptations, 12:12–13, 12:14, 12:43–44, **12:52–61,** 12:*53–59*
Flight membranes, of bats, 12:43, 12:56–58
Flightless dwarf anomalures. *See* Cameroon scaly-tails
Florida Manatee Sanctuary Act, 15:195, 15:209
Florida manatees, 15:191, 15:193, 15:195, 15:209, 15:211–212
  See also West Indian manatees
Florida panthers, 12:*222*
Florida water rats. *See* Round-tailed muskrats
Flower bats, 13:314
  See also specific types of flower bats
Flower-faced bats, 13:409*t*
Fluffy gliders. *See* Yellow-bellied gliders
Fluid dynamics, powered flight and, 12:56
Flukes, whale, 12:*66*
  See also Whales
Flying foxes, **13:319–333,** 13:*323*, 13:*324*
  behavior, 13:319–320
  conservation status, 13:321–322
  crepuscular activity cycle, 12:54
  distribution, 13:*319*
  evolution, 13:319
  feeding ecology, 13:320
  golden-crowned, 13:*338*
  gray-headed, 13:312, 13:320
  habitats, 13:319
  humans and, 13:322
  physical characteristics, 12:*54*, 12:58, 13:307, 13:319
  reproduction, 13:320–321
  species of, 13:*325*–330, 13:331*t*–332*t*
  taxonomy, 13:309, 13:319
Flying jacks. *See* White-faced sakis
Flying lemurs. *See* Colugos
Flying mice, 16:300
  big-eared, 16:300, 16:*301*, 16:*304*–305
  Zenker's, 16:300, 16:*301*, 16:*304*
  See also Pygmy gliders
Flying squirrels, 12:14, 12:131, **16:135–142,** 16:*138*, 16:141*t*–142*t*
Folivores, 14:8
Food and Agriculture Organization of the United Nations (FAO), on livestock, 12:175
Foose, Thomas J., 12:206, 12:207
Foot drumming, 12:83
Foraging
  memory and, 12:152–153
  subterranean mammals, 12:75–76
  See also Feeding ecology
Forbidden tree kangaroos, 13:32
Forebrain, 12:6
  See also Physical characteristics
Foregut fermentation, 12:14–15, 12:47–48
Forelimbs, bat wings and, 12:56
Forest Department (Government of Assam), on pygmy hogs, 15:287
Forest dingos, 12:180
Forest dormice, 16:317–320, 16:*322*, 16:*323*–324
Forest-dwelling bandicoots, 13:3, 13:4
Forest elephants, 15:162–163, 15:*166*
Forest foxes. *See* Crab-eating foxes
Forest giraffes. *See* Okapis
Forest hogs, 15:277, 15:*278*, 15:*283*, 15:*284*
Forest-living guenons, 14:188, 14:192, 14:193
Forest marmots. *See* Woodchucks
Forest mice, 16:266, 16:267, 16:270
  See also Gray climbing mice
Forest musk shrews, 13:197, 13:265, 13:266, 13:*269*, 13:*272*
Forest rats, red, 16:*281*
Forest shrews, 13:*270*, 13:*273*
Forest spiny pocket mice, 16:199
Forest tarpans, 12:175
Forest wallabies, 13:34, 13:35, 13:83, 13:86
Foresters. *See* Eastern gray kangaroos
Forests, 12:218–219
Fork-crowned lemurs. *See* Masoala fork-crowned lemurs
Formosan serows, 16:87, 16:90
Fossa, 12:*98*, 12:*198*, **14:347–355,** 14:*348*, 14:*353*, 14:*354*
Fossa fossana. See Malagasy civets
Fossil fuels, 12:213
  See also Humans
Fossils. *See* Evolution
Fossorial mammals, 12:13, 12:69, 12:82
  See also specific fossorial mammals
Fossorial spiny rats, Bishop's, 16:450, 16:458*t*
Founder species
  bottlenecks and, 12:221–222
  in zoo populations, 12:207–208
  See also Conservation status
Four-eyed opossums, 12:38, 12:254
Four-horned antelopes. *See* Chousinghas
Four-striped grass mice, 12:*117*, 16:*252*
Four-toed hedgehogs. *See* Central African hedgehogs
Four-toed jerboas, 16:212, 16:216, 16:223*t*
Four-toed sengis, 16:518, 16:519, 16:521, 16:*523*, 16:*526*–527
Fox squirrels
  big cypress, 16:*164*
  Delmarva, 16:167
  eastern, 16:*164*, 16:*166*, 16:*168*, 16:*169*, 16:171
  Mexican, 16:175*t*
Foxes, **14:265–273,** 14:*275*, **14:278–285**
  behavior, 14:267–268
  conservation status, 12:192, 14:272, 14:273
  distribution, 14:267
  evolution, 14:265
  feeding ecology, 14:255, 14:269–270, 14:271
  feral cats and, 12:185
  habitats, 14:267
  humans and, 14:273
  as pests, 12:188
  physical characteristics, 14:266–267
  reproduction, 12:107, 14:271, 14:272

Foxes *(continued)*
species of, 14:278–283, 14:283*t*, 14:284*t*
taxonomy, 14:265
Fragmentation, habitat, 12:214, 12:222–223
See also Conservation status
Franciscana dolphins, 15:6, 15:9, **15:23–26,** 15:*24*, 15:*25*
Frank, Morris, 14:293
Frankfurt Zoo, Germany, 12:*210*
Frankham, R., 12:206, 12:208
Franklin Island house-building rats. *See* Greater stick-nest rats
Franklin Island stick nest rats. *See* Greater stick-nest rats
Franquet's epauletted bats. *See* Singing fruit bats
Franz Josef (Emperor of Austria), 12:203
Fraser's dolphins, 15:57*t*
Freckled antechinuses. *See* Southern dibblers
Free-tailed bats, **13:483–496,** 13:*487*, 13:*489*
behavior, 13:312, 13:484–485
conservation status, 13:487
distribution, 13:*483*
evolution, 13:483
feeding ecology, 13:485
habitats, 13:311
humans and, 13:487–488
physical characteristics, 13:307, 13:308, 13:310, 13:483
reproduction, 13:487–488
species of, 13:491–493, 13:493*t*–495*t*
taxonomy, 13:483
French Island, Australia, 12:*223*
Friendly whales. *See* Gray whales
Friends of the Arabuko-Sokoke Forest, 16:525
Fringe-lipped bats, 13:314, 13:*417*, 13:*422*, 13:*423*–424
Frosted hairy dwarf porcupines, 16:365, 16:374*t*
Frugivores, 14:8
Fruit bats, 12:13, 13:*310*, 13:312
Fulk, G. W., 16:435
Fulvous leaf-nosed bats. *See* Fulvus roundleaf bats
Fulvous lemurs. *See* Brown lemurs
Fulvus roundleaf bats, 13:405, 13:*406*, 13:*407*, 13:408
Funambulus spp., 16:163
Funambulus tristriatus, 16:167
Funisciurus spp., 16:163
Funisciurus anerythrus, 16:165
Funisciurus carruthersi, 16:167
Funisciurus congicus. See Striped tree squirrels
Funisciurus isabella, 16:167
Funnel-eared bats, **13:459–465,** 13:*460*, 13:*461*, 13:*462*, 13:*463*
Fur. *See* Hairs
Fur seals, 12:*106*, **14:393–403,** 14:*400*, 14:*401*, **14:407–408**
behavior, 14:396–397
conservation status, 12:*219*, 14:399
distribution, 12:138, 14:394–395
evolution, 14:393
feeding ecology, 14:397–398
habitats, 14:395–396
humans and, 14:400
physical characteristics, 14:393–394
reproduction, 12:*92*, 14:398–399
species of, 14:*402*–403, 14:407*t*–408*t*
taxonomy, 14:256
Fur trade, 12:173, 14:374, 14:376
See also Humans
Furipteridae. *See* Smoky bats
Furipterus spp., 13:467, 13:468
Furipterus horrens. See Smoky bats
Fussion-fission communities, 14:229

G

Gabon Allen's bushbabies, 14:*27*, 14:28–*29*
Gaia: An Atlas of Planet Management, 12:218
Gaindatherium browni, 15:249
Gaits, 12:*39*
See also Physical characteristics
Galadictis spp., 14:347, 14:348
Galagidae. *See* Bushbabies
Galago spp., 14:23
Galago alleni. See Bioko Allen's bushbabies
Galago cameronensis. See Cross River Allen's bushbabies
Galago demidoff. See Demidoff's bushbabies
Galago gabonensis. See Gabon Allen's bushbabies
Galago gallarum. See Somali bushbabies
Galago granti. See Grant's bushbabies
Galago matschiei. See Dusky bushbabies
Galago moholi. See Moholi bushbabies
Galago nyasae. See Malawi bushbabies
Galago orinus. See Uluguru bushbabies
Galago rondoensis. See Rondo bushbabies
Galago senegalensis. See Senegal bushbabies
Galago udzungwensis. See Uzungwa bushbabies
Galago zanzibaricus, 14:26
Galagoides spp., 14:23
Galagoides demidoff. See Demidoff's bushbabies
Galagoides thomasi. See Thomas's bushbabies
Galagoides zanzibaricus. See Zanzibar bushbabies
Galagonidae. *See* Bushbabies
Galagos, vision, 12:79
Galápagos fur seals, 12:*92*, 14:394, 14:395, 14:399, 14:*400*, 14:408*t*
Galápagos sea lions, 12:*64*, 12:*65*, 12:*131*, 14:*394*, 14:*397*, 14:399–400, 14:*401*, 14:*404*, 14:405–406
Galea spp., 16:389, 16:390
Galea flavidens. See Yellow-toothed cavies
Galea musteloides. See Cuis
Galemys spp., 13:279
Galemys pyrenaicus. See Pyrenean desmans
Galericinae, 13:203
Galidia spp., 14:347, 14:348
Galidia elegans. See Ring-tailed mongooses
Galidia elegans elegans. See Eastern Malagasy ring-tailed mongooses
Galidictis grandidieri. See Western Malagasy broad striped mongooses
Galidiinae, 14:347, 14:349, 14:352
Gallardo, M. H., 16:433
Gambarian, P. P., 16:211
Gambian epauletted fruit bats, 13:*336*
Gambian rats, 16:*286*, 16:*289*–290
Gambian slit-faced bats, 13:376*t*
Gametes, 12:101, 12:105
See also Reproduction
Ganges and Indus dolphins, 15:4, 15:5, **15:13–17,** 15:*14*, 15:*15*, 15:*16*
Ganges river dolphins. *See* Ganges and Indus dolphins
Gansu hamsters, 16:239
Gansu moles, 13:280, 13:288*t*
Gansu pikas, 16:491, 16:*496*, 16:*499*, 16:500
Gaoligong pikas, 16:501*t*
Garden dormice, 16:317, 16:318, 16:319, 16:*321*, 16:*322*, 16:*324*–*325*
Gardner, Allen, 12:160
Gardner, Beatrice, 12:160
Garnett's bushbabies. *See* Northern greater bushbabies
Gaurs, 12:178, 15:267, 15:271, 16:2, 16:4, 16:6, 16:24*t*
Gayals, 12:180
Gazella spp., 15:265, 16:1, 16:7
Gazella bennettii, 16:48
Gazella cuvieri, 16:48
Gazella dama. See Dama gazelles
Gazella dama mhorr. See Mhorr gazelles
Gazella dorcas. See Dorcas gazelles
Gazella gazella. See Mountain gazelles
Gazella granti. See Grant's gazelles
Gazella leptoceros. See Slender-horned gazelles
Gazella rufifrons. See Red-fronted gazelles
Gazella rufina. See Red gazelles
Gazella soemmerringii. See Persian gazelles
Gazella spekei, 16:48
Gazella subgutturosa, 16:48
Gazella subgutturosa marica. See Arabian sand gazelles
Gazella thomsonii. See Thomson's gazelles
Gazelles, 15:146, **16:45–58,** 16:*49*
behavior, 16:46, 16:48
conservation status, 16:48
distribution, 16:4, 16:*45*, 16:46
evolution, 15:265, 16:1, 16:45
feeding ecology, 16:48
habitats, 16:46
humans and, 12:139, 16:48
physical characteristics, 16:2, 16:45
reproduction, 16:48, 16:50–55
species of, 16:50–57
taxonomy, 16:1, 16:45
Gebe cuscuses, 13:32, 13:60
Geist, V., 16:89
Gelada baboons. *See* Geladas
Geladas, 14:*195*, 14:*201*, 14:203
behavior, 14:7, 14:192
conservation status, 14:194
distribution, 14:*192*
evolution, 14:189
feeding ecology, 14:193
habitats, 14:191
reproduction, 14:193
taxonomy, 14:188
Gemsboks, 12:*95*, 16:2, 16:4, **16:27–43,** 16:*29*, 16:*31*, 16:*32*, 16:*33*, 16:*34*, 16:41*t*
Gene duplication, 12:30–31
Generation length, 12:207
Genetic diversity. *See* Biodiversity
Genetic drift, 12:222
Genetics
evolution and, 12:26–35
small populations and, 12:221–222
zoo populations and, 12:206–209

Genets, **14:335–345,** 14:*339*
behavior, 14:258, 14:337
conservation status, 14:338
distribution, 14:336
evolution, 14:335
feeding ecology, 14:337
habitats, 14:337
physical characteristics, 14:335–336
reproduction, 14:338
species of, 14:*340–341*, 14:344*t*
taxonomy, 14:256, 14:335
Genetta spp., 14:335
Genetta abyssinica. See Abyssinian genets
Genetta genetta. See Common genets
Genetta johnstoni. See Johnston's genets
Genetta servalina, 14:336
Genetta tigrina. See Blotched genets
Genital displays, squirrel monkeys, 14:104
Genomes, 12:26–27
Gentle jirds. *See* Sundevall's jirds
Gentle lemurs, 14:3, 14:8, 14:10–11, 14:*52*
Geocapromys spp., 16:461, 16:462
Geocapromys brownii. See Jamaican hutias
Geocapromys ingrahami. See Bahamian hutias
Geoffroy's bats, 12:54
Geoffroy's cats, 14:374, 14:*378*, 14:*384*–385
Geoffroy's horseshoe bats, 13:390, 13:392
Geoffroy's ocelots. *See* Geoffroy's cats
Geoffroy's spider monkeys, 12:7, 14:*160*, 14:*161*, 14:*163*
Geoffroy's tailless bats, 13:*422*, 13:427–*428*
Geoffroy's tamarins, 14:*123*, 14:132*t*
Geoffroy's woolly monkeys. *See* Gray woolly monkeys
Geogale spp. *See* Large-eared tenrecs
Geogale aurita. See Large-eared tenrecs
Geogalinae. *See* Large-eared tenrecs
Geological Society of America, 12:65
Geomagnetic fields, 12:84
Geomyidae. *See* Pocket gophers
Geomyoidea, 16:121
Geomys spp., 16:185
Geomys arenarius. See Desert pocket gophers
Geomys bursarius. See Plains pocket gophers
Geomys personatus. See Texas pocket gophers
Geomys pinetis. See Southeastern pocket gophers
Geomys tropicalis. See Tropical pocket gophers
Geophytes, 12:69, 12:76
Georychinae, 16:339
Georychus spp. *See* Cape mole-rats
Georychus capensis. See Cape mole-rats
Geoxus spp. *See* Mole mice
Geoxus valdivianus. See Long-clawed mole mice
Gerbil mice. *See* Gray climbing mice
Gerbillinae, 16:127, 16:283–284, 16:285
Gerbillurus setzeri. See Setzer's hairy-footed gerbils
Gerbillus spp., 12:114
Gerbils, 16:123, 16:127, 16:128, 16:*282*, 16:*283*, 16:*284*, 16:296*t*
See also Sundevall's jirds
Gerenuks, 12:*154*, 15:140, 16:7, 16:*46*, 16:*47*, 16:48, 16:*49*, 16:*55*–56
German shepherds, 14:288–289
Gervais' beaked whales, 15:69*t*
Gervais' funnel-eared bats, 13:461, 13:*462*, 13:*463*, 13:465
Gestation, 12:95–96
See also Reproduction
Ghost bats, 13:308, **13:355–365,** 13:357, 13:*359*
See also Australian false vampire bats
Ghost-faced bats, 13:316–317, 13:435, 13:436, 13:*439*, 13:*440*, 13:441, 13:442*t*
Giant anteaters, 12:*29*, 13:*176*, 13:*177*, 13:178–179
behavior, 13:*173*
conservation status, 13:153
feeding ecology, 13:151, 13:*174*
habitats, 13:150, 13:172
physical characteristics, 13:149
Giant armadillos, 13:148, 13:150, 13:152, 13:*181*, 13:182, 13:*186*, 13:*189*–190
Giant bandicoots, 13:17*t*
Giant beavers, 16:177
Giant bottlenosed whales. *See* Baird's beaked whales
Giant duikers, 16:73
Giant flying squirrels, 16:135, 16:136, 16:137
Giant golden moles. *See* Large golden moles
Giant herbivorous armadillos, 13:181
Giant hutias, **16:469–471,** 16:*470*, 16:471*t*
Giant Indian fruit bats. *See* Indian flying foxes
Giant kangaroo rats, 16:200, 16:201–*202*, 16:203, 16:*204*, 16:*207*
Giant leaf-nosed bats. *See* Commerson's leaf-nosed bats
Giant mastiff bats, 13:*489*, 13:491–*492*
Giant muntjacs, 15:346, 15:*347*, 15:*353*–354
Giant noctules. *See* Greater noctules
Giant otter shrews, 13:196, 13:234*t*
Giant otters, 14:321, 14:324
Giant pandas, 12:*118*, 12:*208*, 14:*301*, 14:303–*304*
behavior, 14:297
captive breeding, 12:224
conservation status, 14:300
distribution, 12:136, 14:296
feeding ecology, 14:255, 14:260, 14:*297*, 14:299
habitats, 14:297
physical characteristics, 14:296
reproduction, 12:110, 14:299
taxonomy, 14:295
Giant pangolins, 16:110, 16:112, 16:*114*, 16:117
Giant pocket gophers. *See* Large pocket gophers
Giant pouched bats, 13:364*t*
Giant rat-kangaroos, 13:70
Giant rats, 16:265, 16:*286*, 16:*289*, 16:291
Giant sable antelopes, 16:33
Giant schnauzers, 14:292
Giant sea cows, 12:15
Giant sengis, 16:517–521
Giant shrews, 13:*254*, 13:*255*, 13:262
Giant squirrels, 16:163, 16:166, 16:174
Giant stag. *See* Irish elk
Giant Zambian mole-rats, 12:73, 16:340, 16:345, 16:349*t*
Giarda, beavers and, 16:127
Gibbons, **14:207–223,** 14:*216*, 14:*217*
behavior, 12:145, 14:6–7, 14:8, 14:210–211
conservation status, 14:11, 14:214–215
distribution, 12:136, 14:5, 14:*207*, 14:210
evolution, 14:207–209
feeding ecology, 14:211–213
habitats, 14:210
humans and, 14:215
locomotion, 12:43
physical characteristics, 14:209–210
reproduction, 12:107, 14:213–214
self recognition, 12:159
sexual dimorphism, 12:99
species of, 14:*218–222*
taxonomy, 14:4, 14:207–208
in zoos, 12:210
Gibb's shrew-moles. *See* American shrew moles
Gibnuts. *See* Pacas
Gidley, J. W., 16:479
Gilbert's potoroos, 13:34, 13:39, 13:74, 13:76, 13:77, 13:*78*
Gilchrist's disease, beavers and, 16:127
Gile's planigales, 12:300*t*
Gilliard's flying foxes, 13:332*t*
Ginkgo-toothed beaked whales, 15:69*t*
Giraffa spp., 15:265
Giraffa camelopardalis. See Giraffes
Giraffa camelopardalis reticulata. See Reticulated giraffes
Giraffa camelopardalis rothschildi. See Rothschild's giraffes
Giraffa camelopardalis tippelskirchi. See Masai giraffes
Giraffe gazelles. *See* Gerenuks
Giraffes, 12:*6*, 12:*82*, 12:*103*, 12:116, **15:399–408,** 15:*401*, 15:*403*, 15:*407*
behavior, 15:401–403, 15:*402*
conservation status, 15:405
distribution, 15:*399*, 15:401
evolution, 15:138, 15:265, 15:399
feeding ecology, 15:403–404, 15:*404*
habitats, 15:401
horns, 12:40, 12:102
humans and, 15:405–406
migrations, 12:167
physical characteristics, 15:138, 15:140, 15:399–401, 15:*402*
reproduction, 15:*135*, 15:143, 15:*400*, 15:404–405
species of, 15:408
taxonomy, 15:138, 15:263, 15:265, 15:399
Giraffidae, 12:135, 15:137, 15:266, 15:267, 15:268, **15:399–409,** 15:*407*
Glacier bears. *See* American black bears
Glaciers, 12:17–18
Glander, K. E., 14:162
Glands, scent, 12:81
See also Physical characteristics
Glaucomys spp., 16:136–137
Glaucomys sabrinus. See Northern flying squirrels
Glaucomys volans. See Southern flying squirrels
Gleaning bats, 13:313
See also specific gleaning bats
Gliders, 13:31, 13:35, 13:39
See also specific types of gliders
Gliding possums, 13:33, **13:125–133,** 13:133*t*
Glires, 16:479–480
See also Lagomorpha; Rodents
Gliridae, 16:283
Glironia spp. *See* Bushy-tailed opossums
Glironia venusta. See Bushy-tailed opossums
Glirulus japonicus. See Japanese dormice
Global positioning systems (GPS), 12:199
Global Strategy for the Management of Farm Animal Genetic Resources, 12:175
Globicephala spp. *See* Pilot whales
Globicephala macrorhynchus. See Short-finned pilot whales

Globicephala melas. See Long-finned pilot whales
Globicephalinae, 15:41
Globin genes, 12:30
Glossophaga longirostris. See Southern long-tongued bats
Glossophaga soricina. See Pallas's long-tongued bats
Glossophaginae, 13:413, 13:414, 13:415, 13:417, 13:418, 13:420
Glover's pikas, 16:491
Glyphotes spp., 16:163
Glyptodonts, 12:11, 13:147–150, 13:152, 13:181, 15:137, 15:138
Gnus, 12:116, 15:140
Goas. *See* Tibetan gazelles
Goats, 12:*5*, 15:146, 15:263, 16:1, **16:87–105**, 16:*96*, 16:*97*
behavior, 12:*151*, 15:271, 16:92–93
bezoar, 12:179
conservation status, 16:94–95
digestive system, 12:122
distribution, 16:4, 16:*87*, 16:90–91
evolution, 16:87–89
feeding ecology, 16:93–94
habitats, 16:91–92
humans and, 15:273, 16:8, 16:9, 16:95
migrations, 12:165, 15:138
neonatal milk, 12:127
physical characteristics, 16:5, 16:89–90
reproduction, 16:94
species of, 16:98–104*t*
taxonomy, 16:87–89
See also Domestic goats
Goeldi's monkeys, **14:115–129**, 14:*121*, 14:*123*, 14:*126*
behavior, 14:6–7, 14:117–120
distribution, 14:116
evolution, 14:115
feeding ecology, 14:120, 14:121
habitats, 14:116
physical characteristics, 14:115, 14:116
reproduction, 14:10, 14:121–124
species of, 14:*127*, 14:129
taxonomy, 14:4, 14:115
Goitered gazelles. *See* Persian gazelles
Golden angwantibos, 14:16, 14:21*t*
Golden anteaters. *See* Silky anteaters
Golden bandicoots, 13:*3*, 13:4, 13:12, 13:16*t*
Golden bats. *See* Old World sucker-footed bats
Golden-bellied tree shrews, 13:297*t*
Golden-bellied water rats, 16:261*t*
Golden-capped fruit bats. *See* Golden-crowned flying foxes
Golden cats, 14:390*t*–391*t*
Golden-cheeked gibbons, 14:*216*, 14:*220*, 14:221–222
Golden-crowned flying foxes, 13:*338*, 13:*341*–342
Golden-crowned sifakas, 14:11, 14:63, 14:*67*
Golden eagles, 12:192
Golden-faced sakis, 14:*146*
Golden hamsters, 16:239–240, 16:*241*, 16:243, 16:*244*, 16:*245*, 16:246–247
Golden-headed lion tamarins, 14:118, 14:*126*, 14:*128*–129
Golden horseshoe bats, 13:*407*, 13:*408*, 13:409
Golden jackals, 14:*268*, 14:271, 14:272
Golden lion tamarins, 14:*120*, 14:*126*
behavior, 14:117, 14:118
conservation status, 12:204, 12:208, 12:224, 14:11
humans and, 14:124
physical characteristics, 14:115–116
reproduction, 14:121–122, 14:124
Golden-mantled ground squirrels, 16:143, 16:159*t*
Golden mice, 16:277*t*
Golden moles, 12:79–80, **13:215–223**, 13:*216*, 13:*219*
behavior, 13:198
distribution, 12:135, 13:197
evolution, 12:29
feeding ecology, 12:84
habitats, 13:197
physical characteristics, 12:29, 13:195–197
species of, 13:220–222, 13:222*t*
taxonomy, 13:193–194
Golden pottos. *See* Calabar angwantibos
Golden ringtail possums, 13:123*t*
Golden-rumped lion tamarins, 14:11, 14:117, 14:132*t*
Golden-rumped sengis, 16:521, 16:522, 16:*523*, 16:*524*–*525*
Golden snub-nosed monkeys, 14:*5*, 14:175, 14:*177*, 14:*182*–183
Golden spiny mice, 16:*251*, 16:261*t*
Golden squirrels. *See* Persian squirrels
Golden takins, 16:2
Golden-tipped bats, 13:*519*, 13:*520*, 13:521, 13:522
Gompotheres, 15:137
Gongshan muntjacs, 15:346, 15:*347*, 15:*350*, 15:352
Goodall, Jane, 12:157
Goodfellow's tree kangaroos, 13:33
Goose-beaked whales. *See* Cuvier's beaked whales
Gorals, 16:87, 16:88, 16:89, 16:90, 16:92, 16:95
Gorgopithecus spp., 14:189
Gorilla spp. *See* Gorillas
Gorilla beringei. See Eastern gorillas
Gorilla beringei beringei. See Mountain gorillas
Gorilla beringei graueri. See Eastern lowland gorillas
Gorilla gorilla. See Western gorillas
Gorillas
behavior, 12:*141*, 14:7, 14:228, 14:232
conservation status, 12:215, 14:235
distribution, 14:5, 14:227
evolution, 12:33, 14:225
feeding ecology, 14:232, 14:233
field studies of, 12:200
habitats, 14:6, 14:227
humans and, 14:11
memory, 12:153
physical characteristics, 14:4, 14:226
population indexes, 12:195
reproduction, 14:234–235
taxonomy, 14:225
in zoos, 12:208, 12:*210*
See also specific types of gorillas
Gould's wattled bats, 12:*59*
Goya tuco-tucos, 16:431*t*
GPS (Global positioning systems), 12:199
Gracile chimpanzees. *See* Bonobos
Gracile opossums, 12:249
Gracilinanus spp. *See* Gracile opossums
Gracilinanus aceramarcae, 12:251, 12:254
Gracilinanus agilis, 12:251
Gracilinanus emiliae, 12:251
Gracilinanus marica. See Northern gracile mouse opossums
Gracilinanus microtarsus, 12:251
Graell's black-mantled tamarins, 14:116
Grahame, Kenneth, 13:200
Grammar. *See* Syntax
Grampus griseus. See Risso's dolphins
Grant's bushbabies, 14:26, 14:32*t*
Grant's desert golden moles, 12:71–72, 12:76, 13:*197*, 13:215, 13:*217*, 13:*219*, 13:*220*–221
Grant's gazelles, 15:271, 16:48, 16:57*t*
Graphiurinae. *See* African dormice
Graphiurus spp. *See* African dormice
Graphiurus crassicaudatus. See Jentnink's dormice
Graphiurus kelleni. See Kellen's dormice
Graphiurus murinus. See Woodland dormice
Graphiurus ocularis. See Spectacled dormice
Graphiurus parvus. See Savanna dormice
Graphiurus rupicola. See Stone dormice
Graphiurus surdus. See Silent dormice
Grasscutters. *See* Cane rats
Grasshopper mice, 16:125
Grauer's shrews, 13:277*t*
Gray-backed sportive lemurs, 14:75, 14:*78*, 14:*79*
Gray-bellied douroucoulis. *See* Gray-bellied night monkeys
Gray-bellied night monkeys, 14:135, 14:138, 14:*139*, 14:*140*
Gray-bellied owl monkeys. *See* Gray-bellied night monkeys
Gray brocket deer, 15:396*t*
Gray-cheeked flying squirrels, 16:141*t*
Gray-cheeked mangabeys, 14:*196*, 14:*201*–202, 14:204*t*
Gray climbing mice, 16:*286*, 16:*290*–291
Gray dorcopsises, 13:*91*, 13:*94*, 13:100
Gray dwarf hamsters, 16:247*t*
Gray four-eyed opossums, 12:250, 12:*256*, 12:263
Gray foxes, 14:265, 14:272, 14:284*t*
Gray fruit bats. *See* Black flying foxes
Gray gentle lemurs, 14:*50*
Gray gibbons. *See* Mueller's gibbons
Gray hamsters, 16:243
Gray-headed flying foxes, 13:312, 13:320
Gray kangaroos, 13:34, 13:84
Gray long-eared bats, 13:516*t*
Gray mouse lemurs, 14:*37*, 14:*38*, 14:39, 14:*40*, 14:*42*–43
Gray myotis, 12:*5*, 13:310–311, 13:316
Gray-neck night monkeys, 14:135
Gray rheboks, 16:27, 16:28, 16:29, 16:30, 16:31, 16:32
Gray sac-winged bats, 13:*359*, 13:360–*361*
Gray seals, 14:417, 14:418, 14:421–422, 14:*426*, 14:*427*–428
Gray shrews, 13:276*t*
Gray slender lorises, 14:*17*, 14:*18*
Gray slender mouse opossums, 12:251, 12:*256*, 12:261
Gray snub-nosed monkeys, 14:175, 14:185*t*
Gray squirrels, 12:*93*, 12:139, 16:163, 16:*164*, 16:*165*, 16:*168*, 16:*170*–171
Arizona, 16:167, 16:*168*, 16:*171*–172
vs. red squirrels, 12:183–184
western, 16:175*t*

Gray tree kangaroos. *See* Bennett's tree kangaroos
Gray whales, **15:93–101**, 15:*94*, 15:*96*, 15:*97*, 15:*98*, 15:*99*, 15:*100*
behavior, 15:95–97
conservation status, 15:9, 15:99–100
distribution, 15:5, 15:*93*, 15:94–95
evolution, 15:3, 15:93
feeding ecology, 15:97–98
habitats, 15:6, 15:95
humans and, 15:100
migrations, 12:87
physical characteristics, 15:4, 15:93–94
reproduction, 15:*95*, 15:98–99
taxonomy, 15:93
Gray wolves, 12:*30*, 14:*271*, 14:*275*, 14:*276–277*
behavior, 14:258, 14:*266*
conservation status, 14:273
distribution, 12:129, 12:131–132
habitats, 14:267
reproduction, 14:*255*, 14:*259*
smell, 12:*83*
Gray woolly monkeys, 14:160, 14:*161*, 14:*165–166*
Grayish mouse opossums, 12:264*t*
Gray's beaked whales, 15:70*t*
Great apes, **14:225–240**, 14:*236*
behavior, 14:228–232
conservation status, 12:215, 14:235
distribution, 12:135, 12:136, 14:*225*, 14:227
evolution, 14:225
feeding ecology, 14:232–234
habitats, 14:227–228
humans and, 14:235
imitation, 12:158
language, 12:162
physical characteristics, 14:225–227
reproduction, 14:234–235
self recognition, 12:159
species of, 14:*237*–240
taxonomy, 14:4, 14:225
tool use, 12:157
See also Apes
Great Basin kangaroo rats. *See* Chisel-toothed kangaroo rats
Great Basin pocket mice, 16:*200*
Great Danes, 14:289
Great fruit-eating bats, 13:434*t*
Great gray kangaroos. *See* Eastern gray kangaroos
Great long-nosed armadillos, 13:191*t*
Great northern sea cows. *See* Steller's sea cows
Great-tailed trioks, 13:133*t*
Great whales. *See* specific great whales
Greater bamboo lemurs, 14:11, 14:*48*
Greater bandicoot rats, 16:249–250
Greater bent-winged bats. *See* Common bentwing bats
Greater bilbies, 13:*2*, 13:4, 13:5, 13:6, 13:19–*20*, 13:*21*, 13:*22*
Greater bulldog bats, 13:313, 13:443, 13:*444*, 13:*445*, 13:*446*, 13:447, 13:*448*, 13:*449*
Greater bushbabies. *See* Brown greater bushbabies; Northern greater bushbabies
Greater cane rats, 16:333, 16:*334*, 16:*336*, 16:*337*
Greater Cuban nesophontes, 13:243
Greater dog-faced bats, 13:*357*, 13:*359*, 13:361–*362*
Greater doglike bats. *See* Greater dog-faced bats
Greater dwarf lemurs, 14:*4*, 14:*36*, 14:*40*, 14:*41–42*
Greater Egyptian jerboas, 16:212, 16:216, 16:223*t*
Greater false vampire bats, 13:313–314, 13:*379*, 13:*381*, 13:*382*
Greater fat-tailed jerboas, 16:212, 16:214, 16:216, 16:223*t*
Greater gliders, 13:33, 13:113–120, 13:*115*, 13:*118*, 13:*119*
Greater gliding possums, **13:113–123**, 13:122*t*–123*t*
Greater guinea pigs, 16:399*t*
Greater hedgehog tenrecs, 13:200, 13:*227*, 13:234*t*
Greater horseshoe bats, 13:*389*, 13:*391*, 13:*395*, 13:*397*–398
behavior, 13:390
conservation status, 13:392–393
distribution, 13:389
reproduction, 13:314, 13:372
Greater house bats, 13:*484*, 13:486, 13:*489*, 13:*490*, 13:492
Greater Japanese shrew moles, 13:280, 13:282, 13:288*t*
Greater kudus, 16:6, 16:11, 16:*12*, 16:*13*, 16:*16*, 16:*19*, 16:20, 16:25*t*
Greater long-tailed hamsters, 16:239, 16:247*t*
Greater long-tongued fruit bats, 13:*339*, 13:*344*, 13:346
Greater long-tongued nectar bats. *See* Greater long-tongued fruit bats
Greater Malagasy bushy-tailed rats, 16:*281*
Greater Malay mouse deer, 15:325, 15:*326*, 15:*327*, 15:*328*, 15:*330*, 15:*332*–333
Greater marsupial moles. *See* Marsupial moles
Greater Mascarene flying foxes, 13:331*t*
Greater mouse-eared bats, 13:497, 13:*503*, 13:515*t*
Greater mouse-tailed bats, 13:352
Greater naked-backed bats. *See* Naked-backed fruit bats
Greater naked-tailed armadillos, 13:185, 13:*186*, 13:*189*
Greater New Zealand short-tailed bats, 13:454, 13:455, 13:*456*, 13:457–458
Greater noctules, 13:313, 13:516*t*
Greater one-horned rhinoceroses. *See* Indian rhinoceroses
Greater rabbit-eared bandicoots. *See* Greater bilbies
Greater sac-winged bats, 12:81, 13:*356–358*, 13:*359*, 13:*360*
Greater short-tailed bats. *See* Greater New Zealand short-tailed bats
Greater shrews, 13:276*t*
Greater sloth lemurs, 14:*68*, 14:*70*, 14:71
Greater spear-nosed bats, 13:*422*, 13:*424*
behavior, 13:312, 13:416, 13:419
echolocation, 12:87
feeding ecology, 13:418
vocalizations, 12:85
Greater stick-nest rats, 16:*254*, 16:*259–260*
Greater thumbless bats. *See* Schnabeli's thumbless bats
Greater two-lined bats. *See* Greater sac-winged bats
Greater white-lined bats. *See* Greater sac-winged bats
Green acouchis, 16:*408*, 16:*410*, 16:*412*, 16:414
Green colobus. *See* Olive colobus
Green monkeys. *See* Grivets
Green ringtails, 13:*114*, 13:*116*, 13:*118*, 13:*119*, 13:120
Greenhalli's dog-faced bats, 13:494*t*
Greenland. *See* Nearctic region
Greenland collared lemmings. *See* Northern collared lemmings
Gregarious behavior, 12:145–147
See also Behavior
Grevy's zebras, 15:*215*, 15:*221*, 15:*227*, 15:*231*, 15:*232–233*
behavior, 15:219, 15:*228*
conservation status, 15:222
evolution, 15:216
physical characteristics, 15:*225*
reproduction, 15:221
Griffin, Donald, 12:54
Grivets, 14:*187*, 14:*196*, 14:199
behavior, 12:143, 14:7
communication, 12:84, 12:160
distribution, 14:*197*
evolution, 14:188
Grizzlies. *See* Brown bears
Grooming behavior. *See* Behavior
Groove-toothed rats. *See* Angoni vlei rats
Groove-toothed swamp rats. *See* Creek rats
Ground cuscuses, 13:*60*, 13:*62*, 13:*64–65*
Ground pangolins, 16:*109*, 16:*110*, 16:*112*, 16:113, 16:*114*, 16:117–*118*
Ground sloths, 12:132, 15:137, 15:138
Ground squirrels, **16:143–161**, 16:*149–150*
behavior, 12:*71*, 12:141, 16:124, 16:144–145
conservation status, 16:147–148
distribution, 12:131, 16:*143*, 16:144
evolution, 16:143
feeding ecology, 16:145, 16:147
habitats, 16:144
humans and, 16:126, 16:148
physical characteristics, 12:113, 16:143–144
reproduction, 16:147
species of, 16:151–158, 16:158*t*–160*t*
taxonomy, 16:143
Groundhogs. *See* Woodchucks
Groves, C. P.
on capuchins, 14:101
on tarsiers, 14:91
Grysboks, **16:59–72**, 16:71*t*
Guadalupe fur seals, 14:394, 14:395, 14:398, 14:399, 14:408*t*
Guadalupe storm-petrels, 14:291
Guadeloupe raccoons, 14:309, 14:310, 14:316*t*
Guam flying foxes, 12:138, 13:332*t*
Guanacaste squirrels. *See* Deppe's squirrels
Guanacos, 15:*134*, **15:313–323**, 15:*315*, 15:*319*, 15:*322*
behavior, 15:316–317, 15:*318*
distribution, 12:132, 15:*313*, 15:315
domestication of, 12:180
evolution, 15:151, 15:313–314
feeding ecology, 15:317
habitats, 15:316
humans and, 15:318

Guanacos *(continued)*
physical characteristics, 15:314
reproduction, 15:*316*, 15:317
taxonomy, 15:313
Guano bats. *See* Brazilian free-tailed bats
Guard dogs, 14:288, 14:289, 14:292–293
Guenons, 14:5
behavior, 12:146, 14:7, 14:192
evolution, 14:188
guilds of, 12:116
habitats, 14:191
humans and, 14:194
physical characteristics, 14:189
reproduction, 14:10, 14:193
taxonomy, 14:188
Guenther's dikdiks, 16:71*t*
Guenther's long-snouted dikdiks. *See* Guenther's dikdiks
Guerezas, 14:174, 14:*175*, 14:*176*, 14:*178*
Guguftos. *See* Gerenuks
Guianan mastiff bats. *See* Greater house bats
Guianan sakis. *See* White-faced sakis
Guiaras, 16:458*t*
Guide dogs, 14:*293*
Guiffra, E., 15:149
Guilds, of species, 12:116–119
Guinea baboons, 14:*188*, 14:194, 14:205*t*
Guinea pigs, 16:*392*, 16:*394*, 16:*395*–396
behavior, 12:148
distribution, 12:132
domestic, 12:180–181, 16:*391*, 16:399*t*
greater, 16:399*t*
humans and, 16:128, 16:393
montane, 16:399*t*
neonatal requirements, 12:126
neonates, 12:125
physical characteristics, 16:390
reproduction, 12:106
shiny, 16:399*t*
taxonomy, 16:121, 16:389
Gulf porpoises. *See* Vaquitas
Gulf Stream beaked whales. *See* Gervais' beaked whales
Gulo spp., 14:321
Gulo gulo. *See* Wolverines
Gummivory, 14:8–9
Gundis, **16:311–315**
Gunning's golden moles, 13:222*t*
Gunnison's prairie dogs, 16:159*t*
Gunn's bandicoots. *See* Eastern barred bandicoots
Gursky, S., 14:98
Gymnobelideus spp., 13:128, 13:129
Gymnobelideus leadbeateri. *See* Leadbeater's possums
Gymnures, 13:193, 13:195, 13:197, 13:199, **13:203–214,** 13:212*t*–213*t*

H

Habitat loss, 12:214
See also Conservation status
Habitats
aardvarks, 15:156
agoutis, 16:407, 16:411–414, 16:415*t*
anteaters, 13:172, 13:177–179
armadillos, 13:183, 13:187–190, 13:190*t*–192*t*
Artiodactyla, 15:269
aye-ayes, 14:86
baijis, 15:20
bandicoots, 13:2–3
dry-country, 13:10, 13:14–16, 13:16*t*–17*t*
rainforest, 13:10, 13:14–17*t*
bats, 13:311
bulldog, 13:445, 13:449–450
disk-winged, 13:474, 13:476–477
Emballonuridae, 13:355, 13:360–363, 13:363*t*–364*t*
false vampire, 13:379, 13:384–385
funnel-eared, 13:460, 13:463–465
horseshoe, 13:389–390, 13:396–400
Kitti's hog-nosed, 13:368
Molossidae, 13:484, 13:490–493, 13:493*t*–495*t*
mouse-tailed, 13:351, 13:353
moustached, 13:435, 13:440–442, 13:442*t*
New Zealand short-tailed, 13:454, 13:457
Old World fruit, 13:319, 13:325–330, 13:331*t*–332*t*, 13:334, 13:340–347, 13:348*t*–349*t*
Old World leaf-nosed, 13:402, 13:407–409, 13:409*t*–410*t*
Old World sucker-footed, 13:480
slit-faced, 13:371, 13:375, 13:376*t*–377*t*
smoky, 13:467–468, 13:470
Vespertilionidae, 13:498, 13:506, 13:509–514, 13:515*t*–516*t*, 13:521, 13:524–525, 13:526*t*
bears, 14:296–297, 14:302–305, 14:306*t*
beavers, 16:179, 16:183–184
bilbies, 13:20
botos, 15:28
Bovidae, 16:4–5
Antilopinae, 16:46, 16:50–55, 16:56*t*–57*t*
Bovinae, 16:13, 16:18–23, 16:24*t*–25*t*
Caprinae, 16:91–92, 16:98–103, 16:103*t*–104*t*
duikers, 16:74, 16:80–84, 16:84*t*–85*t*
Hippotraginae, 16:29–30, 16:37–42, 16:42*t*–43*t*
Neotraginae, 16:60, 16:66–71, 16:71*t*
bushbabies, 14:6, 14:24, 14:28–32, 14:32*t*–33*t*
Camelidae, 15:315–316, 15:320–323
Canidae, 14:267, 14:276–283, 14:283*t*–284*t*
capybaras, 16:402
Carnivora, 14:257
cats, 14:371, 14:379–389, 14:390*t*–391*t*
Caviidae, 16:390–391, 16:395–398, 16:399*t*
Cetacea, 15:5–6
Balaenidae, 15:109, 15:115–117
beaked whales, 15:61, 15:64–68, 15:69*t*–70*t*
dolphins, 15:43–44, 15:53–55, 15:56*t*–57*t*
franciscana dolphins, 15:24
Ganges and Indus dolphins, 15:14
gray whales, 15:95
Monodontidae, 15:83–84, 15:90
porpoises, 15:34–35, 15:38–39, 15:39*t*
pygmy right whales, 15:104
rorquals, 15:6, 15:121, 15:127–130, 15:130*t*
sperm whales, 15:74, 15:79–80
chevrotains, 15:327, 15:331–333
Chinchillidae, 16:378, 16:382–383, 16:383*t*
colugos, 13:302, 13:304
coypus, 16:124, 16:474
dasyurids, 12:288–289, 12:294–298, 12:299*t*–301*t*
Dasyuromorphia, 12:279–280
deer
Chinese water, 15:374
muntjacs, 15:344, 15:349–354
musk, 15:336, 15:340–341
New World, 15:382–384, 15:388–396, 15:396*t*
Old World, 15:359, 15:364–371, 15:371*t*
Dipodidae, 16:213, 16:219–222, 16:223*t*–224*t*
Diprotodontia, 13:34–35
dormice, 16:318, 16:323–326, 16:327*t*–328*t*
duck-billed platypuses, 12:*229*, 12:230, 12:244, 12:*247*
Dugongidae, 15:200, 15:203
echidnas, 12:237, 12:240*t*
elephants, 15:166–167, 15:174–175
Equidae, 15:226, 15:232–235, 15:236*t*
Erinaceidae, 13:204, 13:209–213, 13:212*t*–213*t*
giant hutias, 16:469, 16:471*t*
gibbons, 14:210, 14:218–222
Giraffidae, 15:401, 15:408
great apes, 14:227–228, 14:237–240
gundis, 16:312, 16:314–315
Herpestidae, 14:349, 14:354–356, 14:357*t*
Heteromyidae, 16:200, 16:205–208*t*, 16:209*t*
hippopotamuses, 15:306, 15:310–311
hot spots, 12:218–219
humans, 14:247
hutias, 16:461, 16:467*t*
Hyaenidae, 14:362, 14:364–367
hyraxes, 15:179–180, 15:186–188, 15:189*t*
Indriidae, 14:65, 14:69–72
Insectivora, 13:197
koalas, 13:44–45
Lagomorpha, 16:481–482
lemurs, 14:6, 14:49, 14:55, 14:57–60
Cheirogaleidae, 14:36, 14:41–44
sportive, 14:75, 14:79–83
Leporidae, 16:506, 16:511–5135, 16:515*t*–516*t*
Lorisidae, 14:14, 14:18–20, 14:21*t*
Macropodidae, 13:34, 13:35, 13:85–86, 13:93–100, 13:101*t*–102*t*
manatees, 15:206, 15:211–212
Megalonychidae, 13:156, 13:159
moles
golden, 13:197, 13:216, 13:220–222, 13:222*t*
marsupial, 13:26, 13:28
monitos del monte, 12:274
monkeys
Atelidae, 14:156–157, 14:162–166, 14:166*t*–167*t*
Callitrichidae, 14:116, 14:127–131, 14:132*t*
Cebidae, 14:103, 14:108–112
cheek-pouched, 14:191, 14:197–204, 14:204*t*–205*t*
leaf-monkeys, 14:174, 14:178–183, 14:184*t*–185*t*
night, 14:137, 14:141*t*
Pitheciidae, 14:145, 14:150–152, 14:153*t*
monotremes, 12:230
mountain beavers, 16:132
Muridae, 16:283–284, 16:288–295, 16:296*t*–297*t*

Habitats *(continued)*
Arvicolinae, 16:226, 16:233–238, 16:237*t*–238*t*
hamsters, 16:240, 16:245–246, 16:247*t*
Murinae, 16:250–251, 16:255–260, 16:261*t*–262*t*
Sigmodontinae, 16:266–267, 16:272–276, 16:277*t*–278*t*
musky rat-kangaroos, 13:70
Mustelidae, 14:321, 14:327–331, 14:332*t*–333*t*
numbats, 12:304
octodonts, 16:434, 16:438–440, 16:440*t*
opossums
New World, 12:251, 12:257–263, 12:264*t*–265*t*
shrew, 12:268, 12:270
Otariidae, 14:395–396, 14:402–406
pacaranas, 16:386–387
pacas, 16:418, 16:423–424
pangolins, 16:110, 16:115–119
peccaries, 15:292–293, 15:298–300
Perissodactyla, 15:217–218
Petauridae, 13:125–126, 13:131–133, 13:133*t*
Phalangeridae, 13:59–60, 13:*63–66t*
pigs, 15:277, 15:284–288, 15:289*t*–290*t*
pikas, 16:491, 16:497–501, 16:501*t*–502*t*
pocket gophers, 16:187, 16:192–194, 16:195*t*–197*t*
porcupines
New World, 16:367, 16:371–373, 16:374*t*
Old World, 16:352, 16:357–364
possums
feather-tailed, 13:141, 13:144
honey, 13:136
pygmy, 13:106, 13:110–111
primates, 14:6
Procyonidae, 14:310, 14:313–315, 14:315*t*–316*t*
pronghorns, 15:412
Pseudocheiridae, 13:114–115, 13:119–123, 13:122*t*–123*t*
rat-kangaroos, 13:74, 13:78–81
rats
African mole-rats, 16:342–343, 16:347–349, 16:349*t*–350*t*
cane, 16:334, 16:337–338
chinchilla, 16:444, 16:447–448
dassie, 16:330
spiny, 16:450, 16:453–457, 16:458*t*
rhinoceroses, 15:252–253, 15:257–261
rodents, 16:124
sengis, 16:519, 16:524–530, 16:531*t*
shrews
red-toothed, 13:248, 13:255–263, 13:263*t*
West Indian, 13:244
white-toothed, 13:266–267, 13:271–275, 13:276*t*–277*t*
Sirenia, 15:192
solenodons, 13:238, 13:240
springhares, 16:308
squirrels
flying, 16:135–136, 16:139–140, 16:141*t*–142*t*
ground, 16:144, 16:151–158, 16:158*t*–160*t*
scaly-tailed, 16:300, 16:302–305
tree, 16:164–165, 16:169–174*t*, 16:175*t*
Talpidae, 13:280, 13:284–287, 13:287*t*–288*t*
tapirs, 15:238, 15:245–247
tarsiers, 14:93–94, 14:97–99
tenrecs, 13:226–227, 13:232–234, 13:234*t*
three-toed tree sloths, 13:162, 13:167–169
tree shrews, 13:291, 13:294–296, 13:297*t*–298*t*
true seals, 14:257, 14:418, 14:427–428, 14:430–434, 14:435*t*
tuco-tucos, 16:425–426, 16:429–430, 16:430*t*–431*t*
Viverridae, 14:337, 14:340–343, 14:344*t*–345*t*
walruses, 12:14, 14:412
wombats, 13:52, 13:55–56
Xenartha, 13:150
Hadropithecus spp., 14:63
Haemochorial placentation, 12:94
Hafner, John C., 16:199
Hagenbeck, Carl, 12:204
Hagenbeck exhibits, 12:204–205
Hainan gymnures, 13:207, 13:213*t*
Hainan hares, 16:509
Hainan moonrats. *See* Hainan gymnures
Hairless bats. *See* Naked bats
Hairs, 12:3, 12:7, 12:38–39, 12:*43*
bats, 12:59
domestication and, 12:174
evolution of, 12:19, 12:22
for heat loss prevention, 12:51
longtail weasels, 12:*47*
marine mammals, 12:63, 12:*64*
in sexual selection, 12:24
See also Physical characteristics
Hairy-armed bats. *See* Leisler's bats
Hairy big-eyed bats, 13:434*t*
Hairy dwarf porcupines
Bahía, 16:366, 16:*370*, 16:*371*, 16:373
black-tailed, 16:365, 16:374*t*
brown, 16:366, 16:374*t*
frosted, 16:365, 16:374*t*
orange-spined, 16:366, 16:374*t*
Paraguay, 16:366, 16:367, 16:368, 16:374*t*
white-fronted, 16:365, 16:374*t*
Hairy-eared mouse lemurs, 14:36, 14:38, 14:39, 14:*40*, 14:*41*
Hairy-footed flying squirrels, 16:141*t*
Hairy-footed jerboas, 16:214, 16:215, 16:216, 16:*218*, 16:*221*
Hairy-fronted muntjacs. *See* Black muntjacs
Hairy-legged vampire bats, 13:314, 13:*414*, 13:433*t*
Hairy long-nosed armadillos, 13:185, 13:191*t*
Hairy-nosed otters, 14:332*t*
Hairy-nosed wombats, 13:37, 13:51, 13:52
Hairy slit-faced bats, 13:376*t*
Hairy-tailed moles, 13:*282*, 13:*283*, 13:*284*, 13:286
Hairy-winged bats. *See* Harpy-headed bats
Haitian nesophontes, 13:243, 13:244
Haitian solenodons. *See* Hispaniolan solenodons
Haliaeetus leucocephalus. See Bald eagles
Halichoerus grypus. See Gray seals
Hamadryas baboons, 12:146, 12:*209*, 14:7, 14:192, 14:194, 14:*195*, 14:*202–203*
Hammer-headed fruit bats, 12:84, 13:312, 13:334, 13:*338*, 13:*341*, 13:342
Hamsters, 16:123, 16:128, **16:239–248,** 16:*240*, 16:*241*, 16:*244*, 16:282, 16:283
Hantaviruses
mice and, 16:128
Sigmodontinae, 16:270
Hanuman langurs. *See* Northern plains gray langurs
Hapalemur spp., 14:8
Hapalemur aureus, 14:11
Hapalemur griseus griseus. See Gray gentle lemurs
Hapalemur simus. See Greater bamboo lemurs
Haplonycteris fischeri. See Fischer's pygmy fruit bats
Haplorhine primates
evolution, 12:11
reproduction, 12:94
See also Apes; Humans; Monkeys; Tarsiers
Harbor porpoises, 15:7, 15:9, 15:33, 15:*34*, 15:35, 15:*36*, 15:*37*, 15:*38*
Harbor seals, 14:*261*, 14:418–420, 14:*424*, 14:*426*, 14:*428–429*
Hard head whales. *See* Gray whales
Hardwicke's hedgehogs. *See* Collared hedgehogs
Hardwicke's lesser mouse-tailed bats, 13:*352*, 13:*354*
Hare, Brian, 14:287
Hare-wallabies, 13:83, 13:84, 13:85, 13:86, 13:88
Harem groups. *See* One-male groups
Harems. *See* Polygyny
Hares, **16:505–516,** 16:*510*
behavior, 16:482–483, 16:506–507
conservation status, 12:222, 16:509
distribution, 12:129, 12:135, 16:*505*–506
evolution, 16:480, 16:505
feeding ecology, 16:484–485, 16:507
habitats, 16:481, 16:506
humans and, 16:488, 16:509
milk of, 12:126
physical characteristics, 12:38, 16:505
predation and, 12:115
reproduction, 16:485–487, 16:507–509
species of, 16:*511*–513, 16:515*t*–516*t*
taxonomy, 16:505
Harlequin bats, 13:*503*
Harlow, Harry, 12:151–152
Harney, B. A., 16:200
Harp seals, 14:*421*, 14:*422*, 14:424, 14:*426*, 14:*429*–430
Harpagornis moorei. See Eagles
Harpiocephalus spp., 13:519
Harpiocephalus harpia. See Harpy-headed bats
Harpy fruit bats, 13:*339*, 13:*341*, 13:345
Harpy-headed bats, 13:*523*, 13:524
Harpy-winged bats. *See* Harpy-headed bats
Harpyionycteris whiteheadi. See Harpy fruit bats
Harris's antelope squirrels, 16:*150*, 16:*154*, 16:156
Harris's olingos, 14:310, 14:316*t*
Hartebeests, 15:139, **16:27–43,** 16:*30*, 16:42*t*
Hartenberger, J. L., 16:311
Hartman, Daniel, 15:191
Hartmann's mountain zebras, 15:222
Harvest mice, 16:*254*, 16:*255*, 16:258
plains, 16:*264*
saltmarsh, 16:*265*
Sumichrast's, 16:*269*
Harvey's duikers, 16:85*t*
Hatshepsut (Queen of Egypt), 12:203
Hawaiian hoary bats, 12:188–192
Hawaiian honeycreepers, 12:188
Hawaiian Islands, introduced species, 12:188–192

Hawaiian monk seals, 12:138, 14:*420*, 14:422, 14:*425*, 14:*429*, 14:431–432
Hayes, Cathy, 12:160
Hayes, Keith, 12:160
Hazel dormice, 16:318, 16:*319*, 16:*320*, 16:321, 16:*322*, 16:*323*, 16:326
Health care, in zoos, 12:209–211
Hearing, 12:5–6, 12:8–9, 12:50, 12:72, 12:77, 12:*81*, 12:81–82
See also Physical characteristics
Heart-nosed bats, 13:314, 13:*380*, 13:381–382
Hearts, 12:8, 12:36, 12:45, 12:*46*, 12:60
See also Physical characteristics
Heat loss, 12:51, 12:59–60
See also Thermoregulation
Hector's beaked whales, 15:70*t*
Hector's dolphins, 15:43, 15:*45*, 15:48
Hedgehog tenrecs, 13:199
Hedgehogs, 12:*185*, 13:193, 13:195–198, 13:200, **13:203–214,** 13:212*t*–213*t*
Heermann's kangaroo rats, 16:201, 16:*204*, 16:*207*, 16:208
Helarctos malayanus. See Malayan sun bears
Heliophobius spp. *See* Silvery mole-rats
Heliophobius argenteocinereus. See Silvery mole-rats
Heliosciurus spp., 16:163
Heliosciurus rufobrachium, 16:165
Helogale spp., 14:347
Helogale parvula. See Dwarf mongooses
Helohyus spp., 15:265
Helpers, in social systems, 12:146–147
See also Behavior
Hemibelideus spp., 13:113
Hemibelideus lemuroides. See Lemuroid ringtails
Hemicentetes spp., 13:225, 13:226, 13:227–228
Hemicentetes nigriceps. See White-streaked tenrecs
Hemicentetes semispinosus. See Yellow streaked tenrecs
Hemiechinus auritus. See Long-eared hedgehogs
Hemiechinus collaris. See Collared hedgehogs
Hemiechinus nudiventris, 13:207
Hemigalinae, 14:335
Hemigalus derbyanus. See Banded palm civets
Hemitragus spp., 16:87
Hemitragus hylocrius. See Nilgiri tahrs
Hemitragus jayakari. See Arabian tahrs
Hemoglobin, 12:30
See also Physical characteristics
Heptaxodontidae. *See* Giant hutias
Herbert River ringtails, 13:117, 13:*118*, 13:*121*–122
Herbivores
digestive system, 12:120–124
distribution, 12:135, 12:136
in guilds, 12:118–119
migrations, 12:164–166, 12:164–168
periglacial environments and, 12:25
subterranean, 12:69, 12:76
teeth, 12:14–15, 12:*37*, 12:46
See also Ungulates; specific herbivores
Herding dogs, 14:288–289, 14:292
Hero shrews. *See* Armored shrews
Herodotius pattersoni, 16:517
Herodotus, 14:290
Herpailurus yaguarondi. See Jaguarundis
Herpestes spp., 14:347, 14:348–349
Herpestes auropunctatus. See Indian mongooses
Herpestes edwardsii. See Indian gray mongooses
Herpestes ichneumon. See Egyptian mongooses
Herpestes javanicus. See Small Indian mongooses
Herpestes nyula. See Indian mongooses
Herpestes palustris. See Bengali water mongooses
Herpestes pulverulentus. See Cape gray mongooses
Herpestes sanguinus. See Slender mongooses
Herpestidae. *See* Fossa; Mongooses
Herpestinae, 14:347, 14:349
Herring whales. *See* Fin whales
Hershkovitz, P.
on *Callicebus* spp., 14:143
on capuchins, 14:101
on squirrel monkeys, 14:101
Hershkovitz's night monkeys, 14:135, 14:138, 14:141*t*
Heterocephalus spp. *See* Naked mole-rats
Heterocephalus glaber. See Naked mole-rats
Heterodonty, 12:46
Heterohyrax spp., 15:177
Heterohyrax antineae. See Ahaggar hyraxes
Heterohyrax brucei. See Bush hyraxes
Heterohyrax brucei antineae, 15:179
Heterohyrax chapini. See Yellow-spotted hyraxes
Heteromyid rodents, 12:131
See also specific rodents
Heteromyidae, 16:121, 16:123, **16:199–210,** 16:*204*, 16:208*t*–209*t*
Heteromys spp., 16:199, 16:200, 16:201, 16:203
Heteromys desmarestianus. See Desmarest's spiny pocket mice
Heteromys nelsoni. See Nelson's spiny pocket mice
Heterothermy, 12:50, 12:59–60, 12:73, 12:113
See also Thermoregulation
Hexaprotodon spp., 15:301
Hexaprotodon liberiensis. See Pygmy hippopotamuses
Hexolobodon phenax. See Imposter hutias
Hibernation, 12:113, 12:120, 12:169
See also Behavior
High altitudes, 12:113
High mountain voles, 16:225, 16:228
Higher primates, 14:1, 14:2, 14:3, 14:7
See also specific species
Highland yellow-shouldered bats, 13:433*t*
Highveld mole-rats, 16:350*t*
Hildebrandt's horseshoe bats, 13:387, 13:389, 13:391, 13:*393*
Hildegarde's tomb bats, 13:*357*
Hill, J. E., 16:199, 16:211
Hill long-tongued fruit bats. *See* Greater long-tongued fruit bats
Hill wallaroos. *See* Common wallaroos
Himalayan gorals, 16:87, 16:*96*, 16:*98–99*
Himalayan marmots, 16:*146*
Himalayan musk deer, 15:335, 15:*339*, 15:*340*
Himalayan pikas, 16:*494*, 16:501*t*
Himalayan pipistrelles, 13:*503*
Himalayan snow bears. *See* Brown bears
Himalayan tahrs, 12:137, 16:90, 16:91, 16:92, 16:93, 16:*94*
Himalayan water shrews, 13:*253*, 13:*256*, 13:258–259
Hind limbs, bats, 12:57
See also Bats
Hindgut fermentation, 12:14–15, 12:48
Hinnies, 12:177, 15:223
Hipparionid horses, 15:137
Hippocamelus spp., 15:382
Hippocamelus antisensis. See Peruvian huemuls
Hippocamelus bisulcus. See Chilean huemuls
Hippopotamidae. *See* Hippopotamuses
Hippopotamus spp., 15:301
Hippopotamus amphibius. See Common hippopotamuses
Hippopotamus lemerlei, 12:136, 15:302
Hippopotamuses, 12:*62*, 12:*116*, **15:301–312,** 15:*302*, 15:*304*, 15:*309*
behavior, 15:306
cetaceans and, 12:30
conservation status, 15:308
distribution, 12:135, 15:*301*, 15:306
evolution, 15:136, 15:137, 15:264, 15:267, 15:304
feeding ecology, 15:271, 15:306, 15:308
habitats, 15:269, 15:306
physical characteristics, 12:68, 15:138, 15:139, 15:267, 15:304–306
reproduction, 12:91, 15:308
species of, 15:310–311
taxonomy, 15:301–302, 15:304
Hipposideridae. *See* Old World leaf-nosed bats
Hipposideros spp., 13:401, 13:402, 13:405
Hipposideros ater. See Dusky leaf-nosed bats
Hipposideros bicolor, 13:*404*
Hipposideros caffer. See Sundevall's roundleaf bats
Hipposideros cineraceus. See Ashy roundleaf bats
Hipposideros commersoni. See Commerson's leaf-nosed bats
Hipposideros diadema. See Diadem roundleaf bats
Hipposideros fulvus. See Fulvus roundleaf bats
Hipposideros larvatus. See Roundleaf horseshoe bats
Hipposideros ridleyi. See Ridley's roundleaf bats
Hipposideros ruber. See Noack's roundleaf bats
Hipposideros stenotis. See Northern leaf-nosed bats
Hipposideros terasensis, 13:405
Hippotraginae, 16:1, **16:27–44,** 16:*35–36*, 16:42*t*–43*t*
Hippotragini, 16:27, 16:28, 16:29, 16:31, 16:34
Hippotragus spp., 16:27, 16:28
Hippotragus equinus. See Roan antelopes
Hippotragus leucophaeus. See Bluebucks
Hippotragus niger. See Sable antelopes
Hippotragus niger variani. See Giant sable antelopes
Hirolas. *See* Hunter's hartebeests
Hispaniolan giant hutias, 16:471*t*
Hispaniolan hutias, 16:*463*, 16:*464*, 16:465–*466*
Hispaniolan solenodons, 13:195, 13:198, 13:237, 13:*238*, 13:*239*, 13:240–241
Hispid cotton rats, 16:268, 16:269, 16:270, 16:*271*, 16:*272*, 16:275–276
Hispid hares, 16:482, 16:487, 16:505, 16:506, 16:509, 16:515*t*

INDEX

Hispid pocket gophers, 16:196*t*
Hispid pocket mice, 16:*204*, 16:*205*
Histoplasmosis, 13:316
The History of Animals (Aristotle), 12:203
HIV (Human immunodeficiency virus), 12:222
Hoary bats, 13:310, 13:312, 13:497, 13:499, 13:500, 13:501, 13:*505*, 13:*510*–511
Hoary marmots, 16:*144*, 16:145, 16:158*t*
Hodgson's brown-toothed shrews, 12:*134*, 13:*254*, 13:*256*, 13:259
Hodomys alleni. *See* Allen's woodrats
Hoeck, H. N., 16:391
l'Hoest's guenons, 14:188, 14:194
Hoffman's two-toed sloths, 13:*148*, 13:*156*, 13:*157*, 13:*158*
Hog deer, 12:19, 15:356, 15:*360*, 15:*362*–*363*
Hog-nosed bats. *See* Kitti's hog-nosed bats; Mexican hog-nosed bats
Hoggar hyraxes. *See* Ahaggar hyraxes
Holoarctic true moles, 12:13
Holochilus spp., 16:268
Holochilus chacarius, 16:270
Homeothermy, 12:50
  See also Thermoregulation
Hominidae, **14:225–253,** 14:*236*
  behavior, 14:228–232, 14:249–252
  conservation status, 14:235
  distribution, 14:*225*, 14:227, 14:*241*, 14:247
  evolution, 14:225, 14:241–244
  feeding ecology, 14:232–234, 14:247–248
  habitats, 14:227–228, 14:247
  physical characteristics, 12:23–24, 14:225–227, 14:244–247
  reproduction, 14:234–235, 14:248–249
  species of, 14:*237*–240
  taxonomy, 14:4, 14:225, 14:241–243
Homininae, 14:225, 14:241–243, 14:247, 14:248
Hominoids, 14:4, 14:5
Homo spp., 14:242, 14:243
Homo erectus, 12:20, 14:*243*, 14:244, 14:*249*
Homo habilis, 14:*243*
Homo neanderthalensis, 14:*243*–244
Homo rudolfensis, 14:243
Homo sapiens, **14:241–253,** 14:*245*, 14:*246*, 14:*247*, 14:*248*, 14:*250*
  animal intelligence and, 12:150
  behavior, 14:249–252, 14:*251*, 14:*252*
  body size of, 12:19
  desert adaptation of, 12:21
  domestication and commensals, **12:171–181**
  ecosystems and, 12:213
  evolution, 14:241–244, 14:*243*
  feeding ecology, 14:*245*, 14:247–248
  field studies of mammals, **12:194–202**
  habitats, 14:247
  mammalian invasives and pests, **12:182–193**
  physical characteristics, 14:244–247
  reproduction, 12:94, 12:104, 12:106, 12:109, 14:248–249
  taxonomy, 14:1, 14:3, 14:4, 14:241–243
  zoos, **12:203–212**
  See also Humans
Homo sapiens neanderthalensis, 14:243–244
Homoiodorcas spp., 16:1
Homologous similarities, 12:27–28
Homunculus spp., 14:143
Honey badgers, 14:259, 14:319, 14:*322*, 14:332*t*
Honey bears. *See* Kinkajous
Honey possums, **13:135–138,** 13:*136*, 13:*137*, 13:*138*
  behavior, 13:35, 13:37, 13:136–137
  conservation status, 13:138
  distribution, 13:*135*, 13:136
  evolution, 13:31, 13:135
  feeding ecology, 13:38, 13:137
  habitats, 13:34, 13:35, 13:136
  humans and, 13:138
  physical characteristics, 13:135–136
  reproduction, 13:39, 13:137–138
  taxonomy, 13:135
Honeycutt, R. L.
  on Caviidae, 16:389
  on octodonts, 16:433
Hooded seals, 12:126, 12:127, 14:417, 14:420, 14:424, 14:*425*, 14:432–*433*
Hooded skunks, 14:333*t*
Hoofed mammals. *See* Ungulates
Hook-lipped rhinoceroses. *See* Black rhinoceroses
Hooker's sea lions, 14:394, 14:395, 14:396, 14:399, 14:400, 14:*401*, 14:*403*–404
Hoolock gibbons, 14:207–208, 14:209, 14:210, 14:211, 14:212, 14:213, 14:*216*, 14:*218*
Hooves, 12:39–40, 12:42
  See also Physical characteristics; Ungulates
Hoplomys spp., 16:449
Hoplomys gymnurus. *See* Armored rats
Horns, 12:*5*, 12:10, 12:40, 12:*43*, 12:*47*, 12:102, 15:*132*
  See also Antlers
Horse-faced bats. *See* Hammer-headed fruit bats
Horse-like antelopes. *See* Hippotragini
Horse racing, 12:*173*, 12:174
Horsehead seals. *See* Gray seals
Horses, **15:225–236,** 15:*230*, 15:*231*
  behavior, 15:142, 15:219, 15:226–228
  conservation status, 15:221, 15:222, 15:229–230
  distribution, 15:217, 15:*225*, 15:226
  domestication, 15:145, 15:146, 15:149
  evolution, 15:136, 15:137, 15:138, 15:215, 15:216, 15:225
  feeding ecology, 15:141, 15:220, 15:228
  gaits of, 12:*39*
  habitats, 15:218, 15:226
  humans and, 15:222, 15:223, 15:230
  locomotion, 12:42, 12:*45*
  neonatal milk, 12:126
  numbers and, 12:155
  as pests, 12:182
  physical characteristics, 15:139, 15:140, 15:217
  reproduction, 12:95, 12:106, 15:221, 15:228–229
  species of, 15:232–236*t*
  taxonomy, 15:225
  vision, 12:79
Horseshoe bats, 13:310, **13:387–400,** 13:*394*, 13:*395*
  behavior, 13:312, 13:390
  conservation status, 13:392–393
  distribution, 13:*387*, 13:388–389
  echolocation, 12:86
  evolution, 13:387
  feeding ecology, 13:390–391
  humans and, 13:393
  physical characteristics, 13:387–388
  reproduction, 13:315, 13:392
  species of, 13:*396*–400
  taxonomy, 13:387
Horsfield's shrews, 13:276*t*
Horsfield's tarsiers. *See* Western tarsiers
Hose's leaf-monkeys, 14:175, 14:184*t*
Hose's palm civets, 14:344*t*
Hose's pygmy flying squirrels, 16:*138*, 16:*139*, 16:140
Hot cave bats, 13:415–416
Hotson's mouse-like hamsters, 16:243
Hotspots, biological, 12:218
Hottentot golden moles, 13:215–216, 13:217, 13:*219*, 13:*220*
Hounds, 14:288
House bats, 13:516*t*
House-building rats. *See* Greater stick-nest rats
House mice, 16:*254*, 16:*255*–*256*
  behavior, 12:*192*
  distribution, 12:137, 16:123
  feral cats and, 12:185
  humans and, 12:173, 16:126, 16:128
  reproduction, 16:125, 16:*251*, 16:*252*
  smell, 12:80
House rats. *See* Black rats
Howler monkeys, **14:155–169,** 14:*161*
  behavior, 12:*142*, 14:7, 14:157
  conservation status, 14:159–160
  distribution, 14:156
  evolution, 14:155
  feeding ecology, 14:158
  habitats, 14:156–157
  humans and, 14:160
  physical characteristics, 14:*160*
  population indexes, 12:195
  reproduction, 14:158–159
  species of, 14:*162*–163, 14:166*t*–167*t*
  taxonomy, 14:155
  translocation of, 12:224
Hubb's beaked whales, 15:69*t*
Huemuls. *See* Chilean huemuls; Peruvian huemuls
Hugh's hedgehogs, 13:207, 13:213*t*
Human immunodeficiency virus (HIV), 12:222
Humane Society of the United States, 14:291, 14:294
Humans
  aardvarks and, 15:159
  agoutis and, 16:409, 16:411–414
  anteaters and, 13:175, 13:177–179
  ape enculturation by, 12:162
  armadillos and, 13:185, 13:187–190
  Artiodactyla and, 15:273
  aye-ayes and, 14:88
  baijis and, 15:21–22
  bandicoots and, 13:6
    dry-country, 13:12, 13:14–16
    rainforest, 13:12, 13:14–16
  bats and, 13:316–317
    American leaf-nosed, 13:316–317, 13:420, 13:423–432
    bulldog, 13:447, 13:449–450
    disk-winged, 13:474, 13:476–477
    Emballonuridae, 13:358, 13:360–363
    false vampire, 13:317, 13:382, 13:384–385
    funnel-eared, 13:461, 13:463–465
    horseshoe, 13:393, 13:397–400

Humans *(continued)*
Kitti's hog-nosed, 13:369
Molossidae, 13:487–488, 13:490–493
mouse-tailed, 13:352, 13:353
moustached, 13:438, 13:440–441
New Zealand short-tailed, 13:455, 13:457–458
Old World fruit, 13:322, 13:325–330, 13:337, 13:340–347
Old World leaf-nosed, 13:405, 13:407–409
Old World sucker-footed, 13:480
slit-faced, 13:373, 13:375–376
smoky, 13:468, 13:470–471
Vespertilionidae, 13:502, 13:506–514, 13:522, 13:524–525
bears and, 14:300, 14:303–305
beavers and, 16:180–181, 16:183–184
bilbies and, 13:22
botos and, 15:30
Bovidae and, 16:8–9
Antilopinae, 16:48, 16:50–56
Bovinae, 16:15, 16:18–23
Caprinae, 16:95, 16:98–103
duikers, 16:77, 16:80–84
Hippotraginae, 16:34, 16:37–42
Neotraginae, 16:64, 16:66–70
bushbabies and, 14:26, 14:28–32
Camelidae and, 15:318, 15:320–323
Canidae and, 14:262, 14:273–274, 14:277–278, 14:280–283
capybaras and, 16:405–406
Carnivora and, 14:262–263
cats and, 14:290, 14:291, 14:292–293, 14:375–376, 14:380–389
Caviidae and, 16:393, 16:395–398
Cetacea and, 15:10–11
Balaenidae, 15:112, 15:115–118
beaked whales, 15:62, 15:64–68
dolphins, 15:51
franciscana dolphins, 15:25
Ganges and Indus dolphins, 15:16
gray whales, 15:100
Monodontidae, 15:87–88, 15:90–91
porpoises, 15:36, 15:38–39
pygmy right whales, 15:105
rorquals, 15:125, 15:128–130
sperm whales, 15:78, 15:79–80
chevrotains and, 15:329, 15:331–333
Chinchillidae and, 16:380, 16:382–383
colugos and, 13:302, 13:304
coypus and, 12:184, 16:126–127, 16:477
cross-species diseases, 12:222–223
dasyurids and, 12:291, 12:294–298
Dasyuromorphia and, 12:283–284
deer and
Chinese water, 15:376
muntjacs, 15:346, 15:349–354
musk, 15:338, 15:340–341
New World, 15:384–385, 15:388–396, 15:388–396*t*, 15:396*t*
Old World, 15:361, 15:364–371, 15:371*t*
Dipodidae and, 16:219–222
Diprotodontia and, 13:40
distribution, 14:5, 14:*241*, 14:247
dogs and, 14:288, 14:*289*, 14:292–293
dormice and, 16:321, 16:323–326
duck-billed platypuses and, 12:234, 12:247
Dugongidae and, 15:191, 15:201, 15:203–204
echidnas and, 12:238
elephants and, 12:173, 15:172, 15:174–175
encephalization quotient, 12:149
Equidae and, 15:230, 15:232–235
Erinaceidae and, 13:207, 13:209–212
field studies of mammals, 12:*195–201*
giant hutias and, 16:470
gibbons and, 14:215, 14:218–222
Giraffidae and, 15:405–406, 15:408–409
great apes and, 14:235, 14:237–240
gundis and, 16:313, 16:314–315
hearing, 12:82
Herpestidae and, 14:352, 14:354–356
Heteromyidae and, 16:203, 16:205–208
hippopotamuses and, 15:308, 15:311
hutias and, 16:463
Hyaenidae and, 14:362, 14:365–367
hyraxes and, 15:184, 15:186–188
Ice Age evolution of, 12:20–21
Indriidae and, 14:67, 14:69–72
Insectivora and, 13:200
as invasive species, 12:184–185
koalas and, 13:49
Lagomorpha and, 16:487–488
lemurs and, 14:52, 14:56–60
Cheirogaleidae, 14:39, 14:41–44
sportive, 14:77, 14:79–83
Leporidae and, 16:509, 16:511–515
Lorisidae and, 14:16, 14:18–20
Macropodidae and, 13:89–90, 13:93–100
mammal distribution and, 12:139
mammals and, 12:15
manatees and, 15:209, 15:212
Megalonychidae and, 13:158, 13:159
memory, 12:152, 12:154
milk of, 12:97
mirror self-recognition, 12:159
moles and
golden, 13:218, 13:220–221
marsupial, 13:27, 13:28
monitos del monte and, 12:275
monkeys and
Atelidae, 14:160, 14:162–166
Callitrichidae, 14:124, 14:127–131
Cebidae, 14:106, 14:108–112
cheek-pouched, 14:194, 14:197–204
leaf-monkeys, 14:175, 14:178–183
night, 14:138
Pitheciidae, 14:148, 14:150–153
monotremes and, 12:234
mountain beavers and, 16:133
Muridae and, 16:127–128, 16:285, 16:288–295
Arvicolinae, 16:229–230, 16:233–237
hamsters, 16:243, 16:245–247
Murinae, 16:253, 16:256–260
Sigmodontinae, 16:270, 16:272–276
musky rat-kangaroos and, 13:71
Mustelidae and, 14:324–325, 14:327–331
neonates, 12:125
numbats and, 12:306
octodonts and, 16:436, 16:438–440
opossums
New World, 12:254, 12:257–263
shrew, 12:268, 12:270
Otariidae and, 14:400, 14:403–406
pacaranas and, 16:388
pacas and, 16:420–421, 16:423–424
pangolins and, 16:113, 16:115–120
peccaries and, 15:296, 15:298–300
Perissodactyla and, 15:222–223
Petauridae and, 13:129, 13:131–132
Phalangeridae and, 13:60–61, 13:*63–66t*
pigs and, 15:*150*, 15:273, 15:281–282, 15:282, 15:284–288, 15:289*t*–290*t*
pikas and, 16:495, 16:497–501
pocket gophers and, 16:190, 16:192–195
porcupines and
New World, 16:369, 16:372–373
Old World, 16:354, 16:357–364
possums and
feather-tailed, 13:142, 13:144
honey, 13:138
pygmy, 13:108, 13:110–111
primates and, 14:11
Procyonidae and, 14:311, 14:313–315
pronghorns and, 15:416–417
Pseudocheiridae and, 13:113–123, 13:117
rat-kangaroos and, 13:76, 13:78–81
rats and
African mole-rats, 16:345, 16:347–349
cane, 16:335, 16:337–338
chinchilla, 16:445, 16:447–448
dassie, 16:331
spiny, 16:451, 16:453–457
rhinoceroses and, 15:254–255, 15:257–262
rodents and, 16:126–128
sengis and, 16:522, 16:524–530
shrews and
red-toothed, 13:252, 13:255–263
West Indian, 13:244
white-toothed, 13:268, 13:271–275
Sirenia and, 15:191, 15:195–196
social cognition, 12:160
solenodons and, 13:239, 13:241
spermatogenesis, 12:103
springhares and, 16:310
squirrels and
flying, 16:137
ground, 16:126, 16:148, 16:151–158
scaly-tailed, 16:300, 16:303–305
tree, 16:167, 16:169–174
Talpidae and, 13:282, 13:284–287
tapirs and, 15:242–243, 15:245–247
tarsiers and, 14:95, 14:97–99
tenrecs and, 13:230, 13:232–233
three-toed tree sloths and, 13:165, 13:167–169
tree shrews and, 13:292, 13:294–296
true seals and, 14:423–424, 14:428–434
tuco-tucos and, 16:427, 16:429–430
Viverridae and, 14:338, 14:340–343
walruses and, 14:415
wombats and, 13:53, 13:55–56
Xenartha and, 13:153
See also Conservation status; Domestication; *Homo sapiens*
Humboldt's woolly monkeys, 14:*156*, 14:167*t*
Humidity, subterranean mammals and, 12:73–74
Humpback dolphins, 15:5, 15:6, 15:57*t*
Humpback whales, 15:*9*, 15:*125*, 15:130*t*
behavior, 15:7, 15:123
distribution, 15:5, 15:121
evolution, 12:*66*, 15:119
feeding ecology, 15:8, 15:*123*, 15:*124*
migrations, 12:87, 15:6
physical characteristics, 15:4, 15:120
vocalizations, 12:85
Hunter's hartebeests, 16:27, 16:29, 16:33,

Humpback whales *(continued)*
16:42*t*
Hunting, 12:24–25, 14:248
See also Humans
Hunting dogs, 12:146, 14:288, 14:292
Huon tree kangaroos. *See* Matschie's tree kangaroos
Huskies, Siberian, 14:289, 14:*292*, 14:293
Hussar monkeys. *See* Patas monkeys
Hutias, 16:123, **16:461–467,** 16:*464*, 16:467*t*
See also Giant hutias
Hutterer, R., 16:433
Hutton's tube-nosed bats, 13:526*t*
Hyaena brunnea. See Brown hyenas
Hyaena hyaena. See Striped hyenas
Hyaena hyaena barbara, 14:360
Hyaena hyaena dubbah, 14:360
Hyaena hyaena hyaena, 14:360
Hyaena hyaena sultana, 14:360
Hyaena hyaena syriaca, 14:360
Hyaenidae. *See* Aardwolves; Hyenas
Hybrid gibbons, 14:208–209, 14:210
Hydrochaeridae. *See* Capybaras
Hydrochaeris hydrochaeris. See Capybaras
Hydrodamalinae. *See* Steller's sea cows
Hydrodamalis gigas. See Steller's sea cows
Hydrodamalis stelleri. See Giant sea cows
Hydromys chrysogaster. See Golden-bellied water rats
Hydropotes spp. *See* Water deer
Hydropotes inermis. See Chinese water deer
Hydropotes inermis argyropus. See Siberian water deer
Hydropotes inermis inermis, 15:373
Hydropotinae. *See* Chinese water deer
Hydrurga leptonyx. See Leopard seals
Hyemoschus spp. *See* Water chevrotains
Hyemoschus aquaticus. See Water chevrotains
Hyenas, 12:*142*, 14:256, 14:*259*, 14:260, **14:359–367,** 14:*361*, 14:*363*
Hylobates spp., 14:207–208, 14:209, 14:211
Hylobates agilis. See Agile gibbons
Hylobates agilis albibarbis, 14:208
Hylobates concolor, 14:207
Hylobates gabriellae, 14:207
Hylobates hoolock. See Hoolock gibbons
Hylobates klossi. See Kloss gibbons
Hylobates lar. See Lar gibbons
Hylobates leucogenys, 14:207
Hylobates moloch. See Moloch gibbons
Hylobates muelleri. See Mueller's gibbons
Hylobates pileatus. See Pileated gibbons
Hylobates syndactylus. See Siamangs
Hylobatidae. *See* Gibbons
Hylochoerus spp., 15:275, 15:276
Hylochoerus meinertzhageni. See Forest hogs
Hylomyinae, 13:203, 13:204
Hylomys spp., 13:203
Hylomys hainanensis. See Hainan gymnures
Hylomys megalotis. See Long-eared lesser gymnures
Hylomys sinens, 13:207
Hylomys sinensis. See Shrew gymnures
Hylomys suillus. See Lesser gymnures
Hylomys suillus parvus, 13:207
Hylopetes spp., 16:136
Hylopetes lepidus. See Gray-cheeked flying squirrels
Hyotherium spp., 15:265
Hypercapnia, 12:73
Hyperoodon spp. *See* Bottlenosed whales
Hyperoodon ampullatus. See Northern bottlenosed whales
Hyperoodon planifrons. See Southern bottlenosed whales
Hypertragulidae, 15:265
Hypnomys spp., 16:317–318
Hypogeomys antimena. See Malagasy giant rats
Hypoxia, 12:73, 12:114
Hypsignathus monstrosus. See Hammer-headed fruit bats
Hypsiprimnodontidae. *See* Musky rat-kangaroos
Hypsiprymnodon moschatus. See Musky rat-kangaroos
Hyracoidea. *See* Hyraxes
Hyracotherium spp., 15:132
Hyraxes, **15:177–190,** 15:*178*, 15:*185*
behavior, 15:180–181
conservation status, 15:184
digestive system, 12:14–15
distribution, 15:*177*, 15:179
evolution, 15:131, 15:133, 15:177
feeding ecology, 15:182
habitats, 15:179–180
humans and, 15:184
physical characteristics, 15:178–179
reproduction, 12:94, 12:95, 15:182–183
species of, 15:186–188, 15:189*t*
taxonomy, 15:134, 15:177–178
Hyspiprymnodon bartholomai, 13:69–70
Hystricidae. *See* Old World porcupines
Hystricognathi, 12:11, 12:71, 16:121, 16:122, 16:125
Hystricomorph rodents, 12:92, 12:94
Hystrix spp., 16:351
Hystrix africaeaustralis. See South African porcupines
Hystrix brachyura. See Common porcupines
Hystrix crassispinis. See Thick-spined porcupines
Hystrix cristata. See North African porcupines
Hystrix indica. See Indian crested porcupines
Hystrix javanica. See Javan short-tailed porcupines
Hystrix macroura, 16:*358*
Hystrix pumila. See Indonesian porcupines
Hystrix sumatrae. See Sumatran porcupines
Hysudricus maximus, 15:164

I

Iberian desmans. *See* Pyrenean desmans
Iberian lynx, 14:262, 14:370–371, 14:374, 14:*378*, 14:*387*, 14:389
Ibex, 16:4, 16:8, 16:9, 16:90, 16:92–93, 16:95
Ice Age, **12:17–25**
Ice rats, 16:*283*
Ichneumia spp., 14:347
Ichneumia albicauda. See White-tailed mongooses
Ichnumeons. *See* Egyptian mongooses
Ichthyomyines, 16:264, 16:268, 16:269
Ichthyomys spp., 16:266, 16:269
Ichthyomys pittieri, 16:269
Ictonyx spp. *See* Zorillas
Ictonyx striatus. See Zorillas
Idaho ground squirrels, 16:147, 16:*150*, 16:*153*, 16:157–158
Idionycteris phyllotis. See Allen's big-eared bats
Idiurus macrotis. See Big-eared flying mice
Idiurus zenkeri. See Zenker's flying mice
Ili pikas, 16:487, 16:495, 16:502*t*
Imitation, 12:158
See also Behavior
Immersion exhibits, 12:205
Immunocontraception, 12:188
See also Reproduction
Immunoglobins, neonatal, 12:126
Impalas, 16:*36*, 16:41–42
behavior, 16:5, 16:31, 16:*39*, 16:41–42
black-faced, 16:34
conservation status, 16:31
distribution, 16:29, 16:*39*
evolution, 16:27
habitats, 16:30
physical characteristics, 16:28
predation and, 12:116
reproduction, 16:32
smell, 12:80
taxonomy, 16:27
Implantation, delayed, 12:109–110
See also Reproduction
Imposter hutias, 16:467*t*
Inbreeding, 12:110, 12:222
See also Captive breeding; Domestication
Incisors, 12:46
See also Teeth
Incus, 12:36
Indexes
indirect observations and, 12:199–202
population, 12:195–196
Indian blackbuck antelopes, 15:271
Indian civets, 14:336, 14:345*t*
Indian crested porcupines, 16:*352*, 16:*356*, 16:*358*, 16:359–360
Indian desert hedgehogs, 13:204, 13:206, 13:*208*, 13:*210*
Indian elephants. *See* Asian elephants
Indian flying foxes, 13:307, 13:*320*, 13:322, 13:*324*, 13:*327*, 13:328
Indian fruit bats, 13:*305*, 13:*338*, 13:*344*
Indian giant squirrels, 16:167, 16:174*t*
Indian gray mongooses, 14:351–352
Indian mongooses, 12:190–191, 14:*350*, 14:352
Indian muntjacs, 15:343, 15:*344*, 15:*346*, 15:*347*, 15:*350*, 15:351–352
Indian pangolins, 16:110, 16:113, 16:*114*, 16:*115*, 16:116
Indian porcupines. *See* Indian crested porcupines
Indian rabbits. *See* Central American agoutis
Indian rhinoceroses, 15:*223*, 15:*251*, 15:254, 15:*256*, 15:*258–259*
behavior, 15:218
conservation status, 12:224, 15:222, 15:254
habitats, 15:252–253
humans and, 15:254–255
Indian spotted chevrotains. *See* Spotted mouse deer
Indian tapirs. *See* Malayan tapirs
Indian tree shrews, 13:197, 13:292, 13:*293*, 13:*294*
Indian wild asses. *See* Khurs
Indian Wildlife Protection Act of 1972, 15:280
Indian wolves, 12:180

Indiana bats, 13:310–311, 13:316, 13:502
Indirect observations, 12:199–202
Indo-Burma hotspot, 12:218
Indo-Pacific bottlenosed dolphins, 15:7
Indochinese brush-tailed porcupines. *See* Asiatic brush-tailed porcupines
Indochinese gibbons. *See* Black crested gibbons
Indochinese shrews. *See* Gray shrews
Indonesian cuscuses, 13:34
Indonesian mountain weasels, 14:324
Indonesian porcupines, 16:*356*, 16:*357*
Indonesian stink badgers, 14:333*t*
Indopacetus spp., 15:59
Indopacetus pacificus. *See* Longman's beaked whales
Indopacific beaked whales. *See* Longman's beaked whales
Indri indri. *See* Indris
Indricotherium asiaticum, 15:249
Indricotherium transouralicum, 15:216
Indriidae, 12:136, 14:4, **14:63–72,** 14:*68*
behavior, 14:*64*, 14:65–66
conservation status, 14:67
distribution, 14:*63*, 14:*65*
evolution, 14:63
feeding ecology, 14:66
habitats, 14:65
humans and, 14:67
physical characteristics, 14:63–65
reproduction, 14:66–67
species of, 14:*69*–72
taxonomy, 14:63
Indris, **14:63–72,** 14:*65*, 14:*66*, 14:*68*
behavior, 14:6, 14:8, 14:65, 14:66
conservation status, 14:67
distribution, 14:65
evolution, 14:63
feeding ecology, 14:66
habitats, 14:65
humans and, 14:67
physical characteristics, 14:63–65
reproduction, 14:66
taxonomy, 14:4, 14:63
Induced luteinization, 12:92
Induced ovulation, 12:109
See also Reproduction
Indus dolphins. *See* Ganges and Indus dolphins
Infectious diseases, 12:215–216
Infrared energy, 12:83–84
Inia geoffrensis. *See* Botos
Inia geoffrensis boliviensis, 15:27
Inia geoffrensis geoffrensis, 15:27
Inia geoffrensis humboldtiana, 15:27
Iniidae. *See* Botos
See also River dolphins
Inioidea, 15:23
Insectivora, **13:193–201,** 14:8
field studies of, 12:201
phylogenetic trees and, 12:33
reproduction, 12:90, 12:91, 12:92, 12:94, 12:95
subterranean, 12:71
See also specific insectivores
Insectivorous bats, 13:313
See also specific insectivorous bats
Inserted sequences. *See* Retroposons
Insight, problem solving by, 12:140
See also Cognition
Insular fruit bats. *See* Tongan flying foxes
Insulation, hair for, 12:38
See also Physical characteristics
Integrated pest management (IPM), 12:186–188, 12:193
Integrated rabbit control, 12:186–188
Integumental derivatives, 12:7–8
Integumentary system, 12:36–39
See also Physical characteristics
Intelligence, **12:149–164**
Inter-sexual selection, 12:101–102
See also Reproduction
Interdependence, in ecosystems, 12:216–217
Interindividual discrimination, 12:9–10
Intermediate horseshoe bats, 13:390, 13:392
Intermediate slit-faced bats, 13:373, 13:376*t*
Internal organs, 12:174
See also Hearts; Lungs
International Committee on Zoological Nomenclature
bushbabies, 14:23
primates
Lorisidae, 14:13
International Species Information System (ISIS), 12:205, 12:206
International Union for Conservation of Nature. *See* IUCN Red List of Threatened Species
International Whaling Commission, 15:9
Balaenidae, 15:*111*
gray whales, 15:99
northern bottlenosed whales, 15:62, 15:65
pygmy right whales, 15:105
rorquals, 15:124–125
sperm whales, 15:78
Intestines, 12:47–48, 12:123–124
See also Physical characteristics
Intra-sexual selection, 12:101–102
See also Reproduction
Introduced species, 12:139, 12:215
Australian region, 12:137–138, 12:185–188
competition and, 12:118
Hawaiian Islands, 12:188–192
invasives and pests, 12:182–193
Invasive species, **12:182–193**
Iomys horsfieldii. *See* Javanese flying squirrels
IPM (Integrated pest management), 12:186–188, 12:193
Irenomys spp., 16:264
Irenomys tarsalis. *See* Chilean tree mice
Iriomote cats, 14:369
Irish elk, 12:19, 12:102
Iron, 12:120
Irrawaddy dolphins, 15:6, 15:43, 15:44, 15:50
Ischyrorhyncus spp., 15:27
Isectolophidae, 15:237
ISIS (International Species Information System), 12:205, 12:206
Island flying foxes, 13:*323*, 13:*325*
Island foxes, 12:192, 14:273, 14:284*t*
Island seals. *See* Harbor seals
Island spotted skunks, 12:192
Islands, 12:131, 12:220–221
Isolobodon spp., 16:463
Isolobodon portoricensis. *See* Puerto Rican hutias
Isoodon auratus. *See* Golden bandicoots
Isoodon macrourus. *See* Northern brown bandicoots
Isoodon obesulus. *See* Southern brown bandicoots
Isothrix bistriata. *See* Toros
Isothrix pagurus, 16:450–451
IUCN African Elephant Specialist Group, 15:163
IUCN Caprinae Specialist Group (CSG), 16:87
IUCN Chiroptera Specialist Group, 13:358, 13:420
IUCN Red List of Threatened Species
African elephants, 15:175
African wild dogs, 14:278
agoutis, 16:409
Ahaggar hyraxes, 15:183
Alouatta palliata mexicana, 14:162
Alouatta seniculus insulanus, 14:163
annamite striped rabbits, 16:509
Antilopinae, 16:48
aquatic genets, 14:341
arctic ground squirrels, 16:152
Artiodactyla, 15:272–273
Arvicolinae, 16:229
Asian elephants, 15:174
babirusas, 15:288
bactrian camels, 15:317, 15:320
Baikal seals, 14:430
Baird's beaked whales, 15:65
Balaenidae, 15:112
bald uakaris, 14:151
bandicoots, 13:6
beaked whales, 15:61–62
bears, 14:300
belugas, 15:86
black crested gibbons, 14:221
black muntjacs, 15:346, 15:351
black rhinoceroses, 15:260
black spotted cuscuses, 13:39
black-tailed prairie dogs, 16:155
blackish squirrel monkeys, 14:110
Blanford's foxes, 14:280
Blanville's beaked whales, 15:67
Bovidae, 16:8
Bovinae, 16:14
Brazilian free-tailed bats, 13:493
Brazilian shrew mice, 16:274
brown long-eared bats, 13:514
brush-tailed bettongs, 13:39, 13:76
Bryde's whales, 15:129
Burmeister's porpoises, 15:39
bush pigs, 15:285
California leaf-nosed bats, 13:423
Caprinae, 16:94–95
capuchins, 14:106
Carnivora, 14:262
cats, 14:374
Cebus apella robustus, 14:111
Chacoan peccaries, 15:295, 15:300
cheetahs, 14:382
chevrotains, 15:329
Chilean shrew opossums, 12:268, 12:270
chinchillas, 16:380
Chinese dormice, 16:324
Chinese short-tailed shrews, 13:256
Chinese water deer, 15:376
clouded leopards, 14:387
Colombian woolly monkeys, 14:166
colugos, 13:302
common bentwing bats, 13:525
common bottlenosed dolphins, 15:54
common warthogs, 15:286
Cuvier's beaked whales, 15:66

IUCN Red List of Threatened Species *(continued)*
dassie rats, 16:331
dasyurids, 12:290–291
Dendromus vernayi, 16:291
desert dormice, 16:325
Dipodidae, 16:216
disk-winged bats, 13:473, 13:474
dolphins, 15:48
Dorcas gazelles, 16:51
dormice, 16:321
dugongs, 15:201, 15:204
duikers, 16:77
eastern barred bandicoots, 13:4
eastern tree hyraxes, 15:183, 15:187
edible dormice, 16:326
Ethiopian wolves, 14:277
Eurasian beavers, 16:180, 16:183
Eurasian lynx, 14:388
Eurasian wild pigs, 15:288
European ground squirrels, 16:156
falanoucs, 14:343
false vampire bats, 13:382
Fea's muntjacs, 15:346, 15:351
fin whales, 15:128
five-toed pygmy jerboas, 16:220
flying squirrels, 16:137
forest dormice, 16:324
forest hogs, 15:284
franciscana dolphins, 15:25
funnel-eared bats, 13:461
garden dormice, 16:325
Genetta genetta isabelae, 14:341
Geoffroy's cats, 14:385
giant anteaters, 13:153
giant kangaroo rats, 16:207
giant mastiff bats, 13:492
giant pandas, 14:304
Gilbert's potoroos, 13:39, 13:76
giraffes, 15:408
Giraffidae, 15:405
golden bandicoots, 13:4
golden-cheeked gibbons, 14:222
golden moles, 13:217
Gongshan muntjacs, 15:346, 15:352
Goodfellow's tree kangaroos, 13:39
Grant's desert golden moles, 13:221
gray-backed sportive lemurs, 14:79
gray whales, 15:99
gray woolly monkeys, 14:166
great apes, 14:235
greater bilbies, 13:4, 13:22
greater house bats, 13:491
greater Malay mouse deer, 15:333
guanacos, 15:317–318, 15:322
Hainan moonrats, 13:207
hamsters, 16:243
harbor porpoises, 15:38
hazel dormice, 16:326
Heermann's kangaroo rats, 16:208
Hemiechinus nudiventris, 13:207
Herpestidae, 14:352
Heteromyidae, 16:203
Hispaniolan hutias, 16:466
Hispaniolan solenodons, 13:241
Hispid cotton rats, 16:276
horseshoe bats, 13:392
Hugh's hedgehogs, 13:207
hutias, 16:463
hyenas, 14:362
Hylomys suillus parvus, 13:207
Idaho ground squirrels, 16:158
Indian rhinoceroses, 15:259
Indriidae, 14:67
Insectivora, 13:199
jaguars, 14:386
Jamaican hutias, 16:465
Japanese dormice, 16:325
Javan rhinoceroses, 15:258
killer whales, 15:53
koalas, 13:48
Lagomorpha, 16:487
large golden moles, 13:221
lesser-crested mastiff bats, 13:490
lesser Malay mouse deer, 15:332
lesser mouse-tailed bats, 13:*353*
linh duongs, 16:14
lions, 14:380
little pygmy possums, 13:108
long-beaked echidnas, 12:233, 12:238
long-eared jerboas, 16:219
long-footed potoroos, 13:76
long-nosed potoroos, 13:39
long-snouted bats, 13:429
long-tailed chinchillas, 16:383
Longman's beaked whales, 15:68
Malagasy giant rats, 16:291
manatees, 15:209
maned sloths, 13:165, 13:169
maned wolves, 14:282
marsupial moles, 13:27
masked titis, 14:153
Michoacán pocket gophers, 16:193
Milne-Edwards's sportive lemurs, 14:81
Mindanao gymnures, 13:212
monitos del monte, 12:274
mountain beavers, 16:133
mountain pygmy possums, 13:39
mouse-tailed bats, 13:353
moustached bats, 13:436
Murinae, 16:253
musk deer, 15:338
Mustelidae, 14:324
naked bats, 13:491
North African porcupines, 16:361
northern bettongs, 13:76
northern bottlenosed whales, 15:65
northern hairy-nosed wombats, 13:39
northern minke whales, 15:129
northern muriquis, 14:165
northern sportive lemurs, 14:83
Novae's bald uakaris, 14:151
numbats, 12:305
octodonts, 16:436
Old World fruit bats, 13:321, 13:337
Old World leaf-nosed bats, 13:405
Old World monkeys, 14:175
Old World opossums, 12:254, 12:257–263
Old World porcupines, 16:353–354
Otariidae, 14:399
otter civets, 14:343
pacaranas, 16:388
pangolins, 16:113
Parnell's moustached bats, 13:440
Pel's anomalures, 16:303
Perissodactyla, 15:221–222
Philippine colugos, 13:302
Philippine gymnures, 13:207
picas, 16:495, 16:500
pigs, 15:280–281, 15:289*t*–290*t*
Pitheciidae, 14:148
pocket gophers, 16:190
polar bears, 14:305
porpoises, 15:35–36
primates, 14:11
Procyonidae, 14:310
Proserpine rock wallabies, 13:39
pumas, 14:382
pygmy hippopotamuses, 15:308, 15:311
pygmy hogs, 15:287
Querétaro pocket gophers, 16:194
red-backed squirrel monkeys, 14:109
red bald uakaris, 14:151
red-tailed sportive lemurs, 14:80
red-toothed shrews, 13:252
red viscacha rats, 16:438
Rio de Janeiro rice rats, 16:275
Roach's mouse-tailed dormice, 16:325
rodents, 16:126
rorquals, 15:125
Russian desmans, 13:286
scaly-tailed possums, 13:66
Schnabeli's thumbless bats, 13:471
sengis, 16:522, 16:525–526, 16:529
Shepherd's beaked whales, 15:68
short-beaked echidnas, 12:238
Siberian zokors, 16:294
Sigmodontinae, 16:270
sika deer, 15:369
Sirenia, 15:195
slit-faced bats, 13:373
small-toothed sportive lemurs, 14:83
smoky bats, 13:468
snow leopards, 14:387
solenodons, 13:239
southern African hedgehogs, 13:210
southern bearded sakis, 14:151
southern long-nosed bats, 13:429
southern muriquis, 14:164
spectacled bears, 14:305
spectacled dormice, 16:323
sperm whales, 15:78, 15:79
sportive lemurs, 14:76
squirrel monkeys, 14:105–106
Sumatran rhinoceroses, 15:257
Talpidae, 13:282
tapirs, 15:241–242, 15:243, 15:257–262
tarsiers, 14:95
Tasmanian bettongs, 13:76, 13:78
Tasmanian wolves, 12:309
tenrecs, 13:229
thick-spined porcupines, 16:358
tigers, 14:381
tree squirrels, 16:167
true seals, 14:422
tuco-tucos, 16:427
Ucayali bald uakaris, 14:151
Uta Hick's bearded sakis, 14:151
Vancouver Island marmots, 16:156
Vespertilionidae, 13:502, 13:522
vicuñas, 15:317–318, 15:321
Viverridae, 14:338
volcano rabbits, 16:514
water chevrotains, 15:331
weasel sportive lemurs, 14:82
western barbastelles, 13:507
western barred bandicoots, 13:4–5
white bald uakaris, 14:151
white-cheeked gibbons, 14:221
white-footed sportive lemurs, 14:82

IUCN Red List of Threatened Species *(continued)*
white rhinoceroses, 15:262
white-striped free-tailed bats, 13:492
white-tailed mice, 16:289
white-toothed shrews, 13:268
wood lemmings, 16:236
yellow-breasted capuchins, 14:112
yellow-footed rock wallabies, 13:39
yellow-spotted hyraxes, 15:183
IUCN/SSC Pigs, Peccaries, and Hippos Specialist Group, 15:275
cebu bearded pigs, 15:281
peccaries, 15:295
pygmy hogs, 15:280, 15:281, 15:287
IUCN (World Conservation Union), 12:213–214
Izubrs, 15:358

J

Ja slit-faced bats, 13:373, 13:377*t*
Jackals, 12:*144*, 12:148, 14:*255*, 14:260, **14:265–273, 14:283*t***
Jackrabbits
antelope, 16:487, 16:*506*, 16:515*t*
black-tailed, 16:484, 16:*485*, 16:487, 16:*508*, 16:515*t*
tehuantepec, 16:487
white-sided, 16:516*t*
Jackson's mongooses, 14:352
Jacobson's organ, 12:80, 12:*103*
See also Physical characteristics
Jaculus spp., 16:212
Jaculus blanfordi, 16:216
Jaculus blanfordi margianus, 16:216
Jaculus blanfordi turcmenicus, 16:216
Jaculus jaculus. *See* Lesser Egyptian jerboas
Jaculus orientalis. *See* Greater Egyptian jerboas
Jaguars, 12:*123*, 14:371, 14:374, 14:*377*, 14:*386*
Jaguarundis, 12:116, 14:390*t*
Jamaican fruit-eating bats, 13:*416*, 13:417, 13:419, 13:420, 13:*421*, 13:*428*, 13:432
Jamaican hutias, 16:*462*, 16:*463*, 16:*464*, 16:*465*
Jamaican monkeys, 12:129
Jameson, E. W., 16:143
Janis, Christine, 15:134
Japan. *See* Palaearctic region
Japanese beaked whales. *See* Ginkgo-toothed beaked whales
Japanese dormice, 16:317, 16:319, 16:*322*, 16:*324*, 16:325
Japanese macaques, 12:135, 12:*147*, 12:157, 14:*3*, 14:5, 14:*8*, 14:10, 14:*191*, 14:194
Japanese mountain moles, 13:282
Japanese serows, 16:87, 16:*90*, 16:92, 16:93, 16:95, 16:103*t*
Jar seals. *See* Harp seals
Jardin des Plantes, Paris, France, 12:203
Jarvis, Jennifer U. M., 12:70
Javan gibbons. *See* Moloch gibbons
Javan gold-spotted mongooses. *See* Small Indian mongooses
Javan pigs, 15:280, 15:289*t*
Javan rhinoceroses, 12:218, 15:222, 15:249, 15:251, 15:252, 15:254, 15:*256*, 15:*258*
Javan rusa deer, 15:*360*
Javan short-tailed porcupines, 16:*359*
Javan slit-faced bats, 13:373, 13:376*t*
Javan warty pigs. *See* Javan pigs
Javanese flying squirrels, 16:141*t*
Javanese tree shrews, 13:297*t*
Javelinas. *See* Collared peccaries
Jaws, 12:9, 12:11, 12:*41*
of carnivores, 12:14
hearing and, 12:50
See also Physical characteristics
Jentink's dormice, 16:327*t*
Jentink's duikers, 16:76, 16:*79*, 16:*82*
Jerboa marsupial mice. *See* Kultarrs
Jerboa mice. *See* Australian jumping mice
Jerboas, 16:123, **16:211–224**, 16:*218*, 16:223*t*–224*t*
Jerdon's palm civets, 14:338
Jirds. *See* Sundevall's jirds
Johnston, Sir Harry, 15:406
Johnston's genets, 14:338
Johnston's hyraxes, 15:177, 15:189*t*
Juan Fernández fur seals, 14:394, 14:395, 14:398, 14:399, 14:407*t*
Juliana's golden moles, 13:216, 13:222*t*
Juliomys spp., 16:264
Jumbie bats. *See* Northern ghost bats
Jumping mice, 16:123, 16:125, **16:211–224**, 16:223*t*–224*t*
Jumping shrews. *See* Round-eared sengis
Jungle cats, 14:290, 14:390*t*
Jungle wallabies. *See* Agile wallabies
Junk DNA, 12:29
Juscelinomys spp., 16:266

K

K-strategists, 12:13, 12:97
Kadwell, M., 15:313
Kafue lechwes, 16:7, 16:33, 16:34
Kai horseshoe bats, 13:392
Kalahari mole-rats, 12:74
Kalinowski's agoutis, 16:409, 16:415*t*
Kangaroo mice, **16:199–210**, 16:*204*, 16:208*t*–209*t*
Kangaroo rats, **16:199–210**, 16:*204*
behavior, 16:124, 16:200–202
conservation status, 16:203
distribution, 16:123, 16:*199*, 16:200
evolution, 16:199
feeding ecology, 12:117, 16:202
habitats, 16:124, 16:200
humans and, 16:203
physical characteristics, 16:123, 16:199
reproduction, 16:203
species of, 16:207–208, 16:208*t*–209*t*
taxonomy, 16:121, 16:199
vibrations, 12:83
water balance, 12:113
Kangaroos, 12:*141*, 13:*31*, **13:83–103**, 13:*84*, 13:*85*, 13:*91*, 13:*92*, 13:101*t*–102*t*
behavior, 13:35, 13:*36*, 13:86–87
conservation status, 13:89
digestive systems of, 12:14, 12:123
distribution, 12:137, 13:*83*, 13:84–85
evolution, 13:31–32, 13:83
feeding ecology, 13:87–88
habitats, 13:85–86
humans and, 13:89–90
locomotion, 12:43
physical characteristics, 13:32, 13:33, 13:83–84
reproduction, 12:106, 12:108, 13:88–89
scansorial adaptations, 12:14
species of, 13:93–100, 13:101*t*–102*t*
taxonomy, 13:83
Kannabateomys amblyonyx. *See* Southern bamboo rats
Kaokoveld hyraxes, 15:177
Karanisia spp., 14:13
Karoo rats. *See* Angoni vlei rats
Kelaart's long-clawed shrews, 13:*269*, 13:*271*, 13:275
Kellen's dormice, 16:327*t*
Kelpies, 14:292
Kemp's grass mice, 16:*271*, 16:*272–273*
Kenagy, G. J., 16:202
Kennalestids, 13:193
Keratin, 12:7–8
Kerguelen Island fur seals. *See* Antarctic fur seals
Kerivoula spp., 13:519, 13:520, 13:521
Kerivoula lanosa, 13:*522*
Kerivoula papillosa. *See* Papillose bats
Kerivoula papuensis. *See* Golden-tipped bats
Kerivoula picta. *See* Painted bats
Kerivoulinae. *See Kerivoula* spp.
Kermode bears. *See* American black bears
Kerodon spp. *See* Rock cavies
Kerodon rupestris. *See* Rock cavies
Key deer, 15:*382*
Keystone species, 12:216–217
Khaudum Game Reserves, 12:*221*
Khulans. *See* Kulans
Khurs, 15:222, 15:*229*
See also Asiatic wild asses
Kiangs, 12:134, 15:226, 15:*231*, 15:*234–235*
Killer whales, 12:*204*, 15:*52*, 15:*53*
behavior, 15:7, 15:44, 15:45, 15:46
conservation status, 15:*49*, 15:50
distribution, 12:138, 15:43
echolocation, 12:87
feeding ecology, 15:8, 15:46, 15:*47*
habitats, 15:44
humans and, 15:10–11
migrations, 12:*165*
physical characteristics, 15:*4*, 15:42, 15:43
pygmy, 15:2, 15:56*t*
reproduction, 15:48
See also False killer whales
Kin recognition, 12:110
See also Behavior
King colobus, 14:175, 14:184*t*
King Mahendra Trust, 12:224
Kinkajous, 14:260, 14:309, 14:*310*, 14:311, 14:*312*, 14:313–*314*
Kirk's dikdiks, 16:*61*, 16:*62*, 16:*63*, 16:*65*, 16:*68*, 16:69
Kit foxes. *See* Swift foxes
Kitti's hog-nosed bats, 12:12, 12:58, 12:136, 13:307, **13:369–371**, 13:*370*
Klipdassies. *See* Bush hyraxes; Rock hyraxes
Klipspringers, 12:148, 16:5, 16:7, 16:60, 16:*65*, 16:*67–68*
Kloss gibbons, 14:207–212, 14:212, 14:215, 14:*217*, 14:*218*
Knock-out mice, 16:128

Koala lemurs, 14:73–74, 14:75
Koalas, 12:*215*, 12:*223*, 13:*37*, **13:43–50**, 13:*44*, 13:*45*, 13:*46*, 13:*47*, 13:*48*, 13:*49*
behavior, 13:35, 13:37, 13:45–46
conservation status, 13:48–49
digestive system, 12:15, 12:*127*
distribution, 12:137, 13:*43*, 13:44
evolution, 13:31, 13:43
feeding ecology, 13:38, 13:46–47
habitats, 13:34, 13:44–45
humans and, 13:40, 13:49
physical characteristics, 13:32, 13:43–44
reproduction, 12:*103*, 13:33, 13:38, 13:39, 13:47–48
Kobs, 15:272, 16:5, 16:27, 16:29, 16:30, 16:34, 16:42*t*
Ugandan, 16:7
white-eared, 16:7
Kobus spp., 16:27, 16:28
Kobus defassa, 16:29
Kobus ellipsiprymnus. See Waterbucks
Kobus kob. See Kobs
Kobus kob leucotis. See White-eared kobs
Kobus kob thomasi. See Uganda kobs
Kobus leche. See Kafue lechwes
Kobus leche kafuensis. See Kafue lechwes
Kobus leche robertsi. See Roberts' lechwes
Kobus leche smithemani. See Black lechwes
Kobus megaceros. See Nile lechwes
Kodiak bears, 12:24, 14:*298*
See also Brown bears
Kogia spp., 15:73, 15:74–75, 15:76, 15:78
Kogia breviceps. See Pygmy sperm whales
Kogia sima. See Dwarf sperm whales
Kollokodontidae, 12:228
Komba spp., 14:23
Kongonis. *See* Red hartebeests
Koopman's porcupines, 16:365, 16:374*t*
Korea. *See* Palaearctic region
Korrigums, 16:34
Koshima Island, 12:157
Koslov's pikas, 16:487, 16:495
Koupreys, 12:136, 12:177, 16:24*t*
Kravetz, F. O., 16:267
Kudus, 15:271, 15:273, 16:11, **16:11–25**, 16:*16*, 16:24*t*–25*t*
Kuehneotherium spp., 12:11
Kuhl's pipistrelle bats, 13:516*t*
Kulans, 12:134, 15:222, 15:229
See also Asiatic wild asses
Kultarrs, 12:278, 12:288, 12:*292*, 12:*296*, 12:297
Kunkele, J., 16:391
Kunsia spp., 16:265, 16:266
Kunsia tomentosus, 16:265, 16:266
Kuril seals. *See* Harbor seals

L

La Plata river dolphins. *See* Franciscana dolphins
Laboratory animals, 12:173
See also specific animals
Lacher, T. E., 16:389
Laconi, M. R., 16:269
Lactation, 12:89, 12:125–127
See also Reproduction
Lactose, 12:126–127
Ladakh pikas, 16:502*t*
Lagenodelphis hosei. See Fraser's dolphins
Lagenorhynchus acutus. See Atlantic white-sided dolphins
Lagenorhynchus obliquidens. See Pacific white-sided dolphins
Lagenorhynchus obscurus. See Dusky dolphins
Lagidium spp., 16:377
Lagidium peruanum. See Mountain viscachas
Lagidium viscacia. See Southern viscachas
Lagidium wolffsohni. See Wolffsohn's viscachas
Lagomorpha, **16:479–489**
digestive systems, 12:14–15
neonatal milk, 12:127
reproduction, 12:12, 12:103, 12:107–108
vibrations, 12:83
See also Rabbits
Lagonimico spp., 14:115
Lagorchestes spp. *See* Hare-wallabies
Lagostomus maximus. See Plains viscachas
Lagostrophus spp., 13:85
Lagostrophus fasciatus. See Banded hare-wallabies
Lagostrophus hirsutus. See Rufous hare-wallabies
Lagothrix spp. *See* Woolly monkeys
Lagothrix cana. See Gray woolly monkeys
Lagothrix flavicauda. See Yellow-tailed woolly monkeys
Lagothrix lagotricha. See Humboldt's woolly monkeys
Lagothrix lugens. See Colombian woolly monkeys
Lagothrix poeppigii, 14:160
Lagurus lagurus. See Steppe lemmings
Lama spp., 12:132, 15:265, 15:313, 15:314, 15:317
Lama glama. See Llamas
Lama guanicoe. See Guanacos
Lama pacos. See Alpacas
Lamini, 15:313
Land carnivores, **14:255–263**
See also specific carnivores
Lander's horseshoe bats, 13:389–390
Language, 12:160–162
See also Behavior
Langurs, 12:123, 14:7, 14:11, 14:172
Lankin, K., 15:147
Lappet-brown bats. *See* Allen's big-eared bats
Laptev walruses, 14:409, 14:412
Lar gibbons, 14:207–213, 14:*209*, 14:*217*, 14:*219*
Large-antlered muntjacs. *See* Giant muntjacs
Large bamboo rats, 16:*283*, 16:*287*, 16:*288*, 16:294–295
Large-eared horseshoe bats, 13:*392*
Large-eared hutias. *See* Eared hutias
Large-eared pikas, 16:*496*, 16:*498*
Large-eared slit-faced bats, 13:*372*, 13:*373*, 13:376*t*
Large-eared tenrecs, 13:225, 13:226, 13:227, 13:228, 13:*229*, 13:*231*, 13:*232–233*
Large-footed myotis, 13:313
Large golden moles, 13:215, 13:217, 13:*219*, 13:*220*, 13:221
Large hairy armadillos, 13:183, 13:*184*, 13:191*t*
Large Indian civets. *See* Indian civets
Large intestine, 12:47–48, 12:124
See also Physical characteristics
Large Japanese moles, 13:287*t*
Large Malayan leaf-nosed bats. *See* Diadem roundleaf bats
Large naked-soled gerbils. *See* Bushveld gerbils
Large northern bandicoots. *See* Northern brown bandicoots
Large pocket gophers, 16:*191*, 16:*193*–194
Large slit-faced bats, 13:313–314, 13:*374*, 13:*375*
Large-spotted civets. *See* Oriental civets
Large-toothed bandicoots, 13:4, 13:17*t*
Larynx, 12:8
See also Physical characteristics
Lasionycteris noctivagans. See Silver-haired bats
Lasiorhinus spp. *See* Hairy-nosed wombats
Lasiorhinus krefftii. See Northern hairy-nosed wombats
Lasiorhinus latifrons. See Southern hairy-nosed wombats
Lasiurus borealis. See Red bats
Lasiurus cinereus. See Hoary bats
Lasiurus cinereus semotus. See Hawaiian hoary bats
Lataste's gundis. *See* Mzab gundis
Latidens salimalii. See Salim Ali's fruit bats
Laurasiatheria, 12:26, 12:*30*, 12:33, 15:134
Lavia frons. See Yellow-winged bats
Lawrence, D. H., 12:52
Layard's beaked whales. *See* Strap-toothed whales
Le Carre, John, 13:200
Le Gros Clark, Wilfred E., 13:289
Leadbeater's possums, 13:*126*, 13:*130*, 13:132
behavior, 13:35, 13:37, 13:127
conservation status, 13:39, 13:129
habitats, 13:34
physical characteristics, 13:32, 13:34
reproduction, 13:132
taxonomy, 13:132
Leaf deer. *See* Leaf muntjacs
Leaf-eared mice, 16:269, 16:270
Leaf-monkeys, **14:171–186**, 14:*176*, 14:*177*
Asian, 14:172
behavior, 12:153, 14:7, 14:174
conservation status, 14:175
distribution, 14:5, 14:*171*, 14:173
evolution, 14:171–172
feeding ecology, 14:8, 14:9, 14:174
habitats, 14:174
humans and, 14:175
physical characteristics, 14:172–173
reproduction, 14:174–175
species of, 14:*178*–183, 14:184*t*–185*t*
taxonomy, 14:4, 14:171–172
Leaf muntjacs, 15:346, 15:*348*, 15:*349*, 15:354
Leaf-nosed bats
Commerson's, 13:401, 13:*402*, 13:*404*
dusky, 13:*403*, 13:*404*
East Asian tailless, 13:409*t*
evolution, 12:11
New World, 13:311, 13:315
northern, 13:*403*, 13:*404*
trident, 13:*406*, 13:*407*
Learning, 12:9–10, 12:140–143, 12:151–155
See also Cognition
Learning set, 12:152

Least chipmunks, 16:*149*, 16:*152*
Least shrew-moles. *See* American shrew moles
Least shrews, 13:196
Least weasels, 14:333*t*
distribution, 14:321
habitats, 14:257
physical characteristics, 14:256, 14:321
predation and, 12:115, 12:116
reproduction, 12:107, 14:262, 14:324
Lechwes, 16:27–32, 16:34, 16:42*t*
black, 16:34
Kafue, 16:7, 16:33, 16:34
Nile, 16:4, 16:27, 16:34, 16:*35*, 16:*37*
Roberts', 16:33
Leisler's bats, 13:516*t*
Leithia spp., 16:317
Leithiinae, 16:317
Lemmings, 12:115, 16:124, **16:225–238,** 16:*231*, 16:*232*
Lemmiscus curtatus. See Sagebrush voles
Lemmus spp., 16:225–226, 16:227
Lemmus lemmus. See Norway lemmings
Lemmus sibiricus. See Brown lemmings
Lemniscomys barbarus. See Zebra mice
Lemur catta. See Ringtailed lemurs
Lemur coronatus. See Crowned lemurs
Lemur fulvus. See Brown lemurs
Lemur fulvus rufus. See Red-fronted lemurs
Lemur macaco. See Black lemurs
Lemur mongoz. See Mongoose lemurs
Lemur possums. *See* Lemuroid ringtails
Lemur rubriventer. See Red-bellied lemurs
Lemuridae. *See* Lemurs
Lemuroid ringtails, 13:115, 13:*116*, 13:*118*, 13:*119*
Lemurs, **14:47–61,** 14:*49*, 14:*53*, 14:*54*
baboon, 14:63, 14:*68*, 14:*69*, 14:72
behavior, 12:*142*, 14:6, 14:7–8, 14:*49*, 14:50–51
blue-eyed, 14:*2*
cathemeral, 14:8
conservation status, 14:52
distribution, 12:135–136, 14:5, 14:*47*, 14:49
diurnal, 14:7, 14:8, 14:9
evolution, 14:2, 14:3, 14:47
feeding ecology, 14:9, 14:51
fork-marked, 12:38
gentle, 14:3, 14:8, 14:10–11, 14:*50*
greater bamboo, 14:11, 14:*48*
habitats, 14:6, 14:49
humans and, 14:52
nocturnal, 14:7, 14:8
physical characteristics, 12:79, 14:4, 14:48–49
reproduction, 14:10–11, 14:51–52
sexual dimorphism, 12:99
sloth, 14:63, 14:*68*, 14:*70*, 14:71
species of, 14:55–60
taxonomy, 14:1, 14:3–4, 14:47–48
See also Dwarf lemurs; Mouse lemurs; Sportive lemurs; Strepsirrhines
Leonard, Jennifer, 14:287
Leontopithecus spp. *See* Lion tamarins
Leontopithecus caissara. See Black-faced lion tamarins
Leontopithecus chrysomelas. See Golden-headed lion tamarins
Leontopithecus chrysopygus. See Golden-rumped lion tamarins
Leontopithecus rosalia. See Golden lion tamarins
Leontopithecus rosalia rosalia. See Lion tamarins
Leopard seals, 12:87, 12:138, 14:256, 14:260, 14:417, 14:420, 14:*421*, 14:435*t*
Leopards, 12:*133*, 14:374, 14:*377*, 14:*385*, 14:391*t*
behavior, 12:*146*, 12:*152*, 14:*371*
conservation status, 14:374
distribution, 12:129, 14:370
habitats, 14:257
humans and, 14:376
Leopardus (Felis) pardalis. See Ocelots
Leopardus pardalis. See Ocelots
Leopardus tigrinus. See Little spotted cats
Leopardus wiedii. See Margays
Leopold, Aldo, 12:223
Leopoldamys sabanus, 12:220
Lepilemur spp. *See* Sportive lemurs
Lepilemur dorsalis. See Gray-backed sportive lemurs
Lepilemur edwardsi. See Milne-Edwards's sportive lemurs
Lepilemur leucopus. See White-footed sportive lemurs
Lepilemur microdon. See Small-toothed sportive lemurs
Lepilemur mustelinus. See Weasel sportive lemurs
Lepilemur ruficaudatus. See Red-tailed sportive lemurs
Lepilemur septentrionalis. See Northern sportive lemurs
Lepilemuridae. *See* Sportive lemurs
Leporidae. *See* Hares; Rabbits
Leporinae, 16:505
Leptailurus (Felis) serval. See Servals
Leptobos spp., 15:265, 16:1
Leptochoeridae, 15:264
Leptonychotes weddellii. See Weddell seals
Leptonycteris spp., 13:415
Leptonycteris curasoae. See Southern long-nosed bats
Leptonycteris nivalis, 13:417, 13:420
Lepus spp. *See* Hares
Lepus alleni. See Antelope jackrabbits
Lepus americanus. See Snowshoe hares
Lepus arcticus. See Arctic hares
Lepus brachyurus. See Hares
Lepus californicus. See Black-tailed jackrabbits
Lepus callotis. See White-sided jackrabbits
Lepus capensis. See Cape hares
Lepus europaeus. See European hares
Lepus flavigularis. See Tehuantepec jackrabbits
Lepus mandshuricus. See Manchurian hares
Lepus othus, 16:487
Lepus timidus. See Mountain hares
Lesser anomalures, 16:*301*, 16:*302*, 16:303–304
Lesser anteaters. *See* Northern tamanduas; Southern tamanduas
Lesser apes. *See* Gibbons
Lesser beaked whales. *See* Pygmy beaked whales
Lesser bilbies, 13:2, 13:19, 13:*20*, 13:22
Lesser bulldog bats, 13:443, 13:*444*, 13:*445*, 13:446, 13:*448*, 13:449–*450*
Lesser cane rats, 16:333, 16:334, 16:*336*, 16:337–*338*
Lesser-crested mastiff bats, 12:85, 13:315, 13:486, 13:*489*, 13:*490*
Lesser Cuban nesophontes, 13:243
Lesser dog-faced bats, 13:363*t*
Lesser doglike bats. *See* Lesser dog-faced bats
Lesser Egyptian jerboas, 16:*212*, 16:*213*, 16:223*t*
Lesser false vampire bats, 13:*380*
Lesser fat-tailed jerboas, 16:212, 16:216, 16:223*t*
Lesser flat-headed bats. *See* Bamboo bats
Lesser forest wallabies, 13:102*t*
Lesser gliders, 13:33, 13:37
Lesser gymnures, 12:220, 13:204, 13:*208*, 13:*210*, 13:211
Lesser Haitian ground sloths, 13:*159*
Lesser hedgehog tenrecs, 13:*193*, 13:196, 13:199, 13:*228*, 13:234*t*
Lesser horseshoe bats, 13:316, 13:390, 13:393, 13:*394*, 13:*399*–400
Lesser kudus, 16:4, 16:11
Lesser Malay mouse deer, 15:325, 15:*327*, 15:*328*, 15:*330*, 15:*331*, 15:332
Lesser marsupial moles. *See* Northern marsupial moles
Lesser naked bats, 13:493*t*
Lesser New Zealand short-tailed bats, 13:313, 13:*454*, 13:455, 13:*456*, 13:457
Lesser pandas. *See* Red pandas
Lesser pink fairy armadillos. *See* Pink fairy armadillos
Lesser rorquals. *See* Northern minke whales
Lesser sac-winged bats, 13:363*t*
Lesser sheath-tailed bats, 13:*359*, 13:*362–363*
Lesser short-nosed fruit bats, 13:311, 13:*336*
Lesser short-tailed bats. *See* Lesser New Zealand short-tailed bats
Lesser shrews. *See* Lesser white-toothed shrews
Lesser spectral tarsiers. *See* Pygmy tarsiers
Lesser thumbless bats. *See* Smoky bats
Lesser tree shrews. *See* Pygmy tree shrews
Lesser white-lined bats. *See* Lesser sac-winged bats
Lesser white-nosed monkeys, 14:*192*
Lesser white-tailed shrews, 13:267
Lesser white-toothed shrews, 13:*267*, 13:*268*, 13:276*t*
Lesser woolly horseshoe bats, 13:388, 13:389, 13:*395*, 13:*397*
Lesser yellow bats. *See* House bats
Lestodelphys spp. *See* Patagonian opossums
Lestodelphys halli. See Patagonian opossums
Leumuridae, 12:79
Lexigrams, 12:161–162
l'Hoest's guenons, 14:188, 14:194
Liberian mongooses, 14:349, 14:352, 14:*353*, 14:*354*, 14:356
Liberiictis spp., 14:347
Liberiictis kuhni. See Liberian mongooses
Libyan wild cats, 12:180, 14:290
Libypithecus spp., 14:172
Lichtenstein's hartebeests, 16:27, 16:29, 16:33, 16:*35*, 16:*38*, 16:40
Lichtenstein's jerboas, 16:212, 16:215, 16:223*t*
Lickliter, R., 15:145
Life histories, **12:97–98,** 12:107–108
See also Reproduction
Life of Mammals, 13:217
Life spans, 12:97–98
Lift, for powered flight, 12:56
Limbs, 12:3–4, 12:8, 12:40–41, 12:56–57
See also Physical characteristics

Limnogale spp., 13:225, 13:226, 13:227, 13:228
Limnogale mergulus. See Aquatic tenrecs
Lincoln Park Zoo, Chicago, 12:203
LINEs (Long interspersed nuclear elements), 12:29–30
Linh duongs, 15:263, 16:1–2, 16:14
Linnaeus, Carolus
on Archonta, 13:333
on Cetacea, 15:1
on Insectivora, 13:193
on primates, 14:1
Linsangs, **14:335–341,** 14:*339,* **14:345*t***
Liomys spp., 16:199, 16:200, 16:203
Liomys irroratus. See Mexican spiny pocket mice
Liomys salvini. See Salvin's spiny pocket mice
Lion-tailed macaques. *See* Nilgiri langurs
Lion tamarins, 14:7, 14:8, 14:11, 14:115–121, 14:*119,* 14:123, 14:124
Lions, 14:*369,* 14:*370,* 14:*371,* 14:*374,* 14:*377*
behavior, 12:*9,* 12:*144,* 12:146–147, 12:*153,* 14:258, 14:259, 14:260, 14:371
conservation status, 12:*216*
distribution, 12:129–130, 14:370
feeding ecology, 12:*116,* 14:372
habitats, 12:*136*
humans and, 14:262, 14:376
physical characteristics, 14:370
reproduction, 12:*91,* 12:103, 12:107, 12:*109,* 12:*136,* 14:261, 14:262, 14:372–373
Lipids, 12:120
Lipotes spp., 15:19
Lipotes vexillifer. See Baijis
Lipotidae. *See* Baijis
See also River dolphins
Lipotyphla, 13:227
Lissodelphinae, 15:41
Lissodelphis spp., 15:42
Litocranius walleri. See Gerenuks
Litopterna, 12:11, 12:132, 15:131, 15:133, 15:135–138
Little brown bats, 13:*498,* 13:*499,* 13:*504,* 13:*511*–512
behavior, 13:499
conservation status, 13:316
distribution, 13:310–311
echolocation, 12:85
feeding ecology, 13:313
reproduction, 12:110, 13:314, 13:315, 13:501
Little brown myotis. *See* Little brown bats
Little chipmunks. *See* Least chipmunks
Little collared fruit bats, 13:351*t*
Little dorcopsises. *See* Lesser forest wallabies
Little five-toed jerboas, 16:215, 16:216, 16:217, 16:*218,* 16:*220,* 16:221–222
Little flying cows, 13:350*t*
Little golden-mantled flying foxes, 13:332*t*
Little long-fingered bats, 13:*520*
Little Nepalese horseshoe bats, 13:389
Little northern cats. *See* Northern quolls
Little pikas. *See* Steppe pikas
Little pocket mice, 16:200, 16:*202,* 16:*267*
Little pygmy possums, 13:106, 13:*109,* 13:*110*–111
Little red brocket deer, 15:396*t*
Little red flying foxes, 12:*56,* 13:*321,* 13:*324,* 13:*325,* 13:330
Little red kalutas, 12:282, 12:299*t*
Little rock wallabies. *See* Narbaleks
Little spotted cats, 14:391*t*
Little water opossums. *See* Thick-tailed opossums
Little yellow-shouldered bats, 13:417, 13:418, 13:*421,* 13:*429,* 13:431
Livestock. *See* Domestic cattle
Livingstone's flying foxes. *See* Livingstone's fruit bats
Livingstone's fruit bats, 13:*323,* 13:*326,* 13:327
Llamas, 15:137, **15:313–323,** 15:*314,* 15:*317,* 15:*319,* 15:*322,* 15:323
Llanocetus denticrenatus, 15:3
Llanos long-nosed armadillos, 13:183, 13:184, 13:191*t*
Lobodon carcinophagus. See Crab-eater seals
Locomotion, 12:3–4, 12:8, 12:40–41, 12:41–44
evolution of, 12:14
subterranean mammals, 12:71–72
See also Physical characteristics
Loder's gazelles. *See* Slender-horned gazelles
Loess steppe, 12:17–18
Lonchophylla robusta. See Orange nectar bats
Lonchophyllinae, 13:413, 13:414, 13:415, 13:420
Lonchorhina aurita. See Sword-nosed bats
Lonchothrix emiliae. See Tuft-tailed spiny tree rats
Long-beaked echidnas, 12:108, 12:137, 12:227–238, 12:*237,* 12:*238,* 12:*239,* 12:240*t*
Long-clawed ground squirrels, 16:159*t*
Long-clawed mole mice, 16:265, 16:267
Long-clawed mole voles, 16:227, 16:*232,* 16:*233,* 16:236
Long-clawed squirrels, 16:144
Long-eared bats, 13:497–498
Long-eared hedgehogs, 13:204, 13:*207,* 13:*208,* 13:*209,* 13:210–211
Long-eared jerboas, 16:211–212, 16:215, 16:216, 16:*218,* 16:*219*
Long-eared lesser gymnures, 13:203
Long-eared mice, 16:265
Long-fingered bats, 13:313, 13:515*t*
Long-fingered trioks, 13:133*t*
Long-finned pilot whales, 15:9, 15:56*t*
Long-footed potoroos, 13:32, 13:74, 13:76, 13:77, 13:*79,* 13:80
Long-footed tree shrews. *See* Bornean tree shrews
Long interspersed nuclear elements (LINEs), 12:29–30
Long-legged bandicoots. *See* Raffray's bandicoots
Long-nosed bandicoots, 13:*2,* 13:11, 13:17*t*
Long-nosed bats. *See* Proboscis bats; Southern long-nosed bats
Long-nosed echymiperas. *See* Rufous spiny bandicoots
Long-nosed mice, 16:267
Long-nosed potoroos, 13:39, 13:74, 13:*76,* 13:77, 13:78, 13:*80*
Long-snouted bats, 13:415, 13:417, 13:*421,* 13:*428,* 13:429
Long-snouted dolphins. *See* Spinner dolphins
Long-spined hedgehogs. *See* Brandt's hedgehogs
Long-tailed chinchillas, 16:377, 16:*378,* 16:*379,* 16:380, 16:*381,* 16:*382,* 16:383
Long-tailed dunnarts, 12:*292,* 12:*295,* 12:298
Long-tailed field mice, 16:*250,* 16:261*t*
Long-tailed fruit bats, 13:*335*
Long-tailed gorals, 16:87, 16:91, 16:93, 16:103*t*
Long-tailed macaques, 14:10, 14:193
Long-tailed marsupial mice. *See* Long-tailed dunnarts
Long-tailed mice, 16:262*t,* 16:270
Long-tailed moles, 13:288*t*
Long-tailed pangolins, 16:109, 16:110, 16:112, 16:*114,* 16:*118*–119
Long-tailed planigales, 12:288, 12:*292,* 12:*294,* 12:298
Long-tailed porcupines, 16:*355,* 16:*359,* 16:363–364
Long-tailed pygmy possums, 13:*105,* 13:*106,* 13:*107*
Long-tailed shrews, 13:195, 13:198
Long-tailed shrew tenrecs, 13:*228*
Long-tailed tenrecs, 13:*226*
Long-term memory, 12:152, 12:153–154
See also Cognition
Long-tongued bats. *See* specific types of long-tongued bats
Long-winged bats. *See* Common bentwing bats
Longevity. *See* Life spans
Longman's beaked whales, 15:*63,* 15:*67,* 15:68
Longtail weasels, 12:*47*
Lontra spp., 14:321
Lontra canadensis. See Northern river otters
Lontra felina. See Marine otters
Lontra provocax. See Southern river otters
Lophiomeryx spp., 15:343
Lophiomys imhausi. See Crested rats
Lophocebus spp., 14:188, 14:191
Lophocebus albigena. See Gray-cheeked mangabeys
Lophocebus aterrimus. See Black mangabeys
Lord Derby's anomalures, 16:*301,* 16:*302*–303
Lord Derby's scalytails. *See* Lord Derby's anomalures
Loridae. *See* Lorises
Loris lydekkerianus. See Gray slender lorises
Loris tardigradus. See Slender lorises
Lorises, **14:13–22,** 14:*17*
behavior, 12:116, 14:7–8, 14:14–15
conservation status, 14:16
distribution, 14:5, 14:*13,* 14:14
evolution, 14:2, 14:3, 14:13
feeding ecology, 14:9, 14:15
habitats, 14:14
humans and, 14:16
physical characteristics, 12:79, 14:13–14
reproduction, 14:16
species of, 14:18–19, 14:21*t*
taxonomy, 14:1, 14:3–4, 14:13
See also Strepsirrhines
Lorisidae. *See* Lorises; Pottos
Lorisinae, 14:13
Lower Californian fur seals. *See* Guadalupe fur seals
Lower jaw. *See* Mandible
Lower primates, 14:1
See also specific lower primates
Lowland ringtails, 13:113, 13:114, 13:*115,* 13:*118,* 13:*121*

Lowland tapirs, 15:223, 15:*238*, 15:*239*, 15:241–*242*, 15:*244*, 15:*245*
Loxodonta spp., 15:162, 15:165
Loxodonta adaurora, 15:162
Loxodonta africana. *See* African elephants
Loxodonta africana africana. *See* Savanna elephants
Loxodonta africana cyclotis. *See* Forest elephants
Loxodonta atlantica, 15:162
Loxodonta exoptata, 15:162
Loxodonta pumilio. *See* Pygmy elephants
Luckhart, G., 15:146
Lucy. *See Australopithecus afarensis*
Lumholtz's tree kangaroos, 13:102*t*
Luna, L., 16:266
Lundomys spp., 16:265, 16:268
Lungs, 12:8, 12:*40*, 12:45
bats, 12:60
marine mammals, 12:68
subterranean mammals, 12:73
See also Physical characteristics
Lushilagus, 16:480
Luteinizing hormone, 12:103
Lutra spp., 14:319
Lutra lutra. *See* European otters
Lutra maculicollis. *See* Spotted-necked otters
Lutra sumatrana. *See* Hairy-nosed otters
Lutreolina spp. *See* Lutrine opossums
Lutreolina crassicaudata. *See* Thick-tailed opossums
Lutrinae. *See* Otters
Lutrine opossums, 12:250–253
Lutrogale perspiciallata. *See* Smooth-coated otters
Luxury organs, 12:22–24
See also Physical characteristics
Lycaon pictus. *See* African wild dogs
Lyle's flying foxes, 13:331*t*
Lyme disease, 16:127
Lyncodon patagonicus. *See* Patagonian weasels
Lynx, 12:116, 12:139, 14:370–371, 14:374
Lynx canadensis. *See* Canada lynx
Lynx (Felis) canadensis. *See* Canada lynx
Lynx (Felis) lynx. *See* Eurasian lynx
Lynx (Felis) pardinus. *See* Iberian lynx
Lynx (Felis) rufus. *See* Bobcats
Lynx lynx. *See* Lynx
Lynx pardinus. *See* Iberian lynx
Lynx rufus. *See* Bobcats
Lyons, D. M., 15:145

M

Macaca spp. *See* Macaques
Macaca arctoides. *See* Stump-tailed macaques
Macaca assamensis, 14:194
Macaca cyclopis, 14:194
Macaca fascicularis, 14:194
Macaca fuscata. *See* Japanese macaques
Macaca hecki, 14:194
Macaca leonina, 14:194
Macaca maurus, 14:193–194
Macaca mulatta. *See* Rhesus macaques
Macaca nemestrina, 14:194
Macaca nigra. *See* Celebes macaques
Macaca nigrescens, 14:194
Macaca ochreata, 14:194
Macaca pagensis, 14:11, 14:193
Macaca radiata. *See* Bonnet macaques
Macaca silenus, 14:193–194
Macaca sinica. *See* Toque macaques
Macaca sylvanus. *See* Barbary macaques
Macaca thibetana, 14:194
Macaca tonkeana, 14:194
Macaques, 14:*195*
conservation status, 14:11, 14:193–194
distribution, 12:137, 14:5, 14:191
evolution, 14:189
habitats, 14:191
humans and, 12:173, 14:194
learning and, 12:143, 12:151–153
reproduction, 14:10, 14:193
species of, 14:200–201, 14:205*t*
taxonomy, 14:188
MacArthur Foundation, 12:218
Macdonald, D. W., 16:392
MacInnes's mouse-tailed bats, 13:352
Macleay's dorcopsises. *See* Papuan forest wallabies
Macleay's marsupial mice. *See* Brown antechinuses
Macleay's moustached bats, 13:435, 13:436, 13:442*t*
Macrauchenids, 15:133
Macroderma gigas. *See* Australian false vampire bats
Macrogalidia musschenbroekii. *See* Sulawesi palm civets
Macroglossus sobrinus. *See* Greater long-tongued fruit bats
Macropodidae, **13:83–103**, 13:*91*, 13:*92*, 13:101*t*–102*t*
Macropodinae, 13:83
Macropodoidea, 13:31–33, 13:38, 13:39, 13:69, 13:70, 13:71, 13:73
Macropus spp., 13:33, 13:83, 13:85, 13:86–87
Macropus agilis. *See* Agile wallabies
Macropus antilopinus. *See* Antilopine wallaroos
Macropus bernardus. *See* Black wallaroos
Macropus dorsalis. *See* Black-striped wallabies
Macropus eugenii. *See* Tammar wallabies
Macropus fuliginosus. *See* Western gray kangaroos
Macropus giganteus. *See* Eastern gray kangaroos
Macropus greyi. *See* Toolache wallabies
Macropus parma. *See* Parma wallabies
Macropus parryi. *See* Whiptail wallabies
Macropus robustus. *See* Common wallaroos
Macropus rufogriseus. *See* Red-necked wallabies
Macropus rufogriseus banksianus. *See* Red-necked wallabies
Macropus rufus. *See* Red kangaroos
Macroscelidae. *See* Sengis
Macroscelides spp., 16:518
Macroscelides proboscideus. *See* Round-eared sengis
Macroscelididae. *See* Sengis
Macroscelidinae. *See* Soft-furred sengis
Macrotis spp. *See* Bilbies
Macrotis lagotis. *See* Greater bilbies
Macrotis leucura. *See* Lesser bilbies
Macrotus californicus. *See* California leaf-nosed bats
Madagascan fruit bats, 13:*334*
Madagascar. *See* Ethiopian region
Madagascar flying foxes, 13:*322*, 13:*323*, 13:*325–326*
Madagascar hedgehogs, 12:*145*, 13:196, 13:226
Madagascar slit-faced bats, 13:375
Madoqua spp. *See* Dikdiks
Madoqua guentheri. *See* Guenther's dikdiks
Madoqua kirkii. *See* Kirk's dikdiks
Madoqua piacentinii. *See* Silver dikdiks
Madoqua saltiana. *See* Salt's dikdiks
Madrid Zoo, Spain, 12:203
Magellanic tuco-tucos, 16:427, 16:430*t*
Magnetoreception, in subterranean mammals, 12:75
Maher, C. R., 15:413
Mahogany gliders, 13:40, 13:129, 13:133*t*
Maias. *See* Bornean orangutans
Makalata armata. *See* Armored spiny rats
Makalata occasius, 16:450
Malabar civets, 14:262, 14:338
Malabar spiny dormice, 16:*287*, 16:*290*, 16:293
Malabar spiny mice. *See* Malabar spiny dormice
Malabar squirrels. *See* Indian giant squirrels
Malagasy brown mongooses, 14:357*t*
Malagasy civets, 14:335
Malagasy flying foxes. *See* Madagascar flying foxes
Malagasy giant rats, 16:*121*, 16:*286*, 16:*289*, 16:291
Malagasy mice. *See* Malagasy giant rats
Malagasy mongooses. *See* Falanoucs
Malagasy rats. *See* Malagasy giant rats
Malagasy reed rats, 16:*284*
Malas. *See* Rufous hare-wallabies
Malawi bushbabies, 14:32*t*
Malayan colugos, 13:*299*, 13:*300*, 13:*301*, 13:*302*, 13:*303*, 13:*304*
Malayan flying lemurs. *See* Malayan colugos
Malayan gymnures. *See* Malayan moonrats
Malayan moonrats, 13:194, 13:198, 13:*204*, 13:*208*, 13:212
Malayan pangolins, 16:110, 16:113, 16:*114*, 16:*115*, 16:116
Malayan porcupines. *See* Common porcupines
Malayan sun bears, 14:295, 14:296, 14:*297*, 14:298, 14:299, 14:300, 14:306*t*
Malayan tapirs, 15:222, 15:237, 15:238, 15:*239*, 15:*240*, 15:241, 15:242, 15:*244*, 15:*247*
Malaysian long-tailed porcupines. *See* Long-tailed porcupines
Malaysian short-tailed porcupines. *See* Javan short-tailed porcupines
Malaysian stink badgers. *See* Indonesian stink badgers
Malaysian tapirs. *See* Malayan tapirs
Male reproductive system, 12:91
See also Physical characteristics; Reproduction
Mallee kangaroos. *See* Western gray kangaroos
Malleus, 12:36
See also Physical characteristics
Mammae. *See* Mammary glands
Mammals
aquatic, 12:3, 12:14, 12:44, **12:62–68,** 12:82, 12:91, 12:127, 12:201
arboreal, 12:14, 12:43, 12:79, 12:202
behavior, 12:9–10, **12:140–148**
vs. birds, 12:6

Mammals *(continued)*
conservation status, 12:15, **12:213–225**
dinosaurs and, 12:33
distribution, **12:129–139**
diving, 12:67–68
draft, 12:15
evolution, 12:10–11
feeding ecology, 12:12–15
fossorial, 12:13, 12:69, 12:82
marine, 12:14, 12:53, 12:62–68, 12:82, 12:126–127, 12:138, 12:164
nocturnal, 12:79–80, 12:115
physical characteristics, **12:36–51**
reproduction, 12:12, 12:51, **12:89–112**
reptiles and, 12:36
saltatory, 12:42–43
subterranean, **12:69–78,** 12:83, 12:113–114, 12:201
terrestrial, 12:13–14, 12:79, 12:200–202
volant, 12:200–201
See also specific mammals and classes of mammals
Mammary glands, 12:4–5, 12:6, 12:14, 12:36, 12:38, 12:*44*, 12:51, 12:89
See also Physical characteristics
Mammary nipples. *See* Nipples
Mammoths, 15:138, 15:139
See also Woolly mammoths
Mammuthus spp., 15:162, 15:165
Mammuthus columbi. See Columbian mammoths
Mammuthus primigenius. See Woolly mammoths
Management, zoo, 12:205–212
Manatees, 12:*13*, 12:*67*, 15:*193*, **15:205–213,** 15:*210*
behavior, 15:193
conservation status, 15:194–195, 15:208–209
distribution, 15:192, 15:*205*
evolution, 15:191, 15:205
feeding ecology, 15:193, 15:206, 15:208
field studies of, 12:201
habitats, 15:192, 15:206
humans and, 15:195–196, 15:209
physical characteristics, 15:*192*, 15:205
reproduction, 12:103, 12:*106*, 15:193–194, 15:208
taxonomy, 15:191
Manchurian hares, 16:481, 16:506
Mandible, 12:9, 12:36
See also Physical characteristics
Mandrills, 14:*189*, 14:*195*, 14:*202*, 14:203–204
conservation status, 14:194
habitats, 14:191
physical characteristics, 14:190, 14:191
reproduction, 14:193
sexual dimorphism, 12:99
taxonomy, 14:188
Mandrillus leucophaeus. See Drills
Mandrillus sphinx. See Mandrills
Maned rats. *See* Crested rats
Maned sloths, 13:165, 13:*166*, 13:*167*, 13:169
Maned wolves, 14:265–266, 14:271, 14:272, 14:*275*, 14:*276*, 14:282
Mangabeys, 14:188, 14:191, 14:193
Mangroves, 12:219
See also Habitats
Manidae. *See* Pangolins
Manis spp., 12:135
Manis crassicaudata. See Indian pangolins
Manis gigantea. See Giant pangolins
Manis javanica. See Malayan pangolins
Manis pentadactyla. See Chinese pangolins
Manis temminckii. See Ground pangolins
Manis tetradactyla. See Long-tailed pangolins
Manis tricuspis. See Tree pangolins
Mann, G., 16:270
Mantled baboons. *See* Hamadryas baboons
Mantled black-and-white colobus. *See* Mantled guerezas
Mantled guerezas, 14:*175*, 14:*176*, 14:*178*
Mantled howler monkeys, 14:*161*, 14:*162*
Mantled mangabeys. *See* Gray-cheeked mangabeys
Maras, 16:123, 16:125, **16:389–400,** 16:*390*, 16:*391*, 16:*392*, 16:*394*, 16:*397*
Marbled polecats, 14:323, 14:333*t*
Marcano's solenodons, 13:237
Mares, M. A., 16:269, 16:433–434, 16:436
Margays, 12:116, 14:391*t*
Marianas flying foxes. *See* Marianas fruit bats
Marianas fruit bats, 13:*323*, 13:*326*, 13:327
Marianna flying foxes. *See* Marianas fruit bats
Marine carnivores, **14:255–263**
See also specific animals
Marine Mammal Protection Act (U.S.), 15:35, 15:38, 15:125
Marine mammals, 12:14, 12:138
echolocation, 12:53
evolution, 12:62–68
hearing, 12:82
migrations, 12:164
neonatal milk, 12:126–127
physical characteristics, 12:63, 12:*63*, 12:*64*, 12:*65*, 12:65–67, 12:*66*, 12:*67*
rehabilitation pools for, 12:*204*
water adaptations, 12:62–68
See also specific marine mammals
Marine otters, 14:324
Marion Island, 12:185
Mark/recapture, 12:196–197
Markhors, 16:87, 16:90–94, 16:104*t*
Marmosa spp. *See* Mouse opossums
Marmosa andersoni, 12:254
Marmosa canescens. See Grayish mouse opossums
Marmosa mexicana. See Mexican mouse opossums
Marmosa murina. See Murine mouse opossums
Marmosa robinsoni, 12:252
Marmosa rubra. See Red mouse opossums
Marmosa xerophila. See Orange mouse opossums
Marmosets, **14:115–133,** 14:*125*, 14:*126*
behavior, 14:6–7, 14:8, 14:116–120, 14:*123*
conservation status, 14:124
distribution, 14:116
evolution, 14:115
feeding ecology, 14:8–9, 14:120, 14:121
habitats, 14:116
humans and, 14:11, 14:124
physical characteristics, 14:116, 14:*122*
reproduction, 14:10, 14:121–124
species of, 14:129–131, 14:132*t*
taxonomy, 14:4, 14:115
Marmosops spp. *See* Slender mouse opossums
Marmosops cracens, 12:254
Marmosops dorothea, 12:251
Marmosops handleyi, 12:254
Marmosops incanus. See Gray slender mouse opossums
Marmosops invictus. See Slaty slender mouse opossums
Marmota spp. *See* Marmots
Marmota caligata. See Hoary marmots
Marmota camtschatica. See Black-capped marmots
Marmota caudata. See Himalayan marmots
Marmota flaviventris. See Yellow-bellied marmots
Marmota marmota. See Alpine marmots
Marmota menzbieri, 16:147
Marmota monax. See Woodchucks
Marmota olympus. See Olympic marmots
Marmota vancouverensis. See Vancouver Island marmots
Marmotini, 16:143
Marmots
alpine, 12:*76*, 12:147, 12:148, 16:*144*, 16:*150*, 16:*155*, 16:157
behavior, 12:147, 16:124, 16:145
black-capped, 16:*146*, 16:159*t*
conservation status, 16:147
evolution, 16:143
feeding ecology, 16:145, 16:147
habitats, 16:124
Himalayan, 16:*146*
hoary, 16:*144*, 16:145, 16:158*t*
humans and, 16:148
Olympic, 16:*148*
reproduction, 16:147
taxonomy, 16:143
Vancouver Island, 16:147, 16:*150*, 16:*152*, 16:156
yellow-bellied, 16:*147*, 16:158*t*
Marsh deer, 15:379, 15:383, 15:*384*, 15:*386*, 15:*391*–392
Marsh mongooses, 14:349, 14:*352*
Marsh rabbits, 16:481, 16:*481*, 16:*506*
Marsh rats, 16:265, 16:266, 16:268
Marsh rice rats, 16:268, 16:*271*, 16:*273*–274
Marsupial cats. *See* Dasyurids
Marsupial mice. *See* Dasyurids
Marsupial moles, 12:137, **13:25–29,** 13:*26*, 13:*27*, 13:*28*
behavior, 12:71–72, 12:113
distribution, 12:137, 13:*25*, 13:26
evolution, 12:13, 13:25
physical characteristics, 12:13, 12:71, 13:25–26
Marsupials
brain, 12:49
digestive system, 12:123
distribution, 12:129, 12:136–137
evolution, 12:11, 12:33
locomotion, 12:14
neonatal requirements, 12:126–127
physical characteristics, 13:1
reproduction, 12:12, 12:15, 12:51, 12:89–93, 12:*102*, 12:106, 12:108–109, 12:110
See also specific types of marsupials
Marsupionatia, 12:33
Martens, 14:256, 14:257, 14:319, 14:323, 14:324
American, 14:*320*, 14:*324*
yellow-throated, 14:*320*
Martes spp. *See* Martens
Martes americana. See American martens

Martes flavigula. See Yellow-throated martens
Martes foina. See Stone martens
Martes martes. See European pine martens
Martes pennanti. See Fishers
Martes zibellina. See Sables
Masai giraffes, 15:*402*
Mashona mole-rats, 16:344–345, 16:349*t*
Masked mouse-tailed dormice, 16:327*t*
Masked palm civets, 14:336, 14:344*t*
Masked shrews, 13:*251*
Masked titis, 14:144, 14:145, 14:148, 14:*149*, 14:*150*, 14:152–153
Masoala fork-crowned lemurs, 12:38, 14:35, 14:39, 14:*40*, 14:*41*, 14:44
Massoutiera spp., 16:311
Massoutiera mzabi. See Mzab gundis
Mastiff bats, **13:483–496**, 13:*489*, 14:287
  behavior, 13:485
  conservation status, 13:487
  distribution, 13:*483*
  evolution, 13:483
  feeding ecology, 13:485
  physical characteristics, 13:483
  reproduction, 13:486–487
  species of, 13:*490–492*, 13:493*t*–495*t*
  taxonomy, 13:483
Mastodons, 12:*18*, 15:138
Mastomys natalensis. See Natal multimammate mice
Mate selection, 12:101–102, 12:143–144
  See also Reproduction
Maternal care, 12:96–97, 12:107, 12:143–144
Maternal recognition, 12:84–85
  See also Behavior
Mating, 12:98–99, 12:107
  See also Reproduction
Matschie's bushbabies. *See* Dusky bushbabies
Matschie's tree kangaroos, 13:*37*, 13:*86*, 13:*91*, 13:*98*, 13:100
Matsuzawa, Tetsuro, 12:161–162
Mauritian tomb bats, 13:*359*, 13:*362*, 13:363
Mawas. *See* Sumatran orangutans
Maxilla, 12:9
  See also Physical characteristics
Maxwell's duikers, 16:*74*, 16:*79*, 16:*80*
Mazama spp., 15:382, 15:384
Mazama americana. See Red brocket deer
Mazama americana temana. See Mexican red brocket deer
Mazama bororo. See Brocket deer
Mazama bricenii. See Merioa brocket deer
Mazama chunyi. See Dwarf brocket deer
Mazama gouazoubira. See Gray brocket deer
Mazama pandora. See Yucatán brown brocket deer
Mazama pocket gophers, 16:195*t*
Mazama rufina. See Little red brocket deer
Mazes, in laboratory experiments, 12:153, 12:*154*
McGuire, Tamara L., 15:29
Meadow jumping mice, 16:*212*, 16:216
Meadow voles, 16:125–126, 16:*226*, 16:227, 16:238*t*
Mearns's flying foxes, 13:332*t*
Medem's collared titis, 14:148
Medicine
  conservation, 12:222–223
  zoological, 12:210–211
  See also Diseases
Mediterranean blind mole-rats. *See* Blind mole-rats
Mediterranean horseshoe bats, 13:391, 13:393, 13:*394*, 13:*399*
Mediterranean monk seals, 14:262, 14:418, 14:422, 14:435*t*
Meerkats, 14:259, 14:*347*, 14:357*t*
Meester, J., 16:329
Megachiroptera. *See* Pteropodidae
Megaderma lyra. See Greater false vampire bats
Megaderma spasma. See Lesser false vampire bats
Megadermatidae. *See* False vampire bats
Megaerops niphanae. See Ratanaworabhan's fruit bats
Megaladapis spp. *See* Koala lemurs
Megaladapis edwardsi, 14:73–74
Megaladapis grandidieri, 14:73
Megaladapis madagascariensis, 14:73
Megaloglossus woermanni. See African long-tongued fruit bats
Megalomys spp., 16:266
Megalonychidae, 12:11, 13:148, **13:155–159**
Megamuntiacus spp., 12:129, 15:263
Megamuntiacus vuquangensis. See Giant muntjacs
Megantereon spp. *See* Saber-toothed cats
Megaoryzomys spp., 16:266
Megapedetes spp., 16:307
Megaptera novaeangliae. See Humpback whales
Megaroyzomys spp., 16:270
Megasores gigas. See Mexican giant shrews
Megatapirus spp., 15:237
Megatheriidae, 13:148
Mehely's horseshoe bats, 13:393
Melatonin, subterranean mammals and, 12:74–75
Meles meles. See European badgers
Melinae. *See* Badgers
Mellivora capensis. See Honey badgers
Mellivorinae, 14:319
Melogale moschata. See Chinese ferret badgers
Melon-headed whales, 15:1
Melursus ursinus. See Sloth bears
Melville, Herman, 15:78
Membranes, flight. *See* Flight membranes
Memory, 12:152–155
  See also Cognition
Men. *See* Humans
Menstrual cycles, 12:109
Menstruation, 12:109
Mentawai gibbons. *See* Kloss gibbons
Mentawai Island langurs, 14:172, 14:173, 14:175, 14:*177*, 14:*181*, 14:183
Mentawai Island leaf-monkeys, 14:175, 14:184*t*
Mentawai Islands snub-nosed leaf-monkeys. *See* Mentawai Island langurs
Mentawai langurs, 14:6–7
Menzies' echymiperas, 13:10, 13:12
Mephitinae. *See* Skunks
Mephitis spp., 14:319, 14:321
Mephitis macroura. See Hooded skunks
Mephitis mephitis. See Striped skunks
Mercer, J. M., 16:143
Merck's rhinoceroses, 15:249
Mérida small-eared shrews, 13:*253*
Merioa brocket deer, 15:396*t*
Meriones spp., 16:127
Meriones crassus. See Sundevall's jirds
Meriones unguiculatus. See Gerbils
Merker, St., 14:95
Merriam's desert shrews. *See* Mexican giant shrews
Merriam's kangaroo rats, 16:202
Merriam's pocket gophers, 16:197*t*
Merrins. *See* Bridled nail-tailed wallabies
Merychippus spp., 15:137
Merycodontinae, 15:265–266, 15:411
Merycoidodontidae, 15:264
Mesaxonia, 15:131
Mesechinus dauuricus. See Daurian hedgehogs
Mesechinus hughi. See Hugh's hedgehogs
Mesocapromys spp., 16:462
Mesocapromys angelcabrerai, 16:461
Mesocapromys auritus. See Eared hutias
Mesocapromys nanus. See Dwarf hutias
Mesocricetus spp., 16:239
Mesocricetus auratus. See Golden hamsters
Mesocricetus brandti. See Brandt's hamsters
Mesocricetus newtoni. See Romanian hamsters
Mesomys didelphoides. See Brazilian spiny tree rats
Mesomys obscurus, 16:451
Mesonychidae, 12:30, 15:2, 15:41, 15:131–132, 15:133
Mesopithecus spp., 14:172
Mesoplodon spp., 15:2, 15:59
Mesoplodon bidens. See Sowerby's beaked whales
Mesoplodon bowdoini. See Andrew's beaked whales
Mesoplodon carlhubbsi. See Hubb's beaked whales
Mesoplodon densirostris. See Blainville's beaked whales
Mesoplodon europaeus. See Gervais' beaked whales
Mesoplodon ginkgodens. See Ginkgo-toothed beaked whales
Mesoplodon grayi. See Gray's beaked whales
Mesoplodon hectori. See Hector's beaked whales
Mesoplodon layardii. See Strap-toothed whales
Mesoplodon mirus. See True's beaked whales
Mesoplodon perrini. See Perrin's beaked whales
Mesoplodon peruvianus. See Pygmy beaked whales
Mesoplodon stejnergeri. See Stejneger's beaked whales
Mesoplodon traversii. See Spade-toothed whales
Mesopropithecus spp., 14:63
Metabolism, body size and, 12:21
Metachirus spp. *See* Brown four-eyed opossums
Metachirus nudicaudatus. See Brown four-eyed opossums
Mexican big-eared bats. *See* Allen's big-eared bats
Mexican black agoutis, 16:409, 16:*410*, 16:*412*, 16:413
Mexican black howler monkeys, 14:*158*, 14:166*t*
Mexican fishing bats, 13:313, 13:500
Mexican fox squirrels, 16:175*t*
Mexican free-tailed bats. *See* Brazilian free-tailed bats
Mexican giant shrews, 13:263*t*
Mexican ground squirrels, 12:*188*
Mexican hairless, 14:287

Mexican hairy porcupines, 16:*366*, 16:367, 16:*370*, 16:*372*, 16:373
Mexican hog-nosed bats, 13:*414*, 13:415, 13:418, 13:433*t*
Mexican mouse opossums, 12:*256*, 12:*260*–261
Mexican red brocket deer, 15:*384*
Mexican spiny pocket mice, 16:208*t*
Mexico. *See* Nearctic region; Neotropical region
Mhorr gazelles, 16:*47*
Miacids, 14:256
Mice
 behavior, 12:142–143, 16:124–125
 birch, **16:211–224**
 Delany's swamp, 16:297*t*
 distribution, 12:137, 12:138, 16:123
 dormice, 12:134, 16:122, 16:125, **16:317–328**
 evolution, 16:121
 fat, 16:296*t*
 grasshopper, 16:125
 gray climbing, 16:*286*, 16:*290*–291
 humans and, 12:139, 16:128
 jumping, 16:123, 16:125, **16:211–224**
 kangaroo, **16:199–201,** 16:*204*, 16:208*t*–209*t*
 knock-out, 16:128
 learning and, 12:154
 meadow jumping, 16:*212*, 16:216
 metabolism, 12:21
 old-field, 16:125
 Old World, **16:249–262**
 pocket, 12:117, 16:121, 16:123, 16:124, **16:199–210**
 pouched, 16:297*t*
 pygmy, 16:123
 pygmy rock, 16:*286*, 16:*289*, 16:290
 reproduction, 12:103, 16:125–126
 South American, **16:263–279**
 taxonomy, 16:121
 white-tailed, 16:*286*, 16:*288–289*
 See also Flying mice; Muridae
Michoacán pocket gophers, 16:190, 16:*191*, 16:*192*, 16:193
Mico spp. *See* Amazonian marmosets
Mico intermedius. See Aripuanã marmosets
Micoureus spp. *See* Woolly mouse opossums
Micoureus alstoni. See Alston's woolly mouse opossums
Micoureus constantiae, 12:251
Microbats. *See* Microchiroptera
Microbial fermentation, 12:121–124
Microbiotheridae. *See* Monitos del monte
Microbiotherium spp., 12:273
Microcavia spp. *See* Mountain cavies
Microcavia australis. See Mountain cavies
Microcavia niata. See Andean mountain cavies
Microcebus spp., 14:36, 14:37–38
Microcebus berthae. See Pygmy mouse lemurs
Microcebus griseorufus. See Mouse lemurs
Microcebus murinus. See Gray mouse lemurs
Microcebus ravlobensis, 14:39
Microcebus rufus. See Red mouse lemurs
Microchiroptera, 12:58, 12:79, 12:85–87, 13:309
Microdictyon spp., 12:64
Microdipodops spp., 16:199, 16:200
Microdipodops megacephalus. See Dark kangaroo mice
Microdipodops pallidus. See Pale kangaroo mice
Microgale spp., 13:196, 13:225, 13:226–227, 13:228, 13:229
Microgale brevicaudata, 13:227
Microgale dobsoni. See Dobson's shrew tenrecs
Microgale gracilis, 13:226
Microgale gymnorhyncha, 13:226
Microgale longicaudata. See Madagascar hedgehogs
Microgale nasoloi. See Nasolo's shrew tenrecs
Microgale principula. See Long-tailed shrew tenrecs
Microgale pusilla, 13:227
Micromys minutus. See Harvest mice
Micronesian flying foxes. *See* Marianas fruit bats
Micronycteris microtis. See Common big-eared bats
Microorganisms, Ice Age mammals and, 12:25
Microperoryctes spp., 13:2
Microperoryctes longicauda, 13:10
Microperoryctes murina. See Mouse bandicoots
Microperoryctes papuensis. See Papuan bandicoots
Micropotamogale spp., 13:225
Micropotamogale lamottei. See Nimba otter shrews
Micropotamogale ruwenzorii. See Ruwenzori otter shrews
Micropteropus pusillus. See Dwarf epauletted fruit bats
Microsciurus spp., 16:163
Microtinae. *See* Arvicolinae
Microtus spp. *See* Voles
Microtus agrestis. See Field voles
Microtus cabrearae, 16:270
Microtus californicus. See California meadow voles
Microtus chrotorrhinus. See Rock voles
Microtus evoronensis, 16:229
Microtus mujanensis, 16:229
Microtus ochrogaster. See Prairie voles; Voles
Microtus oeconomus. See Root voles
Microtus pennsylvanicus. See Meadow voles
Microtus pinetorum. See Woodland voles
Midas free-tailed bats, 13:494*t*
Midas tamarins, 14:124, 14:132*t*
Middle ear, 12:8–9, 12:36
 See also Physical characteristics
Migration, 12:87, 12:116, **12:164–170**
 See also Behavior
Migratory squirrels. *See* Gray squirrels
Miles, Lynn, 12:161
Milk, 12:96, 12:106, 12:126–127
 See also Lactation
Miller's mastiff bats, 13:494*t*
Miller's monk sakis, 14:148
Milne-Edwards's sifakas, 14:63, 14:*68*, 14:*70*
Milne-Edwards's sportive lemurs, 14:76, 14:*78*, 14:80–81, 14:*83*
Milton, K., 14:164
Milus. *See* Pere David's deer
Mimolagus, 16:480
Mimotoma, 16:480
Mimotonids. *See* Lagomorpha
Mindanao gymnures, 13:207, 13:*208*, 13:212
Mindanao moonrats. *See* Mindanao gymnures
Mindanao wood-shrews. *See* Mindanao gymnures
Miner's cats. *See* Ringtails
Miniature donkeys, 12:*172*
Miniature horses, 15:*216*
Miniopterinae. *See Miniopterus* spp.
Miniopterus spp., 13:519, 13:520, 13:521, 13:522
Miniopterus australis. See Little long-fingered bats
Miniopterus medius. See Southeast Asian bent-winged bats
Miniopterus schreibersi. See Common bentwing bats
Ministry of Environment and Forests (Government of India), on pygmy hogs, 15:287
Mink, 14:319, 14:321, 14:323, 14:324
 learning set, 12:152
 neonatal requirements, 12:126
 reproduction, 12:107
 See also American mink
Mink (Keeping) Regulations of 1975 (United Kingdom), 12:183
Minke whales, 15:5, 15:8, 15:9, 15:120–121, 15:123–124
 Antarctic, 15:1, 15:120, 15:125, 15:130*t*
 dwarf, 12:*200*
 northern, 15:120, 15:*122*, 15:125, 15:*126*, 15:128–129
Minnesota Zoo, 12:210
Mioeuoticus spp., 14:13
Miopithecus spp. *See* Talapoins
Miopithecus talapoin. See Angolan talapoins
Mirounga spp. *See* Elephant seals
Mirounga angustirostris. See Northern elephant seals
Mirounga leonina. See Southern elephant seals
Mirror self-recognition (MSR), 12:159
Mirza spp., 14:37, 14:39
Mirza coquereli. See Coquerel's mouse lemurs
Miss Waldron's red colobus, 12:214
Mitered leaf-monkeys. *See* Banded leaf-monkeys
Mitochondrial DNA (mtDNA), 12:26–27, 12:28–29, 12:31, 12:33
Mitred horseshoe bats, 13:389
Mixed-species troops, Callitrichidae and, 14:119–120
Moas, 12:*183*, 12:184
Moberg, G. P., 15:145
Moby Dick, 15:78
Mogera spp., 13:279
Mogera etigo. See Echigo Plain moles
Mogera robusta. See Large Japanese moles
Mogera tokudae. See Sado moles
Moholi bushbabies, 14:10, 14:25, 14:27, 14:*29*
Molar teeth, 12:9, 12:14, 12:46
 See also Teeth
Mole-like rice tenrecs, 13:234*t*
Mole mice, 16:265
Mole-rats, 12:70, 12:72–77, 12:115, 12:116
 See also African mole-rats
Mole-shrews, 13:*253*, 13:*255*
Mole voles, 12:71, 16:225, 16:228, 16:230
Molecular clocks, 12:31, 12:33
Molecular genetics, **12:26–35**
Moles, **13:279–288,** 13:*283*
 behavior, 12:113, 13:197–198, 13:280–281
 claws, 12:39
 conservation status, 13:282
 distribution, 12:131, 13:197, 13:279, 13:280
 evolution, 12:13, 13:279

Moles *(continued)*
feeding ecology, 12:75–76, 13:198, 13:281–282
habitats, 13:280
humans and, 13:200, 13:282
physical characteristics, 13:195–196, 13:279–280
reproduction, 13:199, 13:282
species of, 13:284–287, 13:287*t*–288*t*
taxonomy, 13:193, 13:194, 13:279
See also Marsupial moles; specific types of moles
Moloch gibbons, 14:*215*, 14:*217*, 14:*219*, 14:220
behavior, 14:210, 14:211
conservation status, 14:11, 14:215
distribution, 14:208
evolution, 14:207
feeding ecology, 14:213
Molossidae. *See* Free-tailed bats; Mastiff bats
Molossops spp., 13:483
Molossops greenhalli. *See* Greenhalli's dog-faced bats
Molossus ater. *See* Greater house bats
Molossus molossus. *See* Pallas's mastiff bats
Molossus pretiosus. *See* Miller's mastiff bats
Molsher, Robyn, 12:185
Molts, 12:38
Moluccan cuscuses, 13:66*t*
Monachinae, 14:417
Monachus monachus. *See* Mediterranean monk seals
Monachus schauinslandi. *See* Hawaiian monk seals
Monachus tropicalis. *See* West Indian monk seals
Mongolian gazelles, 12:134, 16:48, 16:*49*, 16:*53*–54
Mongolian hamsters, 16:239, 16:247*t*
Mongolian pikas. *See* Pallas's pikas
Mongolian wild horses. *See* Przewalski's horses
Mongoose lemurs, 14:6, 14:8, 14:*48*, 14:50, 14:*53*, 14:*56*–57
Mongooses, **14:347–358,** 14:*353*
behavior, 12:114, 12:115, 12:146–147, 14:259, 14:349–351
in biological control, 12:190–191
conservation status, 14:262, 14:352
distribution, 12:136, 14:348–349
evolution, 14:347
feeding ecology, 14:260, 14:351–352
habitats, 14:349
humans and, 14:352
physical characteristics, 14:347–348
reproduction, 14:261, 14:352
species of, 14:*354*–356, 14:357*t*
taxonomy, 14:256, 14:347
Monitos del monte, 12:132, **12:273–275,** 12:*274*
Monk sakis, 14:143, 14:144–145, 14:147, 14:153*t*
Monk seals, 12:138, 14:256, 14:418
Hawaiian, 12:138, 14:*420*, 14:422, 14:*425*, 14:*429*, 14:431–432
Mediterranean, 14:262, 14:418, 14:422, 14:435*t*
West Indian, 14:422, 14:435*t*
Monk sloths, 13:*166*, 13:*168*
Monkeys, 14:3, 14:8
Allen's swamp, 14:194, 14:*196*, 14:*197*
behavior, 12:148
coloration in, 12:38
encephalization quotient, 12:149
learning set, 12:152
lesser white-nosed, 14:*192*
memory, 12:152–154
numbers and, 12:155
theory of mind, 12:159–160
tool use, 12:157
in zoos, 12:*207*
See also specific types of monkeys
Monodelphis spp. *See* Short-tailed opossums
Monodelphis americana, 12:251
Monodelphis brevicaudata. *See* Red-legged short-tailed opossums
Monodelphis dimidiata. *See* Southern short-tailed opossums
Monodelphis domestica, 12:251
Monodelphis iheringi, 12:251
Monodelphis kunsi. *See* Pygmy short-tailed opossums
Monodelphis osgoodi, 12:250
Monodelphis rubida, 12:250
Monodelphis scallops, 12:250
Monodelphis sorex, 12:250
Monodelphis unistriata, 12:250
Monodon monoceros. *See* Narwhals
Monodontidae, **15:81–91,** 15:*89*
Monogamy, 12:98, 12:107, 14:6–7
Canidae, 14:271
Carnivora, 14:261
gibbons, 14:213, 14:214
See also Behavior
Monophyllus redmani, 13:415
Monotremata. *See* Monotremes
Monotrematum sudamericanum. *See* South American monotremes
Monotremes, **12:227–234,** 15:133
behavior, 12:230–231
conservation status, 12:233–234
development, 12:106
distribution, 12:136–137, 12:230
evolution, 12:11, 12:33, 12:227–228
feeding ecology, 12:231
habitats, 12:230
humans and, 12:234
physical characteristics, 12:38, 12:228–230
reproduction, 12:12, 12:51, 12:89–93, 12:108, 12:231–233
skeletons, 12:41
Montane bamboo rats, 16:450–451, 16:*452*, 16:*455*, 16:457
Montane guinea pigs, 12:181, 16:399*t*
Montane tree shrews, 13:297*t*
Montane woolly flying squirrels, 16:136
Montezuma (Aztec chief), 12:203
Moon-toothed degus, 16:433, 16:440*t*
Moonrats, 13:194, 13:197, 13:198, 13:199, 13:203, 13:*204*, 13:*208*, 13:212
Moose, 15:*133*, 15:145–146, 15:*263*, 15:*269*, 15:*387*
behavior, 15:383, 15:394
conservation status, 15:384, 15:394
distribution, 12:132, 15:*391*, 15:394
evolution, 15:379, 15:380, 15:381
feeding ecology, 15:394
habitats, 15:383, 15:394
humans and, 15:384–385
migrations, 12:164–165
physical characteristics, 15:140, 15:381, 15:394
reproduction, 15:384, 15:394
taxonomy, 15:394
Mops midas. *See* Midas free-tailed bats
Mops niangarae, 13:489
Mops spurelli. *See* Spurelli's free-tailed bats
Moraes, P. L. R., 14:164
Moreno glacier, 12:*20*
Morenocetus parvus, 15:107
Morganucodon spp., 12:11
Mormoopidae. *See* Moustached bats
Mormoops spp. *See* Ghost-faced bats
Mormoops blainvillii. *See* Antillean ghost-faced bats
Mormoops megalophylla. *See* Ghost-faced bats
Mormopterus spp., 13:483
Mormopterus acetabulosus. *See* Natal free-tailed bats
Mormopterus beccarii. *See* Beccari's mastiff bats
Mormopterus phrudus, 13:487
Moschidae. *See* Musk deer
Moschinae. *See* Musk deer
Moschus berezovskii, 15:333
Moschus chrysogaster. *See* Himalayan musk deer
Moschus fuscus, 15:333
Moschus moschiferus. *See* Siberian musk deer
Moss, Cynthia, 15:167
Moss-forest ringtails, 13:114, 13:115, 13:122*t*
Mottle-faced tamarins, 14:132*t*
Mottled tuco-tucos, 16:427, 16:430*t*
Mouflons, 12:178, 15:147, 16:87, 16:91
Mt. Graham red squirrels, 16:167
Mountain beavers, 12:131, 16:122, 16:123, 16:126–127, **16:131–134,** 16:*132*, 16:*133*
Mountain brushtail possums, 13:*59*, 13:60, 13:*66t*
Mountain cavies, 16:389, 16:390, 16:391, 16:392, 16:393, 16:*394*, 16:*395*, 16:399*t*
Mountain cottontails, 16:481, 16:516*t*
Mountain cows. *See* Central American tapirs
Mountain deer, 15:384
Mountain gazelles, 12:139, 16:48, 16:*49*, 16:54–*55*
Mountain goats, 12:*21*, 12:131, 12:165, 15:*134*, 15:138, 16:5, 16:87, 16:*92*, 16:93–94, 16:*96*, 16:*99*
Mountain gorillas, 12:*218*, 14:225, 14:227
See also Eastern gorillas
Mountain hares, 12:129, 16:481, 16:506, 16:*510*, 16:*512*–513
Mountain lions. *See* Pumas
Mountain marmots. *See* Hoary marmots
Mountain nyalas, 16:11, 16:25*t*
Mountain pacas, 16:*422*, 16:*423*–424
Mountain pocket gophers, 16:*185*, 16:*186*, 16:195*t*
Mountain pygmy possums, 13:35, 13:*109*, 13:*110*, 13:111
behavior, 13:37, 13:106–107
conservation status, 13:39, 13:108
evolution, 13:105
feeding ecology, 13:38, 13:107
habitats, 13:34, 13:106
reproduction, 13:38
Mountain reedbucks, 16:27, 16:29, 16:30, 16:32
Mountain sheep, 12:165–167, 15:138, 16:4, 16:95

Mountain squirrels, 16:167, 16:175*t*
Mountain tapirs, 15:*244*, 15:*245*, 15:246–247
behavior, 15:239, 15:240
conservation status, 15:222, 15:*242*
feeding ecology, 15:*241*
habitats, 15:218, 15:238
humans and, 15:*242*
physical characteristics, 15:216, 15:217, 15:*238*
Mountain tarsiers. *See* Pygmy tarsiers
Mountain viscachas, 16:377, 16:378, 16:*379*, 16:380, 16:*381*, 16:*382*–383
Mountain zebras, 15:215, 15:218, 15:220, 15:222, 15:226–227, 15:*231*, 15:*233*–234
behavior, 12:146
Moupin pikas, 16:502*t*
Mouse bandicoots, 13:2, 13:12, 13:17*t*
Mouse deer, 15:138, 15:266, 15:267, 15:325–328, 15:*330*, 15:*332*–333
Mouse-eared bats, 13:497, 13:498, 13:499
Mouse lemurs, 12:*117*, 14:3, 14:4, 14:6, 14:7, 14:*10*, **14:35–45**, 14:*40*
behavior, 14:36–37
conservation status, 14:39
distribution, 14:*35*, 14:36
evolution, 14:35
feeding ecology, 14:37–38
habitats, 14:36
humans and, 14:39
physical characteristics, 14:35
reproduction, 14:38
species of, 14:*41*–44
taxonomy, 14:35
Mouse-like hamsters, 16:241, 16:*286*, 16:*288*
Afghan, 16:243
Tsolov's, 16:243
Urartsk, 16:243
Mouse opossums, 12:250–254, 12:*252*, 12:*256*, 12:*261*–*262*, 12:264*t*–265*t*
Mouse shrews. *See* Forest shrews
Mouse-tailed bats, 13:308, **13:351–354**
Mouse-tailed dormice, 16:317, 16:318
masked, 16:327*t*
Setzer's, 16:328*t*
See also Roach's mouse-tailed dormice
Moustached bats, 13:308, **13:435–442**, 13:*437*, 13:*439*, 13:442*t*
Moustached guenons, 14:*195*, 14:*198*, 14:199–200
Moustached monkeys. *See* Moustached guenons
Moustached tamarins, 14:117, 14:118, 14:119, 14:120, 14:121, 14:122, 14:*123*, 14:124
Mrs. Gray's lechwes. *See* Nile lechwes
MSR (Mirror self-recognition), 12:159
Mt. Graham red squirrels, 16:167
Mt. Kenya shrews, 13:277*t*
mtDNA. *See* Mitochondrial DNA
Mud digger whales. *See* Gray whales
Mueller's gibbons, 14:*214*, 14:*217*, 14:220–*221*
behavior, 14:210, 14:211
distribution, 14:208, 14:*212*
evolution, 14:207, 14:208–209
feeding ecology, 14:213
taxonomy, 14:208
Mule deer. *See* Black-tailed deer
Mules, 12:177, 15:223
Mulgaras, 12:*278*, 12:*293*, 12:*294*
Multituberculata, 12:11, 16:122
Mungos spp., 14:347, 14:348
Mungos mungo. *See* Banded mongooses
Mungotictis spp., 14:347, 14:348
Mungotictis decemlineata. *See* Narrow-striped mongooses
Mungotinae, 14:347, 14:349
Munnings. *See* Banded hare-wallabies
Muntiacinae. *See* Muntjacs
Muntiacus spp. *See* Feral muntjacs
Muntiacus spp. See Muntjacs
Muntiacus atherodes. *See* Bornean yellow muntjacs
Muntiacus crinifrons. *See* Black muntjacs
Muntiacus feae. *See* Fea's muntjacs
Muntiacus gongshanensis. *See* Gongshan muntjacs
Muntiacus muntjak. *See* Indian muntjacs
Muntiacus muntjak vaginalis. *See* North Indian muntjacs
Muntiacus putaoensis. *See* Leaf muntjacs
Muntiacus reevesi. *See* Reeve's muntjacs
Muntiacus rooseveltorum. *See* Roosevelt's muntjacs
Muntiacus truongsonensis. *See* Truong Son muntjacs
Muntjacs, **15:343–355**, 15:*347*–*348*
behavior, 12:118, 15:344–345
conservation status, 15:346
distribution, 15:*343*, 15:344
evolution, 12:19, 15:343
feeding ecology, 15:345
habitats, 15:344
humans and, 15:346
physical characteristics, 15:267, 15:268, 15:269, 15:343–344
reproduction, 15:346
species, 15:349–354
taxonomy, 12:129, 15:343
Murexia spp., 12:279, 12:288, 12:290
Murexia longicaudata. *See* Short-furred dasyures
Murexia rothschildi. *See* Broad-striped dasyures
Muridae, **16:225–298**, 16:*286*–*287*
Arvicolinae, 16:124, **16:225–238**, 16:*231*, 16:*232*, 16:270
behavior, 16:284
color vision, 12:79
distribution, 12:134, 16:123, 16:*281*, 16:283
evolution, 16:122, 16:281–282
feeding ecology, 16:284
habitats, 16:283–284
hamsters, **16:239–248**
humans and, 16:127–128
Murinae, **16:249–262**, 16:*254*, 16:261*t*–262*t*
physical characteristics, 16:123, 16:283
reproduction, 16:125
Sigmodontinae, 16:128, **16:263–279**, 16:*271*, 16:277*t*–278*t*
species of, 16:288–295, 16:296*t*–297*t*
taxonomy, 16:121, 16:282–283
Murina spp., 13:521, 13:523
Murina huttoni. *See* Hutton's tube-nosed bats
Murina leucogaster. *See* Tube-nosed bats, insectivorous
Murina suilla. *See* Brown tube-nosed bats
Murinae, **16:249–262**, 16:*254*, 16:261*t*–262*t*, 16:283
Murine mouse opossums, 12:*252*
Murininae, 13:519, 13:520–521
Muriquis, 14:155, 14:156, 14:157, 14:158, 14:159, 14:160
Murofushi, Kiyoko, 12:161–162
Muroidea, 16:121
Mus musculus. *See* House mice
Mus spicilegus. *See* Steppe harvesting mice
Muscardinus avellanarius. *See* Hazel dormice
Muscles, in bat wings, 12:56–57
See also Physical characteristics
Muscular diaphragm, 12:8
Musk deer, 12:136, 15:265, 15:266, **15:335–341**, 15:*339*
Musk oxen. *See* Muskoxen
Musk shrews, 12:107, 13:198, 13:199, 13:266
Muskoxen, 16:*97*, 16:*98*, 16:102–103
behavior, 16:92, 16:93
distribution, 12:132, 16:*98*
domestication and, 15:145–146
evolution, 15:138, 16:87, 16:88
feeding ecology, 16:93, 16:94
habitats, 16:91
physical characteristics, 15:139, 16:*89*, 16:90
reproduction, 12:102, 16:94
taxonomy, 16:87, 16:88
Muskrats, 16:*231*, 16:*233*, 16:238*t*
behavior, 16:227
distribution, 16:226
feeding ecology, 16:228
habitats, 16:124
humans and, 16:126, 16:230
physical characteristics, 16:123, 16:226
Musky rat-kangaroos, 13:35, 13:*69*, **13:69–72**, 13:*70*, 13:*71*, 13:*80*
Musser, G. G., 16:282
Mustangs. *See* Domestic horses
Mustela erminea. *See* Ermines
Mustela eversmanni. *See* Steppe polecats
Mustela felipei. *See* Colombian weasels
Mustela frenata. *See* Longtail weasels
Mustela lutreola. *See* European mink
Mustela lutreolina. *See* Indonesian mountain weasels
Mustela macrodon. *See* Sea minks
Mustela nigripes. *See* Black-footed ferrets
Mustela nivalis. *See* Least weasels
Mustela putorius. *See* Ferrets
Mustela putorius furo, 12:222
Mustela vison. *See* American mink
Mustelidae, **14:319–334**, 14:*326*
behavior, 12:145, 14:258–259, 14:321–323
conservation status, 14:262, 14:324
distribution, 12:131, 14:*319*, 14:321
evolution, 14:319
feeding ecology, 14:260, 14:323
habitats, 14:321
humans and, 14:324–325
physical characteristics, 12:37, 14:319–321
reproduction, 12:109, 12:110, 14:324
species of, 14:*327*–331, 14:332*t*–333*t*
taxonomy, 14:256, 14:319
Mustelinae, 14:319
Musth, 15:167, 15:168, 15:170
Mydaus spp. *See* Stink badgers
Mydaus javensis. *See* Indonesian stink badgers
Mydaus marchei. *See* Philippine stink badgers
Myers, Norman, 12:218
Mylodontidae, 12:11, 13:148
Myocastor coypus. *See* Coypus
Myocastoridae. *See* Coypus
Myoictis spp., 12:288
Myoictis melas. *See* Three-striped dasyures

Myomimus spp. *See* Mouse-tailed dormice
Myomimus personatus. *See* Masked mouse-tailed dormice
Myomimus roachi. *See* Roach's mouse-tailed dormice
Myomimus setzeri. *See* Setzer's mouse-tailed dormice
Myoncyteris torquata. *See* Little collared fruit bats
Myoprocta spp., 16:407
Myoprocta acouchy. *See* Green acouchis
Myoprocta exilis. *See* Red acouchis
Myopterus whitleyi. *See* Bini free-tailed bats
Myopus schisticolor. *See* Wood lemmings
Myosciurus spp., 16:163
Myosciurus pumilio. *See* West African pygmy squirrels
Myosorex spp., 13:265, 13:266
Myosorex schalleri. *See* Schaller's mouse shrews
Myosorex varius. *See* Forest shrews
Myospalacinae, 16:282, 16:284
Myospalax myospalax. *See* Siberian zokors
Myotis
gray, 13:310–311, 13:316
large-footed, 13:313
Myotis spp. *See* Mouse-eared bats
Myotis adversus. *See* Large-footed myotis
Myotis bechsteinii. *See* Bechstein's bats
Myotis bocagei. *See* Rufous mouse-eared bats
Myotis capaccinii. *See* Long-fingered bats
Myotis dasycneme. *See* Pond bats
Myotis daubentonii. *See* Daubenton's bats
Myotis formosus, 13:497
Myotis grisescens. *See* Gray myotis
Myotis lucifugus. *See* Little brown bats
Myotis myotis. *See* Greater mouse-eared bats
Myotis mystacinus. *See* Whiskered bats
Myotis nattereri. *See* Natterer's bats
Myotis rickettii. *See* Rickett's big-footed bats
Myotis sodalis. *See* Indiana bats
Myotis vivesi. *See* Mexican fishing bats
Myotis welwitschii. *See* Welwitch's hairy bats
Myoxidae. *See* Dormice
Myoxus glis. *See* Edible dormice
Myoxus japonicus. *See* Japanese dormice
Myrmecobiidae. *See* Numbats
Myrmecobius fasciatus. *See* Numbats
Myrmecophaga spp., 13:149, 13:150
Myrmecophaga tridactyla. *See* Giant anteaters
Myrmecophagidae. *See* Anteaters
Mysateles spp., 16:461
Mysateles garridoi, 16:461
Mysateles melanurus. *See* Black-tailed hutias
Mysateles prehensilis. *See* Prehensile-tailed hutias
Mystacina robusta. *See* Greater New Zealand short-tailed bats
Mystacina tuberculata. *See* Lesser New Zealand short-tailed bats
Mystacinidae. *See* New Zealand short-tailed bats
Mysticeti. *See* Baleen whales
Mystromyinae, 16:282
Mystromys albicaudatus. *See* White-tailed mice
Mytonolagus, 16:480
Myxoma virus, 12:187
Myzopoda spp. *See* Old World sucker-footed bats
Myzopoda aurita. *See* Old World sucker-footed bats
Myzopodidae. *See* Old World sucker-footed bats
Mzab gundis, 16:311, 16:312, 16:313, 16:*314*–315

N

Naemorhedus spp., 16:87
Naemorhedus baileyi. *See* Red gorals
Naemorhedus caudatus. *See* Long-tailed gorals
Naemorhedus goral. *See* Himalayan gorals
Nail-tailed wallabies, 13:83, 13:85, 13:86, 13:88
Nails, 12:39
See also Physical characteristics
Naked-backed bats
big, 13:435, 13:442*t*
Davy's, 13:435, 13:*437*, 13:442*t*
Naked-backed fruit bats, 13:307, 13:308, 13:*335*, 13:348*t*
Naked-backed moustached bats, 13:308
Naked bats, 13:*485*, 13:*489*, 13:*490*, 13:491
Naked bulldog bats. *See* Naked bats
Naked mole-rats, 16:*343*, 16:*346*, 16:*348*–349
behavior, 12:74, 12:147, 16:124, 16:343
communication, 12:76, 12:83
distribution, 16:342, 16:*342*
feeding ecology, 16:344
heterothermia, 12:73
physical characteristics, 12:71–74, 12:*72*, 12:78, 12:*86*, 16:340, 16:341
reproduction, 12:107, 16:126, 16:344–345
studies of, 12:72
taxonomy, 16:339
Naked-tailed armadillos, 13:182, 13:183
Nalacetus spp., 15:2
Namaqua dune mole-rats, 16:*344*, 16:345, 16:349*t*
Namdapha flying squirrels, 16:141*t*
Namib golden moles. *See* Grant's desert golden moles
Nancy Ma's douroucoulis. *See* Nancy Ma's night monkeys
Nancy Ma's night monkeys, 14:135, 14:138, 14:*139*, 14:*140*–141
Nancy Ma's owl monkeys. *See* Nancy Ma's night monkeys
Nandinia binotata. *See* African palm civets
Nandiniinae. *See* African palm civets
Nanjaats. *See* Gerenuks
Nannosciurus spp., 16:163
Nannospalax spp., 16:124
Nannospalax ehrenbergi. *See* Palestine mole rats
Nanonycteris veldkampi. *See* Little flying cows
Napaeozapus spp., 16:212, 16:213, 16:216
Napaeozapus insignis. *See* Woodland jumping mice
Napo monk sakis, 14:148
Narbaleks, 13:*92*, 13:*96*
Nares, 12:67
Narrow-headed golden moles. *See* Hottentot golden moles
Narrow-snouted scalytails. *See* Beecroft's anomalures
Narrow-striped marsupial shrews, 12:300*t*
Narrow-striped mongooses, 14:*352*, 14:357*t*
Narwhal-beluga hybrids, 15:81
Narwhals, 15:7, 15:8, 15:9, **15:81–91**, 15:*83*, 15:*86*, 15:*88*, 15:*89*
Nasal cavity, 12:8
See also Physical characteristics
Nasalis spp., 14:172, 14:173
Nasalis concolor. *See* Pig-tailed langurs
Nasalis larvatus. *See* Proboscis monkeys
Nasolo's shrew tenrecs, 13:*231*, 13:*233*
Nasua spp., 14:309, 14:310
Nasua narica. *See* White-nosed coatis
Nasua nasua. *See* Ring-tailed coatis
Nasua nelsoni. *See* Cozumel Island coatis
Nasuella olivacea, 14:310
Natal duikers, 16:84*t*
Natal free-tailed bats, 13:494*t*
Natal multimammate mice, 16:*249*
Natalidae. *See* Funnel-eared bats
Natalus spp. *See* Funnel-eared bats
Natalus lepidus. *See* Gervais' funnel-eared bats
Natalus major, 13:*460*
Natalus micropus. *See* Small-footed funnel-eared bats
Natalus stramineus. *See* Funnel-eared bats
Natalus tumidifrons. *See* Bahamian funnel-eared bats
Natalus tumidirostris. *See* White-bellied funnel-eared bats
Nathusius's pipistrelle bats, 13:516*t*
Nation-states, 14:252
National Invasive Species Council, 12:184
National Park Service, on feral pigs, 12:192
National Zoo (U.S.), 12:205
Natterer's bats, 13:516*t*
Natterer's tuco-tucos, 16:431*t*
Natural history field studies, 12:198–200
Natural selection, 12:97, 12:144–145
See also Evolution
Nature Biotechnology, 15:151
Navahoceros spp., 15:379, 15:380
Navigation, 12:53–55, 12:87
See also Behavior
Nayarit squirrels. *See* Mexican fox squirrels
nDNA. *See* Nuclear DNA
Neacomys spp., 16:265
Neamblysomus spp., 13:216
Neanderthals, 14:243–*244*, 14:249, 14:251
body size of, 12:19, 12:24–25
brains, 12:24
evolution of, 12:21
See also Homo sapiens
Nearctic region, 12:131–132, 12:134
Necrolemur spp., 14:1
Necromys benefactus, 16:270
Necromys lasiurus, 16:267
Nectar bats, 12:13
See also specific types of nectar bats
Nectogale spp., 13:248, 13:249
Nectogale elegans. *See* Elegant water shrews
Nectomys spp., 16:268
Nectomys parvipes, 16:270
Nectomys squamipes, 16:266, 16:267–268, 16:269
Needle-clawed bushbabies, 14:8–9, 14:26, 14:*27*, 14:*28*
Negros shrews, 13:277*t*
Nelson's antelope squirrels, 16:124, 16:145, 16:147, 16:160*t*
Nelson's kangaroo rats, 16:209*t*
Nelson's spiny pocket mice, 16:208*t*
Neobalaenidae. *See* Pygmy right whales
Neocortex, 12:6, 12:36, 12:49
See also Physical characteristics

Neofelis nebulosa. See Clouded leopards
Neofiber alleni. See Round-tailed muskrats
Neohylomys spp., 13:203
Neomyini, 13:247
Neomys spp., 13:249
Neomys fodiens. See Eurasian water shrews
Neonates
altricial, 12:95–96, 12:97
body fat, 12:125
marsupial, 12:108
nutritional requirements, 12:126–127
precocial, 12:12, 12:95–96, 12:97
See also Reproduction
Neopallium. *See* Neocortex
Neophascogale loreatzi. See Speckled dasyures
Neophoca cinerea. See Australian sea lions
Neophocaena phocaenoides. See Finless porpoises
Neoplatymops spp., 13:311
Neopteryx frosti. See Small-toothed fruit bats
Neotamias spp., 16:143
Neotetracus spp., 13:203
Neotoma spp. *See* Woodrats
Neotoma albigula. See White-throated woodrats
Neotoma cinerea. See Bushy-tailed woodrats
Neotoma floridana. See Eastern woodrats
Neotoma fuscipes. See Dusky-footed woodrats
Neotraginae, 16:1, **16:59–72,** 16:*65*, 16:71*t*
Neotragus batesi. See Dwarf antelopes
Neotragus moshcatus. See Sunis
Neotragus pygmeus. See Royal antelopes
Neotropical region, 12:131, 12:132–133
Neri-Arboleda, Irene, 14:95
Nervous system, 12:49–50
See also Physical characteristics
Nesbit, P. H., 16:333
Nesokia indica. See Short-tailed bandicoot rats
Nesolagus spp. *See* Striped rabbits
Nesolagus limminsi. See Annamite striped rabbits
Nesolagus netscheri. See Sumatra short-eared rabbits
Nesolagus timminsi. See Annamite striped rabbits
Nesomyinae, 16:282, 16:285
Nesomys rufus. See Red forest rats
Nesophontes spp. *See* West Indian shrews
Nesophontes edithae. See Puerto Rican nesophontes
Nesophontes hypomicrus. See Atalaye nesophontes
Nesophontes longirostris. See Slender Cuban nesophontes
Nesophontes major. See Greater Cuban nesophontes
Nesophontes micrus. See Western Cuban nesophontes
Nesophontes paramicrus. See St. Michel nesophontes
Nesophontes submicrus. See Lesser Cuban nesophontes
Nesophontes zamicrus. See Haitian nesophontes
Nesophontidae. *See* West Indian shrews
Nesoryzomys spp., 16:266
Nesoryzomys darwini, 16:270
Nesoryzomys indefessus, 16:270
Nesoscaptor uchidai. See Ryukyu moles
Ness, J. W., 15:145
Nests, 12:96
See also Reproduction
Neurotrichus gibbsii. See American shrew moles
Neusticomys oyapocki, 16:265
Nevo, Eviatar, 12:70
New Guinea. *See* Australian region
New Guinea singing dogs, 14:266
New Guinean quolls, 12:299*t*
New Mexico National Heritage Program, 16:156
New World deer, 12:20, 15:343, **15:379–397,** 15:*386*, 15:*387*, 15:396*t*
New World dogs, 14:287
See also Domestic dogs
New World leaf-nosed bats, 13:311, 13:315
New World monkeys, **14:101–133,** 14:*107*, 14:*125*, 14:*126*
behavior, 14:6–7, 14:103–104, 14:116–120
conservation status, 14:105–106, 14:124
distribution, 14:5, 14:*101*, 14:102–103, 14:*115*, 14:116
evolution, 14:2, 14:101, 14:115
feeding ecology, 14:9, 14:104–105, 14:120–121
habitats, 14:6, 14:103, 14:116
humans and, 14:106, 14:124
physical characteristics, 14:101–102, 14:115–116
reproduction, 14:10, 14:105, 14:121–124
species of, 14:*108*–112, 14:*127*–131, 14:132*t*
taxonomy, 14:1, 14:4, 14:101, 14:115
New World opossums, 12:132, **12:249–265,** 12:*255*, 12:*256*
behavior, 12:251–252, 12:257–263, 12:264*t*–265*t*
conservation status, 12:254, 12:257–263, 12:264*t*–265*t*
distribution, 12:*249*, 12:250–251, 12:257–263, 12:264*t*–265*t*
evolution, 12:249–250
feeding ecology, 12:252–253, 12:257–263, 12:264*t*–265*t*
habitats, 12:251, 12:257–263, 12:264*t*–265*t*
humans and, 12:254, 12:257–263
physical characteristics, 12:250, 12:257–263, 12:264*t*–265*t*
reproduction, 12:253–254, 12:257–263
taxonomy, 12:249–250, 12:257–263, 12:264*t*–265*t*
New World peccaries. *See* Peccaries
New World porcupines, 16:125, **16:365–375,** 16:*370*, 16:374*t*
New Zealand. *See* Australian region
New Zealand fur seals, 14:394, 14:*397*, 14:398, 14:407*t*
New Zealand lesser short-tailed bats. *See* Lesser New Zealand short-tailed bats
New Zealand long-eared bats. *See* Lesser New Zealand short-tailed bats
New Zealand Red Data Book, on lesser New Zealand short-tailed bats, 13:457
New Zealand sea lions. *See* Hooker's sea lions
New Zealand short-tailed bats, **13:453–458,** 13:*456*
Newborns. *See* Neonates
Nicaraguan pocket gophers, 16:196*t*
Niches, ecological. *See* Ecological niches
Nicobar tree shrews, 13:292, 13:297*t*
Nigerian mole-rats, 16:350*t*
Night monkeys, 14:6, 14:8, 14:10, **14:135–142,** 14:*139*, 14:141*t*
Nile lechwes, 15:271, 16:4, 16:27, 16:34, 16:*35*, 16:*38*
Nilgais, 12:136, 15:265, 16:4, 16:11, 16:24*t*
Nilgiri ibex. *See* Nilgiri tahrs
Nilgiri langurs, 14:184*t*
Nilgiri tahrs, 16:90, 16:92, 16:*93*, 16:94, 16:95, 16:103*t*
Nimba otter shrews, 13:234*t*
Nine-banded armadillos, 12:111, 12:131, 13:181, 13:*182*, 13:183–184, 13:*185*, 13:*186*, 13:*187*
Ningaui ridei. See Wongai ningauis
Ningaui timealeyi. See Pilbara ningauis
Nipples, 12:5, 12:89, 12:96–97, 12:106
See also Physical characteristics
Nitrogen
diving mammals and, 12:68
from fossil fuels and fertilizers, 12:213
Noack's African leaf-nosed bats. *See* Noack's roundleaf bats
Noack's roundleaf bats, 13:*406*, 13:*407*, 13:408
Noctilio spp. *See* Bulldog bats
Noctilio albiventris. See Lesser bulldog bats
Noctilio leporinus. See Greater bulldog bats
Noctilionidae. *See* Bulldog bats
Noctules, 13:310, 13:498, 13:500, 13:*501*, 13:*504*, 13:*510*, 13:512
giant, 13:313
greater, 13:516*t*
reproduction, 12:110
Nocturnal flight, bats, 12:52–55, 12:59
Nocturnal mammals
body size and, 12:115
color vision, 12:79
vision, 12:79–80
See also specific nocturnal mammals
Noisy night monkeys, 14:135, 14:138, 14:141*t*
Nollman, Jim, 15:93
Nomascus spp. *See* Crested gibbons
Nomascus concolor. See Black crested gibbons
Nomascus gabriellae. See Golden-cheeked gibbons
Nomascus leucogenys. See White-cheeked gibbons
Nomenclature. *See* Taxonomy
Non-declarative memory, 12:154
Non-shivering thermogenesis (NST), 12:113
Non-spiny bandicoots, 13:9
Nonlinearity, of ecosystems, 12:217
Nonsporting dogs, 14:288, 14:289
Noronhomys spp., 16:266
North Africa. *See* Palaearctic region
North African crested porcupines. *See* North African porcupines
North African gundis, 16:312, 16:313
North African porcupines, 16:*353*, 16:354, 16:*355*, 16:*360*–361
North African sengis, 16:*519*, 16:*523*, 16:*527*, 16:529–530
North African striped weasels, 14:333*t*
North America. *See* Nearctic region
North American badgers, 14:321, 14:*322*, 14:333*t*
North American beavers, 16:*178*, 16:*181*, 16:*182*, 16:*183*–184
behavior, 16:*177*, 16:*180*
feeding ecology, 12:*126*, 16:*179*
habitats, 12:*138*
physical characteristics, 12:111

North American black bears. *See* American black bears
North American kangaroo rats, 16:124
North American mountain goats, 12:131
North American plains bison. *See* American bison
North American pocket gophers, 16:126
North American porcupines, 16:*365*, 16:*370*, 16:*371–372*
behavior, 16:367–368
distribution, 16:123, 16:367
feeding ecology, 12:*123*
humans and, 16:127
physical characteristics, 12:38
reproduction, 16:*366*
North American pronghorns, 12:40
North American red squirrels, 16:*127*, 16:*168*, 16:*173*
See also Red squirrels
North American sheep, 16:92
See also Sheep
North Atlantic right whales, 15:10, 15:107, 15:*108*, 15:*109*, 15:*111*, 15:*114*, 15:*115*, 15:116
North Chinese flying squirrels, 16:141*t*
See also Flying squirrels
North Indian muntjacs, 15:*345*
See also Muntjacs
North Moluccan flying foxes, 13:331*t*
North Pacific right whales, 15:107, 15:109, 15:111–112, 15:*114*, 15:116–*117*
See also Right whales
North Sea beaked whales. *See* Sowerby's beaked whales
Northern Bahian blond titis, 14:11, 14:148
Northern bats, 13:500
Northern bettongs, 13:74, 13:76, 13:77, 13:*79*
Northern bog lemmings, 16:*227*, 16:238*t*
Northern bottlenosed whales, 15:*60*, 15:*61*, 15:62, 15:*63*, 15:*64–65*
Northern brown bandicoots, 13:2, 13:4, 13:*6*, 13:*10*, 13:*11*, 13:*13*, 13:*14*–15
Northern brushtail possums, 13:66*t*
Northern cave-lions, 12:24
Northern chamois, 16:6, 16:87, 16:*96*, 16:*100*
Northern collared lemmings, 16:*226*, 16:*227*, 16:237*t*
Northern common cuscuses. *See* Common cuscuses
Northern elephant seals, 14:*425*, 14:431
behavior, 14:*421*, 14:*423*
distribution, 14:*429*
feeding ecology, 14:419
humans and, 14:424
lactation, 12:127
physical characteristics, 12:67
reproduction, 14:421–422
Northern flying squirrels, 16:135
See also Flying squirrels
Northern four-toothed whales. *See* Baird's beaked whales
Northern fur seals, 14:*401*, 14:*402*
conservation status, 12:*219*
distribution, 14:394, 14:395
evolution, 14:393
reproduction, 12:110, 14:396, 14:398, 14:399
Northern ghost bats, 13:*359*, 13:*360*, 13:362
Northern gliders, 13:133*t*
Northern gracile mouse opossums, 12:264*t*
Northern grasshopper mice, 16:*267*
Northern greater bushbabies, 14:25, 14:*25*, 14:27, 14:*29*, 14:31–32
Northern hairy-nosed wombats, 13:39, 13:40, 13:51, 13:52, 13:53, 13:*54*, 13:*55*, 13:56
Northern leaf-nosed bats, 13:*403*, 13:*404*
Northern lesser bushbabies. *See* Senegal bushbabies
Northern long-nosed armadillos. *See* Llanos long-nosed armadillos
Northern marsupial moles, 13:26, 13:*28*
Northern mastiff bats, 13:493*t*
Northern minke whales, 12:*200*, 15:120, 15:*122*, 15:125, 15:*126*, 15:128–129
Northern muriquis, 14:11, 14:156, 14:159, 14:*161*, 14:*164*, 14:165
Northern naked-tailed armadillos, 13:183, 13:190*t*
Northern needle-clawed bushbabies, 14:26, 14:32*t*
Northern pikas, 16:*493*, 16:*496*, 16:*498*–499
Northern plains gray langurs, 14:10, 14:*173*, 14:175, 14:*176*, 14:*180*
Northern planigales. *See* Long-tailed planigales
Northern pocket gophers, 16:*187*, 16:*189*, 16:195*t*
Northern pudus, 15:379, 15:396*t*
Northern quolls, 12:281, 12:282, 12:299*t*
Northern raccoons, 12:*189*, 14:*309*, 14:310, 14:*311*, 14:*312*, 14:*313*
Northern red-legged pademelons. *See* Red-legged pademelons
Northern right whale dolphins, 15:42
Northern river otters, 12:109, 14:*322*, 14:*323*
Northern sea elephants. *See* Northern elephant seals
Northern sea lions. *See* Steller sea lions
Northern short-tailed bats. *See* Lesser New Zealand short-tailed bats
Northern short-tailed shrews, 13:*195*, 13:*250*, 13:*253*, 13:*256–257*
Northern smooth-tailed tree shrews, 13:297*t*
Northern sportive lemurs, 14:*78*, 14:*83*
Northern tamanduas, 13:171, 13:*176*, 13:*178*
See also Anteaters
Northern three-toed jerboas. *See* Hairy-footed jerboas
Northern water shrews. *See* Eurasian water shrews
Northern water voles, 12:70–71, 12:182–183, 16:229, 16:*231*, 16:*233–234*
Norway lemmings, 16:*227*, 16:*231*, 16:*233*, 16:235
Norway rats. *See* Brown rats
Noses, 12:8, 12:80, 12:*81*
See also Physical characteristics
Nostrils, 12:8
See also Physical characteristics
Notharctus spp., 14:1
Nothrotheriops spp., 13:151
Notiomys edwardsii, 16:266
Notiosorex spp., 13:248, 13:250
Notiosorex crawfordi. *See* Desert shrews
Notiosoricini, 13:247
Notomys alexis. *See* Australian jumping mice
Notomys fuscus. *See* Dusky hopping mice
Notonycteris spp., 13:413
Notopteris macdonaldi. *See* Long-tailed fruit bats
Notoryctemorphia. *See* Marsupial moles
Notoryctes spp., 13:25, 13:26
Notoryctes caurinus. *See* Northern marsupial moles
Notoryctes typhlops. *See* Marsupial moles
Notoryctidae. *See* Marsupial moles
Notoungulata, 12:11, 12:132, 15:131, 15:133, 15:134, 15:135, 15:136, 15:137
Nourishment. *See* Feeding ecology
Novae's bald uakaris, 14:148, 14:151
NST (Non-shivering thermogenesis), 12:113
Nubian ibex, 15:146, 16:87, 16:91, 16:92, 16:94, 16:95
Nubian wild asses, 12:176–177
See also African wild asses
Nubra pikas, 16:502*t*
Nuciruptor spp., 14:143
Nuclear DNA (nDNA), 12:26–27, 12:29–30, 12:31, 12:33
Nucleotide bases, 12:26–27
Numbats, 12:137, 12:277–284, **12:303–306**, 12:*304*, 12:*305*, 12:*306*
Numbers, understanding, 12:155–156, 12:162
See also Cognition
Nutrias. *See* Coypus
Nutritional adaptations, **12:120–128**
Nutritional ecology. *See* Feeding ecology
Nyalas, 16:6, 16:11, 16:25*t*
Nyctalus spp. *See* Noctules
Nyctalus lasiopterus. *See* Greater noctules
Nyctalus leisleri. *See* Leisler's bats
Nyctalus noctula. *See* Noctules
Nyctereutes procyonoides. *See* Raccoon dogs
Nycteridae. *See* Slit-faced bats
Nycteris spp. *See* Slit-faced bats
Nycteris arge. *See* Bate's slit-faced bats
Nycteris avrita, 13:373
Nycteris gambiensis. *See* Gambian slit-faced bats
Nycteris grandis. *See* Large slit-faced bats
Nycteris hispida. *See* Hairy slit-faced bats
Nycteris intermedia. *See* Intermediate slit-faced bats
Nycteris javanica. *See* Javan slit-faced bats
Nycteris macrotis. *See* Large-eared slit-faced bats
Nycteris madagascarensis. *See* Madagascar slit-faced bats
Nycteris major. *See* Ja slit-faced bats
Nycteris thebaica. *See* Egyptian slit-faced bats
Nycteris woodi. *See* Wood's slit-faced bats
Nycticebus bengalensis. *See* Bengal slow lorises
Nycticebus coucang. *See* Sunda slow lorises
Nycticebus pygmaeus. *See* Pygmy slow lorises
Nycticeius humeralis. *See* Evening bats
Nyctimene spp., 13:335
Nyctimene rabori. *See* Philippine tube-nosed fruit bats
Nyctimene robinsoni. *See* Queensland tube-nosed bats
Nyctinomops femorosaccus. *See* Pocketed free-tailed bats

O

Obdurodon spp., 12:228
Obligatory vivipary, 12:6
See also Reproduction

Observational learning, 12:158
  See also Cognition
Oceanodroma macrodactyla. See Guadalupe storm-petrels
Ocelots, 14:374, 14:*378*, 14:382–*383*
Ochotona spp. *See* Pikas
Ochotona alpina. See Alpine pikas
Ochotona argentata. See Silver pikas
Ochotona cansus. See Gansu pikas
Ochotona collaris. See Collared pikas
Ochotona curzoniae. See Plateau pikas
Ochotona dauurica. See Daurian pikas
Ochotona erythrotis. See Chinese red pikas
Ochotona gaoligongensis. See Gaoligong pikas
Ochotona gloveri. See Glover's pikas
Ochotona himalayana. See Himalayan pikas
Ochotona hoffmanni, 16:495
Ochotona huangensis, 16:495
Ochotona hyperborea. See Northern pikas
Ochotona iliensis. See Ili pikas
Ochotona koslowi. See Koslov's pikas
Ochotona ladacensis. See Ladakh pikas
Ochotona macrotis. See Large-eared pikas
Ochotona nubrica. See Nubra pikas
Ochotona pallasi. See Pallas's pikas
Ochotona pallasi hamica, 16:500
Ochotona pallasi sunidica, 16:500
Ochotona princeps. See American pikas
Ochotona pusilla. See Steppe pikas
Ochotona rufescens. See Afghan pikas
Ochotona rutila. See Turkestan red pikas
Ochotona thibetana. See Moupin pikas
Ochotonidae. *See* Pikas
Ochre mole-rats, 16:350*t*
Ochrotomys nuttalli. See Golden mice
Ochrotomys torridus. See Southern grasshopper mice
Octodon spp., 16:433
Octodon bridgesi. See Bridges's degus
Octodon degus. See Degus
Octodon lunatus. See Moon-toothed degus
Octodon pacificus, 16:433, 16:436
Octodontidae. *See* Octodonts; Pikas
Octodontomys spp. *See* Chozchozes
Octodontomys gliroides. See Chozchozes
Octodonts, 12:72, **16:433–441,** 16:*437*, 16:440*t*
Octomys spp., 16:433, 16:436
Octomys mimax. See Viscacha rats
Odd-nosed leaf-monkeys, 14:172
Odd-toed ungulates. *See* Perissodactyla
Odobenidae. *See* Walruses
Odobenus rosmarus. See Walruses
Odobenus rosmarus rosmarus. See Atlantic walruses
Odocoileinae, 15:380–381, 15:382
Odocoileus spp., 12:102, 15:269, 15:380, 15:382, 15:383
Odocoileus bezoarticus. See Pampas deer
Odocoileus hemionus. See Black-tailed deer
Odocoileus virginianus. See White-tailed deer
Odocoileus virginianus clavium. See Key deer
Odocoilid deer. *See Odocoileus* spp.
Odontoceti. *See* Toothed whales
Oecomys spp., 16:266
Ogilby's duikers, 16:*78*, 16:*81*–82
Oilbirds, echolocation, 12:53
Oioceros spp., 15:265, 16:1
Okapia spp., 15:265
Okapia johnstoni. See Okapis
Okapis, 15:267, 15:269, **15:399–409,** 15:*404*, 15:*405*, 15:*406*, 15:*407*, 15:*408*
Olalla, Alfonso, 16:457
Olalla, Carlos, 16:457
Olalla, Manuel, 16:457
Olalla, Ramón, 16:457
Olalla, Rosalino, 16:457
Olallamys albicauda. See Montane bamboo rats
Olallamys edax, 16:450–451
Old English sheepdogs, 14:292
Old-field mice, 16:125
Old World bats, 12:13, 12:44
Old World deer, 12:19, **15:357–372,** 15:*362*, 15:*363*, 15:371*t*
Old World fruit bats, 12:13, 13:*310*, 13:311, **13:319–350,** 13:*323–324*, 13:*338*–339
  behavior, 13:312, 13:319–320, 13:335
  conservation status, 13:321–322, 13:337
  distribution, 13:*319*, 13:*333*, 13:334
  evolution, 13:319, 13:333
  feeding ecology, 13:320, 13:335–336
  habitats, 13:319, 13:334
  humans and, 13:322, 13:337
  physical characteristics, 12:54, 12:*54*, 12:58, 13:319, 13:333–334
  reproduction, 13:320–321, 13:336–337
  species of, 13:*325*–330, 13:331*t*–332*t*, 13:*341*–347, 13:348*t*–349*t*
  taxonomy, 13:319, 13:333
Old World harvest mice. *See* Harvest mice
Old World leaf-nosed bats, 13:310, **13:401–411,** 13:*406*
  behavior, 13:402, 13:405
  conservation status, 13:405
  distribution, 13:*401*, 13:402
  echolocation, 12:86
  evolution, 13:401
  feeding ecology, 13:405
  habitats, 13:402
  humans and, 13:405
  physical characteristics, 13:401–402
  reproduction, 13:405
  species of, 13:*407*–409, 13:409*t*–410*t*
  taxonomy, 13:401
Old World mice, **16:249–262,** 16:*254*, 16:261*t*–262*t*
Old World monkeys, **14:171–186,** 14:*176*, 14:*177*, **14:187–206,** 14:*195*, 14:*196*
  behavior, 14:6–7, 14:174, 14:191–192
  conservation status, 14:175, 14:194
  distribution, 12:135, 14:5, 14:*171*, 14:173, 14:*187*, 14:191
  evolution, 14:2, 14:171–172, 14:187–189
  feeding ecology, 14:9, 14:174, 14:192–193
  habitats, 14:6, 14:174, 14:191
  humans and, 14:175, 14:194
  physical characteristics, 14:172–173, 14:189–191
  reproduction, 14:10, 14:174–175, 14:193
  species of, 14:*178*–183, 14:184*t*–185*t*, 14:*197*–204, 14:204*t*–205*t*
  taxonomy, 14:1, 14:4, 14:171–172, 14:187–188
Old World pigs. *See* Pigs
Old World porcupines, 16:121, **16:351–364,** 16:*355*, 16:*356*
Old World rats, **16:249–262,** 16:*254*, 16:261*t*–262*t*
Old World sucker-footed bats, 12:136, 13:311, **13:479–481,** 13:*480*
Old World water shrews, 12:14
Olds, N., 15:177
Olfactory marking. *See* Scent marking
Olfactory system. *See* Smell
Oligoryzomys spp., 16:128, 16:265–266, 16:270
Oligoryzomys longicaudatus, 16:269
Oligoryzomys nigripes, 16:270
Olingos, 14:309, 14:310, 14:*311*, 14:316*t*
Olive baboons, 14:*6*, 14:*190*
Olive-backed pocket mice, 16:208*t*
Olive colobus, 14:173, 14:175, 14:*176*, 14:*178*, 14:179
Olympic marmots, 16:*148*
Omiltimi rabbits, 16:481
Ommatophoca rossii. See Ross seals
Omnivores
  digestive system, 12:124
  periglacial environments and, 12:25
  See also specific omnivores
Omomyiformes, 14:1, 14:3
Onagers. *See* Asiatic wild asses
Oncifelis colocolo. See Pampas cats
Oncifelis (Felis) geoffroyi. See Geoffroy's cats
Oncifelis geoffroyi. See Geoffroy's cats
Ondatra zibethicus. See Muskrats
One-humped camels. *See* Dromedary camels
One-male groups, 14:7
  See also Behavior
Onnoco agoutis, 16:415*t*
Ontogeny, 12:106, 12:141–143, 12:148
Onychogalea spp. *See* Nail-tailed wallabies
Onychogalea fraenata. See Bridled nail-tailed wallabies
Onychomys spp., 16:125
Onychomys leucogaster. *See* Northern grasshopper mice
Opossums, 12:*96*, 12:*143*, 12:*191*
  See also New World opossums
Orange leaf-nosed bats. *See* Golden horseshoe bats
Orange mouse opossums, 12:254, 12:265*t*
Orange nectar bats, 13:433*t*
Orange-rumped agoutis. *See* Red-rumped agoutis
Orange-spined hairy dwarf porcupines, 16:366, 16:374*t*
Orangutans, 14:7, 14:*227*, 14:*232*, 14:*234*
  behavior, 12:*156*, 12:157–158, 14:6, 14:228–229, 14:232
  Bornean, 14:225, 14:*236*, 14:*237*
  conservation status, 14:235
  distribution, 12:136, 14:5, 14:227
  evolution, 12:33, 14:225
  feeding ecology, 14:233
  habitats, 14:227–228
  language, 12:161
  learning, 12:151
  memory, 12:153
  numbers and, 12:155
  physical characteristics, 14:226–227, 14:*233*
  reproduction, 14:234, 14:235
  social cognition, 12:160
  taxonomy, 14:225
  tool use, 12:157
  See also Sumatran orangutans
Orcaella brevirostris. See Irrawaddy dolphins
Orcas. *See* Killer whales
Orchard dormice. *See* Garden dormice
Orcininae, 15:41
Orcinus orca. See Killer whales

Ordinal relationships, 12:155
Ord's kangaroo rats, 16:*200*, 16:*202*, 16:209*t*
Oreailurus jacobita. See Andean cats
Oreamnos spp., 16:87, 16:88
Oreamnos americanus. See Mountain goats
Oreodontae, 15:136, 15:264
Oreonax spp. *See* Yellow-tailed woolly monkeys
Oreonax flavicauda. See Yellow-tailed woolly monkeys
Oreotragus oreotragus. See Klipspringers
Oribis, 16:*60*, 16:62, 16:*65*, 16:*66*
Oriental civets, 14:345*t*
Oriental region, 12:134, 12:135, 12:136
Orizaba squirrels. *See* Deppe's squirrels
Ormalas. *See* Rufous hare-wallabies
Ornate shrews, 13:*251*
Ornate titis, 14:148
Ornate tree shrews. *See* Painted tree shrews
Ornithorhynchidae. *See* Duck-billed platypuses
Ornithorhynchus anatinus. See Duck-billed platypuses
Orrorin tugenensis, 14:241–242
Orthogeomys spp., 16:185
Orthogeomys cavator. See Chiriquá pocket gophers
Orthogeomys cherriei. See Cherrie's pocket gophers
Orthogeomys cuniculus, 16:190
Orthogeomys dariensis. See Darién pocket gophers
Orthogeomys grandis. See Large pocket gophers
Orthogeomys heterodus. See Variable pocket gophers
Orthogeomys hispidus. See Hispid pocket gophers
Orthogeomys lanius. See Big pocket gophers
Orthogeomys matagalpae. See Nicaraguan pocket gophers
Orthogeomys underwoodi. See Underwood's pocket gophers
Orycteropodidae. *See* Aardvarks
Orycteropus afer. See Aardvarks
Oryctolagus spp. *See* European rabbits
Oryctolagus cuniculus. See European rabbits
Oryctolagus domesticus. See Domestic rabbits
Oryx, 12:204, 15:*137*, 16:5, **16:27–43,** 16:41*t*–42*t*
Oryx dammah. See Scimitar-horned oryx
Oryx gazella. See Gemsboks; Oryx
Oryx leucoryx. See Arabian oryx
Oryzomyines, 16:264
Oryzomys spp. *See* Rice rats
Oryzomys bauri, 16:270
Oryzomys couesi, 16:268, 16:270
Oryzomys galapagoensis, 16:270
Oryzomys gorgasi, 16:270
Oryzomys intermedius, 16:267–268, 16:270
Oryzomys nelsoni, 16:270
Oryzomys palustris. See Marsh rice rats
Oryzomys russatus, 16:270
Oryzomys xantheolus, 16:269
Oryzorictes spp. *See* Rice tenrecs
Oryzorictes hova. See Rice tenrecs
Oryzorictes talpoides. See Mole-like rice tenrecs
Oryzorictinae. *See* Shrew tenrecs
os penis. *See* Baculum
Osbornictis spp., 14:335
Osbornictis piscivora. See Aquatic genets
Otaria byronia. See South American sea lions
Otariidae, **14:393–408,** 14:*401*
  behavior, 14:396–397
  conservation status, 14:399–400
  distribution, 14:*393*, 14:394–395
  evolution, 14:393
  feeding ecology, 14:397–398
  habitats, 14:395–396
  humans and, 14:400
  physical characteristics, 14:393–394, 14:*419*
  reproduction, 14:398–399
  species of, 14:*402*–406, 14:407*t*–408*t*
  taxonomy, 14:256, 14:393
Otariinae. *See* Sea lions
Otocolobus manul. See Pallas's cats
Otocyon megalotis. See Bat-eared foxes
Otolemur spp., 14:23
Otolemur crassicaudatus. See Brown greater bushbabies
Otolemur garnettii. See Northern greater bushbabies
Otolemur monteiri. See Silvery greater bushbabies
Otomops martiensseni. See Giant mastiff bats
Otomops papuensis. See Big-eared mastiff bats
Otomops wroughtoni. See Wroughton's free-tailed bats
Otomyinae, 16:283
Otomys angoniensis. See Angoni vlei rats
Otomys sloggetti. See Ice rats
Ototylomys phyllotis. See Big-eared climbing rats
Otter civets, 14:335, 14:337, 14:338, 14:*339*, 14:*342*, 14:343
Otter shrews, 13:194, 13:199, 13:225
Otters, 12:14, **14:319–325,** 14:*326*, **14:330, 14:332–334**
  behavior, 14:321–323
  conservation status, 14:324
  distribution, 14:*319*, 14:321
  evolution, 14:319
  feeding ecology, 14:260, 14:323
  field studies of, 12:201
  habitats, 14:255, 14:257, 14:321
  humans and, 14:324–325
  physical characteristics, 14:319–321
  reproduction, 14:324
  species of, 14:*330*, 14:332*t*
  taxonomy, 14:256, 14:319
Ottoia spp., 12:64
Ounce. *See* Snow leopards
Ourebia ourebi. See Oribis
Ovarian cycles, 12:91–92
Ovibos moschatus. See Muskoxen
Ovibovini, 16:87, 16:89
Oviparous reproduction, 12:106
  See also Reproduction
Ovis spp., 16:3, 16:4, 16:87, 16:88, 16:90, 16:91, 16:95
Ovis ammon. See Argalis
Ovis ammon hodgsoni. See Tibetan argalis
Ovis ammon musimon. See European mouflons
Ovis aries. See Domestic sheep
Ovis canadensis. See Bighorn sheep
Ovis dalli. See Dall's sheep
Ovis musimon. See Mouflons
Ovis nivicola. See Snow sheep
Ovis orientalis. See Asiatic mouflons; Urials
Ovis vignei. See Urials
Ovulation, 12:92, 12:109
Ovum. *See* Eggs, mammalian
Owen, R. D., 13:389
Owl monkeys. *See* Night monkeys
Owston's palm civets, 14:338, 14:344*t*
Oxen, 15:*149*
Oxygen, subterranean mammals and, 12:72–73, 12:114
Oxymycterus josei, 16:263
Oxymycterus nasutus, 16:267, 16:268
Oxytocin, 12:141
Oyans. *See* African linsangs
Ozotocerus spp., 15:382
Ozotocerus bezoarticus. See Pampas deer

P

Pacaranas, 16:123, **16:385–388,** 16:*386*, 16:*387*
Pacas, 16:124, **16:417–424,** 16:*418*, 16:*419*, 16:*420*, 16:*422*, 16:*423*
Pachycrocuta spp., 14:359
Pachyderms, 12:91
  See also Elephants; Rhinoceroses
Pachygazella, 16:88
Pachyporlax spp., 15:265
Pachyuromys dupras. See Fat-tailed gerbils
Pacific flying foxes. *See* Tongan flying foxes
Pacific rats. *See* Polynesian rats
Pacific sheath-tailed bats, 13:358, 13:364*t*
Pacific walruses, 14:*409*–412, 14:*411*, 14:*414*, 14:415
Pacific white-sided dolphins, 12:*63*, 15:*10*, 15:49
Packrats. *See* Woodrats
Pademelons, 13:34, 13:35, 13:83, 13:85, 13:86, 13:92, 13:97–*98*
Paenungulata, 15:131, 15:133
Pagophilus groenlandicus. See Harp seals
Paguma larvata. See Masked palm civets
Painted bats, 13:*523*, 13:*525*
Painted quaggas. *See* Plains zebras
Painted tree shrews, 13:298*t*
Painted wolves. *See* African wild dogs
Pairs, 12:145–146
  See also Behavior
Pakicetidae, 15:2
Pakicetus spp., 15:2–3, 15:119
Palaeanodonts, 15:135
Palaearctic region, 12:131–132, 12:134–135, 12:136
Palaeomerycids, 15:137, 15:265, 15:343
Palaeomastodon, 15:161
Palaeopropithecus ingens. See Greater sloth lemurs
Palaeotragus spp., 15:265
Palaeotupaia spp., 13:291
Palawan porcupines. *See* Indonesian porcupines
Palawan stink badgers. *See* Philippine stink badgers
Palawan tree shrews, 13:292, 13:298*t*
Pale field rats. *See* Tunney's rats
Pale foxes, 14:265
Pale hedgehogs. *See* Indian desert hedgehogs
Pale kangaroo mice, 16:200, 16:*204*, 16:*206*–207
Pale spear-nosed bats, 13:418, 13:433*t*

Pale-throated three-toed sloths, 13:*162*, 13:*164*, 13:*165*, 13:*166*
Paleaopropithecus spp., 14:63
Paleochoerus spp., 15:265
Paleodonta, 15:264
Paleolaginae, 16:505
Paleomeryx spp., 15:265
Paleopropithecinae. *See* Sloth lemurs
Paleotheres, 15:136
Paleotraginae, 15:399
Palestine mole rats, 16:*287*, 16:*289*, 16:294
See also Blind mole-rats
Pallas's cats, 14:391*t*
Pallas's long-tongued bats, 13:*415*, 13:417, 13:*422*, 13:*427*
Pallas's mastiff bats, 13:486, 13:494*t*
Pallas's pikas, 16:488, 16:*492*, 16:*496*, 16:*499*–500
Pallas's squirrels, 16:*168*, 16:*173*–174
Pallid bats, 13:*308*, 13:313, 13:497, 13:*500*, 13:*505*, 13:*506*
Palm civets, 14:258, 14:260, 14:335, 14:337, 14:344*t*
Pampas cats, 14:391*t*
Pampas deer, 15:379, 15:*381*, 15:383, 15:384, 15:396*t*
Pampas foxes, 14:284*t*
Pampatheres. *See* Giant herbivorous armadillos
Pan paniscus. *See* Bonobos
Pan troglodytes. *See* Chimpanzees
Pan troglodytes schweinfurthii, 14:225
Pan troglodytes troglodytes, 14:225
Pan troglodytes vellerosus, 14:225
Pan troglodytes verus, 14:225
Panama Canal, 12:220
Panamanian squirrel monkeys. *See* Red-backed squirrel monkeys
Pandas, red, 12:*28*, 14:309–*311*, 14:*312*, 14:*315*
See also Giant pandas
Pangolins, 12:135, 13:196, **16:107–120**, 16:*108*, 16:*109*, 16:*111*, 16:*114*
Panthera spp., 14:369
Panthera leo. *See* Lions
Panthera onca. *See* Jaguars
Panthera pardus. *See* Leopards
Panthera tigris. *See* Tigers
Panthera tigris altaica. *See* Amur tigers
Pantherinae, 14:369
Panthers. *See* Pumas
Pantholops hodgsonii. *See* Tibetan antelopes
Pantodonta, 12:11, 15:131, 15:134, 15:135, 15:136
Papillose bats, 13:526*t*
Papio spp. *See* Baboons
Papio anubis. *See* Olive baboons
Papio cynocephalus. *See* Yellow baboons
Papio hamadryas. *See* Hamadryas baboons
Papio hamadryas anubis. *See* Olive baboons
Papio papio. *See* Guinea baboons
Papio ursinus. *See* Chacma baboons
Papionini, 14:188
Pappogeomys spp., 16:185
Pappogeomys alcorni, 16:190
Pappogeomys bulleri. *See* Buller's pocket gophers
Papuan bandicoots, 13:10, 13:12, 13:17*t*
Papuan forest wallabies, 13:*91*, 13:*99*, 13:100–101
Parabos spp., 15:265, 16:1
Paraceratherium [Indricotherium] ransouralicum, 15:138
Paracolobus spp., 14:172
Paracosoryx prodromus, 15:265
Paracrocidura spp., 13:265
Paracrocidura graueri. *See* Grauer's shrews
Paracrocidura maxima. *See* Greater shrews
Paracrocidura schoutedeni. *See* Schouteden's shrews
Paracynictis spp., 14:347
Paradiceros mukiri, 15:249
Paradipodinae, 16:211, 16:212
Paradipus spp., 16:211
Paradipus ctenodactylus. *See* Comb-toed jerboas
Paradoxurinae, 14:335
Paradoxurus spp., 14:336
Paradoxurus hermaphroditus. *See* Palm civets
Paradoxurus jerdoni. *See* Jerdon's palm civets
Paraechinus aethiopicus. *See* Ethiopian hedgehogs
Paraechinus hypomelas. *See* Brandt's hedgehogs
Paraechinus micropus. *See* Indian desert hedgehogs
Paraguay hairy dwarf porcupines, 16:366, 16:367, 16:368, 16:374*t*
Parahyaena brunnea. *See* Brown hyenas
Paralouatta spp., 14:143
Paramyidae, 16:122
Parantechinus spp. *See* Dibblers
Parantechinus apicalis. *See* Southern dibblers
Parantechinus bilarni. *See* Sandstone dibblers
Paranyctimene spp., 13:334
Paranyctoides spp., 13:193
Parapedetes gracilis, 16:307
Parapedetes namaquensis, 16:307
Parapontoporia spp., 15:19, 15:23
Parascalops breweri. *See* Hairy-tailed moles
Parascaptor spp., 13:279
Parasites
Ice Age mammals and, 12:25
subterranean mammals and, 12:77–78
Paratopithecus brasiliensis, 14:155
Paraxerus spp., 16:163
Paraxerus alexandri, 16:167
Paraxerus cepapi. *See* Smith's tree squirrels
Paraxerus cooperi, 16:167
Paraxerus palliatus, 16:167
Paraxerus vexillarius, 16:167
Paraxerus vincenti, 16:167
Paraxonia, 15:131
Pardel lynx. *See* Iberian lynx
Pardiñas, U. F. J., 16:264, 16:266
Parental care. *See* Reproduction
Pariah dogs, 12:173
Parka squirrels. *See* Arctic ground squirrels
Parma wallabies, 13:40, 13:*87*, 13:*91*, 13:*94*–95
Parnell's moustached bats, 12:86, 13:437, 13:*438*, 13:*439*, 13:*441*, 13:*442*–443
Parti-colored bats, 13:516*t*
Patagium. *See* Flight membranes
Patagonian cavies. *See* Maras
Patagonian hares. *See* Maras
Patagonian opossums, 12:249–253, 12:*255*, 12:*258*, 12:259
Patagonian weasels, 14:332*t*
Patas monkeys, 12:146, 14:7, 14:10, 14:188, 14:191, 14:192, 14:*196*, 14:197–*198*
Paternal care, 12:96, 12:107
See also Reproduction
Paternity, sperm competition and, 12:105–106
See also Reproduction
Patterns, hair, 12:38
Patterson, B. D., 16:266
Patterson, Francine, 12:162
Patton, J. L., 16:263
Paucituberculata. *See* Shrew opossums
PCR. *See* Polymerase chain reactions
Peak-saddle horseshoe bats. *See* Blasius's horseshoe bats
Pearson, O. P., 16:267, 16:269
Pearsonomys annectens, 16:266
Pearson's long-clawed shrews, 13:*270*, 13:*272*, 13:274
Pearson's tuco-tucos, 16:*428*, 16:*429*
Pecari spp., 15:267
Peccaries, **15:291–300**, 15:*297*
behavior, 15:293–294
conservation status, 15:295–296
distribution, 15:*291*, 15:292
evolution, 15:137, 15:264, 15:265, 15:291
feeding ecology, 15:141, 15:271, 15:294–295
habitats, 15:292–293
humans and, 15:296
physical characteristics, 15:140, 15:291–292
reproduction, 15:295
species of, 15:298–300
taxonomy, 15:291
Pecoran ruminants, 15:136
Pectinator spp., 16:311
Pectinator spekei. *See* Speke's pectinators
Pectoral girdle, 12:8, 12:41
See also Physical characteristics
Pedetes capensis. *See* Springhares
Pedetidae. *See* Springhares
Pediolagus spp., 16:389, 16:390
Pediolagus salinicola. *See* Salt-desert cavies
Pekinese, 14:289
Pelage. *See* Hairs
Pelea spp., 16:28
Pelea capreolus. *See* Gray rheboks
Peleinae, 16:1
Pelomys fallax. *See* Creek rats
Pel's anomalures, 16:300, 16:*301*, 16:303, 16:*305*
Pel's scaletails. *See* Pel's anomalures
Pelvic bones, 12:8
See also Physical characteristics
Pen-tailed tree shrews, 13:291, 13:*293*, 13:*294*, 13:296
Penant's red colobus, 14:185*t*
Peninsular pronghorns, 15:415
Penises, 12:91, 12:104, 12:105, 12:111
See also Physical characteristics
Pentadactyly, 12:42
See also Physical characteristics
Pentalagus spp. *See* Amami rabbits
Pentalagus furnessi. *See* Amami rabbits
Penultimate Glaciation, 12:20–21
People for the Ethical Treatment of Animals (PETA), 14:294
Peponocephala electra. *See* Melon-headed whales
Peramelemorphia. *See* Bandicoots
Perameles spp. *See* Bandicoots
Perameles bougainville. *See* Western barred bandicoots
Perameles eremiana. *See* Desert bandicoots

Perameles gunnii. See Eastern barred bandicoots
Perameles nasuta. See Long-nosed bandicoots
Peramelidae. *See* Dry-country bandicoots
Peramelinae, 13:9
Perameloidea, 13:9
Percheron horses, 12:176
Perchoerus spp., 15:265
Percival's trident bats, 13:*402*, 13:410*t*
Pere David's deer, 15:*267*, 15:358, 15:360, 15:*362*, 15:366–367
Peres, C. A., 14:165
Perikoala palankarinnica, 13:43
Periptychoidea, 15:134
Perissodactyla, 15:131–132, 15:133, 15:135–141, **15:215–224**
  digestive systems of, 12:14–15, 12:121
  distribution, 12:132
  neonatal milk, 12:127
  vision, 12:79
Perodicticus potto. See Pottos
Perodictinae, 14:13
Perognathinae, 16:199
Perognathus spp., 16:199, 16:200, 16:203
Perognathus fasciatus. See Olive-backed pocket mice
Perognathus flavus. See Silky pocket mice
Perognathus inornatus. See San Joaquin pocket mice
Perognathus longimembris. See Little pocket mice
Perognathus longimembris pacificus, 16:203
Perognathus parvus. See Great Basin pocket mice
Peromyscus attwateri. See Texas mice
Peromyscus boylii. See Brush mice
Peromyscus californicus. See California mice
Peromyscus eremicus. See Cactus mice
Peromyscus leucopus. See White-footed mice
Peromyscus maniculatus. See Deer mice
Peromyscus polinotus. See Old-field mice
Peropteryx kappleri. See Greater dog-faced bats
Peropteryx macrotis. See Lesser dog-faced bats
Peroryctes spp., 13:1–2
Peroryctes broadbenti. See Giant bandicoots
Peroryctes longicauda. See Striped bandicoots
Peroryctes papuensis. See Papuan bandicoots
Peroryctes raffrayana. See Raffray's bandicoots
Peroryctidae. *See* Rainforest bandicoots
Perrin's beaked whales, 15:1, 15:61, 15:70*t*
Persian fallow deer. *See* Fallow deer
Persian gazelles, 16:48, 16:57*t*
Persian moles, 13:282
Persian onagers, 15:*219*, 15:222
Persian squirrels, 16:167, 16:175*t*
Persian trident bats, 13:409*t*
Peruvian beaked whales. *See* Stejneger's beaked whales
Peruvian crevice-dwelling bats, 13:519, 13:520, 13:521, 13:526*t*
Peruvian guemals. *See* Peruvian huemuls
Peruvian huemuls, 15:379, 15:384, 15:396*t*
Peruvian mountain viscachas. *See* Mountain viscachas
Peruvian spider monkeys, 14:*161*, 14:*163*–164
Pest control, rabbits, 12:186–188
  See also specific animals
Pests, mammalian, **12:182–193**
PETA (People for the Ethical Treatment of Animals), 14:294
Petauridae, 13:35, **13:125–133,** 13:*130*, 13:133*t*
Petaurillus hosei. See Hose's pygmy flying squirrels
Petaurinae, 13:125, 13:126, 13:127–128
Petaurirodea, 13:113
Petaurista spp., 16:135
Petaurista petaurista. See Red giant flying squirrels
Petauroidea, 13:32, 13:34, 13:38
Petauroides spp., 13:113, 13:114, 13:116
Petauroides volans. See Greater gliders
Petaurus spp., 13:113, 13:125, 13:126–127, 13:128
Petaurus abidi. See Northern gliders
Petaurus australis. See Yellow-bellied gliders
Petaurus breviceps. See Sugar gliders
Petaurus gracilis. See Mahogany gliders
Petaurus norfolcensis. See Squirrel gliders
Peters's disk-winged bats, 13:*475*, 13:*476*–477
Peters's duikers, 16:84*t*
Peters's dwarf epauletted fruit bats. *See* Dwarf epauletted fruit bats
Peters's sac-winged bats. *See* Lesser dog-faced bats
Peters's squirrels, 16:*168*, 16:*171*, 16:172
Petinomys spp., 16:136
Petinomys genibarbis. See Whiskered flying squirrels
Petrodromus spp., 16:518
Petrodromus tetradactylus. See Four-toed sengis
Petrogale spp. *See* Rock wallabies
Petrogale assimilis. See Allied rock-wallabies
Petrogale concinna. See Narbaleks
Petrogale penicillata. See Brush-tailed rock wallabies
Petrogale persephone. See Proserpine rock wallabies
Petrogale xanthopus. See Yellow-footed rock wallabies
Petromuridae. *See* Dassie rats
Petromus typicus. See Dassie rats
Petromyscinae, 16:282, 16:285
Petromyscus collinus. See Pygmy rock mice
Petropseudes spp., 13:113
Petropseudes dahli. See Rock ringtails
Petrosum, 12:9
Pets, 12:15, 12:173
  cats and dogs as, 14:291
  humans and, 14:*291*
  See also specific animals
Pettigrew, Jack, 13:333
Pfungst, Oskar, 12:155
Phacochoerinae, 15:275
Phacochoerus spp., 15:276
Phacochoerus aethiopicus. See Warthogs
Phacochoerus aethiopicus delamerei. See Warthogs
Phacochoerus africanus. See Common warthogs
Phaenomys spp., 16:264
Phaenomys ferrugineus. See Rio de Janeiro rice rats
Phalanger spp., 13:57–58
Phalanger alexandrae. See Gebe cuscuses
Phalanger gymnotis. See Ground cuscuses
Phalanger lullulae. See Woodlark Island cuscuses
Phalanger maculatus. See Spotted cuscuses
Phalanger matanim. See Telefomin cuscuses
Phalanger orientalis. See Common cuscuses
Phalanger ornatus. See Moluccan cuscuses
Phalanger sericeus. See Silky cuscuses
Phalangeridae, 13:31, 13:32, **13:57–67,** 13:*62*, 13:66*t*
Phalangeroidea, 13:31
Phaner spp. *See* Masoala fork-crowned lemurs
Phaner furcifer. See Masoala fork-crowned lemurs
Phascalonus gigas, 13:51
Phascogale calura. See Red-tailed phascogales
Phascogale tapoatafa. See Brush-tailed phascogales
Phascogales, 12:281–282, 12:284, 12:*289*, 12:290
Phascolarctidae. *See* Koalas
Phascolarctos cinereus. See Koalas
Phascolarctos cinereus adustus. See Queensland koalas
Phascolarctos cinereus cinereus, 13:43
Phascolarctos cinereus victor, 13:43
Phascolarctos stirtoni, 13:43
Phascolosorex spp., 12:288, 12:290
Phascolosorex dorsalis. See Narrow-striped marsupial shrews
Phenacodontidae, 15:132
Phenacomys intermedius. See Western heather voles
Pheromones, 12:37, 12:80
  See also Physical characteristics
Philadelphia Zoo, 12:203
Philander spp. *See* Gray four-eyed opossums
Philander andersoni. See Black four-eyed opossums
Philander opossum. See Gray four-eyed opossums
Philantomba spp., 16:73
Philippine badgers. *See* Philippine stink badgers
Philippine brown deer, 15:371*t*
Philippine colugos, 13:302, 13:*303*, 13:*304*
Philippine flying lemurs. *See* Philippine colugos
Philippine gymnures, 13:207
Philippine sambars. *See* Philippine brown deer
Philippine spotted deer. *See* Visayan spotted deer
Philippine stink badgers, 14:333*t*
Philippine tarsiers, 14:*92*, 14:*93*, 14:94, 14:*95*, 14:*96*, 14:*97*
Philippine tree shrews, 13:292, 13:*293*, 13:295–*296*
Philippine tube-nosed fruit bats, 13:*334*
Philippine warty pigs, 12:179, 15:280, 15:290*t*
Philippine wood shrews. *See* Mindanao gymnures
Philippines. *See* Oriental region
Phiomia, 15:161
Phlitrum, 12:8
Phoca caspica. See Caspian seals
Phoca hispida. See Ring seals
Phoca sibirica. See Baikal seals
Phoca vitulina. See Harbor seals
Phocarctos hookeri. See Hooker's sea lions
Phocidae. *See* True seals
Phocinae, 14:417
Phocoena spp. *See* Porpoises
Phocoena dalli. See Dall's porpoises
Phocoena dioptrica. See Spectacled porpoises
Phocoena phocoena. See Harbor porpoises
Phocoena sinus. See Vaquitas

Phocoena spinipinnis. See Burmeister's porpoises
Phocoenidae. *See* Porpoises
Phocoenoides spp., 15:33
Phodopus spp. *See* Dwarf hamsters
Phodopus sungorus. See Dzhungarian hamsters
Phoenix Zoo, 12:*209*
Pholidota. *See* Pangolins
Photoperiodicity, 12:74–75
Phyllonycterinae, 13:413, 13:415
Phyllonycteris aphylla, 13:420
Phyllonycteris poeyi. See Cuban flower bats
Phyllostomidae. *See* American leaf-nosed bats
Phyllostominae, 13:413, 13:414, 13:416, 13:417, 13:420
Phyllostomus discolor. See Pale spear-nosed bats
Phyllostomus hastatus. See Greater spear-nosed bats
Phyllotines, 16:264
Phyllotis spp., 16:265
Phyllotis darwini. See Leaf-eared mice
Phyllotis sublimis, 16:267
Phylogenetics
  cognition and, 12:149–150
  molecular genetics and, **12:26–35**
  social behavior and, 12:147–148
Physeter spp., 15:73, 15:75
Physeter macrocephalus. See Sperm whales
Physeteridae. *See* Sperm whales
Physical characteristics, **12:36–51**
  aardvarks, 15:155–156
  agoutis, 16:407, 16:411–414, 16:415*t*
  anteaters, 12:39, 12:46, 13:171–172, 13:177–179
  armadillos, 13:182, 13:187–190, 13:190*t*–192*t*
  Artiodactyla, 12:40, 12:79, 15:138–140, 15:267–269
  aye-ayes, 14:85, 14:*86*
  baijis, 15:19–20
  bandicoots, 13:1–2
    dry-country, 13:9–10, 13:14–16, 13:16*t*–17*t*
    rainforest, 13:9–10, 13:14–17*t*
  bats, 12:56–60, 13:307–308, 13:310
    American leaf-nosed, 13:413–415, 13:*414*, 13:425–434, 13:433*t*–434*t*
    bulldog, 13:443, 13:445, 13:449–450
    disk-winged, 13:473, 13:476–477
    Emballonuridae, 13:355, 13:360–363, 13:363*t*–364*t*
    false vampire, 13:310, 13:379, 13:384–385
    funnel-eared, 13:459–*460*, 13:*461*, 13:463–465
    horseshoe, 13:387–388, 13:396–399
    Kitti's hog-nosed, 13:367–368
    Molossidae, 13:483, 13:490–493, 13:493*t*–495*t*
    mouse-tailed, 13:351, 13:353
    moustached, 13:435, 13:440–441, 13:442*t*
    New Zealand short-tailed, 13:453, 13:457
    Old World fruit, 13:319, 13:325–330, 13:331*t*–332*t*, 13:333–334, 13:340–347, 13:348*t*–349*t*
    Old World sucker-footed, 13:479
    slit-faced, 13:371, 13:*372*, 13:375, 13:376*t*–377*t*
    smoky, 13:467, 13:470
    Vespertilionidae, 13:497–498, 13:506–514, 13:515*t*–516*t*, 13:519–520, 13:524–525, 13:526*t*
  bats Old World leaf-nosed, 13:401–402, 13:407–409, 13:409*t*–410*t*
  bears, 14:296, 14:302–305, 14:306*t*
  beavers, 12:111, 16:177–179, 16:183
  bilbies, 13:19
  botos, 15:27–28
  Bovidae, 16:2–4
    Antilopinae, 16:45, 16:50–55, 16:56*t*–57*t*
    Bovinae, 16:18–23, 16:24*t*–25*t*
    Caprinae, 16:89–90, 16:98–103, 16:103*t*–104*t*
    duikers, 16:73, 16:80–84, 16:84*t*–85*t*
    Hippotraginae, 16:28, 16:37–42, 16:42*t*–43*t*
    Neotraginae, 16:59, 16:66–71, 16:71*t*
  bushbabies, 14:23–24, 14:28–32, 14:32*t*–33*t*
  Camelidae, 15:314–315, 15:320–323
  Canidae, 14:266–267, 14:276–282, 14:283*t*–284*t*
  capybaras, 16:401–402
  Carnivora, 12:14, 12:79, 12:82, 12:120, 12:123–124, 14:256–257
  cats, 12:84, 12:120, 14:370, 14:379–389, 14:390*t*–391*t*
  Caviidae, 16:390, 16:395–398, 16:399*t*
  Cetacea, 12:44, 12:67, 12:68, 15:4–5
    Balaenidae, 15:4, 15:107, 15:109, 15:115–117
    beaked whales, 15:59, 15:61, 15:64–68, 15:69*t*–70*t*
    bottlenosed dolphins, 12:67–68
    dolphins, 12:66, 15:42–43, 15:53–55, 15:56*t*–57*t*
    franciscana dolphins, 15:23–24
    Ganges and Indus dolphins, 15:13–14
    gray whales, 15:93–94
    Monodontidae, 15:81–82, 15:90
    porpoises, 15:33–34, 15:38–39, 15:39*t*
    pygmy right whales, 15:103–104
    rorquals, 12:12, 12:63, 12:79, 12:82, 12:213, 15:4, 15:119–120, 15:127–130, 15:130*t*
    sperm whales, 15:73, 15:79–80
  chevrotains, 15:326–327, 15:331–333
  Chinchillidae, 16:377, 16:382–383, 16:383*t*
  colugos, 13:*300*–301, 13:304
  coypus, 16:123, 16:473–474
  dasyurids, 12:287–288, 12:294–298, 12:299*t*–301*t*
  Dasyuromorphia, 12:277–279
  deer
    Chinese water, 15:373–374
    muntjacs, 15:343–344, 15:349–354
    musk, 15:335, 15:340–341
    New World, 15:382, 15:388–396, 15:396*t*
    Old World, 15:359, 15:364–371, 15:371*t*
  Dipodidae, 16:211–212, 16:219–222, 16:223*t*–224*t*
  Diprotodontia, 13:32–33
  dormice, 16:317, 16:323–326, 16:327*t*–328*t*
  duck-billed platypuses, 12:228–230, 12:243–244
  Dugongidae, 15:199, 15:203
  echidnas, 12:235–236, 12:240*t*
  elephants, 12:11–12, 12:79, 12:82, 12:83, 15:165–166, 15:174–175
  Equidae, 15:225–226, 15:232–235, 15:236*t*
  Erinaceidae, 13:203, 13:209–213, 13:212*t*–213*t*
  giant hutias, 16:469, 16:471*t*
  gibbons, 14:209–210, 14:218–222
  Giraffidae, 12:40, 12:102, 15:399–401, 15:408
  great apes, 14:225–227, 14:237–239
  gundis, 16:311–*312*, 16:314–315
  Herpestidae, 14:347–348, 14:354–356, 14:357*t*
  Heteromyidae, 16:199, 16:205–209, 16:208*t*–209*t*
  hippopotamuses, 12:68, 15:304–306, 15:310–311
  humans, 12:23–24, 14:244–247
  hutias, 16:461, 16:467*t*
  Hyaenidae, 14:360, 14:364–366
  hyraxes, 15:178–179, 15:186–188, 15:189*t*
  Indriidae, 14:63–65, 14:69–72
  Insectivora, 13:194–197
  koalas, 13:43–44
  Lagomorpha, 16:480–481
  lemurs, 14:48–49, 14:55–60
    Cheirogaleidae, 14:35, 14:41–44
    sportive, 14:75, 14:79–83
  Leporidae, 12:46, 12:121, 12:122, 16:505, 16:511–514, 16:515*t*–516*t*
  Lorisidae, 14:13–14, 14:18–20, 14:21*t*
  Macropodidae, 13:32, 13:33, 13:83–84, 13:93–100, 13:101*t*–102*t*
  manatees, 15:205, 15:211–212
  marine mammals, 12:63, 12:*63*, 12:*64*, 12:*65*, 12:65–67, 12:*66*, 12:*67*
  Megalonychidae, 13:155, 13:159
  moles
    golden, 13:195–197, 13:216, 13:220–222, 13:222*t*
    marsupial, 13:25–26, 13:28
  monitos del monte, 12:273–274
  monkeys
    Atelidae, 14:155–156, 14:162–166, 14:166*t*–167*t*
    Callitrichidae, 12:39, 14:115–116, 14:127–131, 14:132*t*
    Cebidae, 14:101–102, 14:108–112
    cheek-pouched, 14:189–191, 14:197–204, 14:204*t*–205*t*
    leaf-monkeys, 14:172–173, 14:178–183, 14:184*t*–185*t*
    night, 14:135, 14:137, 14:141*t*
    Pitheciidae, 14:143–144, 14:150–152, 14:153*t*
  monotremes, 12:38, 12:228–230
  mountain beavers, 16:131
  Muridae, 12:79, 16:123, 16:283, 16:288–295, 16:296*t*–297*t*
    Arvicolinae, 16:225–226, 16:233–238, 16:237*t*–238*t*
    hamsters, 16:239, 16:245–246, 16:247*t*
    Murinae, 16:249–250, 16:255–260, 16:261*t*–262*t*
    Sigmodontinae, 16:264–265, 16:272–275, 16:277*t*–278*t*
  musky rat-kangaroos, 13:70
  Mustelidae, 12:37, 12:38, 14:319–321, 14:327–331, 14:332*t*–333*t*
  numbats, 12:303
  octodonts, 16:434, 16:438–440, 16:440*t*
  opossums
    New World, 12:250, 12:257–263, 12:264*t*–265*t*
    shrew, 12:267, 12:270

Physical characteristics *(continued)*
Otariidae, 14:393–394, 14:402–406
pacaranas, 16:386
pacas, 16:417–418, 16:423–424
pangolins, 16:107–110, 16:115–119
peccaries, 15:291–292, 15:298–299
Perissodactyla, 15:216–217
Petauridae, 13:125, 13:131–133, 13:133*t*
Phalangeridae, 13:57, 13:*63–66t*
pigs, 15:275–276, 15:284–288, 15:289*t*–290*t*
pikas, 16:491, 16:497–501, 16:501*t*–502*t*
pocket gophers, 16:186–187, 16:192–194, 16:195*t*–197*t*
porcupines
New World, 16:366–*367*, 16:371–373, 16:374*t*
Old World, 16:351–352, 16:357–363
possums
feather-tailed, 13:140, 13:144
honey, 13:135–136
pygmy, 13:106, 13:110–111
primates, 14:4–5
Procyonidae, 14:309, 14:313–315, 14:315*t*–316*t*
pronghorns, 15:411–412
Pseudocheiridae, 13:113–114, 13:113–123, 13:119–123, 13:122*t*–123*t*
rat-kangaroos, 13:73–74, 13:78–80
rats
African mole-rats, 12:71, 12:72, 12:72–75, 12:73, 12:76–77, 12:78, 12:*86*, 16:339–342, 16:*341*, 16:347–349, 16:349*t*–350*t*
cane, 16:333–334, 16:337
chinchilla, 16:443–444, 16:447–448
dassie, 16:329–330
spiny, 16:*449*–450, 16:453–457, 16:458*t*
rhinoceroses, 15:251–252, 15:257–262
rodents, 16:122–123
sengis, 16:518–519, 16:524–529, 16:531*t*
shrews
red-toothed, 13:247–248, 13:255–263, 13:263*t*
West Indian, 13:243–244
white-toothed, 13:265–266, 13:271–275, 13:276*t*–277*t*
Sirenia, 15:192
size, 12:11–12
solenodons, 13:237–238, 13:240
springhares, 16:307–308
squirrels
flying, 16:135, 16:139–140, 16:141*t*–142*t*
ground, 12:113, 16:143–144, 16:151–158, 16:158*t*–160*t*
scaly-tailed, 16:299–300, 16:302–305
tree, 16:163, 16:169–174*t*, 16:175*t*
subterranean mammals, 12:71–78
Talpidae, 13:279–280, 13:284–287, 13:287*t*–288*t*
tapirs, 15:238, 15:245–247
tarsiers, 14:92–93, 14:97–99
tenrecs, 13:225–226, 13:232–234, 13:234*t*
three-toed tree sloths, 13:161–162, 13:167–169
tree shrews, 13:291, 13:294–296, 13:297*t*–298*t*
true seals, 14:256, 14:417, 14:*419*, 14:427–432, 14:434, 14:435*t*
tuco-tucos, 16:425, 16:429–430, 16:430*t*–431*t*
ungulates, 15:138–141, 15:141–142
Viverridae, 14:335–336, 14:340–343, 14:344*t*–345*t*
walruses, 12:46, 12:68, 14:409–*410*, 14:*413*
whales, 12:67, 12:68
wombats, 13:51, 13:55–56
Xenartha, 13:149
Physics, of powered flight, 12:55–56
Piaggo, A. J., 16:143
Pichi armadillos, 13:*149*, 13:153, 13:184, 13:192*t*
Picket pins. *See* Richardson's ground squirrels
Piebald shrews, 13:*269*, 13:*271*, 13:272–273
Pied tamarins, 14:124, 14:132*t*
Pig-footed bandicoots, 13:2, 13:*5*, 13:9–10, 13:11, 13:*13*, 13:*15*
Pig-tailed langurs, 14:184*t*
Pig-tailed snub-nosed langurs. *See* Mentawai Island langurs
Pigs, 15:263, **15:275–290,** 15:*283*
Asian, 15:149
behavior, 15:278
Chinese Meishan, 15:149
conservation status, 15:280–281
distribution, 12:136, 15:*275*, 15:276–277
European, 12:191–192
evolution, 15:137, 15:265, 15:267, 15:275
feeding ecology, 15:141, 15:271, 15:278–279
feral, 12:*190*, 12:191–192
habitats, 15:277
humans and, 15:273, 15:281–282
physical characteristics, 15:140, 15:268, 15:275–276
reproduction, 15:143, 15:272, 15:279–280
species of, 15:284–288, 15:289*t*–290*t*
taxonomy, 15:275
Vietnam warty, 15:263, 15:280, 15:289*t*
vision, 12:79
See also Domestic pigs; Eurasian wild pigs
Pikas, **16:491–503,** 16:*496*
behavior, 16:482–483, 16:491–493
conservation status, 16:487, 16:495
distribution, 12:134, 16:481, 16:*491*
evolution, 16:479–480, 16:491
feeding ecology, 16:484, 16:485, 16:493–494
habitats, 16:124, 16:481, 16:491
humans and, 16:487–488, 16:495
physical characteristics, 16:123, 16:480–481, 16:491
reproduction, 16:486, 16:494–495
species of, 16:497–501, 16:501*t*–502*t*
taxonomy, 16:479–480, 16:491
Piked whales. *See* Northern minke whales
Pilbara ningauis, 12:*292*, 12:*294*, 12:297–298
Pileated gibbons, 14:208, 14:*211*, 14:*212*, 14:*213*, 14:*217*, 14:*218*–219
behavior, 14:210, 14:211
conservation status, 14:215
evolution, 14:207
feeding ecology, 14:213
Piliocolobus spp. *See* Red colobus
Piliocolobus badius waldronae, 14:179
Piliocolobus gordonorum, 14:175
Piliocolobus kirkii, 14:175
Piliocolobus pennantii, 14:175
Piliocolobus rufomitratus, 14:175
Pilot whales, 15:2, 15:*3*, 15:8, 15:43–46, 15:48, 15:50
Pine squirrels. *See* North American red squirrels
Pink fairy armadillos, 13:*185*, 13:*186*, 13:*189*, 13:190
Pink river dolphins. *See* Botos
Pinnae, 12:54, 12:82
Pinnipeds, 12:14, 12:63, 12:67, 12:68
distribution, 12:129, 12:137, 12:138
humans and, 12:173
reproduction, 12:95, 12:98, 12:107, 12:110
vision, 12:79–80
Pinzon, Martín, 15:242
Pipanacoctomys spp., 16:434, 16:436
Pipanacoctomys aureus, 16:433–434
Pipistrelles, 12:*58*, 13:312, 13:499, 13:500, 13:501
banana, 13:311
Bodenheimer's, 13:313
smell, 12:85
Pipistrellus spp. *See* Pipistrelles
Pipistrellus babu. See Himalayan pipistrelles
Pipistrellus bodenheimeri. See Bodenheimer's pipistrelles
Pipistrellus kuhlii. See Kuhl's pipistrelle bats
Pipistrellus nanus. See African banana bats
Pipistrellus nathusii. See Nathusius's pipistrelle bats
Pipistrellus pipistrellus. See Pipistrelles
Pipistrellus subflavus. See Eastern pipistrelles
Pithecia spp. *See* Sakis
Pithecia aequatorialis, 14:143
Pithecia albicans, 14:143, 14:146, 14:147
Pithecia irrorata, 14:143
Pithecia monachus. See Monk sakis
Pithecia monachus milleri. See Miller's monk sakis
Pithecia monachus napiensis. See Napo monk sakis
Pithecia pithecia. See White-faced sakis
Pithecia pithecia chrysocephala. See Golden-faced sakis
Pitheciidae, **14:143–154,** 14:*149*
behavior, 14:145–146
conservation status, 14:148
distribution, 14:*143*, 14:144–145
evolution, 14:143
feeding ecology, 14:146–147
habitats, 14:145
humans and, 14:148
physical characteristics, 14:143–144
reproduction, 14:147
species of, 14:*150*–153, 14:153*t*
taxonomy, 14:143
Pitheciinae, 14:143, 14:146–147
Pizonyx visesi. See Mexican fishing bats
Placentals, 14:255–256
evolution, 12:11, 12:*31*, 12:33
neonatal milk, 12:127
phylogenetic tree of, 12:11, 12:*32*, 12:33
reproduction, 12:51, 12:89–93, 12:*102*, 12:106, 12:109
See also specific placentals
Placentas, 12:93–94, 12:101, 12:106, 12:109
See also Reproduction
Plagiodontia spp., 16:462
Plagiodontia aedium. See Hispaniolan hutias
Plagiodontia ipnaeum. See Samana hutias
Plain-nosed bats, 13:310, 13:311, 13:313, 13:315
See also Vespertilionidae
Plains baboons, 14:7, 14:10
Plains bison. *See* American bison

Plains harvest mice, 16:*264*
Plains kangaroos. *See* Red kangaroos
Plains mice, 16:*252*
Plains pocket gophers, 12:74, 16:*186*, 16:*191*, 16:*192*–193
Plains viscacha rats. *See* Red viscacha rats
Plains viscachas, 16:124, 16:377, 16:378, 16:379–380, 16:*381*, 16:*382*
Plains zebras, 12:*94*, 15:*220*, 15:*221*, 15:*228*, 15:*231*, 15:234
behavior, 12:146, 15:219, 15:*226*, 15:227
coevolution and, 12:216
conservation status, 15:229
distribution, 15:*233*
feeding ecology, 15:234
habitats, 15:218
true, 15:222
Planigale gilesi. See Gile's planigales
Planigale ingrami. See Long-tailed planigales
Planigale maculata. See Pygmy planigales
Planigales, 12:280, 12:281
Plantigrades, 12:*39*
Plasticity, behavioral, 12:140–143
See also Behavior
Platacanthomys lasiurus. See Malabar spiny dormice
Platalina genovensium. See Long-snouted bats
Platanista spp., 15:13
Platanista gangetica. See Ganges and Indus dolphins
Platanista gangetica gangetica, 15:13
Platanista gangetica minor, 15:13
Platanistidae. *See* Ganges and Indus dolphins
See also River dolphins
Platanistoidea, 15:2, 15:13
Plateau pikas, 16:482, 16:488, 16:492, 16:494, 16:*496*, 16:*497*–*498*
Platygonus spp., 15:291
Platymops spp., 13:311
Platypuses. *See* Duck-billed platypuses
Platyrrhinus helleri, 13:*414*
Play behavior, 12:142
See also Behavior
Plecotus auritus. See Brown long-eared bats; Long-eared bats
Plecotus austriacus. See Gray long-eared bats
Pliauchenia spp., 15:265
Pliny, 13:207
Pliopontes spp., 15:23
PM2000 (software), 12:207
Pocket gophers, 12:71, **16:185–197,** 16:*186*, 16:*187*, 16:*191*, 16:283
behavior, 16:124, 16:188–189
conservation status, 16:190
distribution, 12:131, 16:123, 16:*185*, 16:187
evolution, 16:121, 16:185
feeding ecology, 16:189–190
habitats, 16:187
humans and, 16:126, 16:127, 16:190
photoperiodicity, 12:74
physical characteristics, 16:123, 16:186–187
reproduction, 16:125, 16:190
species of, 16:192–195, 16:195*t*–197*t*
taxonomy, 16:121, 16:185
Pocket mice, 12:117, 16:121, 16:123, 16:124, **16:199–210,** 16:*204*, 16:208*t*–209*t*
Pocketed free-tailed bats, 13:495*t*
Podogymnura aureospinula. See Dinagat gymnures
Podogymnura truei. See Mindanao gymnures
Poebrodon spp., 15:265
Poebrotherium spp., 15:265
Poecilictis libyca. See North African striped weasels
Poecilogale albinucha. See White-naped weasels
Poelagus spp. *See* Bunyoro rabbits
Poelagus marjorita. See Bunyoro rabbits
Pohnpei flying foxes, 13:322
Poiana spp., 14:335
Poiana richardsonii. See African linsangs
Point mutations, 12:27, 12:30
Pointers, 14:288
Poisons, in pest control, 12:186
Polar bears, 12:*132*, 14:*301*, 14:*304*–305
behavior, 12:*143*, 14:258, 14:298
conservation status, 14:300
distribution, 12:129, 12:138, 14:296
evolution, 14:295–296
feeding ecology, 14:260, 14:298–299
habitats, 14:257, 14:296
Ice Age, 12:24
migrations, 12:*166*
physical characteristics, 14:296
reproduction, 12:106, 14:*296*, 14:299
taxonomy, 14:295
Polecats, 14:256, 14:319
marbled, 14:323
steppe, 14:321
Politics, conservation and, 12:224–225
Poll's shrews, 13:276*t*
Pollution, 12:213
Polo, Marco, 15:254–255
Polyandry, 12:107, 12:146
Polygyny, 12:98, 12:107, 12:146, 12:147–148
Polymerase chain reactions (PCR), 12:26
Polynesian rats, 12:189
Pomatodelphis spp., 15:13
Pond bats, 13:313, 13:515*t*
Pondaungia spp., 14:2
Pongidae. *See* Great apes
Ponginae, 14:225
Pongo spp. *See* Orangutans
Pongo abelii. See Sumatran orangutans
Pongo pygmaeus. See Bornean orangutans
Pontoporia blainvillei. See Franciscana dolphins
Pontoporiidae. *See* Franciscana dolphins
See also River dolphins
Poodles, 14:289
Populations
growth, 12:97
human, 12:213
indexes, 12:195–196, 12:200–202
size of, genetic diversity and, 12:221–222
in zoos, 12:207–209
Porcupine quills, 16:*367*
Porcupines, 12:38, 12:*123*, 16:121, 16:123, 16:127
New World, 16:125, **16:365–375**
Old World, 16:121, **16:351–364**
Porpoises, 15:1–2, **15:33–40,** 15:*37*
behavior, 15:6, 15:35
conservation status, 15:35–36
distribution, 15:*33*, 15:34
echolocation, 12:87
evolution, 15:3, 15:33
feeding ecology, 15:35
habitats, 15:6, 15:34–35
humans and, 15:36
physical characteristics, 15:4, 15:33–34
reproduction, 15:9, 15:35
species of, 15:38–39, 15:39*t*
taxonomy, 15:33
Portuguese watchdogs, 14:292
Positive reinforcement, in zoo training, 12:211
Possums, 12:137, 12:145, 13:31–35, 13:37–40, **13:57–67,** 13:*59*, 13:*62*, 13:*66t*
See also specific types of possums
Postcanines. *See* Molar teeth; Premolars
Postnatal care, 12:5
See also Reproduction
Postnatal development, 12:96–97
Pot-bellied pigs, 12:191, 15:*278*
Potamochoerus spp., 15:275, 15:276
Potamochoerus larvatus. See Bush pigs
Potamochoerus porcus. See Red river hogs
Potamochoerus porcus nyassae. See Red river hogs
Potamogale spp. *See* Otter shrews
Potamogale velox. See Giant otter shrews
Potomaglinae. *See* Otter shrews
Potoroidae. *See* Rat-kangaroos
Potoroinae, 13:73
Potoroos, 13:32, 13:73
See also specific types of potoroos
Potorous spp. *See* Potoroos
Potorous gilberti. See Gilbert's potoroos
Potorous longipes. See Long-footed potoroos
Potorous tridactlus. See Long-nosed potoroos
Potos spp., 14:309
Potos flavus. See Kinkajous
Potter, Beatrix, 13:200, 13:207
Pottos, 12:116, **14:13–21,** 14:*14*, 14:*15*, 14:*17*, 14:21*t*
Pouched mice, 16:297*t*
Poverty, effect on conservation, 12:220–221
Powered flight, 12:43–44, 12:55–56
Prairie dogs, 12:*115*, 16:124, 16:126, 16:127, 16:*128*, 16:143, 16:144
See also Black-tailed prairie dogs; Gunnison's prairie dogs
Prairie voles, 16:125, 16:*231*, 16:*234*
Precocial offspring, 12:12, 12:95–96, 12:97, 12:106
Predation, 12:21, 12:22–23, 12:64, 12:114–116
See also Behavior
Prehensile-tailed hutias, 16:461, 16:*464*, 16:*466*
Prehensile-tailed porcupines, 16:365, 16:366, 16:*367*, 16:*368*, 16:*369*, 16:*370*, 16:*372*–*373*
Premack, David, 12:161
Premaxilla, 12:9
Premolars, 12:9, 12:14, 12:46
See also Teeth
Presbytis spp., 14:172, 14:173
Presbytis comata, 14:175
Presbytis femoralis, 14:175
Presbytis fredericae, 14:175
Presbytis frontata, 14:175
Presbytis hosei. See Hose's leaf-monkeys
Presbytis johni. See Nilgiri langurs
Presbytis melalophos. See Banded leaf-monkeys
Presbytis potenziani. See Mentawai Island leaf-monkeys
Presbytis sivalensis, 14:172
Presbytis thomasi, 14:175
Prettyface wallabies. *See* Whiptail wallabies
Price, E. O., 15:145
Pricklepigs. *See* North American porcupines

Prides, lion, 14:258
Primates, **14:1–12**
behavior, 14:6–8
body size of, 12:21
color vision, 12:79
conservation status, 14:11
distribution, 12:132, 14:5–6
evolution, 12:33, 14:1–3
feeding ecology, 14:8–9
habitats, 14:6
haplorhine, 12:11
humans and, 14:11
as keystone species, 12:216–217
locomotion, 12:14
memory, 12:154
neonatal requirements, 12:126–127
physical characteristics, 14:4–5
reproduction, 12:90–91, 12:92, 12:95, 12:97, 12:98, 12:104, 12:105–106, 12:109, 14:10–11
sexual dimorphism, 12:99
taxonomy, 14:1–4
touch, 12:83
See also specific types of primates
Primelephas spp., 15:162
Primitive similarities, 12:28
Priodontes maximus. See Giant armadillos
Prionailurus bengalensis. See Leopards
Prionailurus bengalensis iriomotensis. See Iriomote cats
Prionodon spp., 14:335
Prionodon linsang. See Banded linsangs
Problem solving, by insight, 12:140
Proboscidea. *See* Elephants
Proboscis bats, 13:356, 13:357, 13:*359*, 13:*361*
Proboscis monkeys, 12:219, 14:171–*172*, 14:173, 14:175, 14:*177*, 14:*180*, 14:181–182
Procamelus spp., 15:265
Procapra spp. *See* Central Asian gazelles
Procapra gutturosa. See Mongolian gazelles
Procapra picticaudata. See Tibetan gazelles
Procapra przewalskii, 16:48
Procavia spp., 15:177
Procavia capensis. See Rock hyraxes
Procavia habessinica. See Abyssinian hyraxes
Procavia johnstoni. See Johnston's hyraxes
Procavia ruficeps. See Red-headed rock hyraxes
Procavia syriacus. See Syrian hyraxes
Procavia welwitschii. See Kaokoveld hyraxes
Procaviidae. *See* Hyraxes
Procolobus spp., 14:172, 14:173
Procolobus badius. See Western red colobus
Procolobus badius waldroni. See Miss Waldron's red colobus
Procolobus pannantii. See Penant's red colobus
Procolobus preussi, 14:175
Procolobus rufomitratus, 14:11
Procolobus verus. See Olive colobus
Procyon spp. *See* Raccoons
Procyon cancrivorus. See Crab-eating raccoons
Procyon gloveralleni. See Barbados raccoons
Procyon insularis, 14:310
Procyon lotor. See Northern raccoons
Procyon maynardi. See Bahaman raccoons
Procyon minor. See Guadeloupe raccoons
Procyon pygmaeus. See Cozumel Island raccoons
Procyonidae, 14:256, 14:259, 14:262, **14:309–317**, 14:*312*, 14:315*t*–316*t*
Prodendrogale spp., 13:291
Proechimys spp., 16:449, 16:*449*, 16:*451*
Proechimys albispinus, 16:450–451
Proechimys amphichoricus. See Venezuelan spiny rats
Proechimys gorgonae, 16:450–451
Proechimys semispinosus. See Spiny rats
Profelis aurata. See African golden cats
Progalago spp., 14:23
Prohylobates spp., 14:172, 14:188–189
Prolactin, 12:148
Prolagus spp., 16:480, 16:491
Prolagus sardus. See Sardinian pikas
Prometheomys spp., 16:225
Prometheomys schaposchnikowi. See Long-clawed mole voles
Promiscuity, 12:98, 12:103, 12:107
See also Behavior
Promops centralis. See Big crested mastiff bats
Pronghorns, 12:40, **15:411–417**, 15:*412*, 15:*413*, 15:*414*, 15:*415*, 15:*416*
behavior, 15:271, 15:412–414
conservation status, 15:415–416
distribution, 12:131, 12:*132*, 15:*411*, 15:412
evolution, 15:137, 15:138, 15:265, 15:266, 15:267, 15:411
feeding ecology, 15:414
habitats, 15:269, 15:412
humans and, 15:416–417
physical characteristics, 15:268, 15:411–412
reproduction, 15:143, 15:272, 15:414–415
scent marking, 12:80
taxonomy, 15:411
Pronolagus spp. *See* Rockhares
Pronycticeboides spp., 14:13
Propalaeochoerus spp., 15:265
Propithecus spp., 14:8
Propithecus candidus, 14:63, 14:67
Propithecus diadema, 14:63, 14:67
Propithecus diadema diadema. See Diademed sifakas
Propithecus edwardsi. See Milne-Edwards's sifakas
Propithecus perrieri, 14:63, 14:67
Propithecus tattersalli. See Golden-crowned sifakas
Propithecus verreauxi, 14:7, 14:63
Propithecus verreauxi coquereli. See Coquerel's sifakas
Propleopinae. *See* Giant rat-kangaroos
Propleopus oscillans, 13:70
Prosciurillus spp., 16:163
Prosciurillus abstrusus, 16:167
Prosciurillus weberi, 16:167
Proserpine rock wallabies, 13:32, 13:39
Prosigmodon spp., 16:264
Prosimian primates
behavior, 12:145–146, 14:7–8
distribution, 14:5
evolution, 14:1
feeding ecology, 14:9
habitats, 14:6
reproduction, 12:90, 14:10
taxonomy, 14:1, 14:3
See also Lemurs; Lorises; Primates; Tarsiers
Prosqualondotidae, 15:13
Prostate gland, 12:104
Protapirus spp., 15:237
Protein synthesis, 12:26–27, 12:29–31, 12:122–123
Proteles cristatus. See Aardwolves
Proteles cristatus cristatus, 14:359
Proteles cristatus septentrionalis, 14:359
Proteropithecia spp., 14:143
Protocetidae, 15:2
Protomeryx spp., 15:265
Protosciurus spp., 16:135, 16:163
Prototheria. *See* Monotremes
Protoungulata, 15:131, 15:136
Protoxerus spp., 16:163
Protungulatum spp., 15:132
Przewalski's horses, 15:*231*, 15:*234*–235
behavior, 15:226
conservation status, 12:175, 12:222, 15:149, 15:222
distribution, 12:134
domestication and, 12:175, 15:149
feeding ecology, 15:*217*
genetic diversity and, 12:222
physical characteristics, 15:226
Przewalski's wild horses. *See* Przewalski's horses
Pseudailurus spp., 14:369
Pseudalopex culpaeus. See Culpeos
Pseudalopex fulvipes. See Darwin's foxes
Pseudalopex gymnocercus. See Pampas foxes
Pseudantechinus macdonnellensis. See Fat-tailed pseudantechinuses
Pseudocheiridae, 13:34, 13:35, **13:113–123**, 13:*118*, 13:122*t*–123*t*
Pseudocheirus spp., 13:115
Pseudocheirus caroli. See Weyland ringtails
Pseudocheirus forbesi. See Moss-forest ringtails
Pseudocheirus peregrinus. See Common ringtails
Pseudocheirus schlegeli. See Arfak ringtails
Pseudocherius peregrinus. See Eastern ringtail possums
Pseudochirops spp., 13:113
Pseudochirops albertisi. See d'Albertis's ringtail possums
Pseudochirops archeri. See Green ringtails
Pseudochirops corinnae. See Golden ringtail possums
Pseudochirops cupreus. See Coppery ringtail possums
Pseudochirulus spp., 13:113
Pseudochirulus canescens. See Lowland ringtails
Pseudochirulus herbertensis. See Herbert River ringtails
Pseudois spp., 16:87, 16:90, 16:91
Pseudois nayaur. See Blue sheep
Pseudois schaeferi. See Dwarf blue sheep
Pseudomys australis. See Plains mice
Pseudomys higginsi. See Long-tailed mice
Pseudomys nanus. See Western chestnut mice
Pseudonovibos spiralis. See Linh duongs
Pseudopotto martini. See False pottos
Pseudorca crassidens. See False killer whales
Pseudoryx spp., 15:263
Pseudoryx nghetinhensis. See Saolas
Pseudoryzomys spp., 16:266
Pteromyinae. *See* Flying squirrels
Pteromys volans. See Siberian flying squirrels
Pteronotus spp., 13:308, 13:*437*
Pteronotus davyi. See Davy's naked-backed bats
Pteronotus gymnonotus. See Big naked-backed bats
Pteronotus macleayii. See Macleay's moustached bats
Pteronotus parnellii. See Parnell's moustached bats

Pteronotus personatus. See Wagner's moustached bats
Pteronotus quadridens. See Sooty moustached bats
Pteronura spp., 14:319
Pteronura brasiliensis. See Giant otters
Pteropodidae, 12:86, 13:307, 13:308, 13:309, 13:314
See also Flying foxes
Pteropus spp. *See* Flying foxes
Pteropus admiralitatum. See Admiralty flying foxes
Pteropus alecto. See Black flying foxes
Pteropus argentatus. See Ambon flying foxes
Pteropus brunneus. See Dusky flying foxes
Pteropus caniceps. See North Moluccan flying foxes
Pteropus conspicillatus. See Spectacled flying foxes
Pteropus dasymallus. See Ryukyu flying foxes
Pteropus fundatus. See Banks flying foxes
Pteropus giganteus. See Indian flying foxes
Pteropus gilliardi. See Gilliard's flying foxes
Pteropus hypomelanus. See Island flying foxes
Pteropus livingstonii. See Livingstone's fruit bats
Pteropus lylei. See Lyle's flying foxes
Pteropus macrotis. See Big-eared flying foxes
Pteropus mahaganus. See Sanborn's flying foxes
Pteropus mariannus. See Marianas fruit bats
Pteropus mearnsi. See Mearns's flying foxes
Pteropus melanotus. See Blyth's flying foxes
Pteropus molossinus. See Pohnpei flying foxes
Pteropus niger. See Greater Mascarene flying foxes
Pteropus poliocephalus. See Gray-headed flying foxes
Pteropus pumilus. See Little golden-mantled flying foxes
Pteropus rodricensis. See Rodricensis flying foxes
Pteropus rufus. See Madagascar flying foxes
Pteropus samoensis. See Samoan flying foxes
Pteropus scapulatus. See Little red flying foxes
Pteropus tokudae. See Guam flying foxes
Pteropus tonganus. See Tongan flying foxes
Pterosaurs, 12:52
Ptilocercus spp., 13:292
Ptilocercus lowii. See Pen-tailed tree shrews
Pudu spp., 15:382, 15:384
Pudu mephistophiles. See Northern pudus
Pudu pudu. See Southern pudus
Puerto Rican giant hutias, 16:*470*, 16:471*t*
Puerto Rican hutias, 16:467*t*
Puerto Rican nesophontes, 13:243–244
Pukus, 16:27, 16:29, 16:30, 16:32
Puma concolor. See Pumas
Puma (Felis) concolor. See Pumas
Pumas, 14:*378*, 14:382
behavior, 12:*121*
distribution, 12:129, 14:371, 14:*386*
feeding ecology, 14:372
as keystone species, 12:217
Punares, 16:*450*, 16:458*t*
Punomys spp., 16:264
Punomys lemminus, 16:266
Pygathrix spp., 14:172, 14:173
Pygathrix nemaeus. See Red-shanked douc langurs
Pygathrix nigripes, 14:175
Pygathrix roxellana. See Golden snub-nosed monkeys
Pygeretmus spp., 16:215
Pygeretmus platyurus. See Lesser fat-tailed jerboas
Pygeretmus pumilio, 16:213, 16:214, 16:216, 16:217
Pygeretmus shitkovi. See Greater fat-tailed jerboas
Pygmy anteaters. *See* Silky anteaters
Pygmy antelopes. *See* Dwarf antelopes
Pygmy beaked whales, 15:1, 15:59, 15:70*t*
Pygmy Bryde's whales, 15:1
Pygmy chimpanzees. *See* Bonobos
Pygmy dormice, Chinese, 16:297*t*
Pygmy elephants, 15:163
Pygmy fruit bats. *See* Black flying foxes
Pygmy gliders, 13:33, 13:35, 13:38, 13:*139*, 13:*140*, 13:*141*, 13:142, 13:*143*, 13:*144*
Pygmy gliding possums. *See* Pygmy gliders
Pygmy goats, 16:*95*
Pygmy hippopotamuses, 15:301, 15:*305*, 15:306, 15:*307*, 15:308, 15:*309*, 15:*310*, 15:311
Pygmy hogs, 15:275, 15:277, 15:280, 15:281, 15:*283*, 15:286–*287*
Pygmy jerboas, 16:211, 16:212, 16:213, 16:215
Pygmy killer whales, 15:2, 15:56*t*
Pygmy lorises. *See* Pygmy slow lorises
Pygmy marmosets, 14:*120*, 14:*125*, 14:*130*, 14:131
behavior, 14:117, 14:118
distribution, 14:116
feeding ecology, 14:120
habitats, 14:116
physical characteristics, 14:116
reproduction, 14:121, 14:122, 14:123
taxonomy, 14:115
Pygmy mice, 16:123
Pygmy mouse lemurs, 14:4, 14:35, 14:*36*, 14:39
Pygmy phalangers. *See* Pygmy gliders
Pygmy planigales, 12:300*t*
Pygmy possums, 13:31, 13:33, 13:38, 13:39, **13:105–111,** 13:*109*
Pygmy rabbits, 16:481, 16:487, 16:505, 16:515*t*
Pygmy right whales, **15:103–106,** 15:*104*, 15:*105*
Pygmy ringtails, 13:*118*, 13:*121*, 13:122
Pygmy rock mice, 16:*286*, 16:*289*, 16:290
Pygmy short-tailed opossums, 12:251, 12:254, 12:*256*, 12:*262*–263
Pygmy shrews. *See* American pygmy shrews
Pygmy slow lorises, 14:*14*, 14:15, 14:*16*, 14:*17*, 14:*18*, 14:19
Pygmy sperm whales, 15:6, 15:73, 15:74, 15:76, 15:77, 15:*78*, 15:*79*, 15:80
Pygmy squirrels, 16:163
Pygmy tarsiers, 14:93, 14:95, 14:*97*, 14:99
Pygmy three-toed sloths. *See* Monk sloths
Pygmy tree shrews, 13:*290*, 13:*292*, 13:297*t*
Pygmy white-toothed shrews. *See* Savi's pygmy shrews
Pyrenean desmans, 13:281, 13:282, 13:287*t*
Pyrotheria, 12:11, 15:131, 15:136

Q

Quaggas, 15:146, 15:221, 15:236*t*
Quanacaste squirrels. *See* Deppe's squirrels
Queensland blossom bats. *See* Southern blossom bats
Queensland koalas, 13:43, 13:*44*, 13:*45*, 13:*48*
Queensland tube-nosed bats, 13:*335*, 13:*337*, 13:*339*, 13:*346*, 13:347, 13:349*t*
Quemisia spp., 16:469
Quemisia gravis. See Hispaniolan giant hutias
Querétaro pocket gophers, 16:190, 16:*191*, 16:*192*, 16:194
Quillers. *See* North American porcupines
Quillpigs. *See* North American porcupines
Quills. *See* Spines
Quokkas, 12:123, 13:34, 13:*92*, 13:*93*, 13:95–96
Quolls, 12:279–284, 12:288–291, 14:256

R

r-strategists, 12:13, 12:97
Rabbit calicivirus disease (RCD), 12:187–188, 12:223
Rabbit-eared bandicoots. *See* Bilbies
Rabbit Nuisance Bill of 1883 (Australia), 12:185
Rabbit rats, 16:265, 16:266, 16:269, 16:*271*, 16:*272*
Rabbits, 12:180, 16:479–488, **16:505–516,** 16:*510*, 16:515*t*–516*t*
behavior, 16:506–507
conservation status, 16:509
distribution, 12:137, 16:*505*–506
evolution, 16:505
feeding ecology, 16:507
habitats, 16:513–515
humans and, 16:509
as pests, 12:185–186
physical characteristics, 12:46, 12:121, 12:122, 16:505
reproduction, 12:103, 12:126, 16:507–509
species of, 16:513–515, 16:515*t*–516*t*
taxonomy, 16:505
Rabies, 13:316
Raccoon dogs, 14:265, 14:266, 14:267, 14:*275*, 14:*278*, 14:281–282
Raccoons, 12:79, 12:224, 14:256, 14:260, **14:309–317,** 14:*312*, 14:315*t*–316*t*
Racey, Paul A., 13:502
Radiation, mammalian, 12:11, 12:12
Radinsky, Leonard, 15:132
Radio telemetry, 12:199–202
Raffray's bandicoots, 13:*4*, 13:*13*, 13:*14*, 13:15–16
Rainforest bandicoots, 13:1, **13:9–18,** 13:*13*, 13:16*t*–17*t*
Rainforests, 12:218
See also Habitats
Ramus mandibulae, 12:9, 12:11
Randall, Jan A., 16:200–201, 16:203
Random genetic drift, 12:222
Rangifer spp., 15:382, 15:383
Rangifer tarandus. See Reindeer
Rangifer tarandus caribou. See Canadian caribou
Raphicerus campestris. See Steenboks
Raphicerus melanotis. See Cape grysboks

Raphicerus sharpei. See Sharpe's grysboks
Rare Breeds Survival Trust, on pigs, 15:282
Rat-kangaroos, 12:145, 13:31–32, 13:34, 13:35, 13:38, **13:73–81,** 13:77
Rat-like hamsters, 16:239, 16:240
Ratanaworabhan's fruit bats, 13:349*t*
Ratels. *See* Honey badgers
Rats, 12:*94*
behavior, 12:142–143, 12:*186*, 12:*187*, 16:124–125
cane, 16:121, **16:333–338**
chinchilla, **16:443–448**
dassie, 16:121, **16:329–332,** 16:*330*
distribution, 12:137, 16:123
evolution, 16:121
Hawaiian Islands, 12:189–191
humans and, 12:139, 12:173, 16:128
kangaroo, 16:121, 16:123, 16:124, **16:199–210**
learning set, 12:152
memory and, 12:153
mole, 12:70, 12:72–77, 12:115, 12:116
neonatal requirements, 12:126
numbers and, 12:155
Old World, **16:249–262**
Polynesian, 12:189
reproduction, 12:106, 16:125
South American, **16:263–279**
spiny, 16:253, **16:449–459**
taxonomy, 16:121
See also Black rats; Moonrats; Muridae
Rattus exulans. See Polynesian rats
Rattus norvegicus. See Brown rats
Rattus rattus. See Black rats
Rattus tunneyi. See Tunney's rats
Ratufa spp., 16:163
Ratufa affinis. See Giant squirrels
Ratufa indica. See Indian giant squirrels
Ratufa macroura, 16:167
Rayner, J., 12:57
Razorbacks. *See* Fin whales
RBCs. *See* Erythrocytes
RCD (Rabbit calicivirus disease), 12:187–188, 12:223
Recall, 12:152, 12:154
Recognition, 12:152, 12:154
Record keeping, zoo, 12:206–207
Red acouchis, 16:409, 16:*410*, 16:*411*, 16:413–414
Red-and-white uakaris. *See* Bald uakaris
Red-armed bats. *See* Natterer's bats
Red-backed squirrel monkeys, 14:101, 14:*102*, 14:103, 14:*107*, 14:*108*–109
Red-backed voles, 16:226, 16:227, 16:228
Red bald uakaris, 14:*145*, 14:148, 14:151
Red bats, 13:310, 13:312, 13:314, 13:315, 13:*498*, 13:499–500, 13:501
Red-bellied lemurs, 14:6, 14:*54*, 14:*56*, 14:60
Red-bellied squirrels. *See* Pallas's squirrels
Red-bellied tamarins, 14:119, 14:124
Red blood cells. *See* Erythrocytes
Red brocket deer, 15:*384*, 15:*387*, 15:*393*–394
Red-capped mangabeys. *See* Collared mangabeys
Red colobus, 14:7, 14:172–175
Red Data Books, 16:219, 16:220
aardvarks, 15:158
alpine shrews, 13:260
giant shrews, 13:262
Red deer, 15:*359*, 15:*363*, 15:369–370
behavior, 15:*360*
competition, 12:118
distribution, 15:*364*
evolution, 12:19, 15:357
humans and, 15:273, 15:361
predators and, 12:116
reproduction, 15:*360*
Red duikers, 16:73
Red-flanked duikers, 16:77, 16:*79*, 16:82–*83*
Red forest rats, 16:*281*
Red foxes, 14:*269*, 14:*275*, 14:278–280, 14:*279*, 14:287–288
behavior, 12:145–146, 14:258, 14:259, 14:267–268
conservation status, 14:272
distribution, 12:137
evolution, 14:265
feeding ecology, 14:271
habitats, 14:267
humans and, 14:273
as pests, 12:185
reproduction, 12:107, 14:272
taxonomy, 14:265
Red-fronted gazelles, 16:48, 16:57*t*
Red-fronted lemurs, 14:*51*
Red gazelles, 12:129
Red giant flying squirrels, 16:*136*, 16:*138*, 16:*139*
Red gorals, 15:*272*, 16:87, 16:92
Red-handed howler monkeys, 14:166*t*
Red hartebeests, 12:116, 15:*139*
Red-headed rock hyraxes, 15:177, 15:189*t*
Red howler monkeys. *See* Bolivian red howler monkeys; Venezuelan red howler monkeys
Red kangaroos, 12:*4*, 13:*91*, 13:95
distribution, 13:34, 13:*93*
feeding ecology, 13:38
habitats, 13:34
humans and, 13:90
physical characteristics, 13:*33*
reproduction, 12:108, 13:*88*, 13:*89*
taxonomy, 13:31
Red-legged pademelons, 13:*92*, 13:97–*98*
Red-legged short-tailed opossums, 12:*256*, 12:*262*
Red meerkats. *See* Yellow mongooses
Red monkeys. *See* Patas monkeys
Red mouse lemurs, 14:*10*, 14:*36*, 14:*38*, 14:39, 14:*40*, 14:*42*, 14:43
Red mouse opossums, 12:251, 12:*252*
Red-necked night monkeys, 14:135
Red-necked wallabies, 12:*137*, 13:*34*, 13:*87*, 13:102*t*
Red-nosed tree rats, 16:268
Red pandas, 12:*28*, 14:309–*311*, 14:*312*, 14:*315*
Red river hogs, 15:277, 15:279, 15:*281*, 15:289*t*
Red-rumped agoutis, 16:*409*, 16:*410*, 16:411–*412*
Red-shanked douc langurs, 12:218, 14:*172*, 14:175, 14:*177*, 14:*182*
Red squirrels, 12:139, 12:183–184, 16:*127*, 16:166, 16:*167*, 16:*168*, 16:*169*–170
Red-tailed chipmunks, 16:159*t*
Red-tailed monkeys, 12:38
Red-tailed phascogales, 12:*289*, 12:300*t*
Red-tailed sportive lemurs, 14:*74*, 14:*75*, 14:*78*, 14:79–*80*
Red-tailed wambergers. *See* Red-tailed phascogales
Red titis, 14:153*t*
Red-toothed shrews, **13:247–264,** 13:*253*, 13:*254*, 13:263*t*
Red tree voles, 16:*231*, 16:*234*–235
Red viscacha rats, 16:124, 16:433, 16:434, 16:436, 16:*437*, 16:*438*
Red wolves, 12:*214*, 14:262, 14:273
Redford, K. H., 16:434
Redunca spp., 16:5, 16:27, 16:28
Redunca arundinum. See Southern reedbucks
Redunca fulvorufula. See Mountain reedbucks
Redunca fulvorufula adamauae. See Western mountain reedbucks
Redunca redunca. See Bohor reedbucks
Reduncinae, 16:1, 16:7
Reduncini, 16:27–31, 16:34
Reed rats, Malagasy, 16:*284*
Reedbucks, 16:5, **16:27–43,** 16:*35*, 16:41*t*–42*t*
Reeder, D. M., 16:121, 16:389
Reeve's muntjacs, 15:343, 15:*344*, 15:*345*, 15:346, 15:*348*, 15:*349*
Reference memory. *See* Long-term memory
Referential pointing/gazing, 12:160
Rehabilitation pools, 12:*204*
Reichard, U., 14:214
Reig, O. A., 16:265, 16:435
Reindeer, 15:*270*, 15:*387*, 15:394–395
behavior, 15:383
conservation status, 15:384
distribution, 12:131, 12:138, 15:*393*
domestication of, 12:172–173, 12:180, 15:145
evolution, 15:379, 15:380, 15:381–382
feeding ecology, 12:*125*
field studies of, 12:*197*
habitats, 15:383
humans and, 15:361, 15:384
migrations, 12:87, 12:116, 12:164, 12:*165*, 12:*168*, 12:170
physical characteristics, 15:268, 15:*383*
Reintroductions, 12:139, 12:223–224
red wolves, 12:*214*
zoos and, 12:204, 12:208
See also Conservation status
Reithrodon spp., 16:265
Reithrodon auritus. See Rabbit rats
Reithrodon typicus, 16:268
Reithrodontines, 16:263–264
Reithrodontomys montanus. See Plains harvest mice
Reithrodontomys raviventris. See Saltmarsh harvest mice
Reithrodontomys sumichrasti. *See* Sumichrast's harvest mice
Reithrosciurus spp., 16:163
Remingtonocetidae, 15:2
Removal sampling, 12:197, 12:200
Renal system, 12:89–90
See also Physical characteristics
Renewable resources, 12:215
Rensch's Rule, 12:99
Rephicerus campestris. See Steenboks
Reproduction, 12:12, 12:51, **12:89–112**
aardvarks, 15:158
agoutis, 16:409, 16:411–414
anteaters, 12:94, 13:174–175, 13:177–179
armadillos, 12:94, 12:103, 12:110–111, 13:185, 13:187–190

Reproduction *(continued)*
  Artiodactyla, 12:127, 15:272
  aye-ayes, 14:88
  baijis, 15:21
  bandicoots, 13:4
    dry-country, 13:11, 13:14–16
    rainforest, 13:11, 13:14–17*t*
  bats, 13:314–315
    American leaf-nosed, 13:419–420, 13:423–432
    bulldog, 13:447, 13:449–450
    disk-winged, 13:474, 13:476–477
    Emballonuridae, 13:358, 13:360–363
    false vampire, 13:382, 13:384–385
    funnel-eared, 13:461, 13:463–465
    horseshoe, 13:392, 13:396–400
    Kitti's hog-nosed, 13:368
    Molossidae, 13:486–487, 13:490–493
    mouse-tailed, 13:352, 13:353
    moustached, 13:436, 13:440–441
    New Zealand short-tailed, 13:454, 13:457–458
    Old World fruit, 13:320–321, 13:325–330, 13:336–337, 13:340–347
    Old World leaf-nosed, 13:405, 13:407–409
    Old World sucker-footed, 13:480
    slit-faced, 13:373, 13:375–376
    smoky, 13:468, 13:470–471
    Vespertilionidae, 13:500–501, 13:506–510, 13:512–514, 13:522, 13:524–525
  bears, 14:299, 14:302–305
  beavers, 16:180, 16:183–184
  bilbies, 13:22
  body size and, 12:15, 12:22–24
  botos, 15:29
  bottleneck species and, 12:221–222
  Bovidae, 16:7–8
    Antilopinae, 16:48, 16:50–56
    Bovinae, 16:14, 16:18–23
    Caprinae, 16:94, 16:98–103
    duikers, 16:76–77, 16:80–84
    Hippotraginae, 16:32–33, 16:37–42
    Neotraginae, 16:62–63, 16:66–70
  bushbabies, 14:26, 14:28–32
  Camelidae, 15:317, 15:320–323
  Canidae, 14:271–272, 14:277–283
  capybaras, 16:404–405
  Carnivora, 12:90, 12:92, 12:94, 12:95, 12:98, 12:109, 12:110, 12:126–127, 14:261–262
  cats, 12:103, 12:106, 12:109, 12:111, 12:126, 14:372–373, 14:379–389
  Caviidae, 16:393, 16:395–398
  Cetacea, 12:92, 12:94, 12:103–104, 15:9
    Balaenidae, 15:110–111, 15:115–117
    beaked whales, 15:61
    dolphins, 15:47–48, 15:53–55
    franciscana dolphins, 15:24–25
    Ganges and Indus dolphins, 15:15
    gray whales, 15:98–99
    Monodontidae, 15:85–86, 15:90–91
    porpoises, 15:35, 15:38–39
    pygmy right whales, 15:105
    rorquals, 15:124, 15:127–130
    sperm whales, 15:76–77, 15:79–80
  chevrotains, 15:328–329, 15:331–333
  Chinchillidae, 16:380, 16:382–383
  colugos, 13:302, 13:304
  coypus, 16:477
  dasyurids, 12:289–290, 12:294–298
  Dasyuromorphia, 12:281–283
  deer
    Chinese water, 15:375–376
    muntjacs, 15:346, 15:349–354
    musk, 15:337–338, 15:340–341
    New World, 15:384, 15:388–395
    Old World, 15:360, 15:364–371
  Dipodidae, 16:215–216, 16:219–222
  Diprotodontia, 13:38–39
  domestication and, 12:174
  dormice, 16:320–321, 16:323–326
  duck-billed platypuses, 12:101, 12:106, 12:108, 12:*228*, 12:231–233, 12:*232*, 12:245–246
  Dugongidae, 15:201, 15:203
  echidnas, 12:237–238
  elephants, 12:90, 12:91, 12:94, 12:95, 12:98, 12:103, 12:106–108, 12:*108*, 12:209, 15:170–171, 15:174–175
  Equidae, 15:228–229, 15:232–235
  Erinaceidae, 13:206–207, 13:209–212
  giant hutias, 16:470
  gibbons, 14:213–214, 14:218–222
  Giraffidae, 15:404–405, 15:408–409
  great apes, 14:234–235, 14:237–240
  gundis, 16:313, 16:314–315
  Herpestidae, 14:352, 14:354–356
  Heteromyidae, 16:203, 16:205–208
  hippopotamuses, 15:308, 15:310–311
  humans, 14:248–249
  hutias, 16:463
  Hyaenidae, 14:261, 14:362, 14:365–367
  hyraxes, 15:182–183, 15:186–188
  Indriidae, 14:66–67, 14:69–72
  Insectivora, 13:199
  koalas, 13:47–48
  Lagomorpha, 16:485–487
  lemurs, 14:10–11, 14:51–52, 14:56–60
    Cheirogaleidae, 14:38, 14:41–44
    sportive, 14:76, 14:79–83
  Leporidae, 12:126, 16:507–509, 16:511–515
  Lorisidae, 14:16, 14:18–20
  luxury organs and, 12:23–24
  Macropodidae, 13:88–89
  manatees, 15:208, 15:211–212
  Megalonychidae, 13:157, 13:159
  migrations and, 12:170
  moles
    golden, 13:217, 13:220–221
    marsupial, 13:27, 13:28
  monitos del monte, 12:274
  monkeys
    Atelidae, 14:158–159, 14:162–166
    Callitrichidae, 14:121–124, 14:127–131
    Cebidae, 14:105, 14:108–112
    cheek-pouched, 14:193, 14:197–204
    leaf-monkeys, 14:174–175, 14:178–183
    night, 14:138
    Pitheciidae, 14:147, 14:150–153
  monotremes, 12:12, 12:51, 12:89–93, 12:108, 12:231–233
  mountain beavers, 16:133
  Muridae, 16:125, 16:284–285, 16:288–295
    Arvicolinae, 16:228–229, 16:233–237
    hamsters, 16:242–243, 16:245–246
    Murinae, 16:252, 16:256–260
    Sigmodontinae, 16:269–270, 16:272–276
  musky rat-kangaroos, 13:71
  Mustelidae, 14:324, 14:327–331
  numbats, 12:305
  nutritional requirements and, 12:125–127
  octodonts, 16:436, 16:438–440
  opossums
    New World, 12:253–254, 12:257–263
    shrew, 12:268, 12:270
  Otariidae, 14:398–399, 14:402–406
  pacaranas, 16:387–388
  pacas, 16:419–420, 16:423–424
  pangolins, 16:113, 16:115–120
  peccaries, 15:295, 15:298–300
  Perissodactyla, 15:220–221
  Petauridae, 13:128–129, 13:131–132
  Phalangeridae, 13:60, 13:*63–66t*
  photoperiodicity and, 12:75
  pigs, 15:279–280, 15:284–288, 15:289*t*–290*t*
  pikas, 16:494–495, 16:497–501
  pocket gophers, 16:190, 16:192–195
  porcupines
    New World, 16:369, 16:372–373
    Old World, 16:353, 16:357–364
  possums
    feather-tailed, 13:142, 13:144
    honey, 13:137–138
    pygmy, 13:107–108, 13:110–111
  primates, 14:10–11
  Procyonidae, 14:310, 14:313–315
  pronghorns, 15:414–415
  Pseudocheiridae, 13:116–117, 13:119–123
  rat-kangaroos, 13:76, 13:78–81
  rats
    African mole-rats, 12:107, 16:344–345, 16:347–349
    cane, 16:334–335, 16:337–338
    chinchilla, 16:445, 16:447–448
    dassie, 16:331
    spiny, 16:450, 16:453–457
  rhinoceroses, 15:254, 15:257–262
  rodents, 16:125–126
  sengis, 16:521–522, 16:524–530
  shrews
    red-toothed, 13:251–252, 13:255–263
    West Indian, 13:244
    white-toothed, 13:268, 13:271–275
  Sirenia, 15:193–194
  solenodons, 13:239, 13:241
  springhares, 16:*309*
  squirrels
    flying, 16:136–137, 16:139–140
    ground, 16:147, 16:151–158
    scaly-tailed, 16:300, 16:302–305
    tree, 16:166–167, 16:169–174
  subterranean mammals, 12:76
  Talpidae, 13:282, 13:284–287
  tapirs, 15:241, 15:245–247
  tarsiers, 14:95, 14:97–99
  tenrecs, 13:229, 13:232–233
  three-toed tree sloths, 13:165, 13:167–169
  tree shrews, 13:292, 13:294–296
  true seals, 12:126, 12:127, 14:261, 14:262, 14:420–422, 14:427, 14:429–434
  tuco-tucos, 16:427, 16:429–430
  ungulates, 15:142–143
  Viverridae, 14:338, 14:340–343
  viviparous, 12:5
  walruses, 12:111, 14:414–415
  wombats, 13:53, 13:55–56
  Xenartha, 13:152
  in zoo populations, 12:206–209
Reproductive technology, in zoos, 12:208
Reptiles, mammals and, 12:36

Resistance, viral, 12:187
Resource Dispersion Hypothesis, 14:259–260
Resource-holding power (RHP), 12:144
Respiration
bats, 12:60
marine mammals, 12:67–68
See also Physical characteristics
Respiratory system, 12:45–46, 12:73, 12:114
See also Physical characteristics
Reticulated giraffes, 15:*401*, 15:*402*, 15:*405*
Retinas, 12:79
See also Eyes
Retrievers, 14:*288*
Retrieving information, 12:152
Retroposons, 12:29–30
Reverse sexual dimorphism, 12:99
Rhabdomys pumilio. *See* Four-striped grass mice
Rhagomys spp., 16:264, 16:265, 16:266
Rhagomys rufescens, 16:270
Rheboks, 16:31
Rheomys mexicanus, 16:268
Rhesus macaques, 12:151–153, 14:5, 14:194, 14:*195*, 14:*200*
Rhesus monkeys. *See* Rhesus macaques
Rhims. *See* Slender-horned gazelles
Rhinarium, 12:8
Rhinoceros spp., 15:216, 15:217, 15:252
Rhinoceros sondiacus. *See* Javan rhinoceroses
Rhinoceros sondiacus annamiticus, 12:220
Rhinoceros unicornis. *See* Indian rhinoceroses
Rhinoceroses, 15:146, 15:237, **15:249–262**, 15:*256*
behavior, 15:218, 15:219, 15:253
conservation status, 15:221, 15:222, 15:254
distribution, 12:136, 15:217, 15:*249*, 15:252
evolution, 15:136, 15:137, 15:215, 15:216, 15:249, 15:251
feeding ecology, 15:220, 15:253
habitats, 15:218, 15:252–253
horns, 12:40
humans and, 12:184, 15:223, 15:254–255
migrations, 12:167
neonatal milk, 12:126
physical characteristics, 15:138, 15:139, 15:216, 15:217, 15:251–252
population indexes, 12:195
reproduction, 12:91, 15:143, 15:221, 15:254
species of, 15:257–262
taxonomy, 15:249, 15:251
woolly, 15:138, 15:249
in zoos, 12:209
Rhinocerotidae. *See* Rhinoceroses
Rhinocerotinae, 15:249
Rhinolophidae. *See* Horseshoe bats
Rhinolophus spp. *See* Horseshoe bats
Rhinolophus affinis. *See* Intermediate horseshoe bats
Rhinolophus alcyone, 13:388
Rhinolophus beddomei. *See* Lesser woolly horseshoe bats
Rhinolophus blasii. *See* Blasius's horseshoe bats
Rhinolophus capensis. *See* Cape horseshoe bats
Rhinolophus clivosus. *See* Geoffroy's horseshoe bats
Rhinolophus cognatus. *See* Andaman horseshoe bats
Rhinolophus convexus, 13:392
Rhinolophus denti. *See* Dent's horseshoe bats
Rhinolophus euryale. *See* Mediterranean horseshoe bats
Rhinolophus ferrumequinum. *See* Greater horseshoe bats
Rhinolophus fumigatus. *See* Ruppell's horseshoe bats
Rhinolophus hildebrandti. *See* Hildebrandt's horseshoe bats
Rhinolophus hipposideros. *See* Lesser horseshoe bats
Rhinolophus imaizumii, 13:389, 13:392
Rhinolophus keyensis. *See* Kai horseshoe bats
Rhinolophus landeri. *See* Lander's horseshoe bats
Rhinolophus lepidus. *See* Blyth's horseshoe bats
Rhinolophus luctus. *See* Woolly horseshoe bats
Rhinolophus maclaudi, 13:389
Rhinolophus macrotis. *See* Rufous big-eared horseshoe bats
Rhinolophus megaphyllus. *See* Eastern horseshoe bats
Rhinolophus mehelyi. *See* Mehely's horseshoe bats
Rhinolophus mitratus. *See* Mitred horseshoe bats
Rhinolophus monoceros, 13:389
Rhinolophus paradoxolophus, 13:389
Rhinolophus philippinensis. *See* Large-eared horseshoe bats
Rhinolophus rouxii. *See* Rufous horseshoe bats
Rhinolophus simulator. *See* Bushveld horseshoe bats
Rhinolophus subbadius. *See* Little Nepalese horseshoe bats
Rhinolophus trifoliatus. *See* Trefoil horseshoe bats
Rhinolophus yunanensis. *See* Dobson's horseshoe bats
Rhinonicteris aurantia. *See* Golden horseshoe bats
Rhinophylla spp., 13:416
Rhinophylla pumilio. *See* Dwarf little fruit bats
Rhinopithecus spp., 14:172, 14:173
Rhinopithecus avunculus, 14:11, 14:175
Rhinopithecus bieti, 14:175
Rhinopithecus brelichi. *See* Gray snub-nosed monkeys
Rhinopithecus roxellana. *See* Golden snub-nosed monkeys
Rhinopoma hardwickei. *See* Hardwicke's lesser mouse-tailed bats
Rhinopoma macinnesi. *See* MacInnes's mouse-tailed bats
Rhinopoma microphyllum. *See* Greater mouse-tailed bats
Rhinopoma muscatellum. *See* Small mouse-tailed bats
Rhinopomatidae. *See* Mouse-tailed bats
Rhinos. *See* Rhinoceroses
Rhipidomys spp., 16:265, 16:266
Rhizomyinae, 16:123, 16:282, 16:284
Rhizomys sumatrensis. *See* Large bamboo rats
Rhogeessa anaeus. *See* Yellow bats
Rhombomys spp., 16:127
RHP (Resource-holding power), 12:144
Rhynchocyon spp. *See* Giant sengis
Rhynchocyon chrysopygus. *See* Golden-rumped sengis
Rhynchocyon cirnei. *See* Checkered sengis
Rhynchocyon petersi. *See* Black-and-rufous sengis
Rhynchocyoninae. *See* Giant sengis
Rhynchogale spp., 14:347
Rhyncholestes spp., 12:267–268
Rhyncholestes raphanurus. *See* Chilean shrew opossums
Rhynchomeles spp., 13:10
Rhynchomeles prattorum. *See* Seram Island bandicoots
Rhynchonycteris naso. *See* Proboscis bats
Rhytinas. *See* Steller's sea cows
Ribs, 12:41
See also Physical characteristics
Rice rats
behavior, 16:267, 16:268
distribution, 16:265–266
humans and, 16:128, 16:270
marsh, 16:268, 16:*271*, 16:*273–274*
Rio de Janeiro, 16:*271*, 16:*273*, 16:275
taxonomy, 16:263
Rice tenrecs, 13:225, 13:*226*, 13:227
Richardson's ground squirrels, 16:159*t*
Rickets, 12:74
Rickett's big-footed bats, 13:313
Ricochet saltation, 12:43
Ride, David, 13:105–106
Ridley's leaf-nosed bats. *See* Ridley's roundleaf bats
Ridley's roundleaf bats, 13:410*t*
Right whale dolphins, 15:4, 15:46
Right whales, **15:107–118**, 15:*114*
behavior, 15:109–110
conservation status, 15:9, 15:10, 15:111–112
distribution, 15:5, 15:*107*, 15:109
evolution, 15:3, 15:107
feeding ecology, 15:110
habitats, 15:109
humans and, 15:112
migrations, 12:87
physical characteristics, 15:4, 15:107, 15:109
pygmy, **15:103–106**, 15:*104*, 15:*105*
reproduction, 15:110–111
species of, 15:*115*–118
taxonomy, 15:107
Ring seals, 12:129, 14:*418*, 14:435*t*
Ring-tailed cats. *See* Ringtails
Ring-tailed coatis, 14:316*t*
Ring-tailed mongooses, 14:349, 14:350, 14:*353*, 14:*354*
Ring-tailed rock wallabies. *See* Yellow-footed rock wallabies
Ringed fruit bats. *See* Little collared fruit bats
Ringtail possums, 13:31, 13:34, 13:35, 13:39, **13:113–123**, 13:122*t*–123*t*
Ringtailed lemurs, 12:105, 14:6, 14:7, 14:*48*, 14:*49*, 14:50, 14:*51*, 14:*53*, 14:*55–56*
Ringtails (Procyonidae), 14:7, 14:259, 14:*310*, 14:*312*, 14:*313*, 14:314–315
Rio de Janeiro rice rats, 16:*271*, 16:*273*, 16:275
Rio Negro tuco-tucos, 16:*428*, 16:*429*
Rissoam dwarf. *See* Dzhungarian hamsters
Risso's dolphins, 15:1, 15:8, 15:57*t*
River dolphins, 12:79, 15:2, 15:5–6, 15:13, 15:15, 15:19
See also Baijis; Botos; Franciscana dolphins; Ganges and Indus dolphins;
River hippopotamuses. *See* Common hippopotamuses

River otters, 12:80, 12:132
River otters, northern, 14:*322*, 14:*323*
Riverine rabbits, 16:481, 16:487, 16:506, 16:509, 16:515*t*
Riversleigh, Queensland, 12:228
Roach's mouse-tailed dormice, 16:318, 16:321, 16:*322*, 16:*324*, 16:325
Roan antelopes, 16:*2*, 16:6, 16:27, 16:29–32, 16:41*t*
Roatán Island agoutis, 16:409, 16:*410*, 16:*412*–413
Roberts's lechwes, 16:33
Robinson, George Augustus, 12:310
Robinson, Michael H., 12:205
Robust chimpanzees. *See* Chimpanzees
Rock cavies, 16:*390*, 16:*394*, 16:396–*397*, 16:399*t*
  behavior, 16:392
  distribution, 16:390
  evolution, 16:389
  feeding ecology, 16:393–394
  habitats, 16:*391*
  physical characteristics, 16:390
  reproduction, 16:126, 16:393
Rock-haunting possums. *See* Rock ringtails
Rock hyraxes, 15:177–*180*, 15:*181*, 15:*182*, 15:*183*, 15:*184*, 15:*185*, 15:*187*, 15:188
Rock mice, pygmy, 16:*286*, 16:*289*, 16:290
Rock possums. *See* Rock ringtails
Rock rabbits. *See* American pikas
Rock rats, 16:434, 16:435, 16:*437*, 16:*438*–439, 16:440*t*
Rock ringtails, 13:114, 13:115, 13:117, 13:*118*, 13:*119*, 13:120
Rock voles, 16:238*t*
Rock wallabies, 13:34, 13:83, 13:84, 13:85, 13:86
Rockhares, 16:481–482, 16:505
Rocky Mountain goats. *See* Mountain goats
Rocky Mountain spotted fever, 16:148
Rodentia. *See* Rodents
Rodents, 12:13, **16:121–129**
  behavior, 12:141, 12:142–143, 16:124–125
  color vision, 12:79
  conservation status, 16:126
  digestive system, 12:14–15, 12:121
  distribution, 12:129, 12:131, 12:132, 12:134, 12:135, 12:136, 12:137, 12:138, 16:123
  evolution, 16:121–122
  feeding ecology, 16:125
  field studies of, 12:*197*
  geomagnetic fields, 12:84
  habitats, 16:124
  humans and, 16:126–128
  hystricognathe, 12:11
  Lagomorpha and, 16:479–480
  locomotion, 12:14
  as pests, 12:182
  pheromones, 12:80
  physical characteristics, 16:122–123
  reproduction, 12:90, 12:91, 12:92, 12:94, 12:95, 12:107–108, 12:110, 16:125–126
  taxonomy, 16:121–122
  teeth, 12:*37*, 12:46
  See also specific rodents
Rodricensis flying foxes, 13:*324*, 13:*326*, 13:329
Rodrigues flying foxes. *See* Rodricensis flying foxes
Roe deer
  behavior, 12:145
  competition and, 12:118
  European, 15:384, 15:*386*, 15:*388*
  evolution, 15:379, 15:380
  habitats, 15:269, 15:383
  humans and, 15:384–385
  physical characteristics, 15:382
  predation and, 12:116
  Siberian, 15:384, 15:*386*, 15:*388*–389
  taxonomy, 15:381
Rogovin, K. A., 16:216
Roman Empire, zoos in, 12:203
Romanian hamsters, 16:243
Romer, A. S., 16:333
Romerolagus spp. *See* Volcano rabbits
Romerolagus diazi. *See* Volcano rabbits
Rondo bushbabies, 14:26, 14:33*t*
Rood, J. P., 16:391
Roosevelt's muntjacs, 15:343, 15:*348*, 15:*350*
Roosting bats, 13:310, 13:311, 13:312
  See also specific roosting bats
Root rats. *See* Large bamboo rats
Root voles, 16:*229*
Rorquals, **15:119–130**, 15:*126*
  behavior, 15:6, 15:7, 15:122–123
  conservation status, 15:9, 15:10, 15:124–125
  distribution, 15:5, 15:*119*, 15:120–121
  evolution, 12:*66*, 15:3, 15:119
  feeding ecology, 15:8, 15:123–124
  habitats, 15:6, 15:121–122
  humans and, 15:125
  physical characteristics, 12:12, 12:63, 12:79, 12:82, 12:213, 15:4, 15:94, 15:119–120
  reproduction, 15:124
  species of, 15:127–129, 15:130*t*
  taxonomy, 15:1, 15:119
Ross seals, 12:138, 14:435*t*
Rossetti, Dante Gabriel, 13:53
Roth, V. L., 16:143
Rothschild's giraffes, 15:*402*
Rothschild's porcupines, 16:365, 16:366, 16:374*t*
Rottweilers, 14:292
Rough-haired golden moles, 13:222*t*
Rough-legged jerboas. *See* Hairy-footed jerboas
Rough-toothed dolphins, 15:8, 15:49
Roulin, X., 15:242
Roulin's tapirs. *See* Mountain tapirs
Round-eared bats, 13:311
Round-eared sengis, 16:518, 16:519, 16:*520*, 16:522, 16:*523*, 16:*525*–*526*
Round-tailed muskrats, 16:225, 16:238*t*
Roundleaf bats
  Ashy, 13:412*t*
  Noack's, 13:*406*, 13:*407*, 13:408
  Ridley's, 13:410*t*
  Sundevall's, 13:*403*
  See also specific types of roundleaf bats
Roundleaf horseshoe bats, 13:*403*
Rousette bats, 12:55, 13:333
Rousettus spp., 13:334
Rousettus aegyptiacus. *See* Egyptian rousettes
Rowe, D. L., 16:389
Royal antelopes, 16:60, 16:63, 16:71*t*
Rubrisciurus spp., 16:163
Rufescent bandicoots. *See* Rufous spiny bandicoots
Ruffed lemurs. *See* Variegated lemurs
Rufous bettongs, 13:*74*, 13:*75*, 13:*76*, 13:*77*, 13:*79*, 13:80–81,
Rufous big-eared horseshoe bats, 13:390
Rufous elephant shrews. *See* Rufuou sengis
Rufous hare-wallabies, 13:34, 13:*91*, 13:*98*, 13:99
Rufous horseshoe bats, 13:389, 13:391, 13:392, 13:*395*, 13:*396*–397
Rufous leaf bats. *See* Rufous horseshoe bats
Rufous mouse-eared bats, 13:311
Rufous ringtails. *See* Common ringtails
Rufous sengis, 16:*521*, 16:*523*, 16:*528*, 16:529
Rufous spiny bandicoots, 13:2, 13:4, 13:*13*, 13:*15*, 13:16
Rufous-tailed tree shrews, 13:298*t*
Rufous tree rats, 16:458*t*
Rumbaugh, Duane, 12:161
Rumen, 12:122–124
  See also Physical characteristics
Ruminants
  evolution, 12:216, 15:136, 15:263–265
  feeding ecology, 12:14, 12:118–119, 12:120–124, 15:141, 15:271–272
  physical characteristics, 12:10
  taxonomy, 15:263, 15:266–267
  vitamin requirements, 12:120
  See also specific ruminants
Rumination, 12:122–124
  See also Feeding ecology
Rumpler, Y., 14:74–75
Running, 12:42
  See also Behavior
Runwenzorisorex spp., 13:265
Rupicapra spp. *See* Chamois
Rupicapra pyrenaica. *See* Southern chamois
Rupicapra rupicapra. *See* Northern chamois
Rupicaprini, 16:87, 16:88, 16:89, 16:90, 16:94
Ruppell's horseshoe bats, 13:390
Rusa, 12:19
Russell, Charlie, 15:416
Russian brown bears. *See* Brown bears
Russian desmans, 13:200, 13:280, 13:281, 13:282, 13:*283*, 13:*285*, 13:286
Russian flying squirrels. *See* Siberian flying squirrels
Ruwenzori otter shrews, 13:234*t*
Ruwenzori shrews, 13:*269*, 13:*272*, 13:274–275
Ruwenzorisorex suncoides. *See* Ruwenzori shrews
Ryukyu flying foxes, 13:331*t*
Ryukyu moles, 13:282
Ryukyu rabbits. *See* Amami rabbits

S

Saber-toothed cats, 14:256
Saber-toothed tigers, 12:*23*, 12:*24*, 14:369, 14:375
  See also Saber-toothed cats
Sable antelopes, 16:2, 16:5, 16:27, 16:28, 16:29, 16:31, 16:32, 16:*36*, 16:*39*, 16:41
Sables, 14:324
Sac-winged bats, **13:355–365**, 13:*359*, 13:363*t*, 13:364*t*
Saccolaimus peli. *See* Giant pouched bats

Saccopteryx bilineata. See Greater sac-winged bats
Saccopteryx leptura. See Lesser sac-winged bats
Saccostomus campestris. See Pouched mice
Sachser, N., 16:392
Sacred baboons. *See* Hamadryas baboons
Sacred langurs. *See* Northern plains gray langurs
Saddleback tamarins, 14:116–121, 14:*125*, 14:127–*128*
Saddleback tapirs. *See* Malayan tapirs
Sado moles, 13:282
Sagebrush voles, 16:237*t*
Sage's rock rats, 16:440*t*
Saguinus spp. *See* Tamarins
Saguinus bicolor. See Pied tamarins
Saguinus fuscicollis. See Saddleback tamarins
Saguinus geoffroyi. See Geoffroy's tamarins
Saguinus graellsi. See Graell's black-mantled tamarins
Saguinus imperator. See Emperor tamarins
Saguinus inustus. See Mottle-faced tamarins
Saguinus labiatus. See Red-bellied tamarins
Saguinus leucopus. See White-lipped tamarins
Saguinus midas. See Midas tamarins
Saguinus mystax. See Moustached tamarins
Saguinus nigricollis. See Black-mantled tamarins
Saguinus oedipus. See Cotton-top tamarins
Saguinus oedipus oedipus. See Cotton-top tamarins
Sahara gundis. *See* Desert gundis
Saharagalago spp., 14:23
Sahelanthropus chadensis, 14:241–242
Sahlins, Marshall, 14:248
Saiga antelopes, 16:4, 16:9, **16:45–58,** 16:*49*, 16:*53*
Saiga tatarica. See Saiga antelopes
Saigas. *See* Saiga antelopes
Saimiri spp. *See* Squirrel monkeys
Saimiri boliviensis. See Bolivian squirrel monkeys
Saimiri oerstedii. See Red-backed squirrel monkeys
Saimiri oerstedii citrinellus, 14:105
Saimiri oerstedii oerstedii, 14:105
Saimiri sciureus. See Common squirrel monkeys
Saimiri ustus, 14:101, 14:102
Saimiri vanzolinii. See Blackish squirrel monkeys
Saint Bernards, 14:293
St. Michel nesophontes, 13:243
Sakis, **14:143–154,** 14:*149*
 behavior, 14:145–146
 conservation status, 14:148
 distribution, 14:144–145
 evolution, 14:143
 feeding ecology, 14:146–147
 habitats, 14:145
 humans and, 14:148
 physical characteristics, 14:143–144
 reproduction, 14:147
 species of, 14:*150*–151, 14:153*t*
 taxonomy, 14:143
Salanoia spp., 14:347
Salanoia concolor. See Malagasy brown mongooses
Salim Ali's fruit bats, 13:349*t*
Salinoctomys spp., 16:434, 16:436
Salinoctomys loschalchalerosorum, 16:433–434
Saliva, 12:122
Salpingotus spp., 16:217
Salpingotus crassicauda. See Thick-tailed pygmy jerboas
Salpingotus heptneri, 16:212, 16:216
Salpingotus kozlovi, 16:212, 16:215, 16:216
Salpingotus michaelis, 16:217
Salpingotus pallidus, 16:216
Salpingotus thomasi, 16:217
Salt-desert cavies, 16:389, 16:390–391, 16:393, 16:*394*, 16:*397*, 16:398
Salt licks, 12:17–18
Salta tuco-tucos, 16:431*t*
Saltatory mammals, 12:42–43
Saltmarsh harvest mice, 16:*265*
Salt's dikdiks, 16:*65*, 16:*67*, 16:69–70
Salukis, 14:292
Salvin's spiny pocket mice, 16:*204*, 16:*205*–206
Samana hutias, 16:467*t*
Sambars, 12:19, 15:*358*, 15:*362*, 15:*367*
Samoan flying foxes, 13:332*t*
Samotherium spp., 15:265
San Joaquin antelope squirrels. *See* Nelson's antelope squirrels
San Joaquin pocket mice, 16:*204*, 16:*205*
San Joaquin Valley kangaroo rats, 16:209*t*
Sanborn's flying foxes, 13:331*t*
Sanchez-Cordero, V., 16:200
Sand cats, 14:371, 14:390*t*
A Sand County Almanac, 12:223
Sand gazelles. *See* Slender-horned gazelles
Sand rats. *See* Sundevall's jirds
Sand-swimming, 12:71–72
 See also Behavior
Sandhill dunnarts, 12:301*t*
Sandstone dibblers, 12:300*t*
Sandy wallabies. *See* Agile wallabies
Sangihe tarsiers, 14:95, 14:97, 14:99
Saolas, 12:129, 16:4, 16:11
Sarcophilus laniarius. See Tasmanian devils
Sardinian pikas, 16:502*t*
Sasins. *See* Blackbucks
Sassabies, 16:27, 16:32
Satunin's five-toed pgymy jerboas. *See* Five-toed pygmy jerboas
Sauromys spp., 13:311
Savage-Rumbaugh, Sue, 12:162
Savanna baboons, 14:*193*
Savanna dormice, 16:*320*, 16:327*t*
Savanna elephants, 15:162–163, 15:165
Savannas
 Ice Age and, 12:18–19
 tropical, 12:219
 See also Habitats
Savi's pygmy shrews, 12:12, 13:194, 13:266, 13:*267*, 13:*270*, 13:271–*272*
Savolainen, Peter, 14:287
Scala naturae, 12:149
Scalopinae, 13:279
Scalopus aquaticus. See Eastern moles
Scaly anteaters. *See* Ground pangolins; Pangolins
Scaly-tailed possums, 13:57, 13:59, 13:60, 13:*62*, 13:*65*–66
Scaly-tailed squirrels, 16:122, 16:123, **16:299–306,** 16:*301*
Scammon, Charles, 15:96
Scandentia. *See* Tree shrews
Scansorial adaptations, 12:14
Scapanus latimanus. See Broad-footed moles
Scaponulus oweni. See Gansu moles
Scapteromys spp., 16:268
Scapteromys aquaticus, 16:268
Scapteromys tumidus, 16:266, 16:268
Scaptochirus spp., 13:279
Scaptochirus moschatus. See Short-faced moles
Scaptonyx fuscicaudatus. See Long-tailed moles
Scapula, 12:41
Scent glands, 12:37
Scent-marking, 12:80–81, 14:*123*
 See also Behavior
Scenthounds, 14:*292*
Schaller's mouse shrews, 13:276*t*
Schmidly, D. J., 16:200
Schnabeli's thumbless bats, 13:*469*, 13:*470*–471
Schomburgk's deer, 15:360, 15:371*t*
Schonbrunn Zoo, Vienna, Austria, 12:203
Schouteden's shrews, 13:269, 13:*272*, 13:274
Schreiber's bent-winged bats. *See* Common bentwing bats
Schreiber's long-fingered bats. *See* Common bentwing bats
Scientific research, in zoos, 12:204
Scimitar-horned oryx, 16:27–28, 16:29, 16:31, 16:32, 16:33, 16:41*t*
Sciuridae, 12:79, 16:121–122, 16:123, 16:125, 16:126
Sciurillus spp., 16:163
Sciurognathi, 12:71, 16:121, 16:125
Sciurotamias spp., 16:143, 16:144
Sciurotamias forresti, 16:147
Sciurus spp., 16:163, 16:300
Sciurus aberti. See Abert squirrels
Sciurus alleni. See Allen's squirrels
Sciurus anomalus. See Persian squirrels
Sciurus arizonensis. See Arizona gray squirrels
Sciurus carolinensis. See Gray squirrels
Sciurus deppei. See Deppe's squirrels
Sciurus dubius, 16:163
Sciurus griseus. See Western gray squirrels
Sciurus nayaritensis. See Mexican fox squirrels
Sciurus niger. See Eastern fox squirrels
Sciurus niger avicennia. See Big cypress fox squirrels
Sciurus niger cinereus. See Delmarva fox squirrels
Sciurus oculatus. See Peters' squirrels
Sciurus richmondi, 16:167
Sciurus sanborni, 16:167
Sciurus variegatoides. See Variegated squirrels
Sciurus vulgaris. See Red squirrels
Sciurus yucatanensis. See Yucatán squirrels
Sclater's golden moles, 13:198, 13:200
Sclater's lemurs. *See* Black lemurs
Scolomys spp., 16:265
Scoop (Waugh), 12:183
Scotomanes ornatus. See Harlequin bats
Scotonycteris zenkeri. See Zenker's fruit bats
Scotophilus spp., 13:311, 13:497
Scotophilus kuhlii. See House bats
Scrotum, 12:103
 See also Physical characteristics
Scrub wallabies. *See* Black-striped wallabies
Scutisorex spp., 13:265
Scutisorex somereni. See Armored shrews
Sea bears. *See* Northern fur seals; Polar bears
Sea cows. *See* Dugongidae; Steller's sea cows

Sea lions, 12:14, 12:*145*, 14:256, **14:393–408,** 14:*401*
behavior, 14:396–397
conservation status, 14:399–400
distribution, 12:138, 14:394
evolution, 14:393
feeding ecology, 14:397–398
habitats, 14:395–396
learning set, 12:152
locomotion, 12:44
memory, 12:153
physical characteristics, 12:63, 14:393–394
reproduction, 14:398–399
species of, 14:*403*–406
Sea minks, 14:262
Sea otters, 12:14, 14:258, 14:321, 14:324
conservation status, 12:214
distribution, 12:138
physical characteristics, 12:63, 12:*64*
tool use, 12:*158*
Seals, 12:14, 14:255
distribution, 12:129, 12:136, 12:138
eared, **14:393–408,** 14:*401*
locomotion, 12:44
neonatal milk, 12:126–127
neonates, 12:125
physical characteristics, 12:63
reproduction, 12:91
touch, 12:82–83
See also True seals
Search dogs, 14:293
Seasonality
of migrations, 12:164–170
of nutritional requirements, 12:124–125
of reproduction, 12:91–92, 12:103
Seattle Zoo, 12:*210*
Sebaceous glands, 12:37
See also Physical characteristics
Seba's short-tailed bats, 13:*312*, 13:417, 13:418, 13:419, 13:*421*, 13:*429*–430
Sebum, 12:37
Secondary bony palate, 12:8, 12:11
Seeing-eye dogs, 14:*293*
Sei whales, 15:5, 15:8, 15:120, 15:121, 15:125, 15:130*t*
Seismic communication, 12:76
See also Behavior
Selevinia spp. *See* Desert dormice
Selevinia betpakdalaensis. See Desert dormice
Self-directed behaviors, 12:159
See also Behavior
Self recognition, 12:159–160
See also Cognition
Selfish herds, 12:167–168
Semantic memory, 12:154
Semifossorial mammals, 12:13
See also Rodents
Semnopithecus spp., 14:172, 14:173
Semnopithecus entellus. See Northern plains gray langurs
Senegal bushbabies, 12:105, 14:*24*, 14:*25*, 14:*26*, 14:*27*, 14:*28*, 14:30
Senegal galagos. *See* Senegal bushbabies
Sengis, 12:135, **16:517–532,** 16:*523*, 16:531*t*
Sensory systems, 12:5–6, 12:9, 12:49–50, 12:75–77, **12:79–88,** 12:114
See also Physical characteristics
Seram Island bandicoots, 13:10, 13:12
Serial position effect, 12:154
Serotine bats, 13:498, 13:515*t*
Serows, 12:135, 16:2, 16:87, 16:89, 16:90, 16:92, 16:95, 16:*96*, 16:*98*
Servals, 14:*378*, 14:*381*, 14:384
Service dogs, 14:293
Setifer spp., 13:225, 13:226, 13:230
Setifer setosus. See Greater hedgehog tenrecs
Setonix spp., 13:83, 13:85
Setonix brachyurus. See Quokkas
Setzer's hairy-footed gerbils, 16:296*t*
Setzer's mouse-tailed dormice, 16:328*t*
Seven-banded armadillos, 13:183, 13:*184*, 13:191*t*
Sewall Wright effect. *See* Random genetic drift
Sewellels. *See* Mountain beavers
Sex chromosomes, 12:30–31
Sex determination, 12:10
Sexual dimorphism, 12:99
See also Physical characteristics
Sexual maturity, 12:97, 12:107
See also Reproduction
Seychelles sheath-tailed bats, 13:358, 13:364*t*
Shadow chipmunks. *See* Allen's chipmunks
Shaggy-haired bats, 13:364*t*
Shakespeare, William, 13:200, 13:207
Shamolagus, 16:480
Shanxi hedgehogs. *See* Hugh's hedgehogs
Shark-toothed dolphins, 15:3
Sharp-nosed bats. *See* Proboscis bats
Sharpe's grysboks, 16:*61*, 16:71*t*
Sharpe's grysbucks. *See* Sharpe's grysboks
Sharpe's steinboks. *See* Sharpe's grysboks
Shaw, George, 12:227, 12:243
Sheath-tailed bats, 12:81, **13:355–365,** 13:*359*, 13:364*t*
Sheep, 15:*147*, 15:263, 15:265, 16:1, **16:87–105,** 16:*96*, 16:*97*
behavior, 15:271, 16:92–93
conservation status, 16:94–95
digestive system, 12:122
distribution, 16:4, 16:*87*, 16:90–91
evolution, 16:87–89
feeding ecology, 16:93–94
feral cats and, 12:185
habitats, 16:91–92
humans and, 15:273, 16:8, 16:9, 16:95
nutritional requirements, 12:124–125
physical characteristics, 16:3, 16:89–90
reproduction, 16:94
smell, 12:85
species of, 16:98–104*t*
taxonomy, 16:87–89
See also specific types of sheep
Sheepdogs, 14:288, 14:292
Shenbrot, G. I., 16:216
Shepherd's beaked whales, 15:59, 15:*63*, 15:*67*–68
Shetland ponies, 15:*230*
Shiny guinea pigs, 16:399*t*
Ship rats. *See* Black rats
Short-beaked echidnas, 12:*227*, 12:*239*, 12:*240*, 12:240*t*
behavior, 12:*230*, 12:237, 12:*238*
conservation status, 12:233
distribution, 12:230, 12:236
feeding ecology, 12:*228*, 12:231, 12:*236*, 12:237
habitats, 12:230, 12:237
humans and, 12:238
physical characteristics, 12:235–236
reproduction, 12:108, 12:231–232, 12:233, 12:237
taxonomy, 12:227, 12:235
Short-beaked saddleback dolphins, 15:2, 15:3, 15:42, 15:44, 15:47, 15:56*t*
Short-eared dogs, 14:283*t*
Short-eared sengis. *See* Round-eared sengis
Short-faced bears, 12:24
Short-faced moles, 13:288*t*
Short-finned pilot whales, 15:56*t*
Short-furred dasyures, 12:299*t*
Short-horned water buffaloes, 16:1
Short interspersed nuclear elements (SINEs), 12:29–30
Short-nosed bandicoots. *See* Southern brown bandicoots
Short-nosed fruit bats, 13:311
Short-palate fruit bats, 13:348*t*
Short-snouted sengis, 16:*520*, 16:*523*, 16:*527*–528
Short-tailed bandicoot rats, 16:261*t*
Short-tailed bats, chestnut, 13:433*t*
Short-tailed chinchillas, 16:377, 16:380, 16:383*t*
Short-tailed gymnures. *See* Lesser gymnures
Short-tailed macaques, 14:193
Short-tailed opossums, 12:250–254
Short-tailed porcupines, 16:365, 16:374*t*
Short-tailed shrews, 13:195, 13:197, 13:198, 13:200
Short-tailed wallabies. *See* Quokkas
Short-term memory, 12:152
Shoshani, J., 15:177
Shrew gymnures, 13:213*t*
Shrew mice, 13:197, 16:265, 16:*271*, 16:*273*, 16:274
Shrew moles, 13:196, 13:198, **13:279–288,** 13:*283*, 13:288*t*
Shrew opossums, 12:132, **12:267–271,** 12:*269*
Shrew tenrecs, 13:225
Shrews, 13:193–200, 13:247
alpine, 13:*254*, 13:*258*, 13:260
behavior, 12:145
bicolored, 13:*266*
color vision, 12:79
distribution, 12:131
echolocation, 12:53, 12:86
musk, 12:107, 12:115, 13:198, 13:199, 13:266
Nimba otter, 13:234*t*
otter, 13:194, 13:199, 13:225
red-toothed, **13:247–264**
Ruwenzori otter, 13:234*t*
smell, 12:80
West Indian, 12:133, 13:193, 13:197, **13:243–245**
white-toothed, 13:198, 13:199, **13:265–277**
See also Tree shrews
Shumaker, Robert, 12:161
Siamangs, 14:*216*, 14:222
behavior, 14:*210*, 14:210–211
distribution, 14:*221*
evolution, 14:207
feeding ecology, 14:212–213
habitats, 14:210
physical characteristics, 14:209
taxonomy, 14:207
Siamopithecus spp., 14:2
Siberian-American moose, 15:380

Siberian chipmunks, 16:144, 16:158*t*
Siberian-European moose, 15:380
Siberian flying squirrels, 16:135, 16:*138*, 16:139–*140*
Siberian huskies, 14:289, 14:*292*, 14:293
Siberian ibex, 16:87, 16:91–92, 16:93, 16:*97*, 16:*101*–102
Siberian marals, 15:357
Siberian musk deer, 15:335, 15:*336*, 15:*337*, 15:*338*, 15:*339*, 15:*340*–341
Siberian pikas. *See* Northern pikas
Siberian roe deer, 15:384, 15:*386*, 15:*388*–389
Siberian snow sheep, 12:178
Siberian tigers. *See* Amur tigers
Siberian water deer, 15:376
Siberian zokors, 16:*287*, 16:*290*, 16:293–294
Sichuan golden snub-nosed monkeys. *See* Golden snub-nosed monkeys
Sicista spp., 16:216
Sicista armenica, 16:216
Sicista betulina, 16:214, 16:*216*
Sicista caudata, 16:216
Sicista subtilis. *See* Southern birch mice
Side-striped jackals, 12:117, 14:283*t*
Sidneyia, 12:64
Sifakas, **14:63–72,** 14:*68*
behavior, 14:7, 14:8, 14:65–66
conservation status, 14:67
distribution, 14:65
evolution, 14:63
feeding ecology, 14:66
habitats, 14:65
humans and, 14:67
physical characteristics, 14:63–65
reproduction, 14:66–67
species of, 14:*70*
taxonomy, 14:63
Sight dogs, 14:288
Sighthounds, 14:292
Sigmoceros lichtensteinii. *See* Lichtenstein's hartebeests
Sigmodon spp., 16:127, 16:263, 16:264, 16:265–266
Sigmodon alstoni, 16:270
Sigmodon hispidus. *See* Hispid cotton rats
Sigmodontinae, 16:128, **16:263–279,** 16:*271*, 16:277*t*–278*t*, 16:282
Sigmodontini, 16:264
Sigmodontomys aphrastus, 16:270
Sika deer, 12:19, 15:358, 15:361, 15:*363*, 15:*368*, 15:369
Silent dormice, 16:327*t*
Silky anteaters, 13:*172*, 13:*174*, 13:*175*, 13:*176*, 13:*177*
Silky cuscuses, 13:66*t*
Silky pocket mice, 16:199, 16:208*t*
Silky shrew opossums, 12:*269*, 12:*270*
Silt, from glaciers, 12:17
Silver dikdiks, 16:63, 16:71*t*
Silver foxes. *See* Red foxes
Silver-haired bats, 13:310, 13:*503*, 13:*505*, 13:*508*, 13:509
Silver pikas, 16:487, 16:495
Silverback jackals, 12:8, 12:117, 12:216, 14:*256*, 14:*257*, 14:*269*, 14:*272*, 14:283*t*
Silverbacks, 12:*141*, 14:228
Silvered langurs. *See* Silvery leaf-monkeys
Silvery gibbons. *See* Moloch gibbons
Silvery greater bushbabies, 14:33*t*
Silvery leaf-monkeys, 14:*177*, 14:*181*
Silvery lutungs. *See* Silvery leaf-monkeys
Silvery marmosets, 14:*124*, 14:132*t*
Silvery mole-rats, 12:72, 12:74, 12:75, 12:76, 16:339, 16:342, 16:344, 16:349*t*
Silvery mountain voles, 16:*232*, 16:*235*, 16:236
Simakobus. *See* Mentawai Island langurs
Simian primates, 12:90, 14:8
Simias spp. *See* Mentawai Island langurs
Simias concolor. *See* Mentawai Island langurs
Simmons, N. B., 13:333
Simplex uterus, 12:90–91, 12:*102*, 12:110–111
Simplicendentata. *See* Rodents
Simpson, George G.
on deer, 15:380–381
on Dipodidae, 16:211
on Lagomorpha, 16:479
on tree shrews, 13:289
SINEs (Short interspersed nuclear elements), 12:29–30
Singapore roundleaf horseshoe bats. *See* Ridley's roundleaf bats
Singing fruit bats, 13:*338*, 13:*341*, 13:342
Singing rats. *See* Amazon bamboo rats
Sinoconodon spp., 12:11
Sirenia, **15:191–197**
behavior, 15:192–193
conservation status, 15:194–195
distribution, 12:138, 15:192
evolution, 15:133, 15:191
feeding ecology, 15:193
habitats, 12:14, 15:192
humans and, 15:195–196
physical characteristics, 12:63, 12:68, 15:192
reproduction, 12:103, 15:193–194
taxonomy, 15:131, 15:134, 15:191–192
See also Dugongs; Manatees; Steller's sea cows
Sitatungas, 16:2, 16:4, 16:11
Six-banded armadillos. *See* Yellow armadillos
Size. *See* Body size
Skeletal system, 12:40–41, 12:65, 12:174
See also Physical characteristics
Skew-beaked whales. *See* Hector's beaked whales
Skin, 12:7, 12:36, 12:*39*
See also Physical characteristics
Skull, 12:8, 12:10, 12:41, 12:174
See also Physical characteristics
Skunks, **14:319–325,** 14:*320*, 14:*326*, **14:329, 14:332–334**
behavior, 12:*186*, 14:321–323
conservation status, 14:324
distribution, 14:3*19*, 14:321
evolution, 14:319
feeding ecology, 12:192, 14:323
habitats, 14:319–321
humans and, 14:324–325
physical characteristics, 12:38, 14:319–321
reproduction, 14:324
species of, 14:*329*, 14:332*t*, 14:333*t*
taxonomy, 14:256, 14:319
Slaty mouse opossums. *See* Slaty slender mouse opossums
Slaty slender mouse opossums, 12:265*t*
Slender Cuban nesophontes, 13:243
Slender falanoucs. *See* Falanoucs
Slender-horned gazelles, 16:48, 16:57*t*
Slender lorises, 14:13, 14:14, 14:*15*, 14:16, 14:*17*, 14:*18*, 14:21*t*
Slender mongooses, 14:349, 14:357*t*
Slender mouse opossums, 12:250–253
Slender-tailed dunnarts, 12:*290*, 12:301*t*
Slender tree shrews, 13:297*t*
Slit-faced bats, 13:310, **13:371–377,** 13:*372*, 13:*374*, 13:376*t*–377*t*
See also specific types of slit-faced bats
Sloggett's vlei rats. *See* Ice rats
Sloth bears, 14:295, 14:296, 14:297, 14:298, 14:299, 14:300, 14:306*t*
Sloth lemurs, 14:63, 14:*68*, 14:*70*, 14:71
Sloths, 13:147–153, 13:155, 13:156, 13:165
claws, 12:39
digestive systems of, 12:15, 12:123
lesser Haitian ground, 13:*159*
locomotion, 12:43
monk, 13:*166*, 13:*168*
reproduction, 12:94, 12:103
southern two-toed, 13:*148*, 13:*157*
West Indian, **13:155–159**
See also Tree sloths; specific types of sloths
Slow lorises, 14:13, 14:14
See also Pygmy slow lorises; Sunda slow lorises
Small Asian sheath-tailed bats, 13:*357*
Small-eared dogs, 14:266
Small-eared shrews, 13:198
Small fat-tailed opossums, 12:265*t*
Small five-toed jerboas. *See* Little five-toed jerboas
Small flying foxes. *See* Island flying foxes
Small-footed funnel-eared bats, 13:*461*, 13:*462*, 13:*464*–465
Small hairy armadillos, 13:*186*, 13:187–*188*
Small Indian mongooses, 14:349, 14:*353*, 14:*355*
Small intestine, 12:47–48, 12:123
See also Physical characteristics
Small Lander's horseshoe bats, 13:388
Small mouse-tailed bats, 13:352
Small Population Analysis and Record Keeping System (SPARKS), 12:206–207
Small red brocket deer. *See* Brocket deer
Small-scaled pangolins. *See* Tree pangolins
Small screaming armadillos. *See* Small hairy armadillos
Small-spotted genets. *See* Common genets
Small Sulawesi cuscuses, 13:57–58, 13:60, 13:*62*, 13:*63*–64
Small-toothed fruit bats, 13:348*t*
Small-toothed moles, 13:282
Small-toothed mongooses. *See* Falanoucs
Small-toothed palm civets, 14:344*t*
Small-toothed sportive lemurs, 14:*78*, 14:*79*, 14:82–83
Smaller horseshoe bats. *See* Eastern horseshoe bats
Smell, 12:5–6, 12:49–50, 12:80–81, 12:*81*, 12:*83*
See also Physical characteristics
Smilodon californicus. *See* Saber-toothed cats
Sminthopsis archeri, 12:279
Sminthopsis crassicaudata. *See* Fat-tailed dunnarts
Sminthopsis leucopus. *See* White-footed dunnarts
Sminthopsis longicaudata. *See* Long-tailed dunnarts
Sminthopsis murina. *See* Slender-tailed dunnarts

Sminthopsis psammophila. See Sandhill dunnarts
Sminthopsis virginiae, 12:279
Smintinae, 16:211, 16:212
Smith, M. F., 16:263
Smith's rock sengis. *See* Western rock sengis
Smith's tree squirrels, 16:*165*
Smoky bats, 13:308, **13:467–471,** 13:*468*, 13:*469*, 13:*470*
Smoky pocket gophers, 16:197*t*
Smoky shrews, 13:*194*
Smooth-coated otters, 14:325
Smooth-toothed pocket gophers. *See* Valley pocket gophers
Snow leopards, 14:370–371, 14:372, 14:374, 14:377, 14:*380*, 14:386–387
Snow sheep, 16:87, 16:91, 16:92, 16:104*t*
Snow voles, 16:228
Snowshoe hares, 16:*508*, 16:*510*, 16:*511*
feeding ecology, 16:484
habitats, 16:481, 16:506
humans and, 16:488
physical characteristics, 12:38, 16:*479*, 16:*507*
predation and, 12:115
Snub-nosed monkeys, 14:11
See also Golden snub-nosed monkeys
Social behavior, 12:4–6, 12:9–10
See also Behavior
Social cognition, 12:160
See also Cognition
Social complexity, human, 14:251–252
Social grooming, 14:7
See also Behavior
Social learning, 12:157–159
Social organization. *See* Behavior
Social systems, 12:143–148
See also Behavior
Social tuco-tucos, 16:124, 16:426, 16:427, 16:*428*, 16:*429*–430
Sociobiology, 12:148
Sociological techniques, species lists and, 12:195
Sodium monofluoroacetate (1080), 12:186
Soft-furred sengis, 16:518, 16:520–521, 16:525–530
Soft palate, 12:8
See also Physical characteristics
Sokolov, V. E., 16:216
Solenodon spp. *See* Solenodons
Solenodon arredondoi. See Arredondo's solenodons
Solenodon cubanus. See Cuban solenodons
Solenodon marconoi. See Marcano's solenodons
Solenodon paradoxus. See Hispaniolan solenodons
Solenodons, 12:133, 13:193, 13:194, 13:195, 13:197, 13:198, 13:199, **13:237–242**
Solenodontidae. *See* Solenodons
Solisorex spp., 13:265
Solisorex pearsoni. See Pearson's long-clawed shrews
Solitary behavior, 12:145, 12:148
See also Behavior
Solos, 14:208
Somali bushbabies, 14:26, 14:32*t*
Somali sengis, 16:522, 16:531*t*
Somali wild asses, 12:177
See also African wild asses
Somalian hedgehogs, 13:213*t*
Somatosensory perception, 12:77
See also Physical characteristics
Sommer, V., 14:214
Sonar, in field studies, 12:201
Sonoran pronghorns, 15:415, 15:416
Sooty mangabeys, 14:194, 14:204*t*
Sooty moustached bats, 13:435, 13:436, 13:442*t*
Sorex spp., 13:251, 13:268
Sorex alpinus. See Alpine shrews
Sorex araneus. See Common shrews
Sorex bendirii, 13:249
Sorex cinereus. See Masked shrews
Sorex daphaenodon, 13:248
Sorex dispar. See Long-tailed shrews
Sorex fumeus. See Smoky shrews
Sorex hoyi. See American pygmy shrews
Sorex minutissimus, 13:247
Sorex minutus. See Eurasian pygmy shrews
Sorex mirabilis. See Giant shrews
Sorex ornatus. See Ornate shrews
Sorex palustris. See American water shrews
Sorex tundrensis. See Tundra shrews
Soriacebus spp., 14:143
Soricidae. *See* Shrews
Soricini, 13:247
Soriciniae. *See* Red-toothed shrews
Soriculus spp., 13:248
Soriculus caudatus. See Hodgson's brown-toothed shrews
Sotalia fluviatilis. See Tucuxis
Sound receptors, 12:9
Sounds, 12:81–82
Sousa spp. *See* Humpback dolphins
Sousa chinensis. See Humpback dolphins
Sousa teuszi. See Atlantic humpbacked dolphins
South African crested porcupines. *See* South African porcupines
South African ground squirrels, 16:145, 16:*149*, 16:*151*, 16:154
South African pangolins. *See* Ground pangolins
South African porcupines, 16:*352*, 16:*354*, 16:*355*, 16:*361*–362
South America. *See* Neotropical region
South American beavers. *See* Coypus
South American bush rats. *See* Degus
South American deer, 15:380, 15:382, 15:383
South American fur seals, 14:394, 14:399, 14:407*t*
South American maras, 16:125
South American mice, **16:263–279,** 16:*271*, 16:277*t*–278*t*
South American monotremes, 12:228
South American rats, **16:263–279,** 16:*271*, 16:277*t*–278*t*
South American sea lions, 14:394, 14:396, 14:399, 14:*401*, 14:*404*–405
South American tapirs. *See* Lowland tapirs
Southeast Asia. *See* Oriental region
Southeast Asian bent-winged bats, 13:526*t*
Southeast Asian porcupines. *See* Common porcupines
Southeast Asian shrews, 13:276*t*
Southeastern pocket gophers, 16:187, 16:195*t*
Southern African hedgehogs, 13:204, 13:*206*, 13:207, 13:*208*, 13:*209*–210
Southern Bahian masked titis, 14:148
Southern bamboo rats, 16:450, 16:*452*, 16:*454*, 16:456–457
Southern beaked whales. *See* Gray's beaked whales
Southern bearded sakis, 14:*146*, 14:148, 14:151
Southern birch mice, 16:*214*, 16:*218*, 16:*219*
Southern blossom bats, 13:*336*, 13:*339*, 13:*346*–347
Southern bottlenosed whales, 15:69*t*
Southern brown bandicoots, 13:2, 13:5, 13:10–11, 13:*12*, 13:16*t*
Southern chamois, 16:87, 16:103*t*
Southern dibblers, 12:281, 12:*293*, 12:*295*–296
Southern elephant seals, 14:256, 14:421–422, 14:*423*, 14:435*t*
Southern flying squirrels, 12:*55*, 16:135, 16:*136*, 16:*138*, 16:139, 16:*140*
Southern four-toothed whales. *See* Arnoux's beaked whales
Southern fur seals. *See* Galápagos fur seals
Southern giant bottlenosed whales. *See* Arnoux's beaked whales
Southern grasshopper mice, 16:277*t*
Southern hairy-nosed wombats, 13:40, 13:51, 13:52, 13:53, 13:*54*, 13:*55*
Southern koalas, 13:44
Southern lesser bushbabies. *See* Moholi bushbabies
Southern long-nosed armadillos, 13:191*t*
Southern long-nosed bats, 13:415, 13:417, 13:*419*, 13:420, 13:*421*, 13:*428*
Southern long-tongued bats, 13:433*t*
Southern muriquis, 14:11, 14:156, 14:158, 14:159, 14:*161*, 14:*164*
Southern naked-tailed armadillos, 13:*185*, 13:190*t*
Southern needle-clawed bushbabies, 14:26, 14:*27*, 14:*28*
Southern opossums, 12:131, 12:*143*, 12:*251*, 12:*253*, 12:264*t*
Southern pocket gophers, 16:*188*, 16:195*t*
Southern pudus, 15:379, 15:*382*, 15:384, 15:*387*, 15:*389*, 15:392–393
Southern red-backed voles, 16:237*t*
Southern reedbucks, 16:27–30, 16:*33*, 16:*35*, 16:*37*
Southern right whale dolphins, 15:42
Southern right whales, 15:107, 15:*108*, 15:109, 15:*110*, 15:111, 15:*112*, 15:*113*, 15:*114*, 15:*117*–118
See also Right whales
Southern river otters, 14:324
Southern short-tailed bats. *See* Greater New Zealand short-tailed bats
Southern short-tailed opossums, 12:250, 12:251, 12:265*t*
Southern short-tailed shrews, 13:198, 13:*252*
Southern tamanduas, 13:*151*, 13:*172*, 13:*176*, 13:177–*178*
Southern three-banded armadillos, 13:*186*, 13:188–*189*
Southern tree hyraxes, 15:177, 15:*181*, 15:*182*, 15:184, 15:*185*, 15:*186*
Southern tuco-tucos, 16:430*t*
Southern two-toed sloths, 13:*148*, 13:*157*
Southern viscachas, 16:*379*, 16:380, 16:383*t*
Sowerby's beaked whales, 15:69*t*
Spade-toothed whales, 15:1, 15:70*t*

Spalacinae, 16:123, 16:282, 16:284
Spalacopus spp., 16:433
Spalacopus cyanus. See Coruros
Spalacopus cyanus maulinus, 16:439
Spalax spp. *See* Blind mole-rats
Spalax ehrenbergi. See East Mediterranean blind mole-rats
Spalax graecus. See Bukovin mole rats
Spallanzani, Lazzaro, 12:54
Spanish ibex, 16:87, 16:91, 16:104*t*
SPARKS (Small Population Analysis and Record Keeping System), 12:206–207
Spatial memory, 12:140–142, 12:153
Spear-nosed bats, greater, 13:312
Species
 bottleneck, 12:221–222
 eusocial, 12:146–147
 founder, 12:207–208, 12:221–222
 invasive, **12:182–193**
 keystone, 12:216–217
 lists, 12:194–195
 See also Introduced species; specific species
Species Survival Commission (SSC), 14:11
Species Survival Plan (SSP), 12:209
Speckled dasyures, 12:281, 12:288, 12:*293,* 12:*295*
Speckled marsupial mice. *See* Southern dibblers
Spectacled bears, 14:*301,* 14:305
 behavior, 14:297, 14:298
 conservation status, 14:300
 distribution, 14:*302*
 evolution, 14:295–296
 habitats, 14:297
 physical characteristics, 14:296, 14:*299*
Spectacled bushbabies. *See* Dusky bushbabies
Spectacled dormice, 16:318, 16:*322,* 16:*323*
Spectacled flying foxes, 12:*57,* 13:319, 13:*321,* 13:*322,* 13:*324,* 13:*326,* 13:329–330
Spectacled fruit bats. *See* Spectacled flying foxes
Spectacled porpoises, 15:4, 15:33, 15:34, 15:35, 15:36, 15:39*t*
Spectacled squirrels. *See* Peters' squirrels
Spectaled hare wallabies, 13:34
Spectral bats, 13:*310,* 13:313, 13:418, 13:419, 13:420, 13:*422,* 13:*424–425*
Spectral tarsiers, 14:*92,* 14:*93,* 14:94, 14:95, 14:*96,* 14:*97,* 14:98
Speke's pectinators, 16:312, 16:313, 16:*314*
Speothos venaticus. See Bush dogs
Sperm, 12:91, 12:101, 12:104
 competition, 12:98–99, 12:105–106
 delayed fertilization and, 12:110
 See also Reproduction
Sperm whales, **15:73–80,** 15:*74,* 15:*75,* 15:*76,* 15:77
 behavior, 15:6, 15:7, 15:74–75
 conservation status, 15:78, 15:79–80
 distribution, 15:*73,* 15:74
 evolution, 15:3, 15:73
 feeding ecology, 15:75–76
 habitats, 15:74
 humans and, 15:78
 physical characteristics, 15:4, 15:73
 reproduction, 15:9, 15:76–77
 species of, 15:79–80
 taxonomy, 12:14, 15:1, 15:2, 15:73
Spermatogenesis, 12:94–95, 12:102–103
Spermophilopsis spp., 16:143, 16:144, 16:145, 16:147
Spermophilus spp., 16:143
Spermophilus armatus. See Uinta ground squirrels
Spermophilus beldingi. See Belding's ground squirrels
Spermophilus brunneus. See Idaho ground squirrels
Spermophilus citellus. See European ground squirrels
Spermophilus columbianus. See Columbian ground squirrels
Spermophilus lateralis. See Golden-mantled ground squirrels
Spermophilus leptodactylus. See Long-clawed ground squirrels
Spermophilus mexicanus. See Mexican ground squirrels
Spermophilus mohavensis, 16:147–148
Spermophilus parryii. See Arctic ground squirrels
Spermophilus richardsonii. See Richardson's ground squirrels
Spermophilus saturatus, 16:143
Spermophilus tridecemlineatus. See Thirteen-lined ground squirrels
Spermophilus washingtoni, 16:147–148
Sphaerias blanfordi. See Blanford's fruit bats
Sphiggurus spp., 16:366, 16:367
Sphiggurus insidiosus. See Bahía hairy dwarf porcupines
Sphiggurus mexicanus. See Mexican hairy porcupines
Sphiggurus spinosus. See Paraguay hairy dwarf porcupines
Sphiggurus vestitus. See Brown hairy dwarf porcupines
Sphiggurus villosus. See Orange-spined hairy dwarf porcupines
Spicer, G. S., 16:143
Spider monkeys, **14:155–169,** 14:*161*
 behavior, 14:7, 14:157
 conservation status, 14:159–160
 distribution, 14:156
 evolution, 14:155
 feeding ecology, 14:158
 habitats, 14:156–157
 humans and, 14:160
 reproduction, 14:158–159
 species of, 14:*163*–164, 14:167*t*
 taxonomy, 14:155
Spilocuscus spp., 13:57
Spilocuscus maculatus. See Common spotted cuscuses
Spilocuscus papuensis. See Waigeo cuscuses
Spilocuscus rufoniger. See Black-spotted cuscuses
Spilogale spp., 14:319, 14:321
Spilogale gracilis. See Western spotted skunks
Spilogale gracilis amphiala. See Island spotted skunks
Spilogale putorius. See Eastern spotted skunks
Spines, 12:38, 12:*43*
Spinner dolphins, 15:*1,* 15:6, 15:44, 15:46, 15:*48,* 15:49, 15:*50,* 15:*52,* 15:*54*–55
Spiny bandicoots, 13:9, 13:17*t*
 See also Rufous spiny bandicoots
Spiny hedgehogs, 13:199, 13:203
Spiny mice. *See* Egyptian spiny mice
Spiny pocket mice, 16:199, 16:208*t*
 Desmarest's, 16:*204*
 forest, 16:199
 Mexican, 16:208*t*
 Nelson's, 16:208*t*
 Salvin's, 16:*204,* 16:*205*–206
Spiny rats, 16:253, **16:449–459,** 16:*450,* 16:*451,* 16:*452,* 16:*453,* 16:458*t*
Spiny tenrecs, 13:225, 13:226, 13:228, 13:230
Spiny tree rats
 Brazilian, 16:451, 16:458*t*
 tuft-tailed, 16:458*t*
Spix's disk-winged bats, 13:312, 13:*474,* 13:*475,* 13:*476*
Splay-toothed beaked whales. *See* Andrew's beaked whales
Spontaneous ovulation, 12:92
Sporting dogs, 14:288
Sportive lemurs, **14:73–84,** 14:*78*
 behavior, 14:8, 14:75–76
 conservation status, 14:76
 distribution, 14:*73,* 14:75
 evolution, 14:73–75
 feeding ecology, 14:8, 14:76
 habitats, 14:75
 humans and, 14:77
 physical characteristics, 14:75
 reproduction, 14:76
 species of, 14:*79–83*
 taxonomy, 14:3–4, 14:73–75
Spotted bats, 13:*314,* 13:500, 13:*505,* 13:*509*–510
Spotted cuscuses, 13:57, 13:*59*
Spotted deer. *See* Chitals
Spotted dolphins, 12:*164,* 15:49
 See also Atlantic spotted dolphins
Spotted hyenas, 12:*4,* 12:145, 12:*147,* 14:259, 14:260, 14:*359,* 14:360, 14:*361,* 14:*362,* 14:*363,* 14:*364*–365
Spotted mouse deer, 15:325, 15:*327,* 15:*330,* 15:*332,* 15:333
Spotted-necked otters, 14:321
Spotted seals. *See* Harbor seals
Spotted skunks
 eastern, 14:*321,* 14:324
 island, 12:192
 western, 14:324
Spotted-tailed quolls, 12:*279,* 12:280–281, 12:*282,* 12:289, 12:299*t*
Springboks, 13:*139,* 16:7, **16:45–58,** 16:*47,* 16:*48,* 16:*49,* 16:*50,* 16:56*t*–57*t*
Springhares, 16:122, 16:123, **16:307–310,** 16:*308,* 16:*309,* 16:*310*
Spurelli's free-tailed bats, 13:496*t*
Squalodelphinidae, 15:13
Squalodontidae, 15:3, 15:13
Squamous cells, 12:36
 See also Physical characteristics
Square-lipped rhinoceroses. *See* White rhinoceroses
Squirrel gliders, 13:37, 13:*127,* 13:*128,* 13:*129,* 13:133*t*
Squirrel monkeys, 14:4, **14:101–110,** 14:*107*
 behavior, 14:103, 14:104
 conservation status, 14:105–106
 distribution, 14:*101,* 14:102–103
 evolution, 14:101
 feeding ecology, 14:104–105
 habitats, 14:103
 humans and, 14:106

INDEX

Squirrel monkeys *(continued)*
memory, 12:153
physical characteristics, 14:101–102
reproduction, 14:10, 14:105
species of, 14:*108*–110
taxonomy, 14:101
Squirrels
behavior, 16:124
claws, 12:39
complex-footed, 16:136
conservation status, 16:126
distribution, 12:131, 16:123
flying squirrels, **16:135–142**
ground, **16:143–161**
habitats, 16:124
humans and, 16:127
learning set, 12:152
locomotion, 12:43
physical characteristics, 16:123
scaly-tailed, 16:123, **16:299–306**
tree, 12:115, **16:163–176,** 16:*168*, 16:174*t*–175*t*
SSC (Species Survival Commission), 14:11
SSP (Species Survival Plan), 12:209
Stapes, 12:36
Star-nosed moles, 12:*70*, 12:76, 12:83, 13:195, 13:196, 13:*279*, 13:280, 13:*281*, 13:*282*, 13:*283*, 13:*285*, 13:287
States (social organizational structures), 14:252
Steatomys pratensis. See Fat mice
Steenboks, 15:*136*, 16:2, **16:59–72,** 16:*61*, 16:*63*, 16:*64*, 16:*65*, 16:66–*67*, 16:71*t*
Steenbucks. *See* Steenboks
Stegodons, 15:161
Steinboks. *See* Steenboks
Steinbucks. *See* Steenboks
Steiromys spp., 16:366
Stejneger's beaked whales, 15:70*t*
Steller sea lions, 14:*394*, 14:396, 14:400, 14:*401*, 14:*402*, 14:406
Steller's sea cows, **15:199–204,** 15:*202*, 15:*203*
distribution, 15:*199*
evolution, 15:191, 15:199
feeding ecology, 15:193, 15:200
habitats, 12:138, 15:192, 15:200
humans and, 12:215, 15:195, 15:201
physical characteristics, 15:192, 15:199
reproduction, 15:194, 15:201
taxonomy, 15:191–192
Stenella spp., 15:43–44
Stenella attenuata. See Spotted dolphins
Stenella coeruleoalba. See Striped dolphins
Stenella frontalis. See Atlantic spotted dolphins
Stenella longirostris. See Spinner dolphins
Steno bredanensis. See Rough-toothed dolphins
Stenoderma rufum, 13:420
Stenodermatinae, 13:413, 13:414, 13:416–417, 13:418, 13:420
Stenoninae, 15:41
Stenopodon galmani, 12:228
Stephen's kangaroo rats, 16:203, 16:209*t*
Steppe
human evolution and, 12:20
loess, 12:17–18
See also Habitats
Steppe deer, 15:379
Steppe harvesting mice, 12:15
Steppe lemmings, 16:*232*, 16:*233*, 16:237
Steppe pikas, 16:483, 16:491, 16:495, 16:*496*, 16:501
Steppe polecats, 14:321
Steppe rhinoceroses, 15:249
Steppe sicistas. *See* Southern birch mice
Stereoscopic vision. *See* Depth perception
Sternal bones, 12:41
See also Skeletal system
Sternal glands, 12:37
See also Physical characteristics
Steropodontidae, 12:228
Stewart Island short-tailed bats. *See* Greater New Zealand short-tailed bats
Sthenurinae, 13:83
Stick nest rats. *See* Greater stick-nest rats
Stink badgers, 14:321
Stinkers. *See* Swamp wallabies
Stinson, N., 16:215
Stoats. *See* Ermines
Stoker, Bram, 13:317
Stomachs, 12:47–48, 12:121–122, 12:123
See also Physical characteristics
Stone dormice, 16:327*t*
Stone martens, 12:132
Stone tools, 14:242
Storing information, 12:152
Storm-petrels, Guadalupe, 14:291
Strap-toothed whales, 15:70*t*
Straw-colored fruit bats, 13:310, 13:*338*, 13:340–*341*
Strepsirrhines, 12:8, 12:94, 14:3, 14:9
See also Lemurs; Lorises
Strier, K. B., 14:165
Strigocuscus spp., 13:57
Strigocuscus celebensis. See Small Sulawesi cuscuses
Strip sampling, 12:197–198
Striped-backed duikers. *See* Zebra duikers
Striped bandicoots, 13:17*t*
See also Eastern barred bandicoots
Striped dolphins, 15:48, 15:57*t*
Striped dwarf hamsters, 16:247*t*
Striped-faced fruit bats, 13:350*t*
Striped field mice, 16:261*t*
Striped hyenas, 14:359, 14:360, 14:362, 14:*363*, 14:*365*–366
Striped phalangers. *See* Striped possums
Striped polecats. *See* Zorillas
Striped possums, **13:125–133,** 13:*126*, 13:*127*, 13:*130*, 13:*131*, 13:132, 13:133*t*
Striped rabbits, 16:482, 16:487, 16:505, 16:506, 16:509
Striped skunks, 14:*326*, 14:*329*
behavior, 14:321
feeding ecology, 14:*321*
habitats, 14:321
humans and, 12:*186*, 14:325
physical characteristics, 14:321
reproduction, 12:109, 12:110, 14:324
Striped tree squirrels, 16:165, 16:*166*
Stuart's antechinuses. *See* Brown antechinuses
Stuhlmann's golden moles, 13:218, 13:*219*, 13:*220*, 13:221
Stump-eared squirrels. *See* Fox squirrels
Stump-tailed macaques, 14:*192*, 14:194
Stump-tailed porcupines. *See* Short-tailed porcupines
Sturnira spp., 13:418
Sturnira lilium. See Little yellow-shouldered bats
Sturnira ludovici. See Highland yellow-shouldered bats
Sturnira thomasi, 13:420
Styloctenium wallace. See Striped-faced fruit bats
Stylodipus andrewsi. See Andrews's three-toed jerboas
Stylodipus telum, 16:214, 16:215, 16:216, 16:217
Subantarctic fur seals, 14:394, 14:395, 14:398, 14:399, 14:407*t*
Subitizing, 12:155
Subterranean mammals, **12:69–78,** 12:113–114
field studies of, 12:201
vibrations, 12:83
See also specific mammals
Subterranean mice, 16:270
Subunguis, 12:39–40
Suckling, 12:89, 12:96–97
See also Mammary glands
Sudoriferous sweat glands. *See* Apocrine sweat glands
Sugar gliders, 12:148, 13:33, 13:37, 13:126, 13:*128*–129, 13:*130*, 13:*131*
Suidae. *See* Pigs
Suiformes, 15:266, 15:275, 15:292, 15:302
Suinae, 15:136, 15:267, 15:275
Sukumar, Raman, 15:171
Sulawesi. *See* Australian region
Sulawesi bear cuscuses, 13:35, 13:*58*, 13:60, 13:*62*, 13:*63*
Sulawesi palm civets, 14:338, 14:344*t*
Sulawesi tarsiers. *See* Spectral tarsiers
Sumatran orangutans, 14:11, 14:225, 14:235, 14:*236*, 14:*237*–238
Sumatran porcupines, 16:*356*, 16:*357*–358
Sumatran rabbits, 16:487, 16:516*t*
Sumatran rhinoceroses, 15:*250*, 15:*256*, 15:*257*
behavior, 15:218
conservation status, 12:218, 15:222, 15:254
evolution, 15:249
habitats, 15:252
humans and, 15:255
physical characteristics, 15:251
reproduction, 15:220–221, 15:254
See also Rhinoceroses
Sumatran short-tailed porcupines. *See* Sumatran porcupines
Sumatran surilis. *See* Banded leaf-monkeys
Sumatran thick-spined porcupines. *See* Sumatran porcupines
Sumichrast's harvest mice, 16:*269*
Sun bears. *See* Malayan sun bears
Suncus spp., 13:265, 13:266, 13:267
Suncus etruscus. See Savi's pygmy shrews
Suncus murinus. See Musk shrews
Sunda porcupines. *See* Javan short-tailed porcupines
Sunda sambars. *See* Timor deer
Sunda shrews, 13:277*t*
Sunda slow lorises, 14:*15*, 14:*17*, 14:*18*–19
Sunda water shrews, 13:263*t*
Sundasciurus spp., 16:163
Sundasciurus brookei, 16:167
Sundasciurus jentinki, 16:167
Sundasciurus juvencus, 16:167
Sundasciurus moellendorffi, 16:167
Sundasciurus rabori, 16:167

Sundasciurus samarensis, 16:167
Sundasciurus steerii, 16:167
Sundevall's jirds, 16:*286*, 16:291–*292*
See also Gerbils
Sundevall's roundleaf bats, 13:*403*
Sunis, 16:*60*, 16:71*t*
Supply and demand economies, 12:223
Surat Thani Province, Thailand, 12:220
Surdisorex spp., 13:265
Surdisorex norae. See Aberdare shrews
Surdisorex polulus. See Mt. Kenya shrews
Suricata spp., 14:347, 14:348
Suricata suricatta. See Meerkats
Suricates. *See* Meerkats
Surveys, biodiversity, 12:194–198, 12:*195*
Survival of the fittest. *See* Natural selection
Sus spp., 15:265, 15:275, 15:276, 15:278
Sus barbatus. See Bearded pigs
Sus bucculentus. See Vietnam warty pigs
Sus cebifrons. See Cebu bearded pigs
Sus celebensis. See Celebes pigs
Sus domestica. See Domestic pigs
Sus philippensis. See Philippine warty pigs
Sus salvanius. See Pygmy hogs
Sus scrofa. See Eurasian wild pigs; Feral pigs
Sus scrofa riukiuanus, 15:280
Sus scrofa vittatus. See Asian banded boars
Sus timoriensis. See Timor wild boars
Sus verrucosus. See Javan pigs
Sustained-yield harvests, 12:219
Susus. *See* Ganges and Indus dolphins
Swamp deer. *See* Barasinghas
Swamp lynx. *See* Jungle cats
Swamp mice, Delany's, 16:297*t*
Swamp monkeys, 14:193
Swamp rabbits, 16:481, 16:516*t*
Swamp rats. *See* Creek rats
Swamp wallabies, 13:*92*, 13:*93*, 13:95
Sweat glands, 12:7, 12:36–38, 12:89
See also Physical characteristics
Swift foxes, 12:125, 14:265, 14:267, 14:270, 14:271, 14:284*t*
Sword-nosed bats, 13:*414*, 13:*415*
Syconycteris australis. See Southern blossom bats
Sylvicapra spp., 16:73
Sylvicapra grimmia. See Bush duikers
Sylvilagus spp. *See* Cottontails
Sylvilagus aquaticus. See Swamp rabbits
Sylvilagus audubonii. See Desert cottontails
Sylvilagus bachmani, 16:481
Sylvilagus brasiliensis, 16:481
Sylvilagus cunicularius, 16:481
Sylvilagus floridanus. See Eastern cottontails
Sylvilagus graysoni, 16:481
Sylvilagus insonus. See Omiltimi rabbits
Sylvilagus mansuetus, 16:481
Sylvilagus nuttallii. See Mountain cottontails
Sylvilagus palustris. See Marsh rabbits
Sylvilagus transitionalis, 16:481
Sylvisorex spp. *See* Forest musk shrews
Sylvisorex megalura. See Forest musk shrews
Sylvisorex vulcanorum. See Volcano shrews
Symbolic communication, 12:161–162
Symmetrodontomys spp., 16:264
Symmetry, biological success and, 12:23–24
Sympatric occurrence, 12:116
Symphalangus spp., 14:207
Symphalangus syndactylus. See Siamangs
Symplesiomorphies, 12:7
Synapomorphies, 12:7
Synapsids, 12:10
Synaptomys borealis. See Northern bog lemmings
Syncerus caffer. See African buffaloes
Synocnus comes. See Lesser Haitian ground sloths
Syntax, 12:161
Syntheosciurus spp., 16:163
Syntheosciurus brochus. See Mountain squirrels
Syrian bears. *See* Brown bears
Syrian golden hamsters. *See* Golden hamsters
Syrian hamsters. *See* Golden hamsters
Syrian onagers, 15:222
Syrian rock hyrax, 15:189*t*
Systema Naturae, 13:193
Systematics. *See* Taxonomy

T

Taber, A. B., 16:392
Tachyglossidae. *See* Echidnas
Tachyglossus spp. *See* Short-beaked echidnas
Tachyglossus aculeatus. See Short-beaked echidnas
Tachyglossus aculeatus acanthion, 12:236
Tachyglossus aculeatus aculeatus, 12:236
Tachyglossus aculeatus lawesi, 12:237
Tachyglossus aculeatus multiaculeatus, 12:236
Tachyglossus aculeatus setosus, 12:236
Tachyoryctes splendens. See East African mole rats
Tactile hairs, 12:38–39
See also Physical characteristics
Tadarida aegyptiaca. See Egyptian free-tailed bats
Tadarida australis. See White-striped free-tailed bats
Tadarida brasiliensis. See Brazilian free-tailed bats
Tadarida fulminans, 13:*487*
Tadarida teniotis. See European free-tailed bats
Taeniodonta, 12:11
TAG (Felid Taxonomic Advisory Group), 14:369
Tahkis. *See* Przewalski's horses
Tahrs, 16:87, 16:88, 16:92
See also specific types of tahrs
Tailless bats, Geoffroy's, 13:*422*, 13:427–*428*
Tails, 12:*42*
in locomotion, 12:43
subterranean mammals, 12:72
See also Physical characteristics
Tajik markhors, 16:*8*
Takins, 15:*138*, 16:87–90, 16:92, 16:93, 16:94, 16:97, 16:*101*, 16:103
Talapoins, 14:188, 14:191, 14:193, 14:*196*, 14:*198*–199
Talas tuco-tucos, 16:*426*, 16:*427*
The Tale of Mrs. Tiggy-Winkle, 13:200
Talpa spp., 13:279
Talpa europaea. See European moles
Talpa streeti. See Persian moles
Talpidae, **13:279–288,** 13:*283*
behavior, 13:280–281
conservation status, 13:282
distribution, 13:*279*, 13:280
evolution, 13:279
feeding ecology, 13:281–282
habitats, 13:280
physical characteristics, 13:279–280
reproduction, 13:282
species of, 13:*284*–287, 13:287*t*–288*t*
taxonomy, 13:194, 13:279
vision, 12:79
Talpinae, 13:279, 13:280
Tamandua spp. *See* Tamanduas
Tamandua mexicana. See Northern tamanduas
Tamandua tetradactyla. See Southern tamanduas
Tamanduas, 13:151, 13:172
Tamarins, **14:115–132,** 14:*125*
behavior, 14:6–7, 14:8, 14:116–120, 14:*123*
conservation status, 14:11, 14:124
distribution, 14:116
evolution, 14:115
feeding ecology, 14:120–121
habitats, 14:116
humans and, 14:124
physical characteristics, 14:115–116, 14:*122*
reproduction, 14:10, 14:121–124
species of, 14:*127*–129, 14:132*t*
taxonomy, 14:4, 14:115
Tamias spp., 16:143
Tamias minimus. See Least chipmunks
Tamias palmeri, 16:147–148
Tamias ruficaudus. See Red-tailed chipmunks
Tamias senex. See Allen's chipmunks
Tamias sibericus. See Siberian chipmunks
Tamias striatus. See Eastern chipmunks
Tamias townsendii. See Townsend's chipmunks
Tamiasciurus spp., 16:163
Tamiasciurus douglasii. See Douglas's squirrels
Tamiasciurus hudsonicus. See North American red squirrels
Tamiasciurus hudsonicus grahamensis. See Mt. Graham red squirrels
The Taming of the Shrew, 13:200
Tamini, 16:143
Tammar wallabies, 12:185, 13:*89*, 13:101*t*
Tapeta lucida, 12:79–80
Taphozoinae, 13:355
Taphozous spp. *See* Tomb bats
Taphozous achates. See Brown-bearded sheath-tail bats
Taphozous hildegardeae. See Hildegarde's tomb bats
Taphozous mauritianus. See Mauritian tomb bats
Taphozous melanopogon. See Black-bearded tomb bats
Taphozous perforatus. See Egyptian tomb bats
Taphozous troughtoni, 13:358
Tapir Specialist Group, IUCN, 15:242
Tapiridae. *See* Tapirs
Tapirs, 12:131, 15:136, 15:137, 15:215–223, **15:237–248,** 15:*244*
Tapirus spp. *See* Tapirs
Tapirus bairdii. See Central American tapirs
Tapirus indicus. See Malayan tapirs
Tapirus pinchaque. See Mountain tapirs
Tapirus terrestris. See Lowland tapirs
Target training, of zoo animals, 12:211
Tarpans, 12:175, 15:222
Tarsiers, **14:91–100,** 14:*96*
behavior, 14:8, 14:94–95
conservation status, 14:95
distribution, 12:136, 14:5, 14:*91*, 14:93

Tarsiers *(continued)*
evolution, 14:2, 14:91
feeding ecology, 14:8
habitats, 14:6, 14:93–94
humans and, 14:95
locomotion, 12:43
physical characteristics, 14:92–93
reproduction, 14:95
species of, 14:*97–99*
taxonomy, 14:1, 14:3–4, 14:91–92
Tarsiidae. *See* Tarsiers
Tarsipedidae. *See* Honey possums
Tarsipedoidea, 13:32, 13:139
Tarsipes rostratus. See Honey possums
Tarsius spp. *See* Tarsiers
Tarsius bancanus. See Western tarsiers
Tarsius dianae. See Dian's tarsiers
Tarsius eocaenus spp., 14:2
Tarsius pumilus. See Pygmy tarsiers
Tarsius sangirensis. See Sangihe tarsiers
Tarsius spectrum. See Spectral tarsiers
Tarsius syrichta. See Philippine tarsiers
Tasmacetus spp., 15:59
Tasmacetus shepherdi. See Shepherd's beaked whales
Tasman beaked whales. *See* Shepherd's beaked whales
Tasmanian barred bandicoots. *See* Eastern barred bandicoots
Tasmanian bettongs, 13:*73*, 13:74, 13:76, 13:77, 13:*78*
Tasmanian devils, 12:*278*, **12:287–291,** 12:*288*, 12:*293*, 12:*295*, **12:296**
behavior, 12:281, 12:*281*
conservation status, 12:290
distribution, 12:137, 12:279
evolution, 14:256
feeding ecology, 12:289
habitats, 12:280, 12:289
humans and, 12:284
physical characteristics, 12:288
reproduction, 12:282
Tasmanian ringtails. *See* Common ringtails
Tasmanian wolves, 12:*31*, 12:137, 12:277–284, **12:307–310,** 12:*308*, 12:*309*, 12:*310*, 14:256
Tassel-eared marmosets, 14:*118*, 14:132*t*
Tassel-eared squirrels. *See* Abert squirrels
Taste, 12:50, 12:*80*
See also Physical characteristics
Tatera spp., 16:127
Tatera leucogaster. See Bushveld gerbils
Tate's trioks, 13:129, 13:133*t*
Tattersall, Ian
on koala lemurs, 14:73
on *Lepilemur* spp., 14:75
Taurine, 12:127
Taurotragus spp., 16:11
Taurotragus oryx. See Elands
Taxidea taxus. See North American badgers
Taxonomy
aardvarks, 15:155
agoutis, 16:407, 16:411–414
anteaters, 13:171, 13:177–179
armadillos, 13:181–182, 13:187–190
Artiodactyla, 15:263–267
Atelidae, 14:155, 14:162–166
aye-ayes, 14:85
baijis, 15:19
bandicoots, 13:1
dry-country, 13:9, 13:14–16
rainforest, 13:9, 13:14–16
bats, 13:309
American leaf-nosed bats, 13:413, 13:423–432
bulldog bats, 13:443, 13:449
disk-winged, 13:473, 13:476
Emballonuridae, 13:355, 13:360–363
false vampire, 13:379, 13:384
funnel-eared, 13:459, 13:463–465
horseshoe, 13:387, 13:396–399
Kitti's hog-nosed, 13:367
Molossidae, 13:483, 13:490–491, 13:491–493
mouse-tailed, 13:351, 13:353
moustached, 13:435, 13:440–441
New Zealand short-tailed, 13:453, 13:457
Old World fruit, 13:319, 13:325–330, 13:333, 13:340–347
Old World leaf-nosed, 13:401, 13:407–409
Old World sucker-footed, 13:479
slit-faced, 13:371, 13:375
smoky, 13:467, 13:470
Vespertilionidae, 13:497, 13:506–514, 13:519, 13:524–525
bears, 14:256, 14:295, 14:302–305
beavers, 16:177, 16:183
bilbies, 13:19
botos, 15:27
Bovidae, 16:1–2
Antilopinae, 16:45, 16:50–55
Bovinae, 16:11, 16:18–23
Caprinae, 16:87–89, 16:98–103
duikers, 16:73, 16:80–84
Hippotraginae, 16:27–28, 16:37–41
Neotraginae, 16:59, 16:66–70
bushbabies, 14:23, 14:28–31
Camelidae, 15:313–314, 15:320–323
Canidae, 14:265–266, 14:276–282
capybaras, 16:401
Carnivora, 14:255, 14:256
cats, 14:369, 14:379–389
Caviidae, 16:389–390, 16:395–398
Cetacea, 12:14, 15:2–4, 15:133, 15:266
Balaenidae, 15:107, 15:115–117
beaked whales, 15:59, 15:64–68
dolphins, 15:41–42, 15:53–55
franciscana dolphins, 15:23
Ganges and Indus dolphins, 15:13
gray whales, 15:93
Monodontidae, 15:81, 15:90
porpoises, 15:33, 15:38–39
pygmy right, 15:103
rorquals, 15:1, 15:119, 15:127–129
sperm, 15:73, 15:79–80
Cheirogaleidae, 14:35, 14:41–44
chevrotains, 15:325–326, 15:331–333
Chinchillidae, 16:377, 16:382–383
colugos, 13:299, 13:304
coypus, 16:473
dasyurids, 12:287, 12:294–298, 12:299*t*–301*t*
Dasyuromorphia, 12:277
deer
Chinese water, 15:373
muntjacs, 15:343, 15:349–354
musk, 15:335, 15:340–341
New World, 15:379–382, 15:388–395
Old World, 15:357–358, 15:364–371*t*
Dipodidae, 16:211, 16:219–222
Diprotodontia, 13:31–32
dormice, 16:317, 16:323–326
duck-billed platypuses, 12:243
Dugongidae, 15: 199, 15:203
echidnas, 12:235
elephants, 12:*29*, 15:161–165, 15:174–175
Equidae, 15:225, 15:232–235
Erinaceidae, 13:203, 13:209–212
giant hutias, 16:469
gibbons, 14:207–208, 14:218–222
Giraffidae, 15:399, 15:408
great apes, 14:225, 14:237–239
gundis, 16:311, 16:314–315
Herpestidae, 14:347, 14:354–356
Heteromyidae, 16:199, 16:205–208
hippopotamuses, 15:301–302, 15:304, 15:310–311
humans, 14:241–243
hutias, 16:461
Hyaenidae, 14:359–360, 14:364–366
hyraxes, 15:177–178, 15:186–188
Indriidae, 14:63, 14:69–72
Insectivora, 13:193–194
koalas, 13:43
Lagomorpha, 16:479–480
lemurs, 14:47–48, 14:55–60
Cheirogaleidae, 14:35, 14:41–44
sportive, 14:3–4, 14:73–75, 14:79–83
Leporidae, 16:505, 16:511–514
Lorisidae, 14:13, 14:18–20
Macropodidae, 13:83, 13:93–100
manatees, 15: 205, 15: 211–212
Megalonychidae, 13:155, 13:159
moles
golden, 13:215–216, 13:220–221
marsupial, 13:25, 13:28
monitos del monte, 12:273
monkeys
Callitrichidae, 14:115, 14:127–131
Cebidae, 14:101, 14:110–112
cheek-pouched monkeys, 14:187–188, 14:197–203
leaf-monkeys, 14:4, 14:171–172, 14:178–183
night, 14:135
Pitheciidae, 14:143, 14:150–152
squirrel, 14:101, 14:108–110
mountain beavers, 16:131
Muridae, 16:281–283, 16:288–294
Arvicolinae, 16:225, 16:233–237, 16:238*t*
hamsters, 16:239, 16:245–246
Murinae, 16:249, 16:255–260
Sigmodontinae, 16:263–264, 16:272–275
musky rat-kangaroos, 13:69–70
Mustelidae, 14:319, 14:327–331
numbats, 12:303
octodonts, 16:433–434, 16:438–440
opossums
New World, 12:249–250, 12:257–263, 12:264*t*–265*t*
shrew, 12:267, 12:270
Otariidae, 14:256, 14:393, 14:402–406
pacaranas, 16:385–386
pacas, 16:417, 16:423
pangolins, 16:107, 16:115–119
peccaries, 15:291, 15:298–299
Perissodactyla, 15:215–216
Petauridae, 13:125, 13:131–132
Phalangeridae, 13:57, 13:*63–65*
pigs, 15:275, 15:284–288

Taxonomy *(continued)*
pikas, 16:491, 16:497–501
pocket gophers, 16:185, 16:192–194
porcupines
New World, 16:365–366, 16:371–373
Old World, 16:357–363
possums
feather-tailed, 13:139–140, 13:144
honey, 13:135
pygmy, 13:105–106, 13:110–111
primates, 14:1–4
Procyonidae, 14:309, 14:313–315
pronghorns, 15:411
Pseudocheiridae, 13:113, 13:119–122
rat-kangaroos, 13:78–80
rats
African mole-rats, 16:339, 16:347–349
cane rats, 16:333, 16:337
chinchilla rats, 16:443, 16:447–448
dassie rats, 16:329
spiny, 16:449, 16:453–457
rhinoceroses, 15:249, 15:251, 15:257–262
rodents, 16:121–122
sengis, 16:517–518, 16:524–529
shrews
red-toothed, 13:247, 13:255–263
West Indian, 13:243
white-toothed, 13:265, 13:271–275
Sirenia, 15: 191–192
solenodons, 13:237, 13:240
springhares, 16:307
squirrels
flying, 16:135, 16:139–140
ground, 16:143, 16:151–157
scaly-tailed, 16:299, 16:302–305
tree, 16:163, 16:169–174
Talpidae, 13:193, 13:194, 13:279, 13:284–287
tapirs, 15:237–238, 15:245–247
tarsiers, 14:91–92, 14:97–99
tenrecs, 13:225, 13:232–233
three-toed tree sloths, 13:161, 13:167–169
tree shrews, 13:289, 13:294–296
true seals, 14:256, 14:417, 14:427–433
tuco-tucos, 16:425, 16:429
ungulates, 15:131–138
Viverridae, 14:335, 14:340–343
walruses, 14:256, 14:409
wombats, 13:51, 13:55–56
Xenartha, 13:147–149
Tayassu spp., 15:267, 15:291
Tayassu pecari. See White-lipped peccaries
Tayassu tajacu. See Collared peccaries
Tayassuidae. *See* Peccaries
Tayras, 14:323, 14:*326*, 14:*327*, 14:331
Tcharibbeena. *See* Bennett's tree kangaroos
Teaching, mother chimpanzees, 12:157
Teats. *See* Nipples
Tectum mesencephali, 12:6
Teeth, 12:9, 12:11, 12:*37*
digestion and, 12:46
domestication and, 12:174
evolution and, 12:11
feeding habits and, 12:14–15
subterranean mammals, 12:74
suckling and, 12:89
See also Physical characteristics
Tehuantepec jackrabbits, 16:487
Telefomin cuscuses, 13:32, 13:60, 13:66*t*
Telemetry, 12:199–202
Telencephalon. *See* Forebrain
Teleoceros spp., 15:249
Temminck's ground pangolins. *See* Ground pangolins
Temminck's trident bats, 13:410*t*
Templeton, A. R., 14:244
Ten-lined mongooses. *See* Narrow-striped mongooses
Tenrec spp. *See* Tenrecs
Tenrec ecaudatus. See Common tenrecs
Tenrecidae. *See* Tenrecs
Tenrecinae. *See* Spiny tenrecs
Tenrecomorpha, 13:215
Tenrecs, **13:225–235**, 13:*231*
behavior, 13:197–198, 13:227–228
conservation status, 13:229
distribution, 12:135, 12:136, 13:197, 13:*225*, 13:226
echolocation, 12:86
evolution, 13:225
feeding ecology, 13:198, 13:228–229
habitats, 13:197, 13:226–227
humans and, 13:230
physical characteristics, 13:195–196, 13:225–226
reproduction, 13:199, 13:229
species of, 13:232–234, 13:234*t*
taxonomy, 13:193, 13:194, 13:225
Tent-making bats, 13:416, 13:*421*, 13:*425*, 13:431
Terrace, H. S., 12:160–161
Terrestrial mammals
adaptations in, 12:13–14
field studies of, 12:200–202
vision, 12:79
See also specific terrestrial species
Terrestrial pangolins, 16:107–112
Terrestrial tree shrews, 13:*291*, 13:*292*, 13:*293*, 13:295, 13:*296*
Terriers, 14:288, 14:292
Territoriality, 12:*142*
evolution of, 12:18
nutritional requirements and, 12:124
See also Behavior
Testes, 12:91, 12:102–103
See also Physical characteristics
Testosterone, 12:148
Tetanus, subterranean mammals and, 12:77–78
Tethytragus spp., 16:1
Tetonius spp., 14:1
Tetracerus quadricornis. See Chousinghas
Tetraclaenodon spp., 15:132
Texas kangaroo rats, 16:209*t*
Texas mice, 16:278*t*
Texas pocket gophers, 16:195*t*
Thallomys paedulcus. See Tree rats
Thamins. *See* Eld's deer
Thaptomys nigrita, 16:270
Theory of Mind, 12:159–160
Therapsida, 12:10, 12:36
Therians
mammary nipples in, 12:5
reproduction, 12:51
See also Marsupials; Placentals; specific animals
Thermal imaging, in field studies, 12:198–199
Thermal isolation, 12:3
Thermoregulation, 12:3, 12:36, 12:50–51, 12:59–60, 12:65–67, 12:73–74, 12:113
See also Physical characteristics
Theropithecus spp. *See* Geladas
Theropithecus gelada. See Geladas
Thick-spined porcupines, 16:353–354, 16:*356*, 16:*357*, 16:358
Thick-spined rats. *See* Armored rats
Thick-tailed bushbabies. *See* Brown greater bushbabies
Thick-tailed opossums, 12:251, 12:*255*, 12:*260*
Thick-tailed pangolins. *See* Indian pangolins
Thick-tailed pygmy jerboas, 16:215, 16:216, 16:223*t*
Thin-spined porcupines, 16:365, 16:368, 16:369, 16:374*t*, 16:450, 16:*452*, 16:*453*, 16:457
Thinhorn sheep, 12:131, 12:178, 16:87, 16:91, 16:92
Think Tank exhibit (Smithsonian), 12:151
Thinking. *See* Cognition
Thinopithecus avunculus. See Tonkin snub-nosed monkeys
Thirteen-lined ground squirrels, 16:143–144
Thomasomyines, 16:263–264
Thomasomys aureus, 16:266, 16:268
Thomas's bats. *See* Shaggy-haired bats
Thomas's bushbabies, 14:33*t*
Thomas's flying squirrels, 16:141*t*
Thomomys spp., 16:185
Thomomys bottae. See Valley pocket gophers
Thomomys bulbivorus. See Camas pocket gophers
Thomomys mazama. See Mazama pocket gophers
Thomomys monticola. See Mountain pocket gophers
Thomomys talpoides. See Northern pocket gophers
Thomomys umbrinus. See Southern pocket gophers
Thomson's gazelles, 12:*10*, 15:*133*, 16:*46*, 16:*49*, 16:51
coevolution and, 12:216
conservation status, 16:48
distribution, 16:*50*
in guilds, 12:119
physical characteristics, 15:269
predation and, 12:116
Thorold's deer. *See* White-lipped deer
Three-banded armadillos, 13:*182*, 13:*183*, 13:184
Three-pointed pangolins. *See* Tree pangolins
Three-striped dasyures, 12:300*t*
Three-striped douroucoulis. *See* Three-striped night monkeys
Three-striped night monkeys, 14:135, 14:*136*, 14:*137*, 14:*139*, 14:*140*, 14:141
Three-toed jerboas, 16:211, 16:212, 16:213, 16:214, 16:216
Three-toed tree sloths, 13:*150*, **13:161–169**, 13:*162*, 13:*164*, 13:*165*, 13:*166*
behavior, 13:*152*, 13:*153*, 13:162–164
conservation status, 13:153, 13:165
distribution, 13:*161*, 13:162
evolution, 13:161
feeding ecology, 13:164–165
habitats, 13:162
humans and, 13:165
physical characteristics, 13:149, 13:161–162
reproduction, 13:*147*, 13:165
species of, 13:167–169
taxonomy, 13:147

Threshold relationships, in ecosystems, 12:217
Thrichomys apereoides. See Punares
Thrombocytes, 12:8
Thrust, for powered flight, 12:56
Thryonomyidae. *See* Cane rats
Thryonomys spp. *See* Cane rats
Thryonomys arkelli, 16:333
Thryonomys gregorianus. See Lesser cane rats
Thryonomys logani, 16:333
Thryonomys swinderianus. See Greater cane rats
Thumbless bats. *See* Smoky bats
Thylacines. *See* Tasmanian wolves
Thylacinidae. *See* Tasmanian wolves
Thylacinus cynocephalus. See Tasmanian wolves
Thylacinus potens, 12:307
Thylacomyinae. *See* Bilbies
Thylacosmilus spp., 12:11
Thylamys spp. *See* Fat-tailed mouse opossums
Thylamys elegans. See Elegant fat-tailed opossums
Thylamys pusilla. See Small fat-tailed opossums
Thylogale spp. *See* Pademelons
Thylogale stigmatica. See Red-legged pademelons
Thyroptera spp. *See* Disk-winged bats
Thyroptera discifera. See Peter's disk-winged bats
Thyroptera laveli, 13:473, 13:474
Thyroptera tricolor. See Spix's disk-winged bats
Thyropteridae. *See* Disk-winged bats
Thyroxine, 12:141
Tibetan antelopes, 12:134, 16:*9*
Tibetan argalis, 12:134
Tibetan foxes, 14:284*t*
Tibetan gazelles, 12:134, 16:48, 16:57*t*
Tibetan muntjacs. *See* Tufted deer
Tibetan sand foxes. *See* Tibetan foxes
Tibetan water shrews. *See* Elegant water shrews
Tibetan wild asses. *See* Kiangs
Tiger ocelots. *See* Little spotted cats
Tiger quolls. *See* Spotted-tailed quolls
Tigers, 14:*377*, 14:*380*–381
behavior, 14:371, 14:372
conservation status, 12:218, 12:220, 14:374
distribution, 12:136, 14:370
humans and, 12:139, 14:376
physical characteristics, 14:370
reproduction, 14:372–373
stripes, 12:38
in zoos, 12:210
Tillodontia, 12:11, 15:131, 15:133, 15:135, 15:136
Timber wolves. *See* Gray wolves
Timor deer, 15:371*t*
Timor wild boars, 15:290*t*
Tinbergen, N., 12:140–143, 12:147–148, 15:143
Titis, **14:143–154**, 14:*149*
behavior, 14:7, 14:8, 14:145, 14:146
conservation status, 14:11, 14:148
distribution, 14:144–145
evolution, 14:143
feeding ecology, 14:146–147
habitats, 14:145
humans and, 14:148
physical characteristics, 14:143–144
reproduction, 14:147
species of, 14:*150*, 14:151–153, 14:153*t*
taxonomy, 14:143
Togo mole-rats, 16:350*t*
Tolypeutes spp. *See* Three-banded armadillos
Tolypeutes matacus. See Southern three-banded armadillos
Tolypeutes tricinctus. See Brazilian three-banded armadillos
Tolypeutinae, 13:182
Tomb bats, 13:355, 13:356, 13:357
black-bearded, 13:*357*, 13:364*t*
Egyptian, 13:364*t*
Hildegarde's, 13:*357*
Mauritian, 13:*359*, 13:*362*, 13:363
Tommies. *See* Thomson's gazelles
Tomopeas ravus. See Peruvian crevice-dwelling bats
Tomopeatinae. *See* Peruvian crevice-dwelling bats
Tonatia spp. *See* Round-eared bats
Tonatia sylvicola. See White-throated round-eared bats
Tongan flying foxes, 13:*323*, 13:*327*–328
Tongan fruit bats. *See* Tongan flying foxes
Tonkin snub-nosed monkeys, 14:185*t*
Tool use, 12:140–141, 12:156–158
See also Behavior
Toolache wallabies, 13:39
Tooth replacements, 12:9
Toothed whales, 12:14, 15:1
behavior, 15:7
echolocation, 12:85–87
evolution, 15:3, 15:41, 15:59
feeding ecology, 15:8
geomagnetic fields, 12:84
hearing, 12:81–82
humans and, 15:11
physical characteristics, 12:67, 15:4, 15:81
Topis, 15:272, 16:*2*, 16:42*t*
behavior, 16:5, 16:30
distribution, 16:29
evolution, 16:27
habitats, 16:29
humans and, 16:34
predation and, 12:116
reproduction, 16:7, 16:33
taxonomy, 16:27
Toque macaques, 14:194, 14:205*t*
Torontoceros spp., 15:379
Toros, 16:450–451, 16:*452*, 16:*453*, 16:454–455
Torpor, 12:113
Torrens creek rock-wallabies. *See* Allied rock-wallabies
Torres-Mura, J. C., 16:435, 16:436
Tossunnoria spp., 16:88
Touch, 12:82–83
See also Physical characteristics
Townsend's chipmunks, 16:*146*
Toy dogs, 14:288, 14:289
Trace elements, 12:120
Trachops cirrhosus. See Fringe-lipped bats
Trachypithecus spp., 14:172, 14:173
Trachypithecus auratus, 14:175
Trachypithecus auratus auratus, 14:*175*
Trachypithecus cristatus. See Silvery leaf-monkeys
Trachypithecus delacouri, 14:11, 14:175
Trachypithecus francoisi, 14:175
Trachypithecus geei, 14:175
Trachypithecus johnii, 14:175
Trachypithecus laotum, 14:175
Trachypithecus pileatus. See Capped langurs
Trachypithecus poliocephalus, 14:11, 14:175
Trachypithecus vetulus, 14:175
Trachypithecus villosus, 14:175
Tragelaphine antelopes. *See* Tragelaphini
Tragelaphini, 16:2, 16:3, 16:11–14
Tragelaphus angasii. See Nyalas
Tragelaphus buxtoni. See Mountain nyalas
Tragelaphus eurycerus. See Bongos
Tragelaphus imberbis. See Lesser kudus
Tragelaphus scriptus. See Bushbucks
Tragelaphus spekii. See Sitatungas
Tragelaphus strepsiceros. See Greater kudus
Tragocerinae, 15:265
Tragulidae. *See* Chevrotains
Traguloids. *See* Chevrotains
Tragulus spp. *See* Mouse deer
Tragulus javanicus. See Lesser Malay mouse deer
Tragulus meminna. See Spotted mouse deer
Tragulus napu. See Greater Malay mouse deer
Tragus, bats, 12:54
See also Bats
Training, of zoo animals, 12:211
Transgenic mice, 16:128
Translocations, 12:224
Traps, in field studies, 12:194–197, 12:200–202
Trawlers (bats), 13:313–314
Tree dassies. *See* Tree hyraxes
Tree hyraxes, 15:177–184
Tree kangaroos, 13:34, 13:35, 13:36, 13:83–89
Bennett's, 13:*86*, 13:*91*, 13:*99*–100
black, 13:32
Doria's, 13:102*t*
forbidden, 13:32
Goodfellow's, 13:33, 13:39
greater, 13:36
Lumholtz's, 13:102*t*
Matschie's, 13:*37*, 13:*86*, 13:*91*, 13:*98*, 13:100
Tree pangolins, 16:109, 16:110, 16:112, 16:*114*, 16:*117*, 16:119–120
Tree porcupines, 16:365, 16:367, 16:*368*, 16:369
Tree rats, 16:*254*, 16:*257*
Brazilian spiny, 16:451, 16:458*t*
red-nosed, 16:268
rufous, 16:458*t*
tuft-tailed spiny, 16:458*t*
Tree shrews, **13:289–298**, 13:*293*
behavior, 13:291
conservation status, 13:292
distribution, 12:136, 13:*289*, 13:291
evolution, 13:289–291
feeding ecology, 13:291–292
habitats, 13:291
humans and, 13:292
Indian, 13:197, 13:292, 13:*293*, 13:*294*
locomotion, 12:14
physical characteristics, 13:291
reproduction, 12:90, 12:92, 12:94, 12:95, 12:96, 13:292
species of, 13:*294–296*, 13:297*t*–298*t*
taxonomy, 13:289
Tree sloths, 13:147, 13:149, 13:150–151
three-toed, **13:161–169**
two-toed, **13:155–159**

Tree squirrels, 12:115, **16:163–176,** 16:*168*
behavior, 16:165–166
conservation status, 16:167
distribution, 16:*163*–164
evolution, 16:163
feeding ecology, 16:166
habitats, 16:124, 16:164–165
humans and, 16:126, 16:167
physical characteristics, 16:123, 16:163
reproduction, 16:166–167
species of, 16:169–174, 16:174*t*–175*t*
taxonomy, 16:163
Tree voles, 16:225, 16:226, 16:228, 16:229
See also Red tree voles
Trefoil horseshoe bats, 13:389
Tremarctinae, 14:295
Tremarctos ornatus. See Spectacled bears
Tres Marias cottontails, 16:509
Triaenops persicus. See Persian trident bats
Tribes, human, 14:251
Tribosphenic molars, 12:9
See also Teeth
Trichechidae. *See* Manatees
Trichechinae, 15:205
Trichechus spp. See Manatees
Trichechus inunguis. See Amazonian manatees
Trichechus manatus. See West Indian manatees
Trichechus manatus latirostris. See Florida manatees
Trichechus manatus manatus. See Antillean manatees
Trichechus senegalensis. See West African manatees
Trichomys spp., 16:450
Trichosurus arnhemensis. See Northern brushtail possums
Trichosurus caninus. See Mountain brushtail possums
Trichosurus vulpecula. See Common brushtail possums
Trichys spp., 16:351
Trichys fasciculata. See Long-tailed porcupines
Trident bats, 13:*402*, 13:409*t*–410*t*
Trident leaf-nosed bats, 13:*406*, 13:*407*
Trilobites, 12:64
Trinidadian funnel-eared bats. *See* White-bellied funnel-eared bats
Trogontherium spp., 16:177
Trogopterus xanthipes. See Complex-toothed flying squirrels
Tropical deciduous forests, 12:219
Tropical pocket gophers, 16:190, 16:195*t*
Tropical rainforests, 12:218
Tropical savannas, 12:219
Tropical whales. *See* Bryde's whales
Tropiese grysboks. *See* Sharpe's grysboks
True bats. *See* Microchiroptera
True chevrotains. *See* Water chevrotains
True elephants. *See* Elephants
True lemmings, 16:225–226
True lemurs. *See* Lemurs
True plains zebras, 15:222
True ruminants. *See* Ruminants
True seals, **14:417–436,** 14:*425*, 14:*426*
behavior, 12:85, 12:87, 14:418–419
conservation status, 14:262, 14:422
distribution, 12:129, 12:138, 14:*417*–418
evolution, 14:417
feeding ecology, 14:260, 14:*261*, 14:419–420
habitats, 14:257, 14:418
humans and, 14:423–424
physical characteristics, 14:256, 14:417, 14:*419*
reproduction, 12:126, 12:127, 14:261, 14:262, 14:420–422
species of, 14:*427*–434, 14:435*t*
taxonomy, 14:256, 14:417
True's beaked whales, 15:70*t*
Truong Son muntjacs, 15:346, 15:*348*, 15:*351*, 15:353
Tscherskia spp. *See* Greater long-tailed hamsters
Tscherskia triton. See Greater long-tailed hamsters
Tsessebes, 16:32
Tsolov's mouse-like hamsters, 16:243
Tuans. *See* Brush-tailed phascogales
Tube-nosed bats
brown, 13:*523*, 13:524
fruit, 13:333, 13:*334*, 13:*335*, 13:*337*, 13:*339*, 13:*346*, 13:347, 13:349*t*
Hutton's, 13:526*t*
insectivorous, 13:*520*
Tubulidentata. *See* Aardvarks
Tuco-tucos, 16:123, **16:425–431,** 16:*428*, 16:430*t*–431*t*
Tucuxis, 15:6, 15:43, 15:44, 15:46, 15:48, 15:50
Tuft-tailed spiny tree rats, 16:458*t*
Tufted capuchins. *See* Black-capped capuchins
Tufted deer, 15:268, 15:343, 15:346, 15:*347*, 15:352–*353*
See also Muntjacs
Tule elk, 15:*266*
Tundra shrews, 13:263*t*
Tunney's rats, 16:*128*
Tupaia belangeri. See Belanger's tree shrews
Tupaia chrysogaster, 13:292
Tupaia dorsalis. See Golden-bellied tree shrews
Tupaia glis. See Common tree shrews
Tupaia gracilis. See Slender tree shrews
Tupaia javanica. See Javanese tree shrews
Tupaia longipes. See Bornean tree shrews
Tupaia minor. See Pygmy tree shrews
Tupaia montana. See Montane tree shrews
Tupaia nicobarica. See Nicobar tree shrews
Tupaia palawanensis. See Palawan tree shrews
Tupaia picta. See Painted tree shrews
Tupaia splendidula. See Rufous-tailed tree shrews
Tupaia tana. See Terrestrial tree shrews
Tupaiidae. *See* Tree shrews
Tupaiinae, 13:291, 13:292
Turbaries, 15:281
Turkestan red pikas, 16:*496*, 16:*497*, 16:499
Turkish vans, 14:291
Tursiops spp. *See* Bottlenosed dolphins
Tursiops aduncus. See Indo-Pacific bottlenosed dolphins
Tursiops truncatus. See Common bottlenosed dolphins
Twilight bats. *See* Evening bats
Two-humped camels. *See* Bactrian camels
Two-toed anteaters. *See* Silky anteaters
Two-toed tree sloths, 13:150, 13:*152*, 13:153, **13:155–159,** 13:*156*
Tylonycteris spp., 13:311
Tylonycteris pachypus. See Bamboo bats
Tylopoda. *See* Camelidae
Tympanic bone, 12:9
See also Physical characteristics
Tympanic membrane, 12:9
See also Physical characteristics
Tympanoctomys spp., 16:433, 16:434
Tympanoctomys barrerae. See Red viscacha rats
Typhlomys cinereus. See Chinese pygmy dormice

U

Uakaris, **14:143–154,** 14:*149*
behavior, 14:146
conservation status, 14:148
distribution, 14:144–145
evolution, 14:143
feeding ecology, 14:146–147
habitats, 14:145
humans and, 14:148
physical characteristics, 14:143–144
reproduction, 14:147
species of, 14:*150*, 14:151, 14:153*t*
taxonomy, 14:143
Ucayali bald uakaris, 14:148, 14:151
Uganda kobs, 16:7
Uinta ground squirrels, 16:160*t*
Uintatheres, 15:131, 15:135, 15:136
Uintatherium spp., 15:135, 15:136
Ulendo: Travels of a Naturalist in and out of Africa, 12:218
Ultrasound, bats and, 12:54–55
Uluguru bushbabies, 14:26, 14:32*t*
Uncia (Panthera) uncia. See Snow leopards
Uncia uncia. See Snow leopards
Underwood's pocket gophers, 16:196*t*
Unguis, 12:39–40
Ungulata. *See* Ungulates
Ungulates, **15:131–144**
behavior, 12:145, 15:142
coevolution and, 12:216
distribution, 12:137
domestication, 15:145–153
evolution, 15:131–138
feeding ecology, 15:141–142
field studies of, 12:200
guilds, 12:118–119
neonatal requirements, 12:126–127
physical characteristics, 12:40, 15:138–141, 15:142
predation and, 12:116
reproduction, 12:90, 12:92, 12:94, 12:95, 12:97, 12:98, 12:102, 12:107–108, 15:142–143
stomach, 12:123
See also Cattle; specific types of ungulates
Unguligrades, 12:*39*
Unicolored tree kangaroos. *See* Doria's tree kangaroos
United Kingdom
American mink, 12:182–183
gray squirrels, 12:183–184
United States Department of the Interior
babirusas, 15:288
Heermann's kangaroo rats, 16:208
maned sloths, 13:165
San Joaquin pocket mice, 16:205
sportive lemurs, 14:76

United States Fish and Wildlife Service
Aplodontia rufa nigra, 16:133
Arizona gray squirrels, 16:172
babirusas, 15:280
black-tailed prairie dogs, 16:155
Caprinae, 16:95
gray whales, 15:99
Heteromyidae, 16:203
Idaho ground squirrels, 16:158
Indian rhinoceroses, 12:224
peninsular pronghorns, 15:415
pygmy hogs, 15:280, 15:287
Sonoran pronghorns, 15:415–416
Vespertilioninae, 13:502
Unstriped ground squirrels, 16:159*t*
Urartsk mouse-like hamsters, 16:243
Urea, 12:122–123
Ureters, 12:90
See also Physical characteristics
Urials, 12:178, 16:6, 16:87, 16:89, 16:91, 16:92–93, 16:104*t*
Urinary tract, 12:90–91
See also Physical characteristics
Urocyon cinereoargenteus. See Gray foxes
Urocyon littoralis. See Island foxes
Uroderma bilobatum. See Tent-making bats
Urogale spp., 13:292
Urogale everetti. See Philippine tree shrews
Urogenital sinus, 12:90, 12:111
See also Physical characteristics
Uropsilinae, 13:279, 13:280
Uropsilus spp. *See* Shrew moles
Uropsilus investigator. See Yunnan shrew-moles
Uropsilus soricipes. See Chinese shrew-moles
Urotrichus talpoides. See Greater Japanese shrew moles
Ursidae. *See* Bears
Ursinae, 14:295
Ursus americanus. See American black bears
Ursus americanus floridanus, 14:300
Ursus arctos. See Brown bears
Ursus arctos middendorffi. See Kodiak bears
Ursus arctus. See Brown bears
Ursus maritimus. See Polar bears
Ursus thibetanus. See Asiatic black bears
USDI. *See* United States Department of the Interior
Uta Hick's bearded sakis, 14:148, 14:151
Uterine glands, 12:94
See also Physical characteristics
Uterus, 12:90–91, 12:*102*
See also Physical characteristics
Uzungwa bushbabies, 14:33*t*

V

Vaginas, 12:91
See also Physical characteristics; Reproduction
Valais goats, 15:*268*
Valley pocket gophers, 12:74, 12:*74*, 16:*188*, 16:*189*, 16:*190*, 16:*191*, 16:*192*
Val's gundis. *See* Desert gundis
Vampire bats, 13:*308*, 13:*311*, 13:*422*
behavior, 13:419
conservation status, 13:420
distribution, 13:415
feeding ecology, 13:314, 13:418, 13:*420*
hairy-legged, 13:314, 13:434*t*
humans and, 13:316, 13:317, 13:420
infrared energy, 12:83–84
physical characteristics, 13:308, 13:414
reproduction, 13:315, 13:419, 13:420
species of, 13:*425*–426
See also False vampire bats
Vampyrum spectrum. See Spectral bats
Van Zyl's golden moles, 13:222*t*
Vancouver Island marmots, 16:147, 16:*150*, 16:*152*, 16:156
Vaquitas, 15:9, 15:33, 15:34, 15:35–36, 15:39*t*
Varecia spp., 14:51
Varecia variegata. See Variegated lemurs
Variable flying foxes. *See* Island flying foxes
Variable pocket gophers, 16:196*t*
Variegated lemurs, 14:*53*, 14:*57*, 14:58–59
Variegated spider monkeys, 14:*156*
Variegated squirrels, 16:*168*, 16:*170*, 16:172
Varying hares. *See* Mountain hares
Vas deferens, 12:91
Veldkamp's dwarf epauletted bats. *See* Little flying cows
Veldkamp's dwarf fruit bats. *See* Little flying cows
Velvet (antlers), 12:40
See also Physical characteristics
Velvety free-tailed bats. *See* Greater house bats
Venezuelan hemorrhagic fever, 16:270
Venezuelan red howler monkeys, 14:*157*, 14:*158*, 14:*161*, 14:*162*–163
Venezuelan spiny rats, 16:*450*
Venom, platypus, 12:232, 12:244
Venous hearts, bats, 12:60
Vermilingua, 13:171
Verraux's sifakas, 14:66, 14:67
Vertebral column, 12:8
See also Physical characteristics
Vertebrates, flight, 12:52–61
See also specific vertebrates
Vervet monkeys. *See* Grivets
Vesper mice, 16:267, 16:269, 16:270
Vespertilio murinus. See Parti-colored bats
Vespertilionidae, **13:497–526,** 13:*504*, 13:*505*, 13:*523*
behavior, 13:498–500, 13:521
conservation status, 13:501–502, 13:522
distribution, 13:*497*, 13:498, 13:*520*, 13:520–521
evolution, 13:497, 13:519
feeding ecology, 13:313, 13:500, 13:521
habitats, 13:498, 13:521
humans and, 13:502, 13:522
physical characteristics, 13:310, 13:497–498, 13:519–520
reproduction, 13:500–501, 13:522
species of, 13:*506*–514, 13:515*t*–516*t*, 13:524–525, 13:526*t*
taxonomy, 13:497, 13:519
Vespertilioninae, **13:497–517,** 13:*504*, 13:*505*, 13:515*t*–516*t*
Veterinary medicine, in zoos, 12:210–211
Vibrations, 12:83, 12:*83*
See also Physical characteristics
Vibrissae, 12:38–39, 12:*43*, 12:72
See also Physical characteristics
Victoriapithecus spp., 14:172, 14:188–189
Vicugna spp. *See* Vicuñas
Vicugna vicugna. See Vicuñas
Vicuñas, 12:132, 15:142, 15:151, **15:313–323,** 15:*314*, 15:*317*, 15:*319*, 15:*321*
Vietnam warty pigs, 15:263, 15:280, 15:289*t*
Vietnamese pot-bellied pigs, 15:*278*
Virginia opossums, 12:*250*–*253*, 12:*251*, 12:*254*, 12:*255*, 12:*257*, 12:258
Viruses, in rabbit control, 12:187–188
Visayan spotted deer, 15:360, 15:371*t*
Visayan warty pigs. *See* Cebu bearded pigs
Viscacha rats, 16:436, 16:*437*, 16:*438*, 16:439–440
Viscachas, **16:377–384,** 16:*380*, 16:*381*, 16:383*t*
Vision, 12:5–6, 12:50, 12:76–77, 12:79–80
See also Eyes; Physical characteristics
Visual social signals, 12:160
See also Behavior
Vitamin requirements, 12:120, 12:123
See also Feeding ecology
Vitelline sac. *See* Yolk sac
Viverra spp., 14:335, 14:338
Viverra civettina. See Malabar civets
Viverra megaspila. See Oriental civets
Viverra zibetha. See Indian civets
Viverricula spp., 14:335, 14:338
Viverricula indica, 14:336
Viverridae, **14:335–345,** 14:*339*
behavior, 14:258, 14:337
conservation status, 14:262, 14:338
distribution, 12:135, 12:136, 14:*335*, 14:336
evolution, 14:335
feeding ecology, 14:260, 14:337
habitats, 14:337
humans and, 14:338
physical characteristics, 14:256, 14:335–336
reproduction, 14:338
species of, 14:*340*–343, 14:344*t*–345*t*
taxonomy, 14:256, 14:335
Viverrinae, 14:335
Viviparous reproduction, 12:5, 12:7, 12:106
See also Reproduction
Vizcachas. *See* Plains viscachas
Vlei rats. *See* Angoni vlei rats
Vocalizations, 12:85
See also Behavior
Voice organs, 12:8
See also Physical characteristics
Volant mammals, 12:200–201
Volcano rabbits, 16:482, 16:483, 16:487, 16:505, 16:509, 16:*510*, 16:513–*514*
Volcano shrews, 13:277*t*
Voles, **16:225–238,** 16:*231*, 16:*232*
behavior, 12:141–142, 12:148, 16:226–227
conservation status, 16:229–230
distribution, 16:*225*, 16:226
evolution, 16:225
feeding ecology, 16:227–228
habitats, 16:124, 16:226
humans and, 16:127
parasitic infestations, 12:78
physical characteristics, 16:225–226
predation and, 12:115
reproduction, 16:228–229
species of, 16:233–237*t*, 16:238*t*
taxonomy, 16:225
Vombatidae. *See* Wombats
Vombatiformes, 13:31, 13:32
Vombatus ursinus. See Common wombats

Vormela spp., 14:321
Vormela peregusna. See Marbled polecats
Vorontsov, N. N., 16:211
Vortex theory, of animal flight, 12:57
Voss, R., 16:269
Vu Quang oxen. *See* Saolas
Vulpes spp. *See* Foxes
Vulpes cana. See Blanford's foxes
Vulpes corsac. See Corsac foxes
Vulpes ferrilata. See Tibetan foxes
Vulpes pallida. See Pale foxes
Vulpes velox. See Swift foxes
Vulpes vulpes. See Red foxes
Vulpes zerda. See Fennec foxes

W

Wagner's bonneted bats, 13:493*t*
Wagner's moustached bats, 13:435, 13:*436*, 13:442*t*
Wagner's sac-winged bats. *See* Chestnut sac-winged bats
Wahlberg's epauletted fruit bats, 13:*313*, 13:*338*, 13:*343*
Waigeo cuscuses, 13:60
Waipatiidae, 15:13
Walia ibex, 16:87, 16:91, 16:92, 16:94, 16:95, 16:104*t*
Walker, E. P., 16:333
Wallabia spp., 13:83, 13:85
Wallabia bicolor. See Swamp wallabies
Wallabies, 13:*39*, **13:83–103,** 13:*91*, 13:*92*
  behavior, 12:145, 13:36, 13:86–87
  digestive system, 12:123
  distribution, 12:137, 13:*83*, 13:84–85
  evolution, 13:31–32, 13:83
  feeding ecology, 13:87–88
  habitats, 13:34, 13:35, 13:85–86
  humans and, 13:89–90
  physical characteristics, 13:32, 13:33, 13:83–84
  reproduction, 13:88–89
  species of, 13:93–100, 13:101*t*–102*t*
  taxonomy, 13:83
Wallaby-rats. *See* Tasmanian bettongs
Wallaroos, 13:34, 13:101*t*
Waller's gazelles. *See* Gerenuks
Walruses, **14:409–415,** 14:*410*, 14:*411*, 14:*413*
  distribution, 12:138
  feeding ecology, 14:260
  habitats, 12:14
  locomotion, 12:44
  migrations, 12:*170*
  physical characteristics, 12:46, 12:68
  reproduction, 12:111
  taxonomy, 14:256
Wapiti, 12:19, 15:271
Waris. *See* White-lipped peccaries
Warrens, rabbit, 12:186
Warthogs, 15:*276*, 15:277, 15:278–279, 15:280, 15:289*t*
  See also Common warthogs
Watchdogs. *See* Guard dogs
Water balance, 12:113
Water bats. *See* Daubenton's bats
Water buffaloes, 15:145, 15:263, 16:2, 16:4, 16:11, 16:12, 16:*17*, 16:20–21
  Asian, 16:2, 16:4, 16:*17*
  distribution, 16:*18*
  humans and, 16:8, 16:15
  physical characteristics, 16:3, 16:12, 16:13
  reproduction, 16:14
  short-horned, 16:1
Water chevrotains, 15:325, 15:*326*, 15:328, 15:329, 15:*330*, 15:*331*
Water civets. *See* Otter civets
Water deer, 15:268
  See also Chinese water deer
Water mice, 16:*271*, 16:*272*, 16:275
Water mongooses. *See* Marsh mongooses
Water opossums, 12:14, 12:249–253, 12:254, 12:*255*, 12:257–*258*
Water rabbits, 16:516*t*
Water rats, 16:266, 16:267–268, 16:269, 16:270
  false, 16:262*t*
  Florida, 16:225
  golden-bellied, 16:261*t*
Water shrews. *See* American water shrews
Water voles, 16:225, 16:226, 16:227, 16:229, 16:230
Waterbucks, 12:116, 16:*12*, 16:27, 16:29, 16:30, 16:32, 16:*36*, 16:37–38, 16:*41*
Waterman, J. M., 16:145
Watts, P., 12:57
Waugh, Evelyn, 12:183
Weaning, 12:97, 12:126
  See also Reproduction
Weasel lemurs. *See* Sportive lemurs
Weasel sportive lemurs, 14:*78*, 14:*80*, 14:82
Weasels, **14:319–328,** 14:*326*, **14:332–334**
  behavior, 12:114, 14:321–323
  conservation status, 14:324
  distribution, 14:*319*, 14:321
  evolution, 14:319
  feeding ecology, 14:260, 14:323
  humans and, 14:324–325
  physical characteristics, 14:319–321
  reproduction, 14:324
  species of, 14:*327*–328, 14:332*t*, 14:333*t*
  taxonomy, 14:256, 14:319
Weddell seals, 12:85, 12:87, 12:138, 14:*423*, 14:*426*, 14:*428*, 14:430–431
Wedge-capped capuchins. *See* Weeper capuchins
Weeper capuchins, 14:101, 14:102, 14:103, 14:*103*, 14:*107*, 14:*110*, 14:112
Weidomyines, 16:263–264
Weight reduction, in bats, 12:58–59
Welwitch's hairy bats, 13:497
West African brush-tailed porcupines. *See* African brush-tailed porcupines
West African manatees, 15:191–193, 15:*209*, 15:*210*, 15:*211*–212
West African pygmy squirrels, 16:167, 16:174*t*
West African rousettes. *See* Egyptian rousettes
West Australian boodies, 13:35
West Caucasian turs, 16:87, 16:91, 16:94
West Indian manatees, 15:*193*, 15:*196*, 15:*205*, 15:*210*, 15:*211*–212
  behavior, 15:*207*
  conservation status, 15:195
  distribution, 15:192
  feeding ecology, 15:193, 15:*206*
  habitats, 15:192, 15:*194*
  migrations, 12:87
  physical characteristics, 15:192, 15:*209*
  reproduction, 15:*208*
  taxonomy, 15:191
West Indian monk seals, 14:422, 14:435*t*
West Indian shrews, 12:133, 13:193, 13:197, **13:243–245,** 13:*244*
West Indian sloths, **13:155–159**
West Indies. *See* Neotropical region
Western, David, 12:97
Western Australian fur seals. *See* New Zealand fur seals
Western barbastelles, 13:500, 13:*505* 13:506–*507*
Western barred bandicoots, 12:*110*, 13:*1*, 13:*4*–5, 13:10, 13:12, 13:17*t*
Western bonneted bats, 13:493*t*
Western chestnut mice, 16:*252*
Western chipmunks. *See* Least chipmunks
Western Cuban nesophontes, 13:243
Western cusimanses, 14:357*t*
Western European hedgehogs, 12:*185*, 13:*200*, 13:*204*, 13:*205*, 13:*208*, 13:*209*, 13:212*t*–213*t*
Western fat-tailed dwarf lemurs, 14:*37*, 14:*38*, 14:*40*, 14:*41*, 14:42
Western gorillas, 12:222, 14:225, 14:*231*, 14:232, 14:*236*, 14:*238*
Western gray kangaroos, 13:101*t*
Western gray squirrels, 16:175*t*
Western hare-wallabies. *See* Rufous hare-wallabies
Western heather voles, 16:238*t*
Western hog-nosed skunks, 14:332*t*
Western hyraxes. *See* Red-headed rock hyraxes
Western jumping mice, 16:212, 16:*215*, 16:216
Western Malagasy broad striped mongooses, 14:*350*
Western Malagasy bushy-tailed rats, 16:*284*
Western mastiff bats. *See* Western bonneted bats
Western mountain reedbucks, 16:34
Western pocket gophers. *See* Valley pocket gophers
Western pygmy possums, 13:*107*
Western quolls. *See* Chuditches
Western red colobus, 14:*174*, 14:175, 14:*176*, 14:*178*–179
Western ringtails. *See* Common ringtails
Western rock sengis, 16:522, 16:531*t*
Western rock wallabies. *See* Brush-tailed rock wallabies
Western spotted skunks, 14:*320*, 14:324
Western tarsiers, 12:*81*, 14:*92*, 14:93, 14:*94*, 14:95, 14:*96*, 14:*97*–98
Western thumbless bats. *See* Schnabeli's thumbless bats
Western tree hyraxes, 15:177, 15:178, 15:184, 15:*185*, 15:*186*–187
Weyland ringtails, 13:114, 13:122*t*
Weyns's duikers, 16:85*t*
Whales, 12:14, 15:1–11, 15:*84*
  beaked, **15:59–71,** 15:*63*, 15:69*t*–70*t*
  belugas, 12:*85*, 12:87, 15:9, 15:10–11, **15:81–91**
  distribution, 12:138
  gray, 12:87, 15:3, 15:4, 15:5, 15:6, 15:9, **15:93–101**

Whales, *(continued)*
migrations, 12:169
neonatal milk, 12:126
physical characteristics, 12:63, 12:67, 12:68
reproduction, 12:91, 12:94, 12:103, 12:108
right, **15:107–118,** 15:*114*
smell, 12:80
sperm, **15:73–80**
teeth, 12:46
See also Killer whales; Rorquals
Whiptail wallabies, 13:*86*, 13:101*t*
Whiskered bats, 13:515*t*
Whiskered flying squirrels, 16:142*t*
Whiskers. *See* Vibrissae
Whispering bats. *See* Brown long-eared bats
Whistle pigs. *See* Woodchucks
Whistlers. *See* Hoary marmots
Whistling hares. *See* American pikas
Whistling rats. *See* Angoni vlei rats
White bald uakaris, 14:148, 14:151
White bats, 13:*313*, 13:414, 13:416, 13:*418*, 13:*421*, 13:*427*, 13:431–432
See also Northern ghost bats
White-bellied duikers, 16:84*t*
White-bellied fat-tailed mouse opossums. *See* Small fat-tailed opossums
White-bellied funnel-eared bats, 13:*462*, 13:463–*464*
White-bellied hedgehogs. *See* Central African hedgehogs
White-bellied pangolins. *See* Tree pangolins
White-bellied spider monkeys, 14:160, 14:167*t*
White-breasted hedgehogs. *See* Eastern European hedgehogs
White-browed gibbons. *See* Hoolock gibbons
White-capped sea lions. *See* Australian sea lions
White-cheeked gibbons, 14:*208*, 14:*216*, 14:*221*
White-collared mangabeys. *See* Collared mangabeys
White-eared kobs, 16:7
White-faced arboreal spiny rats, 16:450, 16:*452*, 16:*455*
White-faced sakis, 14:143–*144*, 14:147, 14:*149*, 14:*150*
White fin dolphins. *See* Baijis
White-footed dunnarts, 12:289, 12:301*t*
White-footed mice, 16:127, 16:128, 16:*267*, 16:*268*
White-footed sportive lemurs, 14:74, 14:*76*, 14:*78*, 14:*79*, 14:81–82
White-footed voles, 16:237*t*
White-footed weasel lemurs. *See* White-footed sportive lemurs
White-fronted capuchins, 14:101, 14:102, 14:*107*, 14:*110*
White-fronted hairy dwarf porcupines, 16:365, 16:374*t*
White-handed gibbons. *See* Lar gibbons
White-lipped deer, 12:134, 15:358, 15:*363*, 15:*368*, 15:370–371
White-lipped peccaries, 15:291–299, 15:*295*, 15:*297*
White-lipped tamarins, 14:*119*
White-naped weasels, 14:333*t*
White-necked fruit bats. *See* Tongan flying foxes
White-nosed bearded sakis, 14:144, 14:145, 14:147, 14:148, 14:153*t*
White-nosed coatis, 14:*310*, 14:*312*, 14:*314*
White-nosed monkeys, lesser, 14:*192*
White-nosed sakis. *See* White-nosed bearded sakis
White pigs, 15:277
White rhinoceroses, 15:215, 15:*249*, 15:*256*, 15:261
behavior, 12:145, 15:218, 15:253
conservation status, 15:222, 15:*254*
distribution, 15:*260*, 15:261
feeding ecology, 15:220, 15:253
habitats, 15:*253*
physical characteristics, 15:216, 15:217, 15:251, 15:*252*
reproduction, 15:221, 15:222
scent marking, 12:80–81
See also Rhinoceroses
White seals. *See* Crab-eater seals
White-shouldered capuchins. *See* White-throated capuchins
White-sided jackrabbits, 16:516*t*
White-streaked tenrecs, 13:228, 13:*229*
White-striped dorcopsises, 16:102*t*
White-striped free-tailed bats, 13:*487*, 13:*489*, 13:*492*
White-tailed deer, 15:*386*, 15:*389*–390
behavior, 15:271, 15:*381*, 15:*382*
competition, 12:118
ecosystem effects, 12:217
evolution, 15:380
habitats, 15:383
humans and, 15:384–385
reproduction, 15:384
scent marking, 12:80–81, 12:*142*
translocation of, 12:224
White-tailed gnus. *See* Black wildebeests
White-tailed mice, 16:*286*, 16:288–*289*
White-tailed mongooses, 14:259, 14:350
White-throated capuchins, 14:*1*, 14:101, 14:102, 14:*103*, 14:*104*, 14:*107*, 14:*111*
White-throated round-eared bats, 13:435*t*
White-throated wallabies. *See* Parma wallabies
White-throated woodrats, 16:277*t*, 16:*282*
White tigers, 14:*372*
White-toothed shrews, 13:198, 13:199, 13:247, **13:265–277,** 13:*269*, 13:*270*, 13:*276t–277t*
White-whiskered spider monkeys, 14:160, 14:167*t*
White-winged vampires, 13:314
Widdowson, Elsie, 12:126
Widow monkeys. *See* Collared titis
Wiedomys pyrrhorhinos, 16:267
Wild boars. *See* Eurasian wild pigs
Wild cats, 14:257, 14:262, 14:290, 14:292, 14:375, 14:*378*, 14:*383*–384
See also Cats
Wild cattle. *See* Cattle
Wild dogs. *See* Dogs
Wild goats, 15:146, 15:148, 15:149, 16:87, 16:*88*, 16:90–93, 16:95, 16:104*t*
See also Goats
Wild oxen. *See* Aurochs
Wild sheep, 15:269, 15:271, 15:273, 16:87, 16:*88*, 16:95
See also Sheep
Wild yaks. *See* Yaks
Wildcats. *See* Wild cats
Wildebeests, 12:*164*, **16:27–43,** 16:*35*
behavior, 16:5, 16:30–31
conservation status, 16:33–34
distribution, 16:29, 16:*38*
evolution, 16:27–28
feeding ecology, 16:31–32
in guilds, 12:119
habitats, 16:4, 16:29–30
humans and, 16:34
migrations, 12:*164*, 12:*166*
physical characteristics, 16:2, 16:28–29
reproduction, 12:101, 12:106, 16:32–33
species of, 16:37–38, 16:42*t*
taxonomy, 16:27–28
Wilfredomys spp., 16:264
Wilfredomys oenax. *See* Red-nosed tree rats
Wilson, D. E., 16:121, 16:389
The Wind in the Willows, 12:183, 13:200
Winemiller, Kirk O., 15:29
Wing loading (WL), bats, 12:57–58
Wing sacs, 12:81
Wings, bat, 12:56–58, 12:56–60, 12:*58*
See also specific types of bats
Wisents, 16:14
Wiwaxia, 12:64
WL. *See* Wing loading
Woermann's bats. *See* African long-tongued fruit bats
Wolffsohn's viscachas, 16:380, 16:383*t*
Wolverines, 14:*326*, 14:*328*, 14:330–331
behavior, 14:323
conservation status, 12:220
distribution, 12:131
feeding ecology, 14:323
habitats, 14:321
humans and, 14:324
reproduction, 12:105, 12:107, 12:109, 12:110, 14:324
Wolves, **14:265–278,** 14:*275*, **14:282–285**
aardwolves, 14:259, 14:260, **14:359–367**
behavior, 12:*141*, 12:*142*, 14:267–268, 14:269
conservation status, 14:262, 14:263, 14:272–273
domestic dogs and, 12:180
eurasian, 12:180
evolution, 14:265–266
Falkland Island, 14:262, 14:272, 14:284*t*
feeding ecology, 14:269, 14:270–271
habitats, 14:267
humans and, 14:273–274
physical characteristics, 14:266
predation and, 12:116
red, 12:*214*, 14:262, 14:273
reproduction, 14:272
species of, 14:*276*–*278*, 14:282, 14:284*t*
taxonomy, 14:265
See also Gray wolves; Tasmanian wolves
Wombats, **13:51–56,** 13:*52*, 13:*54*
behavior, 13:52, 13:55–56
conservation status, 13:40, 13:53, 13:55–56
distribution, 13:*51*, 13:*55*–56
evolution, 13:31, 13:51
feeding ecology, 13:38, 13:52–53, 13:55–56
habitats, 13:*35*–36, 13:52, 13:55–56
humans and, 13:40, 13:53, 13:55–56
physical characteristics, 13:32, 13:33, 13:51, 13:55–56
reproduction, 13:38, 13:39, 13:53, 13:55–56
taxonomy, 13:51, 13:55–56

Women. *See* Humans
Wongai ningauis, 12:300*t*
Wood lemmings, 16:228, 16:229, 16:*232*, 16:*235*–236, 16:270
Wood rats, white-throated, 16:*282*
Woodchucks, 12:*31*, 16:126, 16:*149*, 16:152–*153*
Woodland dormice, 16:327*t*
Woodland jumping mice, 16:212, 16:213, 16:*214*, 16:215, 16:216, 16:224*t*
Woodland Park Zoo, 12:205
Woodland voles, 16:238*t*
Woodlark Island cuscuses, 13:58, 13:60, 13:66*t*
Woodrats, 12:131, 16:126
  Allen's, 16:277*t*
  bushy-tailed, 16:*124*, 16:277*t*
  dusky-footed, 16:277*t*
  eastern, 16:125, 16:277*t*
  white-throated, 16:277*t*
Woods, C. A., 16:436
Wood's slit-faced bats, 13:375
Woolly dormice, 16:327*t*
Woolly false vampire bats, 13:*414*, 13:416, 13:433*t*
Woolly flying squirrels, 16:136, 16:141*t*
Woolly horseshoe bats, 13:388, 13:389, 13:390, 13:392
Woolly lemurs, 14:63, 14:66
  See also Eastern woolly lemurs
Woolly mammoths, 12:*18*, 12:*19*, 12:*21*, 12:*22*, 15:162
Woolly monkeys, 14:7, 14:11, 14:155, 14:156, 14:157, 14:158–159
Woolly mouse opossums, 12:250–253, 12:*256*, 12:*261*–262
Woolly opossums, 12:249–253, 12:254
Woolly rhinoceroses, 15:138, 15:249
  See also Rhinoceroses
Woolly spider monkeys. *See* Northern muriquis; Southern muriquis
Wooly tapirs. *See* Mountain tapirs
Work dogs, 14:288, 14:289, 14:*289*
Working memory. *See* Short-term memory
World Conservation Union (IUCN), 12:213–214
  See also IUCN Red List of Threatened Species
A World of Wounds: Ecologists and the Human Dilemma, 12:217
World Wildlife Fund for Nature (WWF), 12:224
World Wildlife Fund Global 200 Terrestrial Ecoregions, 16:386–387
Woylies. *See* Brush-tailed bettongs
Wrinkle-faced bats, 12:79, 13:310, 13:*421*, 13:*430*, 13:432
Wrinkle-lipped free-tailed bats, 13:493*t*
Wroughton's free-tailed bats, 13:487, 13:495*t*
Wu Wang (Emperor of China), 12:203
Wurrups. *See* Rufous hare-wallabies
Wyulda spp., 13:57
Wyulda squamicaudata. *See* Scaly-tailed possums

X

Xanthorhysis tabrumi, 14:91
Xenarthra, 12:26, 12:33, 12:132, **13:147–154**, 15:134
Xenothrix spp., 14:143
Xenothrix mcgregori. *See* Jamaican monkeys
Xenungulates, 15:135, 15:136
Xerini, 16:143
Xeromys myoides. *See* False water rats
Xerus spp., 16:143, 16:144
Xerus inauris. *See* South African ground squirrels
Xerus rutilus. *See* Unstriped ground squirrels

Y

Yaks, 12:*48*, 16:*1*, 16:*17*, 16:*18*
  distribution, 16:4
  domestic, 12:177–178
  evolution, 16:11
  high-altitudes and, 12:134
  humans and, 16:15
  physical characteristics, 16:2
Yellow armadillos, 13:184, 13:*186*, 13:*188*
Yellow baboons, 14:*193*
Yellow-backed duikers, 16:*75*, 16:*78*, 16:*82*, 16:83–84
Yellow bats, 13:311, 13:497
Yellow bear cuscuses, 13:60
Yellow-bellied capuchins. *See* Yellow-breasted capuchins
Yellow-bellied gliders, 13:126, 13:*127*, 13:128–129, 13:*130*, 13:*131*–132
Yellow-bellied marmots, 16:*147*, 16:158*t*
Yellow-breasted capuchins, 14:11, 14:101, 14:102, 14:106, 14:*107*, 14:*111*, 14:112
Yellow-cheeked crested gibbons. *See* Golden-cheeked gibbons
Yellow-crowned brush-tailed tree rats. *See* Toros
Yellow-faced pocket gophers, 16:*191*, 16:*193*, 16:194
Yellow-footed antechinuses, 12:277, 12:299*t*
Yellow-footed rock wallabies, 13:33, 13:39, 13:*92*, 13:*96*, 13:97
Yellow fruit bats. *See* Straw-colored fruit bats
Yellow golden moles, 13:198, 13:*222t*
Yellow-handed titis. *See* Collared titis
Yellow house bats. *See* House bats
Yellow mongooses, 14:*348*, 14:350, 14:357*t*
Yellow-sided opossums. *See* Southern short-tailed opossums
Yellow-spotted hyraxes, 15:177, 15:189*t*
  See also Rock hyraxes
Yellow-spotted rock hyraxes. *See* Rock hyraxes
Yellow streaked tenrecs, 13:*199*, 13:*227*, 13:*228*, 13:234*t*
Yellow-tailed woolly monkeys, 14:11, 14:155, 14:156, 14:159, 14:167*t*
Yellow-throated martens, 14:*320*
Yellow-toothed cavies, 16:399*t*
  See also Cuis
Yellow-winged bats, 13:381, 13:*381*, 13:*383*, 13:*384*
Yerkes Regional Primate Research Center, 12:161
Yerkish, 12:161
Yolk sac, 12:93
  See also Reproduction
Yorkshire pigs, 12:*172*
Yorkshire terriers, 14:288, 14:289
Yucatán brown brocket deer, 15:396*t*
Yucatán squirrels, 16:175*t*
Yuma antelope squirrels. *See* Harris's antelope squirrels
Yunnan shrew-moles, 13:282

Z

Zacatuches. *See* Volcano rabbits
Zaedyus pichiy. *See* Pichi armadillos
Zaglossus spp. *See* Long-beaked echidnas
Zaglossus bartoni. *See* Long-beaked echidnas
Zaglossus bruijni. *See* Long-beaked echidnas
Zalambdalestids, 13:193
Zalophus californianus. *See* California sea lions
Zalophus californianus wollebaecki. *See* Galápagos sea lions
Zalophus wollebaeki. *See* Galápagos sea lions
Zambian mole-rats, 12:75
Zanj sengis. *See* Black-and-rufous sengis
Zanzibar bushbabies, 14:27, 14:*30*
Zanzibar leopards, 14:385
Zapodidae, 16:123, 16:125, 16:283
Zapodinae, 16:212
Zapus spp., 16:212, 16:213, 16:216
Zapus hudsonius. *See* Meadow jumping mice
Zapus hudsonius campestris, 16:216
Zapus hudsonius preblei, 16:216
Zapus princeps. *See* Western jumping mice
Zapus trinotatus, 16:212, 16:216
Zapus trinotatus orarius, 16:216
Zarhachis spp., 15:13
Zebra antelopes. *See* Zebra duikers
Zebra duikers, 16:2, 16:*75*, 16:*79*, 16:*80*, 16:81
Zebra mice, 16:*250*
Zebra mongooses. *See* Banded mongooses
Zebras, **15:225–236**, 15:*231*
  behavior, 12:145, 15:142, 15:219, 15:226–228
  conservation status, 15:221, 15:222, 15:229–230
  distribution, 15:217, 15:*225*, 15:226
  domestication and, 15:145, 15:146
  evolution, 15:136, 15:137, 15:215, 15:216, 15:225
  feeding ecology, 12:118–119, 15:220, 15:228
  habitats, 15:218, 15:226
  humans and, 15:222, 15:223, 15:230
  migrations, 12:164, 12:*167*
  physical characteristics, 15:139, 15:140, 15:216, 15:217, 15:225–226
  reproduction, 15:221, 15:228–229
  species of, 15:232–234, 15:236*t*
  taxonomy, 15:225
Zebu cattle, 16:6
Zeitgeber, 12:74
Zenkerella insignis. *See* Cameroon scaly-tails
Zenker's flying mice, 16:300, 16:*301*, 16:*304*
Zenker's fruit bats, 13:349*t*
Zhou, K., 15:19
Ziegler, T. E., 14:165
Zinser's pocket gophers, 16:197*t*
Ziphiidae. *See* Beaked whales
Ziphius spp., 15:59
Ziphius cavirostris. *See* Cuvier's beaked whales
Zohary, D., 15:147
Zokors. *See* Siberian zokors

Zoogeography, 12:129–139
See also Distribution
Zoological Society of London, 12:204
Zoos, **12:203–212,** 12:*205*, 12:*206*, 12:*207*
animal husbandry, 12:209–211
ethics in, 12:212
exhibit designs, 12:204–205
history, 12:203
inbreeding in, 12:222
management of, 12:205–212
purpose of, 12:203–204
Zorillas, 14:321, 14:323, 14:332*t*
Zorros, 14:265
Zulu golden moles, 13:222*t*
Zygodontomys brevicauda, 16:268, 16:269, 16:270
Zygogeomys spp., 16:185
Zygogeomys trichopus. See Michoacán pocket gophers
Zygomatic arches, 12:8
Zygotes, 12:92, 12:101
See also Reproduction